Prof Dr Artur Goldschmidt and Dr Hans-Joachim Streitberger

BASF Handbook
Basics of Coating Technology

3rd Revised Edition

Published in cooperation with
BASF We create chemistry
VINCENTZ Providing Professional Information

Cover: © pleinpicture/Biwa Inc.

Bibliographische Information der Deutschen Bibliothek

Die Deutsche Bibliothek verzeichnet diese Publikation in der Deutschen Nationalbibliographie; detaillierte bibliographische Daten sind im Internet über http://dnb.ddb.de abrufbar.

Prof Dr Artur Goldschmidt and Dr Hans-Joachim Streitberger

Vincentz Network, Plathnerstraße 4c, 30175 Hannover, Germany
T +49 511 9910-033, F +49 511 9910-029, books@european-coatings.com
ISBN 3-86630-336-X
ISBN 978-3-86630-336-2

Layout: Siegfried Urbich, Mediendesign, Celle, Germany
Printed by: Bonifatius Druck/Buch/Verlag, Paderborn, Germany

Prof Dr Artur Goldschmidt and Dr Hans-Joachim Streitberger

BASF Handbook

Basics of Coating Technology

3rd Revised Edition

Foreword to the 3rd Revised Edition

Glasurit's "Paints and Coatings" handbook has been a standard work in the field of coating materials since it was first published in 1934. The latest German edition appeared in 2014 and serves as the basis for this third English edition, which is sponsored by BASF Coatings. Compared with its 2007 predecessor, this third edition contains more up-to-date market data and incorporates new raw materials, new application technologies, new legislative developments, mainly in Europe, as well as modern coating processes.

The main new developments covered in the raw materials and coatings sector are nanotechnology, polyurethane dispersions for water-borne coatings and smart coatings. As for pretreatment processes, application methods and recovery techniques, the latest advances here include workpiece rotation during electrocoating, the integrated paint process and new methods for handling overspray in the automotive industry. More than 200 new references have been added and some 500 others updated.

The proven arrangement of chapters has been left unchanged.

The handbook starts logically with a brief historical summary of painting, its economic and technical significance and the social framework underpinning the industry. Next comes a chapter on raw materials, product composition, the various production processes and the principles behind product formulation. Descriptions are then given of the properties of the liquid and solid coating materials employed in the various applications. This is followed by a chapter dealing with the theoretical aspects of coating compositions. Applications are then explained for the reader against the background of the current major industrial coating processes. This chapter clearly shows just how important the substrate is because of the impact it has on the coating outcome. A chapter follows dealing with the demands on environmental compliance, health, safety and quality as specified in current legislation, mainly in Europe and Germany. After that, the reader is taken on a journey through various sectors of the paint and coatings industry, accompanied by numerous examples and illustrations. This chapter will be of particular interest for newcomers to the industry and practically-minded readers alike. The book concludes with tables of selected standards, explanations of physical symbols, useful websites, definitions of terms and a comprehensive index round.

Literature references are provided for readers seeking more in-depth information. These are to be found at the end of each chapter and are repeated on occasion. Many of the comments and scenarios are based on experience gained by BASF Coatings and are not directly attributable.

We would like to thank BASF Coatings for making this edition possible, along with countless persons for their support and invaluable tips. Special thanks are due to Dr Walter Jouck, Dr Peter Betz, Sabine Rüttgers, Rolf Döring, Jürgen Book, Heike Thöne and Dr Peter Bachhausen from BASF Coatings. We are indebted to Bernd Reinmüller,

Director at the German DIN Institute for the trouble he took to ensure that the standards listed are up to date.

Finally, we would like to express our gratitude to the many companies which kindly provided us with pictures and charts for publication.

Prof Dr Artur Goldschmidt and Dr Achim Streitberger

Paderborn and Münster, November 2017

Contents

1 Introduction

2 Coating Materials

3 Coatings

4 Coating Technology

5 Safety, Environmental Protection and Health

6 Principles of Quality Management

7 Coating Industries

8 Industrial Standards

9 Appendix

10 Index

Contents

1	**Introduction**	**15**
1.1	Definition, Tasks and Economic Importance	15
1.2	Retrospective	21
1.3	References	25
2	**Coating Materials**	**27**
2.1	Raw Materials	27
2.1.1	Film Forming Agents	30
2.1.1.1	Natural Materials	34
2.1.1.2	Synthetic Resins	49
2.1.2	Solvents	127
2.1.2.1	Theory of Solubility	129
2.1.2.2	Physical Properties	131
2.1.2.3	Chemical and Physiological Properties	134
2.1.2.4	Important Solvents for Paints	134
2.1.3	Pigments and Extenders	138
2.1.3.1	Physical Principles behind Chroma and Hiding Power	142
2.1.3.2	The most Important Pigments for Coating Materials	145
2.1.3.3	Extenders	165
2.1.4	Plasticisers and Additives	167
2.1.4.1	Plasticisers	168
2.1.4.2	Additives	171
2.1.5	Nanoproducts	191
2.1.5.1	Description and Definition	191
2.1.5.2	Synthesis of Nanoproducts	192
2.1.5.3	Sample Applications	194
2.1.5.4	Outlook	198
2.1.6	Summary	198
2.2	From Raw Material to Coating Material	199
2.2.1	General Rules on Drawing up Formulations	200
2.2.1.1	Purpose and Quality	200
2.2.1.2	Production Resources, Application Systems and Object to be Painted	200
2.2.1.3	Function in the Paint System	202
2.2.1.4	Cost Effectiveness and Availability	203
2.2.1.5	Occupational Health, Safety and Environmental Protection Regulations	203
2.2.2	Material Flow in a Paint Factory	204
2.2.3	Theory of Dispersion	206
2.2.3.1	Pigment-Specific Properties for Dispersion Processes	206
2.2.3.2	Wetting and Separation of Agglomerates	210
2.2.3.3	Stabilisation of Pigment Dispersions	215
2.2.3.4	Optimum Mill Base Formulation	223
2.2.4	Production of Coating Materials	226
2.2.4.1	Processing Sequences in a Paint Plant	227
2.2.4.2	Agitation and Agitators	228

2.2.4.3	Dispersing and Dispersers	234
2.2.4.4	Separation Processes in the Manufacture of Paint	250
2.2.5	Summary	258
2.3	Characterisation of Coating Materials	258
2.3.1	Measurement Accuracy	260
2.3.2	Testing Raw Materials and Coating Materials	263
2.3.2.1	Safety and Environment-Related Performance Indicators	264
2.3.2.2	Chemical Characterisation	268
2.3.2.3	Physical Indicators	279
2.3.3	Pigment-Specific Tests	305
2.3.3.1	Indicators of the Pigment as a Raw Material	305
2.3.3.2	Tests with the Pigmented Coating Material	314
2.3.4	Processability of Coating Materials	319
2.3.5	Summary	322
2.4	Literature	322
3	**Coatings**	**327**
3.1	From Coating Material to Coating: Film Formation	327
3.1.1	Wetting and Levelling	327
3.1.2	Solidification of the Film	329
3.1.2.1	Physical Drying	330
3.1.2.2	Chemical Curing	335
3.1.2.3	Flow Patterns in the Solidifying Film	336
3.1.3	Film Shrinkage	336
3.1.4	Special Features of High Solids Paints and Water-Borne Paints	337
3.1.5	Monitoring the Film Forming Process with Measuring Instruments	340
3.1.5.1	Levelling and Sagging	340
3.1.5.2	Film Formation by Air Drying Paints	342
3.1.5.3	Indirect Methods	343
3.1.6	Summary	343
3.2	Properties and Testing of Coatings	344
3.2.1	Film Thickness	344
3.2.1.1	Wet Films	345
3.2.1.2	Dry Films	346
3.2.2	Dry Film Density	352
3.2.3	Measurement of Voids in Coating Films	354
3.2.4	Visual Properties	355
3.2.4.1	Gloss	355
3.2.4.2	Colour and Colourimetry	363
3.2.5	Mechanical Properties	375
3.2.5.1	Adhesion	376
3.2.5.2	Elasticity	387
3.2.5.3	Hardness	398
3.2.5.4	Abrasion Resistance and Mar Resistance	404
3.2.5.5	Other Tests	407
3.2.6	Summary	408
3.3	Durability of Coatings	408

3.3.1	Fundamental Features of Ageing	408
3.3.2	Ageing Tests	416
3.3.2.1	Methods for Testing Surface Durability	417
3.3.2.2	Corrosion Protection Tests	426
3.3.2.3	Chemical Resistance	433
3.3.3	Summary	435
3.4	Literature	435
4	**Coating Technology**	**439**
4.1	Influence of the Substrate	441
4.1.1	Wood and Wood Materials	442
4.1.1.1	Wood as a Material	442
4.1.1.2	Wood as a Workpiece	444
4.1.1.3	Pretreatment of Wood	446
4.1.1.4	Surface Finishing of Wood Panels	447
4.1.2	Metallic Materials	448
4.1.2.1	Properties of Metals	448
4.1.2.2	From the Material to the Workpiece	453
4.1.2.3	Pretreatments	454
4.1.3	Plastics	465
4.1.3.1	Types of Plastics and their Properties	465
4.1.3.2	Pretreatments	468
4.1.4	Mineral Substrates	471
4.1.4.1	Concrete, Mortar, Plaster	471
4.1.4.2	Glass	473
4.1.5	Other Substrates	473
4.1.5.1	Rubber	473
4.1.5.2	Leather	474
4.1.6	Workpiece Design and Coatability	474
4.1.7	Summary	476
4.2	Paint Processing	477
4.2.1	Processing of Wet Paints	478
4.2.1.1	Dip Coating (Object Taken to Paint)	479
4.2.1.2	Brushing, Manual and Mechanical Roller, Flow and Curtain Coating Methods (Paint Applied Directly to the Object)	500
4.2.1.3	Spray Process: Paint Applied Indirectly to the Object	506
4.2.1.4	Paint Supply Systems	539
4.2.1.5	Booth Conditioning and Overspray Elimination	556
4.2.1.6	Automated Systems and Robots for Paint Processing	562
4.2.1.7	Conveyors	566
4.2.1.8	Paint Removal	567
4.2.2	Processing of Powder Coatings	569
4.2.2.1	Retrospective	569
4.2.2.2	Plant and Equipment Details	570
4.2.2.3	Powder Coating Materials	580
4.2.2.4	Powder-Specific Test Methods	582
4.2.3	Summary	584

4.3	Drying and Curing	584
4.3.1	Film Formation by Heat Transfer	585
4.3.1.1	Theory	585
4.3.1.2	Oven Design	597
4.3.2	Curing by UV and Electron-Beam Radiation	603
4.3.3	Summary	610
4.4	Literature	611
5	**Safety, Environmental Protection and Health**	**615**
5.1	Legislative Framework	615
5.2	Safety	619
5.2.1	Manufacturing	620
5.2.2	Storage	622
5.2.3	Transportation	622
5.2.4	Application	623
5.3	Environmental Protection	625
5.3.1	Exhaust Air	625
5.3.2	Wastewater	632
5.3.3	Recycling and Disposal	633
5.4	Health	635
5.5	Ecobalances	638
5.6	Environment-Compatible Paints and Coatings	642
5.6.1	Low-Emissions Coatings	643
5.6.1.1	High Solids	643
5.6.1.2	Water-Borne Paints	644
5.6.1.3	Powder Coatings	647
5.6.1.4	Radiation Curable Paints and Coatings	648
5.6.1.5	Other Coating Systems	651
5.6.2	Biobased Coating Systems	652
5.6.3	Foil Coating	653
5.7	Economics of Coating Processes	654
5.8	Literature	656
6	**Principles of Quality Management**	**659**
6.1	Evolution of Quality Concepts over Time	659
6.2	Defects in Coating Processes and Applied Coatings	661
6.2.1	Identifying Defects and their Causes	662
6.2.2	The Most Frequent Causes of Surface Defects	666
6.2.3	Summary	674
6.3	Material Control	674
6.4	Defect Prevention by Process Control and Control Loops	676
6.5	Quality Management	677
6.6	Literature	683
7	**Coating Industries**	**685**
7.1	Automotive OEM Coating	686
7.1.1	Pretreatment	691

7.1.2	Electrocoating	694
7.1.3	Seam Sealant and Underbody Protection	698
7.1.4	Primer Surfacer	698
7.1.5	Topcoat Application	702
7.1.6	Repair in Automotive OEM Lines	707
7.1.7	Outlook	709
7.2	Automotive Refinishing	711
7.3	Automotive Supply Industry	718
7.4	Coil Coating	725
7.5	Commercial Vehicles	730
7.6	Mechanical Engineering	733
7.7	White Goods	734
7.8	Building Supplies	735
7.9	Rail Vehicles	737
7.10	Wood Coating	738
7.11	Other Fields of Application	742
7.11.1	Protection of Structures	742
7.11.2	Steel Furniture	743
7.11.3	Aviation Industry	744
7.11.4	Electrically Insulating Coatings	746
7.11.5	Communications Equipment	747
7.11.6	Road Marking paints	748
7.11.7	Shipbuilding	749
7.11.8	Wind Energy	750
7.11.9	Smart Coatings	752
7.11.10	Miscellaneous Coating Applications	753
7.12	Literature	755
8	**Standards**	**759**
8.1	General Information on Standardisation Work	759
8.2	Selection of DIN Standards for the Paints and Coatings Industry and their Circle of Users	761
8.2.1	Standards by Numerical Order	761
8.2.1.1	Standards on Terminology	761
8.2.1.2	Standards on Coating Materials	761
8.2.1.3	Standards on Coatings	762
8.2.1.4	Standards on Aluminium	766
8.2.1.5	Standards on Coil Coatings	766
8.2.1.6	Standards on Galvanised Steel	767
8.2.1.7	Standards on Wood Outdoors	767
8.2.1.8	Standards on Anticorrosion Protection of Steel Structures	767
8.2.1.9	Standards on Nuclear Facilities	768
8.2.1.10	Standards on Aerospace	768
8.2.1.11	Standards on Exterior Masonry and Concrete	768
8.2.1.12	Standards on Shipbuilding	769
8.2.1.13	Dispersion Coatings for Indoors	769
8.2.1.14	Synthetic Resin Plasters	769

8.2.1.15 Powder Coatings 769
8.2.1.16 Road Marking Paints 770
8.2.1.17 Radiator Coatings 770
8.2.1.18 Surface Preparation 770
8.2.1.19 External Thermal Insulation Composite Systems (ETICS) 771
8.2.1.20 Colourimetry 771
8.2.1.21 DIN Reports 771
8.2.2 Standards by Keyword 772
8.3 Literature 781

9 Appendix 783
9.1 List of Physical Constants 783
9.1.1 Latin Symbols 783
9.1.2 Greek Symbols 786
9.2 Internet Addresses of Interest to the Coatings World 787
9.2.1 Organisations/Associations/Authorities 787
9.2.2 Publications/Information 788
9.2.3 Institutes/Universities 789
9.2.4 Miscellaneous 790
9.3 Acronyms/Abbreviations 790

Authors 799

Index 801

1 Introduction

1.1 Definition, Tasks and Economic Importance

The task of coating technology is to provide surface protection, decorative finishes and numerous special functions for commodities and merchandise by means of organic coatings. Many everyday products are only made usable and thus saleable because of their surface treatment. For this, the right coating formulations, production plant, coating material and suitable coating processes for the product must be available. However, the level of quality attainable via the coating process is not solely a function of the coating material. The object to be coated, along with its design, specific material treatment and an appropriate application method, are further variables which play a significant role. In addressing the ongoing tasks of quality optimisation and rationalisation while minimising the impact on humans and the environment, it is vital that the dependencies mentioned above be not only recognized but also taken into account as the framework shaping the conditions in which work advances from development through to application. Coating technology, therefore, is an interdisciplinary subject.

Figure 1.1.1: Coated everyday objects *Source: VdL*

Paints and coating materials are not end products, but merely starter or intermediate products which, for the above-mentioned reasons, require a skilled and conscientious user if they are to be converted into the actual end product, namely the coating itself. Only the cured coating, which in many cases is a system of several individual coats, can meet the wishes invested in and the requirements demanded of the coated products.

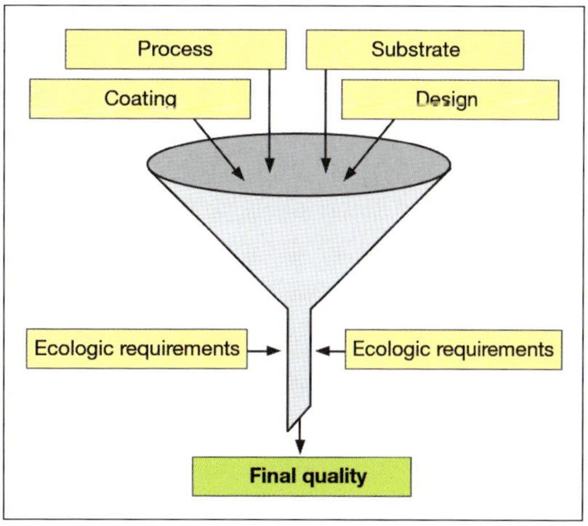

Figure 1.1.2: Factors determining the quality of coatings

Two of the most important of the many functions which coatings have to meet are **protection** and **decoration**. Others worth noting are the informative tasks and the achievement of special physical effects. The conspicuousness of emergency service vehicles, the camouflaging of military equipment, and road or airport markings are just some of the informative tasks required of coatings. Colour markings enable areas or spaces to be clearly signed or divided. Colour coding helps to indicate the contents of containers and the materials being conveyed in pipes. Optical effects induced by coloured or metallic pigments lend a coating a particular optical attraction. Deliberately generated surface textures such as scars or wrinkles expand the range of effects which can be achieved. The use of colour schemes based on the known physiological and psychological effects of colours also contributes in various ways to improved working conditions and enhanced safety wherever work is done in rooms and on machines. Functional pigments produce colours that vary with temperature, for example as a result of their thermochromic properties, and therefore afford an indirect way of measuring the temperatures of objects.

The most important task of coatings, in economic terms, is surface protection. Thus, coatings help to retain value and improve the usability properties of almost all products and so are of huge economic significance. Particular mention should be made of the protection of metal goods that acquire lasting anticorrosive protection only when they have been painted.

It is vital in this regard, e.g. in the automotive sector, for the resistance of the coating system to external, sometimes aggressive natural and anthropogenic atmospheric influences such as tree resins, bird droppings, acids, alkalis, salts and organic solvents, to be guaranteed.

The protective function of paint on cars must not be impaired even under extreme mechanical impact, such as stone chippings thrown up from the road by traffic or by brush action in carwashes.

Furthermore, coatings must withstand combined, i.e. physical and chemical, effects to which objects are subjected, for example, in different weather conditions. The interaction of sunshine, rain, heat and frost, combined with emissions from heating systems and internal combustion engines, ozone and salt fog makes great demands on a coating's resistance and protective properties.

However, a surface protection coating can also be applied in order to meet quite different requirements. Floors and steps can be made non-slip, and their utility value enhanced, by means of rough or high-grip coatings. By contrast, surface friction can be reduced by using smooth coatings to produce a high degree of slip. Flammable materials can be rendered safe by means of flame retardant coatings. Antibacterial coatings help keep surfaces sterile in production and storage facilities in dairies and breweries and prevent the growth of barnacles and algae on ships' hulls. In the electrical engineering sector, insulating coatings provide effective and lasting insulation for wire, windings and condenser materials. On the other hand, conductive coatings can be used to make insulating substrates electrically conductive or even to print electrical circuits. Furthermore, organic coatings can help to reduce noise pollution. Acoustic insulation coatings for machines and underbody protection coatings for passenger cars are examples of this.

This broad spectrum of requirements explains why no single coating material can satisfy every wish simultaneously and in the same way. The goal of providing coating materials for the lasting protection, decoration and improvement of objects made of wood, metal, plastic or mineral materials at reasonable prices can be met only by adopting different formulations based on a range of materials and material combinations. Each of these combinations targets a limited field of substrates, a selected application method and a specific profile of film properties.

Coating technology is used in metal processing, in the manufacture of plant and machinery and in the electrical engineering industry. All kinds of road and rail vehicles, ships and aircraft are important objects which require painting or coating. Effective surface protection by means of paints and coatings is also indispensable in the civil engineering sector, for steel and concrete structures and in wood processing. Even plastics and leather require coating in many cases. Modern paper, plastic and sheet-metal packaging materials would be inconceivable without the protection and decoration afforded by coatings.

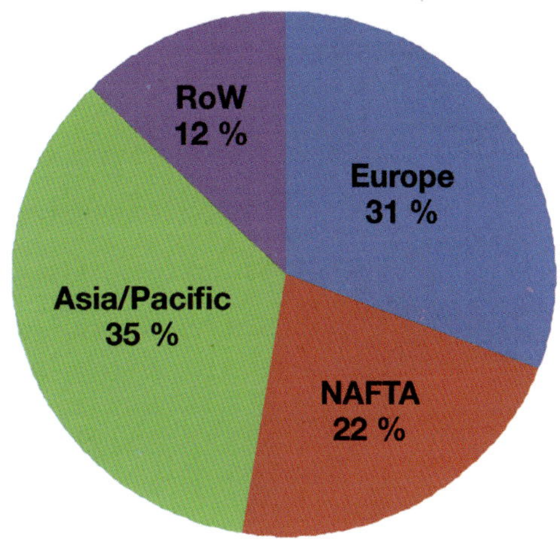

Figure 1.1.3: Regional division of the market for paint and coatings in 2010

The global paint and coatings market reflects economic developments in the regions. It is most highly developed, for example, in the triad of North America, Europe, and South East Asia. Per capita consumption of paints and coatings in these regions is approx. 4.5 kg. Growth in coatings consumption is determined by economic development in the various regions and countries [1.3.1].

The broad field of applications for coatings and their widespread use are explained by the high value and great benefits which they offer. The fact that there are few objects which do not require coating is an indication of the enormous importance of coating technology. It would present an incomplete picture to evaluate this importance merely in terms of the 30.5 million tonnes of coating materials manufactured worldwide in 2010 with a value of some 80 billion euros (Figure 1.1.3).

Although quoting the quantities of coating materials is not a direct indication of the added value of industrial commodities, it does permit that area to be calculated which can be protected or decorated by means of coating materials, with due allowance for the film thickness to be applied. An annual production quantity of 30.5 million tonnes (see above) is enough to coat a surface area of some 340,000 km^2 in a wet film 100 µm (0.1 mm) thick. That is equivalent to about 3/4 of the surface area of Germany. The same quantity of paint applied to a 10 m wide strip in a film thickness of 100 µm, on the other hand, would stretch from the earth to the moon about 100 times or go around the world 650 times.

A more meaningful evaluation would consider the value of the effectively protected and improved products. On the assumption that painting or coating the goods produced generates an added value of 20 % in the form of extended service life and increased attractiveness, this translates to 140 billion euros for the German market in 2012, or 70 times the value of the paints and coatings sold.

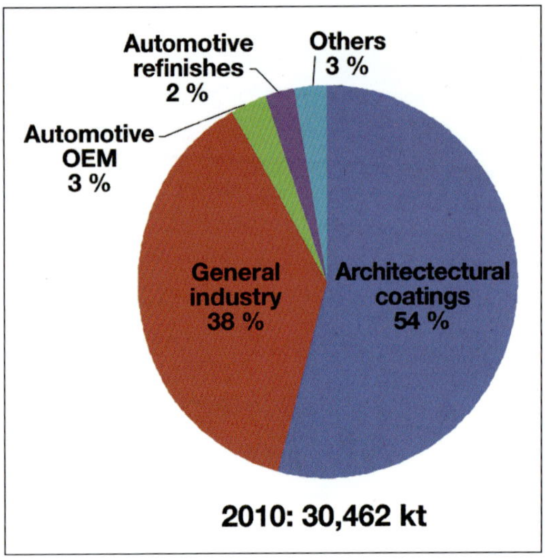

Figure 1.1.4: Worldwide coatings market in 2010 by segments

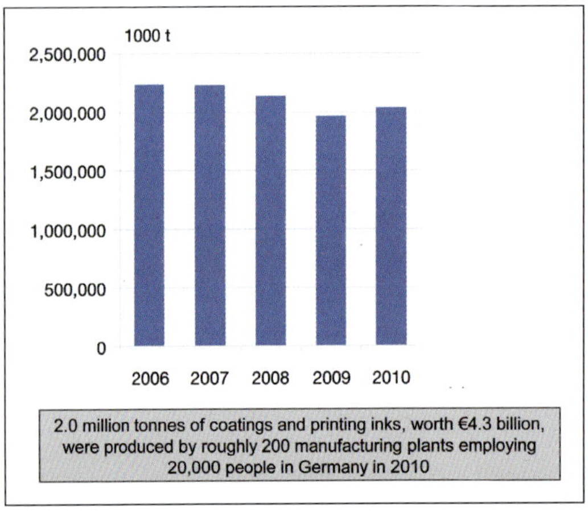

Figure 1.1.5: Production of coatings in Germany from 2006 to 2010

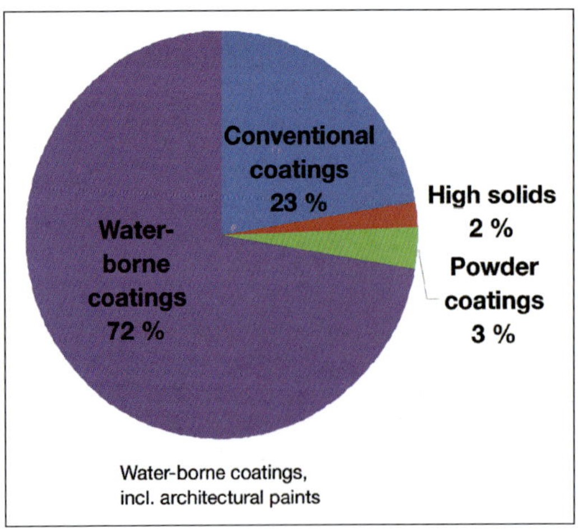

Figure 1.1.6: German coatings market in 2010 by product classes *Source VdL*

The division of the market into branches and segments is not uniform around the world. A number of such divisions, however, seem to agree on certain segments, such as decorative paints, general industrial paints, automotive paints, and printing inks.

The largest market for paints and coatings, at 54 %, is that of decorative paints. This is followed, at 38 %, by the market for the industrial coating of a huge range of objects, from compact discs via plastic bumpers for cars to rail vehicles. Automotive coating lines and refinishing body shops are each clearly defined segments with a high technological value, though of less significance in terms of volume sales. Although not shown in Figure 1.1.4, printing inks represent approx. 4 % of global demand for coating materials and are a separate segment in technological and marketing terms, though not from the point of view of their composition.

The size of the European market in 2011 was 9.4 million tonnes. It has undergone a slight shift towards industrial coatings and printing inks compared with the sectoral division in the rest of the world. Germany leads with a consumption of approx. 1.5 million tonnes, ahead of Italy, France, the UK and Spain, which are all in the range between 0.8 and 1.0 million tonnes. The North American market (NAFTA) had a volume of 6.6 million tonnes in 2010 and is served by about 800 companies.

Figure 1.1.5 gives an overview of the economic development of the paint and coatings industry in Germany since 2006 which is closely linked to the country's overall economic development.

One of the characteristic features of coatings technology in addition to coatings consumption is the ongoing high energy consumption for processing coatings which is estimated at approx. 200 billion kWh annually worldwide. This figure is the equivalent of the energy content of approx. 30 million tonnes of crude oil. If the raw materials required for paint production are also added to this figure in the form of crude oil equivalents, the result is a total crude oil requirement of approx. 120 million tonnes for the global manufacture and processing of coating materials, or some 3 % of global annual crude oil extraction of approx. 4 billion tonnes in 2012.

The legal requirements imposed on environmentally compatible coating processes have led to the greater use of appropriate coating materials over the last 20 years. These include, in particular, solvent-free powder coatings, water-borne paints, in which organic solvents are replaced in whole or in part by water, high solid paints and radiation-curable paints, which are processed either in aqueous solution or completely without conventional solvents but with the aid of low-molecular reactive thinners. Statistics from the Association of the German Paint Industry (VdL) show that these coatings have achieved the greatest growth over the last 10 years.

As a result of stricter legislation and ongoing improvements in our knowledge of their toxicology, the raw materials are having to be replaced regularly, with all the attendant development costs, if the quality standard achieved is to be maintained.

As far as energy consumption is concerned, there is still a need to be more economical in the use of raw materials and energy. A proportion of the material is lost en route to the finished coating. Spray application, which is specified for many objects to be painted because of its optical attractiveness and range of colours, is prominent in this regard. As far as coating of wood and plastics is concerned, the more effective electrostatic spray methods have not yet found universal acceptance. In addition, paint lines lose substantial quantities of heat energy. In recent years a number of developments have increased the efficiency of coating processes to such an extent that growth in the paint and coatings market in the industrialized countries has only just been below the growth level of the gross domestic product (GDP).

As a result of the use of solvents as an application aid for coatings with an average organic solvent content of 50 %, it has been estimated that additional hydrocarbon emissions of approx. 200,000 tonnes occurred in Germany alone in 2007 [1.3.2].

Whereas organic emissions from motor vehicles have been successfully reduced to less than 1/3 of their previous level in the last 30 years by the introduction of the catalytic converter, successes in coating technology have been much more modest so far by comparison. Despite the use of water-borne paints and powder coatings, including by small and medium-sized enterprises after the enactment of the EU VOC (volatile organic compounds) directive of 1999 and its implementation in Germany under 31 BImSchG (BundesImmissionsSchutzGesetz = Federal Immission Control Act), the German Ministry of the Environment estimates that 30 % of all solvent emissions in 2020 will stem from the paints and coatings industry [1.3.3].

Manufacturers and processors of paints and coatings have substantial gains in terms of occupational health and safety. For some time, the chemical industry, for example, has led the accident statistics in Germany for the industry in having the least number of incidents. Health risks posed by paints and coatings are being kept as low as possible

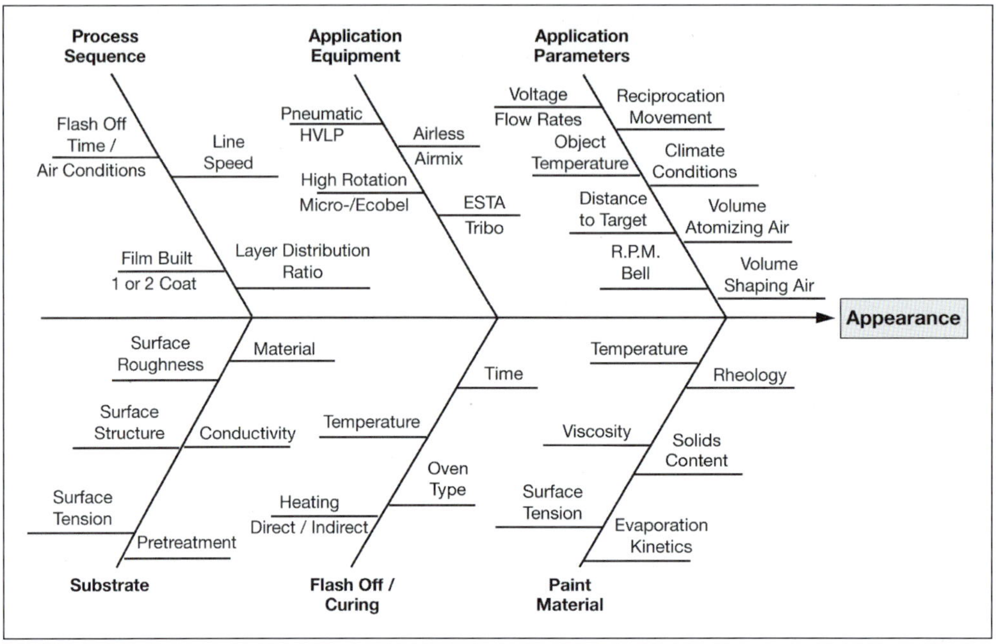

Figure 1.1.7: Various factors determining the final coating result of the spray application process

through REACH [1.3.4] and the Biocides Ordinance [1.3.5] in Europe and through the Clean Air Act (CAA) and the Toxic Substances Control Act (TSCA), legislation and civic engagement in North America, and these have brought about substantial changes in coatings technologies and formulations.

An analysis of this situation reveals that manufacturers and consumers of paints and coatings, though occupying different value added stages, are highly interconnected. Manufacturers develop and supply coatings materials to the consumers who, by applying the products, induce physical, physico-chemical and chemical processes that convert them into an adhering, mechanically solid and, at the same time, flexible coating.

The chemical path by means of which the raw material becomes a finished coating starts at the raw materials or paint manufacturer and is then deliberately interrupted before being taken up again during application by the paint consumer. Although the performance profile of a coating is initially shaped by the paint and thus by the paint manufacturer, it is the processor who actually generates the finished properties. The industrial scale coating of consumer goods is therefore a joint effort between paint and coatings manufacturers and processors.

Paint manufacturers who really know their job not only are responsible nowadays for developing, manufacturing and selling paint. Their task also includes creating the conditions for successful painting through the provision of a permanent technical presence and support. This relates primarily to materials and processes, but also extends to detailed environmental protection and occupational safety issues. Paint manufacturers offer a package, as it were, in which the material is just one component among many (see Figure 1.1.7). A consequence of this in past years has been that the responsibility for producing a coating outcome in the required quality has increasingly been transferred into the hands of the paint supplier (see Chapter 7.1.7).

Apart from the technical tasks of manufacturing and processing coating materials, particular attention needs to be paid to quality assurance methods. Quality assurance provides the link between production and R&D and sales within a company. Production must be capable of reproducibly providing the quality demanded by the customer, while sales must identify the total costs in order that appropriate prices may be obtained.

However, paint manufacturers face specific problems because they are expected to produce material of constant quality and, at the same time, paints of constant processability. This is the only way to create the best conditions for achieving a uniform result in the painted article. This means that there is more to the production of paint and coatings than merely manufacturing a product whose composition is identical with a defined standard. Rather, since physical variables can only rarely serve as criteria for the practical properties of coatings, paint testing of necessity must include simulating the application method used by the processor of these materials. This gives rise to a large number of different test methods because of the very wide range of specification conditions and the different requirements on the coating process resulting from them. Standardising these tests and reducing their overall number are also a priority task for all concerned.

Quality and costs of a coating are defined, as mentioned earlier, not only by the paint material and an appropriate application method but also and significantly by the substrate, i.e. by the material and the design of the object to be painted. It is therefore clear that it is extremely important to address surface treatment and the selection and design properties of the material during the product design stage and to incorporate these features into the overall planning.

Coating technology is therefore a complex marriage of chemical, physical, process-engineering, environmental, toxic and economic variables. This discipline is in a constant state of change as a result of technical progress and is further accelerated by legislative requirements. It is therefore incumbent on companies to link well known features with new knowledge. Industrial coating technology can only be fully understood if, in addition to detailed knowledge of paint processing, the properties of the coating material and of the object to be coated are known and also if all the quality-shaping variables within the range of economic and environmental requirements are addressed.

1.2 Retrospective

The aim of the following retrospective of materials and painting methods in the past is to highlight the entangled paths which coating technology has taken down to the present day by illustrating a few key events that will make its progress more transparent.

The early coating materials were natural resins . *Dioscorides* and *Plinius* report, among other things, on the use of countless exotic natural resins from the time of the Ancient Greeks and Romans. Later papers describe the importance of colophony, copals, shellac and dammar. Later still, in the 12th century, come reports of the combination of hardened natural resins such as amber with resinifying, i.e. chemically hardening, natural oils [1.3.6].

The application methods up to that point consisted solely of brushing or wiping techniques, without the generation of significant emissions, waste water or paint waste.

Rodgerus von Helmarshausen, also known as *Theophilus*, describes how to make coatings and provides a detailed recipe In chapter 21 of his book *Schedula Diversarium Artium*, which dates from around 1000 AD:

1* Put **linseed oil** into a small pot and add the **gum** which is called **fornis** ...

1* Place together four stones ... place a common pot upon them and put into it the above mentioned gum fornis", which in Latin is called **"glassa"** ...

2* Then place fire carefully underneath until this gum liquefy.

3* Have also a third pot nigh, placed upon the coals, in which is hot **linseed oil** ...

4* Pour the warm oil into it and stir it with the iron ...

5* And take care in this, that in weight there are **two parts of oil** and the third part of gum...

Figure 1.2.1: First recipe of a coatings material by Theophilius

Around the end of the first millennium, the monk *Rodgerus of Helmarshausen* reports in his book *Schedula Diversarium Artium* the first details of the composition and manufacture of the then standard paint [1.3.7]. He is therefore, regarded as the creator of the first binding paint formulation and of the detailed instructions on how to make it.

Linseed oil and amber (called "rubber" in German back then), are boiled together in a ratio of 2:1, with the hardened resin acting as a nonvolatile film former, and the linseed oil as a chemically crosslinking reactive thinner. Solvents to regulate viscosity were not used at that time because they were not available in sufficient quantities.

In the 10th century it was the Arabs in the guise of the doctor *Abu Mansur* who taught the Europeans the art of distillation [1.3.8]. This art was then used in Europe for, among other things, extracting turpentine oil. When the *van Eyck brothers* took *Rodgerus of Helmarshausen's* formulation, which was the first to be systematically described, and extended it by adding turpentine oil in the early 15th century, physical drying and thus emissions had been invented [1.3.9]. This outcome is noteworthy on one hand because it greatly expanded the use of painting and on the other because the associated environmental problems are still tying up a considerable part of paint-specific research capacity to this day.

The memorable "invention" of emissions was followed very much later by the "invention" of waste. By this is meant the introduction of the continuous production line by *Henry Ford* in the early 20th century and consequently the start of industrial-scale painting technology.

The production line has brought the world huge benefits in terms of speedy and economical production systems. However, the necessary faster coating processes could only be fully exploited by employing quick-drying paints and a new application method. Consequently, the oil paints which took several weeks to cure were superseded by quick-drying cellulose paints with solvent contents of up to 80 %. At the same time the waste-free brush method which had been standard up to that point was replaced by the spray method. However, the introduction of the spray gun, which had been invented by *de Vilbiss* back in the 19th century, also ushered in a new problem because it had an extremely poor material application efficiency of less than 50 %: the generation of paint waste and, at the same time, a dramatic increase in organic emissions because of the considerably higher solvent contents of the cellulose paints.

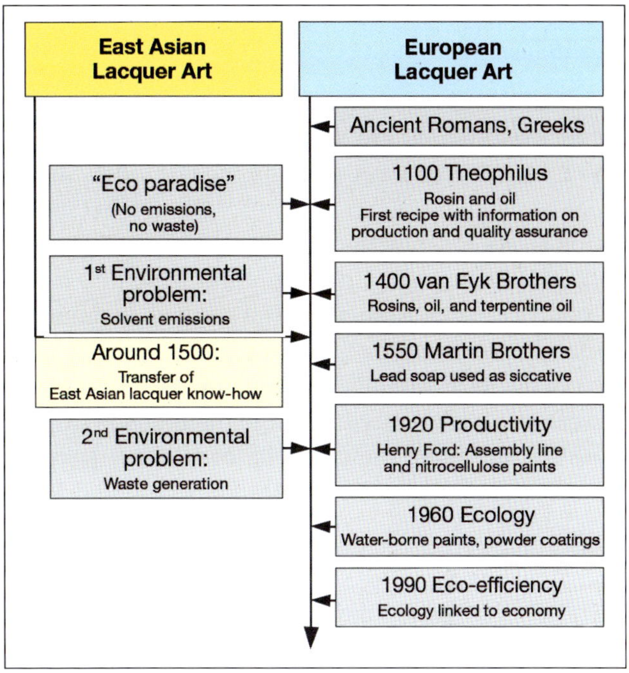

Figure 1.2.2: Historical highlights in the evolution of coating technology

So, 1920 introduced a completely new situation into coating technology, but failed to provide an appropriate approach for the new problem. This approach had to be developed in the course of the following decades. The production line meant technical progress in the form of increased coating quality as a result of improved application and a more economical production system, though at the expense of the environment in the form of waste and emissions.

The introduction of the production line therefore shared the problems of emissions and waste which had been known since the time of the *van Eyck brothers* and also signalled the start of the metamorphosis of craftsmanship into a segment of industrial production technology.

It took government action in the form of environmental legislation for people finally to become aware of this situation and for themselves to take appropriate concerted action. The innovations which subsequently occurred were no longer the outcome of individual ideas but were increasingly achieved thanks to cooperation among all those involved in the finished coating.

The first goal was to rationalise the processes and optimise quality on the basis of conventional, i.e. solvent-borne, coating technology. Later, from the 1960s onwards, research started in a new direction from several different locations simultaneously, initially separated but later with a common goal in view: the development of low-solvent and low-waste painting methods [1.3.9].

The results of this work are new water-borne materials, radiation-curable systems and waste-free powder coatings in conjunction with new processing methods. Electrocoat painting, roller and curtain coating, e.g. for coil coating, and the application of solid powder coatings are the most noteworthy innovations of the last 50 years. Environmental protection on one hand and occupational safety issues on the other are beginning to become much more structured. With the passage of Germany's *Bundesimissionsschutzgesetz* (Federal Immission Control Act) in March 1974 and the oil crises, research into environmentally compatible coating technologies was expanded to include the aspect of resource conservation.

The various trade associations and technical bodies of the American and European paint and coatings industries were also involved in this. Efforts in this direction were strengthened by the voluntary commitment ("responsible care") of the paint and coatings industry in almost every region of the world.

Successful examples of such a concerted approach to dealing with a problem by a number of different groups and bodies include the development of the new paint process for the Mercedes A-Class in the late 1990s in which a paint manufacturer, a paint processor and a plant manufacturer cooperated, and also the "Expansion of the application fields for powder coating technology" research project funded by Germany's Federal Ministry for Education and Research (i.e. BMBF) and coordinated by the German Research Association for Surface Treatment (i.e. DFO) together with a total of 20 companies from the fields of paint production, paint processing, plant engineering, the raw materials industry and countless research institutes. The same approach can be seen in North America, where the industry-supported LEPC (Low Emission Paint Consortium) under the umbrella of USCAR (United States Council for Automotive Research), invested a great deal of money in the 1990s to develop low-emissions paint technologies for automotive coating [1.3.10].

The globalization of the world economy, particularly the manufacturing industry, is currently confronting the paint and coatings industry with new challenges. Since quality requirements have been standardised worldwide as a result of consolidation, in the automotive industry for example, the major paint suppliers have positioned themselves globally and now guarantee the same quality standard for coatings everywhere. No end to this development is presently in sight. In addition, sales channels have expanded for many paint suppliers through the introduction of e-commerce, the electronic transaction of purchases over the internet. Of necessity, all aspects are steps in the direction of consolidation.

The quality requirements, particularly those imposed by the automotive industry in the form of ISO standards, VDA standards and the QS9000 series, mean that the entire process from the product idea, via development, scale-up, production and sale to the final application is falling more and more completely within the sphere of responsibility of paint and coatings companies. It now seems that the entire process chain, starting with the production of the paint materials to the finished (coated) article, will become a key competency of the major paint companies. It is therefore certain that the active players in this field will continue to grow closer together. Only in this way will it be possible to develop efficient processes for the respective sectors or applications.

The conditions for coating technologies that have zero environmental impact and make the most efficient use of resources are largely in place. Future success will depend on whether holistic concepts can be found [1.3.11]. Product life-cycle or ecoefficiency audits [1.3.12] aimed at achieving sustainable development will therefore provide the principles underpinning research and development programmes more and more frequently [1.3.13] (see Chapter 5.5).

In-depth knowledge is an essential factor for the successful introduction of environmentally compatible coating technologies in conjunction with the implementation of new ideas in processing technologies. This includes details on correlations and interdependencies between the individual scientific disciplines and thus of the various sectors involved in the paint process. This book has been written not least for this reason.

1.3 References

1.3.1 A. Mehta, PPCJ 194 (4481), p. 20 (2004)
1.3.2 T. May, JOT 12/2000, p. 34; N. J. Sarginson, ECJ 11/2000, p. 27
1.3.3 H. Schrübbers, JOT 48 (2), 30 (2008)
1.3.4 T. May, MO 58 (4), p. 44 (2004), M. Carrol, PPCJ 202 (4570), p. 29 (2012)
1.3.5 D. Leroy, ECJ 04/2006, p. 16
1.3.6 Kittel, Lehrbuch der Lacke und Beschichtungen, Vol. 1 (2nd Edition.), p. 25ff, Hirzel Verlag, Stuttgart, 1998
1.3.7 Rodgerus von Helmershausen, schedula diversarium artium, Hrsg. V. Albert Ilg, Verlag W. Braumüller, Wien 1874
1.3.8 O. Krätz, 7000 Jahre Chemie, Nikol p. 15ff, Verlagsgesellschaft Hamburg 1990
1.3.9 Die Lack Story, 100 Jahre Farbigkeit zwischen Schutz, Schönheit und Umwelt, Verband der Lackindustrie aus Anlass seines 100-jährigen Bestehens
1.3.10 T. Meschievitz, Y. Rahangdale, P. Pearson, Proceedings XVII Surcar, Cannes, 1995
1.3.11 P. Svejda, Coat.World 16 (6), p. 34 (2011)
1.3.12 D. Engel, T. May, R.R. Matheson, P.Q. Nguyen, ECJ 01-02/2004, p. 40
1.3.13 G. Kirchner, C. Kraft, ECJ 07-08/2011, p. 22

2 Coating Materials

In accordance with the international standard ISO 4618-2 as well ISO 4618-3 and the supplementary German standard DIN 55 945, and in general agreement with the English speaking experts, the correct designation for materials for organic coat-ings that provide surface protection is coating materials. Paints, on the other hand, is the standard term for coating materials that possess particular properties. In the English speaking world, the term "coating" is often synonymous with paint, the layer on the coated product, and the painting process.

The standard defines paints as liquid, paste or powder coating materials which enable optically opaque coatings to be produced that have decorative, protective and, if necessary, specific technical properties. It should be noted that many terms in general use, such as varnish and vehicle, are different from the terms specified in international standards (see Chapter 8). Paints are therefore just one part of the large class of coating materials, but nevertheless have to meet a broad spectrum of extremely different requirements. They can be sub-divided in accordance with various principles: by the function of the coating (clearcoat, metallic paint, solid paint), by the particular layer in the coating system (primer, primer surfacer, topcoat), by the purpose of the paint (car paint, decorative paint, industrial paint), by the degree of environmental compatibility (water-borne paint, high solid paint, radiation-curable system, powder coating), by the chemistry of film forming agent (alkyd resin paint, acrylic resin paint, cellulose nitrate lacquer) and by the processing conditions (stoving enamel, oxidatively curing coating material), to name the major classification criteria.

The selection of the right coating materials the optimum performance profile therefore depends equally on the quality requirements, the specified application conditions, the curing process, the design features and the various constituent materials of the object to be painted. All coating materials, however varied their use may be, must have the same properties of wetting substrates, transforming into a closed film, flowing and then solidifying so that the desired mechanical and chemical protection of the object is achieved.

The performance profile of the paints therefore has to be tailored not only to application but also to subsequent film formation. A fresh paint finish applied by spray gun must exhibit sufficient flowability for the droplets to seal all pores. At the same time or subsequently, it must level out all the peaks and troughs in the film surface that are caused by application. Surface tension and viscosity are the most important paint parameters in this regard. Furthermore, special effects such as high-gloss finish, chroma and oriented reflection (metallic effect) may have to be achieved. Clearcoats for wooden panelling, for example, therefore have a completely different composition from anticorrosive paints and from coating materials for applying colour designs to leather.

2.1 Raw Materials

Resins in their capacity as film forming agents (often incorrectly termed binders), along with additives, solvents, pigments and extenders are the usual ingredients of liquid coating materials. Binders, according to ISO 4618, are the pigment-free and extender-free

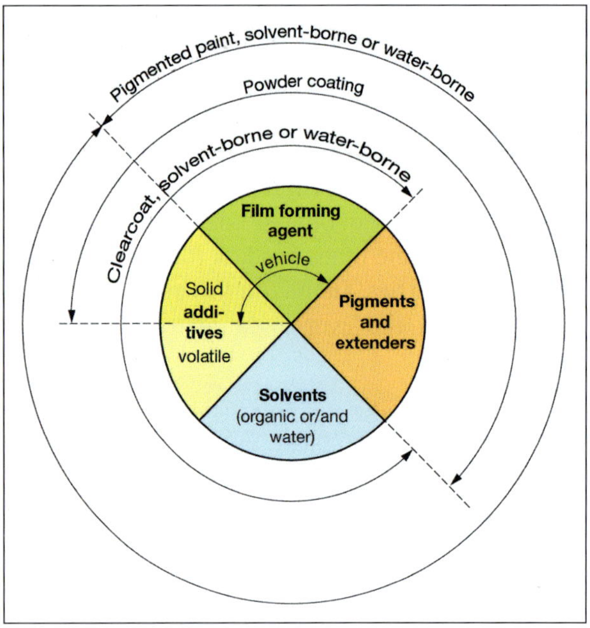

Figure 2.1.1: Typical composition of coating materials

fractions of the dried or cured coating. They are, therefore, composed of the film forming agent and the nonvolatile fraction of the additives.

If there are no pigments and extenders, the material is a clearcoat, and if there are no solvents when higher molecular film forming agents are used, the coating material can be ground to produce a powder coating. If the film forming agents are lower molecular liquid products, it is also possible to eliminate some or all of the organic solvent. Where the resins or dispersions are miscible with water, the solvents can be replaced by water.

Irrespective of the intended application, a **film forming agent** which solidifies as a result of physical or chemical processes is the one indispensable component of a coating material. Traditional film forming agents are exclusively oligomeric to polymeric organic materials because of the range of requirements imposed on adhesion to the substrate, mechanical strength combined with elasticity, and resistance to ambient effects. They can be manufactured from natural substances which have been chemically modified or by industrial synthesis.

Despite the existence of a broad range of synthetic resins, the natural substances or modified natural substances have lost little of their importance. Cyclorubber or chlorinated rubber, cellulose esters and alkyd resins based on natural fatty acids are just a few examples of the technically important group of renewable paint raw materials. A foreseeable shortage of raw materials derived from crude oil has spurred activity aimed at finding a new basis for synthetic chemical raw materials and intermediates (see Chapter 5.6.2) [2.4.1].

Film forming agents can be divided into two basic groups, irrespective of their origin. The first group consists of the higher molecular, physically curing film forming agents such as cellulose nitrate, cellulose acetobutyrate, thermoplastic acrylates and also PVC copolymers. These resins set by releasing solvents, but do not undergo a chemical change during film formation.

The second group comprises chemically reactive film forming agents or combinations thereof and are used for high-grade decorative and industrial paints. A crosslinking reaction initiated by heat or catalysts after application enables lower molecular and thus high solid base resins to be used. As an example, consider stoving enamels based on acrylic-amino resin, alkyd-amino resin or phenolic resin combinations. Oxidatively curing alkyd resins (incorrectly termed "air drying") start to crosslink at room temperature, though much more slowly.

Modern polyurethane and epoxy resin paints cure very quickly, even at room temperature. The latter can generally only be used in the form of 2-component materials because of their high chemical reactivity.

A coating material formulation usually contains **solvents** in addition to a film forming agent. Trouble-free processing and film formation are possible only if the right solvents are selected and combined. The most common organic solvents are aromatic hydrocarbons such as xylene and solvent naphtha, aliphatic hydrocarbons such as mineral spirit or vernice, esters such as ethyl acetate or butyl acetate, alcohols such as propanol or butanol, ethers such as butyl glycol, ether esters such as ethyl and butyl glycol acetate, and various ketones. It is important that the combination of solvents chosen from the huge range available matches the resin type.

It is critical, for the sake of the application process, to choose solvent combinations that have the correct evaporation properties. This is the only way to strike a balance between the levelling and the non-sag properties of fresh paint on vertical surfaces.

Traditionally, solvents have served merely as viscosity regulators, but they can also react chemically with the paint resins during film formation, i.e. they can do more than function as thinner. Such reactive thinners remain in the solidifying film and thus contribute to a reduction in organic emissions. Paints which can be cured by means of UV radiation or electron beams also deserve to be mentioned in this context, as do unsaturated polyesters cured by organic peroxides.

The **pigments** used in coating materials provide chroma and hiding power. Extenders influence technical properties, such as sandability and hardness, but also increase the nonvolatile content of the coating material. In many cases, by absorbing UV radiation, pigments and extenders prevent premature degradation both of the substrate to be protected by painting and of the polymeric protective coating itself. A further important task for certain special pigments and extenders is to optimise protection against corrosion.

The various types of pigment are divided into organic and inorganic absorption pigments, metallic and interference pigments, and anticorrosive pigments. The first two groups consist of extremely finely dispersed, generally crystalline solids. The most important among the inorganic pigments are titanium and iron oxides and the mixed-phase pigments. Organic pigments vary hugely in their composition. Modern, synthetically manufactured organic pigments also offer excellent lightfastness and long-term outdoor exposure properties. Their particular advantage lies in their brilliance and high tinting strength. At elevated temperatures, however, and with certain exceptions, they are inferior to the inorganic pigments in terms of heat stability.

Special optical effects are achieved by means of pearlescent, interference or metallic pigments. Coatings whose colour or brightness varies with the viewing angle are of particular interest in the automotive sector because of their attractiveness.

The final coating material components to be mentioned are the **additives** and **plasticisers**. With the exception of the plasticisers, these substances are added to the resin/solvent or pigment mixture in sometimes minute quantities. Despite this low concentration, they have a significant impact on the properties of coating materials and coatings.

Additives facilitate dispersion of the pigments during production. In the ready-to-use paint they suppress the tendency for the pigments to settle, enhance flowability when the paint is applied, improve levelling during film formation and improve technical properties by influencing the surface smoothness or roughness. They prevent the pigments from floating to the surface, accelerate curing, while also exerting an influence on the gloss, increasing resistance to harmful UV radiation and reducing premature degradation by bacteria or mildew which attack paint.

Whereas in the past raw materials were mostly chosen on the basis of their chemical and physico-chemical properties, recent insights have shifted the focus increasingly onto particle morphology, dimensions and the associated interaction with other surfaces. Particles or surface structures with dimensions on the nanometre scale exhibit interesting structure-related properties. The behaviour of such nanoparticles can differ quite substantially from that of individual molecules as well as that of particles of larger dimensions. This brings an additional variable into play for many technical areas, including the composition and formulation of coating materials.

The size, shape and surface structure of a wide range of established and new raw materials are therefore being adjusted for the purpose of trimming the size-dependent properties to the optimum extent for the intended application. For instance, it is possible to modify the dimensions of TiO_2 and SiO_2 particles such that they lose the ability to scatter light yet retain their UV stability and immobility in the coating. As a result, the use of micronized, i.e. nanoscale, TiO_2 to formulate clearcoats makes it possible to achieve stable UV absorption without adversely affecting permeability to visible light.

Numerous other applications of nanoparticles and nanostructured surfaces have opened up new problem-solving approaches in paint research (see Chapter 2.1.5).

2.1.1 Film Forming Agents

Whereas it is possible for a coating material to dispense with the raw materials such as solvents, additives or pigments mentioned in the first section, a film forming agent or a combination of such agents is absolutely essential if the specified tasks as a paint are to be met. The film forming agent must meet numerous criteria. These essentially concern adhesion to the substrate and the film's mechanical strength (cohesion) in conjunction with retention of elasticity. Such a range of requirements makes it necessary to use macromolecular polymeric materials. Their molecular structure enables them to meet many requirements through mechanical entanglement of the molecular chains. It is just as feasible to achieve these properties with synthetically manufactured oligomeric or polymeric materials as with appropriate natural polymers, although the last of these must be chemically modified if they are to find practical application. However, optimum coating properties can only be obtained if the film forming agent can wet the substrate as a resin solution, as a dispersion or as a molten powder coating and can be transformed into a visually attractive, smooth film by means of subsequent levelling.

The mechanical strength can therefore be developed through the release of solvents (physical drying) or through molecular enlargement with simultaneous crosslinking of the molecules of the film forming agent (chemical hardening) .

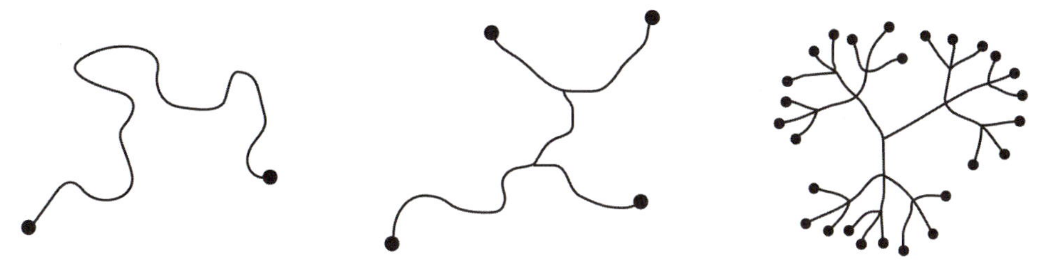

Figure 2.1.2: Various forms of film forming molecules

All film forming agents which solidify by physical drying, i.e. purely by releasing solvents, therefore remain sensitive to solvents and can be liquefied again by heat treatment. These film forming agents are thermoplastic. Consequently, coatings made with them are also thermoplastic.

Chemically reactive film forming agents, on the other hand, perform differently. These are chemically crosslinked with reactive groups of the resin molecules during film formation and are thereby transformed into an insoluble thermosetting film which cannot be reliquefied.

Unlike purely physical drying, this supplementary chemical hardening also permits the use of low molecular film forming agents. This means that the amount of solvents which are needed for processing can be substantially reduced in the chemically reactive systems. Chemically reactive systems therefore help to lower solvent emissions, unlike thermoplastic materials.

The varied shapes and sizes of the film forming agent molecules (see Figure 2.1.2), and of those functional groups at the points of linkage within the polymer chains and of those at the end of the molecular chains, determine the film forming agents' properties and thus the application for which they can be used.

Linearly structured molecules behave differently from branched ones in many respects. The large number of chain ends in dendrimers (tree-like polymers) affords much higher functionality despite the low molecular weight compared with the former, thereby providing the necessary conditions for high crosslinking density.

The size or size distribution of these resin molecules has a particular impact on the processing properties, though also, of course, on the quality of the end product. This applies as much to the processing state as to the structure of the molecular bond in the finished coating. In thermoplastic materials, i.e. where nonreactive film forming agents are used, the molecule size remains unchanged even after processing. Chemically reactive coating materials, on the other hand, undergo considerable changes in molecule size as a result of chemical crosslinking. When paint resins and their usability are being categorised, therefore, the molecular weight and molecular weight distribution serve as important indicators of the molecule size.

Polymer chemists have developed a number of methods for determining these indicators. Resin solution viscosity (viscometry), diffusability and the osmotic pressure (osmometry) which occurs when semi-permeable membranes are employed act as indicators of molecular weight. More up to date and informative are methods involving chromatographic separation of the individual film forming agent molecules which can vary considerably in size. With the appropriate equipment, different dwell times in the cavities of polystyrene gels can be used to determine average molecular weights and the molecular weight distribution concurrently (see Chapter 2.3.2). The molecular weight distribution is particularly important because the use properties cannot be specified by means of the average molecular weight alone.

Reactive end groups and the type of chemical link between the basic structural elements also help to shape properties. Information on the crosslinking capabilities of the resins and their polarity can be obtained by experimentally determining the number of free carboxyl, hydroxyl or isocyanate groups in the form of the acid, hydroxyl or isocyanate values. Similarly, the iodine absorption value provides information on the presence of carbon double bonds which in turn can be used to describe the oxidative curing of alkyd resins (see Chapter 2.3.2).

To understand the nature of the film forming agents' basic structural elements, wet chemical degradation reactions must first be conducted. Only then can the functional groups created be identified. This can be done either by determining the aforementioned resin indicators or by identifying and quantifying functional groups or characteristic molecular moieties with the aid of analytical instruments. These are divided into spectroscopic, thermal and chromatographic techniques (see Chapter 2.3.2).

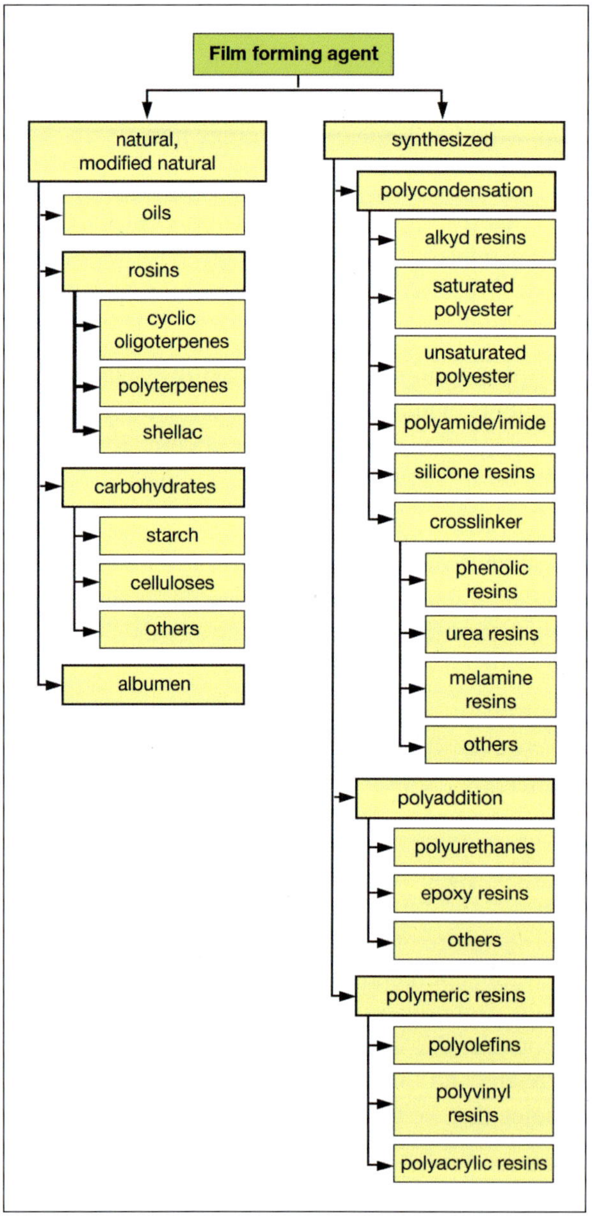

Figure 2.1.3: Classification of important film forming agents

Film forming agents, which have proved their value over thousands of years for protecting and decorating objects, are extremely varied and belong to a huge range of classes of chemical substances. A number of different approaches can be used to classify them systematically. One is to classify them by the chemical principle of the molecular structure; another consists in dividing them into film forming agents which either remain thermoplastic during film formation or are transformed into thermosets. A further meaningful classification method is to distinguish between natural materials, modified natural materials and materials which are obtained synthetically. As division into natural and synthetic products also reflects the history of paint, this chapter will also be structured in this way.

Accordingly, natural oils which have been important for thousands of years should be mentioned first. They were combined at an early stage with hard resins such as shellac to increase the initial hardness because solidification was a slow, purely chemical process. Other important representatives of such hard resins are colophony, copal, dammar and amber. They belong to the large group of natural isoprene derivatives. Latex milk which is produced from certain plants is also an isoprene derivative in the form of 1,4-cis-polyisoprene. It was used centuries ago as natural rubber to make water-repellent coatings. Later modifications to produce cyclorubber and chlorinated rubber gave access to those film forming agents which are still in use today for heavy-duty anticorrosive protection and the protection of damp rooms because of their hydrophobic properties.

Carbohydrates such as cellulose and starch which are produced in nature to excess by photosynthesis proved to be useful raw materials for paints in the form of organic or inorganic esters and ethers more than 100 years ago. Even in the 21st century, cellulose nitrate and cellulose acetobutyrate are still technically important base resins for the woodworking and furniture sectors on one hand and the automotive sector on the other.

Overall, with the invention or synthesis of artificial resins, natural materials have lost their pre-eminence because of their restricted availability and the limited adaptability of material and processing methods.

The economic breakthrough for synthetic resins came with the invention of phenolic resins and combinations of amino resins of enhanced yellowing resistance with alkyd resins and saturated polyesters. The first synthetic resins also included the unsaturated

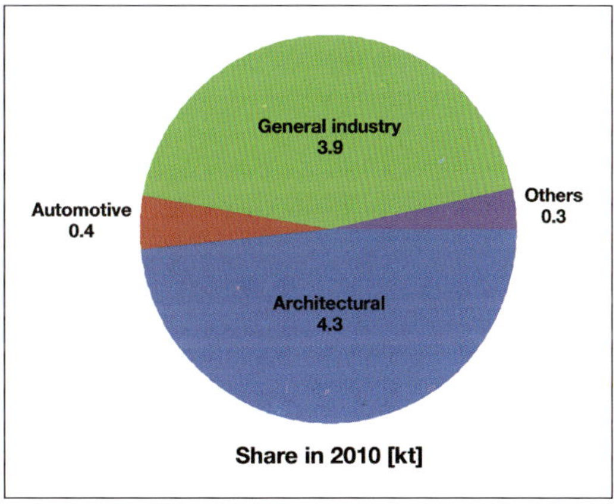

Figure 2.1.4: Global resins market for coating materials in 2010 *Source: authors' own estimate*

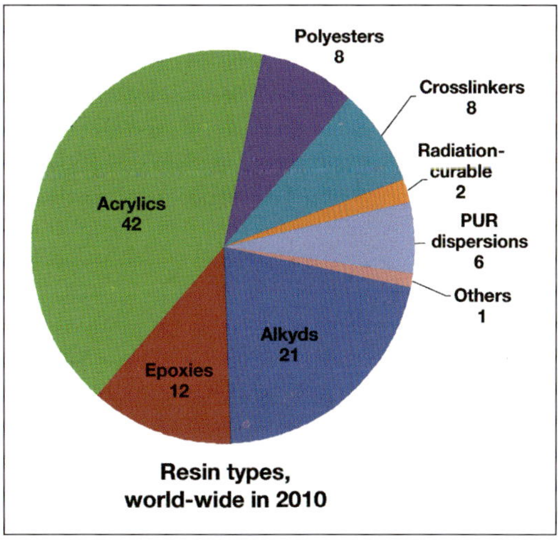

Figure 2.1.5: Film forming agents' share of the overall market in 2010 *Source: authors' own estimate*

polyesters which, when combined with chemically reactive solvents, enabled paints to be developed that offered a combination of high film thickness and low emissions because of the chemical bond that formed between the film forming agent and the solvent. A degree of long-term outdoor weatherability and chemical resistance which had not been possible previously was made possible by the invention of polyurethane and epoxy resins. Thanks to the resistance conferred by silicone resins to heat and long-term outdoor exposure, coating materials can be designed that feature correspondingly excellent properties.

Polymer resins such as polyacrylics, polyvinyl chloride, polyvinyl esters and ethers, polystyrene and copolymers are now used in all kinds of paint systems, e.g. as aqueous dispersions, radiation curable systems and powder coatings (see below).

Film forming agents are by far the largest component in a coatings formulation, in terms of quantity. In 2010 they had a global market volume of around 8.9 million tonnes. In first place came emulsion polymers, including acrylic resins, followed by alkyd resins and then epoxies. Polyesters, the mostly water-borne polyurethane dispersions and the crosslinkers for the various resin types have roughly the same market share.

The film forming agent, in being responsible for coating functionality, stability and application properties, performs roughly 70 % of the tasks of the various paint components found in all coating materials.

2.1.1.1 Natural Materials

By far the most important natural materials used as film forming agents are the **natural oils** which are also known as fatty oils. Being tri-esters of glycerine with polyunsaturated fatty acids, they became extremely important as film forming agents for surface protection a considerable time ago because of their ability to transform from a low viscosity liquid into a solid by interaction with atmospheric oxygen. This capability distinguishes them from fats based on saturated fatty acid esters. Fatty oils are pure natural materials and are available in sufficient quantity as inexpensive raw materials for paints.

Even though natural oils have increasingly diminished in importance with the development of synthetic film forming agents, their fatty acid components are still used by raw materials suppliers for modern alkyd resins.

Numerous types of **fatty oils** are produced naturally by various plants and certain marine animals. If they are to be used successfully in technical applications, they need to be isolated from seeds and fruits by cold- or warm-pressing and to be freed of impurities. These are mainly carbohydrates, proteins, and lecithin, which is of interest as a paint additive. Due to its surfactant character, lecithin is used to improve the dispersibility of coloured pigments in stoving enamels. It has an inhibiting effect on the oxidative crosslinking of fatty oils and so must be quantitatively removed from them.

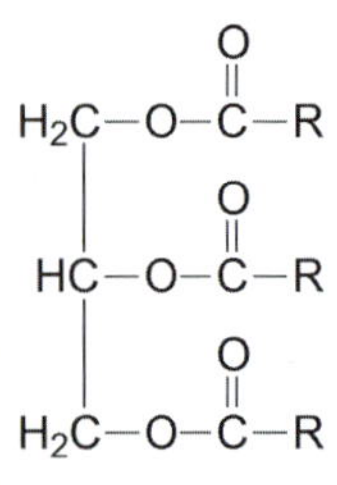

R = alkyl chain of fatty acids (see Figure 2.1.7)

Figure 2.1.6: Fatty oils as triglycerides of unsaturated fatty acids

The huge diversity of oils which can be used in paint production, without exception esters of glycerine, stems from the

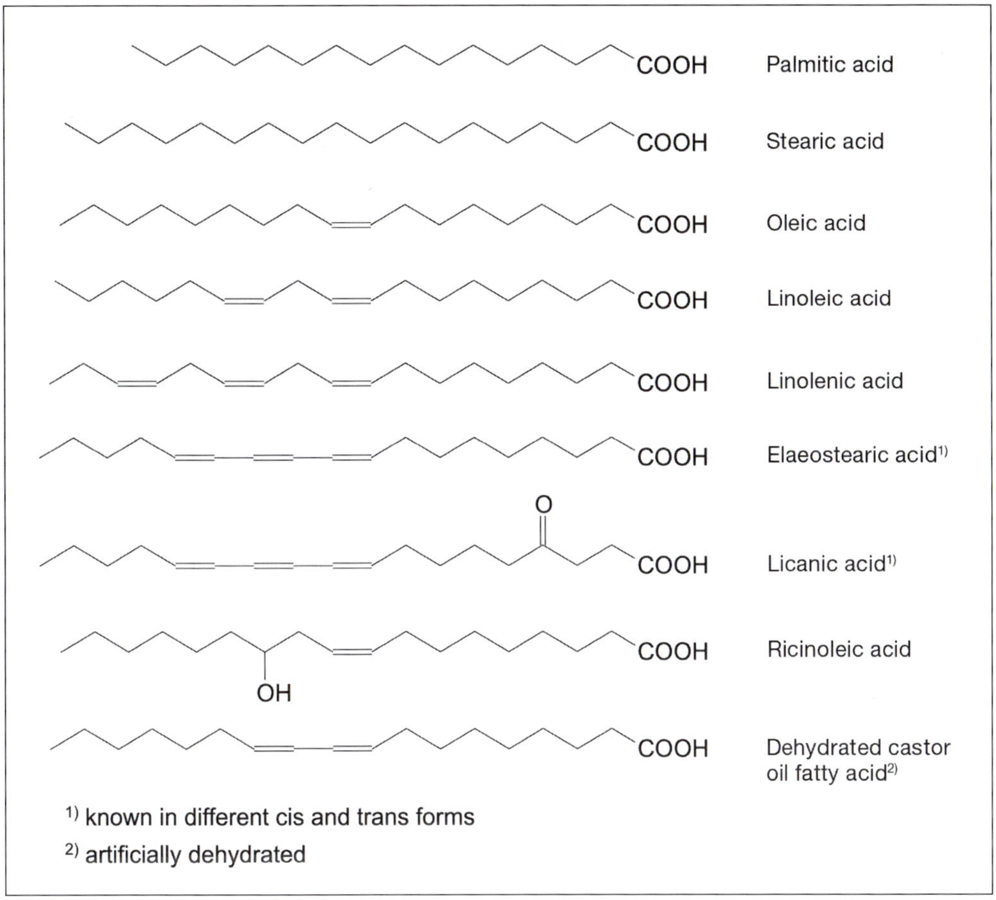

Figure 2.1.7: Important natural fatty acids as the constituents of fatty oils

wide availability of **fatty acids**, the acidic esterification component. Nature produces these in varying chain lengths and, at the same time, a fluctuating number of double bonds and different cis/trans forms.

The fatty acids shown in Figure 2.1.7, i.e. stearic and palmitic acid, cannot be cured by oxygen. Their analogue with a double bond in the cis position is oleic acid. Its ability to be crosslinked by oxidation is poor, and so it cannot be used by itself in practice as a paint raw material. The C_{18} fatty acids containing two or three double bonds are much more effective in terms of paint curing by means of atmospheric oxygen. In the former group, the most important role is taken by the 9,12- and 9,11-octadecanoic acids. Unlike the 9,12- variant, linoleic acid, 9,11-octadecanoic acid does not occur naturally. It is obtained by artificial transformation (isomerisation) of natural linoleic acid or by dehydration from castor oil or ricinoleic acid. Both double bonds occur proportionately in conjugated form and consequently confer increased reactivity on the film forming agents into which they are chemically incorporated.

Fatty acids containing three isolated double bonds are even more reactive than those with two double bonds. Linoleic acid, which has three isolated double bonds and is the main component of linseed oil, gives the oil an extremely pronounced oxidative curing capability which has been used since time immemorial. Elaeostearic acid, which is contained in tung oil and also has three double bonds, albeit in a conjugated configuration, is among the most reactive of all fatty acids. Unfortunately, the highly unsaturated oils are less resistant to yellowing and therefore unsuitable for white coatings.

Among the large group of oils which are of importance in paint and coating technology, special mention should be made of linseed oil, soy oil, tung oil, oiticica oil, groundnut oil, sunflower oil, cottonseed oil, castor oil and fish oil.

It is also worth noting that tall oil, an important raw material for synthetic paint resins, is not actually an oil. It is not, as would be anticipated from its name, a triester of glycerine, but rather is a mixture of various fatty acids formed during cellulose production by the alkaline saponification of oil-containing byproducts.

A proven indicator for the oxidative hardening of oils is the **iodine value**. This indicates the quantity of iodine which is added to the double bonds for every 100 g substance. An iodine value > 150 is a pointer to rapidly hardening oils, a figure of 150 to 100 to "semi-drying", i.e. less reactive oils, and < 100 to non-drying oils. Although the last of these

	C_8	C_{10}	C_{12}	C_{14}	C_{16}	C_{18}	$C_{18'}$	$C_{18''}$	$C_{18'''}$	$C_{20'}$	$C_{20''''}$	$C_{22'}$	$C_{20''''}$
Palm kernel oil	3–5	3–7	40–50	14–18	7–9	1–3	10–20	–	–	–	–	–	–
Peanut oil	–	–	–	–	6–16	1–7	35–70	15–50	–	–	–	–	–
Soya bean oil	–	–	–	–	7–11	2–6	15–30	40–55	4–10	–	–	–	
Sunflower oil	–	–	–	–	3–7	1–5	15–45	45–70	–	–	–	–	–
Linseed oil	–	–	–	–	4–7	1	12–35	20–25	35–60	–	–	–	–
Fish oil				6–8	15–20	2–4	10–18	1–3	0–2	2–8	15–30	1–8	5–20
Tung oil					3–4	1–3	5–10	8–10	75–85*	–	–	–	–
Oiticica oil					6–8	3–5	5–15	10–25	75–85**				
Castor oil					1–2	1–2	85–90***	2–7	–	–	–	–	–

* α-eleostearic acid (conjugated double bonds)
** α-licanic acid (conjugated double bonds)
*** ricinoleic acid

Figure 2.1.8: Selection of important natural oils and the composition of their fatty acids

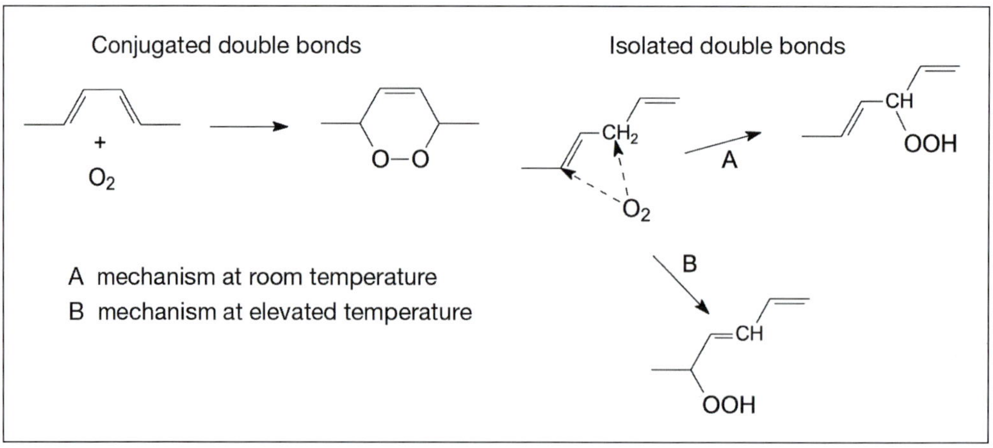

Figure 2.1.9: First steps in the oxidative hardening of oils by forming hydroperoxides and peroxides

react with atmospheric oxygen, they solidify only incompletely such that mechanical hardness cannot be achieved even after an extended reaction period.

Oxidative Drying of Oils

The solidification caused by the action of atmospheric oxygen has been described by numerous authors [2.4.2, 2.4.3]. It is now known that there is no single mechanism by means of which all oils react in the same way [2.4.4]. Various reaction mechanisms have been identified that vary with the type of fatty acid and the number and configuration of the double bonds, as well as the temperature. Whereas it has been shown that the formation of cyclic peroxides is the first stage when conjugated double bonds are present, hydroperoxides are formed initially in the case of isolated double bonds. The atmospheric oxygen acts either directly on the double bonds or on the activated methylene group between the double bonds as a function of temperature. Whereas at low temperatures the action primarily occurs the activated methylene group, at higher temperatures the oxygen adds to a double bond and forms new conjugated double bonds at the same time.

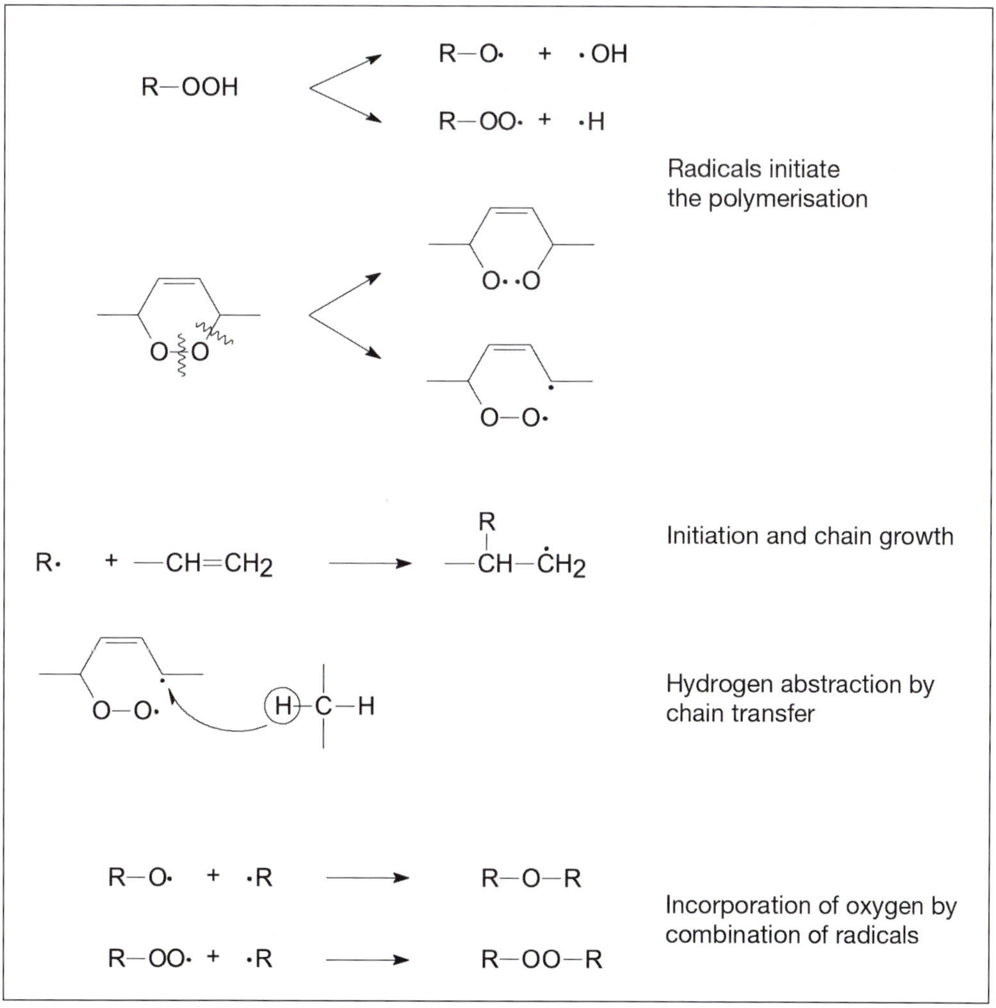

Figure 2.1.10: Mechanism of oxidative curing of unsaturated fatty acids

When the peroxides and hydroperoxides decompose into radicals, the molecule-enlarging crosslinking process begins which then initiates radical polymerisation. At the same time, additional chain-transfer and recombination reactions start because of the high radical density. This extremely complex process is also subject to numerous secondary reactions.

The chemical crosslinking process can be accelerated substantially by the use of specific catalysts, known as siccatives. Metallic soaps, whose cations accelerate peroxide decomposition by reversible valence change, are suitable as siccatives.

The hydrogen and hydroxyl ions formed by hydroperoxide or peroxide decomposition combine to form water. They shift the equilibrium towards the radicals and are therefore the driving force behind the catalytic action.

The natural "drying" oils, with a molecular weight of approx. 880, are relatively low molecular substances. In chemical terms, therefore, it is a long journey until the crosslinking caused by oxidative hardening produces a hard and, at the same time, resistant coating.

The natural oils' low molecular weight leads to a low viscosity and consequently a high tendency to run on vertical surfaces. To remedy these shortcomings, various methods of preliminary, targeted molecular enlargement of the oils have been used for years to increase the viscosity. One such method involves heating the oils in the absence of oxygen at temperatures of between 260 °C and 300 °C over a period of a few hours. This leads to thermally induced polymerisation reactions of the double bonds to form so-called **stand oils** of higher viscosity. The goal of molecular enlargement can be achieved in the same time-frame at temperatures of between 100 °C and 150 °C by heating in the presence of oxygen. By contrast with the stand oils mentioned above, the increase in the molecular weight in the case of the so-called **blown oils** is achieved by molecular enlargement via oxygen bridges.

$$
\left.\begin{array}{l}
Me^{2+} + ROOH \longrightarrow Me^{3+} + RO\cdot + OH^{-} \\
Me^{3+} + ROOH \longrightarrow Me^{2+} + ROO\cdot + H^{+}
\end{array}\right\} H_2O
$$

Figure 2.1.11: Acceleration of peroxide decomposition by siccatives

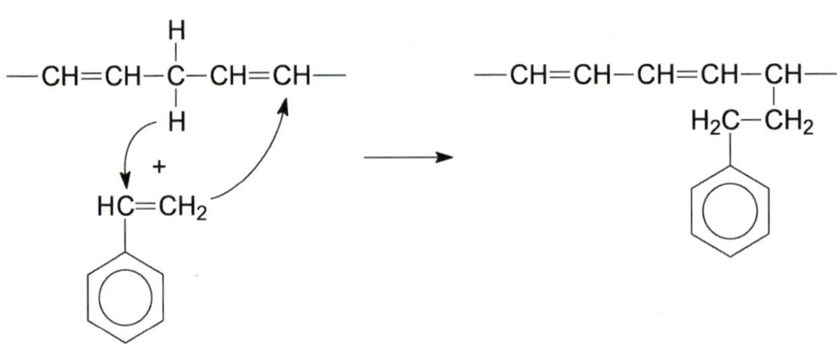

Figure 2.1.12: Modification of oils by reaction with styrene

The reaction of oils with styrene or cyclopentadiene also leads to an increase in viscosity and thus to the faster formation of mechanically solid films.

One way of making natural oils more reactive is to increase the number of double bonds and simultaneously raise the content of conjugated double bonds. The number of double bonds can be increased by epoxidation of the double bonds in the fatty acids, followed by ring opening and dehydration.

Fatty oils, particularly **linseed oil**, can be also converted into candidate raw materials for coatings by reaction with maleic anhydride. Addition of the maleic anhydride to the fatty acids' double bonds, followed by treatment with butanol to split open the ring and then neutralisation with amines yields anionic paint raw materials that are soluble in water. Such modification products have been highly successful for many years as film forming agents for anodic electrocoating (see Chapter 4.2.1).

Another way to improve the processing properties of oxidatively curing oils is to blend them with natural **hardened resins**. However, the resulting increased viscosity has to be lowered again by the addition of appropriate solvents. The hardened resins accelerate physical drying, while the oils ensure subsequent chemical crosslinking.

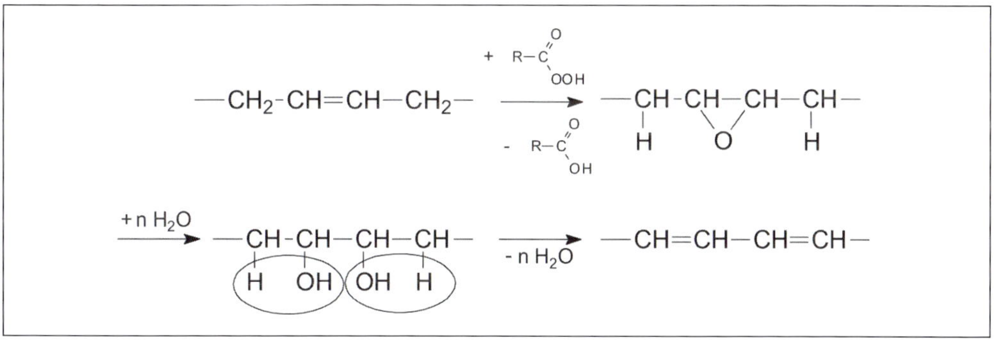

Figure 2.1.13: Increasing the number of conjugated double bonds of oils by epoxidation

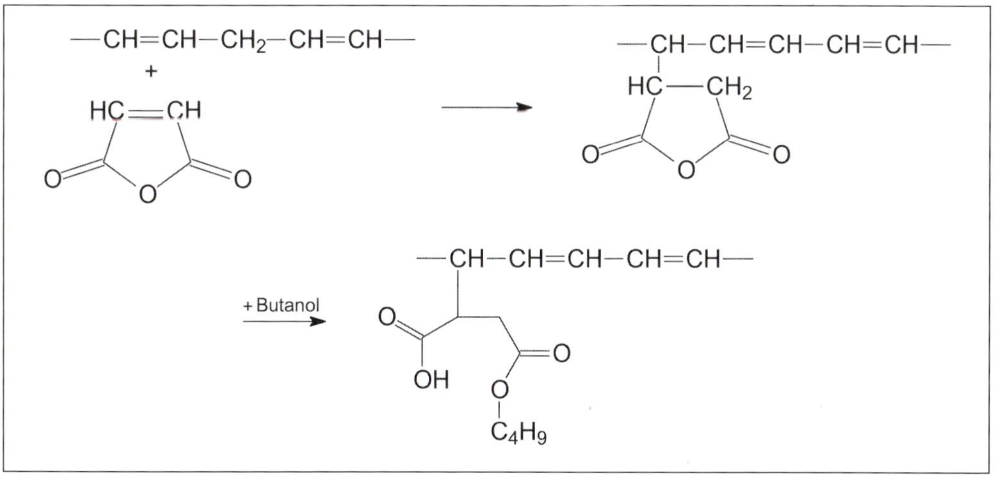

Figure 2.1.14: Water-soluble resins obtained by reaction of oils with maleic anhydride

Natural Resins

Other important natural materials for paint formulations are the cyclic diterpenes and triterpenes. These include copals, amber, dammar and colophony. Since colophony still plays a very important technical role in printing inks, it is discussed here as a representative example of the whole range of natural resins. It should not be forgotten that amber, which is now known solely as a material for ornamental objects, used to serve as a raw material for particularly wear-resistant coatings. Whereas dammar and the copals have also become less important in coating technology, shellac, a natural polyester obtained from cycloaliphatic, hydroxy-functional polycarboxylic acids still serves as a raw material for the surface treatment of furniture.

Colophony

As just mentioned, colophony is by far the most important and is consumed in by far the most notable quantities of all the natural resins. Colophony is a constituent of conifers and is extracted from them as balsam by tapping. This involves cutting notches in the trees which exude the balsam as a sealing compound. The raw material extracted in this way is a mixture of mono- and diterpenes, along with numerous impurities.

The balsam is separated into volatile turpentine oil, a blend of various monoterpenes including limonene and camphene, and the residue, colophony. This was previously done with steam distillation, but fractional distillation is now the preferred method. Its main constituent is abietic acid (see Figure 2.1.15). Other sources of natural resins, corresponding to colophony in terms of composition, are tree stumps from cleared forests and the waste water from cellulose production.

Despite colophony's good compatibility with fatty oils, its softening point, which is too low at 65 °C and its high content of acid groups are performance shortcomings, though these can be eradicated by chemical modification.

The softening point can be raised and the degree of acidity lowered at the same time either by neutralisation or by esterification with polyfunctional alcohols. Calcium oxide and zinc oxide have proved successful for the salt formation, and glycerine and pentaerythritol for the molecular enlargement by esterification.

Esterification of abietic acid can be carried out stoichiometrically on one hand to produce non-polar combination resins for oils (see Figure 2.1.16) or can be carried out with excess alcohol on the other to produce polar modifying agents for cellulose nitrate.

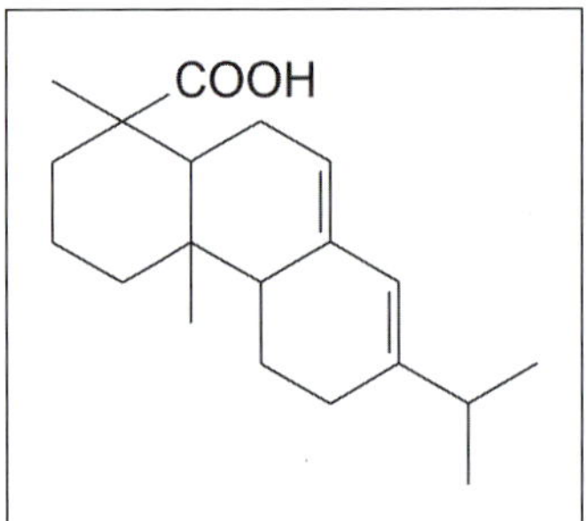

Figure 2.1.15: Abietic acid as the main constituent of colophony

While the salts of colophony have practically no technical significance any more, the products modified with maleic acid are still important in the printing ink sector (see Figure 2.1.17). After thermal rearrangement of abietic acid to laevopimaric acid, the latter reacts with maleic anhydride in the form of a 1,4-Diels-Alder addition to form the intermediate stage of a tricarboxylic acid. This then reacts with polyols to yield maleic resins.

These **maleic resins** are light-coloured, hard, high-melting, oxidation-resistant and readily soluble and may be combined advantageously with many other film

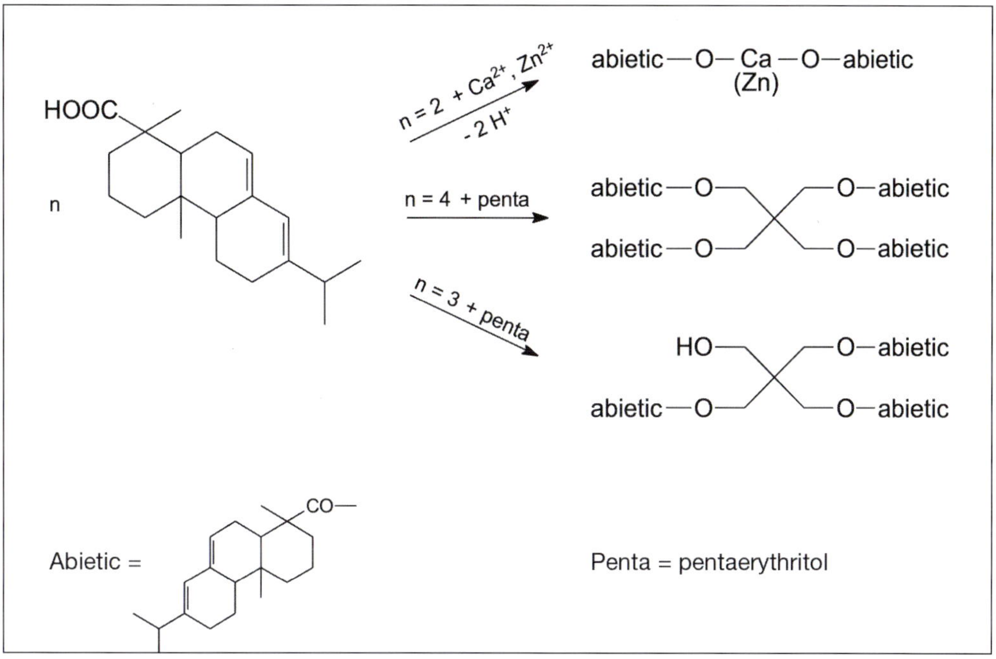

Figure 2.1.16: Modification of colophony by salt formation or esterification

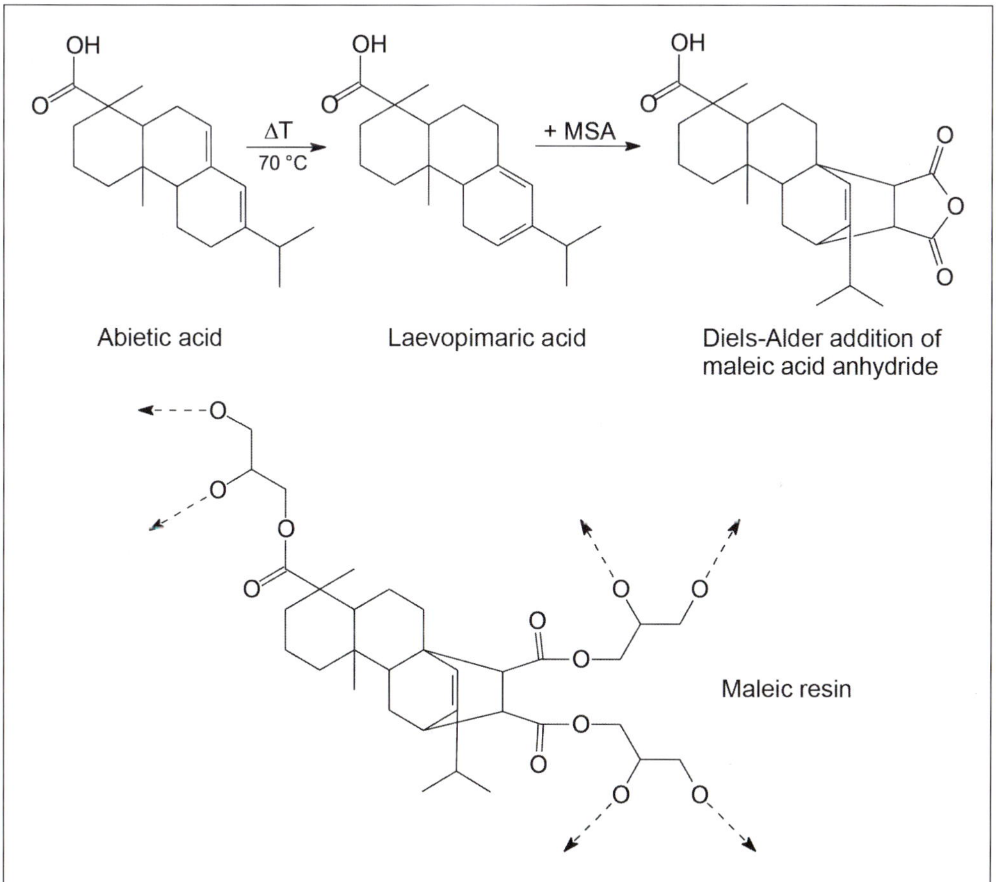

Figure 2.1.17: Production of maleic resins from colophony, maleic anhydride and polyols

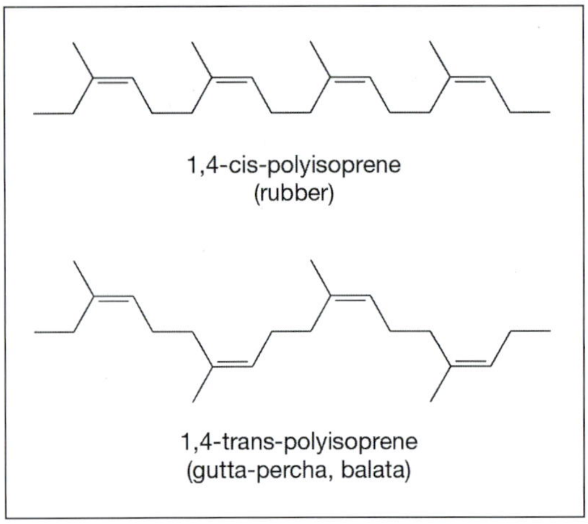

Figure 2.1.18: Rubber and gutta-percha as isomeric forms of 1,4-polyisoprene

forming agents. They still play a significant role, particularly as additives for physically drying and oxidatively curing coating materials.

Natural Rubber

Nature is not just capable of synthesising cyclic terpenes or terpenoids. In certain plants it can also join activated isoprene in a 1,4-polymerisation stereospecifically to form macromolecules. 1,4-trans-Polyisoprene has achieved less technical importance as gutta-percha or balata than the isomeric 1,4-cis-polyisoprene, known as natural rubber, i.e. caoutchouc. The latter is extracted as latex from notches cut in *Hevea Brasiliensis* which is grown in plantations specifically created to harvest this raw material. After treatment of the aqueous dispersion, the crude rubber is then isolated.

Although 1,4-cis-polyisoprene is now manufactured synthetically, there is still a high proportion of natural rubber on the market. While unmodified, natural rubber plays a major role as the raw material for rubber, natural rubber needs to be chemically modified before it can serve as a film forming agent, because of its insolubility in all commercial paint solvents and its lack of resistance to atmospheric oxygen.

The very high molecular weight and the double bonds, which become capable of cross-linking due to atmospheric oxygen, can be effectively reduced by selected heat treatment in the presence of catalysts. New valency bonds are then formed by the initiation of hydrogen ions when the double bonds are split. As a result of the new bonds, 6-membered intramolecular rings are formed from the originally aliphatic molecular chain. This cyclisation eliminates more than 80 % of the previously existing double bonds. If the reaction is controlled properly, the molecular weight of the resulting **cyclorubber** is also substantially reduced. The molecular weight of cyclorubber intended for use in coating technology is only 3,000 to 10,000 g/mol. This guarantees adequate solubility in aliphatic, cycloaliphatic, aromatic and terpene hydrocarbons. Cyclorubber is a hard resin which has a melting point in the range between 130 °C and 140 °C, and exhibits good compatibility

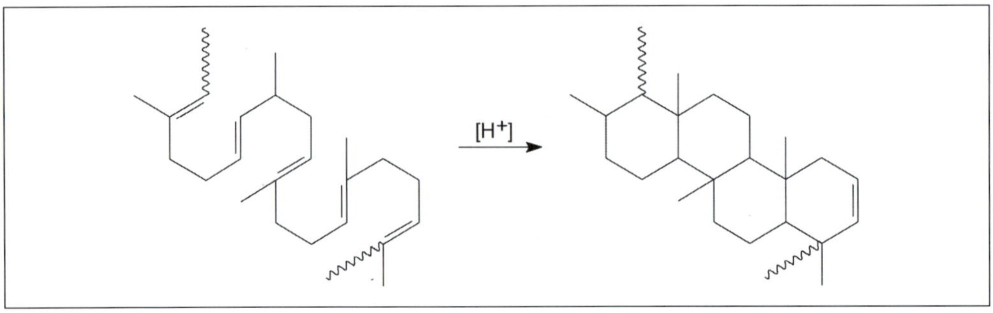

Figure 2.1.19: Cyclisation of rubber

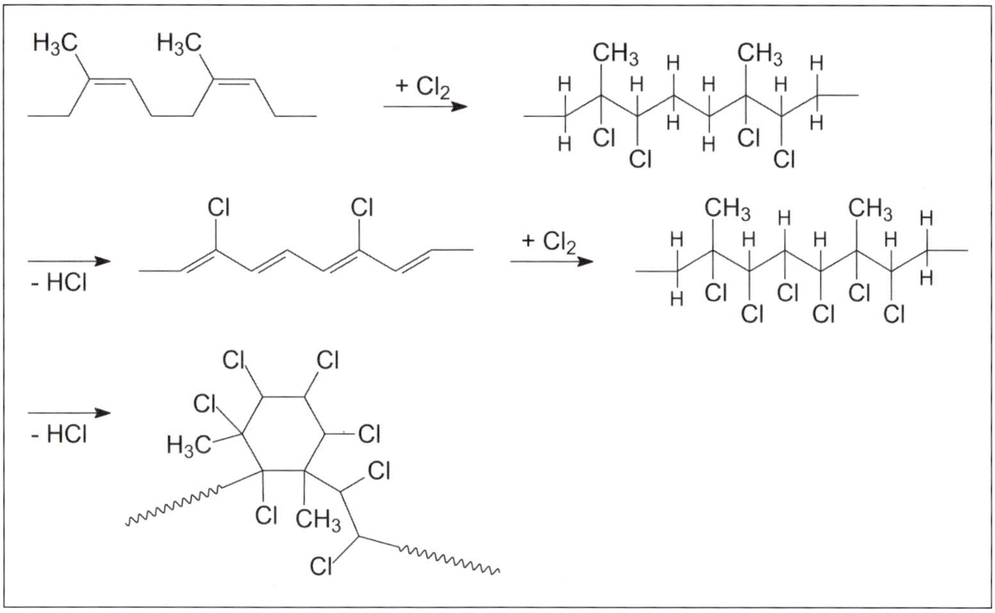

Figure 2.1.20: Chlorination of rubber

with drying oils, alkyd resins and bituminous film forming agents. It has excellent chemical resistance due to the absence of saponifiable groups and also good compatibility with alkaline pigments due to the lack of acid groups. Its relatively good heat resistance enables the manufacture of coatings which will be subjected to thermal stress. Cyclorubber is preferred for applications in heavy-duty anticorrosive protection, for underwater coatings and for coating vessels used in dairies, breweries and other branches of the food industry. The low solid content of the coating formulations and the low resistance to light and long-term outdoor exposure have prevented it from being used more widely.

Another option for using caoutchouc for coating applications is to chlorinate it under the influence of UV light. A synchronous reaction occurs in which the chlorine is first attached to the double bonds and then is removed as HCl, leaving behind new double bonds. Then chlorine then becomes attached again to these new double bonds, and HCl is again removed and so on until complete chlorination of all the C atoms has been achieved in this way. The chlorination reaction is halted at a chlorine content of 66 % and a density of 1.6 g/cm^3.

Unlike cyclorubber, **chlorinated rubber** is soluble in polar solvents because of its higher polarity. Leaving its lower heat resistance aside, chlorinated rubber has a similar performance profile to cyclorubber.

Carbohydrates

Carbohydrates are also natural materials which have significantly influenced industrial coating technology in the form of numerous modifications. **Cellulose nitrate** and **cellulose acetobutyrate** still serve today as film forming agents for wood and automotive paints. Because of their high availability in nature, the unmodified hydrocarbons are also presented below to increase understanding of them.

Carbohydrates are termed mono-, di- or polysaccharides in accordance with the number of basic structural elements which are chemically linked to each other. The mono-saccharides

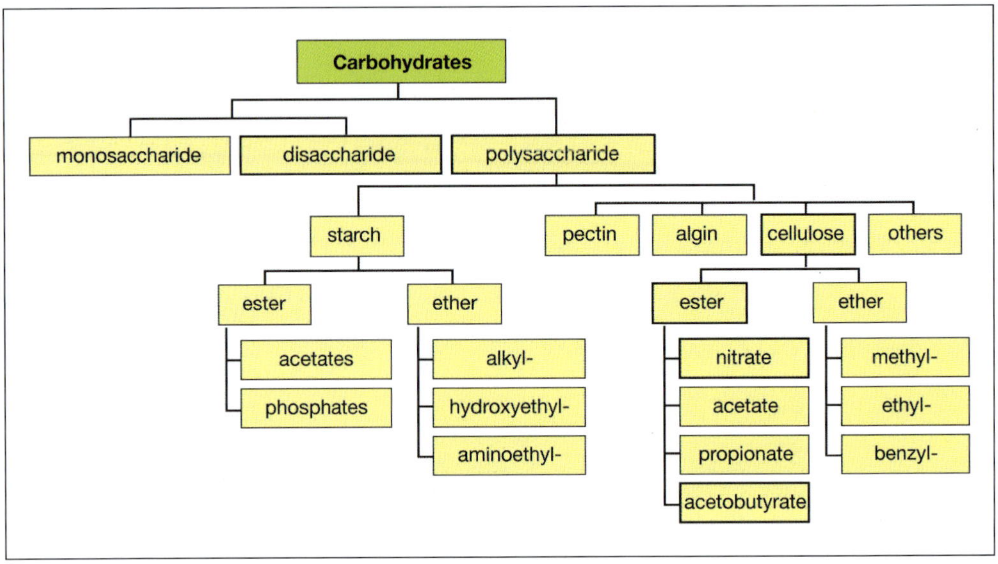

Figure 2.1.21: Important carbohydrates for use in coatings

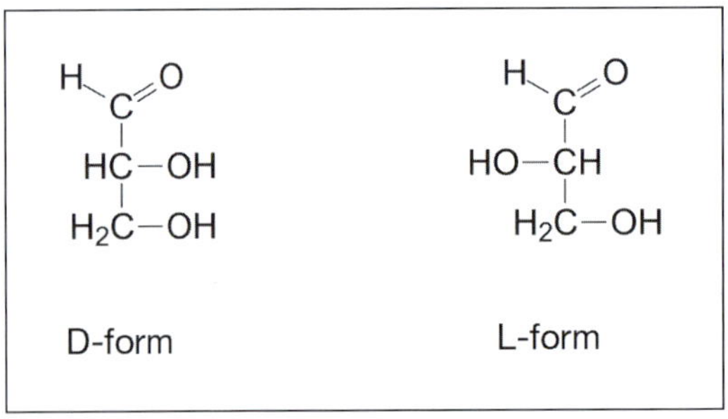

Figure 2.1.22: The two enantiomeric (D-, L-) forms of glycerine aldehyde as the base substances of all carbohydrates

or sugars as the basic structural elements of this class of materials have the empirical chemical formula $C_nH_{2n}O_n$ or, using the older notation, $[C(H_2O)]_n$. This also explains the name carbohydrates.

The **monosaccharides** synthesised by nature, principally with 5 or 6 C atoms in the aliphatic parent chain, are derived from glycerine aldehyde, the smallest base material meeting the definition of carbohydrates.

Glycerine aldehyde contains one aldehyde and two hydroxyl groups in the presence of an asymmetric carbon atom which gives the molecules stereoisomerism in the form of enantiomeric constituents, i.e. an isomer with its mirror-image isomer.

Apart having a longer parent chain, the longer-chained pentoses which contain 5 C atoms, and the hexoses, which contain 6, also differ from glycerine aldehyde in the number of asymmetric C atoms. The existence of enantiomers and diastereoisomers means that there are many individual forms, the number of which is determined by the number of asymmetric atoms. If there are *n* asymmetric C atoms, there will be 2^n different variants in the form of enantiomers and diastereoisomers. In the case of the hexoses, there are thus 2^4, i.e. sixteen, variants comprising eight D- and eight L-forms. As far as coating technology is concerned, however, only the mannoses, the galactoses and the glucoses are important.

The degree of variety among the monosaccharides is increased still further by the position of the carbonyl group, but also by the ability to form intramolecular hemiacetals that can vary in ring size at the same time. Since the cyclising formation of hemiacetal gives rise to a further asymmetric C atom, the individual hexoses differ not only in their D- and L-forms, but also in their α- and β-forms, depending on the steric configuration of the

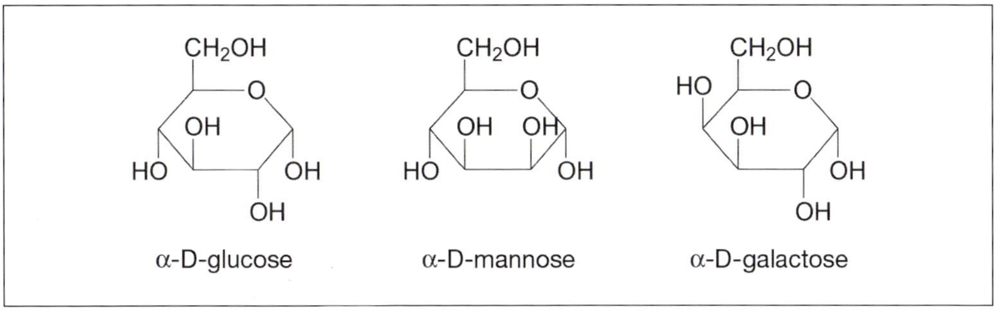

Figure 2.1.23: Glucose, mannose and galactose as raw materials for resins

hydroxyl group of the hemiacetal. All monosaccharides are soluble in water, colourless, odourless and sweet-tasting because of their large number of hydroxyl groups.

In the context of their importance in coating technology, it is worth noting that the terminal aldehyde and OH groups can easily be oxidised, separately or together, to hydroxy-functional carboxylic acids. Reduction of the carbonyl group is also easily possible. When glucose is reduced, for example, the result is sorbitol, a hexavalent alcohol which became important, among other things, as a polyol for the production of long-oiled alkyd resins.

Of the **disaccharides**, cane sugar, i.e. sucrose, is the most important substance of relevance to coating technology. The total of eight hydroxyl groups, 3 primary and 5 secondary, can be converted into resin-like esters with the aid of organic acids such as acetic acid, benzoic acid or mixtures of acetic and butyric acids. These are compatible with a large number of film forming agents and exhibit good solubility. As additives they increase hardness and increase the dependence of viscosity on temperature.

Monosaccharides, the basic structural elements of carbohydrates, are capable of self-condensation to form long-chain **polysaccharides**.

Starch and cellulose are produced naturally in huge quantities as highly polymerised forms of glucose and are therefore available for use as paint raw materials in sufficient volume. Despite their similar chemical structure, their range of properties is extremely different. Both starch and cellulose consist of D-glucose units linked via the glycosidic hydroxyl group with the fourth C atom of the next structural element. The mere difference in the

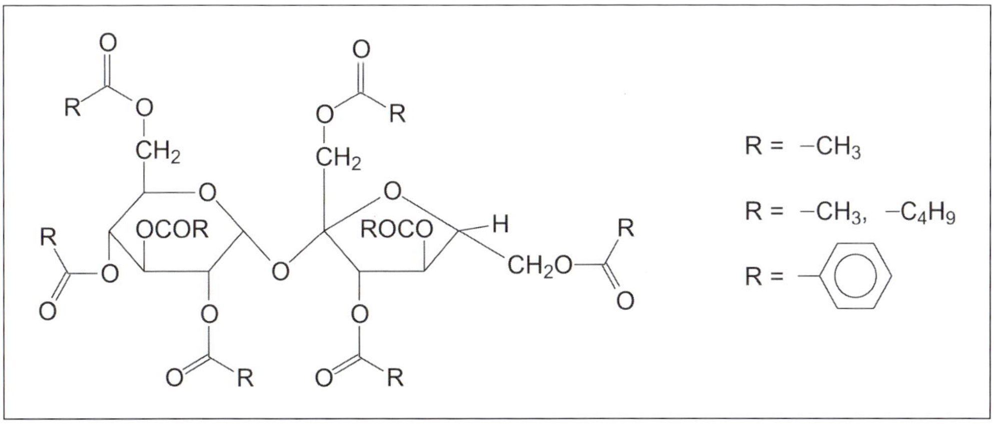

Figure 2.1.24: Sucrose esters as raw materials for coatings

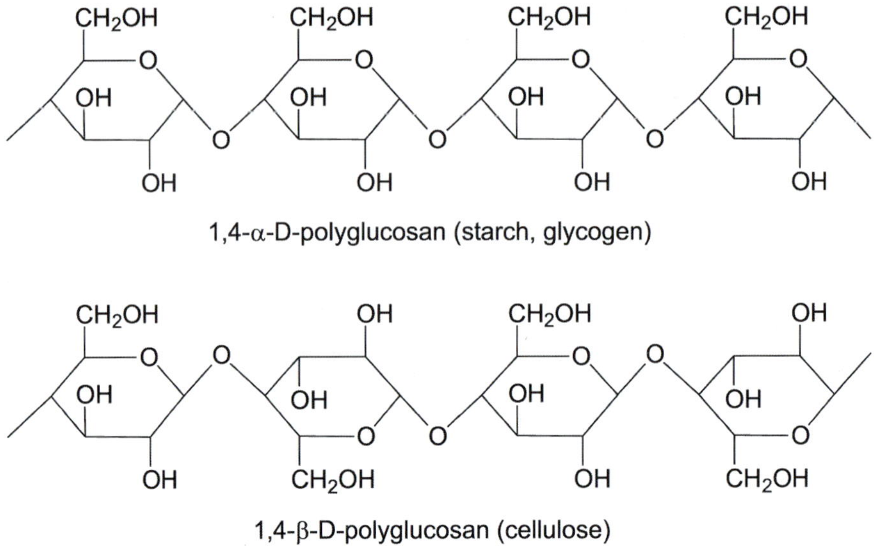

Figure 2.1.25: Chemical structure of starch and cellulose

steric configuration of the glycosidic hydroxyl group is responsible for the major differences in their solubility and resistance to saponifying agents. While starch serves as an energy-storage medium for plants and must, therefore, exhibit a certain solubility in water, cellulose as a skeletal material is absolutely insoluble in all usual solvents, especially in water.

These different requirements are met by the different ways in which intermolecular hydrogen bridging bonds can be formed during α- und β- linking of the structural elements.

The two products are only of any significance to coating technology in chemically modified form. Dextrins, i.e. partially hydrolysed starch, and also etherified and esterified starch have acquired importance as physiologically harmless adhesives and film forming agents for paper coatings. **Aldehyde starch** occupies a special position because of its water resistance. It is obtained from starch by oxidation with per-iodic acid. The glucose structural elements are transformed into aldehyde groups without any significant reduction in molecular weight by splitting their parent chains between C_2 and C_3, with subsequent oxidation.

The derivatives of **cellulose** are of much greater technical importance. As with starch, the ethers and esters merit particular attention. In this regard the three hydroxyl groups per glucose unit can be made to react partially or completely.

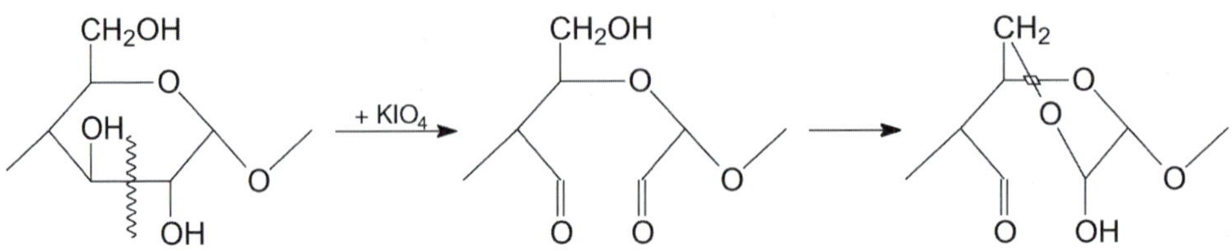

Figure 2.1.26: Synthesis of aldehyde starch by degradation of starch by per-iodic acid

Of the ethers, benzyl cellulose and ethyl cellulose are worth mentioning. Benzyl cellulose served in the past as a film forming agent for heavy duty anticorrosive protection products because of its non-saponifiable ether groups and its low water vapour permeability. Cyclorubber or chlorinated rubber coatings have succeeded in ousting benzyl cellulose in terms of quantity because of their lower cost.

Ethyl cellulose suffered a similar fate. The commercial product, which has an ethoxyl content of 45 to 50 % and is readily soluble in alcohols and aromatic hydrocarbons, was unable to match modern synthetic resins because of its brittleness on exposure to UV light. Ethyl cellulose is still used to a limited extent for paper films and printing inks for food packaging.

The water-soluble variants of methyl or hydroxyethyl cellulose, however, continue to be successfully used as thickening agents for dispersions, as film forming agents for distemper and as wallpaper paste.

Cellulose can also be modified with organic and inorganic acids at different esterification levels. The most important examples are cellulose nitrate, commonly known as **nitrocellulose**, collodion cotton or soluble guncotton, and the esters of acetic, propionic and butyric acids, though the latter are only used as mixed esters of acetic and butyric acids because of the decreasing hardness with increasing chain length of the esterification component.

Cellulose nitrate is produced from cellulose through the action of a mixture of nitric and sulphuric acids (nitrating acid). Different esterification levels and a varying reduction in the original molecular weight are produced, depending on the reaction conditions. The target esterification level is achieved at a nitrogen content of between 10.6 and 12.2 % in the end product. Products with a lower nitrogen content lack the necessary solubility. The lower end of the nitrogen content range, from 10.6 % to 11.2 %, is made up of the alcohol-soluble products, known as alcohol-soluble cellulose nitrate, while the higher end, i.e. 11.8 and 12.3 % nitrogen, comprises the ester-soluble products, known as ester-soluble cellulose nitrate. In the case of products with a higher nitrogen content, the blendability

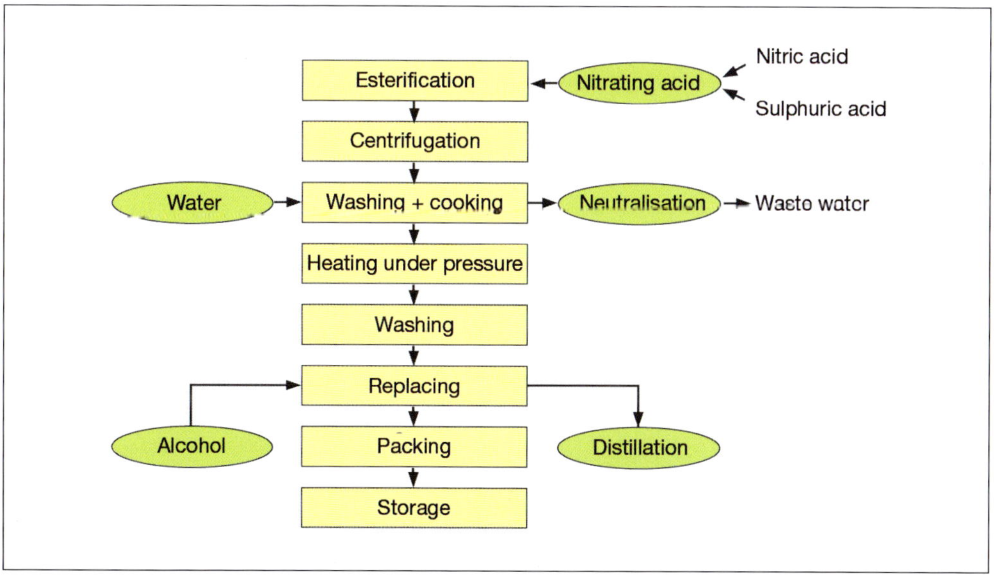

Figure 2.1.27: Synthesis path of cellulose nitrate

of the solutions with aromatic hydrocarbons causes problems. Furthermore, the reaction product has a pronounced thermal instability. All cellulose nitrate types used for the production of coating materials require special precautionary measures during handling, because they are easy to ignite and highly flammable. Thus, they are never supplied as dry goods, but instead are usually delivered wetted in butanol in a ratio of 2:1. They can also be reliably transported in the form of chips processed with plasticisers or concentrated solutions in suitable solvents.

The properties which make cellulose nitrate stand out as an exclusively physically drying film forming agent are its extremely rapid drying and its ability to completely liberate its solvent. With the advent of coating materials based on cellulose nitrate, for the first time paints were available which were dry and could be sanded and polished within a few minutes of application, and could therefore keep pace with the short cycle times of modern industrial production.

Cellulose nitrate is combined with many other film forming agents to regulate the desired application properties. In addition to shellac and modified colophony, amino, amido and ketone resins can be used (see elsewhere in this chapter). Alkaline pigments may not be combined with cellulose nitrate due to the latter's sensitivity towards saponifying agents.

The most important differences between the cellulose esters of organic acids compared with cellulose nitrate are their improved lightfastness and reduced flammability. The favourable performance profile of cellulose esters made with organic acids is rounded out by improved resistance to extremes of temperature, good electrical insulation ability and resistance to water, oil, grease and sweat. Particular disadvantages of acetic acid esters are their limited solubility in organic solvents and restricted compatibility with other resins. This disadvantage is not so marked in mixed esters of propionic and butyric acids.

The cellulose acetate types used for paint manufacture have acetyl contents of between 38 and 40 %. Cellulose propionate contains 2.5 to 3.5 % acetyl and 45 % propionyl groups. In the case of cellulose acetobutyrate, the acetyl content is between 2 and 30 %, and the butyryl content between 15 and 53 %. In all products, the residual hydroxyl group content is between 1.6 and 5 %.

The esters of cellulose produced with organic acids can be used, like cellulose nitrate, to manufacture physically drying coating materials for a wide range of applications. The advantages are to be found in greatly improved resistance to light, water and heat, as

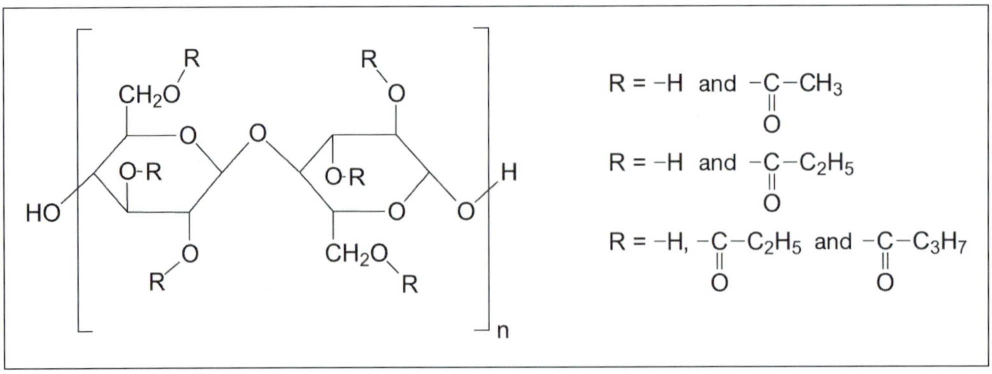

Figure 2.1.28: Cellulose esters of organic acids

already mentioned. The last of these enables cellulose esters from organic acids to be used as a rapidly drying component in stoving enamels. During film formation by metallic paints, for example, the position of bronzes in the film can be stabilised effectively.

For environmental reasons, however, and despite the aforementioned advantages in terms of application technology, no great future can be forecast for solvent-containing cellulose derivatives in view of their solvent levels of up to 80 %.

Proteins

Apart from natural resins, oils and hydrocarbons, nature also provides proteins by way of macromolecular substances. Based as they are on amino acids, these are essentially polyamides. They can occur either in isolation or in association with other polymeric substances, e.g. polysaccharides. Products which merit attention for practical operations include collagens, casein, which is extracted from milk, and zein, which is obtained from corn (maize). Collagens, which are connective tissue proteins, are transformed into bone glue via the intermediate stage of gelatin by boiling them in aqueous alkaline solution. This product, known as animal glue, is compatible with fatty oils and is used for artists' paints, particularly water-colours.

Casein, a calcium-containing phosphoprotein, is insoluble in neutral water and can be converted into highly viscous fillers, casein fillers, in reactions with calcium oxide. Zein has a similar performance profile, though with a different solubility. Unlike casein, it is readily soluble in alcohol.

2.1.1.2 Synthetic Resins

The path from natural materials and chemically modified natural materials to the synthetic resins which are widely available today started in the early years of the 20th century. The most important reasons for the success of synthetic resins are on one hand the scope for tailoring products to a huge range of application fields with much higher reproducibility and on the other their unlimited availability.

In Germany the starting point in the systematic search for new resins, apart from the improved technical possibilities, was the political situation at the turn of the last century but one. Hard resins such as copals, colophony, dammar and particularly shellac, all of which were foreign-sourced raw materials, were products which it was imperative to replace in the medium term by synthetic alternatives because their availability could not be guaranteed.

The invention of glyptals as hard resins and the more fortuitous discovery of phenolic resins during the search for means of synthesising shellac are just a couple of examples of noteworthy inventions from the pioneering phase of synthetic resin chemistry [2.4.4].

Whereas at the start of the new era, a great deal had to be left to chance due to a lack of knowledge surrounding the production of large molecules and to the absence of a sound knowledge of the structure/property correlations of resins, the situation changed, at the latest, when scientists were in the position to comment more precisely on the construction and structure of macromolecules. Systematic developments for producing specific product properties in tailor-made resins only became possible when scientists learned how to produce linear, branched and even crosslinked macromolecules specifically by chemically bonding polyfunctional structural elements and when it became known that these could become entangled with each other because of the free mobility of their chains, a fact which guarantees elasticity and mechanical strength. It was the scientists *Staudinger, Carothers* and *Kuhn* who laid the foundations of modern polymer chemistry [2.4.5] through their detailed knowledge of the construction of macromolecules.

Thanks to a very wide range of analytical techniques, we now have a detailed picture of the structure and properties of macromolecular materials. The desire at that time to develop or synthesise paint resins to meet the varied range of practical requirements has now become conventional technology.

Synthetic production of oligomeric and macromolecular materials is possible via chemical conversions of monomers, provided that these are at least bifunctional. Classifying reactions as step and chain reactions to reflect the type of molecular growth has proved useful. The first group includes **polycondensation**, which is accompanied by the formation of low molecular cleavage products, and **polyaddition**, which produces no cleavage products. The second group includes polymerisation, with the sub-groups of radical and ionic growth of the molecules during their production.

The synthesis of polyesters from dicarboxylic acids and diols by elimination of water is an example of a typical polycondensation reaction, while the formation of polyurethanes from polyfunctional isocyanates and OH-functional reactants without the cleavage of molecular elements of low molecular weight is an example of polyaddition.

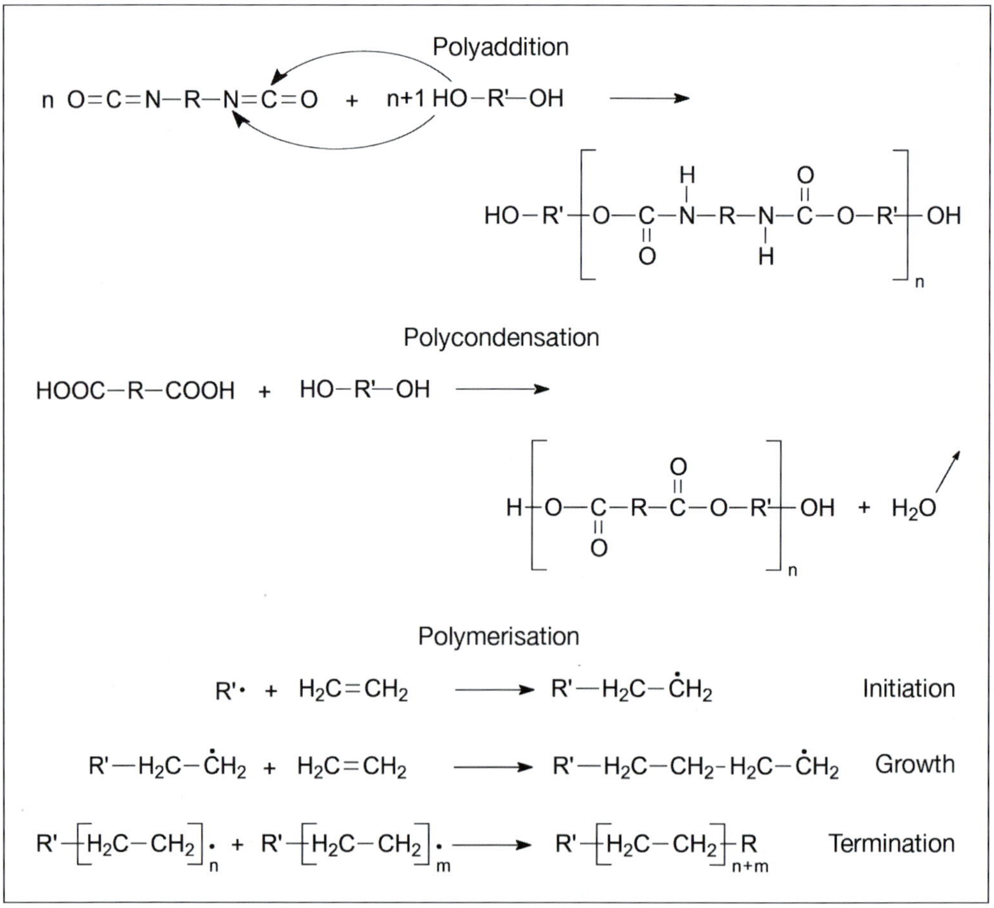

Figure 2.1.29: Basic reactions for synthesising film forming agents

Polymerisation as a chain reaction operates in accordance with a quite different mechanism. With the aid of suitable starting radicals or ions, monomers – which in the vast majority of cases are basic structural elements containing C=C bonds – become activated and thereby enabled to attach further monomers to the growing chain. In radical polymerisation, molecular enlargement does not stop until one activated chain combines with another activated chain.

The different reaction mechanisms behind polycondensation and polyaddition reactions are used to produce simple reactive resins of low to medium molecular weight. Both types of reaction are important in synthesising film forming agents for low-solvent industrial coatings. Polymers, with the exception of the reactive acrylic resins, are generally high molecular, non-reactive thermoplastic materials with a relatively high content of organic solvent or are dispersed in water in the form of emulsion polymers.

For further details on the principles of macromolecular chemistry, please refer to [2.4.6, 2.4.7].

Polycondensation Resins

In the 1920s, the esterification of glycerine with phthalic anhydride was achieved for the first time, and led to the first fairly useful substitutes for natural hardened resins. However, these **glyptal resins**, as they were known from the abbreviation for gly-

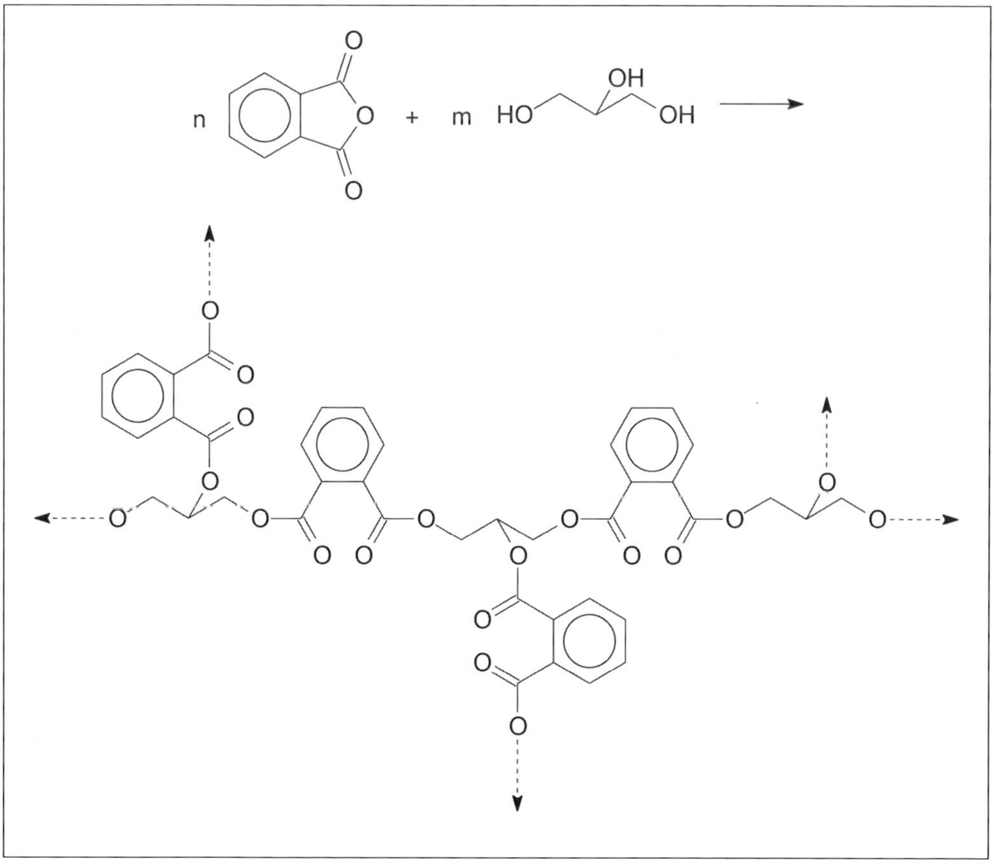

Figure 2.1.30: Glyptal resins as replacements for natural resins

cerine and phthalic acid, tended to gel prematurely or caused incompatibilities with chemically curing oils, to an extent depending on the ratio in which the raw materials were mixed. Only later was it discovered that, because of their bi- and tri-functional structural elements, reaction mixtures of phthalic acid or phthalic anhydride and glycerine – as a function of the mixing ratio – either crosslink during the synthesis process itself, and thus become unusable as film forming agents, or lose their compatibility with fatty oils due to the excessive polarity that results from the chain-breaking excess of glycerine.

This knowledge led *Patton* to formulate the so-called P_{gel} value as an easily calculated coefficient for reliably predicting the "cookability" of raw materials mixtures for polyester [2.4.8]:

$$P_{gel} = m_0/e_a$$

m_0 = sum of moles
e_a = sum of acid equivalents

The P_{gel} value, the quotient of the sum of the moles weighed out and the sum of the acid equivalents (see Chapter 2.3.2) should be as near to 1 as possible. Useful resins can only be synthesised if this is so. If the P_{gel} value is too high, higher molecular resins cannot be made; if it is < 1, higher conversion rates as a function of the raw material composition lead to premature crosslinking and thus gelling. The P_{gel} value represents the maximum permissible conversion. It enables a critical acid value to be calculated, below which gelling of the reaction mixture will occur.

Without this new knowledge, *Kienle* successfully managed as early as 1927 to cut the Gordian knot represented by the tightrope walk between tendency to gel and compati-

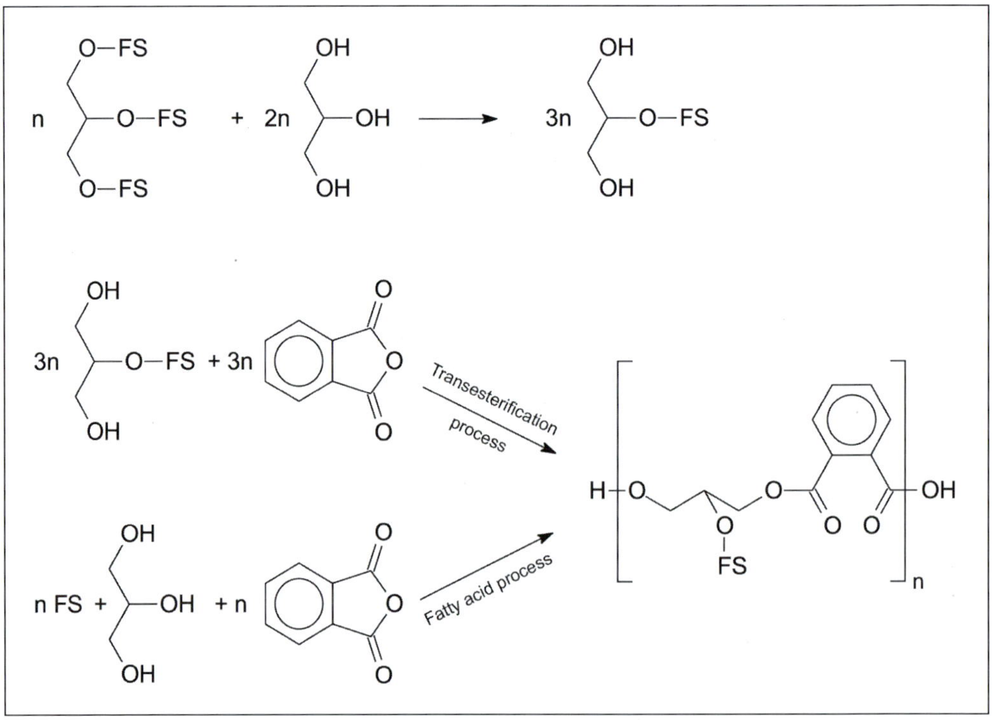

Figure 2.1.31: Synthesis of alkyd resins by fatty acid and transesterification methods

bility with oils. Once people learned to bond oils to glyptals chemically, the lack of oil compatibility and the premature gelling of glyptals were consigned to history at a stroke.

This marked the birth of **alkyd resins**, a class of substances which still heads the list of film forming agents for industrial coatings and decorative paints because of its balanced film properties and its price.

Two means of chemically linking the physically drying, hard polyesters with the chemically crosslinking oils were developed: the oils are either separated into their starting materials of glycerine, the polyol and fatty acids, i.e. monocarboxylic acids, by saponification, and then re-esterified jointly in the presence of phthalic anhydride and another polyol (fatty acid method), or the oils are converted by transesterification in the presence of polyols into a hydroxy-functional intermediate stage which then reacts with phthalic anhydride to form the alkyd resins.

The essential reasons for the success of this class of substances are the simplicity of the synthesis methods, the availability of the raw materials, namely oil as a natural material

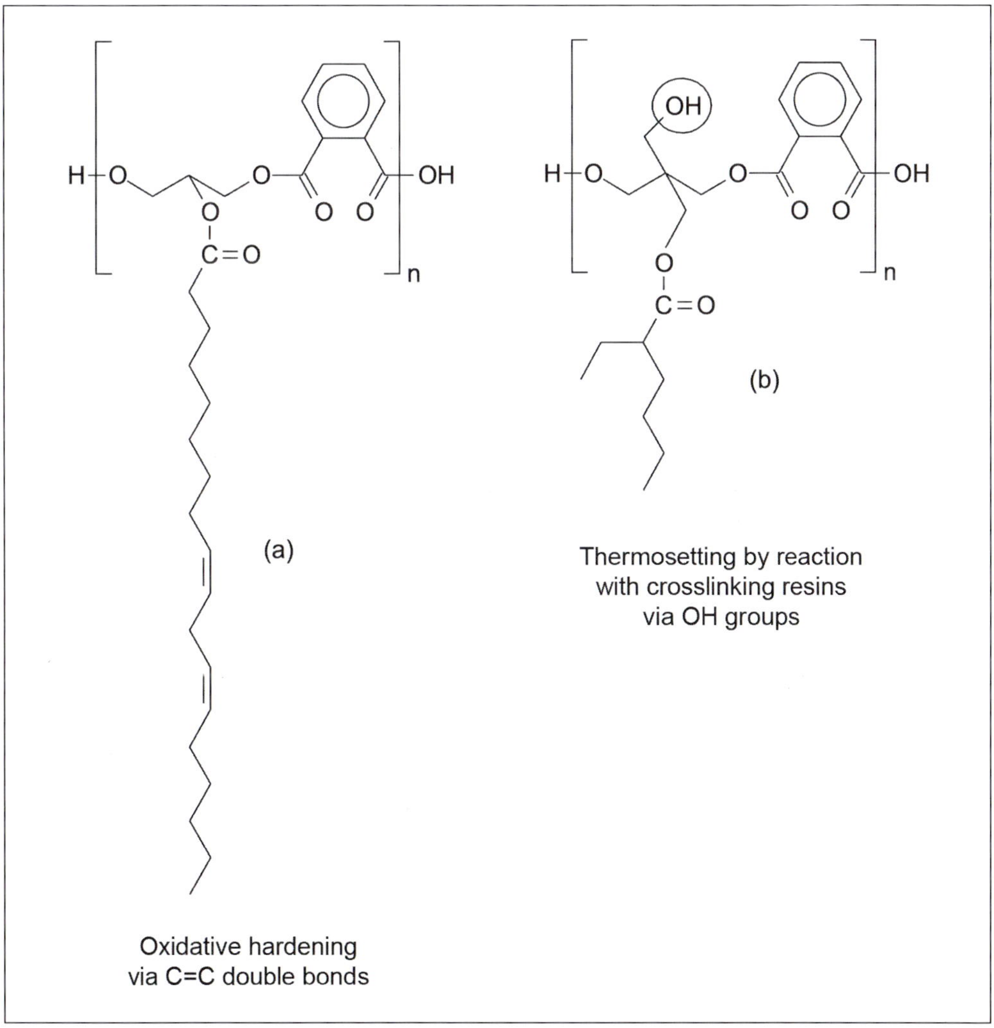

Figure 2.1.32: Alkyd resins: suitable for oxidative hardening (a), for hardening with reactive partners (b)

and phthalic anhydride as an inexpensive product from the petrochemical sector, and the scope for meeting many practical demands by varying the mixing ratios and using different types of oil.

It should be mentioned, to complete the picture that, after the incorporation of hydroxyl groups, the process of crosslinking, which originally occurred via the double bonds of the fatty acids only, could also be carried out with corresponding complementary resins such as phenol and amino resins. The class of substances comprising the previously oxidatively curing alkyd resins therefore found it had an important new member in the guise of film forming agents for stoving enamels.

The term alkyd is a modification of the original name “alcid”, which is derived from the words *alcohol* and *acid*. Although the term alkyd resin makes no reference to the fatty acid component, alkyd resins are always understood to be oil-modified polyesters which are sub-divided into long-, medium- and short-oiled alkyd resins, in accordance with the proportion of fatty acids or oils. The percentage of fatty acid which is esterified in imaginary fashion with glycerine is termed the oil length, irrespective of the polyol actually used by the resin manufacturers. The oil ratio and oil length are therefore only identical if glycerine serves exclusively as the polyol.

Alkyd resins, as mixed esters of polyols, fatty acids and generally o-phthalic acid, are in turn given a merely imaginary division into a physically drying, hard phthalate part and an initially soft, but chemically reactive oil part in order that their characteristics may be described. In line with the above definition, **long-oiled alkyd resins** have an oil length of more than 60 %, **medium-oiled** 40 to 60 % and **short-oiled** less than 40 %.

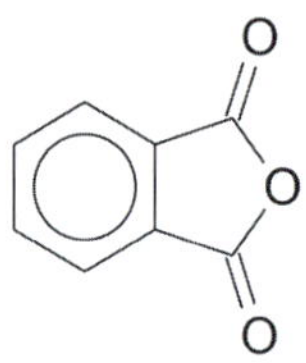

Figure 2.1.33: o-Phthalic anhydride

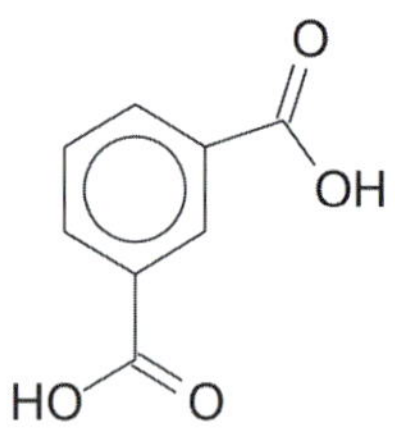

Figure 2.1.34: Isophthalic acid

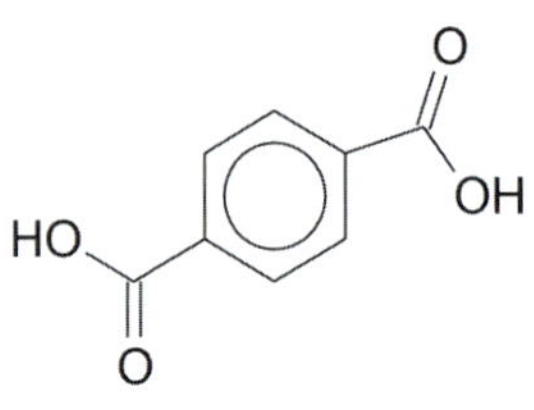

Figure 2.1.35: Terephthalic acid

In addition to the basic structural elements already mentioned, i.e. fatty acids, glycerine and phthalic anhydride, numerous mono- and dicarboxylic acids and polyols have now become accepted as supplementary raw materials or as alternatives for modifying, i.e. adapting, properties.

The most popular of the dicarboxylic acids, measured in terms of quantities used, for the production of alkyd resins, is still **o-phthalic acid** or phthalic anhydride. It is manufactured by oxidising o-xylene and has a relatively low melting point of 131 °C.

Isophthalic acid does not form an internal anhydride and, because of the resulting higher polarity, does not melt below 348 °C. Preliminary esterification with polyols is therefore recommended. Because it has less of a tendency to form cyclic byproducts, isophthalic acid yields resins of higher viscosity and is therefore of particular interest for long-oiled alkyd resins. Isophthalic acid alkyds have greater heat resistance because of their inability to form internal anhydrides.

Terephthalic acid does not form an internal anhydride either and is primarily used, because of its even higher melting point and sublimation point, as a dimethyl ester which is liquid at room temperature. Its incorporation into the alkyd resin proceeds via transesterification. Such alkyd resins are also thermally stable and are therefore important film forming agents for electrically insulating coatings.

Trimellitic anhydride is trifunctional. If only two carboxyl groups are esterified, and the remaining free acid group is neutralised with amines, water-soluble alkyd resins result.

Tetrahydrophthalic acid plays merely a subordinate role for alkyd resins. It has more important uses in unsaturated polyesters. It is employed to mitigate the inhibiting effect of atmospheric oxygen on film formation (see Chapter 2.1.4 and 4.3.4).

Hexahydrophthalic acid confers increased elasticity on alkyd resins, along with a simultaneous improvement in UV transmittance.

Endomethylene tetrahydrophthalic acid, a 1,4-Diels-Alder adduct made from cyclopentadiene and maleic acid, mitigates the inhibiting effect of oxygen, just like tetrahydrophthalic acid.

Maleic anhydride is an essential raw material for unsaturated polyesters (see below). Maleic anhydride reduces the discolouration of alkyd resins during their manufacture, and so is added to the raw material mixture in small quantities (< 2 %).

A proven means of elasticising alkyd resins is to replace the aromatic or cycloaliphatic dicarboxylic acids proportionally with aliphatic dicarboxylic acids.

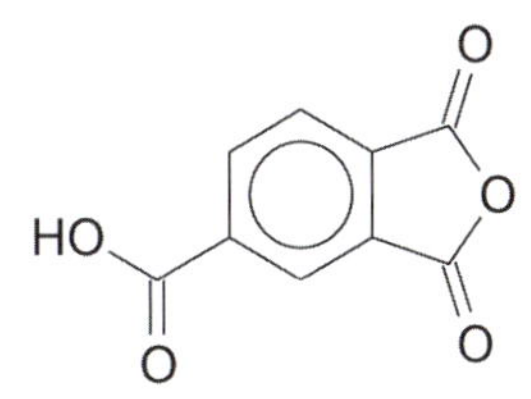

Figure 2.1.36: Trimellitic anhydride (TMA)

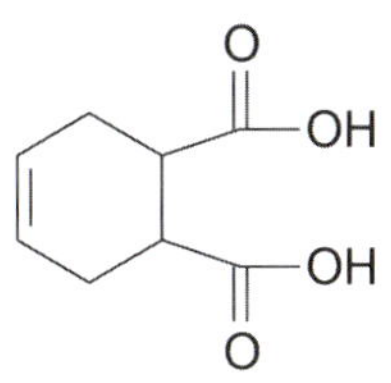

Figure 2.1.37: Tetrahydrophthalic acid

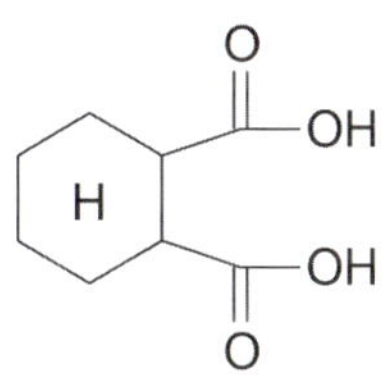

Figure 2.1.38: Hexahydrophthalic acid

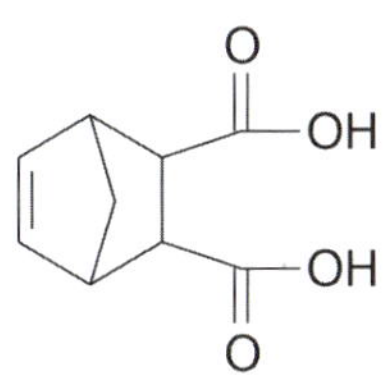

Figure 2.1.39: Endomethylene tetrahydrophthalic acid

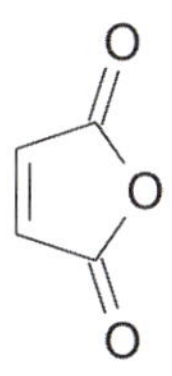

Figure 2.1.40: Maleic anhydride

Coating Material

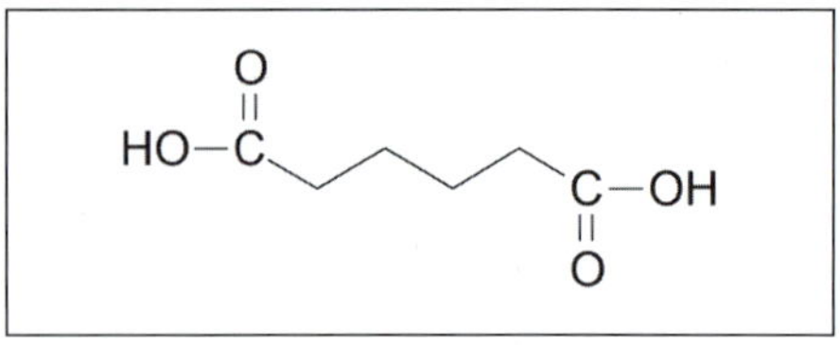

Figure 2.1.41: Adipic acid

Adipic acid (C_6) and the longer-chain azelaic acid (C_9), which is accessed by oxidative cleavage of oleic acid, are among the most important aliphatic dicarboxylic acids.

Apart from glycerine, other polyols are important esterification components for alkyd resins and polyesters:

Diols such as ethanediol (ethylene glycol), 1,2-propanediol (propylene glycol), 1,3-propanediol, 1,4-butanediol und 2,2-dimethyl-1,3-propanediol (neopentyl glycol) mainly serve to construct the basic chain of polyesters and less that of short-oiled alkyd resins.

Glycerine is a triol. However, it is only used nowadays in short- and medium-oiled, oxidatively and thermally curing alkyd resins.

Trimethylolpropane (TMP) and trimethylolethane are interesting alternatives to glycerine because of their three equivalent hydroxyl groups. Compared with glycerine, TMP-based alkyd resins lead to improved compatibility with non-polar solvents such as benzine, for the same oil length.

Pentaerythritol, with its four hydroxyl groups, is an important structural element for long-oiled alkyd resins. Dipentaerythritol, a byproduct of the manufacture of pentaerythritol, enables the oil length to be extended even further.

The range of natural fatty acids has now been expanded by a large number of **synthetic monocarboxylic acids**. While monounsaturated or polyunsaturated C_{18} acids are of

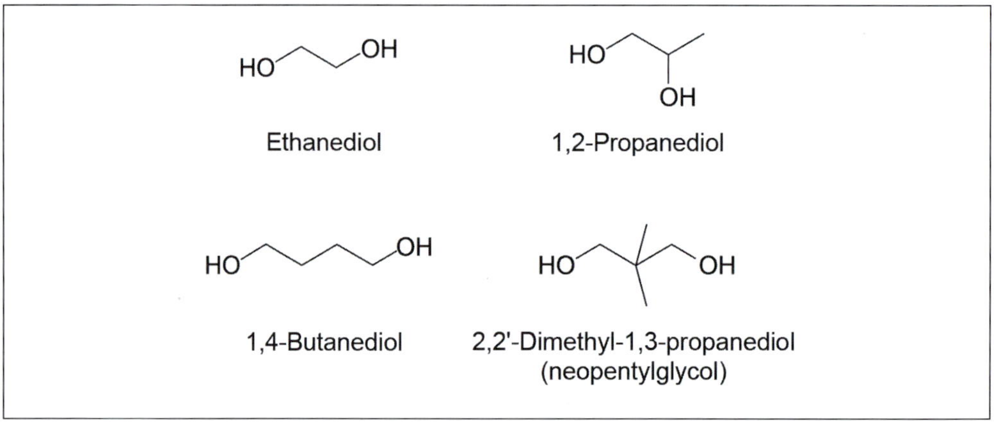

Figure 2.1.42: Diols as components for esterification of alkyd resins and unsaturated polyester resins

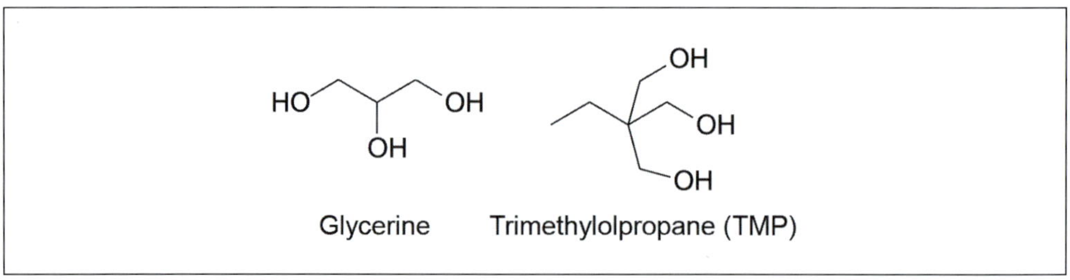

Figure 2.1.43: Triols as components for esterification of alkyd resins and saturated polyester resins

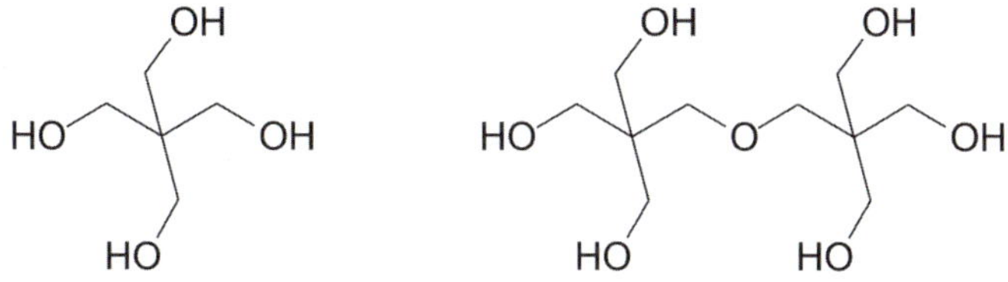

Pentaerythritol Dipentaerythritol

Figure 2.1.44: Tetrols and higher polyols for alkyd resins and saturated polyester resins

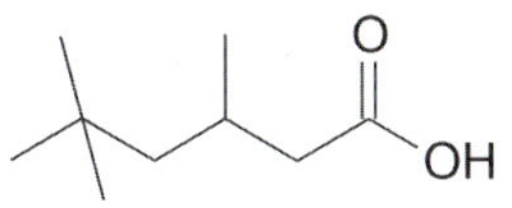

Figure 2.1.45: Isononanoic acid as an example of branched primary and synthetic fatty acids

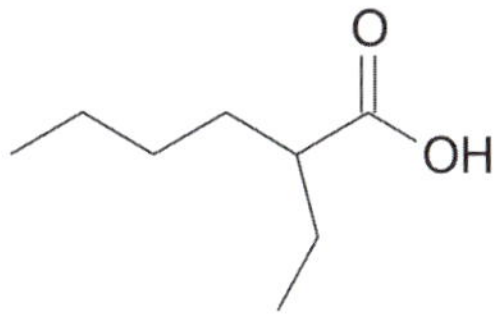

Figure 2.1.46: α-Ethylhexanoic acid as an example of branched, secondary synthetic fatty acids

particular interest for the oxidatively curing alkyd resins, thermally curing alkyd resins are generally manufactured with the aid of saturated, short-chain natural or synthetic monocarboxylic acids. By contrast with natural fatty acids, synthetic monocarboxylic acids feature greater purity and thus lead to more consistent quality in the resins. The saturated monofunctional structural elements which are important for thermally curing alkyd resins are either unbranched short-chain C_4 to C_{12} acids obtained from natural materials or synthetic fatty acids which are branched to a greater or lesser extent.

Isononanoic acid is highly branched in the alkyl group, but has a primary carboxyl group, nonetheless. Like the natural fatty acids, therefore, it is readily esterified, though equally readily saponified.

α-Ethylhexanoic acid, with a branch in the α-position to the carboxyl group, is harder to esterify, but it is at the same time also harder to saponify. This means that coatings have a greater resistance to chemicals.

As tert.-carboxylic acids, **versatic acids** with alkyl chains (C_9 to C_n) of different lengths have virtually no esterification capability due to pronounced steric hindrance. They are therefore generally provided as glycidic esters bearing a reactive epoxy group. This behaves like a diol under the conditions prevailing during alkyd resin production.

As a supplement to the aliphatic fatty acids, the aromatic **benzoic acid** and p-tert.-butyl benzoic acid as monofunctional structural elements serve as modifying agents for improving compatibility with other resins, in addition to their role as viscosity regulators.

There are two different methods for **manufacturing** the **alkyd resins**. Either free fatty acids, dicarboxylic acids and polyols are made to react together in an appropriate mixing ratio (fatty acid process) or the actual esterification is preceded by a transesterification of the natural oils with appropriate quantities of polyols into bifunctional starter products. Only after this is esterification carried out with phthalic anhydride (see Figure 2.1.31).

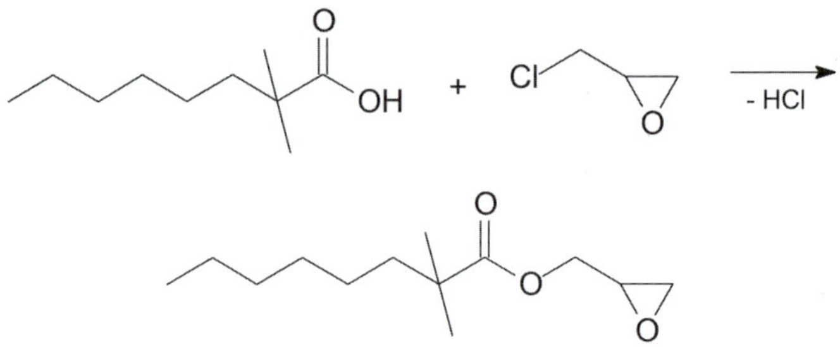

Figure 2.1.47: Epoxy-modified versatic acid as raw material for alkyd resins

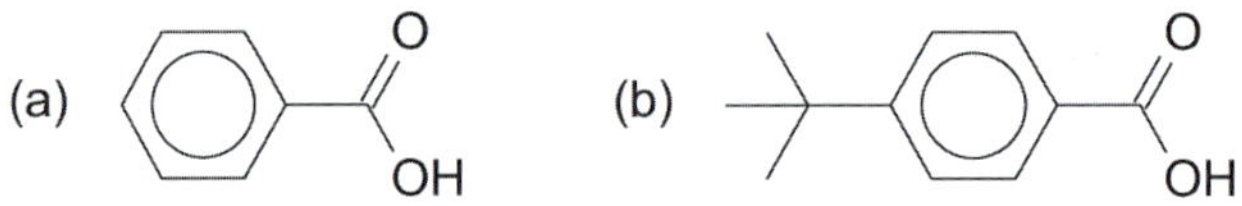

Figure 2.1.48: (a) Benzoic acid and (b) p-tert.-butyl benzoic acid

Figure 2.1.49: Vessel for manufacturing alkyd resins

Source: BASF Coatings

Unfortunately transesterification of the starting oils into bifunctional structural elements is not quantitative. In addition to measurable quantities of non-transesterified triglycerides, monofunctional and thus chain-breaking fractions are found.

If mention is also made of the fact that the relatively high transesterification temperature of up to 260 °C permits molecule-enlarging preliminary reactions via the fatty acids' double bonds, it will be clear that the various production processes cannot yield identical products. The molecule sizes and, in particular, the molecular weight distributions differ substantially at times, even when the formula ingredients have the same composition. Nevertheless, the two processes have not lost their technical importance to this day. Whereas in some countries the transesterification process is preferred, the fatty acid process is the more sensible method because it allows greater flexibility in the selection of raw materials for customised film forming agents.

If alkyd resins are to be manufactured in uniform quality, it is imperative in the current standard batch process that appropriate measurement methods be used to monitor molecular growth during the synthesis. This is usually done by conducting the polycondensation reaction over a specified temperature curve and measuring the cleavage products removed by distillation. Corresponding performance indicators are a rise in viscosity and a reduction in free acid groups (acid value). In the case of thermally curing alkyd resins, the number of free hydroxyl groups (hydroxyl value) must also be specified because of the later crosslinking with complementary resins (see Chapter 2.3.2).

The **properties of alkyd resins** are mainly determined by the composition of the raw materials, the molecular weights achieved and the molecular weight distributions. In this context, the molecular weight distribution is influenced by the production process and the reaction conditions. The latter give rise to various secondary reactions which impact quality to an extent depending on the production temperature and the reactivity of the raw materials. Unwanted preliminary reactions of the double bonds, etherification, dehydration with simultaneous attachment of hydroxy- and carboxy-functional molecule groups and unwanted cyclisation are just some of the reactions which affect the molecular size and form.

Preparatory fractionation of the alkyd resin and selective combination of the individual fractions enable the effect of the different sizes of the structural elements on the application properties to be studied. It turns out that, with the exception of unwanted residual monomers, all fractions contribute positively to the overall quality. While the high molecular fractions ensure good, rapid physical hardening during film formation, it is the polar, low molecular structural elements which promote wetting of the pigments during incorporation and thus uniform distribution in the liquid matrix [2.4.9].

In **long-oiled alkyd resins,** the incorporated fatty acids are primarily unsaturated, i.e. they render the resin amenable to oxidative crosslinking. Reactivity is closely correlated with the properties of the oil from which the relevant fatty acid mixture is derived and with the subsequent treatment (isomerisation, dehydration) which it has undergone. A high oil content delivers good levelling, leads to high elasticity and easy processability using workshop application methods. However, the slow curing of the film, and thus the mechanical sensitivity of the coatings shortly after processing, is a disadvantage. The most important alkyd resins in this class are based on linseed oil. As these yellow very markedly, resins based on soya oil and tall oil fatty acids are preferred in the decorative paints sector for light coatings which have to resist yellowing (see Figure 2.1.8).

Long-oiled alkyd resins should be used as sole film forming agents. **Medium-oiled alkyds** feature universal solubility and good compatibility with modified natural resins, cellulose nitrate and amino and amido resins. **Short-oiled alkyds**, both with oxidative crosslinking and with saturated monocarboxylic acids, are among the most widely used film forming agents for stoving enamels when used in combination with amino or amido resins, and frequently also with epoxy and phenolic resins. The performance profile is largely determined by the oil length and the type of fatty or monocarboxylic acids.

If saturated or only weakly unsaturated fatty acid mixtures such as those of coconut oil, castor oil or groundnut oil are added, the resulting alkyd resins do not crosslink oxidatively or do so only negligibly. They can, therefore, serve as plasticising components in cellulose nitrate and acid-curing coatings.

The choice of suitable fatty acid or monocarboxylic acid mixture for particular applications therefore plays an important role for stoving enamels, too. Stoving enamel coatings with low stoving temperatures of < 100 °C require oxidative secondary crosslinking.

This is achieved with high levels of fatty acids that have good resistance to yellowing and a high number of conjugated double bonds. Ricinoleic fatty acid, which is derived from castor oil fatty acid by dehydratation, is ideal because it absorbs less atmospheric oxygen and tends to yellow less than the analogous isolenic acids. Hardness and chemical resistance are achieved by using branched synthetic fatty acids. The aromatic monocarboxylic acids which are sometimes also used in the fatty acid component of alkyd resins make it easier for solvents to be released from the liquid coating film and promote physical drying.

By contrast with the long- and medium-oiled alkyd resins, the emphasis of the cross-linking function in short-oiled stoving enamel alkyd resins is transferred to conversion of the hydroxyl groups and in water-soluble alkyd resins is transferred to conversion of the carboxyl groups.

Despite the wide range of properties of alkyd resins, as a result of the type of oil, oil length, molecular weight, molecular weight distribution and free content of carboxyl and hydroxyl groups, numerous attempts have been made to adapt alkyd resins better to practical requirements by chemical modification.

The **modification of alkyd resins** with long-chain polyaminoamides made from dimerised fatty acids and aliphatic polyamines leads to **thixotropic alkyd resins** that possess the special feature of time- and shear-dependent flow (see Chapter 2.3.2).

The originally incompatible components can be rendered compatible by means of transesterification and transamidation. At the same time, because of the formation of numerous intermolecular hydrogen bridging bonds, viscosity in the form of pseudoplasticity and thixotropy is a function of the shear load. Since the thixotropy is also time-dependent, such resins open up new routes for formulating run-resistant coatings. Thixotropy is responsible for a temporary reduction in viscosity in response to the high level of shear exerted during spraying or brush application; this reduction is then reversed at a speed that depends on the level and type of thixotropy. Thixotropy is an important phenomenon and aid for trouble-free film formation, particularly on vertical surfaces. Alongside curing via solvent release or via chemical crosslinking, it is the third physical variable that affects film formation and it harmonises the levelling and running of coating materials, particularly in the case of water-borne paints (see Chapter 5.6.2).

By analogy with the addition reactions familiar from the fatty oils, **styrene-modified alkyd resins** result from the conversion of styrene or vinyl toluene with oxidatively curing alkyd resins. The chemical linking of the styrene to the double bonds of unsaturated fatty acids can represent up to 50 % of the oil component. Physical drying is substantially improved as a result. The good resistance that coatings made of styrenated alkyds present to water and alkalis is offset by a permanent sensitivity towards solvents.

The incorporation of acrylic acid or methacrylic acid esters leads to **acrylated alkyd resins**. They also dry very quickly and yield coatings with good adhesion and excellent long-term outdoor exposure properties.

The partial replacement of the o-phthalic acid by diisocyanates yields the **urethane alkyds** group. The basic chain of these materials exhibits increased stability against saponifying agents, with the increase depending on the degree of substitution. Quantitative substitution of the dicarboxylic acid leads to the **urethane oils** which are parti-

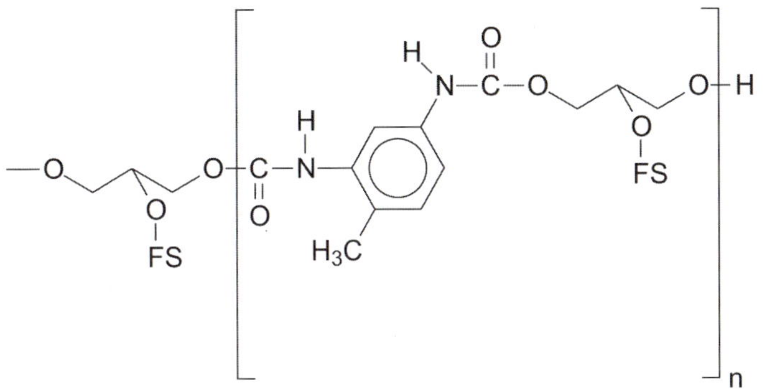

Figure 2.1.50: Structural formula of urethane oils based on toluylene diisocyanate (TDI) and fatty acid (FS)

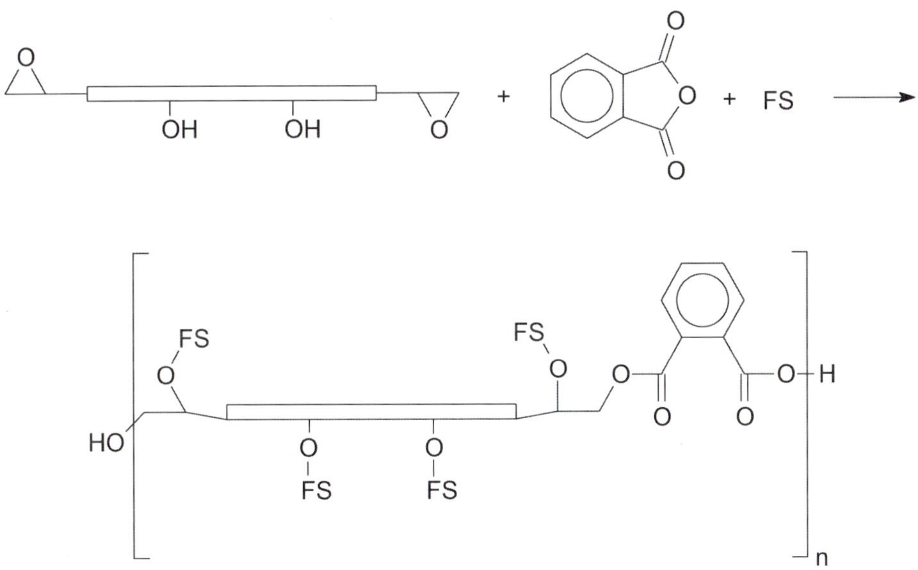

Figure 2.1.51: Synthesis of an epoxy-modified alkyd resin (FS = fatty acid)

cularly wear-resistant and, at the same time, have better resistance to acids and alkalis. Aromatic diisocyanates impair the resistance to yellowing. To ensure the production of resins which are resistant to yellowing, aliphatic modification agents, such as hexamethylene diisocyanate (HDI) or isophorone diisocyanate (IPDI), must be used.

Modification with epoxy resins is also possible. This is based on the partial replacement of the polyol, which is necessary for the manufacture of the alkyd resin, by a suitable epoxy resin (see Figure 2.1.51). During this process the epoxy groups and the hydroxyl groups of the epoxy resin act as the polyol component for the mono- and dicarboxylic acids.

Another way of modifying resins is the chemical bonding of epoxy resins with an alkyd resin. As expected, the result is a shift in properties towards the resin used for the modification. Improvements in the adhesion, resistance to saponification and hardness on one hand but also impairment of the lightfastness on the other are the outcome if aromatic ether types serve as the epoxy resins.

Hydroxy- or methoxy-functional silicone resins are also suitable for chemical bonding with hydroxy-functional alkyd resins. Such **silicone-modified alkyd resins** effectively improve the heat resistance and long-term outdoor exposure properties of the corresponding coatings, though only if the silicone resin is present in a proportion of 20 % or more. Despite the associated high price, silicone-modified alkyd resins have created an important role for themselves, particularly where iso- or terephthalic acid is used.

Coating Material

Through-hardening of oxidatively curing alkyd resins can be improved substantially by chemically modifying them with diketenes to enable them to form complexes with organic aluminium compounds. When unmodified alkyd resins undergo oxidative curing, molecular enlargement starts at the surface of the forming film, as a result of the crosslinking. This is accompanied by a reduction in oxygen permeability. Good through-hardening therefore requires only slow surface hardening. Thus if both properties need to be combined, further crosslinking mechanisms must be sought. The aforementioned use of diketenes to convert the alkyd resins' hydroxyl groups leads to acetoacetic ester groups which cause additional through-hardening of **metal supported crosslinked alkyd resins** during film formation by trans-complexation of an aluminium-acetylacetone complex added as a liquid component. The acetylacetone is released and then escapes.

The search for coating materials which can be processed in an environment-compatible manner has shown **water-soluble** and **water-emulsifiable** alkyd resins to be inexpensive film forming agents.

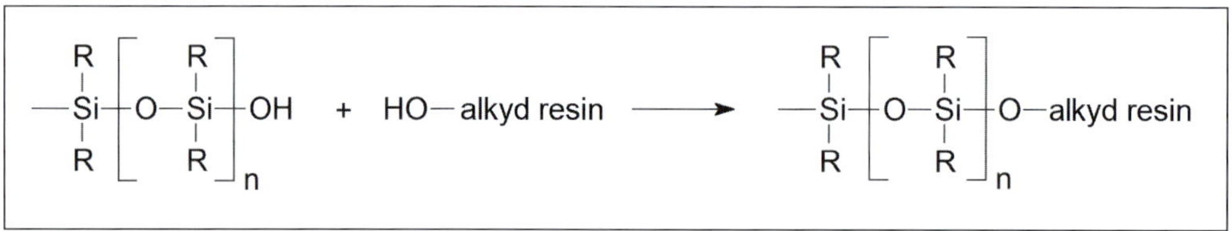

Figure 2.1.52: Structural formula of silicone-modified alkyd resins

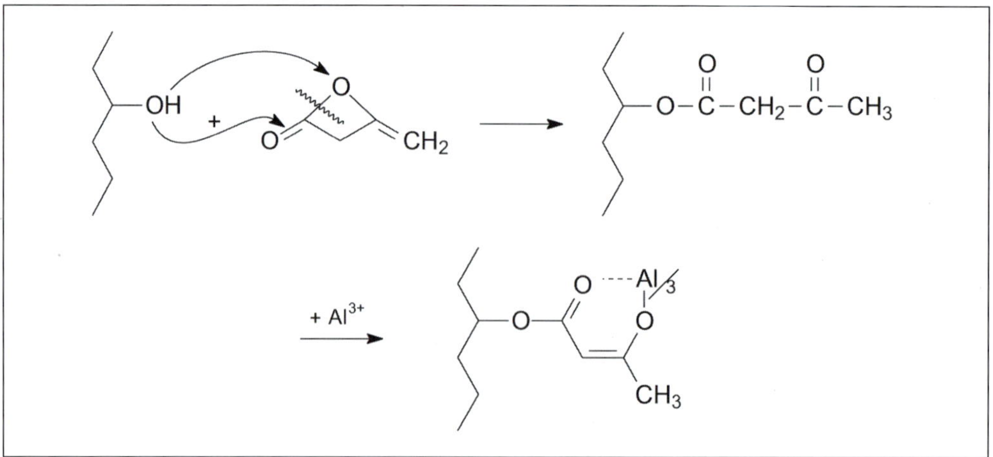

Figure 2.1.53: Film forming of metal-supported alkyd resins

Solubility in water or the ability to form stable emulsions in water is achieved by integrating polar functional groups corresponding to the polar nature of the water into the film forming agent. Free carboxyl and carboxylate groups have proved suitable in practice. The number of carboxyl and carboxylate groups determines whether the outcome is either a homogeneous solution or a dispersion or emulsion, i.e. a heterogeneous distribution of extremely fine resin droplets in water.

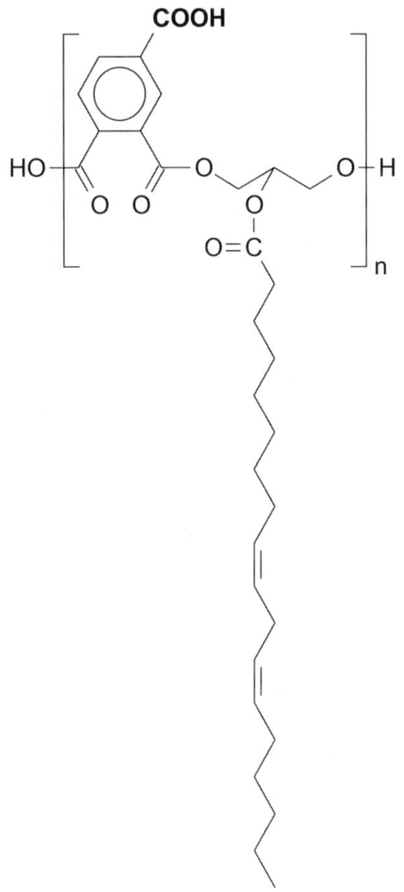

Figure 2.1.54: Typical structural element of alkyd resins for water-soluble resins

Neutralisation of the polar carboxyl groups with amines increases their polarity and enables the manufacture of stable emulsions and solutions as a function of the quantity of amines and the supplementary solvents used, i.e. the cosolvents, which can be mixed with water. Glycol ethers and classic alcohols in volumes ranging from 5 % to 15 % have proved successful as cosolvents for aqueous solutions. The water-soluble alkyd resins obtain the number of carboxyl groups needed for hydrophilisation through substitution of trimellitic acid for some or all of the customary o-phthalic acid. This is the only way in which the demands for a sufficiently high molecular weight and a sufficient number of carboxyl groups to ensure solubility in water can be met. Compared with solutions, emulsions offer the advantage of being able to cope with a smaller number of polar groups and significantly lower levels of cosolvents.

The environmental benefit of reduced emissions during the processing of oxidatively curing alkyd resins dissolved or emulsified in water must be offset in the form described thus far against the disadvantages of the polar carboxyl or hydroxyl groups which impair the film resistance. If, as happens in the film formation of conventional alkyd resins, physical curing is followed by oxidative curing, the carboxyl groups remain in the cured film. By contrast, the carboxyl groups in the thermally curing alkyd resins react with complementary resins. Initial problems with the resistance of emulsions, which was compromised by saponifiable ester groups, were overcome by displacement into the interior of the emulsion droplets.

Saturated Polyester Resins

The starting point for the success of oilfree **saturated polyesters**, which are now becoming more and more important, is the glyptals which were controversial at the time of their invention. In the early days they could only be manufactured with difficulty and were often incompatible with non-polar partners. In addition, the alkyd resins first synthetically manufactured by *Kienle* with their short-oiled variants exhibited a certain degree of yellowing when subjected to thermal loads, light or long-term outdoor exposure. Only when fatty acids were completely dispensed with did it prove possible to manufacture non-yellowing resins.

Usable resins were produced by replacing the glycerine with neopentyl glycol, trimethylol propane or pentaerythritol and substituting isophthalic acid or terephthalic acid for phthalic anhydride. The elasticising which now becomes necessary due to the lack of fatty acids is achieved by internal plasticising by incorporating adipic acid or sebacic acid into the resin. Any solubility in water which is desired is obtained, as in the case of the conventional alkyd resins, by replacing the dicarboxylic acids with the trifunctional trimellitic acid or tetrafunctional pyromellitic acid. The molecular form and the molecular weight, and therefore the solubility and compatibility with other resins, are affected by the overall functionality of the raw materials and the degree of conversion.

The success of saturated polyesters is due in part to their ability to provide specific properties for powder coatings and coil coatings. In both cases, thermal resistance in conjunction with high requirements imposed on application properties may be mentioned. This has led to the establishment of saturated polyesters as the base resins for sophisticated polyurethanes in addition to their standard applications as stoving enamels.

Saturated polyesters, whether they are purely hydroxy-functional or, as specialty resins for water-borne paints and powder coatings, also carboxy-functional, are exclusively polar representatives, compared with their fatty acid-containing rivals, and are therefore also only compatible with correspondingly polar complementary resins. Low-molecular melamine resins highly etherified with methanol have proved successful. The high degree of etherification promotes mutual crosslinking of the melamine resin and the saturated polyester. Quality-diminishing self-curing, as occurs in melamine resins with a low etherification level, is thus prevented (see below).

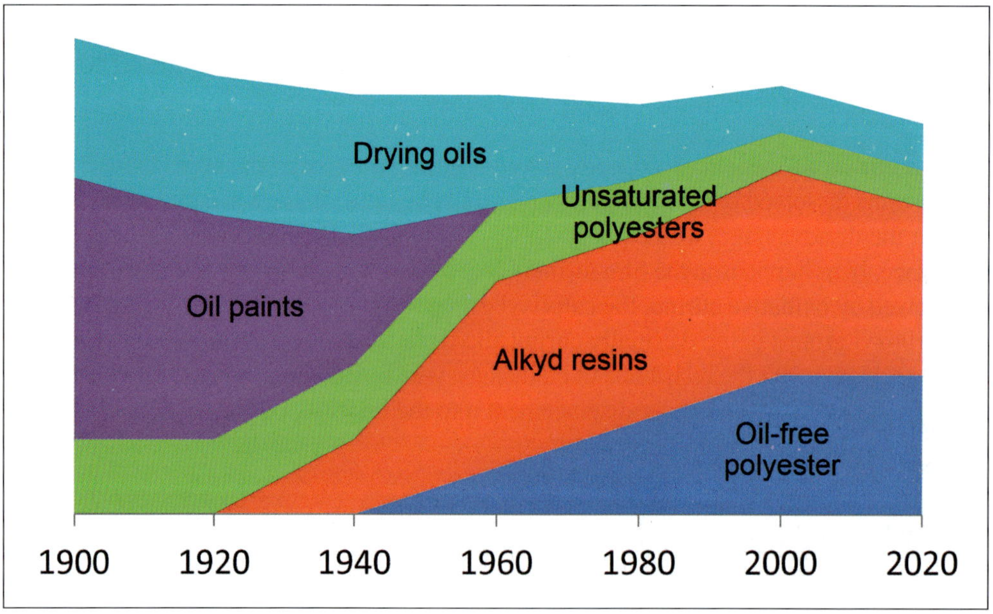

Figure 2.1.55: Qualitative chart of the development of the saturated polyester market in comparison to other resin types *Source: authors' estimate*

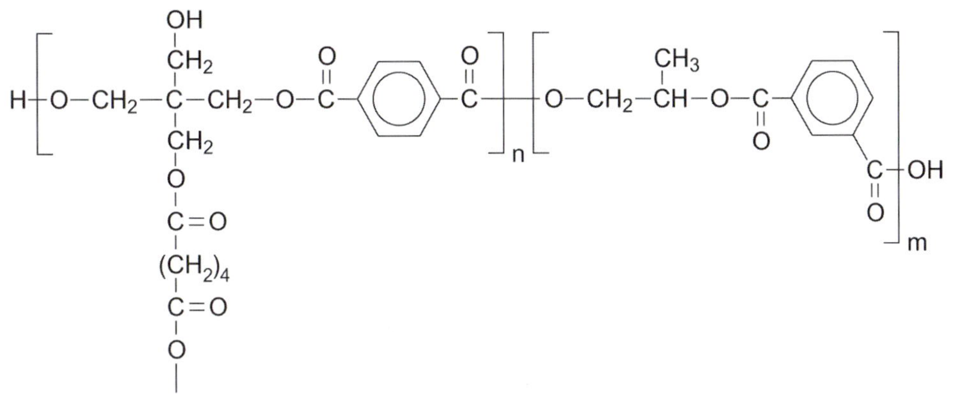

Figure 2.1.56: Typical structural unit and building blocks of a saturated polyester

Despite having a lower number of basic structural elements by comparison with alkyd resins, saturated polyester products are available in a wide range. The solvent-borne polyesters with molecular weights up to 25,000 g/mol should be mentioned first. As linear or branched products they are OH-functional and therefore suitable for crosslinking with melamine or isocyanate resins.

Solvent-free polyesters with similarly high molecular weights are used as solid resins, with OH- or carboxy-functional again being used for powder coatings. Those representatives which have hydroxyl groups are combined with polyisocyanates, while the carboxy-functional types were previously combined with triglycidyl isocyanurate (TGIC), though this crosslinking component has now been replaced by isophthalic acid, trimellitic acid glycidyl ester or N,N,N',N'-tetrakis-(2-hydroxyethyl)adipamide (Primid) or similar ß-hydroxyl alkylamides because of its proven mutagenic effect [2.4.10].

The basic reactions b) and c) in Figure 2.1.57 are polycondensation reactions and suffer from the disadvantage that, in the event of relatively large film thicknesses, removal of the cleavage products generated during crosslinking can give rise to flaws in the film.

Hydroxy- and at the same time carboxy-functional resins with acid values of 40 to 80 mg KOH/g (see Chapter 2.3.2) become water-soluble after neutralisation with amines, although cosolvents are required for acid values in the lower range quoted. Water-borne stoving enamels can be formulated for the automotive and coil-coating sectors and for industry in general by combining them with water-soluble, highly etherified melamine resins.

Low molecular hydroxy-functional polyesters with molecular weights of 1,500 to 4,000 g/mol can also be processed into 1-component high solid coatings with solid contents of up to 70 %. Even higher solid contents are possible if the molecular weights are reduced still further and crosslinking is carried out with low molecular isocyanate hardeners. The merely slight increase in viscosity which is a result of physical drying after application has to be compensated by pseudoplasticising or thixotroping.

A further application area for saturated hydroxy-functional polyesters opens up if the hydroxyl groups at the ends of the polyester chains are esterified with acrylic acid. This makes it possible to use high-energy UV or electron radiation for curing (see Chapter

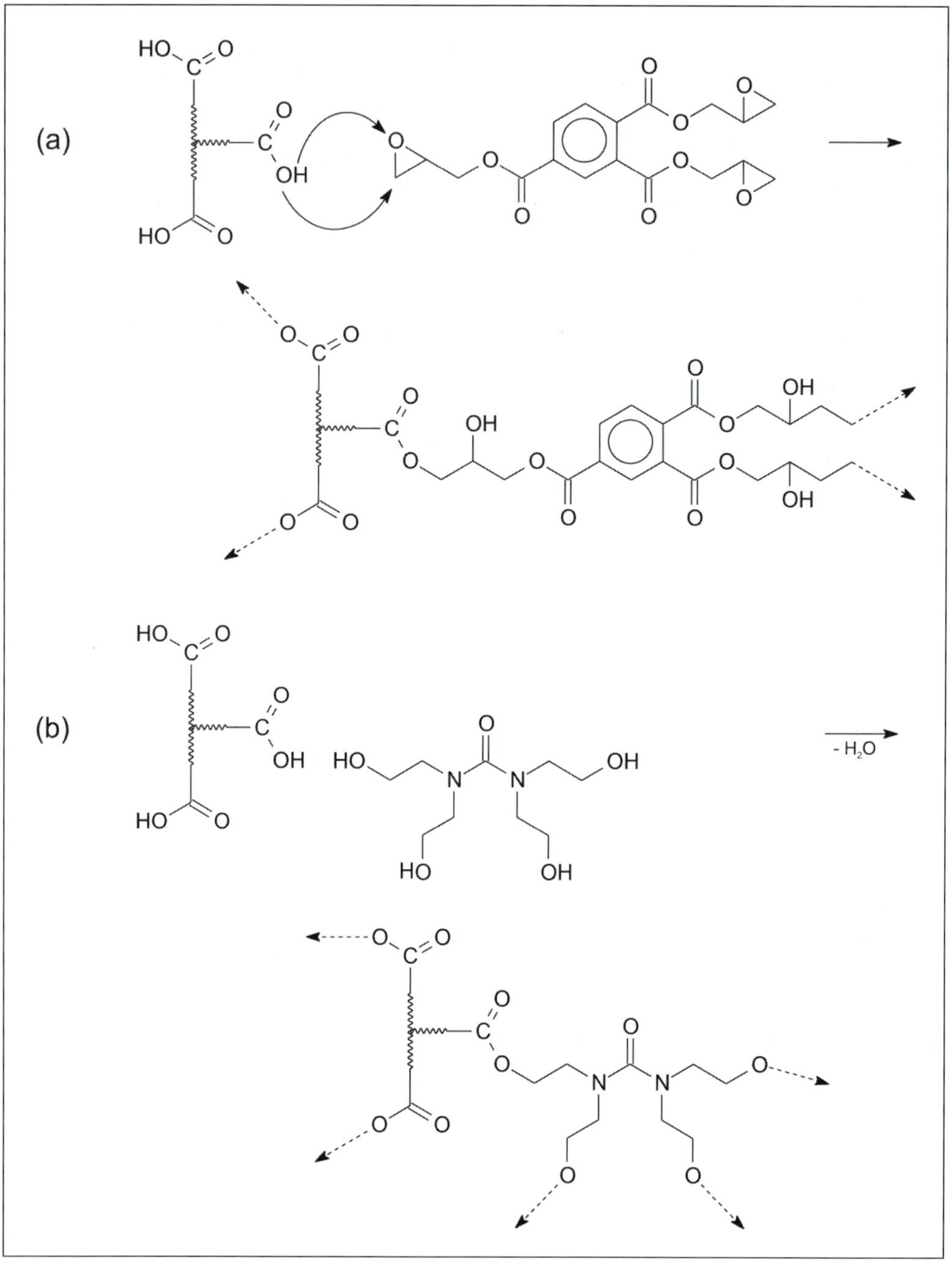

Figure 2.1.57: Crosslinking reactions of saturated polyester resins for powder coatings (a) with trimellitic glycidyl ester, (b) with ß-hydroxyl alkylamides, (c) with blocked isocyanates (continued on page 67)

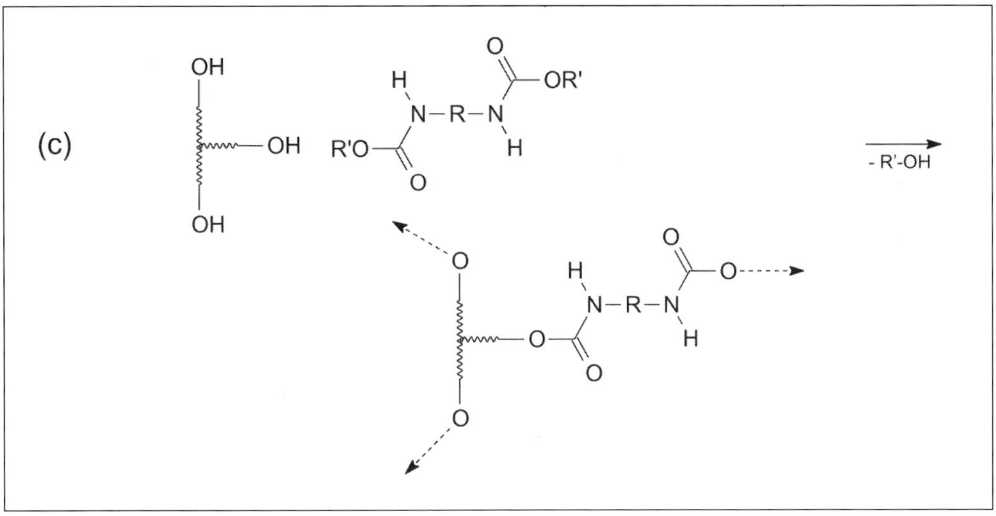

Figure 2.1.57 (continued from page 66): Crosslinking reactions of saturated polyester resins for powder coatings (a) with trimellitic glycidyl ester, (b) with β-hydroxyl alkylamides, (c) with blocked isocyanates

4.3.2). Such resins can also be copolymerised with monomers of acrylic acid or methacrylic acid, a fact which overcomes the general incompatibility of polyesters with acrylic resins.

Modifications of saturated polyesters with silicone resins also yield interesting film forming agents for coatings that possess extreme long-term outdoor exposure properties if they are made to react with melamine resins in polycondensation reactions or with aliphatic/cycloaliphatic polyisocyanates in polyaddition reactions.

Saturated polyesters generally feature a high level of hardness combined with pronounced elasticity. High crosslinking densities and mobile segments in the basic chains are the reasons that these are suitable for coil coatings which have to withstand extreme deformation without suffering damage. Good gloss, excellent film build, resistance to yellowing when over-stoved and good chemical resistance are other noteworthy properties.

Unsaturated Polyester Resins

Unsaturated polyester resins are semi-finished products which are transformed into a thermosetting coating through copolymerisation of their double bonds with styrene or vinyl toluene acting as chemically reactive solvent. Unsaturated polyesters are members of the class of polycondensates or condensation polymers because they are manufactured from dicarboxylic acids and diols in polycondensing reactions.

The special feature of this class of substances, when compared with the saturated polyesters, lies in the linear structure of the resin molecules and the proportion of maleic anhydride used. This makes crosslinking with the styrene solvent possible. In this process the ratio of maleic acid to phthalic acid in conjunction with the quantity of styrene determines the crosslink density of the cured film and thus the solvent resistance and the mechanical/technological properties. Since the styrene as a reactive diluent does not leave the paint film on curing and, in addition, no volume-reducing cleavage products are formed, unsaturated polyesters can be applied in high film thicknesses without causing any problem. They therefore have important applications for thick-film coatings in the wood and furniture sectors, as well as resins for automotive repair fillers.

In addition to the ratio of maleic anhydride to phthalic anhydride, elasticising basic structural elements/units can also affect the mechanical/technological properties. On the dicarboxylic acid side these are adipic acid or sebacic acid. On the polyol side the standard components ethanediol and 1,2-propanediol are partially or wholly replaced by 1,3-propanediol, neopentyl glycol, diethylene glycol, 1,4-butanediol or triethylene glycol.

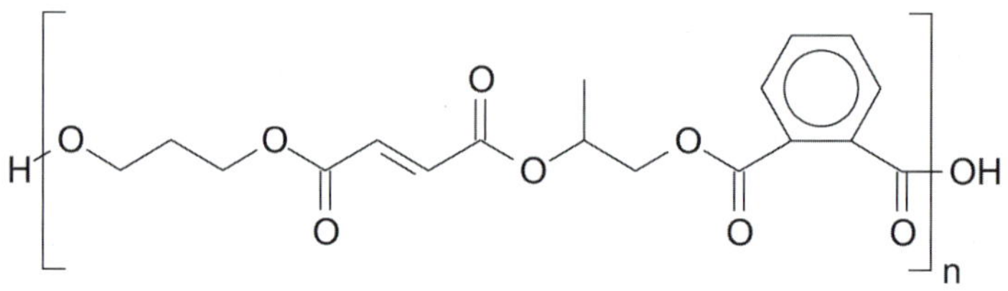

Figure 2.1.58: Structural unit of an unsaturated polyester

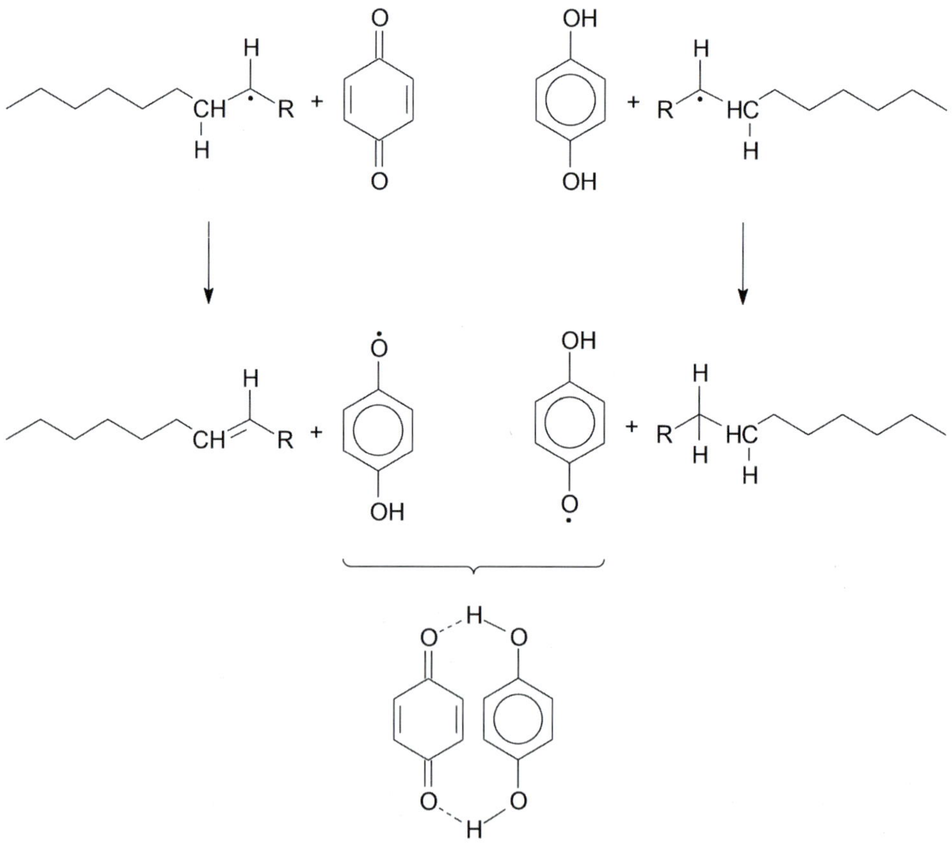

Figure 2.1.59: Chemical effect of the inhibitors quinone and hydroquinone

When unsaturated polyesters are being synthesised, trouble-free esterification is only possible if atmospheric oxygen is excluded at reaction temperatures of 190 to 210 °C and with the addition of inhibitors because of the risk of premature crosslinking.

Radical scavengers such as hydroquinone and quinone in volumes of 0.001 to 0.1 % have proved to be successful inhibitors. If unwanted curing has started, they can halt it again by chain transfer. The ready-to-use unsaturated polyester is a mixture of the actual polyester resin, styrene, and hydroquinone which serves as a stabiliser to increase storage stability.

Since the inhibitor retains its inhibiting properties when initiators are added to harden the coating, it affects the starting point of the curing process and the crosslinking rate. Manufacturers make use of this fact by varying the inhibitor quantity to adjust the pot life to operating conditions.

Not all application techniques can be used to apply unsaturated polyesters to the object to be coated. Spraying is not an option because of the huge increase in the material surface area during application and the resulting acceleration in evaporation of the styrene reactant. This leaves roller and curtain coating as usable processing methods. Both are used solely for planar objects such as coils, and wooden and plastic panels.

Since the styrene solvent is also a reactant in the curing process and therefore must not leave the forming film, thermal processes cannot be used to accelerate crosslinking of the film. The mutual bonding of styrene to the double bonds of the basic chain of the unsaturated polyester resin can therefore only be induced by "cold" formation of starting radicals. These can be obtained from catalytic decomposition of peroxides and hydroperoxides or from photoinitiators exposed to high-energy radiation (see Chapter 2.1.4). Whereas the first option yields 2-component paints with a limited pot life, the second

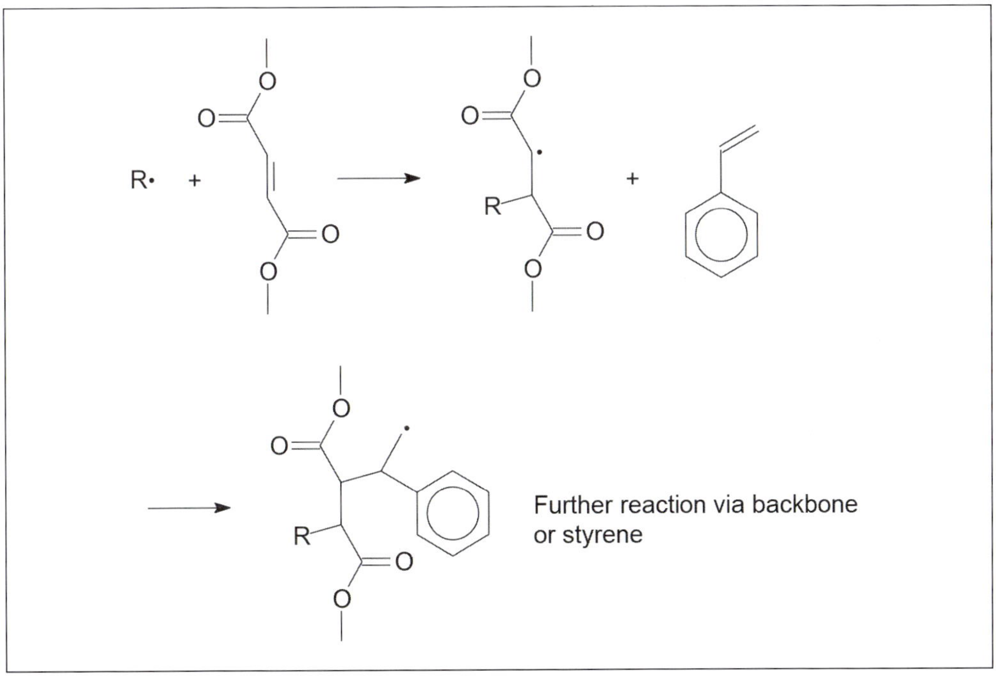

Figure 2.1.60: Crosslinking reaction of unsaturated polyesters with styrene

opens up the possibility of 1-component paints which are UV-curable. 2-Component paints consist of the base containing the unsaturated polyester, styrene, inhibitor, and accelerator, and the hardener, which contains the peroxide.

Decomposition of peroxides or hydroperoxides is achieved by the catalytic action of certain metallic soaps. Cobalt salts can be successfully used to accelerate radical formation because of their ability to reversibly change valency. Commercial radical initiators are available in a wide ranged, including dibenzoyl peroxide, cumyl hydroperoxide and t-butyl perbenzoate.

Tertiary amines are alternatives to cobalt salts. Their effect is based on radical-forming hydrogen transfer.

Both product groups of accelerators have an established track record. Whereas the highly reactive amine is gradually consumed by the curing process and its effect thereby diminishes, the cobalt salt acts purely catalytically, with relatively low reactivity initially but better through-hardening. Blends of the two product groups are therefore used to ensure the best processing properties. The amines' tendency to yellow is a disadvantage to note in light coatings.

The chemical destruction of the inhibitors that sets in after decomposition of the initiators and the subsequent exothermic crosslinking reaction give rise to a restricted pot life. Users have two performance indicators to enable them to make a quantitative evaluation. The **gel time** is defined in DIN 53 184 as the time which elapses until the temperature of the reaction mixture has risen by 10 °C as a result of the heat of the crosslinking reaction. The **curing time** is the time taken to reach the maximum temperature.

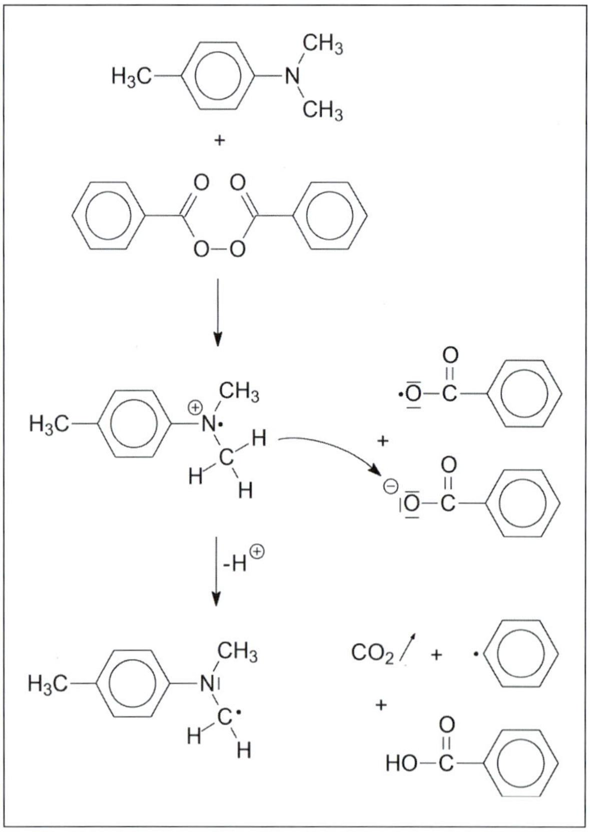

Figure 2.1.61: Mechanism of catalysis by amines for the decomposition of peroxides

Because of the limited pot life and at the same time the relatively slow curing of unsaturated polyester resins, additional curing methods have developed. The process of generating radicals by means of **photoinitiators** under the simultaneous effect of UV radiation has been in use since the 1950s. The pot life was now no longer limited and this advantage was then supplemented by a significant acceleration in the curing process. Photoinitiators or sensitisers are benzoin derivatives, benzil ketals, α-halogen ketones and benzophenone derivatives. The last of these do not decompose into radicals, but rather abstract hydrogen atoms from suitable synergists which are thereby transformed into radicals (see Chapter 2.1.4).

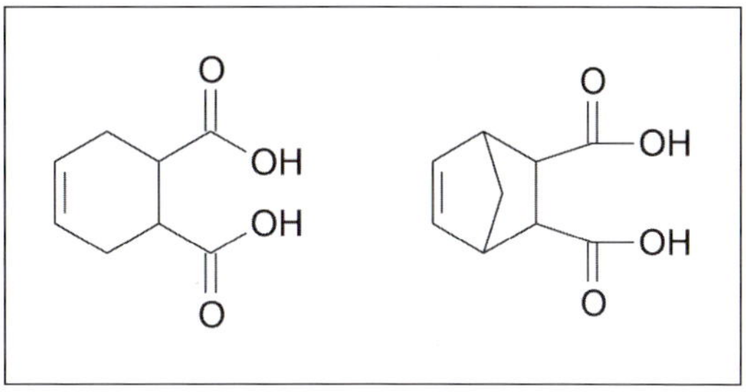

Figure 2.1.62: Dicarboxylic acids for unsaturated polyester coatings that are not inhibited by oxygen

Mercury vapour lamps serve as the radiation source for the curing process, supplying the necessary energy in the form of short electromagnetic waves from low pressure and high pressure lamps. Electron beam curing further shortens the crosslinking time while also enabling photosensitisers to be dispensed with. Accelerated electrons are capable of forming radicals or radical ions by direct interaction with the coating resins (see Chapter 4.3.2).

Intensive research in the field of radiation curing brought new understanding of the mechanism behind photopolymerising curing. The ratio of the rate of chain growth to chain termination is crucial for a high degree of crosslinking and thus high hardness in the coatings. Out of the huge range of different classes of substances, special acrylic resins proved advantageous in this context. Consequently, more than 60 % of radiation curable coatings contained acrylic resins as their base resin by 1995 (see below).

No matter which processing method is used, crosslinking is inhibited by atmospheric oxygen. As a diradical present in high concentrations, it intervenes in the molecule-enlarging chemical reactions to terminate radical chains. All the standard resins described thus far therefore have the drawback of failing to achieve optimum surface cure, coupled with permanent stickiness. There are two different approaches to eradicating this problem. Either the interaction of the oxygen with the forming coating must be stopped by adding floating waxes, or the chain terminating effect of the oxygen must be prevented by the chemical insertion of special groups into the polyester resin, i.e. by incorporating these groups into the coating formulation via additives. The first method requires an additional working stage for removing the floating layer of wax by polishing.

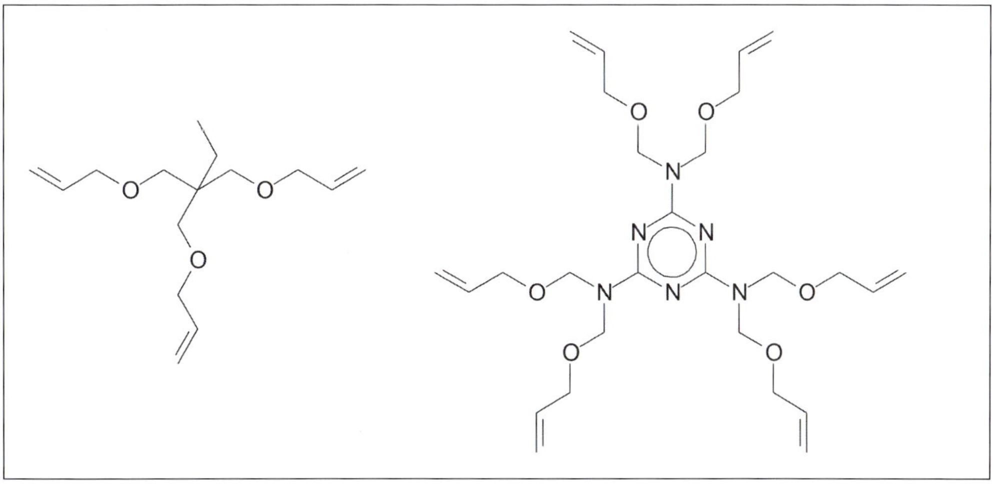

Figure 2.1.63: Examples of additives for suppressing inhibition by oxygen

The chemical route proceeds either by adding hydroperoxide-forming and thus oxygen-deactivating structural elements during production of the polyester or by subsequently including additives which contain allyl groups. Possible substances for the first case are tetrahydrophthalic acid and endomethylene tetrahydrophthalic acid.

The paints formulated on these principles can be safely applied in high film thicknesses by means of the low-waste-generating processes of curtain and roller coating. As a result, these methods are still widely used in the wood and furniture sector to this day.

Polyamides

Polyamides are polymeric compounds whose basic structural groups are bonded to each other via acid amide groups. Despite being related to polyesters; polyamides are only of negligible importance in coating technology terms.

Synthesised from aliphatic dicarboxylic acids and aliphatic diamines or ω-aminocaprolactam by *Carothers* in the USA in 1929 and by *Schlack* in Germany in 1935, they originally served as synthetic replacements for the then expensive Japan silk because of their chemical similarity to natural silk [2.4.11, 2.4.12].

It was only later that methods of manufacturing polyamides from other basic structural elements were developed. The greatest significance in terms of coating technology may be ascribed to polycondensates of dimerised fatty acids and aliphatic polyamines.

Polyaminoamides have proved successful as thixotropic agents for alkyd resins and as hardeners for epoxy resins, the application depending on the mixing ratio of the starting components. If an excess of dicarboxylic acids is chosen, starting materials for providing thixotropy are obtained. In the presence of excess amines, the then liquid polycondensates become raw materials for flexible epoxy resin coatings.

The polyamide manufactured from ω-aminoundecanoic acid merits mention as a base resin for thermoplastic powder coatings processed by the fluidised-bed method. The amino acid is made available by thermal decomposition of ricinoleic acid. followed by attachment of amino groups to the resulting product.

Like the related polyimides, polybenzimidazoles and polybenzimidazopyrrolones, polyamides are proven starter products for thermally stable plastics and coatings.

On account of the low mobility of the rigid monomers which are unable to form coils, the entropy changes only minimally during synthesis or decomposition. Consequently, the ratio of the change in enthalpy to the change in entropy has a relatively high value.

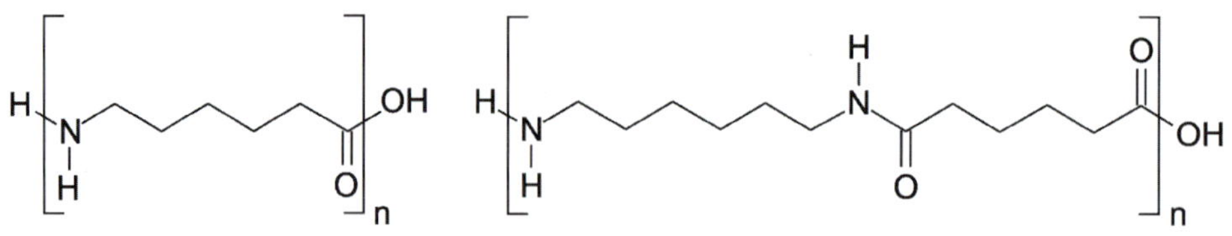

Figure 2.1.64: Structural units of polyamide 6 (from ε-aminocaprolactam) and polyamide 66 (from hexamethylene-diamine and adipic acid)

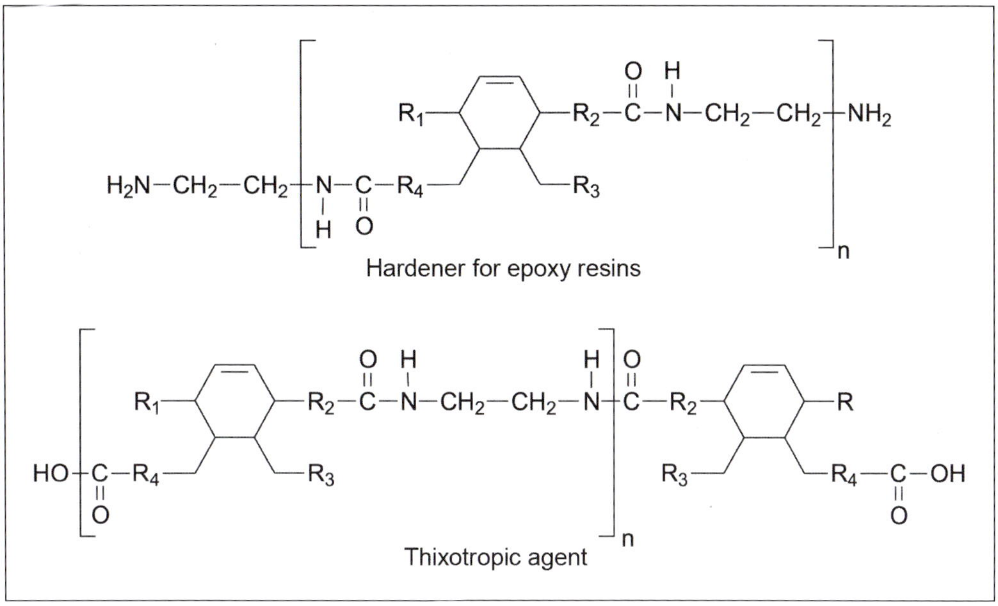

Figure 2.1.65: Polyaminoamides as thixotropic agents and hardeners for epoxy resins

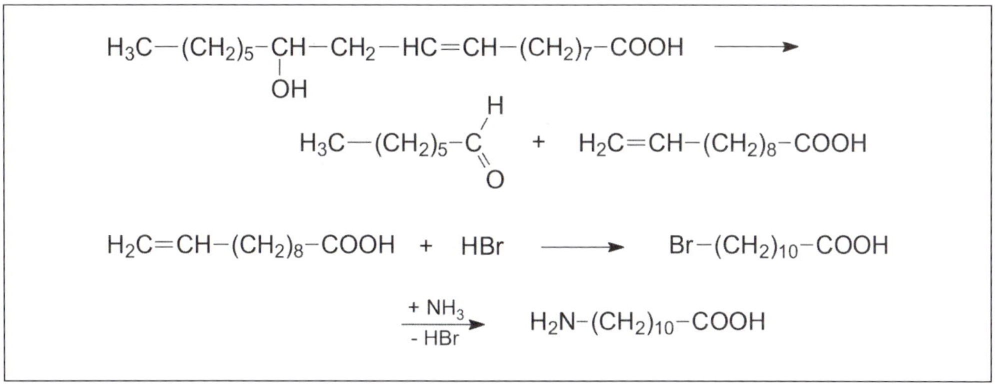

Figure 2.1.66: Synthesis of ω-aminoundecanoic acid from ricinoleic acid

$$T_c = \Delta H / \Delta S$$

T_c = ceiling temperature
ΔH = reaction enthalpy
ΔS = reaction entropy

The variable stated as the ceiling temperature T_c is the temperature at which the rates of polymer formation and polymer decomposition become the same. As ΔS is relatively small for the polyamide derivatives mentioned above, they have exceptional heat resistance.

Polyimides

Polyimides are manufactured from aromatic tetracarboxylic acids and aromatic diamines via the intermediate stage of acid polyamides. Heat resistances of more than 250 °C for continuous exposure and up to 600 °C for transient heat exposure are obtained with this class of substances.

The solubility which is necessary for use as coating materials is obtained by substituting diphenyl endomethylene cyclohexane tetracarboxylic acid for the pyromellitic acid, though this leads to a gradual reduction in the heat resistance.

If the tetracarboxylic acid is replaced by the bifunctional isophthalic acid phenyl ester and if tetraaminobiphenyl, a 4-functional structural element, is simultaneously substituted for the diaminophenyl ether, the result is the polybenzimidazole class of substances via the intermediate stage of a polyamide.

The poor resistance of polybenzimidazoles to atmospheric oxygen can be eliminated by partially replacing the amino groups with mercaptan groups, i.e. by converting the

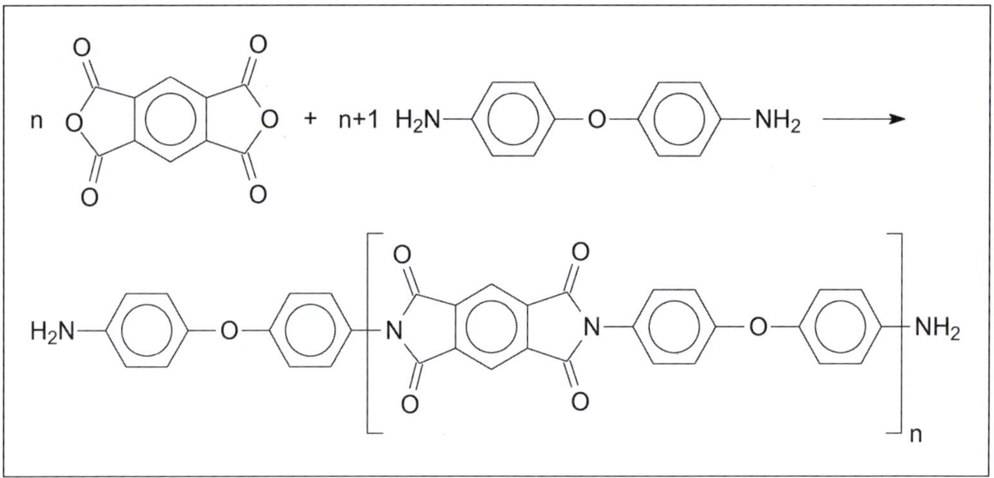

Figure 2.1.67: Synthesis of polyimides from pyromellitic acid and diaminodiphenyl ether

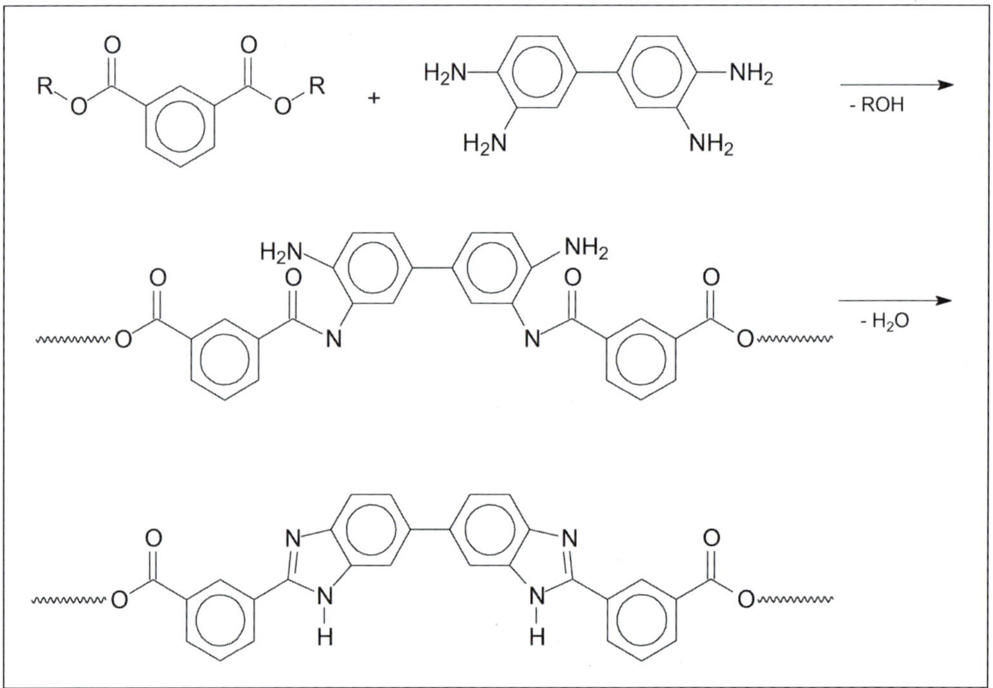

Fig.2.1.68: Polybenzimidazoles from isophthalic acid phenyl ester and tetraaminobiphenyl

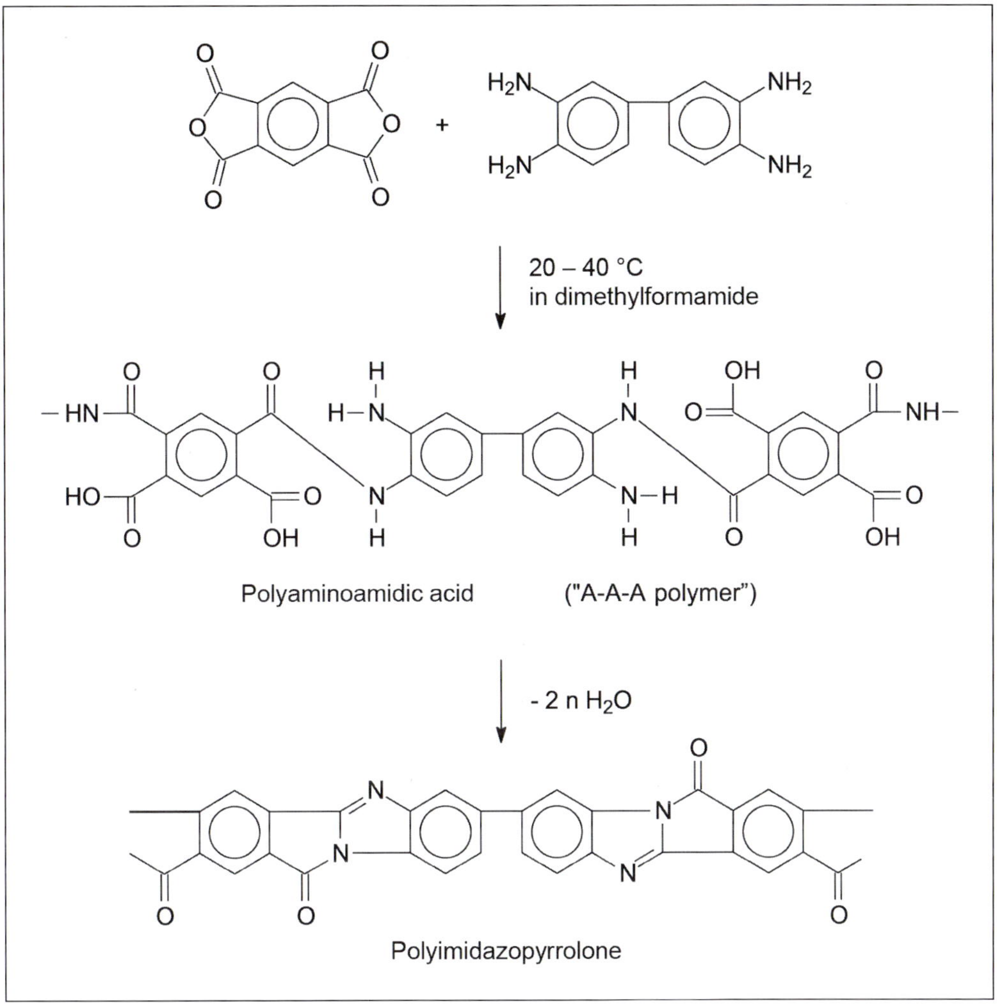

Figure 2.1.69: Polyimide azopyrrolones from pyromellitic acid and tetraaminobiphenyl

imidazoles into thiazoles. Only exotic solvents can be used for processing as coating materials with dimethyl sulphoxide.

A further increase in the molecular stability to heat can be achieved by polycondensation of the 4-functional pyromellitic acid with the equally 4-functional tetraaminobiphenyl to form the **polyimidazopyrrolones**. Apart from further-improved heat resistance, this class of substances also has excellent stability against the effect of high-energy radiation. The polymeric azopyrrolones therefore serve as resins for special aerospace applications.

As polyimides cannot be crosslinked and are difficult to dissolve, they are only employed as a raw material for special coatings, particularly wherever high heat resistance is called for.

Silicone Resins

By contrast with the film forming agents which have been discussed so far and which essentially consist of carbon chains, the elements that make up the backbone of **silicone**

resins (polysiloxanes) are silicon and oxygen. Hydrocarbon groups are found only as substituents in the side chains. Greater stability is achieved because of the higher bond energy of 375 kJ/mol in the alternating silicon/oxygen bonds compared with 340 kJ/mol for the carbon/carbon single bond. Any organic alkyl or aryl groups attached to the backbone are always directly connected to the silicon. Different groups of product can be synthesised in accordance with the type of basic structural elements employed. These groups range from liquid to paste-like silicone oils or silicone greases via elastomeric silicone rubbers to solid silicone resins.

Due to their good resistance to long-term outdoor exposure, chemicals and damaging atmospheric effects and to their excellent mechanical/technological properties, silicone resins have always proved successful wherever particularly high demands are imposed on the coating. They are important components for coil coatings. This application is particularly demanding in terms of heat resistance because of the high coil speeds and the associated high stoving temperatures.

Polysiloxanes are now manufactured exclusively by the *Müller-Rochow* synthesis method [2.4.13]. This entails making extremely finely-ground elemental silicon react with alkyl halides over a cuprous catalyst to form mono-, bi- and trifunctional organosilanes. The individual components are then separated by distillation, before being joined back together again in defined mixtures for molecule-enlarging hydrolysis, in accordance with the chemical goal. The mixing ratio of the mono-, bi- and trifunctional basic structural elements determine the molecular size and shape. Monofunctional chlorosilane terminates chains and thus limits molecular growth, whereas trifunctional silanes cause molecules to branch.

The polysiloxanes can be divided into various types on the basis of the organic groups bonded to the parent chain. If only chloromethanes are used for reaction with silicon, extremely non-polar, macromolecular **silicone oils** are obtained. These

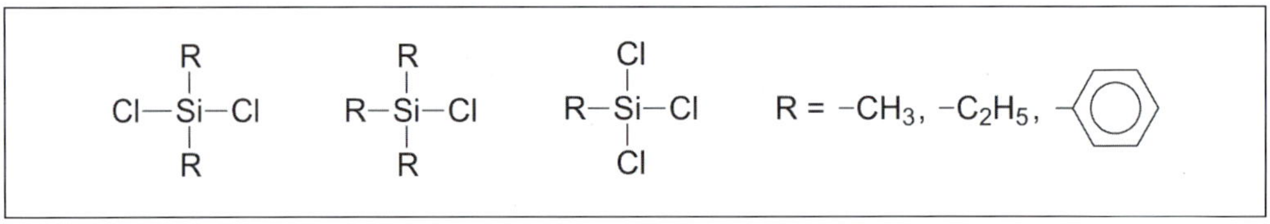

Figure 2.1.70: Basic components for the synthesis of polysiloxanes

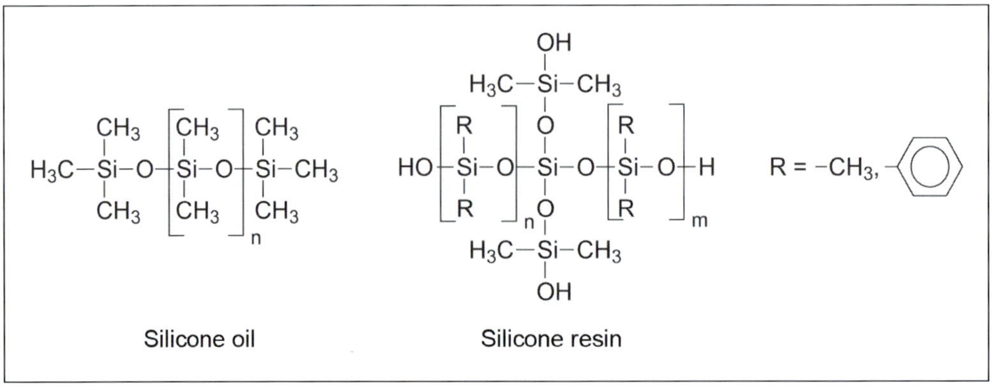

Figure 2.1.71: Structural units of silicone oils and silicone resins

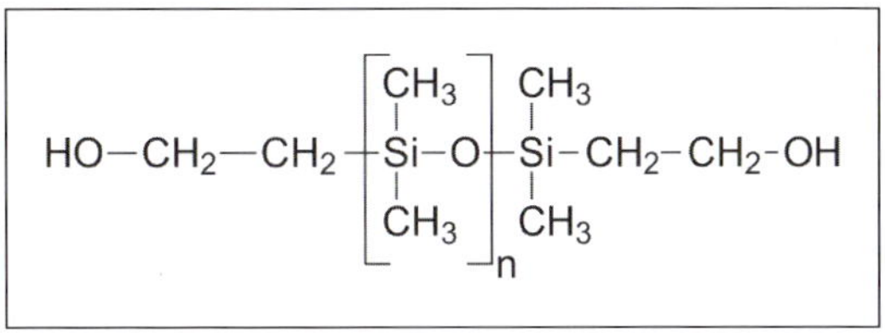

Figure 2.1.72: Polysiloxanes for hardening with isocyanate crosslinkers

are not chemically reactive, nor do they develop mechanical strength, regardless of their molecular weight. On account of their very low surface tension of 20 mN/m silicone oils and silicone pastes are important additives for wetting, for improving levelling and for defoaming.

Their incompatibility with almost all surface coating resins is also deliberately exploited to create surface textures, such as hammer finishes.

Compatibility with the usual surface coating resins is improved by replacing some of the methyl groups with bulky, polar phenyl groups and by simultaneously ensuring that the ends of the chains bear reactive groups. The hydroxy-functional chain ends can undergo co-condensation with other hydroxy-functional reactants either directly or after etherification with methanol. Precondensation with saturated polyesters or OH functional acrylic resins is worth mentioning in this context. Final hardening can then be carried out with the usual crosslinking resins.

Apart from the good properties already described, silicone resins have a continuous heat resistance of between 180 and 200 °C. Even transitory peak temperatures of up to 600 °C can be withstood without any quality impairment.

A further increase in the resistance of polysiloxanes can be achieved by crosslinking with isocyanates. This, however, entails replacing the acidic SiOH groups of the conventional silicone resins by neutral Si-CH_2-CH_2-OH (see below).

As they are expensive, silicone resins are also frequently modified with resins from other classes of substance. The mutual chemical bonding between the silicone resins and the modification resins takes place at an early stage in the manufacturing process. Important examples of the resultant modified products are **silicone polyesters, silicone alkyds, silicone-reinforced epoxy resins** and silicone **acrylates**.

The properties of the modified resins are determined by the starter resins and the proportions in the resin combination. Where the silicone ratio exceeds 50 %, the character of the silicone resin is dominant. In these cases, the modification leads to improved application properties relative to those of pure silicone resins, with only a slight deterioration in heat resistance. Where the silicone resin proportion is below 50 %, the partner's character prevails. Thus, oxidative crosslinking of a paint film consisting of an alkyd resin modified with 30 % silicone takes hardly any longer, assuming standard drying, than in the case of an unmodified alkyd resin.

The price of silicone resins limits the use of pure polysiloxanes to applications in which high resistance to heat and/or long-term outdoor exposure is indispensable. With the exception of certain branches of the electrical industry, such applications include the manufacture of ovens, stoves and other goods subjected to extreme temperatures. Recently a further application field has been added, namely the coating of cladding elements which are mounted on façades of high-rise buildings which are difficult to access.

Coating Material

Crosslinking Resins for Polycondensation Reactions

Many of the paint resins described require a reactant in order that the desired coating properties may be obtained via crosslinking. **Crosslinking agents** are generally low molecular, high-functionality resins. Important crosslinking resins for polycondensation reactions include **phenolic resins** and **amino resins**, with the subgroups of urea resins and melamine resins. Another group of crosslinking agents is the polyaddition isocyanate resins which are described in greater detail elsewhere.

When the chemist *C. H. Meyer,* an employee at *Louis Blumer,* received an order from his brother-in-law, a merchant, to manufacture shellac synthetically, but at the same time also received the incorrect information that shellac, being a vegetable product, would also contain phenol and formaldehyde, he could not possibly have known that, despite being given inaccurate information, he would succeed in manufacturing a resin that was extraordinarily similar to shellac, i.e. a secretion from an insect known as a lac and thus demonstrably an animal product. Even though the product was synthesised on a completely incorrect premise, the supposed imitation shellac was to go down in history as the first synthetic coating resin [2.4] – not by *C.H. Meier,* however, but by the Belgian *Baekeland,* who, after refining the synthesis of the phenolic resin class of substances under the name *Bakelite,* was able to introduce the world to the first mass-produced plastic [2.4.4].

Although a few decades earlier, under different circumstances and with a different objective, *Adolf von Beyer* had failed to synthesise phenolic resins [2.4.14], *C. H. Meier* succeeded in isolating a chemically inert synthetic resin which was soluble in polar solvents by mixing phenol with a mistakenly low volume of formaldehyde. The new coating was named **Novolak**. Today it is regarded as the original resin in a now large class of resins with important descendants in coating technology.

Novolaks

The synthesis of novolaks from phenol and formaldehyde succeeded despite the high reactivity of the reaction mixture because the reactants were mixed in a ratio that only permitted a limited molecular size.

The electrophilic addition of formaldehyde to the phenol body is continued by a subsequent polycondensation stage to yield the novolak. If the proportion of formaldehyde is increased during the synthesis process, this leads to premature crosslinking and the product's therefore becoming unusable because of the trifunctionality and high reactivity of the phenol in the acid environment.

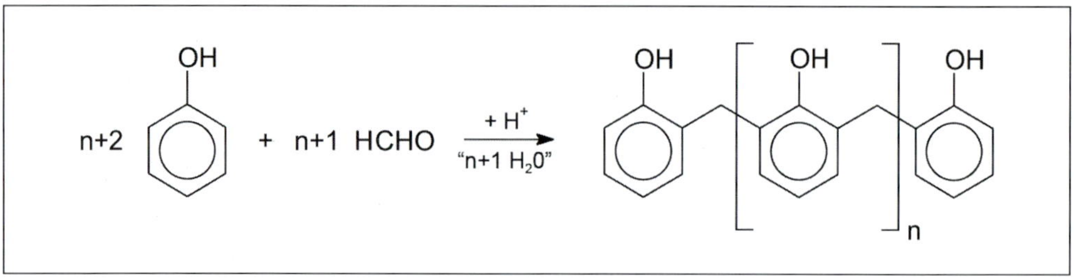

Figure 2.1.73: Synthesis of novolak from phenol and formaldehyde

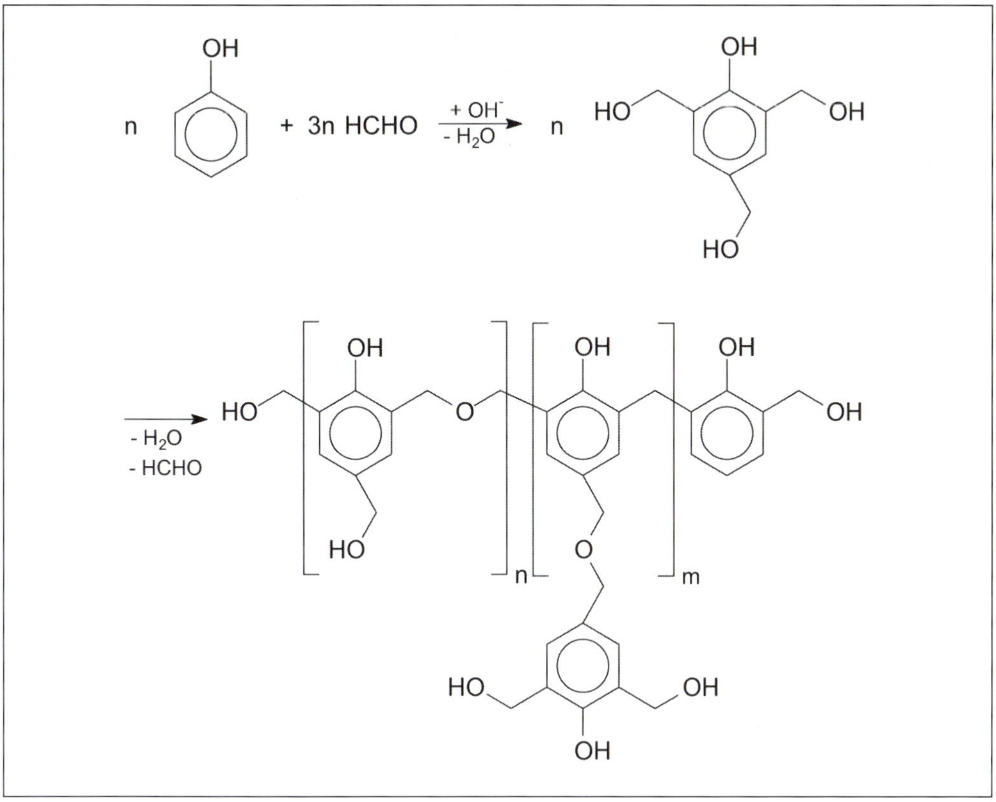

Figure 2.1.74: Synthesis of resols from phenol and excess formaldehyde

Only when scientists learned to slow the reaction rate down by increasing the pH value such that intermediate stages of the polycondensation processes could be stabilised by premature cooling was the now more important resols subgroup invented.

Resols contain hydroxymethylene groups and are therefore reactive resins which can self-cure to become resites by heating or acid catalysis. Phenolic resins are less suitable for use as sole film forming agents because of their poor elasticity. However, an additional modification is needed if they are to be mixed with elasticising OH-functional alkyd resins or saturated polyesters.

To improve compatibility, reduce the tendency to self-condensation and to achieve additional elasticity, phenolic resins must be etherified with aliphatic alcohols.

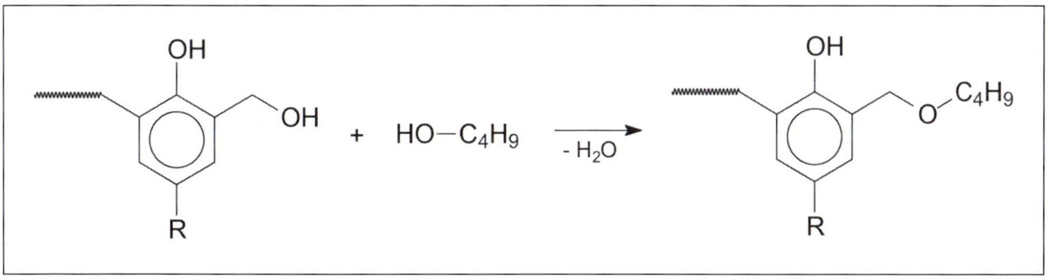

Figure 2.1.75: Etherification of resols with butanol

The variety of phenolic resins commercially available today partly results from the different types and levels of etherification. These determine reactivity, compatibility with other resins and solubility. Moreover, there has been no shortage of attempts to manufacture further variants by using additional raw materials and modifying the base resin.

Replacing phenol with alkylphenol produces a reduction of functionality and, in accordance with the type, position and length of the alkyl groups, permits the polarity to be lowered. The improvement in compatibility with non-polar oils should also be noted in this regard.

Phenolic Resins

An opposite effect is achieved by incorporating diphenolic acid (DPA), a conversion product from laevulinic acid and phenol. Such products are converted into phenolic resins which are water-soluble but, at the same time, are stable when stored after neutralisation of their carboxyl groups.

Through modification with cardanol, a natural material obtained from cashew nuts (CNSL or cashew nut shell liquid), phenolic resins even acquire oxidatively curing properties. The conversion products of resols with colophony or maleic resins, termed artificial copals, which are modifications of already manufactured phenolic resins, are worth noting [2.4.15].

During the production and processing of phenolic resins, the acidic phenol group remains chemically unchanged. Despite chemical crosslinking, therefore, films made of crosslinked phenolic resins are unstable in alkalis. Stability in alkalis can be generated by chemical inactivation of the phenol group. Transesterifications with acetyl chloride or trans-etherifications with epichlorohydrin have proved successful. The latter yield either epoxy novolaks that contain epoxy groups or hydroxy-functional glycerobenzyl resins, with the outcome being determined by the stoichiometric ratio of the reactants.

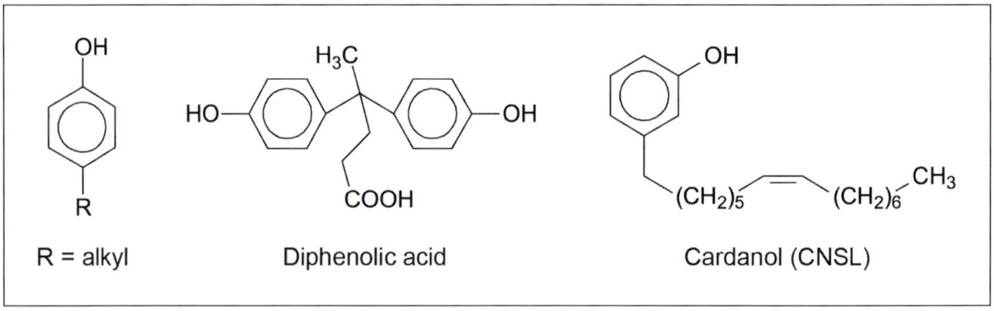

Figure 2.1.76: Phenol derivatives as components for phenolic resins

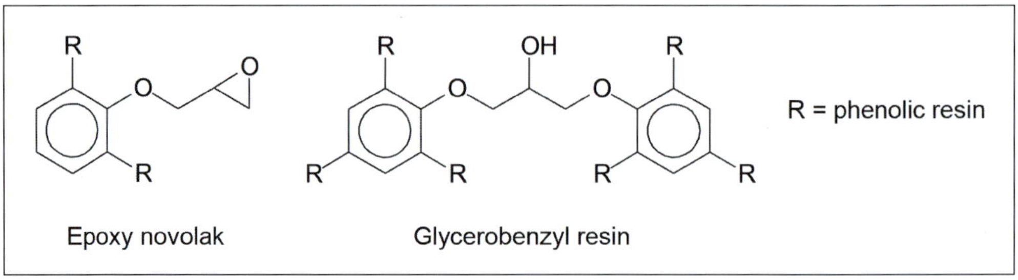

Figure 2.1.77: Alkali-resistant phenolic resins obtained by modification of the phenolic group

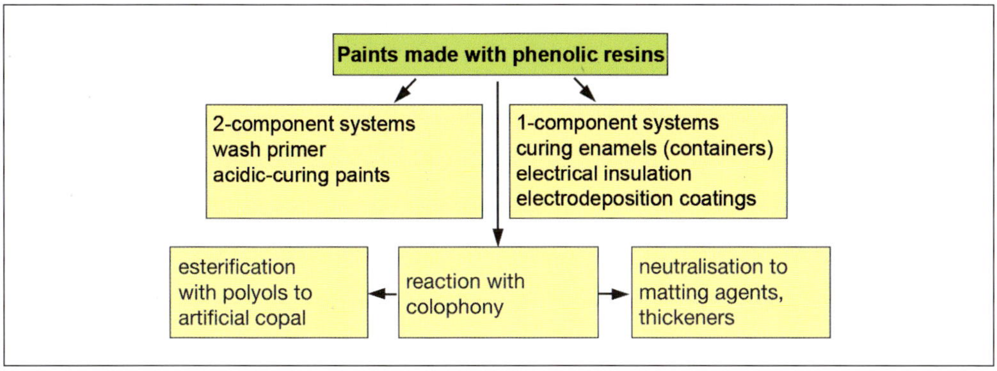

Figure 2.1.78: Applications of phenolic resins

In general, phenolic resins are found in numerous applications today because of their good acid stability, hardness, adhesion and electrical insulation capability. They deserve mention as coatings for receptacles ("gold coatings") and, in for their water-soluble variant, as electrocoating paints.

When combined with polyvinyl butyral (see below) and phosphoric acid they are known as wash primers and offer effective anticorrosive protection as primer coats for untreated steel or ground areas. In combination with alkyd resins and epoxy resins they are also important film forming agents for stoving enamel primers and primer surfacers processed on an industrial scale.

Since phenolic resins require relatively high stoving temperatures of 180 °C and they display marked yellowing at the same time, there has been no shortage of attempts to develop more reactive and, at the same time, more colour-stable alternatives. Urea, melamine, benzoguanamine, various urethanes and recently also glycoluril are the preferred materials from the broad range of products tested to replace phenol in reaction with formaldehyde. They are all nitrogen-containing compounds with reactive amino or amido groups and can readily be converted to corresponding methylol compounds with formaldehyde. Molecular enlargement can be achieved, if required, by subsequent self-condensation. The reactivity of aniline and alkyl-substituted benzenes with formaldehyde has also been investigated, but these materials remain only suitable for exotic niche markets because of their different chemical properties.

Urea Resins

Urea resins, which are etherified to various degrees with mostly butanol and methylolated to various degrees with formaldehyde, generally occur in solution in butanol. Commercially available urea resins are relatively low molecular products with molecular weights < 1000 g/mol.

Intramolecular dehydration and subsequent ring formation leading to hexahydrotriazine as a secondary reaction during manufacture is the reason that the functionality of the urea resins is lower than would be anticipated from the chemical structure of the urea.

Urea resins do not serve as sole film forming agents owing to the brittleness of their films and a lack of resistance to water. These properties, together with lightfastness and sandability, are improved in combinations with soft resins and cellulose nitrate.

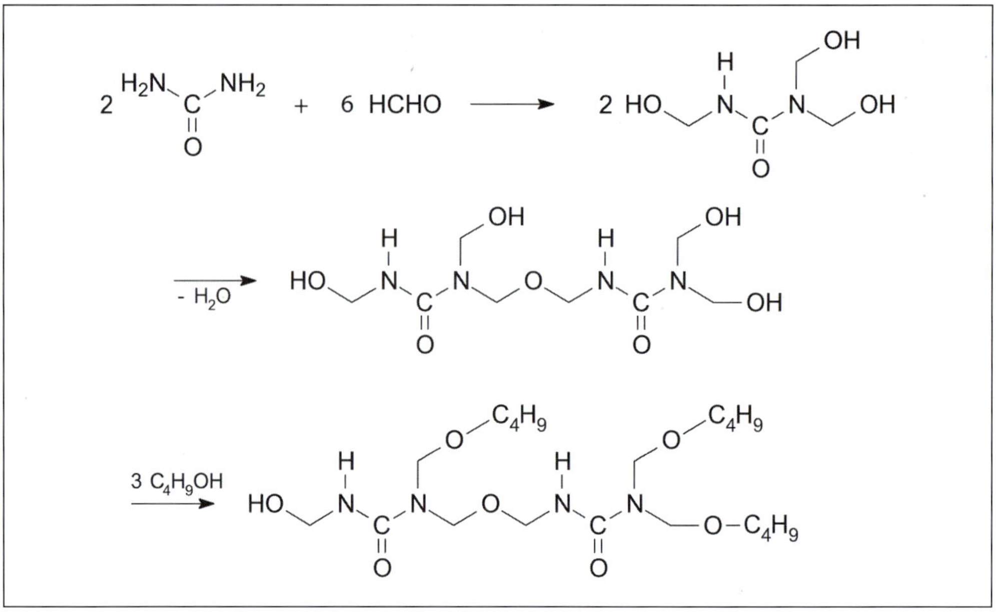

Figure 2.1.79: Synthesis of partially etherified urea resins

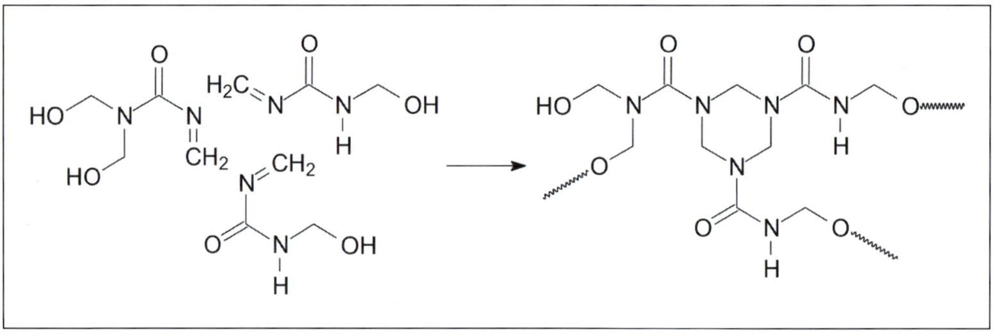

Figure 2.1.80: Side reactions during the synthesis of urea resins

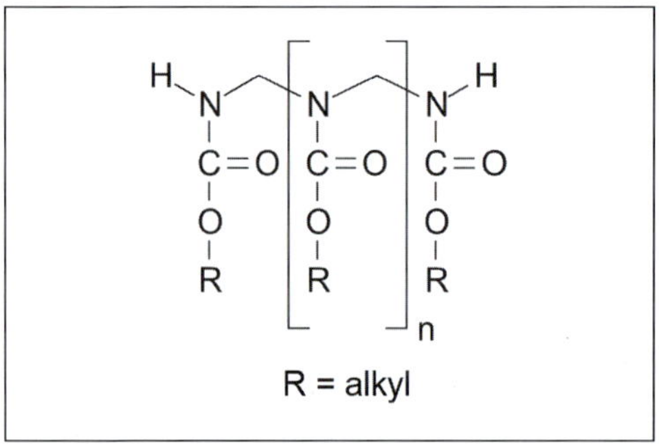

Figure 2.1.81: Chemical structure of urethane resins

When the resins are used in acid-curing paints, plasticising is generally carried out with alkyd or polyester resins. The most common partners of etherified urea resins in stoving enamels are lean to semi-fatty hydroxy-functional alkyd resins.

Urea resins can be elasticised and their compatibility enhanced at the same time by precondensation with saturated polyesters. This generates plasticised urea resins which can be used in all relevant application fields which require acid-curing or thermally reactive products, and they can also serve as sole binders.

Urethane resins, which are similar to urea resins, are reaction products of esters of carbamic acid. Because of the bifunctionality of the starter materials only linear structures are possible. They remain soft and elastic up to high molecular weights. Their plasticity increases with the chain length of the carbamic acid ester. Urethane resins are employed only as modifying agents in combination with other resins, for example as elasticising agents for acid-curing urea resins.

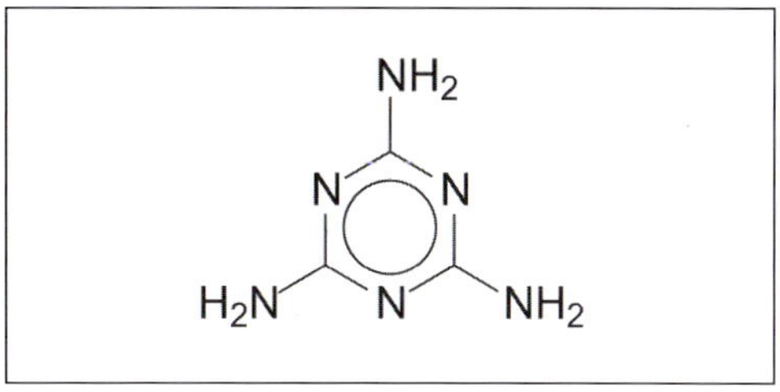

Figure 2.1.82: Melamine (2,4,6-triamino-1,3,5-triazine)

Coating Material

Melamine Resins

A higher crosslink density compared with the urea resins can be obtained by replacing urea with its maximum functionality of four by the hexafunctional melamine. This is readily obtained by conversion of urea with carbon dioxide and is thus an inexpensive raw material.

The target properties during the production of melamine resins are influenced by the mixing ratio of melamine to formaldehyde which is usually between 1:3 and 1:6, the etherification alcohol and the degree of etherification. Different molecular weights occur as a function of the reaction conditions.

Whereas traditional, polynuclear melamine resins are condensation products which consist of 5 to 6 melamine structural elements and which, for the purpose of imbuing them with sufficient reactivity and compatibility with standard solvents, are only partially etherified with butanol, a different route is followed as regards the molecular structure of today's melamine resins. A new generation of low molecular, in the most extreme case even mononuclear, products which are highly etherified with methanol has been created to improve compatibility with polar saturated polyesters through to water-soluble reactants. Due to their low molecular mass they are liquid and at the same time are soluble in both organic solvents and water.

For this reason, high solid and water-borne paints are formulated with these melamine resin variants. If we add that a high etherification level lowers undesirable self-condensation during crosslinking with hydroxy-functional reactants, the importance of the new melamine resin class, such as hexamethoxymethyl melamine (HMMM), is underlined from a quality perspective, too.

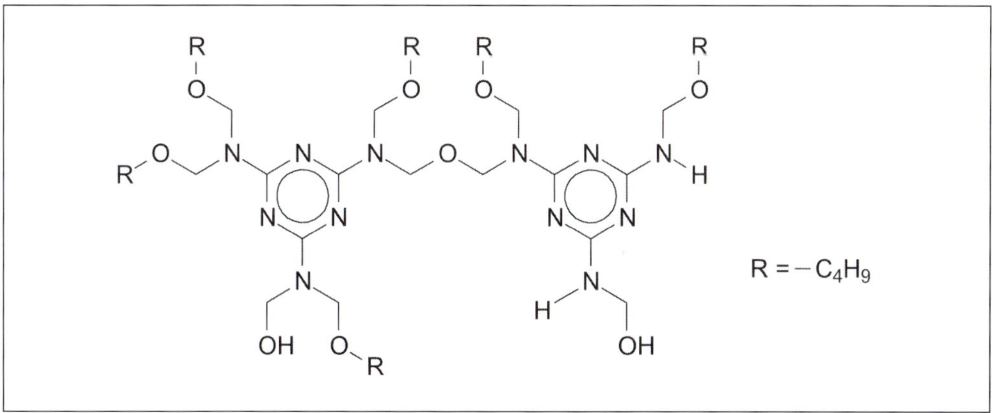

Figure 2.1.83: Traditional condensed melamine resin for crosslinking OH-functional reaction partners

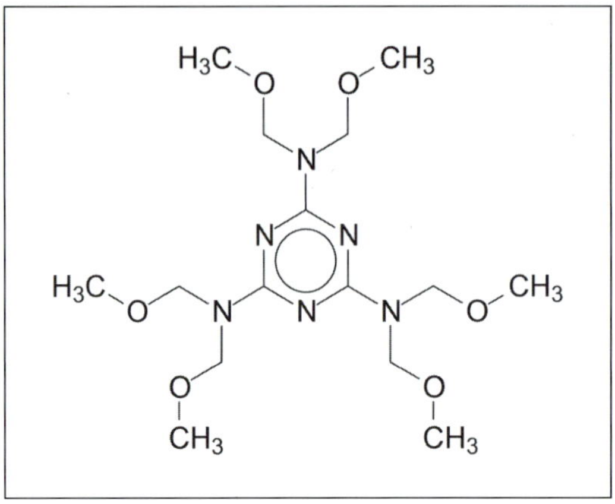

Figure 2.1.84: Hexamethoxymethyl melamine (HMMM)

The lower reactivity which is a consequence of the high etherification level must be compensated for by acid catalysis using carboxyl groups of the reactant or by separately added acids.

Since the important stoving enamels processed on an industrial scale are exclusively melamine resin combinations with short-oiled alkyd resins, saturated polyesters or OH-functional acrylic resins whose crosslink density determines the quality, it is important to pay particular attention to the etherification level. While the capability for external crosslinking increases with rising etherification level, the thermal reactivity shows a marked deterioration.

A low etherification level leads to stability problems during storage because of the tendency towards self-condensation.

If water-soluble reactants are used as the reactants for melamine resins, it is important to ensure that the carboxyl groups which confer solubility in water react with the melamine resins during the stoving process to improve long-term outdoor exposure properties. Only if this is the case will it be possible to obtain coatings whose quality level is comparable to that of traditional solvent-borne coatings.

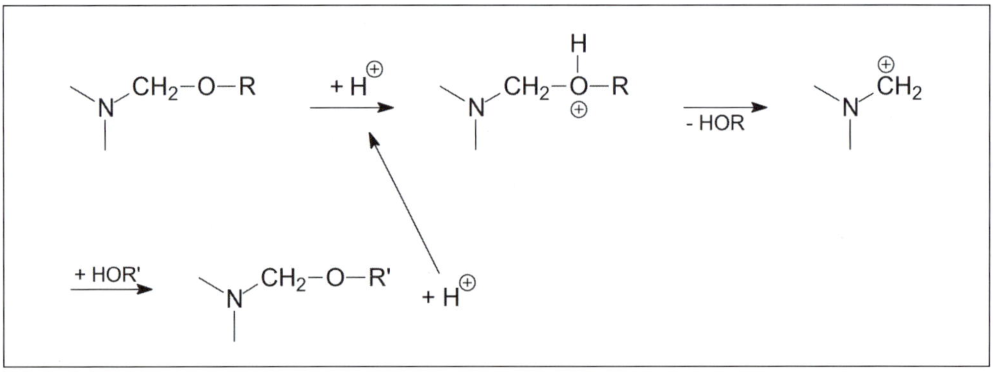

Figure 2.1.85: Reaction scheme for crosslinking highly etherified melamine resins with OH-functional resins

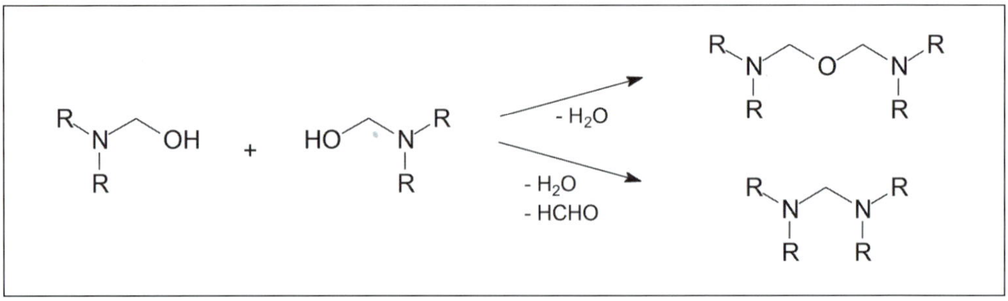

Figure 2.1.86: Self-condensation reaction of melamine resins

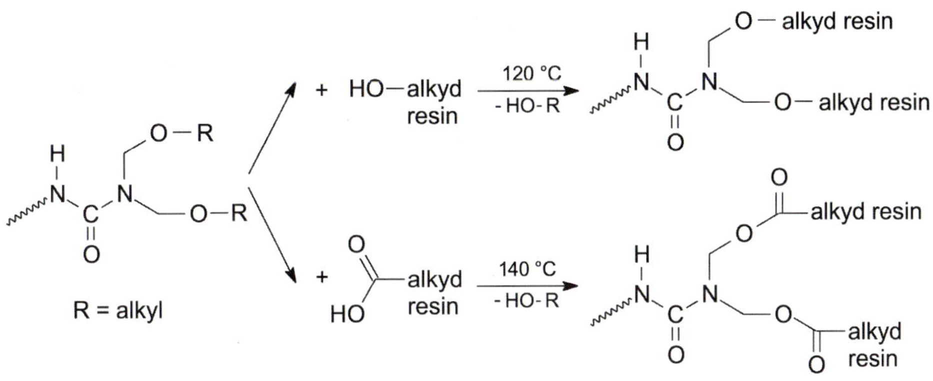

Figure 2.1.87: Crosslinking of melamine resins with OH-functional resins and acidic resins

Since, however, the activation energy for the conversion of melamine resins with carboxyl groups is significantly higher than for conversion with OH groups, higher stoving temperatures must be set when stoving water-borne materials.

Measurable quantities of formaldehyde are released during the thermal crosslinking of melamine resins. In partially etherified melamine resins, the free formaldehyde formed can amount to up to two per cent of the melamine resin content. With the more highly etherified alternatives it is now possible to reduce this formaldehyde formation to less than 0.1 %. A concomitant feature of this is the reduction in the overall quantity of cleavage products during the stoving process, which in turn leads to a fall in the formation of gloss-reducing microstructures in the surface of the coating. The outcome is improved film build.

In general, coatings produced with melamine resins offer improved long-term outdoor exposure, water resistance, gloss retention, resistance to salt spray and excellent hardness because they have a higher crosslink density and lower stoving temperature than urea resins.

Combinations with medium- and short-oiled alkyd resins, saturated polyesters and reactive acrylic and epoxy paints span the wide range of applications found in automobiles, domestic appliances, machine and coil coating of steel and aluminium.

Usable resins for coatings can also be synthesised from **benzoguanamine** or **glycoluril** by conversion with formaldehyde in a manner similar to the manufacture of phenolic resins.

Compared with traditional melamine resins, **benzoguanomine resins** improve the levelling, gloss and adhesion of the coating, although this comes at the expense of a deterioration in resistance to long-term outdoor exposure.

Glycoluril, a reaction product of glyoxal and urea, permits the synthesis of mononuclear, highly etherified crosslinking resins that possess good storage stability and excellent adhesion to metallic substrates. Less formaldehyde is released during processing, while anticorrosive protection improves at the same time.

Other, admittedly exotic, reactants for formaldehyde are sulphonamides and alkyl-substituted benzenes. While, for example, the reaction between toluene-p-sulphonamide

Coating Material

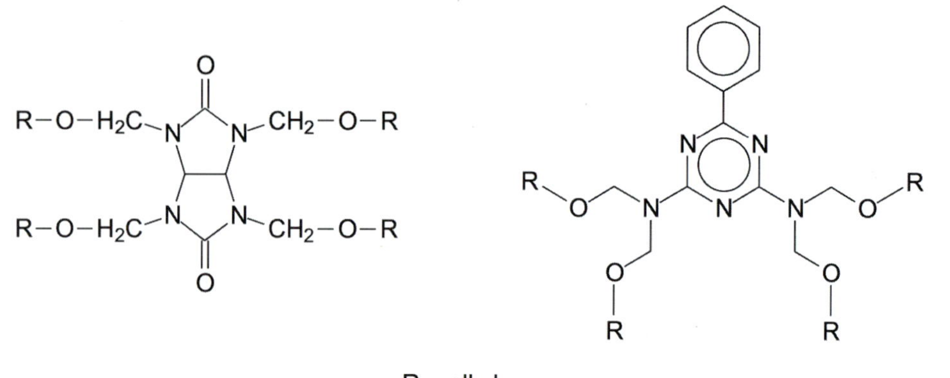

Figure 2.1.88: Glycoluril and benzoguanamine resins

and formaldehyde proceeds in accordance with expectations in the first reaction stage, there then follows practically quantitative trimerisation to a stable **sulphonamide resin** which principally serves as a modifying resin because of its lightfastness, solubility and compatibility with other paint resins.

The reactivity scheme of aniline is similar. Here, too, triazine rings are initially formed as a result of the action of formaldehyde. Unlike the sulphonamide resins these can be opened by a thermal load above 140 °C, with formation of long chains. The reaction product remains thermoplastic, is lightfast and displays interesting dielectric properties. These products are therefore mainly used for electrical insulation.

If, on the other hand, substituted benzenes such as toluene or xylene are converted with formaldehyde under extreme reaction conditions, the outcome is ring-substituted reaction products. These have properties similar to those of phenolic resins. They serve almost exclusively as modifying agents in the printing ink sector.

Also, those ketones which contain activated methylene groups next to the carbonyl group can be converted with formaldehyde. These resins primarily serve as binders for adhesives and printing inks.

While cyclohexanone can be converted with formaldehyde, it can also be transformed into usable paint resins by self-condensation. Converting the oxygen in the carbonyl group with the reactive hydrogens in α-position also gives rise to lightfast, nonsaponifiable resins that offer good compatibility.

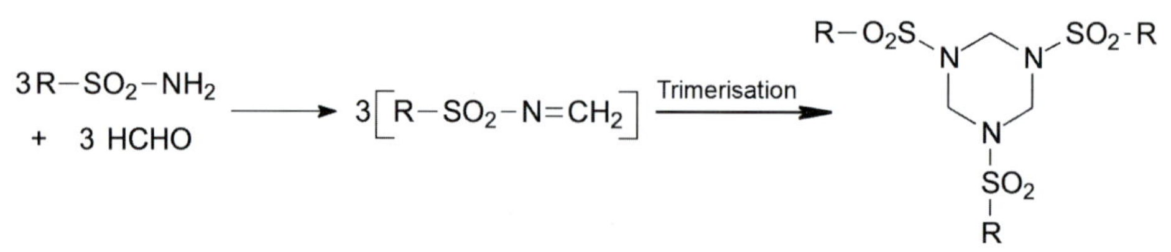

Figure 2.1.89: Trimerisation of methylolated sulphonamide resins

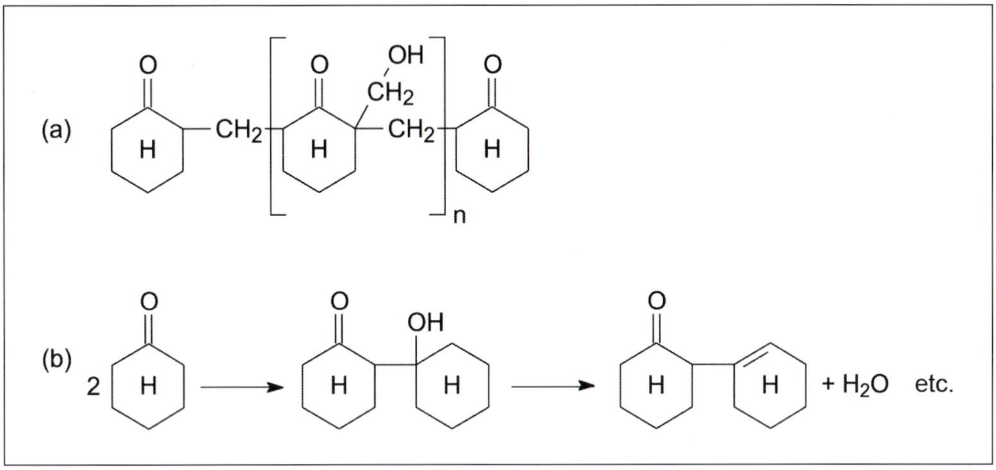

Figure 2.1.90: Ketone resins from cyclohexanone (a) with and (b) without formaldehyde

Polyaddition Resins

Polyaddition is a step reaction in which the reactants join to form macromolecules without, in contrast to polycondensation, releasing low molecular fragments. Molecular enlargement is achieved by transferring a hydrogen atom to the reactant while forming new bonds at the same time. Such polyaddition resins have only achieved any degree of technical importance for two classes of substance. These are the **polyurethanes**, which can be formed from OH-functional and isocyanate-containing structural elements, and the **epoxy resins**, which react with amines or carboxylic acids.

After molecular mixing of the two components, both classes of substance can react and enlarge their molecules at room temperature. Coating materials made from them are mostly formulated as 2-component systems.

Polyurethanes

When *O. Bayer* transferred the reaction of ethanol with ethyl isocyanate – described by *Wurtz* as long ago as 1849 – to polyfunctional structural elements in 1940, he laid the

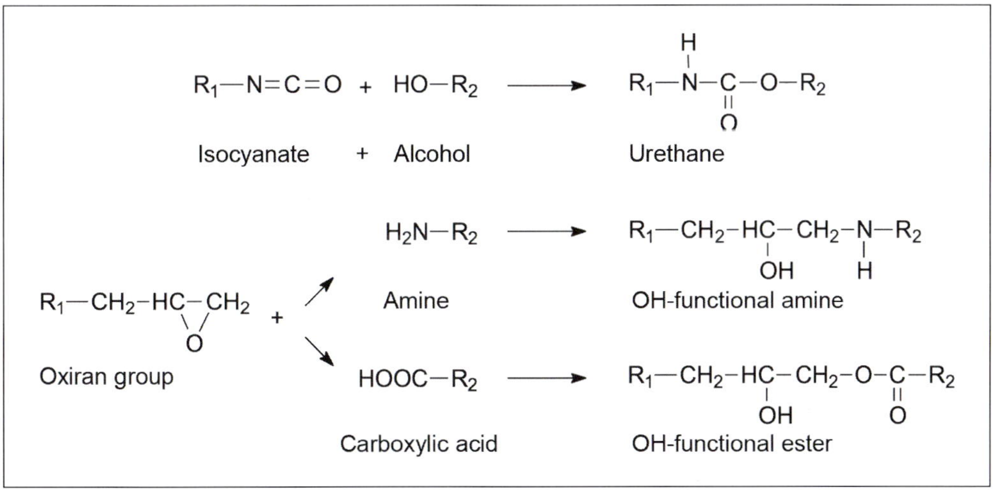

Figure 2.1.91: Basic reaction of the most important polyaddition resins

foundation for the new class of substances known as the **polyurethanes** [2.4.16]. They outperformed the polymers known at that time in terms of elasticity and resistance to saponification and consequently quickly established themselves for use in plastics, adhesives and coatings.

The conversion of resins containing hydroxyl groups with an **isocyanate** hardener leads to the formation of urethane groups. Because isocyanates are highly reactive, the manufacture of polyurethanes also features numerous secondary reactions which vary with the reaction conditions and the choice of raw materials.

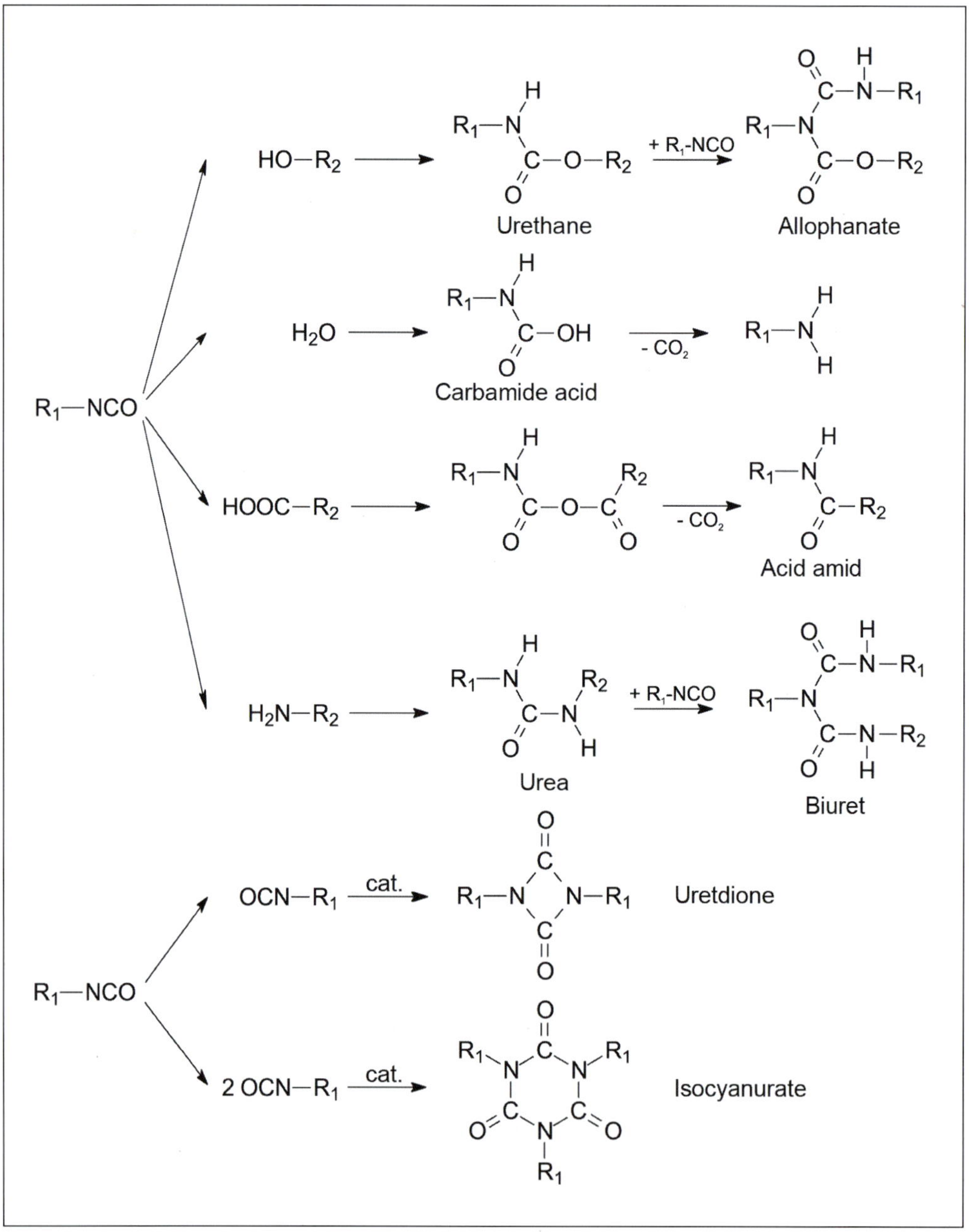

Figure 2.1.92: Principal and secondary reactions in the manufacture of polyurethanes

If the isocyanate hardener is present in excess, this can continue to react to form allophanates because of the active hydrogen in the urethane group. Isocyanates also react with ambient humidity. The resulting unstable carbamic acid undergoes decarboxylation to a primary amine. This in turn is able to form urea or biuret structures if there is an excess of isocyanate hardeners. When acid resins, such as polyester resins, are used, account must also be taken of the carboxyl groups which will also enter into reaction with isocyanates. The adduct formed initially soon decarboxylates to an acid amide. It is therefore recommended that polyesters with the lowest-possible acid value be used. Two further isocyanate reactions afford commercially available hardeners. This is because, under certain reaction conditions and the right choice of catalyst, isocyanates can dimerise to uretdiones or trimerise to isocyanurates. Whereas the isocyanurate ring is only split back into its starter materials at temperatures > 300 °C, the uretdione is split at temperatures > 140 °C. Uretdiones are therefore proven thermoreactive reactants for OH-functional resins (see Figure 2.1.92).

Large-scale synthesis of isocyanate hardeners is generally performed by converting primary amines with phosgene in the presence of CO_2 as catalyst.

Because of moves to replace chlorine processes, phosgene-free syntheses are gaining in importance. They are based on the thermal degradation of suitable derivatives of carbamic acid or urea [2.4.17].

The rate at which polyurethane coatings cure is essentially determined by the reactivity of the isocyanate group. Here, the electrophilic property of the carbon in the isocyanate group is the deciding factor. The first attack is by the electrophilic carbon of the isocyanate at the OH group of the reactant. The hydrogen atom from the hydroxyl group is then transferred to the nitrogen in the isocyanate group, thereby giving rise to the urethane group.

As aromatic rings on the nitrogen of the isocyanate group increase the carbon's electrophilic property through electron withdrawal, aromatic isocyanates are substantially more reactive than aliphatic ones.

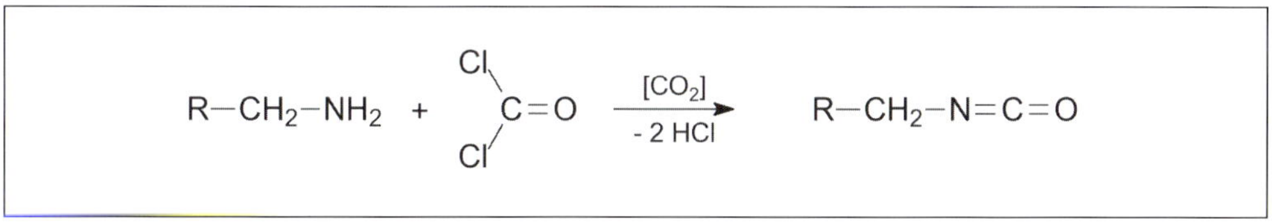

Figure 2.1.93: Synthesis of isocyanates from phosgene and amines

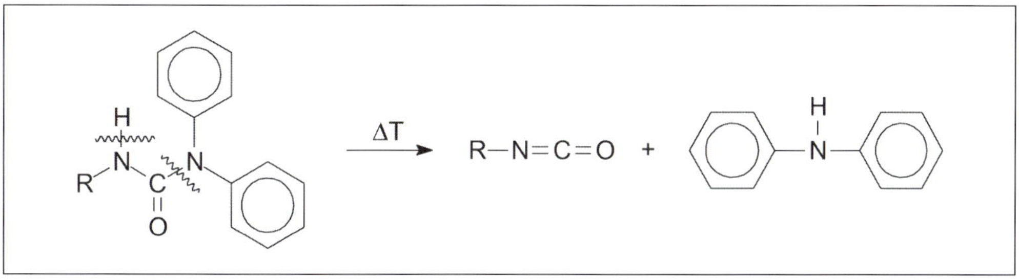

Figure 2.1.94: Synthesis of isocyanates by thermal degradation of derivatives of urea

By contrast, the influence of the OH-functional resin component on the reaction rate can be rated as low. Although secondary hydroxyl groups are basically more reactive than their primary counterparts, steric hindrance frequently overcompensates for this.

Countless types of hardener are available commercially because of the various requirements imposed on polyurethane paints. These in turn are manufactured from 2-functional basic structural elements by molecular enlargement, with simultaneous increase

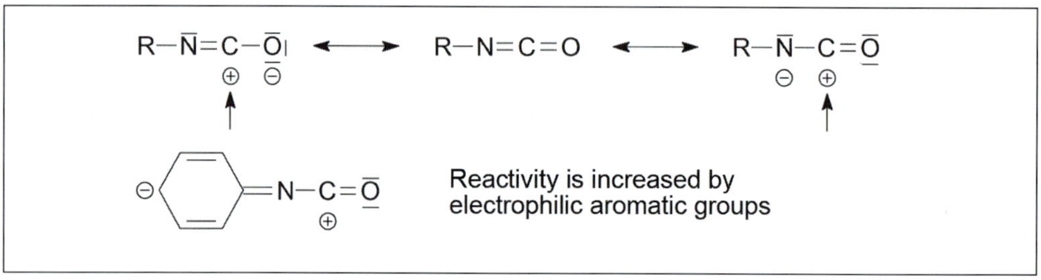

Figure 2.1.95: Impact of structure on the reactivity of isocyanates

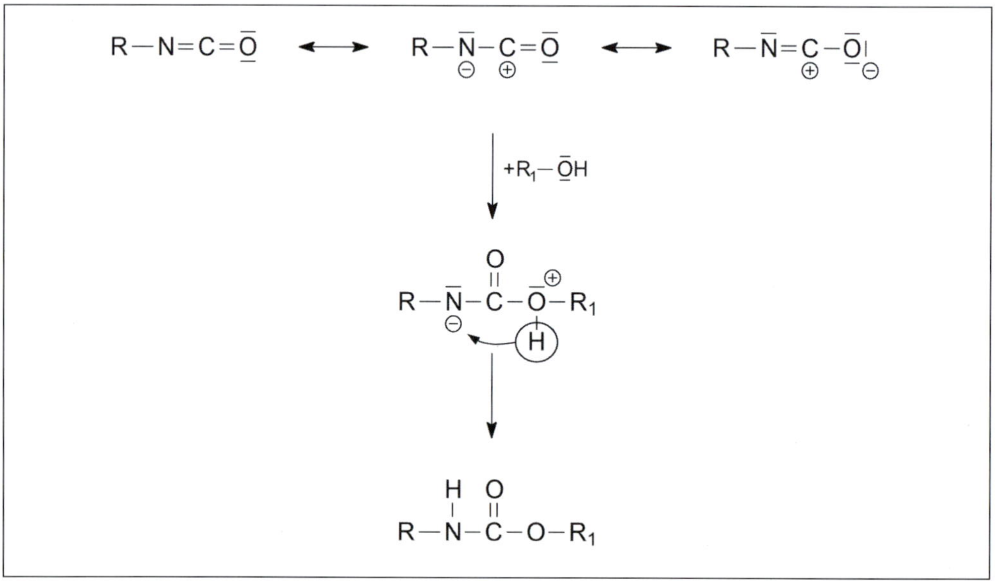

Figure 2.1.96: Reaction scheme for crosslinking isocyanates with OH groups

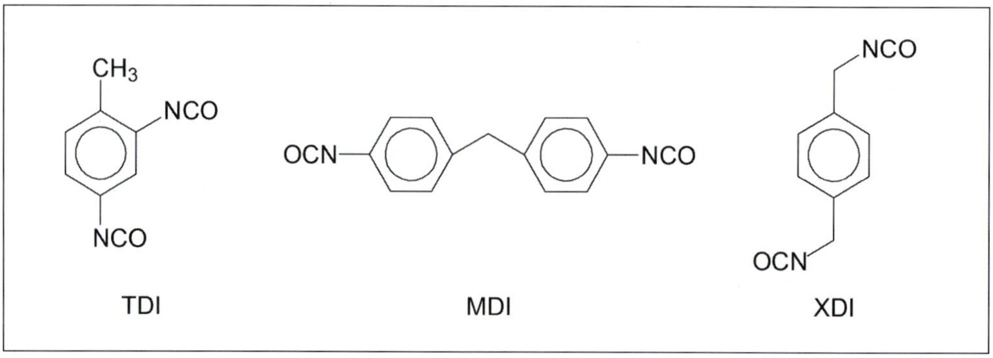

Figure 2.1.97: Aromatic diisocyanates as basic structural elements for hardeners

in functionality. Important basic materials for aromatic hardeners are toluylene diisocyanate (TDI), diphenylmethane diisocyanate (MDI) and xylylene diisocyanate (XDI).

The manufacture of aliphatic or cycloaliphatic hardeners usually involves hexamethylene diisocyanate (HDI), isophorone diisocyanate (IPDI) or $H_{12}MDI$, a hydrogenated variant of MDI.

The disadvantage of the lower reactivity of aliphatic isocyanates can be more than offset by suitable catalysts.

Coating Material

The isocyanates described cannot be used as raw materials for paints because of their low molecular weight, high vapour pressure and the associated potential physiological hazard, e.g. an occupational exposure limit (OEL) for HDI of 0.01 ppm (see Chapter 5.4). An additional molecular enlargement step should therefore be carried out. This can be done using appropriate reactants or by adduct formation with itself. Practical examples include the conversion products of TDI with trimethylolpropane (TMP) and the triadduct of HDI after its partial hydrolysis with water. Important adducts from a coating perspective also include the isocyanurates of HDI or IPDI as well as those formed from mixtures of HDI and TDI.

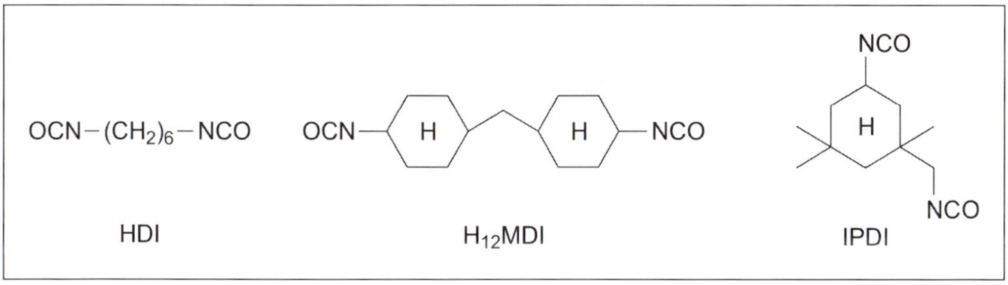

Figure 2.1.98: Aliphatic diisocyanates for non-yellowing and weather resistant polyurethane coatings

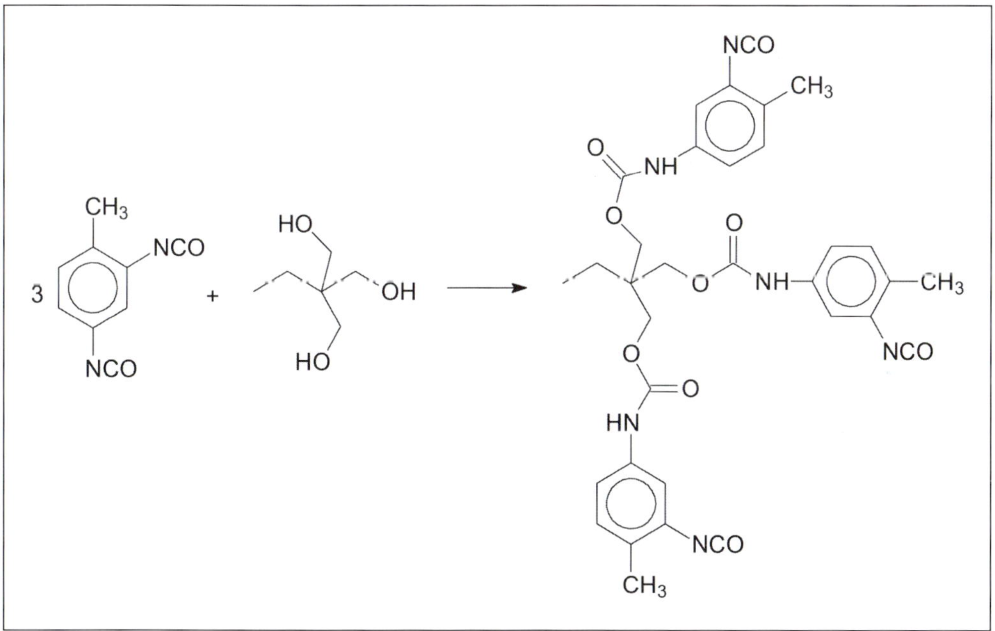

Figure 2.1.99: Synthesis of high molecular isocyanate crosslinker by forming adducts with triols

Saturated polyesters and polyacrylates are of similar technical importance as resins for the OH-functional parts of the generally 2-component paint formulations known as base paints. To further improve the chemical and weathering resistance of the already excellent polyurethanes, the polyols required to manufacture the hydroxy-functional base resins can be equipped with fluorine groups. Provided that no extreme demands are placed on the resistance to yellowing, even alkyd resins, as oil modified polyesters, make appropriate partners.

Owing to the reactivity of the isocyanates and, consequently, the incipient chemical reaction between mixtures of the base-paints and hardener at room temperature, common polyurethane coatings have only a limited pot life. A proven performance indicator is the time taken for the processing viscosity to double.

Both components in 2-component polyurethane coating systems must be metered extremely accurately and homogenised prior to processing. So, if large quantities of materials are being processed, suitable metering and homogenisation systems are required (see Chapter 4.2.1). This additional expense is more than compensated for in many appli-

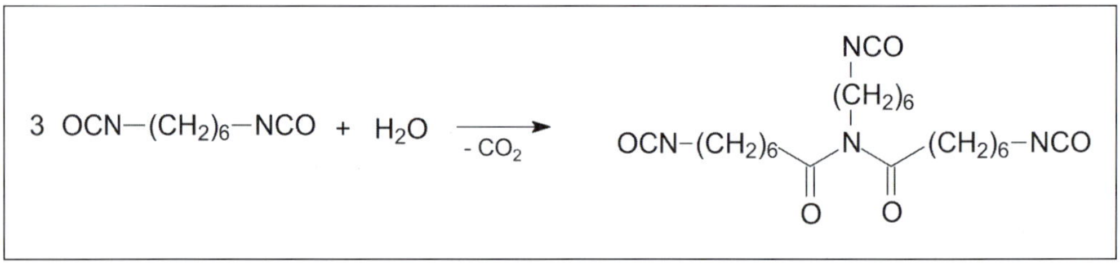

Figure 2.1.100: Molecular enlargement by partial saponification of hexamethylene diisocyanate (HDI) with water

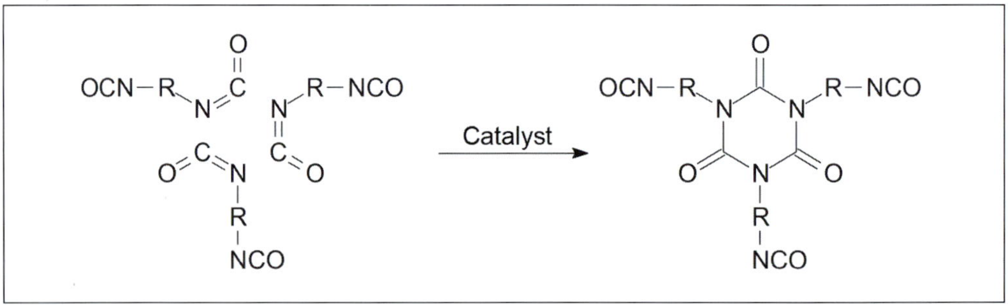

Figure 2.1.101: Molecular enlargement of isocyanates by trimerisation

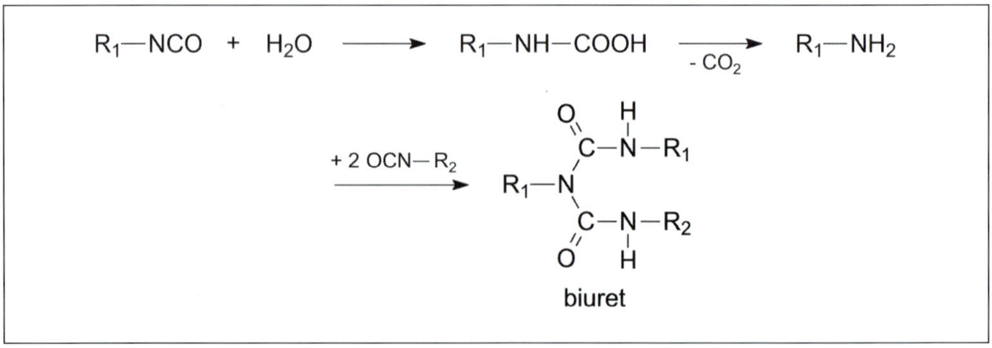

Figure 2.1.102: Traditional hardening of polyurethane resin with water

cations, e.g. for automotive refinish paints or aircraft paints, by the advantage of rapid curing at room temperature. Two-component paint systems are conquering the market for many applications in the transport sector, wood and plastics, and very large objects such as ships, pipelines, and buildings [2.4.18].

Due to the technical outlay entailed and the additional sources of errors during metering and homogenising, countless attempts have been made to process polyurethanes as 1-component coatings too. One means of achieving storage-stable 1-component systems consists in involving ambient humidity in the formation of films by resins which have no OH groups. The isocyanate groups are partially hydrolysed by the water and converted through decarboxylation into amines with which excess isocyanate crosslinks.

In addition to these conventional systems a new generation of humidity-hardening polyurethane coatings has been developed. Since ketimines, aldimines and also oxazolidines react more quickly with ambient humidity than do the isocyanates, but will not react with isocyanates due to a lack of active hydrogen atoms, corresponding paints can be processed as 1-component systems. The ambient humidity which is permitted access to the film after processing does not now cause any hydrolysis of isocyanates and thus does not form CO_2 either; instead it opens the oxazolidine ring or splits the imines. Both cases give rise to the necessary hydrogen for crosslinking in the form of OH or NH_2 groups.

Another method of designing 1-component coatings consists in carrying out a preliminary conversion of the isocyanates into chemically inert intermediate stages which can,

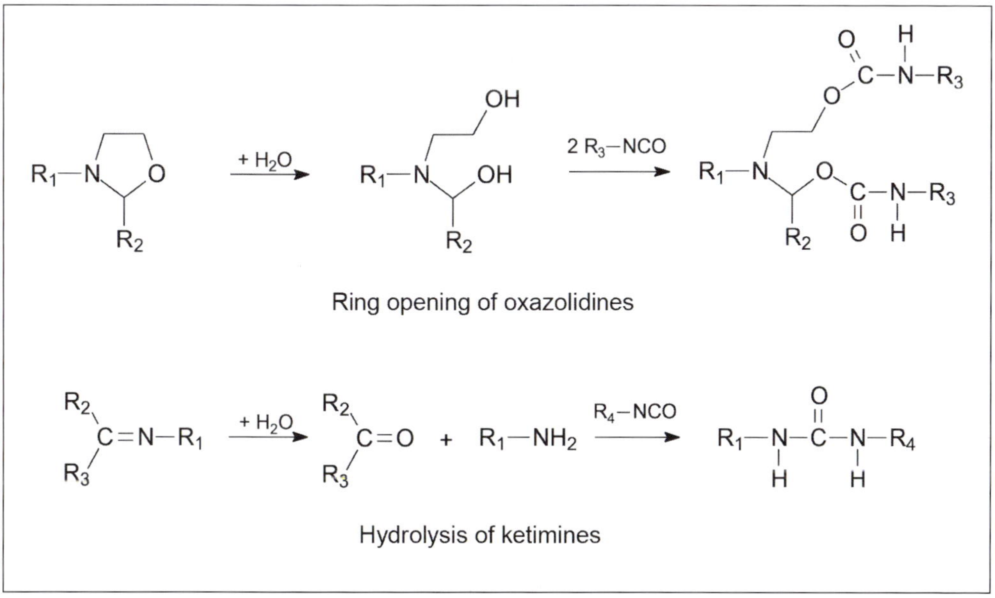

Figure 2.1.103: Hardening reaction of polyurethane resins with water in the presence of oxazolidines or ketimines

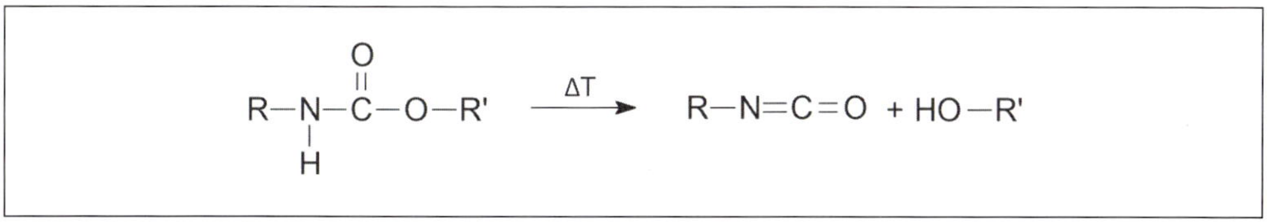

Figure 2.1.104: Deblocking reaction of capped isocyanate hardeners

however, be reactivated by heat action. As already mentioned, such blocking can be carried out with the material itself or with external materials. These are cleaved again, i.e. unblocked, during thermal reactivation and then have an impact on air flow and consumption in the drier as volatile cleavage products (see Chapter 4.3.1).

Traditional blocking agents for isocyanate groups are phenols and ε-aminocaprolactam. Because of the relatively high cleavage temperature of 160 to 180 °C, oximes or CH-acid compounds such as acetoacetic ester, acetylacetone or malonic ester have been successfully tested as blocking agents. Other blocking agents which also resplit materials at around 140 °C are 1,2,4-triazole and dimethylpyrazole [2.4.19].

The blocked, thermally curing 1-component polyurethane paints occupy a special position in that they exhibit better chemical resistance and elasticity combined with greater hardness than the traditional stoving enamels containing amino resin crosslinking agents. Representatives blocked with ε-aminocaprolactam are used to meet special demands for coil coating and can coating, for sterilisation-proof coatings on the insides

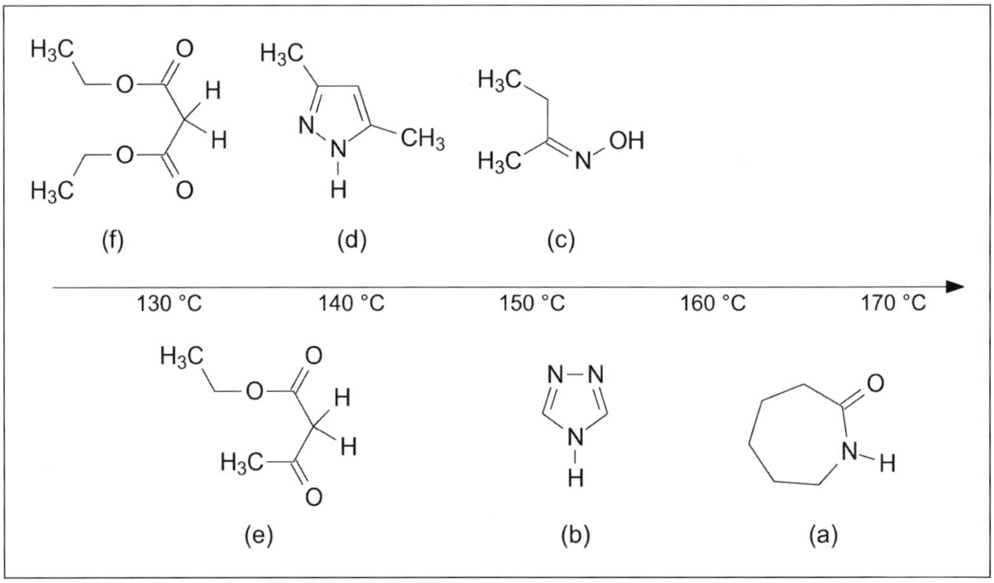

Figure 2.1.105: Blocking agents for isocyanate crosslinkers and their respective deblocking temperature

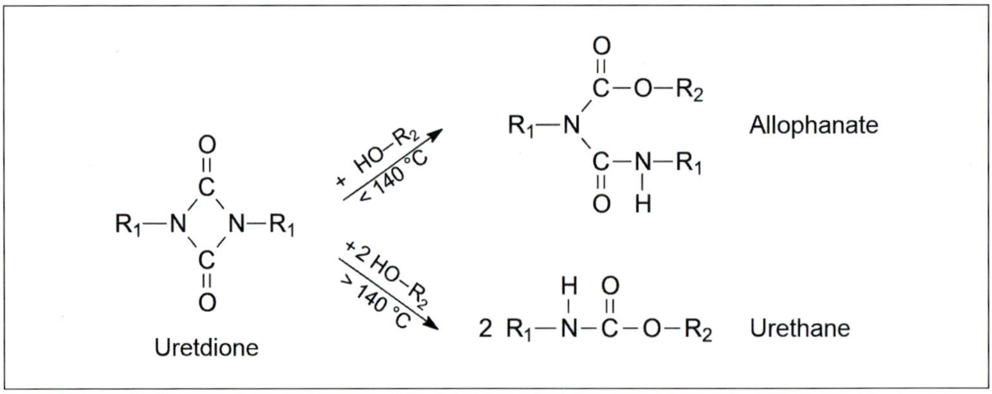

Figure 2.1.106: Uretdiones: Hardeners for OH-functionalised resins by thermolysis

of cans intended for aggressive foodstuffs and on cans for fuels and oils. Blocked isocyanate crosslinking agents are also used in electrical insulation coatings because of their high heat resistance and their constant electrical properties.

In powder coatings the blocking agent ε-aminocaprolactam (a) is being replaced more and more by the more reactive 1,2,4-triazole (b). Furthermore, isocyanates of the trimeric HDI and IPDI blocked with methyl ethyl ketoxime (c) are being incorporated into stoving primer surfacers for automotive paints. Isocyanate hardeners blocked with ethyl acetoacetate (e) or diethyl malonate (f) have proved successful for reactive topcoats and clearcoats because of their low stoving temperatures. 3,5-dimethyl-l,2-pyrazole (d) also recently became available as a blocking agent for similarly low reaction temperatures (see Figure 2.1.105).

Blocked isocyanates are also used in cationic electrocoat paints to achieve a high crosslink density.

The formation of cleavage products of blocked isocyanates is a disadvantage which does not occur when the already mentioned uretdiones are used. After heating, the uretdiones are activated by ring opening or cleavage, as a function of temperature, without the formation of cleavage products. In the case of ring opening, allophanates are formed after addition of the OH-functional reactant, while with cleavage, traditional urethanes are formed. This has enabled **1-component polyurethane paints** to be formulated which have already proved successful as powder coatings [2.4.20].

In view of the reaction of isocyanates with water and the associated destruction of the crosslinking functions, precautions must be taken when designing water-borne polyurethane paints to prevent this preliminary reaction of the isocyanate groups. This can involve either, as just described, blocking the isocyanates or using the reactive groups before film formation with the specific intention of building up the basic chains to high-molecular and physically drying resins.

The latter method is used with water-borne **1-component polyurethane dispersions** or PUDs [2.4.21]. These predominantly anionically stabilised dispersions are manufactured by converting diisocyanates, diols and sterically hindered dihydroxycarboxylic acids into prepolymers, followed by neutralisation with amines and then transformation into a fine dispersion. Steric hindrance of the carboxylic acids prevents them from reacting with the isocyanates. This preserves the carboxylate groups for anionic stabilisation. Subsequently, during the emulsification, the NCO end-groups are transformed into further hydroxyl groups by adding further diols. These polyols can be polyester diols, polycarbonates or polyethers [2.4.22]. The polar groups of such resins permit full-scale conversion into a stable emulsion if adequate shear forces are applied.

For trouble-free dispersing, the prepolymers need to be dissolved in appropriate solvents, such as acetone, which then have to be stripped after the emulsification. Owing to the low boiling point of acetone, N-methylpyrrolidone (NMP) and N-ethylpyrrolidone (NEP) have been able to establish themselves as alternatives. However, these substitutes for acetone cannot be removed completely after the emulsification. Thus undesirable residues of the not entirely physiologically safe solvents NMP and NEP remain in the dispersions.

The prepolymer and melt processing methods are two further production variants that render the use of NMP and NEP obsolete. Whereas in the acetone process, polyols are

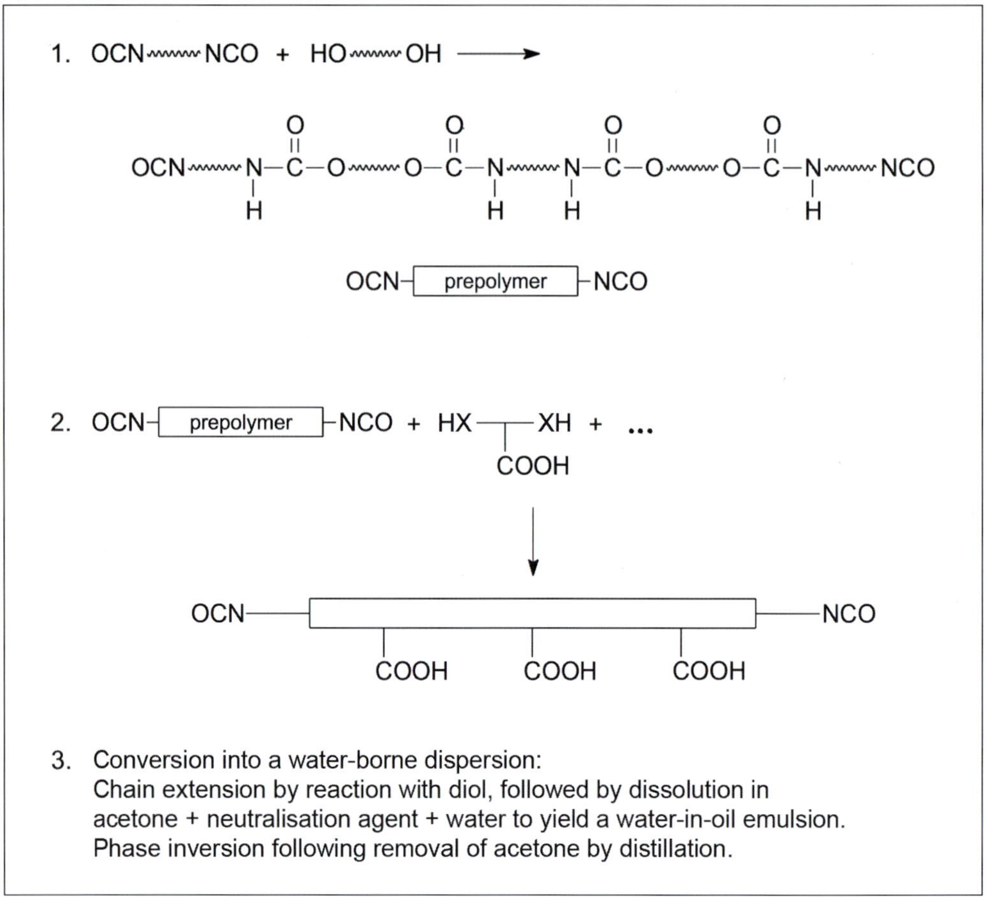

Figure 2.1.107: Chemical steps for manufacturing water-borne polyurethane dispersions

used irrespective of the type of solvent employed for the purpose of transforming the NCO groups at the chain ends, the prepolymer process utilises water. The amino groups formed by partial hydrolysis of the terminal NCO groups serve the function of chain extension with excess isocyanate.

In melt processing, the third manufacturing process, chain extension of the prepolymers occurs in the aqueous phase. Figure 2.1.108 uses the example of a cationic, urea-blocked prepolymer to show that subsequent conversion with formaldehyde leads to the formation of reactive hydroxymethylene groups. During film formation via acid catalysis or heating, these are capable of inducing molecular enlargement by crosslinking. For ready-to-use coatings, plasticisers are usually also incorporated into the dispersions.

Reactive PUR dispersions can be synthesised by modifying the prepolymer with unsaturated fatty acids [2.4.23]. Aside from the familiar capability of the traditional urethane oils to engage in oxidative crosslinking, the presence of fatty acid groups also has a positive effect on emulsifiability.

Polyurethane dispersions can also be converted into reactive systems by incorporation of OH groups. Thus, by crosslinking with melamine resins dissolved in the aqueous phase, they can be transformed into thermosetting films. HMMM resins have proved ideal for

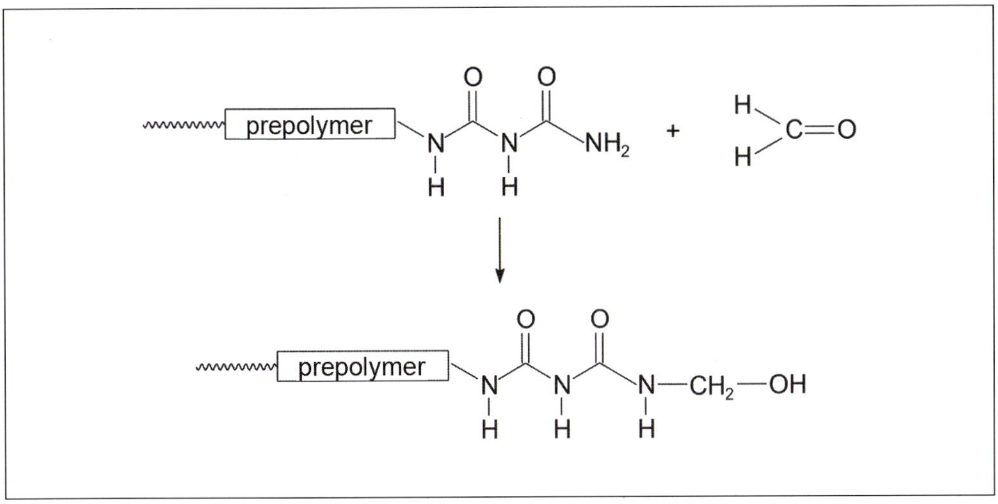

Figure 2.1.108: Production of reactive polyurethane dispersions

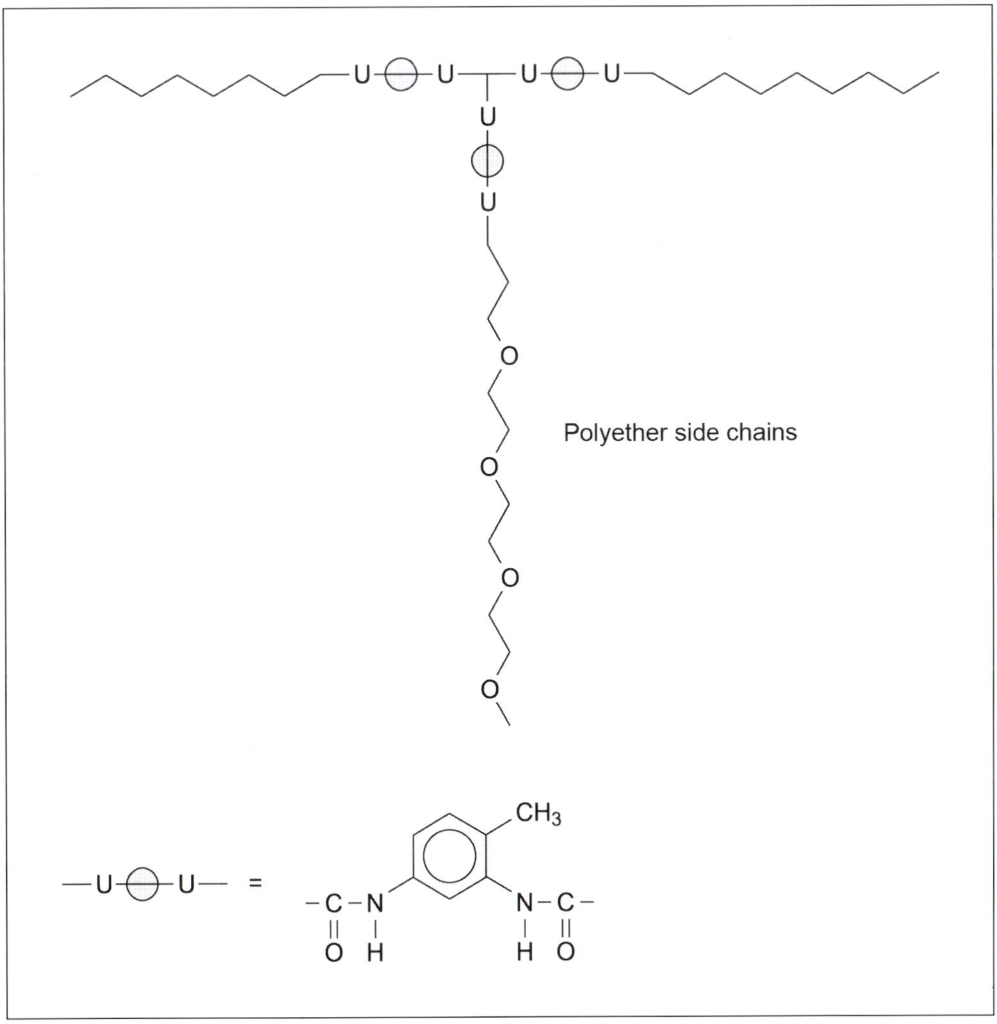

Figure 2.1.109: Typical structure of a nonionic polyurethane dispersion

this group of water-borne polyurethane stoving systems. Conversion with dispersible isocyanates is increasingly being used for producing high-quality coatings [2.4.24].

In the meanwhile, analogous cationic and **nonionic polyurethane dispersions** have been developed. The resin molecules containing amino groups are converted with acids, usually acetic acid, into the water-soluble form. It should be noted that, e.g. OH-functional acrylic dispersions made to react with water-dispersible isocyanate crosslinkers are also very commonly referred to in the literature as 2-component PUR systems [2.4.25, 2.4.26].

Generally polyurethane dispersions have the advantage over emulsion polymers that, if subjected to high enough shear forces, they can be produced in a particularly fine form that will dry physically to yield glossy, dense, homogeneous films. The films are tough and elastic due to the elastomeric character and the resulting hydrogen bridging bonds. The dispersions can also be combined with other water-dilutable binders. For this reason, they are particularly suitable for water-thinnable primer surfacers, one-coat topcoats, plastic coatings, and wood and film coatings. They can also be combined with primary acrylic dispersions to form water-thinnable decorative paints. The result is diffusion-resistant, glossy films that can be used both for high-quality applications, such as automotive base coats, automotive primer surfacers and for painting windows.

Contrary to the assumption that OH-functional water-borne polyurethane dispersions made with isocyanates cannot be processed as 2-component coatings, both chemical and process engineering methods have been developed to improve the resistance of the hardener to the aqueous environment [2.4.25]. Improved emulsifiability involving the incorporation of polyether groups, reduced reactivity and a suitable dispersing system have led to attractive water-borne 2-component coating systems. Mixing with such hardener components takes place directly prior to processing by means of jet dispersion of the hardener in the base paint, with the result that hydrolysis can only take place very slowly. Acrylic or polyester polyols containing OH-groups are combined with an isocyanate-containing hardener using appropriate metering and emulsifying equipment. A relatively coarse primary emulsion is first obtained; this is then so finely distributed in a downstream dispersing machine that the necessary molecular blending occurs by diffusion during film formation [2.4.27].

Despite the heterogeneous distribution of the components and the reduced reactivity, the possibility that some of the isocyanate groups are split hydrolytically due to the influence of the water at the dispersion interface cannot be excluded. To ensure optimum crosslinking, therefore, it is necessary to remove the water quickly from the film and also to provide excess hardener.

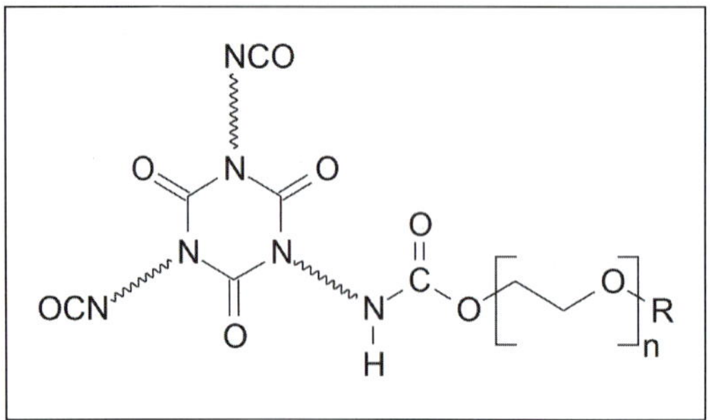

Figure 2.1.110: Isocyanate crosslinker for water-borne 2-component paints

Consideration is being given to the use of allophanate- modified polyethers as new hardeners to further improve the emulsifiability of the isocyanate hardener and the film properties (see Figure 2.1.111) [2.4.25].

The range of possible applications for polyurethane paints is as wide as the

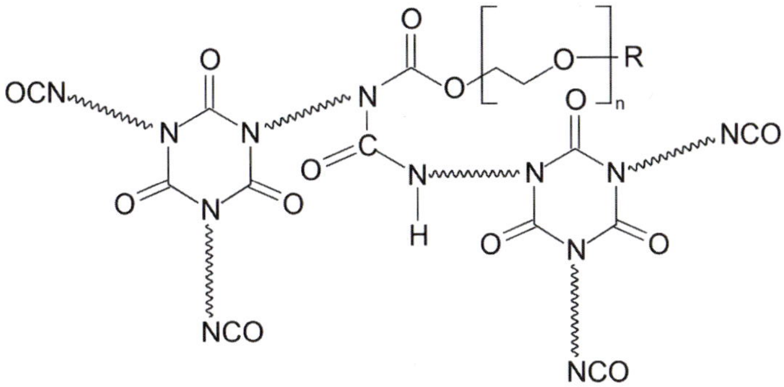

Figure 2.1.111: Allophanate-modified hardener for 2-component dispersions

available variety of raw materials. Pigmented and unpigmented polyurethanes for a huge range of substrates with an impressive list of benefits are in use in the individual industries. These benefits include their good adhesion, wear resistance, long-term outdoor exposure properties, and hardness combined with a high degree of elasticity and a high resistance to chemical agents such as acids, alkalis, household chemicals, solvents and fuels. 2-Component PUR systems are very successful as automotive refinish paints. Their use on automotive coating lines has also advanced greatly in recent years for the above mentioned reasons. Polyurethanes have had a dominant role for quite some time now when it comes to coatings for large vehicles such as buses, commercial vehicles, road tankers, passenger and freight carriages on the railways, and also aircraft. They have also proved successful for coating highly stressed parts in the wood and furniture industry. Other applications include coatings on masonry, concrete, asbestos cement, plastics and even on rubber.

Humidity-hardening polyurethane paints yield finishes with excellent mechanical properties and good resistance to the effects of water, acids, alkalis and solvents. They are therefore used on highly stressed wooden parts such as parquet and hall floors, plastics, fabrics and often also on concrete surfaces. In the last of these, they largely protect concrete structures against penetration by moisture, road salt, oil contamination and other critical active substances.

Polyurethane paints are now used in all environment-compatible coating systems, such as powder coatings, water-borne, solvent-free, liquid and radiation-curable coatings.

In view of the fact that intermediate stages involving isocyanate groups are not unproblematic from the point of view of industrial health and are an integral part of the product chain extending from raw material to finished polyurethane coating, there have been many attempts to arrive at, from a quality perspective, the attractive end stage of a polyurethane structure by other **reaction chains** whose every intermediate stage is **isocyanate-free**.

One promising route to the desired urethane groups is the modification of hydroxyl groups with methyl carbamate. Suitable base products here are the hydroxy-functional acrylic resins, which have been known for a long time. The two components are chemically linked by transesterification in conjunction with thermal cleavage of methanol from the methyl ester of carbamic acid (see Figure 2.1.112).

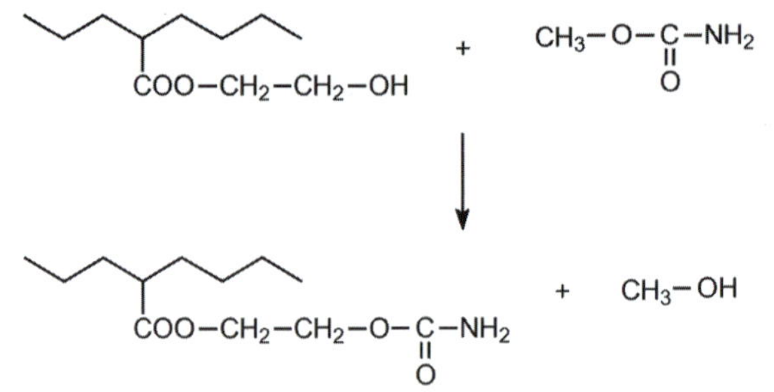

Figure 2.1.112: Conversion of hydroxy-functional acrylic resin to the carbamide-functional ester, with release of methanol

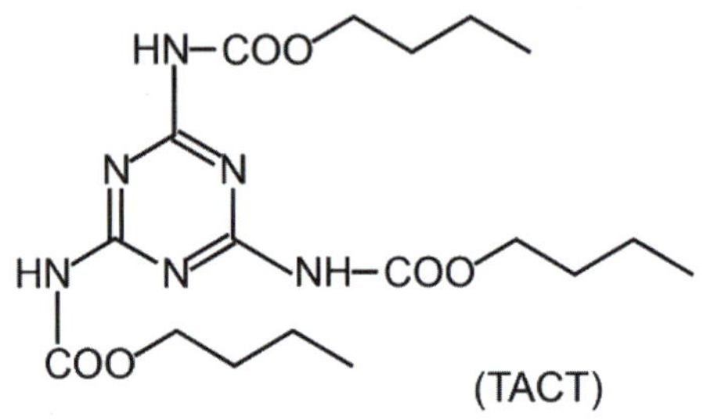

Figure 2.1.113: Trisbutoxycarbonylaminotriazine

As described in Chapter 2.1.1.2 on acrylate resins, carbamic acid esters of preferably hydroxy-functional acrylic resins can be crosslinked with etherified melamine resins. Hexamethoxymethyl melamine (HMMM) fully etherified with methanol proves to be particularly suitable for a co-condensation. Thermal initiation during the hardening of such 1-component systems cleaves methanol and leads to the formation of coatings containing urethane groups (see Chapter 2.1.1.2 and Figure 2.1.145b).

A further possibility of obtaining polyurethanes without the isocyanate-containing intermediate stages is afforded by the reaction of trisalkoxycarbonylaminotriazine (TACT) with OH-functional reaction partners [2.4.29]. Hydroxy-functional acrylic resins are also suitable. for designing high-quality industrial paints Chemical linkage occurs during thermal crosslinking with TACT, with elimination of the esterification alcohol butanol.

Both concepts, be they the reaction of carbamate-containing acrylic resins with HMMM resins or the reaction of TACT with unmodified hydroxy-functional acrylic resins, yield urethane-containing crosslinking sites and thus films possessing the outstanding properties of the traditional PUR coatings. In particular, these compounds are characterised by excellent long-term outdoor exposure properties and chemical resistance combined with a high degree of hardness and elasticity, with the triazine component additionally producing excellent surface smoothness along with high gloss and film build.

Epoxy Resins

The second important group of polyaddition resins from a coatings perspective is that of the **epoxy resins**. Even though the first successful attempts at manufacturing epoxy compounds took place more than 100 years ago, it nevertheless took more than 35 years for the first technically usable paint raw materials to become available [2.4.30].

Today epoxy resins have a secure place among the high-performance polymers because of their excellent adhesion to metallic substrates and good mechanical/technological properties as adhesives, plastics and coating materials. As with the polyurethanes group, the epoxy resins are generally processed as 2-component paints.

The oxirane or epoxy group is characteristic of epoxy resins. It consists of a three-membered ring containing oxygen with a high ring strain. The reactivity is further enhanced by the electronegativity of the oxygen and the resulting polarity of the epoxy groups. The carbon atoms of the three-membered ring which are starved of electrons therefore become electrophilic and can take part in addition reactions with numerous nucleophilic partners. The reactions with amines and carboxylic acids are of particular importance for practical operations. The oxirane group can also be polymerised such that the ring is opened even without reactants. The reaction, initiated by Lewis acids, such as BF_3, or ferrocene complexes that split under UV radiation, affords polyethers of industrial importance.

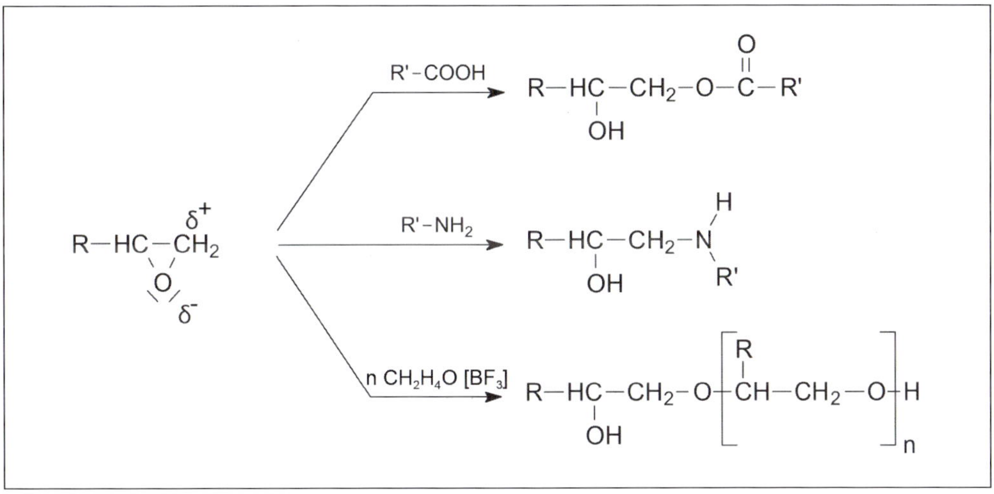

Figure 2.1.114: Crosslinking reactions of epoxy resins

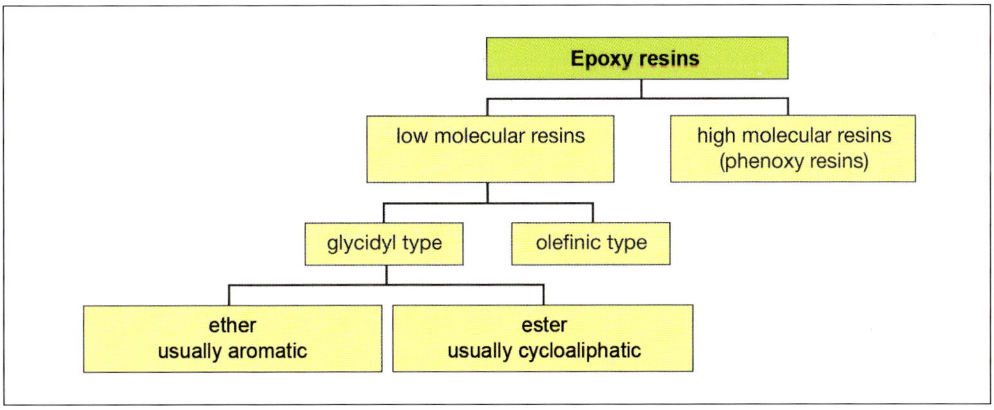

Figure 2.1.115: Classification of epoxy resins

By far the most important group among the epoxy resins is formed by the aromatic glycidyl ether types. They are manufactured by polycondensation of epichlorohydrin and bisphenol A. The molar mass of the individual representatives of this group is determined solely by the mixing ratio of the reactants. If a molar ratio for epichlorohydrin to bisphenol A of 2:1 is chosen, the outcome is the smallest representative of this resin class. If the ratio is increased to 3:2 or even further towards 1:1, higher molecular products will be synthesised. They differ from the first representative described in having additional hydroxyl groups. As a result, in addition to the reactivity of the epoxy groups, the epoxy resins become of interest for all partners capable of reacting with OH groups.

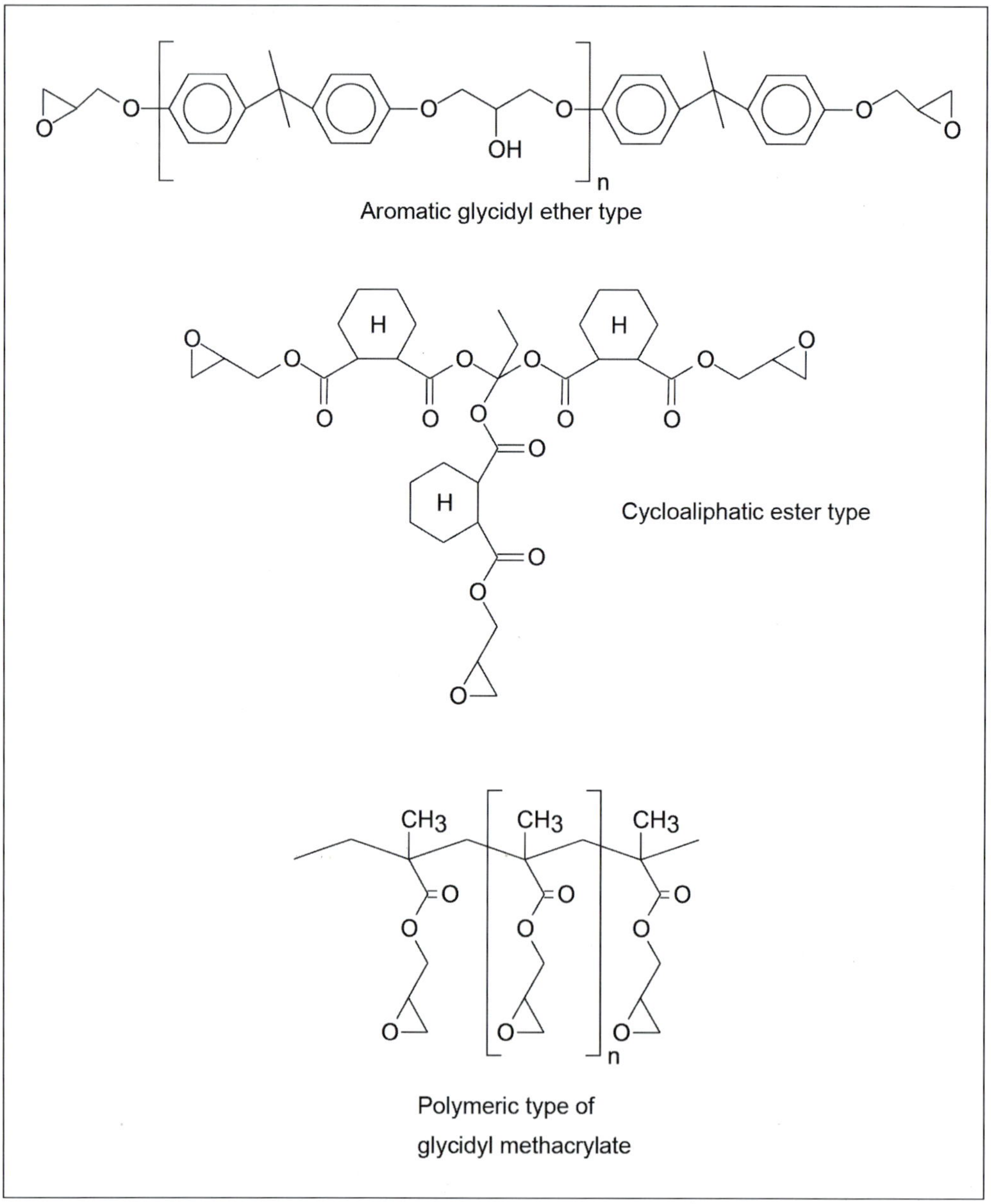

Figure 2.1.116: Chemical structure of important epoxy resins

If a mixing ratio close to 1:1 is chosen, the resin molecules become so large that the terminal oxirane groups lose more and more importance in respect of crosslinking. Such high molecular phenoxy resins therefore only have a role as hydroxy-functional resins.

The light-unstable ether groups formed by the reaction of epichlorohydrin with bisphenol A restrict the use of this resin group to primers and primer surfacers. Ester types must continue to be used for lightfast topcoats. Technically important examples are

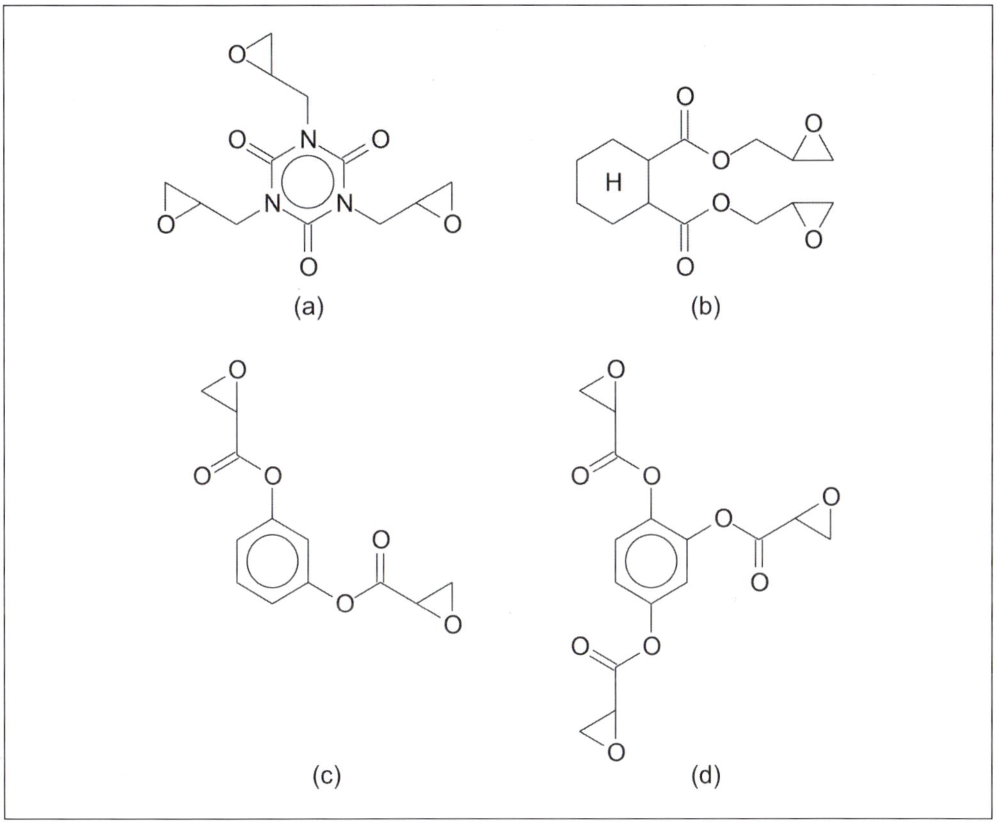

Figure 2.1.117: (a) Triglycidyl isocyanurate TGIC and (b-d) alternatives

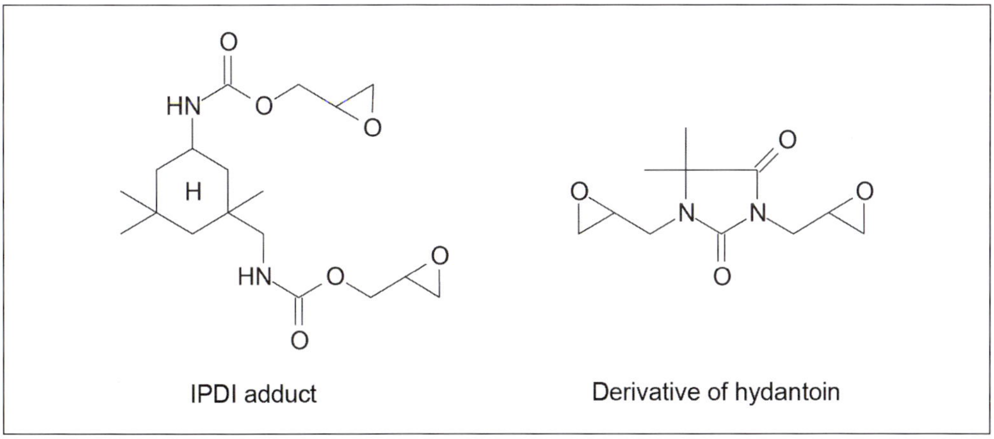

Figure 2.1.118: Heat-resistant and non-yellowing epoxy crosslinkers

the glycidyl esters of hexahydrophthalic acid and isophthalic acid or trimellitic acid. **Triglycidyl isocyanurate** (TGIC), formerly the standard hardener for acid reactants in powder coatings, may now only be used if appropriately labelled because of its toxicity.

The resulting decision to dispense with TGIC has caused problems in the field of powder coatings. Today, in addition to the isophthalic acid and trimellitic acid esters of **epichlorohydrin**, there are other alternative partners, albeit ones which crosslink in polycondensing reactions, for the acid acrylic or polyester resins. Noteworthy glycidyl group-free crosslinking agents include hydroxy-functional adipic acid amides which have been largely able to supersede TGIC (see above).

Other yellowing-resistant epoxy crosslinkers with good long-term outdoor exposure properties are the 2- and 3-functional reactants of epichlorohydrin with isophorone diisocyanate or hydantoin, with hydantoin derivatives also having the benefit of higher heat resistance.

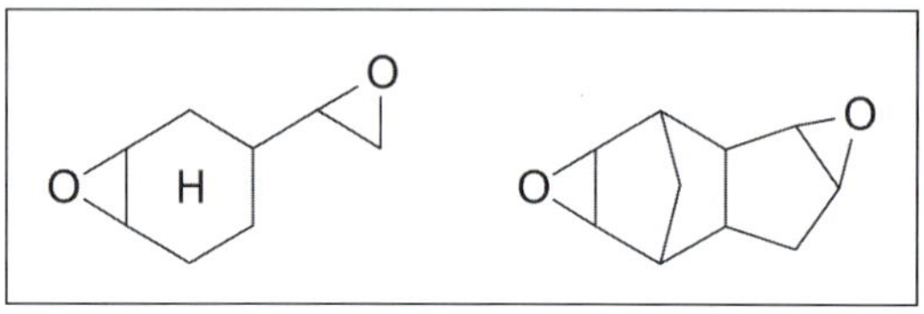

Figure 2.1.119: Olefin types of epoxy components as reactive diluents

A completely different approach to epoxy resins involves the epoxidation of carbon double bonds. This yields so-called olefin types. The reaction of

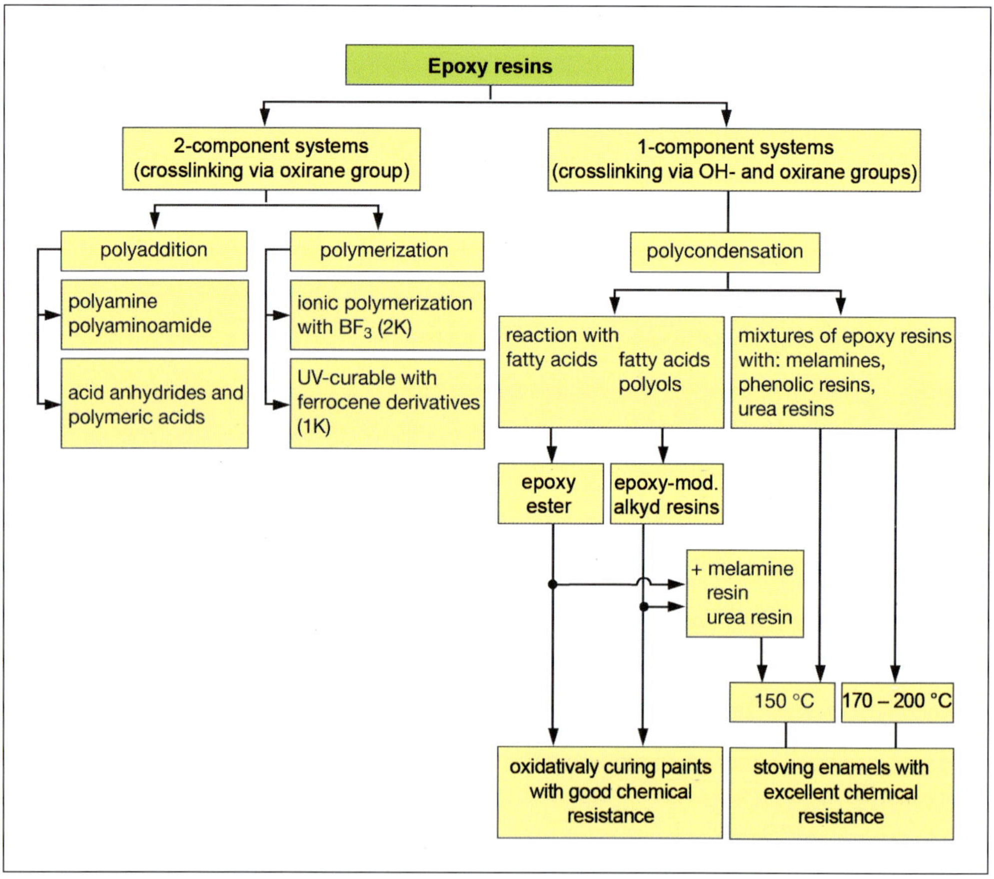

Figure 2.1.120: Summary of applications of epoxy resins in coating formulations

tetrahydrostyrene with peracetic acid, for example, leads to a bifunctional cycloaliphatic, yet low molecular epoxy resin. These solvent-free, low viscosity structural elements have proved their worth as **reactive diluents**. As solvents they have a pronounced viscosity-lowering effect, but are also involved in the polyaddition process of crosslinking during hardening. As they therefore do not leave the film, they help to lower organic emissions.

As outlined in principle above, the **crosslinking of epoxy resins** can take place, in accordance with the reactant, either exclusively at the oxirane groups or additionally at the less reactive hydroxyl groups. Consequently, epoxy resins can be cured as 1-component or 2-component coatings. The usual method of curing epoxy resins involves amines, acid anhydrides or carboxylic acids. In addition, mercaptan hardeners also serve as reactants for special applications.

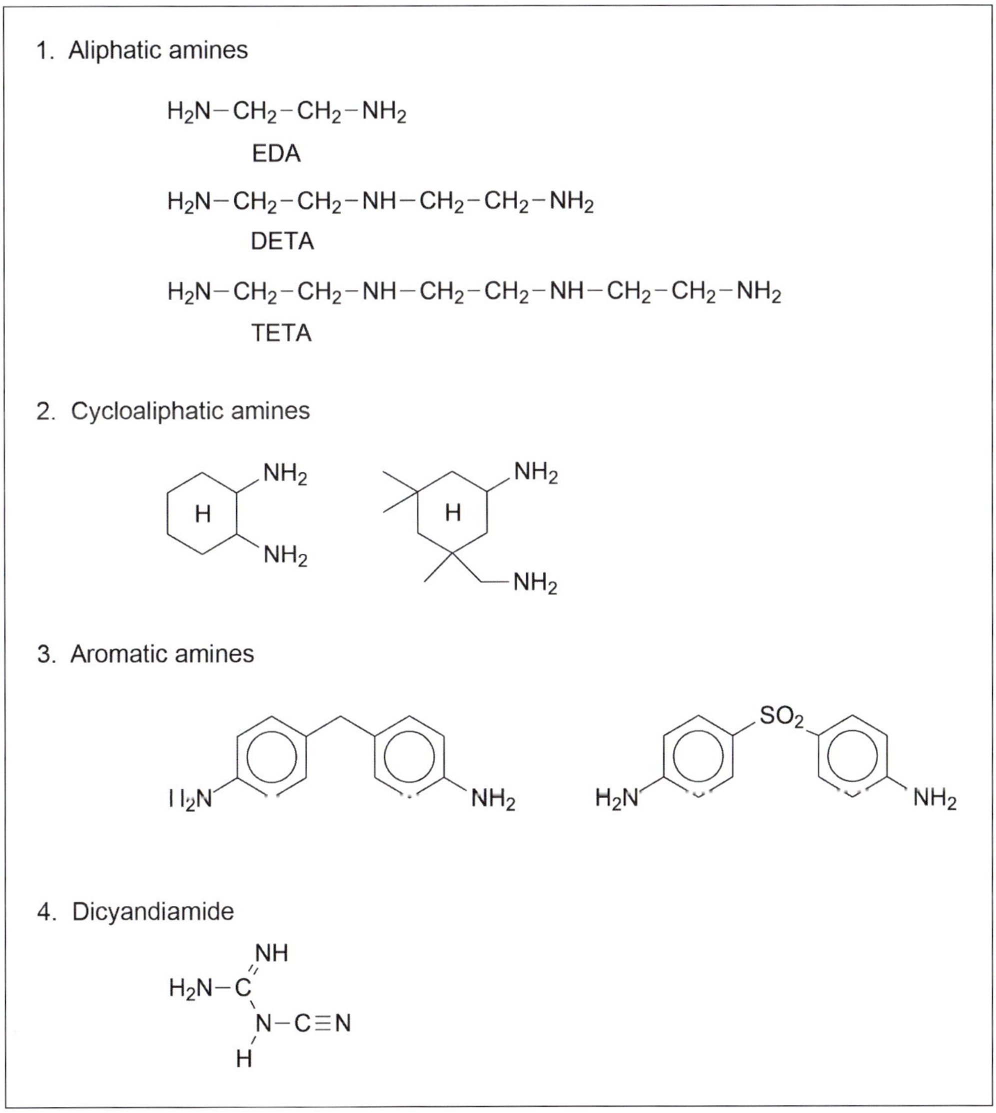

Figure 2.1.121: Amine hardeners for epoxy resins (EDA = ethylenediamine, DETA = diethylenetriamine, TETA = triethylenetetramine)

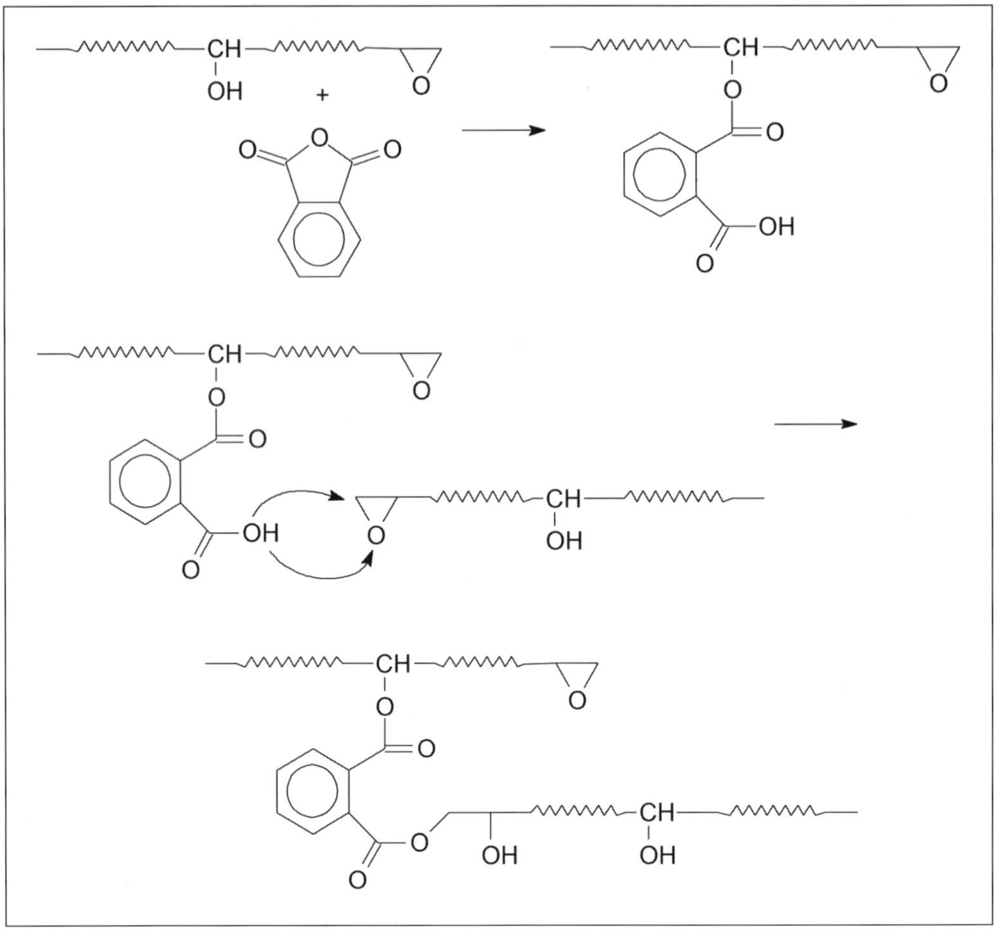

Figure 2.1.122: Crosslinking of epoxy resins with anhydrides, for example phthalic anhydride

The reaction rate of amine curing depends on the type of amino group. The processing or curing times can be varied via the basicity of the amine hardener in association with the steric configuration of the amino group. The most reactive with a pot life of 15 to 30 minutes are the short-chain aliphatic polyamines (ethylenediamine, diethylenetriamine etc). The longer chain polyaminoamides are less reactive and yield more elastic films. Mixtures of epoxy resins with cycloaliphatic or aromatic amines must even be heated briefly to 120 to 140 °C because of their inadequate reactivity. Dicyandiamide occupies a special position among the amine hardeners. This product, which is solid at room temperature, is incompatible with epoxy resins and forms two phases from homogeneous melts when solidifying by cooling. The reaction takes place only at the common interface and is therefore so slow that the mixture is ground and can be used as a powder coating which is stable in storage. The reaction of dicyandiamide with epoxy groups proceeds differently from that of traditional amines. Primary adducts are rearranged into acid amides and urethanes, among other things, by secondary reactions. Dicyandiamide is therefore not 4-functional, as would be expected from its chemical structure, but only 3-functional [2.4.32].

The second, no less important route for industrial coating technology to highly crosslinked and adhesive coatings involves the reaction of epoxy resins with acid anhydrides.

Due to the lack of active hydrogen atoms, acid anhydrides cannot react directly with epoxy resins. Only once the reaction of the anhydride ring with the hydroxyl groups of the epoxy resin chain has been initiated by heat action does this lead to ring opening and the simultaneous formation of a carboxyl group. If the ring-opening OH groups are absent, reaction of the acid anhydride with the epoxy resin due to the addition of catalytic quantities of butanol is possible because a new OH group is formed for each bond between a carboxyl group and an epoxy ring.

If the performance profile of anhydride-cured and amine-cured epoxy resins is compared, it will be noted that the systems differ significantly in their acid and alkali resistances. The amine-hardened films are sensitive to acids, while the acid-hardened esters are more susceptible to saponification by alkaline compounds.

If aliphatic or cycloaliphatic epoxy resins are used in place of aromatic ones during acid curing, the resulting coatings feature improved electrical insulation properties, even when subjected to continuous heat treatment.

On account of the ring tension in the epoxy group, epoxy resins can be polymerised to polyethers with the assistance of suitable starter ions such that the ring can even be opened without supplementary reactants. The reaction is particularly rapid if the starter ions are provided by the *Lewis* acid, boron fluoride. One method of successfully extending the pot life is to form a complex of boron fluoride and amines. The boron fluoride can then be released again by means of thermal decomposition at the time of film formation.

This reaction sequence is used on a large scale for the production of polyethylene glycols from ethylene oxide. The degree of polymerisation determines whether the polymer is liquid or wax-like (carbowaxes); it serves as a plasticiser or dispersant in water-borne paints.

In the search for high-quality 1-component systems the experience of blocking gained with the polyurethanes was successfully transferred to epoxy resins. Inactivated amine hardeners in the form of ketimines or aldimines achieve a pot life of more than 200 days if stored correctly, for a curing time of less than 16 hours. The ketimines and aldimines hydrolyse as a result of the ambient humidity, and release the amines required for crosslinking (see Figure 2.1.103).

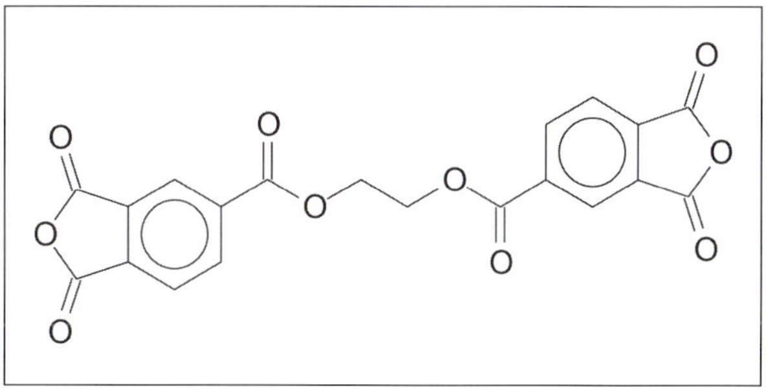

Figure 2.1.123: Example of an anhydride hardener produced by reaction of trimellitic anhydride with 1,2-ethanediol

The simple mixing of epoxy resins with amino resins or phenolic resins also yields usable, thermally curable 1-component coatings. Such combinations feature excellent resistance to a wide range of aggressive chemicals.

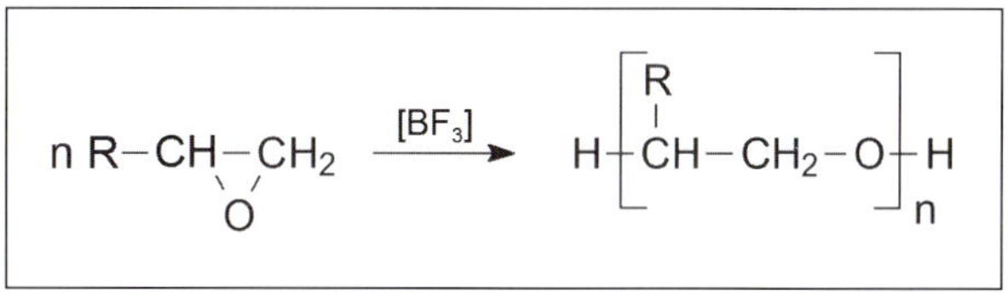

Figure 2.1.124: Polymerisation via ring opening of epoxy resins with BF_3

The reaction of epoxy resins with fatty acids to form epoxy resin esters yields a resin class

Coating Material

in which the excellent adhesion and hardness values of the epoxy resins are combined with the elasticity and good levelling properties of the oils. Furthermore, it is possible via the choice of fatty acid type to engineer purely oxidative curing or, in conjunction with amino resins, thermal curing.

Epoxy resin esters are therefore important basic structural elements for high-quality primers and primer surfacers for industrial coatings.

Due to good adhesion to many different substrates, high chemical resistance and the associated good corrosion protection afforded to steel structures, construction machines and vehicles as metallic objects, as well as the sealing of concrete and mineral substrates, epoxy resin paints have evolved into an economically important class of products. The special properties of the solvent-borne 2-component and 1-component materials led to their being accorded priority when it came to adapting to low-solvent, mainly water-borne systems and powder coatings [2.4.33].

If the epoxy alkyd resin esters are modified via the double bonds in the fatty acids with maleic anhydride and if the incorporated anhydride ring is opened in similar fashion to the maleinised oils with butanol and is then neutralised with amines, the result is water-soluble resins for anodic electrocoating.

Bisphenol A epoxy resins are also used for the manufacture of resins for cathodic electrocoat paints, which are significantly more important today (see Figure 2.1.127). The cations which are required for dispersion in water are obtained by adduct formation with amines and subsequent neutralisation with acetic acid.

The important electrocoat primers in the automotive and subcontractor industries are based on resins which consist of polyester core segments and added epoxy resins. The epoxy end groups are then converted with amines as described. Crosslinking takes place mainly with admixed and dispersed blocked isocyanates which then react with the OH and NH groups of the base resin after thermal unblocking from about 150 °C [2.4.34].

The changeover to cathodically depositable coatings that was achieved in a very short period of time in automotive coating, despite the necessity for substantial plant outlay (see Chapter 4.2.1.1), is a good example of the technical advances made in corrosion protection (see Chapter 7.2).

Initially attempts were made to produce **water-thinnable epoxy resins** for spray or manual application by simply converting conventional systems into a dispersion with

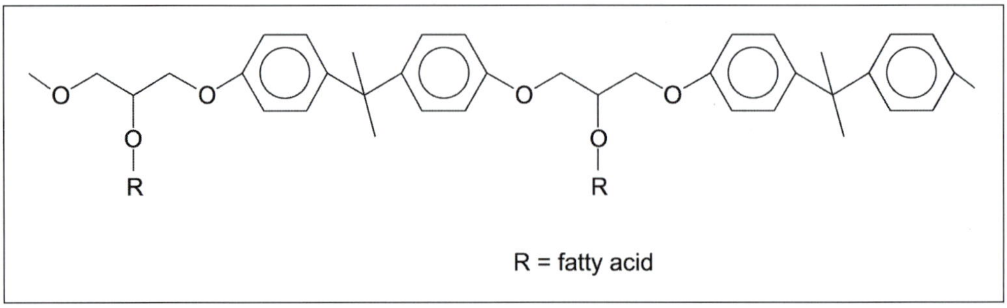

Figure 2.1.125: Structure of epoxy resins modified with fatty acids

the aid of suitable agitators but without resorting to any chemical changes, and then to simultaneously or subsequently combine them with the traditional water-soluble hardeners. The type of crosslinking partner would then produce hot- or cold-curing paints. However, the vagaries of the base-paint material led to specific problems: epoxy resins, even at relatively low molecular weights, are solids and so are not emulsifiable. It was therefore necessary either to resort to very low molecular weight base resins or to add unwanted organic solvents. It was found that low molecular weight epoxy resins cannot be converted into stable dispersions in the absence of stabilisers. Conventional emulsifiers, such as nonylphenol ethoxylate, may increase the stability and, with drop sizes in the range 5 to 10 µm, allow the production of sufficiently fine dispersions needed for subsequent film formation. As plasticisers, however, they degrade the mechanical-technological properties of the films, and through migration and evaporation over the long-term, lead to a reduction in quality which was already moderate to start with, ranging from embrittlement through to partial detachment.

A further weak point of epoxy group-containing dispersions is the interaction with the surrounding water. The finer the emulsion, the more extensive is the interaction. However, a fine emulsion is desirable from the point of view of film forming and levelling properties.

As described in numerous patents, suitable additives have been developed that eliminate these weak points and enable the quality to be adapted to the traditional role models that were dissolved in solvents. In addition to permanently effective stabilisation of the emulsion, it proved possible to reduce the droplet sizes to 1 to 5 µm and, at the same time, form an effective barrier layer against the influence of water, without adversely affecting the application properties.

The replacement of conventional emulsifiers by chemically reactive alternatives, such as epoxy-modified polyalkyl ether polyols or polyamidoamine-modified polyethers [2.4.35] in conjunction with novel hardeners obtained from the reaction of diamines with aliphatic diglycidyl ethers [2.4.36] paved the way for substantial progress with regard to the emulsifiability and fineness of the dispersion. The mobility conferred by the aliphatic chains of the polyethers additionally boosts the elasticity and maximum attainable film thicknesses, in association with a wider range of application.

Epoxy resin emulsions optimised by such modification are limited to liquid, i.e. low molecular, resin types which yield films with a correspondingly high degree of crosslinking. If lower levels of crosslinking are intended to induce further elasticisation, higher molecular base resins are indispensable. Since these are only available as solid resins, organic solvents or reactive diluents must be used as viscosity regulators. This has less to do with the process viscosity of the dispersion than with the fact that the resin components in the binder droplets must have a low viscosity which is necessary for uniform mixing.

The quality of current water-borne epoxy paints compares favourably with that conventional epoxy paints dissolved in solvents. Recent studies have even reported that the use of active pigments such as zinc phosphate which is classified as environmentally harmful can be dispensed with when water-borne anticorrosion primers are being designed [2.4.37].

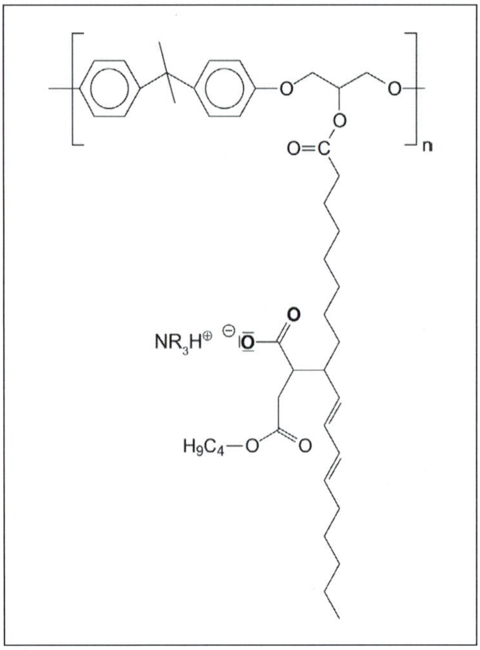

Figure 2.1.126: Structural element of maleinised epoxy resin esters for anodic electrocoating

In addition to their importance for conventional 2-component and 1-component systems as well as their use as base resins for anodic and cathodic electrocoat paints, epoxy resins have also found a broad field of application as film formers for powder coatings, as primers and, in mixtures with polyesters, as so-called hybrid powders for topcoats [2.4.38].

The good film properties in conjunction with the processing-specific advantage of the lower stoving temperatures without elimination of condensation products have made polyaddition coating resins particularly attractive. The search for further cross-linking possibilities through polyaddition has thrown up several product groups.

The addition of active methylene and amino groups to vinyl ketones or acrylic

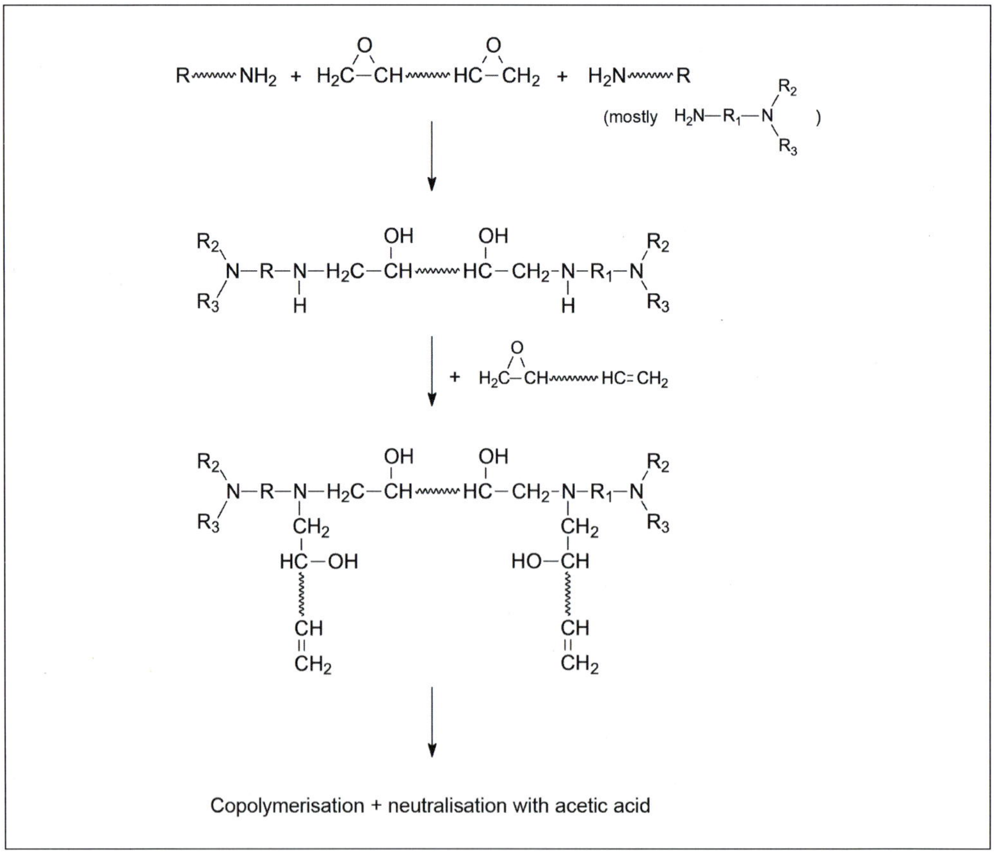

Figure 2.1.127: Example of the chemical structure of cathodic electrocoat epoxy-based resins

esters, the reaction of carboxylic acids with carbodiimides and the varied reaction options for oxazolines with carboxylic acids, carboxylic acid anhydrides and with itself are just some of the novel crosslinking reactions which are worth noting. An example which merits a special mention is the Michael addition of malonic acid esters or polyamines with resins containing acrylic ester. The discovery of catalysts means that the crosslinking temperatures can be reduced to as low as 70 °C, as a result of which new options are opening up even for thermally sensitive substrates and for the automotive refinishing sector [2.4.39].

Carbodiimides, which have been known for many decades as a mild condensing agent in the chemistry of natural products, can be employed as paint resins if they are made to react with carboxyl-group-containing film forming agents such as acid polyesters or polyacrylates.

Oxazolines, which have been known for their role in the emulsification of alkyd resins for decades, are experiencing renewed interest. Their synthesis works by the intramolecular dehydration of hydroxy-functional acid amides or by removal of NH_3 from the products of the reaction of ethanolamine with nitriles.

Oxazolines, made from the analogous unsaturated fatty acid amides, are eminently suitable for the synthesis of oxidatively curing polyamides if they made to undergo ring-opening polymerisation in the presence of appropriate catalysts.

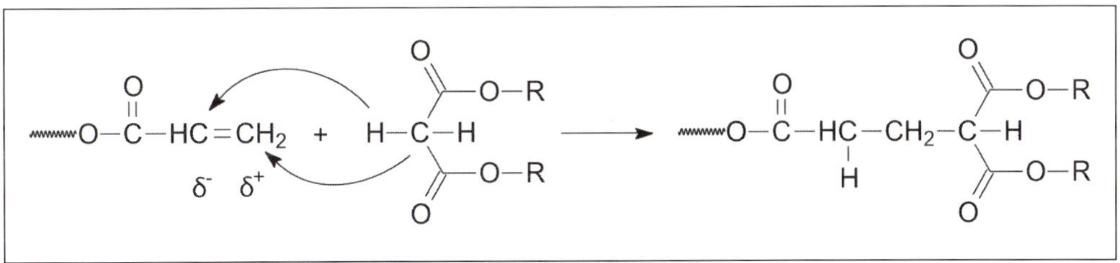

Figure 2.1.128: Michael addition of malonates to acrylic acid derivatives

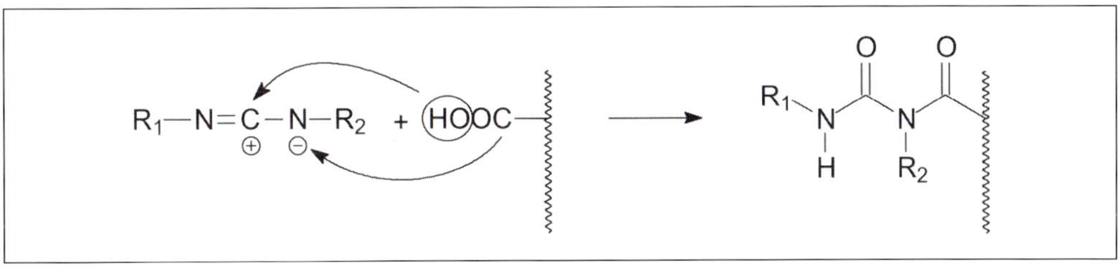

Figure 2.1.129: Reaction scheme for the addition of carboxylic acids to carbodiimides

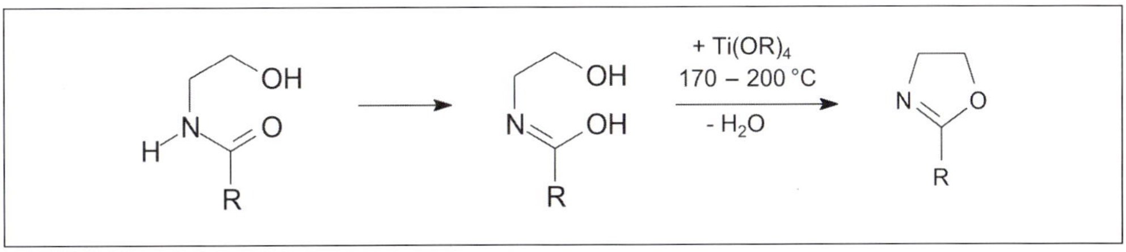

Figure 2.1.130: Synthesis of oxazolines from ß-hydroxy acid amides

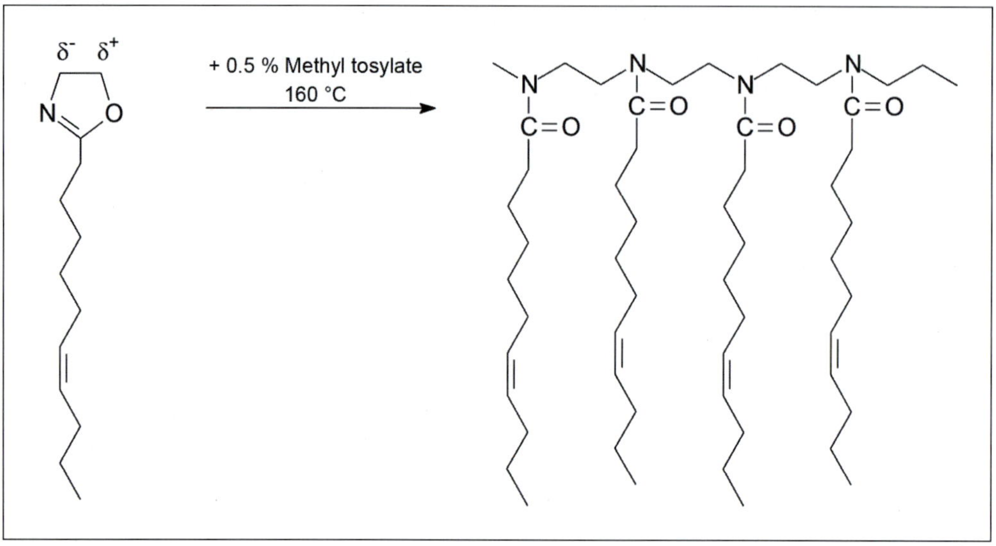

Figure 2.1.1131: Ring opening polymerisation of oxidative-hardening oxazolines

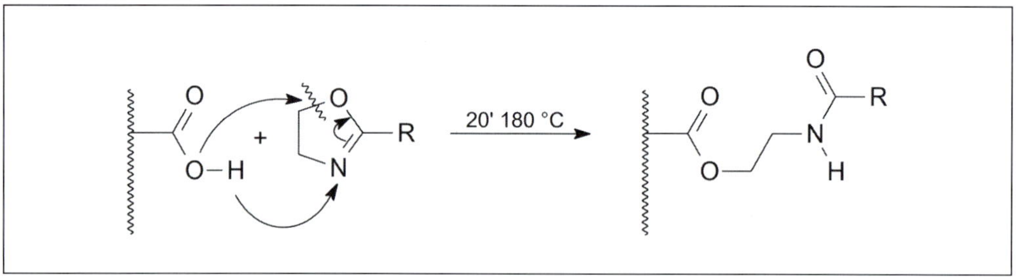

Figure 2.1.132: Crosslinking reaction of oxazolines with organic acids

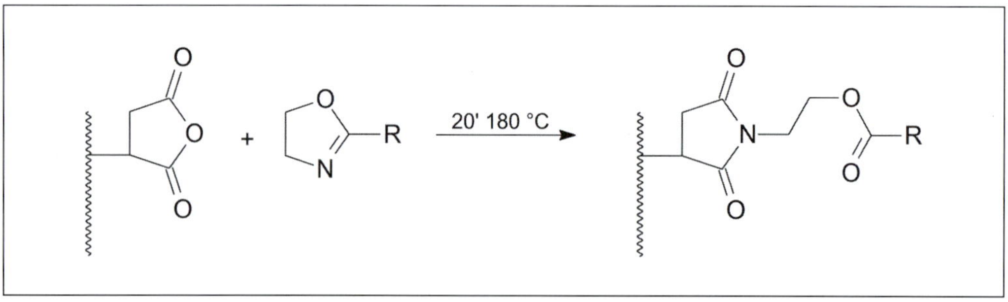

Figure 2.1.133: Crosslinking reaction of oxazolines with acid anhydrides

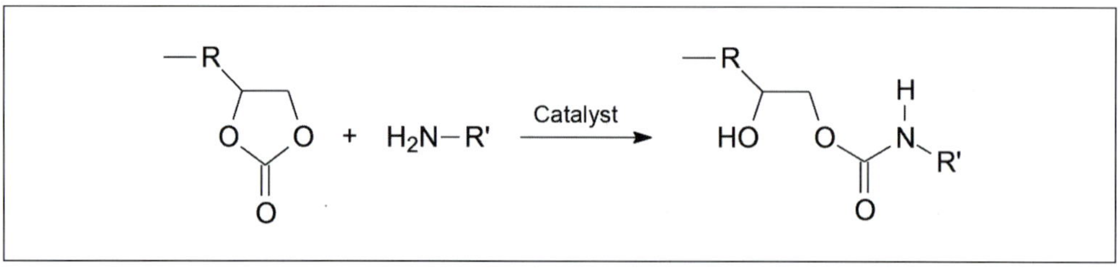

Figure 2.1.134: Scheme of crosslinking reaction of carbonic acid esters with polyamines

Film forming agents containing oxazoline groups are also partners for acid polymers. They react with carboxyl groups at temperatures above 180 °C, leading to ring opening, formation of esters and, at the same time, formation of acid amide groups.

A reaction with acid anhydrides is also possible. For example, maleinised epoxy resin esters can be converted with oxazolines to stable ester imides.

Cyclic carbonic esters can also be used as crosslinking resins. Cyclic propylene carbonate reacts with primary and secondary amines at low temperatures to form polyurethanes.

Resins by Polymerisation

Polymerisation has become by far the most important type of reaction for the manufacture of synthetic materials. As a result of the mechanism and kinetics of the reaction, polymers generally have substantially higher molecular weights than condensation resins, which are more suitable for paint purposes. Since a high molecular weight in a film forming agent is associated with reduced solubility, increased viscosity in the paint solutions and thus a low solid content, the use of high molecular polymers in dissolved form faces considerable environmental problems. As solvent-borne paints, they are therefore waning in importance.

A way out of this problem is afforded by processing as a dispersion, or as a powder coating. The aqueous emulsion polymers play a very major role. They serve as the basis for dispersion coatings, fillers and synthetic resin plasters. Acrylic or polyvinyl acetate dispersions for coating external façades or interior surfaces can be particularly mentioned in this regard. Dispersions, in which solvents or plasticisers act as the fluid phase instead of water, include particularly the organosols and plastisols which are used as

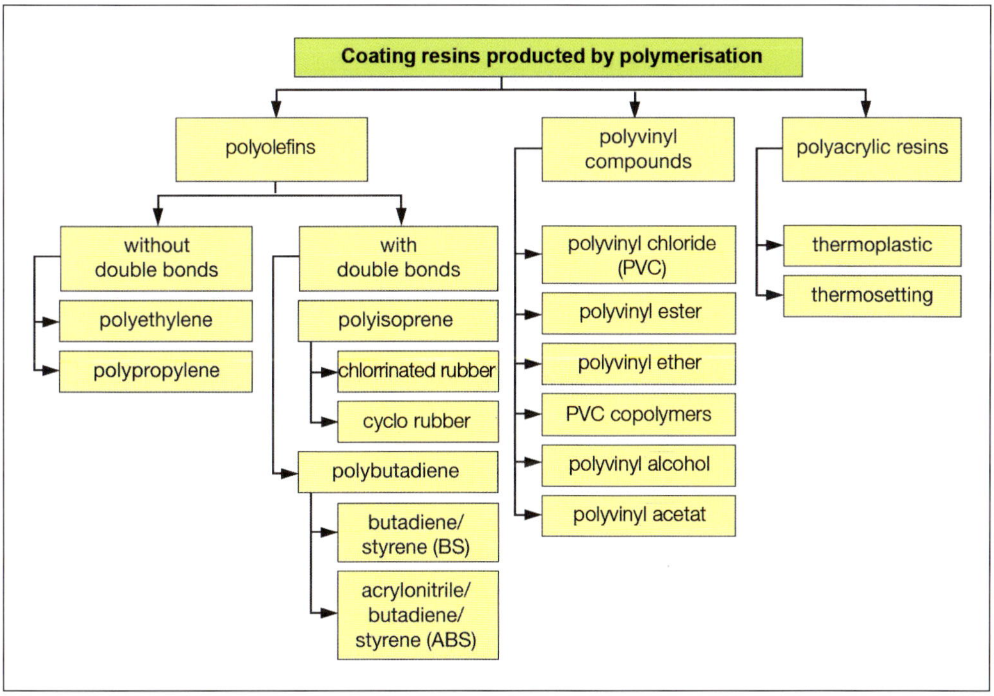

Figure 2.1.135: Groups of important coating resins produced by polymerisation

stoving enamels. The plastisols contain plasticisers, while the organosols also contain nonsolvents which are still volatile. Those polymers which are important in coating technology are divided into polyolefins, polyvinyl compounds and polyacrylates.

The classic **polyolefins**, such as polyethylene, polypropylene, i.e. those which are no longer amenable to chemical crosslinking during film formation, have failed to gain any technical importance as film forming agents because of their insolubility in standard solvents and their tendency to crystallise. Polyisobutylene, obtained by ionic polymerisation, partially soluble in hydrocarbons and possessing excellent cold resistance, has proved a successful elasticising component for bitumen and for flexible coatings of food wrappings and textiles.

Low molecular polyethylene wax dispersions are additives for influencing the surface smoothness and effect formation of metal effect coatings. Following sulphochlorination they perform special tasks as film forming agents. The situation with polyolefins is different, as these still contain reactive double bonds after their manufacture. Polybutadiene and polyisoprene belong to this product group. Low molecular polybutadiene oil, like fatty oils or oxidatively curable epoxy resin esters, is maleinisable and thus a proven film forming agent for anodic electrocoating [2.4.40].

The most important members of the **polyvinyl resins** group are polystyrene, polyvinyl chloride (PVC) and various polyvinyl esters and polyvinyl ethers.

Polystyrene, a homopolymer, is not suitable for use in coatings because of its inability to wet pigments and its high degree of brittleness. Only following copolymerisation with acrylonitrile (better colour absorption) and butadiene (elastic properties) is the resulting **acrylonitrile-butadiene-styrene copolymer** (ABS), which is processed from aqueous dispersion, of interest for coating, among other things, surfaces at risk of chipping.

Copolymers of butadiene with styrene in the form of aqueous dispersions have acquired another application field. On account of their residual C=C double bonds the films are capable of oxidative secondary crosslinking. This helps to increase hardness and resistance to water, chemicals and detergents. Latex paints are manufactured with styrene-butadiene copolymers. They are suitable for internal and external application, particularly on mineral substrates and, in some cases, also on metals.

Unless it is modified, **PVC** too is not usable as a film forming agent. Its poor solubility and sensitivity to light are the most serious shortcomings from the coatings perspective. Only post-chlorinating the raw material, which already contains 56 % chlorine, to a

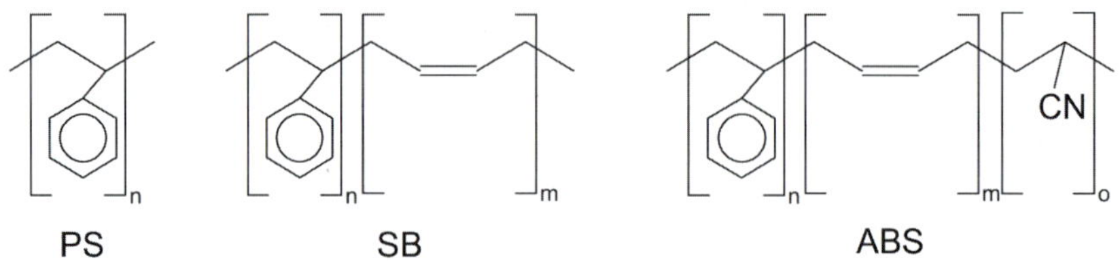

Figure 2.1.136: Polystyrene (PS) as a homopolymeric compound as well as styrene-butadiene (SB) and acrylonitrile-butadiene-styrene (ABS) as copolymeric compounds

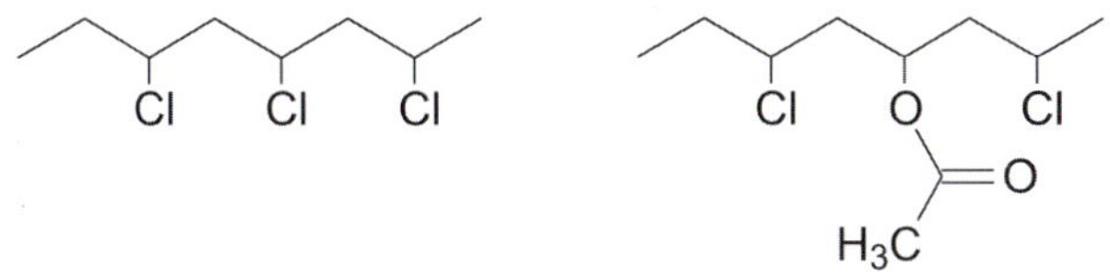

Figure 2.1.137: Molecular segments of polyvinyl chloride (PVC) and a copolymer with vinyl acetate

chlorine content of approx. 65 % improves solubility and thus permits the manufacture of physically drying film forming agents. These were used in the past, in combination with tar and bitumen, for heavy-duty anticorrosive protection.

PVC attains much greater importance when used as a dispersion in plasticisers. Such plastisols can be processed without emissions. They then gel under the action of heat to form a highly elastic compound which is processed in thick films. Underbody protection and seam sealing compounds in the automotive sector are important applications for PVC plastisols. PVC plastisols are also processed as topcoats in the coil-coating sector.

Coatings made of nonplasticised polyvinyl chloride have a high surface hardness and wear resistance and above-average resistance to water, solvents, chemicals and fuels. As a result, they are particularly suitable for cladding containers, storage tanks and similar objects.

Their poor lightfastness and heat resistance can be overcome by the use of stabilisers. Good adhesion to iron can only be achieved by the use of special wash primers.

Water-borne PVC dispersions are also used to coat textiles, leather and synthetic leather. PVC-based powder coatings are only of interest for the fluidised-bed process (see Chapter 4.2.2). A widely used method of obtaining film forming agents with better solubility, adhesion and compatibility properties is copolymerisation of the vinyl chloride with vinyl acetate, vinyl propionate, vinyl isobutyl ether or in smaller proportions with free acrylic and maleic acids. The amount of ester- and ether-type comonomers ranges from 5 % to 50 %, while that of the adhesion-enhancing acid components is of the order of 1 %. Apart from the molecular weight, the type and proportion of comonomers used have a defining effect on the solubility and compatibility of the resins, as well as on the properties of the coatings made with them.

With few exceptions, the copolymers of vinyl chloride are physically drying film forming agents. Important applications for PVC copolymers include internal coatings of beverage cans, heat-sealable coatings on aluminium foils for lamination and bonding, high-adhesion coatings on hot-galvanised substrates and aluminium sheets. PVC copolymers are also used to protect concrete floors and for road-marking paints. When combined with alkyd resins they serve as primers and high-build anticorrosive coatings.

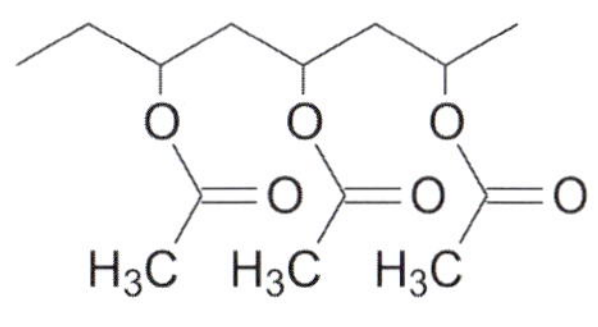

Figure 2.1.138: Molecular segment of polyvinyl acetate

The homopolymer **polyvinyl acetate** is a crystal-clear, lightfast, odourless and tasteless coating resin. It is available in a wide range of different molecular weights. The molecular weight is not only important for the viscosity of solutions, but also for compatibility with other paint raw materials and for the mechanical properties of the dried films. In all cases polyvinyl acetate has good adhesion, even on difficult substrates. The preferred solvents for polyvinyl acetate are esters and ketones. There is good compatibility with cellulose nitrate, acetobutyrate, chlorinated rubber and polyacrylates.

Carefully coordinated combinations increase the bonding strength and build. Despite their good barrier effect against gases, polyvinyl acetate films are extremely permeable to moisture. This permits the exchange of moisture between the coated substrate and the atmosphere.

Water-borne polyvinyl acetate dispersions with solid contents of 50 to 60 % have incomparably more important uses than the solvent-borne types. For the homopolymers the minimum processing temperature, i.e. the minimum film forming temperature (MFT), is 15 °C which is too high for central European climate conditions. To lower this and thus to guarantee film formation even at lower temperatures, the material has to be plasticised. Plasticisation can be achieved by adding plasticisers or by copolymerising vinyl acetate with plasticising monomers. Materials which can be considered for use as plasticising monomers are esters of vinyl alcohol with lauric or stearic acids, as well as alkyl esters of maleic and fumaric acids.

Polymers of vinyl propionate have greater resistance to alkalis. The high inherent elasticity of **polyvinyl propionate** means that for many applications it is necessary to raise the product's hardness by copolymerisation, e.g. with vinyl chloride.

Aqueous dispersions (see below) of polymers of vinyl acetate and propionate prepared by emulsion polymerisation have been the foundation of emulsion paints for decades. They are very widely employed in the manufacture of coatings for buildings, wood coatings, coatings for wallpaper and textiles and of fillers and synthetic plasters. The numerous advantages of good compatibility with many other film forming agents, physiological safety, lightfastness and film gloss that can be achieved with polyvinyl acetate must be weighed against the disadvantages of their tendency to be swelled by water and the saponifiability of the ester groups.

Polyvinyl alcohol cannot be manufactured from its monomer because of the instability of the vinyl alcohol, which immediately rearranges into acetaldehyde. Polyvinyl alcohol therefore has to be obtained indirectly by saponification of polyvinyl acetate. As a technical material it is used as a thickening agent and as the starting material for polyvinyl acetals.

Polyvinyl acetals are products of the reaction of polyvinyl alcohol with aldehydes or ketones. Polyvinyl formal is produced by the reaction of formaldehyde with polyvinyl alcohol, while polyvinyl butyral, which is more important from a coatings perspective, is from reaction with butyraldehyde. As polyvinyl alcohol is manufactured from polyvinyl acetate, residual amounts of unsaponified ester groups remain. There are also hydroxyl groups which escaped acetalation. Apart from the molecular weight, the defining factors for the products' solubility properties and compatibility are the respective rations of residual acetate and hydroxyl groups to the acetal groups formed.

Polyvinyl butyral is compatible with all phenolic and urea resins, and with some alkyd resins and cellulose nitrate. Low molecular types serve as plasticising components in

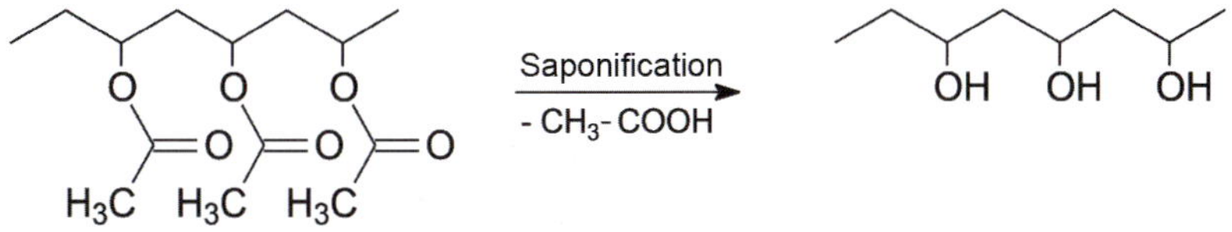

Figure 2.1.139 Saponification of polyvinyl acetate

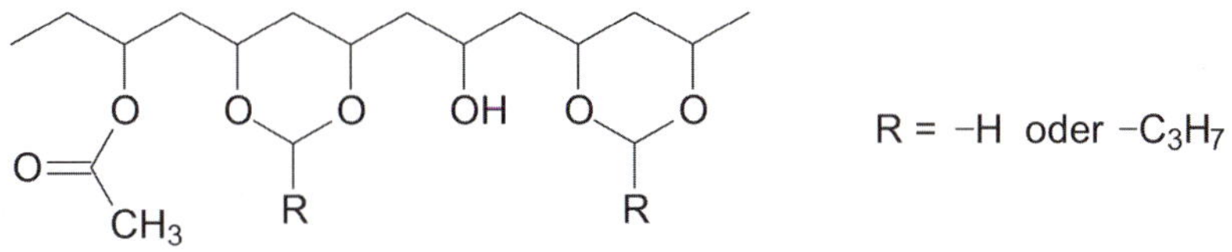

Figure 2.1.140: Molecular segment of polyvinyl acetal

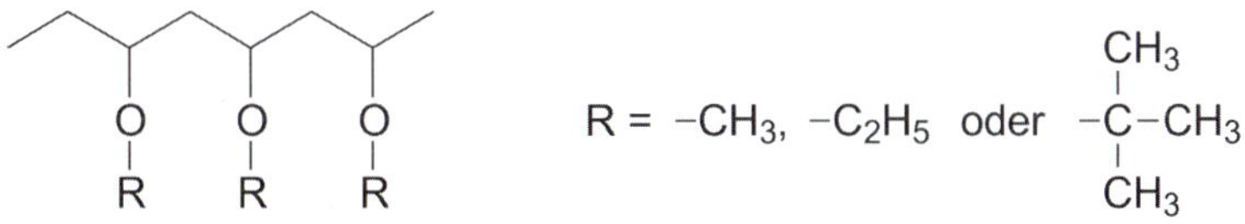

Figure 2.1.141: Molecular segment of polyvinyl ether

stoving enamels based on phenolic resins and in acid-curing coatings based on urea or melamine resins. Medium-molecular types are often blended with epoxy and phenolic resins and have an important role in corrosion control in reactive mixtures with phosphoric acid as wash primers.

Other noteworthy polyvinyl compounds are the **polyvinyl ethers** which are more resistant to saponification than polyvinyl esters, but are not lightfast. The copolymer of polyvinyl isobutyl ether and PVC is the most important commercial product in this class of substances and is ideal for physically drying primers because of its good adhesion to nonferrous metals.

Most of the polymers described thus far are processed as aqueous dispersions when used for coatings. They are mostly produced by emulsion polymerisation [2.4.41].

In this manufacturing process, the monomeric structural elements to be polymerised are first transformed into a coarse aqueous emulsion stabilised by surfactants. Excess surfactants aggregate together in the polar aqueous phase to form micelles. These are small nests composed of polar moieties of surfactant facing outwards and less polar moieties facing inwards.

The hydrophobic monomers, which have poor solubility in water, ultimately migrate into the hydrophobic cores of the micelles. Water soluble starting radicals R· initiate a polymerisation reaction in the micelles. Via of the surfactant and radical concentration, it is possible to control the growth of the particles in the micelles such that the desired particle size is reached when the monomer has been fully used up.

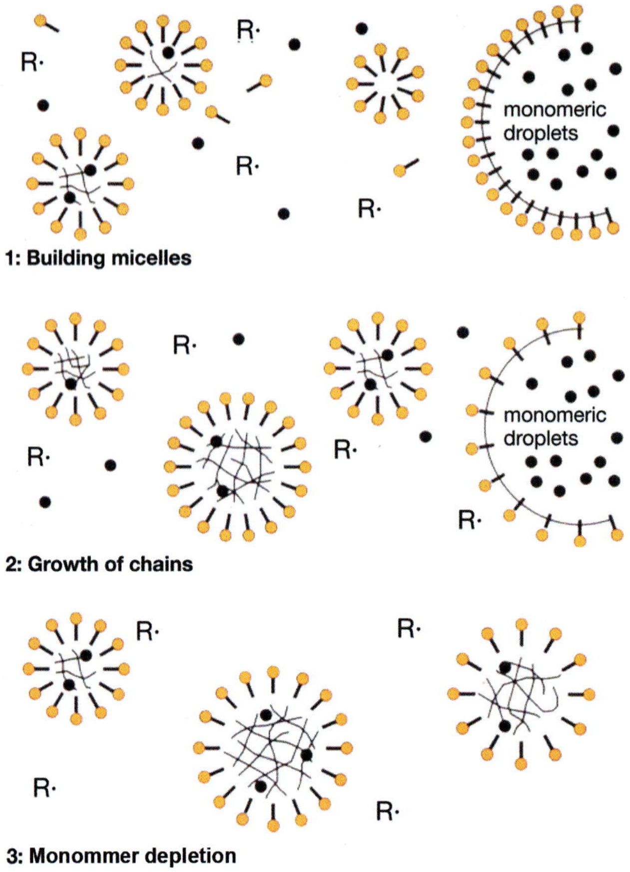

Figure 2.1.142: Schematic diagram of emulsion polymerisation Source: BASF SE

Acrylic Resins

The term acrylic (acrylate) resins, which are also called polyacrylates or polyacrylics, refers generally not just to the polymers of the pure acrylic acid derivatives, but rather to the copolymers of the esters of acrylic and methacrylic acids and styrene. The various monomers can be randomly incorporated into the polymer chains in almost any mixing ratio.

The character and properties of the acrylic resins are defined by the monomer mixture employed and also by the molecular weight. Thermoplastic acrylic resins having molecular weights of more than 100,000 g/mol and no reactive groups have found their niche in coating technology just as much as low molecular chemically reactive types having molecular weights of less than 20,000 g/mol. Thermoplastic polyacrylates are mainly processed in aqueous dispersions nowadays (see above). For environmental and economic reasons, the proportions of genuine organic partial solutions are low these days. The chemically reactive variants are used in conjunction with crosslinking agents such as melamine resins as 1-component paints or with isocyanate hardeners as 2-component materials. Important monomers for thermoplastic coating materials include ethyl, methyl and butyl acrylates and analogous esters of methacrylic acid and styrene.

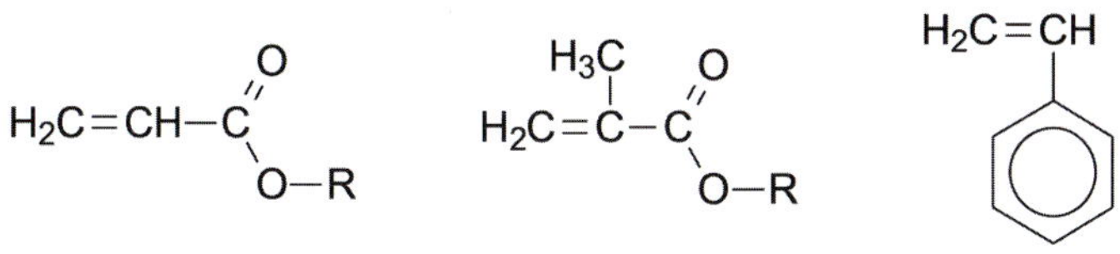

Ester of acrylic acid   Ester of methacrylic acid   Styrene

R = – H, $-CH_3$, $-C_2H_5$, $-C_3H_7$ and $-C_4H_9$

Figure 2.1.143: Basic components of thermoplastic acrylic resins

Thermally curing coating paints, known as "thermosetting paints", use resins which also contain proportions of free acrylic acid, hydroxy-functional acrylic esters, acrylamides, glycidic esters of acrylic acid and corresponding methacrylic acid derivatives (see Figure 2.1.145b).

The film properties and chemical reactivity can be controlled by the number and type of functional groups. In addition to hydroxyl and carboxyl groups, amino or amido groups also merit a mention in this regard. The latter can occur free, mixed with formaldehyde and also etherified.

The mixing ratio of the randomly copolymerising basic structural elements affects the softening point, also termed the glass transition temperature (see Chapter 3.2.5) and thus also the minimum film forming temperature for aqueous dispersions. The methyl side groups of the methacrylic esters stiffen the molecular chains in similar fashion to the aromatic groups in styrene. Polymethacrylates or styrene-containing polyacrylates are therefore harder than the corresponding pure polyacrylates. The softening point of the polymers is also affected by the length of the alcohol group of the acrylic or methacrylic esters. Given comparable molecular weights of the polymers, the butyl esters are relatively flexible whereas the methyl esters are rigid and hard. The length of the alcohol group affects the mechanical properties, solubility in solvents and compatibility with other resins.

A characteristic feature of the performance profile of acrylic resins, irrespective of whether they are thermoplastic or chemically reactive, is the very high degree of saponification of the polymeric acrylic and methacrylic esters. They are therefore proven film forming agents for coating concrete, plastics and nonferrous metals such as copper, silver, brass or chromium-plated iron. Aqueous thermoplastic acrylic resin dispersions have now also gained access to the wood coatings field due to the relatively rapid evaporation of the water.

However, the high molecular weight and associated poor pigment absorption prevents them from achieving peak performance in terms of gloss. Despite such difficulties thermoplastic acrylic resins have achieved a certain importance as rapidly drying automotive refinishing paints for spray cans because of their excellent resistance combined with good polishing properties. They allow flaws and contamination to be sanded away during refinishing work without the need for any additional secondary coating, since the coating merely needs to be heated to enable it to flow again and create a smooth, glossy film. This method, known as the "reflow" process was used on automotive coating lines in the US for a time in the 1970s.

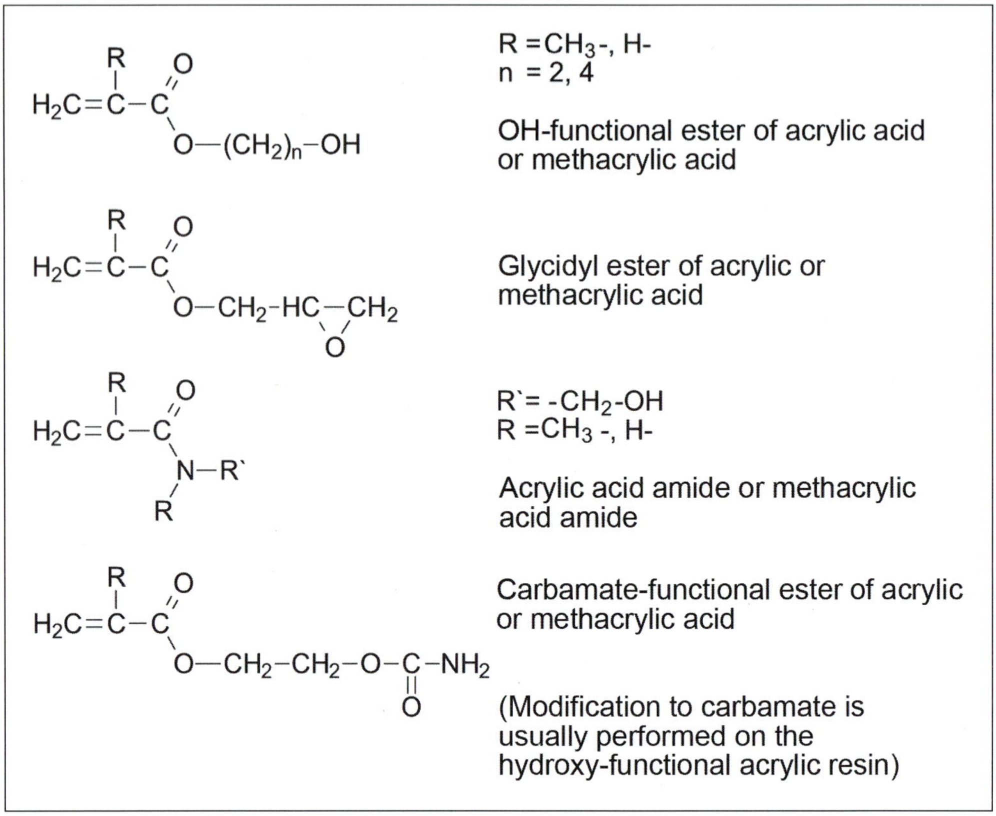

Figure 2.1.144: Basic components for thermosetting acrylic resins

Functional polyacrylates, combined with melamine resins or isocyanate hardeners to achieve crosslinking, are widely used on industrial coating lines and for refinishing. Because of the benefits they bring, they have largely superseded the topcoats based on alkyd/melamine resin combinations which previously served as standard. To ensure high cracking resistance (see Chapter 3.3.2) under extreme climatic conditions, clearcoats for metallic base coats are formulated primarily with externally crosslinking acrylic resins.

For reactive **OH-functional polyacrylates**, which are usually crosslinked with melamine resins or isocyanate hardeners, new reaction partners have been found. Trisalkoxycarbonylaminotriazine, or TACT (see Figure 2.1.113), is a crosslinking component which, due to the formation of urethane groups in the film, rivals melamine resin and isocyanate-crosslinked variants in terms of the properties of the finished coating. Reactive acrylic resins which can be crosslinked with melamine resins, polyisocyanates and the newer TACT hardeners have become widely accepted as high-quality industrial and automotive finishes.

Mention should also be made of the previously described **carbamate-modified acrylic resins** in combination with HMMM resins as a new variant for heat curable polyurethane acrylates (see Figure 2.1.145a and b).

If glycidyl acrylate or glycidyl methacrylate is used proportionately in the manufacture of acrylic resins, such special resins can also be made to react with acid polyesters or

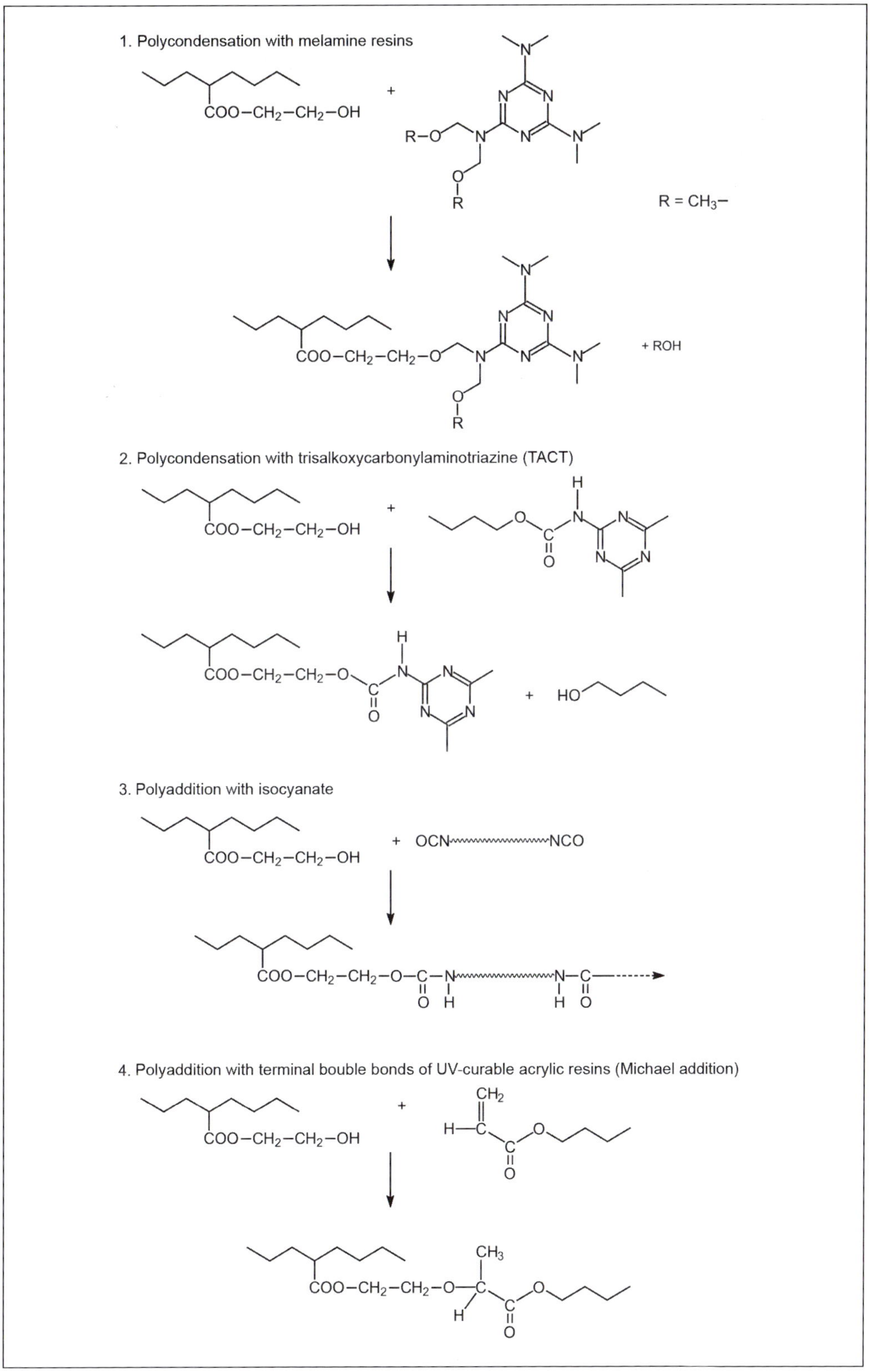

Figure 2.1.145a: Reaction schemes for external crosslinking of OH-functional acrylic resins

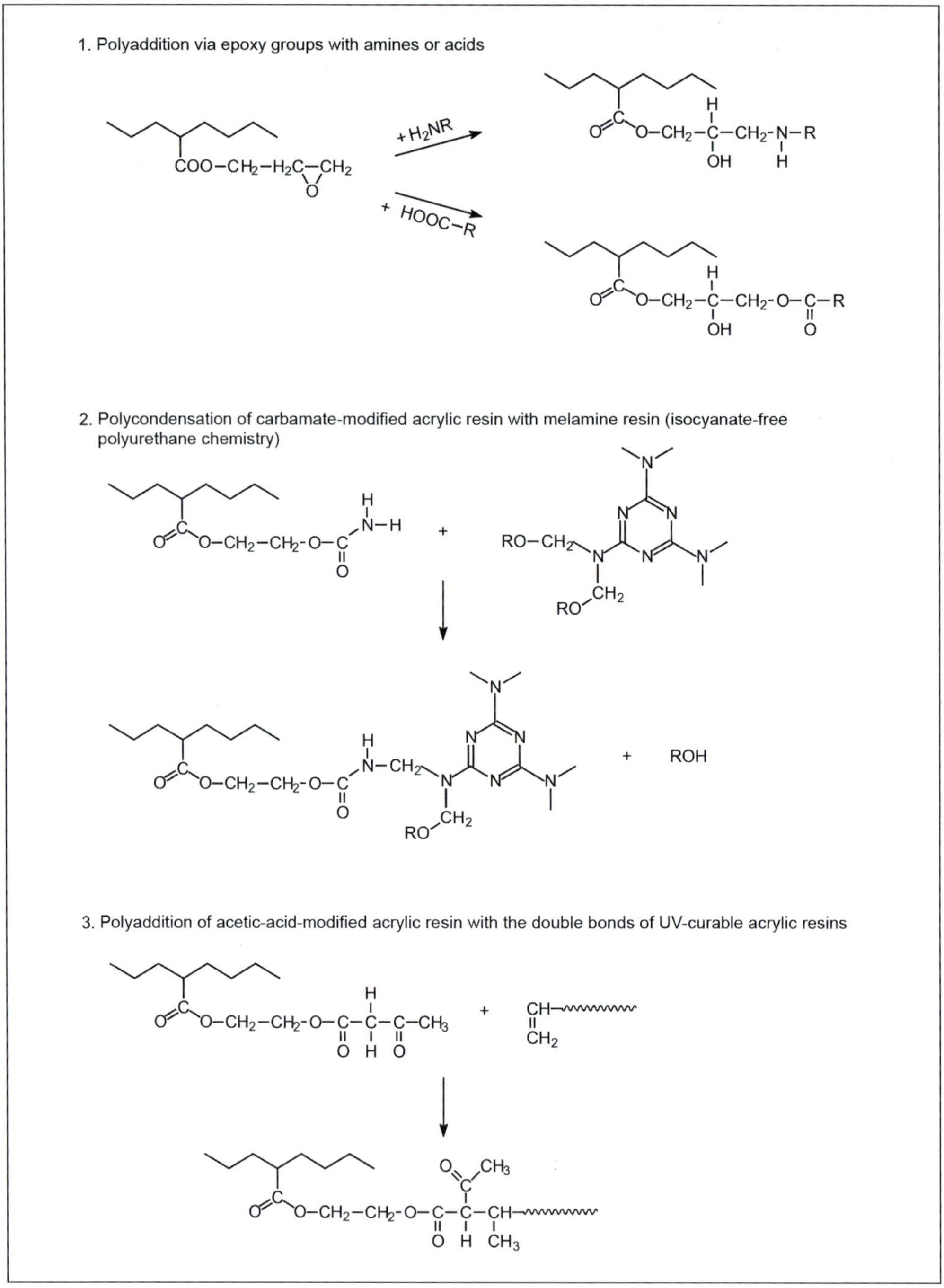

Figure 2.1.145b: Reaction schemes for external crosslinking of modified acrylic resins

acid acrylic resins via the epoxy groups which are now present in the molecule. This crosslinking reaction is used in powder coating technology, among others.

If a specific modification is carried out, it is possible to produce **self-curing acrylic resins**. One of the many options that these resins contain in their chain molecules, in addition to the hydroxyl groups, is amino groups formed by copolymerisation with

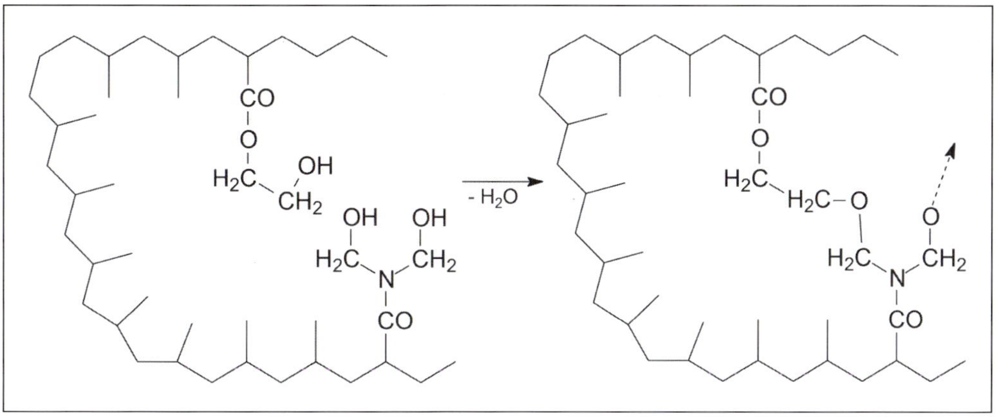

Figure 2.1.146: Typical reaction scheme for self-curing acrylic resins

acrylamide. These can be methylolated and additionally etherified with butanol. Starting from the modified amino groups, a thermosetting network is formed with the hydroxyl groups during stoving, without the need for the presence of an external resin. These are the base resins for detergent-proof single-film coatings for household appliances, dishwashers and refrigerators.

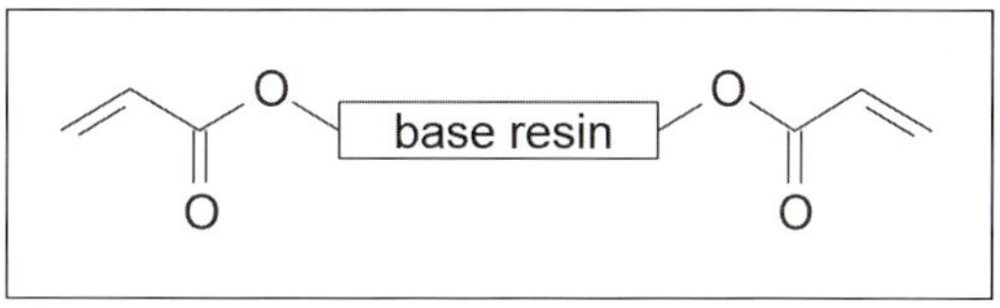

Figure 2.1.147: Acrylic acid-modified resin for radiation curing

Increasing the proportion of the hydroxy- and particularly the carboxy-functional monomers in the monomer mixture leads to special products which, after the addition of amines as polymer anions, become soluble in water and which can therefore also serve as resins for anodic electrocoating. Cathodically depositable acrylic resins can also be manufactured by polymerisation of acrylate esters that contain amino groups in the esters. After addition of blocked isocyanates and neutralisation with organic acids the mixture can be dispersed and used for topcoats that possess acceptable long-term outdoor exposure properties [2.4.42].

The **radiation curable acrylic resins** have a completely different chemical constitution from the film forming agents described thus far (see Chapter 5.6.1.4). Unlike classic acrylic resins, radiation curable resins still contain carbon double bonds. These are inserted by the incorporation of terminal acrylic ester groups. They are, therefore, not genuine acrylic resins, but rather acryl-modified film forming agents from a huge variety of origins.

Such resins contain terminal active double bonds. Consequently, their suitability as film forming agents for radiation curable coatings differs from that of the styrene-containing unsaturated polyesters which previously served as standard because they have a better ratio of growth to termination during curing. They lead to higher molecular networks and as a result are more resistant to chemical and mechanical stresses.

Various prepolymers, which are enabled to bond chemically with acrylic acid via either the terminal epoxy or hydroxyl groups, are used to manufacture the acrylic acid-containing film forming agents for UV-curing.

The first commercially available resins were acrylic acid-modified epoxy resins. Since one OH group is created per reaction stage by the addition to the epoxy groups, this can also be made to react with acrylic acid chloride, with the release of HCl. The resulting increased functionality leads to even higher degrees of crosslinking and thus to even greater resistance to chemicals and improved hardness.

The manufacture of modified polyester or urethane resins is similar to that of acrylic acid-modified epoxy resins. Polyesters are bound to acrylic acid via terminal OH groups in a condensation reaction, while isocyanate-containing resins are bound to OH-functional acrylic esters in a polyaddition reaction

In keeping with the characteristic properties of the prepolymers, the radiation curable systems dominate specific application fields. The reactive epoxy acrylates confer high chemical resistance, while polyester acrylates are extremely lightfast and therefore suitable for packaging material coatings and for wood treatment materials. Urethane acrylates are particularly suitable for coating plastics and wood because of their toughness, good adhesion and wear resistance and are also suitable for coating leather and for printing inks.

If the coating has to meet special demands on heat resistance combined with good long-term outdoor exposure and mechanical/technological properties, silicone acrylates are the ideal starting products. A particularly noteworthy property is the outstanding scratch resistance, which stems from the very high surface hardness. Fluorinated acrylic resins additionally increase the lightfastness [2.4.43].

Coating materials for UV curing made purely of modified acrylic resins are generally too viscous to be processed without difficulty. They are therefore dissolved in solvents. However, the alternative practice of using lower molecular reactive diluents that

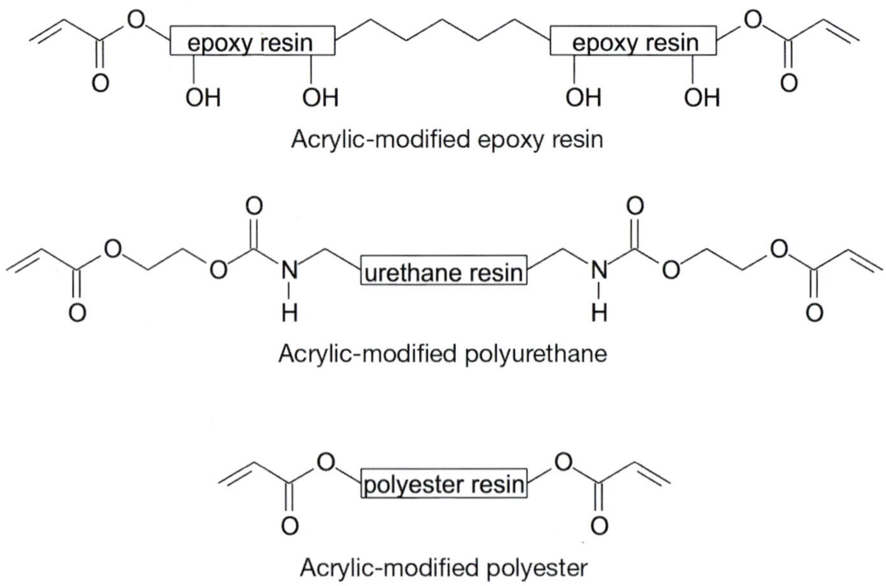

Figure 2.1.148: Examples of radiation curable, acrylic acid-modified film forming agents

copolymerise with the base resin has become widely accepted. Such formulations are practically emissions-free. The acrylic esters of trimethylolpropane and pentaerythritol have been superseded by propoxylated glycerine triacrylate or tris-(2-hydroxyethyl) iso-cyanatotriacrylate because of their irritation on contact with the skin. New reactive thinners which are regarded as safer include acrylic acid-modified carbonic esters and oxazolidones.

One of the other ways of influencing the rheological properties during processing consists in transforming the prepolymers into a water-soluble or water-dispersible form. However, radiation curing is only possible once the water has completely evaporated. Although interest in this relatively new variant of radiation curable coatings is growing, it has yet to find an important market.

As is the case for unsaturated polyesters and coatings made from them, photoinitiators are absolutely essential for initiating hardening by UV radiation, despite the higher reactivity of the acrylate functions (see Chapter 2.1.4 and Chapter 4.3.3).

The special performance profile and rapid or cost-effective curing of flat coating films led to attempts to extend the application spectrum of the radiation curable coatings [2.4.44]. Difficulties arise when the objects to be coated have a complicated shape and, as a result, not all the areas are equally accessible to the radiation that triggers curing. Such is the case, for example, with the folds and various undercuts found in vehicle interiors. In order to enable uniform curing and thus uniform quality on the outer skin and so-called shadow areas, coatings were developed which complemented radiation curing by offering scope for conventional film crosslinking. As a result, these **dual-cure coatings** reach the necessary crosslinking density in their "shadow areas" as well [2.4.45]. Apart from thermally initiated reaction of hydroxyl groups with melamine resins, pro-

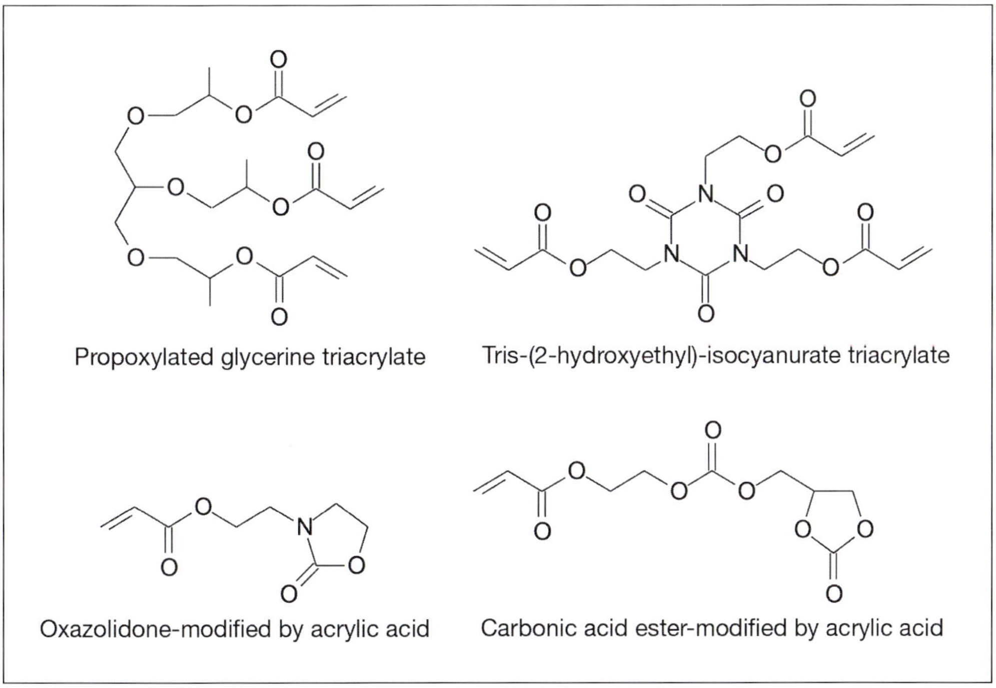

Figure 2.1.149: Reactive diluents for radiation curing and subsequent thermal crosslinking

ven linkage mechanisms for this purpose include polyaddition with isocyanate groups, which proceeds without the need for additional heating. Suitable isocyanate-modified acrylic resins can also find application as moisture-curing 1-component systems, or with corresponding OH-functional complementary resins as 2-component systems. More recent dual-cure systems utilise the possibility afforded by thermal crosslinking of ß-keto acid-modified acrylic resins with the terminal double bonds of the radiation curable base resins (see Figure 2.1.145b).

Coating Material

On account of the greater reactivity of primary hydroxyl groups in the presence of catalytically active acids, OH-functional acrylic resins are amenable to linkage with terminal double bonds, without the need for prior modification with acetic acid ester.

A new variant of radiation curable film forming agents are resins which are manufactured from **dendritic polyesters** and which bear terminal acrylic acid groups. Crosslinking requires only minimal quantities of photoinitiators. It is even possible to dispense with them completely if adequate radiation energy is available and amine synergists are present. The synergists, i.e. amines with a methylene group close to the nitrogen, generate radicals and reduce the oxygen inhibition. They are then incorporated into the film forming agent.

The special molecular form of the dendritic polyesters leads to low viscosities, even at high molecular weights. The high functionality yields extremely high crosslink densi-

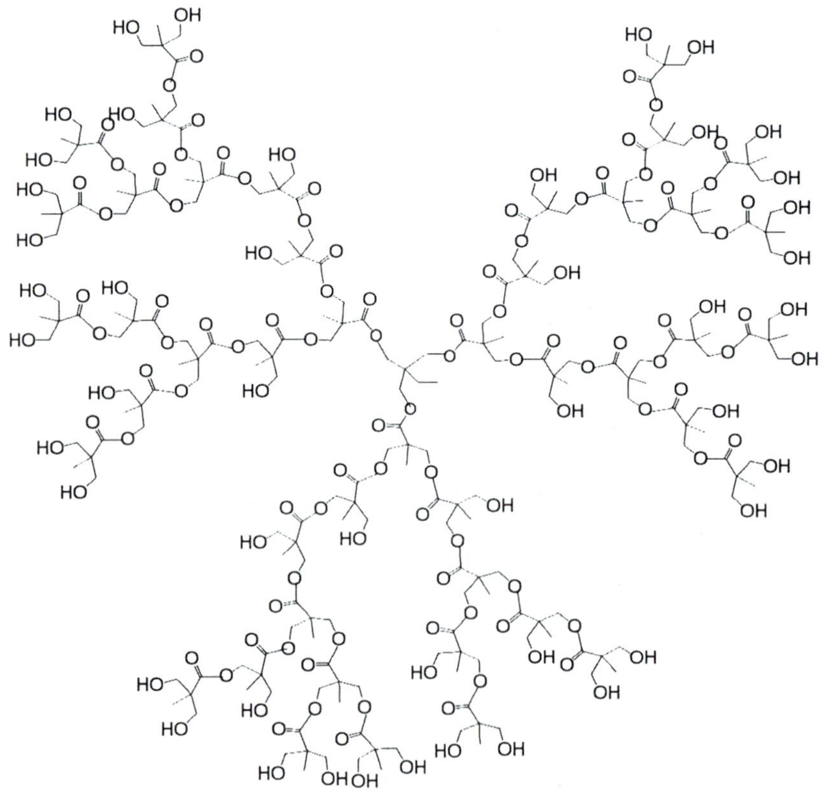

Figure 2.1.150: Structure of dendritic polyesters (dendrimers)

ties and thus to scratch and chemical resistance on a level which could not otherwise be achieved [2.4.46]. Other interesting properties of dendrimeric film forming agents are their extraordinary curing rate and also extremely low shrinkage during crosslinking because of their particular molecular form.

2.1.2 Solvents

Solvents, as defined in ISO 4618, are liquids comprising one or more components that are volatile under the specified drying conditions and can dissolve film forming agents purely physically without chemical reactions.

As low-viscosity substances, they are the exceptions among all the raw materials for coatings. They perform a special task in that they merely provide assistance during the processing of coating materials. When this task is done, they should remove themselves as quickly and completely as possible. They are thus unlike the reactive thinners and plasticisers which also affect the consistency of the material. Reactive thinners combine chemically with the resin, while plasticisers are chemically inert, very low volatility substances that remain in the film.

Ideally the solvents should produce a uniform liquid phase composed of all resin components. Colloids, gels, emulsions and dispersions, in contrast, are formulated by different methods. Notwithstanding the different formulating approaches, the aim is the same, i.e. to ensure that the coatings can be processed from the liquid phase.

In the case of solutions and to an extent depending on their compatibility with the solvents, the molecules of resin adopt various coil conformations, and the macromolecular chains become entangled to a greater or lesser extent during film formation. Variations in the mechanical strength of the film are a reflection of this. The latter impacts on the crack resistance and impact strength. This effect is particularly pronounced in physically drying coating materials. The choice of solvents for a particular coating system is therefore limited, not just by the application method, but also for mechanical/technological reasons.

However, the effect of the solvents on the film forming process and thus also on the properties of the coating goes well beyond expansion of the coils of resin molecules. The surface tension of solvents which is generally lower than that of the more polar coating resins promotes the wetting of the substrate. To this can be added the fact that the lowering of the viscosity by the solvents also aids improved processing, i.e. better sprayability, coatability, rollability or also dippability. The solvent composition should enable the rheological properties during the entire transition from the liquid coating material via the application to the viscoelastic coating including the drying and hardening phases to be modified such that good levelling is achievable without runs and dewetting from edges, that setting proceeds without blistering and cratering, that good overspray absorption is possible and at the same time that optimum effect formation is obtained with metallic or pearlescent effect coatings (see Chapter 3.1 and 6.2).

Nor should the influence of the solvents on the substrates be forgotten. In the case of plastics, for example, the ability of the substrates to swell and the associated impairment of the visual properties during solvent release and bonding must be taken into account.

Coating	Typical solvent content [%]
Powder coating	–
Silicone paints and plasters	–
Interior emulsion paints	0 – 2
Exterior emulsion paints	1 – 3
Synthetic resin plasters	3 – 5
Water-borne primers	3 – 5
Electrocoating paints	1 – 3
Water-borne paints	5 – 20
High solid paints	25 – 30
Conventional paints	45 – 70

Figure 2.1.151: Typical solvent content of different paints as supplied

Mixtures of solvents are often used because of the need to coordinate the interaction with the resin molecules and the associated evaporation rate during processing and film formation. Solvent mixtures should be capable at any time, under permanently changing circumstances, of guaranteeing the optimum film properties. Despite their environmental problems, therefore, solvents are necessary for coatings.

In recent decades considerable efforts have been made to replace organic solvents by the physiologically safe alternative of water or, as in powder coatings, to dispense with them completely. Given the specific requirements of complex coating processes, the use of organic solvents will undoubtedly remain necessary to some extent for the foreseeable future.

Since current legislation has severely restricted the permissible levels of solvent emissions, operators of coating plants are compelled to reduce emissions by employing secondary measures or by using low-solvent coatings (see Chapter 5.3.1).

Solvents can also be hazardous materials for those people involved in the manufacture and processing of coatings. They are, therefore, substances which need to be evaluated just as critically from the points of view of occupational health and the environment. Numerous toxicological studies over the last few decades have provided comprehensive information on the hazards of handling solvents. Research projects aimed at replacing or reducing the use of solvents are in progress all around the world. Furthermore, classification and labelling schemes have been put in place for standard solvents employed in coatings to simplify their selection and promote correct handling (see Chapter 5.4).

The colloidal solutions which were mentioned earlier and in which the film forming agents are no longer uniformly distributed are clear to slightly opalescent liquids. One indicator of their increasing inhomogeneity is that they scatter light. This is caused by the aggregation of a number of individual molecules which nevertheless are fully acted upon by solvents. Dispersions, which have grown more and more popular in recent years, are heterogeneous systems that exhibit a high degree of light scatter. They are milky liquids that consist of a continuous liquid phase and a discrete phase. The latter may be a solid or a liquid. The discrete phase needs to be stabilised by either electrostatic or entropic means (see Chapter 2.2.3).

2.1.2.1 Theory of Solubility

During the dissolving process solvent molecules penetrate the matrix of the film forming agent and surround its individual molecules with solvent shells. They solvate the resin. The effect of the solvent shells is to diminish the interaction between the resin molecules themselves to such a degree that molecular distribution of the dissolved materials occurs.

The strength of the solvation is dependent in equal measure on the intermolecular interactions of the film forming agent molecules and of the solvents themselves.

The cohesion of the solvent, quantified by its cohesion energy ΔE, can be calculated from the enthalpy of vapourisation ΔH_V. The cohesive energy expressed in terms of the molar volume, namely the cohesive energy density $\Delta E/V$, has a specific value for each solvent. It can be used to estimate the solvent power, provided that corresponding values for the film forming agent to be dissolved are also available. In ideal solvent mixtures, the joint value for the mixture is the geometric mean of the cohesive energy densities of the individual components. The square root of the cohesive energy density is called the

$$\Delta E = \Delta H_V - RT$$

$$\frac{\Delta E}{V} = \frac{\Delta H_V - RT}{V}$$

$$\sqrt{\frac{\Delta E}{V}} = \delta = \sqrt{\frac{\Delta H_V - RT}{V}}$$

ΔE = molar cohesion energy

ΔH_V = molar enthalpy of evaporation

$\frac{\Delta E}{V}$ = molar cohesive energy expressed in terms of volume

δ = solubility parameter

Figure 2.1.152: Definition and calculation of solubility parameters

Component 1	Component 2	Solubility	$\Delta\delta$
n-Hexane	methanol	poor	14.1
	polydimethylsiloxane	good	0.4
	polyisobutylene	good	1.4
n-Heptane	diethyl ether	good	0.2
	benzene	good	3.5
	ethylene glycol	poor	13.2
	water	poor	30.6
n-Octane	methylene chloride	good	4.7
Toluene	polyvinyl acetate	good	0.4
Propanol	polydimethylsiloxane	poor	9.2
Ethanol	polyvinyl acetate	good	6.9
	polyisobutylene	poor	9.8
Methanol	polyacrylate	moderate	8.4
	polystyrene	poor	11.0
Ethyl acetate	polystyrene	good	0.8
Acetone	polymethyl methacrylate	good	0.6

Figure 2.1.153: Solubility potential of two components according to the difference thesis $\Delta\delta < 6\ [(J/cm^3)^{1/2}]$

	δ	δ_d	δ_p	δ_h
n-Hexane	14.9	14.7	0	0
Cyclohexane	16.8	16.8	0	0
Toluene	18.2	18.0	1.4	2.0
Xylene	18.0	17.8	1.0	3.1
Ethyl benzene	18.0	17.8	0.6	1.4
Styrene	19.0	18.6	1.0	4.1
n-Propanol	24.3	15.1	6.1	17.6
2-Propanol	23.5	15.3	6.1	17.2
n-Butanol	23.3	16.0	6.1	15.8
Isobutanol	21.9	15.3	5.7	15.8
2-Ethylhexanol	19.4	16.0	3.3	11.9
Cyclohexanol	23.3	17.4	4.1	13.5
Diacetone alcohol	18.8	15.8	8.2	10.8
Acetone	20.5	15.6	11.7	4.1
Methyl ethyl ketone	19.0	16.0	9.0	5.1
Methyl isobutyl ketone	17.2	15.3	6.1	4.1
Cyclohexanone	19.8	17.8	7.0	7.0
Ethyl acetate	18.6	15.1	5.3	9.2
Butyl acetate	17.4	15.1	3.7	7.6
Butyl glycol	18.2	16.0	6.3	12.1
Ethyl diglycol	19.6	16.2	7.6	12.3
Butyl diglycol	18.2	16.0	7.0	10.6
Water	47.8	14.3	16.3	42.6
Polystyrene	19.0	17.5	6.1	4.0
Polyvinyl acetate	23.0	18.9	10.1	8.1
Polymethyl methylacrylate	22.0	18.7	10.1	8.5
Polyvinyl chloride	22.4	19.1	9.1	7.1
Epoxy resins	23.4	17.3	11.2	11.2

Figure 2.1.154: Solubility parameter δ of different solvents and their disperse contribution δ_D, polar contribution δ_P and hydrogen bridging contribution δ_H

solubility parameter δ. Solvents with similar δ values have comparable molecular interaction forces and thus also a similar solvent power. They therefore mix together well. Mutual solubility should therefore always be guaranteed if the differences do not exceed 6 $[J/cm^3]^{1/2}$. In a good fit with this thesis, the Figure 2.1.153 for several pairs shows that the actual solubility properties depend on the differences in the solubility parameters. Unfortunately, this fit is not guaranteed in all cases.

To eradicate this shortcoming, the δ value is divided into three individual components. The indices relate to the non-polar dispersion forces, the dipole forces and the interactions caused by hydrogen bridges. The solubility-related properties of the solvents are now expanded by three individual values to yield a 3-dimensional solubility parameter [2.4.47].

Figure 2.1.154 shows that it is quite possible for the same or similar δ_D values to have different δ_P and δ_H values. Despite this optimisation in the calculation of solubility parameters, some inconsistencies remain. Formulators will find it useful but should exercise caution.

2.1.2.2 Physical Properties

With regard to their action on the various film forming agents, solvents or solvent groups can turn out to be genuine active solvents, latent solvents, but also nonsolvents. The genuine or active solvents dissolve a substance readily at room temperature. The latent solvents, which are incapable of dissolving a particular substance by themselves, can have their solvent power activated by addition of a genuine solvent, or by addition of a nonsolvent. Supplementary solvents that exert a high activation effect are alcohols, ketones, higher-molecular esters and sometimes even water. The activation is based on a shift in the solubility parameters which is achieved by targeted mixing. Nonsolvents are incapable by themselves of dissolving the film forming agent in question. The definition of a solvent or a solvent mixture as a genuine solvent, a latent solvent or a nonsolvent is specific to a particular film forming agent only. With a different film forming agent, conditions may be completely different and other solvents may act as a solvent or nonsolvent.

As already mentioned above, the solvents in coating materials only have a temporary function which relates to the coating material while it is still liquid: this starts with the manufacture of the liquid coating material, continues with application and finishes with evaporation from the liquid film. The type and quantity of solvents or solvent mixtures depend particularly on the type of film forming agent and the application conditions. During their dwell time in the coating material the solvents fulfil a range of tasks and exert influences in a number of directions.

They dissolve solid or liquid film forming agents and also other components in the coating formulation and convert them into solutions that possess particular flow properties. In dispersion-based paints they are the liquid phase in which the solids are distributed. During the dispersion process, they assist with wetting of the pigments and extenders via the dissolved binders. In the case of dispersion paints, the solvents or the liquid phase are generally the sole medium for the dispersion. They regulate the paint viscosity during both manufacture and processing.

Thanks to their selective evaporation properties, they intervene to regulate the film forming process and they can improve levelling, increase the gloss and prevent the formation of blisters, craters etc. by transporting volatile reaction products. In electrostatic spraying, they are also used for adjusting the conductivity.

Important physical performance indicators of the solvents used for the coating materials are the density, refractive index, boiling temperature or boiling ranges and the vapour pressure or evaporation time. For safety reasons other values such as the flash point, ignition point and explosion limits must also be added (see Chapter 5.2).

With reference to their boiling properties, solvents are divided into low boilers (boiling range < 100 °C), middle boilers (boiling range 100 to 150 °C) and high boilers (boiling range > 150 °C). Interestingly there are no real correlations between the boiling points or boiling ranges of the solvents and the time they need for evaporation from the paint

Solvent	Molecular weight	Boiling point/ range [°C]	Desity at 20 °C [g/cm³]	Refractive index at 20 °C	Evapouration number	Vapour pressure at 20 °C [hPa]	Flash point [°C]	Ignition temperature [°C]	Explosion limits [Vol.%]
Aliphatic hydrocarbons and their mixtures									
n-Hexane	86.2	65–70	0.675	1.372	1.4	190	–22	240	1.2/7.4
Solvesso 100	123	155–181	0.877	1.502	40–45	3	41	>450	0.8/7.0
Solvesso 150	135	178–209	0.889	1.515	120	1	62	>450	0.6/7.0
Solvent naphtha	123	150–195	0.870	1.500	40–45	3	41	>450	0.8/7.0
White spirit 135/180	131	135–175	0.766	1.428	25–30	70	~22	210	0.6/7.0
Cyclohexane	84,2	80.5–81.5	0.778	1.426	3.5	104	–17	260	1.2/8.3
Aromatic hydrocarbons									
Xylene	106.2	137–142	0.874	1.498	17	90	25	562	1.0/7.6
Toluene	92.1	110–111	0.873	1.499	6.1	290	6	569	1.2/7.0
Styrene	104.2	145	0.907	1.547	16	60	31	490	1.0/6.3
Alcohols									
Propanol	60.1	97.2	0.804	1.386	16	19	23	360	2.1/13.5
n-Butanol	74.1	117.7	0.811	1.399	33	6.6	34	360	1.4/11.3
Iso-Butanol	74.1	107.7	0.802	1.396	25	12	28	410	1.5/12
Ethylhexanol	130.2	183.5–185	0.833	1.432	600	0.5	76	250	1.1/12.7
Isotridecyl alcohol	200.2	242–262	0.845	1.448	>2000	<0.01	115	250	0.6/4.5
Glycolether									
Butyl glycol	118.2	167–173	0.901	1.419	163	1	67	240	1.1/10.6
Propyl glycol	104.2	150.5	0.911	1.414	75	2	51	235	1.3/15.8
Hexyl glycol	146.2	208	0.887	1.429	ca. 1200	0.08	91	220	1.2/8.4
Methyl diglycol	120.2	194.2	1.021	1.424	576	0.3	90	215	1.6/16.1
Butyl diglycol	162.2	224–234	0.956	1.431	>1200	0.1	98	225	0.7/5.3
Methoxy propanol	90.1	122.8	0.934	1.403	25		38	270	1.7/11.5
Ethoxy propanol	104.1	132	0.896		33	<10	40	255	1.3/12

Figure 2.1.155: Physical and safety related data of important solvents (selected from different resources without claiming to be complete and accurate)

Continued on page 133

film as it forms. Consequently, the evaporation time of a solvent is expressed as an evaporation number which indicates how fast it evaporates relative to diethyl ether. With diethyl ether assigned an evaporation number of 1, solvents can be subdivided into the following volatility groups: high volatility (evaporation number < 10), medium volatility (evaporation number 10 to 35), low volatility (evaporation number 35 to 50) and very low volatility (evaporation number > 50) (see Chapter 2.3.2.3).

If macromolecular film forming agents are dissolved in a solvent, the evaporation rate is lowered due to the interaction of the solvent molecules with the film forming agent. The more soluble the polymer is in the relevant solvent, the greater is this effect. The actual conditions are more complicated, however, since paints always contain combinations of solvents whose solubility and evaporation properties have to be coordinated with each other.

Solvent	Molecular weight	Boiling point/ range [°C]	Desity at 20 °C [g/cm³]	Refractive index at 20 °C	Evapouration number	Vapour pressure at 20 °C [hPa]	Flash point [°C]	Ignition temperature [°C]	Explosion limits [Vol.%]
Esters									
Butyl acetate	116.2	123–127	0.880	1.394	11	13	25	400	1.2/7.5
Ethylethoxy-proprionate	146.2	170	0.943		96	2	59	327	1.05/
Ethyl acetate	88.1	76–78	0.900	1.372	2.9	97	–4	460	2.1/11.5
Isobutyl acetate	116.2	114–118	0.871	1.390	8	18	19	400	1.6/10.5
Ethoxypropyl acetate	146.2	158	0.941	1.405		2.27	54	325	1.0/9.8
Methoxypropyl acetate	132.2	143–149	0.965	1.402	33	4.2	45	315	1.5/10.8
Pentyl acetate	130.2	146	0.876	1.405	14	6	23	380	1.1/
Butylglycol acetate	160.2	190–198	0.945	1.415	250	0.4	75	280	1.7/8.4
Ketones									
Methy isobutyl ketone	100.2	115.9	0.800	1.396	7	20	14	475	1.7/9.0
Acetone	58.1	56.2	0.792	1.359	2	245	–19	540	2.1/13
Cyclohexanone	98.2	155	0.945	1.451	40	3.5	44	455	1.1/7.9
Methyl ethyl ketone	72.1	79.6	0.805	1.379	2.6	96	–14	514	1.8/11.5
Methyl n-amyl ketone	114.2	150.5	0.816	1.411	25	5	49	393	1.1/7.9
Methyl isoamyl ketone	114.2	141-148	0.813	1.407	55	6	41	425	1.1/8.2
Isophorone	138.2	215	0.922	1.478	230	3	96	460	0.8/3.8
Others									
Water	18	100	1.000				-	-	-
Propylene carbonate	102.2	242	1.208		>1000		123		
N-Methyl-pyrrolidone	99.1	202.0	1.0260	1.4604		0.32	86.0		1.3/9.5
N-Ethylpyrrolidone	113.1	212.5	0.9974	1.4665		0.18	90.5	245	1.3/7.7

Figure 2.1.155: Physical and safety related data of important solvents (selected from different resources without claiming to be complete and accurate) *Continued from page 132*

Solvents with a low evaporation number cause rapid initial drying, while medium-volatility solvents ensure that the coating is kept open for a certain time so that any entrained air or gaseous reaction products can escape. High boilers leave the coating film only slowly and positively influence levelling, smoothness and gloss of the coating in the final phase of film forming. It is important for the slowest-evaporating component of the solvent combination to be a genuine solvent for the particular film forming agent combination. This will ensure that film faults are avoided (see Chapter 6.2).

Where solvent mixtures consist of two or more components it is possible at certain mixing ratios for systems to be formed which have a constant boiling point which is lower or even higher than that of the individual components. The occurrence of such azeotropic solvent mixtures is due to intermolecular forces or molecular associations between the solvents themselves. As a consequence of the formation of an azeotrope, therefore, it is possible for a solvent mixture to exhibit properties which, during evaporation from a liquid coating film, are different from those that would be anticipated on the basis of the individual components' evaporation times. This can also affect the level of the flash point (see Chapter 2.3.2).

2.1.2.3 Chemical and Physiological Properties

An important prerequisite for the technical usability of a solvent is its chemical stability. Almost all aliphatic and aromatic hydrocarbons meet this requirement. Alcohols also exhibit good stability. However, alcohols can be oxidised to aldehydes or ketones or even to carboxylic acid by strong oxidisers, though not by atmospheric oxygen. Because of their ready reaction with isocyanates, alcohols (and glycols and glycol monoalkyl ethers) must never be mixed with polyurethane hardeners. Ether and glycol ether react with atmospheric oxygen to form peroxides which in turn can lead to the formation of acids. Esters and ketones also display high resistance. Under unfavourable conditions, however, ketones can resinify, and esters can be split by saponification.

Solvents in liquid and particularly in gaseous form can be absorbed by human, animal and, should the situation arise, by plant organisms and cause damage in the event of long-term exposure. The hazard and the degree of hazard depend on the nature of the particular solvent. An important measure of hazards for humans is the threshold concentration of a solvent in the air, below which, even upon repeated and long-term exposure, no harm to health is to be expected. The Figures for the **maximum admissible concentration** (MAC level; in German: MAK, today AGW)) in the workplace, otherwise known as the occupational exposure limit, are given in mg material per m^3 air (mg/m^3) and for solvents also in ml solvent vapour per m^3 air (ml/m^3) (see Chapter 2.3.2). Comparison figures obtained from animal experiments are also often used as guidelines for evaluating the harmfulness or toxicity (see Chapter 5.4). The usual method is to determine the concentration of the harmful substances which is lethal to a given proportion of the animal population studied (usually rats). The figure is quoted in the form of a solid concentration (LD = **lethal dose**) or gas phase concentration (LC = **lethal concentration**), with an additional index denoting the percentage of animals which died after exposure.

The odour emitted by solvents is not directly linked to their harmfulness. Odours can cause significant annoyance in certain cases. The measure of odour intensity is the **odour threshold** and is determined by olfactometric methods (see Chapter 5.3.1). It indicates the smallest concentration of the solvent concerned, expressed in mg/m^3, that can just be perceived in the air by people with an average sense of smell and without prior training. MAC (MAK, today AGW) figures and other occupational health and safety information for the most important solvents are summarised in Figure 2.1.156. The corresponding risk phrases classify the solvents as R 10 "Flammable" and R 11 "Highly Flammable" (in 2015 these were replaced by H and P phrases; see Chapter 5.1).

2.1.2.4 Important Solvents for Paints

The many organic solvents which are important for coatings can be divided by their chemical character into groups that have common features. The aliphatic, cycloaliphatic and aromatic hydrocarbons, esters, ethers, alcohols, glycol ethers and ketones are particularly common. The range of applications in which these solvents are used and their solvent power are evaluated below.

Solvent	**MAC* [ppm]**	**MAC* [mg/m³]**	**LD_{50} rats, oral [mg/kg]**	**LC_{50} rats, inhalation [mg/l]**	**Class according to GefStoffV**
Aliphatic hydrocarbons and their mixtures					
n-Hexane	50	180	>2000	>5	R 11
Solvesso 100	50	200	>2000	>5	R 10
Solvesso 150	50	200	>2000	>5	
Solvent naphta	50	200	>2000	>5	R 10
Petroleum benzine 135/180	100	500	>2000	>5	R 10
Cyclohexane	200	700	>2000	1800 ppm	R 11
Aromatic hydrocarbons					
Xylene	100	440	>2000	>5	R 10
Toluene	50	190	>2000	>5	R 11
Styrene	20	86	>2000	11.9	R 10
Alcohols					
Propanol			>2000	9.8	R 10
n-Butanol	100	300	700	>17.7	R 10
Iso-Butanol	100	310	>2000	>6.5	R 10
Methoxy propanol	100	370	>2000	>6	R 10
Ethoxy propanol			>2000	>5	
Ethylhexanol			>2000		
Isotridecyl alcohol			>2000		
Glycols					
Butyl glycol	20	98	1480		
Propyl glycol	20	86	>2000	2132 ppm	
Hexyl glycol			1480		
Methyl diglycol			>2000		
Butyl diglycol		100	>2000		
Esters					
Butyl acetate	100	480	>2000		R 10
Ethylethoxy proprionate			>2000		
Ethyl acetate	400	1500			R 11
Isobutyl acetate	100	480	>2000		R 11
Ethoxypropyl acetate			>2000	7	R 10
Methoxypropyl acetate	50	270	>2000	>23.8	R 10
Pentyl acetate	50	270	>2000		R 10
Butylglycol acetate	20	130	>2000	2.7	
Ketones					
Methyl isobutyl ketone	20	83	>2000	8.3	R 11
Acetone	500	1200	>2000	>5	R 11
Cyclohexanone	20	80	1900	10.2	R 10
Methyl ethyl ketone	200	600	>2000	12	R 11
Methyl isoamyl ketone	50	230	>2000	3813 ppm	R 10
Methyl-n-amyl ketone	50	238	1600	>2000 ppm	R 10
Isophorone	2	11	>2000	>7	
Others					
Water	–	–	–	–	–
N-Methypyrrolidone	20	82	>2000	>2000	
N-Ethylpyrrolidone			1350	>5.1	

** MAK, today AGW*

Figure 2.1.156: Physical and safety related data of important solvents (selected from different resources without claiming to be complete and accurate)

Aliphatic Hydrocarbons

$$H_3C-(CH_2)_x-CH_3$$

Aliphatic hydrocarbons, also termed gasolines, benzines or paraffin hydrocarbons, combine good stability with good solvent power for non-polar to slightly polar film forming agents. They are mixtures consisting of a large number of similar individual components. Mixtures are generally categorised by their boiling ranges into the light petroleum benzines and the significantly heavier mineral spirits. The petroleum benzines, which are mainly used for rapidly drying coatings, have flash points ranging from below 21 °C to < 0 °C, which means that the coatings manufactured from them may only be processed under stringent safety precautions. Mineral spirits, all with flash points above 21 °C, are good solvents for chlorinated rubber, long-oiled alkyd resins, some vinyl chloride copolymers and polyvinyl ethers. Polar film forming agents such as cellulose derivatives, epoxy, polyester, PUR, phenol, amino and amido resins are not dissolved.

Cycloaliphatic Hydrocarbons

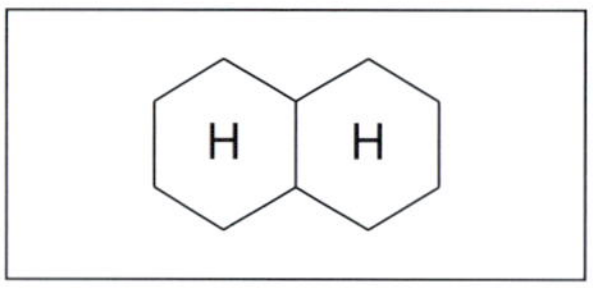

The cycloaliphatic hydrocarbons, like the example of hydrogenated naphthalene shown, lie between the aliphatic and the aromatic hydrocarbons in terms of solvent power. The range of resins soluble in cycloaliphatic hydrocarbons corresponds largely to that of the aliphatic hydrocarbons. The cycloaliphatics are less common in the paint and coatings industry than the paraffin hydrocarbons. The main examples are hydrogenated naphthalene products such as tetrahydronaphthalene (tetralin) and decahydronaphthalene (decalin), both high boilers with good solvent power. The hydrocarbons are exclusively of petrochemical origin nowadays.

Terpene Hydrocarbons

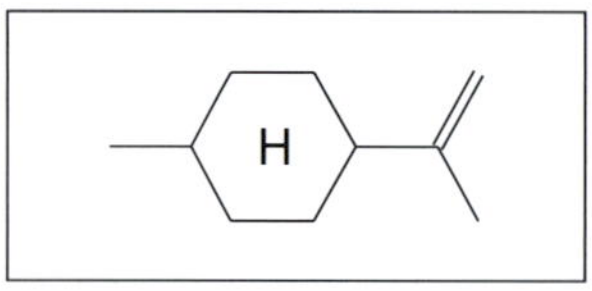

The terpene hydrocarbons, headed by turpentine oil, are of plant origin. They are obtained by distilling the balsam of various pine species. They were mainly used in combination with mineral spirit in oxidatively crosslinking oil and alkyd resin coatings and have now largely been replaced by aliphatic hydrocarbons ("turpentine substitute").

Aromatic Hydrocarbons

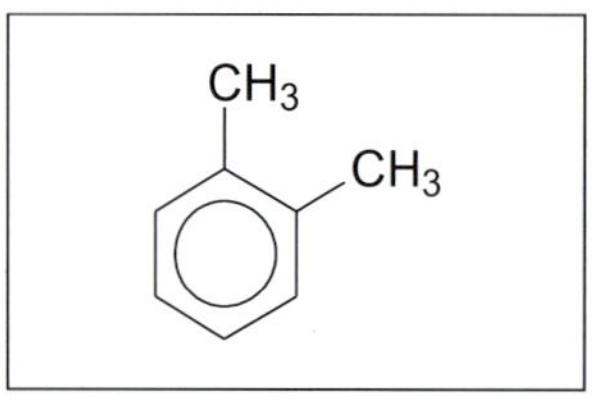

Compared with petroleum benzines, the aromatic hydrocarbons have much broader solvent power for alkyd resins of all types, saturated polyesters, polystyrene, polyvinyl ether, polyacrylates and polymethacrylates, polyvinyl acetate, vinyl chloride and vinyl acetate copolymers. Aromatic hydrocarbons can be used as diluents for solutions of cellulose nitrate and other cellulose esters in ester and ketone mixtures. Apart from toluene and commercially available technical xylene, a mixture of changing quantities of o- (see above), m- and p-xylene, higher-substituted benzene derivatives known as solvent naphtha are in widespread use. These are primarily alkylbenzenes, including tri- and also tetramethylbenzene, isopropylbenzene (cumene), methyl ethyl benzene and diethylbenzene. These high-boiling aromatic mixtures, like toluene and xylene themselves, do not contain the potentially physiologically hazardous benzene which has not been used in the paint and coatings industry for quite some time because of its carcinogenic effect.

Styrene (vinyl benzene) and vinyl toluene, also aromatic hydrocarbons, act simultaneously as solvents and reactants for chemical crosslinking with unsaturated polyester resins.

Chlorinated Hydrocarbons

$Cl_3C{-}CH_3$

Despite their excellent solvent power for a very broad range of film forming agents, chlorinated hydrocarbons are hardly used at all now in paint formulations in Germany for environmental and physiological reasons. However, they are still found in adhesion primers for plastics coatings in North America and Asia.

Alcohols

$R{-}OH$

Because of their hydroxyl groups the alcohols and the glycols and glycol ethers differ from the hydrocarbons in having a high dipole moment and a marked tendency to form hydrogen bridging bonds. The alcohols' solvent power is influenced by the mass ratio of polar hydroxyl groups to the non-polar organic group. The lower alcohols with a pronounced polarity are good solvents for highly polar film forming agents, such as phenolic resins, amino resins, cellulose nitrate, ethyl cellulose and polyvinyl acetate. As the carbon chain lengthens, the alcohols' solvent power for polar film forming agents diminishes, as a result of which they can only serve as a diluent with them. Higher alcohols can also be used to improve the storage stability of stoving enamels with etherified crosslinking resins. They prevent the cleavage of alcohols from the ether groups.

Ketones

$R{-}C({=}O){-}R'$

Like the hydroxyl group, the carbonyl groups introduce high dipole moments and a definite tendency towards hydrogen bridging bonds into the solvent molecules. The lower ketones are also excellent solvents for polar film forming agents such as cellulose ether, ketone resins, copolymers of vinyl chloride, for polystyrene, polyacrylates or chlorinated rubber, as well as for the low molecular film forming agents of high solid paints. The higher ketones with their long carbon chains shift the solvent power towards less polar film forming agents.

Esters

$R{-}C({=}O){-}O{-}R'$

The esters resemble the ketones in the size of their dipole moments and their tendency to form hydrogen bridges and also have excellent solvent power for most film forming agents. Acetic acid and propionic acid predominantly serve as the acid components for the esters used in the paint and coatings industry. Esters based on butyric and carbonic acids, on the other hand, are of less importance. In this class of solvent, the chain length of the alcohol component influences the solvent power.

Glycol Ethers

$R{-}O{-}CH_2{-}CH_2{-}OH$

Glycol ethers, particularly ethers of ethylene glycol, diethylene glycol and propylene glycol, also feature high dipole moments and a marked tendency to form hydrogen bridges. They are excellent solvents for a broad range of polar and semi-polar film forming agents and have a stabilising effect in alkyd resin paints because of their alcoholic hydroxyl group. They are gaining in significance as important cosolvents for waterborne paints. The ethers of propylene glycol are coming to the fore, as more information becomes available about the possible harmfulness to human health of ethers of ethylene glycol.

Special Solvents

Solvents from groups other than those listed are used only rarely and in very limited fashion in the paint and coatings industry. Cresols serve as special solvents for wire enamels. Methyl formamide is also occasionally used for wire enamels. The use of N-methylpyrrolidone and N-ethylpyrrolidone is restricted to just a few special coatings based on polyurethanes. Propylene carbonate and members of this class of substance are still fairly rare but are gaining in importance.

2.1.3 Pigments and Extenders

Combination of film forming agents with solvents and possibly with additives affords transparent coating materials, the clearcoats. Despite the varied use of clearcoats to seal wood and plastics or to cover metallic coatings, the most frequently employed coating materials are actually not transparent. Pigments, effect-inducing materials and extenders are added to provide chroma, hiding power, and special optical effects, as well as functional properties such as sandability and hardness. The chroma is generated by selective absorption or interference, and the hiding power by light scattering. Pigments have the ability to provide hiding power and chroma simultaneously. They are fine substances which are insoluble in the film forming agent matrix. The term colorant is a generic term for all additives, including soluble dyes, which lend colour. Dyes play only a subordinate role in the coatings field compared with insoluble pigments. Although soluble dyes have good tinting strength and high colour purity, they do not have any scattering or hiding power. Paints made with soluble dyes are therefore only suitable for transparent coatings for paper or metal foils or for coating signal lamps made of transparent plastic. With a few exceptions, soluble dyes are inferior to pigments in terms of lightfastness.

Pigments are divided into organic and inorganic colorants in accordance with their chemical structure. In addition to the tasks already mentioned of giving colour, creating effects and providing hiding power, pigments are frequently expected to take on special tasks such as stabilising coatings against the influence of UV light or even affording supplementary anticorrosive protection. Certain conditions must be met if such tasks are to be achieved. The pigment particles must be evenly distributed in the liquid environment, be well wetted by the film forming agent solution, permit trouble-free coating with stabilising resin molecules and enable rheological properties which are compatible with the processing method. Further demands that pigments have to meet include complete insolubility in the surrounding matrix, good lightfastness, resistance to long-term outdoor exposure and heat, lack of sensitivity to the effects of chemicals, as well as no physiological risk.

Extenders are not primarily concerned with optical performance. Rather, their purpose is to raise the solid content of the paint and improve the mechanical/technological properties of the coating.

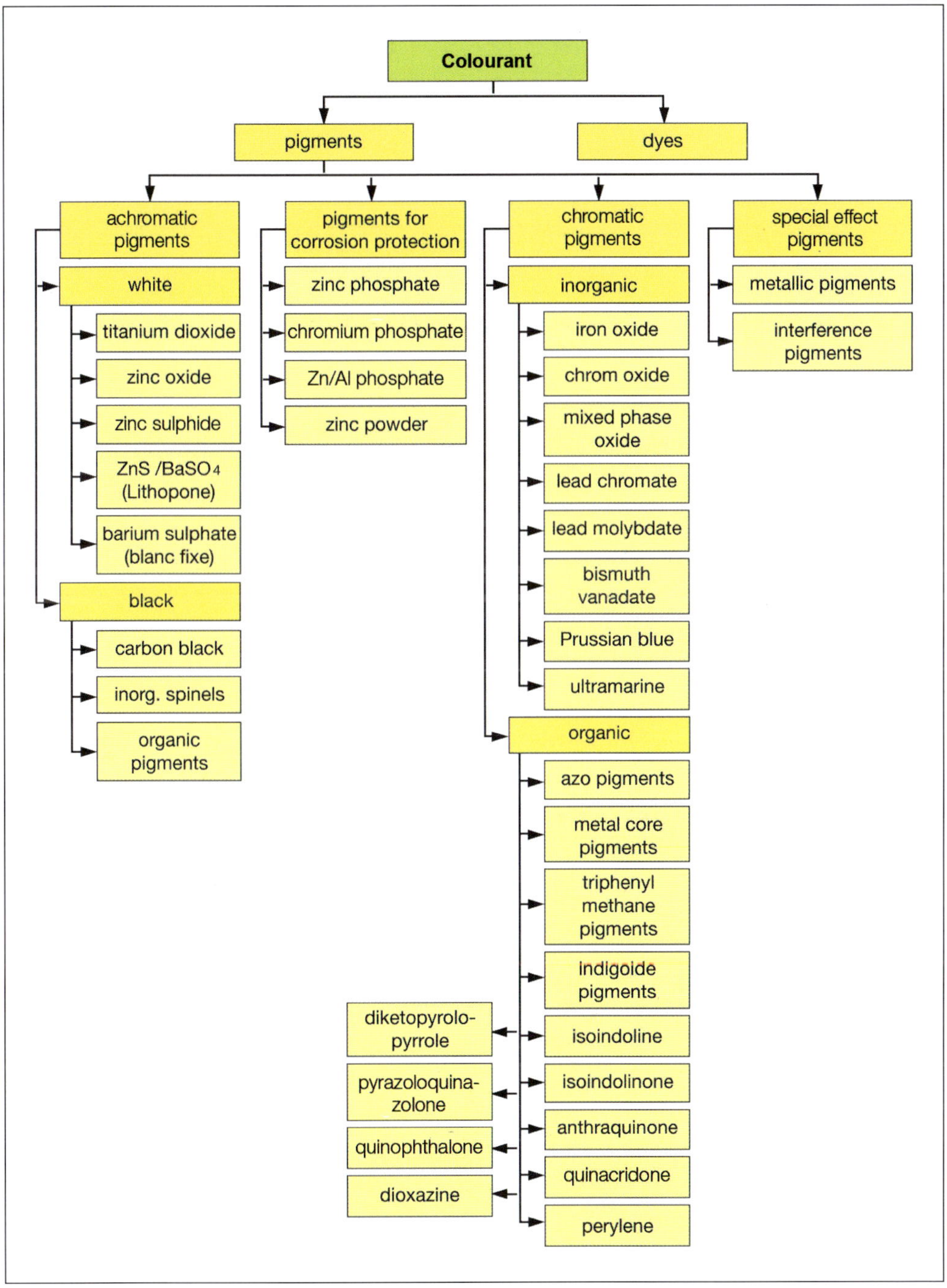

Figure 2.1.157: Classification of pigments according to function

Tasks of pigments	Requirements for pigments and extenders	Special tasks of extenders
• selective absorption • light scattering • optical effects by oriented reflection or interference • UV-protection • corrosion protection	• dispersibility • unsoluble • lightfast and weather resistant • heat resistant • chemical resistant • physiological compatible	• "filling" • sandability • improvement of mechanicaltechnological coating properties

Figure 2.1.158: Tasks and performance requirements imposed on pigments and extenders

Economic Importance

Because of the optical attractiveness which can be achieved by the use of pigments, coatings act as an enticement for consumers to buy goods. The countless colour nuances specifically created to accentuate car designs in conjunction with effects and high-gloss finishes are an example of this. Colour-psychological studies provide an insight into the interactions of colour and form and into the aesthetic perceptions which they generate in the observer. At regular intervals, carmakers devise new colour schemes for solid and effect hues for model series that are intended to appeal to customers' tastes and encourage them to buy (see Chapter 7.1.5). Whether or not changes in the buyer's taste come from within or are attributable to the effects of advertising, it is an accepted fact that the percentage shares of car colours vary greatly with time and region. Thus, a 2007 study by one paint-maker revealed huge differences in the popularity of colours in the three regions of North America, Japan and Europe. More recently, in 2012, more than 70 % of new car registrations in Germany concerned non-chromatic colours (see Figure 7.1.17).

The pigments which are constantly being developed and produced by the pigments industry and had a global volume of 8.5 million tonnes in 2012 are not just used to colour coating materials. Only somewhat more than half of the largest inorganic product group, which accounted for roughly 7 million tonnes, is employed in the form of pigments, especially titanium dioxide, for coating materials, especially architectural paints, on account of the price [2.4.49]. Other materials also coloured with pigments are plastics, construction materials, paper and ceramics. With organic colorants, the situation is different. Only one quarter of the world total of about 1.5 million tonnes is destined for the paint and coatings sector. By far the largest proportion is used as dyes, i.e. as soluble colorants for colouring textiles. The roughly 400,000 tonnes accounted for by insoluble pigments, goes to printing inks, paints and plastics, but is the most interesting product line in terms of value.

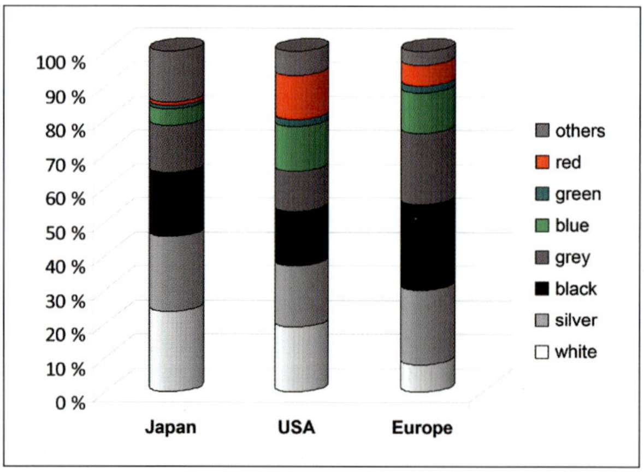

Figure 2.1.159: Regional differences in the popularity of car colours in 2007 Source: [2.4.48]

After titanium dioxide, the most important class of substances, with a share of 25 % of world production, is that of the iron oxides. Germany produces about 15 % of the inorganic and 25 % of the organic pigments manufactured in the world. On average, the proportion of pigment in a coating formula is 32 % [2.4.50].

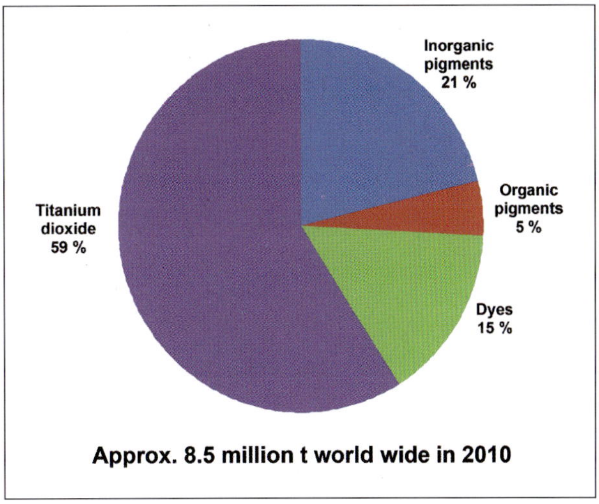

Figure 2.1.160: The colorants market by segment in 2010

Source: authors` own estimate

History

The use of colorants goes back over thousands of years. Natural iron oxides, carbon black and organic natural materials such as egg whites and tree resins form the basis for the materials used in the cave paintings at Lascaux and Altamira which are more than 40,000 years old.

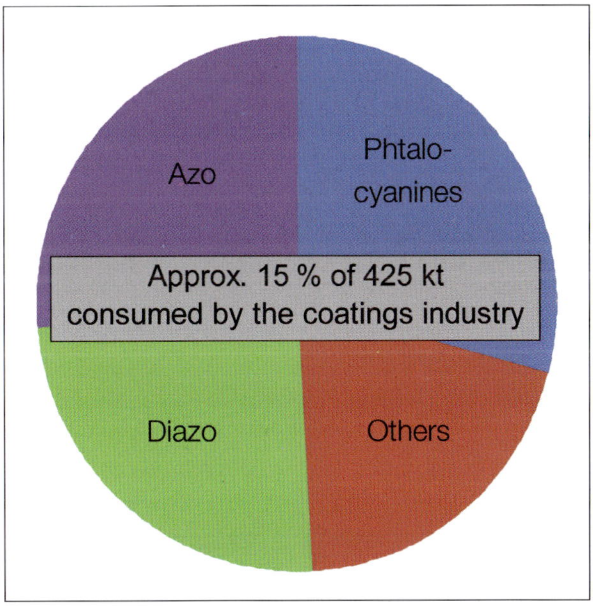

Figure 2.1.161: Global consumption of organic pigments by product class

Carbon black, copper-calcium silicate and ultramarine, the pigment from the purple snail, the plant-derived indigo plus kermes and crimson travelled along with age-old cultures from China to Egypt via India and then finally also to Europe. In his book "Germania" *Tacitus* reports in detail on dye works north of the Alps in which a colourless substance, dyer's woad extracted from a meadow plant, was transformed by atmospheric oxygen into a deep blue, insoluble pigment: indigo. The discovery of America enriched the colour palette with the addition of red cochineal, an extract from particular insects.

The following centuries were given over to chemical synthesis of nature's colour originals. For obvious reasons the first successes were obtained with inorganic substances which were copied from nature or were the byproducts of technical processes. Examples include zinc oxide, chromium yellow, cadmium sulphide, chromium oxide and the wide range of iron oxides. In the 19th century the more complicated structure of ultramarine was successfully synthesised.

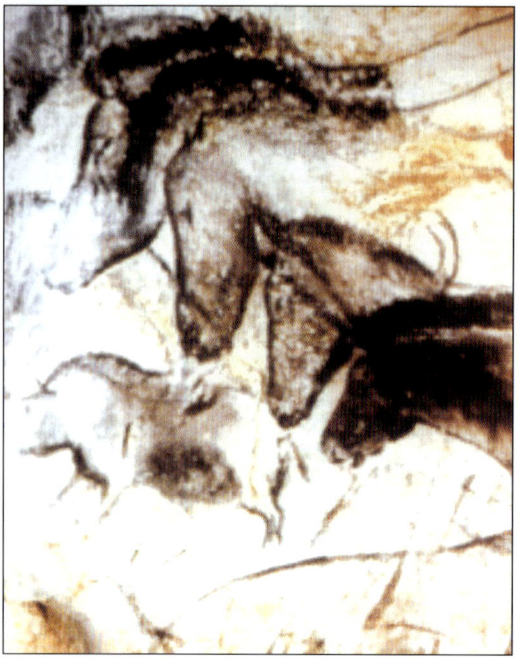

Figure 2.1.162: Cave paintings in Lascaux

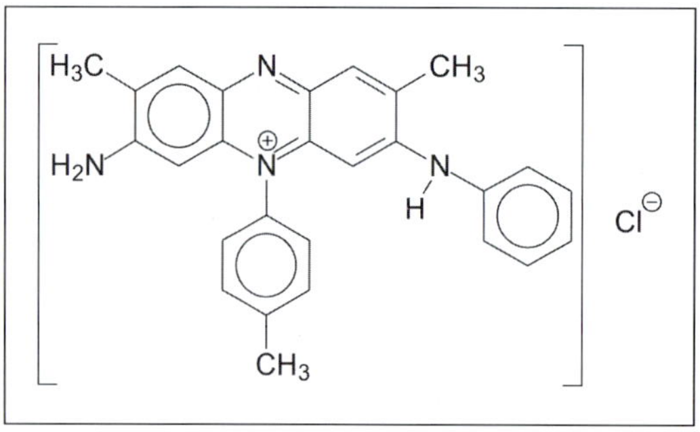

Figure 2.1.163: Mauveine as the first fully synthetic organic dyestuff

The foundation stone for the successful synthesis of organic dyestuff was laid in the middle of the 19th century when aniline was isolated from coal. *Perkin* achieved the breakthrough to synthetic dyestuffs in 1858 with the deliberate oxidation of aniline to yield a deep violet dyestuff, mauveine, also known as *Perkin's* mauve.

The availability of large quantities of organic dyestuffs which were also inexpensive made the world a more colourful place. Triphenylmethane dyes, azo dyes, synthetic indigo blue or red alizarin are just a few of the subsequent discoveries which revolutionised the world of colours. Today's chemists are still developing attractive, new colorants and combining them with new optical effects. The metallic effect, caused by oriented reflections from mini aluminium mirrors, or the enormous attraction of interference pigments whose colour impressions vary with the angle of view are undoubtedly the most noteworthy inventions of recent decades in this field.

2.1.3.1 Physical Principles behind Chroma and Hiding Power

Colour as a sensory impression is generated when light interacts with the pigments and part of the spectrum is removed from it by absorption or interference. People perceive colour via the cones on the retina. These comprise three groups of sensory cells with different sensitivities to light. If part of the spectrum is removed from the white light by interaction with a selectively absorbent material, the individual sensors register changes in the intensities of the reflected light. These are transmitted to the brain where they are transformed into a colour impression (see Chapter 3.2.4).

The second visual requirement imposed on coatings is hiding power. Leaving aside black pigments which absorb light from the whole range of the visible spectrum, this effect is only achieved by those pigments which are capable of scattering light. Since the scattering power, just like the tinting strength, depends on differences in the refractive indices of the pigment and the binder and also on the particle size, pigment manufacturers make every effort to achieve the optimum particle size for tinting strength and hiding power in their commercial products to ensure economical use.

Owing to the small particle size, the resulting proximity, and the polarity of the pigments, forces of attraction are generated which are strong enough to make the pigment particles combine to form agglomerates or aggregates. Whereas in agglomerates the original particles (primary particles) are only loosely connected to each other by a few points of contact, in aggregates the particles are rigidly joined together via large areas of contact. The coating manufacturer's aim of restoring agglomerates and aggregates to their primary particles to optimise tinting strength and hiding power is only achieved, however, in the case of the agglomerates. Aggregates in pigment powders therefore signify a quality-shortcoming.

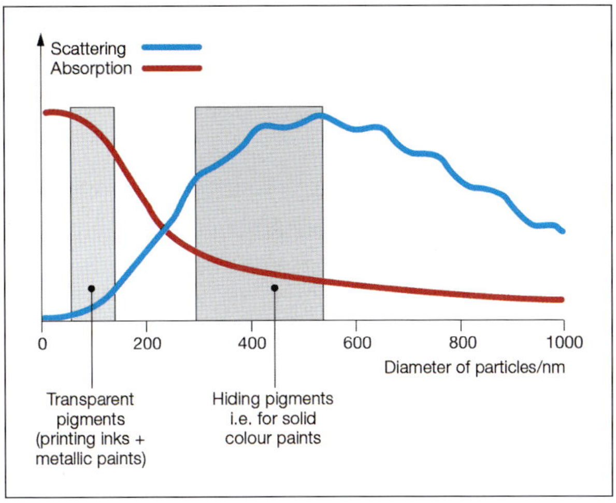

Figure 2.1.164: Absorption and scattering of light as a function of pigment particle size

To ensure lasting stabilisation, separation of agglomerates into their primary particles must be accompanied by total envelopment in suitable resin solutions. Such stabilisation prevents later reagglomeration and leads to consistent quality, even when the coating materials are stored for extended periods (see Chapter 2.2.4).

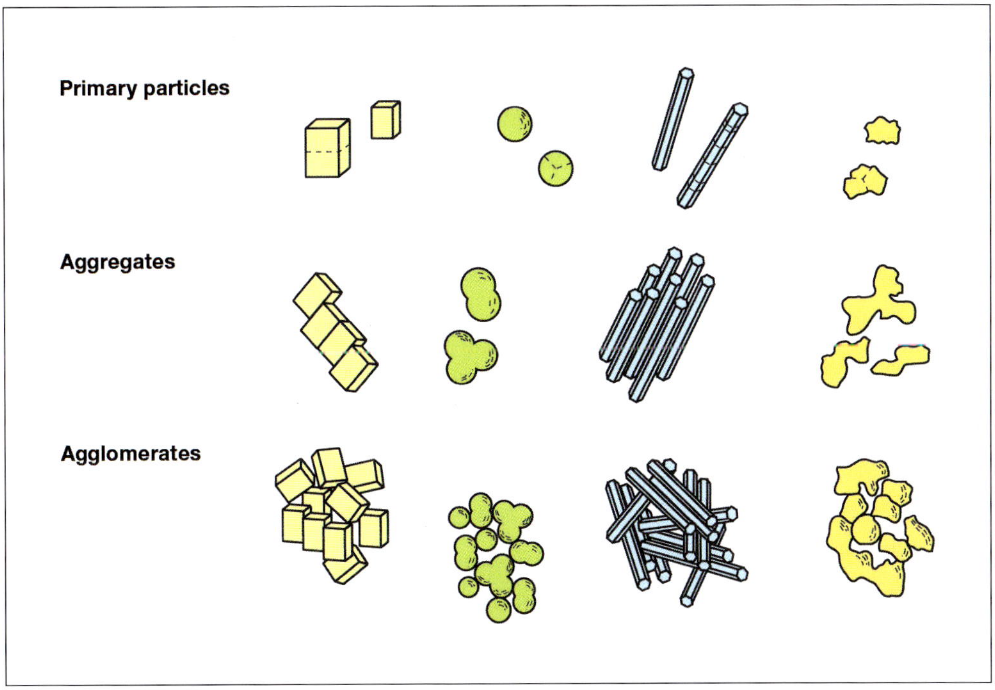

Figure 2.1.165: Primary particles, agglomerates and aggregates of pigments

Coating Material

Pigment Performance Indicators

The particle size and particle size distribution of pigments affect the colour, tinting strength and hiding power to a large degree. They are therefore subjected to precise measurement.

Direct measurement using light-optical methods is not possible due to the small size of the particles. Proven methods utilise electron microscope images, measurements of accelerated sedimentation with disc centrifuges, determinations of particle size with the *Coulter* counter and evaluations of the diffraction or scattering of laser light (see Chapter 2.3.3). The **particle size distribution curves** which can be recorded by means of these measuring methods determine the colour and coating-specific properties as a function of the position of the maximum and the width of the distribution.

A proven practical method of mathematically processing and characterising the particle size distribution was developed empirically by *Rosin, Remmler Sperling* and *Bennet* [2.4.51] (see Chapter 2.3.3).

A property which is functionally linked to the **particle size** and is also important for pigment dispersion is the mass- or volume-specific **pigment surface area**. The mathematical processing of vapour pressure measurements after *Brunauer, Emmet* and *Teller* (BET method) permits the specific surface area of the pigments under test to be determined from the pressure drop caused by the adsorption of nitrogen (see Chapter 2.3.3).

Apart from the particle size and the specific surface area, the mass per unit volume, i.e. the **density**, is a property which is of interest both technically and economically. It determines the level of paint consumption at specified film thicknesses.

Likewise, the bulk density indicates the amount of cavities in the pigment powder. It also provides information about the cavities to be filled when the pigments are being dispersed.

In solvent-borne paints the **oil absorption value** and in water-borne paints the **water absorption value** yield supplementary information about the amount of binder needed to ensure thorough wetting (see Chapter 2.3.3).

A particularly important means of evaluating the colour of dyestuffs is the **Colour Index**, issued by the British Society of Dyers and Colourists and the American Association of Textile Chemists and Colourists, which documents all standard colorants in a multivolume publication. It uniquely characterises each material by two or three letters and a combination of numbers. The letters come first, followed by a sequential number. The first letter indicates the type of colorant, while the second and third letters, if any, indicate the colour, together with the sequential number. The description of the composition and chemical structure then follows in the second part of the Colour Index. The Colour Index supplements the manufacturers' information on tradenames, fastness properties and application fields [2.4.52].

During the manufacture of pigmented paints, the change in particle size and tinting strength over time serves as a measure of the extent of dispersing. The stability of the pigment dispersion can be seen in the tendency towards settling and the results of the rub-out test on freshly applied coatings (see Chapter 2.3.3). Changes in tinting strength under specifically applied shear are indicative of particle enlargement and thus poor stabilisation of the pigment dispersion.

2.1.3.2 The most Important Pigments for Coating Materials

The large class of insoluble colorants which are important to coating technology can be divided into achromatic pigments, anticorrosive pigments and coloured pigments by analogy with DIN 53944. The first group contains the "white" scattering pigments, and the black pigments which absorb light from the entire visible spectrum, the second contains special inorganic pigments, and the third consists of the organic and inorganic coloured pigments and the interference pigments.

Designation of colorant groups in the Color Index (selection)

1. letter	2. letter
A = Acid Dye	B = blue
B = Basic Dye	G = green
P = Pigment	BK = black
	R = red
	V = violet
	W = white
	Y = yellow

Characterisation or chemical structure

11000 – 36999	azo pigments
41000 – 41999	diphenylmethane pigments
58000 – 72980	anthraquinone pigments
74000 – 74999	phthalocyanine pigments
77000 – 77999	inorganic colorants

Figure 2.1.166: Keys to the Colour Index

Achromatic Pigments

Among the scattering pigments or **white pigments, titanium dioxide** has successfully captured the lion's share of the market because of its high refractive index and its extraordinary chemical resistance [2.4.53]. Of the three known crystal forms, brookite is unimportant, and anatase is only of subordinate importance. With a share of about 90 % of all coating pigments, rutile is by far the most important. It is extracted from ilmenite, a naturally occurring iron titanate, by the sulphate process, or from natural rutile by the chloride process. The sulphate process was perceived as being less environmentally friendly, but the introduction of acid recycling has meant that it has regained its share, with the result that the two processes now enjoy equal status. Irrespective of the production process, however, the raw products fail to meet the in part rigorous requirements imposed on pigment dispersion and outdoor exposure properties unless they are chemically modified and the pigment surface is given additional secondary treatment.

a) Sulphate process — Ilmenite

$$FeTiO_3 + 2\,H_2SO_4 \longrightarrow TiOSO_4 + 2\,H_2O + FeSO_4$$

$$TiOSO_4 + 2\,H_2O \longrightarrow TiO(OH)_2 + 2\,H_2SO_4$$

$$TiO(OH)_2 \xrightarrow[-\,H_2O]{} TiO_2$$

b) Chloride process — Nat. rutile

$$TiO_2 + C + 2\,Cl_2 \longrightarrow TiCl_4 + CO_2$$

$$TiCl_4 + O_2 \longrightarrow TiO_2 + 2\,Cl_2$$

Figure 2.1.167: Manufacturing methods for titanium dioxide

The necessary protection against the influence of UV light is achieved by doping the rutile crystal with aluminium or zinc, while the dispersion properties are achieved by coating the pigment surface with oxides of aluminium and silicon. Even organic adsorption films have proved successful for grades that are particularly easy to disperse. Titanium dioxide in the modification described has incomparable long-term outdoor exposure properties and lightfastness combined with excellent hiding power. Its resistance to chemical action and its insolubility in acids and alkalis mean that it can also be regarded as a safe product from a physiological perspective. Formerly important white pigments such as white lead (basic lead carbonate), zinc white (zinc oxide), zinc sulphide and lithopone (zinc sulphide with barium sulphate) have either been replaced for occupational health reasons or have largely been redeployed to primer coats or ground coats because their properties in topcoats are inferior to those of titanium dioxide.

To counter the frequently tight market situation and associated price pressure, the effectiveness of pigmentation with titanium dioxide is enhanced by recourse to inter alia extenders whose particle size has been optimised to prevent reagglomeration and to boost light scattering or at least retain it at the same level in the corresponding production processes [2.4.54].

Black pigments are mainly derived from **carbon blacks**. The fineness of the pigment particles, and consequently their dispersion properties, varies with the production process and aftertreatment. It is now possible to produce extremely fine particles, in the nanometre range, by incompletely combusting a mixture of natural gas and oil. With specific surface areas of up to 1000 m^2/g and thus an extremely high tinting strength, they achieve a high hiding power even at extremely low concentrations [2.4.55].

Less important are carbon blacks obtained from incomplete combustion of plant or animal materials. They are coarser and are sometimes contaminated with quality-impairing mineral materials. Another pigment, which is substantially easier to disperse than carbon black, is **black iron oxide**, a magnetite modified with copper and chromium. Black iron oxide is inferior to carbon back in terms of tinting strength and chemical resistance.

Anticorrosive Pigments

As opposed to the pigments which give coatings hiding power and chroma and therefore primarily serve aesthetic purposes, anticorrosive pigments are purely functional and have achieved extraordinary economic importance for very different reasons. The active, anticorrosive pigments intervene in the corrosion processes occurring at the interface between the metallic substrate and the coating by entering into chemical reactions. Their corrosion-inhibiting effect is based on valency changes in the soluble pigment components. In the past, red lead and chromium-containing pigments were by far the most effective and therefore also the most commonly employed anticorrosive pigments. However, the proven physiological hazards of these pigments, the necessary labelling and the restrictions imposed on processing them mean that they have been almost completely superseded in Europe and North America by **zinc phosphates** and **zinc aluminium phosphates**. These substitutes are also among the important active anticorrosive pigments. For optimum effectiveness, zinc phosphates are crystallised in such a way that they are particularly fine and so have a correspondingly large surface area. As it takes a low pH to guarantee the effectiveness of the zinc phosphate pigment, acid film forming agents and extenders need to be used.

Zinc dust occupies a special position as an anticorrosive pigment. The protective effect of zinc dust primers, always formulated at high pigment volume concentrations (PVC), can be explained electrochemically by the large difference between zinc and iron in the electrochemical series of the elements. Zinc dust is used both in organic and in waterborne film forming agents. Zinc dust primers are weldable at high PVC, i.e. their related high electrical conductivity.

Anticorrosive protection is also possible without chemical intervention. It is achieved by significantly reducing the diffusibility and permeability of corrosive agents such as oxygen, water and salts through appropriate formulation. Lamellar pigments such as aluminium silicates, lamellar iron oxide or mica are best suited to influence this property positively.

Coloured Pigments

Among the **inorganic coloured pigments** there are major applications for numerous yellow, red and brown variants of iron oxide, in addition to the already mentioned black iron oxide. The most important basic types are **red iron oxide** (α-Fe_2O_3 as haematite), **yellow iron oxide** (α-FeOOH as goethite) and brown iron oxide, which is a mixed pigment consisting of red, yellow and black iron oxides. Over the course of time the natural or earth pigments, because of their quality-impairing impurities, have been replaced more and more by synthetic products, the mineral pigments. These are manufactured either by oxidation of scrap iron with nitrobenzene (aniline process) or by oxidising hydrolysis of iron (II) sulphate (*Penniman* process).

Iron oxide pigments have excellent hiding power and long-term outdoor exposure properties, are lightfast and have excellent chemical resistance. Unlike red iron oxide, which is stable up to 1000 °C, the yellow iron oxide pigments and their brown mixtures have poor heat resistance. The hydrated oxides dehydrate above temperatures of as low as 160 °C and are thereby converted into the red form, haematite.

A large proportion of iron oxide pigments is used in primers. In topcoats, the surfaces of iron oxide pigments are frequently covered with organic pigments to increase the

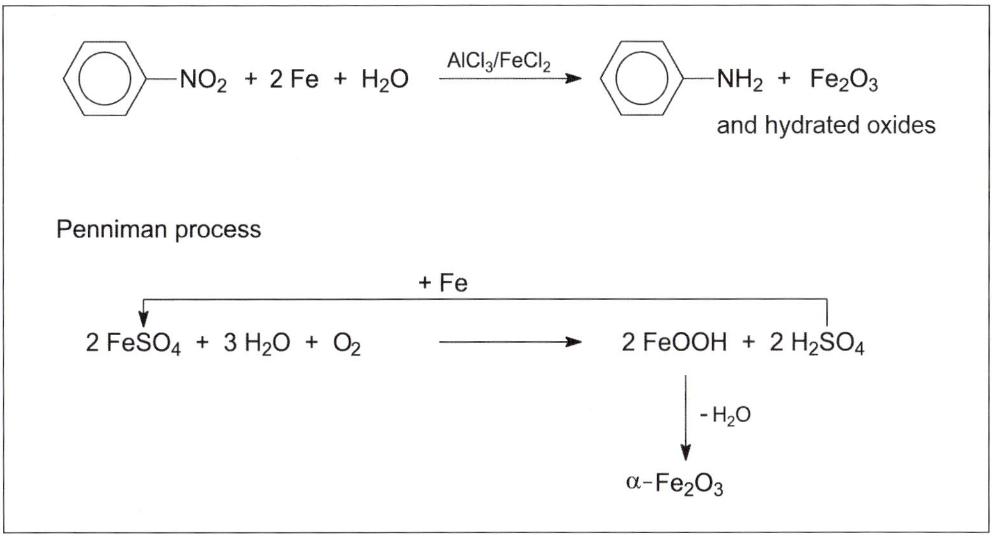

Figure 2.1.168: Manufacturing methods for iron oxide pigments

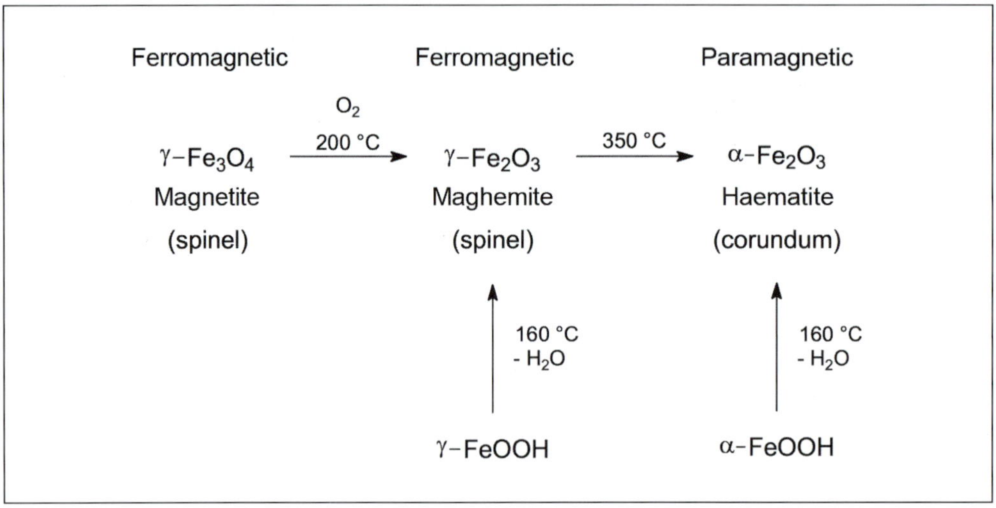

Figure 2.1.169: The most important iron oxides

brilliance. Iron oxide pigments with extremely small particle diameters serve as colorants in metallic effect coatings and transparent paints.

Chromium oxide pigments are used far less commonly. Chromium oxide (Cr_2O_3) and hydrated chromium oxide ($Cr_2O_3 \cdot 3H_2O$) are limited in terms of colour to greens. Their fastness properties and the thermal instability of the hydrated form match those of iron oxide pigments. They are manufactured by reduction of the hexavalent chromate to the trivalent chromium oxide. It crystallises like α-Fe_2O_3 in the corundum matrix and therefore has a high hardness. Due to its good chemical and thermal resistance, chrome oxide green's main application fields are chemically resistant coatings for concrete and cement, emulsion paints and a few industrial coatings. Chromium oxide green is used in conjunction with micaceous iron ore for heavy-duty anticorrosive protection. Its characteristic reflection properties in the infrared range have also made it well-known as a pigment for camouflage coatings.

Other metal oxides which are technically important for yellows include the **mixed-phase pigments**. These consist of antimony-containing titanium dioxide of the rutile type which is doped with small quantities of nickel during manufacture to produce the green-tinged nickel titanium yellow, and with chromium to produce the red-tinged chromium titanium yellow. This group of pigments provides a suitable replacement for lead chromate because it has good heat and chemical resistance while also being regarded as physiologically safe. The mixed-phase pigment group also contains a number of spinels such as cobalt blue, cobalt green as inverse spinel and the already mentioned black spinel.

Rutile grid with hosts	Spinels
• Nickel titanium yellow $(Ti,Ni,Sb)O_2$	• Spinel blue ($CoAl_2O_4$)
• Chromium titanium yellow $(Ti,Cr,Sb)O_2$	• Cobalt blue $(Co(Al,Cr)_2O_4$
	• Spinel black $(Cu(Cr,Fe)_2O_4$
	• Zink iron brown $(Zn(Fe))Fe_2O_4$
	Inverse spinels
	• Cobalt green $(Co,Ni,Zn)_2TiAlO_4$

Figure 2.1.170: Mixed-phase pigments

Although containing both hexavalent chromium and lead, lead chromate pigments are still of technical importance as mixed pigments of chromate and sulphate. Their toxicity is low because of the extremely poor solubility of the lead chromate and the resulting lack of biological availability. Their brilliant colours are limited to yellow. Partial replacement of the lead by molybdenum yields molybdate orange and molybdate red. All lead chromate-containing pigments are naturally sensitive to acids, reducing chemicals and the effect of sunlight. However, suitable stabilisation processes permit chromate pigments of increased resistance to be made. Each grain of the lead chromate pigment is coated with a protective layer of a chemically resistant material.

Lead-free and chromium-free, brilliant and, at the same time, highly opaque inorganic yellow pigments consisting of **bismuth vanadate** and **bismuth molybdate** have gained in importance. As oxidic 2-phase pigments they are generally green-tinged, with particularly high colour purity and also a good gloss finish. For the yellow range, a new pigment based on niobium and tin is being propagated that is primarily intended to replace the lead chromate pigments [2.4.56].

Pigments discovered by chance include **Prussian blue** or mineral blue. While attempting to make cochineal red lake, *Diesbach* was surprised to isolate a blue substance. Cyanide in the potash that was needed to adjust the pH had caused a deep blue iron-cyanide complex to form in the presence of iron ions and overdye the cochineal paint. Industrial production nowadays involves the incomplete oxidation of the colourless, iron (II)-containing intermediate stage. The presence of both iron (II) and iron (III) in conjunction with the water of crystallisation gives the cyanide complex the required pigment properties [2.4.57]. The water of crystallisation is particularly important with regard to the colour. Prussian blue pigments are pyrophoric, i.e. they can spontaneously combust in the presence of oxidising agents. This fact must therefore be borne in mind, as special fire protection precautions have to be put in place because of the simultaneous grinding of Prussian blue with lead chromate, which acts as an oxidising agent, to produce the chromium green mixed pigment. Prussian blue is an extremely light inorganic pigment which has an unusually low density of 1.75 g/cm^3 and has poor acid resistance and only limited stability when exposed to alkalis, despite supplementary stabilisation. Its tinting strength is high, as a result of which it can be used for industrial paints with particularly deep blue tones because of the very good lightfastness. One disadvantage is its metallic, copper-coloured shimmer after lengthy exposure to the weather. This bronzing effect can be reduced by subsequent treatment.

Ultramarine occupies a special place among inorganic pigments. This semi-precious stone, which was used in Ancient Egypt to decorate the burial sites of the pharaohs, even fascinated Goethe. It is no surprise, therefore, that successful attempts at synthesising this natural product were reported as long ago as the beginning of the

$$BiVO_4 \cdot x\,Bi_2MO_6 \qquad x = 0.2 - 2 \quad \text{common: } 0.75$$

Figure 2.1.171: Bismuth vanadate as a multi-phase oxidic pigment

$$Fe^{(III)}\,R\,Fe^{(II)}\,(CN)_6\;x\,H_2O \qquad R = Na^+,\,K^+ \text{ or } NH_4^+$$

Figure 2.1.172: Chemical composition of Prussian blue (mineral blue)

$Na_8[Al_6Si_6O_{24}] \cdot S_{2-4}$

Figure 2.1.173: Chemical formula of ultramarine

19th century. The first factory for the manufacture of artificial ultramarine was built in 1830 by *Leverkus* in a town which was then named after him and is thus known today as Leverkusen [2.4.58].

Ultramarine is an aluminium silicate with a zeolite-like crystal matrix; the unusually warm blue tone comes from the incorporation of sodium and sulphur. On account of its poor hiding power and its sensitivity to acids, ultramarine has proved more suitable for artists' paints and translucent painting on porcelain than for coating consumer goods. Violet and red ultramarines can also be manufactured by varying the proportion of sulphur and by incorporating amines.

In addition to the pigments already mentioned, there are numerous other inorganic oxides which are used for colouring glass or ceramics because they have high thermal resistance. Blue Wo_3O_8 or yellow WoO_3, MnO_2, mixed oxides of silicon with calcium, magnesium and iron in the form of hornblende or green earth merit a special mention.

Organic Coloured Pigments

Organic coloured pigments feature extensive, selective light absorption and therefore high colour purity, cover very wide colour ranges, yet are so close-meshed in their colour range that they are available in a wide variety of colour nuances. Given the high absorption capacity, the scattering and hiding powers are lower than in most inorganic coloured pigments. To achieve satisfactory coverage of the substrate, the organic pigments are often combined with opaque inorganic white or coloured pigments. Long-term outdoor exposure and lightfastness are often less good. Weaknesses in lightfastness occur especially at high degrees of brightening, i.e. with increasing levels of added white pigment. Another disadvantage which can affect certain organic pigments despite their other benefits is inadequate heat resistance. Their use in stoving enamels is therefore limited. As developments have occurred, however, organic pigments have come onto the market which are significantly better at meeting the requirements imposed on heat resistance. These are frequently associated with high manufacturing costs and high outlay on the dispersion process.

All organic pigments have chromophore structures, which create a colour impression by absorbing light in the visible spectrum. Chromophore groups generally comprise a number of conjugated double bonds. These need not necessarily be carbon-carbon double bonds. Nitrogen-nitrogen double bonds, the azo groups or double bonds between carbon and nitrogen or oxygen are also very effective. The chromophore structures are generally also supplemented by auxochrome groups, which lead to colour intensification and colour shifts through inductive effects. The degree of electronegativity determines whether such shifts in absorption occur towards the longer (bathochromic shift) or the shorter wavelengths (hypsochromic shift).

At 0.01 to 0.1 µm the particle size of organic pigments is smaller than that of inorganic pigments by about one power of 10. The smaller particle sizes of the organic pigments require particularly effective dispersion processes to ensure complete and even distribution within the coating material.

Classification of the many organic pigments mostly follows the typical functional groups in their chemical structure.

Azo pigments are the most important group among the organic pigments in terms of quantity (see Figure 2.1.161). They derive their name from the azo group, which acts a link between the individual parts of the molecule.

The structure and chemical functionality of the ligands which are linked by azo groups determine the performance profile of the pigments. A distinction is drawn between monoazo and diazo pigments in accordance with the number of azo groups in the molecule.

The first synthetic azo product to be manufactured commercially was BASF's Lithol Red. Azo pigments are now manufactured in an extremely large variety. Despite weaknesses when subjected to thermal stresses above approx. 140 °C, their mediocre lightfastness and their limited solvent resistance, the standard products in this class of substances represent by far the largest proportion of the organic pigments used in the paint and coatings industry.

Azo pigments
• Monoazo pigments
• Diazo pigments
• (azo condensation pigments)
• BONS pigments
Non-azo pigments
• Metal-complex pigments
• Indigoid pigments
• Triphenylmethane pigments
• Isoindolinone, isoindoline
• Anthraquinones
• Quinacrinones
• Perylene, perinone
• Dioxazines
• Quinaphthalones, pyrazoloquinazolone
• Diketo pyrrolopyrole (DPP)

Figure 2.1.174: Classification of organic coloured pigments by chemical class

Synthesis of the azo pigments requires a primary aromatic amine component which is diazotised with nitrous acid. This is usually aniline with a very wide range of possible substituents. Coupling of the diazotised amine on one hand with acetoacetanilide and phenylmethylpyrazolone affords yellow-orange pigments and, on the other, with ß-naphthol and naphthol AS leads to orange to red pigments.

The weaknesses in heat resistance, lightfastness and solubility of the monoazo pigments can be reduced by molecular enlargement. Either a diazotised diamine is bound to two conventional couplers or a 2-functional coupler is used with two diazotised amines. The first case yields the diaryl pigment class (benzidine type), and the second that of the bis-acetoacetic arylides.

Other proven methods of stabilising azo pigments consist in performing an azo condensation in which diamines are used to couple carboxyl group-containing monoazo pigments via the formation of acidamide groups.

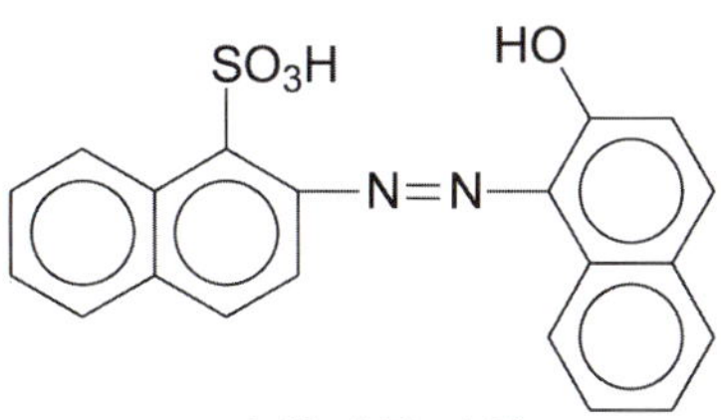

Figure 2.1.175: Lithol Red: the first azo pigment of industrial importance

A further reduction in solubility together with an improvement in lightfastness can result from strong intermolecular interactions created by additional acidamide groups in the coupling components. Prime examples of these modifications are the variants of the classic azo pigments containing benzimidazolone groups.

Strong intermolecular interactions arising from numerous hydrogen bridges lead to heat resistances of up to 220 °C, combined with good lightfastness and excellent solvent resistance. Similar improvements can be achieved with coupling components synthesised from cyanuric chloride and acetoacetic anilide. The molecules in this

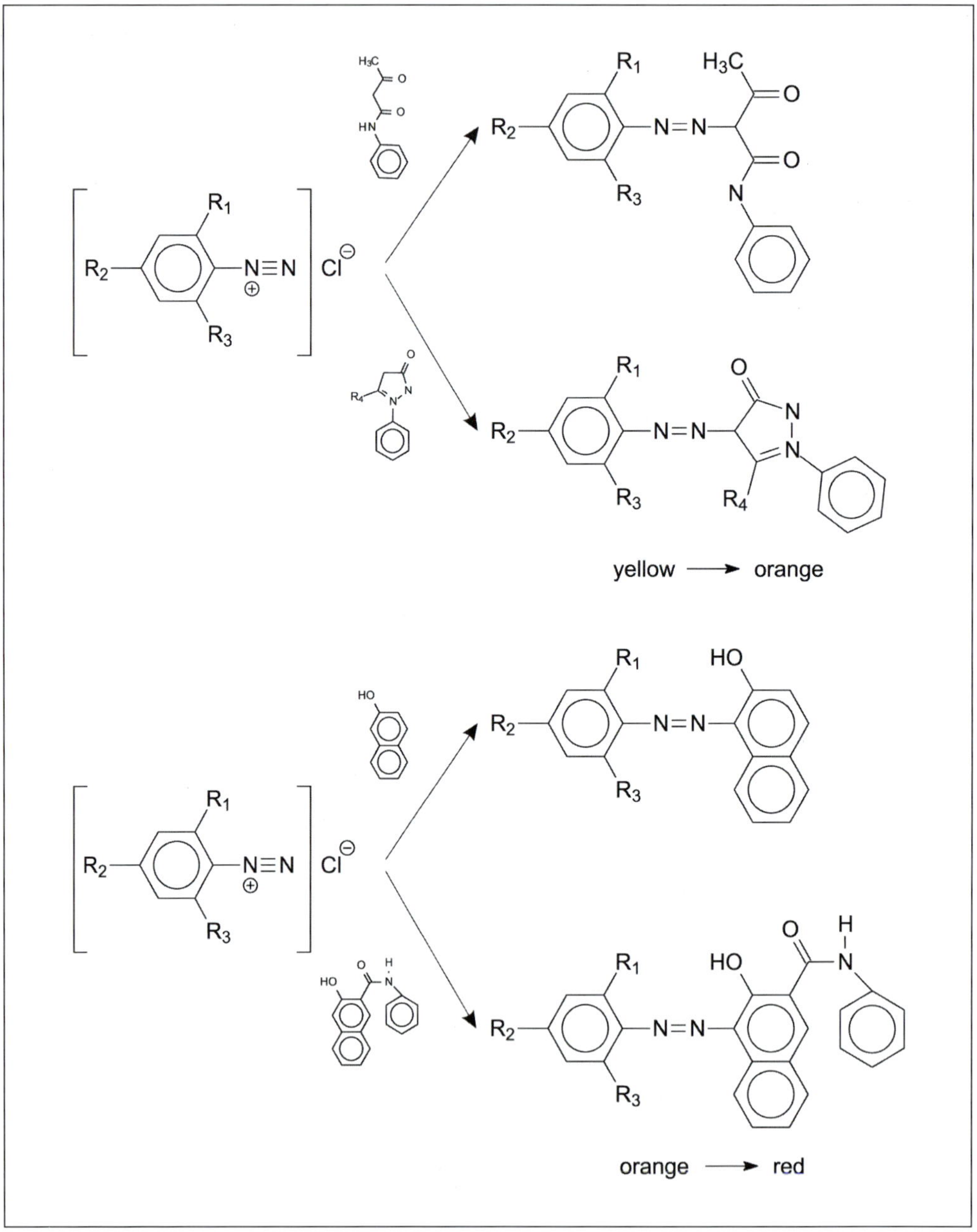

Figure 2.1.176: Chemical formulas of monoazo pigments (yellow, orange, red)

pigment class are planar and therefore permit the formation of a large number of hydrogen bridging bonds.

Another method, known as lacquering, for manufacturing azo pigments which are insoluble in organic media, proceeds via salt and complex formation with metal cations. This is only possible if there are acid groups in the molecule. If ß-oxy-naphthoic acid (German acronym: BONS) serves as the coupler, the carboxyl group can be converted to sparingly soluble metal salts with the aid of aluminium, iron or manganese ions. When these BONS pigments are used for pigmenting water-borne paints, their insolubility needs to be tested.

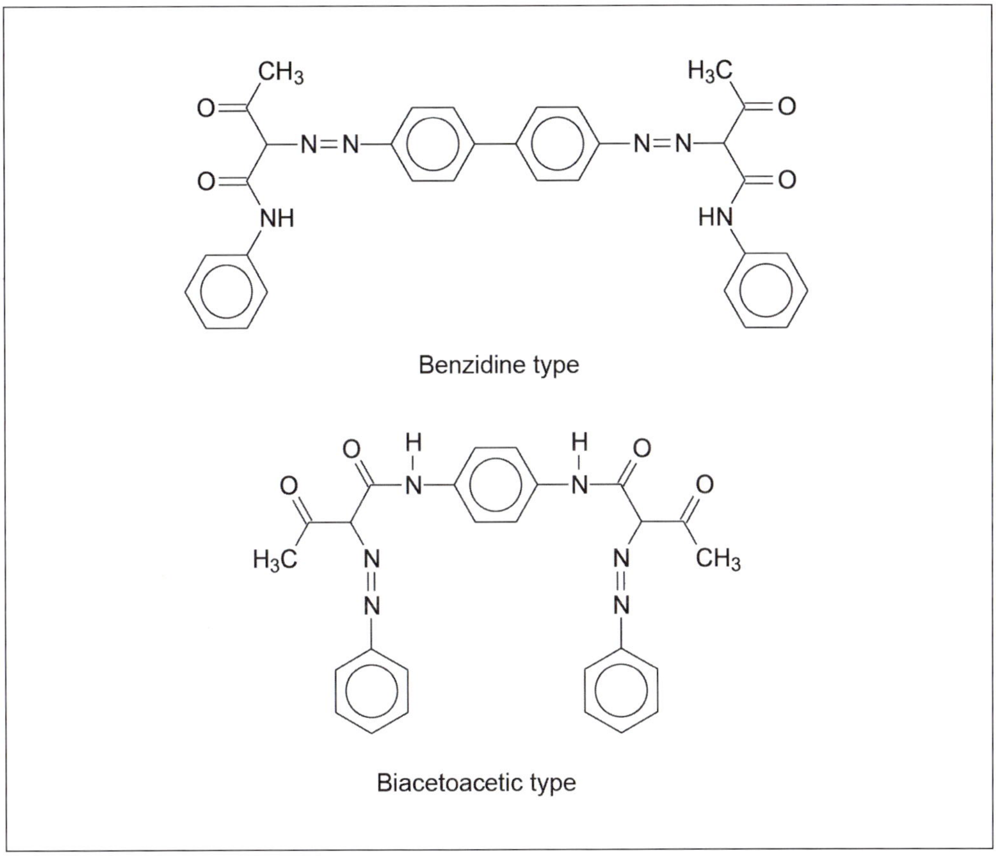

Figure 2.1.177: Chemical structure of diazo pigments

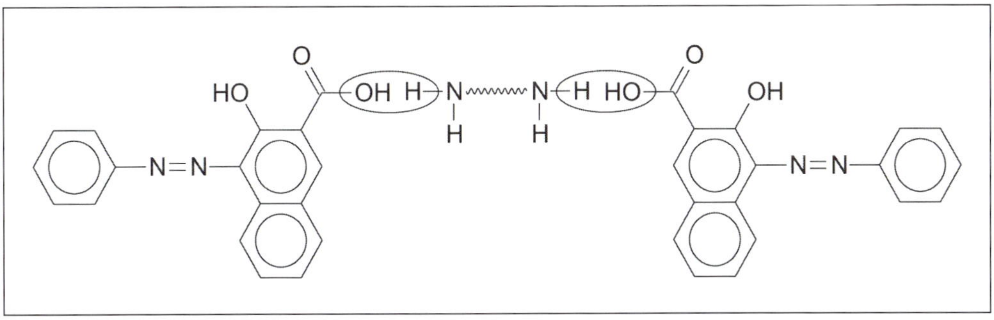

Figure 2.1.178: Synthesis of azo condensation pigments

The **metal complex pigments**, which are of most interest because of their enormous resistance to the effects of chemicals, acids, alkalis, light and long-term outdoor exposure, cover the blue and green ranges of the colour spectrum. The most important representative is chemically related to blood pigment and leaf green and consists of a polycyclic, continuously conjugated ring system, in which the magnesium of chlorophyll or the iron of haemin has been replaced by copper. This class of substances therefore has the name **copper phthalocyanines**. They resist concentrated sulphuric acid and remain unchanged even at temperatures of up to 500 °C. Copper phthalocyanine pigments exist in several crystal forms, of which the red-hued α-form is less resistant than the green-hued β-form.

The incorporation of halogen into the basic skeleton of the copper phthalocyanines yields green pigments which also have a high resistance.

Other far less important metal complex pigments are the nickel complexes of certain hydroxy-functional azo pigments and the iron complex of l-nitroso-2-naphthol.

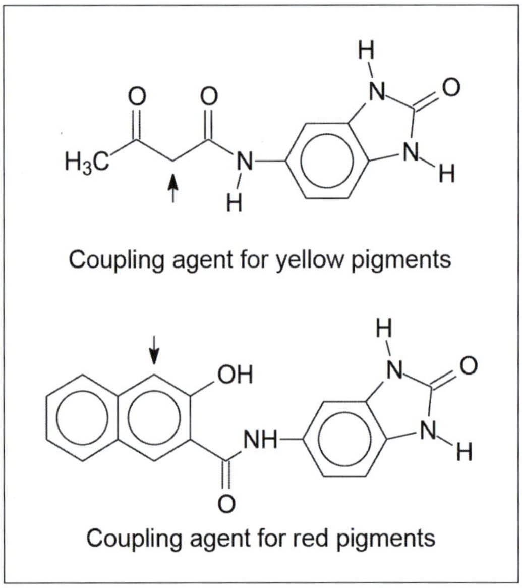

Figure 2.1.179: Benzimidazolone-containing coupling agents for azo pigments

The alkyl-substituted amine derivatives of triphenylmethane, the **tri-phenylmethane pigments**, are of almost unparalleled brilliance and colour purity. They are converted into the insoluble form by salt formation with complex poly acids. Triphenylmethane pigments are mainly used in the printing inks sector. Their fastness properties are inad-equate for use in external coatings subjected to long-term outdoor exposure.

The **carbonyl pigments** are a technically important and diverse group of pigments in terms of the number of variants. The group takes its name from the presence of at least 2 carbonyl groups which are in a conjugated

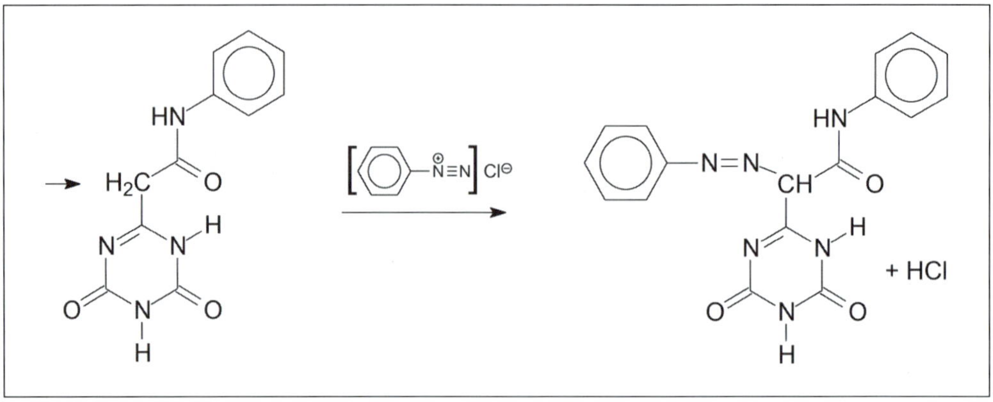

Figure 2.1.180: Synthesis of azo pigments with coupling agents containing cyanuric chloride

configuration to each other or to other double bonds. Compared with the azo and phthalocyanine pigments the quantity, though not the number, of carbonyl pigments used in the paint and coatings industry is comparatively low. The reasons for this are the sometimes very intricate and thus expensive synthesis processes. However, these pigments are particularly valuable when it comes to the manufacture of yellow, red or violet colours that offer the highest standards of fastness and resistance. The first of the frequently complex carbonyl pigments to be considered here is the indigo pigments. Up until a few years ago, only derivatives of thioindigo had any practical importance for the paint and coatings industry. A specific example to mention is **tetrachlorothioindigo**. This is resistant to solvents, heat resistant up to 200 °C, and lightfast and stable in long-term outdoor exposure, even when blended with white pigments.

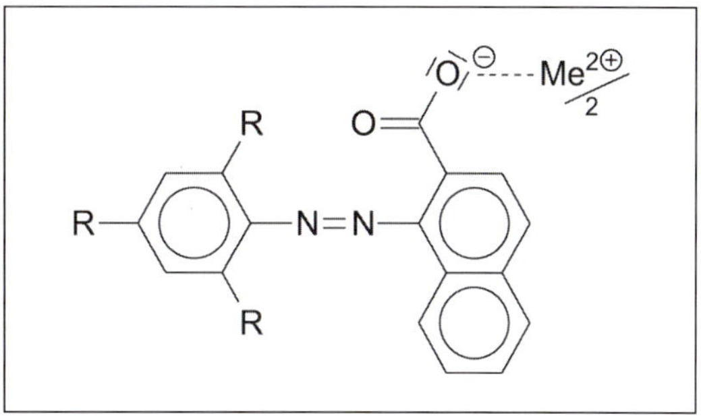

Figure 2.1.181: Azo pigments on the basis of β-oxy-naphthoic acid (BONS pigments)

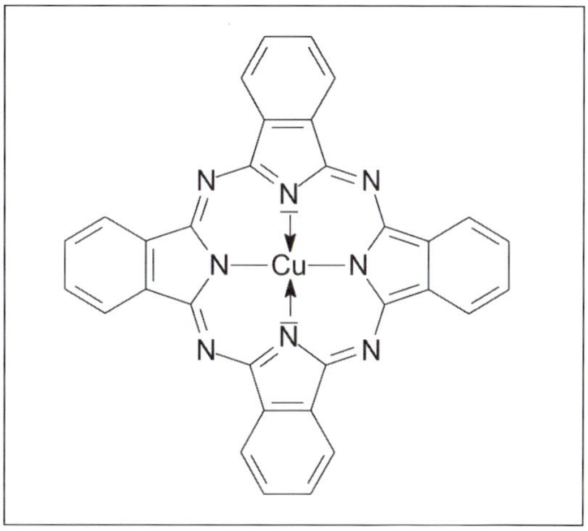

Figure 2.1.182: Structural formula of copper phthalocyanine

The carbonyl pigments also include the **isoindolinones** which are used for plastics and high-grade industrial coatings because of their good lightfastness and long-term outdoor exposure properties. They span the yellow, orange and red range and are formally accessible by condensation of phthalimides with diamines. Colour shifts from green to red via yellow and orange are achieved merely by varying the diamine component.

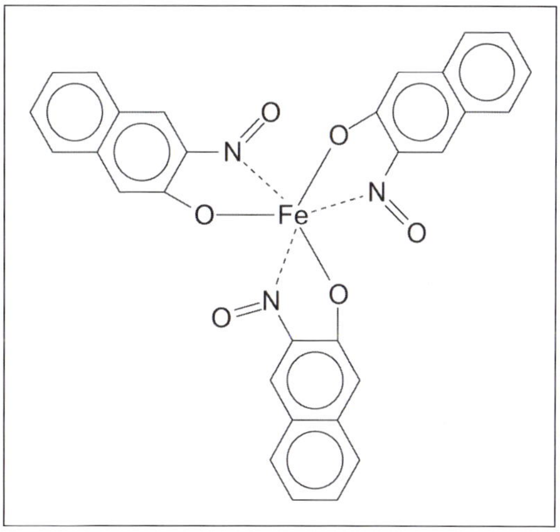

Figure 2.1.183: Chemical structure of the iron complex of 1-nitroso-2-naphthol

By contrast with the isoindolinones, which have been known for some time, the **isoindolines**, i.e. compounds in which both carbonyl groups of the phthalimides have been replaced by double bonds with suitable substituents, have even better long-term outdoor exposure properties and excellent hiding power.

Natural pigments which have been known for centuries, such as madder lake, kermes, and cochineal, provided the blueprint for constructing the **anthraquinone pigment** group.

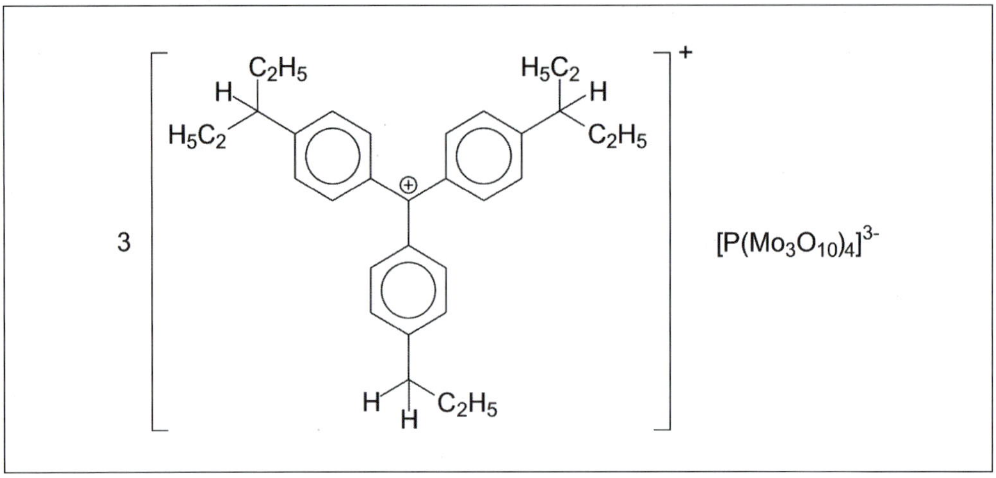

Figure 2.1.184: Example of a triphenylmethane pigment

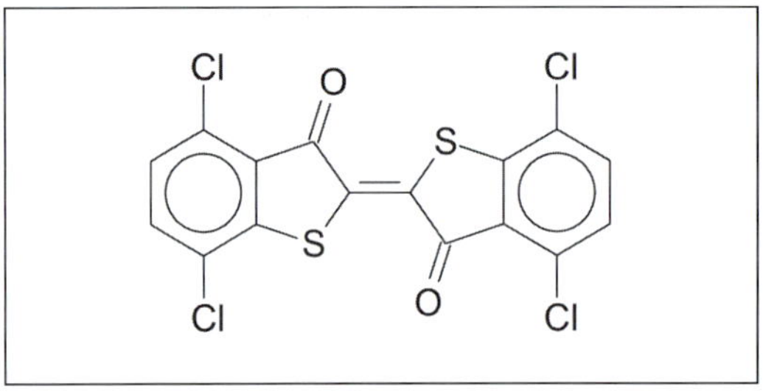

Figure 2.1.185: Chemical structure of tetrachlorothioindigo

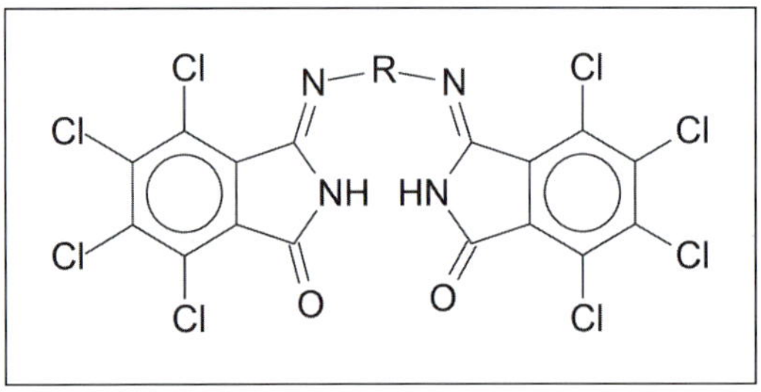

Figure 2.1.186: Chemical structure of isoindolinone pigments

The basic product, formerly manufactured by oxidation of anthracene and more recently by *Friedel-Crafts* acylation of substituted aromatics by phthalic anhydride, has no pigment properties itself, although it is inherently coloured. The molecule is too small to meet the requirements imposed on exposure to light, outdoor conditions and chem-icals in combination with heat resistance. Attempts at molecular enlargement have been successful, and numerous high-grade products are available today as a result. The mere chemical bonding of two aminoanthraquinone molecules with a dicarboxylic acid affords excellent yellow pigments. Highly stable pigments are also obtained by condensing anthrapyrimidine carbonic acid with 1-aminoanthraquinone.

Molecular enlargement by linking two anthraquinone molecules is also achieved by reducing or oxidising dimerisation.

Superior pigment properties are obtained by the formation of further aromatic rings. The highest-grade, polycyclic anthraquinone pigments numbers **flavanthrone** (red-tinged yellow), **pyranthrone** (red to orange) and **anthanthrone** (scarlet) in their group. Although they are very expensive because of their multistage synthesis, they also rank among the most lightfast organic pigments of all. Despite its excellent properties, production of flavanthrone had to be halted for cost reasons.

The quinacridones too are high-performance red to violet pigments. The colour of this pigment class is also determined by the crystal form. Of the five modifications now

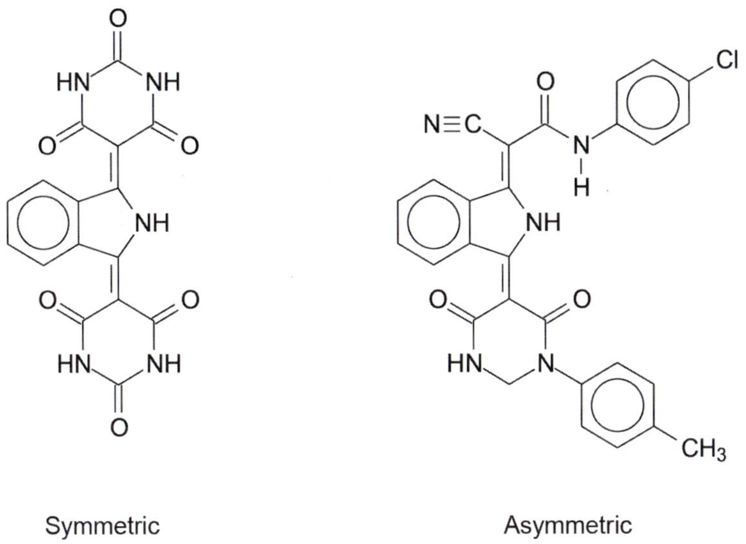

Figure 2.1.187: Typical chemical structures of isoindoline pigments

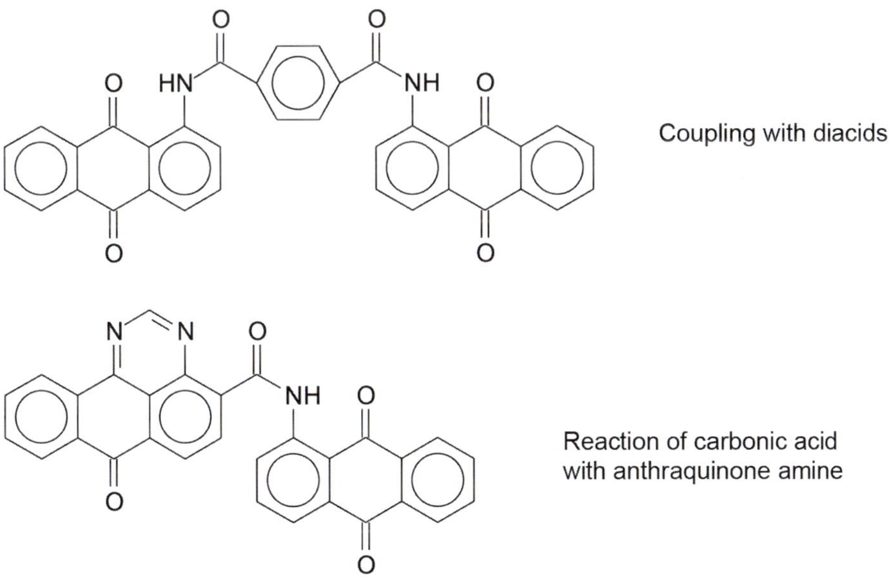

Figure 2.1.188: Anthraquinone pigments by condensation of anthrachinone amine with different compounds

known, the red γ- and the violet β-forms are the most important. They receive top marks for lightfastness and solvent resistance, but are sensitive to alkalis.

The **perylene** and **perinone pigments** are closely related to each other. Starting from aromatic, polynuclear tetracarboxylic acids, the synthesis process involves condensation with monoamines or diamines. The first case yields the perylene pigment group, while

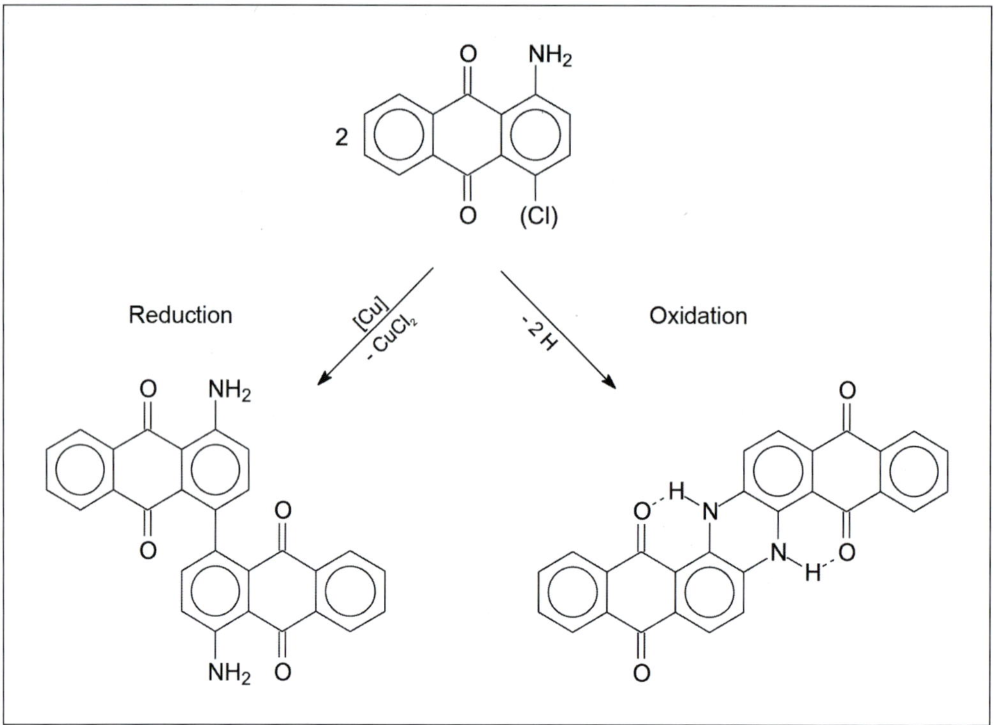

Figure 2.1.189: Synthesis of dimerised anthraquinone pigments by oxidising and reducing coupling reaction

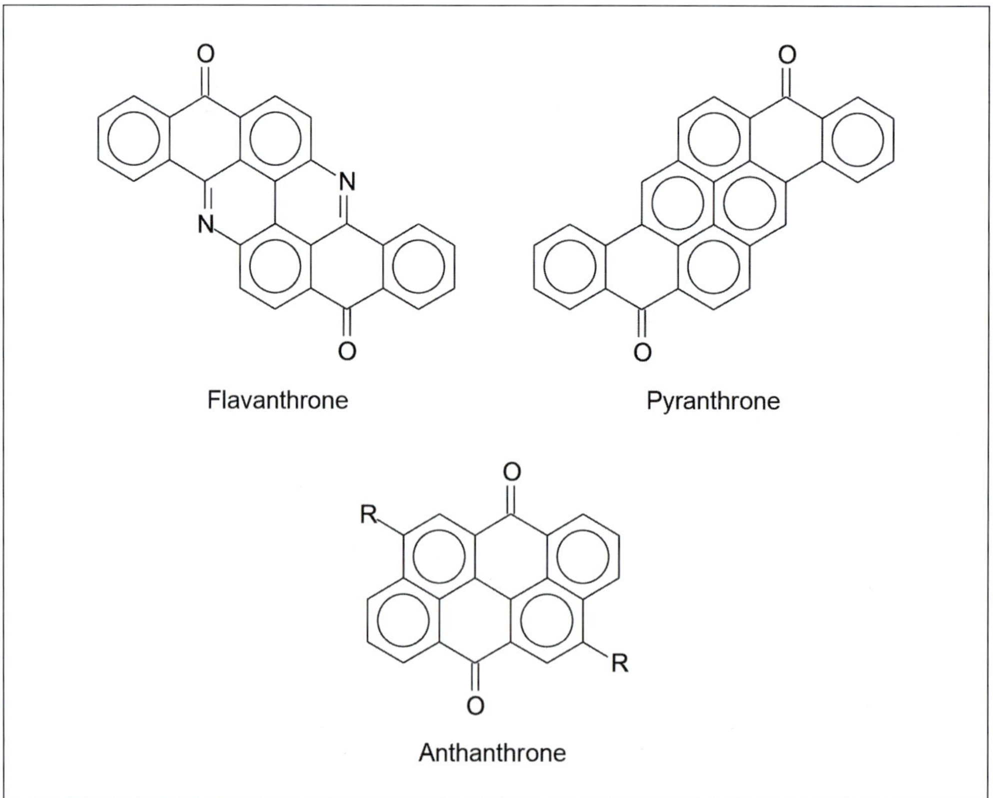

Figure 2.1.190: Chemical structures of condensed anthraquinone pigments

the second leads to the perinones. Because of their symmetrical and asymmetrical forms (cis/trans) the latter have different colour properties.

Perylene pigments, though this also applies to perinones, exhibit excellent lightfastness and long-term outdoor exposure properties, but also high heat resistance because of their chemical structure. Like the above mentioned anthraquinones and quinacridones, therefore, they are also used in high-grade industrial paints, particularly powder coatings. The fine-particle types are particularly suitable for metallic coatings because of the less pronounced scattering power, and the coarser types for opaque solid colours, i.e. without any added white pigments, generally combined with inorganic pigments.

The violet **dioxazine pigments** do not belong to the carbonyl pigment class, yet they also deliver brilliant colours and are extremely lightfast in suitable concentration. Because of the violet colour and their high lightfastness, even in tiny concentrations, they are used, among other things, to "clean up" the white tone of the slightly yellowish rutile form of titanium dioxide.

The dioxazine pigments have even better resistance properties if imidazolone groups are chemically incorporated [2.4.59].

Despite these numerous optically attractive pigment classes which are also suitable for paints in terms of application engineering and long-term outdoor exposure proper-

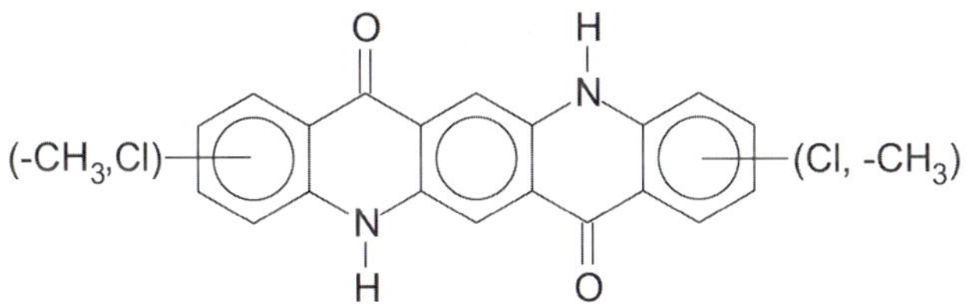

Figure 2.1.191: Typical chemical structures of quinacridones

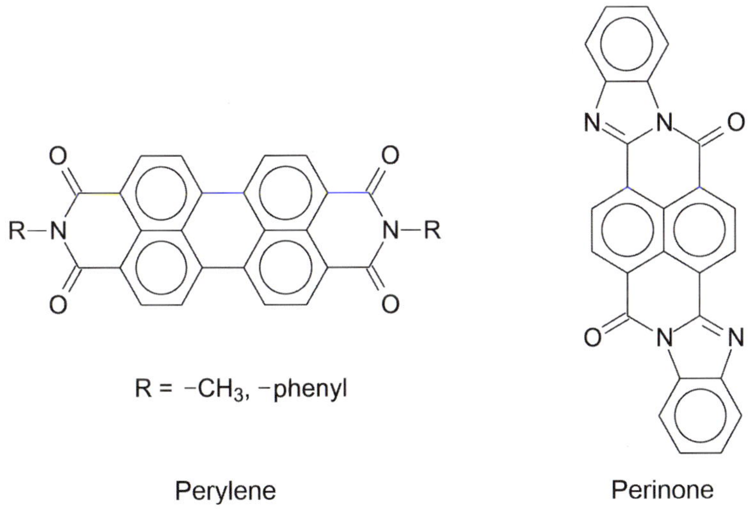

Figure 2.1.192: Perylene and perinone pigments

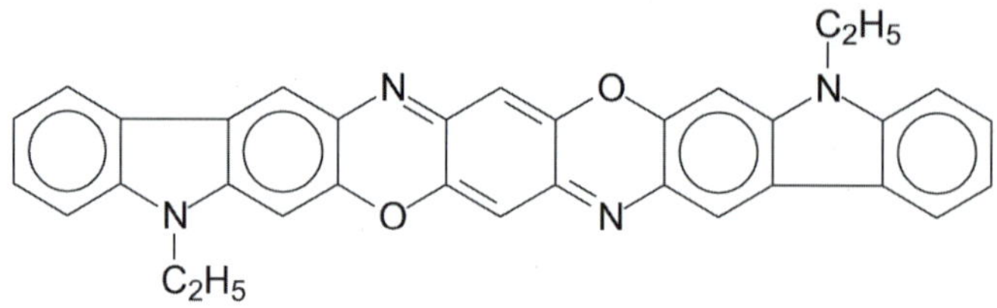

Figure 2.1.193: Chemical structure of a dioxazine pigment

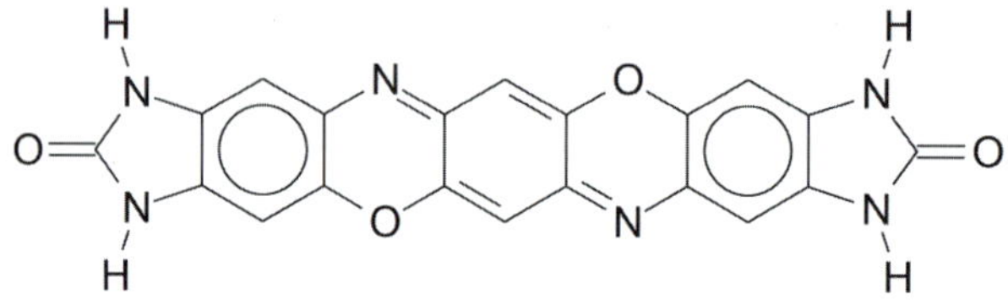

Figure 2.1.194: Structure of an imidazolone-modified dioxazine pigment

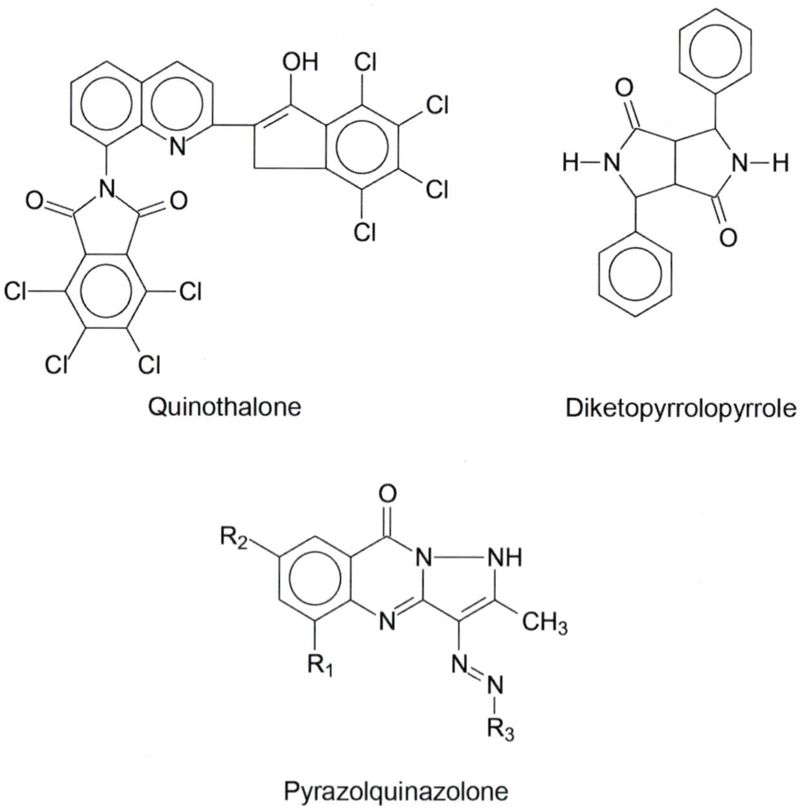

Figure 2.1.195: Structure elements of newly developed pigments

ties, researchers have continued their attempts to develop new knowledge based on the structure/property correlations of familiar pigments. **Quinophthalones, diketopyrrolopyrrole (DPP)** and **pyrazoloquinazolone pigments** are undoubtedly the most important outcome of pigment research in recent decades to find technical use. Of these, the red DPP has gained an important place in automotive paints because of its high colour purity.

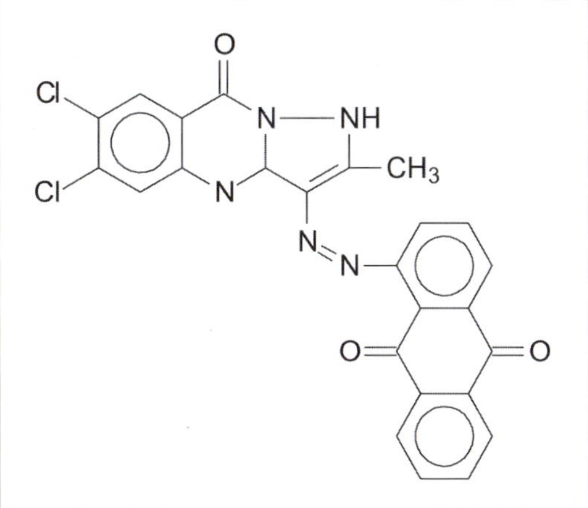

Figure 2.1.196: Chemical formula of a pigment made of azo and anthraquinone structures

The example of the pyrazoloquinazolone pigments shows that rigorous implementation of the theoretical and practical knowledge of structure/property correlations can lead to further improvements, even though quality standards are already high. Planar molecules with as many carbonyl and NH groups as possible in the right configuration lead to good crystallisation and strong intermolecular interaction, both of which are important prerequisites for lightfastness and resistance to heat and solvents. The knowledge gained with acetanilide, pyrazolones and benzimidazolones among the azo pigments resurfaces in optimised form in the pyrazolone matrix of quinazolone. To this must be added the fact that the properties of pyrazoloquinazolone can be further improved by azo coupling with anthraquinone. They currently represent the leading edge in pigment development with a range of brilliant and, at the same time, highly opaque products.

Effect pigments

Metal pigments, also termed "bronzes", are platelet-like metal particles which are generally composed of aluminium and are up to 20 µm long and up to 1 µm thick. Their task is to settle in a plane-parallel configuration in the paint film such that the brightness of the reflected light varies with the angle of view; this is also known as "flip-flop". The coatings thereby gain a metallic appearance which is particularly attractive. The properties of the aluminium platelets in the freshly applied (still liquid) film vary with the surface treatment already received. Those which were modified with stearic acid float to the surface of the coating and form a glossy silver interface between the coating and its

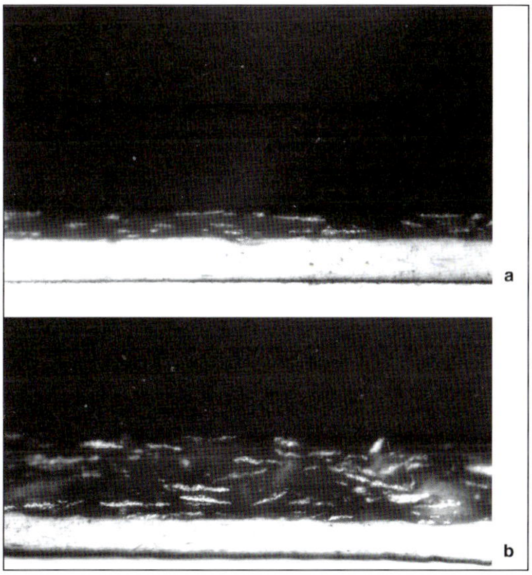

Figure 2.1.197: Metal pigments in a cross section view of a coating with (a) good and (b) bad metallic effect ("flip flop")
Source: BASF Coatings

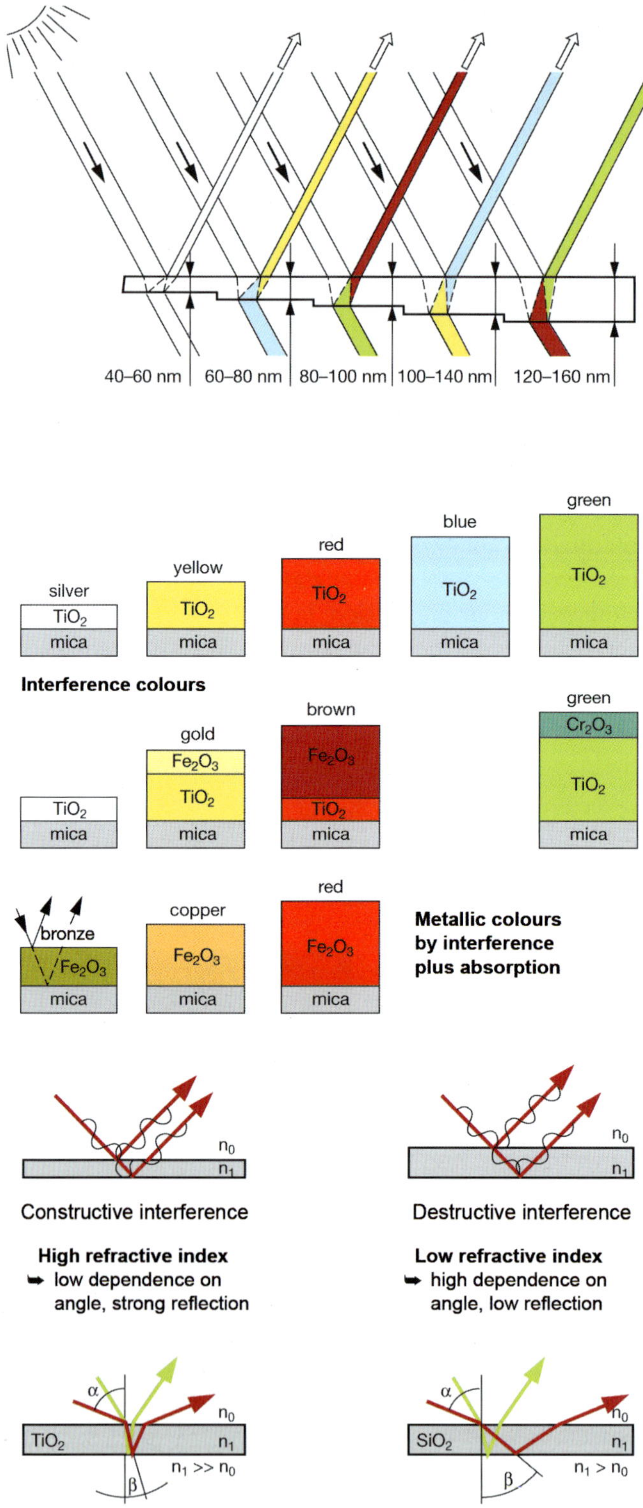

Figure 2.1.198: Path followed by light rays through interference pigments

surroundings. The pigment scales are like leaves floating on the surface of water, and this phenomenon is therefore termed "leafing". They are in parallel alignment and are fixed in this position when the film hardens. The extensive reflection of IR and UV light by aluminium particles is put to use to confer long-term protection on topcoats for industrial steel structures, bridges, lighting masts, large containers, refineries or large chemical plants. This enables effective protection of the objects against light and thermal radiation.

In the case of "non-leafing" types stabilised by oleic acid this floating to the interface of the coating does not occur. If the tiny aluminium mirrors which are now evenly distributed in the film can also be aligned parallel with both each other and the substrate, the outcome is the metallic finish which looks so attractive on cars, once it has been sealed with a weatherproof clearcoat [2.4.60].

For use in water-borne basecoats, the aluminium platelets on the surfaces must be specially stabilised to suppress reaction with water and the associated formation of hydrogen. This can be achieved with a chromate conversion coating, an organic coating [2.4.61] or silicate conversion coating [2.4.62].

The **interference and pearlescent pigments** cause colour shifts that change with the angle of view. This effect, which is based on physical principles of light interference, has been copied from pearls, fish scales, peacock feathers, and beetle carapaces.

The cause of the selective cancelling of portions of the visible light lies in the double reflection at the surface and on the underside of the thin pigment platelets. The pigments are always structured in principle such that thin films of oxides with varying refractive indices are applied to a platelet like, reflective carrier [2.4.63]. If the film is thin enough, the light rays which are offset from each other by half a wavelength are cancelled out, whereas those offset by a whole wavelength are amplified. The colour of the reflected light is complementary to the transmitted light and varies with the angle of view for different angles of incidence because of the changing differences in path lengths of the separated rays (see Chapter 3.2.4).

As the visible light has short wavelengths the interference pigments are applied in thin film thicknesses. The pigments were originally manufactured exclusively by precipitating titanium dioxide films onto mica platelets, but then underwent numerous modifications to include both the colour-providing component and the carrier. Additional application of layers possessing selective absorption properties enables interference coatings to be transformed into further attractive colours by absorption coatings.

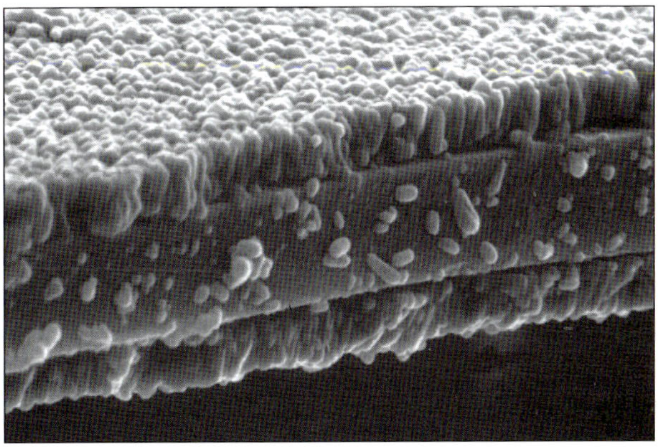

Figure 2.1.199: SEM picture of a cross-section of an interference pigment on an SiO_2 carrier Source: BASF SE

In addition to mica as the carrier material, aluminium platelets, aluminium oxide [2.4.64], sili-

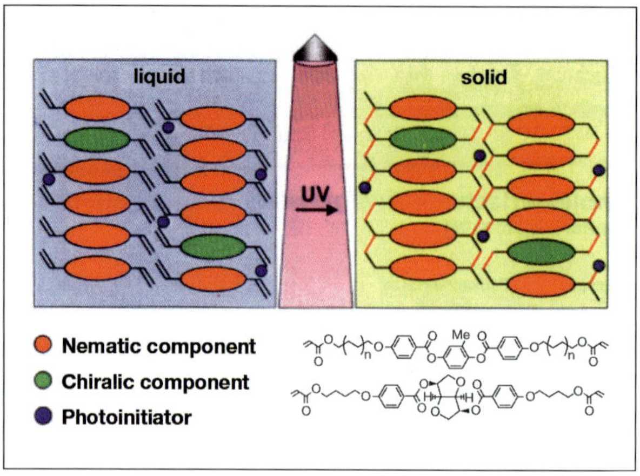

Figure 2.1.200: Liquid crystal pigments fixed in a polymer matrix by polymerisation *Source: BASF Coatings*

con dioxide, and even lamellar iron oxide are now also used for downstream application of the λ/2 films. The latter can be applied from the gaseous phase by means of CVD (chemical vapour deposition).

Another group of organically based interference pigments is the **liquid crystal (LC) pigments**. The liquid crystals which are oriented in a cholesteric phase in the form of helices and then fixed by copolymerisation also give rise to interference effects with light because of their molecular order and thus represent an extension of the already wide range of interference pigments.

However, the high transparency of LC pigments demands that carbon black or opalescent pigments also be added. Only in this way is it possible to achieve both a good effect and hiding power at the same time [2.4.65].

Current Pigment Research

Today's paint and coatings industry offers a large range of organic and inorganic pigments with a broad spectrum of colours and effects. Suitable products are available for meeting various requirements relating to heat resistance, insolubility in an aqueous or organic environment, and lightfastness, as well as hiding power and gloss. In practice, mixtures of organic and inorganic pigments are frequently used in order to combine the strengths of the individual classes of substances with each other cost-effectively.

Modern pigment research uses not only the practical experience of structure/property correlations, but also the modern possibilities of molecular modelling to gain information on optically attractive pigments, without having to synthesise them first. It should be noted in this regard that the colour properties depend not only on the structure of the individual pigment molecule, but also on its configuration within the crystal matrix. Predicting the right crystal packing out of the large number of possibilities is a problem that still has not been fully resolved to this day, despite a far-reaching understanding of the ways of controlling crystal growth.

Whenever the selection of suitable colorants is being considered, it is essential to bear in mind that the performance profile of a coating finish is not determined solely by the pigments. Particle shape, distribution, configuration and interaction with the polymer environment on one hand and the interaction with additives on the other ultimately determine the colour, effect, hiding power and the long-term outdoor exposure properties. Pigments alone, therefore, though important, are but one of the elements which determine the quality of a coating finish.

2.1.3.3 Extenders

Extenders generally consist of inorganic substances of variable chemical composition and, consequently, different physical properties. Their task is to fill the coatings spatially with skeletal material and therefore to influence the mechanical properties in a particular direction that reflects the function of the relevant paint film. The refractive index is frequently used to demarcate them from pigments. For extenders this is less than 1.7 which is why this family of substances does not attain any noteworthy scattering or hiding power. However, for coating formulation principles, they and the pigments must be taken into account when the pigment volume concentration (PVC) is being calculated (see Chapter 2.3.3). The PVC is one of the principal factors governing the mechanical/technological properties of coatings.

When the film volume is being increased with extenders, their grain size distribution plays an important role. Correctly chosen combinations of small and larger particles allow the greatest possible packing density. The remaining volume fraction available for the binder in the cavities is then only small, and the PVC has a correspondingly high numerical value. Each coating system has a critical PVC (CPVC) which, if exceeded, prevents the desired film properties from being achieved since the binder fraction is too low to constitute a continuum.

Impact strength, swelling power and water vapour permeability are film properties which can be influenced by the addition of extenders. However, the effect of extenders is not limited to the interior of a coating; it also extends to their interfaces. In the film surface they can exert a matting effect by increasing the surface area, thereby enhancing the adhesion of the following coating film. Wear resistance and sandability are further surface properties of films which are influenced by extenders.

In volume terms, extenders obtained from minerals and rocks top the product range. After large-scale extraction, the natural raw materials are comminuted in grinding processes. The degree of comminution and treatment determines how the materials are classified:

- **Coarse powders** with particle sizes over 250 µm
- **Medium powders** with particle sizes between 50 µm and 250 µm
- **Fine powders** with particle sizes between 10 µm and 50 µm
- **Pulverised powders** with particle sizes < 10 µm.

Synthetically manufactured extenders usually consist of the same chemical compounds that are used as natural extenders. However, they are purer, and their properties are more reliably reproducible.

Extenders are mainly used for primers, fillers and primer surfacers. Large quantities are also contained in emulsion paints. They are used only minimally, on the other hand, in solvent-borne topcoats.

The following extenders or groups of extenders are worthy of particular note:

Kaolin, also termed China clay, is a natural, very white aluminium silicate and is ideal for use in emulsion paints. Its lamellar structure imparts thixotropic properties to coatings. The hardened films are mechanically tough and exhibit good adhesion to the substrate.

Silicates	
Natural, plate-like magnesium silicate hydrates	Talc
Natural aluminium silicate hydrates	China clay
Natural potassium aluminium silicate hydrates	Muscovite (mica)
Carbonates	
Natural calcium carbonates from shell residues	Chalk
Natural calcium carbonate, crystalline	Calcite, marble
Precipitated calcium carbonates	
Natural calcium magnesium carbonates	Dolomite
Sulphates	
Natural barium salts	Barytes, heavy spar
Synthetic barium salts	Blanc fixe
Silicic acids	
Natural silicic acids (SiO_2)	Quartz
Synthetic silicic acids	Aerosils
Other extenders	
	Graphite, micaceous iron

Figure 2.1.201: Important extenders for coatings

Talc, a natural magnesium silicate, also has a lamellar structure. With a hardness of 1 on the *Mohs'* scale of hardness (see Chapter 3.2.5) talc is extremely soft and promotes the sandability of fillers and primers by reducing the cohesive strength. **Chalk flint,** despite its misleading name, is not a type of chalk at all; instead, it is kaolin contaminated with quartz and dolomite.

Thanks to their hardness **silicon carbide** and **quartz powder** reduce abrasion of coatings for, e.g., road marking and floor covering.

The silicate extenders hornblende, feldspar and kieselgur are only of minor importance. Whereas hornblende and feldspar are complex silicates with Ca, Mg, Al and Ba content, kieselgur (diatomaceous earth) consists of the ground-up siliceous shells of fossilised diatoms.

Carbonates in the form of chalk and calcite powders are important extenders in the paint and coatings industry and account for roughly ¾ of the total quantities employed there. The chalk comes from marine deposits of animal and plant skeletal substances. Microscopic examination of the starting material reveals residues of the shells of small marine creatures. Calcite, on the other hand, is crystalline calcium carbonate which does not contain any biogenic residues. It is finer and brighter than chalk. Also in frequent use are calcite powders which have undergone secondary treatment as well as synthetically manufactured calcium carbonate or precipitated chalk. Extenders made of calcium carbonate allow fillers, primer coats and primer surfacers with high packing densities and excellent application properties to be produced. Another important application field for calcium carbonate is in emulsion paints. The sensitivity of this raw material to acids is a disadvantage which excludes it from many application fields. Dolomite, i.e. calcium magnesium carbonate, is demonstrably less acid-sensitive and harder at the same time.

By far the most important **sulphate extender** is barium sulphate. A major advantage of this extender is its high resistance to acids and other chemicals. The paint and coatings industry uses both naturally occurring and also synthetically manufactured (precipitated) barium sulphate. The former consists of ground heavy spar or baryte, which is often contaminated by impurities because of its natural origin. It is not very bright and is characterised by different crystal forms and grain sizes within the individual particles. This helps a high packing density to be obtained for fillers and primer coats. Synthetically produced barium sulphate, under the tradename "blanc fixe", differs from ground heavy spar in its high degree of purity, its fineness and the uniformity of its grain size. Consequently, it is also suitable for more demanding applications. Barium sulphate has the highest density of all extenders, at 4.5 g/cm^3. This must be borne in mind when it is used in water-borne paints.

Calcium sulphate, also known as lightspar, is generally in the monohydrate form. Due to its sensitivity to water it is only used for less demanding coatings on wood, paper and plaster.

Textured, particularly **fibrous extenders**, such as fibreglass, substantially boost the tensile and tear strengths by forming a fine, matted network.

Other extenders with special roles include hydrous **aluminium oxide powders**, which have proved successful as flame-retardent extenders in conjunction with other additives which release inert gas, such as urea or melamine.

Lamellar extenders are also worth noting, especially micaceous iron ore in the form of lamellar haematite and mica itself. If such extenders are successfully aligned plane-parallel with the substrate, a water-vapour barrier is created and anticorrosive protection is improved at the same time.

Should a coating need to be electrically conductive, this can be achieved by adding graphite.

Synthetically produced **hollow microspheres** of SiO_2 or organic material are particularly light and simultaneously bring about an increase in the scattering power and thus in the hiding power due to the increase in relative refractive index that results from the air-filled cavities. An improvement in impact resilience is also achieved.

2.1.4 Plasticisers and Additives

While natural or synthetically produced resins as film forming agents are responsible for the protective effect, though it is solvents which allow them to be processed in the first place, and while pigments generate optical attractiveness through chroma and specific effects, it falls to plasticisers and additives acting as coating auxiliaries to fulfil the varied and important task of optimising the paint and paint film properties. Plasticisers serve to lower the film forming temperature and to elasticise the coating. They are largely chemically inert and therefore mostly function by physical means only.

Additives are generally added to the raw materials or the coating material in small and even minute quantities, but they affect coating manufacture, processing, film formation and ultimately also the optical and mechanical/technological properties of the film itself. Furthermore, they are capable of improving the film resistance and thereby to delay the ageing process.

Additives are often multifunctional and so influence several properties of paints and the finished coatings at once. Thus numerous wetting agents and auxiliary materials for the pigment dispersion process are exclusively substances which have a surfactant structure but can also be found in the same or only slightly modified form as defoamers or antisettling agents.

Silicone oils are also polyfunctional. They influence levelling and the floating of pigments, they reduce the drag of coated surfaces and thus indirectly improve scratch resistance. They simultaneously optimise substrate wetting and make coatings water-repellent by lowering the surface tension.

Whether additives are always able to achieve their tasks of modifying and optimising properties depends not least on their dose.

2.1.4.1 Plasticisers

As already mentioned a number of times, in addition to good adhesion to the substrate coatings are also expected to combine a high level of hardness and good elasticity. Many film forming agents do not meet these requirements because their chain mobility is restricted by molecular interaction, i.e. by strong *van der Waals* forces. The task of the plasticiser, therefore, is to reduce or entirely prevent the formation of such bridging bonds. This can be achieved by incorporating elasticising segments or monomers which sterically hinder such interaction into the film forming agents during manufacture. Such a system of influencing the elasticity in advance, i.e. during the manufacture of resins, is termed internal plasticisation. If internally plasticised resins cannot be used for reasons of poor processability or nonoptimum film properties, secondary elasticising of "hard" resins must be carried out. This can be achieved in two different ways. Either the hard film forming agent is "diluted" with highly elastic resins, such as aliphatic polyesters or traditional plasticisers in the form of low molecular auxiliary materials such as phthalic esters are added to halt intermolecular interactions.

The aliphatic polyesters mentioned first are nonvolatile and function more or less as oligomeric solvents. As they do not migrate readily, they offer the advantage of good long-term effects. However, if such plasticising resins or soft resins are added to excess, they give rise to a sticky surface on the coating. Important plasticising resins are polyesters of adipic, azelaic or sebacic acids with aliphatic 1,2- or 1,4-diols.

All plasticising resins have good resistance to leaching and migration and will only slightly lower the resistance of the coatings to petrol, oils, greases and soaps. They are resistant to UV light and are regarded as physiologically safe.

The disadvantage of the stickiness which occurs at high doses is avoided by recourse to traditional plasticisers. Such plasticisers are, according to ISO 4618, liquid or solid, inert

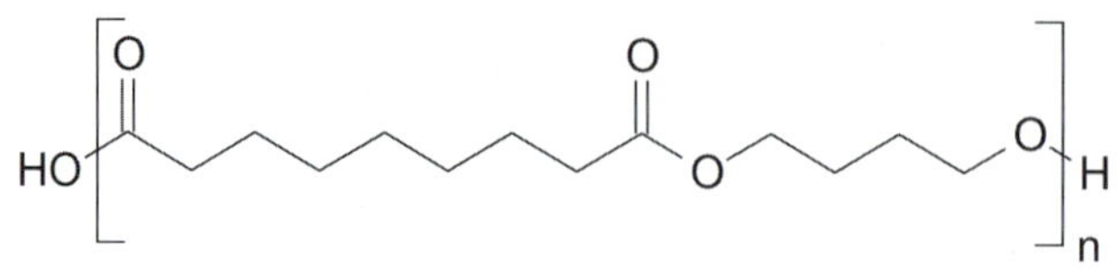

Figure 2.1.202: Plasticiser resin from azelaic acid and 1,4-butanediol

organic substances that have a low vapour pressure, particularly those of an ester-like nature. Mainly by virtue of their swelling properties, they can interact physically and non-chemically with high polymers to form a homogeneous system. This interaction has its origins in their special molecular structure. Plasticiser molecules always contain polar and non-polar moieties. The former enable them to interact with the polar groups of the film forming agents to be plasticised. Steric hindrance by the latter, non-polar moieties prevents the resin molecules from interacting with each other and so promotes their mobility.

The outcome is that the glass transition temperature T_g is lowered (see Chapter 3.2.5). This is a concentration-dependent effect and serves as an important criterion for gauging the effectiveness of plasticisers.

Effective plasticisation is therefore contingent on the use of substances which upon being heated penetrate the resin to be plasticised. Traditional plasticisers have a certain degree of volatility because of their low molecular weight. This makes them inferior to the plasticising resins in terms of lasting effectiveness.

Other demands placed on plasticisers include physiological safety, effectiveness on exposure to low temperatures, heat and light, and resistance to leaching and saponifi-

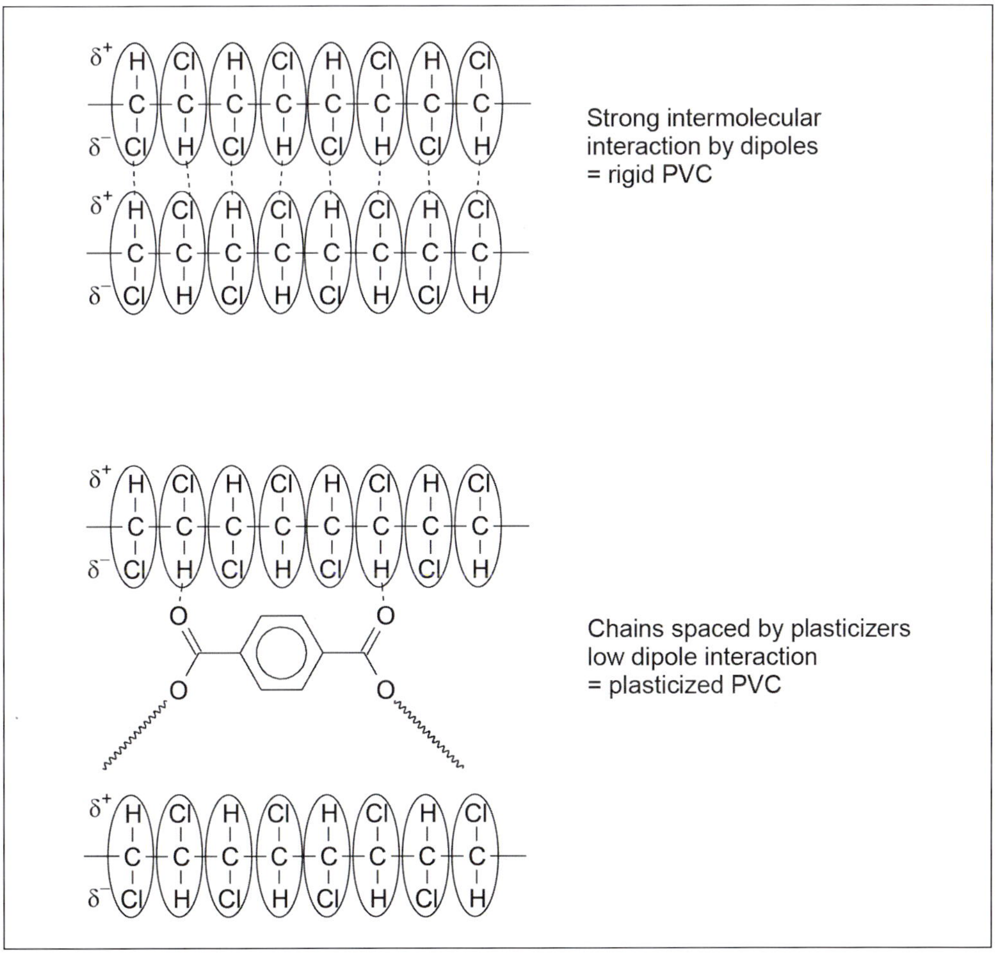

Figure 2.1.203: Effect of classic plasticiser on polyvinyl chloride (PVC, schematic)

cation. Materials which meet these demands include esters of phthalic acid, phosphoric acid, adipic acid and sebacic acid with aliphatic alcohols of different chain length and various degrees of branching.

Epoxidised oils and various fatty acid esters are employed as secondary plasticisers in conjunction with traditional plasticisers to boost the elasticising effect.

The widely employed phthalate plasticisers, usually esters of α-ethyl hexanol, are colourless and odourless, lightfast general-purpose plasticisers that possess moderate cold resistance. Although phosphoric esters are inferior to phthalic esters in their lightfastness, they gel rapidly and have better leaching resistance. By comparison, the acyclic esters of aliphatic dicarboxylic acids, generally adipic and sebacic acids, imbue the coatings with excellent elasticity, even at low temperatures. Due to their high price they are often mixed with phthalate plasticisers.

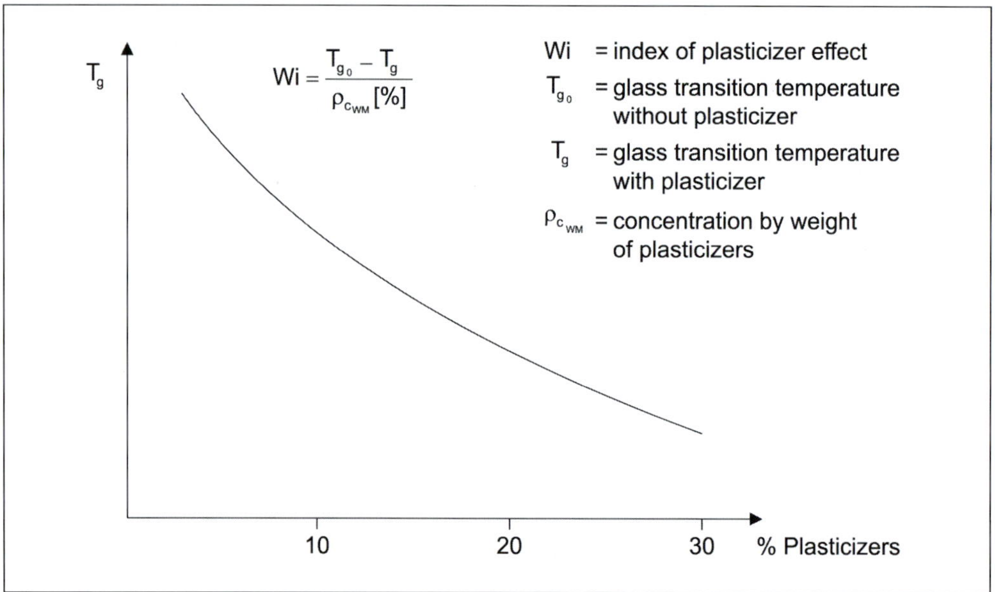

Figure 2.1.204: Lowering the glass transition temperature (T_g) of a polymer with plasticisers

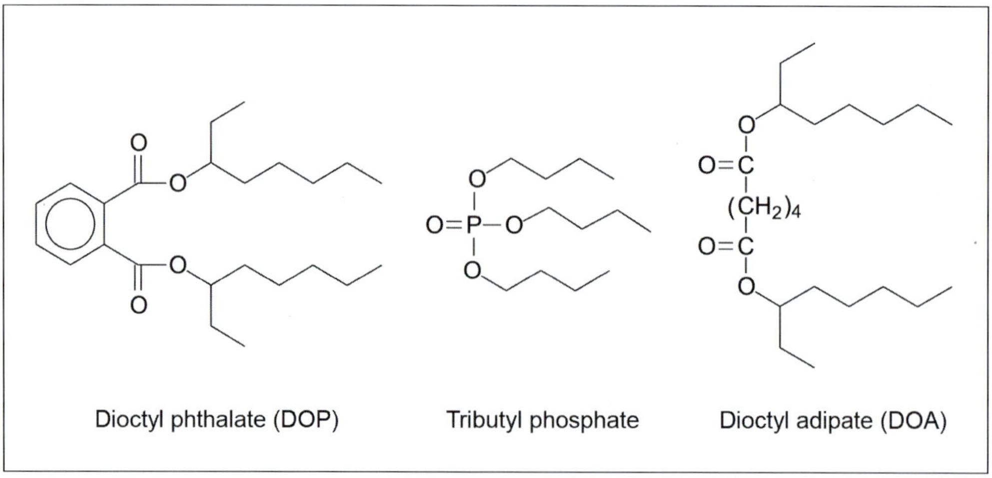

Figure 2.1.205: Chemical structures of traditional plasticisers

	Phosphoric acid ester	Adipic and sebacic acid ester	Phthalic acid ester	Ricinus oil acid ester
1. Drying oils	–	–	–	+
2. Asphalts	–	–	–	–
3. Bitumen	–	–	–	–
4. Maleic resins	+	+	+	+
5. Oil-free polyols	+	+	+	–
6. Alkyd resins	+	+	+	+
7. Phenolic resins				–
8. Urea resins	+	+	+	+
9. Melamine resins	+	+	+	+
10. Polyvinyl acetals	+	+	+	+
11. Polyvinyl chloride	(+)	+	+	+
12. Polyvinyl acetate	+	+	+	
13. Polyvinyl ester dispersions	+	+	+	

Figure 2.1.206: Hint for using plasticisers in film forming agents

Figure 2.1.206 presents a selection of resins which require plasticising and shows their compatibility with the most important plasticisers or plasticising resins.

2.1.4.2 Additives

When formulations are being drawn up it is often assumed that the use of appropriate film forming agents and pigments will produce the desired properties for manufacturing and film performance. If this approach fails for qualitative and economic reasons, the next step is to resort to additives. It should be noted that determining the correct dosage of the correct additive requires meticulous attention because of the tiny quantities added and the physical effects they frequently exert at interfaces. If, for example, a polar surface is to be made water-repellent by means of an additive, such an effect can only be achieved with a monomolecular covering of surfactant. If the dosage is too small, the goal will not be achieved. If it is too high, double layers will be formed which will ultimately restore the status quo.

The coating additives usually employed to optimise and modify properties can be classified in accordance with various criteria. Although various additives often are multi-functional, classification by intended use is the established method. For example, it will be found that surfactants of the same chemical nature are used for dispersing of pigments, defoaming, and combating floating effects, as well as for modifying the processing properties of water-borne paints.

The sections below deal first with additives for manufacturing, storage and processing and then discuss the additives used to form and stabilise the film properties.

Additives for Optimising Coating Manufacture and Storage Stability

If the technical process of coating manufacture is examined to identify scope for optimisation by means of paint additives, the first areas to focus on are the **rheological properties** and the physical phenomena that occur at interfaces. The rheological properties determine, among other things, the extent to which the pigment agglomerates penetrate into the cavities and permit the shear forces transferred to the pigments in the dispersing machines to be modified. The ability to influence flow properties is of critical importance during paint processing and subsequent film formation (see Chapter 3.1.1). Additives which influence surface tension are of particular interest in coatings manufacture and storage because pigments often need to be dispersed rapidly and stabilised better.

The forces of attraction which exist between molecules give rise to cohesion and a tendency to minimise the surface area for a given volume. This property, known as surface tension, can cause problems when different materials interact in the coating formulation. Thus fine particle pigments will agglomerate to form larger units and will resist wetting by resin solutions unless the surface tensions of the substances at the interfaces are coordinated (see Chapter 2.3.3). Compatibilisers in the form of **surfactants** are then needed to ensure that the pigment particles are completely coated in resin solutions, that all of the air is eliminated from the agglomerates, and that any water adsorbed on the polar surfaces of many pigments is removed. In short, they promote wetting of the agglomerates.

Surfactant molecules have a characteristic structure which enables them to fulfil their tasks. It is obtained by linking extremely hydrophilic and extremely hydrophobic molecular segments together chemically. These gravitate towards interfaces of different polarity, such that the polar moiety turns towards the polar substance and the non-polar towards the less polar substance. For example, the carboxylate groups as salts of saturated fatty acids, along with their non-polar alkyl groups, bring such disparate substances as oil and water together and so support the production of stable emulsions.

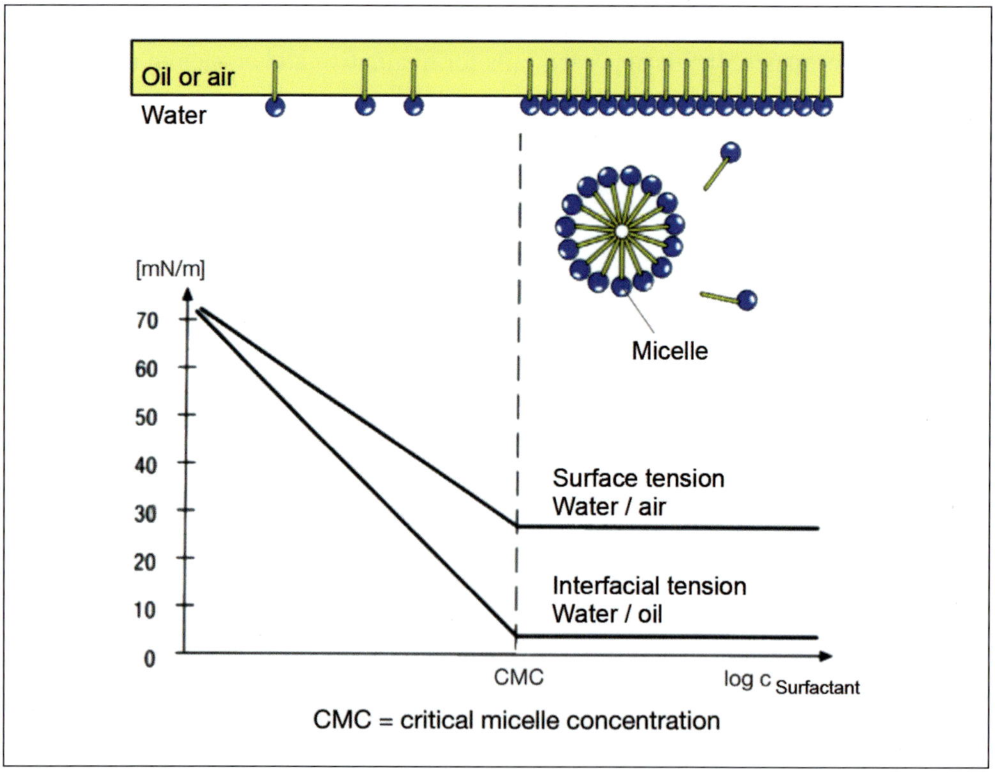

Figure 2.1.207: Dependence of surface tension and interfacial tension on surfactant concentration *Source: BASF SE*

Because surfactants accumulate in the interfaces, the direct relationship between their effectiveness and their concentration holds only up to a certain limit. An excess of most surfactants leads to the formation of surfactant aggregates in the aqueous phase, termed micelles, without any further lowering of interfacial tension. The concentration at which the interfacial tension is a minimum is known as the critical micelle concentration (CMC). It is the concentration at which surfactants have theoretically reached their optimum performance from an application engineering viewpoint.

In accordance with their charge or charge distribution, surfactants are divided into ionic, nonionic and amphoteric types, while the ionic types are further subdivided into anionic and cationic types. Important anionic surfactants include the sulphuric acid-based semi-esters of longchain fatty alcohols and the salts of alkylbenzenesulphonic acids.

Examples of commonly used cationic surfactants are quaternary ammonium bases that contain alkyl groups.

Where ionic surfactants are being used in a formulation with ionically stabilised waterborne paints, the electric charges must be of the same type. Otherwise, substantial stability problems will occur.

As ampholytes, amphoteric surfactants contain both positive and negative charges in addition to a non-polar group. Lecithin, a companion substance in vegetable oils, is important in coating technology.

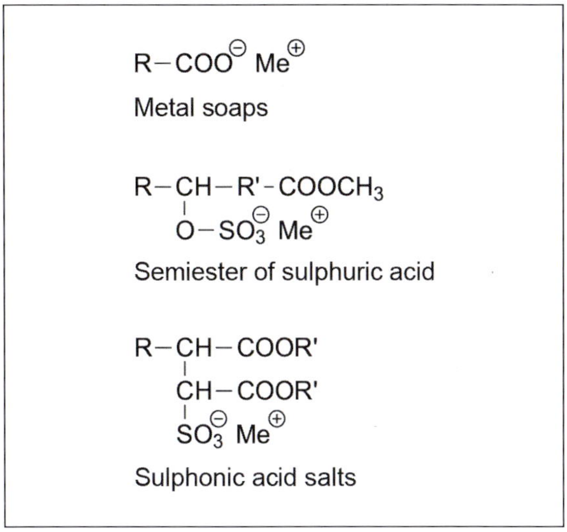

Figure 2.1.208: Structures and examples of anionic surfactants

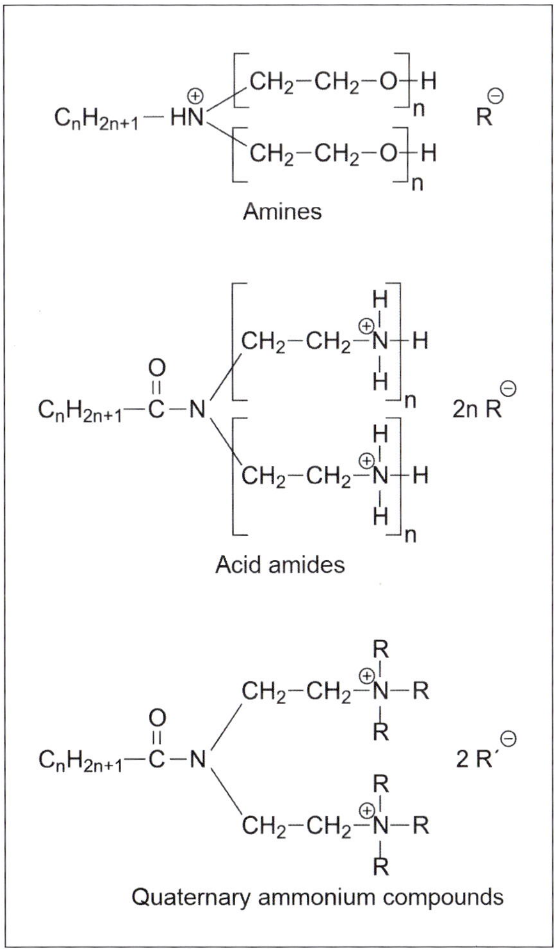

Figure 2.1.209: Structures and examples of cationic surfactants

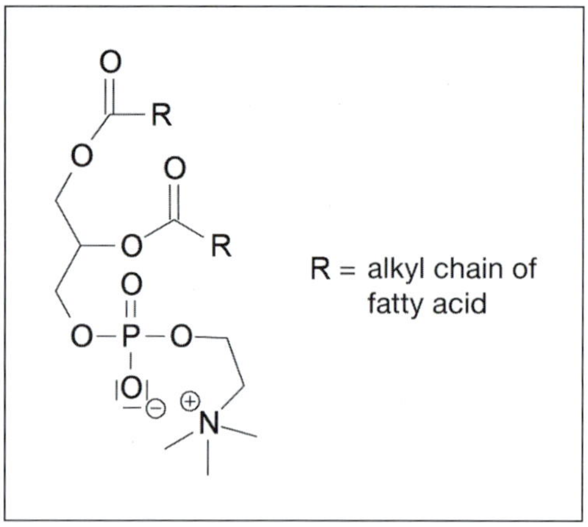

Figure 2.1.210: Chemical structure of lecithin

Lecithins should not be used in the formulation of "air-drying" coatings as they inhibit oxidative curing of alkyd resins.

Nonionic surfactants also have excellent interfacial properties. A distinction must be drawn between the alkylpolyalkylene glycol ethers or ether esters on one hand and butynediol derivatives on the other. The latter orient themselves in the interface with their hydroxyl groups pointing towards the polar side and have a variable space requirement because of the special structure of the non-polar alkyl groups. As a result, there is also no clearly defined critical micelle concentration for them.

The **HLB** value (hydrophilic lipophilic balance) can aid the correct selection of surfactants. It is defined as the ratio of polar groups to the molecular weight, multiplied by 20.

The HLB value varies from 0 to 20 in accordance with the ratio of polar to non-polar groups. Surfactants with low HLB values are compatible with oils and become water-soluble as the values increase. Wetting agents, which act between pigment surfaces and resin solutions, lie in the range 7 to 9. Lower values are targeted for defoaming agents, while higher ones are needed for wash-active substances and solubilisers for water-borne systems.

If the task is to combat **foam** effectively, the causes of the foaming must first be identified. Air bubbles entrapped in liquid coating materials escape relatively quickly as a function

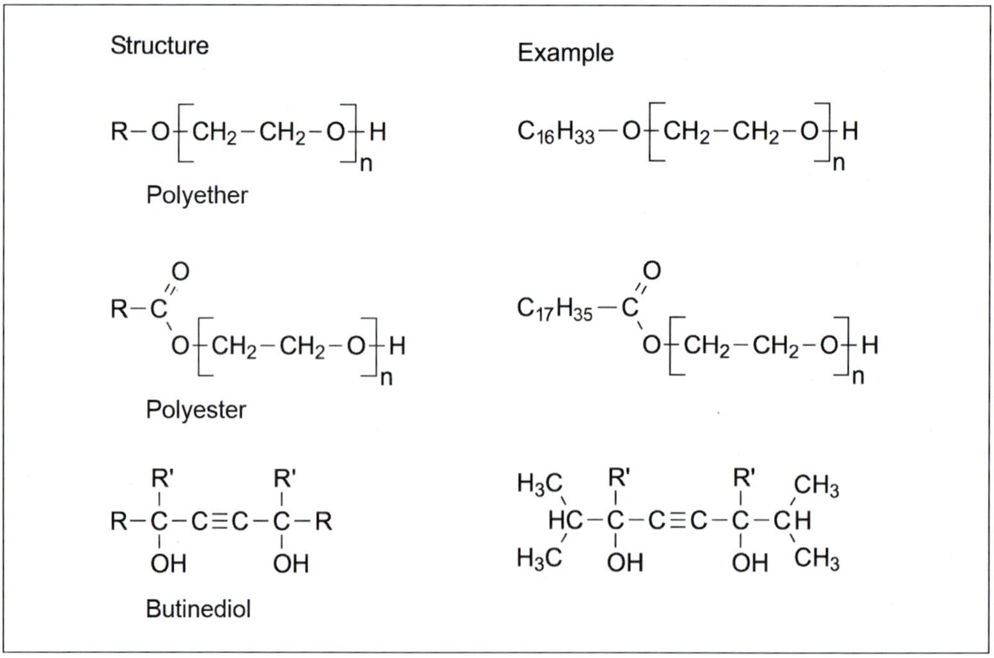

Figure 2.1.211: Typical chemical structures of nonionic surfactants

of their size from surfactant-free, low viscosity liquids and therefore are not a problem. However, at higher viscosities and in the presence of surfactant additives, two aspects of the deaeration process become more difficult. High viscosities slow down the ascent of bubbles while surfactants stabilise them. The first phenomenon causes air to be trapped in the liquid paint in the form of **microfoam**, while the second manifests itself as large **spherical** or **polyhedral bubbles** on the surface. Both phenomena must be combated by various means.

$$\text{HLB} = \frac{\text{mass of polar groups}}{\text{molecular mass}} \cdot 20$$

Figure 2.1.212: HLB value for characterising surfactants

As a bubble of gas rises from a surfactant-free liquid rise to the interface, the liquid lamella surrounding the gas becomes increasingly thinner and, due to the acting surface tension, eventually ruptures, allowing the gas to escape.

In the case of a rising surfactant-stabilised bubble of gas, the physicochemical processes are more complicated. Once this reaches the interface, a membrane which is stabilised on both sides by surfactants forms on the surface of the liquid. Liquid is unable to drain quickly from the membrane, and rupturing of the lamella, which now contains surfactant, does not take place. On account of the mutual repulsion between the surfactant molecules on the lower and upper sides of the thin membrane, the membrane remains around the gas bubble in a finite film thickness. Supplementary stabilisation is achieved in that, if the lamella is stretched, the resultant shortage of surfactant molecules gives rise to an increase in the surface tension, and this is equivalent to an increase in the restoring force. The lamella becomes elastic and stable as a result of this *Gibbs-Marangoni*-effect [2.4.66]. Effective combating of foam therefore requires the use of additives to create inhomogeneities in the surface tension of the foam lamella and thus to induce instability and rupture. Such an effect can only be achieved if the defoaming additives are virtually insoluble in the foaming substance.

The **defoaming agent**, in the form of tiny droplets or solid particles, must therefore enter the foam lamella and then penetrate and destabilise the surfactant film.

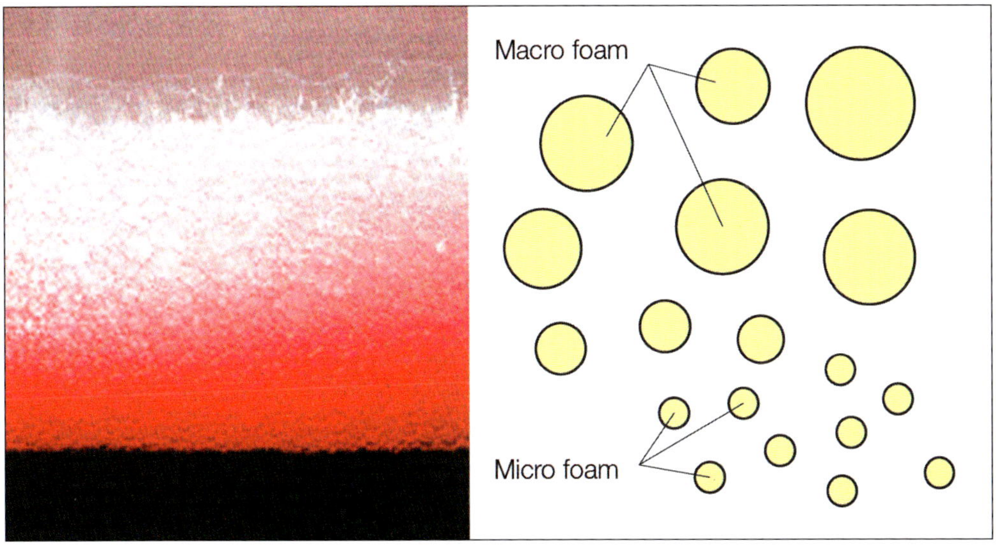

Figure 2.1.213: Surface defects caused by micro- and macro-foaming *Source: Evonik Industries*

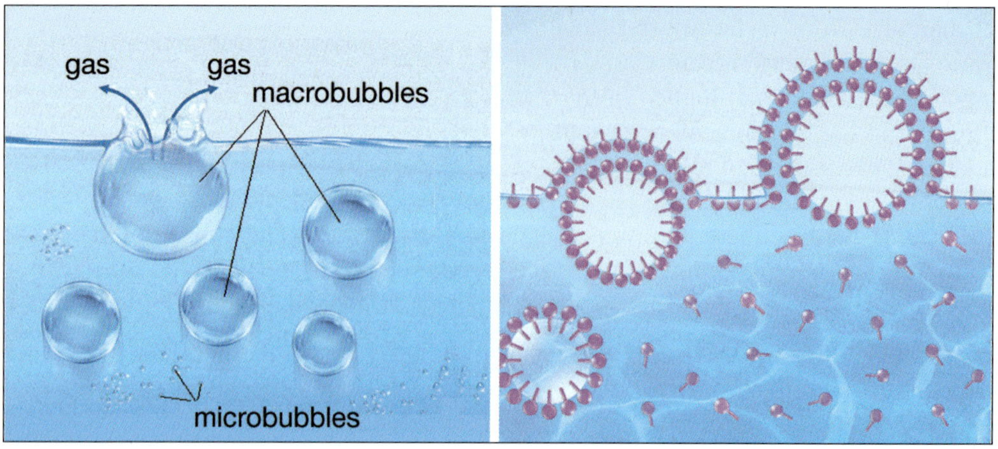

Figure 2.1.214: Foam in liquids free of surfactant (left) and containing surfactant (right) *Source: Krüss*

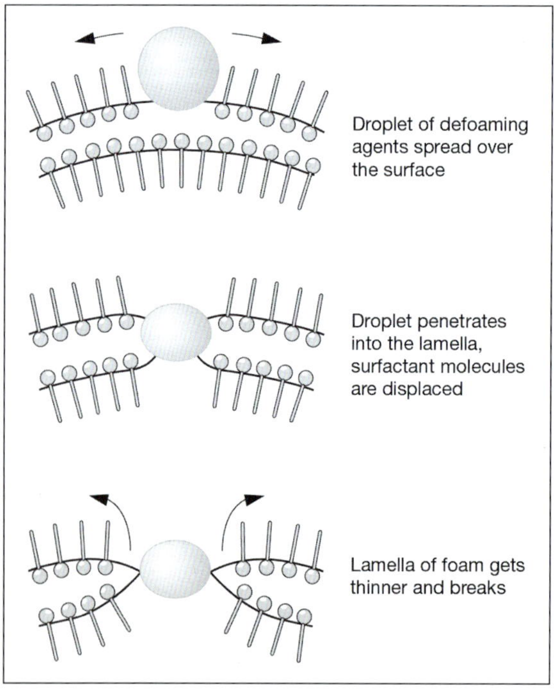

Figure 2.1.215: Mechanism behind defoaming

Mineral oils or polyether-modified polysiloxanes will initiate the spreading effect necessary for destabilising the foam lamella if they are heterogeneously distributed in the interface. Heterogeneous distribution will frequently be enhanced further if they are adsorptively attached to extremely finely distributed SiO_2 particles.

Predictions about the defoaming effect of additives can be made by calculating the spreading coefficient S and the penetration coefficient E (Figures 2.1.216 and 2.1.217). Both variables must be positive in order for defoaming to be induced.

Both the manufacture and processing of coating materials can give rise to extremely fine inclusions. This may be due in the former case to air entrainment during stirring and release of air inclusions from the pigment agglomerates and, in the latter, to air entrapment during application and bubble formation as the coating cures. As small bubbles will have a low rate of ascent, total escape from the liquid phase is not possible. The consequences are air inclusions in the coating or fine craters on its surface. Such film faults should be eradicated by adding **deaerators** to increase the rate at which the gaseous inclusions ascend. Sensitive variables here include the viscosity, the size of the bubble, and surfactant-induced stabilisation of the bubble in the surrounding resin solution. Since the viscosity has a specified magnitude in many cases, the remedial action must concentrate on destabilising and then combining the bubbles with a view to making them larger. As a doubling of bubble

$$S = \sigma_F - \sigma_{EF} - \sigma_F$$
$$E = \sigma_F + \sigma_{EF} - \sigma_E$$

σ_F = surface tension of liquid
σ_{EF} = surface tension of defoaming agent
σ_E = interfacial tension between defoaming agent and liquid

Figure 2.1.216: Equations for calculating the spreading and the penetration coefficients, S and E

size causes a squaring of the rise of ascent, and thus of the deaeration process, additives which cause approaching bubbles to coalesce and thus to become larger prove to be particularly effective. Deaerators, therefore, are additives which, because of their deliberately chosen incompatibility, accumulate in the interfaces of the air bubbles, which they then destabilise. A corresponding effect is eradicating the flow-inhibiting surfactant molecules in the phase interface [2.4.67].

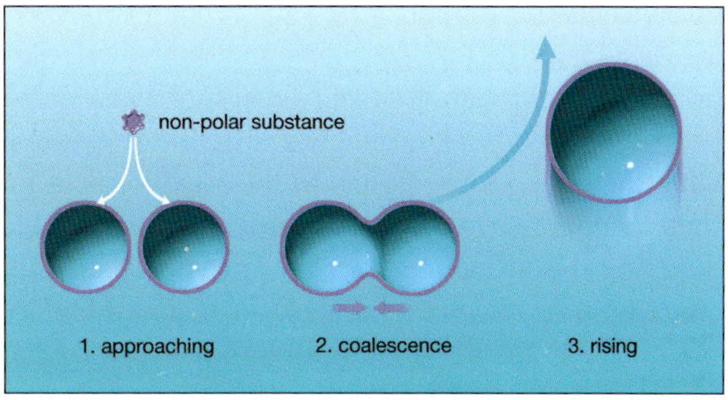

Figure 2.1.217: Mechanism behind defoaming *Source: Krüss*

$$u = k \cdot \frac{\Delta\rho \cdot g}{\eta} \cdot d^2$$

Figure 2.1.218: Equation for calculating the sedimentation rate of pigments

The use of **inhibitors** is necessary in many cases for improving the storage stability of coatings. Their role is to prevent premature crosslinking of the film forming agent combinations. For example, the unsaturated polyesters dissolved in styrene are prevented from crosslinking prematurely by hydroquinone or quinone in exactly the same way as are oxidatively curing alkyd resins by phenol bodies or oximes. In the first case the hydroquinone or quinone molecules inactivate any unwanted starting radicals which may have been generated, and in the second sterically hindered phenol bodies functioning as antioxidants slow down the formation of peroxides by atmospheric oxygen. The oximes' effect is based on the formation of complexes to deactivate siccatives which promote crosslinking; this prevents skinning of decorative paints in the can. Stoving enamels which crosslink with melamine resin are inhibited by the use of monoalcohols to shift the reaction equilibrium towards the starting products.

There are two fundamentally different ways to reduce the tendency of pigments to settle. The rate u at which pigment particles settle in the earth's gravitational field is a function of the difference in density $\Delta\varrho$, the acceleration due to gravity g, the viscosity at rest η and the particle size d. **Sedimentation inhibitors** are additives which have a negative effect on the settling of pigments by stabilising them and increasing the viscosity η at rest. If reagglomeration is not to occur after lengthy storage and thus the particle size is not to change, the only effective approach is to use additives that lead to the formation of intrinsic viscosity and thixotropy. The latter creates a yield point during storage, but does not impair the flow properties under the shear conditions encountered during processing.

Coating Material

Additives for Optimising Processing Properties

Pore-free films with good adhesion are a precondition for the protective function exerted by coatings. Uniform, thorough wetting of the entire substrate is absolutely essential for this. Even the object to be coated has been thoroughly cleaned and pretreated it is often still necessary to add **wetting agents** to the coating material to coordinate the interfacial energy between the substrate and the coating (see Chapter 2.2.3 and Chapter 3.1.1). These are usually surfactant-like substances in which the polar moiety consists of hydrophilic polyether groups, and the non-polar moiety of hydrocarbon chains or methyl-substituted polysiloxanes. Low molecular silicone polyethers have proved adept at effectively adapting water-borne paints to the conditions of the substrate. They are more effective than comparable hydrocarbon-based additives. Fluorosurfactants are highly effective at lowering the surface tension of water-borne coatings to below 20 mN/m. Their fluorine containing chains are kept short to ensure a low PBT (persistent, bioaccumulative and toxic) effect, and are gaining in importance [2.4.68].

As a result of the application process, fresh coating films exhibit a more or less pronounced texture. Before physical or chemical hardening, wavy surface profiles need to be levelled by lowering the peaks and raising the troughs (see Chapter 3.1.1). At the same time action must be taken to prevent the still liquid paint from running off vertical surfaces or, as a result of its surface tension, pulling away from the edges of the object to be coated. Runs impair the coating's appearance and dewetting from edges impairs its function. If no acceptable compromise can be found between levelling and running, even after the solvent composition has been matched to the chemical reactivity of the developing film, the assistance of **levelling additives** must be sought. The cause of the levelling problems must be considered if the right additives are to be selected.

Figure 2.1.219: Effect of wetting agents on coatings

Source: Evonik Industries

Figure 2.1.220: Effect of levelling agents on the surface of coatings

Source: Evonik Industries

On closer analysis of the film forming processes it becomes clear that levelling of the originally textured surface can be counteracted by currents within the film. Solvent evaporation causes inhomogeneities in the film's surface tension and the density. The generally lower surface tension or lower density of the evaporating solvents causes the surface tension and density in the surface of the forming film to rise. If the viscosity is not yet high enough at this point, vertical currents ensue. The consequence is the formation of a texture in the surface which opposes levelling. These film faults, known as *Bénard's* cells, cause separation of the pigments and thus give rise to colour shifts due to differences in mobility of the pigments. Such effects can only be combated with means other than those used for the classic levelling faults described above. The latter must be eliminated by means of viscosity-lowering additives, The formation of *Bénard's* cells, on the other hand, is countered with additives that raise the viscosity and lower the surface tension.

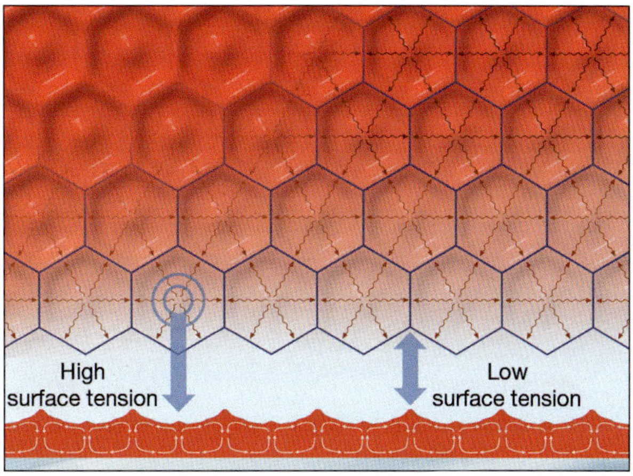

Figure 2.1.221: Bénard's cells: seen from (a) above, (b) the side

Source: Evonik Industries

Since both phenomena occur together in practice, targeted and effective optimisation of the levelling properties is contingent on using the correct dose of additives. Suitable levelling agents are high-boiling solvents (high boilers) to eliminate the classic levelling faults, or silicone oils which have been rendered compatible with the coating by means of side chains to avoid *Bénard's* cells. The latter are similar in chemical composition and effect to the substrate-wetting agents. Consequently, they also promote wetting, while additionally improving surface smoothness and so reducing the drag of coatings. This manifests itself in improved scratch resistance, even though the physical hardness remains unchanged.

Modern additives for increasing the intrinsic viscosity of the film as it forms and thus for avoiding *Bénard's* cells include linear oligomeric polyurethanes, polyacrylates, cellulose ethers, polyvinyl pyrrolidones and special polyethers. Pyrogenic silicas (for example "Aerosil") or montmorillonites (bentonites) have also proved effective.

Other film faults include cavities in the form of **craters** (see Chapter 6.2). These are punctiform depressions in coatings which can sometimes extend as far as the substrate. The reasons for such film faults always lie in a local inhomogeneity in the surface tension of the coating. Incompatible impurities on the substrate, in the applied coating material or in the form of oversprays, dust particles or oil droplets are the commonest causes of craters. Other causes of craters are gel particles from the film forming agents. These are high-molecular fractions of the film forming agent which are at the limit of compatibility and which can cause craters because of their lower polarity and the associated varying surface tension. The negative effect on the surface is frequently amplified by the accumulation of non-polar additives in the gel particles.

In all cases the effect is the same. The paint, which is still liquid after application, cannot flow evenly. It is drawn away from the area of low surface tension and accumulates around it. The result is a cavity which can be either just a blemish or a weakness in the protection afforded to the substrate.

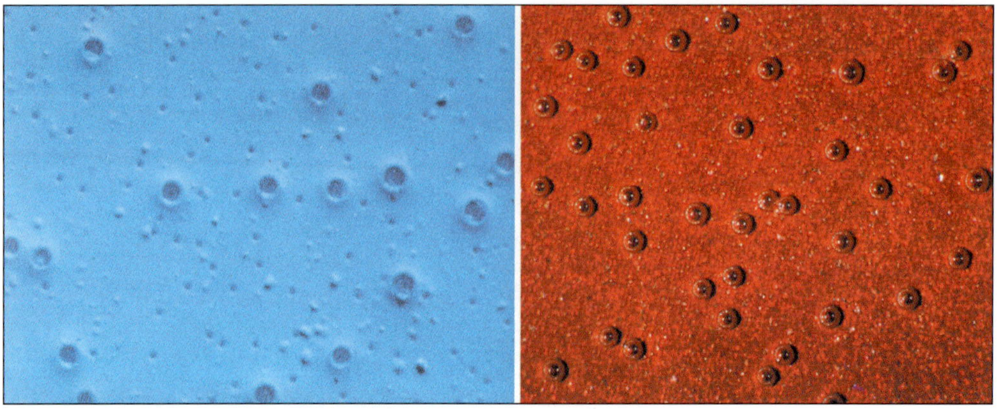

Figure 2.1.222: Typical craters and pinholes in a coating film *Source: Evonik Industries*

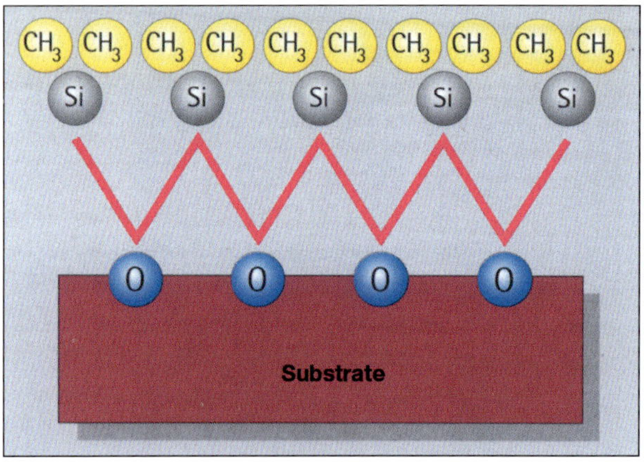

Figure 2.1.223: Schematic diagram showing how polymeric surfaces are rendered water repellent by silicone oils *Source: Evonik Industries*

Before craters are combated with additives it is essential to attempt to identify and eliminate the cause of the craters, if possible (see Chapter 6.2). Gel particles, for example, should be removed from the film forming agent by filtration or centrifuging. If straying overspray is causing craters in the fresh coating, the causes should be eliminated by modifying the surface tensions of the interacting paints. Fluorine-containing surfactants or low molecular silicone oils are the most effective additives because of their low surface tension.

Pinholes are similar in appearance to the film faults caused by escaping air bubbles or by craters (see Figure 2.1.222). These are bubbles which are still intact or have already ruptured and are caused by excessively fast solvent evaporation or by cleavage products of chemical crosslinking. If the resin combinations have been set such that they are too reactive, when the coating is heated the viscosity increases so fast in the early stage that solvents still contained in the film become congested at the polymer skin while they evaporate, as a result of which intact or ruptured bubbles form.

The aim of combating pinholes with additives is always to retain the solvents' ability to diffuse throughout the film forming process. This can be achieved by adding low volatility solvents or by reducing the reactivity of the film forming agents. Longchain alcohols such as dodecyl alcohol have proved successful for most stoving enamels. In industrial coating processes pinholes can also be avoided by making appropriate adjustments to the heating gradient.

Protection of the coated substrate is definitely guaranteed if the coating keeps all harmful agents away from the substrate. Water plays a special role in many chemical reactions which occur at the interface to the substrate. Additives which reduce the influence of water are therefore helpful in many cases. Particular success has been achieved with additives which accumulate on the surface of the coating because of their chemical structure where they attach themselves via their polar molecule groups. The non-polar groups then create a **water repellent state**. Polyolefin waxes, salts of stearic acid, and polysiloxanes make suitable additives.

Apart from modifying wetting properties, levelling and running, preventing surface faults or adjusting the water-repellent finish of organic protective films, additives perform tasks in which they **adapt the material to the type of application**. Adjustments to the electrical conductivity and the pH are two noteworthy examples of how coating additives can accomplish this. The paints' electrical conductivity influences how they are processed by electrostatic methods. Maximum coating yield for solvent-borne paints requires electrical resistances of 10^5 bis 10^8 Ω cm. Such ranges can be set without difficulty, e.g. for stoving enamels dissolved in xylene and butanol, by varying the mixing ratio of the two solvents. If non-polar coatings containing petroleum benzine are being processed, the resistance cannot be adjusted via the solvent compositions alone. The remedy in such cases is to use ionic additives which are compatible with the system. Proven additives include tetraalkylammonium sulphate, ethoxylated ammonium sulphate and monobutyl phosphate.

The stability of aqueous polymer solutions or dispersions is influenced by the pH. It is regulated with acidic or alkaline **neutralising agents**. Amines such as dimethylamine, diethanolamine and 2-methyl-2-aminopropanol are the best known pH regulators for anionic paints. The last of these has become a proven neutralising agent by virtue of its mild odour and high basicity (see Chapter 5.6.2).

The **rheological profile** is adjusted to the processing method for coating materials by adding cellulose ether or urea crystals in a specific precipitated form in addition to the selected solvents [2.4.69]. A far more important role is played by those additives for water-borne paints which influence the rheological properties of the coating material. Such products, which are grouped under the popular terms **"thixotroping agents"** or "thickeners", vary substantially in their nature. The bentonites, i.e. inorganic compounds based on lamellar silicates, are important members of this group (see Chapter 2.1.3). Their effect is based on the construction of tertiary structures by means of dipoles and electrostatic effects [2.4.70]. The platelets are negative on the flat side and positive on the edge. This means that it is possible to build structures which are similar to a house of cards under no-load conditions and which are destroyed by shear. As re-construction is a slow process, thixotropy is the outcome (see Chapter 2.3.2). The bentonites unfold their effect in formulations for water-borne coating materials in weight proportions of 0.2 to 5 %.

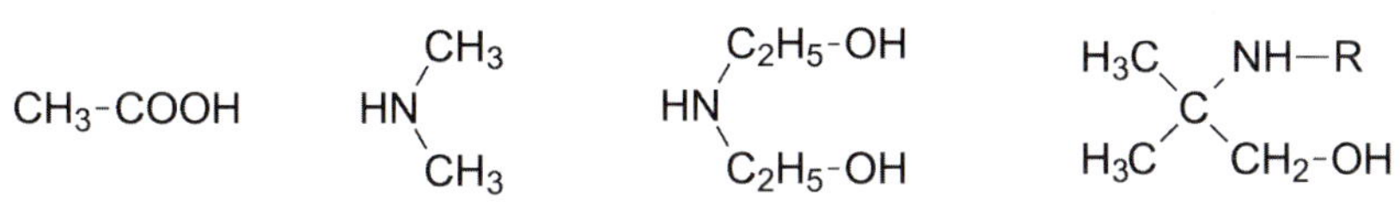

Acetic acid | Dimethylamine | Diethanolamine | 2-Methyl-2-amino-propanol

Figure 2.1.224: Important neutralising agents

A further class of thickeners is represented by the polyacrylic acids. Their effect is based on expanding the molecular coil via electrostatic repulsion of the carboxylate ions. They are effective in the alkaline range and, used in similar concentrations to the bentonites, lead to pseudoplastic formulations at corresponding pH values. The high carboxyl group concentration has an impact on the film properties in that the resistance to water is reduced.

Specifically developed high molecular surfactants, e.g. on the basis of polyethylene oxide with terminal polyurethane groups, are being used with increasing success as **associative thickeners** for aqueous dispersions. They generally feature a very low critical micelle concentration and at concentrations usually in the range 0.05 to 0.5 % lead to inherently viscous behaviour. Their rheological effectiveness can be readily regulated at a molecular level via the polyethylene and polyurethane group [2.4.71].

Additives for Influencing Crosslinking and Film Properties

The mechanical-technological properties of coatings are influenced by **physical drying** and, in the case of industrial coatings, also by **chemical curing**. Wherever thermally unstable substrates are coated and to save process energy, **catalysts** are frequently used to lower the crosslinking temperature. Organic acids such as p-toluenesulphonic acid have proved effective for polycondensation reactions.

Crosslinking of isocyanates can be speeded up by adding tertiary amines, metallic soaps and organometallic compounds. N-Methyl morpholine, diazabicyclooctane (DABCO) and dibutyltin dilaurate are worth mentioning in this context.

In the case of epoxy resins, *Mannich* bases accelerate curing with amines, while boron fluoride activates ring-opening polymerisation of the epoxy groups to afford polyethers. In order to damp the very strong reactivity of boron fluoride, it is common practice to use its amine adducts which are then split again by heating, thereby reactivating the boron fluoride (see Chapter 2.1.1).

The catalysts used for initiating and accelerating polymerisation reactions, termed **initiators**, form a large number of subgroups. Foremost among these are the peroxides and hydroperoxides which have an established role in the radical curing of unsaturated polyesters. Dibenzoyl peroxide, cumene hydroperoxide and tertiary-butyl perbenzoate are but a few examples from the large group of peroxide catalysts. They can be separated into radicals by means of heat or accelerators. The latter are metallic soaps which, when interacting with the peroxides, transform them at room temperature by means of a reversible valency change into radicals that are capable of polymerisation (see Chapter 2.1.1).

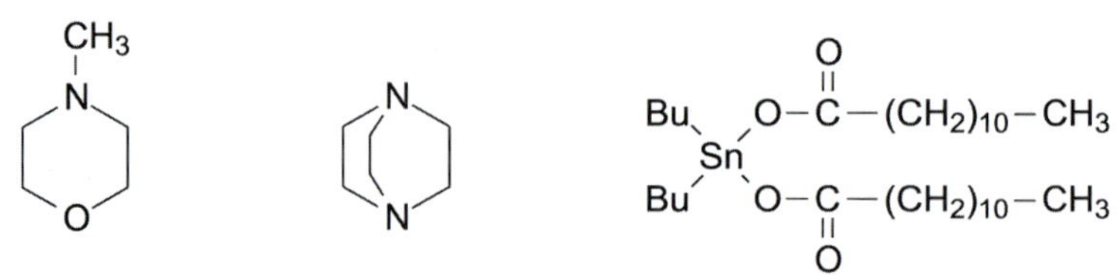

Morpholine | Diazabicyclooctane (DABCO) | Dibutyl tin dilaurate (DBTL)

Figure 2.1.225: Morpholine, diazabicyclooctane, dibutyltin dilaurate as catalysts for hardening polyurethane coatings

Other radical formers are boron trialkyls. They decompose into radicals in the presence of atmospheric oxygen, with formation of trialkylboroxine.

Examples of modern coating materials characterised by extremely short crosslinking times are the UV-curable and electron-beam-curable coating systems. Whereas electron-beam-curable coatings crosslink under the influence of electron beams without the need for additives, film forming agents activated by UV radiation require **sensitisers** or **photoinitiators** (see Chapters 4.3.2 and 5.6.5). These are additives which are split into radicals either homolytically by the high-energy UV radiation or by hydrogen abstraction. The latter case requires synergists to assist radical formation. The synergists are amines which make a hydrogen atom from their alkyl group available for abstraction and radical formation. The first group contains benzoin derivatives, benzil ketals and hydroxycyclohexyl phenyl ketone, and the second benzophenone (see Chapter 2.1.1).

2,4,6-Trimethylbenzoyl phosphine oxide (TPO) is of particular technical interest because it acts in the longwave UV range. TPO enables even coatings opaquely pigmented with titanium dioxide to be cured by UV radiation. However, appropriate longwave emitters must be used for this (see Chapter 4.3.2).

Examples of chemical crosslinking by ionic polymerisation are rare in coating technology. The ring-opening polymerisation of oxazolines to polyamides by means of methyl tosylate and the UV-initiated polymerisation of epoxy resins by catalysis with ferrocene-complexes or sulphonium salts (AB) are the most important reactions to mention in this context.

A further group of crosslinking catalysts for accelerated curing is that of the **siccatives** which are used to oxidatively cure alkyd resins. They are chemically identical with the accelerators for unsaturated polyesters. The peroxides required for crosslinking the unsaturated polyesters can be dispensed with when alkyd resins are cured because they are formed in the unsaturated side chains of the alkyd resins by the atmospheric oxygen during film formation (see Chapter 2.1.1).

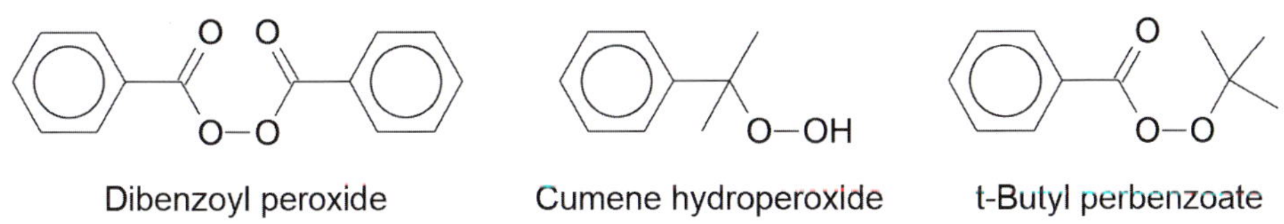

Figure 2.1.226: Peroxides and hydroperoxides as initiators for the radical polymerisation reaction

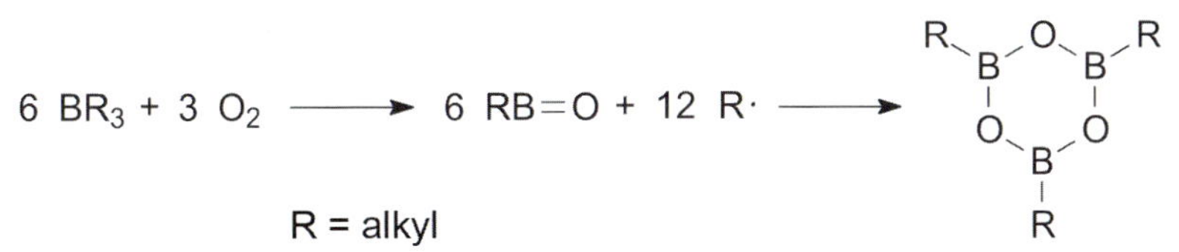

Figure 2.1.227: Decomposition of boron trialkyls in the presence of atmospheric oxygen

The most important siccatives are cobalt, manganese, zinc and zirconium soaps. The more hydrophobic anions which therefore promote solubility for the siccative are octoates and naphthenates or, more rarely, abietates.

Additives also help to improve **film properties** such as the visual and mechanical-technological properties, thermal resistance and long-term outdoor exposure.

Visual properties are those which cause a sensory stimulus in the viewer after light has interacted with a coating. The physical effects responsible for this can be divided into two categories. The first is **gloss** caused by the reflection of light on the surface and the second is the **colour**, caused by absorption, scattering or interference after the light has interacted with the coloured pigments and **effect** materials inside the coating (see Chapter 3.2.4). The task of gloss-influencing additives is either to smooth the film surface in order to increase the gloss or to create microstructures to reduce the gloss. Increasing the gloss by means of additives with a specified film forming agent and pigment mixture

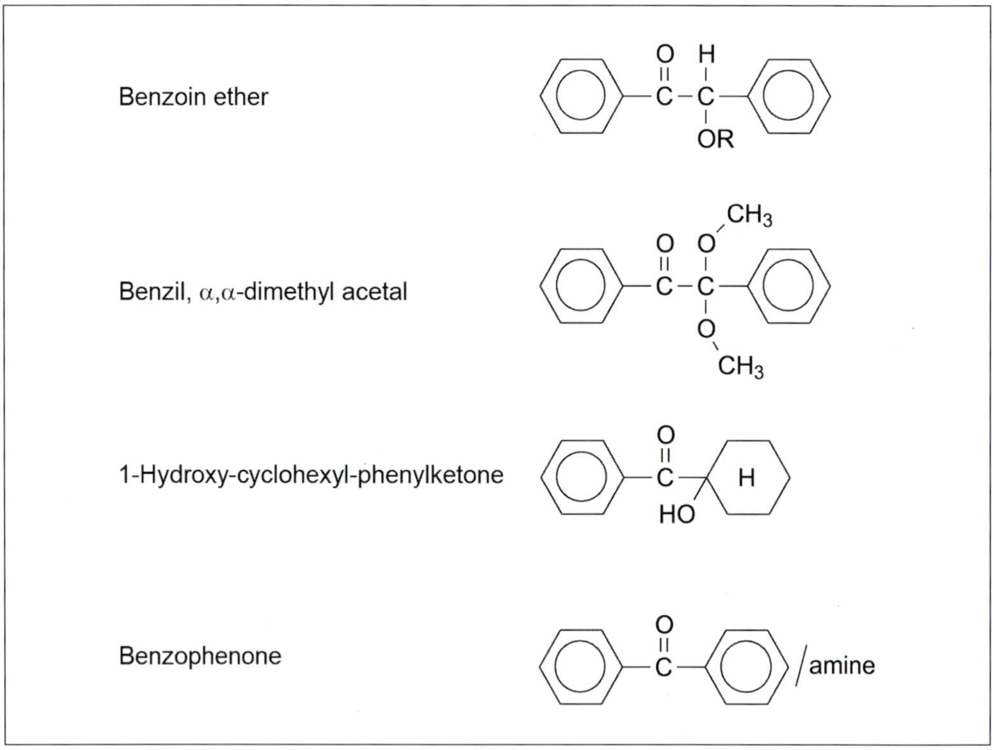

Figure 2.1.228: Photoinitiators for activating crosslinking of UV-curable coatings

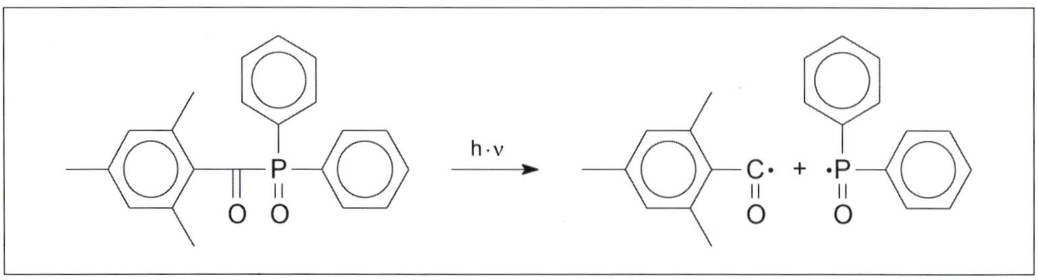

Figure 2.1.229: Splitting of trimethylbenzoyl phosphine oxide (TPO) into radicals

Coating Material

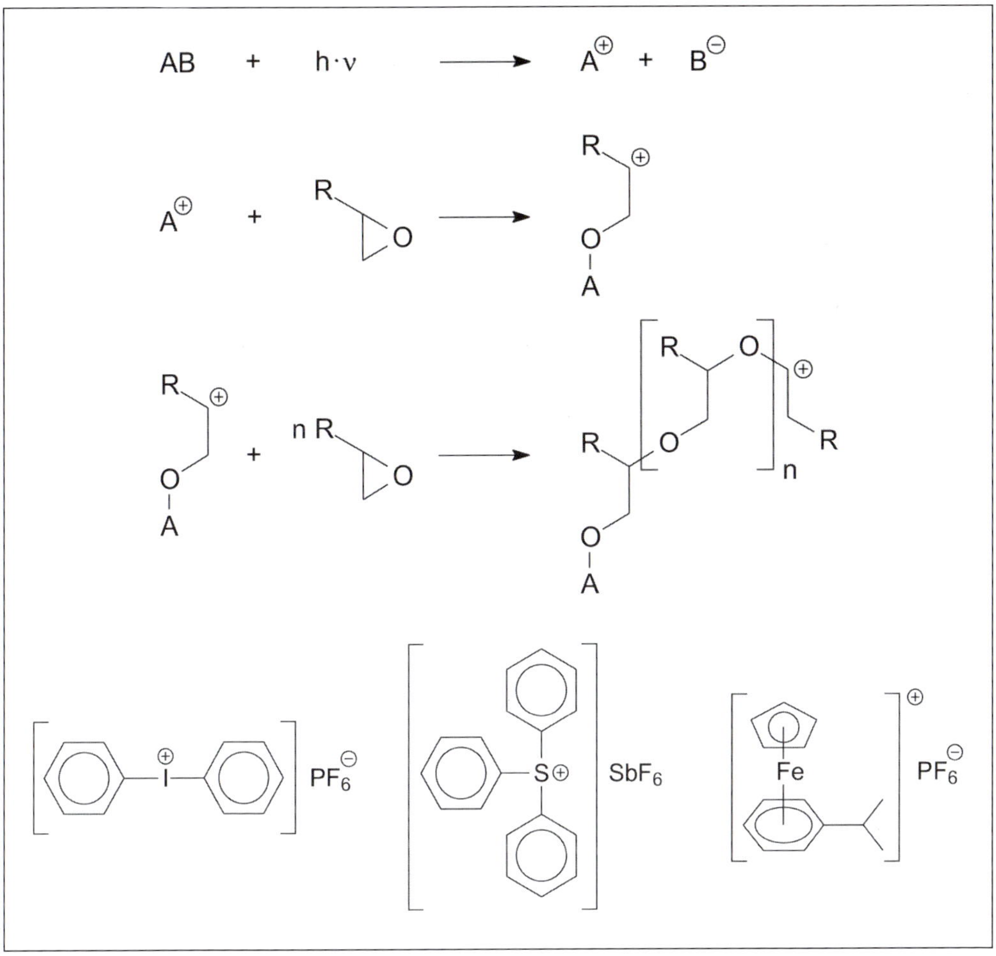

Figure 2.1.230: Mechanism behind UV curing of epoxy resins and examples of initiators on the basis f (a) iodonium, (b) sulphonium and (c) ferrocenium salts

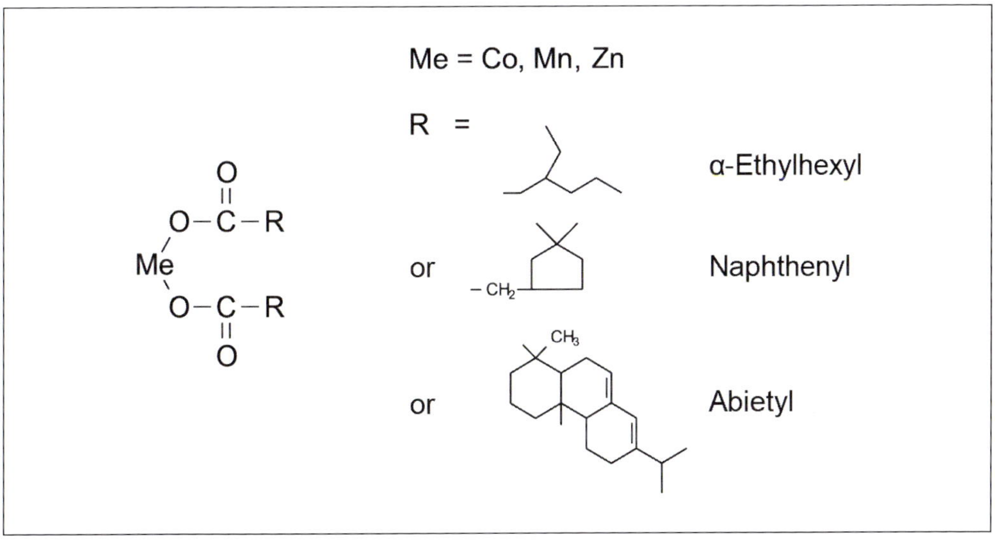

Figure 2.1.231: Siccatives for hardening alkyd resins

is only possible via the degree of dispersion and by means of levelling agents. A reduction in gloss is achieved by means of micronized extenders, such as pyrogenic silicas, floating wax dispersions, micronized polyethylene, natural waxes and aluminium or zinc soaps.

Other additives improve the alignment of the platelet-like effect substances either, for example, directly by means of polyethylene wax dispersions or indirectly by influencing the flow properties during **formation of the effect**. The viscosity, which is still low when alignment of the platelet-like pigments starts, must subsequently be raised as quickly as possible by means of thixotroping additives. Additives already mentioned above in connection with combating floating effects in the form of *Bénard's* cells can also be used successfully in this context.

The role of **optical brighteners** is to transform UV light which is invisible into longer wavelengths and thus into visible light by means of fluorescence. The reduction in yellowing achieved through increased reflection in the blue range and the high conspicuity of pigmentations used on emergency services vehicles are attributable to fluorescence effects. In accordance with the task in hand, either stilbene derivatives, benzimidazoles or pyrazoline derivatives are mixed with the coating material to serve as additives or special fluorescing and thus optically brightening pigments.

Manufacturers wishing to achieve very special film properties cannot do without additives. Flammable objects made of plastic and wood are coated with a flame retardent material for some applications. Although chlorine-containing film forming agents are flame retardent, they form toxic products at high temperatures due to chemical decomposition. **Flameproofing agents** in the form of bromine-containing additives are in use today whose task is to react and combine with the radicals released during combustion and so halt the exothermic chain reaction of combustion. They are also used in plastics [2.4.72]. The most important formulations are intumescent coatings, which represent the optimum combination of attractive surface and effect [2.4.73] (see Chapter 7.11). Brominated cycloaliphatic and aromatic hydrocarbons are flame retardants which work synergistically with borates or phosphites. The dense smoke generated by the volatile reaction products cuts off the supply of oxygen to the fire. Another effective flameproofing method is achieved by the corresponding effect of organic and inorganic extenders. Hydrated aluminium oxides in conjunction with compounds that form nitrogen and carbon dioxide have proved particularly effective when residual carbon remains in the form of expanded crusts under the thermal load. These compounds include urea, methylol urea, melamine, and ammonium hydrogencarbonate which decomposes entirely to gases.

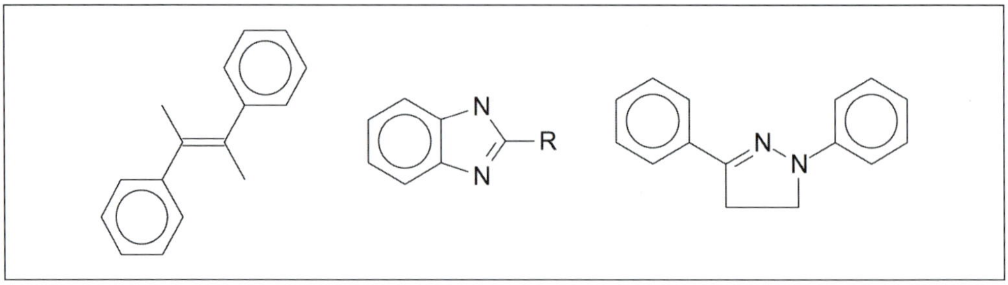

Figure 2.1.232: Derivatives of stilbene (on the left), benzimidazole (in the middle) and pyrazoline (on the right hand side) as optical brighteners

A particularly important task of coatings is to protect the substrate against the harmful influence of UV radiation. The latter is present in the spectrum of sunlight at wavelength from 295 to 400 nm. Because of its high energy content it is capable of degrading macromolecular materials such as wood, textiles, plastic, and coatings by radical formation. Some of the radicals form new bonds. In conjunction with the shortened chains this means a loss of mechanical strength combined with a reduction in elasticity. Poor cohesion and brittleness are the consequences. Supplementary changes in the colour of the substrate, such as in wood, are also attributable to photochemical reactions (see Chapter 3.3.1). Such damage only occurs in the substrate if the high energy radiation is able to penetrate through the coating as far as the substrate. This is only the case, however, if the pigments or film forming agents are incapable of UV absorption. For this reason, the use of **UV absorbers**, eminently for clearcoats, is sensible and necessary.

Although their protective effect increases with the depth of penetration of the UV radiation into the coating because of the increasing probability of interaction, there is less protection on the surface. Effective protection against light is therefore only possible if the UV absorbers are also supplemented by radical scavengers to protect the surface.

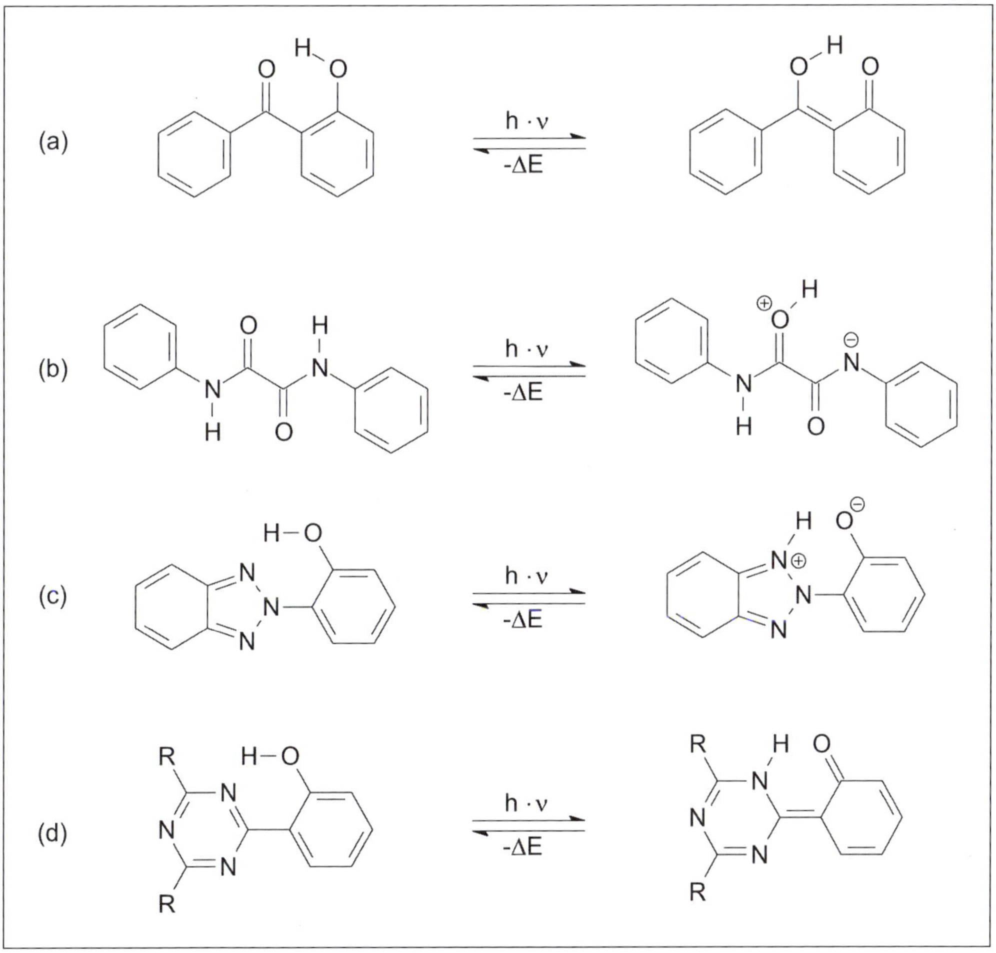

Figure 2.1.233: Absorption of UV light by mesomeric structures of the important light stabilisers a) hydroxybenzophenone, b) oxalanilide, c) benzotriazole und d) triazine

These transform the radicals formed by the influence of light into stable products. High quality clearcoats therefore always contain combinations of radical scavengers and UV absorbers to ensure long-term protection.

A few inorganic substances, such as micronized iron oxides, ferrocene derivatives and organic nickel complexes make effective and, at the same time, chemically stable UV absorbers. However, all these active substances cannot be used in most cases as they have an inherent colour.

Where colourless and, at the same time, colour stable light stabilisers are demanded, (a) hydroxybenzophenone, (b) oxalanilides, (c) benzotriazoles and (d) triazines have proven track record (see Figure 2.1.233).

Their effect is based on the ability to induce mesomeric structural changes which are reversible by UV light. The different energy levels of the interface structures lead to selective absorption of UV light, with simultaneous transformation of the absorbed energy into heat and restoration of the UV absorber to its default state.

The UV absorbers presented in Figure 2.1.233 are merely basic substances. To reduce their volatility they must be provided with larger functional groups or be chemically bound to the molecules of the polymeric matrix. Particularly in the case of 2-coat effect paints, it is recommended that UV absorbers even be added to the base coat because of the tendency to migrate throughout the layers.

Hindered amine light stabilisers (HALS) are used as radical scavengers. In the presence of atmospheric oxygen they are transformed into stable nitrosyl radicals by high-energy radiation. These are capable, according to *Denisov*, of trapping the radicals which destroy the film and of permanently inactivating them by later combining them with other radicals [2.4.74].

Coordinated combinations of HALS enable 2-coat metallic coatings to achieve colour stability, gloss retention and crack resistance of more than 10 years.

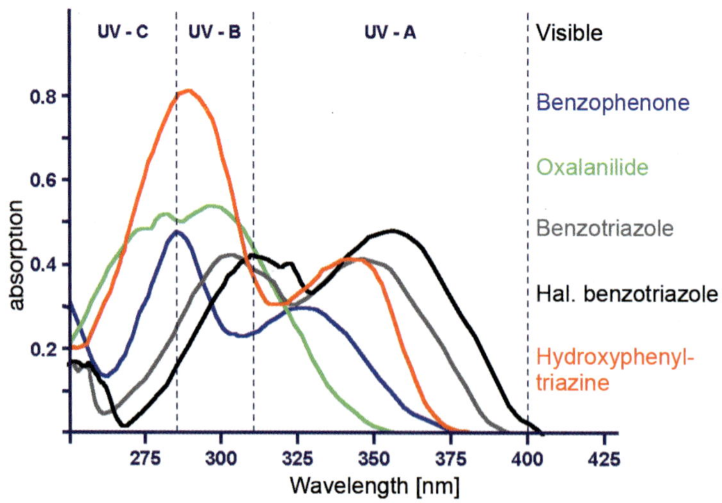

Figure 2.1.234: Absorption spectra of the common UV absorbers *Source: Clariant*

Paints and coatings, being organic substances, can be attacked and sometimes wholly degraded by micro-organisms in the proper environment (warmth, moisture). **Preservatives** are intended to help prevent the destructive actions of bacteria, moulds, and algae. Bactericidal, fungicidal or algicidal additives serve to protect not only the paint and the coating, but also the substrate where this may be biodegradable, such as wood. When biocidal additives are being selected, therefore, it is important to distinguish between **preservation during storage, protection of the coating itself and protection of the substrate**. The protective agents should be chosen with the intended application in mind and be used in the correct dosage. Low-solvent water-borne paints that are free of heavy metals pose particular problems in this context. They offer the microorganisms the best growth conditions of all coating materials. Plant and vessels for the manufacture and storage of aqueous dispersions are sterilised by regular flushing with 0.5 % formaldehyde solution or the addition of barium metaborate as a preservative. A 10 % solution of butyl glycol in water is frequently used for cleaning and sterilising.

An excellent long-term effect combined with a balanced influence on moulds and bacteria is achieved by adding chloromethyl isothiazolinone (CMI), methyl isothiazolinone, loosely bound formaldehyde in the form of N-methylol chloroacetamide, tetramethylol acetylene diurea (tetramethylol glycoluril) and nanoscale silver as depot substances for the coating materials [2.4.75].

Coatings manufacturers have removed the highly effective organo-mercury compounds and chlorophenols from their range of raw materials because of their potential

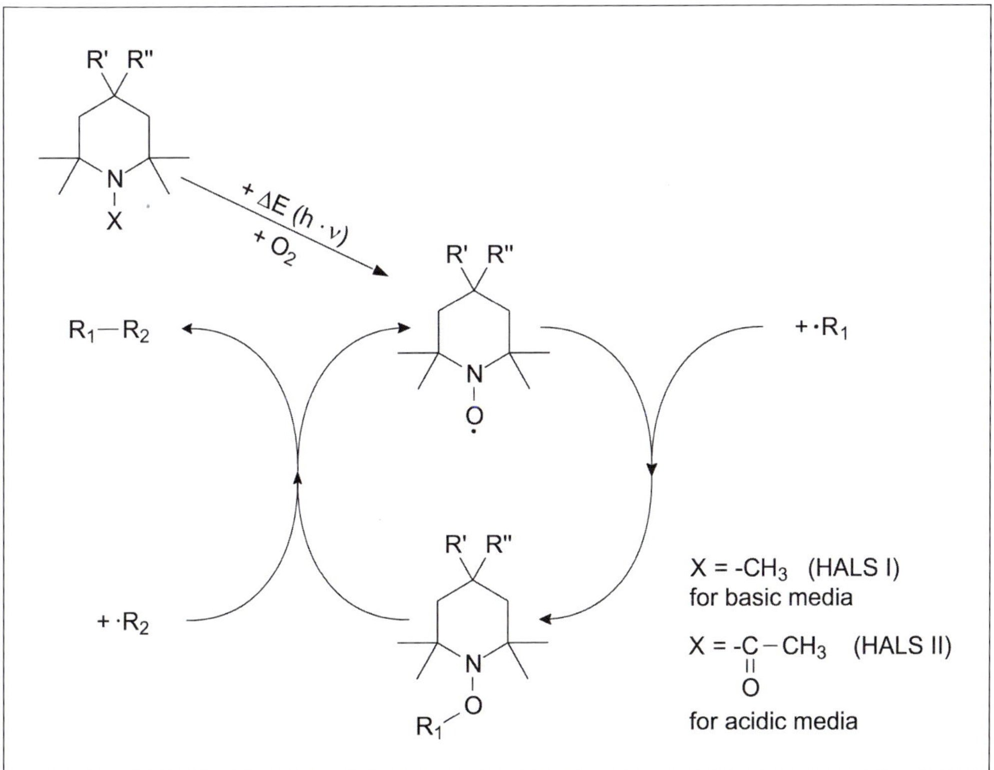

Figure 2.1.235: Mechanism of radical scavenging by hindered amine light stabilisers (HALS) according to Denisov

physiological harmfulness. Organo-tin compounds will also diminish in importance in the future on account of such labelling and the associated legislative restrictions.

Recent developments are following two lines of approach with regard to toxicological aspects. One consists in impregnating crystalline carriers with biocidal metal ions. This quasi ion-exchange process allows the dosage to be precisely controlled so that hazards to human health can be ruled out. Zeolite impregnated with silver ions, for example, have proved very successful. The other, a completely different means of achieving effective preservation which is also safe for humans, is the development of biocidal substances which target the micro-organisms' specific metabolism. The thiocarbamates and compounds containing halogenated methyl thioether groups (R-S-CX_3) are worth mentioning. The latter work by opening the R-S bond and creating new bonds with the micro-organisms' protein molecules.

Apart from serving as preservatives for paints, biocides are also used to help maintain the durability of coatings, e.g. by protecting facades against algal colonisation. Anti-fouling paints can also contain biocides in various forms for preventing the growth of barnacles and algae (see Chapter 7.11).

The EU Biocidal Products Directive (BPD) and the current regulation passed in 2012 seek to reclassify the impact of biocides on the environment and humans in the context of a product and application evaluation and to eliminate critical products (see Chapter 5.4) [2.4.76].

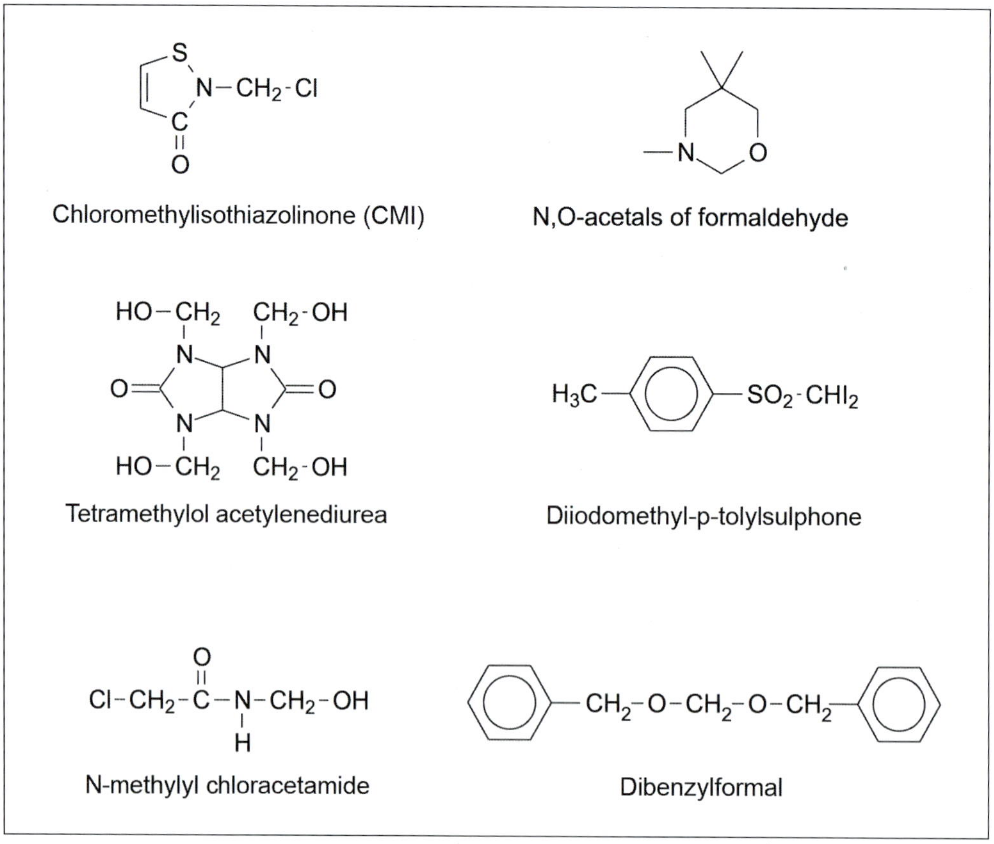

Figure 2.1.236: Biocides for increasing the shelf life of paints

2.1.5 Nanoproducts

2.1.5.1 Description and Definition

Just a few decades ago, it was customary to characterise and classify the individual ingredients for coating materials according to their chemical nature. In the meanwhile, a complementary classification principle which is largely independent of the chemical composition has become established.

This ordering principle which is relevant not only to coating technology takes less account of the chemical properties than of the effective surface interacting with the environment. Since this is directly linked to the particle size and shape, the characterisation of the interaction for the purpose of quantitatively describing the properties is supplemented by morphological aspects.

As particle size decreases, the interacting surfaces and interface properties grow more important. The particles or surface structures of interest to coating technology are in the order of magnitude of 50 to 300 nanometres, which is significantly greater than the dimensions of molecules, but still small enough for some of the usual material properties to be lost. Such particle-size-dependent peculiarities do not only apply to isometric particles. Textured surfaces exhibit a similar behaviour when the structural features are of the same order of magnitude. In addition to 3-dimensional particles, quasi 2-dimensional nano-leaves and 1-dimensional nano-fibres are equally of interest. The range between 50 and 300 nanometres is referred to as the tolerance range in which surface properties with increasing quantum-physical effects dominate volume properties. The European Commission's definition of nanoparticles goes down even further, defining the range as being 1 to 100 nm [2.4.77].

The definition of the BMBF (Federal Ministry of Education and Research) states: nanotechnology describes the production, study and application of structures, molecular materials, internal interfaces and surfaces that have at least one critical dimension <100 nm. The nanoscale dimensions of the system components alone lead to new functionalities and properties for improving existing or new products and application areas [2.4.78]

In the 1970s, *Norio Taniguchi* coined the term nanotechnology for dealing with such small particles. Derived from the Greek word **nanos**, or dwarf, it alludes to the extremely small particle dimensions involved. The term nanotechnology, as applied in coating technology, is interpreted differently from *E. Drexler's* identical term for a branch of technology which refers exclusively to the construction of miniaturised machines. That term primarily concerns tiny devices, motors and sensors for studying and controlling the microworld in organisms.

In the coatings world, nanotechnology in the sense presented above is really a new term for what is largely an old concept. Given that the dust all around us is nanoscale in its dimensions, it becomes clear that nanotechnology is as old as the universe itself and not just an invention of recent decades.

Thus, more than 100 years ago, *W. Ostwald* described a state of matter, termed colloidal, which belongs to the realms of the nano world in view of the size of particles in colloids. He described colloids back then as "the world of neglected dimensions".

Thus, much of what had been happening in the coatings world for a considerable time would very much later come to be recognised as being nano-specific. Examples are organic pigments, certain extenders and additives, as well as the finely divided polymers in a paint dispersion.

The need to stabilise pigments in the liquid matrix of polymeric resin solutions is indicative of the interactions occurring at the nanoscale level. Nanoscale means that due to the size of the particles or to surface structures at levels bordering on the dimensions of individual molecules *zwitters* are created which combine the properties of individual molecules and those of macroscopic crystalline or amorphous solids, thereby giving rise to new performance properties which are less material-dependent and more size-dependent.

One of many examples of the zwitter-like behaviour of nanoparticles manifests itself in micronized titanium dioxide. While ensuring particle-specific immobilisation within the polymeric environment of a coating, micronisation leads to an enormous increase in surface area and thus to increased absorption of UV light, but without the light scattering that typically occurs with larger particles. It is thus possible to provide permanent light protection for clearcoats without impairing their transparency.

Nanoparticles are chameleons which fulfil their tasks by combining properties which are compatible in this discrete size range only. Further examples are the influencing and selective adjusting of the rheological properties of coating materials during storage, application and film formation by means of pyrogenic silicas and other nanoparticles that act in a similar manner.

2.1.5.2 Synthesis of Nanoproducts

In principle, nanoparticles can be produced by **comminution** (grinding or emulsifying), by **growth** (crystallisation) or by **sol-gel technology**. The first case calls for laborious comminution techniques on account of the particles' small dimensions. The increase in surface energy that accompanies the decrease in particle size leads to the formation of agglomerates and aggregates, the extent of which is a function of the polarity of the substance to be comminuted, with the result that subsequent or simultaneous stabilisation is necessary for the optimal formation of the particle-size-dependent properties. This stabilisation can be electrical, steric or entropic in nature (see also Chapter 2.2).

Other ways of forming particles are hydrolysis in a liquid environment and combustion in the gas phase. These are the techniques by which, for example, tiny SiO_2 particles or finely divided carbon blacks are produced.

The opposite approach to particle formation is crystallisation or precipitation from suitable solutions. For example, the extender "blanc fixe" is converted into an extremely finely divided precipitate of $BaSO_4$ by precipitation from a $BaCl_2$ solution. The concentration and the precipitation temperature are used to adjust the fineness down to the nanoscale.

Precipitation from solution also affords a way of obtaining amorphous gel-like dispersions or emulsions. It is known that compact nanoparticles are formed from gel-like hydroxides by hydrolysis of certain salts followed by drying. A further widely known example of the production of nanoparticles by hydrolysis is the reaction of $TiCl_4$ in aqueous phase (TiO_2 production by the chloride method). Another established method is the production of ultra-fine pyrogenic silica by gas-phase hydrolysis from $SiCl_4$.

A similar route to nanoparticles is afforded by gel-sol technology. The property of water-soluble silicates to transform themselves into colloidal silica in the presence of acids has long been known. Elimination of water from the orthosilicic acid leads to the formation of higher molecular weight nuclei which consist of insoluble SiO_2 and which, termed "water glass", grow to the size of nanoparticles, and will form stable sols when the pH is raised.

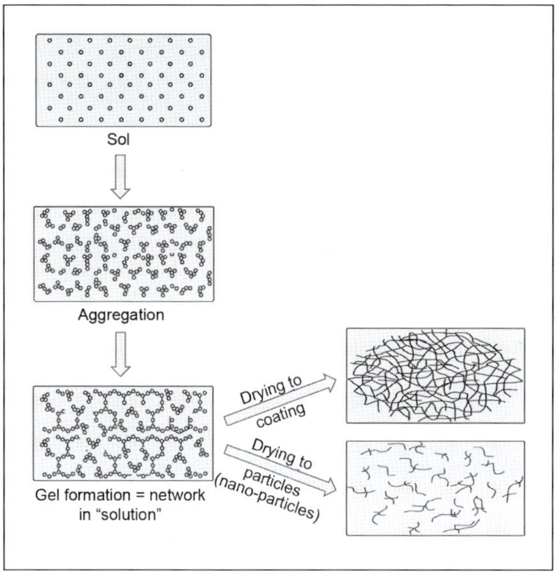

Figure 2.1.237: Principle behind the production of nanoparticles by sol-gel technology [2.4.79]

Colloidal silicas can be converted into solid films by drying in the presence of CO_2, with formation of SiO_2 and corresponding carbonates. Because of their brittleness, gas permeability and extremely poor stability to atmospheric influences, their sole use is as silicate paints in masonry protection. Improvements from the point of view of coating technology are obtained by replacing inorganic silicates with orthosilicates. With tetraethyl orthosilicate as the starting product, highly branched polyanhydrides are produced by partial hydrolysis and subsequent polycondensation.

$$Si(OR)_4 + H_2O \rightarrow HO\text{-}Si(OR)_3 + ROH$$

$$(RO)_3Si\text{-}OH + HO\text{-}Si(OR)_3 \rightarrow (RO)_3Si\text{-}O\text{-}Si(OR)_3$$

etc. until condensation is complete.

Like the water glasses, inorganic polysiloxanes synthesised in this way are unsuitable for general purpose use as film formers due to a lack of stability and elasticity. It was only with the direct bonding of an alkyl group to the central silicon atom, combined with the incorporation of reactive groups at the end of the carbon chain that the breakthrough was made to products which could be used in paint technology.

```
      R
      |
      O
      |
RO-Si-(CH2)n-R'
      O
      |
      R
```

where R′= -H, -Phenyl, $-NH_2$, $-O-CH_2-CH-CH_2$ (epoxide: CH and CH_2 bridged by O), $-OCO-C(R'')=CH_2$

where R″ = -H, $-CH_3$

The saturated aliphatic side chain confers elasticity combined with water repellency while the reactive end groups offer scope for subsequent chemical bonds to conventional coating resins. Additional crosslinking by UV radiation is possible via the double bonds of a terminal acrylic acid in ester form.

The organic/inorganic **hybrid polymers** which are accessible in this way serve as the building blocks for polycondensation leading to the formation of sols or gels. These contain alcohol (organosol) or water (hydrosol) by way of solvent. In solution, the primary particles grow by aggregation, thereby assuming particle sizes in the nano range. For further processing, the "solutions" now present as a gel are either converted into solid nanoparticles or ready-made coating materials. The large number of film forming agents accessible via sol-gel technology can be further extended by replacing the central silicon atom with aluminium, titanium or zirconium.

Hybrid formation with organic and inorganic components boosts the coating properties considerably. The lower network density reduces the brittleness while improving the elasticity and adhesion to the substrate.

The wide range of nanoscale paint raw materials available today offers plenty of scope for devising tailor-made solutions for all kinds of surface problems. Increased scratch resistance, self-cleaning surfaces, anti-fogging properties and lasting UV protection are just some of these.

2.1.5.3 Sample Applications

Rheological Properties

Seemingly contradictory requirements are often imposed on coating materials, such as a reduction in the tendency of pigments to settle during storage on one hand and optimisation of the rheological properties during processing and film formation on the other. High viscosity and, if possible, a yield point are needed during storage, while extremely low viscosity under shear is desirable during processing, and good levelling must not lead to runs on application to vertical surfaces. It is not possible to achieve all this satisfactorily with film forming solutions on their own. Frequently the only way to meet these paradoxical requirements is to employ additives drawn from the large group of nanoproducts. Pyrogenic silicas (aerosils) and bentonites as prepared layer silicates which have at least one dimension in the nano range are suitable for adjusting the rheological properties due to their ability to reversibly build up superstructures (see also Chapters 2.1.3 and 2.1.4).

Mechanical and Technological Properties

Mechanical and technological properties such as elasticity, hardness and adhesion, which are determined by the polymeric matrix of conventional coating materials, can be improved by modification with nanobased raw materials. To this end, high crosslinking densities are used to create more cohesive systems with hard domains in the matrix by means of inorganic nanoparticles or hybrid polymers. This can be achieved either by mixing alone with the hard nanoscale silicon polycondensates or by additionally bonding the chemically reactive hybrid variants to the other film forming agents.

The mix of hard fillers based on silicates and the viscoelastic network of the film forming agent provides a combination of high elasticity with hardness and/or internal strength. Given proper embedding or chemical bonding, the last of these brings about improvements in scratch, abrasion and scrubbing resistance, without impairment of gloss or film build.

When the performance properties are being adapted to the practical requirements a distinction must be made between scratch resistance and abrasion resistance. There are huge differences between these two different types of stress. While scratches are caused by only a few impacts involving high mechanical stress, abrasion is characterised by frequent, but usually only slight, mechanical action. This means that different mechanical-technological properties need to be optimised when the quality is being adjusted. For good abrasion resistance, surface properties such as slip and microstructures are more important than high hardness or internal strength (cohesion), which however have to be given priority in the case of scratch resistant coatings. Different approaches based on the application of nanoparticles have been proposed or are already being implemented in the products (see Chapter 3.2.5.4). They are based on the generation of very thin, hard surfaces by floating of the nanoparticles, on anisotropic dosing in the film, and on the production of hard matrix segments by silane groups in the polymeric film network [2.4.80, 2.4.81].

Dirt Adhesion, Cleaning Effects, Anti-Fogging and Anti-Reflection Coatings

The search for simple, effective cleaning of soiled surfaces is far older than any discussion about the opportunities and limitations of nanotechnology. As far back as the 1970s one of the declared objectives of research was the development of automotive paints which would allow cleaning with little effort and without mechanical stress on the coating. However, a coating which shakes off dirt simply by means of engine vibrations has still not been developed to this day.

The situation is different for other objects, e.g. in the sanitary and hygiene fields. Dirt- and water-repellent specialty products already abound here. **Easy-to-clean** is a catch-all term to summarise all surface finishes whose low surface tension inhibits wetting, rendering them water repellent and dirt repellent to water-borne agents (wetting: see Chapter 2.3.2.3)

In the meantime, nanoscale polysiloxane hydrophobing agents for lowering the surface energy are now available. Hydroxy-functional alkoxysilanes bearing non-polar, optionally fluorinated organic groups bond to the polar groups of the substrate via the hydroxyl groups. The extremely thin layers may be water-repellent, but have only a limited lifetime. Thicker layers with less interaction with dirt can be obtained by using low-energy additives in conventional coating materials. However, these need to penetrate the interface during film formation so that they can unfold their effect without impairing the bond to the substrate at the same time.

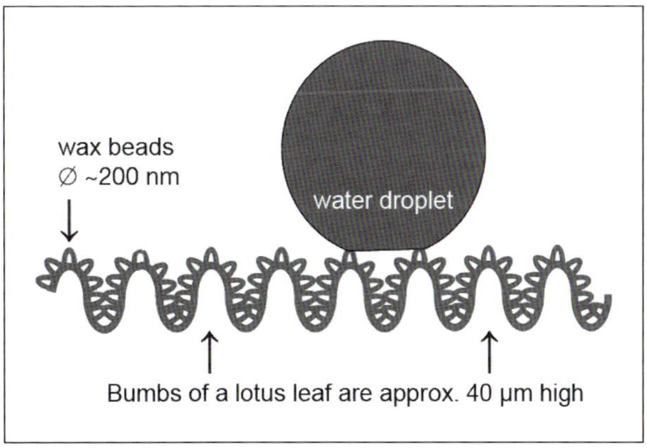

Figure 2.1.238: Dirt-repellent surfaces of the lotus plant (lotus effect) *Source: S. Sepeur*

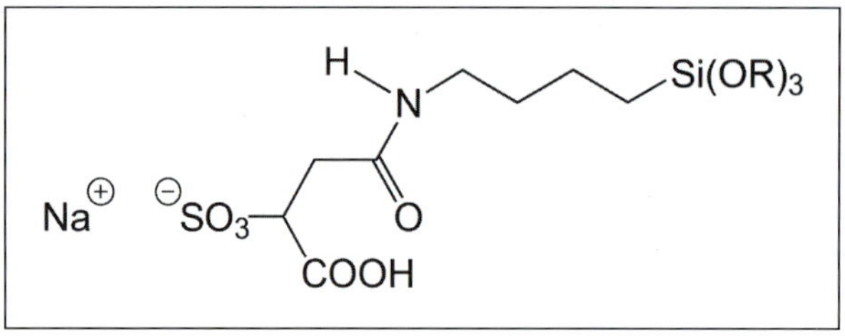

Figure 2.1.239: Silanes bearing hydrophilic acid groups for anti-fogging effects

Given their technical and economic importance the studies on the development of dirt-repellent surfaces were intensified. The breakthrough in self-cleaning surfaces came when, in the 1970s, *W. Barthlott* used a scanning electron microscope to identify structures some 30 to 50 micrometres high on the leaves of lotus plants which in turn were coated with a layer of ultrafine wax beads less than 100 nm in diameter. The extraordinarily low surface tension of the micro-textured wax layer in combination with the structure of the surface lowers the leaves' wettability by water droplets. The filigree structure and the low surface tension of the lotus leaves reduce the area of contact be-tween the water droplets and the lotus leaf to only a few percent. The energy necessary to allow the water droplet to penetrate into the interstices of the wax beads is too great for the as the surface of the leaf could apply. The water droplet rolls off, picking up dirt particles because of the smaller difference between its surface tension and the dirt particle's surface tension.

The effectiveness of the self-cleaning method "developed" by nature prompted researchers to use nanoparticles to re-create nature's model.

Today, various commercial façade paints are available which contain mineral ultrafine extenders in combination with silicone resins and which have a surface analogous to the role model found in nature [2.4.82]. That said, critical comparison of the technical alternative with its natural counterpart reveals rather modest success.

In a manner akin to the self-cleaning lotus effect, nanotechnology is also proving instrumental in special developments in **anti-fogging** and **anti-reflex coatings**. The extremely fine drops of mist responsible for fogging can be removed by adjusting the surface tension of the coating over the substrate and using a suitable microstructure. The fogging of mirrors or car windscreens which is caused by light scattering from the drops of mist can be eliminated by a combination of improved wettability and hydrophilic nanoparticles of defined pore structure. The water penetrates into the pores, causing the drops responsible for scattering the light to coalesce to a continuous phase. Surfactant-like aminosilanes bearing additional polar acid groups make suitable additives for creating the anti-fogging effect.

Antimicrobial nanoparticles such as vanadium pentoxide have been proposed for long-lasting **antifouling coatings** on ships' hulls [2.4.83]. For reduced absorption of dirt by architectural paints, modified silica particles in the nanometre range have been proposed which migrate to the film surface because of their low surface tension, where they form a thin protective layer [2.4.84].

Optical Properties

Nanoparticles enhance not only the mechanical-technological properties of coatings they also permit the optical properties to be optimised to specification. Hiding power and tinting strength are properties of pigmented coating materials which depend on the scattering

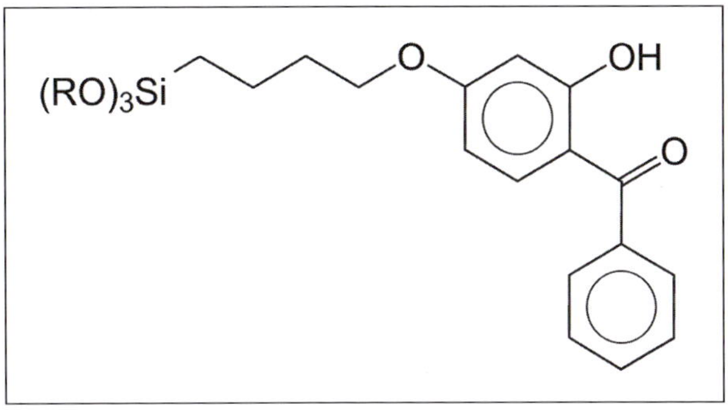

Figure 2.1.240: Migration-proof silyl esters with chemically bound UV absorbers

and absorption of the light interacting with the pigments. **Hiding power** increases with decrease in particle size until it reaches an optimum, and then decreases again. **Tinting strength** also increases with decrease in particle size but due to equally decreasing light stability also reaches an optimum value in the nanometre range, at an order of magnitude at which scattering pigments have already almost completely lost their ability to scatter light. Micronized TiO_2 or ZnO having the dimensions of nanoparticles achieve high transparency and can therefore serve as permanent UV absorbers because of their good absorption capacity for UV light [2.4.85].

Oligomeric silane nanoparticles increase their migration resistance and stability to aggressive UV light by chemically bonding to organic UV absorbers.

Among the optical effects of nanostructures is the colouring bestowed on thin platelets as a result of **interference**. Less than 250 nm thick, they bring about cancellation of portions of the visible light that has been doubly reflected at the surface and on the underside, thus generating a colour that varies with the angle of incidence. The interference pigments which have long been standard in terms of colour properties are therefore also a member of the class of substances known as nanoproducts (see Chapter 2.1.3).

Pretreatment with Nanoscale Adjuvants

The lasting protection afforded by coatings is contingent on good anchorage to the substrate. This applies in particular to anticorrosive protection. Where a substrate needs to be protected against corrosion, the first step is to carry out intensive, thorough cleaning. This is followed by a step intended to optimise the surface to receive the coating. The usual methods here are mechanical and chemical treatment by acids, alkalis or solvents in the case of metallic substrates and, mainly for plastic substrates, **corona pretreatment** and flame treatment. In the case of plastics, the flames act as oxidants, helping with the formation of polar groups (see Chapter 4.1.3.2). The simultaneous incorporation of **silicon-containing nanoparticles**, such as the polycondensates of R_3-Si-O-CH_3, into the heated and therefore soft surface during flame treatment creates a good basis for anchoring the subsequent coating. Especially for anticorrosive protection of metals, silicon-containing sol-gel products are now a good substitute or additive for chromium-containing preparations because they form a strong chemical bond with the surface. Silanes bond to the substrate via the formation of oxygen bridges or covalent bonds between their free hydroxyl groups and the polar metal atoms. Additional functional groups in the organic group then bond to the subsequent coating. This process has been employed in the automotive industry for a few years now [2.4.86].

Other Functions

There are numerous other areas which could be mentioned in addition to those above. Troublesome **electrostatic charging** of plastics and non-conducting coatings can be eliminated by means of electrically conducting nanoparticles either through mutual contact or via silane ammonium ions with hydrophobic groups floating out towards the surface.

Bactericidal effects are also feasible with nanostructured surfaces of suitably low surface tension. These hydrophobic surfaces present the pathogens with only small areas of attack where they can permanently anchor. Bactericidal substances in the form of nanoparticles facilitate lasting bactericidal finishing of surfaces [2.4.87].

2.1.5.4 Outlook

Nanoscale coating raw materials have existed much longer than the term nanotechnology, which was coined in the 1970s. While their effects were known in principle, the details of the underlying cause were not. It was only later that systematic study and research into the dependence on particle size and surface structure, overlaid with chemical influences, began. While numerous new fields of application are being found, however, there are also unanswered questions about the physiological interactions and about the respirability of such small particles [2.4.77, 2.4.88]. Nonetheless there is no disputing the fact that future findings on the toxicological issues will be accompanied by the discovery of new possibilities in the world of "nano dwarfs".

2.1.6 Summary

Close inspection shows that the group of basic raw materials is transformed into a large quantity of individual substances. The **film forming agents** alone comprise a large and constantly growing number of synthetic products in addition to the natural products which are still important for coating technology in some sectors. Particular mention has to be made of the new film forming agents for emissions- and waste-free powder coatings which cure by polyaddition and polymerisation, and of the radiation curable systems. Nor should the trend towards water-dilutable coating materials be forgotten which are proving their value for both conventional processing and in the field of electrocoating.

The situation is similar with colour and effect **pigments** and **extenders**, with the trend heading away from natural raw materials and towards defined synthetic products. The market offers a large and yet still increasing variety of organic and inorganic pigments of known chemical structure which can be reproducibly manufactured. Particular focuses of development work include the synthesis of new effect substances in conjunction with penetration into new areas of colour space in strict compliance with all technical and occupational health information for handling these materials in the production plant, during processing and later reprocessing and recycling.

Physiological issues relating to humans and the environment are of very particular interest in the field of **additives, plasticisers** and **solvents**. Whereas in the 1920s solvents for coatings were regarded as an indicator of technical and economic progress, the perspective changed fundamentally at the latest with the passing of many laws in North America and Europe in 1974. The reduction or replacement of solvents by physiologically safe water or demonstrably safer solvents is the watchword for current and future activities. Solvents which can be incorporated into the film like the reactive thinners in the case of radiation curable coatings and the systematic development and use of nanoproducts are another focus of future developments in coatings.

High quality, with due observance of all occupational health and safety and environment-related requirements, is underpinning the development of raw materials. However, lasting success can only be expected if the costs in a global economy also translate to a positive bottom line on the environmental balance sheet.

2.2 From Raw Material to Coating Material

The paint and coating raw materials described in Chapter 2.1 form the individual modules which are pieced together to produce coating materials which meet the varied requirements of practical applications. The formulations which have to be developed for this contain not only detailed information on the raw materials to be used, but also provide precise details on the proportions, the sequence in which the formulation ingredients are to be combined and the settings for the machines to be used in production. In the case of coating materials and specifically of paints, the binder or coating resin is the most important component, followed by the pigments and additives. The last of these, in turn, are crucial from a costs point of view and are of much greater importance for water-borne and high solid products than for systems dissolved in conventional solvents.

The correct selection of the raw materials must be based on accurate knowledge of their interaction in the product when it is used for its intended purpose, knowledge of the processing methods and the available plant and machinery at the paint manufacturer, and the economics of the coating material itself.

The various requirements imposed by the processor or the end user on the final product quality, in conjunction with the environmental demands and the prices to be achieved in the market place, are variables which have to be covered by a diverse range of raw materials.

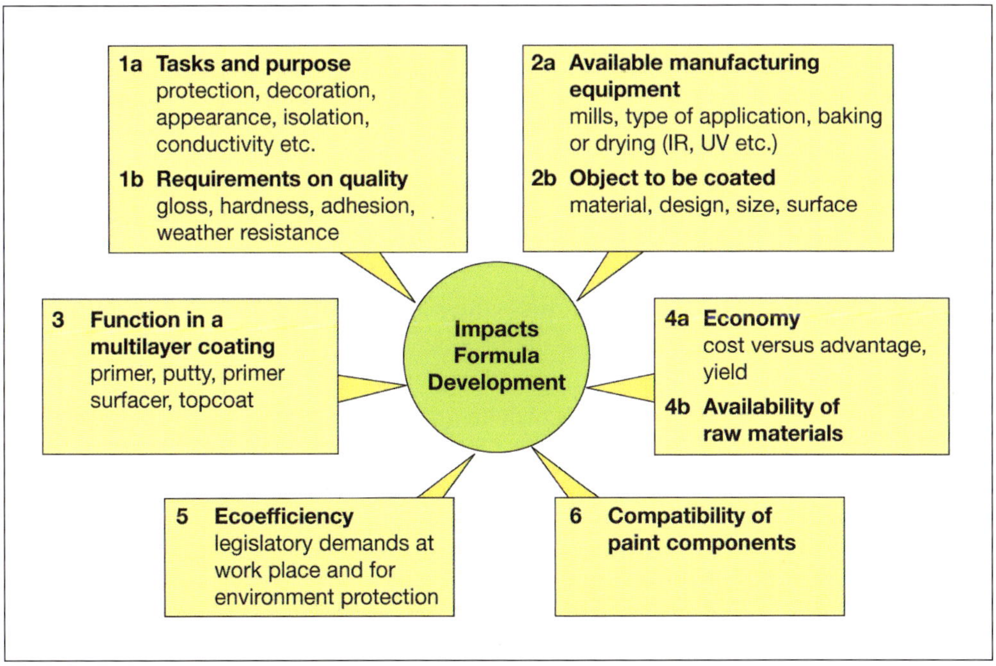

Figure 2.2.1: Overview of the factors influencing coatings formulations

The chosen composition for a coating material is therefore subject to preconditions, demands and external influences. It is thus obvious that universal solutions to problems cannot exist in the field of paints and coatings. The formulator or paint manufacturer is forced to find the optimum solution under the given conditions. Bearing in mind, furthermore, that in the automotive sector alone the number of producible formulations together with the necessary raw materials is hugely increased as a result of the desired variety of colours and effects, it is no wonder that major paint factories have to have many thousands of formulations on hand. Efforts have been underway for years to arrive at a product range which contains as few intermediates and masterbatches as possible.

2.2.1 General Rules on Drawing up Formulations

2.2.1.1 Purpose and Quality

If the customer merely wishes decorative goals to be met, more attention will have to be paid to the pigmentation and the selection of the effect materials than to the other raw materials. A high gloss requires a low pigmentation level, which leads to inadequate hiding power. In order to achieve a large amount of room for manoeuvre in terms of colour and gloss a certain separation of functions in the form of two coats has become accepted, and is standard practice in automotive paints, where the two coats are applied wet on wet, with intermediate flash off. The hiding power is provided by a base coat, irrespective of gloss requirements, while the gloss itself is delivered by an additional clearcoat on top of the base coat.

If the purpose of the paint or coating is the protection of commodities, the selection criteria are aimed initially at the film forming agent. Special demands are imposed on low water vapour and ion permeability of coatings for the protection of metal objects subject to long term outdoor exposure. Further selection criteria include excellent adhesion under extreme climatic conditions combined with the physical and chemical effect exerted by anticorrosive pigments. If, for example, zinc dust is used for active corrosion control, the pigmentation level is a decisive factor. Weight concentrations of more than 90 % are needed in order that the metal particles may touch each other and so develop their protective effect.

But the functional tasks of coating materials go far beyond corrosion control. Antichipping materials for cars, strippable coatings with a defined poor adhesion and coatings to increase or reduce electrical conductivity require significantly different raw material compositions. For sound insulation purposes high-build coatings are used which have a high degree of solvent retention because of the thickness of their coat and thus require new formulation principles. Sound insulating coating materials are therefore solely aqueous dispersions (hydrosols) or dispersions which are distributed in plasticisers (plastisols).

2.2.1.2 Production Resources, Application Systems and Object to be Painted

When the formulation is being drawn up and the raw materials are being assembled the technical resources for the manufacture of the paints, the customer's processing systems and the particular features of the object to be painted must also be addressed.

When it comes to pigment dispersion, formulations for manufacturing coloured coating materials with the aid of dissolvers (disc agitators) differ considerably from those produced in agitator mills. Pigments such as titanium dioxide (TiO_2) or those which are prepared with additives for simple dispersing [2.4.89] can be incorporated into many resins with the aid of dissolvers, which are energy efficient. The pigment-binder ratio and the rheological properties have to be adapted to the specific machines through the incorporation of appropriate solvents and additives [2.4.90]. In particular, the additives are important for producing low-viscosity pastes for the manufacture of low-solvent and especially water-borne coating materials [2.4.91].

The processing systems in operation at the paint customer also have a significant impact on the formulation. It is obvious that a DIY enthusiast's brush or roller requires different material properties from the spray guns or dip baths of industrial scale paint processors. The influencing of flow properties during film formation by means of solvents and additives such that the painter can produce run-free coatings which nonetheless are still fluid enough to yield a smooth finish when applied with a brush requires a different product composition from that designed for quick-drying spray paints that do not require post-finishing treatment.

Even paint films applied by the same processing method can differ significantly in the choice of film forming agents and solvents, with the choice depending on whether the subsequent curing is instigated by tempered air (convection drying) or infrared (IR) radiation. Where curing is solely by means of IR radiation, the solvent composition depends on the radiation's wavelength. If the electromagnetic waves are not absorbed by the paint and if the film therefore heats up from below, different solvents have to be used because of the lower pinholing tendency than in the case of radiation absorption in the upper sections of the setting coating (see Chapter 4.3).

If the high energy IR rays are replaced by "cold" UV radiation or electron beam, completely different types of film forming agent have to be used. The main constituents of the formulation then include resins which cure solely by polymerisation without any heat action and which may be combined with reactive diluents. It should be noted in passing that UV-absorbing pigments inhibit curing with UV radiation and therefore must not be used in that application.

In many cases the object to be coated is also the determining factor in the choice of raw materials. An already installed wooden window frame is difficult to heat up in a drier after coating. The same applies to thermally unstable plastics. The paint manufacturer cannot therefore rely on using stoving enamels based on alkyd or acrylic resins, which are established materials for metal coatings. Oxidatively curing alkyd resins or, where there are particularly high quality demands on the films, 2-component paints would be the appropriate base materials.

If more complex objects incorporating cavities with difficult access have to be coated, dip-coating processes are the only alternative to the standard spray application. If the latest electrocoating method is chosen, all the formulation ingredients have to be matched to the electrochemical deposition process. Water-dispersible resins can be made to coagulate on the objects to be painted, which have been configured as electrodes, via changes in the pH value, and these determine the formulation (see Chapter 4.2.1.1).

If flat items such as endless metal sheets are to be coated, coil coating methods can be used. Although higher viscosities and the associated higher solid contents mean lower solvent contents, totally different solvents have to be employed because of the high coil speed and the associated high temperature during drying. The raw materials of choice here would be long alkyl ether esters as solvents and silicone-modified polyesters as resins.

2.2.1.3 Function in the Paint System

In many cases the demands made of the coating cannot be met with a single coat of paint. Multiple coats with discrete, clearly defined tasks then replace the single coat (see Chapter 7.1 and 7.2). Three- and four-coat systems have proved highly adept at meeting particularly stringent demands on the quality of coatings where protective and decorative functions enjoy equal priority. This is the only way, for example, that the demands imposed on automotive paint finishes can be met.

Primers whose main task is corrosion protection are made with resins which feature excellent adhesion even under extreme climatic conditions. Aromatic epoxy resins, aromatic polyurethanes or modifications of these can therefore be used on account of their good adhesion and despite their lack of lightfastness, as the lack of lightfastness in the aromatic structural elements can be ignored. These are always optically covered by a further functional layer, the primer surfacer. This second layer provides everything needed to make the topcoat attractive while at the same time meeting all the demands on mechanical-technological properties. As indicated by the name, primer surfacers have to prime the surface, i.e. to even out any unevenness and to be readily sandable so that their surface can be smoothed, if required. As an intermediate coat acting in conjunction with the topcoat, primer surfacers also take on the function of protecting against stone chipping. With this purpose in mind, the chosen resins will have a glass transition temperature in the operating temperature range. Furthermore, extenders have to be used which create predetermined breaking points under extreme impact loads. This prevents the coating from separating from the substrate. The lightfastness requirements imposed on the primer surfacer coat are not as rigorous as those on topcoats. As with the primer coat, epoxy resin esters and polyurethanes are the standard film forming agents for the functional intermediate coating.

The tasks of the last coat, the topcoat, are quite different. The emphasis here is on the visual properties, provided that the mechanical-technological properties have been addressed. A high level of gloss and pronounced optical colour effects are the usual requirements imposed on the topcoat. The functionality of the coating depends also on good adhesion to the primer surfacer and a high degree of hardness combined with good resistance to wear and scratching. When it comes to selection of the raw materials for topcoats, the primary considerations are resistance to sunlight and outdoor exposure, as well as to chemicals from industrial emissions and natural atmospheric effects such as tree resins and bird droppings, in addition to the above-mentioned factors. Topcoat formulations differ significantly, therefore, from the above-mentioned items. Weather resistant alkyd-melamine resin or acrylic-melamine resin paints or polyurethanes based on polyester or acrylic resins are the standard film forming agents. By contrast with primer surfacers, the principle governing the use of single coat topcoats is to employ the minimum necessary amount of pigment.

As already mentioned, two-layer coatings have become the accepted standard in many applications as a result of the greater demands on the durability of topcoats. Formulations for two-layer topcoats differ significantly from the single coat solid colour coats which used to be standard. The base coat contains quick drying film forming agents which are frequently soluble or dispersible in water. The solid content of metallic and pearlescent effect paints is set particularly low in order that a good effect may be obtained. For environmental reasons, therefore, the conventional solvent-borne base coats are currently being replaced by water-soluble alternatives. The last coat, which provides the overall seal, is pigment-free. Film forming agents such as alkyd resins and polyesters, which are standard for pigmented paints, last only a few years as clearcoats on outdoor exposure. Loss of gloss and crazing are the usual signs of degradation. Today's clearcoats are usually acrylic resin combinations. Their high durability of more than 10 years stems from the addition of UV absorbers and stabilisers.

2.2.1.4 Cost Effectiveness and Availability

The demands of cost effective production have a particular impact on formulation design. In the manufacture of pigmented coatings, it is the incorporation of the pigments into the formulation which requires the greatest input during production. The goal, therefore, is to identify that combination of pigment, film forming agent and solvent which offers the highest possible pigment content and thus the smallest possible volume, yet supports rapid dispersion of the pigment [2.4.89, 2.4.92]. The resulting concentrates are generally termed pastes and are used for manufacturing paint formulations by simple mixing of the necessary components. Such pastes are initially only intended for cost effective dispersion with a particularly high efficiency in dispersers. The high pigment content of the mill base means that it is essential to apply a subsequent clearcoat which is appropriate to the requirements. Consequently, formulations for the manufacture of coating materials from colour concentrates come in two parts. First a paste formulation with the highest possible pigment proportion has to be found, so that in the second stage instructions can be given on processing the colour concentrates. The completion or "make-up" stage ensures that the demands on the coating and on processability are met.

In addition to manufacturers' need for cost effective production, processors require coating materials which enable large areas to be covered at a low consumption rate. Paints always have a high yield if high hiding power combined with low dry film density makes it possible to apply thin film thicknesses (see Chapter 5.7).

2.2.1.5 Occupational Health, Safety and Environmental Protection Regulations

Paint raw materials are regularly replaced by others for reasons of occupational health, safety and environmental protection as new knowledge becomes available. Nonetheless, the goal of developing emissions-free, physiologically safe paint technology still requires a great deal of effort, which is reflected in new types of paint and their formulations.

The replacement of volatile organic solvents by water, the reduction in solvent use through higher solid contents, the replacement of volatile solvents by reactive diluents, as well as the complete omission of solvents in the use of powder coatings have brought new formulation principles to the process of developing paints. Novel film forming agents, new additives and physiologically less harmful pigments are superseding the existing range of raw materials continuously. Environmentally polluting emissions of organic solvents and additives containing heavy metals are being avoided by means of new environmentally compatible painting technologies (see Chapter 5.6).

Coating Material

2.2.2 Material Flow in a Paint Factory

The stages from raw material supply to production of the semifinished and finished products through to supply of the final product are supported by numerous organisational units that provide control and assistance. These units include the laboratories of Research & Development, Technical Service, Quality Testing, Production Planning and Order Processing.

In addition to those production runs by which a company restocks its inventory, especially of branded materials, and which it can plan and control itself, there will be a high proportion of toll runs for general industrial customers who themselves engage in mass coating. The first point of contact for such customers is the sales department, the reason being that, in the event of a production run having to be initiated, the order can be given

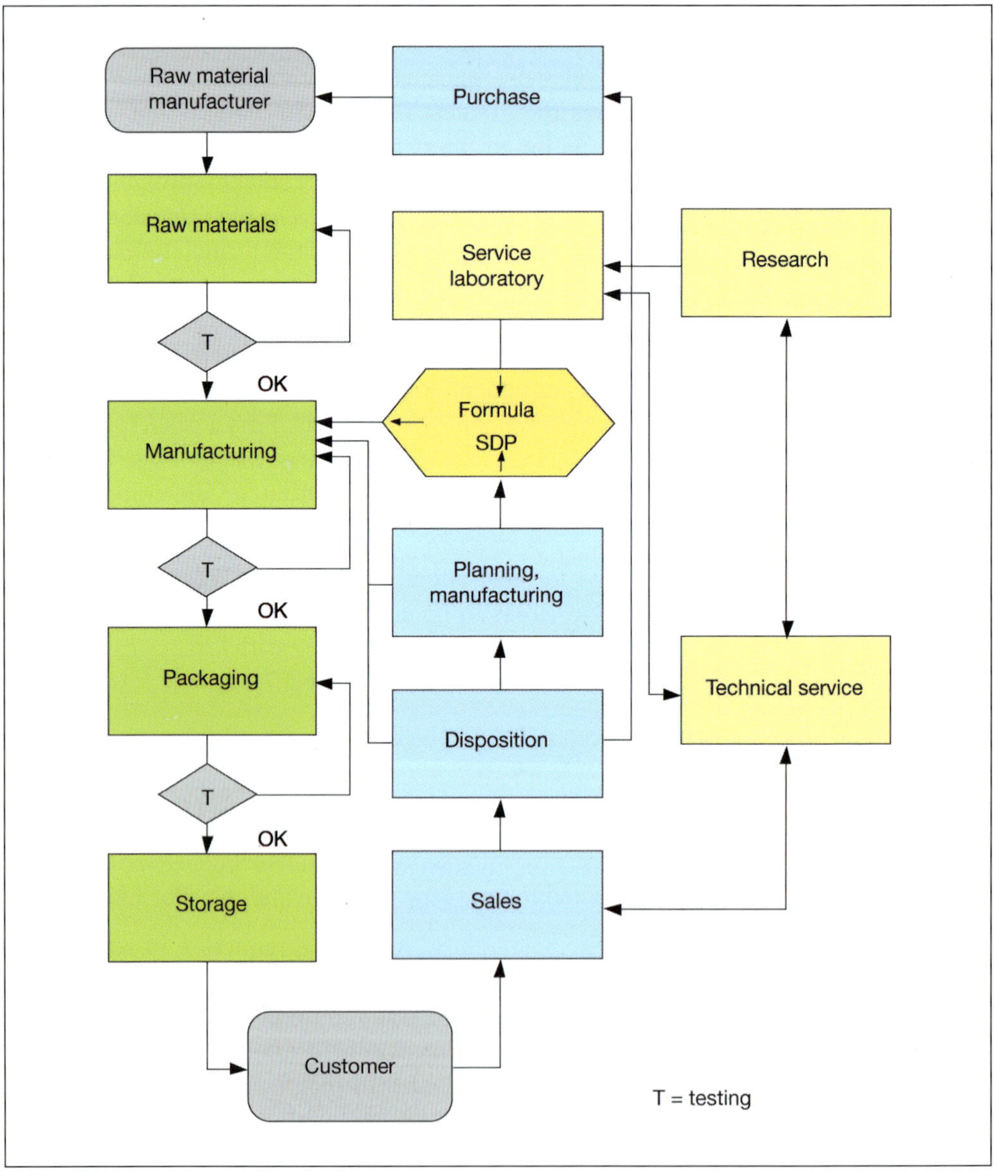

Figure 2.2.2: Organisation, processes and supporting functions in a paint factory

for the various necessary steps, viz. scheduling and the production planning responsible for the issuing of production orders, right through to the technical facilities charged with the development and maintenance of the formulations.

Since large paint plants with an extensive product range will hold anywhere up to 20,000 "live", i.e. continually updated, formulations they are compelled to maintain comprehensive service and product development laboratories. They need to have excellent market knowledge and to provide good technical backup for the systems they present to consumers. The plants have to be fast and versatile to cope with the production demands arising from an extremely large number of raw materials and formulations. Unlike continuous-production plants the paint and coatings industry has to operate in batch production mode as this is the only way to handle the production demands of the above-mentioned diversity of products. Since customer orders differ in terms not only of the formulation itself but also of the order quantity, the plants must have an appropriate range of production systems. Toll manufacturing leads to a huge range of batch sizes, from just a few kilogrammes to 20 to 30 tonnes.

Two fundamental strategies are possible when manufacturing the batch of coating material before the dispersing with the production of factory packed paints ("from scratch") or colour concentrates ("from pastes") with subsequent completion ("made up"). Production from scratch starts with all the raw materials. It is also common for blends of pigments to be dispersed in a batch. Manufacturing "from pastes" (also termed mixed production), on the other hand, is based on premixed pigment pastes which come in series. Generally, each paste contains only a single pigment. The paste series differ from one another in the type of film forming agent needed for obtaining specific properties.

In a stage prior to the actual paint production, the major paint makers manufacture substantial volumes of the film forming agents for their key products in their own resin factories using formulations developed in-house. Unlike smaller paint operations which mainly buy-in their film forming agents and then convert them into other types of solution if necessary, they have the capability to develop and manufacture customised film forming agents. These are optimally adapted to the formulations, the production conditions, the coating system and the intended use of the particular coating material.

A very complex logistical system operates in paint plants because of the various production modes already mentioned and as a result of the different consistencies of the raw materials, intermediate products, final products and packaging materials used.

Figure 2.2.3 shows a simplified block diagram of the material flow in the production of paints and coatings. The material flow ranges from the stocking-in of the raw materials to the provision of the finished products which are ready for dispatch. The deciding factors in terms of material flow and feed are the raw material type, its consistency, the consumption quantities and the packaging form for the product when it leaves the plant. Large plants have to install large tank storage facilities for resin solutions and solvents which are equipped to the latest safety standards. The preferred delivery method here is by road tanker. In addition to the liquid raw materials which are conveyed in large quantities via pipe bridges to the locations where they are processed, a large number of raw materials in paste, solid and powder form are delivered and made ready for use in containers, drums or bags, which may or may not be palletised. Depending on the specified storage conditions, indoor and outdoor storage and

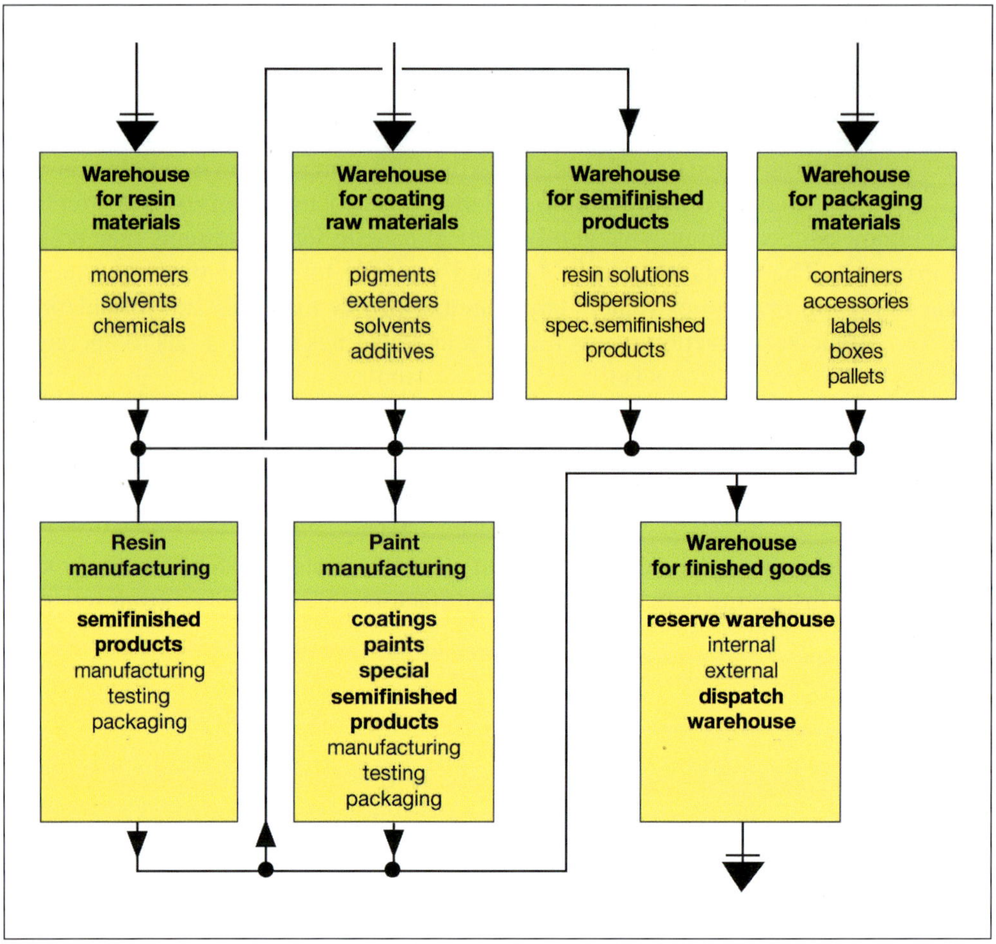

Figure 2.2.3: Block layout of the flow of materials in the production of coating materials

stacking facilities are provided for this. Computer-controlled high bay warehouses are commonly used for large unit quantities and transfer volumes. Material movement is accomplished in a wide variety of different ways, the choice depending on the paint plant's structure and facilities.

The system outlined is much more complex if a company has several sites and each site in turn has several production units. This calls for a highly sophisticated data and information flow system which can only be achieved by means of computer centres.

Logistical and quality decisions are constantly required during the flow of material and information. This necessitates an integrated organisation which pulls together sales, logistics, purchasing, production, quality related test units and application centres.

2.2.3 Theory of Dispersion

2.2.3.1 Pigment-Specific Properties for Dispersion Processes

Whereas the manufacture of clearcoats is limited to the mixing of various film forming agents with solvents and additives, the production of pigmented systems is based on

$$I = I_0 \cdot k \cdot \left[\frac{n_{Pi}^2 + n_{Bi}^2}{n_{Pi}^2 - 2 \cdot n_{Bi}^2}\right]^2 \cdot \frac{V^2}{\lambda^4}$$

valid, if $\frac{d}{\lambda} << 1$

I = intensity of scattered light
I_0 = intensity of incoming light
k = constant
n_{Pi} = refractive index of pigment
n_{Bi} = refractive index of binder
V = volume of pigment particles
λ = wavelength of incoming light

Figure 2.2.4: Equation for intensity of scattered light

complex wetting and stabilisation processes. For a better understanding of the physicochemical background, it is worth recalling the specific properties of the pigments and their interaction with the film forming agent solutions (see Chapter 2.1.3).

The interaction of pigments with incident white light leads to selective removal of wavelengths from the light, with remaining wavelengths being scattered as widely as possible. The first property gives rise to chroma, while the second is the prerequisite for good hiding power. The ability to produce chroma is proportional to the area of the pigment surface which can interact with the light.

This means that for a specified pigment quantity the degree of selective absorption increases with decrease in particle size up to a discrete stage. This stage is reached when the particles are completely penetrated by light.

The **scattering power** increases with the number of dipoles in the pigments which can be excited by light. For this reason, the scattering power increases initially with increase in particle size (see Chapter 2.1.3).

However, this increase only continues as long as all the dipoles in each pigment particle can interact with the incident light. But this is only possible up to a specific particle size for each pigment. Above this critical size the dipoles inside the pigment become optically inactive. Furthermore, interference occurs and ultimately leads to a decrease in scattering power. There is, therefore, an optimum particle size or particle diameter for the scattering power.

For titanium dioxide, the most important light scattering pigment, this optimum size is about 0.25 µm. Experience shows us that this is in the general range of the absorbed wavelengths in the light spectrum. According to *Weber* [2.4.93] the following applies empirically:

$$D \approx \frac{\lambda_{max}}{(n_{Pi} - n_{Bi}) \cdot 2.1}$$

D = ideal diameter of particles for light scattering
λ_{max} = max. wavelength of absorbed light
n_{Pi} = refractive index of pigment
n_{Bi} = refractive index of binder

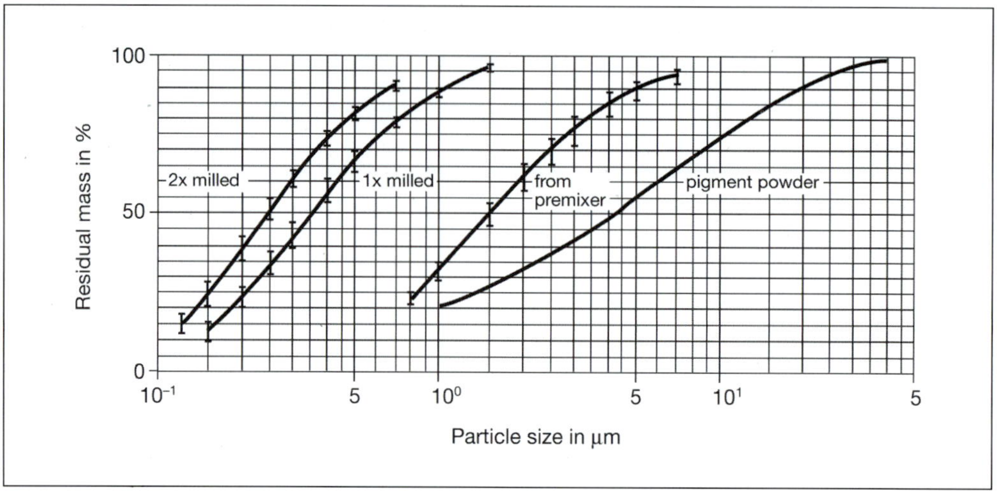

Figure 2.2.5: Pigment particle size as a function of agitation and dispersion processes

Such small particles tend to combine to form larger units in accordance with their shape and polarity. If the primary particles come into contact with each other along edges and at corners, the conglomerations can become very large with diameters of more than 100 µm, though the forces of attraction remain relatively low. Such particles are termed **agglomerates** (see Chapter 2.1.3).

If the primary particles become connected via common surfaces, the forces of attraction increase so much that separation by mechanical means is difficult. However, such **aggregates** have a much lower volume.

The cavities in the agglomerates and the cavities in the pigment powder in the delivery form can represent up to 75 % of the overall volume, the precise percentage depending on the pigment form and size. The pigment volume concentration (PVC) in that event is only 25 %.

$$PVC = \frac{V_{Pi}}{V_{Pi} + V_{Bi}} \cdot 100$$

V_{Pi} = volume of pigment

V_{Bi} = volume of binder

The forces of attraction among the pigments themselves are the cumulative total of *Coulomb* forces, *van der Waals* forces and hydrogen bridging bonds within the chemical structure. The range and strength of the forces of attraction are very greatly influenced by the type of interaction. Whereas *Coulomb's* law states that electrostatic forces decrease with the square of the distance d between electric charges, they decrease by d^{-4} in the case of permanent dipoles and by d^{-7} in the case of induced dipoles (see below). The degree of agglomeration and the strength of the agglomerates are therefore defined by the size, form and type of pigment.

The greater or lesser degree of pigment agglomeration means that incorporation, i.e. uniform separation into primary particles and thorough wetting of the latter, requires the use of special techniques. Simply admixing materials does not yield technically usable products in the majority of cases, with smooth coating finishes being just as difficult to achieve as an intact, pore-free coating. What is more, the optical properties, such as gloss and film build, and the protective effect are completely inadequate in such coatings

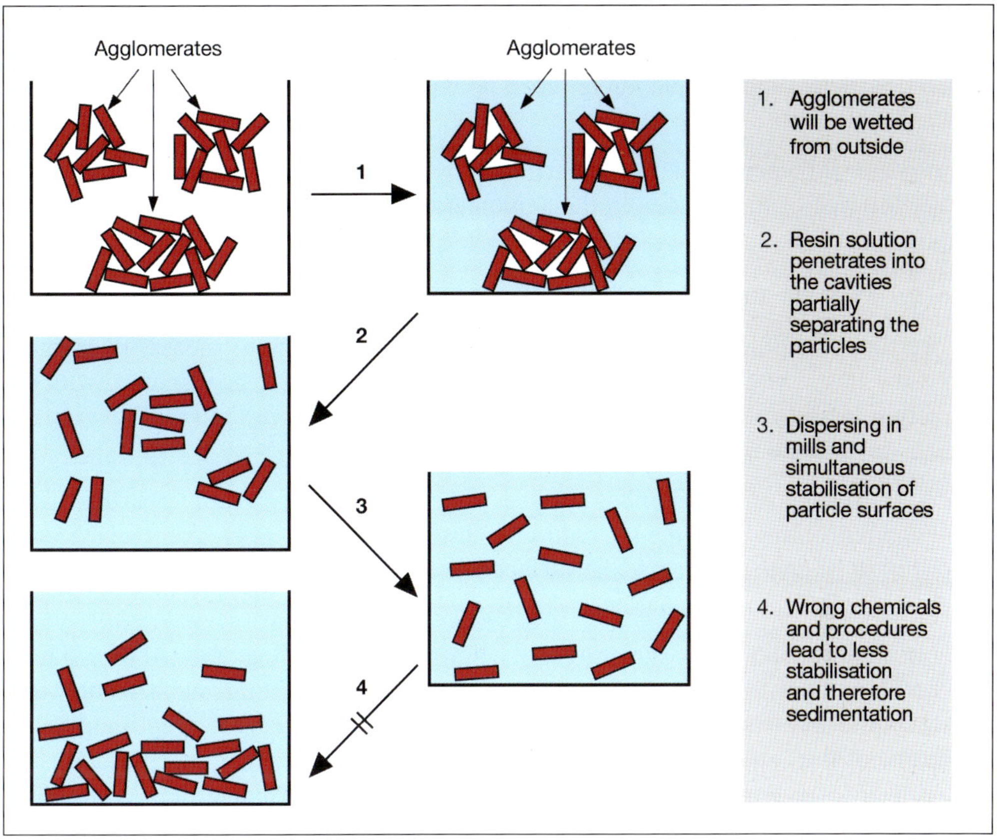

Figure 2.2.6: Schematic diagram of the dispersion and stabilisation of pigments

because of the high gas and water vapour permeability. Dispersion techniques then have to be used. Chroma and hiding power also would only partially develop, as not all of the pigment surfaces would be available to the light. Given that pigment prices can sometimes be high, incomplete development of pigment optical properties constitutes poor economics.

To eliminate these technical and economic deficiencies, therefore, specific expertise is necessary in the selection of resin and solvent, the right ratio of raw materials (formulation) and supplementary mechanical separation of the agglomerates.

Once dispersion has taken place, collisions occur because of Brownian motion. Reagglomeration sets in because the forces of attraction are still active. Stabilisation of the dispersed pigments is therefore absolutely essential.

The manufacture of pigmented coating materials thus concentrates both on separating agglomerates while retaining the primary particles and on completely covering the entire pigment surface with a stabilising shell which is intended to prevent reagglomeration during storage, processing or film formation. The latter phenomenon, which is also termed flocculation, gives rise to colour shifts and a more pronounced tendency to sedimentation.

The dispersion process is therefore divided into the individual stages of: wetting the pigment powder which consists of agglomerates, aggregates and primary particles with the resin solution; penetrating the cavities of the agglomerates; their subsequent conversion into the primary particles by means of mechanical separation; stabilisation of the primary particles; and completion or make-up with resin solutions for the manufacture of saleable products.

The process of dispersion is thus determined by the ease with which surfaces can be wetted, the size of the cavities in the agglomerates, and the depth to which the resin solution penetrates.

2.2.3.2 Wetting and Separation of Agglomerates

The **wetting** of pigments is a surface energy process. The key to successful pigment processing lies in matching the surface tension of the pigment to be dispersed with the interacting binder solution. The surface tension is the work required per new unit of surface area formed by wetting of the pigment.

Thorough wetting of the pigment surfaces by resin solutions is only possible if the energy gain resulting from the interaction of fluid and pigment is greater than the work which has to be done to increase the surface area of the liquid. The ratios are described quantitatively by *Young's* equation [2.4.94].

Good wetting of solid surfaces manifests itself in visible spreading of the wetting liquid. The wetting properties can be measured by determining the wetting (contact) angle α of various liquid droplets applied to the surface of the solid. Wetting is at its optimum when the cosine of the contact angle approaches 1. At the same time the interfacial tension $\sigma_{Pi/Pa}$ converges on zero. This state is reached if the surface tension of the wetting paint σ_{Pa} is equal to or lower than that of the solid pigment surface σ_{Pi} to be wetted. These relations also hold for the interaction of pigments and resin solutions (see Chapter 2.3.2).

The surface tension of solvents is generally lower than that of film forming agents. Polymer solutions therefore also show a decrease in surface tension with decrease in solid content. This means that the wettability of pigments can also be influenced by the resin-solvent ratio.

Wetting of the outer agglomerate shell is therefore followed by **penetration of the resin solution into the cavities of the agglomerates**. Thorough moistening of the pigment powder is facilitated by a low solid content, i.e. a low viscosity, but it is simultaneously slowed down because the surface tension is then lower. The *Washburn* equation, which is derived from the *Hagen-Poiseuille* equation, states that the penetration rate $\dot{V}$ of resin

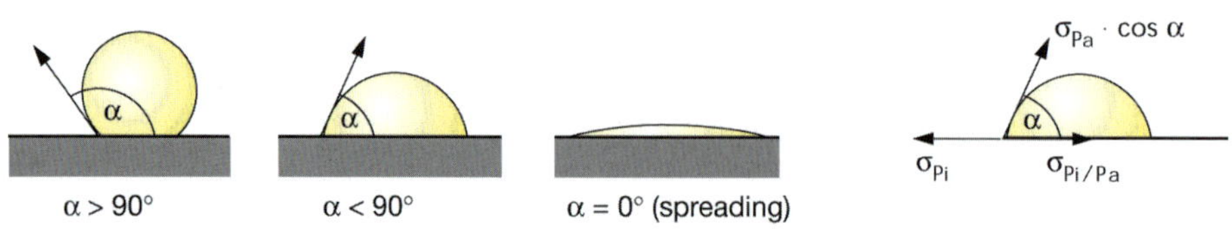

$$\sigma_{Pi} = \sigma_{Pa} \cdot \cos\alpha + \sigma_{Pi/Pa} \qquad \cos\alpha = \frac{\sigma_{Pi} - \sigma_{Pi/Pa}}{\sigma_{Pa}}$$

Figure 2.2.7: Wetting of surfaces as a function of interfacial tension

$$\dot{V} = \frac{\pi \cdot r^3 \cdot \sigma_{Pa} \cdot \cos\alpha}{k \cdot \eta \cdot l}$$

$\dot{V}$ = penetration rate
r = average radius of capillary tube
σ_{Pa} = surface tension of paint
α = wetting angle
η = viscosity
l = length of capillary tubes

Figure 2.2.8: Washburn equation for calculating penetration into pigment agglomerates

solutions into the agglomerate cavities depends very clearly on the radius of the cavities r (see Chapter 2.3.2). Surface tensions σ_{Pa} which are as high as possible promote wetting of the cavity walls, provided they are still smaller than those of the pigment. High viscosities η slow down the penetration rate.

To attain the simultaneous goals of lower surface tension in the resin solution employed for wetting and high surface tension for rapid penetration into the cavities, it is necessary to find the optimum resin solid content and thus the most dispersion friendly mixture for each pigment.

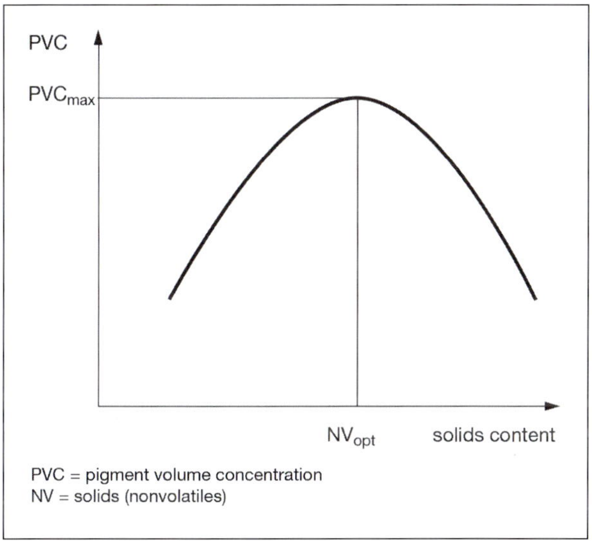

Figure 2.2.9: Dependence of optimum mill base formulation on solid content for the dispersion process

The pigment absorption capability of polymer solutions and the wetting speed are thus variables which are a function of solid content. The ideal ratios exist whenever wetting and thorough moistening are matched to the specific conditions by adjusting the viscosity and the surface tension. Mixtures having a minimum interfacial tension therefore enable pigments to be changed into a flowable and thus dispersible state with the aid of relatively little resin solution. The attainable pigment volume concentration of the mixture of resin solution and pigment is then at a maximum. The composition of the raw materials at the maximum pigment volume concentration for a flowable dispersion mixture is termed the optimum mill base formulation, though it is optimum only inasmuch as surface features have been addressed, without allowance for the material specific and mechanical influences which will be described later.

Various features are used to describe the **material-specific dispersion properties** of pigments. These include the **oil absorption value** and the **wetting volume** (see Chapter 2.3.3). The oil absorption value enables certain predictions to be made about the pigments' wetting properties. If a resin solution is used instead of linseed oil, the oil absorption value becomes the **wetting point**. With the aid of wetting points determined for binder-solvent mixtures of different percentage composition, the PVC can be used to establish the optimum mill base composition.

Apart from the shape and size of the pigment particles the **particle size distribution** also has a particular bearing on the dispersion rate. Large particles with a narrow distribution band have better dispersing properties than smaller ones with a broad distribution band.

A full description of the dispersion process requires that the chemical character of the pigment surface also be taken into account. Although polar surfaces are easier to wet than nonpolar ones, the forces of attraction among the primary particles themselves are higher. The pigment agglomerates are more solid and therefore more difficult to separate.

It has already been mentioned that the dispersion rate is also a function of the resin-solvent ratio. The dispersion rate falls with increase in solid content, i.e. with increase in viscosity. Although high viscosities are desirable for transmitting shear forces to the material to be dispersed in dispersers, they are a disadvantage because of the resulting poorer wetting and slower cavity penetration.

Apart from the viscosity and surface tension of the film forming agent solution already mentioned, the **size of the film forming agent molecules** influences the dispersion properties. Resins can be particularly well dispersed if they have a high solid content and a relatively low viscosity. Small resin molecules are more polar and can also penetrate the cavities better. On account of their high solid content they also offer a large number of film forming agent molecules for covering the pigment surfaces completely.

Both the molecule size and the **molecule shape** have an impact on the depth of penetration into the agglomerate cavities. Rigid and long molecule chains such as those of cellulose nitrate are less capable of initiating thorough wetting simply for steric reasons, despite their good affinity for the pigment surface.

Of the many resin types (see Chapter 2.1.1) low molecular **alkyd resins** have proved to be ideal film forming agents for ensuring optimum dispersion of pigments. They have polar groups, yet have good wetting properties and a high solid content.

The type and quantity of **solvents** and solvent combinations also have a bearing on viscosity and surface tension, while also determining the configuration of the resin molecules which are in the form of coils. Since the low molecular solvent molecules have much greater mobility than the film forming agents, they also penetrate the agglomerate cavities more quickly. They induce preliminary wetting of the pigment surfaces, termed **pseudowetting**, during dispersion [2.4.95]. The type of solvent and the polarity of the film forming agent must therefore be matched to each other such that the pseudowetting by solvents will evolve into real stabilisation by replacing them with film forming agent molecules.

Only some of the agglomerates can be destroyed by physicochemical interaction of the pigments with the resin solution. Complete conversion into the primary particle in tandem with best possible pigment wetting can only be achieved by the additional **transmission of mechanical forces to the pigments**. Of the usual methods (see Figure

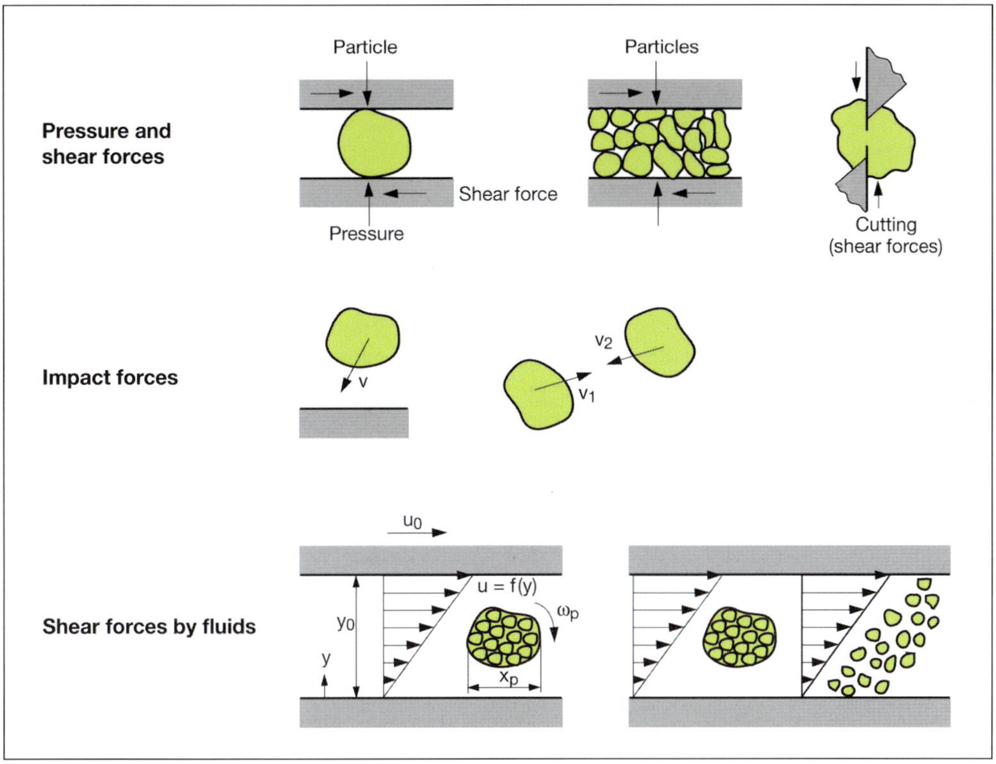

Figure 2.2.10: Methods of particle comminution

2.2.10), only the gentle method involving the transmission of shear forces in laminar flows of viscous media can be used where the specification is that the primary particle must be retained. Viscosity in this context does not refer to the measurable viscosity of the ternary binder/pigment/solvent mixture. The significantly lower viscosity of the pigment-free resin solution is the sole determining factor in the transmission of shear forces [2.4.96].

It has also already been mentioned that although higher shear forces can be transmitted by increasing the viscosity, the overall dispersion rate can fall on the other hand because of the slower penetration into the agglomerate capillaries (see above). Care should therefore be taken to ensure a balanced match of viscosity and shear gradient when formulations and operating procedures are being devised.

Numerous dispersers have been developed to transmit shear forces to pigment agglomerates by laminar flows of viscous fluids. The most important are **dissolvers**, which are suitable for easily dispersible pigments, **three roll mills** for highly viscous pastes and the very effective **agitator mills** for universal use. A small number of **ball mills** are also in use, though these are uneconomic by comparison with agitator mills and dissolvers. Even though these units differ greatly in design, they all share the same operating principle. They permit the transmission of shear forces in laminar flow gradients under tensile-compressive loads which promote dispersion (see Chapter 2.3.3).

Dissolvers have a toothed disc which moves at high speed in a matching smooth walled container. This disc transmits such large quantities of energy to the radially spun mill

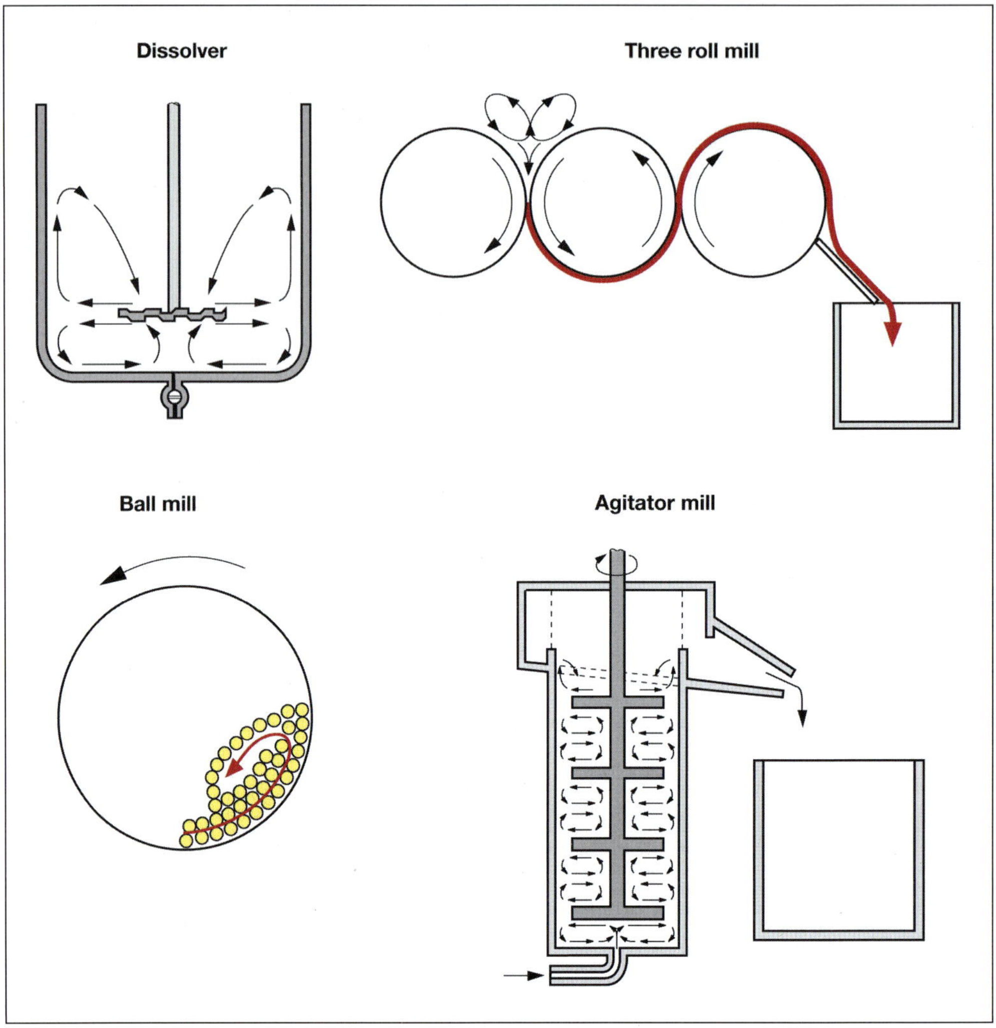

Figure 2.2.11: Schematic drawings of the key dispersing devices: dissolver, three roll mill, ball mill, and agitator mill

base by centrifugal force that separation of the agglomerate is assured. Another type of dissolver is the rotor-stator unit which is capable of dispersing in a small area by means of centrifugal and shear forces.

Ball mills are hollow cylinders which rotate about a horizontal axis. They are filled to a particular level with grinding media, generally steatite or steel balls. The movement of the grinding media relative to each other and to the drum wall generates friction surfaces which slide over each other and in which dispersion takes place.

Agitator mills come in many different variants and, like ball mills, employ moving grinding media. However, the grinding media in this case are substantially smaller than the balls in ball mills which measure a few centimetres in size. The balls in agitator mills are only a few millimetres in size and have to be artificially moved to create a high degree of relative motion.

Three-roll mills have a centre roll which is mounted in fixed bearings and which is flanked in the same plane by rolls pressing against it under great pressure. The three rolls are driven in opposite directions at different speeds, which increase progressively from the feed roll to the centre roll to the apron roll. The mill base is fed into the wedge-shaped space formed by the feed and centre rolls. It then passes through the nip, before being conveyed via the centre roll to the apron roll. Here there is a doctor blade which transfers the dispersed material to a separate container.

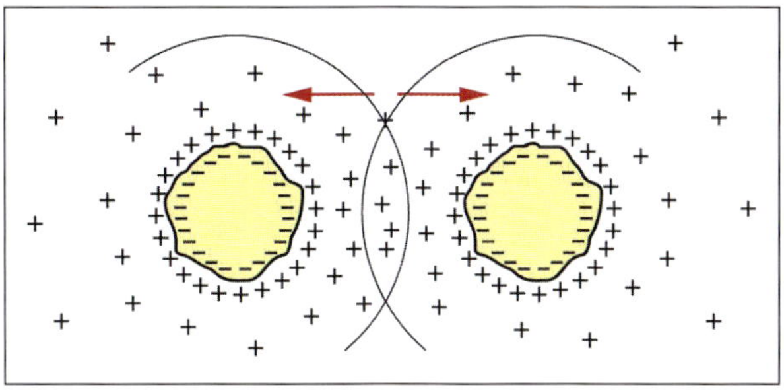

Figure 2.2.12: Repulsion of particles of same electrical charge

2.2.3.3 Stabilisation of Pigment Dispersions

The uniform distribution of the pigments brought about by the dispersion process is only a feasible technical solution if the distribution state is also retained during formulation, storage, processing and the subsequent film formation. Reagglomeration of the primary particles would lead to a reduction in gloss and a change in the tinting strength. Other side effects would be colour shifts in pigment mixtures, specks and sedimentation.

If the surfaces of the pigments are changed by the dispersion process to such an extent that, in the event of a collision, no adhesion and thus no agglomeration takes place, the necessary stabilisation has been attained. In the manufacture of pigmented coating materials any consideration of the pigment wetting and homogenisation process must therefore also address the stability of the resulting product. For technical and economic reasons, the dispersion effort and method must always be related to the covering of the pigment surface with substances which contribute to stabilisation of the product. There are two ways to achieve this, in principle. In polar media electrical double layers are the predominant contributors to stabilisation, whereas in nonpolar environments steric, osmotic and entropic phenomena are the main contributors. All these mechanisms can work in isolation or together.

Thermodynamically, according to the *Gibbs-Helmholtz* equation

$$\Delta G = \Delta H - T\Delta S$$

the change in free enthalpy ΔG must be positive if the approaching particles are already stabilised. Such a case occurs where the entropy ΔS diminishes on approach or the enthalpy ΔH increases. If one of the two variables has a different algebraic sign from the other, stabilisation still exists if the difference between the two is positive. If the entropy changes positively and at the same time the enthalpy negatively, no stabilisation can be achieved.

Stabilisation by Means of Electrical Double Layers

One of the first – still current – theories on the interpretation of the stability of dispersions, be they resin particles or pigments, was developed by the Russians *Derjaguin* and *Landau* [2.4.97] and, independently, also by *Verwey* and *Overbeek* [2.4.98] in the Netherlands. This is a relatively simple method of describing surface energies when there is a polar, i.e. ion-forming environment, which makes it a useful method for

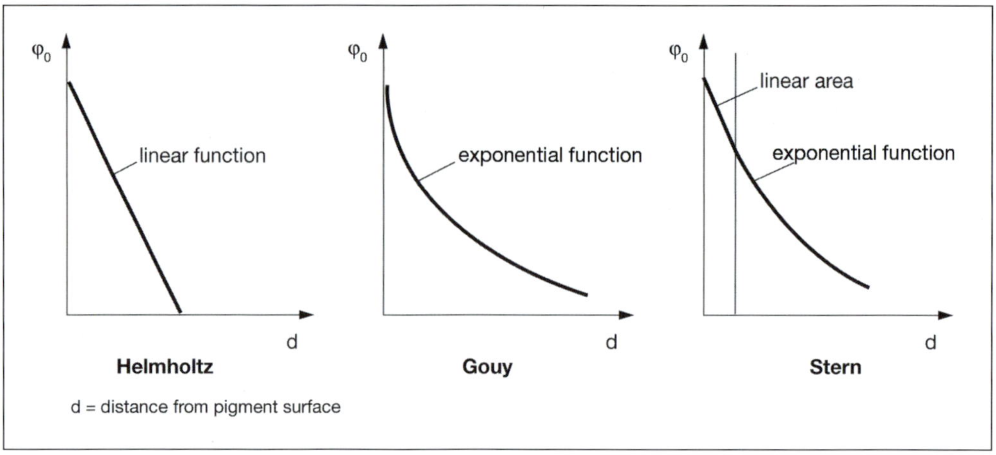

Figure 2.2.13: Curve of the electrical potential φ of electrostatically stabilised pigments

practical operations. The mathematical correlations which are known by the term DLVO theory show that a pigment dispersion is stable if the repulsive forces acting on the surface as a result of ion adsorption exceed the forces of attraction.

$$V_T = V_A + V_R$$

The total potential surface energy V_T is composed of the terms for attraction V_A and electrically induced repulsion V_R.

Electrical double layers are formed by the adsorption of charge carriers at the pigment surface which attract the moving countercharges in the environment disproportionately in the form of an ion cloud.

The repulsive forces are activated when ion clouds of two particles with the same charge approach and mingle. The models proposed by *Helmholtz, Gouy* and *Stern* provide a precise description of the conditions in the vicinity of stabilised pigment surfaces [2.4.99]. All the authors agree that an electrical potential φ is generated from the pigment surface to the freely mobile counterions in the environment as a result of the adsorption of charge carriers. However, opinions diverge when it comes to the exact potential profile. Whereas *Helmholtz* assumes a linear fall in the potential to the counterions which are a short distance from the environment, as in a bulb condenser, *Gouy* and *Stern* and see the potential profile as being more complex.

Gouy postulates a reduction in potential in the form of an exponential function. *Stern*, on the other hand, combines the *Helmholtz* model with that of *Gouy*. There is an initial potential drop in the vicinity of the pigment or dispersion particles, followed by a potential profile with an exponential function. All three models can be applied in practice. Concentrated solutions are largely described by the *Helmholtz* model while extremely dilute solutions follow the model of *Gouy*. The *Stern* theory is consequently the most suitable for real solutions. Because of the easier applicability of *Gouy's* model the comments below refer to this model only.

If φ_0 is the potential at the interface to the pigment, the following applies for the potential φ at a distance d.

$$\varphi = \varphi_{0\,e}^{-\chi \cdot d}$$

The variable χ assumes a practical significance in that it gives an indirect indication of the thickness of the electrical double layer. $1/\chi$ is the film thickness at which the potential φ has fallen to φ_0/e. According to *Debye* and *Hückel's* theory, where

$$J = \frac{1}{2}\sum_{i=1}^{n}(c_i \cdot z_i^2)$$

$$\chi = \sqrt{\frac{8 \cdot \pi \cdot e_0 \cdot 1000}{\varepsilon \cdot k \cdot T}} \cdot \sqrt{J}$$

c_i = concentration
z_i = valency
e_0 = elementary quantum of electricity
ε = dielectric constant
k = Boltzmann constant
T = absolute temperature

χ is proportional to the square root of the ionic strength J. J is composed of the product of the ion concentration c_i and the valency z_i^2 of all the ions. The thickness of the electrical double layer is therefore also determined by the type and quantity of electrolyte.

The stabilisation of pigment dispersions is particularly good if the electrical double layer has a large sphere of influence. As defined in the formulae above, however, the double layer decreases with increase in electrolyte concentration. This enables the particles to come closer together and causes a decrease in pigment stabilisation. If, on the other hand, the double layer consisting of freely mobile ions has a larger sphere of influence at a low concentration, penetration of the double layers occurs as two pigment particles approach, even at an extended distance. A repulsive effect is therefore activated even at extended distances.

In accordance with the **DLVO theory** the following applies to the decrease in the *van der Waals* forces of attraction V_A as a function of distance from the surface:

$$V_A = \frac{A}{6} \cdot \left[\frac{2}{s^2 - 4} + \frac{2}{s^2} + \ln\left[\frac{s^2 - 4}{s^2}\right]\right]$$

with $s = \frac{R}{r}$

R = distance of particle clusters
r = radius of particles
A = Hamaker constant

for short distances applies:

$$V_A = -\frac{A}{12 \cdot \pi \cdot d^2} \quad \text{where} \quad d = R - 2 \cdot r$$

The DLVO theory permits also the estimation of the repulsive forces

for large particles applies ($\chi \cdot r >> 1$)

$$V_R^{el} = \frac{\varepsilon \cdot r \cdot \varphi_0^2}{d + 2 \cdot r} \cdot \ln(1 + e^{-\chi \cdot d})$$

for small particles applies ($\chi \cdot r << 1$)

$$V_R^{el} = \frac{\varepsilon \cdot r \cdot \varphi_0^2}{d + 2 \cdot r} \cdot e^{-\chi \cdot d}$$

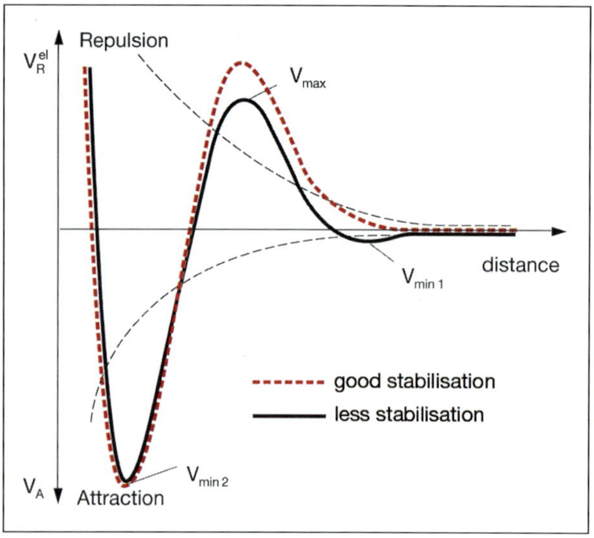

Figure 2.2.14: Forces of attraction and repulsion of electrostatically stabilised pigments as a function of distance from the particle surface

and with infinite dilution

$$(\chi \rightarrow 0)$$

$$V_R^{el} = \frac{\varepsilon \cdot r \cdot \varphi_0^2}{R}$$

An overall balance for the potential energy as a function of distance from the pigment surface can be drawn up with the aid of the equations for V_A and V^{el}_R, the result of which is the curves shown in Figure 2.2.14.

The overall curve which is calculated by adding the individual potentials has three extreme values. The outer minimum V_{min1} is particularly flat and, although it leads to the formation of flocs in the event of pigment collisions, these are extremely soft and can be separated from each other again quantitatively by means of simple shear action during agitation processes. This minimum can be influenced by the concentration of the electrolyte. It can disappear completely with the correct stabilisation (see red curve).

Stabilisation of the pigment dispersion is determined by V_{max}. According to the DLVO theory, energy barriers can be calculated which cannot be overcome even if the particles have a relatively high kinetic energy. The distance and the level of Vmax can also be influenced as described by the concentration and valency of the electrolytes and also by the pH value. Charge reversal occurs at the isoelectric point when the pH reaches a certain value. This leads to destabilisation of the system, such that $\varphi_0 = 0$. There is then no longer a repulsive barrier.

Practical Implications of DLVO Theory for Coatings

It is particularly important in practice to be able to measure the stability of a dispersion by simple means or to record the influence of additives or formulation modifications. The DLVO theory states that the electrical potential of the boundary surface between the pigment and the solution φ_0 leads to the energy barrier being raised by the power of two. Knowing φ_0 or being able to influence it is therefore important for an objective evaluation of the stabilisation of pigment dispersions. Unfortunately, φ_0 cannot be measured because of the greater or lesser degree of shielding by the electrical double layer. However, it is possible to comment indirectly on φ_0 by determining the migration rate of the stabilised pigment particles in an electrical field.

If dispersions of ionically stabilised pigments are introduced into an electrical field E, a force acts on the particles which causes them to move towards the electrode of opposite charge. As, in accordance with *Stokes'* law, a quantifiable counterforce is formed due to the viscosity η during flow, a steady state is quickly attained. The particles will then move in the dielectric material of dielectric constant ε in accordance with their charge at a uniform speed u. Since the transported charged particles are partly shielded by counterions due to a firmly attached liquid shell which moves with them, the calculation of the electrokinetic processes need only make allowance for the net charges of the pigment particles, which act rather like a bulb condenser. By measuring the migration rate of the pigment particles which are enveloped with charges in the electrical field it is possible to determine the so-called **zeta potential** (**ζ potential**) [2.4.100].

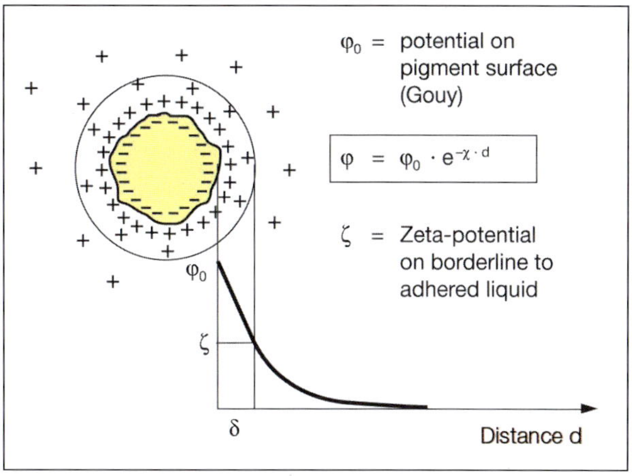

Figure 2.2.15: Schematic representation of the zeta potential as a measure of the degree of stabilisation of dispersed particles

$$u = k \cdot \frac{\varepsilon \cdot \zeta \cdot E}{\eta} \qquad \zeta = \text{electrokinetic potential}$$

This has a functional correlation with the potential φ_0 and therefore enables the repulsion barrier V_{max} to be calculated.

The migration rate of the particles is measured with the aid of laser doppler anemometry, as follows. Two coherent laser beams are passed through the sample. The particles in the measurement range cause fluctuating scatter of the laser light because of movement by the particles. This effect enables the migration rate and thus the ζ potential to be calculated.

It may be noted in conclusion that actual conditions are more complicated because, on one hand, the thickness of the adhering and thus shielding layer is a function of the speed of the pigments in the electrical field and, on the other, deformations associated with the removal and reestablishment of the double layer must be taken into account. It can be demonstrated, however, that such relaxation phenomena occur so quickly that they can generally be ignored.

Steric Stabilisation

While the DLVO theory provides a good approximation of the situation regarding pigment stabilisation in aqueous media, it is ineffective at describing the corresponding situation in a nonpolar environment. It has been shown in practice, however, that pigment dispersions can also be transformed into a stable state without electrical charges.

The first noteworthy attempt at entropic (steric) stabilisation of pigments in a liquid, nonionic environment was the model proposed by *Mackor* in 1950 [2.4.101]. To be stabilised, the pigment must be surrounded by polymer chains which are fixed at one point but which otherwise can move freely. When two particles stabilised in this way approach, the mobility of the polymer chains and thus the entropy are reduced. In accordance with the *Gibbs-Helmholtz* equation ΔS becomes negative and ΔG therefore positive. No further approach therefore occurs. As this type of stabilisation only refers to the entropic part, it is termed **entropic stabilisation**. Another term for it is steric stabilisation.

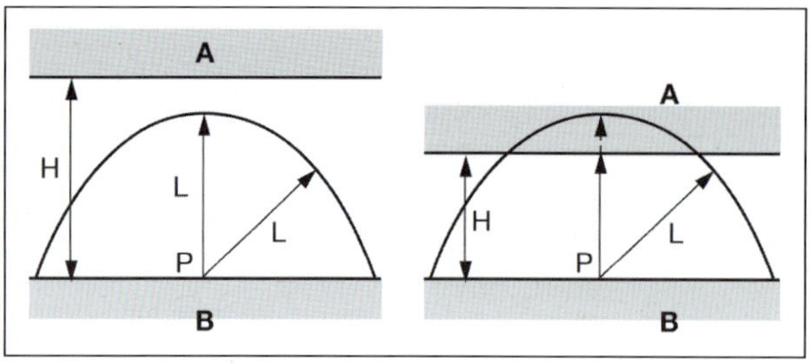

Figure 2.2.16: Model for explaining entropic stabilisation

Figure 2.2.16 shows a simple model for explaining entropic stabilisation. Given a sufficiently large distance between the pigment surfaces A and B the polymer chain which is fixed at point P but can otherwise move freely and has a length L is capable of adopting all the positions within the semi-circle of radius L. If the surfaces A and B approach to within a distance of $H < L$, the free mobility of the chains is impaired. The probability W_∞ is reduced to W_H, and the change in entropy ΔS is then

$$\Delta S = S_H - S_\infty, \qquad \text{because of} \quad S = k \cdot \ln W \quad \text{follows}$$

$$\Delta S = k \cdot \ln \frac{W_H}{W_\infty}$$

ΔS = change of entropy during approaching of particles
W = probability of replacement

If it is assumed that the free mobility of the adsorbed polymer threads of length L decreases linearly upon approach of the two solid surfaces, the following therefore holds for the entropically induced repulsion energy V^S_R if n_s molecules are present per unit surface area:

$$V_R^S = n_S \cdot k \cdot T \cdot \Theta_\infty \cdot \left(1 - \frac{H}{L}\right)$$

H = distance of particles
L = length of chain
Θ = coverage of surface

Θ_∞ stands for a completely covered surface. This equation, which applies only to planar surfaces, was expanded by *Sato* to include spherical particles. It then permits the energy barriers to be calculated for polymer chains of different mobility and of varying covering density and temperature [2.4.102].

A different approach attempts to interpret entropic stabilisation in terms of **osmotic effects**. These manifest themselves in an increase in concentration of the adsorbed molecules and thus in the osmotic pressure in the overlapping area upon approach and mingling of the movable polymer shells [2.4.103]. Comparison of the interpretations of stabilisation by means of osmotic and steric effects reveals that both models can be influenced in the same direction by the same variables. A high surface covering of freely movable polymer chains, uncoiled if at all possible, i.e. well solvated and of adequate chain length, yields the optimum pigment stabilisation.

The choice of solvent is also important when it comes to pigment stabilisation. Owing to the greater mobility of the solvent molecules, it is these which initially effect primary

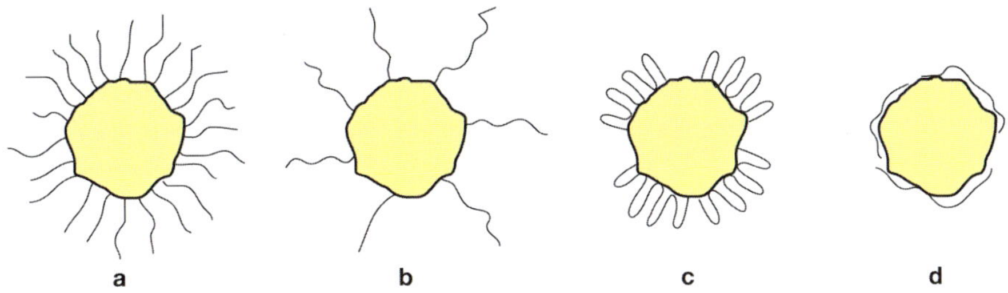

Figure 2.2.17: Schematic representation of different types of entropic stabilisation

Coating Material

wetting of the pigment cavities. Subsequent complete wetting conversion is only achieved if the wetting process of an adequate concentration of film forming agent molecules leads to a gain in free enthalpy. The use of relatively nonpolar solvents enhances instantaneous wetting with polar binder molecules and also wetting conversion of the already adsorbed solvent components.

In practice reagglomerations are occasionally observed which are not in line with the theory. Even where there is adequate electrical or steric stabilisation so-called bridging flocculation of pigment particles can occur. This effect is observed if resin molecules have several polar groups and are consequently capable of combining several pigment particles. Such destabilisation is also observed when either the chain length of the protective polymer shell exceeds an optimum level or a combination of elevated concentration and very polar solvents gives rise to twisting or looping of the chains in a manner similar to the formation of micelles [2.4.104].

Mixed Stabilisation

The process of correlating theoretical considerations on the stabilisation of pigment surfaces with practical operations is made more difficult by the fact that different stabilisation mechanisms can overlap. Mixed forms of particle stabilisation were discovered during calculations of electrostatically stabilised gold sols after the addition of nonionic polyethylene glycol. Thus, *Heller* and *Pugh* discovered that addition of polyethylene glycol to electrostatically stabilised gold sols enhances the already existing stabilisation [2.4.105]. This stabilisation can be further increased by raising the concentration and chain length of the hydrophilic but nonionogenic polyether. This effect is attributable to the mobility of the macromolecular chains of the polyether adsorbed on the gold surface and therefore constitutes supplementary steric stabilisation. Such combinations of **steric and electrostatic stabilisation** are also frequently observed in pigmented coating materials.

The zeta potential ζ, which is an indicator of the stability of pigment distributions stabilised exclusively by electrostatic charges, cannot be employed as the main stabilisation criterion in the case of combined stabilisation mechanisms because the thicker, immo-

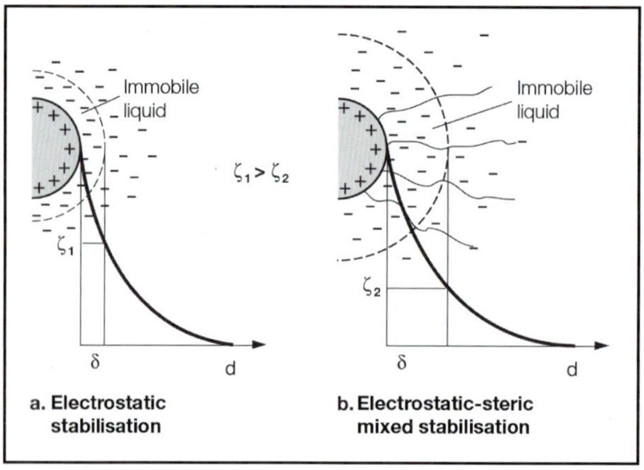

Figure 2.2.18: Mixed stabilisation of pigments and their effect on the zeta potential

bile liquid shell δ cannot be estimated. The zeta potential is lowered without any subsequent adverse effect on stabilisation. On the contrary, the presence of freely mobile molecule chains in two converging pigment particles leads to a reduction in entropy and thus to additional stabilisation.

Stabilisation Limits

Despite the best possible stabilisation, it is possible for the primary particles to converge on each other to within range of the forces of attraction if the kinetic energy is sufficiently high, for example at relatively high temperatures. The outcome of this is pigment flocculation. The flocs can frequently be separated from each other again by applying relatively low shear forces. However, under certain conditions, materials are forced together and irreversible material changes occur. Despite optimum dispersion and the best possible stabilisation, flocculation can occur as a result of solvent loss and inadequate covering of the pigment surfaces if the material to be dispersed is wrongly handled when the pigment concentrates are let down. Specific examples of this from daily coating practice are **pigment shock** and **solvent shock**.

While the mill base composition is based on the best conditions for dispersion (e.g. low solid content due to low viscosity and surface tension), the binder, solvent and pigment components in the finished coating material have to meet the requirements imposed on the coating to be produced and so the mill base frequently has a different composition from the end product.

During the let-down stage, the film forming agents employed will generally have a higher solid content than the mill base and can draw solvents from it, reducing the paste volume and causing the pigments to approach each other. This leads to **depletion flocculation** during the make-up process. The extensive destruction of the pigment dispersion is caused by a combination of osmotic processes induced by differences in concentration and differences in the mobility of the solvent and film forming agent molecules.

Pigment shock can be avoided by ensuring merely small differences in the solid contents of both the film forming agent in the dispersion and the let-down solution and by ensuring rapid homogenisation through intensive agitation.

The same problem can occur when operating with a lower rather than a higher binder concentration or, in an extreme case, only with solvents. Paint consumers experience this on a daily basis when adjusting the coating material to spray viscosity. Due to their high mobility the solvent molecules penetrate the pigmented coating material quickly and in large quantities. In the process of this it is possible for the stabilising film forming agent layer to be replaced locally by solvents to a greater or lesser degree. **Solvent shocks** should be expected whenever pigment dispersions are mixed with excessively

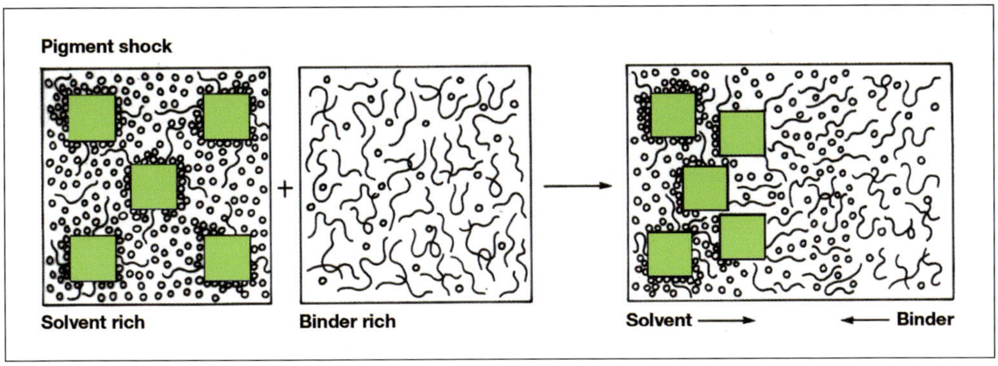

Figure 2.2.19: Pigment shock caused by differences in concentrations of the solvent in the paste and in the let-down resin

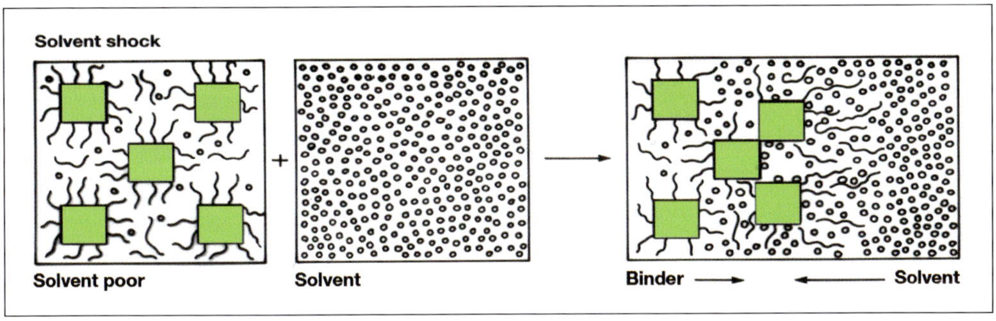

Figure 2.2.20: Solvent shock caused by inexpert addition of solvents

polar solvents. Adjustment to processing viscosity should therefore only be carried out at the dilution specified by the paint manufacturer in conjunction with thorough agitation.

The **reagglomeration** resulting from inexpert handling during let-down, mixing or adjusting leads to an increase in the average particle size and thus a reduction in tinting strength. Both phenomena are used to evaluate the degree of flocculation. The particle size, increased by flocculation, leads to accelerated sedimentation, and this can be used as an indicator (see Chapter 2.3.4). A further highly informative method for determining the degree of flocculation of coloured paints is the **rub-out test**. This consists in nothing more than manually rubbing a freshly applied paint film and subsequently evaluating the resulting colour change (see Chapter 2.3.4).

2.2.3.4 Optimum Mill Base Formulation

Since the paint manufacturer's goal is to convert as much pigment powder into stabilised primary particles within a short period of time at the lowest possible cost, mill base formulations are sometimes significantly different from those of the saleable coating materials. A proven method of determining the minimum quantity of resin solution for a given pigment and at the same time shortening the dispersion time for paints whose colour can only be adjusted with a number of pigments is to disperse these individually. The pigment pastes produced in this way are mixed later in accordance with a specified formulation and adjusted to the required final composition. Paint makers therefore produce numerous pastes and series of pastes with maximum pigment content. These are then adjusted to a standardised tinting strength and stocked. The handling of a customer

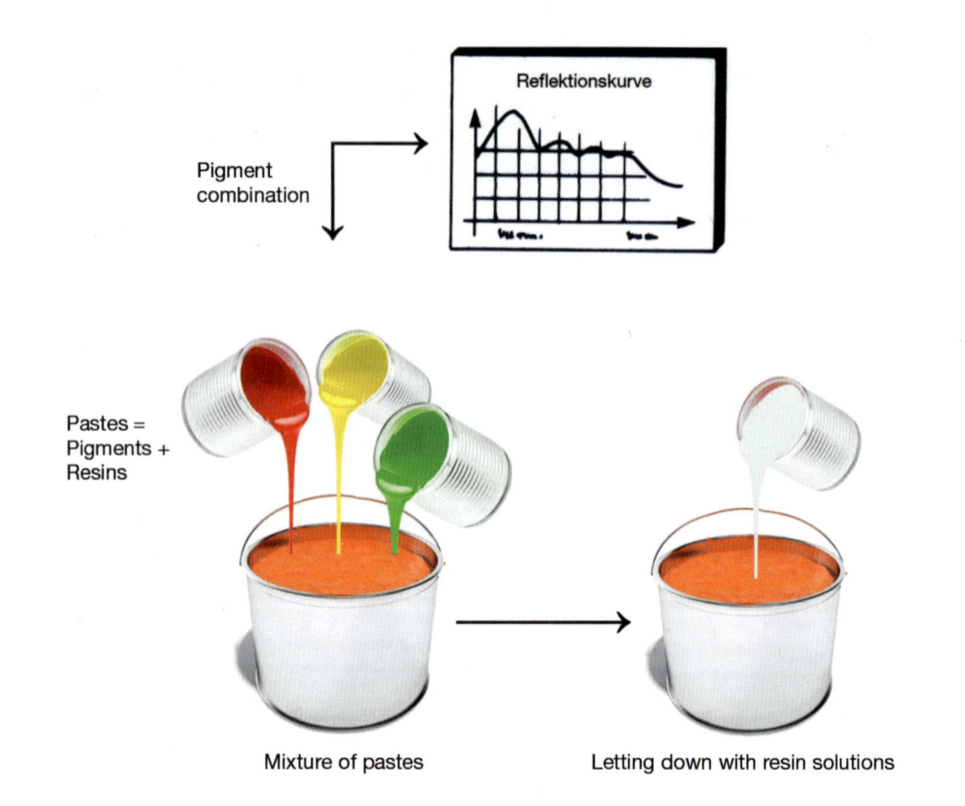

Figure 2.2.21: Paint-making with colour pastes *Source: Andrey Kuzmin/Magnum – www.Fotolia.com*

order is thus limited to mixing and make-up processes. The advantages are a major reduction in time input and a fast delivery capability, but these must be set against the costs of stocking sometimes more than 30 such pastes per quality series.

For the paint manufacturer it is particularly important for economic reasons and for the qualitative reasons described in the previous Chapter to determine the optimum mill base composition and to use this when drawing up the formulation. This is done by determining that solid content of a film forming agent solution at which a specified pigment quantity can be converted into a flowable mixture with the minimum amount of resin solution.

For better evaluation the pigment absorption capability is quantified with the aid of indicators. An important indicator for determining the optimum mill base composition is the **yield value point**. This is defined as the volume of a film forming agent solution which is required to convert a specified pigment quantity into a flowable state. The characterising feature of this property is that the mixture flows in a continuous thread from a glass rod (see Chapter 2.2.3.2).

If the compositions of the ternary mixtures of film forming agent, solvent and pigment determined from the individual yield points are plotted in a concentration triangle, the ranges of the flowable mixtures can be attractively presented in chart form. If the two ends of the **yield point curve** are extended to the triangle sides and the area of the supercritical pigmentation is hatched, the area enclosed by the curve with the exception of the hatched, supercritical area contains all the flowable binder/solvent/pigment mixtures. Supercritically pigmented mill base compositions lie above the critical pigment volume concentration (CPVC), at which the volume of the film forming agent is lower than that of the spaces between the pigments (see above). The CPVC can be calculated from the bulk volume of the pigments and must be converted into the critical pigment weight concentrations (CPWC) for entering in the concentration triangle.

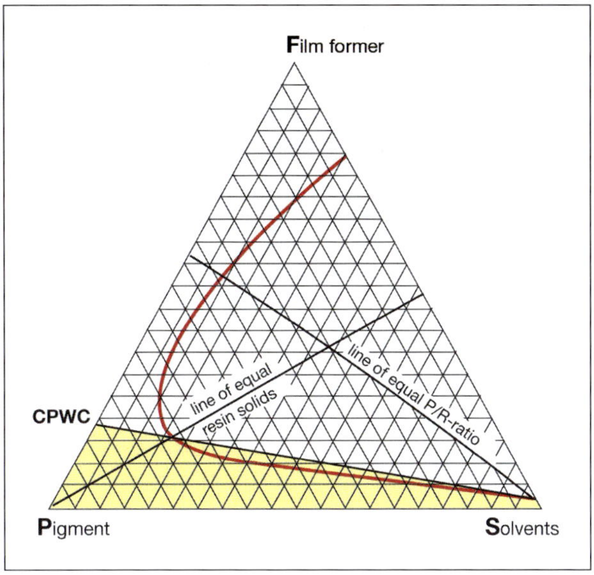

Figure 2.2.22: Concentration triangle showing how a flow curve varies with film forming agent, solvent and pigment

In most cases further corrections to the mill base formulation have to be made to achieve cost effective production. Reasons for this include the fact that the method of flow point determination fails to account for the rheological behaviour of the paste after the dispersion process, and the unique properties of the disperser. Supplementary experiments with varying raw material mixtures are therefore essential in the search to identify the optimum composition for cost effective production under realistic conditions. The rheological properties of the mill base change significantly with decrease in particle size. This frequently involves an increase in the viscosity and the degree of pseudoplasticity. However, the yield point method simplifies work in that a large proportion of the theoretically possible mill base mixtures can be excluded.

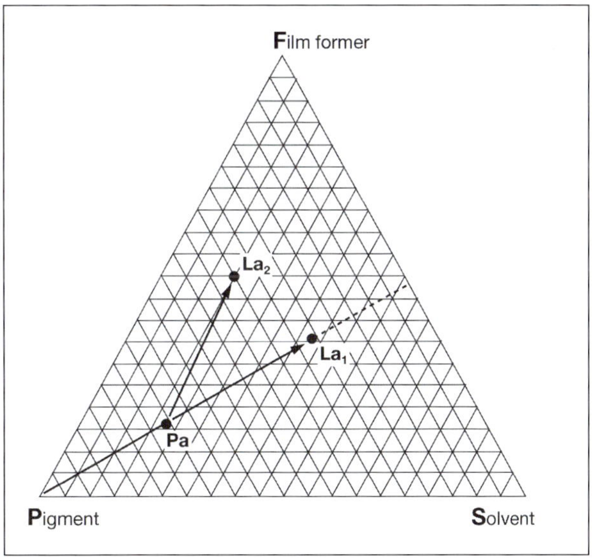

Figure 2.2.23: Using the concentration triangle to explain pigment shock during let-down

Owing to the very different **stresses exerted on the mill base** in the different dispersers the formulations must be adapted to the particular machines. Dissolver formulations differ substantially from those designed for agitator mills or three-roll mills. Dissolvers achieve only moderate shear gradients. Optimum dispersion of pigments is therefore contingent on paying particular attention to the instantaneous viscosity under shear. It is essential to prevent the mill base from developing thixotropic properties.

Roll mills whose rolls are operated at different specified speeds feature a constant shear gradient because the gap between the rolls is constant. Mill base mixtures can acquire a relatively high viscosity because of forced conveyance through the gap. The resin solid content and the binder proportion or the pigmentation level in the mill base are correspondingly high in roll mill formulations.

The high rotational speed of agitator mills requires lower viscosities and frequently, therefore, higher solvent contents. The adaptation of the rheological properties in formulations for agitator mills often leads to raw material mixtures with extremely low levels of film forming agent. The latter should be borne in mind in order that pigment shocks may be avoided when pigment pastes are let down. The risk of a pigment shock can be identified graphically in the concentration triangle of pigment, film forming agent and solvent when the compositions of the pigment concentrate Pa and the saleable coating material La do not lie on the extension of the line formed by connecting the point of 100 % pigment concentration with the point of position of the paste. For La_2 there is a serious risk of a shock whereas La_1 proves to be noncritical during let-down (see Figure 2.2.23).

2.2.4 Production of Coating Materials

The formulations drawn up by the laboratory must be supplemented by further detailed standard operating procedures and test specifications for large scale production. The latter refer equally to technical processing requirements and to quality. Manufacturing instructions contain information about the sequence in which raw materials are added, machine settings and temperatures for homogenization, dispersion and filtration systems, information on process monitoring and interim tests. Information on approval tests, instructions for filling methods, the container type and size to be chosen, together with safety information and container labelling must be clearly specified in the manufacturing instructions.

Modern make-to-order production can then proceed by one of three methods: **Direct dispersion**, production of pastes with downstream let-down and the production of mixing paints. While the mixing paint principle has won through for the very varied colour requirements of decorative and automotive refinishing paints, large scale users tend to prefer **paste production** and direct dispersion as their method of production. Mixing paints are coating materials with a standardised colour and tinting strength which mostly contain only one pigment and are turned into the final product by simply stirring together, without the addition of a binder. They are stored by paint users such as bodyshops and dealers in purpose-built mixing machines and converted into the end products in accordance with given **mixing formulations**.

Direct dispersions and products to be made from pastes are produced by the paint manufacturer and sold as ready-to-use paint.

Paste production of individual pigments has proved its worth, despite the expense of storing the standardised pastes, because dispersing grinding batches is more effective and thus more economical. The individual pigment pastes are combined in particular mixing ratios as per the colour requested by the processor, with make-up to coating material specification simply by admixing of just film forming agent solutions and additives.

2.2.4.1 Processing Sequences in a Paint Plant

The sequence of steps from raw material supply, pigment dispersion and paste production, paint make-up and transfer to containers is shown in the overview in Figure 2.2.24.

The material and production flow passes through four floors in a vertical direction. The top floor houses the **interim storage** and feed systems of the materials required for a batch before the dispersing process. Solid materials, such as pigments and extenders, are moved in bags, drums or cartons, while materials in liquid and paste form are transported in drums, cans or pails. The film forming agent solutions and solvents, which are

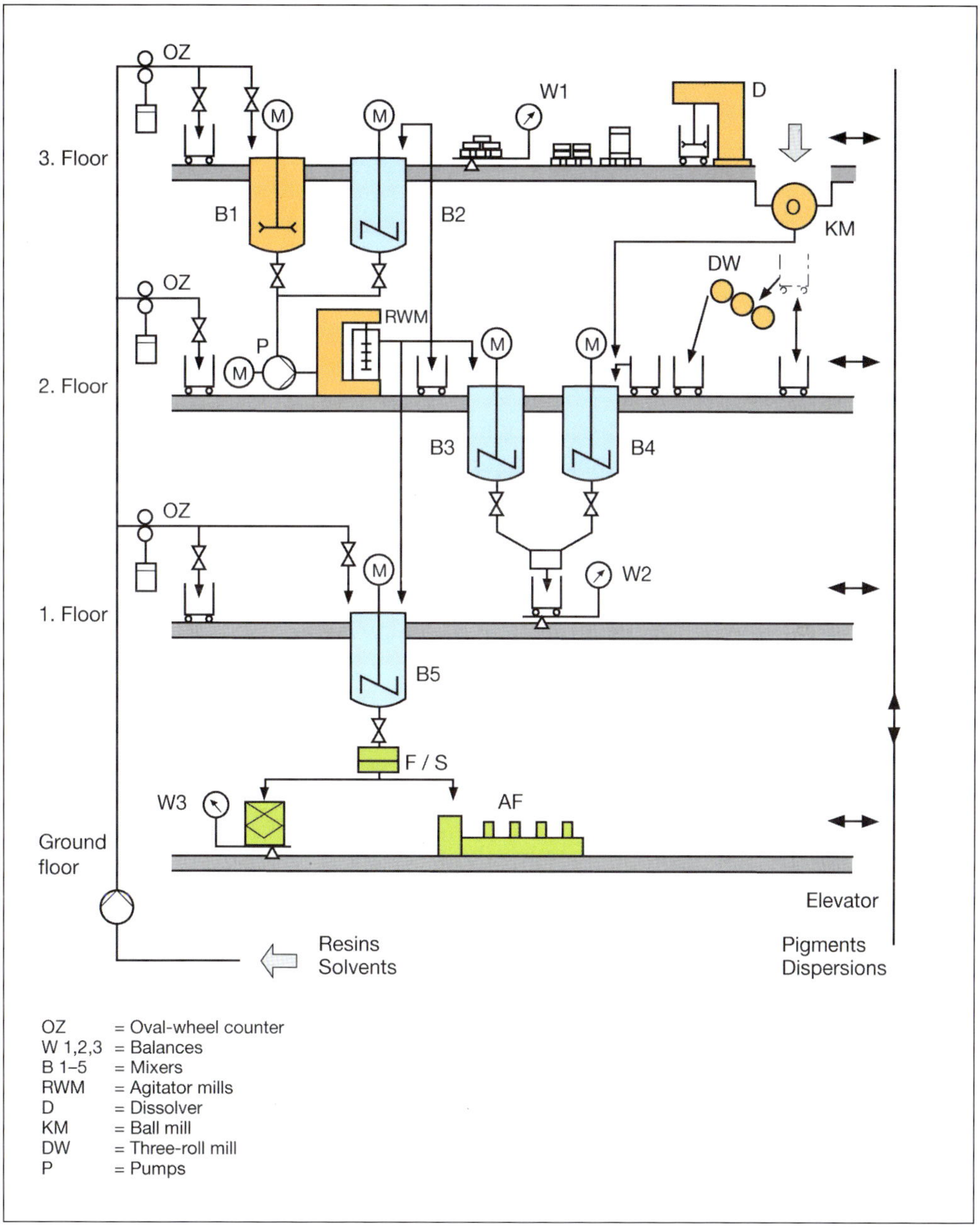

Figure 2.2.24: Production sequence in a paint plant

often stored in tanks and transported through pipelines, are supplied to mobile containers, dissolvers or premixers via supplementary units fitted with flow meters.

The third floor is where **fine dispersion** in mill units takes place. The material to be dispersed is stored as a pigment paste in mixers, and its tinting strength is standardised and approved.

On the next floor down can be found an appropriate number of completing mixers of different sizes corresponding to the range of formulations processed. In mixed production, material is conveyed from the paste mixers into the final mixers via specific dispensing units. For production from scratch the material to be dispersed is conveyed directly from the dispersers into the final mixers. Samples are then taken from the final mixers for testing and approval.

On the ground floor the material undergoes cleaning processes involving screening and filtration, sometimes also including centrifuging, and approval and is then conveyed to the **filling lines**. The material is then transferred into tankers, containers, drums or small receptacles.

Various means of **checking** the material flow to achieve reproducible product quality are possible. On one hand, the raw materials can be prebatched and supplied in this form to the relevant production facilities. On the other, the raw materials are weighed together with the individual production containers. This process involves placing these containers on load cells.

Since the mixers and machines frequently have to be changed to suit the product type and batch size, this leads to a complex material flow system. A sensible approach is to assign machine and instrument groups to particular product groups by compatibility features based on colour or chemical properties. This enables the cleaning input for the machines to be minimised. **Quality tests** are carried out in almost all functional areas where there is ongoing production. The correction strategy, particularly where colour settings are concerned, is designed such that the target quality value is achieved incrementally from one side only. Correction here does not mean the elimination of faults but, in most cases, the addition of amounts of selected raw materials which were deliberately underdispensed initially.

The essential basic process engineering operations in the manufacture of pigmented coating materials are the dispersion, dispensing (see Chapter 4.2.1) and conveying of liquids and solids, homogenisation generally by means of stirring and kneading and the subsequent screening, filtration or centrifuging stages. The theoretical principles and equipment for these basic process engineering operations are explained in greater detail below.

2.2.4.2 Agitation and Agitators

The basic mechanical process engineering operations for combining materials by means of **agitation** consist essentially in the dissolving of resins, the mixing of liquids, and the premixing of material to be dispersed or the tinting of colour pastes. In all cases the aim is to eliminate inhomogeneities in material mixtures. These can manifest themselves in differences in concentration, density or colour. Differences in temperature or refractive indices can also be used to characterise the mixing quality.

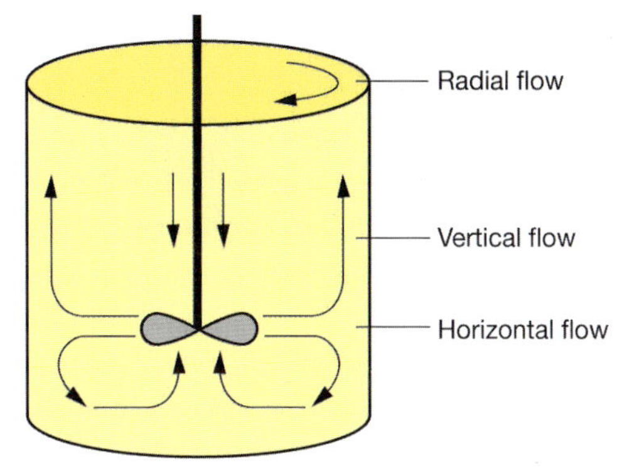

Figure 2.2.25: Characteristic flow patterns in stirred vessels

Mixing quality is measured in terms of local deviations from the mean in the abovementioned variables. If a sufficient number of individual readings are taken at various locations in the material to be homogenised, the degree of homogeneity can be quantified by determining the mean and the standard deviation. The standard deviation is indirectly proportional to the mixing quality and converges on zero when the individual components have achieved the optimum level of dispersion.

Homogenisation of liquids or liquid-solid mixtures is carried out by generating the most irregular flow patterns possible, so that the desired degree of homogenisation can be achieved as quickly as possible with minimum energy input. Where random shifts in position through free product flow are prevented by excessively high viscosities and are attainable only by means of pressure or shear forces, this is termed **kneading**.

Homogenisation of free flowing material mixtures during agitation is achieved by creating vertical and horizontal flow patterns. These are generated by rotating agitator blades. There are numerous **types of agitator**, and their choice depends on the type of material and the desired degree of homogenisation.

Homogenisation which has to be carried out under economic conditions is best achieved when horizontal and vertical circular motions are present in the mixing vessel at the same time. When materials of different density are being homogenised, vertical flow patterns are recommended because the heavier components have a tendency to settle. Movement of the entire amount of material together in the mixing vessel, which occurs frequently during the agitation process, should be avoided wherever possible because of the unwanted separation caused by centrifugal forces.

The mixing time is short if the components to be mixed undergo a large number of changes of location. The latter can occur at the agitator itself or within the flow patterns generated by it. They can be achieved by impact, flow around obstacles, the intersection of flow patterns, and speed differences at the interfaces of parallel flow patterns.

Two different agitator designs have become established. Either the entire contents of the vessel are circulated by means of a large agitator surface area and slow material motion or small partial quantities are transported at high speed by means of impact and mixing eddies with a small agitator surface area and a high rotational speed. The first concept is represented by blade agitators, and the second by propeller agitators.

At high agitation speeds the occurrence of conical eddies can be observed which entrain air into the mixture. To avoid this, either baffles are fitted in the agitators or the agitator shafts are configured eccentrically.

A broad range of different consistencies has to be covered in the paint and coatings industry when homogenising materials. The viscosity of the material to be agitated

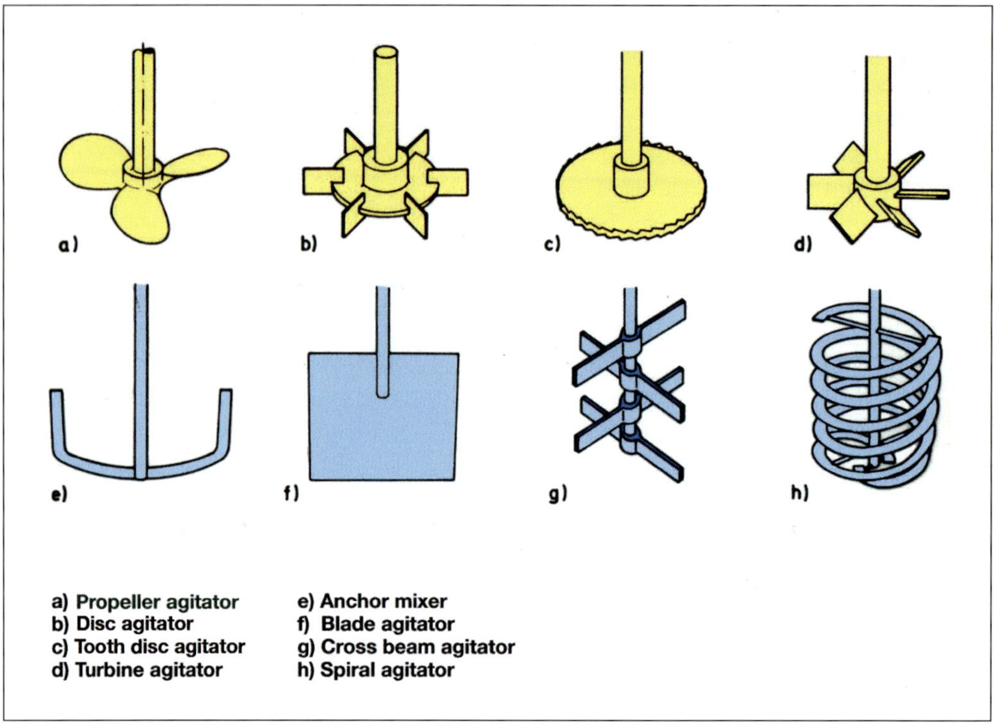

Figure 2.2.26: Different types of agitator

is therefore adjusted continually by altering the rotational speed or by means of gear mechanisms and interchangeable agitator blades.

Figure 2.2.26 shows a small selection of technically important agitator types. The **anchor mixer** is an old design which is unsuitable from an economic perspective. It rotates slowly and is now only used in resin reactors to achieve better heat transfer and eliminate vapour bubbles. It is unsuitable for homogenisation tasks because it lacks vertical flow.

The blade agitator is also a simple agitator design. It is mainly used to maintain mixtures which have already been homogenised. Due to the size of the agitator blades there is the risk of radial motion being induced in the agitated material. Such motion can be prevented or reduced by means of baffles. Vertical flow patterns are not very pronounced in blade agitators.

Flow is different in the beater or its modern variant, the **multistage impulse agitator**. Both achieve intensive vertical motion.

The **propeller agitator** moves part of the mill base at high speed. It entrains the material axially and discharges it in turn axially. This guarantees good vertical flow. The undesired circular motion can be prevented by rotating rings. The propeller agitator copes with a broad range of consistencies and has therefore become accepted in practical paint production operations as a universal machine. Its disadvantage is that the generated flow pattern is often not turbulent enough, as a result of which long mixing times have to be accepted.

The **agitator shaft** and its seal are important when it comes to avoiding contamination of the paint material in all stationary open or closed agitators. The shaft seal can be of a noncontact design if no pressure differences exist or are generated between the interior of the vessel and the atmosphere.

Shaft seals can be made of various materials. If the temperature remains below 100 °C during mixing of the components, simple ring seals made of elastomeric plastics are used as standard. For overpressures of up to 2 bar or a vacuum of down to 0.1 bar moulded rings with lubricants are used for sealing to temperatures of up to 130 °C. For extreme sealing requirements, high pressure differences and elevated thermal loads it makes sense to use mechanical seals with a supplementary sealing liquid and heat discharge via an air cooling system. It is essential that the use of silicone-containing sealing liquids and lubricants be avoided for the reasons mentioned previously.

Safety regulations must be observed when agitators and other mixing equipment are operated. Explosion protection systems, rigorous earthing of all agitators and vessels, and the fitting of grilles in vessel openings are important measures for ensuring accident-free operation.

Useful data relating to the correct design of agitators and their adaptation to the material to be mixed can be obtained from mathematical equations employed in **fluid mechanics**. A number of preliminary considerations must be borne in mind.

Firstly, the *Reynolds number* Re, which is important in calculating flow processes in pipes, must be redefined for agitation processes. *Büche* replaced the flow rate u by the blade tip speed ω r and the pipe diameter d by the diameter of the agitator blade in Re [2.4.106]. Dividing by π yields the *Reynolds' number* for agitation processes Re_M.

$$Re = \frac{\rho \cdot d \cdot u}{\eta} \quad \text{because} \quad u = \omega \cdot r \quad \text{and} \quad \omega = 2 \cdot \pi \cdot n \quad \text{becomes} \quad Re' = \frac{\rho \cdot d^2 \cdot \pi \cdot n}{\eta}$$

$$Re_M = \frac{Re'}{\pi} = \frac{\rho \cdot d^2 \cdot n}{\eta}$$

When $Re_M < 30$, laminar flow is present. A figure between 30 and 100,000 represents a transitional range, culminating in turbulence at $Re_M > 100{,}000$.

Important practical information on the power consumption and mixing time of agitator mills can be obtained from the generally applicable flow resistance law.

$$F_W = C_W(Re) \cdot \frac{\rho_L}{2} \cdot u^2 \cdot A$$

The resistance F_w acting on the agitator blade is composed of the drag coefficient C_w, the density of the liquid ρ_L, the surface area A subject to flow and the square of the path speed u. The value for u is the path speed at the outermost point of the agitator blade. Since the power L corresponds to the work per unit time, so that F_W s/t = F_W u, the following results:

$$L = C_W(Re) \cdot \frac{\rho_L}{2} \cdot u^3 \cdot A$$

If the speed u is replaced by ωr = n π d where n = rotational speed, d = diameter and r = radius of the agitator, and if it is agreed that the height and diameter of an agitator

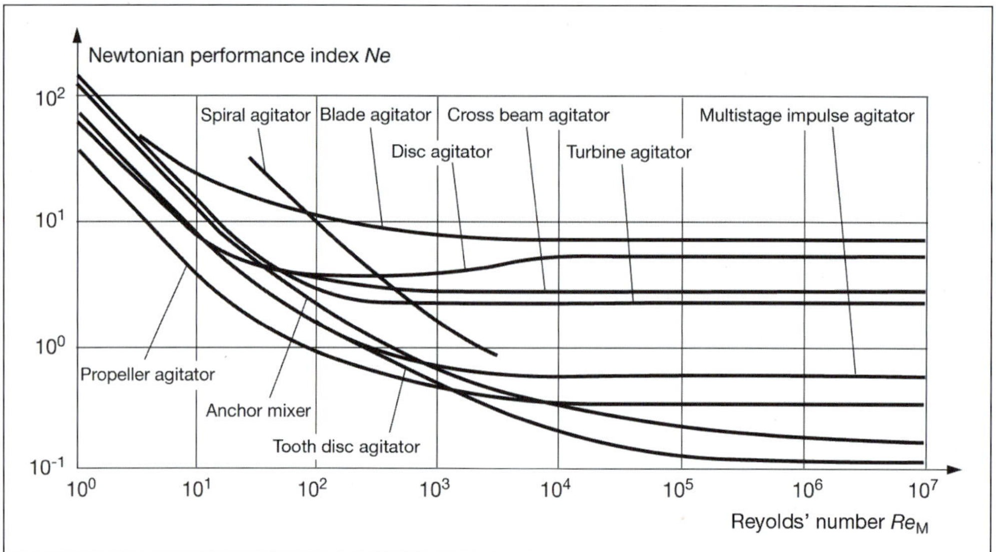

Figure 2.2.27: Diagram for determining the performance index Ne of agitator mills

are always in the same ratio to each other (h/d = k), irrespective of the total surface area and thus A = k d^2, because A = h d, it follows that

$$L = C_W(Re) \cdot \frac{\pi^3}{2} \cdot k \cdot \rho_L \cdot u^3 \cdot d^5$$

If the constant values and the variables which are a function of Re are aggregated, the result is the *Newtonian* performance index $Ne(Re_M)$.

$$L = Ne(Re_M) \cdot \rho_L \cdot n^3 \cdot d^5$$

Ne has to be established when the power consumption of model agitators is being determined as a function of the rotational speed. A geometrical similarity between the model and production system is essential for transfer of the results to large scale production. In order to calculate the power consumption, Re_M has to be determined for the relevant mixing process and Ne has to be located in tables or graphs so that the power consumption L can then be established.

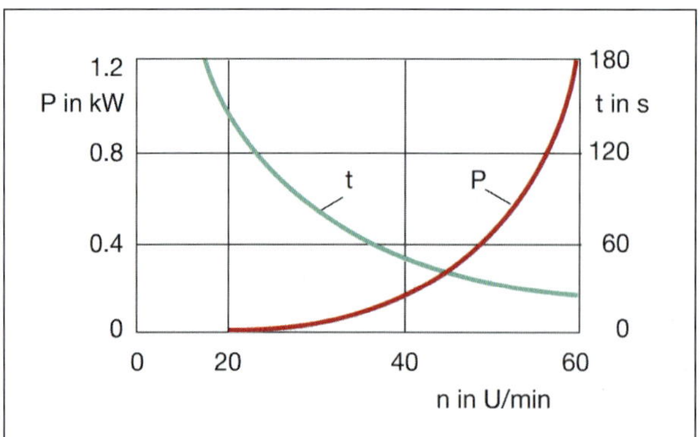

Figure 2.2.28: Grinding power and time as a function of the rotational speed n of an agitator

Cost effective operation of mixing systems requires low-energy homogenisation and also the shortest possible **mixing time**. The energy required for mixing is made up of the product of power and time. Since, however, the power consumption of an agitator mill increases to the power of three with the rotational speed, but the reduction in the agitation time only

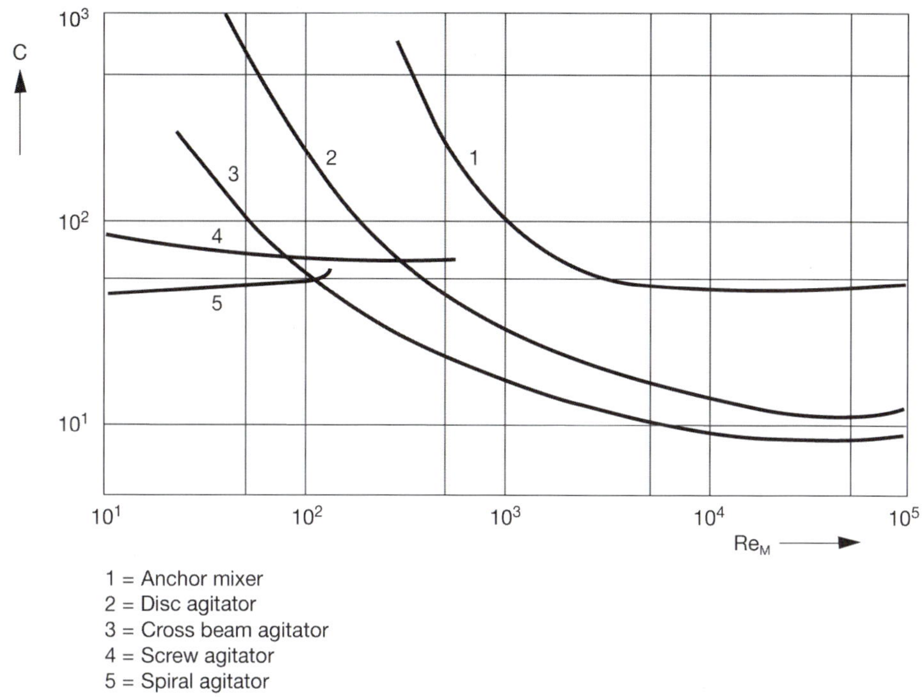

Figure 2.2.29: Mixing coefficient C as a function of the Reynolds number Re for various agitator types

decreases by the power of one, agitation times can only be reduced by means of high energy input.

The mixing coefficient C, which is a function of the *Reynolds'* number, is used to calculate the mixing time required for complete homogenisation. This has been determined experimentally for all the major agitator types and is transferable to larger agitators, provided their geometry is similar. Figure 2.2.29 shows the curve of the mixing coefficient as a function of the number for the relevant flow state. The mixing time t_M itself is defined as $C(Re_M)$.

As shown, an increase in the agitation speed causes an increase to the power of three in the power consumption in the energy calculation. Nonetheless, almost negligibly shorter agitator utilisation with a corresponding added value and order book situation can be more sensible than operating more slowly to save energy. The cost of the agitation energy relative to the total costs of the agitation process is also a decisive factor.

It may be stated in conclusion that mixing and agitation are the commonest process engineering operations in a paint plant. Numerous types of agitation equipment, operated either in stationary or mobile mode, are used for all kinds of homogenisation. Aside from the attainable mixing quality, the cost effectiveness of the process is of special interest.

A problematic feature is that the theoretical considerations are limited in general to Newtonian flow properties. Changes in rheological properties during the manufacturing process of coating materials with non-Newtonian mixtures cannot generally be fully calculated.

2.2.4.3 Dispersing and Dispersers

Machines for Wet Paints

The task of dispersers is to separate pigment agglomerates in viscous fluids from each other by the transmission of shear forces. Judicious raw material selection and formulation simultaneously creates the conditions additionally required for stabilising coverage of the pigment surfaces. The shear forces generated in the machines by flow are enhanced by supplementary rotation of the particles which is induced by centrifugal forces. In addition, there are also tensile and compressive loads at work which help in breaking down the agglomerates. Since the transmission of shear forces right down to the level of the primary particles is a necessity, flow states must be ensured which support the transmission of forces irrespective of particle size.

The following relation shows that such a requirement cannot be met under turbulent flow. The transmissible separation forces τ per unit surface area are proportional to the particle size x_{Pi}, the energy dissipation density N/V and to the density of the fluid phase ρ_{Fl} and are inversely proportional to the pigment density $\rho_{Pi.}$ Consequently, the transmissible separation forces diminish with decrease in particle size during the dispersion process, with the result that it is not possible to go below a certain particle size.

$$\tau = 2 \cdot \rho_{Fl} \cdot \left(\frac{N}{V}\right)^{\frac{2}{3}} \cdot \left(\frac{x_{Pi}}{\rho_{Pi}}\right)^{\frac{2}{3}}$$

$\frac{N}{V}$ = energy dissipation density
ρ_{Fl} = density of liquid
ρ_{Pi} = density of pigment
x_{Pi} = particle size of agglomerates

If the maximum attainable energy dissipation density of 10^3 to 10^4 kW/m^3 is used in this formula, the maximum attainable decomposition will deliver particle sizes ranging from 1 to 10 µm, with the precise figure depending on the agglomerate strength. Consequently, primary particles with sizes of less than 1 µm can only be achieved with a power input of >10^5 kW/m^3.

Under laminar flow, on the other hand, the situation is different. In accordance with the Newtonian equation $\tau = \eta$ D, the shear stress τ is not a function of particle size. The energy introduced by viscous resin solutions is then independent of particle size.

$$\tau = 2.5 \cdot \eta_{Fl} \cdot D$$

In this equation the velocity gradient D and the viscosity η_{Fl} are the only variables with an impact on the dispersion process. Since the shear energy is converted quantitatively into heat (dissipated), and rises by the power of two with increase in veloc-ity gradient D (see Chapter 2.3.3), steps must be taken to ensure an effective means of cooling for high power dispersers.

The machines used for dispersing pigments are varied, and come in a wide range of designs. Nonetheless, all have the same operating principle. They transmit high shear forces in laminar flow gradients to separate the agglomerates. The most important machines used commonly in large scale production are dissolvers, agitator mills and roll mills.

Dissolvers are disc agitators which under certain conditions can transmit enough shear stress to pigments to effect dispersion. The conditions for this are met if the agitator disc and the agitator vessel are matched in size and this is complemented by appropriate agitator disc sizes and shapes, an adequate rotational speed, an optimised fill level and adapted mill base formulations. Even when all these conditions are met, the shear forces transmitted by dissolvers are not high. Dissolvers are therefore only suitable for the dispersion of easily dispersed pigments, such as for emulsion paints and decorative paints. Different dispersion units must be used if requirements are more demanding. Dissolvers are then only used for preliminary homogenisation or preliminary dispersion. This enables the throughputs of the actual dispersion systems to be significantly increased (see below).

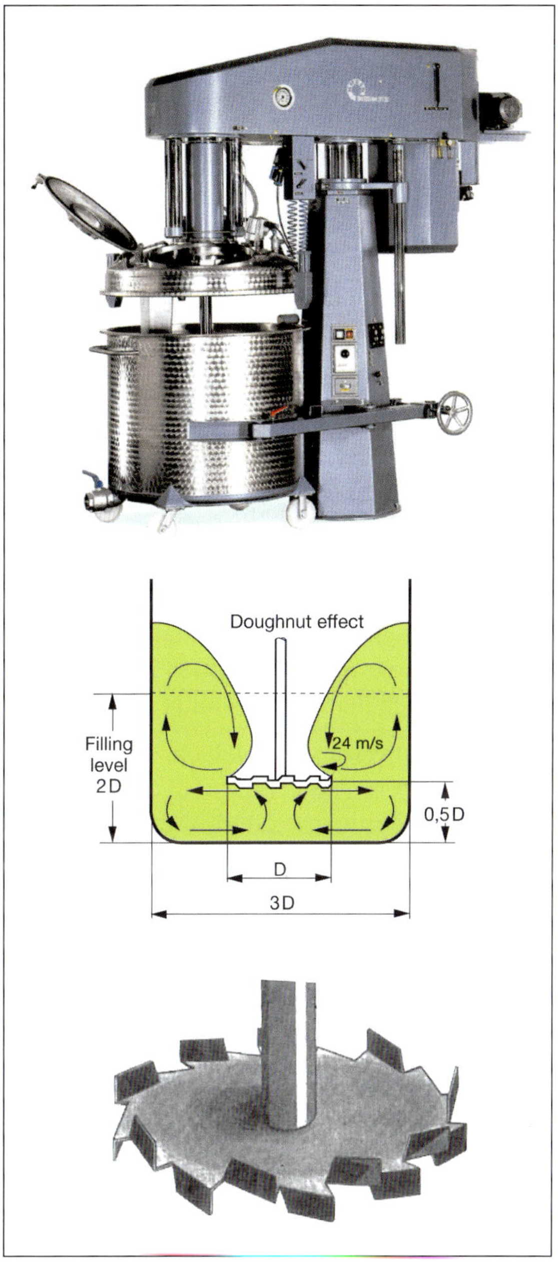

Figure 2.2.30: Schematic diagram of flow in dissolvers and pictures of a dissolver and dissolver disc Source: Netzsch

To generate the necessary laminar flow the sizes of the vessels and agitator discs and the fill level and disc configuration must, as mentioned above, be matched to each other. Optimum dispersion results are achieved if the vessel diameter is 2.5 to 3 times the size of the disc diameter and if the distance from the agitator disc to the base corresponds to 0.5 and the fill level to 2 disc diameters. Under such conditions dispersions are possible once the tangential speed u_A at the periphery of the disc has reached 24 m/s. The limit speed is therefore a function of the disc size.

$$u_A = \omega \cdot r \geq 24\ \text{m/s} \quad \text{i.e.}$$

$$24 = 2 \cdot \pi \cdot n \cdot r \quad \text{or} \quad n \geq \frac{7.5}{D}\ s^{-1}$$

The design of the agitator disc (see Figure 2.2.30) also has an impact on the dispersal effect. The agitator disc is similar to the blade of a circular saw and is fitted with teeth

Coating Material

which are set at 90°. This toothing is only of minor importance for the dispersion process. It merely ensures better circulation of the mill base and additional alternation between tensile and compressive loading of the material to be dispersed. For laminar flow gradients the disc must effect horizontal discharge of the mill base. It is therefore essential for the agitator shaft to be absolutely vertical.

The main part of the dispersal work is not performed, as might at first be thought, in the region of the disc but rather horizontally behind the disc. As a result of the varying lengths of the flow paths and the different braking forces in the upper and lower circulation regions sufficiently high speed differentials are generated behind the disc to create dispersion. Furthermore, the rapid alternation of brief overpressure and underpressure zones promotes the thorough moistening, wetting and mechanical separation of the agglomerates.

The dispersal effect is also influenced by the mill base composition. At a binder solution viscosity of only 100 mPa s the viscosity of the mill base increases, for example, to 3 to 4 Pa s upon addition of the pigment. The dissolver reacts sensitively to the rheological properties of the mill base batches. While dilatancy is a desirable flow characteristic, there is the danger with thixotropic and very pseudoplastic mill base mixtures of incomplete circulation and thus incomplete dispersion of the pigments. With dilatant flow, however, the power consumption can rise dramatically when the rotational speed is increased, such that the possibility of the agitator shaft fracturing cannot be excluded.

If the shear forces and rheological properties are conducive to pigment dispersion, dissolvers represent a rapid type of disperser. Generally, the dispersion process is complete in less than 15 minutes. No further reduction in the fineness of the mill base will take place, even if agitation is continued for an extremely long period.

In operation, a dissolver is a stationary machine which can be loaded with mobile mill base containers. If the mixing vessel is also permanently installed, the dissolver will usually have a volume of between 1,000 and 5,000 litres. Dissolvers with mobile mixing vessels can be moved hydraulically upwards and downwards for charging and for replacement of the agitator discs. The rotational speed is generally infinitely variable. The size and thus the power of small, medium and large dissolvers varies from less than 25 kW right up to 200 kW.

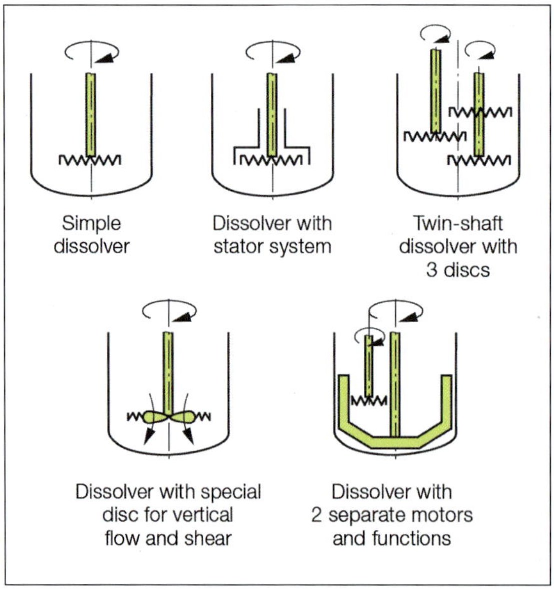

Figure 2.2.31: Schematic drawings of different types of dissolver

Apart from conventional dissolvers numerous special designs will be found in paint plants. They differ in the type and number of agitator discs and in the configuration of the agitator vessels. Of the multishaft machines, the **twin-shaft dissolver** is the most commonly encountered

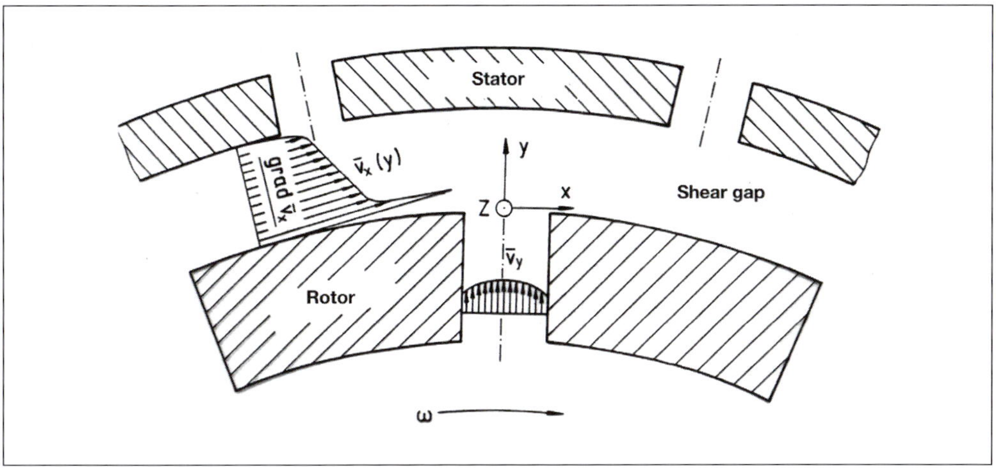

Figure 2.2.32: Schematic drawing and function of an in-line dissolver

disperser. It has agitator discs which are offset in height and mounted on two separate shafts, and so offers better agitation properties (see Figure 2.2.31).

Figure 2.2.33: Three roll mill

Rotor-stator agitators (inline dissolvers) [2.4.107] also have a successful track record for high speed homogenisation, particularly for the manufacture of water-borne paints. The high rotational speed of the rotor makes for rapid, effective distribution of the formulation ingredients, particularly when systems of high pseudoplasticity have to be processed. They feature small dispersion chambers and are mainly used for continuous processes.

The principle of dispersion on **friction rolls** is implemented in single-roll, two-roll, three-roll and multi-roll mills. The only one of any significance for the paint and coatings industry is the three-roll mill which has become less important because its throughput is low compared with other types of dispersing systems. Its use is now limited to the manufacture of fillers, highly viscous printing inks and, because of its gentle treatment of the pigment surfaces, to the dispersion of sensitive surface treated pigments and heat sensitive pastes.

Three-roll mills consist essentially of three metal or plastic cylinders arranged one after the other. The centre roll has a fixed mount, while the two outer ones are pressed against it under pressures of up to 1,000 bar. Rotation of the individual rolls at different speeds generates high shear forces on viscous liquids in the two nips. A speed ratio of 1:3:9 has repeatedly proved in numerous experiments to yield the best and most cost effective dispersing results.

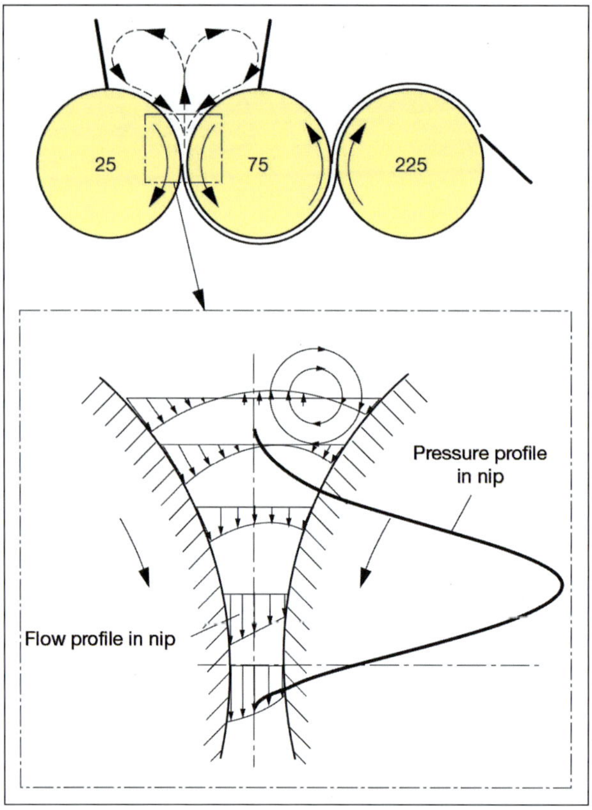

Figure 2.2.34: Functional principle of the three roll mill and pressure profile in the nip

In modern systems a gear mechanism can be used to set varying peripheral speeds. The mill base throughput $\dot{V}$ is determined by the roll width b, rotational speed u_m and nip width s.

$$\dot{V} = 1.225 \cdot u_m \cdot s \cdot b$$

where u_m is the arithmetic mean of the rotational speeds of the two rolls.

The rolls themselves generally have an extremely hard outer lining which is 1 cm thick. Beneath this there are softer elements on the inside. These enable better heat transfer to the cooling hoses inside the rolls. The contact pressure for the rolls is provided via spindles or by motors with a corresponding hydraulic system. Under operating conditions, a nip of approximately 100 µm is left which is many times larger than the diameter of the pigments to be dispersed.

Before a three-roll mill is used, the mill base should undergo preliminary dispersion in either a dissolver or a kneader, the choice between the two depending on the consistency. The mill base is then transferred to a feed device between the first two rolls. The mill base is drawn into the nip because it adheres to the wall of the rolls.

As a result of the shear stress in the direction of flow and the diminishing nip gap, pressure builds up in addition to the shear load and helps to promote dispersion. This external pressure also causes pressure rises inside the agglomerates which may still contain air and promotes the comminution processes. The width and height of the pressure profile depend on the diameter of the rolls and the viscosity of the mill base. The thinner the mill base and the smaller the roll diameter, the narrower the profile is. The effect is enhanced by the shear of the material, caused by the different roll speeds. The pressure is at its maximum shortly before the narrowest part of the nip. This is followed directly by a zone of underpressure, as a result of which a sudden pressure drop causes pigment wetting which is advantageous for the mill base. The mill base is then separated at the nip exit and distributed onto both rolls in a ratio corresponding to the rotational speeds. At a speed ratio of 1:3, the amount of material transferred to the second roll is in the same proportion. This also applies to transfer from the second to the third rolls.

The dispersed mill base adhering to the third roll is then removed from the roll by a doctor blade and transferred to a storage container.

As with the other disperser types, the viscosity of the material to be dispersed needs to be adjusted to the three-roll process. If it is too low, this leads to spraying in the filling gap and to inadequate shear forces; high viscosities, on the other hand, do not cause problems by comparison. However, because of the high energy dissipation and the corresponding heat build up, appropriate measures must be taken to ensure good cooling, or supplementary high boiling liquids must be added to prevent the rolls from drying out. The latter would cause irrevocable damage in the form of grooves in the roll surface.

Another, albeit less important, application nowadays for the three-roll mill is the **flushing of pigments**. In flushing, the pigment cake, which is generally aqueous, still wet and therefore not yet or only slightly agglomerated during pigment production, is not dried by the usual process, but rather is transferred from the aqueous to the organic phase by treatment with organic binder solutions in kneaders. The purpose of subsequent roll dispersion is not so much to separate the agglomerates as to completely remove the residual water. Consequently, the rolls are not cooled, but are heated with steam to ensure faster removal of the water. This method enables high tinting strengths and good fineness levels to be achieved with pigments that are hard to disperse. Carbon black pastes manufactured in this way, for example, are used to produce haze free, glossy, deep black topcoats.

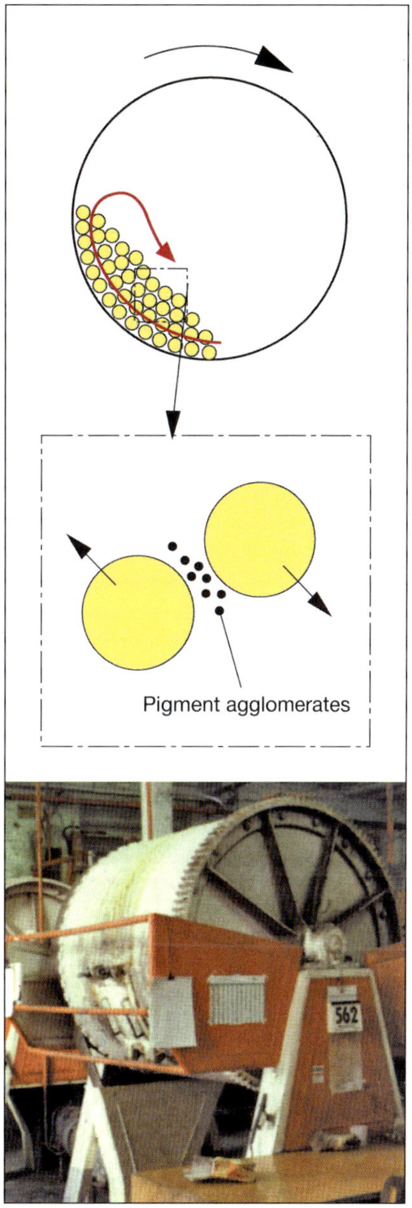

Figure 2.2.35: Working principle and picture of a ball mill

Agitator mills are refinements of **ball mills**, which were encountered fairly frequently in the past. The latter have been displaced more and more by agitator mills because of their poor economy. Their use is now limited to just a few applications in dispersing. Since agitator mills, which have now taken on such significance, were ultimately derived from them, ball mills will also be discussed to enable better understanding of their mechanical development.

Ball mills are hollow cylinders which rotate about a horizontal axis in operation. They are filled to about 40 % of their volume with mill base and grinding media, generally porcelain or steel balls with a diameter of 20 to 30 mm, in a given ratio. The relative motion of the grinding balls passing each other at different speeds generates friction surfaces which support dispersion of the pigments. Above a certain rotational speed, the moving grinding media are lifted up and cause an additional unwanted impact load

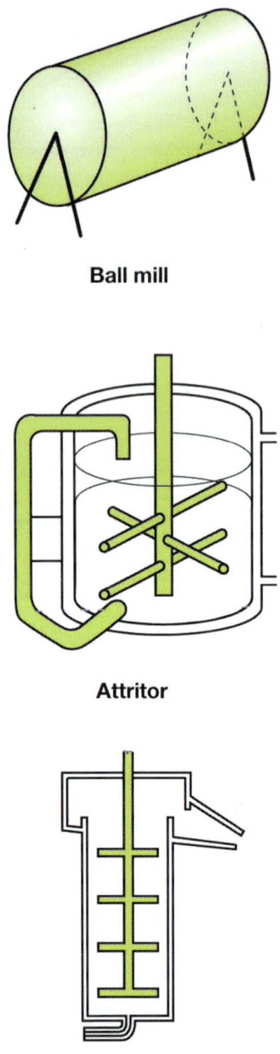

Figure 2.2.36: From ball mills to agitator mills via attritors

on the pigment agglomerates when they fall back down. This leads to comminution of the primary particle and a deterioration in the optical power of the pigments. The drum is therefore set to rotate more slowly for rolling friction.

The actual dispersion process occurs whenever grinding media pass each other at different speeds and generate the shear stress necessary for dispersion at the point where they are closest together. Upper layers roll downwards, while lower layers are entrained upwards by the mill jacket. Since the movement of the grinding media is caused solely by gravity, but this is a mass dependent variable, grinding media in ball mills must have a certain minimum size. The consequence of this is that the ratio of the effective grinding volume to the total volume is extremely low in ball mills. Noteworthy disadvantages include the associated long dispersal times of up to 5 days, extensive cleaning work and a not inconsiderable noise level.

Owing to their closed design, however, ball mills have the benefit of being able to use low boilers as solvents. A further advantage worth mentioning is their low maintenance during the dispersion process.

If it is desired to improve the unfavourable ratio of the dispersion volume to the total volume, the only variable is the size of the grinding media. However, as the grinding media volume decreases, so their mass also decreases. The resulting lower gravitational force and, at the same time, the decreasing kinetic energy therefore have an adverse effect on the dispersion rate.

It was, therefore, an obvious step to artificially generate the energy necessary for dispersion by installing additional agitators for use with relatively small grinding media. One of the first dispersers developed in accordance with this principle in the 1920s in the US was the **attritor**. This consists of an upright milling vessel in which balls 3 to 5 mm thick are artificially moved by means of an agitator.

The mill base is fed into the mill chamber at the bottom and drawn off again at the top. The cycle carries on continually until the required particle fineness or tinting strength has been achieved. The original designs suffered from a high level of wear and high power consumption but these drawbacks were largely eradicated by changing the shape and the type of agitators and by making the grinding media even smaller. The attritor became slimmer and taller, with the outcome that it became possible to achieve adequate dispersion after one pass. In principle, therefore, the **agitator mill** had been invented.

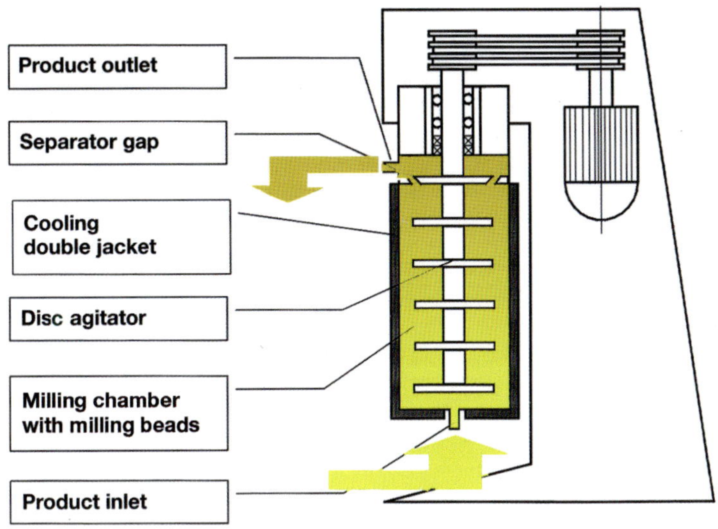

Figure 2.2.37: Schematic drawing of an agitator mill

The grinding media, consisting in the past solely of sand particles with diameters of approx. 1 mm, were moved artificially by adapted agitator units; this permitted the transmission of substantially higher shear stresses than was previously possible with conventional equipment. The separation of the grinding media from the mill base was and still is to some extent achieved by means of wedge wire sieves at the open upper outlet.

However, a one-off pass through the disperser presupposes preliminary homogenisation in a separate mixer or dissolver. Such refinements of the milling process brought about a considerable improvement in effectiveness and product quality.

Further optimisations were directed primarily at transmitting even higher shear forces and influencing the residence time distribution of the pigments. Agitator mills which were usually open and fitted with wedge wire sieves were increasingly replaced by closed systems. By a closed system is meant a disperser which enables the separation of the mill base from the grinding media by means of friction gaps. Such closed agitator mills bring numerous additional benefits. Dispersion can take place at increased pressure, thereby making higher throughputs possible. Material feed or material throughput can be varied by means of infinitely adjustable pumps in order that the disperser may be better adapted to material specific features in many areas. As with ball mills, steps must be taken to provide an effective cooling system in agitator mills too.

The grinding media and the mill base lie in the dispersing chamber of the agitator mill as a mixture which is then set in intensive, defined motion by the agitators. The grinding media take on varying levels of kinetic energy as a result of friction and impact at the energy-bearing surfaces. In the presence of viscous fluids, high shear forces can then be transmitted at the points where the grinding media interact.

Larger grinding media are also suitable for larger pigments because of the better entrainment of the pigment agglomerates into the areas of high shear. With progressive break-

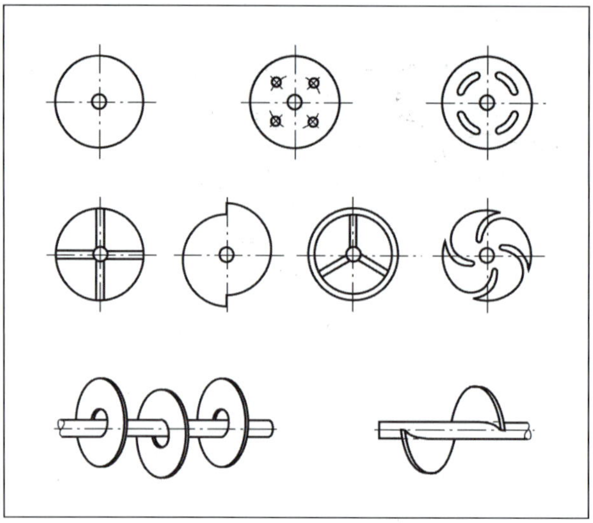

Fig. 2.2.38 : Agitator discs for agitator mills

down, however, the number of smaller particles greatly increases. The number of effective dispersal areas is then no longer sufficient in the case of larger grinding media and faster fine dispersion is not possible. Optimum conditions prevail, therefore, when larger grinding media with a larger entrainment area ensure preliminary dispersion in the first part of the agitator mill, while the small grinding balls, which have by far the larger total frictional surface area, handle the fine dispersion of the smaller agglomerates. Smaller beads are also generally used for the smaller organic pigments, while correspondingly larger ones are used for agglomerates of inorganic pigments which are sometimes rather large. For those involved in practical operations it is also important that the grinding media and the pigments be of sufficiently different sizes so that they may be separated without difficulty.

The density of the grinding media also has a bearing on the dispersal effect. Under otherwise identical conditions, zirconium oxide beads with a density of 5.6 g/cm^3 transmit higher shear stresses to the pigments than conventional sand or plastic particles with densities of only 2.6 g/cm^3 and less.

It has already been mentioned that the shear stresses can be increased still further by increasing the rotational speed of the agitators. However, disproportionate wear on the agitator mills and the grinding media places imposes limits on the amount by which the speed can be increased.

This desire to intensify the milling process led to refinements of the agitator tools. Solid discs, which used to be standard and in which shear was caused solely by the adhesion of

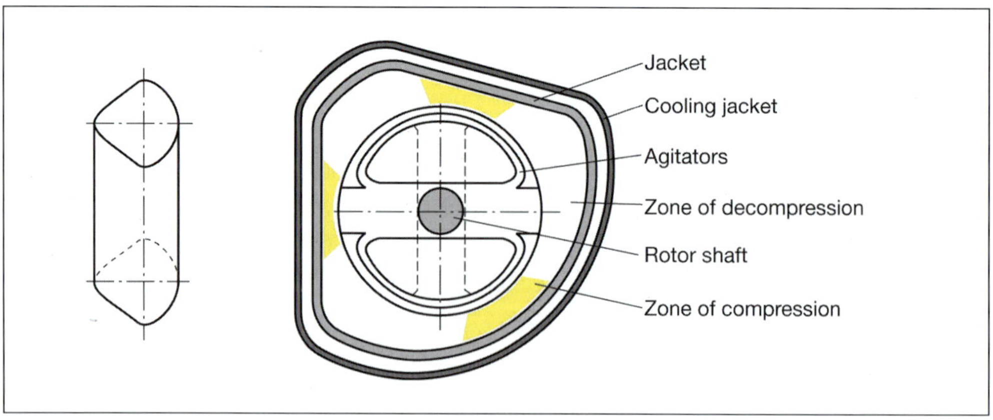

Figure 2.2.39: Cross-section of a chamber of an agitator mill with zones of underpressure and overpressure

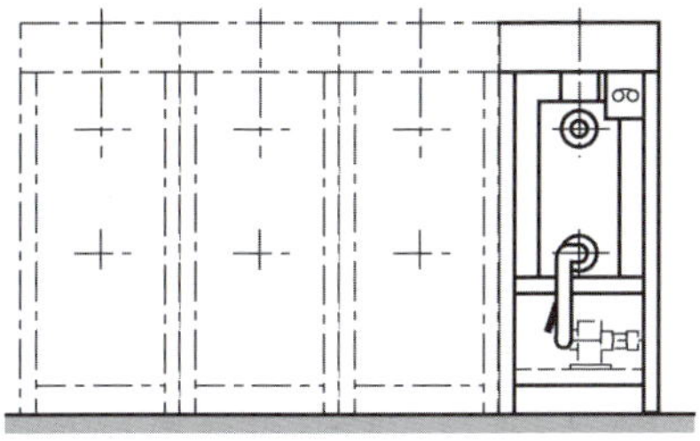

Figure 2.2.40: Schematic drawing of a multi-compartment mill

the mill base to the disc and the centrifugation of the layers lying above this, are being replaced by other designs including spoked ring discs and punched discs.

Punched discs have dispersing properties similar to those of solid discs. However, the apertures offer the advantage of better removal of the air released by the agglomerates and other materials and of better pressure equalisation. Cam discs fitted with projecting impactors have a greater influence on the dispersal effect. Screw discs also deserve mention. They exert an additional pressure on the mill base against its direction of flow, thereby also enhancing the dispersion outcome. The negative aspect is the broadening of the residence time distribution associated with the return transport.

Since it is not only shear forces in laminar flow, but also tensile/compressive effects which promote the dispersion process, grinding trough shapes have been developed which permit an alternation between areas of underpressure and overpressure. The sectional drawing in Figure 2.2.39 illustrates the regular passing of material through compression and decompression zones in a corresponding milling unit.

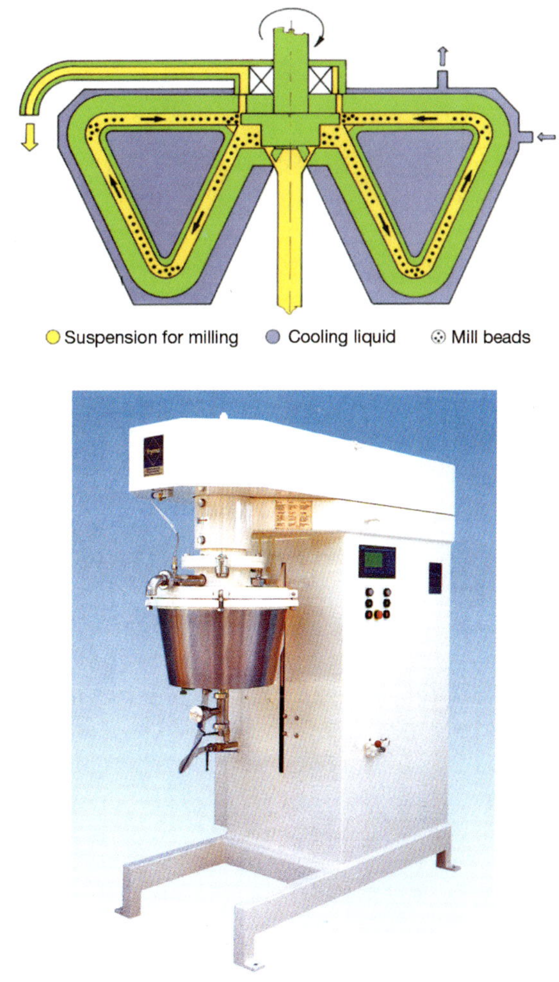

Figure 2.2.41: Schematic diagram and picture of a annular-gap mill

Source: Fryma

Figure 2.2.42: Basket mill with integrated disperser *Source: Getzmann*

The grinding trough shape also affects the residence time distribution of the mill base. High grinding troughs with only a small diameter lead to narrower residence time distributions than short and correspondingly wider ones. A different and better way to have a positive impact on the residence time range is to use **multi-compartment mills**.

Faster passes with short residence times through several smaller mills configured in series lead to closer residence time ranges and thus more uniform pigment loading.

Apart from the design and the rotational speed of the agitator mill the quantity of grinding media also has a decisive effect on the dispersion rate and quality. Open agitator mills have a grinding media volume of 40 to 70 %, and closed types up to 90 % in terms of the trough volume. A fundamental principle is that a high quantity of grinding media increases the power consumption and wear of the entire mill unit, while reducing the dispersion duration and generating a high level of heat at the same time. **Annular-gap mills** were developed to enable high cooling performance and a narrow residence time distribution in combination with uniform grinding media loading.

The milling chamber has the shape of an inverted conical envelope whose tip has been folded inwards. The outer wall of the cone is rigidly mounted, while the inner cone rotates. The relatively narrow gap between the rotor and the stator enables high shear forces to be transmitted. Since the flow rate falls with increase in cone diameter, but at the same time the shear stress rises because of high blade tip speeds, the dispersion process becomes more intensive with decrease in particle size. This effect is highly advantageous in the comminution of smaller and thus more solid agglomerates. The mill base is then removed via a separation device on the shaft, and the grinding media are separated axially downwards via recirculation ducts.

The **centrifugal fluidised bed mill** (see Figure 2.2.42) uses a different geometrical configuration. This consists of a cylindrical, horizontally configured mill chamber and a rotor which is fitted with 8 rows of blades and 8 wedge wire sieve tubes. The grinding media contained in the mill are moved by the blades. The material to be dispersed flows by means of three feed pipes from the outside to the inside radially through the rotating grinding media bed to the wedge wire sieve tubes which are also rotating. At this point the material passes through a relatively large separating plane and then exits the milling chamber via a hollow shaft. As a result of centrifugal forces the pressure of the grinding media is adjusted independently of the product throughput. High specific throughputs are possible as a result [2.4.108, 2.4.109]. A high level of wear of the constituent material, however, has meant that this concept did not remain in use for very long.

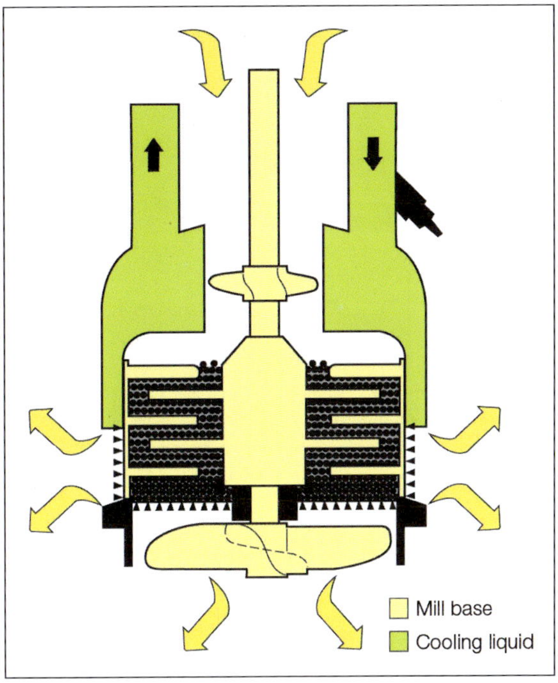

Figure 2.2.43: Schematic diagram of an immersion mill

Medium and small paint plants in particular use **basket** and **immersion mills** in addition to high performance agitator mills. Basket and immersion mills are versatile vessels in which a mobile dispersion unit is immersed after filling with mill base. This unit consists of a sieve-like grinding trough holding the grinding media which are intensively agitated by appropriate equipment. At the same time the mill base is entrained into the dispersion chamber and discharged again after dispersion.

A combination of a basket mill with a dissolver represents an interesting variant. This enables preliminary dispersion to be carried out in the dissolver, with subsequent fine dispersion in an agitator mill in a closed system.

In immersion mills the rotor is also responsible for circulating the liquid. The absence of a pump limits the viscosity spectrum of the mill base. The mill base composition must be determined and assessed for this type of agitator mill in accordance with criteria similar to those adopted for other agitator mills (see above).

The optimum energy input for operating agitator mills is achieved when the mill base has a relatively low viscosity. There are limits to the degree to which the viscosity can be reduced because of the increasing wear of the grinding media and the mill. Furthermore, it is worth noting that pigment shock can be expected during let-down when mixtures with high solvent levels are used.

Agitator mills are still the most economical and thus technically important type of disperser. The most important criterion for the technical evaluation of agitator mills is the ratio of throughput to power consumption with the narrowest possible residence time distribution for the pigments.

In addition to the type of agitator mill the milling process sequence or operational method plays an important role in production output and dispersion quality. For easily dispersible pigments a single pass by the **pass method** is sufficient. In this the homogenised mill base as per (A) in Figure 2.2.44 is conveyed from one vessel via the agitator

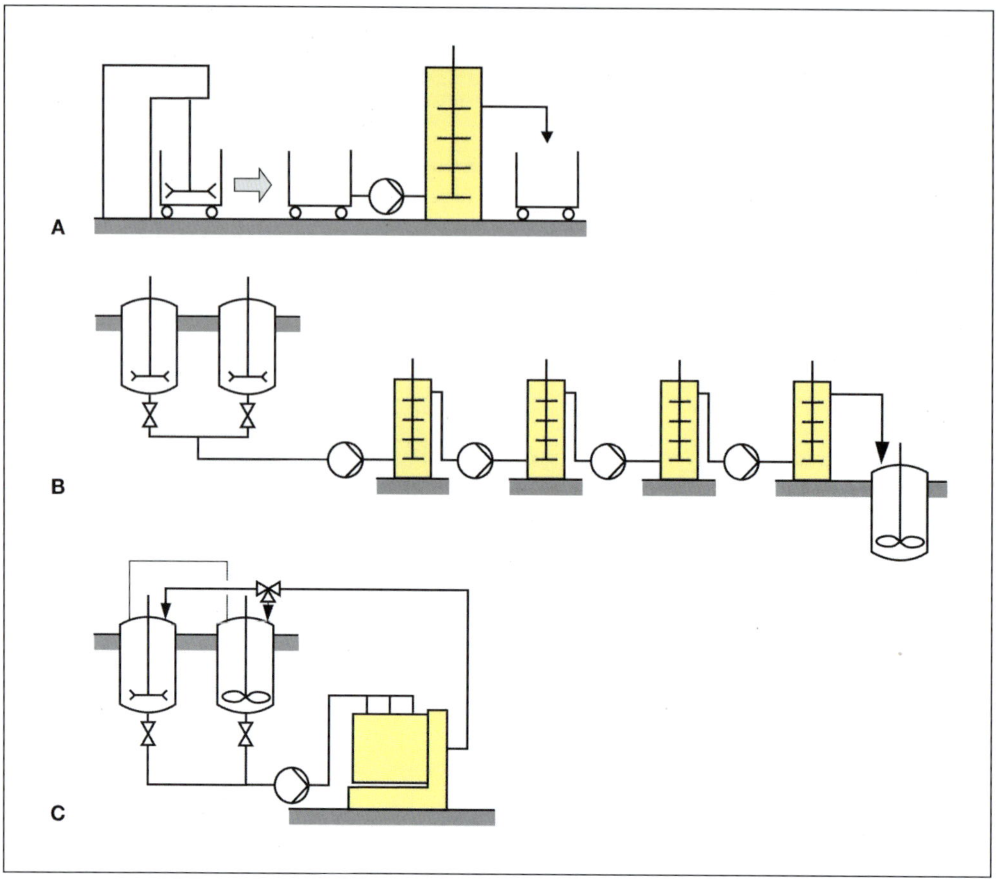

Figure 2.2.44: Process flows for dispersing by the (A) pass, (B) cascade and (C) cycling methods

mill into a second vessel. The milling process is repeated if the quality is below standard. Generally, feed or receiving containers are swapped or, in the case of stationary vessels, the feed and discharge lines are changed over. Most pigments require 2 to 3 passes, though this shuttle process may have to be repeated 5 to 10 times for pigments that are harder to disperse. The **cascade method** represents a continuously operating alternative in which the products pass through several mills in succession as in example (B).

This configuration can also be used for the **cycle method** (C) which requires only one vessel which feeds the mill base continually and then receives it via the agitator mill. To ensure that each volume element of the feed vessel has passed through the mill, a statistical approach would require that 5 to 10 times the vessel capacity be pumped through the system. The volumetric rate of flow should therefore be about 10 times higher than in the pass method. A critical aspect of the cycle method is the necessity for constant and thorough mixing of the material in the feed vessel. The possibility of using a process monitoring system is an advantage. Milling progress can, for example, be correlated with and checked by means of the energy consumption (see Figure 2.2.44).

If dispersion in conventional dispersing equipment is not possible because the viscosity of the mill base is excessively high, **kneaders** must be used [2.4.110]. Their use is essential when fillers or flush pastes are to be manufactured. Kneaders must be sturdy enough to cope with the high compressive and shear stresses to which they are subjected

Figure 2.2.45: Schematic drawing of a paddle kneader and a planetary kneader

and they are therefore generally closed systems. Kneading systems are divided into **paddle and screw type kneaders**. The first group includes **trough, double trough** and **planetary kneaders**, while **single** and **double screw kneaders** are examples of the second and are used particularly for powder production. The first type, particularly double trough kneaders, transmits the shear forces necessary for dispersion of the highly viscous mill base with counter rotating scoops matched to the housing. As a result of the special design, the material is conveyed in regular succession from one section of the kneader to the other. Planetary kneaders are similar to conventional mixers. Several kneading arms which project into the mixing vessel and are fitted with small but stable agitator blades describe a circular path while rotating themselves at the same time in a mixing trough rugged enough to withstand the high shear forces. The shear forces which can be transmitted in planetary kneaders are lower in general than those in double trough kneaders.

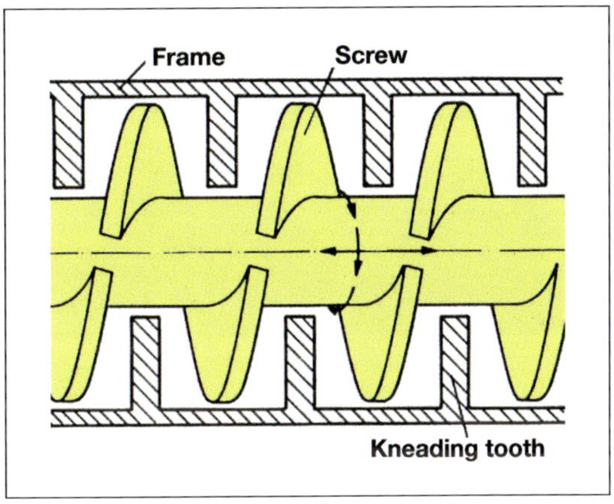

Figure 2.2.46 Schematic diagram of a screw-type kneader

Screw-type kneaders operate in accordance with a different principle. Simple types consist of a tube in which a cylindrical shaft with a screw-shaped metal band rotates. This conveys the mill base to the kneader outlet under shear load. The height of the shear forces can be increased substantially by means of kneading teeth mounted on the housing and the oscillating motion of the agitator shaft.

Machines for Manufacturing Powder Coatings

Heatable kneaders are used in the paint and coatings industry for dispersing powder coatings. The material to be dispersed, which is solid at normal temperatures, is melted and then used as a fluid to transmit the shear forces necessary to crush agglomerates.

Owing to the lack of solvents and the associated high viscosity of powder coatings, conventional production techniques are not suitable for the homogenisation of the resin

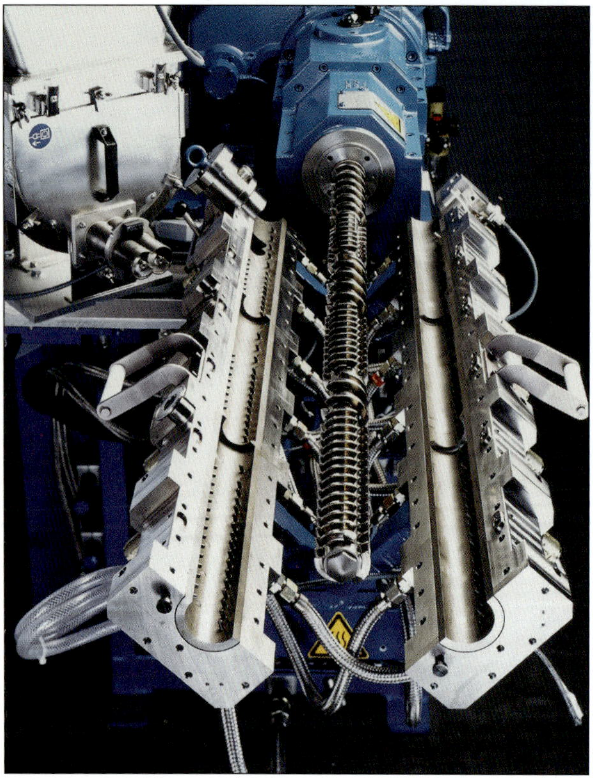

Figure 2.2.47: Screw extruder for manufacturing powder coatings

components and the incorporation of pigments and extenders. The individual components are therefore homogenised or dispersed in the extruder after melting. In preparation for the melting process steps should be taken to premix the formulation ingredients as effectively as possible. This works especially well if the individual solid components are suitably well crushed in advance. Since the solid resins are supplied in coarse form by the manufacturer, homogenisation should be preceded by decomposition of the brittle raw materials by crushing and milling to particle sizes of just a few millimetres. Once weighing and adding of all the formulation ingredients have been completed, components premixing takes place in **tumble** or **plough mixers**. In the case of thermoplastic nonreactive systems, the actual dispersion of the pigments in the polymer matrix can be carried out without difficulty in a kneader after melting has taken place. For reactive systems, on the other hand, conventional batch kneaders are unsuitable because of the excessive residence time of the thermally stressed melt. Continuously operating **extruders** are used for this application. They are capable of homogenising the melt just a few degrees above the melting point, i.e. when it still has a relatively high viscosity, with residence times of only a few minutes. The shear forces are high enough to crush the pigment agglomerates and achieve uniform distribution in the continuous resin phase.

Action needs to be taken to ensure the best possible preliminary homogenisation and cooling in the feed zone in order that raw materials may be prevented from forming lumps. Precise control of the temperature curve over the entire extruder range is also important.

Oscillating **single screw extruders, planetary roller extruders** and **twin screw extruders** have proved to be suitable extruder types in practical operations.

The shear forces to be applied for dispersion are generated in the single screw extruder by a screw oscillating in the direction of material flow. This enables extremely high shear forces to be transmitted between the screw and the matched screw liner. A similar effect can be observed when planetary rollers move about a central screw. High shear forces are also generated by permanently changing the gaps and by the rotary motion.

In twin screw extruders the shear gradient required for homogenisation and dispersion is generated by two meshed, counter-rotating screws. An appropriate configuration of the

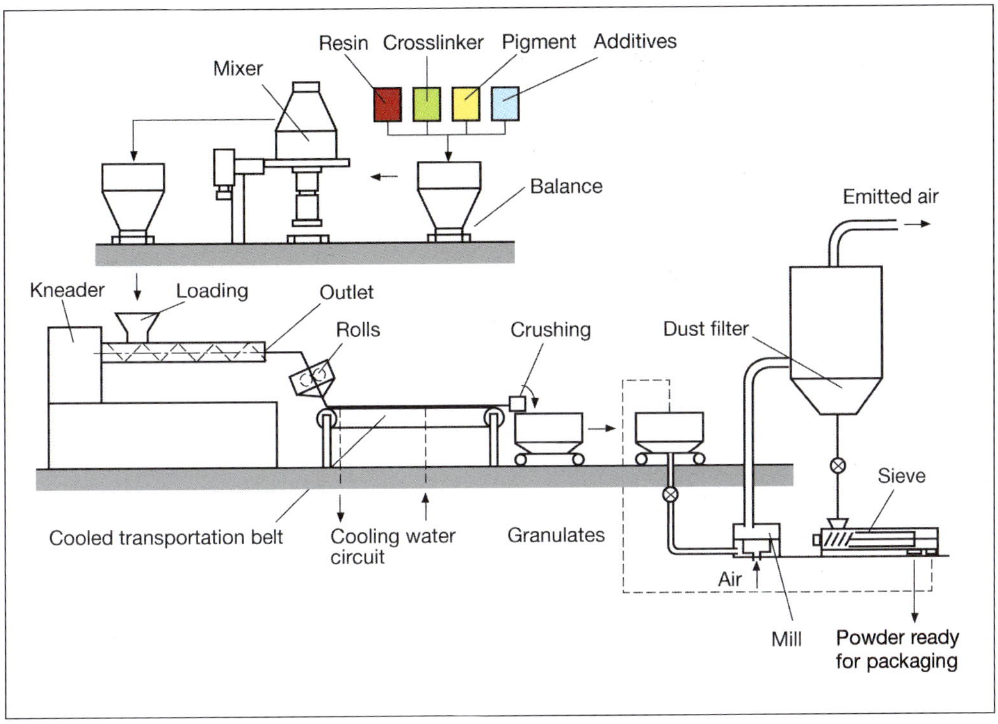

Figure 2.2.48: Traditional process flow for manufacturing powder coatings

shafts enables alternating compression and decompression to produce the required shear forces. Twin-screw extruders are ideal for reactive melts because they have a melt residence time of less than 30 seconds.

A suitable system must be in place to ensure rapid cooling of the reactive melt immediately on exiting the extruder. The outcome of this is that the still liquid mill base is drawn by two cooled rolls into a wide sheet that is only a few millimetres thick.

Figure 2.2.49: Pin mill for grinding powder pellets

Source: BASF Coatings

Conveying the product on a conveyor belt which is also cooled accelerates the temperature drop further. On leaving the cooling section the extrudate is converted back into a pellet material in suitable crushing systems. In order to make a processable powder coating from the material manufactured in this way which is still relatively coarse, fine grinding and pulverising stages must follow. The usual target particle sizes are < 100 µm for relatively thick and < 50 µm for relatively thin film layers. Grinding of the powder coating pellets is generally carried out in **pin, impact disc** or **cone mills**.

All the systems work on a similar principle. The particles to be ground are accelerated to a great speed and then crushed by means of impact or friction against the sides of

Figure 2.2.50: Manufacture of powder coatings

Source: BASF Coatings

the vessel. The air required for this acceleration is cooled to prevent the powder coating particles from sticking together.

When the powder particles exit the milling units they are separated according to their levels of fineness by means of **cyclones, vibrating** and **rotary sieves** or **filter systems**. Selection of the separating system is based on the desired particle size or particle size distribution (see Chapter 4.2.2).

In light of the large energy input for the homogenisation and decomposition of the powder coating particles, there has been no shortage of attempts to develop alternative production processes. Of the possibilities investigated on a laboratory scale, spraying the polymer melts with supercritical carbon dioxide [2.4.111] or ultrasonic atomisation are undoubtedly the most interesting [2.4.112].

2.2.4.4 Separation Processes in the Manufacture of Paint

In the paint manufacturing industry numerous separation processes also have to be carried out during the production of pigmented and nonpigmented coating materials. Examples include the partial or complete removal of solid particles from liquids, the separation of grinding media in the wedge wire sieves of agitator mills, the recovery of hard agglomerates or the removal of foreign matter before the coating materials which are ready for dispatch are transferred to their relevant containers. The separating operations are carried out in sieving and filtration units or by accelerated sedimentation in centrifuges.

Sieving and filtration

Sieving or screening is a sizing process and is used for fractionating mixtures of solids. This can take place in the gaseous phase or in a liquid phase.

The principle of **separation by sieves** lies in offering up the solid particles to be separated as frequently as possible to the openings of the sieve surface. The aim is to permit smaller particles to pass through and to hold back larger ones. The sieves have defined openings and are made of metal, organic fibres or plastics. They are divided on the basis of their design into perforated, woven or clamped sieve surfaces.

To ensure a consistent longterm separating effect it is important to ensure that the sieve design does not permit any shift, i.e. any enlargement of the openings. The necessary

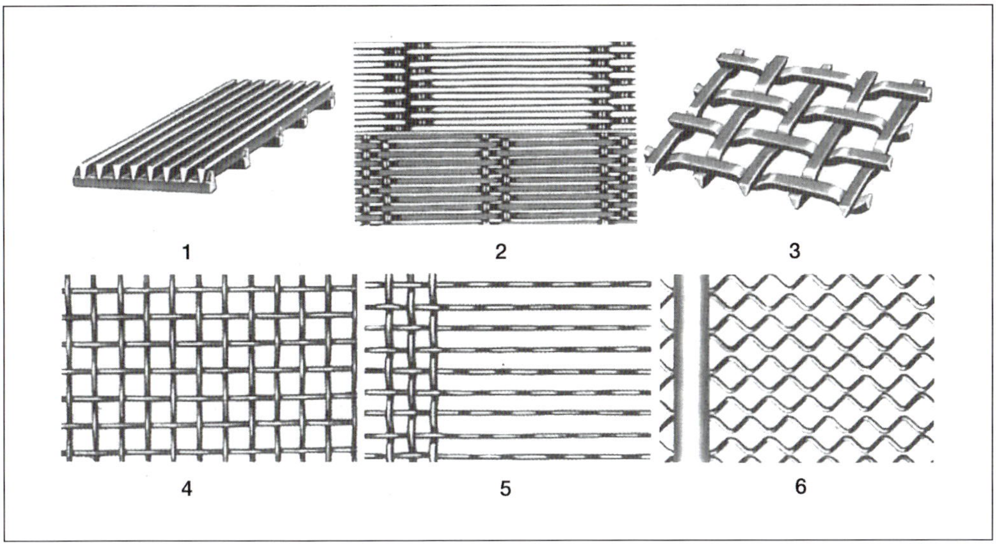

Figure 2.2.51: Different forms of sieves

stability in the sieve surface is achieved by the use of monofilament yarn, a fixed weave in the fabric and nonswelling materials. It is also important in practice to use materials which are not attacked by the material being separated. The latter determines whether solvent, acid, alkali or bacteria resistant sieve fabrics should be used. When hot substances are being dealt with, the heat resistance of the sieve materials must be adequate.

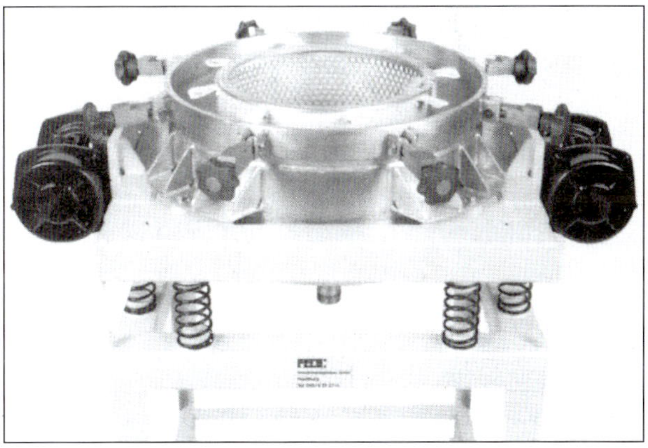

Figure 2.2.52: Vibrating sieve *Source: Netzsch*

High specific separation performance is achieved when all particles are offered up to the sieve openings as frequently as possible by inducing movement in the material to be separated. The frequency with which material is offered up can be increased by horizontal or vertical motion and by vibration.

One proven form of sieve system consists of a vertically configured sieve device through which the material to be cleaned is pressed. The sieving residue is continuously released by vibrations and then removed via gravity sedimentation.

Rigidly fixed, nonmoving sieves can only be used where there are extremely large particle size differences, such as those that occur when the mill base is being separated from the grinding media. The separation of foreign matter when coating materials are being transferred to their storage containers by passing the materials through sieve bags is also impossible without the use of artificial sieve movements.

Filtration is the separation of solid particles from suspensions with the aid of porous filter media. Strictly speaking, therefore, the filtration process is the complete separation

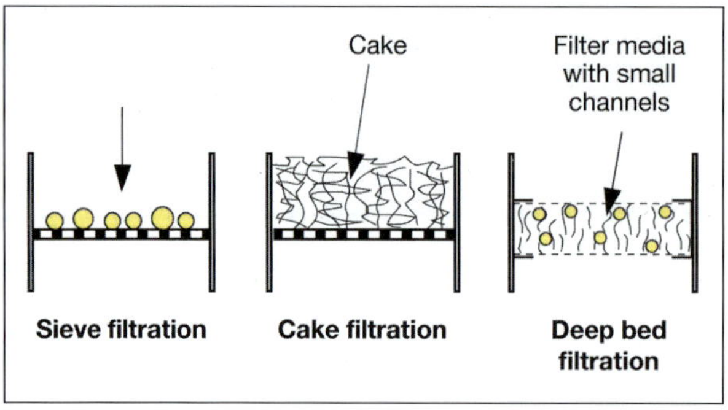

Figure 2.2.53: Different filtration techniques

of solids (filter cake) from the liquid phase (filtrate). In accordance with the purpose of the filtration process, the treatment of the liquid is termed clarification, while separation of the solids is termed cake filtration.

Strictly speaking, the treatment of pigmented coating materials to separate foreign matter and any agglomerates still present is a sieving process. Because of the small size of the pigments the term “filtration” is employed when fine and extremely fine sieves are used, even though the dispersed pigment is still in the filtrate. Whether such processes are termed sieving or filtration in the paint and coatings industry depends merely on the degree of fineness of the separation system.

Filtration processes are divided into surface filtration, cake filtration and deep bed filtration in accordance with the separation method employed. Surface or sieve filtration is a method of solids separation which obeys the laws of sieving. It is used when only small quantities of solids have to be separated by means of fine sieve bags with pore sizes of up to just a few micrometres (clarification).

Cake filtration is a separation process in which the solid cake accumulating on the partition also serves as a filter aid at the same time. It enables the isolation of large quantities of solids. In a pure surface filtration system all the solid particles are larger than the openings in the perforated plate. In the case of cake filtration this is only partially true. In the course of the filtration process the solids form a cake which then enables the separation of finer particles. In both processes the solids are deposited above the filter medium.

The principle of **deep bed filtration** is quite different. This uses thicker filter media whose pores are larger than the diameter of the solids to be separated. As a result these penetrate the labyrinth of the channels in the filter medium where they are lodged either mechanically in relatively narrow side passages or adsorptively on the internal walls of the capillaries.

Due to the fineness of the pore labyrinth, deep bed filtration is not generally carried out with pigmented paints. Deep bed filtration is mainly used for purifying clearcoats and binder solutions. Mechanical choking of the channels, as opposed to purely adsorptive separation, leads to a decrease in filter performance. On the other hand, if filtration is performed for long enough, adsorptively retained particles can be transported more or less quickly through the channels at a rate depending on the magnitude of the attractive forces, with the result that deep bed filters can “blow” after excessively long use.

Deep bed filtration allows even relatively small gel particles, i.e. deformable particles, to be separated if their diameter is small enough. Larger gel particles, on the other hand, choke the pores of the filter medium and lead to a rapid drop in filter performance.

To prevent blockages in the fine filter channels, supplementary filter aids are frequently used to clear critical suspensions. These include diatomaceous earth, fuller’s earth,

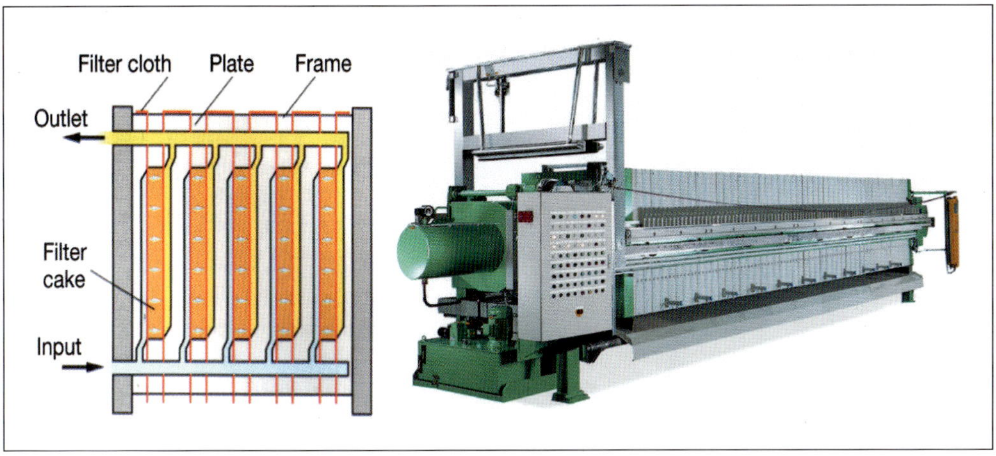

Figure 2.2.54: Schematic diagram and picture of a plate-and-frame filter press *Source: Netzsch*

activated carbon or fine plastic grains with particle sizes of 5 to 50 µm. These are either stirred into the suspension or coated onto the filter beforehand. As a result of adsorption on the filter aid the especially critical elements of the material being filtered are practically retained in the artificially created filter of the filter aid. Constant filtration entails only a small rise in pressure because fixation on the filter aid is overwhelmingly adsorptive. The advantages of filter aids can be increased still further if only part of the filter aid is added to the material being filtered after the rest has been used to form a coating on the filter medium. Filter aids ultimately transform deep bed filtration into cake filtration.

Filtration devices are important for the separation processes in the paint and coatings industry, and there is an enormous variety available. Particular mention may be made of **plate-and-frame** (see Figure 2.2.54) and **chamber filter presses** for cake filtration, and **bag** and **plate filters** for the usual cleaning processes during the manufacture and use of paints, because of their importance in these processes.

The main differences between the **plate-and-frame filter press** and the chamber filter press are the larger chamber volume of the former and the option of carrying out flushing and washing operations. It is therefore used for filtering suspensions with a high solid content, as in the case of cake filtration to recover pigment powder during the manufacture of pigments, for example.

Chamber filter presses are used for the clarification of suspensions with lower solid contents (for example, contaminated clearcoats or binder solutions). Both filter presses comprise numerous modules made of iron or aluminium, connected in parallel. The filter surface area can be varied at will by changing the number of plates.

The filter medium determines whether the filter material consists of cotton or synthetic fabric. Gear pumps are preferred to piston pumps when it comes to achieving uniform filtration quality because there is no pulsating pressure build up.

A variant which is of interest for paint and resin manufacturers is the plate filter. By contrast with plate-and-frame or chamber filter presses, these do not have a fabric pressed between the metal segments, but rather cellulose or plastic reinforced plates. The film

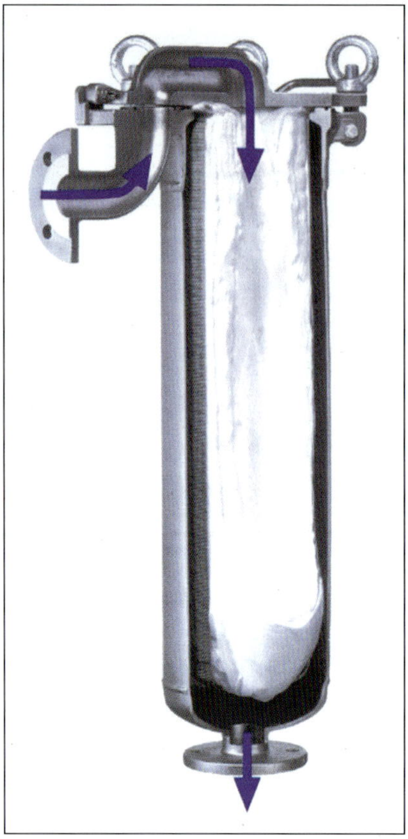

Figure 2.2.55: Bag filter being used in paint production *Source: Netzsch*

thicknesses of these filter media lie between 2 and 6 mm, and the pore size is approx. 0.8 to 15 µm.

Particularly good filter performance is achieved by **double filtration** using different pore sizes.

Cartridge or tubular filters, which are relatively small and easy to install, are also used for cleaning a very wide variety of coating materials. The filter elements are cylindrical cartridges made of paper, felt, synthetic fibres, porous clay or ceramic materials. The synthetic fibre cartridges have proved particularly successful because of their good depth effect.

In recent times the importance of filter cartridges has waned in favour of bag filters. While the first filter bags were practically nothing more than bag shaped sieves, there are now numerous different types that offer a level of depth action which can otherwise only be matched by plate filters. The reasons for the increasing importance of bag filters are essentially commercial. Foremost among these are higher throughputs, lower costs for rigging and derigging, lower material costs and less waste. Certain bag filter materials even allow the type of foreign matter which causes craters to be removed.

Sedimentation

Solid particles of different sizes can be separated from the liquid phase selectively or completely if there is a difference in density between the liquid and dispersed phases. In **gravity sedimentation** the acceleration due to gravity g and, in the opposite direction, the frictional forces of the fluids act on the solid particles. In the stationary state the following applies to the sedimentation rate u.

$$u = \frac{\Delta\rho \cdot g \cdot d^2}{18 \cdot \eta}$$

This relation shows that high sedimentation rates and thus short separation times are only possible with high differences in density $\Delta\rho$ between the particles and the fluid phase, large particles and low viscosity η. The particle size d exerts a quadratic effect on the result. Because of the very low deposition rate of small particles and thus the long separation process, methods are used which achieve sedimentation via centrifugal acceleration $4\,\pi^2\,n^2\,R$, where R is the distance from the axis of rotation.

$$u = \frac{\Delta\rho \cdot 4\pi^2 \cdot n^2 \cdot R \cdot d^2}{18 \cdot \eta}$$

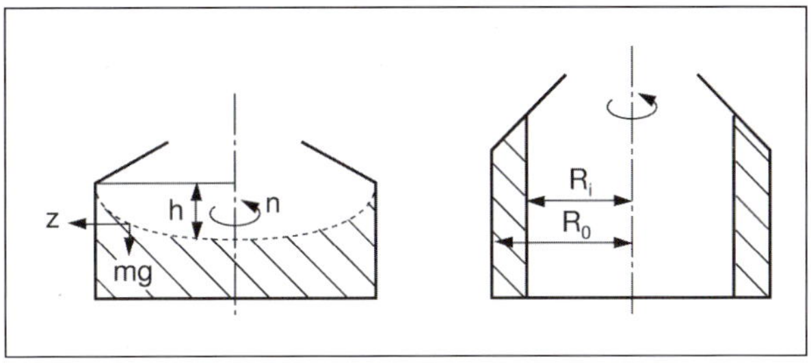

Figure 2.2.56: Schematic diagram of fluid distribution in a centrifuge

The acceleration forces are generated in rotating drums. In the course of this process the mixture to be separated is spun outwards, thereby forming a liquid ring on the drum wall. The relative speed of the liquid to the wall is zero when stationary. The centrifugal force acting during the accelerated separation of the solid particles is quantified by means of the relative centrifugal force C. This is defined as the ratio of the effective centrifugal acceleration to the acceleration due to gravity g.

$$C = \frac{4 \cdot \pi^2 \cdot n^2 \cdot R}{g}$$

To an extent depending on the drum diameter, sufficiently high migration rates are achieved at rotational speeds of 5,000 to 20,000 rpm.

The relative centrifugal force is between 200 and 4,000 for normal centrifuges and between 4,000 and 50,000 for super-centrifuges. Thanks to the enormous increase in the migration rate of small particles (gel particles or agglomerates) in only averagely fast technical centrifuges, it is possible to separate hard residual agglomerates from already dispersed primary particles. The flow rate determines the critical particle diameter. Even gel particles can be separated completely in this way, provided the density differential to the polymer solution is large enough.

The centrifugal separation method has been implemented in various designs. Because of their different separation principles, a distinction is drawn between solid bowl and **sieve centrifuges**. Sieve centrifuges are ultimately merely an accelerated form of sieve filtration. Just like **decanters** as simple solid bowl centrifuges or the very high speed, yet very small tubular centrifuges, they are of virtually no importance for the paint and coatings industry.

By contrast, separators are important devices for the aftertreatment of resin solutions, clearcoats and pigmented coating materials. With a blade tip speed of 160 m/s they reach 6,000 to 8,000 times the acceleration due to gravity.

Nowadays, however, centrifuges are only used where other separation processes are unsuitable. The reasons for this are their lower power, the high operating input for a discontinuous process and safety aspects.

Separators or **disc centrifuges** in the form of solid bowl centrifuges have a large number of conical discs configured in parallel. The paths travelled by the materials to be separated are very short in such systems. The solid particles are pressed against the disc wall and, because of centrifugal force, are transported against the liquid flow to the solids discharge point. The flow rate of the suspension determines the critical particle diameter. Separators are therefore ideal for separating expensive, reusable pigment agglomerates.

Hydrocyclones are particularly simple but not very effective separators for suspensions. By contrast with centrifuges it is not the housing but the liquid which rotates. The rotary flow is induced by the tangential feed of the suspension into a housing which is

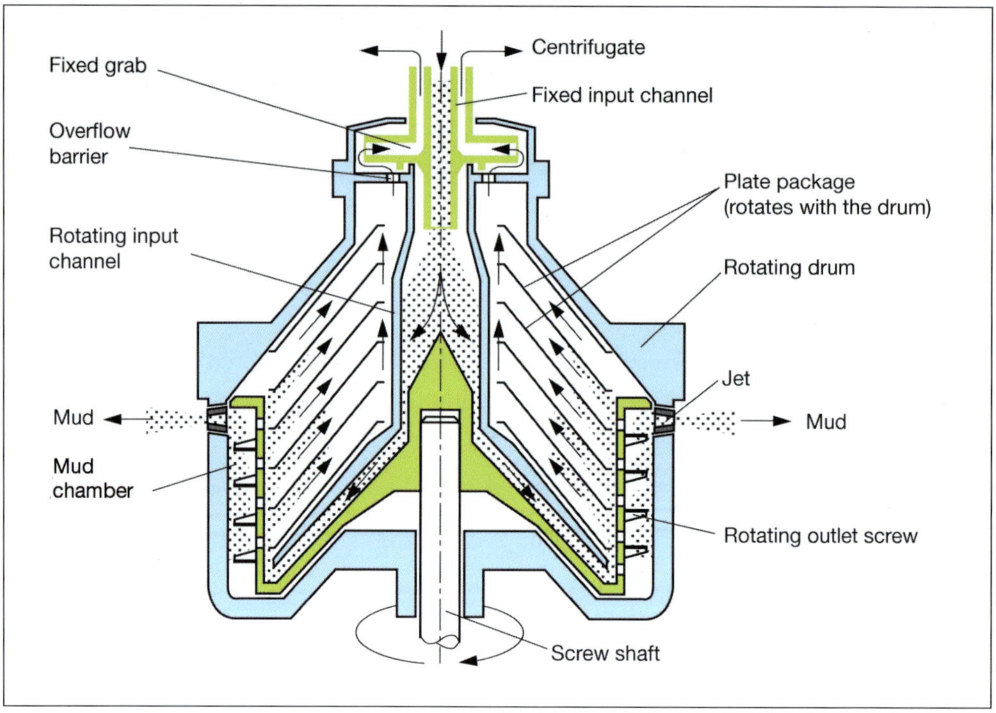

Figure 2.2.57: Cross-section of a disc centrifuge

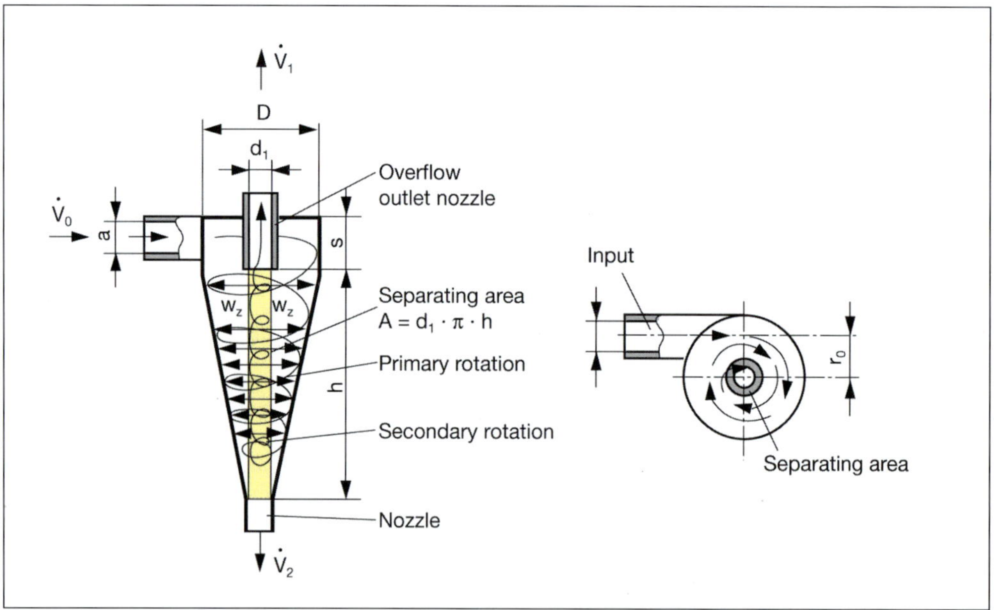

Figure 2.2.58: Schematic drawing of a hydrocyclone

cylindrical at the top but tapers towards the bottom. Centrifugal forces are generated as a function of the feed rate and the diameter of the cyclone and cause the solid particles to move towards the side walls. This principle is of great technical importance in the form of the aerocyclone which uses air as the medium for the classification and recovery of powder coatings (see Chapter 4.2.2).

Practical Examples

Separation processes are useful for almost all stages throughout the manufacture of paints. Cleaning operations are required as far back as the manufacture of polymer solutions by chemical synthesis and the subsequent creation of solutions. In paint manufacture a distinction has to be drawn between the production of clearcoats and of pigmented systems or effect paints. While the aftertreatment of clearcoats serves merely to eliminate foreign matter, the treatment of pigmented systems and effect paints is intended to achieve partial separation of foreign matter and oversize material.

There are a few cases in which it makes more sense to separate hard agglomerates and reuse them as raw materials in other production stages than to damage the products by excessively frequent and aggressive dispersion. The separation processes therefore relate to formulation ingredients and to foreign matter which is not part of the formulation. Since it is not possible to draw a hard and fast line between agglomerates, aggregates and primary particles, the separation measures change important paint properties. For example, the separation of pigments at critical particle diameters which are set too low can alter the hiding power, colour or effect to such an extent that material corrections become necessary.

Foreign matter or formulation ingredients which are to be separated can be rigid (agglomerates, aggregates) or have a changeable outer form. On account of their mobility, length and very small diameter, fibres and fluff must be separated by methods different from those used for gel particles, which are deformable almost at will.

The process and filter medium to be used are essentially determined by the nature of the material to be separated. In the case of metal fabrics and filter cloths which act in two dimensions, mesh sizes of 10 to 60 µm are usual. The elimination of foreign matter from solid colour paints is generally achieved by means of bag filters with mesh sizes of 5 to 50 µm. It is much more difficult to remove foreign matter from metallic paints because the materials creating the effect are particularly large, in one plane at least, on account of their platelet shape. Vibrating sieves with slit-type openings are generally used to enable the bronzes to pass without problem.

Three-dimensional deep bed filters with their fibrous filter plates have far smaller pore diameters. However, it is not only the ratio between the particle diameter and the mesh size which is important for the ability to pass and for the clarification effect. The rheological properties and the surface tension of the suspension are deciding factors in addition to the particle shape and structure. Further influences are exerted by the solvent and the charge of the dispersed particles.

A further important variable for optimising filter performance is reducing the viscosity by raising the temperature of the material being filtered. If, for example, the temperature of an alkyd resin solution is increased by 50 °C, the viscosity drops to such a level that the filter performance is increased 5-fold. Excessive heating can, however, also have a detrimental impact on the filtration effect, for example if foreign matter or dirt particles pass into solution when heated up and then precipitate again after filtration upon cooling down.

In many cases it is expedient, if very fine foreign matter is present, to precipitate this material first by adding suitable flocculants that will cause this foreign material to form larger aggregates. This process is divided into bridging flocculation by bipolar resin molecule chains and flocculation by direct deposition on solid particles (see Chapter 2.2.3). The phenomenon of flocculation, which is undesirable during paint manufacture, is deliberately induced in this particular case. Polymer compounds or metal salts are used as flocculants.

2.2.5 Summary

The manufacture of coating materials is a technical process which has been optimised from economic and environmental perspectives and which is composed of numerous basic operations. Transportation, metering, homogenisation in agitators and dispersers with subsequent separation processes by sieving, filtration and centrifugation are the most important processes during the journey from raw material to coating material. All the production stages are subject to scrutiny by relevant laboratories. Formulations developed by the labs address not only the economic aspects but also the quality standards and environmental regulations. The specific task for a paint maker – manufacturing a semi-finished product which will be converted into a coating at a later date by technical equipment or people – requires special coordination between technical practice in production and application on one hand and theoretical background on the other

The central task of pigment dispersion forms part of the complex sequence of wetting, thorough moistening of the agglomerates with the film forming agent solution and the subsequent mechanical stress in viscous laminar flow gradients. When it comes to rapid and complete dispersion, the type, size and form of pigment are variables in just the same way as polarity, size and form of the polymer molecules. The latter play a fundamental role in shaping the rheological properties and the surface tension of the film forming agent solution.

The pigment stabilisation necessary for rendering usable pigmented coating materials can be achieved by coating with ions and the associated formation of electrical double layers or with sufficiently long mobile and well solvated polymer chains with as complete a coating as possible.

Optimum conditions can only be identified to a limited extent by theoretical considerations. Experimentation is therefore still indispensable for determining the optimum grinding conditions.

2.3 Characterisation of Coating Materials

Coating materials are semifinished products which have to meet a large number of requirements relating to storage stability, processing and properties. It is absolutely essential for the paint maker to manufacture individual products with a consistent quality level which conforms to the agreements made between the manufacturer and consumer. Numerous tests have to be carried out so that the desired property profiles can also be guaranteed. The demand for consistency in processing conditions must always be linked to the requirement for consistent coating quality.

It must be borne in mind that the term "quality" is a variable which cannot necessarily be forecast on the basis of the composition of the coating materials. Although well defined **physical indicators** such as density, viscosity or surface tension, for example, are necessary in evaluating property profiles, they are far from enough to permit binding statements to be made about the processability and final properties of a coating.

All attempts to control the raw material property fluctuations, which cannot be avoided completely, and the absolute reproducibility of the technical process of paint production, which is just as impossible to guarantee, have led to a situation in which the paint manufacturer's task is seen as less about the identical combination of raw materials and more about the creation of an identical property profile. The term "property" refers equally to processing and film properties. Fluctuating product quality is tolerated less and less in industrial scale painting processes because of increasing automation. Achieving identical properties in two batches therefore takes a certain degree of priority over achieving an identical material composition. Processing performance on the paint processor's individual coating lines cannot be forecast solely by reference to physical performance indicators. Only when processing has been carried out under the processor's own conditions with all the tiny variations that this may include or when a comprehensive audit has been made of the application properties by statistical methods is it possible to make a binding assessment of the processing properties of coating materials.

An overview of the physical and processing properties of coating materials can best be drawn up if the requirements and demands are discussed in detail in advance. Issues to be addressed in particular should include **storage stability**, consistency of **application properties** with the various processes and, not least, the **film properties**. The permanent shear stress in the circulation systems or the circulation in dip tanks, for example, should not cause any changes in colour, effect or rheological properties of the material. The last of these is also a significant variable with regard to flow during the film forming process.

In view of the numerous requirements relating to properties and processing performance, the parties involved draw up agreements specifying every detail relating to the quality level and the permitted tolerances. Such specifications or **technical terms and conditions of sale** contain an exact description of properties but also information on relevant measurement and test methods (see Chapter 8).

For the liquid coating material which is still to be processed the deciding criteria are the stability of the product, environment-related aspects and suitability for application. A distinction must be drawn here between application properties which are determined directly and general physical indicators which can only provide indirect information on application properties.

The first group includes all the practical tests for determining the tendency to sag, levelling, and the formation of film faults in the form of pinholes or craters. The indirect test group involving the measurement of physical indicators includes viscosity, surface tension, density and, with special reference to pigmented paints, hiding power, tinting strength and fineness of dispersion.

Quality assurance procedures are not limited, of course, to the use of measuring equipment. Only a comparison of measurement results with the agreed property and production tolerances provides the desired information on acceptance or rejection. In this way, measuring evolves into testing and thus becomes an important element of quality assurance (see Chapter 6). Furthermore, not all test methods yield binding, precisely defined indicators. Many property profiles simply have to be assigned a rating. The results are frequently rated on a scale of 0 (best) to 5 (worst); in North America this scale ranges from 10 to 1. Creating standardised images and using these as a guide during the evaluation process has proved a successful method of objectifying the test results in many cases.

Test methods and evaluation criteria originally drawn up by the individual paint users and specified in writing as in-house standards cause paint processors a large amount of work because of the number of test instruments they have to use and test methods which have to be employed.

Attempts to bring about uniformity through the drafting of international standards has undoubtedly achieved a degree of success. Unfortunately, the various industrial countries have carried out these welcome activities independently of each other, as a result of which it is still necessary to consult a variety of standards. Efforts to specify international standards as a binding requirement have certainly advanced matters, but the goal of binding global standards has not yet been met (see Chapter 8).

2.3.1 Measurement Accuracy

As mentioned above, the manufacturers and consumers of coating materials negotiate **tolerances** in the performance indicators for processing and final properties. Product compliance with specification must therefore be determined by testing to establish whether the performance indicator in question is within permissible tolerances of the ideal value or not.

In order that binding declarations of acceptance or rejection of a product batch may be achieved, the measurement accuracy of the test equipment itself must also be regularly checked when measurements are carried out. Measurement errors cause a loss of confidence in the evaluation of product quality. They are therefore a significant factor in quality assurance. Analyses of the measurement accuracy of all measuring instruments are thus essential aids when measurement results are being processed.

In particular, it is necessary to distinguish between systematic and random deviations from the correct reading.

Figure 2.3.1 shows two targets with numerous hits. In the second case the marksman has fired his shots with low accuracy but good precision, while the hits are scattered over the entire target in the first case. Case 2 is an example of systematic error which can be eliminated via a correction factor. No system can be identified in the first case. These are random deviations.

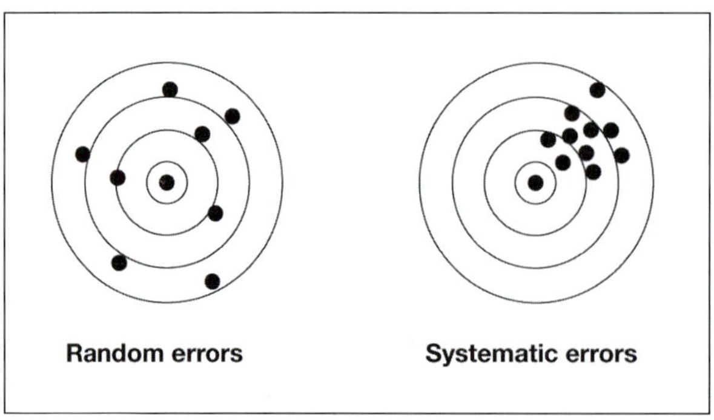

Figure 2.3.1: Visualisation of random and systematic measurement errors

Such random measurement uncertainties are amenable to mathematical analysis. Appropriate treatment of the results of multiple measurements makes it possible to substantially increase the reliability of a statement. In the case of **random errors** calculating the mean is a simple way to obtain a more precise statement. A definition which is meaningful in practice is the **arithmetic mean** $\bar{x}$, i.e. the sum of all the individual readings x_i divided by the number n of individual readings.

$$\bar{x} = \frac{\sum_{i=1}^{n} x_i}{n}$$

Calculating the mean may be meaningful, but it does not permit any comment to be made about the gap between the individual readings and the true value. For this, the deviations of the individual readings from the mean must be included in these considerations.

Since the sum of the errors is zero in the case of random errors because of the uniform distribution of positive and negative deviations from the true value, the sum of the squares $(x_i - \bar{x})^2$ of the individual errors is calculated in order that a so-called confidence range may be arrived at. The square root of the squared mean of the individual errors divided by the number of readings less 1 is termed the **standard deviation** s or mean error.

$$s = \sqrt{\frac{\sum_{i=1}^{n} (x_i - \bar{x})^2}{n-1}}$$

If the standard deviation is expressed in terms of the arithmetic mean of the readings, this yields the coefficient of variation or relative error. The smaller the standard deviation, the more accurate is the measurement process.

Experience shows that the frequency of readings decreases with increase in distance from the mean or from the unknown true reading. As the number of readings increases, the frequency distribution more and more resembles a curve which can be mathematically described by the Gaussian distribution curve J.

$$J = \frac{1}{\sqrt{2\pi} \cdot \sigma} \cdot e^{-\frac{(\bar{x}-\mu)^2}{2 \cdot \sigma^2}}$$

For an infinite number of individual readings, the standard deviation s moves towards σ, and the value $\bar{x}$ approaches the theoretically true value μ. The *Gaussian* curve is clearly defined by σ and μ and permits a distinct statement to be made about the probability of errors in individual readings. If, therefore, a measurement process has been described in a one-off action in terms of its standard deviation based on a sufficiently high number of individual readings, quantitative statements can be made about the probability of individual errors. Thus 68.3 % of all readings will lie within ± 1 σ of the maximum of the *Gaussian* curve, as many as 95.4 % within ± 2 σ and no fewer than 99.7 % of all readings within ±3 σ.

Agreements about the quality that can be achieved by painting can therefore only ever be expressed in terms of percentage certainty. The manufacturer and consumer are left with the residual risk, despite an extremely accurate testing regime based on individual readings, of mistakenly selling merchandise which does not conform to specification as "OK" or of rejecting products which do conform to specification. Taking the example of tinting strength, Figure 2.3.3 shows that a paste A with a tinting strength of 102.5 (and thus a product within the agreed tolerance of 100 ± 5) and a standard deviation of

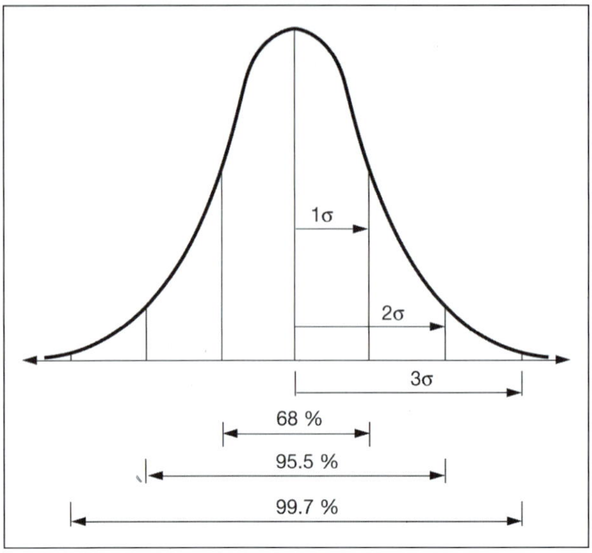

Figure 2.3.2: Error distribution curve showing error probabilities

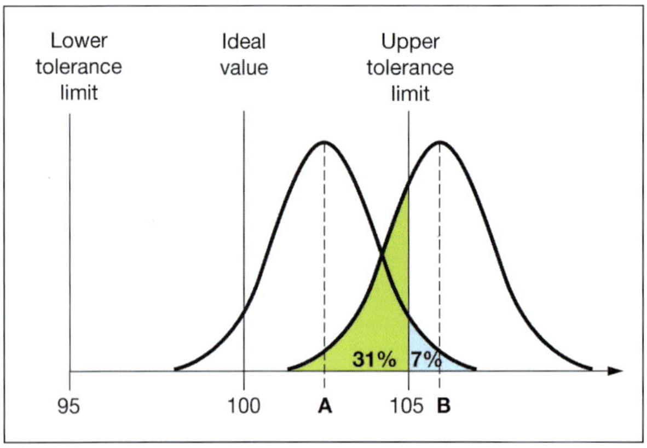

Figure 2.3.3: Quality risks associated with specifications based on single measurements for manufacturer and customer

2 σ has a 7 % probability of being outside the permissible tolerance whereas in the case of product B with a value of 106, and thus clearly outside the agreed limits, 31 % of all readings would nevertheless be identified as being within the specification. The first case represents the manufacturer's risk, and the second case the consumer's risk.

The consequence of this is a need to comply with supplementary safety margins for the agreed tolerances. These are calculated on the basis of the agreements on the reliability of the correctness of the readings. If the standard deviation of the test method is known, a reading is acceptable if it is within the agreed tolerances with the agreed level of confidence (for example 95 %). In other words, the reading may not exceed a deviation of 1.96 s, which represents 95 % of the area below the *Gaussian* curve, from the agreed limit. The standard deviation s must be determined experimentally on an adequate number of individual readings for each measurement method. The threshold value or **significance threshold** is:

$$\frac{s \cdot t_{S,N}}{\sqrt{n}}$$

s = standard deviation
$t_{S,N}$ = Student factor
S = factor for relative accuracy
N = number of measurements to determine the standard deviation

The lower and upper limit y for a significant statement for rejection or acceptance is then:

$$y = \text{tolerance limit} \pm \frac{s \cdot t_{S,N}}{\sqrt{n}}$$

The margins to be met are reduced in the event of multiple readings in the form of $1/\sqrt{n}$, where n is the number of readings for the approval test.

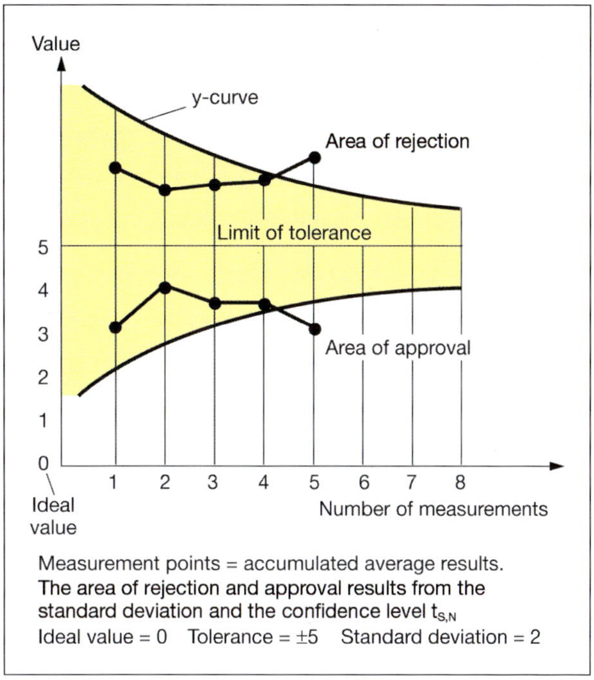

Figure 2.3.4: Method behind the sequential test strategy

These correlations are used when a significant statement has to be made about acceptance or rejection for a minimum number of readings. Tests are therefore not carried out until the "right" outcome is achieved, but until a decision is possible with a calculable risk [2.4.113].

This method of processing readings, termed the **sequential test strategy**, is based initially on the determination of one reading. The position of this reading relative to the **tolerance threshold** is verified. If it lies significantly outside or within an agreed limit, a decision can be made. If it is in a range which does not permit a significant statement, a second reading must be taken, the arithmetic mean of the two calculated and its position relative to the significance with a now narrower range of uncertainty determined.

Testing continues until the result calculated in this way leaves the zone of uncertainty, either into the rejection range or the acceptance range.

It may be noted in conclusion that errors must be expected at every stage in the evaluation of process chains involving tests. The consequence is error propagation over the individual stages of the production and testing sequence. The squares of the standard deviation are then added to the variance s_{total}

$$s_{total} = \sum_{i=1}^{n} s_i^2$$

if only the results of individual readings are available. In the event of multiple readings, the total variance is reduced to

$$s_{total} = \sum_{i=1}^{n} \frac{s_i^2}{n_i}$$

These explanations of error calculation show that readings can only be used with confidence for test purposes and thus for the evaluation of quality if they have undergone appropriate mathematical processing.

2.3.2 Testing Raw Materials and Coating Materials

Numerous test methods have proved successful for evaluating incoming raw materials and performing final inspections on coating materials in paint plants. They can be divided into safety and environment-related indicators, analytical chemistry data, and

Coating Material

physical material and application properties. Due to their special properties and the resulting specific nature of the tests, pigments and pigmented coating materials have been given their own section at the end of this chapter.

2.3.2.1 Safety and Environment-Related Performance Indicators

- **Ordinance on Industrial Safety and Health (BetrSichV)**
  - Flash point, hazard classes, ignition point, ignition limits, ignition classes, explosion limits
- **Hazard Substances Ordinance (GefStV)**
  - MAC, BTLV, TGC values, LD_{50}, LC_{50}, cmr classification
- **Federal Pollution Contol Act (BlmSchG)**
  - VOC limits
- **Waste water**
  - BOD_5, COD

Figure 2.3.5: List of safety and environmental criteria according to German law

Paints are complex mixtures of a range of raw materials which, if their chemical and physical properties are not known, represent a potential hazard to the people in the paint plant and to the environment. Legislators have therefore passed a range of laws and directives to regulate the handling of such materials (see Chapter 5.1 to 5.4).

The most important material properties are flammability, the formation of explosive mixtures and physiological hazards in the event of incorrect handling.

In order to assess the hazard, indicators are defined in the individual laws which are used to assign products to **risk classifications** or to specify limits. Safety variables such as the **flash point**, the **ignition temperature** and the **explosion limits** provide information on the conditions to be met when flammable formulations are handled. The maximum admissible concentration in the workplace (MAC values), biological threshold limit values and the lethal dose (LD) and lethal concentration of gaseous substances (LC values) and the classification as carcinogenic, mutagenic and reproduction-toxic substances give information on the physiological hazard to people and the maximum concentration of immissions. The BOD_5 and COD values, representing the biological and chemical oxygen demands in waste waters, are limits that have to be met to ensure the safe handling of paints and paint raw materials (see Chapter 5).

Flash Point

Above a certain temperature every flammable liquid develops so much vapour, quantifiable by its vapour pressure, that it ignites on approaching an ignition source. However, the initially flammable vapour/air mixture does not necessarily continue to burn for any length of time. The temperature, above which the material is flammable, is termed the flash point. It is therefore the lowest temperature, expressed in °C, at which vapours build up under specified conditions from the liquid under test in such quantities that they form a flammable mixture in combination with atmospheric air.

Statutory hazard categories have been defined with reference to the flash point. Formulations are classified according to the value of the flash point and miscibility with water. It should be noted that flash points of mixtures of substances cannot be calculated from the flash points of the individual components. Azeotropic mixtures of solvents or their interaction with the film forming agents sometimes have a significant impact on the value of the flash point.

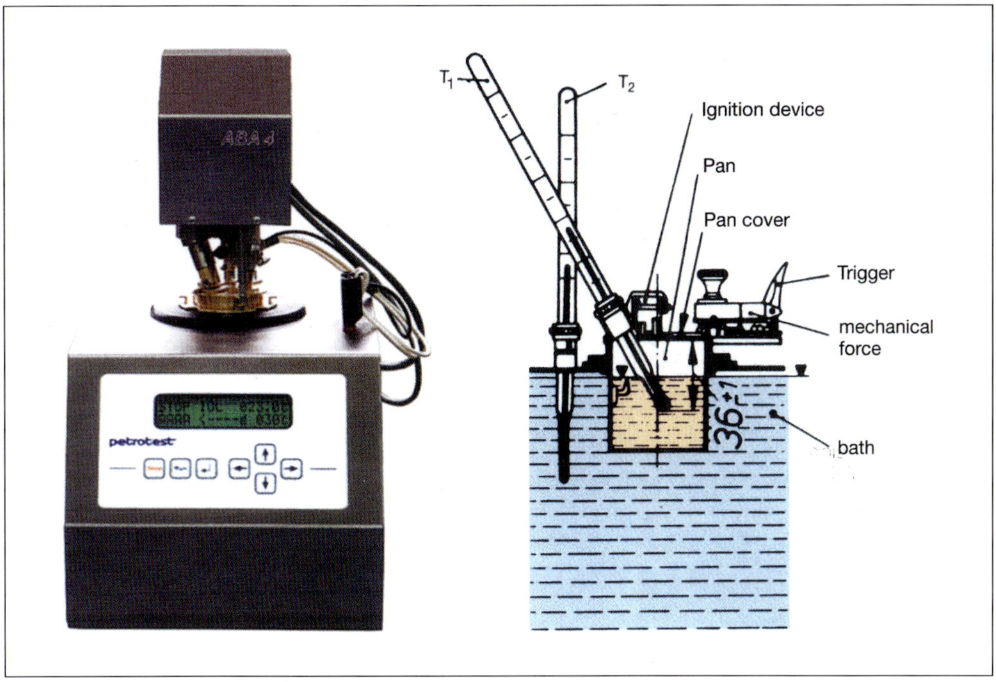

Figure 2.3.6: Schematic diagram and picture of a device for determining flash points according to Abel-Pensky

Flash points are determined according to *Abel-Pensky* (ISO 2719/ISO 3679), *Pensky-Martens* or *Marcusson* [2.4.114].

The main difference between the individual test systems is the heat transfer medium. Whereas *Abel-Pensky* apparatus is fitted with a water bath, *Pensky-Martens* apparatus has an air bath, and *Marcusson* has a sand bath. The actual test procedure is the same for all the systems. At regular intervals an ignition device carries out a 2-second check on the flammability of the liquid which is being heated. The temperature at which a burning liquid, following artificial ignition, is not extinguished after a certain time is defined as the flash point, after conversion to standard temperature and pressure.

$$T_{F_n} = T_{F_B} - \frac{(b - 1013)\,\text{mbar}}{40\,\text{mbar}/°\text{C}}$$

T_{F_n} = standardized flash point

T_{F_B} = measured flash point

b = air pressure in mbar

Explosion Limits

If solvents enter a vapour chamber by means of evaporation, they are capable, above a minimum concentration and after external ignition, of developing so much energy that the ensuing chemical reaction is extremely rapid and is perceived as an explosion. This always occurs therefore whenever, at a sufficiently high temperature, there is a sufficiently high solvent concentration in the vapour chamber for propagation of the artificially initiated oxidation to occur automatically.

With rise in temperature the solvent concentration in the gaseous phase increases further and can become so high that the volume of atmospheric oxygen still available in the environment becomes too small to permit complete oxidation. The mixture then becomes too

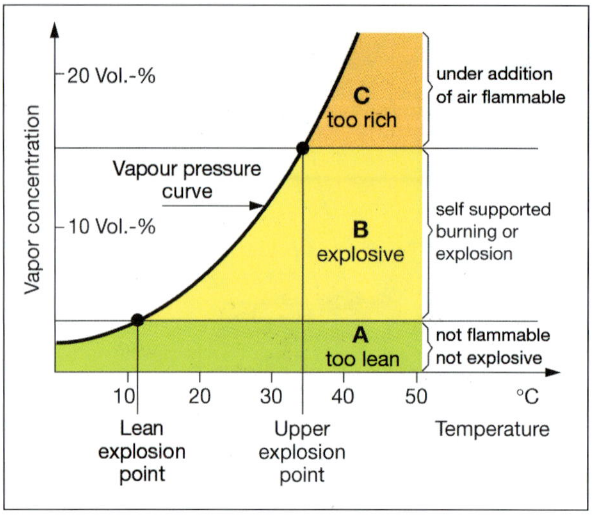

Figure 2.3.7: Generalized vapour pressure curve of a liquid with lower and upper explosion limits

"rich". An explosion is not possible after an **upper explosion limit** has been reached.

Explosive mixtures are thus present when the **lower explosion limit** has been exceeded, but at the same time the upper limit has not yet been reached. However, even then, mixtures with such readiness to ignite must still be supplied with so much energy that the activation energy needed for oxidation of the solvent molecules is exceed-ed. The lower and upper explosion limits are well defined values for most substances. In some materials, such as xylene, whose lower limit is 1 % v/v and upper limit is 6 % v/v, the explosion limits are very close together. Other materials on the other hand, such as hydrogen, have very widely spaced limits (4 % v/v and 76 % v/v).

So if, at manufacturers or processors of coating materials, solvents are released in the form of vapours, these must be diluted to below the explosion limits by supplying pure air. Regulations issued by the appropriate German employers' association and accident prevention regulations, therefore, require that solvent vapours be diluted to less than 0.8 % v/v.

Ignition Temperature

Above a certain temperature an external ignition is no longer required for an explosion to occur. If the activation energy of the reaction mixture is exceeded, a spontaneous exothermic chemical oxidation occurs. The temperature at which a readily ignitable mixture can explode without any external intervention is the ignition temperature.

Ignition temperatures play an important role in defining the hazard potential of explosive gas/air mixtures. For standard paint solvents, the ignition temperatures, divided into temperature categories from Tl to T6, range between 250 °C and 500 °C. To simplify the selection of the right electrical equipment from a safety perspective, binding statutory classifications for solvents have also been specified (see Chapter 5.2).

Health and Safety Indicators

With regard to the protection of people handling hazardous materials in the workplace there is the danger from fire and explosion and also a potential hazard involved in handling toxic, corrosive or irritating formulations.

Germany's Ordinance on Hazardous Substances (GefStV) regulates the handling and labelling of hazardous materials by means of hazard symbols. Supplementary indicators such as the occupational exposure limit value (Arbeitsplatzgrenzwert; AGW), Technical Guide Concentration (technische Richtkonzentration; TRK), Biological Limit Value (biologische Arbeitsplatztoleranz; BAT) and the toxicological terms LD_{50} and LC_{50} are additional tools in evaluating hazards.

AGW (formerly MAK) values correspond to the concentrations of materials which would not be injurious to health even upon eight hours' daily exposure. The figures are stated in units of either mg/m^3 or ml/m^3 (ppm).

For some hazardous materials, e.g. carcinogenic substances, no toxicologically verified limits can be cited. Instead, technical guide concentrations have been defined which are intended to give an indication of the necessary protective measures.

The AGW and TRK values are determined by evaluating toxicological studies on laboratory animals. The toxicity of materials is quantified by the LD_{50} level. The LD_{50} (lethal dose 50 %) is the quantity of a hazardous material (generally taken orally) at which 50 % of the treated laboratory animals die. The LD_{50} is quoted in mg/kg body weight of the laboratory animal. If it is in the range 0 to 25 mg/kg, it is very toxic material. Figures of 25 to 200 mg/kg indicate a toxic material, while LD_{50} values of 200 to 2000 mg/kg are defined as a chemical which is harmful to health. The LC_{50} values permit the inhalation toxicity to be assessed. They indicate the maximum concentration of solvent vapours or gases in the atmosphere at which 50 % of the laboratory survive.

In Germany a list of scientific data for defining maximum admissible concentrations in the workplace published by the Federal Ministry of Labour and Social Affairs is revised each calendar year by the Senate Commission of the Deutsche Forschungsgemeinschaft [DFG – German Research Association] to test materials which are harmful to health. The European Commission then specifies the relevant figures which are subsequently implemented in national regulations or lists in the individual member states.

Apart from occupational health and safety regulations and the associated indicators, coating technology has significant consequences for the environment. Emissions in the form of solvents, noise or odour have to be discussed in this regard, as does the pollution of waste water. The **chemical** (COD) and the **biological oxygen demands** (BOD_5) are important indicators of waste water pollution. Both figures provide information on the level of oxidisable organic pollutants. In the case of the BOD_5 value, samples of the waste water are mixed with bacterial strains. The biological oxygen demand is the volume of oxygen dissolved in 1 litre of water which the microorganisms consume to break down the pollutants. Since this breakdown is associated with a reduction in concentration that follows an exponential function and since the breakdown rate falls, the test period is limited to 5 days. This time limit is indicated by the index 5. The analysis of the chemical oxygen demand COD is a supplementary method. Potassium dichromate solutions serve as the oxidising agent. The volume consumed is determined by titration and converted into oxygen equivalents. The result is quoted in mg O_2/l waste water. The COD method covers all materials, i.e. including inorganic materials, which are oxidisable with dichromates. These include iron (II) ions, sulphides and nitrites as the commonest pollutants in water. For this reason, BOD and COD values are not identical. The two values are a rating basis for calculating discharges into public waste water treatment systems [2.4.115].

Various methods can be used to measure **organic emissions**, termed volatile organic compounds (**VOC**s). To an extent depending on their type and quantity, the volatile compounds are indicated as a mass concentration in mg/m^3 or as a volume concentration in ml/m^3 or ppm. In the event of continuous emissions, it is possible to use the mass flow in g/h. To enable comparative evaluations, the readings must be converted to standard temperature and pressure (air pressure of 1013 hPa and temperature of 0 °C). In cases involving low concentrations and known composition, test tubes are used in which the material to be tested is determined quantitatively in a specific colour reaction. In other measurement methods the test tubes are filled with carrier materials which adsorb the volatile compounds. They are desorbed and quantitatively determined, generally by means of gas chromatography. The adsorption stage can be dispensed with by sampling directly from the emissions in the air using so called gas collectors. This type of sample can then be measured direct with the aid of headspace gas chromatography.

If continuous measurement is necessary, for example in the waste air flows of coating systems, instruments are used which measure changes in the composition of the air flow by means of physical properties such as electrical conductivity in flame ionisation detectors (FID) [2.4.116]. Another type of detector is positioned in the waste air and takes the form of a heated platinum wire which, by catalytically oxidising the organic compounds, is heated still further in proportion to the volume of emissions. The change in its electrical resistance is recorded. Once calibration curves have been recorded, the VOC values can be determined.

If the consumption quantities of coating materials and the air flows in a plant are known, and if the volatile fractions of the coating materials used are known, the material flows and thus the VOC values of a system can be calculated. To determine the volatile components of the coating materials it is enough to base the calculation on the determination of the solid content and the solvent components, with subsequent conversion into volumetric data. This calculation is handled differently in North America from Europe (see Chapter 5.3.1).

2.3.2.2 Chemical Characterisation

Chemical analysis as a means of characterising materials in the paint and coatings industry is mainly used for film forming agents, reactive diluents and additives.

If **film forming agents** are to be developed and produced, and their identity is to be determined with a retained sample, resin-specific indicators are mainly used. The **saponification value, acid value, hydroxyl value, number of NCO groups, iodine value** and the **amine** or **acid equivalents** (MEQ_B, MEQ_S) provide information on the composition of resins and resin solutions.

Saponification Value

Polycondensation resins are identifiable by a finite saponification value (SV). This is determined by mixing the weighed object E of the sample under analysis with a large excess b of 0.1 n KOH solution. After heating and complete saponification, the sample is back-titrated against acid so that the amount of alkaline a used may be determined. The saponification value is calculated from the difference between the acid consumption before and after the saponification of the resins.

$$SV = \frac{(b - a) \cdot 5.61}{E}$$

Acid Value

If resins contain carboxyl or hydroxyl groups, these are also easy to quantify. According to ISO 2114 the acid value AV of a resin is the quantity of KOH in mg which is used to neutralise 1 g substance under specified conditions (indicator: phenolphthalein). If, for example, a resin has an acid value of 56, it contains 1 mmol carboxyl groups per 1 g; or, to express it a different way: an acid value of 56 indicates that there is one carboxyl group in 1000 g substance.

Hydroxyl Value

Determination of the hydroxyl value OHV is a little more complex. ISO 4629 specifies acetylating the hydroxyl groups with an excess of acetic anhydride before saponifying the excess acetic anhydride and, together with the resulting acetic acid, back-titrating it against KOH.

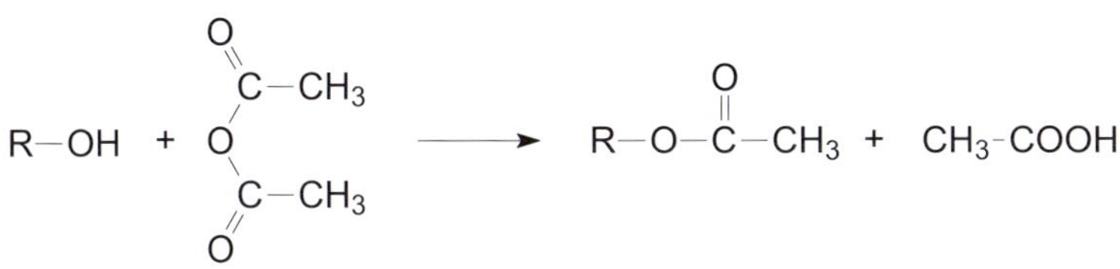

If the resin under analysis also contains carboxyl groups, these are determined during the back-titration process as well. The reading is therefore too low. Consequently, the acid value must be added as a correction term.

$$OHV = \frac{(b - a) \cdot 5.61}{E} + AV$$

NCO and Iodine Values

Other important functional groups such as isocyanates or double bonds can also be quantitatively analysed by simple chemical reactions. Thus the NCO group can be converted into urea with the aid of dibutyl amine. The excess dibutyl amine is back-titrated against hydrochloric acid in the presence of bromophenol blue as an indicator. The process to determine the iodine value is equally straightforward. Halogen adding to double bonds is converted into the analogous quantity of iodine and, in contrast with all other tests specified so far, is expressed in terms not of 1 g, but of 100 g test substance. Of the processes described in ISO 3961 among others, *Hanus's* method may be cited as an example.

Überschuss an IBr mit $I^{\ominus}$ in I_2 überführen

Iodine bromide as the titrating solution b is made to react with the sample. Regardless of whether iodine or bromine is added, the remainder of the titrating solution is quantitatively converted into iodine by the addition of iodide a. The iodine is then titrated against thiosulphate. The iodine value IV is calculated as follows:

$$IV = \frac{(b - a) \cdot 1.26}{E} \qquad 1.26 = \frac{1}{100} \text{ atomic weight of iodine}$$

MEQ_B and MEQ_S Values

The straightforward indicators which apply particularly to water-soluble film forming agents, in addition to the OH and acid values, include the amine and acid values which can be used together as a measure of the degree of neutralisation of anionic or cationic resins or paints.

The amine value in the familiar form of the MEQ_B value (H^+ milliequivalents) indicates how much hydrochloric acid is equivalent to the amine component. To calculate this, the hydrochloric acid consumption is multiplied by the molecular weight of the amine, applied to 100 grams substance and then converted to the solid content.

$$MEQ_B = \frac{\text{ml 1 n HCl} \cdot \text{molecular mass of amine}}{\text{\%FK} \cdot \text{probe mass in grams}} \cdot 100$$

The MEQ_S is calculated by titration against potassium hydroxide solution. Mathematical treatment involves replacing the molecular weight of the amine in the above formula by that of the acid used. Both values are determined for water-borne coating materials in accordance with ISO 15880. The MEQ value is an important complementary test to the pH because, if the concentration of the neutralising agent changes, the pH often only changes insignificantly due to the buffering action of water-soluble paints.

The pH value can be measured quickly and simply with paper strips impregnated with indicators. However, this method is not very accurate. Glass electrodes are used to establish more precise figures.

Instrumental Analysis

Instrumental analytical techniques, some of them complex, are increasingly gaining ground to complement the chemical identification methods introduced in the previous chapters.

Spectroscopic detection of molecular elements, for example by measuring characteristic vibrations, different levels of mobility or adsorptiveness, or the determination of changes induced by heating under defined conditions may be cited as examples. On the other hand, processes which take account of the influence of the molecular weight on the viscosity, vapour pressure, light scatter, and migration rate under gravity have also proved valuable. The following deserve particular mention in this context:

Spectroscopic methods of instrumental analysis are based on the interaction of electromagnetic radiation with the analyte molecules. Changes in the radiation enable inferences to be made about the structures of the molecules.

UV Spectroscopy

UV rays, with wavelengths of between 200 and 400 nm, are capable of exciting electrons. The absorbed energy is used for electronic transitions from the ground to the excited state. Molecular forms, molecular structures and chemical bonds can be identified from the spectra generated in this way. UV spectroscopy is also useful for analysing additives which are used for UV protection of coatings. UV-VIS spectra are increasingly being used to measure the transmission of layers or coatings to check, for example, the barrier effect against UV radiation of the base coats for the integrated paint process (see Chapter 7.1).

This is vital wherever an extender is not used in the integrated paint process, as the electrocoat primer is degraded by UV radiation and adhesion of the base coat is diminished, even to the point of delamination.

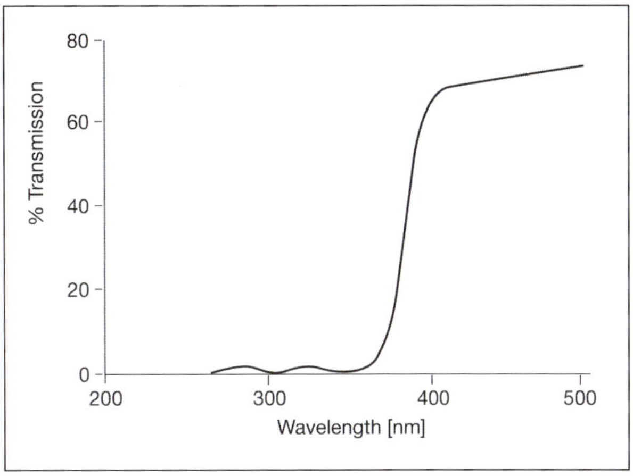

Figure 2.3.8: UV spectrum of a clearcoat based on acrylic resins

Infrared Spectroscopy

The rays of longer wavelengths of 1 to 20 µm, i.e. in the infrared range, are suitable for exciting molecular vibrations which can then be used to identify the chemical structures. Infrared spectroscopy is more universally applicable and much more widespread. Unlike UV spectroscopy, it is not electrons which are excited to rotational, horizontal and bending vibrations, but rather molecular parts because of their significantly longer wavelength, albeit only when the absorption of infrared rays is possible as a result of unequal charge distributions (dipoles) in the vibrating mass. Strong dipoles are therefore particularly prominent in the IR spectrum. The individual parts of the molecules give rise to absorption bands, the position and intensity of which allows conclusions to be drawn about the composition and arrangement of the atoms in the molecule. Infrared spectra are therefore ideal for identifying substances by comparison with reference substances. OH groups, NH groups, carbonyl groups and CH groups exhibit characteristic bands whose exact position in turn is influenced by their immediate environment in the molecule.

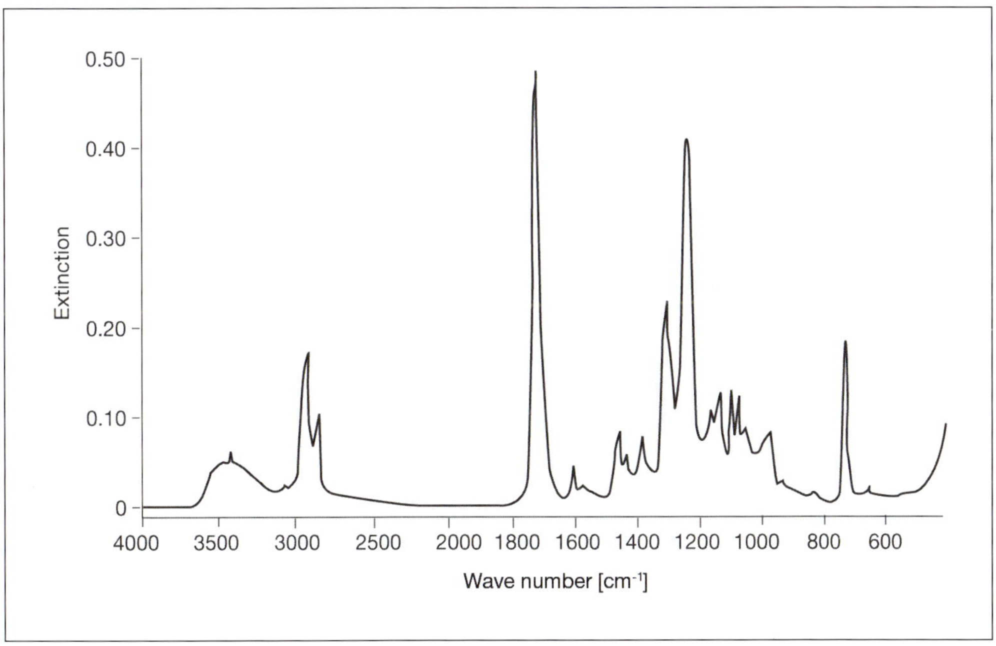

Figure 2.3.9: IR spectrum of an alkyd resin

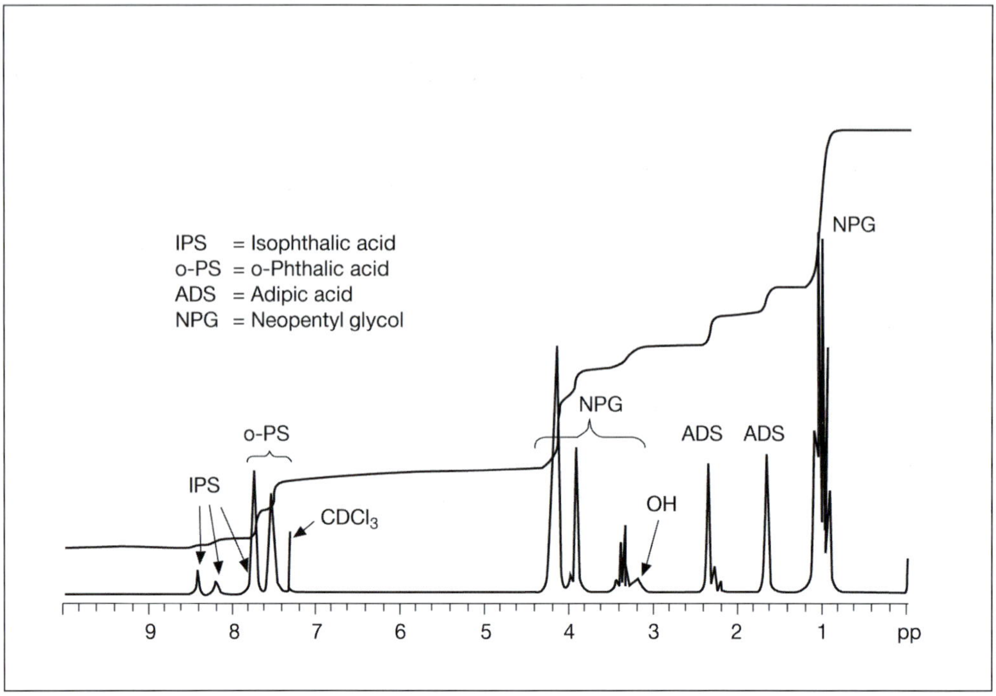

Figure 2.3.10: ^{1}H-NMR spectrum of a polyester resin

This enables conclusions to be drawn about the functional groups, their molecular neighbours and the bond structure.

Modern instruments work extremely quickly and with excellent reproducibility. Using the transmitted light method, they are equally capable of recording spectra of film forming agents from solutions and from solids compressed into pellets with potassium bromide. IR spectra can also be recorded of IR radiation reflected from films.

Infrared spectroscopy also enables mixtures of substances to be identified and the proportions within the mixture to be estimated by comparing the ratios of the characteristic absorption bands of the individual components.

Infrared spectroscopy is particularly successful for analysing PUR paints. The NCO absorption band at approx. 2260 cm^{-1} (approx. 4 µm wavelength) is overlapped by only a few functional groups and is also extremely narrow and therefore suitable for kinetic studies of crosslinking processes in association with isocyanate groups. Tracking the hardening of epoxy resins and the oxidative crosslinking of alkyd resins are further examples of the successful use of IR spectroscopy. A great many standard spectra are now stored in digital form and can be accessed without difficulty and evaluated electronically [2.4.117].

Nuclear Resonance Spectroscopy

Instruments based on nuclear magnetic resonance (NMR) as the measured variable operate on completely differently. The principle here is based on the fact that nuclei with an odd number of protons (^{1}H, ^{13}C) in a homogeneous magnetic field undergo frequency-dependent spin reversal in an irradiated magnetic field because of their magnetic

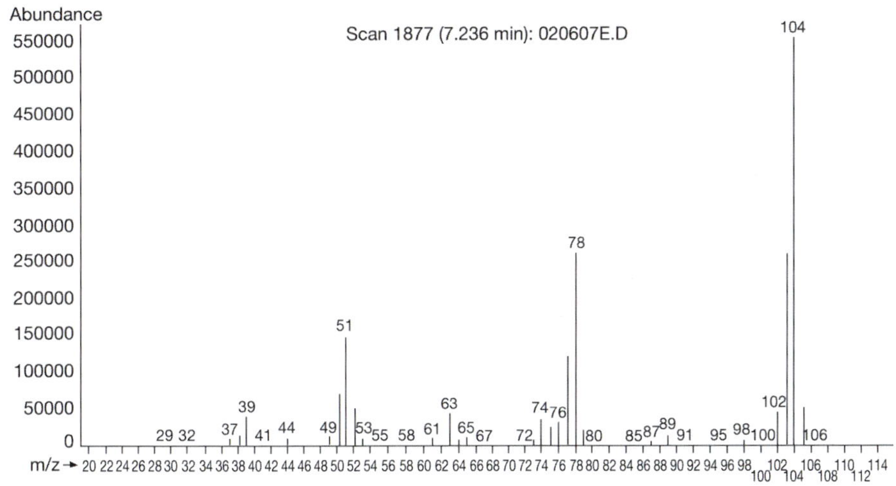

Figure 2.3.11: Mass spectrum of styrene

moment and because of the nuclear spin resulting from the rotation of the nuclei. When nuclear magnetic resonance spectroscopy is used, the sample is first introduced into a strong, linear magnetic field to orient the rotating nuclei. The nuclei, which act as mini magnets, are aligned in parallel and antiparallel form in two different energy levels. If a second magnetic field with a variable frequency acts on the oriented mini magnets, for example of the hydrogen or ^{13}C nucleus, spin reversal takes place in the nuclei at the lower level by absorption of energy at the nucleus's resonant frequency. The frequency of the spin reversal is bond-specific insofar as the electrons shielding the nucleus lead to a change in the resonance frequency and the chemical shift as a function of the chemical bond among the atoms and also of the spatial environment. The shift in the resonant frequencies is determined by reference to an internal standard. Tetramethylsilane serves as a standard particularly for proton resonance spectra. The intensities of the signals are used for quantitative evaluations. The use of ^{1}H-NMR, for example, makes it possible to determine whether protons are assigned to a carboxylic acid, a hydroxyl group or just to a CH bond. This enables molecular structures to be identified [2.4.118].

Intermolecular interactions or cis-/trans-isomerisation can be detected by nuclear magnetic resonance spectra, a fact which ultimately means that secondary structures can also be identified.

Mass Spectroscopy

A common method for structurally identifying molecules is mass spectroscopy. It makes use of the mass-dependent inertia of accelerated charge-bearing molecules during deflection in a magnetic field.

The design of a mass spectrometer consists in principle of a sample reservoir from which the sample is vaporised and then fed to an ionisation chamber. Highly accelerated electrons in the chamber cause ionisation of the substances under analysis. In accordance with the energy input, the molecules may either remain intact or be fragmented. The resulting ions are accelerated by electrical fields, collimated and then deflected from their flight path by an electromagnetic field. The flight path is determined by the charge, particle speed and mass. Where the charge and the speed are the same, deflection is then only a function of mass, and this can therefore be used as a measurement variable for the molecules under analysis or their fragments.

The analysis of macromolecular materials requires fragmentation of the original molecules, as there is no guarantee of volatility. A secondary task thus consists in the imaginary piecing together of the identified molecular fragments.

Mass spectroscopy has largely been superseded as the sole test method for quantitative analyses. However, it has proved to be a useful complementary method for identifying mixtures of substances which have already been separated by gas chromatography or by chemical reactions, in a process termed GC/MS.

Chromatographic Methods

A common feature of all chromatographic separation and identification methods is that the materials or material mixtures are routed as a mobile phase past a stationary phase where material-specific interactions occur. These influence the migration rate in the mobile phase. The retention times measured are then used to identify the material.

Gas Chromatography

Since **gas chromatography** is a process for separating gaseous materials, only materials which can be vaporised are amenable to it.

For macromolecular materials, therefore, it is sensible to produce volatile oligomeric or monomeric elements in advance by means of defined wet chemical decomposition reactions. It is often possible to transform these products into nonpolar forms and thus into more volatile variants through subsequent esterification with methanol or through silylation with trimethylchlorosilane. A simple method of obtaining volatile substances consists in upstream pyrolysis of the test substances in the gas chromatograph itself.

The gaseous initial product in isolation or also as a mixture is carried by an inert carrier gas past a solid material or a liquid absorbed on a support in capillaries. This interaction separates material mixtures into their components, which therefore exit the separation unit at different times. The retention times, measured by means of material specific properties such as thermal conductivity, refractive index and the like, permit conclusions to be drawn about the individual components. A downstream mass spectrometer enables the materials to be identified. Quantitative analysis can then be achieved by comparing the intensities of the signals with those of reference substances (see below).

Liquid Chromatography

Compared with gas chromatography **high performance liquid chromatography** (HPLC) offers the advantage of being able to analyse solutions. The liquid sample in the mobile phase is carried past the solid carrier material in the capillaries (columns) and, due to different interactions there, is transported at varying speeds to the detector. Porous SiO_2 particles modified with amino, cyano or hydroxyl groups serve as polar column material.

HPLC is primarily employed for low volatility substances. It is an established method for trace analysis because it is so sensitive. HPLC is also suitable for analysing polymers. It can separate, for example, acrylic resin mixtures, aliphatic and aromatic epoxy resins, as well as various melamine resins, to mention just a few of the wealth of possibilities [2.4.119].

For the low molecular additives, such as light stabilisers, surfactants and rheological aids, simple column chromatography (LC or liquid chromatography) in combination with mass spectroscopy can be employed for identification purposes.

Gel Permeation Chromatography

An even better method, because it can be used both for nonpolar polymers and for polymer fractions, is **gel permeation chromatography** (GPC). Unlike HPLC, GPC eschews the use of differential polar interactions of material and carrier in the stationary phase and instead utilises cavities of different size in the particles of the stationary phase for separation and identification of polymer mixtures.

Styrene/divinyl benzene copolymers generally serve as gels that have cavities of different size. Their pore diameters lie between 5 and 500 nm. As a result, large molecules penetrate the pores less deeply than the smaller ones and so are also transported more rapidly. Quantitative evaluation must be preceded by calibration against known substances. The concentration in the eluate, plotted against the elution time, provides information on the molecules of different size in the material mixtures. Since macromeric and oligomeric materials, as present in film forming agents, are always mixtures of molecules of different size, gel permeation chromatography is excellent at determining the molecular weight distribution of paint resins and after mathematical treatment also enables mean values to be stated [2.4.120]. GPC is superseding fractional distillation or fractional precipitation methods which were standard in the past but which are expensive to carry out and require particular experimental skills [2.4.121].

Special Methods for Determining Molecular Weights and their Mean Values

Analytical chemists employ a range of methods specifically for characterising the size of molecules and obtaining their mean values where molecules of different size are present. These return either the number-average molecular weight or the weight-average molecular weight.

The **number-average molecular weight** M_n is the sum of the molecular weights of the individual size fractions multiplied by the number of molecules contained in the size fractions, divided by the total number of molecules. The **weight average molecular weight** M_w, by contrast, expresses the molecular weights of the individual size fractions in terms of the sum of the weight fractions, not the number of molecules.

$$M_n = \frac{\sum n_i \cdot M_i}{n_i} \qquad M_w = \frac{\sum w_i \cdot M_i}{w_i}$$

where $w_i = n_i \cdot M_i$

Since M_w must always be greater than M_n wherever there is a **molecular weight distribution**, the quotient M_w/M_n is a measure of the spread of the distribution. This can be stated as the dispersity D or nonuniformity U:

$$D = \frac{M_w}{M_n} \quad \text{and} \quad U = \frac{M_w - M_n}{M_n} \quad \text{or} \quad U = D - 1$$

Membrane and **vapour pressure osmometry** are important methods for measuring the number-average molecular weight. In a resin solution separated from the pure solvent by a semi-permeable wall, diffusion of the solvent into the resin solution will generate osmotic pressure. This pressure is determined by the molecular weight M_n and the concentration. The vapour pressure of the solvent, which has been reduced in relation to the pure solvent by the presence of resins, also serves as a measurement variable. As above, the reduction is a function of the molecular weight M_n and the polymer concentration. MW is calculated by measuring **light scattering** of macromolecules in dilute solutions [2.4.122].

M_w can be determined, approximately rather than exactly, by measuring the migration rate of the polymer in the centrifugal field of an **ultracentrifuge** or its viscosity in dilute polymer solutions. Because of their simple equipment requirements **viscometric methods** have largely superseded the ultracentrifuge. In these, correlations with the molecular weight are established by measuring the different concentrations of solutions and calculating the concentration-dependent rise in viscosity relative to the viscosity of the solvent.

Thermal Analysis

Thermal analysis is a term covering a range of techniques for studying transformations that accompany changes in temperature. If the transformation is also associated with the formation of volatile cleavage products, the gravimetric effect can be used for **thermogravimetric analysis** (TGA). **Differential thermal analysis** (DTA) and **differential scanning calorimetry** (DSC) are common methods of thermal analysis.

In DTA the sample is heated under defined conditions together with a reference material which is inert in the measurement range and is compared with it. The temperature difference between the sample and the reference is zero as long as no thermally induced changes, which are material specific, take place. If, however, changes in the free energy relative to the reference sample occur as a result of changes in the degree of crystal-linity, intermolecular bridging bonds or even chemical reactions, a finite ΔT is measured. The size of the effect and its position on the temperature axis provide important information on the material under analysis. DSC, by contrast, does not measure the temperature difference as a means of obtaining information, but rather determines the extra heat flow needed to equalise the resulting temperature difference.

The position of the glass transition temperature T_g, the crystallinity and the chemical reactivity are some of the material properties which can be identified. For those involved in the paint and coatings industry analyses of already cured coating materials are also of interest, firstly because the heat resistance can be estimated, secondly because undercuring can be highlighted by identifying residual activities, and thirdly because structural change can be determined, for example via a change in the position of T_g (see Chapter 3.2.5).

Dynamic mechanical analysis (DMA) which measures mechanical properties of the test material as a function of time, temperature and frequency of the applied force primarily serves to determine the degree of curing during stoving or over a period of time. The gradients of the temperature rises can be adjusted to match the operating conditions of the coating equipment. Variables determined are T_g, modulus of elasticity E and the shear modulus G of the coatings [2.4.123].

Example of a Full Analysis of a Coating Material

Analysis of the paint begins by determining the nonvolatile components (**solids**). An exact analysis of the volatile components is not possible in every case. Particularly with amino resins there are fluid transitions since fragmentation of resin components can easily occur. The presence of plasticisers also blurs the dividing lines between volatile and nonvolatile components.

To **separate the pigment and the binder** the ready-to-use paint is diluted with a solvent and centrifuged. Highly volatile solvents, such as ethyl acetate, acetone or tetrahydrofuran are preferred. In general centrifuges with speeds of 500 to 6000 rpm are adequate for separating pigments and extenders from the binder. If carbon black or finely distributed organic pigments are present, speeds of up to 20,000 rpm are required to ensure complete separation. Once the pigments have been separated, the diluted binder is decanted, and the pigment is agitated again with a fresh solvent and centrifuged again. This process is repeated, the pigment is finally dried, and a quantitative determination is carried out. The binder proportion is generally calculated from the difference between the solids and pigment components. However, it can also be determined directly by combining the decanted binder portions and distilling off the solvent. The sum of the directly determined binder and pigment components must be equal to the proportion of the nonvolatile components.

The composition of the **volatile components** in solvent-borne systems is determined by means of gas chromatography. In the case of aqueous samples, a supplementary determination of water content is also necessary which can be performed titrimetrically by the *Karl Fischer* method in the case of low concentrations, or by means of the calcium carbide method for higher concentrations.

Before gas chromatography is used, the volatile components can be separated by distillation in a micro rotary evaporator. The separation process can produce certain errors for both high boilers and low boilers. These errors can be avoided by direct injection into the gas chromatograph. The mass spectrum of the individual components can be used in addition to the retention time to identify the GC peaks.

Analysis of the separated **pigment/extender mixture** begins with conventional preliminary tests. The substance under analysis is then treated with both acetic and hydrochloric acids. The proportions soluble in each case are determined, among other ways, polarographically and with the aid of complexometric titration. Infrared spectroscopy is often an indispensable aid in pigment analysis. It can enable various silicates such as kaolin, talcum, barium sulphate and the titanium dioxide modifications rutile and anatase to be distinguished and organic pigments to be identified. It may be noted that IR spectroscopy during pigment analysis can also provide information on film forming agents which have remained on the surface of the pigments.

Synthetic waxes, such as copolymers of ethylene and vinyl acetate, are separated along with the pigments and extenders. The waxes can be dissolved out of the centrifugate by refluxing in tetrahydrofuran and identified using IR spectroscopy.

The bulk of the elements can also be analysed qualitatively and quantitatively by means of atomic absorption spectroscopy (AAS). This method is particularly useful when the elements under analysis are present only in low concentrations.

Apart from simple preliminary tests for nitrogen and chlorine, **binder analysis** starts by recording the IR spectrum of the binder mixture in the solvent-free state. This enables at least the principal components of the mixture to be identified. If esters are present, saponification with ethanolic KOH is useful. When alkyd resins are being analysed, for example, this leads to four fractions, viz. the di- and oligocarboxylic acids, nonsaponifiable components and the fatty acids and polyols. Examples of di- and oligocarboxylic acids that can be analysed polarographically are phthalic, maleic, fumaric and trimellitic acids. Conversion into methyl ester is necessary before gas chromatographic analysis can be conducted. The nonsaponifiable components are analysed by IR spectroscopy. Among other things they contain fragments of amino resins and epoxy resins. HPLC is often ideal for characterising this fraction more accurately.

The **fatty acids** and other higher monocarboxylic acids can be separated on a gas chromatograph if they are methylated beforehand. Capillary columns are available for the identification of linolenic acid, conjugated linoleic acids, ricinoleic acid, abietic acid etc. Since the fatty acid fraction can also contain modified fatty acids such as dimerised, maleinised, styrenated and epoxidised fatty acids, the use of HPLC is advised. It should be noted that escort substances, such as epoxy phenol resin components, can be present in the fatty acid fraction.

The **polyols** occur in free form and can be separated by means of gas chromatography after silylation.

However, saponification is not always a suitable means of characterising a binder mixture completely. Apart from the case of melamine resins, which can be analysed by direct titration in an anhydrous medium or by means of UV spectroscopy after acid catalysed hydrolysis, saponification yields insufficient information, especially where **polyacrylics** are present. In this case pyrolysis in association with gas chromatography is the most appropriate analytical method.

Even when **urethane bonds** are present, normal saponification is inadequate. In this case cleavage can be achieved by means of KOH fusion. The diisocyanates incorporated as urethanes are converted into the corresponding diamines and then identified by means of thin layer chromatography. Thin layer chromatography, a simple, fast and straightforward process, also permits the reference detection of organic peroxides, UV absorbers and plasticisers, but also, after appropriate preparation, of di- and oligocarboxylic acids, lower fatty and aromatic monocarboxylic acids, and finally also polyols which are commonly used in paints.

Another method which can be used to assist in the analysis of binders is extraction of the dried binder. The situation can occur where **phthalic acid plasticisers** are present in addition to **alkyd resins**. While saponification provides a sum for the phthalic acid component, the phthalic acid plasticiser can be quantitatively separated by prior extraction with pentane. Other plasticisers, for example soft resins, can be extracted with aliphatic hydrocarbons. The extract can be identified with the aid of IR spectroscopy. Where a mixture is present, gas chromatography or high performance liquid chromatography must be used.

It is often advisable to continue the extraction process by gradually moving to polar solvents. The binder for a combination cellulose nitrate paint, for example, should first be extracted with pentane to separate the monomeric plasticisers. The extraction should be continued with dichloromethane which permits all synthetic and natural resins with the exception of cellulose nitrate to be identified.

2.3.2.3 Physical Indicators

The most important physical indicators for coating materials from a processing point of view include their solid content, viscosity, surface tension and density. Of these, the solid content and density are the most significant economic variables. They determine the yield (see Chapter 5.7). They also impact on a range of application properties in the same way. Like the surface tension, the viscosity, which varies with the different loads to which coating materials are subjected, is of great relevance to the process engineering aspects of paint manufacture and processing. Since the kinetic energy of the viscous components is converted into heat when the material is flowing, whether during conveying in pipes or homogenisation in agitator vessels, the viscosity is also an economic variable. Atomisation, wetting, levelling, sagging, dewetting from edges and the adhesion of coatings are phenomena which are also affected by the surface tension.

The indicators described below and their measurement methods relate to liquid coating materials. The characteristic indicators for powder coatings and the ways of determining them are described in Chapter 4.2.2.

Solid Content

The ISO 23811 defines the solid content as nonvolatile matter determined as a percentage under clearly defined (i.e. agreed between the customer and the supplier) drying temperatures and times (generally 60 min at 130 °C). Other requirements are a dish diameter of 75 mm and a test portion weighing 2.00 g ± 0.02 g.

Since numerous paints and film forming agent solutions not only lose their solvents on heating, but can also undergo chemical reactions which are associated with the formation of volatile cleavage products, differences in the oven temperature and sample preparation can lead to significant differences in the results. Bearing in mind that the solid content, even leaving its economic importance aside, is also an important environmental variable for calculating environment-related data, measurement errors can have serious consequences. Furthermore, many paint-based indicators are expressed in terms of the solid content.

Role of Viscosity in Evaluating the Flow Properties of Coating Materials

The flow properties of paints and solutions are of huge importance in evaluating the overall manufacturing process of coating materials, starting from the pumping of the raw materials into the storage vessels, via numerous intermediate stages, to the finished products. Moreover, the application properties of the coating materials such as brushability, levelling and tendency to run are closely connected with the flow properties.

Extremely different demands are placed on the flow properties of paints as they progress from manufacture to the finished coating. When conveyed in circulation systems paints should be as liquid as possible whereas, for application to the object,

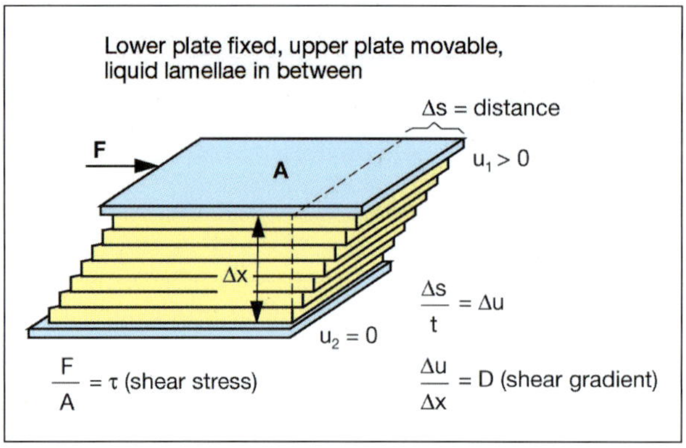

Figure 2.3.12: Newtonian twin-plate model

a smooth film should be produced without running on vertical surfaces. These apparently contradictory demands can only be understood and implemented by those with the expertise to evaluate and influence the flow properties of polymers, polymer solutions and paints in detail.

In the explanations below, "flow" is understood to mean the irreversible deformation of bodies in which the energy required for said deformation is converted into motion and into heat as a result of internal resistance (= viscosity). Another extreme is the automatic elastic recovery of the deformation to the initial state. Between the extremes marked by the purely viscous liquids at one end and the elastic materials which can store all the free deformation energy to ensure recovery at the other lies the wide range of viscoelasticity. Viscoelastic materials combine the properties of viscous flow and reversible storage of deformation energy [2.4.124].

To better understand the phenomenon of viscosity, consider the force which must be applied to stir a thin or a thick liquid. A thick paint poses a higher resistance for the agitator tool than a solvent and therefore requires more energy. The reason for this lies in the higher internal friction of the larger molecules of the film forming agents by comparison with the small structural elements of the solvent.

The physical measurement variable for the quantitative description of this internal friction is viscosity. The physical correlations can be illustrated by means of the *Newtonian* twin plate model.

The gap Δx between two parallel plates of surface area A is filled with a liquid. One of the two plates is moved with a force F and a speed u_1 parallel with the stationary plate. The liquid, which can be imagined to consist of an infinitely large number of liquid lamellae overlying each other in a parallel configuration, adheres to the plates and is therefore also moved at its upper edge at a speed u1, while remaining stationary at the lower, unmoved plate.

Because of the friction between the layers or lamellae a counterforce now acts on the imaginary liquid layers (laminar flow) which are moving parallel with and opposed to each other in the time t in a direction along the path Δs.

The degree of displacement of the individual lamellae relative to each other depends on the distance between the plates, in addition to the resulting speed difference between the two cover plates. The shear force per unit area, the shear stress τ, is thus proportional to $\Delta u/\Delta x$. This variable is termed the speed or shear gradient D. In general, there is a direct proportionality between τ and D for laminar flow in accordance with the *Newtonian* twin plate model, where the proportionality factor η is the viscosity.

$$F = \eta \cdot A \frac{\Delta u}{\Delta x}$$

because of $\frac{F}{A} = \tau$ and $\frac{\Delta u}{\Delta x} = D$ follows

$$\tau = \eta \cdot D$$

In accordance with the definition the viscosity η has the dimension Pa s.

$$\eta = \frac{\tau}{D} \quad [\mathrm{Pa \cdot s}]$$

A liquid consequently has a viscosity of 1 Pa s if there is a velocity gradient of 1 s^{-1} under a shear stress of 1 Pa.

Apart from the thus defined **dynamic viscosity** η it is also usual to apply the **kinematic viscosity** ν per unit density to characterise liquids and paints.

$$\nu = \frac{\eta}{\rho} \left[\frac{m^2}{s}\right] \qquad \varphi = \frac{1}{\eta} \left[\frac{1}{\mathrm{Pa \cdot s}}\right]$$

The reciprocal value of the dynamic viscosity is the fluidity φ.

The dynamic viscosity defined by *Newton* in the twin plate model can be quantified by measuring the resulting velocity gradient under varying shear stress. As defined in $\tau = \eta$ D the viscosity then corresponds to the increase in the straight lines determined.

Examination of the flow properties of a range of liquids shows that not all obey *Newton's* fundamental law. Water, mineral oils and also solvents, i.e. without exception low molecular substances, largely demonstrate the expected linear dependence in a τ/D graph. Viscosity measurements on liquid coating materials, i.e. solutions of higher molecular substances, obey the described linear correlations only at high dilution levels or in exceptional cases. The viscosity of paint preparations, by contrast with *Newtonian* liquids, is a function more or less of the degree of shear. In many cases the viscosity decreases with increase in velocity gradient. In such pseudoplastic liquids the viscosity is not a constant of the material. The τ/D graph is identical under increasing and decreasing shear stress. Binding information about the viscosity must therefore always refer to the shear set during measurement for those pseudoplastic materials.

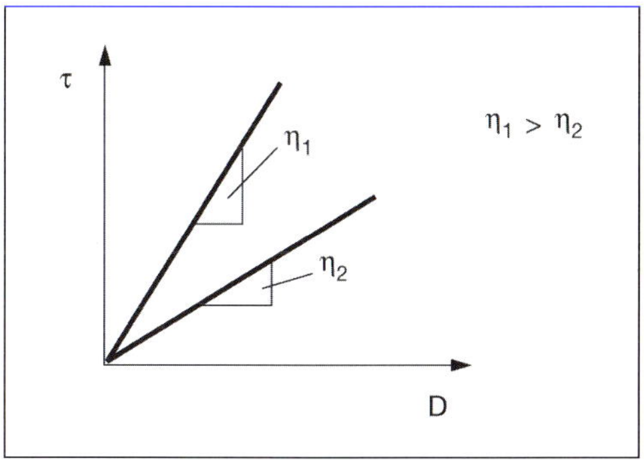

Figure 2.3.13: Flow curves of Newtonian liquids

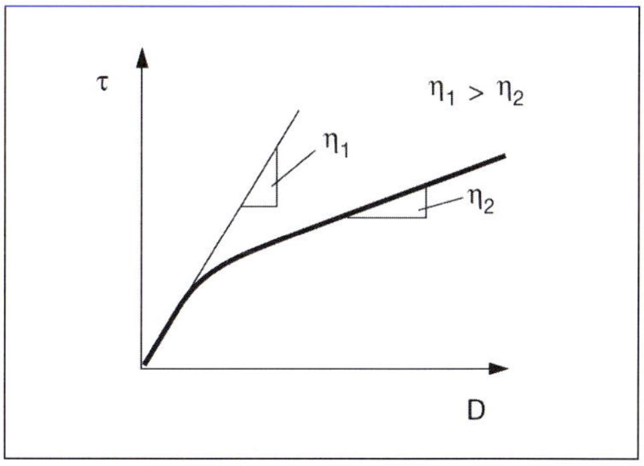

Figure 2.3.14: Flow curve of a pseudoplastic liquid

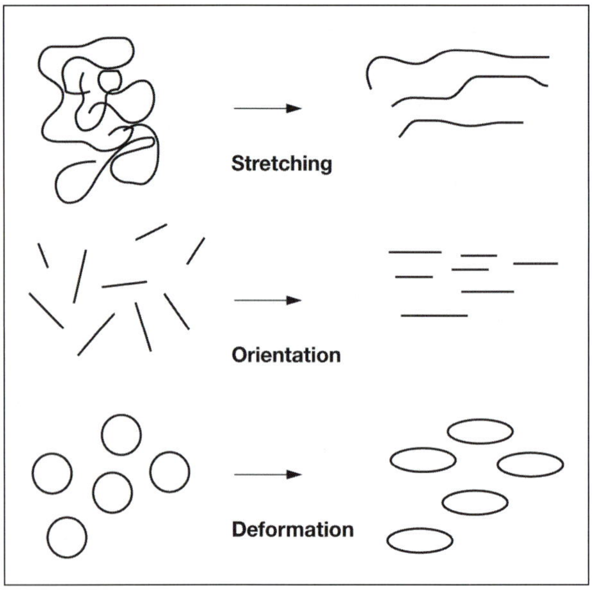

Figure 2.3.15: Explanation of pseudoplasticity as the stretching, orienting and deforming of molecules

A reduction in viscosity with increase in shear stress is a frequent, but also desirable property in paints and pigmented coating systems. It is based on the fact that the flow-inhibiting interaction within the polymer solution decreases with increase in shear stress. This can occur by stretching of macromolecules which are coiled when at rest. Pigments and extenders can also become aligned during flow, thereby reducing internal friction. And finally, deformation of particles has the effect of reducing the viscosity. Pseudoplasticity is the basis for the low and high viscosity of one and the same coating system during processing and subsequent film formation.

Substances are also known in coating technology which cause a rise in viscosity with increase in shear. Sand/water mixtures, starch suspensions and certain types of dispersions and plastisols are examples of such liquids, which are termed **dilatant liquids**. The increase in frictional forces with increase in shear which underlies the phenomenon of dilatancy can be caused by an induced charge exchange at the friction surfaces and an increase in the electrical interaction.

Every coating material has a tendency to run on vertical surfaces under its own weight; this tendency is all the more pronounced, the thicker the freshly applied coat. The paint and coatings industry has succeeded in its endeavours to eradicate these unwanted concomitants by manufacturing corresponding coating materials with a yield point. The weight of the coating material per unit area must then be less than the force which could cause it to run.

Two other phenomena are responsible for the occurrence of yield points, either in isolation or in association with pseudoplasticity: **plasticity** and **thixotropy**. Plasticity is a feature

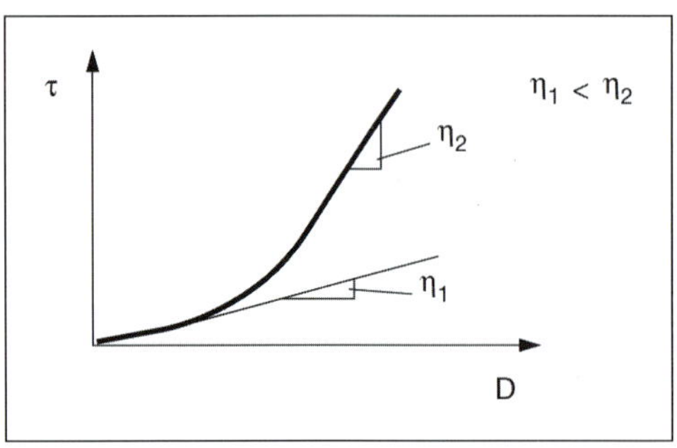

Figure 2.3.16: Rheological profile of dilatant materials

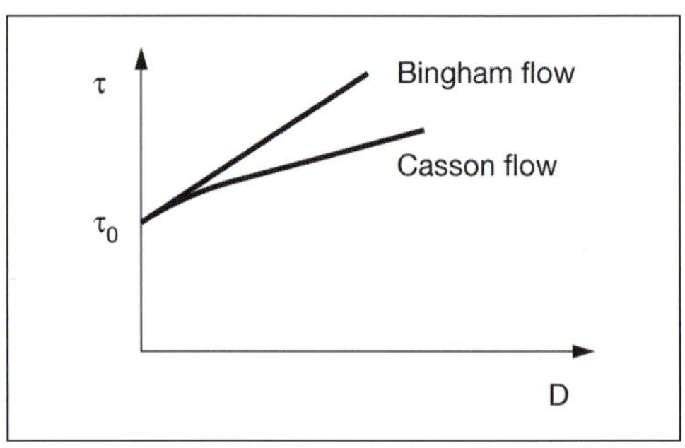

Figure 2.3.17: Yield point of materials exhibiting plastic behaviour

of all liquid substances, which can only flow once a minimum shear stress τ_0 has been exceeded, viz. the yield point (Figure 2.3.17). Many pastes and fillers belong to this group of materials. After starting to flow they can exhibit *Newtonian* flow (*Bingham* liquid) or pseudoplastic properties (*Casson* liquid).

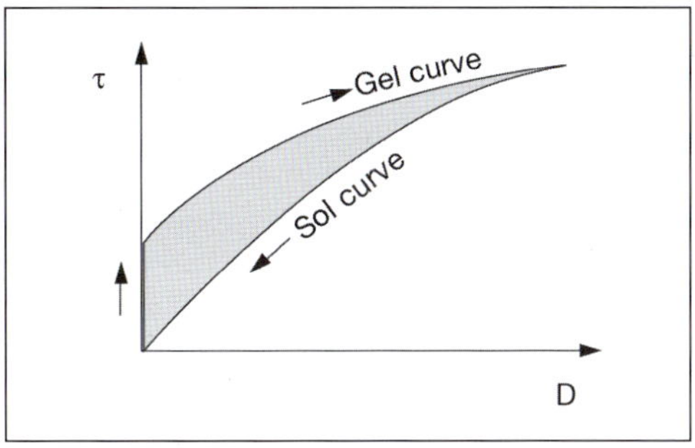

Figure 2.3.18: Flow curve of a thixotropic liquid

In respect of their flow properties, therefore, paints and coating materials cannot generally be described with the linear function formulated by *Newton*. A mathematical relation of more practical relevance was developed by *Ostwald* [2.4.125].

$$(\tau - \tau_0)^n = \eta \cdot D$$

If n = 1 and $\tau_0 = 0$, these are *Newtonian* liquids. If n > 1 or n < 1 and τ_0 in turn is 0, the liquids are pseudoplastic or dilatant. For substances with a yield point, $\tau_0 > 0$.

All rheological phenomena described thus far are reversible processes without any time delays. By contrast, changes in the viscosity of **thixotropic liquids** are additionally dependent on time. The flow properties of coating materials are therefore a function of their history. Starting from the viscosity at rest, the viscosity decreases with increase in shear stress, but then initially reverts to a lower viscosity at rest when the shear rate decreases. Only after a time which is specific to each material is the substance regenerated and the original viscosity reestablished. In the τ/D graph there are, therefore, two different flow curves, for increasing shear stress from rest and falling stresses thereafter. The area enclosed by the curves in Figure 2.3.18 is a measure of the degree of thixotropy.

The term thixotropy is derived from Greek and means change as a result of touch. In this phenomenon a gel liquefies under mechanical action such as vibration, agitation or brushing (sols) and gradually solidifies into a gel-like material again once the action ceases (gels). This reversible process is therefore termed a sol-gel conversion. It is encountered in paints and numerous other materials. Some types of clay, bentonites or hydrated metal oxides have a pronounced tendency when in static concentrated solutions with relatively large volumes of liquid to accumulate as honeycomb-shaped, net-like or matrix-like structures. Consequently, the viscosity of such a solution rises rapidly. However, the intermolecular forces are not large enough to keep the structure intact during agitation or vibration. This process negates the yield point. On reverting to the material's rest state, the structure is reestablished within a certain period of time. It should be noted that the thixotropy of a system decreases with rise in temperature and decreases entirely at temperatures around 80 °C.

Although thixotropy is problematic in the manufacture of paints and coating materials, it is a welcome property in processing. A feature of thixotropic paints, like paints with a set degree of pseudoplasticity, is that they have less tendency to run on vertical surfaces.

In rare cases formulations exhibit the property of **rheopexy**, which is the opposite of thixotropy. In this the viscosity at rest is lower than the viscosity after the action of shear.

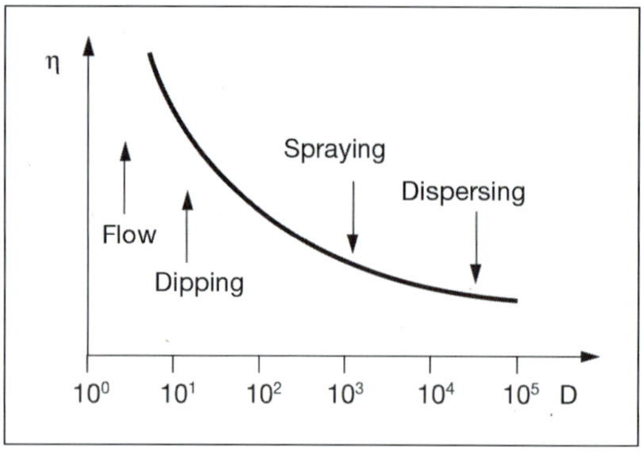

Figure 2.3.19: Shear stress D and viscosity η of ideal pseudoplastic paints from manufacture to application

Evaluating the complex flow properties of paints in practice requires accurate knowledge of the shear stress at every stage of manufacture and processing. A broad band of varying shear stress is covered, from the extremely high shear stress generated when paints are dispersed or sprayed, to the lower shear gradient of dip coating, to the extremely low shear that arises when freshly applied films flow.

The spectrum of D values ranges from 10^{8} to $> 10^{5}\ s^{-1}$ when paints are used.

The ideal pseudoplastic paint for a given application will always exhibit the desired viscosity within this shear gradient spectrum. This is the only way of explaining how coating materials can meet the seemingly contradictory demands outlined above, namely easy brushability or sprayability without dripping, while achieving good levelling at the same time.

Adjusting the pseudoplasticity and the plastic flow properties of a liquid coating material as a way of meeting the sometimes highly diverse requirements of the entire feed and processing stage can only succeed if the temperature dependence of the viscosity is factored into the characterisation of the rheological properties.

Of the numerous empirical and scientific attempts to quantify the impact of temperature on the viscosity of resin solutions and liquid coating materials, the *Andrade* equation has proved useful in practical operations. The *Andrade* equation is based on work by *Arrhenius, Prandtl* and *Eyring*. This work explains the temperature dependence of liquids in terms of the energy exchanged by individual molecules when changing position during the flow process.

$$\eta = A \cdot e^{\frac{B}{T}} \qquad \text{A, B = material specific constants}$$

The temperature dependence of the viscosity of solvent-free polymer melts, as involved in the manufacture and processing of powder coatings, is better explained by the *Williams, Landel* and *Ferry* equation (WLF). Admittedly, this describes the change in viscosity only within a relatively narrow range above the melting point, but with a good level of accuracy [2.4.126].

$$\lg \frac{\eta}{\eta_0} = \frac{-B \cdot (T - T_0)}{C + (T - T_0)}$$

B,C = material constants

η_0 = melt viscosity

T_0 = reference temperature, e.g. melting temperature

For resin solutions and melts the reduction in viscosity accompanying an increase in temperature can be considerable, particularly at higher viscosities. Viscosity reductions of 4 to 10 % are measured per degree of temperature rise, with the exact figure depending on the chemical structure. It is therefore important to ensure good temperature control in dispersers and agitators, and particularly in sensitive measuring instruments for determining viscosity.

Under the known condition that the deformation work done during the flow of viscous liquids is quantitatively converted into heat, the following applies with regard to estimating the heat increase. An imaginary cube of side length x is sheared with a force F along a path s at a shear angle γ. The quantity of heat $\dot{q}$ relative to the volume V arising in the time t is then

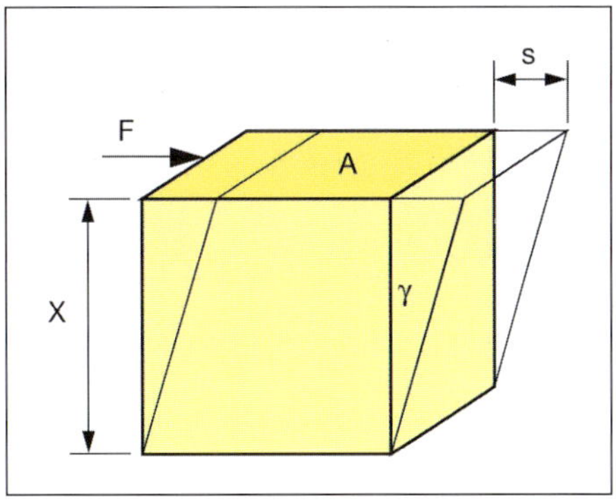

Figure 2.3.20: Model for estimating heat generation during flow of a liquid

Coating Material

$$\dot{q} = \frac{F \cdot s}{V \cdot t} \quad \text{where} \quad F = \tau \cdot A \quad \text{and} \quad V = A \cdot x$$

$$\text{follows} \quad \dot{q} = \frac{\tau \cdot A \cdot s}{A \cdot x \cdot t} = \tau \cdot D = \eta \cdot D^2$$

Heat generation during flow is thus directly proportional to viscosity, but proportional to the square of the shear gradient. This fact is important, among other things, for the design of the temperature control system of sensitive measuring instruments and for the development of manufacturing procedures for paint formulas (see below).

Measurement Methods

A range of instruments is available for measuring viscosity which locate the desired indicators by directly determining the shear stress at a specified shear rate (gap flow), by measuring the discharge rates of flow through pipes (capillary flow) and by determining the sedimentation rate of a defined solid content in liquids (creeping flow).

In every case it is important to ensure that laminar flow conditions obtain during measurement. Only then can *Newton's* fundamental law be used to calculate the viscosity. If capillaries or pipes are used in the measurement process, the flow conditions in pipes can be characterised by calculating the *Reynolds* number as a nondimensional link for all the physical variables that influence flow. These are viscosity, density, flow rate and diameter of the piping.

If *Reynolds* numbers are below 2000, the flow is laminar and suitable for conducting measurements. Above 2000 the liquid threads start to develop mixed flows, while if the *Reynolds* number continues to rise, turbulent flow sets in completely.

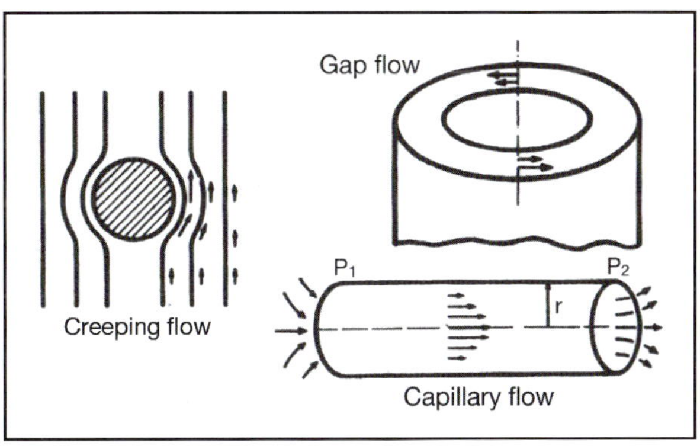

Figure 2.3.21: Characteristic laminar flows for different viscometric techniques

The measurement principle behind **capillary viscometers**, which are widely encountered in coating technology in the form of flow cups, involves the flow of a defined volume of the liquid under analysis through a nozzle of known diameter and length. The measured flow-out time is converted to a viscosity reading with the aid of *Newton's* fundamental law. To this end the law needs to be adapted to the flow conditions in pipes. If the force necessary for flow in pipes is F_2, the counterforce which occurs as a result of friction during flow is F_1 in accordance with *Newton's* law.

$$F_1 = \eta \cdot A \cdot \frac{du}{dr} \qquad F_2 = \Delta p \cdot r^2 \cdot \pi$$

Under steady-state flow,

$$F_1 + F_2 = 0 \quad \text{or} \quad \Delta p \cdot r^2 \cdot \pi = - \eta \cdot A \cdot \frac{du}{dr}$$

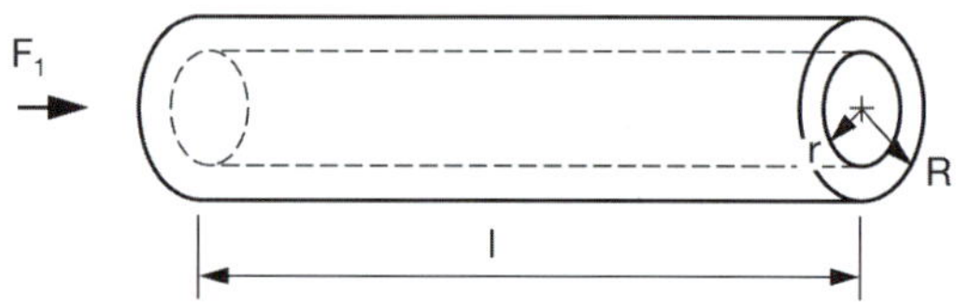

The flow profile u(r) corresponds to a parabolic function. Preparing, inserting the boundary conditions and solving the differential equation yields the following, with r = 0 to R = radius of the capillaries of length 1:

$$u = \frac{\Delta p}{4 \cdot \eta \cdot l} \cdot (R^2 - r^2)$$

The liquid adheres to the side walls with $u_{min.} = 0$, while attaining a maximum u_{max} in the centre.

The conveyed or flow-out volume dV = 2π dr 1 is calculated by using the velocity equation u = 1/t.

The result yielded by solving the new differential equation is

$$V = \frac{\Delta p \cdot \pi \cdot R^4}{8 \cdot \eta \cdot l} \cdot t$$

This equation, which is known as the *Hagen-Poiseuille* equation, enables the viscosity to be calculated for a specified flow rate by measuring the flow-out time. Since the effective pressure Δp is defined by gravity and is a function of time with ρ g h(t) because of the reducing height of the liquid column, this dependence must also be determined. The mean $\bar{h}$ for h(t) is calculated from this.

None of our considerations to date has addressed the fact that the effective static pressure ρ g $\bar{h}$ is reduced during flow for reasons of energy conservation.

The consequence of this effect, first described by *Bernoulli*, is that the dynamic pressure $\rho u^2/2$ of the now flowing liquid has to be deducted from the total pressure ρ g $\bar{h}$ as a so-called *Hagenbach* correction. When all the variables affecting the measurement process have been taken into account, we get:

$$V = \frac{\left(\rho \cdot g \cdot \bar{h} - \frac{\rho}{2} \cdot u^2\right) \cdot \pi \cdot R^4}{8 \cdot \eta \cdot l} \cdot t$$

Mathematical treatment leads to:

$$\frac{\eta}{\rho} = \nu = \frac{g \cdot \bar{h} \cdot \pi \cdot R^4}{8 \cdot V \cdot l} \cdot t - \frac{V}{16 \cdot \pi \cdot l \cdot t}$$

or

$$\nu = A \cdot t - \frac{B}{t}$$

After the constant values have been calculated and aggregated into A and B the kinematic viscosity for *Newtonian* liquids can be calculated by substituting the results for the flow-out times.

Certain restrictions apply to use of the flow cups in the case of pseudoplastic materials. Due to gravity-induced continuous changes in flow rate, it is not possible to set a defined shear gradient, let alone maintain it at a constant level. τ/D graphs for characterising the flow properties as a function of defined

Figure 2.3.22: Temperature-controlled flow cup as per ISO 2431
Source: Erichsen

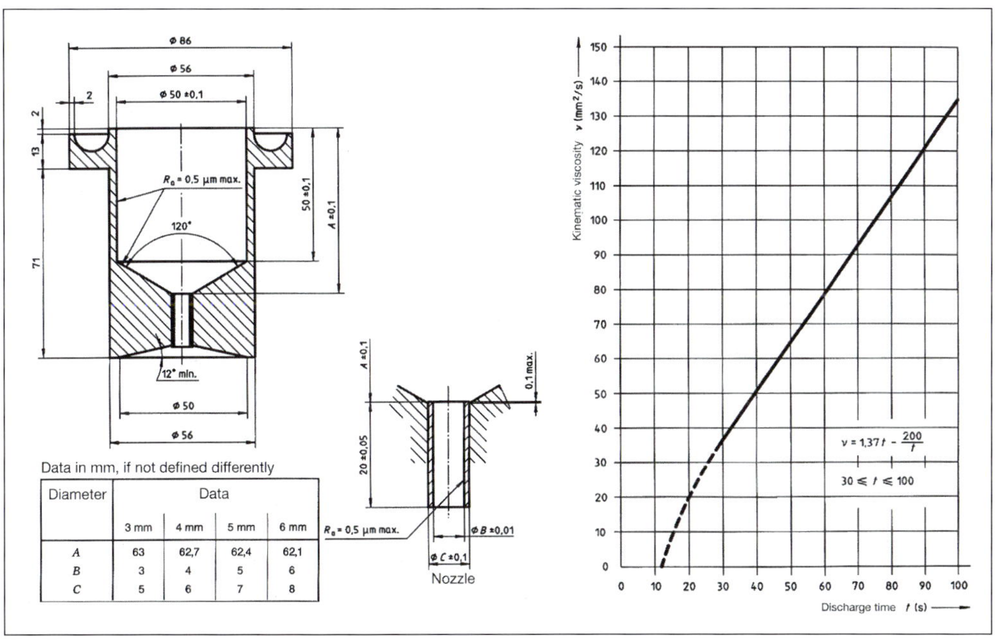

Diameter	Data			
	3 mm	4 mm	5 mm	6 mm
A	63	62,7	62,4	62,1
B	3	4	5	6
C	5	6	7	8

Figure 2.3.23: Cross-section through an ISO-flow cup and a corresponding calibration curve

shear cannot, therefore, be plotted with flow cups. In such cases flow **cups** merely serve ultimately to check the identity of coating materials, and even then only where there is the same level of pseudoplasticity. Flow cups are totally unsuitable for thixotropic coating materials.

The relevant standards specify geometries and calibration curves along with the permissible viscosity ranges for the various cups. The DIN cup, which was the standard in the past, has now been superseded by the ISO cup. The flow-out nozzle has been lengthened five-fold compared with the DIN cup, slows down the flow rate and increases the accuracy and reproducibility of the results.

In view of the temperature dependence of viscosity it is important to ensure good temperature control over the samples under analysis (see Figure 2.3.22).

ISO 2431 describes four flow cups that the same dimensions but different nozzle diameters (3, 4, 5 and 6 mm). The variation in nozzle width makes it possible to cover a broad viscosity spectrum. The permissible working ranges for all cups use flow times of between 30 s and 100 s.

Rotary Viscometers

Rotary viscometers are the appropriate measuring instruments for analysing non-*Newtonian* liquids with a high pseudoplasticity, dilatancy or thixotropy under defined shear gradients. The flow states in the measuring part of the instrument are similar to the laminar flow between the two plates of the model. Cylinder and cone-and-plate viscometers have proved themselves in practical operations. In the first case two concentrically configured cylinders with a narrow gap between them move relative to each other. Either the outer cylinder or the inner cylinder is stationary. In the second case a flat cone rotates on a fixed plate. The liquid under analysis in the space between the cylinders or between the cone and the plate can be sheared in a defined manner by varying the rotational speed. The torque which is generated for a given rotational speed can be measured by the power consumption of the driving motor. Both instrument types permit rheograms (flow curves) to be plotted in every stress range to which coating materials are subjected in practice.

If the outer cylinder of a **cylinder viscometer** is driven, this is termed a *Couette* viscometer, while the design featuring a driven inner cylinder is known as a *Searle* viscometer. In both instrument types the velocity gradient is specified, and the resulting shear stress is then measured.

In the case of substances with time-dependent flow properties care must be taken to allow adequate rest time before starting the measurement process; this allows the gel state to fully develop. The sample must also be checked to ensure that it has reached the specified temperature before measurements begin. Once all the requirements for reproducible readings have been met, flow curves can be plotted, from which the viscosity corresponding to the particular D-value can be calculated. If the flow curves are identical at increasing and decreasing speeds, this is an indication of dilatancy or pseudoplasticity, depending on whether the viscosity is increasing or decreasing. If the curves have different shapes when the speed is increased (gel curve) and subsequently decreased (sol curve), this indicates thixotropy or rheopexy. Newton's fundamental equation is again used to process the readings.

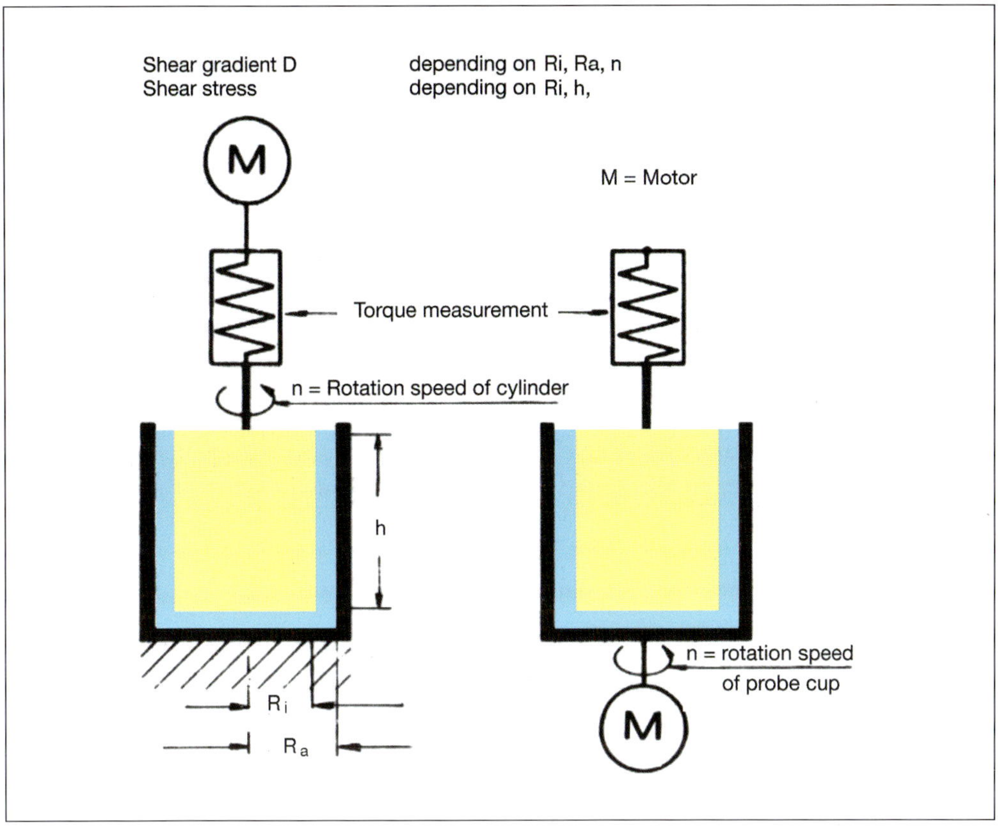

Figure 2.3.24: Cross-sections through cylindrical viscometers

In this

$$\tau = \frac{F}{A} = \frac{F \cdot r}{A \cdot r} = \frac{M}{2 \cdot \pi \cdot r^2 \cdot h} \qquad M = \text{torque}$$

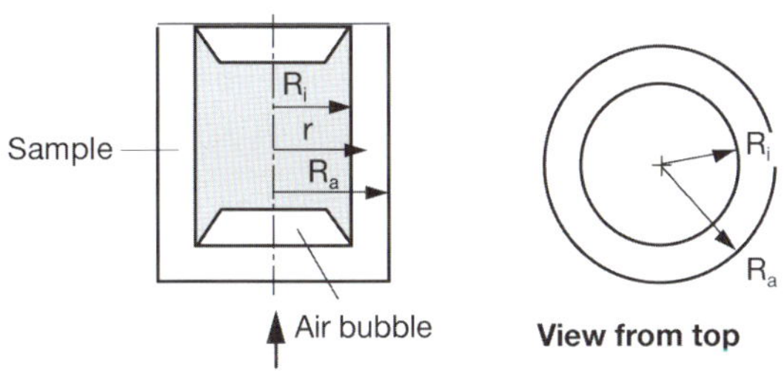

where the generated surface of the liquid $A = 2\,\pi\,r\,h$ and h is the height of the measuring cylinder. If F/A is expanded is expanded to any radius r of the liquid cylinder, a correlation is found between the shear stress τ and the torque M. D is then calculated as follows:

$$D = \frac{du}{dr} = \frac{d\omega \cdot r}{dr}$$

When incorporated into the Newtonian fundamental equation, this yields

$$\frac{M}{2 \cdot \pi \cdot r^2 \cdot h} = \eta \cdot \frac{d\omega \cdot r}{dr}$$

Solving the differential equation, with allowance for the boundary conditions (external radius R_a/internal radius R_i or max. angular velocity $\omega = 2\,\pi\,n$), yields

$$M = \frac{8 \cdot \pi^2 \cdot R_a^2 \cdot R_i^2 \cdot h}{R_a^2 + R_i^2} \cdot \eta \cdot n$$

This equation no longer holds once the gap to be sheared exceeds a certain width. The ratio between the radius of the outer and inner cylinders should not be greater than 1.1 in order that a linear shear gradient may be maintained and the necessary conditions for rapid heat dissipation created.

Apart from cylinder type rotary viscometers **cone-and-plate rotary viscometers** are also in common use. The coating material to be measured is located in these between the rotating cone and the stationary plate. The advantages of these instruments over cylinder viscometers are the very small sample quantity, the rapid attainment of the test temperature and the short measuring and cleaning times. The very small angle of the cone envelope enables high shear gradients to be achieved even at low speeds. Calculation of the viscosity is similar to the method used for cylinder viscometers.

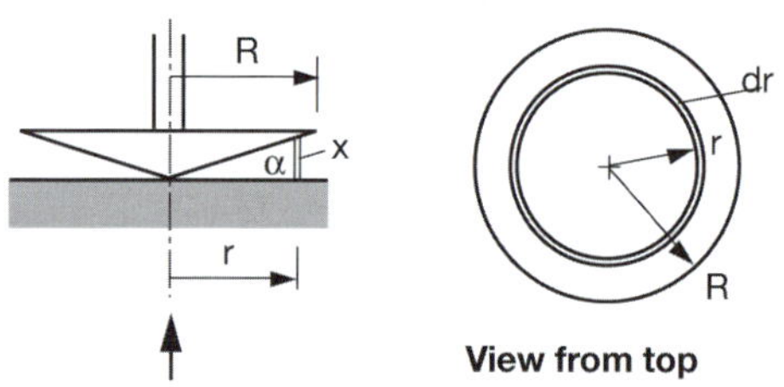

$$D = \frac{\omega \cdot r}{x} = \frac{\omega}{\tan\alpha} \approx \frac{\omega}{\alpha} = \frac{2 \cdot \pi \cdot n}{\alpha}$$

The D-value is a function of the rotational speed ωr and x, the gap between the cone and the plate at the location r. x/r can also be described by $\tan\alpha$ which corresponds to α at very small angles. Since $\omega = 2\,\pi\,n$, the equation for calculating the shear gradient becomes a function of the rotational speed.

Calculation of the torque is somewhat more complicated than is the case for cylinder viscometers. A differential segment dM of the torque M must first be taken into account in order that a usable relation may be obtained by solving the differential equation.

$$dM = \tau \cdot dA \cdot r$$

due to $dA = 2 \cdot \pi \cdot r \cdot dr$

follows after solution of differential equation $\rightarrow \quad \tau = \dfrac{3 \cdot M}{2 \cdot \pi \cdot R^3}$

Substitution into the Newtonian fundamental equation $\tau = \eta\,D$ yields

$$M = \frac{4 \cdot \pi^2 \cdot R^3}{3 \cdot \alpha} \cdot \eta \cdot n$$

Apart from the cone-and-plate instruments which permit a continuous change in the velocity gradient, there are simple viscometers which only operate with a fixed shear gradient. Their simplicity makes them suitable for routine tests both in production monitoring and in R&D laboratories. They have thus become a serious competitor for flow cups which are frequently employed where not permitted, i.e. with non-Newtonian liquids. Cone-and-plate instruments are therefore used not least because they also allow viscosity measurements at elevated temperatures. Resin melts and powder coatings can be measured without difficulty at temperatures of up to 200 °C.

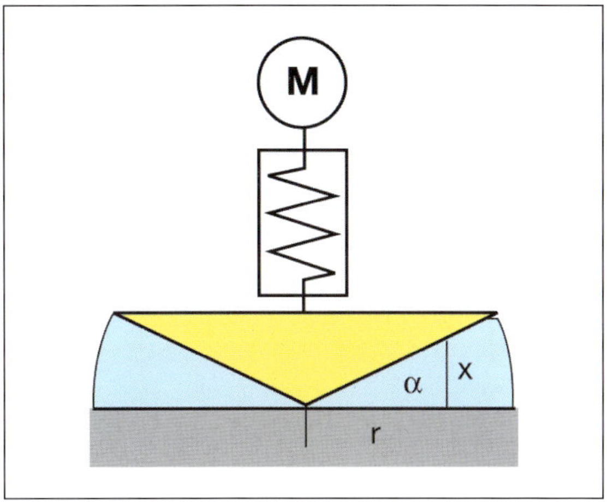

Figure 2.3.25: Schematic diagram of a cone-and-plate viscometer

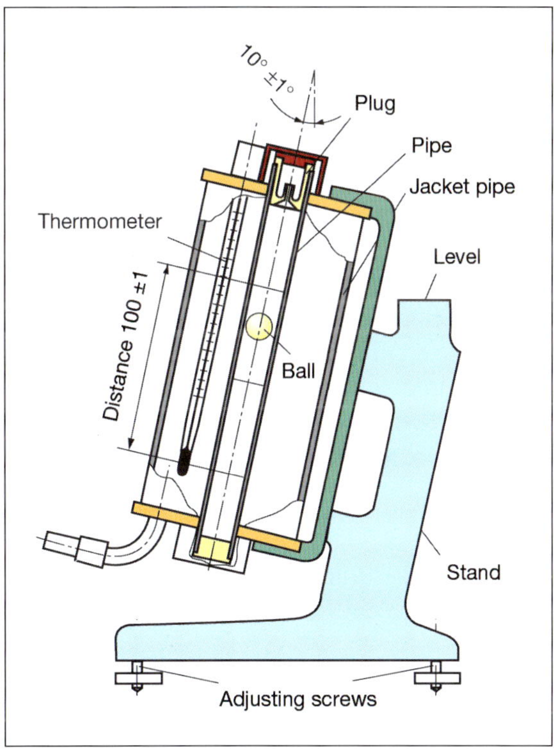

Figure 2.3.26: Schematic diagram of a Höppler viscometer

A further method of measuring viscosity is based on observing balls moving under the force of gravity in a pipe filled with the test liquid. If the diameter of the ball is only slightly less than that of the pipe and the ball moves downwards ($\Delta\rho$ positive) or upwards ($\Delta\rho$ negative) because of the difference in density between itself and the test liquid, the viscosity can be calculated from the time taken to cover a specified distance. These instruments include the **falling ball viscometer** after *Höppler* and the **bubble viscometer**. The viscosity is determined from the resulting rate of fall or rise of the test body in the form of a ball or air bubble. The stationary state for the ball travel u is given by

$$m \cdot g = 6 \cdot \pi \cdot \eta \cdot r \cdot u$$

Solving for this and making the substitution u = s/t, it follows that

$$s = \frac{k}{\eta} \cdot t \quad \text{where} \quad k = \frac{m \cdot g}{6 \cdot \pi \cdot r}$$

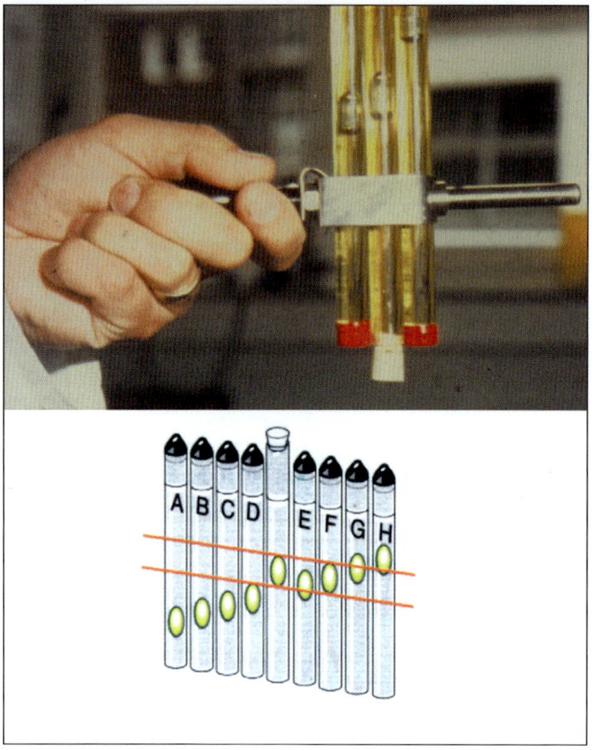

Figure 2.3.27: Bubble viscometer according to ASTM D 1545

Different measurement ranges can be covered by employing balls of different density (glass, steel etc.). Viscosity measurements are also possible by means of a simple comparison of the travel speed of a bubble in liquids of known viscosity. In the case of the bubble viscometers specified in ASTM D 1545 test tubes are filled with liquids of rising viscosity (A–Z10). The liquid under test is then poured into a corresponding empty tube and compared with the rate of rise of the bubbles in the reference standards. The test substance is labelled with the letter which corresponds to its viscosity.

All the instruments discussed so far are suitable to a greater or lesser extent for viscosity measurements aimed at evaluating the molecular structure of resins and their interaction with solvents, as well as for describing pigment dispersions and transport in pipelines and for evaluating atomisation processes. They are not suitable, however, for evaluating the film forming process and the resulting high viscosities. The transition from the liquid to the solid phase, with release of solvents and overlaying with chemical reactions, cannot be simulated in the instruments described.

Since levelling, sagging and surface flaws such as pinholes and craters are very greatly affected by the time profile of the flow properties right up to the set film, there is great interest in measurement methods which can record the rheological properties of paints throughout the entire film forming process.

A very old, yet very simple method of semiquantitatively determining changing viscosity, is afforded by the **"rolling ball method"** developed by *Wolff* and *Zeidler* in 1938 [2.4.126]. Immediately after application of the coating material to be tested, steel balls with a diameter of 2.5 mm are placed on the test panel which is inclined at an angle of 60°. The speed at which the balls roll down is a measure of the viscosity at the time of the measurement. If balls are then dropped onto the curing coating at brief, regular intervals and the travel of the last ball is subtracted from that of the preced-ing one, the distances s determined in this way provide the basis for calculating the instantaneous viscosity η. When s is plotted against t, the viscosity can be calculated from the proportionality factor. The proportionality factor is formed from the density and size of the ball, the density of the film, the acceleration due to gravity g and two empirical constants. This method offers a simple way of determining the change in viscosity. A more elegant solution is provided by DMA analyses (see below).

$$s = \frac{(\rho_b - k_1 \cdot \rho_p) \cdot g \cdot r^2}{k_2 \cdot \eta} \cdot t$$

ρ_b = density of rolling ball

ρ_p = density of paint

k_1, k_2 = empirical constants

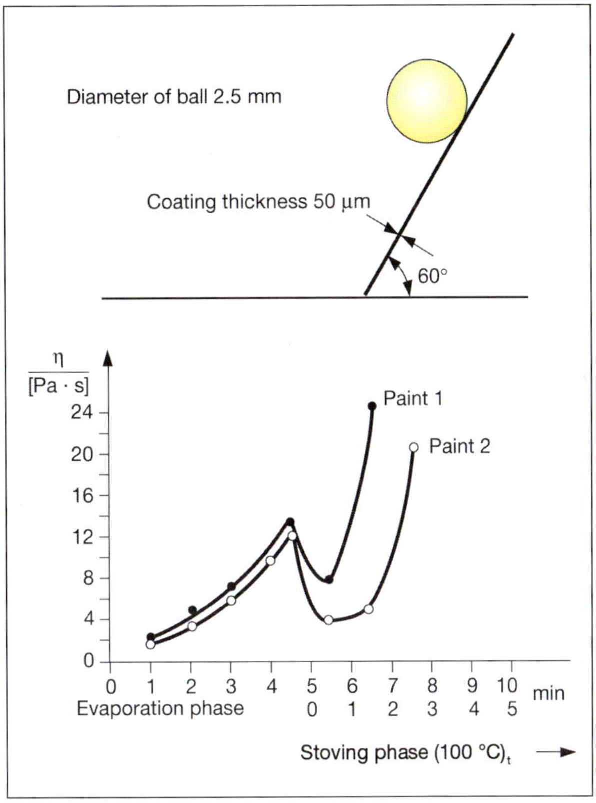

Figure 2.3.28: Generalised curves of the viscosity of paint samples during film formation

Figure 2.3.28 shows the results for two different paints. Due to its lower reactivity in the stoving phase, paint 2, which has a tendency to run, falls to a lower viscosity level than the chemically more reactive and therefore faster curing paint 1. The poor levelling of paint 1 and the tendency to run of paint 2 are clearly confirmed by the flow curves of the "rolling ball method".

Accurate viscosity measurements throughout the curing period of reactive paints can be made with rheometers whose shear travel is shorter than the maximum possible elastic deformation after curing of the film. This can be achieved by replacing the rotation of the rheometer by oscillation. Usually sine-shaped oscillations, whose amplitude must be below the film's yield point, are imposed on the film, which is located between the oscillation plates.

$$\tau = \tau_0 \cdot \sin(\omega \cdot t)$$
$$\gamma = \gamma_0 \cdot \sin(\omega \cdot t - \delta)$$

Such measurement systems allow not only the viscous but also the elastic components to be measured as a result of the temporal shift in the deforming shear stress and the resulting deformation. It is therefore possible to record the complex viscosities of visco-elastic substances (see Chapter 3.2.5).

Surface Tension

Many coating technology processes, e.g. the wetting of pigments, atomisation of coating films and levelling of fresh films, the emulsification process, capillary action during penetration of pigment agglomerates, floating effects, dewetting from edges, crater formation and the like, are affected by the higher potential energy of molecules at interfaces and the resulting surface tension.

The surface tension, in its interaction with the rheological properties, is the most important physical variable in the evaluation of coating materials.

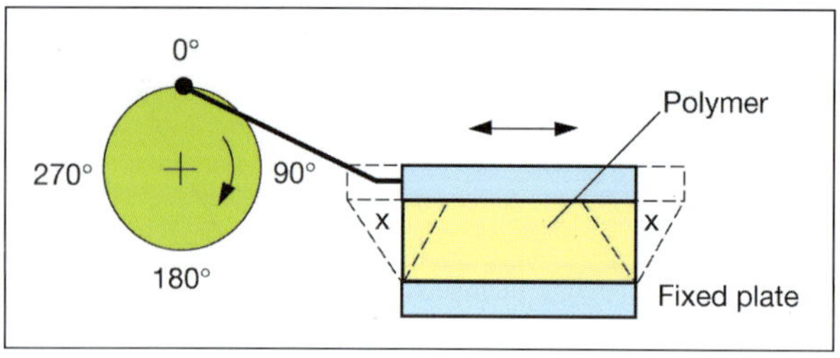

Figure 2.3.29: Schematic diagram of an oscillating rheometer

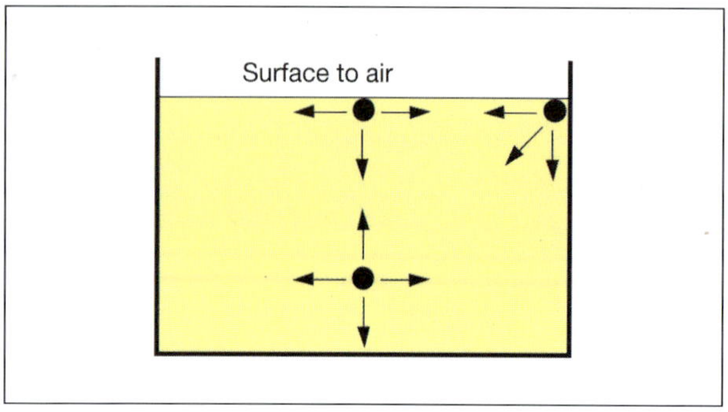

Figure 2.3.30: Schematic diagram of surface tension

Materials with finite extension properties, such as liquids and solids, exhibit a higher energy level at their interfaces than in the rest of their volume. As a result, if gravity effects are excluded, liquids always occupy the minimum surface area. Cohesive forces are generated by the *van der Waals* forces as a consequence of dipole and dispersion forces in conjunction with hydrogen bridges. These cohesive forces are equal in all spatial directions within the liquid. The force resultant assumes the value 0. The molecules at the interface to an imag-inary vacuum or vapour space with comparatively few molecules, on the other hand, behave differently. In this case the sum of all the forces acting is not equal to 0. All the molecules at the surface have an attraction to the inner component of the liquid and consequently a higher potential energy. This enables them to move into the inner part of the liquid if space is or becomes available to them. If, therefore, a change of form leading to a reduction in surface area is possible, that change will occur because of the effective surface tension. Thus, if the influence of external factors such as gravity are excluded, liquids always have the minimum possible surface areas. They transform themselves into spheres because these occupy the least possible surface area for a given volume. Conversely energy has to be expended to increase surface areas. Surface tension a is consequently defined as the work which has to be performed per unit area to increase the surface area.

$$\sigma = \frac{F \cdot s}{A}$$

The dimension of surface tension is N/m. It acts tangentially to the surface. The force counteracting expansion is directed towards the inner part of the liquid.

From a thermodynamic perspective the surface tension σ is equal to the change in the free enthalpy G of the system if surface areas A are changed without material change under isothermal and isobaric conditions.

$$dG = \left(\frac{\delta G}{\delta A}\right)_{p,T} dA \qquad \text{d.h.} \quad \sigma = \left(\frac{\delta G}{\delta A}\right)_{p,T}$$

Surface expansions dA are thus also only possible from a thermodynamic perspective if the free enthalpy in the system increases. The system must therefore perform work to expand surface areas. The work which has to be performed per unit area to expand the surface area varies enormously for individual liquids. It is determined by the molecular interactions described above. These consist of the polar interactions σ^p as the sum of the effect of permanent dipoles (orientation forces), induced dipoles (induction forces), hydrogen bridge bond forces and the non-polar dispersion forces σ^d.

$$\sigma = \sigma^d + \sigma^p$$

σ^d = disperse contribution

σ^p = polar contribution

Substances	σ^d	σ^p	σ_{total}
Water	21.8	51.0	72.8
Ethanol	17.5	4.6	22.1
Ethylenglykol	29.1	20.3	49.4
n-Hexane	18.4	0	18.4
n-Decane	24.0	0	24.0
Glycerine	37.2	27.3	64.5
Toluene	26.1	2.3	28.4
Alkyd resin	26.9	6.3	33.2
Acrylic resin	25.8	7.2	33.0

Figure 2.3.31: Surface tension of resins and liquids in mN/m

Since the surface tensions of liquids of relevance to coating technology are relatively low, the surface tension is given in mN/m rather than N/m. Thus water, which is polar, has a surface tension of 73 mN/m. Other liquids, such as xylene and benzine, have substantially lower values of 30 and 24 mN/m, respectively. The table below (Figure 2.3.31) gives an overview of the surface tensions of the most important resins and solvents.

Surface tension is a temperature-dependent variable. *R. von Etvös, W. Ramsay* and *I. Shields* formulated the following semiempirical relation which holds for almost all liquids:

$$\sigma(T) = A - B \cdot T$$

A and B are constants which have to be determined empirically.

By contrast with viscosity which responds to temperature changes by 4 to 10 %/°C, surface tension is less sensitive, with a response of only 1 %/°C.

Measurement Methods

The surface tension of liquid materials is lowered by surfactants, but these can only act if they can be deposited at the interface, i.e. establish a state of equilibrium, Consequently, a distinction must be drawn in the measurement methods between static and dynamic surface tension.

In **static measurement methods** the surface area is expanded in various ways so slowly that an equilibrium can always be established for additives at the interface. The work needed to achieve this or the potential energy which is increased as a result is measured directly or indirectly. The pull-force methods of *du Nouy* and *Wilhelmy*, and the capillary, drop and rotary methods, have proved particularly successful.

In the **pull-force method** a platinum ring to DIN EN 14210 (*du Nouy*) or a platinum rod (*Wilhelmy*) is immersed in the liquid under analysis whose viscosity should be < 2 Pa s. The test bodies are slowly moved to the interface. As it is raised further, the surface area of the adhering liquid expands. The force expended to achieve this reaches a maximum before falling again as the lamella constricts and then breaks. The maximum force enables the surface tension to be calculated.

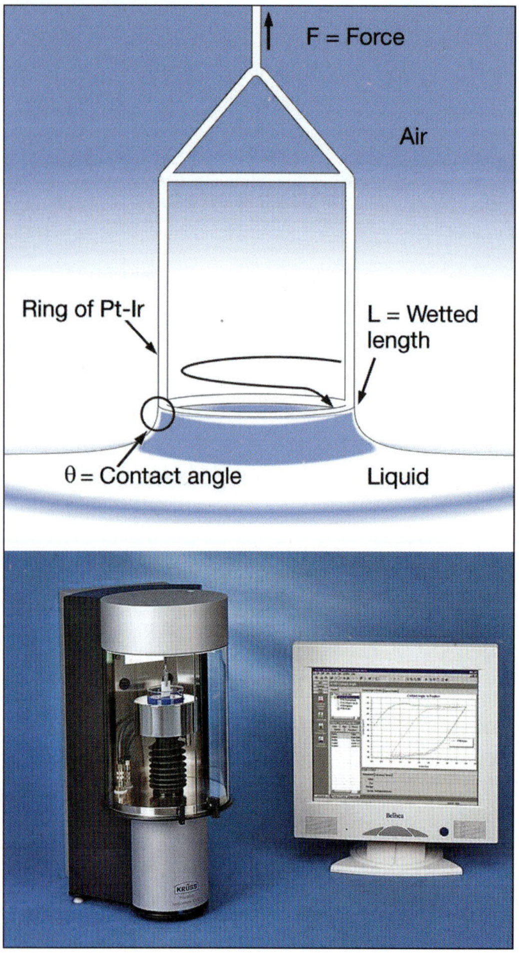

Figure 2.3.32: Tensiometer according to DIN EN 14210

Source: Küss

The work done is F h. This corresponds to the surface tension multiplied by the surface area generated. In the case of a ring, because of the doubled surface area internally and externally, this is $\sigma\ 4\ \pi\ r\ h$ while for the *Wilhelmy* method it is $\sigma\ 2\ l\ h$, where l is the length of the platinum rod. The following applies with regard to the maximum:

$$\sigma = \frac{F}{4 \cdot \pi \cdot r} \quad \textit{de Nouy}$$

$$\sigma = \frac{F}{2 \cdot l \cdot h} \quad \textit{Wilhelmy}$$

The pull-force properties of the lamellae provide supplementary information about the foam persistence of the tested liquid. If the lamella breaks quickly once the maximum has been exceeded, there is no risk of foaming. Liquids with a tendency to foam, on the other hand, form stable lamellae which only break much later.

In the **capillary** or **drop method** a correlation is assumed between the cohesive pressure of the liquid and the surface tension. This can be established by means of a simple theoretical method.

At one end of a liquid filled pipe is a liquid hemisphere of surface area $O = 2\ \pi\ r^2$. The influence of the pressure p leads to an amount of work being done which corresponds to the work required to create the new surface area.

$$\sigma \cdot dO = p \cdot dV$$

dO = change of surface area

dV = change of volume

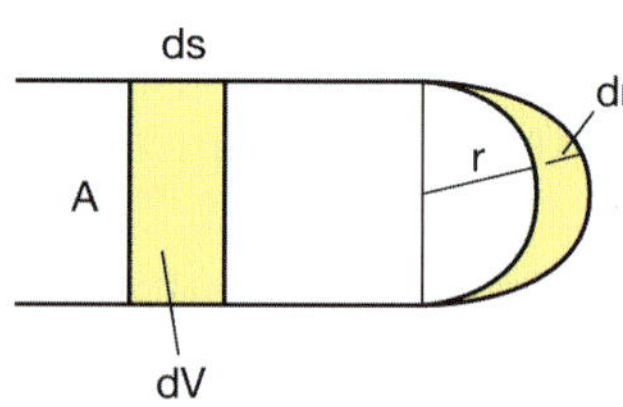

or

$$\sigma \cdot 2 \cdot \pi \cdot \left[(r + dr)^2 - r^2\right] = p \cdot \frac{2}{3} \cdot \pi \cdot \left[(r + dr)^3 - r^3\right]$$

Ignoring dr^2 and dr^3, the equation reduces to

$$p \approx \frac{2 \cdot \sigma}{r}$$

The cohesive pressure p is thus directly proportional to the surface tension σ and indirectly proportional to the radius of the liquid hemisphere r. Decreasing diameters of liquid drops consequently lead to increased cohesive pressure.

The cohesive pressure p = 2σ/r acting in the capillary method draws a wetting liquid into the capillaries. With increasing rise height, gravity also increases and so a steady state is established. For a known capillary radius and wetting angle, the rise height is then a measure of the surface tension.

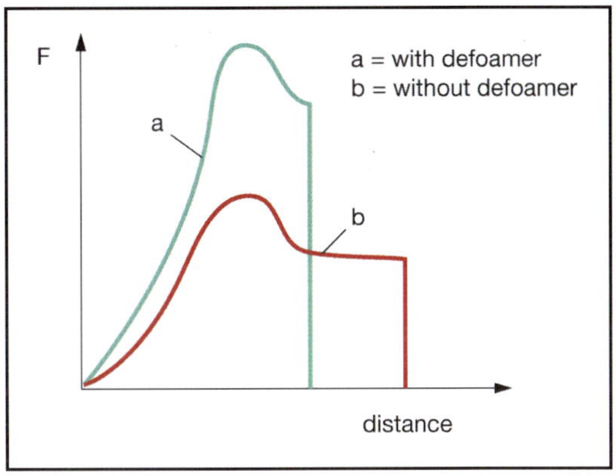

Figure 2.3.33: Pull-force behaviour of stable and unstable lamellae during measurement of static surface tension according to Wilhelmy or de Nouy

if $p = \frac{2 \cdot \sigma}{r'}$ follows $p = \frac{2 \cdot \sigma}{r} \cdot \cos \alpha$

because of $p = \rho \cdot g \cdot h$ follows $\frac{2 \cdot \sigma}{r} \cdot \cos \alpha = \rho \cdot g \cdot h$

$$\sigma = \frac{\rho \cdot g \cdot h \cdot r}{2 \cdot \cos \alpha}$$

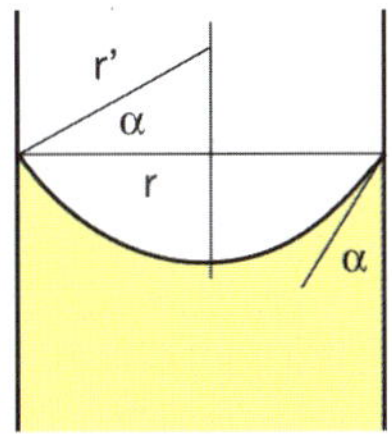

If the capillaries are wetted completely, the relation simplifies to

$$\sigma = \frac{1}{2} \cdot \rho \cdot g \cdot h \cdot r$$

In the drop method, drops whose size or mass m is determined by the surface tension σ are formed by gravity m g acting on a glass tube of radius r. In this process the cohesive pressure p_l counteracts the gravity per unit area p_2. Measurements are made at $p_l = p_2$.

$$p_1 = \frac{2 \cdot \sigma}{r} \qquad p_2 = \frac{m \cdot g}{A}$$

$$\sigma = \frac{m \cdot g}{2 \cdot \pi \cdot r}$$

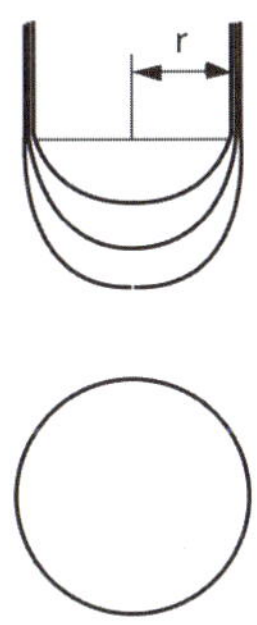

To improve the accuracy of the measurement several drops are generated and the total mass is then divided by the number of drops.

Another means of determining the surface tension of liquids that takes account of gravity involves analysing the geometry of a drop hanging from a needle (pending-drop method). The drop, which is spherical in a gravity-free state, is transformed into an ellipsoid under the effect of gravity. Since the pressure difference between the exterior and interior of an interface is indirectly proportional to the radii of curvature, the

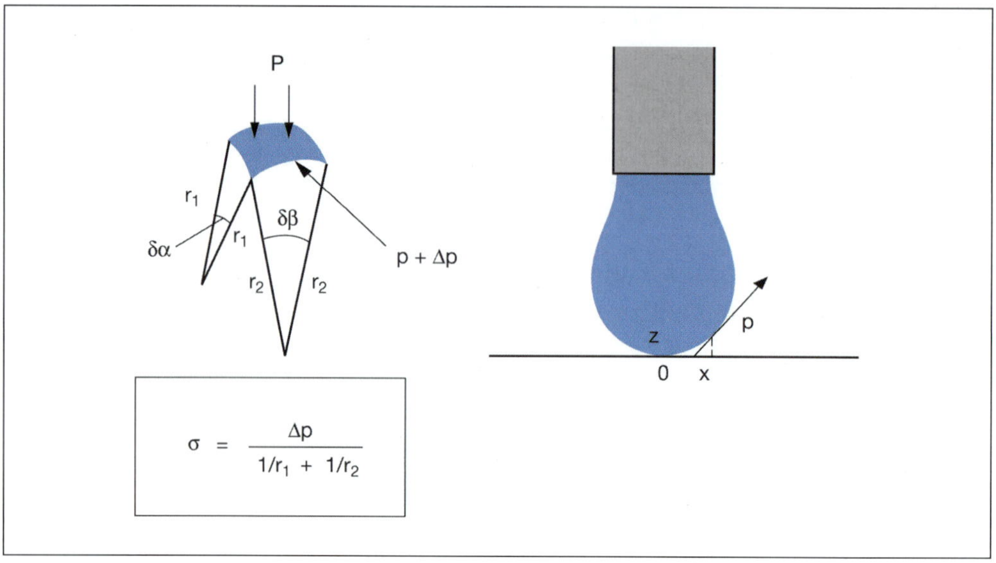

Figure 2.3.34: Measurement of surface tension by the pending-drop method

surface tension of the liquid can be calculated from the changing geometry of the ellipsoid, where Δp = pressure difference between the interior and exterior of the interface.

$$\Delta p = \sigma \cdot \left(\frac{1}{r_1} + \frac{1}{r_2} \right)$$

r_1 and r_2 are the main radii of curvature of the ellipsoid. They can be calculated by image analysis of the drop. This entails measuring the pressure difference at the summit point Δp_s and at all the coordinates of the drop profile. The difference is

$$\Delta p_s - \Delta s = z \cdot \Delta \rho \cdot g$$

where z represents the vertical coordinate as per Figure 2.3.34.

The methods presented here are unsuitable for measuring extremely low surface tensions or interfacial tensions between two liquids. Instead, recourse is made to the spinning-drop method. This consists in injecting a drop of the sample of density r_{Pr}. into a heavier, continuous liquid of density r_s which is immiscible with the test sample and which is located in a tube. This is then rotated at an angular velocity ω. The test drop which is spherical at rest, is first transformed into an ellipsoid by the centrifugal forces. With increase in rotational frequency the ellipsoid is then converted to a cylinder with two hemispherical caps. If the length exceeds the radius r at least four-fold, the interfacial tension can be calculated [2.4.127].

$$\sigma = k \cdot r^3 \cdot \omega^2 \cdot (\rho_s - \rho_{Pr})$$

Dynamic Methods

Paints are always complex mixtures with ingredients which have an effect on surface tension. These have to be deposited at the interface in order for their full impact to develop. However, many processes in paint manufacture and processing operate at such

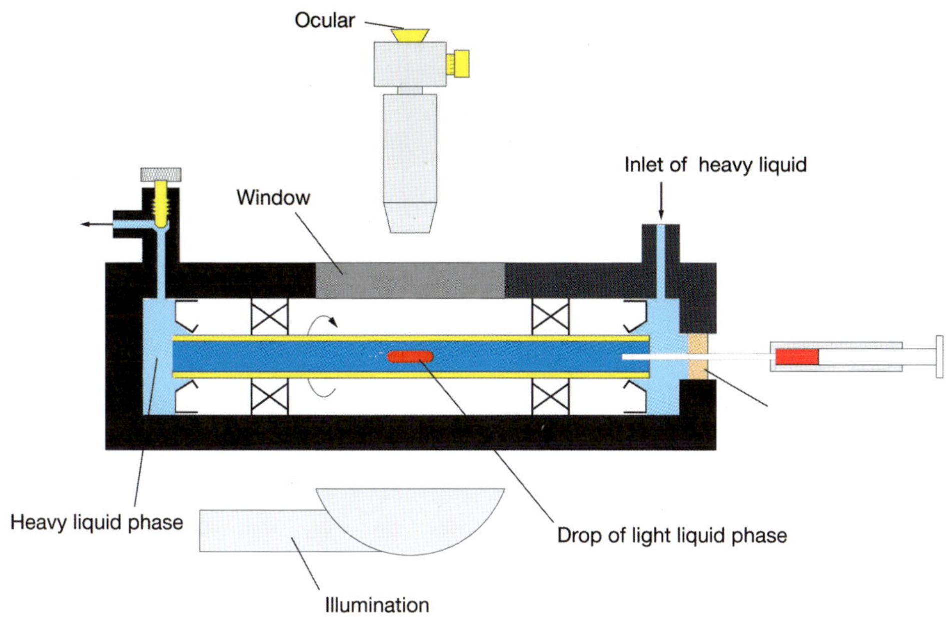

Figure 2.3.35: Measurement of the surface tension of liquids by the spinning-drop method

high speeds that it is impossible to attain an equilibrium and thus the maximum possible effect of surfactant additives. In order to achieve a quantitative analysis of surface tensions, but also to measure the time required to establish equilibrium, **dynamic measurements** have to be conducted on rapidly changing surface areas. The **bubble pressure method** enables the surface tensions of newly generated surfaces to be measured. A capillary of radius r is immersed in the test solution of density ρ. A specified number of bubbles per unit time is ejected from this capillary into the solution by means of a pressure p. The measured pressure required to release the bubble is p_{max}. At that moment the pressure is composed of the hydrostatic pressure $p_0 = \rho\, g\, h$ and the pressure required for bubble formation $2\,\sigma/r_{min}$.

$$p_{max} = \frac{2 \cdot \sigma}{r_{min}} + \rho \cdot g \cdot h$$

The minimum radius r_{min} of the forming bubble is equal to the radius r of the capillaries.

$$\sigma_{dyn} = \frac{(p_{max} - \rho \cdot g \cdot h) \cdot r}{2}$$

If the bubble generation frequency is increased, p_{max} rises and so does the surface tension. Such a rise is an indicator of the kinetics underlying the establishment of equilibrium in surfactants.

Above a certain volumetric flow rate, orientation of the surfactant molecules in the interface cannot take place synchronously. If time-dependent surfactant effects are present, therefore, the surface tension is also a function of the number of bubbles formed per unit time.

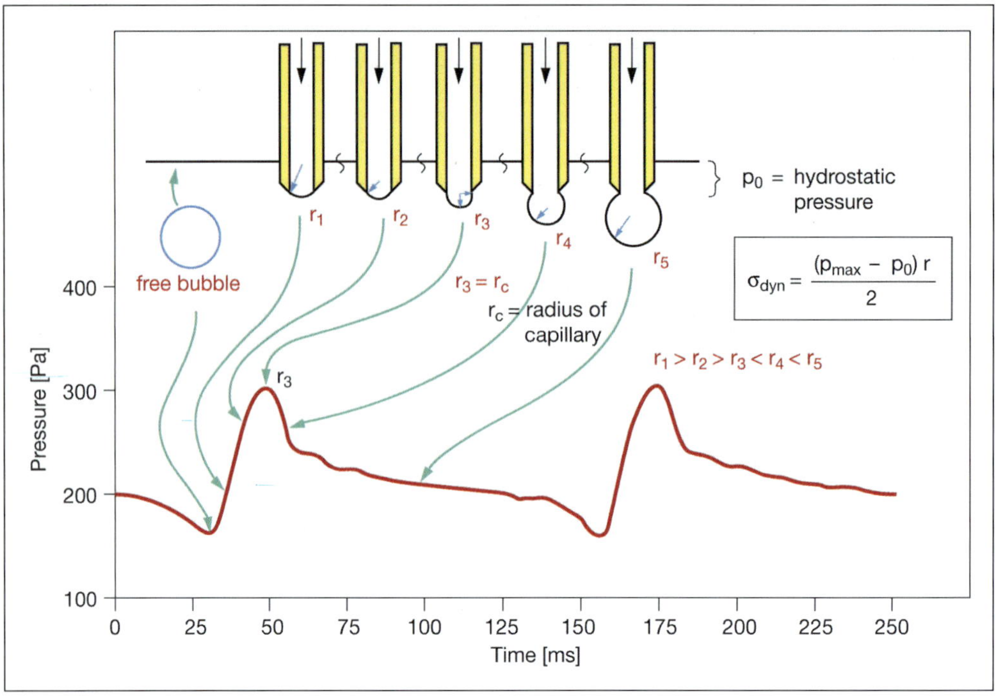

Figure 2.3.36: Measuring the dynamic surface tension of substances that contain surfactants *Source: Krüss*

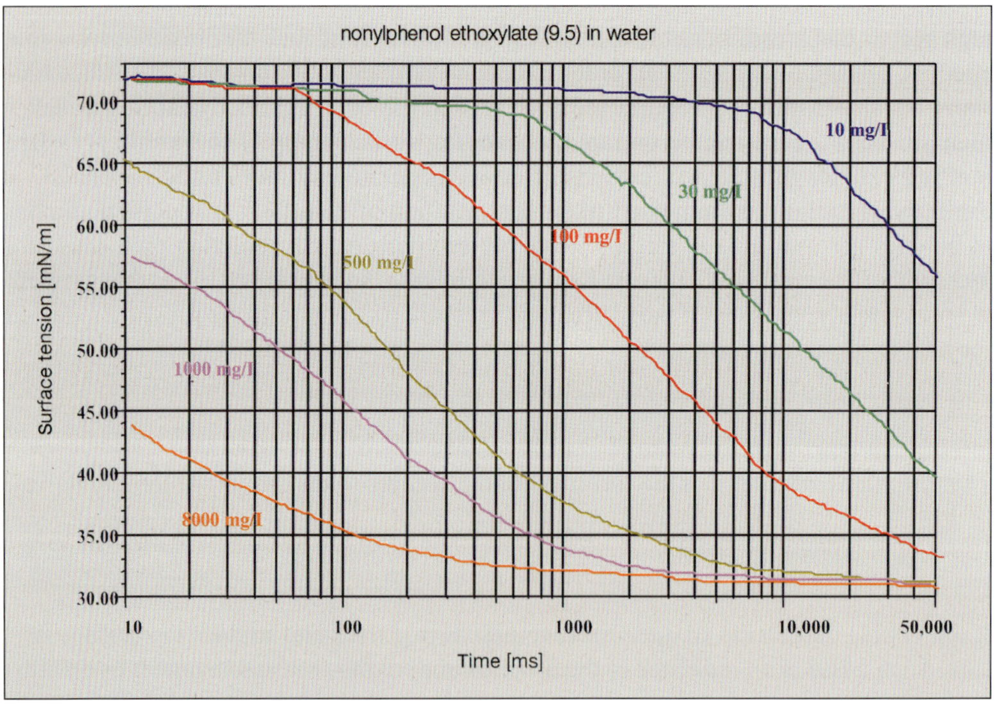

Figure 2.3.37: Time dependence of incipient surface tension *Source: Krüss*

The brief time available to the molecules to reach the interface leads to incomplete coverage of the interface and thus to a higher surface tension. The temporal range of surface ages which can be set in this way extends from 5 ms to 60 s.

The chart in Figure 2.3.37 shows the time dependence of the surface tension generated by the addition of various quantities of surfactant. The smaller the slope of the curves, the faster the surfactant reaches the interface. Assuming comparable molecular weights, equilibrium is established fastest with anionic surfactants, followed by cationic ones. Nonionic surfactants are generally the slowest products to be oriented.

Density

The density of coating materials and coatings is a significant variable with regard to the storage, the processing and, in particular, the yield of paints. Since the protective effect is mainly determined by the film thickness and thus, for a specified surface area, a certain volume ultimately has to be applied, a low density leads to lower material consumption and thus to better yield (see Chapter 5.7). The density ρ is defined in ISO 2811 as the mass per unit volume in kg/m^3.

$$\rho = \frac{m}{V} \quad \left[\frac{kg}{m^3}\right]$$

It is also a temperature-dependent variable because volume is a function of temperature. The degree of temperature dependence is determined by the coefficient of expansion γ

$$\rho(T) = \frac{m}{V_0 \cdot (1 + \gamma \cdot T)}$$

It is therefore essential when density is being measured to ensure that the temperature is also measured or set to a specified value.

Direct and indirect processes are used to determine the density of liquids. Pycnometers belong to the direct type, while the indirect type includes immersed bodies, areometers and oscillators.

Pycnometers as defined in ISO 2811-1 are various shapes of vessel of known volume. Metal pyknometers are preferred in the paint and coatings industry because of their ruggedness, ease of handling and good cleanability.

For accurate readings, the uplift provided by air must be taken into account when the mass is being measured. It is also usual to eliminate the pycnometer volume as a possible source of errors by means of comparative measurements with water, a medium of known density ρ_w. Due to the uplift of the ambient

Figure 2.3.38: Pycnometers according to ISO 2811-1

Source: BASF Coatings

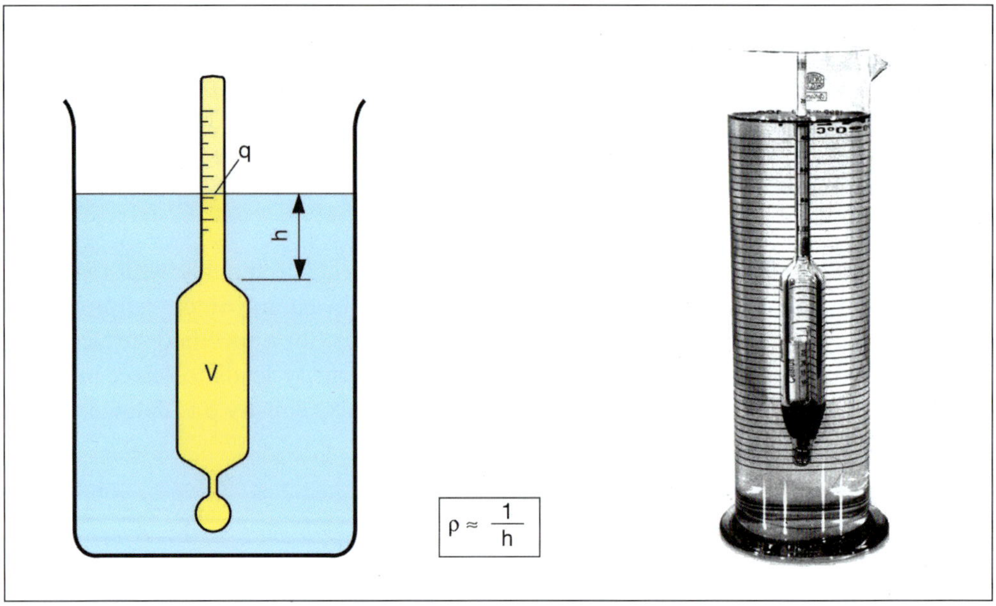

Figure 2.3.39: Schematic diagram and picture of an areometer

air, the density established by weighing is not the density of the paint, but the difference between the densities of paint ρ_{Pa} and air ρ_{Ai}.

$$\rho_{La} - \rho_{Lu} = \frac{m_{La}}{V}$$

If this figure is divided by the result of the comparative reading for water, the outcome is

$$\frac{\rho_{La} - \rho_{Lu}}{\rho_W - \rho_{Lu}} = \frac{m_{La}}{m_W}$$

Solving for ρ_{Pa} leads to an equation which permits readings with accuracy down to the per mill level.

$$\rho_{La} = \frac{m_{La}}{m_W} \cdot \rho_W + \rho_{Lu} \cdot \left(1 - \frac{m_{La}}{m_W}\right)$$

Among the measuring instruments which use the Archimedean principle of uplift is the **areometer**. The glass spindle of the areometer (see Figure 2.3.39) is immersed in the test liquid. Since the immersed body of volume V + h q of the part to be immersed loses the amount of weight equivalent to the weight of the displaced quantity of liquid m, the following applies to the equilibrium

$$m \cdot g = \rho_{La} \cdot g \cdot (V + h \cdot q)$$

$$\rho_{La} = \frac{m}{V + h \cdot q}$$

The distance h on the aerometer scale is a measure of the liquid's density. The aerometer is, of course, unsuitable for use with highly viscous and especially with pigmented and thus opaque test liquids. In such cases recourse can be made to the more precise method of **immersed balls** in accordance with ISO 2811-2. The material under analysis

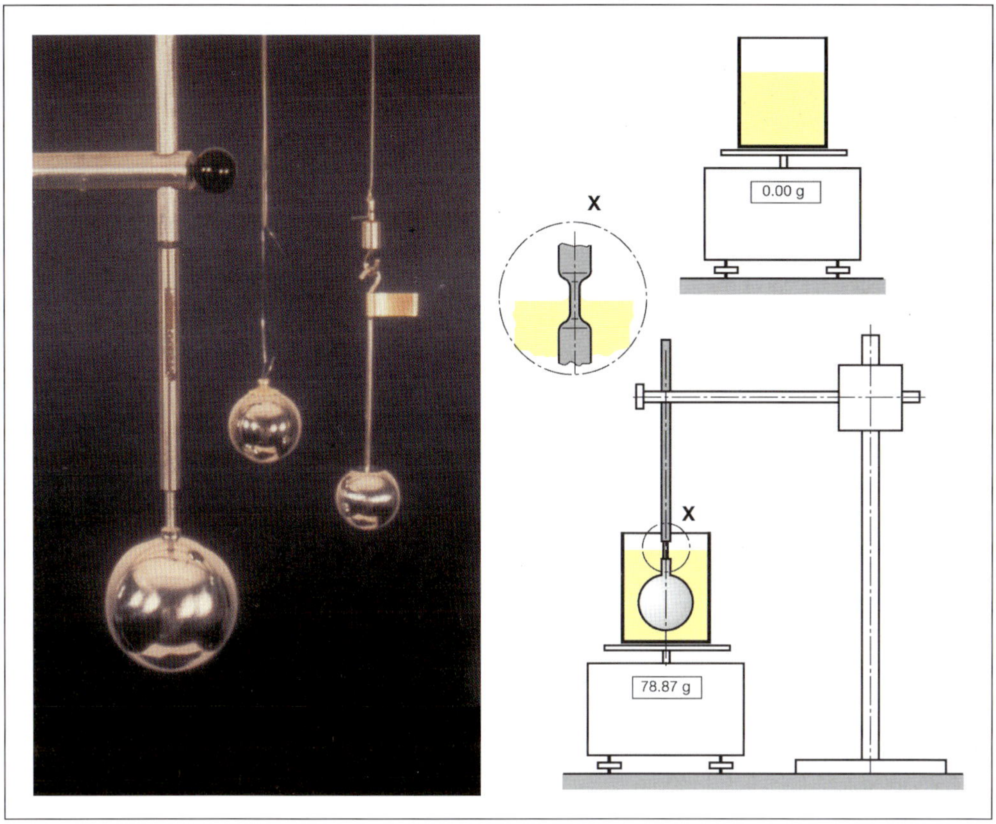

Figure 2.3.40: Picture and schematic diagram of density measurement with immersed balls (Gamma balls)

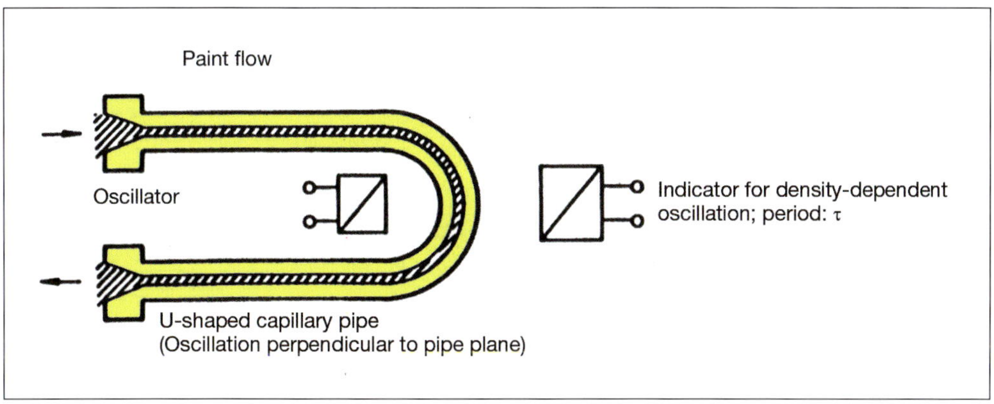

Figure 2.3.41: Measuring principle behind the oscillator method

is located in a test vessel on a balance. After immersion and fastening of a metal ball with a volume of exactly 100 ml the increase in weight caused by the displacement of the paint is expressed in terms of the volume and in this way the density is calculated. The measurement error is less than 0.1 % and is especially low for an immersion process. The immersed body method is universally applicable and easy to use. It is ideal for rapid testing in the plant.

Another indirect method, albeit one which operates by a completely different principle, is the **bending vibration** method. The method according to ISO 2811-3 exploits the fact that the generated frequency of a tuning fork-like glass tube is a function of the densities of its contents. If a ratio is formed of the period of vibration T_1 of the empty resonator tube to that of the resonator tube filled with test liquid T_2, the density ρ can be determined when the mass m_1 and the volume V of the measuring tube are known. The oscillator as a physical pendulum permits density measurements that are accurate to fractions of a per mill.

$$T_1 = 2 \cdot \pi \cdot \sqrt{\frac{m_1}{D}} \qquad T_2 = 2 \cdot \pi \cdot \sqrt{\frac{m_1 + \rho \cdot V}{D}}$$

$$\rho = \frac{m_1 \cdot \left(\frac{T_2^2}{T_1^2} - 1 \right)}{V}$$

The density of coatings (dry film density) and also that of the liquid coating materials are important technical and economic variables (see Chapter 5.7). Means of determining the dry film density are described in Chapter 3.2.2.

Other Indicators

When solvents and resins are being delivered in tankers, the necessary identification checks are frequently carried out in the form of rapid tests to reduce waiting times. In a standards-compliant distillation system the distillate quantities are determined as a function of temperature and compared with those of a reference standard. This formerly common standard method of **fractional analysis** of solvents has now been largely superseded by gas chromatography.

An even easier method involves determining the **evaporation number VDZ** according to DIN 53170. A defined quantity of the solvent under analysis is dripped onto a filter paper at a test temperature of 23 °C. The evaporation time t is determined by comparison with the evaporation time for the same quantity of diethyl ether t_{Ether}.

$$VDZ = \frac{t_X}{t_{Ether}}$$

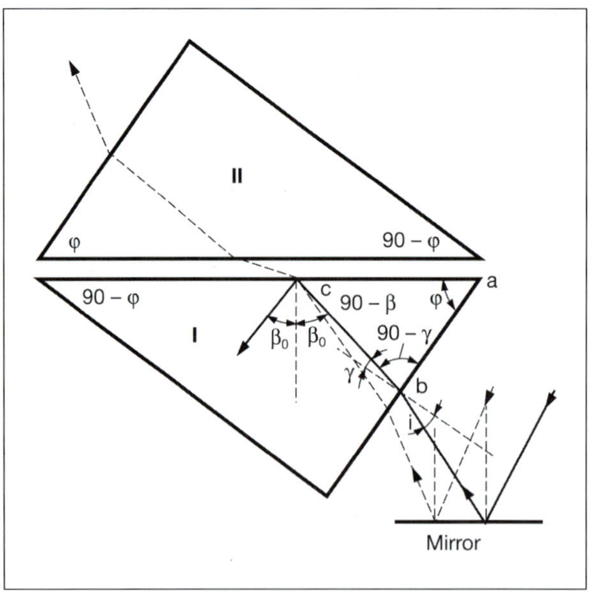

Figure 2.3.42: The path followed by rays inside the prisms of an Abbé refractometer

Determining the **refractive index**, (see DIN 51423-2) is the simplest and one of the most accurate methods of checking the identity of raw material mixtures. It is usually determined with the aid of the Abbé refractometer and the total reflection principle. A beam of light passes through a glass prism after being deflected by a rotatable mirror and then, after exiting and being deflected by an angle corresponding to the refractive index of the liquid, it enters a second prism. From here the light beam enters the observer's eye. Changing the incident angle enables the light beam to be totally reflected in the test liquid. Consequently, the observer can no longer see any light. Since sine $\alpha = 1$ in accordance with Figure 2.3.42, the refractive index of the sample can be calculated by application of *Snellius* law.

2.3.3 Pigment-Specific Tests

The most important requirements imposed on pigments fall into two categories: first, the ageing resistance under the influence of light, outdoor exposure and chemicals as discussed in Chapter 3.3 and, second, the processing properties in coating materials. They are characterised by pigment indicators such as the average particle size, their size distribution, specific surface area, weight per unit volume (bulk density) or volume per unit weight, as well as the moisture content, water-soluble components, specific resistance, pH, acid or alkali number and abrasiveness (hardness).

A complete description requires that the interactions of the pigments with the film forming agent also be included in the characterisation. The flow, softening and wetting points, in the form of the oil absorption value or water equivalent, fineness, tinting strength, tinting strength development and dispersion hardness of the formulations are useful variables in this regard.

2.3.3.1 Indicators of the Pigment as a Raw Material

Particle Size /Particle Size Distribution

Pigment particles are adjusted to their optimum particle size by crystallisation or grind-ing processes. The random processes of growth and size reduction almost always lead to size distributions. If ideal growth is achieved, the particle distributions almost obey a Gaussian curve, but for size reduction, on the other hand, they obey a logarithmic distribution. Of the various measurement methods for describing pigment sizes in practice the distribution curve established empirically by *Rosin, Remmler, Sperling* and *Bennet* (RRSB distribution) has proved the most successful. This permits the distributions to be characterised with two indicators to DIN 66145. The variables d' and n can be calculated by plotting the cumulative frequency as a residual sum R as a function of particle size d and subsequent mathematical treatment involving double logarithms.

$$R = e^{-\left(\frac{d}{d'}\right)^n}$$

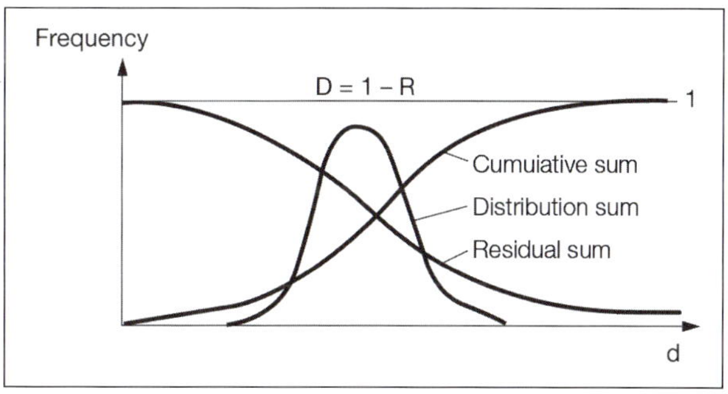

Figure 2.3.43: Definition of mean particle size d' and particle size distribution n of pigments according to Rosin, Remmler, Sperling and Bennet

On account of the exponential function 36.8 % of all particles are > d'. The distribution curve is clearly defined in association with the particle parameter n which characterises the distribution range. In a double logarithmic plot the RRSB curve is

Averaged diameter	a	b	Characterized by	Equation
d_{10}	1	0	Particle diameter expressed in terms of particle number	$d_{10} = \frac{\sum_{i=1}^{n} n_i\, d_i}{\sum_{i=1}^{n} n_i}$
d_{20}	2	0	Particle surface expressed in terms of particle number	$d_{20} = \sqrt[2]{\frac{\sum_{i=1}^{n} n_i\, d_i^2}{\sum_{i=1}^{n} n_i}}$
d_{21}	2	1	Particle surface expressed in terms of sum of diameters	$d_{21} = \frac{\sum_{i=1}^{n} n_i\, d_i^2}{\sum_{i=1}^{n} n_i\, d_i}$
d_{30}	3	0	Particle volume expressed in terms of particle number	$d_{30} = \sqrt[3]{\frac{\sum_{i=1}^{n} n_i\, d_i^3}{\sum_{i=1}^{n} n_i}}$
d_{31}	3	1	Particle volume expressed in terms of sum of diameters	$d_{31} = \sqrt[2]{\frac{\sum_{i=1}^{n} n_i\, d_i^3}{\sum_{i=1}^{n} n_i\, d_i}}$
d_{32} (Sauter diameter)	3	2	Particle volume expressed in terms of particle surface	$d_{32} = \frac{\sum_{i=1}^{n} n_i\, d_i^3}{\sum_{i=1}^{n} n_i\, d_i^2}$

Figure 2.3.44: List of common mean particle sizes according to Megele and Evans

a straight line if the distribution is ideal. Irregularities in the distribution can be seen as jumps (clusters of equal particle sizes) or plateaus (paucity of particles).

$$\underbrace{\lg\lg\left(\frac{1}{R}\right)}_{y} = \underbrace{n}_{a} \cdot \underbrace{\lg d}_{x} + \underbrace{(\lg\lg e - n \lg d')}_{b}$$

$$y = a \cdot x + b$$

In addition to the calculation of d' as the mean particle distribution described above, there are other mathematical methods of determining the mean particle diameters. *Megele* and *Evans* have proposed a general equation for calculating these [2.4.128]:

$$d_{ab} = \left(\frac{\sum_{i=1}^{n} n_i \cdot d_i^a}{\sum_{i=1}^{n} n_i \cdot d_i^b} \right)^{\frac{1}{a-b}}$$

where n_i the number of particles in the categories of diameter d_i, and a and b are positive integers which characterise the type of distribution by volume (3), surface area (2), diameter (1) and number (0). The mean values d_{10}, d_{30} and the diameter d_{32} from the table in Figure 2.3.44 have proved successful in practical operations.

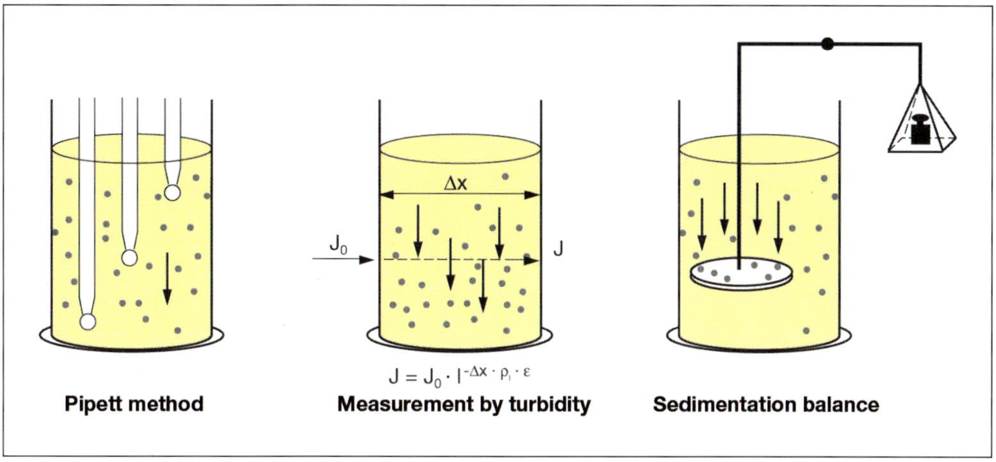

Figure 2.3.45: Measurement of particle sizes by sedimentation and different indication methods

Various processes for measuring particle sizes underpin these mathematical methods of characterisation. Since the pigment particle sizes or size distributions exert a special influence on the colour and rheological properties, particle measurement methods are extremely important. The same applies to powder coatings. The selection of the methods described is also important in determining the drop size and distribution where liquid coating materials are processed by spray application methods.

Mechanical, laser diffraction, electrical and optical methods are used in practice, the choice depending on particle size. The following merit particular mention:

- Dry sieving
- Wet sieving
- Sedimentation (including sedimentation balance)
- Accelerated sedimentation (disc centrifuge)
- Coulter counter
- Electron microscope and image analysis
- Laser scatter (if < 1 μm)
- Laser diffraction (if > 1 μm)

Sieving processes are only successful if relatively coarse pigment or extender particles are to be analysed. Standardised dry sieve sets are used for coarse extender particles.

The substance under analysis is placed in the uppermost of 16 sieving drums arranged one above the other. As a result of intensive oscillation (vibrating sieve) the particles are offered up to the openings in the top sieve and, if they are smaller than the openings, drop into the next sieve chamber. Here the particles are again offered up to the finer sieve tray, and the process is repeated a further 14 times. The mesh sizes start at 1500 μm and end at 50 μm. Weighing the individual sieve fractions gives an overview of the particle size distribution. The residual sum function obtained in this way can be converted after RRSB into a distribution function for determination of d' and n.

The sieve separation process becomes more sensitive and thus feasible for analysing finer particles if the sieving is carried out in suspension. Particles with sizes of up to one micrometre can be analysed in this way.

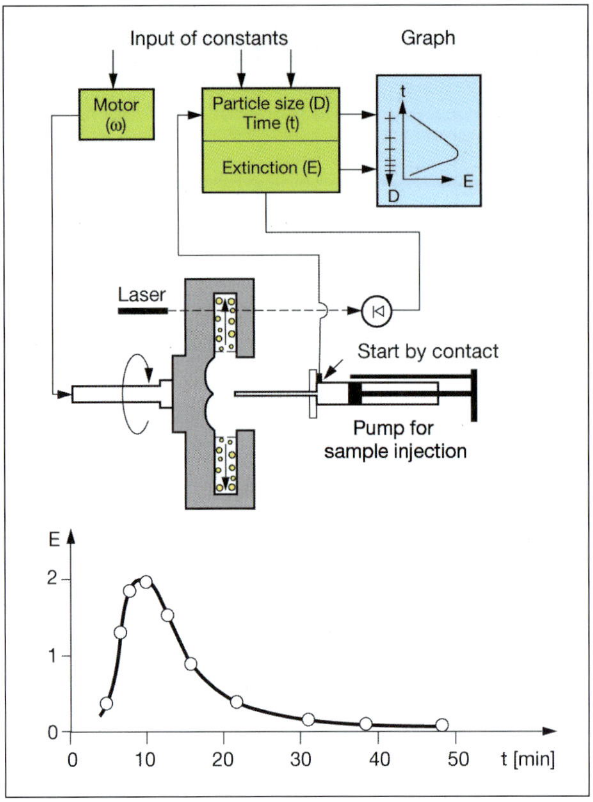

Figure 2.3.46: Schematic drawing of a disc centrifuge

Another measurement process is **sedimentation**. In gravity sedimentation a particle suspension layer is added to a less dense liquid. The particles which are deposited under gravity migrate to the bottom at a sedimentation rate u which is a function of the density difference $\Delta\rho$ between the particle and liquid, the square of the particle diameter d^2 and the viscosity η. In addition, for accurate calculations, a specific form factor f and a concentration-dependent interaction factor b must be included for the particles.

$$u = \frac{\Delta\rho \cdot g \cdot d^2}{18 \cdot \eta} \cdot f \cdot b$$

Measuring the particle speeds during their downward travel or weighing the particle quantities being deposited on a sedimentation balance as a function of time enables the particle sizes or particle size distributions to be calculated. The particle travel can be followed directly after sampling or at a certain time at different levels. The simplest way is to measure the intensity of scattered light in the test configuration itself.

For particles with a diameter of less than 1 µm and also with a small density difference between the fluid and the migrating particles the sedimentation rates under the earth's gravity are so low that excessive test times and results distorted by Brownian movement are inevitable. Accelerated sedimentation processes in ultracentrifuges have proved successful for such particles. A variant of interest to the pigment industry is the disc centrifuge. Practically all pigment particles in the range 0.01 to 1 µm can be analysed accurately by setting appropriately high speeds.

The measurement principle corresponds to that used in gravity sedimentation. It differs only in that it has a centrifuge to increase acceleration g. In the sedimentation equation, the acceleration due to gravity must be replaced by the centrifugal acceleration $\omega^2 r$ or $4\,\pi\,n^2\,r$ (see also Chapter 2.2.4 Sedimentation). The migration rate u is then

$$u = \frac{\Delta\rho \cdot 4\pi^2 \cdot n^2 \cdot R}{18 \cdot \eta} \cdot d^2$$

where R is the radius of the centrifuge. The suspension under analysis is injected into the centre of the rotating disc. The migration rate can be determined in this test configuration by measuring the scattered light of laser beams.

When the results are being evaluated, it should be borne in mind that the calculations are valid initially only for spherical particles. Where form factors are known, however, the mean diameters of irregularly shaped particles can also be calculated.

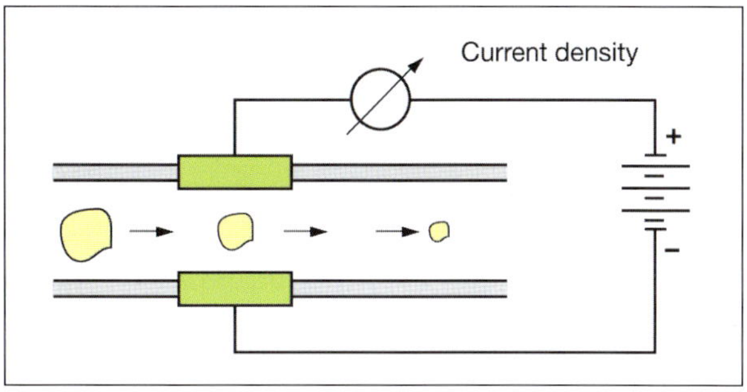

Figure 2.3.47: Schematic drawing of a Coulter counter

A **particle counter** developed by Coulter uses a completely different process for analysing particle distributions. The **Coulter counter** utilises the low electrical conductivity of the particles relative to that of the liquid. If a highly diluted suspension passes through a capillary containing a measuring zone for determining the electrical conductivity, the electrical conductivity falls when a pigment particle passes through. The change in the conductivity is proportional to the particle size.

Type	α	Functionality	proportionality to diameter
Rayleigh scatter	<<1	$I = f(\alpha, n, \varphi)$	$I \sim \frac{D^6}{\lambda^4}$
Mie scatter	≈ 1	$I = f(\alpha, n, \varphi)$	$I \sim D^2$
Fraunhofer diffraction	>> 1	$I = f(\alpha, \varphi)$	$I \sim D^4$ (if $\varphi = 0$)
there is $\alpha = \frac{\pi \cdot D}{\lambda}$		I = intensity of diffraction φ = angle of diffraction	

Figure 2.3.48: Measuring range of calculation functions for various particle sizes

Assuming a sufficiently high dilution, the *Coulter* counter is capable of registering each individual particle and converting the results into a particle size distribution curve with the aid of software.

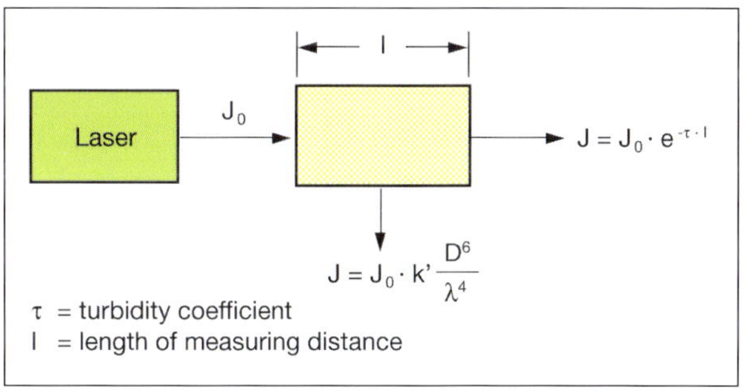

Figure 2.3.49: Methods of measuring particles by light scattering

Optical methods are suitable for the direct characterisation of extremely fine particles, such as are present in dispersions of resins and organic pigments. They can detect particle sizes down to the molecular range. Two phenomena serve as the basis of the analysis: either light scattering or light diffraction, the choice depending on the particle size. In light scattering the interaction of a particle with the light beams is subject to different laws in accordance with the size of the particle. The deciding factor is $\alpha = D\,\pi/\lambda$, which can be calculated from the particle size D and the wavelength λ of the incident light. I is the intensity of the diffusion effect, φ is the diffusion angle and n is the refractive index.

If particles which are smaller than λ/20 (α much smaller than 1) are analysed, either the light scattered by a laser beam is determined perpendicularly to the beam direction or the attenuation of the primary beam itself is measured. In the former case, the scattering density I is converted into the particle size D in accordance with

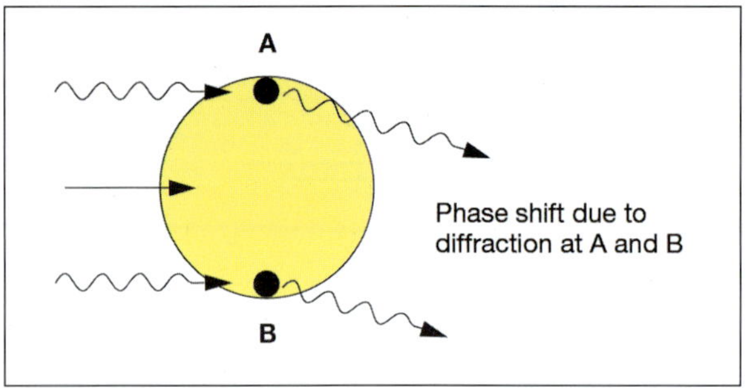

Figure 2.3.50: Shift of wavelength and interference of scattered light on particles

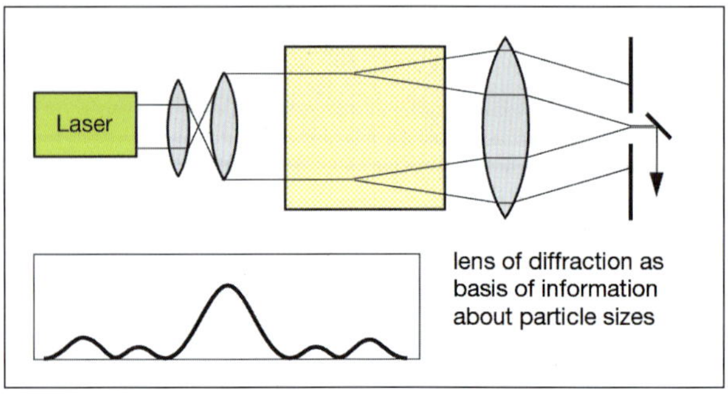

Figure 2.3.51: Diffraction of light as source of information about particle size

$$I = I_0 \cdot k \cdot \frac{D^6}{\lambda^4}$$

In the latter case *Lambert's* law permits the attenuation of the primary beam intensity to be converted into a turbidity factor which is correlated with the particle size [2.4.129].

At the magnitude of the particles described to date other measurement methods record *Brownian* motion. The effect generated by the self-motion of the particles leads to an expansion of the diffusion frequency, as a result of which the incident light is overlapped by higher and lower frequencies. It is then not the intensity of the diffused light but its frequency expansion which is the relevant measurement variable. This method is termed quasielastic light scattering (QELS) or dynamic light scattering (DLS) in the literature [2.4.130].

If the particles under analysis are larger than $\lambda/20$ and α is approx. 1, interference (Mie scattering) occurs in the range of the scattered light.

If two different scattering centres of one and the same particle are observed and there is a sufficiently large distance between points A and B with discrete angles of diffusion, shifts in the wavelengths occur which lead to interference, and this in turn permits calculation of the particles in accordance with *Rayleigh's* equation [2.4.131].

If α significantly exceeds a value of 1, the evaluation of diffraction spectra is the most suitable way of determining particle sizes. By diffraction of light is meant the deviation of the light propagation from the laws of geometric optics. It occurs when free propagation of the wave is changed by objects such as openings or particles [2.4.132]. If a light beam passes a solid surface tangentially, it departs from its original direction all the more, the closer it passes to the particle. This effect is a function of particle size. To meet the conditions for the *Fraunhofer* theory and enable relevant readings, the light source is beamed through the focal point of a convex lens, as a result of which the light source is focused at infinity, and parallel beams of light are formed. By observing the diffraction phenomenon again in the focal plane of a second lens, the diffraction image is also focused at infinity.

Under these conditions the simplified *Fraunhofer* integral can be used for quantitative discussion of the diffraction phenomena [2.4.133, 2.4.134].

$$I = I_0 \cdot \left[\frac{2 \cdot J_1(x)}{x}\right]^2$$

$$x = \frac{2 \cdot \pi \cdot r \cdot s}{\lambda \cdot f}$$

I_0 = intensity of incoming light
J_1 = Bessel equation
r = radius of particle
s = radiant distance to the optical axes
λ = wavelength of laser
f = focal distance of lens

The diffraction spectra of particles can be calculated with the aid of this equation or the particle size of spherical particles can be determined from diffraction images on the basis of this theory.

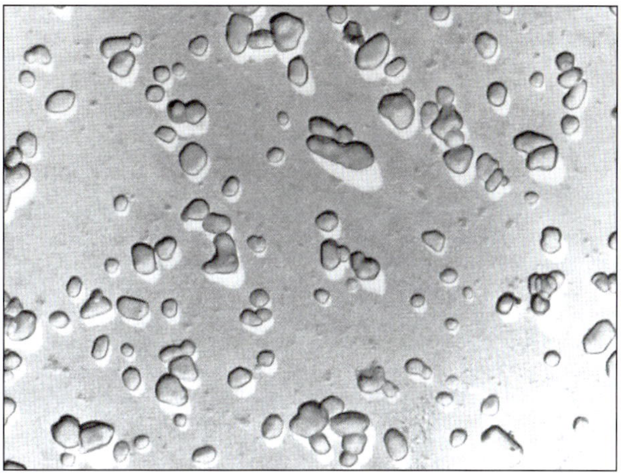

Figure 2.3.52: Electron micrograph of titanium dioxide powder in a resin matrix

Particle sizes can also be determined by means of **electron microscopy**. If the concentration is sufficiently low, the size, size distribution and shape of pigments or "frozen" dispersion particles can be established with the aid digital image analysis (see Chapter 6.2). At the same time, it is possible to derive information about the distribution state of primary particles, agglomerates and aggregates. This requires specific sample preparation which changes the original state of the sample as little as possible [2.4.135].

As outlined in the descriptions of the individual methods for measuring particle sizes, the methods each have specific ranges in which they deliver the best results. It is essential that this be taken into account when methods are selected for a particular measurement problem. The measurement ranges are shown in 2.3.53.

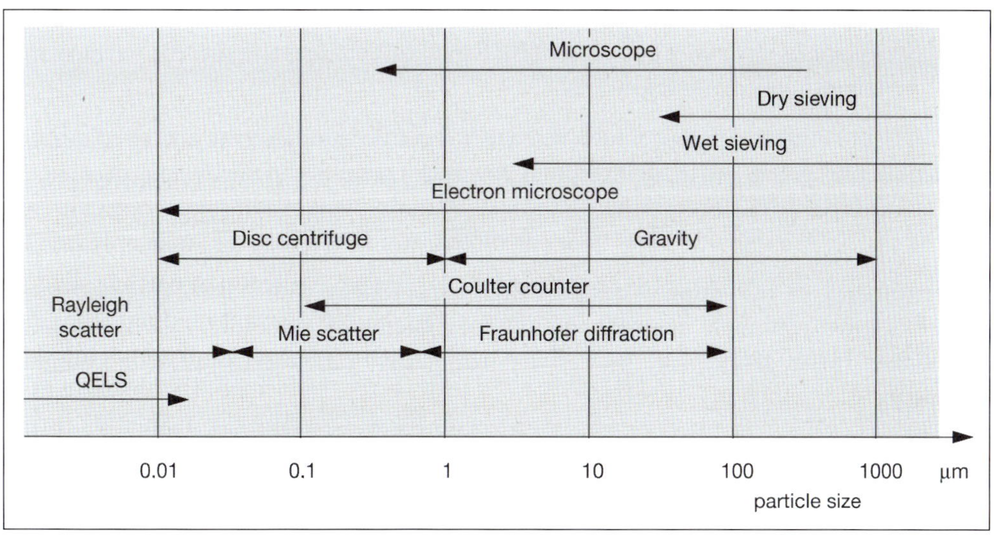

Figure 2.3.53: Optimum measurement ranges for different methods of particle size determination

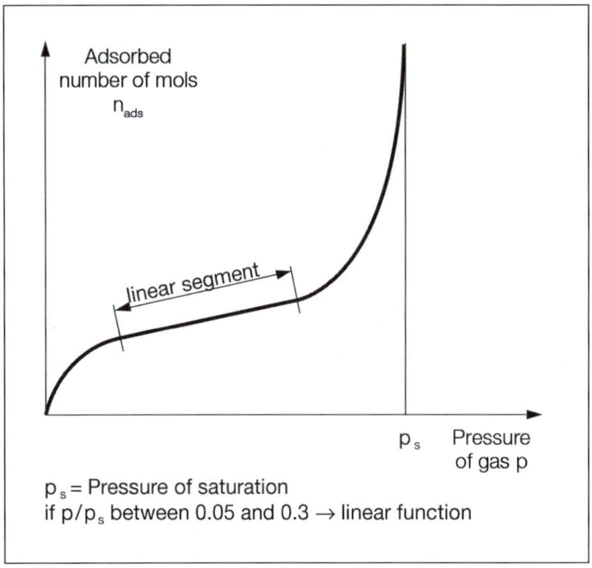

Figure 2.3.54: Isothermal gas adsorption n_{ads} as a function of gas pressure p

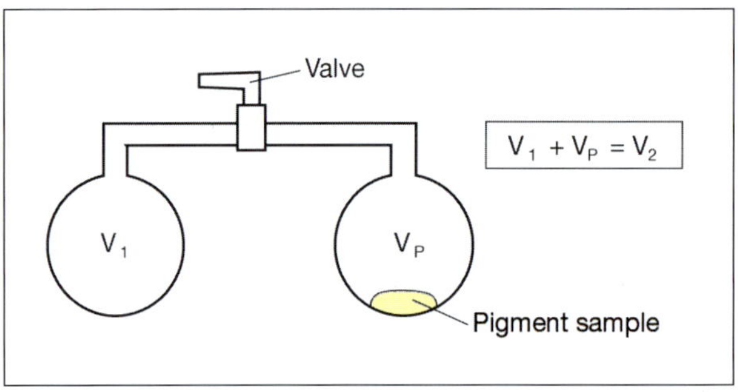

Figure 2.3.55: Simplified diagram of apparatus for measuring gas adsorption on pigments

Specific Pigment Surface Area

An indicator which is functionally linked to the particle size and is important for pigment dispersion is the specific surface area. This is defined as the pigment surface area either per unit mass or per unit volume. Direct determination is not possible because of the small size of the particles. For this reason, the surface area is determined indirectly by means of interaction with gases or liquids.

A measurement method which is also known from other spheres utilises the quantity of gaseous nitrogen adsorbed by the pigment at the temperature of liquid nitrogen. Comparative pressure measurements with helium enable conclusions to be drawn about the number of nitrogen moles adsorbed n_{ads} from the reduced quantity of gaseous nitrogen calculated in accordance with the gas law. The adsorbed number of moles at different pressures and constant temperature yields a typical curve as shown in Figure 2.3.54. Mathematical treatment after *Brunauer, Emmet* and *Teller* (BET method) enables the specific surface area of the pigment to be calculated.

The BET function, which represents the basis for conversion from the adsorbed number of moles of nitrogen n_{ads} to monomolecular coverage n_0, applies to the linear section in the p/p_s range of 0.05 to 0.3. Knowledge of the space requirement of a nitrogen molecule enables the surface area per unit mass to be calculated, after allowance for the sample weight.

To measure n_{ads} as a function of gas pressure in accordance with ISO 9277 closed apparatus of known volume V_1 is used which is correlated with a volume V_p. This contains the sample. If the gas pressure in V_1 is determined with a closed valve after complete evacuation of V_p, then $p_1V_1 = n_1RT$.

On opening of the valve, the volume increases from V_1 to V_2. However, the pressure p_2 measured then does not correspond to the value calculated according to the gas law. Instead, it is lower than expected because of the surface-area-specific gas adsorption and associated reduction in the number of moles from n_1 to n_2. Consequently

$$\frac{p_1 \cdot V_1 - p_2 \cdot V_2}{RT} = (n_1 - n_2) = n_{ads}$$

The number of moles n_0 necessary for monomolecular coverage can be calculated with the BET formula for a known saturation vapour pressure p_s by measuring p and the corresponding number of moles adsorbed n_{ads}.

$$\underbrace{\frac{p}{n_{ads} \cdot (p_s - p)}}_{y} = \underbrace{\frac{c-1}{n_0 \cdot c}}_{a} \cdot \underbrace{\frac{p}{p_s}}_{x} + \underbrace{\frac{1}{n_0 \cdot c}}_{b}$$

If $p/[n_{ads} (p_s - p)]$ is plotted against p/p_s, n_0 is determined from the gradient and the intersection with the ordinate. If n_0 is multiplied by the space requirement for 1 millimol nitrogen (= 97.5 m²) and this figure is expressed in terms of the weighed pigment sample in g, the result is the mass specific surface area O_m in m²/g.

$$O_m = \frac{n_0 \cdot \vartheta}{E}$$

An alternative for determining n consists of single point measurements of the changing thermal conductivity of a helium/nitrogen mixture as a result of nitrogen adsorption.

In the flow apparatus shown in Figure 2.3.56 the difference in the thermal conductivity of a defined gas mixture of helium and nitrogen before and after nitrogen adsorption at the pigment surface is measured in the test vessel and converted into n_{ads}. Mathematical treatment is carried out again as for the gas pressure method [2.4.136].

The determination of the specific surface area can be used to calculate the particle size if the pigment shape is known. To put it another way: the joint determination of particle size and specific surface area also enables form factors to be calculated.

Particle sizes, shapes and distributions also determine the space requirement of pigments in bulk. The **tamped volume** or **tamped density** to ISO 787-11 is used to describe this. Both variables depend on the orientation of the particles in the bed. Thus they are not material-specific properties. The type of transport and storage influence the result. The pigment powder must be precompressed before the test so that comparable and thus reliable data can nonetheless be

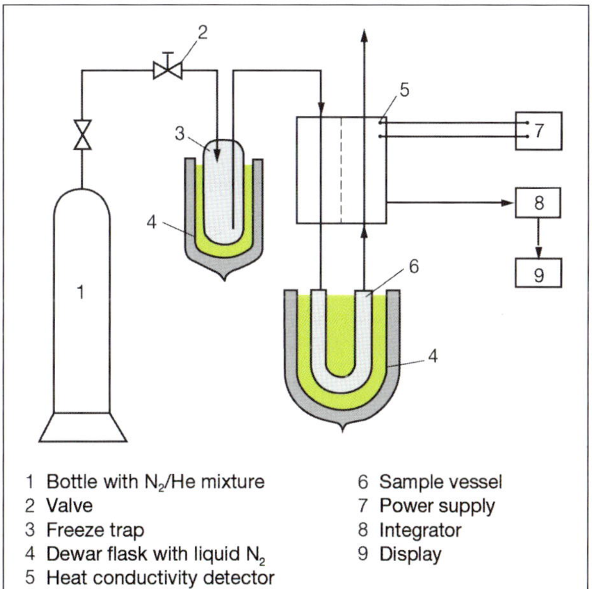

Figure 2.1.56: Determination of the specific surface area by measuring the thermal conductivity

obtained. The mass per unit volume is termed the tamped density, while the tamped volume is the reciprocal of this.

On account of their large and frequently polar surface area, pigments contain a certain quantity of adsorptively bound **moisture**. To an extent depending on the pigment and the storage conditions (temperature, relative humidity) up to 1 % water can be retained. Since water can have a significant adverse effect on pigment dispersion in solvent-borne paints, and CO_2 gas is generated from the interaction with isocyanates, the moisture contents must be determined before the pigments are processed. To this end, the powder is dried in a desiccator until a constant weight is achieved, and the percentage mass loss is recorded. In order that only adsorptively bound water will be recorded if water of crystallisation is also present, the drying temperature must be exactly 105 °C.

A different goal underlies the determination of the **loss on ignition** to ISO 3262-1. In this test, which is primarily important for analytical purposes, the pigment under analysis is heated to 1000 °C in a porcelain crucible and the residue in percentage by weight is termed the loss on ignition.

Since electrolytes in coatings contribute to bubble formation under the action of water, the content of osmotically active salts in pigments must be analysed as they are potential sources of damage. This is achieved by measuring the **electrical conductivity** of an aqueous extract of the pigment powder generally in accordance with ISO 787-14.

If pigments contain soluble acid or alkali components, these can easily be registered in the form of an **acid or alkali value** by mixing them with alkalis or acids. In accordance with ISO 784-4 100 g pigment must be titrated after extraction in the suspension against 0.1 n NaOH or 0.1 n HCl.

Determining the **pH** of pigment suspensions to ISO 787-9 can provide similar information. The pigment powder is first boiled in water and then analysed with a pH meter after removal of any still dissolved carbon dioxide.

Coating materials are sometimes exposed to high shear forces during their manufacture and processing. Hard pigments can cause a high level of material wear in system pumps and valves because of the high flow rates induced by constrictions. Any complete description of pigments and extenders requires data on **abrasiveness**. This property is tested by incorporating a defined pigment quantity into a standard paint and agitating with steel balls of a defined type and quantity in a planetary ball mill. The loss in weight of the steel balls measurable after an agreed test period is stated as a percentage by weight.

2.3.3.2 Tests with the Pigmented Coating Material

The incorporation of pigments into the liquid resin matrix leads to new test demands on the pigments because of the interaction with the binder matrix. Compliance with the entire specification profile to which the coating material is subjected, such as the binder requirement during dispersing, the necessary pigment distribution fineness, the desired tinting strength or the required hiding power, yield, and storage without flocculation or sedimentation can only be obtained by pigment-specific tests in an overall system.

The binder requirement for pigments depends on their form, size and size distribution if the intention is to convert them into a flowable, homogeneous mixture. If this property is to be described in a way which is as pigment-specific as possible, the various influences of the film forming agents and solvents must be eliminated. ISO 787-5, therefore, specifies linseed oil having standardised properties for the determination of the **oil absorption value**. A precisely weighed quantity of pigment is therefore mixed drop-by-drop from a burette with standardised linseed oil and homogenised with a spatula until a coatable, paste-like mass has been produced. The quantity of oil used is expressed in terms of 100 g pigment and then used as an oil absorption value to evaluate the dispersion properties. To characterise pigments for processing in water-borne paints the hydrophobic linseed oil is replaced by water as specified in DIN 55608.

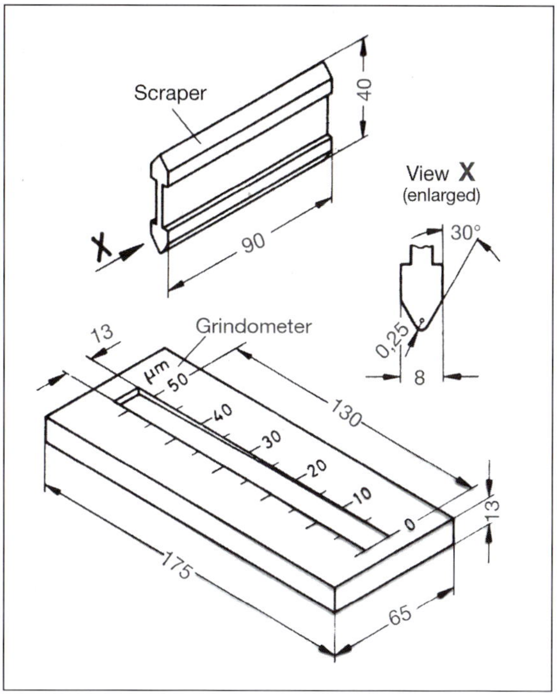

Figure 2.3.57: Device for determining the fineness of grind according to ISO 1524

The oil absorption value is a pigment-specific indicator which should only be used, with certain reservations, for determining the optimum mill base formulation. In many cases the standardised linseed oil should be replaced by the film forming agent solution specified in the formulation (see Chapter 2.2.3).

Important indicators for pigment properties during the dispersion process relate to the degree to which the agglomerates have been broken down. This can be carried out directly or indirectly. In practice, statistical methods are used for optimisation [2.4.137].

The standard test in everyday use involves determining the **fineness of grind**. In this process the material to be dispersed which is under analysis is tested at regular intervals using the grindometer as specified in ISO 1524. In the test a small quantity is taken from the batch for grinding and spread with an applicator over the wedge shaped indentation of a grindometer from the deepest to the shallowest part. In areas in which the particle size exceeds the wedge depth, stripes and grooves are created. The wedge depth, which can be read off on a scale for this range, is a measure of the largest agglomerates still present.

Because assessment is subjective, the evaluation of grindometer measurements leads to a great divergence of results. To objectify and thus to reduce the error range, instruments have been developed which fix the grindometer profile using a high resolution camera and process it objectively after digitisation [2.4.138]. This very simple and therefore commonly applied measurement method is particularly used to determine the progress made in the dispersion process. It only permits an approximate estimate of the maximum possible tinting strength of the individual pigments, especially if fine grain organic pigments have to be dispersed.

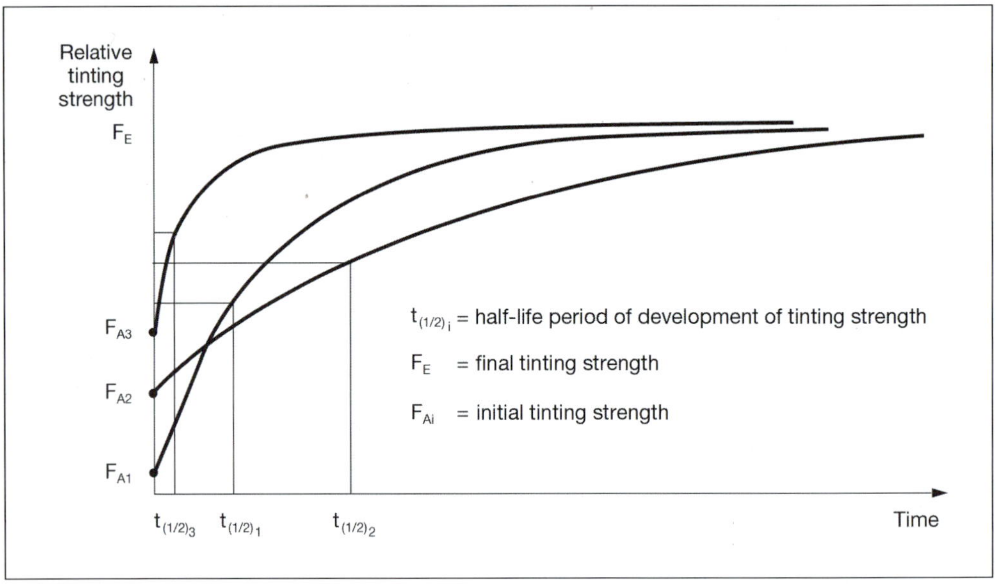

Figure 2.3.58: Development of the relative tinting strength of different pigment dispersions

In such cases and when primary batches are tested, the more complex but therefore more precise measurement method of determining the tinting strength which involves test panels manufactured with the material to be dispersed is chosen. The tinting strength, which increases with increase in dispersion level, cannot, however, be measured directly because of the scattering power S, which changes at the same time. According to *Kubelka* and *Munk* the quotient of the absorption coefficient K and the scattering coefficient S can be calculated from the reflections R as a function of the wavelength λ of the test samples.

$$\frac{K}{S} = \frac{(1 - R(\lambda))^2}{2 \cdot R(\lambda)}$$

K = absorption coefficient
S = scattering coefficient
R = share of reflection

In order that, when the progress of the dispersion process is being tested, the test equipment needed to determine increases in tinting strength may be kept within reasonable bounds despite the varying scattering coefficient, testing is limited to measuring the relative tinting strength. To this end the batches for grinding under analysis are mixed with a sufficient quantity of white paste to keep the scattering coefficient constant, irrespective of the progress of the dispersion process.

If the K/S values determined in accordance with *Kubelka* and *Munk* are then expressed in terms of standard values K(St) and S(St) determined by means of complete dispersion, curves of changes in tinting strength F can be plotted as a function of time.

$$F_{rel} = \frac{K(t)}{S(t)} \cdot \frac{S(St)}{K(St)} \cdot 100$$

if $S(t) = S(St)$ follows

$$F_{rel} = \frac{K(t)}{K(St)} \cdot 100$$

It is possible to derive theoretically from this that curves of changes in tinting strength obey either a hyperbolic or an exponential function, depending on the pigment concentration [2.4.139]. The constants of such functions such as the half-life period provide information on the rate and degree of dispersion quality already achieved. The half-life period is defined as the time that elapses until the colour strength of the starting point has been increased to 50 % of the difference F_E - F_{Ai} (see Figure 2.3.58).

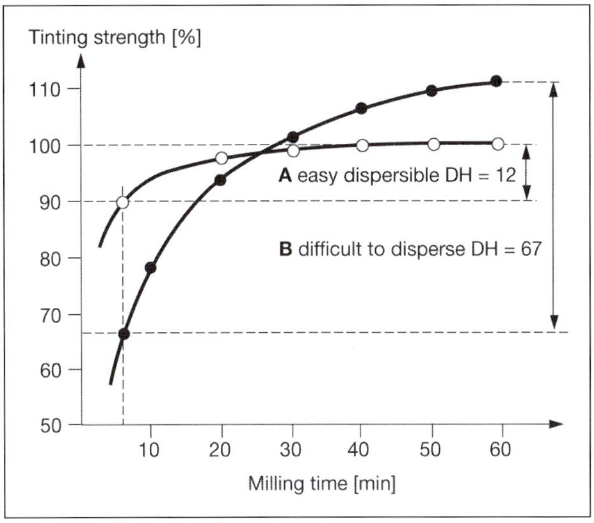

Figure 2.3.59: Development of relative tinting strengths of two copper phthalocyanine pigments

ISO 8781-1 describes dispersion properties in terms of a simple indicator. The **dispersion hardness** DH in % is calculated according to

$$DH = \frac{F_{60} - F_6}{F_6} \cdot 100$$

from the tinting strengths F_6 and F_{60} for dispersion times of 6 and 60 minutes. A comparison of the maximum tinting strength of several pigments which are identical in colour terms can be made using relative measurements by defining the final tinting strength of one of them as 100 % and relating the readings of the others to this. Figure 2.3.59 shows the curves of changes in tinting strength of two blue pigments. They differ from each other only in their pigment shape and size. The needle shaped, coarser pigment has a higher initial tinting strength and low dispersion hardness (DH = 12). The isometric, finer alternative product has a lower tinting strength initially because of the high degree of agglomeration but, having a substantially higher dispersion hardness (DH = 67), develops a higher final tinting strength.

When evaluating the value of a coating, the quantity of paint used in its production is also a deciding criterion, in addition to the quality achieved. Since the protective effect is already achieved in many cases before the substrate is optically covered, the **hiding power** becomes a variable which determines the paint consumption of coatings that provide colour. The hiding power is the reciprocal of the film thickness of a pigmented coating, in which the ratio of the intensity of the reflected light on a white R_w and black substrate R_B has taken on a value > 0.98.

In many cases testing involves the visual evaluation of a wedge-shaped paint film in the wet or coalesced state on a chequered black and white substrate. The two values are not identical because of the changing relative refractive index during the film forming process.

Other processes for determining the hiding power use the possibilities of colorimetry (see Chapter 3.2.4) by plotting reflection curves (R_∞ and R_B) of a hiding and a non-hiding coat, respectively, of thickness h and calculating the scattering coefficient according to

$$S_V \cdot h = \frac{1}{b} \cdot \operatorname{arccoth} \frac{1 - a \cdot R_B}{b \cdot R_B} \qquad \text{where} \quad a = \frac{1}{2} \cdot \left(\frac{1}{R_\infty} + R_\infty \right) \quad \text{and} \quad b = a - R_\infty$$

The volume-related scattering coefficient S_v can be converted into the hiding power Dv as specified in DIN 55987 by means of a correction factor α which is a function of the reflection from the hiding coat.

$$D_V = \alpha \cdot S_V$$

The hiding power specifies the minimum film thickness and, in conjunction with the solid content and the dry film density (see Chapter 3.2.1), largely determines the material consumption and thus the yield per quantity of paint (see Chapter 5.7).

Despite carefully developed formulations and precise knowledge of the important physical processes in the dispersion of pigments it is not always possible to prevent reagglomeration of the pigments that occurs during the storage and processing of coating materials. These conglomerations, which are known as flocculation, give rise to increased **sedimentation**. Sediment in stored paint tins can be qualitatively detected by simple manual stirring. Better, but also more complex measurement processes have proved successful in providing more accurate information on the sedimentation properties of pigmented coating materials.

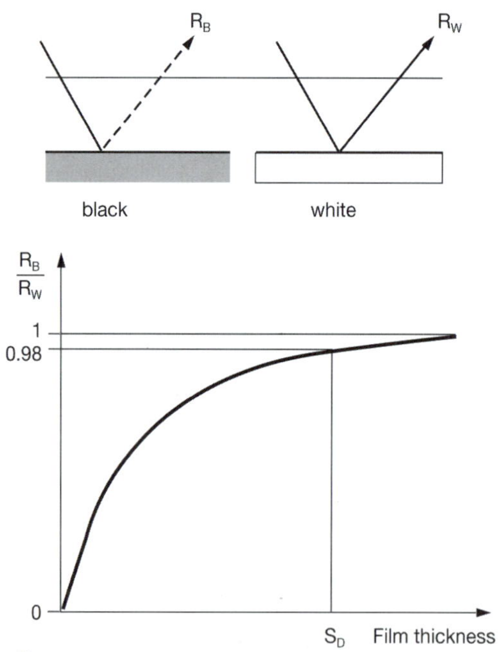

if

$$\frac{R_S}{R_W} > 0.98$$

then the coating covers the underground

Figure 2.3.60: Definition of hiding power according to DIN 55987

The most important method of determining the tendency towards sedimentation consists of immersing a test body into the liquid coating material at a defined rate. The resistance to the penetration of the test body is measured and expressed in terms of the depth of penetration. Whereas only a slight rise in the counterforce is recorded in the upper part of the test vessel, it increases dramatically on penetrating the sedimentation area. This measurement method permits sedimentation profiles to be recorded with a high degree of accuracy [2.4.140].

Another measuring principle involves the attenuation of X-rays passing through the paint sample. The presence of a nonhomogeneous pigment distribution is detected by means of differing levels of absorption of the X-rays.

The degree of flocculation can also be determined indirectly by mechanically rubbing the paint film of a freshly applied paint. This test, known as the **rub out test**, visually evaluates the change in tinting strength caused by rubbing or measures it colorimetrically. If flocs are present, these are converted back into primary particles again. If the rub out process is carried out under defined conditions and if changes in tinting strength are determined colorimetrically by measuring the rubbed out F_r and the nonrubbed out surface areas F_{nr}, it is possible to obtain very accurate information on flocculation during storage or film formation. Quantitative values relating to the degree of flocculation can be derived by determining the rub out number RBN in %.

$$RBZ = \frac{(F_r - F_{nr})}{F_{nr}} \cdot 100$$

2.3.4 Processability of Coating Materials

Coating materials are processed by different methods in accordance with the field of application. Even if brush application plays only a subordinate role in industrial scale paint processing, this application method is still extremely important in the housepaint and DIY fields. Brushes and rollers are important tools for painting doors, windows, facades and similar surfaces. This is revealed by the global breakdown of types of application (see Figure 2.3.61).

For the housepainter, therefore, an analysis of the criteria for good brushability is just as important as the analysis of the variables for sprayability and ease of dip coating in the industrial sector. The different methods of spraying and dipping must be viewed separately. For example, coating materials for conventional spraying must be adjusted and evaluated differently from materials which are to be processed hydraulically, pneumatically or electrostatically. Electrocoating requires different paints from conventional dip coating with solvent-borne or water-borne coating materials.

En route from the paint to the finished coating different material settings have to be drawn up for the coating materials, despite the same quality demands imposed on the coating. In view of the variety of objects to be painted and the resulting different application methods, this is the only way to meet the set target.

The processability test is intended to ensure that the coating material can be processed without problems. In addition to the surface finish of the material under analysis, a range of external conditions exert a great influence on processability. Particular mention may be made of

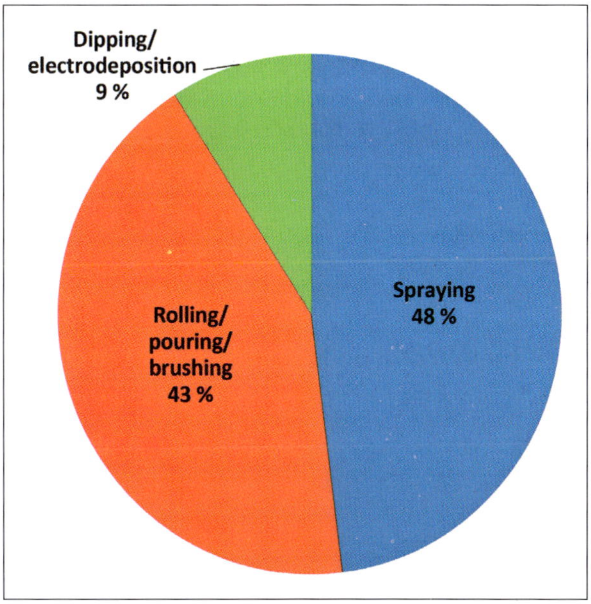

Figure 2.3.61: Global application of paints Source: authors' esiimate

the substrate, the adjustment of the coating material to the processing form; the processing sequence; in the case of manual processing the individual working method; for automatic processing, the setting and condition of the application equipment; and the environmental conditions (temperature, air pressure, relative humidity).

For all paints and all application methods the relevant variables are the viscosity profile as a function of temperature and shear stress, surface tension, density, evaporation properties of the solvents and, from case to case, other specific properties.

An important indicator for the processability of coating materials is the shelf life and the associated time of application. Once the shelf life has expired the application parameters need to be reviewed. However, particularly in the case of water-borne products, the application parameters may fall out of specification if, e.g. the low-solvent, heavy-metal-free products are attacked by bacteria. Such attack is not always visible and very quickly leads to processing and film forming problems but can be tested for [2.4.141].

Brushability

Brushability is the term used to designate a material profile which enables optimum processing by brushing. Thixotropy in conjunction with an appropriately "long" solvent setting is a key property. Only in this way is it possible to level out the applied coating still further by applying another coat. The solvent composition of a brush-applied paint therefore differs substantially from that of a spray paint. The question of the substrate also has to be addressed if good brushability is to be achieved. When absorbent timber is painted, absorption by the substrate can be reduced by choosing a high processing viscosity and a correspondingly high degree of thixotropy. Other features used in evaluating brushability include levelling, sagging, drying properties, and hiding power. These properties are most usefully tested by means of special wedge-shaped applicators. The evaluation of brushability depends very much on subjective variables and so in practical terms is only a relative variable that references a standard which also has to be tested.

Brush-applied paints are subject to high shear stress. The resulting viscosity at any given time determines the application work involved. If, for example, shear gradients of 10^4 s^{-1} are specified, reliable forecasts of brushability can be obtained. An additional brush-applied coat is then superfluous. Simple, inexpensive cone-and-plate viscometers have proved suitable for such forecasts.

Ease of Dipping

The requirements of coating materials are completely different if a dip-application process is to be used. Rigorous demands are imposed on material stability that depend on bath turnover, quantity of components to be coated per unit time and bath size.

Good ease of dipping also means achieving optimum wetting of all areas of the object to be painted in the shortest possible time. A low viscosity is desirable for ensuring that the cavities of complex objects are also reached by paint. At the same time, it is also a precondition for rapid drying of the excess material after removal from the bath. In addition, fast drying prevents the paint from dewetting from the edges. In order to achieve a good compromise between flow and the avoidance of dewetting from edges, the aim must be to strike a balance between viscosity, evaporation properties and surface tension.

Quite different criteria have to be applied in turn when evaluating the ease of dipping of paints for the electrocoating process. The electrochemical deposition equivalent, pH value, degree of neutralisation, electrical conductivity and throwing power, to mention only some of the numerous indicators, are data which need to be precisely determined. They permit more binding forecasts of the anticipated coating result than can be achieved, for example, for manual processing by brush or roller. When pigmented dip-coating paints are processed, pigments with a high specific mass should not be used, regardless of whether or not a conventional or electrochemical method is being used. Lasting homogeneity of the bath material can only be guaranteed and costs kept to a reasonable level by using light to medium-weight pigments. A low tendency towards sedimentation is thus also an important criterion for dip-coating paints. Regular testing of the pigment/ binder ratio and monitoring of the solvent components and their composition are further important criteria for constant processing properties.

Sprayability

Account must be taken of the atomisation methods when evaluating sprayability. Spraying with compressed air with and without supplementary heating of the material, airless atomisation at extreme pressure, electrostatic assistance or the high speed rotary method are based on very different physical principles. One feature they all have in common is that the liquid paint is broken down into more or less small droplets by generating free surface energy and that these droplets are then given kinetic energy to transport them to the object to be painted where they form an intact film in contact with the object. The fundamental difference from the dipping and brushing processes lies in the interim atomisation of the paint. The brief but significant enlargement of the total surface area of the material to be processed and the associated faster solvent release must be taken into account when the formulation for this application method is being drawn up and tested.

For hot application, generally carried out at 60 to 80 °C, the material must be adapted to the changed conditions by appropriate solvent selection. Checking that the formulation is right to ensure applicability is then carried out at these temperatures.

Because of the very small nozzle diameter used for airless atomisation (0.1 to 0.8 mm) special demands are also made of the fineness of grind. At the same time hard and thus abrasive extenders should not be used because of the extremely high tangential speed of the paint of 150 to 200 m/s. Only in this way is it possible to limit nozzle wear.

Very different factors come to the fore in the formulation of paints for electrostatic processing. The electrical conductivity very much determines the degree of charging and therefore determines the atomisation. Physical indicators which influence the spray angle and the droplet size and therefore also the yield and quality of the coating are the surface tension and the dielectric constant.

At this stage it is apparent that in all the important processing stages for liquid paints, whether for spray, brush or dip application, material related properties such as flow properties, wettability, surface tension, evaporation properties and chemical reactivity are interlinked in such a complex fashion that it is impossible to make forecasts about the processability of paints based purely on theoretical considerations. Predictions relating to the properties of paints with a specified processing system are therefore only ventured after extensive simulations and experiments. The more accurately the practical conditions can be simulated, the more authoritative the comment.

Naturally other factors play a role in the processability of powder coatings. These are primarily the particle size and their distribution [2.4.142]. Both parameters are determined not only by the manufacturing process but also by recycling of overspray (see Chapter 5.6.1.3).

2.3.5 Summary

Coating materials and their raw materials are subjected to various basic mechanical and thermal process engineering operations during manufacture and processing. This places certain demands on the individual components and on the coating material. The physical properties of viscosity and surface tension are without doubt the most important fundamental variables as regards conveying in pipes, dispersing the pigments, and application and subsequent film formation. The solid content, density and hiding power then round out the material characterisation.

A complete evaluation of the property profile of coating materials also includes occupational health and safety indicators and systems for measuring them. The flash point, solvent content in the emissions from spray booths and driers, ignition temperatures of explosive gas/air mixtures, as well as the concentration of hazardous work materials must be known if coating plant is to be operated safely.

Where pigmented coating materials are to be evaluated, attention must be paid to specific pigment properties of the raw material both in isolation and in its interaction with the film forming agent. Particle size, particle size distribution, oil absorption value, bulk density, moisture content, water-soluble components and agglomerate strength on one hand and the reduction in fineness of grind, development of tinting strength, as well as stability of the dispersion achieved on the other are just some of the noteworthy properties of coating materials.

In order to provide a comprehensive quality guarantee, in-depth application tests also have to be carried out, including processability, sprayability and the specific properties during pouring or rolling.

If, in addition to identifying the properties, there is a need to comment on the chemical composition, recourse must also be made to the varied methods of instrumental analysis.

2.4 Literature

References

2.4.1 B. Kamm, M. Gerhardt, S. Leiß, Nachr. Chemie 57 (12), p. 1182 (2009)
2.4.2 Mikusch, Öle, Fette, Wachse, 5, p. 113 (1956)
2.4.3 W. J. Muizebelt et al. J. Coatings Techn. 70, 876, p. 83 (1998)
2.4.4 E. Schwenk, 80 Jahre vergessene Erfinder, Firmenschrift der Hoechst AG, Frankfurt Main, p. 4ff
2.4.5 H. G. Elias, Makromoleküle, Vol. 1, Hüthig und Wepf Verlag, Basel, p. 5–12 (1989)
2.4.6 M. D. Lechner, K. Gehrke u. E. H. Nordmeier, Makromolekulare Chemie, Birkhäuser Verlag Basel (1993)
2.4.7 B. Vollmert, Grundriss der Makromolekularen Chemie, E. Vollmert Verlag, Karlsruhe (1988)
2.4.8 M. Dyck, F+L, 71 (11), p. 902 (1965)
2.4.9 H. Neises, F+L, 77 (9), p. 850 (1971)
2.4.10 D. Maetens et al., ECJ 05/1999, p. 26

2.4.11 B. Vollmert, Grundriss der Makromolekularen Chemie, E. Vollmert Verlag, Karlsruhe (1988)
2.4.12 DP 748 253 (1943) IG Farben, P. Schlack
2.4.13 AP 2380995 (1941) General Electric, E. G. Rochow
2.4.14 A. v. Bayer, Ber. Dtsch. Chem. Ges. 5, 25, 280, p. 1094 (1872)
2.4.15 Kunststoff Handbuch, Vol. 10, Hanser Verlag, Munich (1968)
2.4.16 O. Bayer, Angew. Chemie 59, p. 257 (1947); U. Meier-Westhues; Polyurethane, p. 14, Vincentz, Hanover, 2007
2.4.17 Kittel – Lehrbuch der Lacke und Beschichtungen (2nd Edn.), Vol. 2, Bindemittel für lösemittelhaltige und -freie Systeme (Hrsg. W. Krauß), Hirzel, Stuttgart, 2001; U. Meier-Westhues; Polyurethane, p. 25, Vincentz, Hanover, 2007
2.4.18 S. Milmo, ECJ Special Edition 7-8/2007, p. 2
2.4.19 Kittel – Lehrbuch der Lacke und Beschichtungen (2nd Edn.), Vol. 3, Bindemittel für wasserverdünnbare Systeme (Hrsg. K. D. Ledwoch), Hirzel, Stuttgart, 2003
2.4.20 W. Grenda, J.-V. Weiß, F+L 106 (6), p. 97 (2000)
2.4.21 H.-P. Rink, Automotive Paints and Coatings, p. 235, (ed.. Streitberger/Dössel), Wiley-VCH, Weinheim (2008); W. Zhao, PCI 26(11), p. 36 (2010); S. Friebel, C. Philipp, F+L 116(10), p. 25 (2010)
2.4.22 G. Costa, I. Arnici-Kroutilova, M. de Luca, D. Traverso, E. Sironi, J. Furgione, F. Frederici, ECJ 07-08/2011, p. 26
2.4.23 R. S. Stewart, R. Subramanian, JCT Coat. Tech. 6 (10), p. 26 (2009)
2.4.24 R. Harrer, A. Arzt, M. Schafheutle, F+L 118 (3), p. 26 (2012)
2.4.25 M. Shaffer, R. Stewart, K. Allen, A. Wylie, A. Lockhart, JCT Coat. Tech. 6(1), p. 50 (2009)
2.4.26 P. Barbeau, R. Klucker, J.-L. Loubet, S. Pavan, PCI 27 (2), p. 34 (2011)
2.4.27 L. Kahle et al., F+L 102 (3), p. 88 (1996); M. Shaffer, R. Stewart, K. Allen, A. Wylie, A. Lockhart, JCT Coat. Tech. 6 (1), p. 50 (2009)
2.4.28 W. Kubitzka, F+L 97 (3), p. 201 (1991); P. Barbeau, R. Klucker, J.-L. Loubet, S. Pavan, PCI 27 (29), p. 34 (2011)
2.4.29 H.-P. Rink, Automotive Paints and Coatings, p. 248, (ed. Streitberger/Dössel), Wiley-VCH, Weinheim (2007)
2.4.30 P. Castan, CH 211 116 (1938), DE 749512 (1938)
2.4.31 R. Schuler, F+L 118 (3), p. 40 (2012)
2.4.32 H. Batzer, S. A. Zahir, J. Appl. Polym. Sci 19, 609 (1975)
2.4.33 S. Monoghan, M. Lohe, M. Cook, ECJ 11/2005, p.20
2.4.34 US-PS 3947339 (1975) PPG Inc.
2.4.35 EP 69917792 T2 (2005)
2.4.36 EP 102005029792 A1 (2006)
2.4.37 R. Grasböck, M. Geisberger, F+L 113(12), p.36 (2007)
2.4.38 T. Hermann, JOT Spezial 2009, p. 14 (2009); Anonymous, MO 62 (4), p. 22 (2008); Anonymous, JOT 50 (9), p. 70 (2010)
2.4.39 Kittel, Lehrbuch der Lacke und Beschichtungen (2nd Edn.), Vol. 3, Bindemittel für wasserverdünnbare Systeme (ed. K. D. Ledwoch), Hirzel, Stuttgart, 2003
2.4.40 R. Kita, A. Kuni, J. Paint Techn. 48, p. 55 (1976)
2.4.41 H. Fikentscher, Angew. Chemie 51, p. 433 (1938);H. Gerrens, Fortschr. Hocholym. Forschung 1, p. 234 (1959)
2.4.42 US-PS 3799854 (1970) PPG Inc.
2.4.43 M. Durali, K. Wood, D. Battorcchi, M. Hatzenbeller, ECJ 10/2010, p.30
2.444 E. Beck, ECJ 04/2006, p. 32
2.4.45 H. Friedrich, Welt d. Farben 5 (5), p. 8 (2006)
2.4.46 B. Pettersson, F+L 107 (9), p. 123 (2001)
2.4.47 C. M. Hansen, J. Paint Techn. 39, p. 104 (1967); J. Schröder, F+L 104 (3), p. 26 (1998)
2.4.48 K. F. Dössel, JOT 48 (4), p. 6 (2008)
2.4.49 H. Teleng-van Oorschot, PPCJ 202 (4570), p. 22 (2012)
2.4.50 Anonymous, ECJ 01/2013, p. 14
2.4.51 R. Rosin, E. Rammler, K. Sperling, Berichte des Reichskohlenrates C52, Berlin, VDI-Verlag (1933) und I. G. Bennet, J. Inst. Fuel 10, 10 (1936)
2.4.52 Römpp, Lexikon der Lacke und Druckfarben, p. 117, Thieme Verlag, 1997
2.4.53 S. Fitzgerald, PPCJ 202(4576), p. 34 (2012)
2.4.54 M. Diebold, R. Kwoka, D. Mukoda, JCT Coat. Tech. 10(2), p. 30 (2013)
2.4.55 H. Ferch, Pigmente, p. 46ff, Vincentz Verlag, Hanover, 1995
2.4.56 M. Ryan, ECJ 03/2013, p. 74
2.4.57 Kittel – Lehrbuch der Lacke und Beschichtungen (2nd Edn.), Vol. 5, Pigmente, Füllstoffe, Farbmetrik (ed. J. Spille), Hirzel, Stuttgart, 2003
2.4.58 Kittel – Lehrbuch der Lacke und Beschichtungen (2nd Edn.), Vol. 2, Bindemittel für lösemittelhaltige und -frei Systeme (ed. W. Krauß), Hirzel, Stuttgart, 2001
2.4.59 P. Kempter, G. Wilker, F+L 107 (11), p. 29ff (2001)
2.4.60 R. Besold, Pigmente und Farbstoffe für die Lackindustrie, Expert Verlag, Ehningen, 1988

2.4.61 B. Müller, C. Oughourlian, D. Triantafillidis, JCT 73(917), p. 81 (2001)
2.4.62 A .Kreutzer, W. Sauseng, PPCJ 202(4577), p. 18 (2012)
2.4.63 W. Liu, J.B. Carroll, JCT Coat. Tech. 3(10), p. 82 (2006)
2.4.64 G. Pfaff, A. Huber, Welt d. Farben 09/2005, p. 14
2.4.65 F. Meyer, F+L, 104 (8), p. 28 (1998);R. Schmid, N. Mronga, V. Radtke, O. Seeger, ECJ 7–8/97, p. 702; S. R. Sharrock, N. Schuel, ECJ 1–2/00, p. 20
2.4.66 Handbook of Coatings Additives, p. 150, J. Calbo (Ed.), Marcel Dekker Verlag, N.Y., 1987; Anonymous, Welt d. Farben 11, p. 12 (1997)
2.4.67 H. Semmler, Phänomen Farbe 20 (6), p. 20 (2000)
2.4.68 S. Schellenberger, J. Pahnke, R. Friedrich, G. Jonschker, F+L 119 (3), p. 92 (2013)
2.4.69 NL 7613121 (1976) Akzo
2.4.70 W. Huff, PCI 27(2), p. 20 (2011)
2.4.71 H. Reimann et al., F+L 108 (5), p. 44 (2002); H. Wehrens, M. Paczkowski, PCI 23 (4), p. 50 (2007)
2.4.72 S. Narayan, M. Moore, PPCJ 202(4572), p. 26 (2012)
2.4.73 V. Thewes, Kittel – Lehrbuch der Lacke und Beschichtungen, Vol. 6, 2nd Edn. (ed. H.-J. Streitberger), p. 504, Hirzel, Stuttgart, 2008
2.4.74 E. T. Denisov, Y. B. Shilov, Vysokomol. Soyed. A 16, p. 2313 (1974)
2.4.75 H. Zitt, W. Anker, W. Meyer, F+L 108 (6), p. 22 (2002)
2.4.76 B. Kossmann, P. Ottersbach, Phänomen Farbe 22 (2), p. 19 (2002)
2.4.77 V. M. Rowe, PPCJ 202(4571), p. 35 (2012)
2.4.78 G. Jonschker, Praxis der Sol-Gel-Technologie, p. 13, Vincentz, Hanover, 2012; M. Veith, M. Kölbel, JOT 47(7), p. 16 (2007)
2.4.79 G. Jonschker, Praxis der Sol-Gel-Technologie, p. 95, Vincentz, Hanover, 2012
2.4.80 EP 1204701, PPG (2005); EP 2102263B1, BASF (2008)
2.4.81 R. Alezzi, R. Ou, K. Eberts, D. Eberly, S.-P. Lee, PPCJ 202 (4568), p. 41 (2012)
2.4.82 D. Vollmer, L. Mammen, X. Deng, H.-J. Butt, F+L 117 (11), p. 26 (2011)
2.4.83 R. E. Schneider, MO 66 (10), p. 22 (2012)
2.4.84 C. de Lame, J.-M. Claeys, PPCJ 203 (4580), p. 47 (2013)
2.4.85 T. Sawitowski, W. Wessels, Welt d. Farben 3/2006, p. 10
2.4.86 H. Gehmecker, M. Hueber, Oxsilan – Innovative Pretreatment Technology for Multi-Metal Applications, Proceedings Surcar, Barcelona, 2005; J. Rudolph, R. Moyle, JOT 48 (4), p. 16 (2008)
2.4.87 D. Oertle, JOT 49 (12), p. 36 (2009)
2.4.88 M. Vorbau, L. Hillemann, P. Fiala, M. Stintz, A. Rommert, D. Eichstädt, ECJ 02/2011, p. 29
2.4.89 J. A. Gonzalez-Gomez, F+L 111 (4), p. 158 (2005)
2.4.90 F. Lang, F+L 116 (7), p. 18 (2010)
2.4.91 A. Freytag, X. Lien, J. Hajas, F+L 116 (5), p. 25 (2010)
2.4.92 B. Stalder, F+L 113 (12), p. 22 (2007)
2.4.93 H. H. Weber, Kolloid-Zeitschrift, Vol. 174 (1), p. 66 (1961)
2.4.94 J. U. Zilles, Welt der Farben 6, p. 18 (2001); K. Schulz, E. Reuter, P. Glöckner, F+L 116 (6), p. 21 (2010)
2.4.95 K. M. Oesterle et.al, Proceedings XII. Fatipec-Kongress, p. 255 (1974)
2.4.96 U. Kaluza, Physikalisch chemische Grundlagen der Pigmentverarbeitung für Lacke und Druckfarben, Edition Lack+Chemie, Filderstadt, 1980
2.4.97 B. V. Deryaguin, L. D. Landau, Acta Phisiochim URSS 114, 633 (1941)
2.4.98 E. J. W. Verwey, J. G. Overbeek, Theory of the Stability of Liphobic Colloids, Elsevier, Amsterdam, 1948
2.4.99 G. Brezesinski, H. J. Mögel, Grenzflächen und Kolloide, p. 76ff, Spektrum Verlag Heidelberg, 1993
2.4.100 G. Brezesinski, H. J. Mögel, Grenzflächen und Kolloide, p. 161–163, Spektrum Verlag Heidelberg, 1993; L. Frischmann, F+L 112 (1), p. 32 (2006)
2.4.101 E. L. Mackor, J. Colloid Sci. 6, p. 492 (1951)
2.4.102 T. Sato, J. Coat. Techn. 65, 825, p. 113 (1993)
2.4.103 E. W. Fischer, Kolloid-Zeitschrift 160, p. 120 (1958)
2.4.104 C. D. Eisenbach, Ch. Schaller, F+L 107 (11), p. 58 (2001)
2.4.105 W. Heller, T. L. Pugh, J. Polymer Sci. 47, p. 203 (1960)
2.4.106 Vauck, Müller, Grundoperationen der chemischen Verfahrenstechnik, p. 349, VEB Deutscher Verlag für die Grundstoffindustrie Leipzig, 1962
2.4.107 J. Sneeringer, Modernpaintandcoatings 92, p. 30 (2002)
2.4.108 F. Vock, K. Warnke, K. Sollik, F+L 93 (12), p. 995 (1987)
2.4.109 F. Vock, Lackherstellung, p. 272ff, CC Press AG, Thermen, 1977
2.4.110 W. Hemming, Verfahrenstechnik kurz und bündig, p. 112, Vogel Verlag, Würzburg 1975
2.4.111 WO 1994009913 A (1994) Ferro Corp.
2.4.112 R. Mattern, F+L 105 (11), p. 52 (1999)
2.4.113 L. Gall, F+L 81 (11), p. 1015 (1975)

2.4.114 DIN-Taschenbuch 399, Ausgabe 2007-08-01, Beuth Verlag, Berlin
2.4.115 B. Böhnke, Abwassertechnologie, Springer Verlag, Berlin, 1993
2.4.116 Feßmann, Orth, Angew. Chemie und Umwelttechnik für Ingenieure, Ecomed Verlag, Landsberg, 1999
2.4.117 T. Scherzer, S. Müller, R. Mehnert, A. Volland, JCT Coat. Tech. 3 (8), p. 30 (2006)
2.4.118 J. Rudolph, CHIUZ, p. 77 (1967)
2.4.119 Stoye, Freitag, Lackharze, p. 420, Hanser Verlag, Munich, 1996
2.4.120 Stoye, Freitag, Lackharze, p. 422, Hanser Verlag, Munich, 1996
2.4.121 P. Clarke, PCI 27 (2), p. 26 (2011)
2.4.122 H. G. Elias, Makromoleküle, Vol. 1, p. 85ff , Hüthig, Wepf Verlag, Basel, 1990
2.4.123 W. Schlesing, M. Buhk, M. Osterhold, Progr. Org. Coat. 49, p. 197 (2004)
2.4.124 T. Mattke, ECJ 09/2000, p. 83
2.4.125 M. Hediger, Firmenschrift Contraves, Bulletin 6704–652
2.4.126 W. Göring, N. Dingerdissen, C. Hartmann, F+L 89 (4), p. 270 (1977)
2.4.127 J. L. Cayias, R. S. Schechter und W. H. Wade, ACS Symposium Ser. 8, p. 234 (1975)
2.4.128 R. Megele, M. D. Evans, Ind. Eng.Chem. 43 (6), p. 1317 (1951)
2.4.129 G. Brezesinski, H. J. Mögel, Grenzflächen und Kolloide, p. 146, Spektrum Verlag Heidelberg, 1993
2.4.130 R. Pecora, Dynamic Light Scattering, p. 181, Plenum Pr., N.Y., 1985
2.4.131 G. Brezesinski, H. J. Mögel, Grenzflächen und Kolloide, p. 124, Spektrum Verlag, Heidelberg, 1993
2.4.132 Bergmann, Schäfer, Lehrbuch der Experimentalphysik, Vol. 3, p. 388ff, Walter de Gruyter & Co, Berlin, 1993
2.4.133 M. Born, E. Wolf, Principles of Optics – Electromagnetic Theory of Propagation, Interference and Diffraction of Light, p. 446ff, Pergamon Press, London, 1987
2.4.134 R. H. Müller, R. Schumann, Teilchengrößenmessung in der Laborpraxis, Vol. 38, p. 55, Wissenschaftl. Verlagsgesellschaft mbH, Stuttgart, 1996
2.4.135 Reimer, Pfefferkorn, Rasterelektronenmikroskopie, p. 237ff, Springer Verlag, Berlin, 1977
2.4.136 G. Wedler, Adsorption, p. 47ff, Verlag Chemie, Weinheim, 1970
2.4.137 A. Joch, W.-A. Jung, J. Lohmann, F+L 111 (12), p. 26 (2005)
2.4.138 E. Reck, Welt d. Farben, 03/2001, p. 14
2.4.139 U. Zorll, F+L, 80 (1), p. 17 (1974)
2.4.140 Kittel, Lehrbuch der Lacke und Beschichtungen (2nd Edn.), Vol. 8, ed. W. Ritter, W. Zöllner, Hirzel Verlag, Stuttgart, 2004
2.4.141 J. Coulburn, PPCJ 202 (4572), p. 16 (2012)
2.4.142 S. Bryan, PPCJ 202 (4574), p. 27 (2012)

General Literature

H.-J. Adler, K. Potje-Kamloth (ed.), Quo Vadis – Coatings?, XXVI. Fatipec-Kongress, Dresden, Wiley-VCH Verlagsges., Weinheim, 2002
H. Coyard, P. Deligny, N. Tuck, Resins for Surface Coatings, Vol. 1: Acrylics and Epoxies, Wiley-VCH, Weinheim, 2001
U. Poth, R. Schwalm, M. Schwarz, Acrylic Resins, Vincentz, Hanover, 2011
I. D. Morrison, R. Ross, Colloidal Dispersions, John Wiley, New York, 2002
J. Goodwin (ed.), Colloids and Interfaces with Surfactants and Polymers, Wiley, New York, 2009
T. Brock, M. Groteklaes, P. Mischke, European Coatings Handbook (2nd Edn.), Vincentz, Hanover, 2010
H.-J. Streitberger, K.F. Dössel (ed.), Automotive Paints and Coatings, Wiley-VCH, Weinheim, 2008
J. Winkler, Dispersing Pigments and Fillers, Vincentz, Hanover, 2012
T. F. Tadros, Emulsion Science and Technology, Wiley-VCH, Weinheim, 2009
D. Gysau, Fillers for Paints, Vincentz, Hanover, 2017
Kittel – Lehrbuch der Lacke und Beschichtungen (2nd Edn.), Vol. 2, Bindemittel für lösemittelhaltige und -frei Systeme (ed. W. Krauß), Hirzel, Stuttgart, 2001
Kittel – Lehrbuch der Lacke und Beschichtungen (2nd Edn.), Vol. 3, Bindemittel für wasserverdünnbare Systeme (ed. K. D. Ledwoch), Hirzel, Stuttgart, 2003
Kittel – Lehrbuch der Lacke und Beschichtungen (2nd Edn.), Vol. 5, Pigmente, Füllstoffe und Farbmetrik (ed. J. Spille), Hirzel, Stuttgart 2003
J. J. Florio, L. J. Calbo, D. J. Miller (eds.), Handbook of Coating Additives(2nd ed.), Wiley-VCH, Weinheim, 2007
B. Bushan (ed.), Handbook of Nanotechnology, Springer, Berlin, 2004
A. Öchsner, W. Ahmed, N. Ali (Eds.), Nanocomposite Coatings and Nanocomposit Materials, Trans Tech Publications, Zürich, 2009
Z. W. Wicks, F. N. Jones, S. P. Pappas, D. A. Wicks, Organic Coatings (3rd Edn.), Wiley&Sons, Hoboken, 2007
D. Stoye, W. Freitag (eds.), Paints, Coatings and Solvents, Wiley-VCH, Weinheim, 1998
U. Meier-Westhues, Polyurethanes, Vincentz, Hanover, 2007
G. Pfaff (ed.), Spezial Effect Pigments (2nd Edn.), Vincentz, Hanover, 2008

J. Winkler, Titanium Dioxide, Vincentz, (2nd Edn.), Hanover, 2013
C.-S. Chern, Principles and Applications of Emulsion Polymerisation, Wiley, New York, 2008
W. Heilen, S. Herrwerth, Silicone Resins and their Combinations, (2nd Ed.), Vincentz, Hanover, 2015
P. Nanetti, Lackrohstoffkunde (4th Ed.), Vincentz, Hanover, 2013
R. H. Fernando, L.-P. Sing (eds.), Nanotechnology Application in Coatings, ACS Symposium Series 1008, ACS, 2006
B. Müller, Understanding Additives, Vincentz, Hanover, 2010
U. Poth, Automotive Coatings Formulation, Vincentz, Hanover, 2008
T. F. Tadros (ed.), Colloids in Paints (Vol. 6), Wiley-VCH, Weinheim, 2010
T. F. Tadros (ed.), Emulsion Science and Technology, Wiley-VCH, Weinheim, 2009
S. Sepeur (ed.), Nanotechnology, Vincentz, Hanover, 2008
U. Poth, Polyester und Alkydharze, Vincentz, Hanover, 2005
R. T. Mathers, M. A. R. Meier (ed.), Green Polymerization Methods, Wiley-VCH, Weinheim, 2011
U. Schumacher, Th. Feist, D. Lehmann, Das Lernbuch der Lackiertechnik, 4th Edn., Vincentz Network, Hanover, 2013

3 Coatings

3.1 From Coating Material to Coating: Film Formation

Coating material, i.e. paint, is a semi-finished product. Only by coordinating the application process, the subsequent curing, the design requirements and various materials used in the item being coated can the surface properties expected of the coating be achieved. However, there is no doubt that, regardless of how different the demands may be, coating materials will always have to be transformed into a continuous, smooth film with good adhesion in order that they can meet the requirements made of them in the long term after the drying and curing processes.

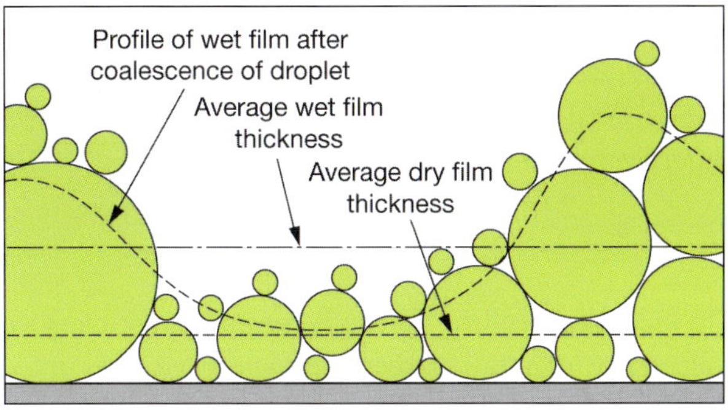

Figure 3.1.1: Schematic texture of a film freshly applied by spray coating as a starting point for film formation

The rough texture of droplets in a fresh coating which has been created by spray application, for example, must be sufficiently flowable to seal any remaining voids and level out the application-related peaks and valleys. In addition to viscosity, the degree of levelling and the rate of levelling are essentially dependent on surface tension and wet film thickness. The temperature characteristic and the rheological behaviour affected by it, as well as evaporation of the solvents, overlie the physical and chemical processes involved in the transition from coating material to coating. Evaporation of solvents, and formation of cleavage products in the case of polycondensation curing systems, give rise additionally to film shrinkage. Reproduction of the substrate surface is therefore a function of not only the solid content but also the quantity of cleavage products and the reactivity of the system. If local changes in density and in surface tension also occur during solvent flash-off, flows can also be expected within the film as it forms.

Owing to the importance of the film forming process to the desired properties of the coating we ought to provide a detailed description of the metamorphosis from coating material to coating.

3.1.1 Wetting and Levelling

Wetting of the substrate is subject to the laws described in Chapter 2.1.4. The liquid coating material or the melting powder paint wets the substrate if the surface tension of the coating material is lower than that of the substrate. The individual droplets created during spray application unite after wetting and, after release of enclosed bubbles of gas, form a continuous film, provided that the viscosity is sufficiently low.

Brush application or various sizes of droplet created by atomisation with the spray gun yields a wet film of pronounced unevenness in the first stage. After wetting, an integral part of film formation is the levelling, smoothing and solidification of surface textures. Eliminating the wet film contour and also preventing sagging on vertical surfaces thus constitute the major task in the second stage of the film forming process.

Although levelling is facilitated by low viscosity, the actual driving force required to create a smooth surface has its origin in surface tension and gravity. Whereas surface tension works to reduce the size of surfaces on both vertical and horizontal surfaces, gravity only enhances levelling on horizontal surfaces. On vertical surfaces it can cause runs. Although the highest-possible surface tension is desirable for levelling, it causes increased dewetting of the still liquid paint away from the edges of the article on account of cohesive pressure p.

$$p = \frac{2 \cdot \sigma}{r} \qquad \sigma = \text{surface tension} \quad r = \text{radius of edge}$$

To facilitate levelling and at the same time prevent sagging and dewetting from edges, it is necessary to factor another parameter, viscosity, into the considerations. Levelling without sagging can be achieved only by striking the right balance of viscosity, with due allowance for viscous or thixotropic behaviour, and surface tension at all stages of film formation.

Detailed investigations into fresh films with defined primary textures were conducted by *Orchard* [3.4.1]. Surface textures made under defined conditions with an initial height h_0 lose height as a function of time t, whilst the minimums increase in the same manner. Levelling down to height h as a function of time follows an exponential function.

$$h = h_0 \cdot e^{-\frac{t}{\tau}} \qquad \text{thereof} \qquad \tau = k \cdot \frac{\eta}{\sigma} \cdot \frac{\lambda^4}{s^3}$$

The initial fast levelling of the peaks and valleys to form a smooth film decelerates as time progresses. Information about the absolute rate is provided by parameter τ in the exponent of the function. The goal of the experiment is to prove that in addition to the material properties mentioned the levelling process is also influenced by application-related variables. The levelling-determining value τ is composed of the material variables of viscosity η, surface tension σ and process-related variables: the distance between the excess heights of primary roughness λ and wet film thickness s. It is worth noting that this influence is much greater than that of material parameters. Bearing in mind that the parameter λ in the case of spray application is determined by the droplet size or the droplet size distribution, the marked dependence of levelling on droplet size and film thickness, which will be familiar to experienced paint formulators, is substantiated by theory.

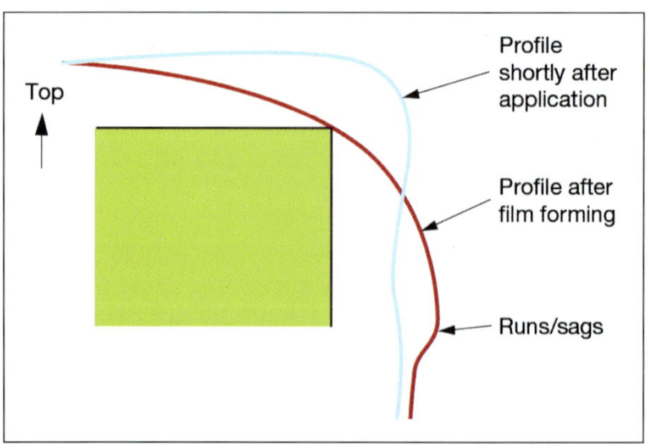

Figure 3.1.2: Dewetting from edges

3.1.2 Solidification of the Film

The essential stages in the transition from a liquid paint to a physically firm coating occur during the fundamental changes in rheological properties. The originally thin paint solidifies more and more and passes through various intermediate stages before it ultimately turns into an elastic body.

The transformation from coating material to coating can be described by the change in viscosity using the following mathematical relationship for resin solutions [3.4.2].

$$\lg \eta = k \cdot \sqrt{M_n} \cdot NV$$

The equation can be used to describe with sufficient accuracy the film forming processes that occur in high-viscosity solutions and melts. Accordingly, viscosity η depends on molecular weight M_n and the solid content NV. After a coating material has been applied, there are two ways in which the viscosity can be boosted to the extent that it acquires the physical strength needed for film formation. First of all, evaporation of the solvents brings about an increase in the concentration of film forming agent. The molecular chains approach each other, increasing the mutual interaction and physical entanglement, and hence the viscosity. If reactive resins are present, chemical curing reactions overlie physical drying. These increase the average molecular weight through crosslinking of the molecular chains. Solidification is therefore the outcome of the combined influences of physical and chemical processes.

Since crosslinking of the resin molecules, other than to raise the viscosity, ultimately has to render the film resistant to solvents and weathering, in the chart shown in Figure 3.1.3 the range of physical hardness is supplemented with that of chemical resistance.

The two fundamentally different formulation principles become easier to understand by tracing the transition from low molecular monomer via paint resin through to finished coating and all the intermediate stages for the equation above (see Figure 3.1.4). The monomeric basic building blocks for manufacturing the resin enter into chemical reactions that give rise to so much molecular enlargement that the substance solidifies (1 → 2). For application to a surface this raw material has to be liquefied again by adding solvents that will convert it into a suitable form. These solvents are released again after application and during film formation. Physical drying therefore means a vertical displacement of the coordinates into the range of physical hardness. No change in molecular weight occurs (2 → 3, top).

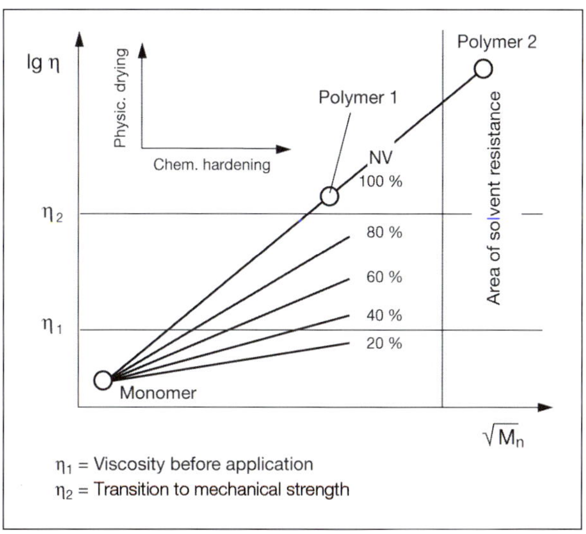

Figure 3.1.3: Dependence of viscosity on molecular weight Mn and solid content (nonvolatiles) in film formation

The situation is different for film forming agents intended for chemical curing. In this case the

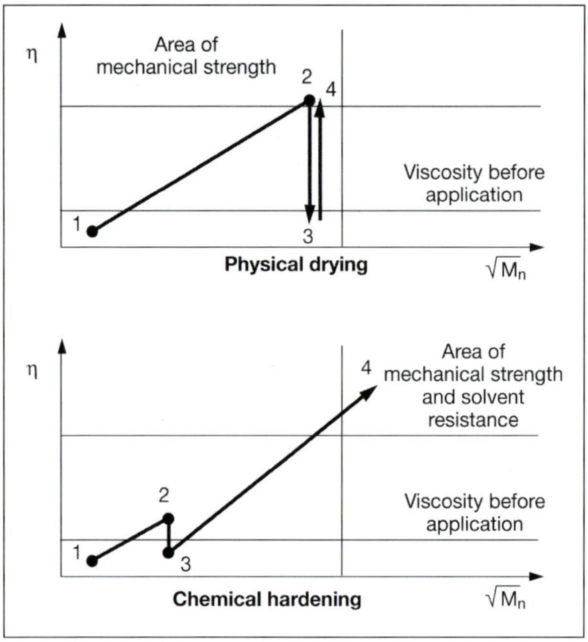

Figure 3.1.4: Principles behind film formation: (A) physical drying and (B) chemical hardening

starting point is assumed to be small resin molecules which are not yet physically stable. Molecular enlargement, which only starts to takes place after application, and an associated increase in molecular weight combined with crosslinking, cause horizontal displacement as shown (3 → 4, bottom). If the coating has to be resistant to solvents and aggressive chemicals, crosslinking of the growing film-forming molecules must yield a dense network.

With the aid of Figure 3.1.4 it is possible to characterise all the paint systems schematically in terms of their **film forming behaviour. Physically drying paints** (A), that is, those which do not undergo any chemical changes during film formation, contain film forming agents of relatively high molecular weight. Owing to the absence of crosslinking these types of paints offer limited solvent resistance. Another disadvantage of physically drying paint systems, also due to the high molecular weight of the film forming agents, consists in a relatively high proportion of solvents in the application state. Physically drying paints therefore always have a low solid content. Due to the associated high emissions during application, physically drying paints containing dissolved film forming agents do not have a great future for ecological reasons. The only advantage which can be attributed to them is that they dry at room temperature. Examples of physically drying paints are thermoplastic acrylates, cellulose derivatives such as cellulose nitrate, cellulose aceto-butyrate, chlorinated rubber and cyclo rubber.

The situation is different for **chemically hardening paints** (B). Their film forming agents have lower molecular weights and so they have a much higher solid content in the application state owing to their better solubility. The lower level of solidification by means of physical drying is more than compensated for, however, by chemical processes during the film forming phase. The molecular weight, and hence the viscosity, can be raised by any amount and the resulting film can be transformed by chemical crosslinking into an insoluble, thermosetting, high molecular network which cannot be melted. Thus, it should be clear that chemically curing reactive paints without exception yield coatings of maximum physical and chemical stability. These advantages come at the cost of extra capital investment needed in the curing process. The attainable quality of coating justifies this approach in many cases, however.

3.1.2.1 Physical Drying

In the case of physical film forming processes a distinction needs to be made between solidification of the polymer due to solvent emissions from polymeric solutions and

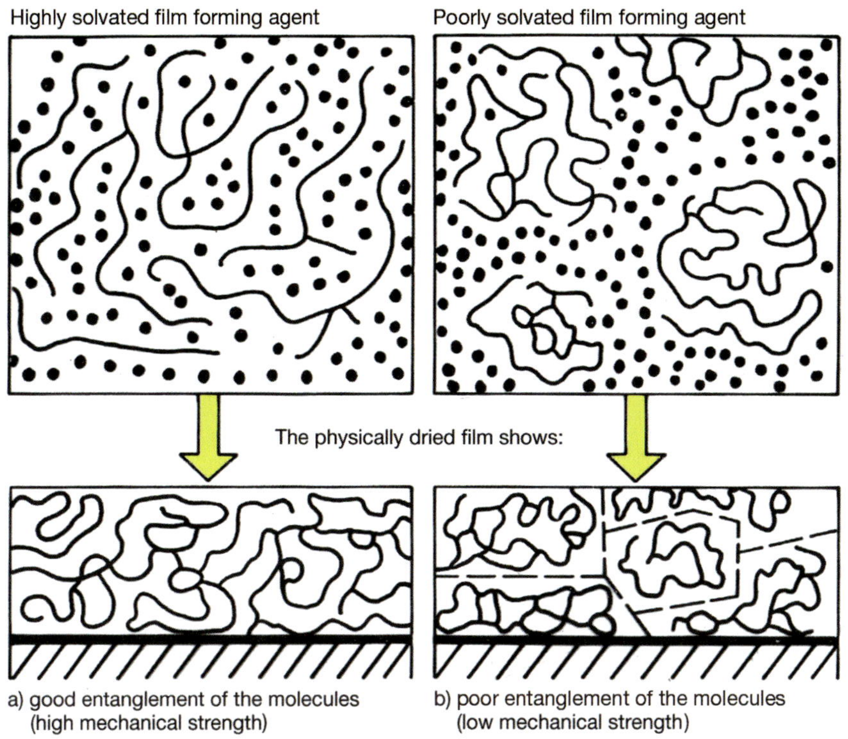

Figure 3.1.5: Molecular orientation of extensively and poorly solvated paints before and after film formation

solvent emissions from dispersions. The last of these can take the form of organosols and hydrosols, as well as plasticiser-containing plastisols. In the case of plastisols the plasticiser penetrates the polymer being plasticised during film formation, but in the case of organosols and hydrosols the liquid constituents, i.e. solvent or water, leave the resulting film in the same way as from solvent-borne paints.

The **film forming process by resin solutions** as a homogeneous mixture of film forming agents and solvents with freely mobile resin molecules in their liquid environment seems to be simpler than for dispersions, but only upon superficial examination. In a moleculary dissolved liquid paint, solvents flow around virtually all the resin molecules. As a result, the polar groups of the film forming agent are saturated via bridging bonds with the similarly polar groups of the solvent molecules. The macromolecules highly coiled in the solvent-free resin are stretched (solvated) to a greater or lesser extent depending on their interaction with the solvent molecules. Interaction of the resin molecules with each other is consequently reduced. As more solvent evaporates, this state changes continuously. Mutual influence of the resin molecules increases, the macromolecules become more and more entangled and solidify with the formation of spatial structures. The degree of uncoiling caused by solvation thus has an influence on subsequent entwinement and therefore on the physical resistance of the coating. Although solvents are used for applying coating materials and they leave the film almost entirely, they can exert different influences on the physical/technological properties via

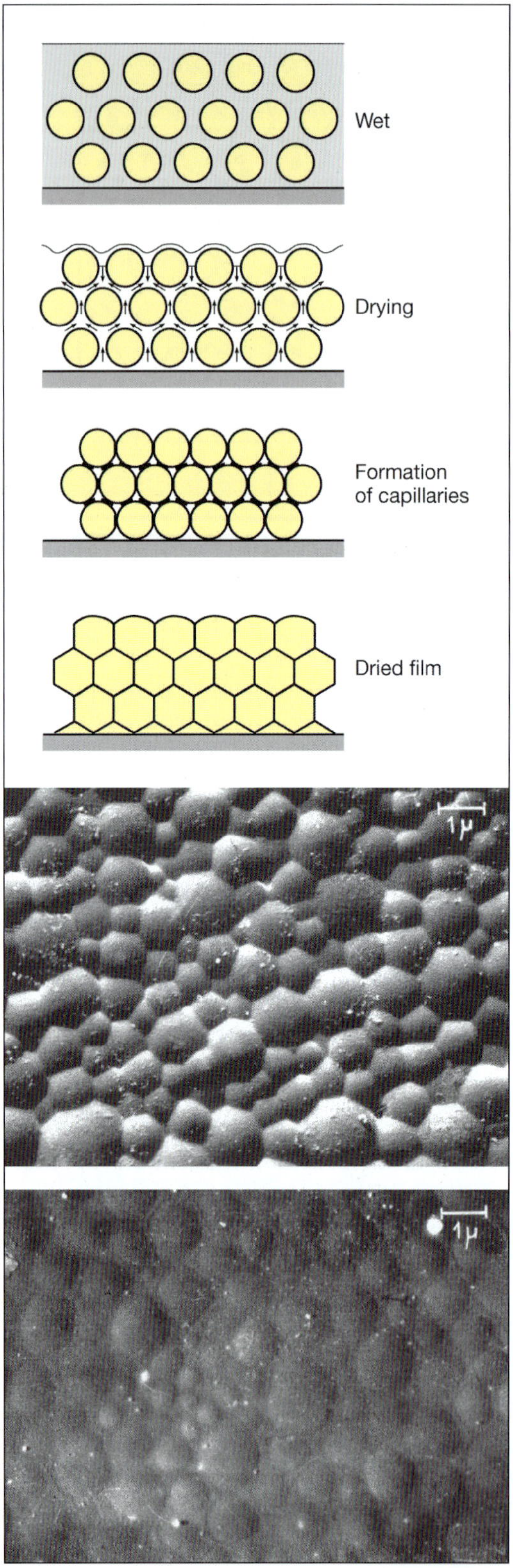

Figure 3.1.6: Schematic drawing and scanning electron micrograph (SEM) of film formation by dispersions

the resulting tertiary structures, even though the coating composition may be the same. This fact applies particularly if the systems dry merely by physical means and without chemical crosslinking.

Film formation by dispersions takes place according to completely different physicochemical principles. In this case, the polymer particles are dispersed in the water and stabilised, and film formation commences with evaporation and associated shrinkage in volume as the discrete dispersion particles come into touch with each other. This enforced convergence overcomes the repulsion barrier, with the result that the forces of attraction prevail. At the moment of contact the surface forces are so strong that complete unification of the individual particles occurs. In the capillaries of the film, which still exist at the beginning, convergence is even supported by capillary forces. Complete fusion of dispersions occurs through deformation of the originally spherical particles by the forces of attraction to such an extent that the intervening spaces disappear completely. The thermodynamic driving force behind film formation in dispersions is therefore the gain in free energy that arises from a reduction in the total surface area of the particles originally dispersed as individuals. The gain in free energy per unit area corresponds to the cohesive pressure $2\sigma/r$ and is therefore inversely proportional to the particle diameter. Consequently, film forming forces are high whenever the dispersed particles are particularly small. At a particle size of 0.05 to 0.1 μm cohesive pressures are calculated to be more than 100 bar. Under such high pressures the solid polymer particles are deformed through hexagonal stacking

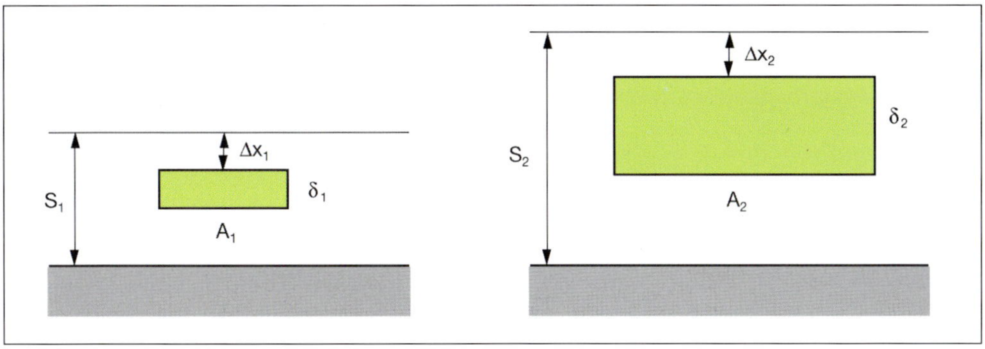

Figure 3.1.7: Model for estimating material transport during film formation

and form rhombic dodecahedrons. This theory is confirmed by electron microscope images.

Formation of the textures just described is only possible if the viscosity of the polymers in the particles permits the necessary flow processes. As viscosity is dependent on temperature there is a characteristic minimum film forming temperature (MFT) for each dispersion. If necessary, it can be lowered by adding organic solvents. For this purpose, only water-miscible solvents which have a higher evaporation number than that of water are suitable. When all the water has evaporated, they act as genuine solvents and soften the mini-spheres of the dispersion, which were previously too hard for deformation. Such solvents create conditions that will allow complete fusion to create a homogeneous film. Another way to lower the minimum film forming temperature is to reduce the molecular weight of the film forming agent. The dispersion becomes softer and thus forms films without solvent, although some properties such as abrasion resistance are compromised.

Film formation by **plasticiser-containing plastisols** (see Chapter 2.1.4) is different again. As a result of brief heating, the plasticiser penetrates the hard dispersion particles, thereby separating intermolecular forces in the polymers, and thus causes elastification and simultaneous solidification of the liquid, which was previously viscous.

Any attempt to quantify film forming processes in terms of an exchange of materials with the gaseous phase reveals a highly complex situation. Known physical laws of kinetics offer no satisfactory basis for calculating the evaporation behaviour of the coating materials. Two of the reasons for this are the constantly changing composition of the complex solvent mixtures in the transition from the coating film to the headspace and the dependence of the evaporation rate on film thickness. If it is assumed that the solvent mixture has a consistent composition over the entire period of the evaporation process, semi-quantitative statements can be made about the dependence of the exchange of materials on film thickness provided that the calculations are expressed isothermally in such short time intervals that *Fick's* first diffusion law holds. A simple thought experiment then leads to the result.

Consider two films of different thickness S_1 and S_2, both of which still contain solvent. Now compare two geometrically similar, cuboid volume elements V_1 and V_2, calculated from the base area A and height δ within the two films. In view of the geometric similarity of the two volume elements with diffusion distances Δx_1 and Δx_2 as well as heights δ_1 and δ_2 it follows that:

$$\frac{\delta_1}{\delta_2} = \frac{\Delta x_1}{\Delta x_2} = \frac{S_1}{S_2}$$

The next step is to look for the times that must elapse in order for the same reduction in concentration Δc to occur in both layers. Δc is the ratio of the mass of the evaporating solvents, Δm, to the volume V.

The requirement is therefore $\Delta c_1 = \Delta c_2$.

Since $\Delta c_1 = \frac{\Delta m_1}{V_1}$ and $\Delta c_2 = \frac{\Delta m_2}{V_2}$ and $V_i = A_i \cdot \delta_i$ follows that

$$\Delta m_1 \cdot A_2 \cdot \delta_2 = \Delta m_2 \cdot A_1 \cdot \delta_1$$

The flow of material per unit area is given by

$$\dot{m}_i = \frac{\Delta m_i}{\Delta t_i} \cdot \frac{1}{A_i}$$

Expressed in terms of the present case it follows that:

$$\dot{m}_1 \cdot A_1 \cdot A_2 \cdot \Delta t_1 \cdot \delta_2 = \dot{m}_2 \cdot A_2 \cdot A_1 \cdot \Delta t_2 \cdot \delta_1$$

In addition, according to *Fick's* first law:

$$\dot{m}_1 = - D \cdot \frac{\Delta c}{\Delta x}$$

$$D_1 \cdot \frac{\Delta c_1}{\Delta x_1} \cdot \Delta t_1 \cdot \delta_2 = D_2 \cdot \frac{\Delta c_2}{\Delta x_2} \cdot \Delta t_2 \cdot \delta_1$$

Since $D_1 = D_2$ (same paint, same solvent composition), it follows that:

$$\frac{\Delta t_1 \cdot \delta_2}{\Delta x_1} = \frac{\Delta t_2 \cdot \delta_1}{\Delta x_2} \quad \text{or} \quad \frac{\Delta t_1}{\Delta t_2} = \frac{\Delta x_1 \cdot \delta_1}{\Delta x_2 \cdot \delta_2}$$

On account of the requisite geometric similarity outlined above

$$\frac{\delta_1}{\delta_2} = \frac{\Delta x_1}{\Delta x_2} = \frac{S_1}{S_2} \quad \text{it follows that} \quad \frac{\Delta t_1}{\Delta t_2} = \left(\frac{S_1}{S_2}\right)^2$$

The ratio of the evaporation times of two paint layers of different thickness is thus equal to the square of the ratio of the film thicknesses. This corroborates the old painter's adage that "two thin coats are better than one thick one" [3.4.3].

A comprehensive analysis of the exchange of materials during the film forming process the temperature dependence of the diffusion coefficient and the vapour pressure of the solvents must include the viscosity of the coating material and the crosslinking reactions in the calculations. Since the material transfer processes are complex and cannot be calculated accurately, practical experience on the part of paint formulators will be crucial for avoiding incorrect coatings, pinholing, runs, craters, lack of gloss and poor levelling arising from inadequate coordination of the paint components.

3.1.2.2 Chemical Curing

Whilst physical drying covers the change in film state due to solvent evaporation, chemical curing relates to changes in the molecular structure of the resins or resin combinations through the creation of new chemical bonds. This involves an enlargement and crosslinking of the film forming agent constituents, which were previously single individuals. In the case of chemically curing paints, the addition of energy serves to speed up the emission of solvent, overcome activation energies and accelerate the ensuing crosslinking reactions. In accordance with the number of reactive groups per molecule these bring about more or less close-knit chemical molecular bonding with a substantial increase in molecular weight.

The bonding of the individual molecules via functional groups can take place by **polyaddition, polycondensation** and **polymerisation** (see Chapter 2.1.1). In the case of polycondensation high levels of activation energies have to be overcome. Above the minimum temperature defined by the activation energy the crosslinking reaction proceeds automatically, with the formation of low molecular cleavage products. This has consequences for the adjustment of air flows in the stoving ovens (see Chapter 4.3.1). Examples of coating materials which cure by polycondensation are stoving enamels based on alkyd resins or acrylic resins in conjunction with phenolic resins, urea resins or melamine resins.

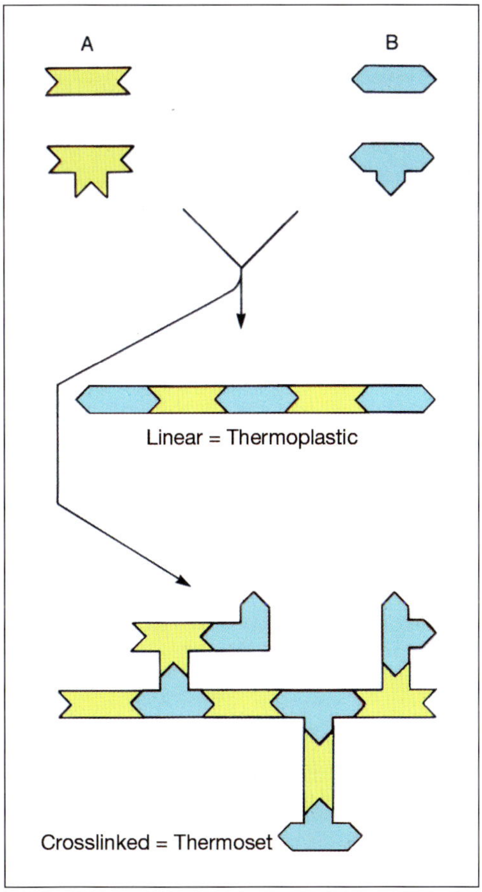

Figure 3.1.8: Schematic diagram of the bonding and crosslinking of film forming agents during chemical curing

Crosslinking stoving enamels require a larger supply of reactive groups than is consumed during crosslinking because the rise in viscosity hampers the crosslinking activity. As the activation energy for the polycondensation reaction is higher than in other crosslinking reactions and if polycondensation is to be performed as completely as possible within the usual set curing time of 30 minutes, the curing temperature exerts a particular influence over the film properties. Overly high temperatures and overly long curing times cause the coating to become brittle due to excessive crosslinking, whilst overly low temperatures and overly short curing times fail to achieve the hardness necessary for coating performance [3.4.4].

This drawback is much less pronounced in curing reactions based on polyaddition or polymerisation. These two mechanisms permit chemical curing without the need for additional heating. Polyaddition systems react as soon as the reactive components have been mixed, without splitting off low molecular reaction products. The normal practice of heating during film formation merely serves to accelerate a crosslinking reaction that is already occurring. The greater reactivity relative to polycondensation means that fewer reactive residual groups are present in the cured film. Two-component paints such as polyurethanes and epoxy resin paints are consequently much more resistant to over-stoving. In the case of polyurethane coatings this property is further reinforced by rapid hydrolysis and the associated inactivation of the remaining isocyanate groups. By comparison with polyaddition, polymerisation-curing coating materials require extra initiation. This can be achieved by adding suitable catalysts or by means of high-energy radiation. The crosslinking reaction is then exceptionally fast and therefore leads to very hard, chemically resistant films. Major examples of polymerisation-curing film forming agents are unsaturated polyesters and acrylic acid derivatives that have been specially developed for radiation curing.

3.1.2.3 Flow Patterns in the Solidifying Film

As the volatile solvents have a lower density than the dried film, the density at the surface which is losing solvent rises during film formation. The heavier but still mobile paint components sink in the substrate which has a lower density. To the same extent, light-weight components containing more solvent are moved upwards. Surface tension, which therefore rises in the areas close to the boundary surface, reinforces flow patterns in the film being formed. Local vertical and horizontal flow patterns take material from the top downwards and transport it upwards again at a different point. This process creates elevations and depressions in the coating as it solidifies.

These surface textures known as Bénard cells run counter to the conventional levelling of elevations created by application. If a paint contains several pigments of different mobility, cell formation is accompanied by floating, pigment separation and flocculation. The effect is particularly pronounced if the resulting films have a low viscosity, if film thicknesses are large and if there are considerable differences in the surface tension and density between upper areas of the film losing solvents and the lower areas of the film richer in solvent. If the solvents are particularly “short”, a temperature gradient additionally occurs because they evaporate rapidly. This reinforces the effect even further (see Chapter 2.1.4). A decrease in tinting strength and therefore changes in colour are the result. It is standard practice to use the reciprocal product of the density and the square of the particle size as a measure of pigment mobility.

3.1.3 Film Shrinkage

Solidification of the film, be it physically due to the emission of solvents or chemically due to the formation of low molecular cleavage products from the crosslinking reaction, brings about a reduction in film volume. The outcome is that irregularities in the substrate to be painted are reproduced to an extent depending on the degree of shrinkage. Such irregularities in automotive steel panels result from deep-drawing and forming in the press shops and give rise to elevations of several micrometres. Even sanding processes cause irregularities in the form of sanding grooves.

Figure 3.1.9 shows two wet films of different solid contents. At the beginning of film formation, the two paints fill both grooves in the substrate completely and create a smooth

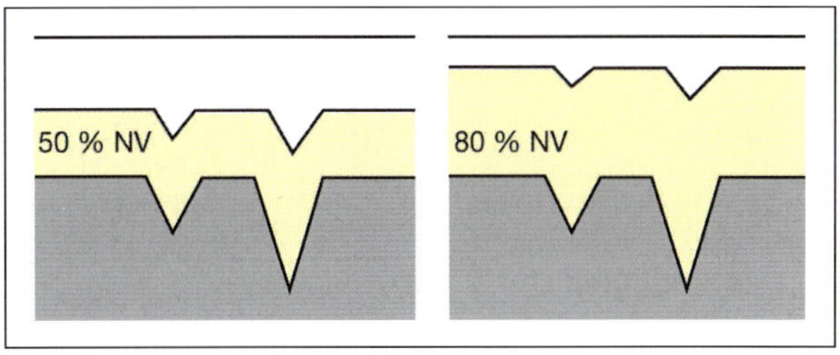

Figure 3.1.9: Surface irregularities and their reproduction in the surface of coatings related to NV

surface. The film shrinks owing to the emission of solvent and the cleavage products produced by chemical crosslinking. As the solid content increases, shrinkage declines and so the substrate is less faithfully reproduced. Some of these irregularities in the film, which is still not crosslinked, will tend to level out as a function of the viscosity of the wet film.

The roughly 10 % share accounted for by cleavage products in polycondensation-curing paints, expressed in terms of the quantity of solvent, seems to be a negligibly small contribution to shrinkage. However, cleavage products are formed only during chemical curing, that is, at temperatures at which the solvents have already largely evaporated. On account of the high film viscosity at that point in time it becomes more difficult to level out the depressions caused by cleavage products. From this it is evident that the relatively small quantities of cleavage products can make a substantially greater contribution to the reproduction of substrate irregularities than the roughly 10-times larger quantity of evaporating solvent.

The deterioration in visual attractiveness associated with the reproduction of micro-textures must therefore be regarded as an interaction between the formation and levelling out of textured surfaces. A high solid content, in conjunction with small quantities of cleavage products, is therefore a good basis for a smooth and hence high-gloss coating of good build. This is particularly evident in the case of radiation-curable paints.

3.1.4 Special Features of High Solids Paints and Water-Borne Paints

Film formation on the part of environmentally friendly alternative materials, which are different by nature, involves certain special features that need to be taken into account (see Chapter 5.6).

High solids paints, i.e., coating materials rich in solids, undergo less film shrinkage than solvent-borne paints, and so have a less pronounced tendency to reproduce surface irregularities because they contain less solvent during film formation. Given that high solids paints drop to a much lower level of viscosity during film formation on account of the lower molecular weight of their resins, it is easy to understand that the less pronounced micro-textures are smoothed out by the excellent levelling of high solids paints, although this is an effect which goes hand in hand with an increase in the tendency to sag. The fact must not be overlooked that polycondensation-curing high solids paints tend to have a higher content of cleaved condensation products because they have a larger number of crosslinking points than do solvent-borne paints. The resulting larger degree of volume shrinkage can negate the advantages that accrue from raising the solid content.

The situation is better for 2-component high solids paints. Here there are absolutely no cleavage products, as crosslinking proceeds by polyaddition. Moreover, by comparison

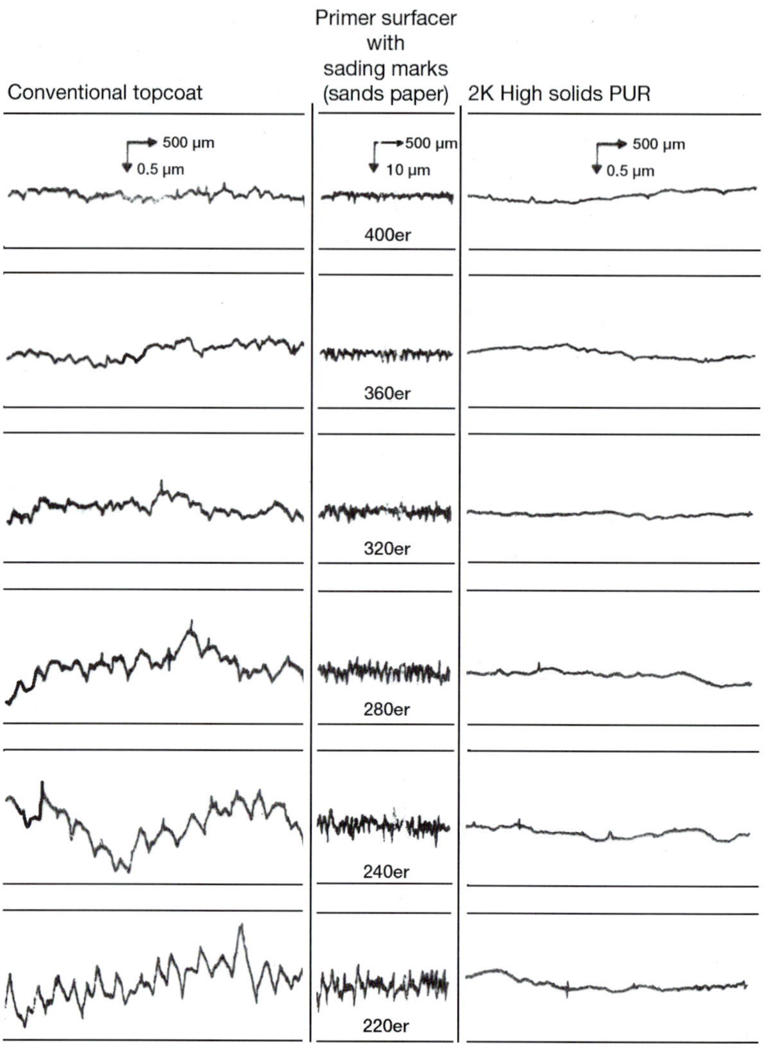

Figure 3.1.10: Surface profiles of conventional 1-component and 2-component high solids coatings as a function of substrate irregularities in the form of defined sanding marks

with 1-component paints a higher solid content can be achieved and so the formation of macro-textures and micro-textures is largely prevented. Also, through optimum matching of solvent release, reactivity and thixotropy it is easier to control the tendency to sag and dewetting from edges that is characteristic of 1-component high solids paints.

The excellent build of 2-component high solids coatings compared to that of their solvent-borne counterparts is illustrated in Figure 3.1.10. The surface textures scanned physically with a profilometer (see Chapter 3.2.4) vary enormously. Whilst solvent-borne stoving enamels have textures which correlate directly with the depth of sanding grooves in the primer, there are virtually no gloss-reducing micro-textures or long-wave irregularities in high solids coatings.

Water-borne paints, too, manifest specific features during film formation. On one hand, there is a dependence of evaporation behaviour on the level of moisture in the ambient air, which thus makes air-conditioning systems a necessarily integral part of coating lines if water-borne paints are to be applied without any problems (see Chapter 4.2.4 and Chapter 5.6). An additional variable that comes into play in the case of physical drying is the degree of neutralisation and the resulting pH. The amines necessary for the solubility of anionic film forming agents influence the viscosity to an extent that depends on the type and quantity of the neutralising agent.

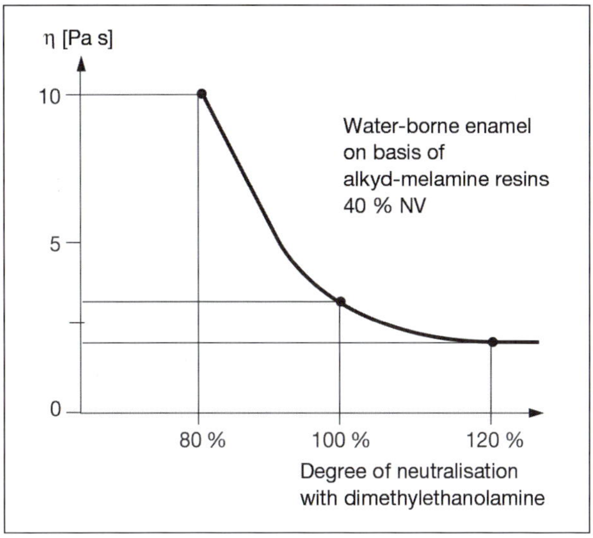

Figure 3.1.11: Viscosity of water-borne paints as a function of the degree of neutralisation

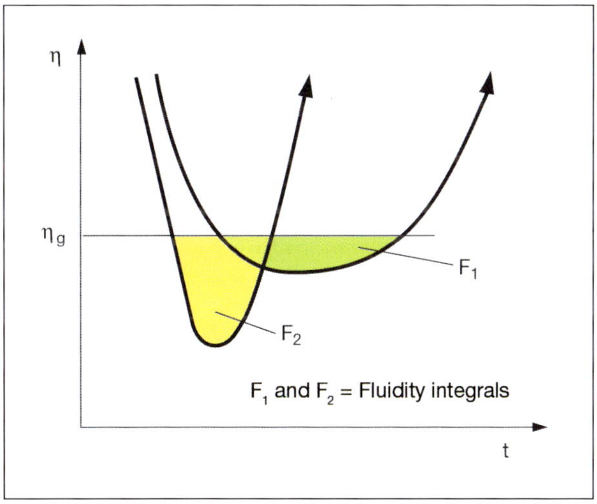

Figure 3.1.12: Fluidity integrals during stoving of powder coatings possessing good (F_1) and poor (F_2) levelling properties

Figure 3.1.11 shows how the viscosity of a dissolved water-borne paint varies with the degree of neutralisation effected by various additions of amine. Since during film formation it is not only water and organic cosolvents that evaporate but also the amine neutralising agent, the rheological behaviour and hence the levelling properties of water-borne paints, relative to those of solvent-borne coating materials, cannot be controlled solely by altering the composition of the solvents.

The situation is completely different for film formation by powder coatings. Here, solid particles made by grinding need to be transformed into a smooth, continuous film after they have melted on the sprayed object. Owing to the absence of solvents and the molecular weight being higher than that of solvent-borne paints the stove-on viscosities are relatively high. Consequently, the problems which need to be overcome have more to do with levelling and ventilation than with sagging. Coatings possessing good levelling properties at film thicknesses equalling those of liquid paints are still undergoing development. The research work is aimed not only at adjusting the flow properties but also at optimising particle size and particle shape. Spherical and also finer particles with a narrow size distribution might improve levelling, even where thin layers are applied [3.4.5].

Viscosity-time curves can assist with characterising the flow properties of powder coatings by using them to calculate dimensions known as fluidity integrals. Here, the time intervals over which an agreed viscosity limit is undershot play a key role. After material-specific constants have been determined experimentally, the fluidity integrals are calculated with computers, as a result of which the levelling properties for any heating gradients can be calculated with considerable accuracy [3.4.6].

3.1.5 Monitoring the Film Forming Process with Measuring Instruments

Quantitative descriptions of film forming processes are obtained with the aid of instrumental techniques which are capable of detecting physical drying, chemical curing and, as is usual in most cases, both phenomena, one overlying the other. The transition from liquid film to a viscoelastic protective layer is monitored by accurately replicating the practical conditions using the important film properties for the case at hand. This can be achieved either by lengthy conditioning at room temperature or by brief exposure to thermal stress. If the second method has to be adopted, defined conditions of film formation have to be set so that the entire range of stoving conditions occurring in practice is covered with as few individual tests as possible. Success in this regard has been achieved with instruments that enable temperature gradients to be set which create different temperature conditions on one and the same test panel. The gradient bar oven used most frequently (Byk-Gardner) consists of a long narrow hot-plate on which an appropriate metal panel is placed. The chosen temperature gradient can then be adjusted from one end of the panel to the other. It is thus possible to implement the film states relevant to quality assurance on just a single test panel. The subsequent measurements of visual and physical/technological properties are then performed at various points on the test piece.

Recourse is frequently also made to dynamic mechanical analysis (DMA) (see Chapter 3.2.5.2), especially for establishing the degree of crosslinking.

3.1.5.1 Levelling and Sagging

Before reliable forecasts can be made about the levelling properties of coating materials, it is essential to ensure that the film is in the same state at the time of measurement as it would be under practical conditions during manual application and in industrial paint shops. Levelling and sag applicators specially developed for manual application are consequently only suitable for the dip and brush coating methods of application employed by contractors. The devices for measuring levelling each have five dual apertures with heights ranging from 250 μm to 4 mm.

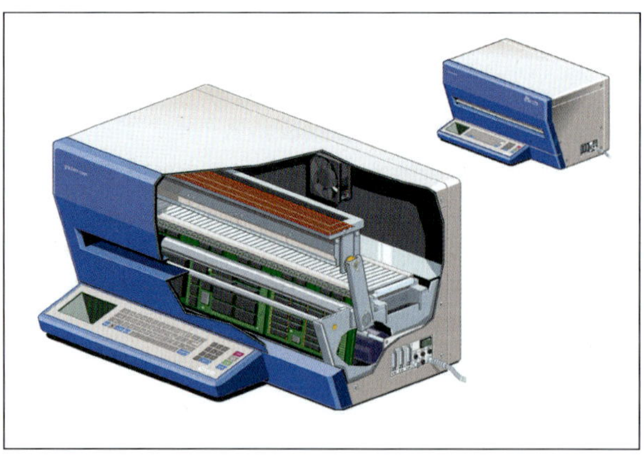

Figure 3.1.13: Gradient bar oven from Byk-Gardner

The horizontally placed test panels (A) are assessed by assigning grades from 1 to 10,

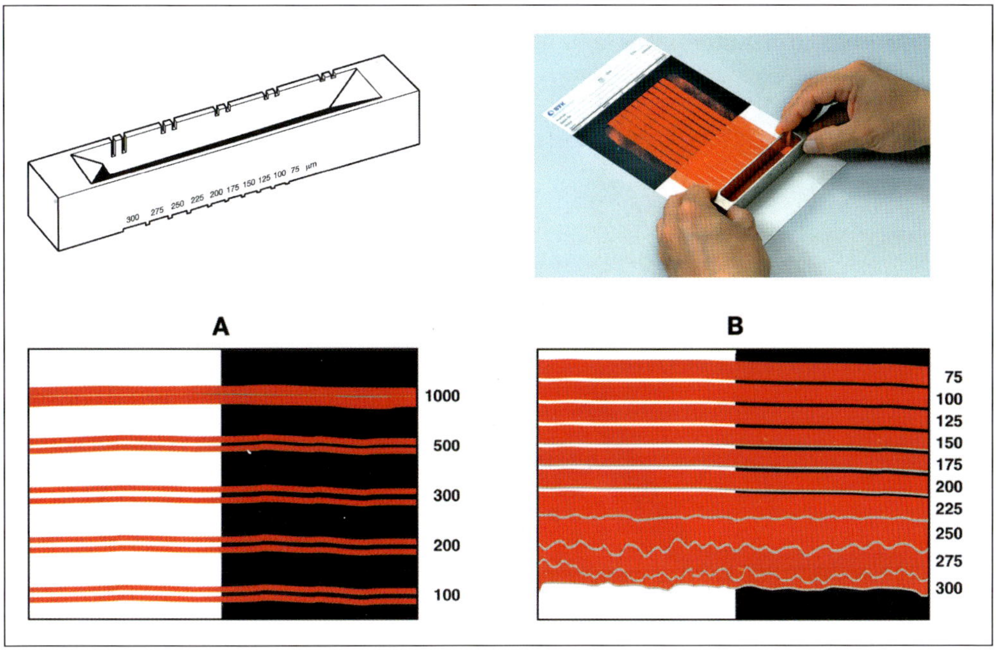

Figure 3.1.14: Levelling and sagging applicators from Stieg

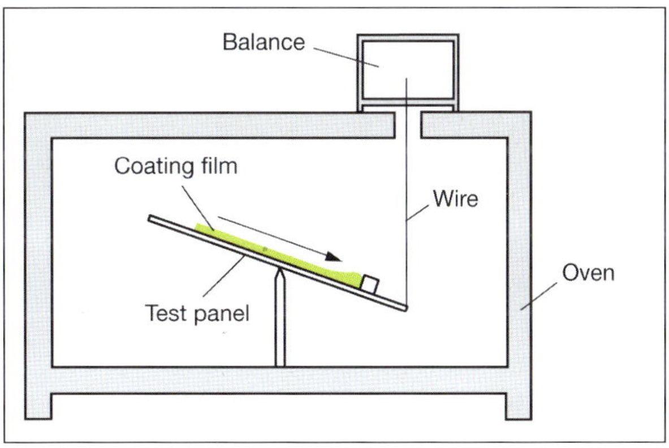

Figure 3.1.15: Schematic drawing of the sag balance from TNO

whereby grade 1 means that no pair of double strips has flowed together and grade 10 means that all the pairs of double strips have flowed together. The sag applicator is of similar design. Ten apertures gradually increasing in height from 75 µm to 300 µm make it possible to apply film thicknesses of different defined heights. If the test panel is then placed upright, the sagging limit is determined by measuring the film thickness at which the runs in the wet film extend to the next strip (B).

A different technique with a similar effect on measurement is to use wedge applicators. The wet film to be tested should exceed the sagging limit at its thickest point. Defined depressions are then made in the fresh paint and the test panel is placed upright so that the sagging tendency transverse to the wedge can be determined.

The film thickness at which sagging into or out of the depression can be observed is given as the sagging tendency. Such a measuring technique is useful for measuring levelling on films made by spray application if the test panels are produced in a manner that reflects application methods in practice.

A different test instrument for quantifying the sagging tendency is the TNO sag balance [3.4.7]. On a test panel placed at an angle the shift in weight associated with sags can be

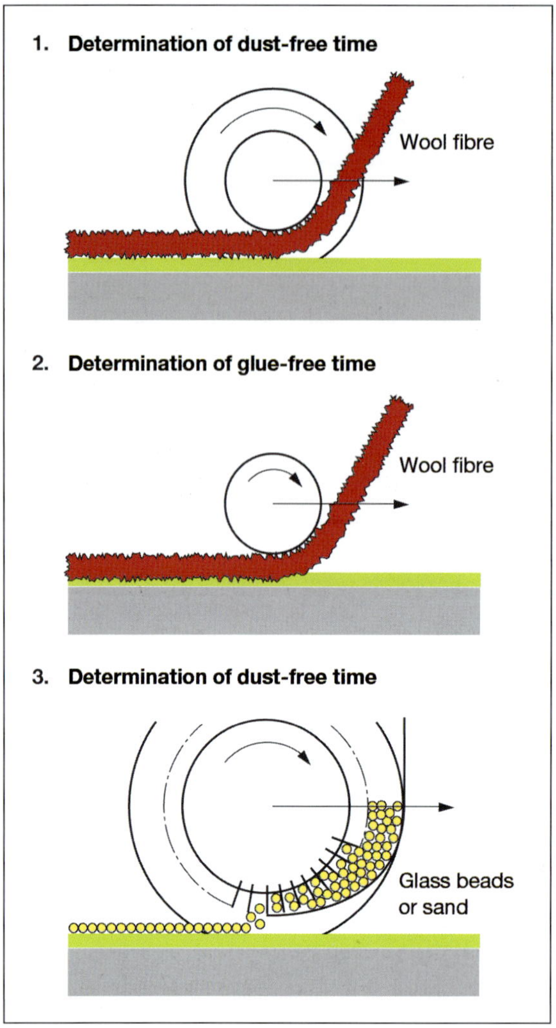

Figure 3.1.16: Methods for measuring film formation by air drying and oxidatively curing paints

read off on a scale connected to the test panel. This technique also permits measurements in a dryer and is therefore also suitable for characterising stoving enamels.

3.1.5.2 Film Formation by Air Drying Paints

Irrespective of whether the paint forms films by physical drying, chemical curing or a combination of the two, woollen threads or sand are placed on or pressed into solidifying films by machine under a defined thrust. At the start of the test the threads and the sand sink into the film while it is still liquid. They then remain on the surface, sticking to the film until it has solidified sufficiently for them to lie loosely on it.

A different testing tool that yields detailed information about the entire process from paint to cured coating is a steel hoop which is drawn through the solidifying layer at a defined velocity and under an applied, adjustable weight. Whilst paint which is still liquid will fill in the traces of the hoop again, the traces will remain in the second stage of drying and will break up

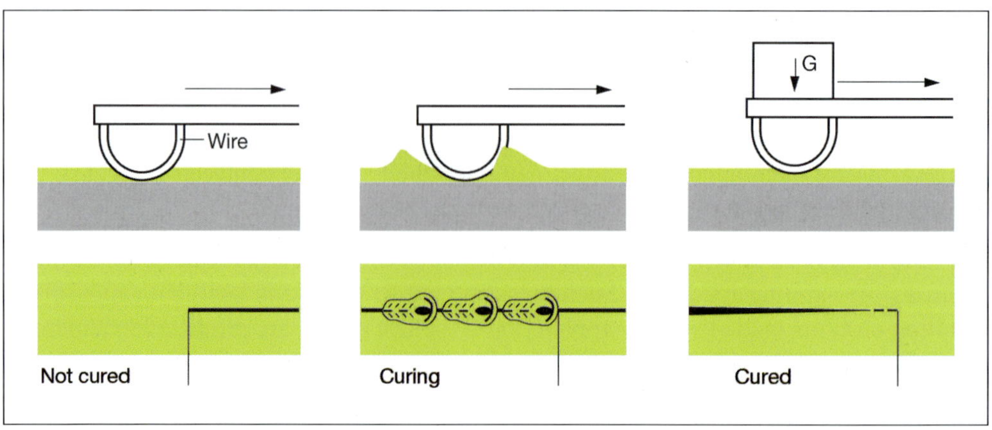

Figure 3.1.17: Method for measuring the process by which coatings solidify

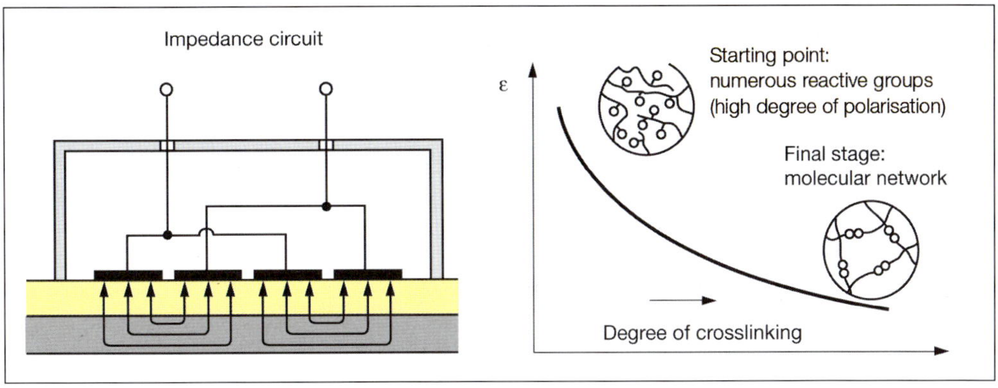

Figure 3.1.18: Correlation between dielectric constant and degree of crosslinking

the surface of the film again after it has solidified further. In the next stage of solidification, the hoop will not leave any traces at all (see Figures 3.1.16 and 3.1.17).

3.1.5.3 Indirect Methods

Numerous indirect methods have proved adept at describing film formation by industrial stoving enamels. Apart from the "rolling ball" technique presented in section 2.3.2 all methods which quantitatively detect the chemical changes in the film forming agent during film formation are to be recommended. Worth mentioning are instrumental methods of analysis such as nuclear magnetic resonance, infrared and Raman spectroscopy, measurements of dielectric constant or dielectric loss, and differential thermal analysis (DTA) and dynamic mechanical analysis (DMA) [3.4.8], which determine caloric effects. For detecting the degree of crosslinking, determination of the glass transition temperature is just as important as that of the still reactive residual groups (see Chapter 2.3.2).

Monitoring the decrease in dielectric constant merits a special mention because it yields reliable readings with simple measuring probes quickly and without any damage. What is measured is the crosslinking-induced reduction in the capacitance of a capacitor composed of the coated metal item, the film of paint and the measuring probe applied. If the film thickness and the area of the probe are known, the dielectric constant can be calculated. The service properties, such as roughness and gloss, can also be used to monitor the progress of film formation [3.4.9].

Furthermore, dissolution tests with aggressive solvents can provide information about the degree of crosslinking. Different degrees of softening provide qualitative information about the degree of crosslinking. If the solvent being used is diluted with a soluble dye, the latter will penetrate the coating if there is insufficient crosslinking and will create colour changes that reflect the degree of crosslinking. To make the test result more objective, effects can be measured by colourimetry and stated in absolute colour differences.

3.1.6 Summary

The transition from coating material to coating is determined by numerous physical and chemical changes. Wetting, levelling, solidification due to solvent release, and chemical crosslinking reactions are the most important individual steps in transition from coating material to a coating suitable for use. Coating materials themselves can be used in various forms: molecularly dissolved, dispersed or solid powder. The desired quality of the coating as manifested by good visual properties, good adhesion and excellent physical/technological properties is defined by the interplay of solvent release, changes in rheological properties and the reactivity of the film forming agents. The last of these leads to chemical crosslinking reactions by polycondensation, polyaddition and polymerisation. The process of solidification is overlaid by other physical phenomena such as flow in the solidifying film and volume shrinkage.

Numerous test methods have been developed which facilitate the reproducibility of film production by quantitatively monitoring the progress of solidification and smoothing. Thus, it is possible not only to arrive at a description of the film forming processes, but also to perform targeted interventions aimed at enhancing quality.

3.2 Properties and Testing of Coatings

By the time film formation is complete, the coating should be capable of delivering the imposed performance properties. Apart from film thickness and physical/technological properties such as adhesion, hardness and elasticity, the visual properties and durability under actual conditions play an important role in the assessment of a coating's performance.

3.2.1 Film Thickness

In addition to an attractive appearance, the choice of coating materials and method of application are vital to long-lasting protection of commodities. For all requirements imposed on the coating there is an optimum layer thickness. High film thicknesses are desirable to ensure good levelling, visual covering of the substrate and good protective action against gases and water vapour. However, the downside here is a deterioration in physical/technological properties with increase in film thickness and not least in the economics of the coating process. The determination and checking of the optimum film thickness and film thickness distribution therefore constitute important preliminary tests in the quality assurance of coatings.

It is frequently difficult to make an exact statement about film thickness because substrate roughness varies to greater and lesser extents. The term film thickness can therefore only be defined as a mean value which depends on the method of measurement and the roughness. For this reason, industrial standards assume that film thickness is to be regarded as the result of a measurement which has been obtained by an agreed, precisely defined method. Figure 3.2.1 demonstrates that the common integral-based measurement of film thicknesses fails to detect local deviations.

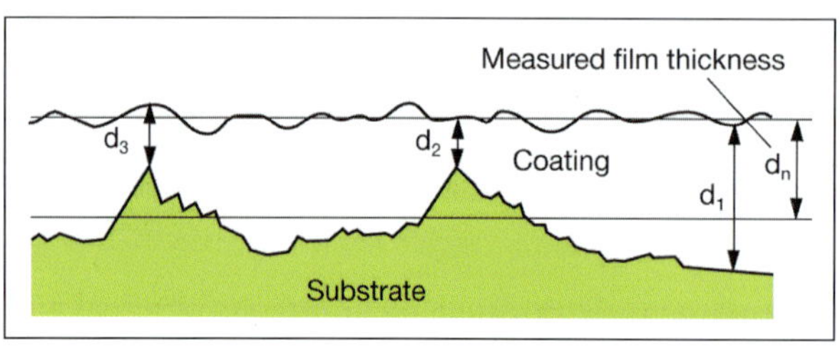

Figure 3.2.1: Determination of film thickness d_n as the mean value of coatings on rough substrates

Any assessment of the protective action of a coating must therefore factor in other criteria. Essentially these are the defini-

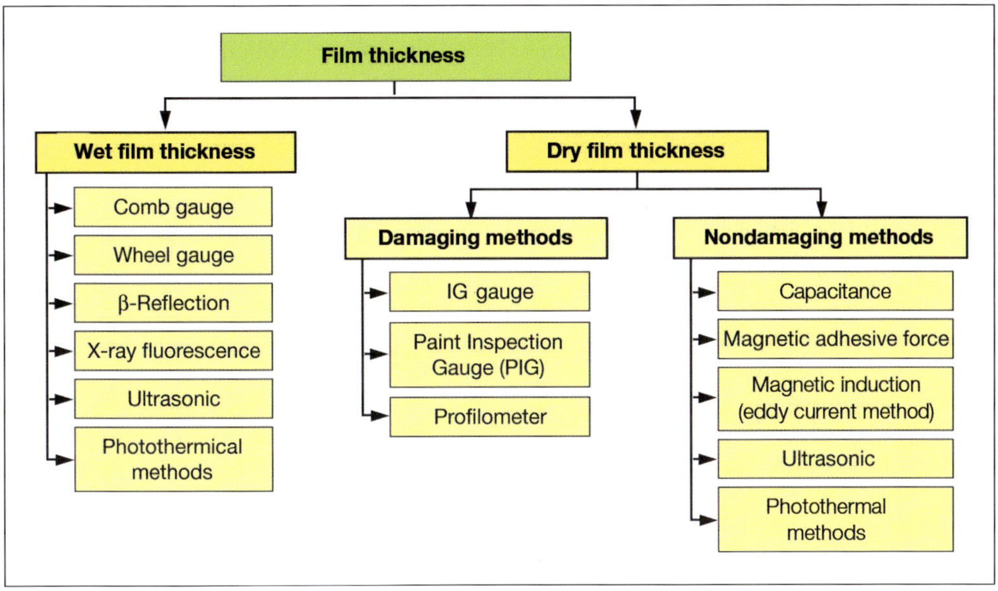

Figure 3.2.2: Overview of standard methods for determining film thickness

tion of substrate roughness, and the voids and irregularities in the organic protective layer that arise during application.

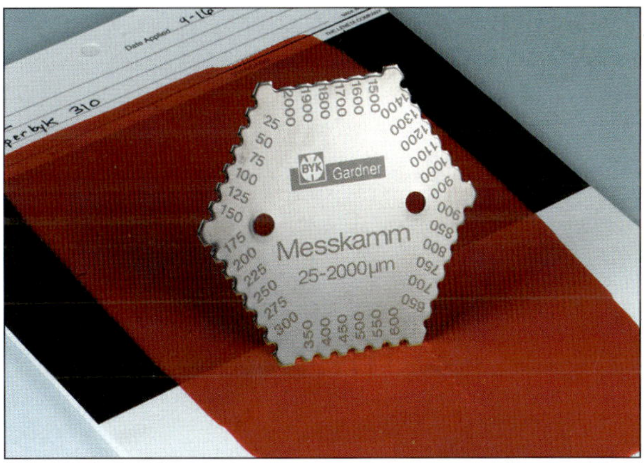

Figure 3.2.3: Measuring comb for determining the wet film thickness
Source: Byk Gardner

Most methods of determining film thickness are based on measurements of the usable or cured coating. In many cases the thickness of the film needs to be determined as it is developing in order that it may still be corrected during application. This also applies to brush or roller application by contractors and to high-speed industrial application of paints by coil coating.

The measurement of film thickness is defined in industrial standards such as ISO 2808. The methods employed can be divided into those for dry films and those for wet films, and some can be used for both types. Determinations of dry film thickness can be split up into damaging and nondamaging methods.

3.2.1.1 Wet Films

For the measurement of wet film thickness there are devices available which are simple and inaccurate and others which are more sophisticated and more accurate. One of the simplest devices is the **measuring comb**.

The measuring comb, which has teeth of different length, is pressed into the fresh coating at right angles. The longest tooth not touched by the paint indicates the wet film thickness.

A **wheel gauge** works on the same principle. It has three parallel wheels, the central one is smaller and eccentric to the outer ones. When the gauge is rolled over the wet paint, the eccentric central rim shows a position at which it is just touching the wet surface. The wet paint thickness at this point is read off from a calibrated scale engraved on the outer wheel.

The **β backscatter method** as per ISO 3543 and **X-ray fluorescence** as per ISO 3497 are more convenient and have measuring errors of ± 1 to 4 %. They are much more accurate and what is more, there is no contact. Both methods can be used for fresh and cured films. Owing to their accuracy and particular suitability for online film thickness measurement these methods will be explained later (see below).

3.2.1.2 Dry Films

Film-Damaging Methods

Film-damaging methods entail removing the coating down to the bare substrate in a defined manner. In the **wedge cut method**, e.g. using the paint inspection gauge (PIG) as per ISO 2808/DIN 50986, a wedge-shaped cut is made in the coating down to the substrate.

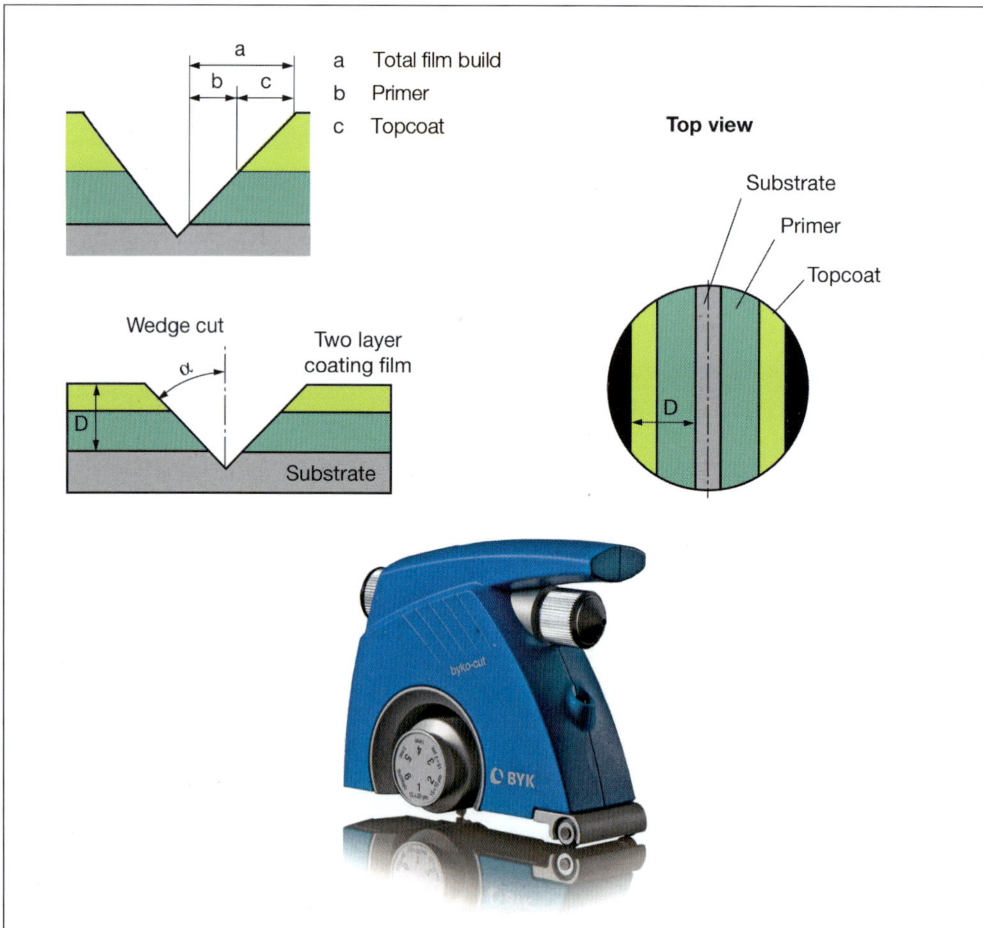

Figure 3.2.4: Measuring principle and instrument for the wedge-cut method of determining film thickness with the paint inspection gauge (PIG) as per ISO 2808/DIN 50986 *Source: Byk Gardner*

With the aid of a measuring microscope, the distance a for the total layer as well as distances b and c for the individual layers are measured and the readings are converted to corresponding film thicknesses, the angle of inclination α, of the resulting edge being known. The wedge cut method is suitable for simultaneously determining individual layers of multilayer systems. It also reliably measures the film thicknesses of coatings on various nonmetallic substrates such as wood and plastics.

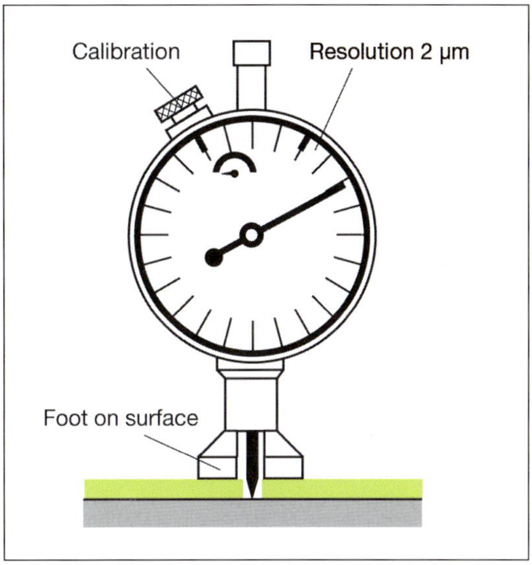

Figure 3.2.5: Schematic drawing of an IG gauge

Another device, albeit one which is only suitable for measuring total film thickness, is the **IG gauge** as per DIN 50933 or ASTM D 1005. It has three feet, two of which are fixed while the other is connected to a gearwheel via a rack. This probe is placed on the coating with its point at a distance from the substrate so that the fixed feet are on the coating and the sensing foot is in the cavity on the substrate. The probe now transmits the height difference to the rotating gearwheel in proportion to the film thickness.

Profilometers (see Chapter 3.2.4) are modern, more accurate devices. They are generally used for measuring surface texture but if the sample is properly prepared they can also be used for measuring film thickness. A diamond or laser beam scans the surface and converts the detected height differences into electronic information about the surface texture. If the coating is damaged before measurement and part of the coating is removed down to the substrate, the profilometer is able to detect this difference in height and thus supply information about film thickness.

Individual layers in multilayer systems can be measured extremely accurately by examining cross-sections under the microscope. Modern **image analysis techniques** permit

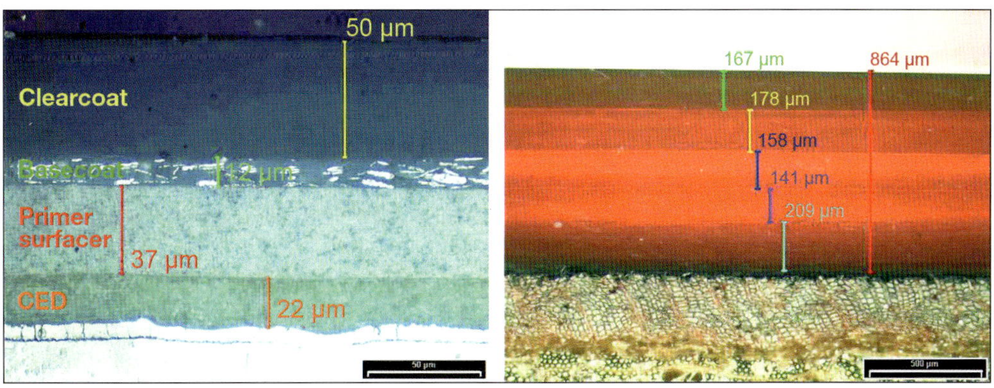

Figure 3.2.6: Evaluating cross-sections of (left) an automotive OEM coating and (right) a wood coating under the microscope *Source: BASF Coatings*

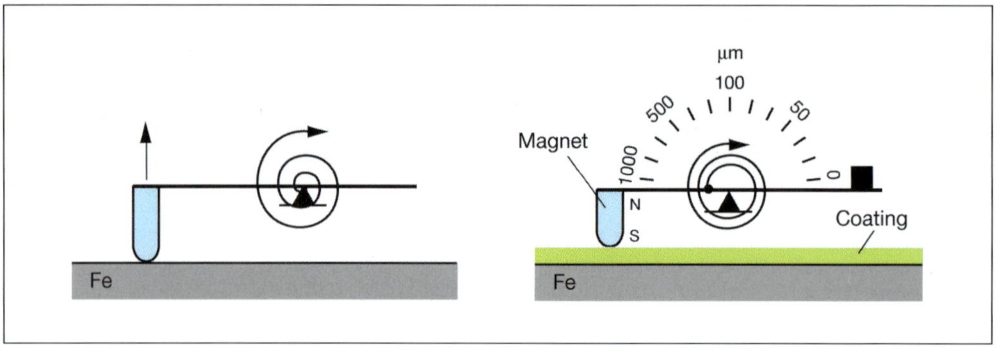

Figure 3.2.7: Use of a permanent magnet for calibration purposes and measurement of film thickness on a ferromagnetic substrate

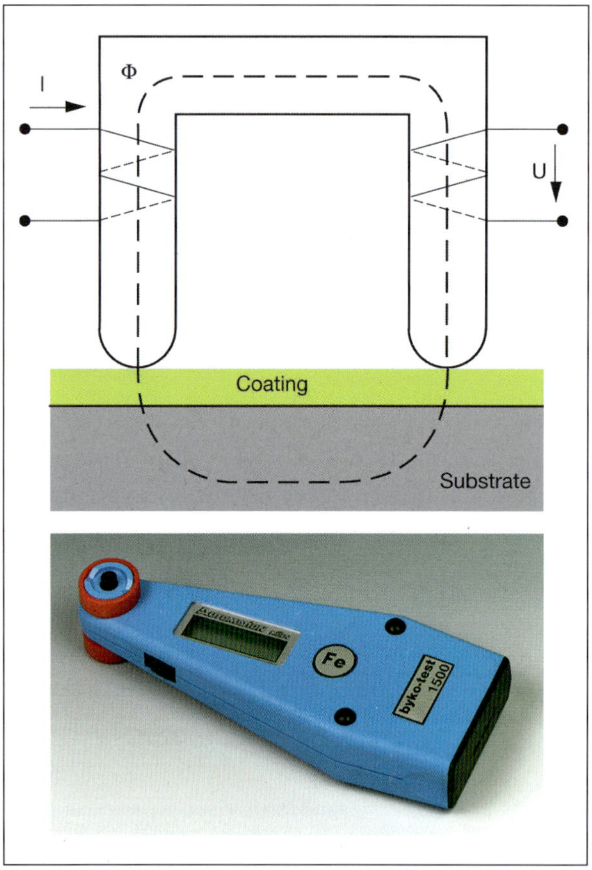

Figure 3.2.8: Film thickness measurements based on the principle of magnetic induction

precise digital processing of the microscopic images which are captured with a video camera.

Nondamaging Methods

If the coatings to be measured are on metal substrates, there are several nondestructive methods for determining film thickness available. The devices employed utilise principles which exploit the material properties of the films and their dependence on film thickness. Three physically different methods have become established.

Two measuring principles can be applied to ferromagnetic substrates, i.e. iron or steel. ISO 2178 describes the measurement of the **magnetic attractive force** of a permanent magnet and the influencing of the magnetic flux of an alternating electromag-netic field by induction. The former utilises the layer-thickness-dependent force of attraction of a permanent magnet. It is measured by continuously increasing a counteracting force. The instantaneous applied force at the moment of pull-off is a measure of the film thickness. The underlying principle is shown in Figure 3.2.7.

Film thickness gauges which exploit the inductive influence of a magnetic substrate on an alternating electromagnetic field are more sophisticated, more sensitive and yield more accurate results.

Current I generates magnetic flux Φ in the primary coil of an electromagnet, which in turn induces a dependent voltage U. In this set-up the thickness of the coating influences the magnetic flux Φ and hence the measured induced voltage U (Figure 3.2.8).

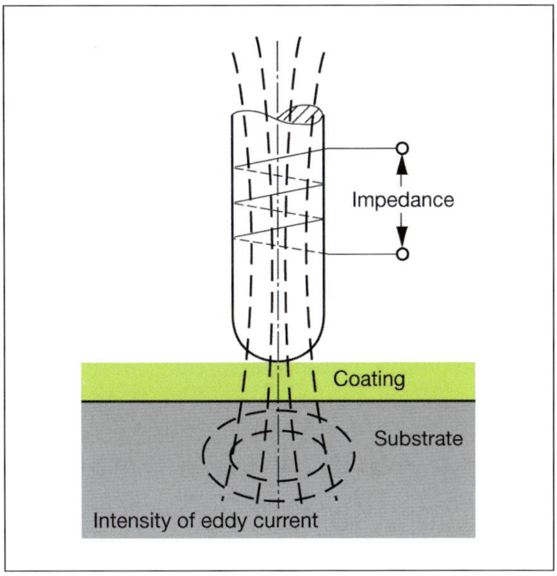

Figure 3.2.9: Determination of film thickness on conductive metallic substrates by the eddy current method

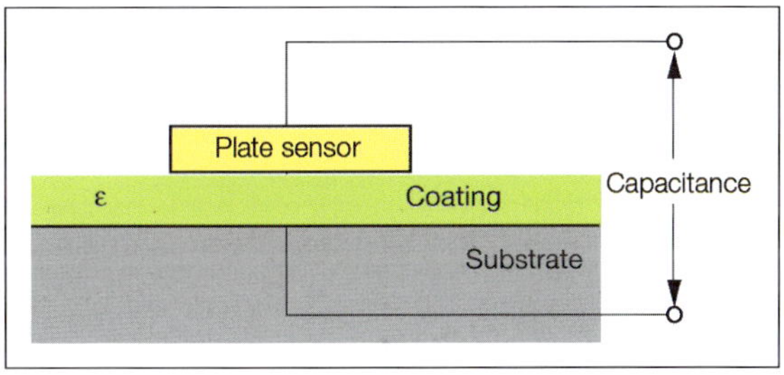

Figure 3.2.10: Principle behind the capacitative method for determining film thickness

A method which is not restricted to magnetic substrates but can be employed with all metallic materials, i.e. aluminium, zinc, copper and magnesium in addition to iron, consists in measuring the change which occurs in the inductive reactance or impedance of a measuring coil through which current is flowing, the change being the result of the feedback effect of inductive eddy currents in the electrically conductive substrate. In the case of the **eddy current method** it is the change in inductive reactance of the measuring coil which serves as the layer-thickness-dependent variable. Details are described in ISO 2360. The substrate dependence leads to deviations, especially in the case of galvanised metals, and so special probes are needed in certain cases [3.4.10].

Capacitative methods of measurement are not quite so common but in principle they can also be used for all electrically conductive substrates. A plate-shaped measuring probe is placed on a flat, painted metal substrate and is used as an electrode of a plate capacitor, the dielectric of which is the coating.

Measuring the capacitance of the capacitor allows the film thickness to be calculated provided that the dielectric constant ε and the measured area are known. Since the dielectric constant is substantially influenced by the film forming agents, the pigments and the moisture content of the coat, this method cannot be used universally. Moreover, measuring errors can be expected if there are irregularities in the coated surface. If, however, identical conditions can be created and measurement proper can be preceded by calibration measurements, film thickness measurements by the capacitative method are by no means inferior to the methods described above.

The path towards automated paint application requires the development of new, nondestructive, contactless measuring techniques.

Coating Film

Contactless Methods

Ultrasound, X-ray fluorescence, β backscatter and photothermometry are now established methods for measuring film thickness. Although they are all elaborate and expensive, they permit noncured films to be measured with a very high degree of accuracy.

Given an ultrasonic source of sufficiently high frequency, film thicknesses can be determined by means of interference effects as described in ISO 2808. The **ultrasound** hitting the coating system is partially reflected from the surface. The remainder penetrates the coating and is split up again at all other boundary surfaces into a reflected component and a component that continues to penetrate. The waves now emerging from the coating with a phase shift, R_2 to R_4, superimpose themselves on the previously reflected ones and also exhibit layer-thickness-specific interference phenomena owing to the dependence of phase shift on film thickness. One major advantage of this technique is that, as opposed to the nondamaging methods described above, it opens up the possibility of measuring the individual layers of multilayer systems on non-metallic substrates. Devices fitted with contact probes can be used to conduct measurements on plastic substrates.

Characteristic secondary radiation emitted from the substrate or from the film of paint upon exposure to beta- or to **X-rays** can also be used to make accurate measurements of film thickness. Two layer-thickness-dependent measuring principles are employed here.

In both the **β backscatter** and **X-ray fluorescence** methods, the radiation penetrates the coating film and is scattered back by the substrate. Where suitable pigments are employed, backscatter will occur from the film of paint itself. The scattered radiation or the emitted secondary radiation is measured with a detector. Provided that the atomic numbers of the elements in the paint and in the substrate are sufficiently different, there are two ways to use the backscattered beta radiation and the X-ray fluorescence to determine the film thickness. If the determination is based on the backscattered radiation or the fluorescence emitted from the film, the intensity is directly proportional to the film thickness. If it is based on the secondary radiation emitted from the substrate, the intensity is indirectly proportional to the film thickness. Measurement of beta backscatter

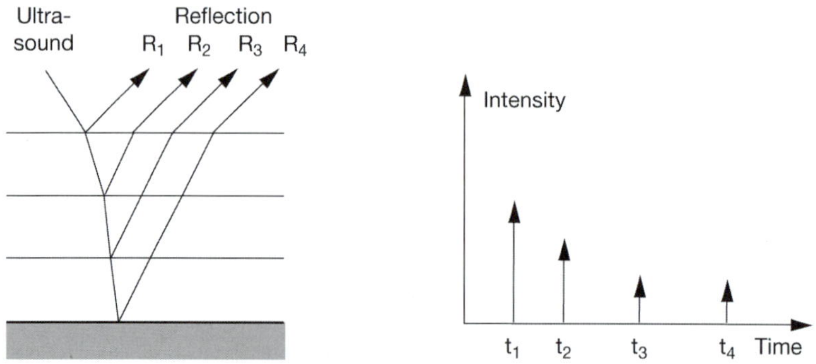

Figure 3.2.11: Principle behind the measurement of film thickness via the reflection of ultrasound from a multilayer coating as per ISO 2808

as per ISO 3882 lends itself particularly to high-speed continuous coating lines for inorganic coatings and coatings that have a high inorganic pigment fraction because the method is not critically dependent on distance, although it does have the drawback of also being less sensitive. The radiation sources are usually radioisotopes.

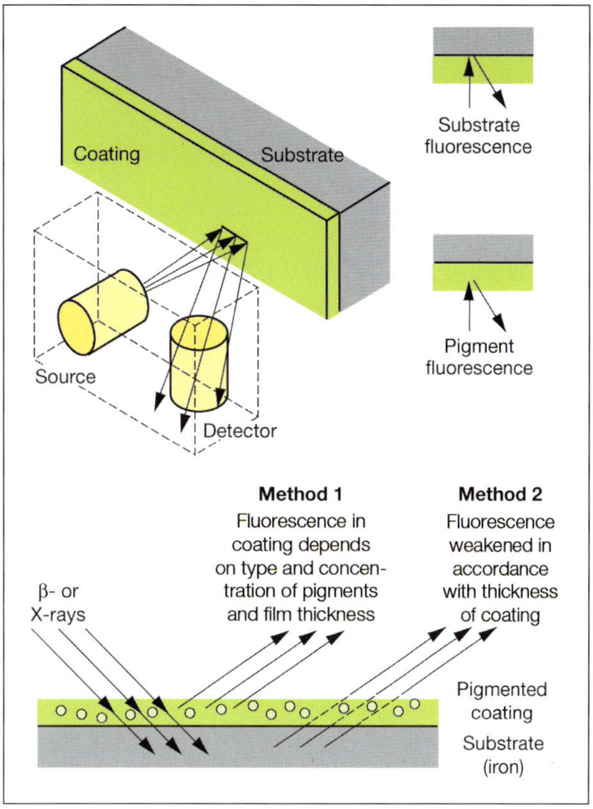

Figure 3.2.12: Reflection of back-scattered beta radiation and X-ray fluorescence by organic coatings

Since both methods provide rapid feedback to the coating process, they have become established ways of making continuous measurements in coil coating which takes place at an extremely high rate of up to 200 m/min.

One method which is suitable both for powder coatings and for liquid paints that have not yet dried is **photothermometry** [3.4.11]. Intermittent pulses of IR frequencies emitted by a radiation source, e.g. a laser, heat the near-surface zones briefly by a few degrees. The temperature rise, the transport of thermal waves to the substrate and the ensuing reflection from the surface of the substrate generate a temperature-time profile in the coating which varies with the film thickness.

The measuring probes are fast-response thermometers, so-called photothermal detectors. Because the IR beam is sensitive to the presence of molecules of water in the film,

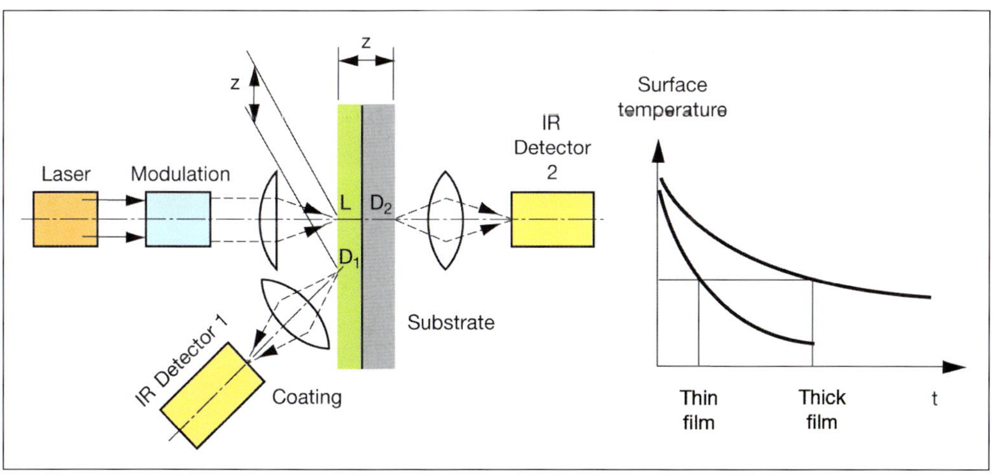

Figure 3.2.13: Photothermometric measurement of film thickness

the thermal wave technique provides detailed information about the drying process, in addition to rapid film thickness measurement. Special infrared spectrometers are also used, particularly for measuring film thickness and monitoring film formation in water-borne paints. The residual water content is indicated by the intensity of the absorption bands characteristic of water and is proportional to the film thickness [3.4.12].

Comparison of Film Thickness Gauges

A comparison of the film thickness gauges normally encountered in practice as regards accuracy, expediency in use and prices or costs always leads to the same result. The reliability of the test result and the speed with which the test results are obtained are always found to be directly proportional to the cost of the method.

The established, no doubt expedient but also low-cost devices, irrespective of whether they operate on the principle of magnetic attraction or magnetic induction, capacitance or the eddy current method, are all relatively inaccurate at ± >5 %, as are the damaging techniques involving the use of the IG gauge and the Paint Inspection Gauge. It is therefore necessary to minimise the errors by conducting multiple measurements and mathematical treatment. On a completely different level are the sophisticated methods of X-ray fluorescence, beta-backscatter and laser thermometry. Given careful device setup and when additional sources of error are taken into account and eliminated, levels of reproducibility of less than 1 % can be achieved.

Other, different methods involving the microscopic examination of cross-sections are also accurate, but have the drawback of damaging the film by sample preparation and so they are restricted to the analysis and determination of the causes of coating defects.

3.2.2 Dry Film Density

Whilst determination of the density of liquids in a pycnometer is straightforward, density measurements in solids have to be performed indirectly, and hence are more complicated, because the volume is seldom known.

One way of measuring dry film density is to determine the **solids volume** as per DIN 53219. For this purpose, a coating is applied to a plate of known mass m_1 and calculated volume V_1. Calculation of volume V_1 proceeds by determining the weight in air m_1 and water m_2 in accordance with

$$V_1 = \frac{(m_1 - m_2)}{\rho_1}$$

m_1 = weight of the test panel in air
m_2 = weighht of the test panel in water
ρ_1 = density of water

Then the panel is coated and weighed again to determine m_3 and, after immersion in water, to determine m_4. The volume of the coated plate, V_2, is then:

$$V_2 = \frac{(m_3 - m_4)}{\rho_1}$$

m_3 = weight of coated test panel in air
m_4 = weight of coated test panel in water

The volume of the dried film V_3, is then given by $V_2 - V_1$. The value $m_3 - m_1$ equals the initial weight of the film. Consequently, dry film density ρ_T is

$$\rho_T = \frac{m_3 - m_1}{V_2 - V_1}$$

Furthermore, it is possible to calculate **dry film density** from the material data of the liquid paint if the density of the wet coat, the density of the solvents and the percentage quantity of the solids (% NV) are known. Since the volume of the liquid paint, V_{La}, is the sum of the volume of the dried film V_T and volatile fraction V_{fl}, it follows that:

$$V_{Pa} = V_T + V_{fl} \quad \text{or} \quad \frac{m_{Pa}}{\rho_{Pa}} = \frac{m_T}{\rho_T} + \frac{m_{fl}}{\rho_{fl}}$$

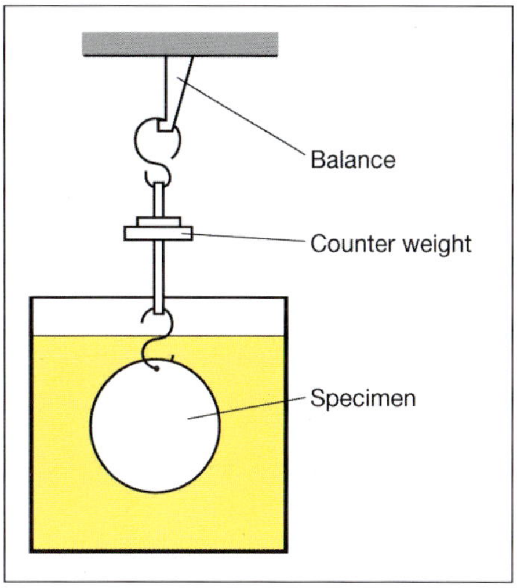

Figure 3.2.14: Measuring the dry film density as per ISO 23811

Since the mass of the paint constituents m_T and m_{fl} is also equal to the mass of the liquid paint m_{La}, it follows that:

$$m_{Pa} = m_T + m_{fl}$$

After solving the equation for ρ_T and substituting the solids (% NV) for m_T/m_{La} x 100 it follows that:

$$\rho_T = \frac{NV \cdot \rho_{Pa} \cdot \rho_{fl}}{100 \cdot \rho_{fl} - (100 - NV) \cdot \rho_{Pa}}$$

On closer analysis this simple relationship proves to be highly sensitive to small errors in the determination of the nonvolatile fraction. This becomes noticeable when stoving temperatures in the oven during laboratory determinations differ from those in industrial application – which is frequently the case. To an initial approximation the density of the volatile fraction can be equated with the density of the solvents in the paint.

On coating lines it is often necessary to determine densities of solids for which there is no relevant data concerning the liquid paint. Pigment powders and powder paints are examples where density calculations fail. In such cases recourse is made to a modified pycnometer method. As described in ISO 787-10 the density of solid powders is calculated from the weights of the pycnometer containing the pigment powder being examined, the pycnometer topped up with petroleum ether, and finally the pycnometer filled with just petroleum ether. The density is determined from

$$\rho_T = \frac{m_{Pi} \cdot \rho_{Pe}}{m_{Pe} - m_{PiPe} + m_{Pi}}$$

ρ_{Pe} = density of petroleum ether
m_{Pi} = weight of pigment
m_{Pe} = weight of petroleum ether
m_{PiPe} = weight of pigment and petroleum ether

For accurate test results it is essential that the powder be completely wetted with the petroleum ether.

3.2.3 Measurement of Voids in Coating Films

Film thickness is an important criterion for the protection of substrates susceptible to corrosion but it is not the only one. Adhesion to the substrate and the integrity of the protective layer are additional factors that govern quality. To an extent depending on the rheological properties of the paint extremely fine air inclusions or wetting flaws introduced by the application process may leave voids in the coating. These may stay in one part of the coating only or may extend in the form of continuous capillaries from the substrate into the surface of the coating, with the likelihood of the latter depending on the number of individual layers applied. Such voids tend to weaken the protection. They allow water and electrolytes to advance down to the substrate virtually without hindrance. Quantitative determination of voids is absolutely vital for assessing the protective action of anticorrosive coatings.

Coating Film

Channels in the coating that extend down to the steel substrate can be rendered visible relatively easily by immersing the item in a copper sulphate solution containing surfactant. The non-noble iron substrate in the area of the voids dissolves and elemental, clearly visible copper is deposited.

$$Fe + Cu^{2+} \longrightarrow Fe^{2+} + Cu\downarrow$$

The number of voids per dm^2 is converted into a **void ratio**. P1 means no continuous voids, P2 fewer than 20, P3 20 to 100, P4 101 to 400 and P5 more than 400.

If the voids are not continuous, they cannot be identified by the simple method of marking with copper. Weak points in the coating then need to be detected by other methods. Continuous voids and latent voids can be reliably located by means of high-voltage electrical spark discharges. As described in DIN 55670, wire brushes or electrically conductive rubber wipers connected to high voltage are passed over the item being examined. The presence of a void is signalled by a spark discharge and a visual or an audible alarm, the latter depending on the depth at which the void occurs and on the applied voltage. This method is also used to determine edge coverage and the extent of protection on highly textured substrates.

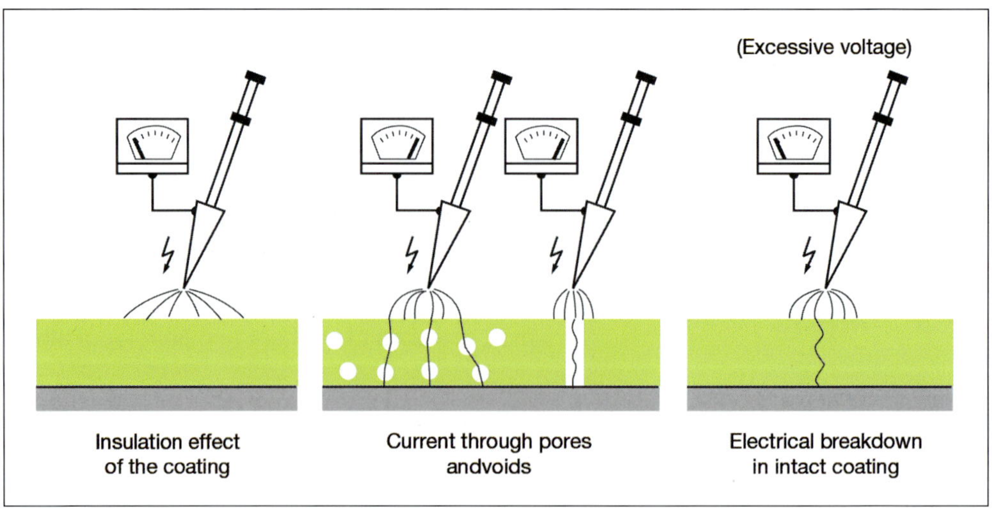

Figure 3.2.15: Locating voids with the aid of high voltage

3.2.4 Visual Properties

The **colour, colour effects and gloss** of surfaces are sensory impressions and hence subjective perceptions of light which has been altered by interaction with coatings.

Studies by colour stylists of the subjective perceptions of colour, gloss and effect in relation to the shapes of coated items are helping to establish a link between aesthetic and functional properties. New visual effects are repeatedly transformed into beauty and elegance with a view to offering incentives and arousing desires, and not just in the automotive sector.

A person's overall visual impression is essentially created by two different physical phenomena. A portion of the light hitting the coating is reflected from the surface and passes without further interaction with the constituents of the coating – irrespective of its colour – into the observer's eye. If this is a case of specular reflection from a smooth surface, the latter is perceived to be glossy. The vast bulk of the light penetrates the coating where, after interacting with the pigments and effect substances, it creates colour, hiding power and additional effects.

3.2.4.1 Gloss

Traditional Definition of Gloss

The traditional definition of gloss considers only the light reflected from the surface of a coating. The reflection of light varies with the smoothness of the surface, and the illuminated environment will be reproduced more or less faithfully. The sensory impression perceived by the eye is recognised as gloss.

As described in ISO 2813 the proportion of the light R reflected from the surface of a coating that does not contain effect pigments is a function of the incident angle and may be calculated with the aid of *Fresnel's* equation.

$$R(n,\varepsilon) = \left[\frac{n^2 \cdot \cos\varepsilon - (n^2 - \sin^2\varepsilon)^{\frac{1}{2}}}{n^2 \cdot \cos\varepsilon + (n^2 - \sin^2\varepsilon)^{\frac{1}{2}}}\right]^2 + \left[\frac{(n^2 - \sin^2\varepsilon)^{\frac{1}{2}} - \cos\varepsilon}{(n^2 - \sin^2\varepsilon)^{\frac{1}{2}} + \cos\varepsilon}\right]^2$$

A plot of the intensity of reflected light as a function of the relative refractive index n, calculated from the ratio of the refractive indices of the coating and air and the angle of the incident light ε, is shown in Figure 3.2.17.

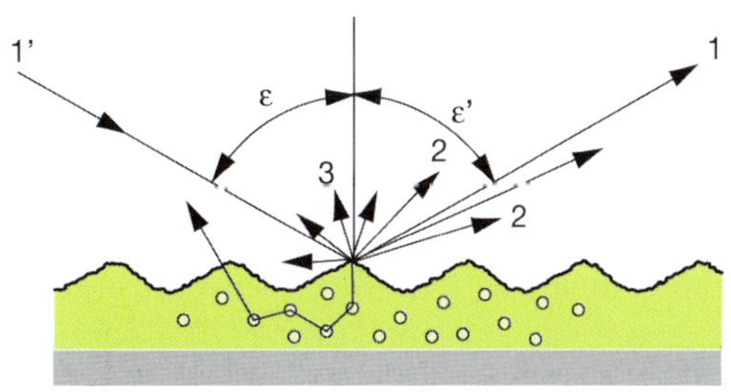

1' Incident light
1 Specular reflection
2 Deviation in path of reflected light due to levelling disturbances
3 Diffuse light caused by scattering from microstructure

Figure 3.2.16: Reflection pattern of light incident on coated surfaces

Surface reflection is found to be relatively low at just 4 % to 8 % for a wide range of incident light angles. Nevertheless, this is perceived by the eye as being intense. Unlike the light that penetrates smooth surfaces it is dispersed either not at all or only diffusely. Consequently, if

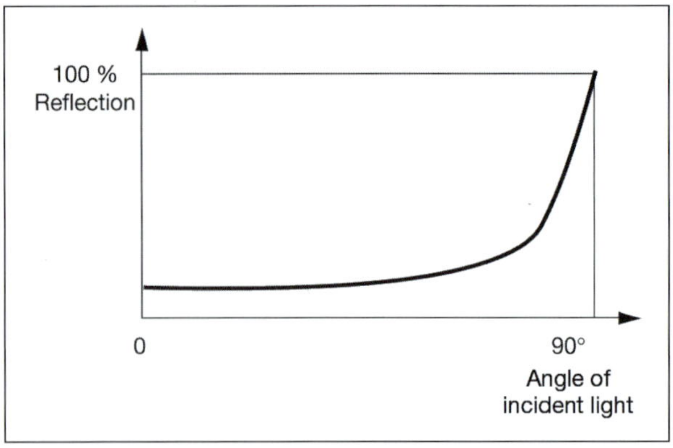

Figure 3.2.17: Reflection as a function of angle of incidence

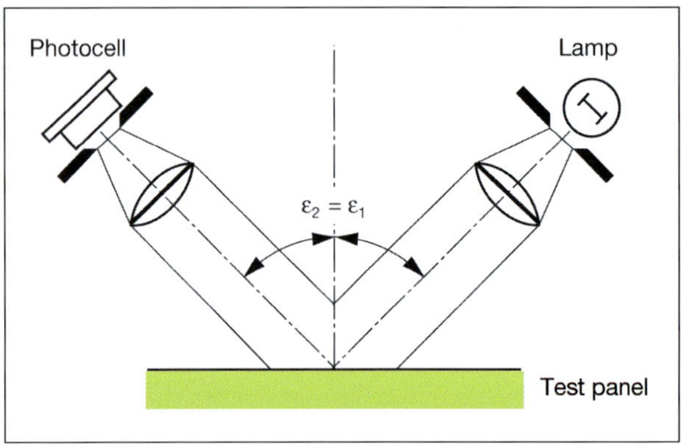

Figure 3.2.18: Schematic drawing of a reflectometer as per ISO 2813

the surface texture is less pronounced, the light mostly undergoes specular reflection, thus giving rise to sharp reproduction of the light source.

Since gloss is clearly perceived as being particularly intense when an illuminated object on a coating of paint is reflected brightly with little distortion and high contrast, it seemed logical when the term ‚gloss' was being defined to use the portions of the specularly reflected light as a measure of gloss. This is the method specified by ISO 2813.

Direct gloss measurements

Gloss measurements are perfored with reflectometers. These instruments have a defined light source which emits light with an area of several square centimetres onto the coated item at an angle ε_1. The intensity of the reflected light ($\varepsilon_1 = \varepsilon_2$) is measured at the equiv-alent angle of reflection ε_2. A smooth surface gives rise to higher reflection values than a rough or wavy surface. High reflectometer values are therefore indicative of high gloss. Simple instruments have become established in practical use that can measure specular reflection on a repro-ducible basis and convert it into reflectometer values. The measurements are always relative and are expressed in terms of the reflected flux of light from a black, polished glass plate with a refractive index of 1.567.

To increase the measuring accuracy and improve the reproducibility of the measurements it has become standard procedure with samples of different gloss to also set various angles of incidence and reflection for the measuring beam. As a higher proportion of light is reflected at larger angles of incidence matt surfaces are measured at a large angle of incident light (85°), semi-gloss ones at a medium angle (60°) and high-gloss ones at only a very small angle (20°). Modern instruments contain all three measuring geometries and can be used universally by switching to the various ranges.

What is certain is that if the specular reflection is of relatively low intensity the sharpness of reproduction will be impaired in different ways as the proportions of nonspecular and diffuse light increase. The nonspecular and diffuse portions are not detected by the reflectometer. However, they exert an influence on the observer's visual perception of surfaces.

To detect the light/dark effects of metallic coatings, which contain 3 to 4 % aluminium flakes in the case of full-shade metallics and 0.2 to 0.5 % in the case of dark metallics,

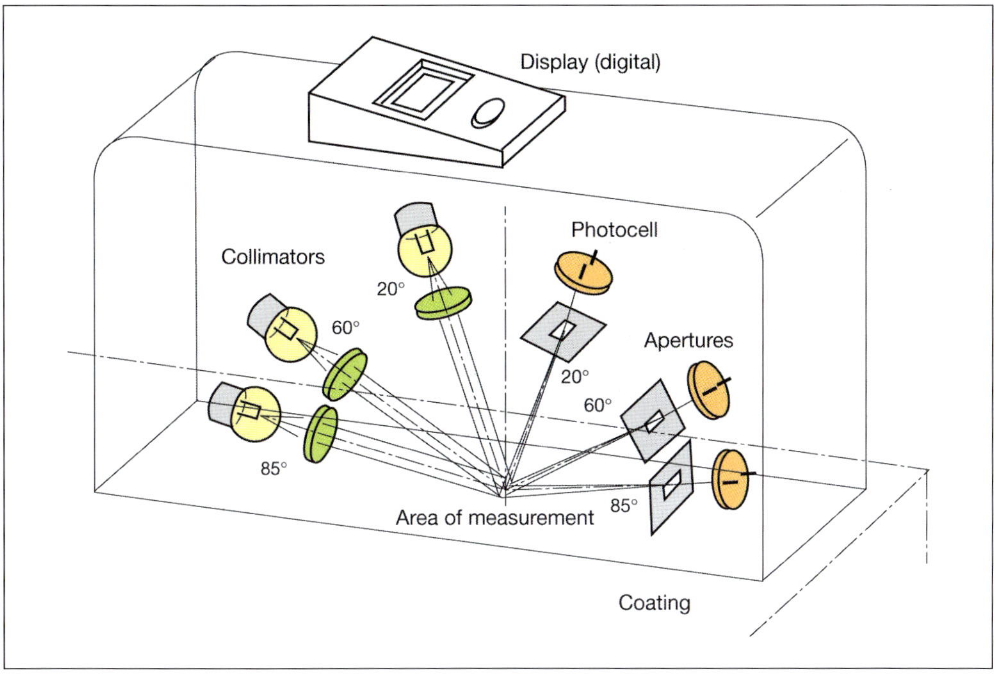

Figure 3.2.19: Schematic diagram of an instrument for measuring gloss at different angles

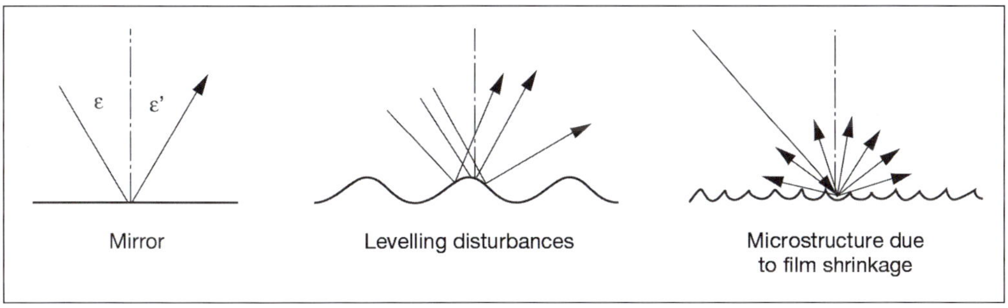

Figure 3.2.20: Reflection of light from surfaces of different textures

the reflectometer measurement is performed at different angles. One way to capture this effect is to measure the light reflected at an angle of 0° from a surface illuminated at 45° with incident light.

Extended Consideration of the Term Gloss

Long-wave structural deviations from an absolutely flawless planar surface give rise to slight local changes in the angle of incident light and therefore in the angle of reflection as well. Short-wave surface disturbances superimposed on these create larger local changes in the angle of incident light. Fine and ultra-fine textures, on the other hand, scatter the light in all directions. Correlations between a sensory impression and physical effects can therefore only be found if the nonspecular reflected light is also detected accurately and quantified with figures.

If the measuring options of the **reflectometer** are extended to include angles of reflection other than the setting $\varepsilon_1 = \varepsilon_2$ statements can be made about the whereabouts of the light flux for nonspecular reflection as well.

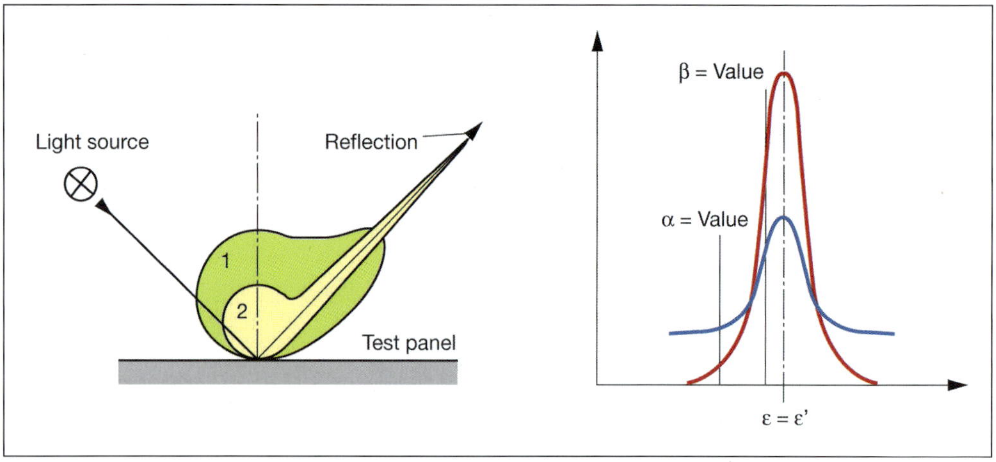

Figure 3.2.21: Distribution curves for light reflected from two different coatings (gloss indicatrix)

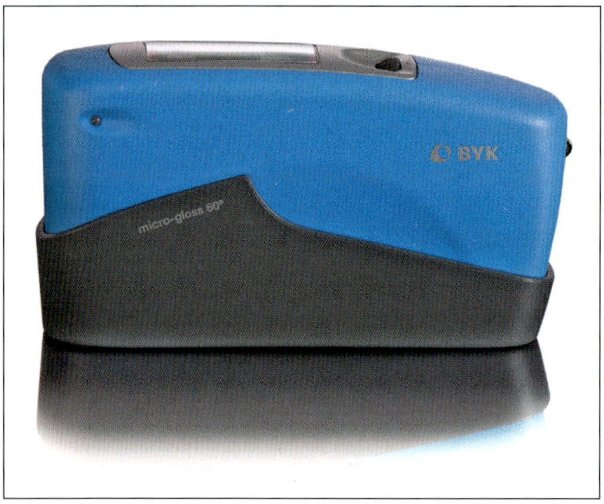

Figure 3.2.22: Glossmeter

Such instruments, known as **goniophotometers**, have a mobile measuring probe and can thus cover all angles of reflection from 0° to 90°. Consequently, the gloss peak is no longer the sole reflectometer value measured but rather an entire reflection curve, the gloss indicatrix.

The gloss indicatrix, with its information about the angle-dependent distribution of all the reflected light, allows of a much more detailed analysis of the integral sensory impression of gloss using numerical values for light flux components. As in simple reflectometer measurements, the curve maximum at the angle of reflection $\varepsilon_1 = \varepsilon_2$ provides information about specular gloss. In addition, owing to the different shape of the gloss indicatrix, statements can be made about the reflected components near the angle of reflection (small angle scatter) and at a larger distance (wide angle scatter). The former indicates the coarse-wave textures with only slight deviation from the horizontal whilst the latter indicates the fine textures which largely create nonspecular scattering effects. This affords a way of differentiating between long-wave levelling disturbances and microstructures in the coating.

For fast comparative measurements instruments have been developed in which the mobile measuring sensor has been replaced by four permanent photosensors. The first is positioned at an angle of reflection $\varepsilon_2 = \varepsilon_1$ the second at $\varepsilon_2 = \varepsilon_1 + 0.3°$, the third at $\varepsilon_2 = \varepsilon_1 + 2°$ and the fourth at $\varepsilon_2 = \varepsilon_1 + 5°$. Mathematical treatment of the individual reflection values enables quantitative statements to be made about specular gloss R_s, levelling disturbances in the form of distinctness of image (DOI) and the fine textures that can

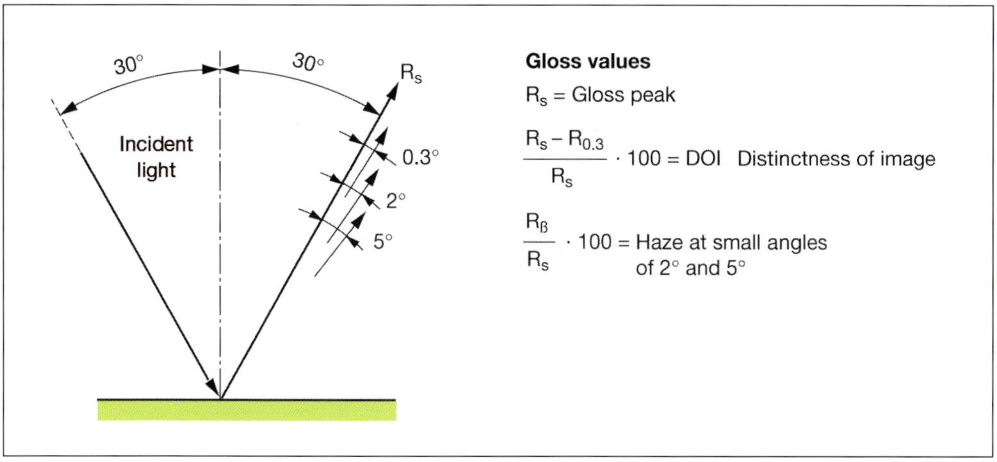

Figure 3.2.23: Measuring principle behind a reflectometer for measuring gloss at several fixed angles of reflection

be perceived as haze H. Haze is a special feature of specular gloss. It is created by near-surface disturbances (e.g. pigment agglomerates) having dimensions in the range of the light's wavelength.

There is no doubt that the goniophotometer as an extension of the reflectometer made it possible to arrive at a much better explanation of the discrepancies between sensory impression and test result. Nonetheless, a reliable forecast of the subjective perception of complex textured coatings still cannot be achieved with it. The physiological impression is more complex because unlike the measuring instrument the observer does not process the sensually perceived images with a sensor but with two eyes. Another reason is that the constant movement of the eyes generates an integrated overall impression that is ultimately the outcome of a large number of individual items of information. Moreover, unlike the measuring instrument, the eyes adapt subconsciously to either the surface or the environment reproduced in a glossy surface – to an extent depending on gloss level and distance from the object.

Indirect Methods

In the search for an even better explanation of the causes of gloss and possibly also more accurate forecasts of the observer's impression, efforts have been undertaken which completely dispense with studies of light reflection and devote themselves to the direct measurement of surface textures. The underlying idea is that a geometric description of the surface structures might be used to find ways of identifying the causes of texture formation above and beyond the relationships involved in subjective perception. If the influences exerted by the formula, dispersion, application or the substrate are known, it becomes possible to not only describe phenomena but also adopt useful measures to optimise product quality.

The surface texture of a coated object is the outcome of an accumulation of complex reproduction of substrate irregularities on one hand and textures resulting from the application and the material on the other. The reproduction of the substrate irregularities is overlaid with (usually) coarse-waved textures induced by the application. These have their origin in the interplay of the levelling of primary structures and the formation of elevations induced by flow patterns in the forming film (see Chapter 3.1.1). In all cases the

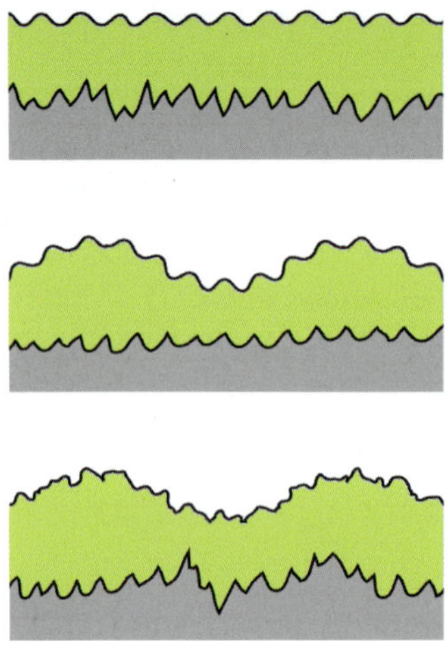

Figure 3.2.24: Superimposed surface structures in films

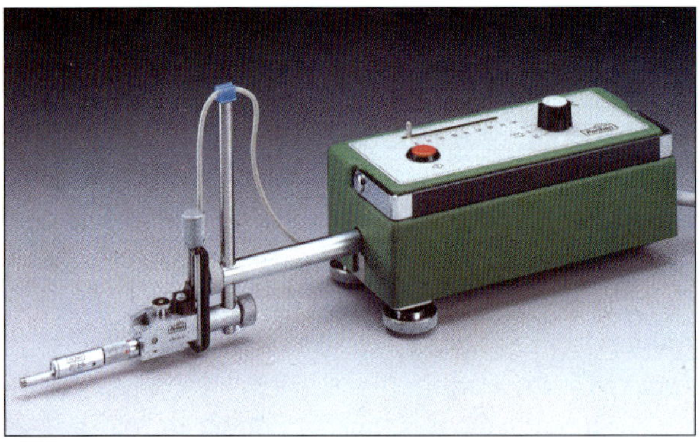

Figure 3.2.25: Profilometer *Source: Perth*

substrate irregularities, overlaid with levelling and film forming disturbances, receive an additional ultra-fine texture that results from film shrinkage. The wavelength and amplitude of the ultra-fine structure are governed just as much by the solid content and the quantity of the cleavage products released during curing as by the fineness of the pigments and the extenders (see Chapter 3.2.3).

It is the measurement of such complex textures and the assignment of indicators to them that permit correlation with the sensory impression of gloss, which is the purpose of the indirect techniques. A proven method for measuring surface textures consists in **scanning the surfaces mechanically** with a diamond tip which can be moved vertically (profilometer). Capable of accommodating deflections of up to 50 µm, a diamond tip having a curvature of approx. 5 µm scans the surface mechanically and then transforms the data electronically into a height profile afterwards. Surface structures are divided up into a long-wave component under the term 'waviness' and a medium-wave and short-wave component under the term 'roughness.'

The long-wave component above 0.24 mm is caused by levelling defects and can be assessed by measuring the DOI (see above). The short-wave component between 0.02 mm and 0.06 mm reproduces the microtexture caused by shrinkage-induced reproduction of the substrate and of the pigments and can be detected by haze measurement [3.2.4].

Better fine resolution is provided by measurement with a laser beam which is focused on the height of the microscopic roughness in question. Such a method can be used successfully with pigmented systems but it fails in the case of metallic coatings, where reflection from the surface is only partial. However, if suitable measuring equipment is used, penetration into the basecoat and additional reflection from the effect substances permits objective quantitative analysis of the metallic effect (flip-flop).

Diagrams obtained by scanning the morphological features of surfaces for the purpose of analysis are assigned various values in accordance with the cataloguing of roughness profiles proposed in ISO 4287. They include the mean roughness value R_a as the arithmetic mean of all the profile values in the **roughness profile** and R_q as the geometric mean.

Supplementary information is supplied by the profile depths R_p and R_m. These are the distances of the highest and lowest profile points from the reference line. The sum of the two variables is R_y, the vertical distance between the highest profile value and the lowest one within test section l. For an even better description of surface profiles the test section l may be divided up into a larger number of individual sections. In the individual subsections the individual peak-to-valley height Z_i corresponding to the R_y value and the peak-to-valley height R_z obtained by averaging the individual peak-to-valley heights of consecutive individual test sections can then be stated and calculated.

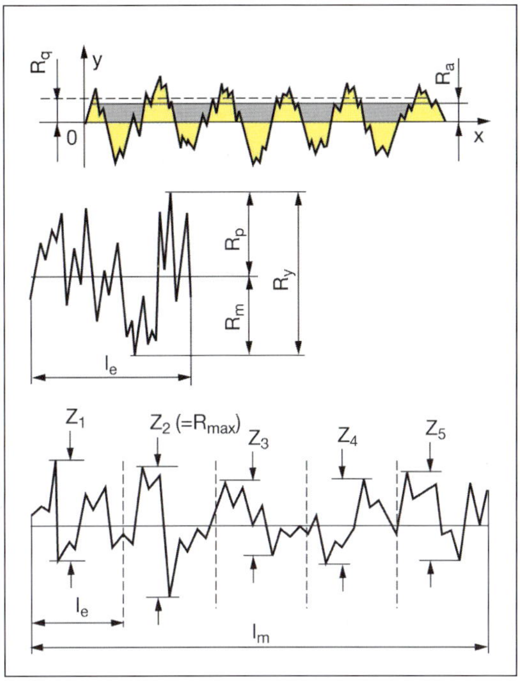

Figure 3.2.26: Definition of measurable data in surface texture

A different principle for measuring surface textures with laser beams is to determine the angle of reflection as it changes when the textured surface is being scanned. This principle is embodied in the **"Wavescan"** instrument which bears a superficial resemblance to a reflectometer. The fundamental difference is the measured area. Whilst the light source of the reflectometer scans the surface integrally with a measured area of several square centimetres to produce an average, the "Wavescan" uses an ultra-fine laser beam ($D < 1\ \mu m$) under an angle of 60° to scan a section of 10 centimetres and thus detects the intensity of the beam reflected as a function of the distance covered. Peaks and valleys appear light-coloured whilst the slopes are shown darker and darker with increase in inclination or rise [3.4.13].

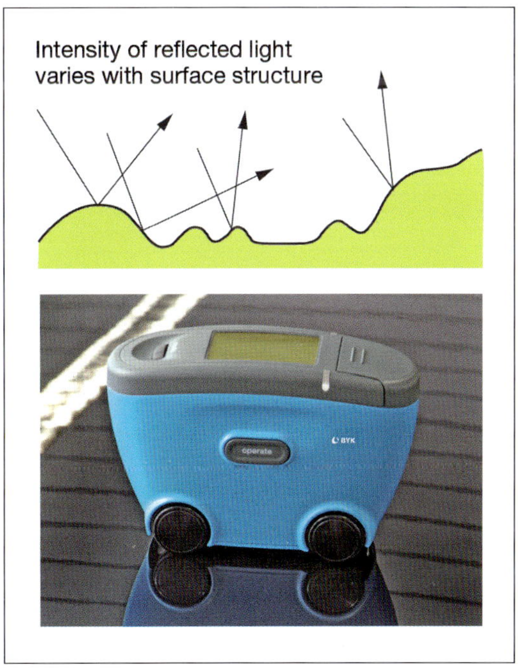

Figure 3.2.27: Measuring principle behind the "Wavescan" instrument Source: Byk Gardner

Mathematical treatment makes it possible to classify the components of the individual structures according to wavelength ranges and amplitudes and to state them separately. All curves can be converted to a multitude of sine and cosine curves of different

Coating Film

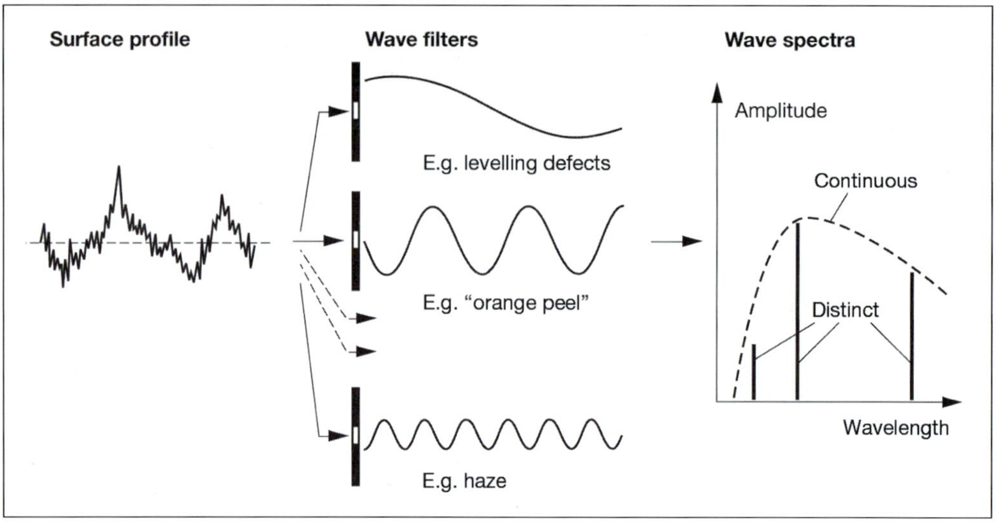

Figure 3.2.28: Fourier analysis of waviness profiles

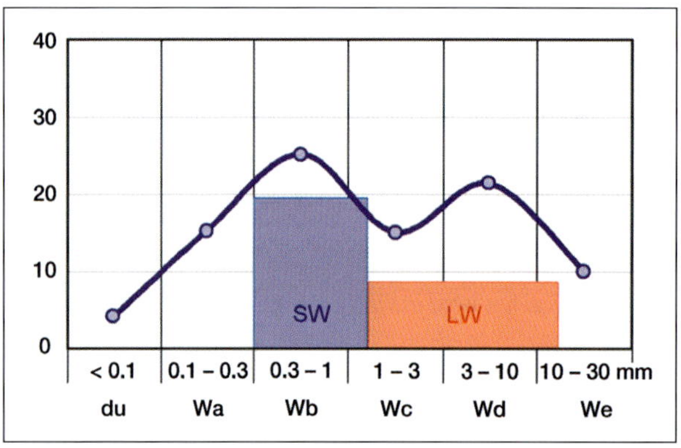

Figure 3.2.29: "Wavescan" values for evaluating surface textures
Source: Byk

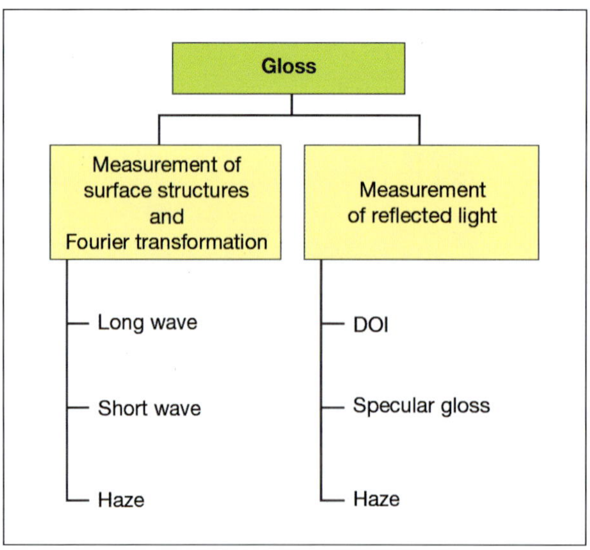

Figure 3.2.30: Variables for describing surfaces

wavelengths and amplitudes, no matter how complex they may be. This operation, known as Fast-Fourier transformation, produces discrete indicators for the short-wave, medium-wave and long-wave components of textured surfaces. The surface profiles obtained primarily by mechanical and optical profilographs or by the "Wavescan" are transformed by means of these conversions into waviness spectra, through the application of waviness filters [3.4.14].

The "Wavescan" is restricted to six wavelength ranges relevant to the assessment of surfaces: du, Wa, Wb, Wc, Wd, We [3.4.15]. These variables, known as short wave and long wave, provide information about fine and coarse textures measuring 0.3 to 1.2 mm and 1.2 to 12 mm (see Figure 3.2.29).

In the meantime, an additional measuring system has been developed for the "Wavescan" which also records matt surfaces, i.e. ones which are highly textured, with satisfactory accuracy. This makes it possible to objectively assess the surface texture of less glossy fillers and primers as well. Optimisation of the ranges for fine and coarse textures has been studied with a view to obtaining better agreement between reading and visual impression [3.4.16].

In the search for other simple instruments for measuring surface textures another device has become established which is referred to as the **Autospect Quality Measurement System** (AQMS). It uses a digital camera to capture a test image of the coating and mathematically evaluates distortion and attenuation of light/dark transitions. In this way it becomes possible, in the manner of a Goniophotometer, to obtain values for the gloss peak, distinctness of image (DOI), appearance and haze.

However, all the technical and physical variables for measuring gloss do not always adequately cover the subjective sensation of gloss. Another approximation of subjective overall impression can be achieved by combining several methods, whereby each one describes partial aspects which can be applied to specific comparisons.

Unambiguous assignment of the readings to the observer's subjective perception continues to be the subject of research activities.

3.2.4.2 Colour and Colourimetry

As opposed to gloss, which is a sensory impression created by the reflection of light from the surface, **colour impression** is the result of a physiological effect arising from the interaction between the light inside the coating and the coating's constituents. Chroma occurs whenever wavelength components are removed from the visible spectrum of available light in the range 380 to 720 nm. Within the spectrum of sunlight, that range represents its maximum radiation intensity (see Figure 3.2.32).

To the extent that colour pigments remove individual wavelength components from the white light through selective absorption or interference, reflection spectra occur which cause the observer to perceive different colour impressions

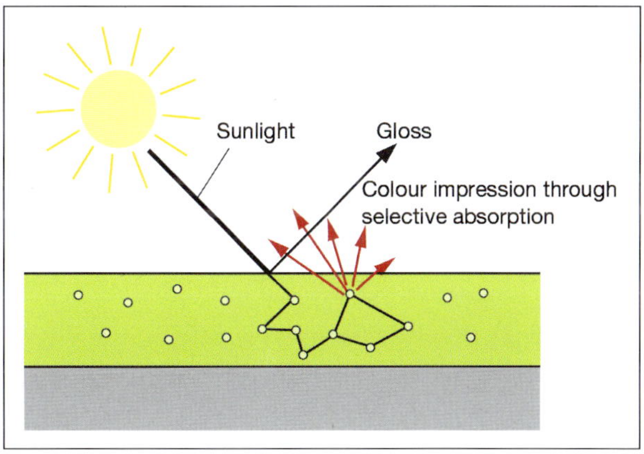

Figure 3.2.31: Schematic representation of the differences between gloss and colour

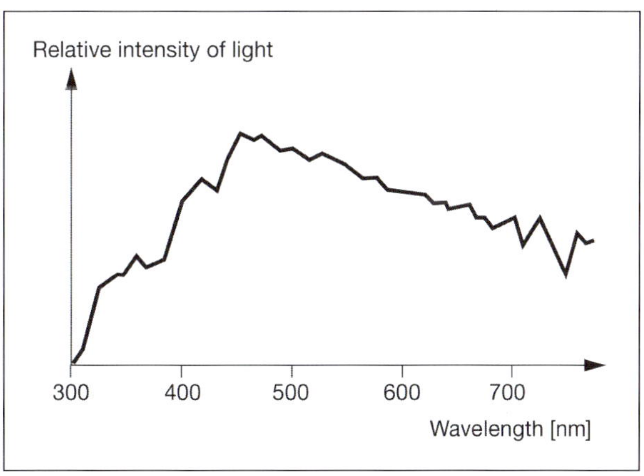

Figure 3.2.32: Spectral distribution of sunlight

with regard to colour, chroma and luminosity. The human eye is so sensitive that it can differentiate between several million individual colours.

Unfortunately, despite this extremely high sensitivity to colour perception, colour impression is subjective, not only among different observers but also in one and the same person at different times or in a different environment. Any objective assessment of colours therefore needs numerical values and measuring methods that can describe colours.

Causes of Chroma

The requirement for perception of colours is a light source with wavelength components in the range from 380 nm to 720 nm, an object with which the light can interact, a coating containing colour pigments in this case, and an observer who perceives with their eyes the light which has been changed and reflected by the coating.

As with a camera, the human eye reproduces the environment on the retina and, on account of sensors located there in the form of rods and cones, it is able to sense different levels of luminosity and colour impressions. The rods are responsible for seeing light and dark and the cones are responsible for detecting colour. They exist in three variants with different degrees of sensitivity to the individual wavelengths and they create the conditions required for sensing colour. The three sensor groups are sensitive in the long-wave, medium-wave and short-wave ranges of the visible spectrum. If a coating is able to completely absorb white light over the entire visible wavelength range, the coating is seen as being black. Conversely, the observer perceives a coating to be white if there is no selective absorption.

If, however, individual ranges are absorbed from the spectrum of incident white light more than other ones, the body appears to be coloured. If the body is perceived to be blue, the long-wave yellow and red radiation is absorbed more than the short-wave blue radiation. A green body reflects green, and absorbs blue, yellow, orange and red entirely. A yellow body reflects yellow and red whilst it absorbs green and blue to varying degrees. A red body reflects red, with blue, green and yellow being absorbed to varying degrees.

A plot of the different degrees of reflection R against the wavelength λ yields a reflection curve which is characteristic of a specific **body colour**. Figure 3.2.33 shows the reflection curves for some colours.

When illuminated with identical light sources, two bodies that have identical reflection curves will have the same colour, physically speaking. The reflection curve of a coating thus permits an objectively accurate description of the physical property of colour

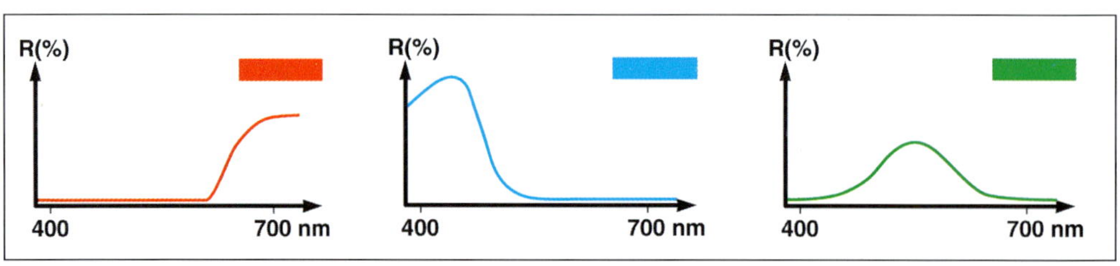

Figure 3.2.33: Examples of reflection curves of red, blue and green coatings

without any subjective influences, as are bound to occur when the human eye perceives a colour impression.

The number of colours can be increased as required by mixing colour pigments. The individual pigment components extract their specific wavelength components from the light and combine the individual absorptions from the incident light to create a new colour impression, which can be seen as a new reflection curve. This principle, known as **subtractive colour mixing**, makes it possible to adjust reflection curves, and hence colours, systematically by mixing pigments.

By contrast with subtractive colour mixing, the superimposition of coloured light sources is termed **additive colour mixing**, a phenomenon which, for example, is used to create television pictures. Here addition of several light sources leads to addition of the individual light components, ultimately through to white light.

Colourimetry

The new spectral distribution which remains after interaction between incident light and coating and which ultimately creates the colour impression is naturally dependent on the **light source**. Different light sources and different coatings lead to different colour impressions. As a result, when exposed to light from an incandescent bulb with a high proportion of radiation in the long-wave red range a coating with reflection in the red range appears to be redder than when exposed to the white light of the sun.

Consequently, if it is desired to make general statements about chroma, the "subjective" cones of the human eye and transmission of impulses to the brain as well as the processing to create a colour impression have to be replaced by photosensors and appropriate mathematical transformation of the signals. This presupposes that the sensitivity of the cones of the human eye is known in full detail. Since human beings perceive colour subjectively, it was necessary to define the colour sensation capacity of a fictional **standard observer**. For this purpose, the International Commission of Illumination (CIE) defined the spectral sensitivity functions of the human being with standard colour eyesight back in 1931. The standard colour matching functions shown in Figure 3.2.34 reflect the averages of the so-called standard eyesight observer. As stated in ISO 11664-1 they represent the viewing ability of over 90 % of all human beings.

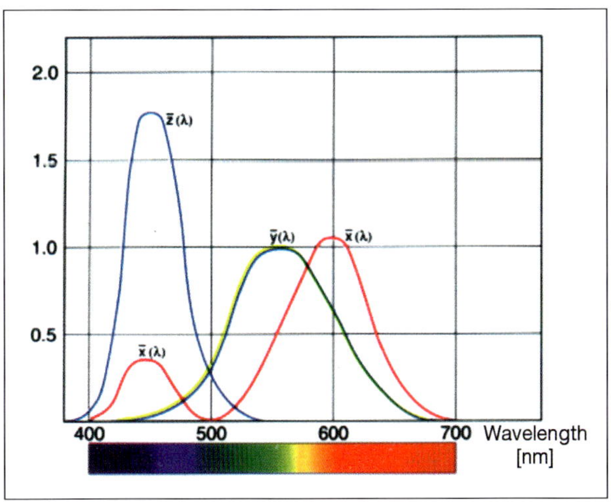

Figure 3.2.34: Sensitivity curves for the eyes of a 2° observer

Source: Minolta

An objective and hence general statement about chroma can only be made if standardised light sources provide a constant supply of light. In order to meet the various conditions encountered in practice the CIE (see Chapter 8) has, among other things, standardised the following types of light in ISO 11664-2:

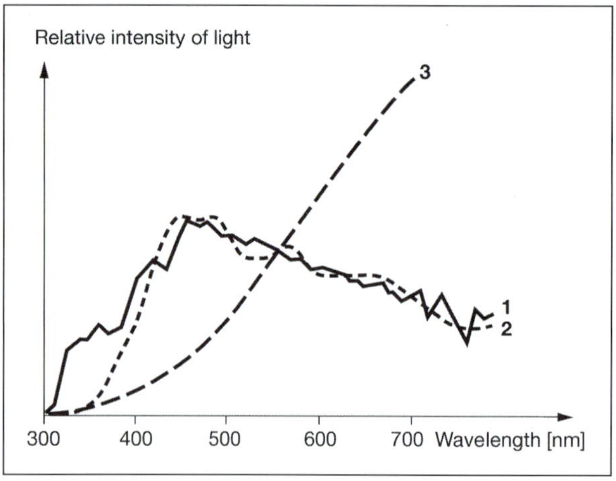

Figure 3.2.35: Distribution of spectral intensity of standardised lights (3) A and (2) D 65 compared with that of (1) sunlight as per CIE

- **Standard light A** for viewing in incandescent bulb light with high proportion of components in the long-wave range
- **Standard light D 65** for daylight under an overcast northern sky
- **Standard light B** for sunlight
- **Standard light TL 84** for department stores

Once account has been taken of the standardised light source, the reflective and absorptive behaviour of the coating and the defined sensitivities of the standard observer, it is then possible to describe colour impressions unambiguously. The most common types of light for coatings are standard lights A and D 65. Transportable light sources for application on site are available for the latter [3.4.17].

When standardised light of intensity S(λ) strikes the coating, it is changed by absorption and scattered to an extent depending on the wavelength. The reflected light R(λ) arrives at a measuring probe where it is detected by sensors and converted to the standard trichromatic values X, Y and Z in accordance with human eye sensitivities $\bar{x}(\lambda)$, $\bar{y}(\lambda)$ and $\bar{z}(\lambda)$.

The standard trichromatic values of body colours are therefore the product of spectral radiation distribution as a function of the type of light S(λ), the reflectance R(λ) and the standard spectral values of the observer $\bar{x}(\lambda)$, $\bar{y}(\lambda)$ and $\bar{z}(\lambda)$.

$$X = \sum_{380}^{720} S(\lambda) \cdot R(\lambda) \cdot \bar{x}(\lambda)$$

$$Y = \sum_{380}^{720} S(\lambda) \cdot R(\lambda) \cdot \bar{y}(\lambda)$$

$$Z = \sum_{380}^{720} S(\lambda) \cdot R(\lambda) \cdot \bar{z}(\lambda)$$

Since the spectral sensitivity of the eye depends on the angle of view and therefore on the size of the object, the CIE originally defined a field of measurement and field of view of 2°. The sensitivity curves of the so-called 2° standard observer were supplemented in 1964 by those of the 10° (wide-field) observer.

The measured standard trichromatic values X, Y and Z are converted into the standard trichromatic value components x, y and z so that further useful indicators may be determined:

$$x = \frac{X}{X+Y+Z}$$

$$y = \frac{Y}{X+Y+Z}$$

$$z = \frac{Z}{X+Y+Z} = 1 - (x+y)$$

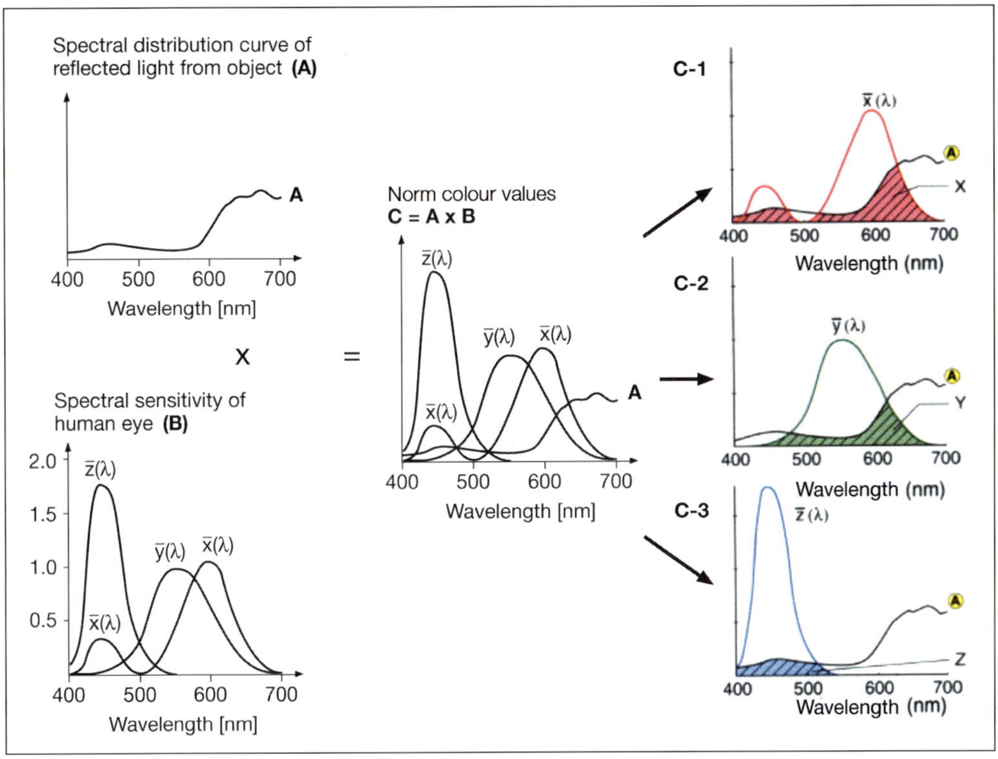

Figure 3.2.36: Determination of X, Y and Z

The value for z is given by z = 1 - (x + y). The standard trichromatic value component x represents the relative red component, y the relative green component. The standard trichromatic value Y is stated as the third numerical value of colour and serves to characterise the luminosity. All colours with the same standard value components x and y have the same type of colour. They differ only in the luminosity Y.

This method permits a **standard chromaticity diagram** to be calculated from the determined x and y values. Achromatic colours such as white, grey and black have a so-called equal energy spectrum with standard trichromatic value components x = y = 0.333.

All naturally occurring colours lie within the spectral locus presented. The pure spectral colours from 380 nm to 720 nm correspond to the points on the spectral locus. The straight line connecting points 380 nm and 720 nm is referred to as the purple straight. It cannot be represented by spectral colours. At the centre of the standard chromaticity diagram is the achromatic point. The more colourful, i.e. the more saturated, a colour is, the further away it will be from the achromatic point and the closer it will be to the spectral locus. If the numerical colour values x, y and Y have each been defined for two colours, it is possible to describe the difference in colour quantitatively.

One drawback of the system is that the calculated colour differences do not always coincide with the subjective sensation of colour difference. Many experiments have therefore been performed to eliminate this drawback of the **CIE system** and to develop a system with equal differences in accordance with perception. To strike a compromise, in 1976 the CIE recommended a transformation formula based on the standard trichromatic values X, Y and Z, which is commonly used nowadays and recognised internationally. In practice a

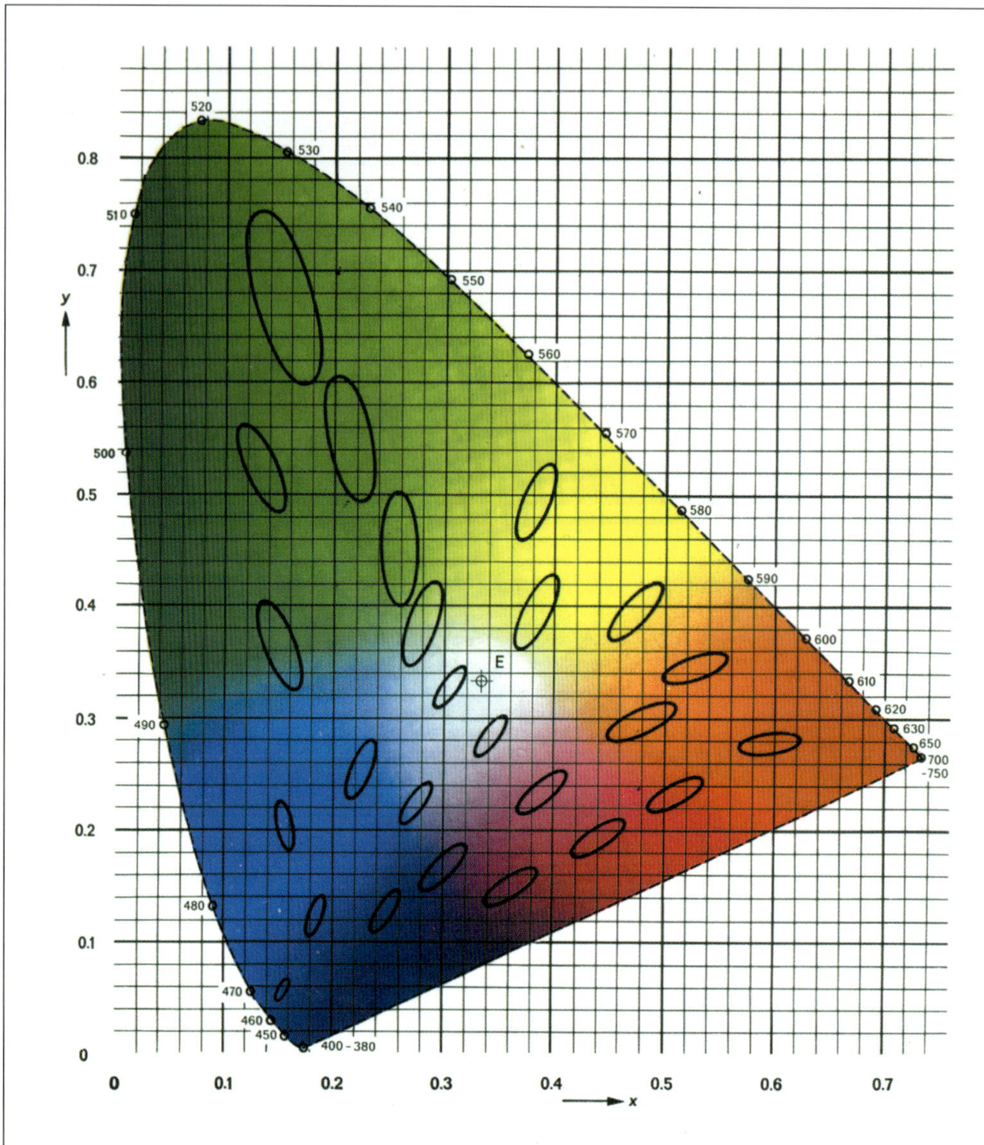

Figure 3.2.37: CIE standard chromaticity diagram from 1931

reliable procedure has been to apply the transformed numerical colour values L*, a* and b* as coordinates of a colour space (CIELAB system as per ISO 11664-4).

$$L^* = 116 \cdot \left(\frac{Y}{Y_n}\right)^{\frac{1}{3}} - 16$$

$$a^* = 500 \cdot \left[\left(\frac{X}{X_n}\right)^{\frac{1}{3}} - \left(\frac{Y}{Y_n}\right)^{\frac{1}{3}}\right]$$

$$b^* = 200 \cdot \left[\left(\frac{Y}{Y_n}\right)^{\frac{1}{3}} - \left(\frac{Z}{Z_n}\right)^{\frac{1}{3}}\right]$$

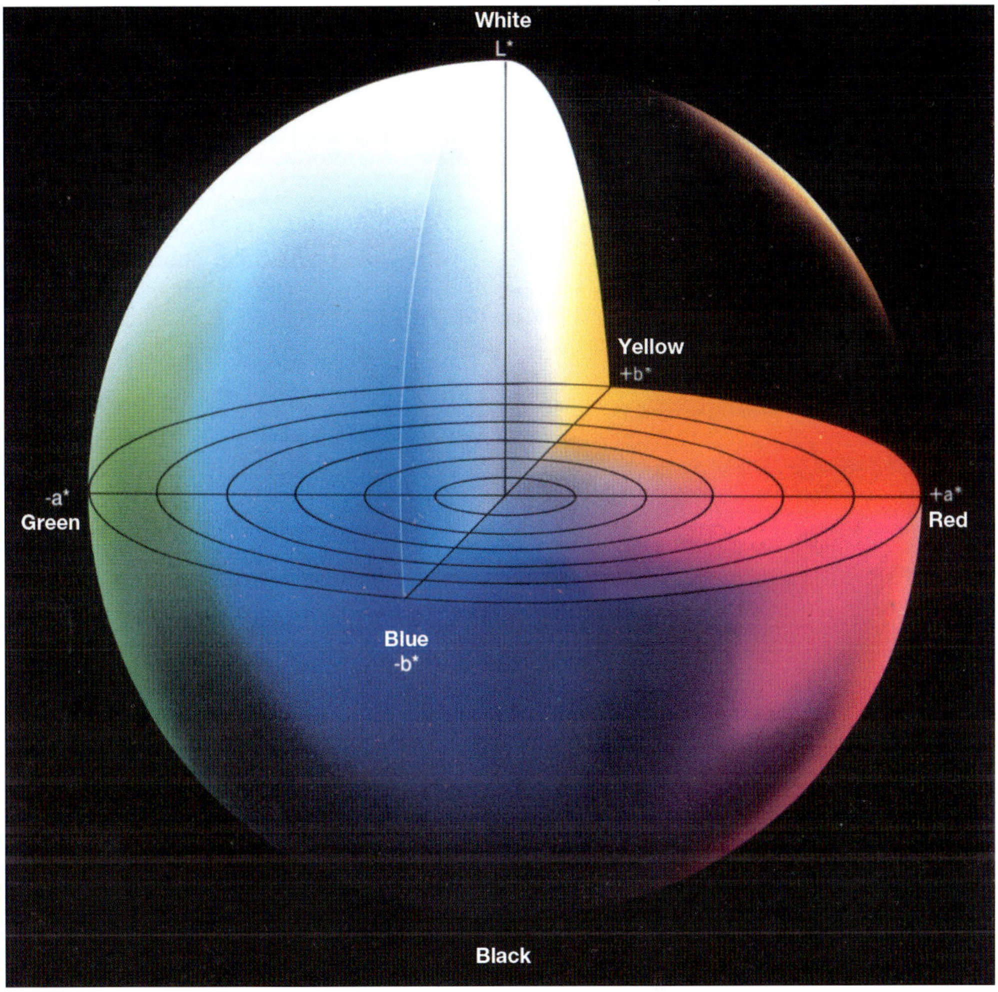

*Figure 3.2.38: Three-dimensional plot of L*a*b* colour space* *Source Minolta*

In these transformation equations X_n, Y_n and Z_n, specific to each type of light and each standard observer, are the standard trichromatic values of the perfectly matt white standard of barium sulphate. The transformation of the standard trichromatic values X, Y and Z to the **CIELAB numerical values** L*a*b* proves highly beneficial in practice because differences in hue and chroma can be represented more clearly on the a*b* plane.

The graphic representation of the colour loci of measured colours L*, a* and b* is three-dimensional. As shown in Figure 3.2.38 the positive and negative directions of the perpendicular a* and b* axes describe the red-green and yellow-blue colour range. The individual quadrants contain orange and brown colours, yellow-green, blue-green and violet ones, whilst the L* axis is perpendicular to axes a* and b*. The latter is calculated from Y and is the measure of luminosity.

The achromatic colours of white, grey and black are on the L* axis. In their case, a* and b* = 0, which corresponds to the achromatic axis. The farther a colour is from the achromatic axis, the more chromatic (saturated) it will be.

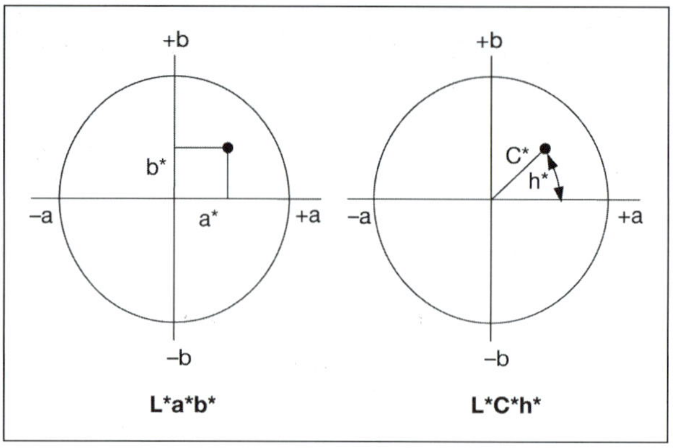

*Figure 3.2.39: Comparison of L*a*b* and L*C*h* coordinates*

Source Minolta

A different representation of the colours in the colour space described is afforded by the **L* C* h* chromatic system**. By contrast with the L*a*b* system, definition of colours is conducted not with cartesian coordinates but with polar coordinates in the form of an angle for the hue h and a radius vector from the centre point for chroma C*. In this

$$C^* = \sqrt{a^{*2} + b^{*2}} \quad \text{and} \quad \tan h^* = \frac{b^*}{a^*}$$

C* stands for chroma and h* for hue. The luminosity L* is identical in both systems. Colours which are on a radius ray have, by definition, the same hue h. Colours with the same L* all have the same luminosity; they are in the same plane parallel to the a*b* plane.

Calculation of Colour Difference

Although it is useful to be able to classify a number of colours in a colour space systematically, it is much more expedient when defining colour specifications to be able to calculate colour differences, as can occur between a standard (original colour) and a delivery (sample).

In practice nowadays the L*a*b* numerical colour values for a standard type of light and a standard observer, e.g. D 65 and 10° standard observer, are determined and assessed to verify the colour consistency between the original colour and deliveries. DIN 6174 quotes two different distance formulas for the L*a*b* system and the L*C*h* system, both of which lead to the same result.

$$\Delta E^*_{ab} = \left[(\Delta L^*)^2 + (\Delta a^*)^2 + (\Delta b^*)^2\right]^{1/2} \quad \text{for} \quad L^*a^*b^*$$

$$\Delta E^*_{ab} = \left[(\Delta L^*)^2 + (\Delta C^*_{ab})^2 + (\Delta h^*_{ab})^2\right]^{1/2} \quad \text{for} \quad L^*C^*h^*$$

ΔL* means the difference in luminosity, ΔC* the difference in chroma and Δh* the difference in hue, whilst Δa* refers to the difference towards the red-green axis and Δb* indicates the difference towards the yellow-blue axis.

One aim of describing colour differences is to perform instrumental analysis on the product batches released by a colourist and to define specifications in terms of numerical values. Achromatic colours with a chroma $C^* < 10$ are better represented by the L*a*b* system whilst more chromatic colours are better represented by the L*C*h* system. A further proposal calculates the colour distances on the basis of L*S*h* colour space, where S* stands for saturation. This colour space is much more symmetrical than the others and so measures the colour distances in better agreement with that perceived by the eyes [3.4.18].

The automotive industry imposes extremely high demands on colour matching owing to the large number of separately coated parts that are often supplied by other companies.

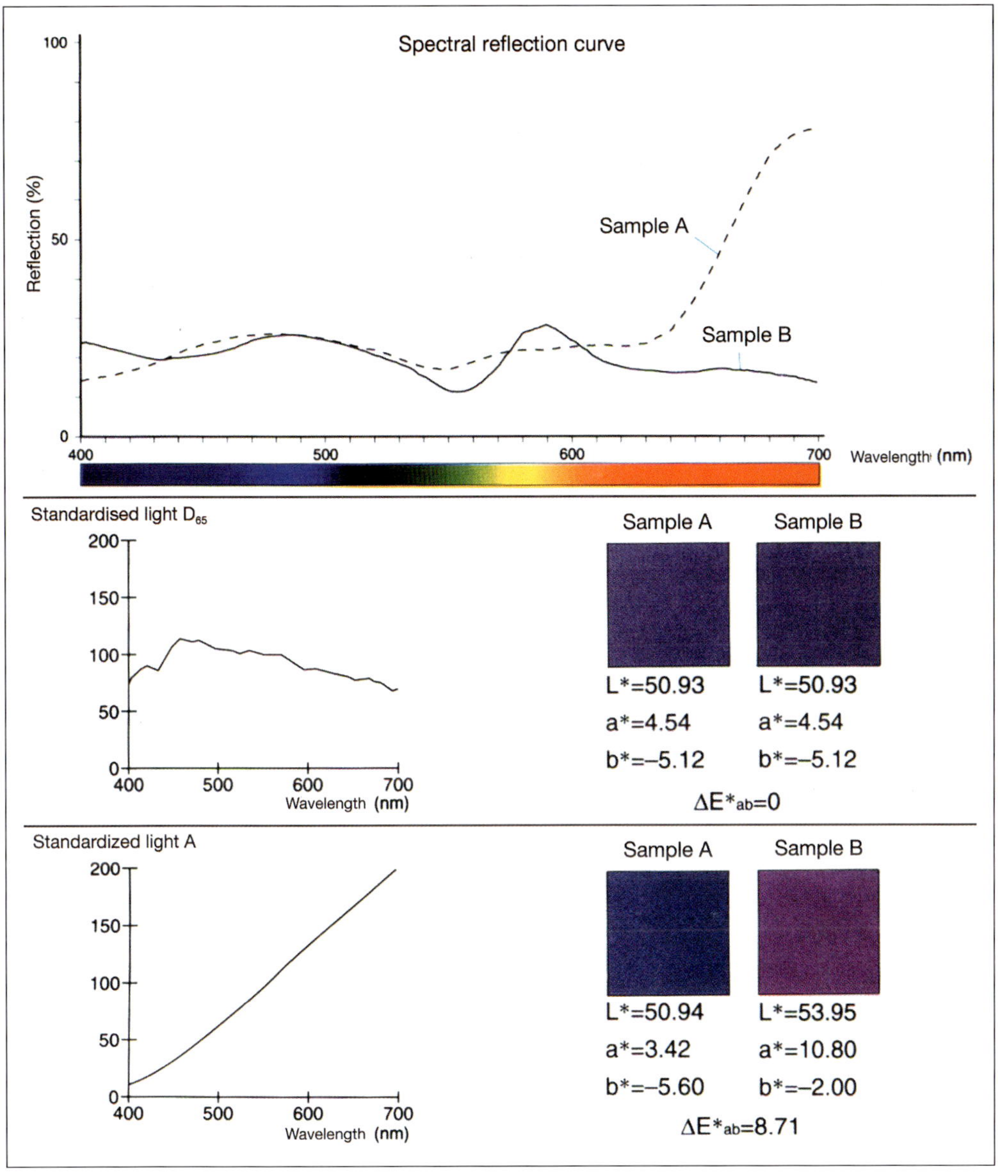

Figure 3.2.40: Example of measurements performed on two metameric coating films

In many cases, the colours are checked by inspection. Increasingly or additionally a colourimetric tolerance value ΔE is specified, based either on BS 6923 (ΔE_{CMC}) or on DIN 6175-1 and -2 (ΔE'). These values are independent of colour and use variably weighted factors from the CIELAB system for each colour. Usually ΔE' values < 1.4 are within the tolerance and so the parts concerned are approved for delivery [3.4.19].

Metamerism

A special feature of human colour perception leads to the fact that, despite different reflection curves, coatings exposed to certain illuminations can create the same colour impression. This phenomenon is termed metamerism. Two coats are metamers whenever they look the same in one type of illumination but look different in at least

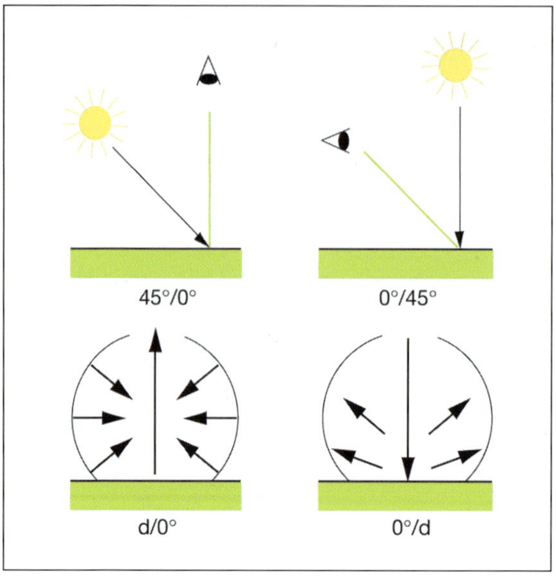

Figure 3.2.41: Geometries recommended by CIE for measuring heads

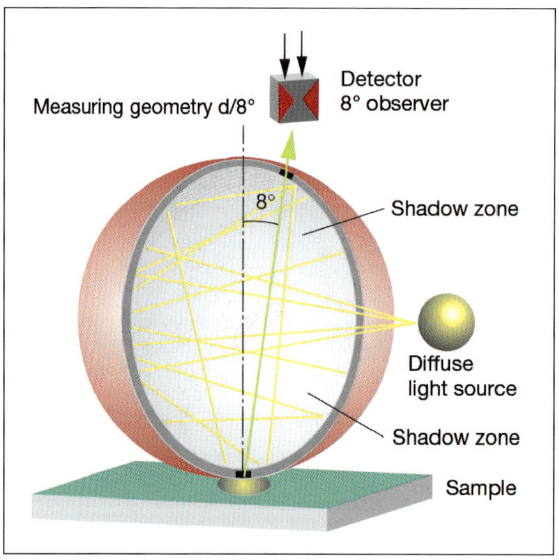

Figure 3.2.42: Principle behind a d/8° measuring head

one other type. This phenomenon assumes that two samples have different reflection curves, integrals of which add up to the same colour values in one type of light but not in the case of another type. The result is identical colours in the first type of light and different colours in the other (see Figure 3.2.40).

Measuring Instruments

As already indicated, the sensory impression complex resulting from light consists of the interactions of the light reflected from the surface (gloss) and of the component which penetrates and is then scattered back diffusely (colour). This reflectance R(λ) is measured with the aid of a **spectrophotometer**. To exclude the gloss factor, the CIE recommends four different measuring geometries for the various coatings:

1. Illumination at 45°/measurement at 0° (45/0)
2. Illumination at 0°/measurement at 45° (0/45)
3. Diffuse illumination/measurement at 0° (d/0)
4. Illumination at 0°/measurement diffuse (0/d)

An additional d/8° arrangement permits simultaneous measurement of gloss and colour in a single instrument that comes close to the sensory impression [3.4.19]. The spectrophotometers mentioned above have established track-records for determining standard trichromatic values.

A spectrophotometer measures the entire spectrum of reflectance R(λ) and uses the individual measurements to calculate the standard colour values X, Y and Z. The reflectance R(λ) needs to be continually calibrated against white and black standards, irrespective of the source of illumination used in the spectrophotometer. For measuring purposes polychromatic light is split into monochromatic light by a prismatic or grid monochromator or by narrow-band filters. A photoelectric cell determines the quantities of light reflected by the standard and the sample and evaluates their ratio. Since two beams are measured, i.e. the sample beam and the reference beam, instruments are referred to as

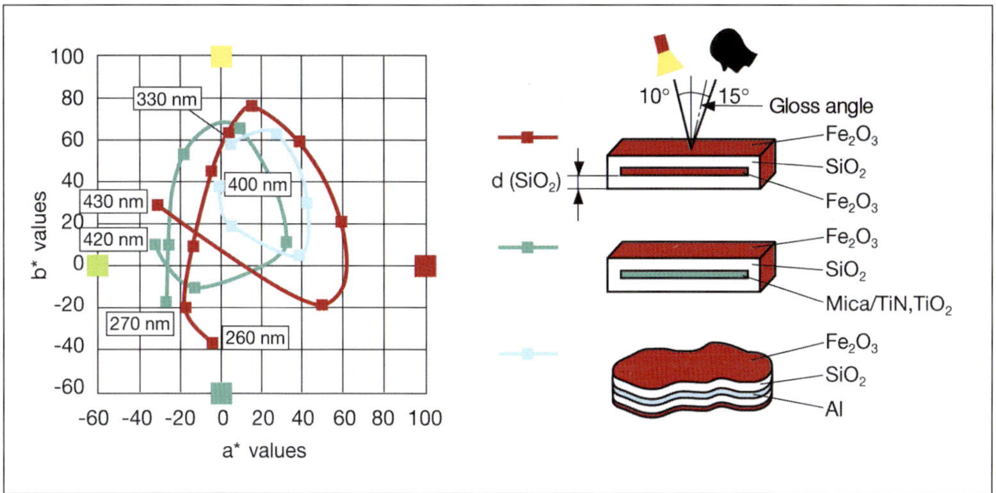

Figure 3.2.43: Colour shift in interference pigments as a function of viewing angle *Source: BASF SE*

dual-beam spectrophotometers. Understandably they have the advantage of compensating for device and power fluctuations and thus eliminating errors. The most accurate measuring instruments known nowadays are therefore based on the dual-beam method.

Although spectrophotometers have one type of light only, they can perform conversions into various types of light because they register the entire spectrum and in so doing cover metamerism.

Measuring the Colour of Effect Coatings

The colour effects of traditional pigments in coatings described so far are largely independent of the angle of observation on account of diffuse backscatter and can be measured by means of the colourimetric instruments described above.

The situation is different, however, in the case of metallic paints and paints containing interference pigments (see Chapter 2.1.3). The former show angle-dependent luminosity differences whilst the latter manifest angle-dependent chroma owing to path differences between the rays of light reflected from the surfaces and from the undersides of the pigment particles. Both require new test methods for characterising colour coatings. The crucial element here is **multiangle measurement**. The market has instruments of different measuring geometry to offer for this [3.4.20]. Normally the angle of incident light is constant at 45°, with five deviations from the angle of reflection (effect angle) ranging from 15° to 110°. Other instruments permit variable adjustment of the angle of illumination, with measurement at fixed effect angles. Even continuous increments of 1° in the illumination and the effect angles are possible nowadays.

Figure 3.2.43 shows an example of colour shift as a function of the angle of observation and illumination. The curves indicate the colour loci of interference pigments in plan view as a function of SiO_2 film thickness. The measurement was conducted with a Zeiss spectrometer, with both the angle of illumination and the angle of measurement being altered independently of one another. The angle of illumination was set at 10° and the angle of measurement was set at 15°, so measurement took place at 5° outside the gloss angle. This 10°/15° setting is roughly equivalent to the colour impression in plan view.

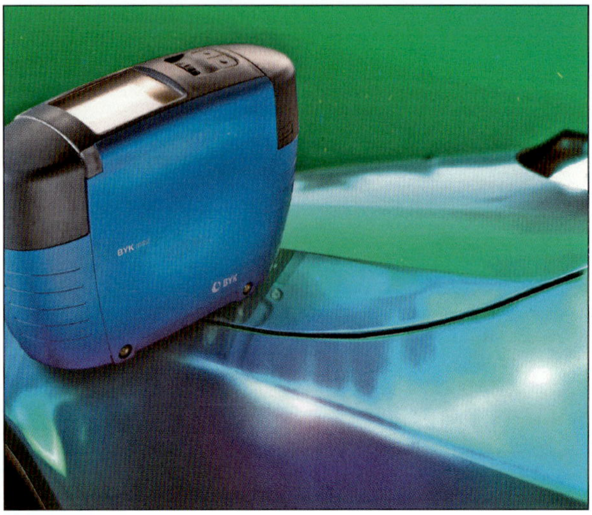

Figure 3.2.44: Multiangle spectrophotometer *Source: Byk Gardner*

The diagram shows three curves. They can be assigned to three chemical systems: the red curve represents the $Fe_2O_3/SiO_2/Fe_2O_3$ system, the silver one the $Al/SiO_2/ Fe_2O_3$ system and finally the blue one the structure of mica/TiO_2/ SiO_2/Fe_2O_3. The colour loci on a curve result solely from variations in the SiO_2 film thickness of a given layer system. As the SiO_2 film thickness increases the colour shifts from blue to green to gold and red of the first interference order and finally to the second and higher interference orders.

New effects in automotive coatings include "sparkle" which is caused by large-area aluminium flakes of low pigmentation or by other effect pigments such as Xirallic pigments [3.4.21]. The corresponding measuring instruments employ three angles of illumination with strong LED lamps. The instruments capture the corresponding images which are then subjected to image analysis to arrive at a sparkle parameter.

Studies of effect coatings comprising the latest generation of interference pigments and conducted with all kinds of measuring instruments are becoming particularly important in the automotive industry for just-in-time delivery of components whose colour must meet colourimetric specifications. Portable instruments are available which can help to gauge colour acceptance based on colour distance measurements and to verify specification compliance of delivered coloured components [3.4.21].

Colour Classification Systems

The concept behind colour classification arose from the need of designers and other people interested in colour to be able to select colours usefully. It consists in compiling **colour sample catalogues** with the closest possible arrangement of those individual colours which are perceived physiologically to have the same colour differences in relation to one another. Colour classification systems enjoy a certain amount of importance for classifying pigments and dyes as well, despite the existence of sophisticated colourimetric techniques (see Chapter 2.1.3). All colour classification approaches make distinctions between the empirical, unsystematic colour collections and the colour classification systems which are based on colourimetric facts [3.4.23].

One of the most important empirical collections is the American **Munsell catalogue**. This largely fulfils the requirements for approximately equidistant individual samples. *Munsell* colour collections are up to date and available in the trade. In addition, mention should be made of a large number of proprietary colour files indexes. The colours represented are selected according to subjective, arbitrary aspects, often based on marketing policy, and are assigned numbers and in some cases colour names. The **RAL register 8040 HR** is an example of a nonproprietary colour chart which is widespread in

Germany and neighbouring countries. Among the matt RAL samples there is a selection of high-gloss RAL charts (RAL 841 GL) whose pigmentation reflects the state of the art. The government institutions such as local authorities, the police and the armed forces have reserved some of these colours for their own use.

The German DIN system is regarded as the most important of the systematic scales. The **DIN colour charts** are based on the criteria of colour saturation and darkness in relation to luminosity. The colours have been selected so that the resulting colour saturation and darkness series each have perceptually equidistant subdivisions. The system of matt samples additionally published from 1960 on is based on the standard C light-type customary at that time and the 2° standard observer. In the meanwhile, a conversion to the 10° observer with light-type D 65 and high-gloss samples with modern pigments has become available.

The **Uniform Colour Scale System** issued by the Optical Society of America (OSA-UCSS) should be mentioned as a systematic colour collection. Based on the CIE it provides a collection of formulas and samples but its use is restricted to the USA.

3.2.5 Mechanical Properties

If a coating is to fulfil its requirements over a lengthy period from the outset, a number of criteria need to be met. Each is contingent on good bonding of the organic layer with the substrate. That is the only way to prevent the destructive influence of water, atmospheric oxygen or chemicals on the substrate to be protected. Furthermore, physical stresses in the form of expansion, compression or torsion must not diminish the bond with the substrate or affect the cohesive strength of the coating. The coating must also offer longterm resistance to abrasion and impact.

The corresponding parameters for adhesion, for elasticity during slow or rapid application of stress, and for physical hardness, all which fall under the term physical/technological properties, must reach a certain minimum level sufficient to cover the extremely temperature range that spans the world's various climate zones.

Specification and testing of the quality level can be conducted in two different ways. In the first, the coatings are exposed to the usual stresses encountered in practice under various degrees of intensity. Examples of test methods which have been used for decades are physical elongation by bending coated metal sheets, blasting with metal balls or road chippings, and artificially damaging and peeling off with defined tools. Their aim is to characterise the coatings' performance. Indicators determined in this way often involve complex links between various physical properties, the assessment of which can often requires the use of a ranking system.

In the second, modern paint research is endeavouring to describe coatings' performance in terms of physical parameters such as modulus of elasticity, adhesive force, relaxation times and retardation times. Such indicators, which need to be measured accurately, are beneficial for making test results objective and for conducting necessary online production tests. They prove to be a great help wherever there is a clear relationship between the indicators and the performance of the coating. The objective of wholly objective analysis, however, has not been achieved yet. Test methods such as *Erichsen* cupping, *Buchholz* and pendulum hardness, cross hatch and mandrel bending, are all still standard tests which are used for quality assurance purposes.

3.2.5.1 Adhesion

Theory of Adhesion

Even if a coating becomes only partially detached from its substrate, it is no longer fit for purpose. Firm attachment to the substrate which also remains intact under the effect of moisture, light, salts and gases, i.e. under the influences encountered in practice, is a necessary requirement for long-lasting protection. On metal substrates good attachment suppresses the formation of corrosive products even if the coating is physically damaged.

The adhesion of a coating is not determined solely by the paint. The special characteristics of the coated object also need to be taken into account. For optimum adhesion, the coating process has to be preceded by cleaning processes or layer-forming pretreatments. Good levels of adhesion are therefore not due solely to the paint but also result from judicious selection and adaptation of paint materials and substrate (see Chapter 4.1).

Terms other than adhesion which are also used in connection with adhesion of coatings include adhesive strength, adhesive power and **bond strength**. Problems arise wherever these are taken to be synonyms for the force per unit area which has to be applied to detach the coating, because the reference parameter, area, needs to be defined more precisely (see Figure 3.2.45). Parameters that need to be considered are the measurable surface area of coating A1, the surface area of the partially highly textured substrate with all its irregularities A2 or only the effective common boundary surface area A3. The last two may be identical if wetting of the substrate is good, but they are always very different from the macroscopically measurable area A1.

For a scientific study of adhesion mechanisms it is useful to consider only the real common surface area and to make allowance for surface texture. However, this calculated indicator is of little interest when it comes to characterising performance properties. Instead, a more pragmatic reference parameter is the macroscopically measurable substrate surface area or coated surface area.

Any attempt to elucidate the causes of interactions by examining the literature will reveal a host of theories which can vary considerably [3.4.24, 3.4.25]. The explanations range from mechanical attachment of the coating in the cavities and fissures of the

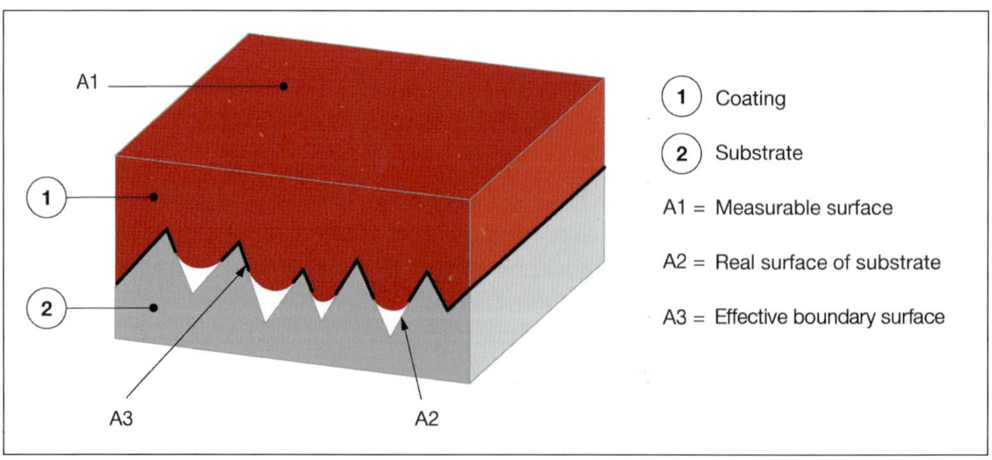

Figure 3.2.45: Schematic drawing of surface areas to be considered in the determination of adhesion

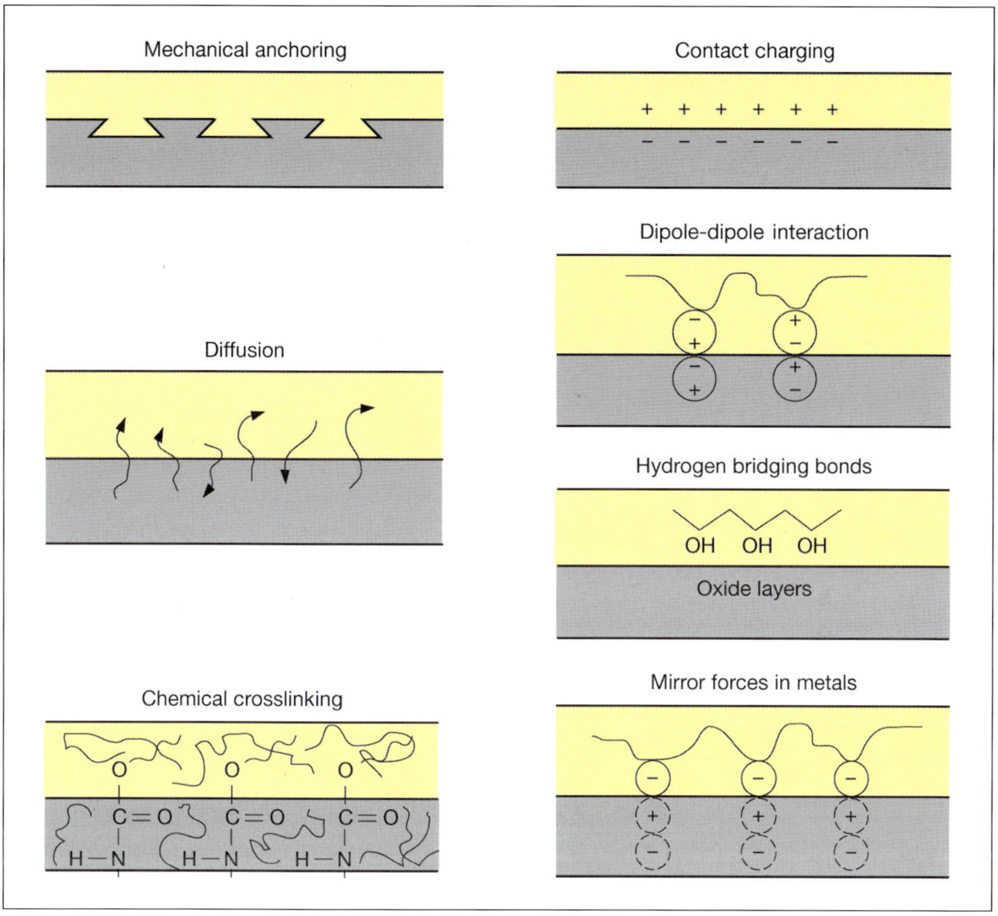

Figure 3.2.46: The mechanisms behind the adhesion of coatings to the substrate

substrate (press-stud theory) or mechanical anchoring and attachment of film forming agent molecules by diffusion or contact charges and the creation of mirror forces through to interactions of polar functional groups, hydrogen bridging bonds, and chemical links between the coating and the substrate.

There is no question that all those mechanisms can affect bond strength and adhesion. There is equally no disputing the fact that the individual mechanisms of adhesion can only come to bear significantly if the necessary criteria have been met. Mechanical attachment to rough metals and wood is without doubt beneficial for bond strength. The same applies to bonds formed when a film penetrates plastics, primers and fillers. However, to hold the press stud and diffusion theories chiefly responsible for the bond formed between a coating and glass or polished metals would be wrong. The same cannot be said of the interactions summed up in the term *van der Waals* forces. Regardless of the type and morphological features of the substrate these generally are the main causes of **adhesion**.

Van der Waals forces are orientation forces (dipole-dipole), induction forces (dipole/induced dipole) and dispersion forces. Furthermore, given a suitable chemical structure and an appropriate substrate, hydrogen bridging bonds also take effect.

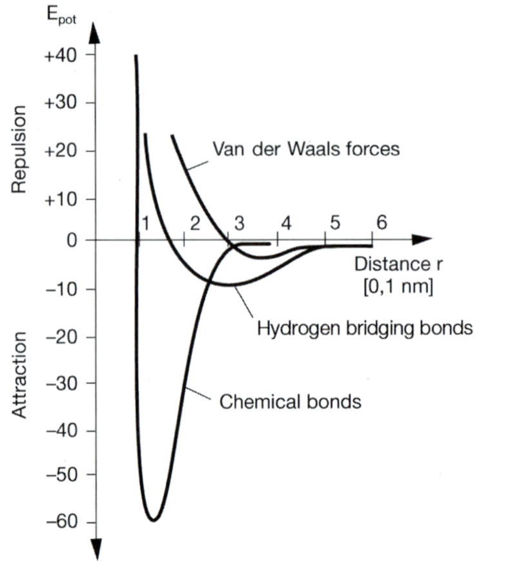

Type of van der Waals forces	Average energy	Range
Orientation forces Dipole-dipole (Keesom forces)	25 kJ / mol	$E = \frac{2\,\mu^4}{3kTr^6}$
Induction forces Dipole-induced dipole (Debye forces)	15 – 20 kJ / mol	$E = \frac{2\alpha \cdot \mu^2}{r^6}$
Dispersion forces (London forces)	5 – 10 kJ / mol	$E = \frac{3h\nu_0 \cdot \alpha}{4r^6}$
Hydrogen bridging bonds	40 – 50 kJ / mol	$E = e^{-kr}$

Figure 3.2.47: Curves and magnitudes of the potential energies of van der Waals forces and hydrogen bonds as causes of adhesion

At fractions of nanometres the range of *van der Waals* forces and hydrogen bridging bonds is extremely short. Optimum adhesion is therefore contingent on good wetting by the material applied, so that the ideal conditions are created for inducing the film forming agent molecules to approach the substrate. The conditions for good wetting are always fulfilled whenever the surface tension of the substrate is higher than that of the still liquid coating material (see Chapters 2.3.2 and 3.1.1). Such a requirement is easily met when metals are coated, due to high **surface tension**. With various nonpolar plastics such as polyethylene or polypropylene having values less than 30 mN/m, the inferior wettability prevents strongly adhering coatings from being obtained, unless polarising surface treatments are performed.

Even so, merely satisfying all the physical boundary surface requirements is not enough to achieve the best possible bond. In addition, the viscosity has to be sufficiently low in order that those functional groups capable of bonding may be given an opportunity for orientation towards the substrate. Furthermore, the rise in surface tension due to evaporation of the solvents, which normally have a lower surface tension, needs to be adjusted in such a way that the wetting behaviour is not impaired by film formation. Should it happen that, at a still relatively low viscosity, the surface tension of the paint exceeds that of the substrate, the ensuing dewetting will give rise to defects in the coating. It is therefore useful to adjust the values of surface tension in paint such that the surface tension in the cured film is similar to or preferably the same as the surface tension in the substrate.

Formulations capable of meeting these requirements can only be prepared if the surface tensions of solids are known and measurable. Whilst the surface tension of liquids can be determined by numerous methods (see Chapter 2.3.2), direct measurement in solids is not possible. It can only be performed indirectly by measuring the contact angle of droplets of liquid which have been applied to the solid surface being examined. If the surface tension of the liquid σ_L is higher than that of the substrate σ_S, a commensurately different contact angle α is created.

If the contact angle α approaches 0, the boundary surface tension $\sigma_{S,L} = 0$ and hence $\sigma_L = \sigma_S$.

Since small **contact angles** α cannot be determined with sufficient accuracy, recourse is made to the *Zisman* method whereby several liquids a, b and c with different but higher surface tensions than the substrate are measured for their resultant contact angle and each calculated cosine is plotted against the surface tension of the liquids. The curve is then extrapolated to the value $\cos \alpha = 1$ [3.4.26]. The measurement is performed on optical benches which allow precise recording of the wetting angles.

Substrate	σ [mN/m]
Stainless steel	1000
Phosphated steel	34
Mercury	480
Glass	74
Polypropylene	30–35
Polyamide	40–43
PVC	36
Polystyrene	32
Polycarbonate	37
Polyester	40–45

Figure 3.2.48: Surface tensions of selected substrates

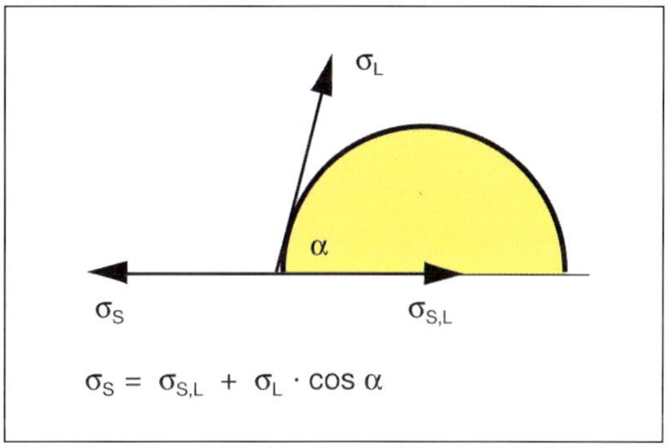

Figure 3.2.49: Definition of contact angle according to DIN 55660-2

Experience of experimentally determined critical surface tensions σ_c shows that deviations from the surface tension of the substrate can occur in many cases. However, the critical surface tension σ_c can be converted into the surface tension of the substrate σ_s by applying a correction in the form of *Good's* interaction parameter Φ.

$$\sigma_c = \Phi^2 \sigma_s$$

The **interaction parameter** provides information about the spatial interaction of groups and hence about the quantity of the maximum number of bonding sites. It is difficult to calculate wherever polymer interaction is involved and so is usually determined by experiment. The *Good* factor assumes a value of 1 if all the polar groups of coating material and substrate can interact. However, it can also drop to values of around 0.3 [3.4.27].

With regard to adhesion, experience teaches that even if the surface tensions of coating and substrate are optimally adapted the expected optimal values are not always found. Better agreement between experiment and theory are obtained if the surface tension is not entered into the calculation as the total value of all the *van der Waals* and hydrogen bridging bond forces. Splitting the surface tension into polar and disperse components σ^p and σ^d

$$\sigma = \sigma^d + \sigma^p$$

enables better forecasts to be made about the maximum possible level of adhesion. The best-possible adhesion is therefore achieved if the polar components of substrate and coating assume the same value:

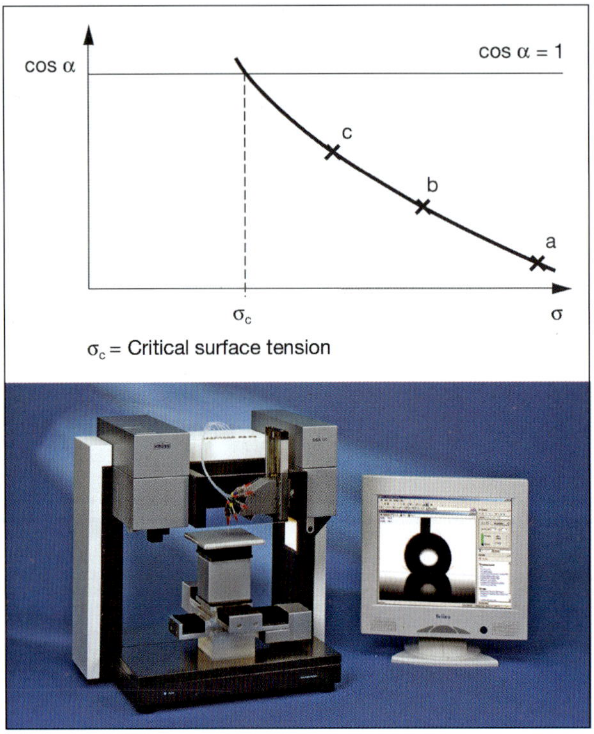

Figure 3.2.50: Zisman diagram for determining the critical surface tension and picture of an instrument for measuring contact angles
Source: Krüss

$$\frac{\sigma_L^p}{\sigma_S^p} \rightarrow 1$$

It must be borne in mind that the best-possible adhesion of the system says nothing about the absolute level of the adhesive forces. These are pre-determined by the polarity of the interacting surfaces and are merely optimised by the adaptation of the surface tensions.

The extent to which adhesion is affected by the ratio of the polar components of surface tension in the substrate to those in the coating has been verified for various polymers [3.4.28]. The examples of acrylic resins and alkyd resins shown in Figure 3.2.51 reveal systematic interrelationships.

Several teams have devised practicable approaches for determining the polar and disperse components. According to *Owens* the boundary surface tension is the geometric mean of the individual constituents [3.4.29]. If Φ = 1, it follows that:

$$\sigma_{S,L} = \sigma_S + \sigma_L - 2 \cdot \sqrt{\sigma_S^d \cdot \sigma_L^d} - 2 \cdot \sqrt{\sigma_S^p \cdot \sigma_L^p}$$

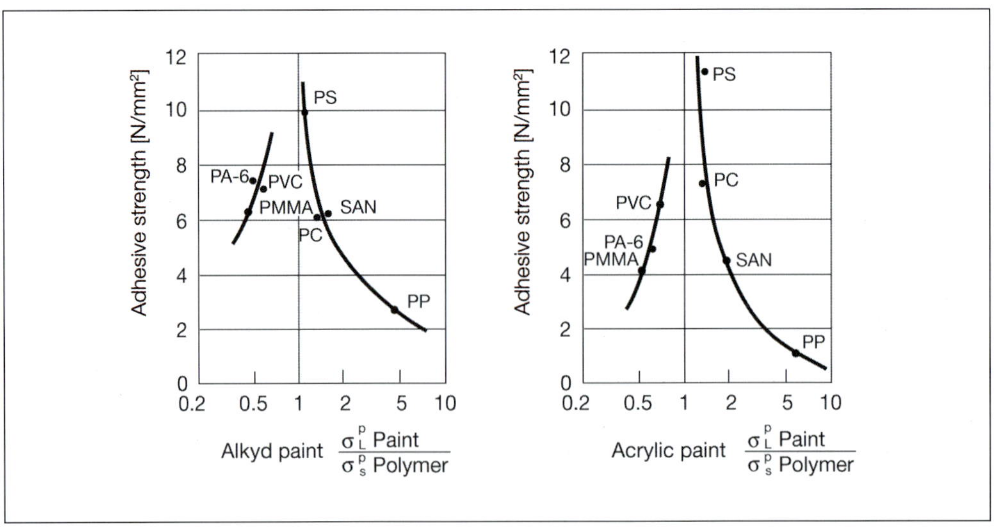

Figure 3.2.51: Adhesive strength of alkyd and acrylic paints on different polymers as a function of the ratio of the polar surface tension components

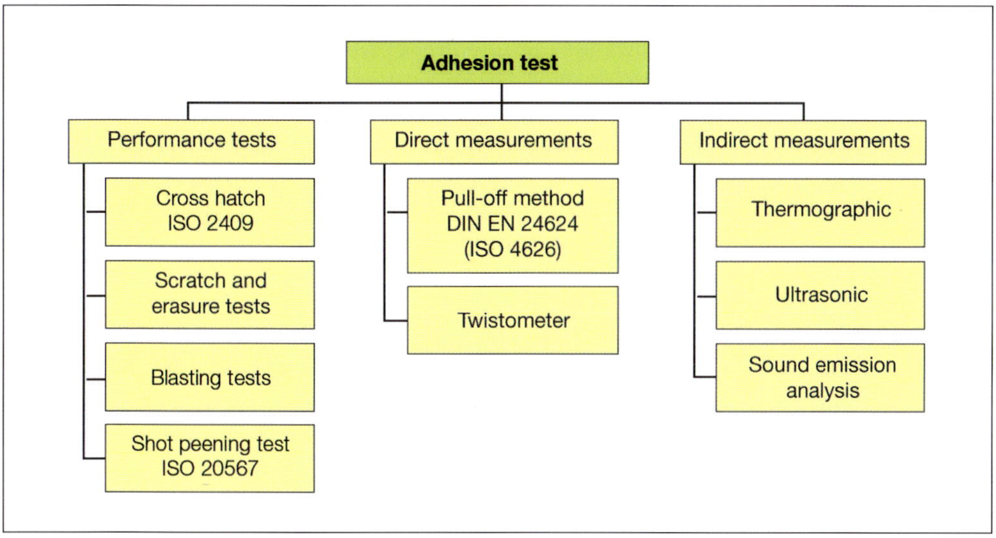

Figure 3.2.52: Overview of adhesion tests

After allowance has been made for *Young's* equation (see Chapter 2.3.3) and after contact angles of liquids of known disperse components σ^d_L have been measured in wetting tests it is possible upon further mathematical treatment to calculate disperse and polar components of the surface tension of the substrate σ^d_S and σ^p_s.

$$\frac{\sigma_L \cdot (1+\cos\alpha)}{2 \cdot \sqrt{\sigma_L^d}} = \sqrt{\sigma_S^p} \cdot \sqrt{\frac{(\sigma_L - \sigma_L^d)}{\sigma_L^d}} + \sqrt{\sigma_S^d}$$

$$y = a \cdot x + b$$

Despite all optimisations, the use of model substances frequently yields only satisfactory agreement between theory and practice. The reasons for this include undefined changes in the boundary surfaces due to water vapour adsorption and cross-contractions of coating materials as they cure on textured surfaces.

Since bonding forces have an extremely short range, the substrate needs to be freed, with the utmost care, of all impurities that could impair wetting. Pretreatments to remove dirt, dust and residues of oil are therefore standard procedures which must precede the coating process. Layer-forming surface pretreatments such as phosphating and chromating have proved reliable, especially for coating metals (see Chapter 4.1.2). The systematic formation of insoluble, crystalline, tenacious salts or oxides bolsters the bond strength through additional attachment, one of the reasons being that the surface area is substantially enlarged.

The influence of solvents on swellable substrates, as in the case of plastics and multilayer coatings, can positively or negatively affect adhesion to an extent depending on the degree of crosslinking (see Chapter 4.1.3). For instance, overstoved coatings feature reduced bond strength owing to a reduced ability to swell.

Studies of the dependence of adhesion on temperature when the real common boundary surface is known and interfering factors on the surface are eliminated permit the activation energies required for adhesion to be calculated. In this way it is then possible to find details about the nature of the bonding forces [3.4.30]. Nevertheless, the theories on the adhesive strength of polymer coatings are still not far enough advanced to allow reliable forecasts of bond strength or adhesion in all cases.

Methods of Measuring the Adhesion of Coatings

Empirical tests are the customary way to assess bond strength and adhesion. Although they are not based on any physically unambiguous principles, they are relatively easy to perform and yield largely reproducible test results.

The **cross hatch test** per ISO 2409 and **scratch and erasure tests** are used for obtaining quick results because they are easy to perform. Although the test result does not provide direct readings, the degree of damage can be rated by comparison with reference images.

Numerical values for the bond strength of coating systems are obtained by determining the minimum **tensile stress or torsional stress** needed to separate the coating from the substrate. **Ultrasonic and thermal flux measurements** are further indirect methods for characterising the adhesion of coating systems.

Pull-Off Method as per ISO 4624

This test consists in applying the coating material to a flat test panel of the substrate being examined. When film formation is complete a dolly is stuck onto the test coating, which is allowed to cure. The dolly is then pulled off vertically. The tensile force must be distributed perpendicularly over the full test area and be increased until failure occurs within 90 seconds of application of the stress. In the case of thin substrates, any risk of deformation which could falsify the result is mitigated by fixing the specimen between two dollies with the aid of a centring device. Care must be taken to ensure that the two dollies are positioned coaxially one above the other.

Special attention must also be paid to the choice of adhesives. It is important for the adhesive bond to be stronger than that of the test coating. Moreover, the adhesives must not cause any swelling in the test coating. The measurement is then conducted with the aid of a tensile testing machine.

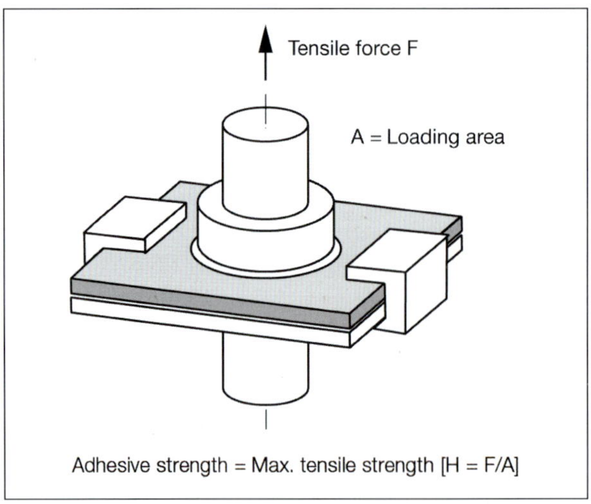

Figure 3.2.53: Measuring adhesion by the pull-off method as per ISO 4624

The parameter determined is the adhesion value H resulting from the force F, per measured area A, which has to be applied until separation occurs. At least three measurements are performed. Analysis of the failure pattern, i.e. position and size of the detached area, yields further information about the adhesion of the coating, especially of multilayer coatings.

Torsion Method with the Epprecht Twistometer

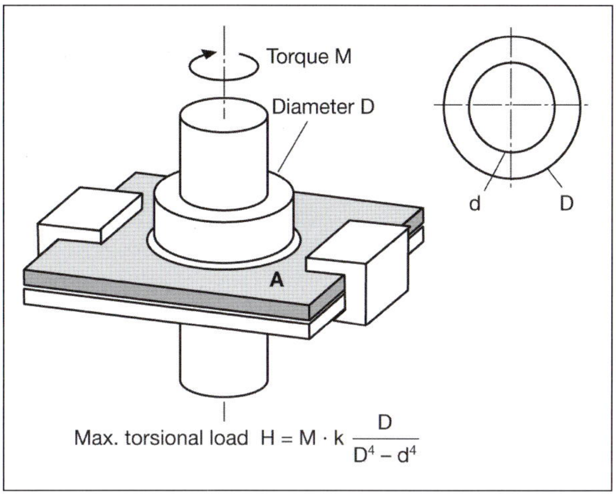

Figure 3.2.54: Experimental set-up for an Epprecht twistometer

The *Epprecht* twistometer is used to determine the bond strength of coatings under torsion. For this purpose, the dolly bonded to the panel as in the previous test set-up is subjected to a torsional load. In this set-up the bonded area is a ring with an outside diameter D and inside diameter d. The torque M which is effective at the moment of failure is converted into the separation force H per unit area. Analysis of the failure patterns makes it relatively easy to locate the point of lowest cohesion within the coating system.

$$H = M \cdot \frac{16}{\pi} \cdot \frac{D}{D^4 - d^4}$$

If the coating becomes entirely detached from the substrate due to adhesive failure, the readings obtained are representative of the bond strength. By contrast, **cohesive failure** (i.e. within the coating) cannot be used to draw direct inferences about adhesion. The result merely indicates that adhesion to the substrate or between two layers of paint exceeds the cohesive capabilities of the paint.

Cross Hatch Method as per ISO 2409

The cross hatch test is a simple empirical test for assessing the bond between single-layer and multilayer coatings and the substrate and between the individual layers. Comparative interpretations of the test results are only possible if the coatings have the same viscoelastic properties. Highly elastic films are easier to deform, as a result of which lower delamination forces are needed during mechanical loading.

On anisotropic substrates such as wood the test results measured at various points may show so much scatter that no absolutely reliable statements can be made about adhesion. The cross hatch test is unsuitable for such substrates.

The test instrument specified by the standard is a cutting tool of precise shape and dimensions. The cutting tool is chosen to suit the film thickness and the type of coating. Consequently, **multicutter instruments** with more than one cutting blade require considerable force application on very hard films and so it is often not possible to make all the cuts down to the substrate. In such cases an **instrument with a single cutting** blade is to be recommended.

Measurements with the multicutter instrument are done by making a continuous longitudinal band of cuts comprising six parallel notches in the sample coating down to the substrate. Another band of cuts is then made at right angles to them. This yields a grid of 25 squares, the so-called cross cuts.

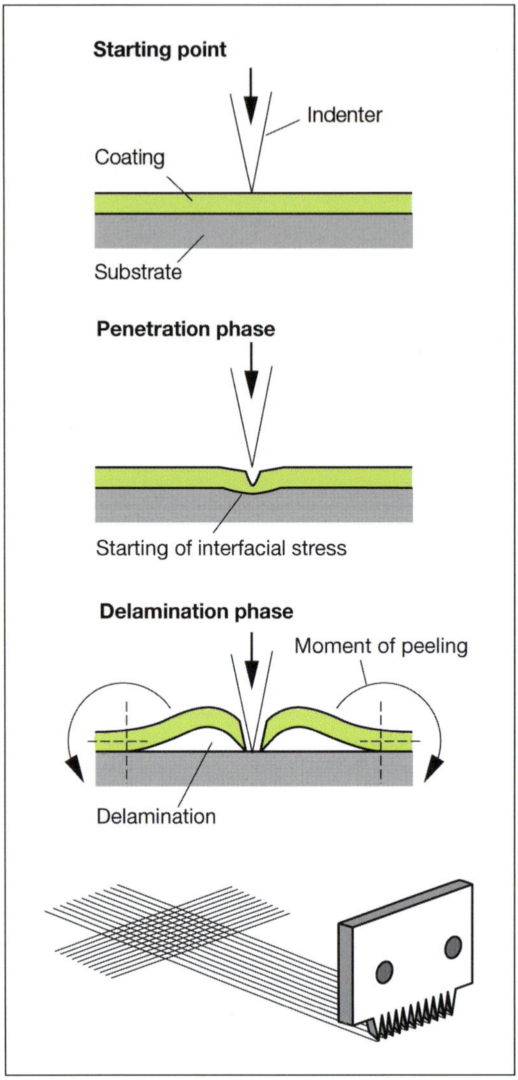

Figure 3.2.55: Schematic drawing of delamination in the cross hatch test as per ISO 2409 and the multicutter device

The single-cutter instrument is best used with a folding ruler which guarantees the necessary cut spacing and parallel cutting.

Evaluation consists in pressing an adhesive tape of defined adhesive force onto the stressed area and then pulling it off the surface at as constant a rate as possible. Cross hatch index numbers ranging from Gt 0 to Gt 5 are assigned by reference to the standard patterns. To increase reproducibility of measurements and make the results more objective the cuts are made in the coating by automatic devices. These are able to adapt the cutting depth to the film thickness. Image analysis may also be used to evaluate the test result [3.4.31].

It is customary in the automotive and other sectors to apply a cross hatch or just a St. Andrew's cross in the centre of the test panel and then to test the adhesion with a jet of water in imitation of real-life conditions, e.g. manual pre-cleaning with a water jet in a carwash. While major differentiation is not possible, the coating must nonetheless pass the test without coming away. This "pressure-water jetting" test is described in ISO 16925.

Peel-Off Method

Pull-off and torsional methods aside, recourse is occasionally made to methods based on determining the work required to peel off the coating and then converting it into adhesion values.

In the **wedge feed method (hesiometer)** a wedge is positioned in the boundary surface between the coating and the substrate and moved parallel to the substrate so that the coating becomes detached. The work done is the product of the feed force F and travel distance s divided by the peeled area A and serves as a direct measure of the bond strength of the two layers. Since A = b s, the peeling work W_K per unit area as a measure of adhesion is given by:

$$W_K = \frac{F \cdot s}{A} = \frac{F}{b}$$

The wedge feed method not only permits statements to be made about adhesion via the failure pattern but also provides information about the hardness and elasticity of the coating.

Cross hatch index	Description	Reference picture
Gt. 0	The cut edges are completely smooth; no partial delamination	——
Gt. 1	No fragments seen at intersections of cross cuts; delaminated area: approx. 5 %	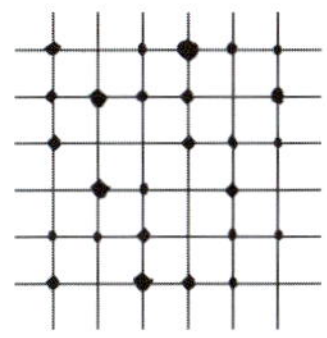
Gt. 2	The coating is delaminated along the cutting lines and/or the cross sections; delaminated area: approx. 15 %	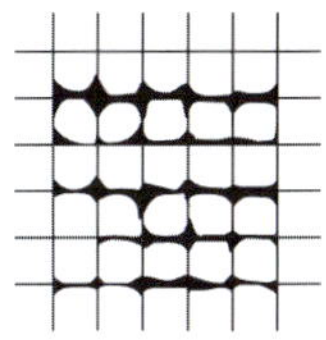
Gt. 3	The coating is partially or wholly delaminated in strips along the cutting lines; delaminated area: approx. 35 %	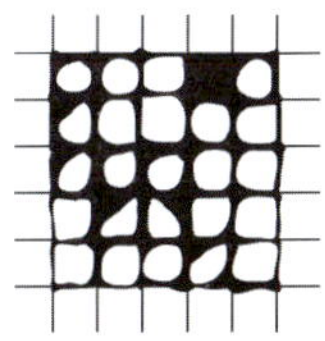
Gt. 4	The coating is delaminated along broad strips and/or individual segments; delaminated area: approx. 65 %	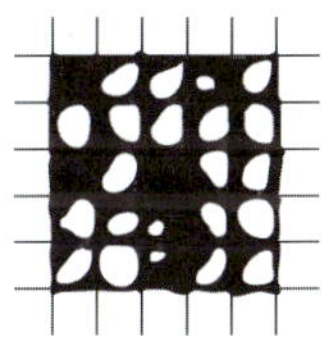
Gt. 5	Delaminated area: more than 65 %	——

Figure 3.2.56: Cross hatch index numbers for evaluating the findings as per ISO 2409

The **peel test** has a similar measuring effect but employs a different measuring technique. It consists in peeling a strip of coating of specified width b from the substrate at an angle β. To ensure that the film does not tear if it has insufficient cohesive strength, a reinforcing wire is placed in the centre of the coating of paint when the test pieces are made. A tensile force F normally in the form of a weight acts at the end of the wire. The weight at which delamination commences is determined. The subsequent calculation of adhesive strength takes account of the angle β at which delamination occurs.

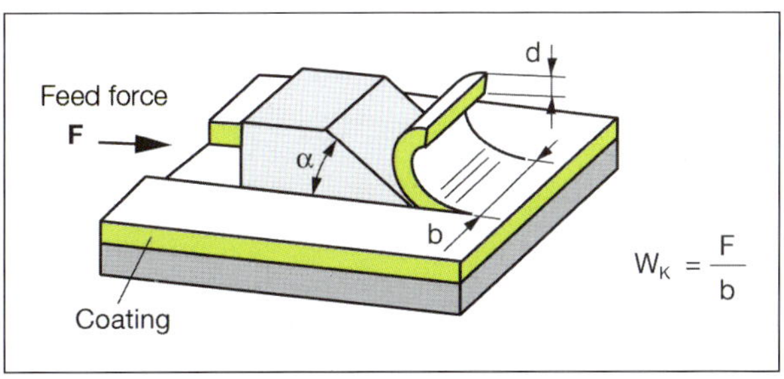

Figure 3.2.57: Measuring principle behind the wedge feed method

Adhesive strength tests based on **scratch and erasure methods** in which the coating is stressed by rubbing or scratching with a knife are of little interest although they are still used by

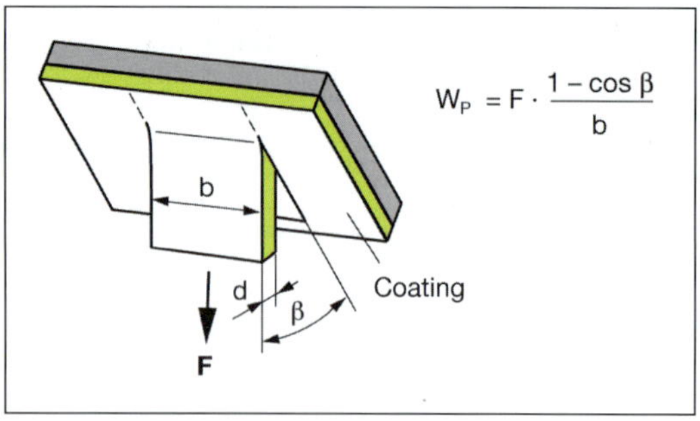

Figure 3.2.58: Measuring principle behind the peel test and calculation of adhesive strength

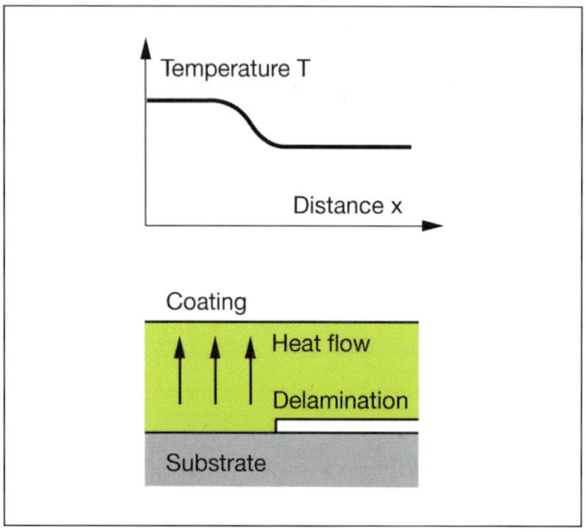

Figure 3.2.59: Heat flow as a basis for measuring adhesion

experienced experts. Even though it is difficult to award wholly reliable grades, the failure pattern provides the experts with information about the adhesion, elasticity and hardness of the coating.

Another variant is the ***Peters* chip test** in which a razor blade is angled so as to cut shavings from the coating. The coating is rated with a number which varies with the length of the shavings removed and which allows a statement to be made about the extent of adhesion.

Another less used method of determining adhesion consists in exerting defined impact stress by means of the **shot peening test** as per EN 2829. Since delamination due to steel shot hitting the surface is influenced by elasticity in addition to bond strength, this is more of a combined test of performance.

So-called **thermographic methods** for determining adhesion forces work on completely different principles. Heat applied to the substrate by convection or IR rays passes through the boundary surface with the coating as a function of the bond and is recorded by an infrared camera in the form of a temperature profile T(x) on the surface of the coating. The thermal flux is impaired at any point where there

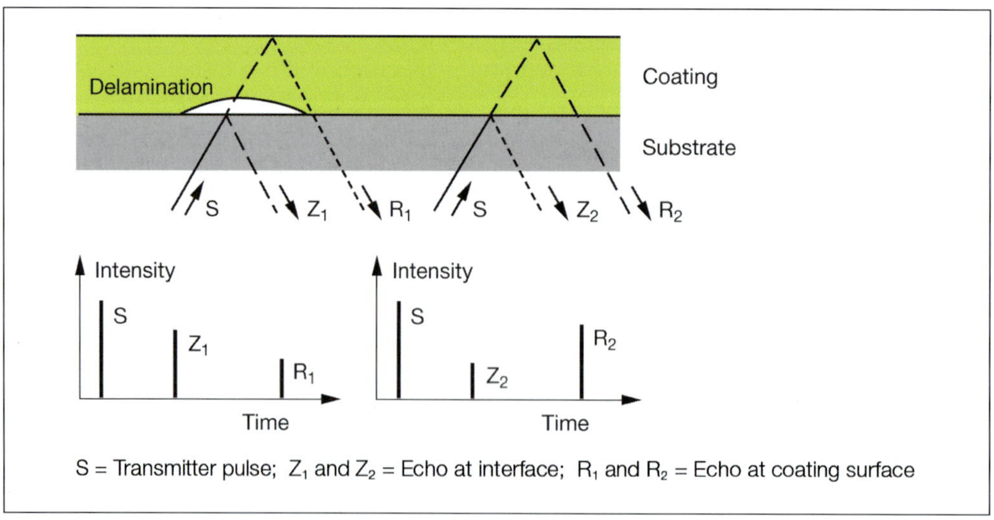

Figure 3.2.60: Schematic drawing of the use of ultrasonic reflection to measure adhesion

is insufficient adhesion so that areas of inferior bond strength are indicated by a low surface temperature.

The drawback of this test method is the lack of differentiation. However, thermal flux measurements are ideal for conducting a non-destructive search for weak points in the bond between layers [3.4.32].

Bond weaknesses in multilayer systems can also be found by acoustic methods. The **ultrasonic pulse echo method** with frequencies above 1 MHz records the sound, Z and R, reflected from the individual layers and the surface. The arrival time of the reflected pulses at the measuring instrument varies with individual film thickness while the intensity of the pulses is a function of the defects in the bond.

The method of **sound emission analysis** is based on the fact that a coating subjected to mechanical stress can emit sound signals. When stress limits are exceeded, the energy stored during elastic deformation can spontaneously transition into new energy states. The energy released is manifested as a sound pulse [3.4.33]. The measurement is performed by elongating a coated steel panel at a defined rate. Incipient damage to the bond is 'heard' before it can be 'seen'. The sound emission is measured, amplified and analysed. This measuring method does not permit any fine differentiation of bond strength and so only rough estimates can be made between good and not so good adhesion.

3.2.5.2 Elasticity

Theory

Coatings are subjected to all types of different deformation throughout their service life which have to be withstood without physical damage, loss of performance or loss of adhesion. The usual physical types of deformation are elongation, compression, shear, bending and torsion. In practice, automotive coatings, for example, can experience brief

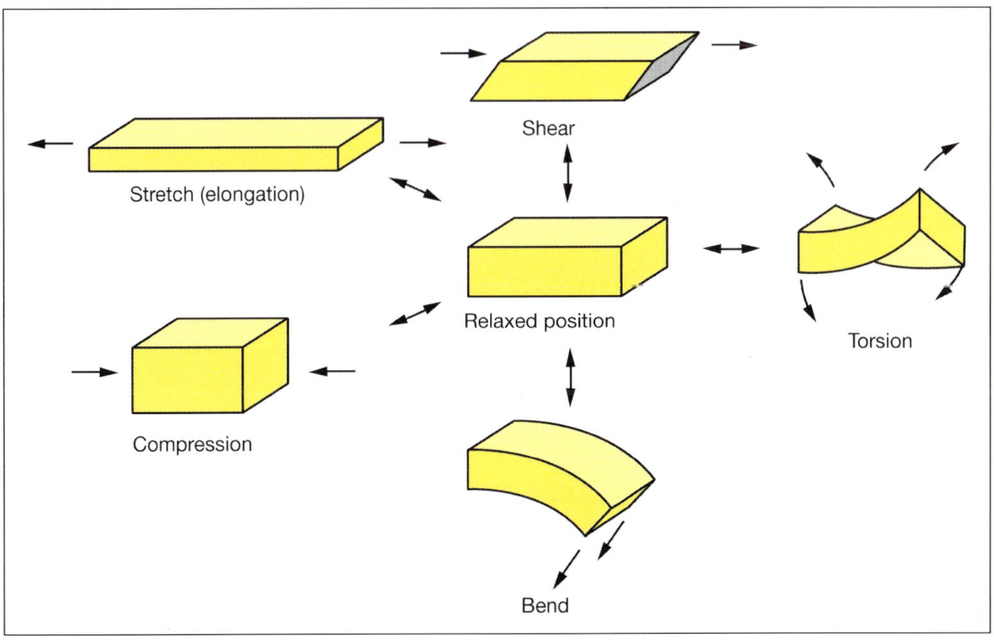

Figure 3.2.61: Types of mechanical deformation

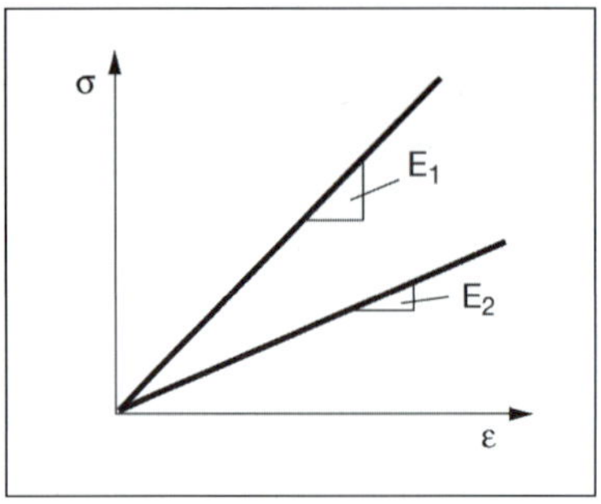

Figure 3.2.62: Stress-strain diagram of elastic materials (Hooke's law)

deformation stresses due to stone-chip impact and engine vibration. Permanent deformation, whether caused unintentionally when a wing is dented or intentionally when coated panels are drawn during the manufacture of precoated beverage cans, refrigerators or other building materials, must not affect the coating. Nor must any thermally induced differential volume changes experienced by the coating and substrate lead to irreversible damage as a result of the internal stresses and strains generated.

Materials with pronounced elastic properties can satisfy most of these requirements. Ideal elasticity is obtained when molecules and molecular chains of the organic film forming agents are capable of reversible reorientation and storing all of the deformation energy applied. This creates the condition for full restoration when the deforming force no longer acts.

The interrelationships of ideal elasticity can be described by *Hooke's* law in terms of the tensile force F acting on a body in the case of elongation and the resultant deformation.

The tensile stress σ acting as a force on an area is directly proportional to the ε-value, which is the elongation Δ l divided by the initial length l. The proportionality factor E in the corresponding stress-strain diagram is the modulus of elasticity by definition.

$$\sigma = E \cdot \varepsilon \qquad \text{where} \quad \sigma = \frac{F}{A} \quad \text{and} \quad \varepsilon = \frac{\Delta l}{l}$$

The other types of deformation such as **shear** and **torsion** can be described in simple terms for ideal-elastic bodies, with the individual moduli for describing the elastic properties being functionally related to one another and interconvertible.

Consider shear as an example: if shear stress τ is acting tangentially on a test body there will be a shear angle γ. The quotient is the shear modulus G.

$$\tau = G \cdot \gamma$$

τ = shear stress
G = shear modul
γ = shear angle

The moduli E from the elongation test and G from the shear test can be converted into one another as follows:

$$G = \frac{E}{2 \cdot (1 + \mu)}$$

Since *Poisson's* ratio μ, the quotient of the relative transverse and longitudinal contractions, assumes values within the region of 0.5 in practice, the modulus of elasticity E is approximately three times the value of the shear modulus G.

The type of **elasticity** described so far, as it applies to metals for example, is independent of **deformation duration** and, within certain temperature intervals, also independent of temperature. Such simple manifestations of elasticity are not found in polymers and therefore not in coatings either. Bending tests reveal linear dependencies only if deformations are small or if the applied force is low. High shear and tensile stress leads to considerable deviations from linearity. Dependencies on time and temperature are also observed. Slow deformation due to bending and fast deformation due to knocks or impact can produce different results on one and the same coating. Any determination of the elastic properties of coatings must therefore differentiate between bending elasticity as a slow deformation and impact resilience as a fast deformation.

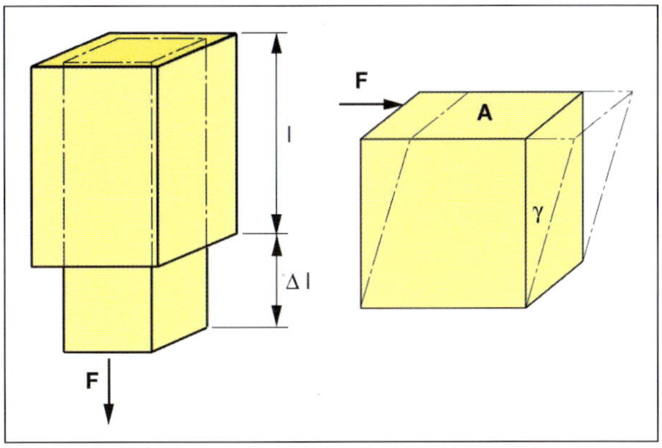

Figure 3.2.63: Elongation and shear

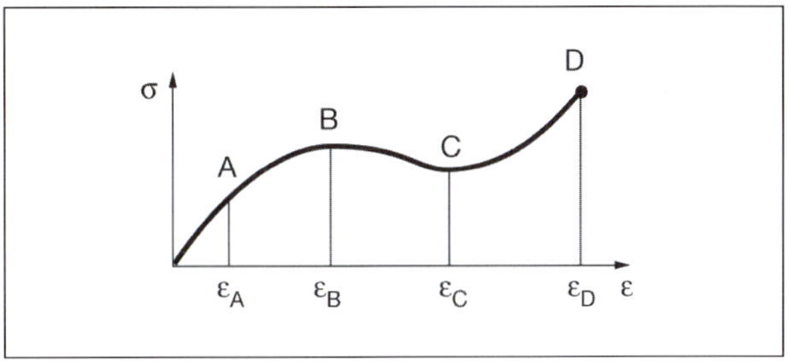

Figure 3.2.64: Stress-strain diagram of viscoelastic films

A stress-strain diagram of slow deformation (see Figure 3.2.64) shows that *Hooke's* law is only followed over a small range of elongation up to point A. This short linear section is then immediately followed by the range of disproportionate elongation A to C. The modulus of elasticity, the gradient of the locus, falls. The product consequently becomes more elastic. The causes are to be found in the separation of intermolecular interactions that stabilise the texture. The chains become more mobile and find it easier to glide past one another. In the case of fixed loops in thermoplastics or chemical linkage of individual molecule segments in thermosets, the orientating of the previously randomly coiled macromolecules entails storage of free energy. Order increases and so the entropy of the system declines. The source of the restoration forces acting on deformed bodies is not only the exploitation of the stored energy (see above) but also the striving of polymeric material to assume the previous random arrangement of molecular chains because of the accompanying gain in entropy.

Elastic deformation of coatings, also by analogy with the *Gibbs-Helmholtz* equation (see Chapter 2.2.3), entails entropy-induced storage of free energy due to "reversible" flow. Such properties are therefore fittingly described by the term **viscoelasticity**.

After full orientation of the molecular chains due to incipient deformation of bond angles the stress-strain diagram resumes a steeper rise as of point C. The curve ends at point D because the tensile forces exceed the cohesive forces of the polymers. Cracking or complete failure is the result.

The deformation of polymers is thus a superposition of energy and entropic changes of state that is dependent on the mobility of the molecular chains. Viscoelasticity is therefore said to have energy- and entropy-elastic components. This provides a molecular explanation for the marked dependence of the elastic properties of coatings on temperature.

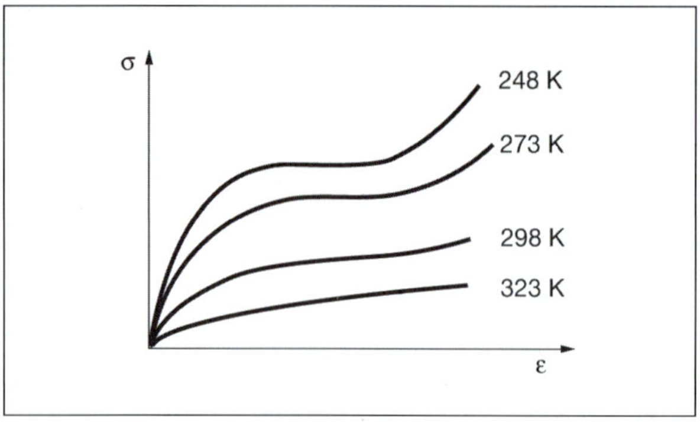

Figure 3.2.65: Schematic drawing of the stress-strain response of coatings as a function of temperature

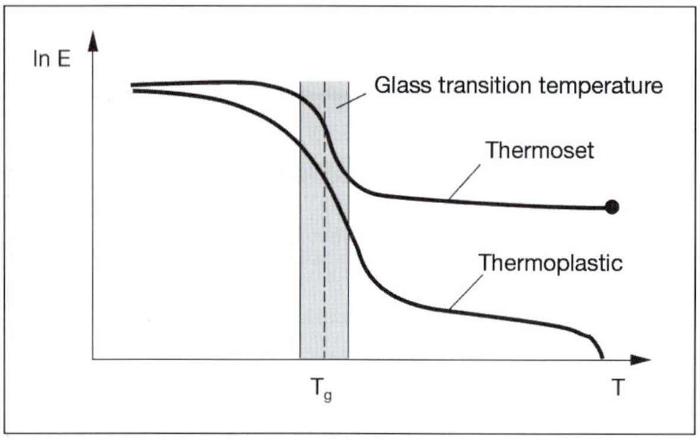

Figure 3.2.66: Logarithmic modulus of elasticity of polymeric films under constant stress as a function of temperature

A knowledge of how the modulus of elasticity varies with temperature is essential as regards the practical behaviour of coatings. In the low temperature range, segment mobility in thermosets and thermoplastic materials is considerably restricted by the fixation of the molecular structure. Deformations are only possible at close range through changes in bond angles. Changes in shape at the macroscopic level are minimal. Consequently, the moduli are high and the polymer is within the region of energy elasticity. As the temperature increases, the modulus of elasticity undergoes only a slight drop at first but after that it plunges at a temperature characteristic of each polymer to a much lower plateau. Below this temperature the stabilising intermolecular bridging bonds, acting as fixed points, cannot be overcome. It takes a further rise in temperature to cancel out the stabilisation, with entropy-elastic deformation with greater effects now becoming possible due to the free mobility of the molecule segments. The turning point in this segment of the curve is referred to as the **glass transition temperature** T_g and it has an established material characteristic in coating technology.

The glass transition temperature can fluctuate between -100 °C and +100 °C in accordance with the structure and molecular weight of the film forming agents. For stoving enamels it is normally between 20 °C and 40 °C. Owing to the enormous differences in viscoelastic behaviour, the testing and stressing temperatures of coatings should always lie on the same side of the glass temperature and not on one side and the other. A correct forecast of behaviour in practice would otherwise only be a matter of chance.

With thermoplastics, if the temperature continues to rise there comes a temperature which is determined by the molecular weight and the polar components and at which the mobility increases enough for long chain segments to glide past one another irreversibly. The material then acquires the properties of a liquid. A corresponding drop does not happen in the case of thermosets. They decompose at a correspondingly high temperature (ceiling temperature), without previously having become flowable.

It is worth noting that with thermosetting coatings the difference in modulus of elasticity declines in the entropy-elastic range. The dependence of the modulus of elasticity on temperature can therefore also serve to indicate the degree of crosslinking in reactive paints.

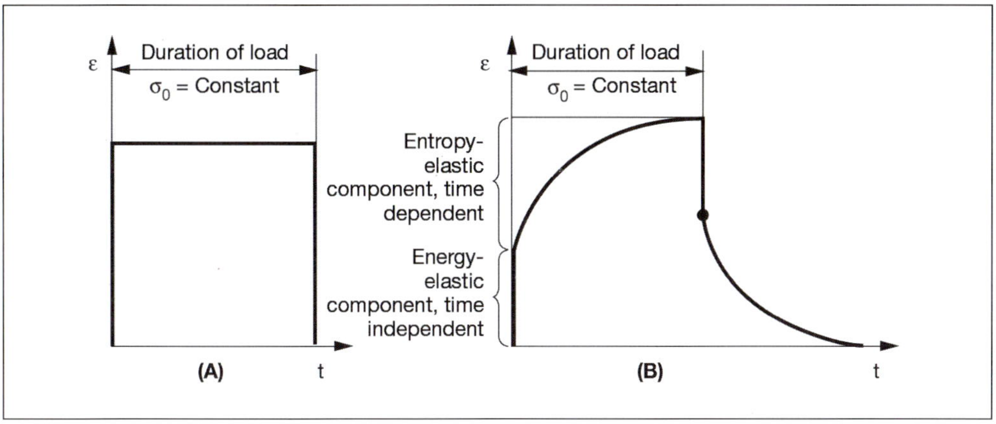

Figure 3.2.67: Strain undergone by (A) energy-elastic and (B) viscoelastic materials as a function of time

The difference in the slow bending elasticity and fast impact resilience of identical coatings can be explained by the length of time it takes for molecule displacement, which is responsible for the viscoelastic behaviour, to occur over long distances.

Figure 3.2.67 shows the time-dependence of the deformation of an energy-elastic steel spring and of a viscoelastic polymer film under constant tensile stress σ_0. Whilst the spring spontaneously undergoes constant elongation, the stressed polymer film exhibits the same time-independent elongation in the first phase of loading only and on account of energy-elastic deformation. The larger component of the elongation attributable to displacement of molecule segments is complete only after a finite amount of time has elapsed. This behaviour, which is known as retardation, can be described quantitatively by the **retardation time**. It can be determined from the dependence of the elongation on time and is defined as the time which elapses before $t = t_{ret}$. By that stage the elongation $\varepsilon(t)$, which follows an exponential function, has reached 63.2 % of its maximum value.

$$\varepsilon(t) = \frac{\sigma_0}{E} \cdot (1 - e^{-\frac{t}{t_{ret}}})$$

Another viscoelastic parameter becomes experimentally accessible by determining the stress induced by a constant, predefined strain as a function of time.

Like that for retardation, the curves for **thermoplastics** and **thermosets** follow an exponential function.

$$\sigma = \sigma_0 \cdot e^{-\frac{t}{t_{rel}}}$$

t_{rel} or **relaxation time** is the time required for re-orientation of the molecular chains until σ has dropped to 36.8 % of σ_0. Ideal thermoplastics can drop to 0 if sufficient time is allowed. Thermosets always retain a crosslinking-induced residual stress.

Viscoelastic behaviour can be described with the aid of a highly descriptive model. The application of a force and its effect on change of shape can be represented by connecting steel springs and dampers in parallel and in series. The latter are similar to cylinders filled with a viscous liquid and a suitable plunger inside [3.4.34].

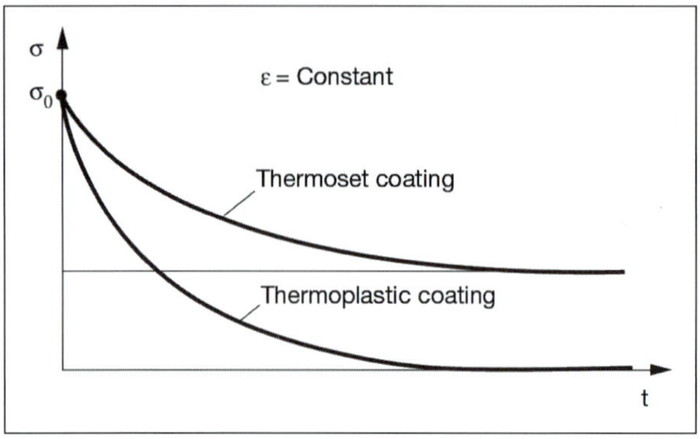

Figure 3.2.68: Relaxation of internal stress in thermoplastic and thermosetting coatings

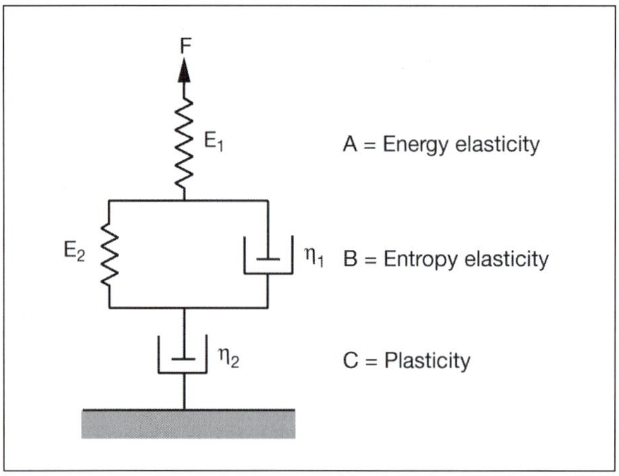

Figure 3.2.69: Schematic diagram for describing viscoelastic properties of materials according to Burger

The three rheologically relevant components of the model are the spring with modulus E_1 which serves to represent energy elasticity A, a combination of the spring with modulus E_2 and the damper with η_1 to represent entropy elasticity B, and the damper with η_2 to represent plasticity C. Whilst energy-elastic component A is independent of time and temperature, the entropy-elastic and plastic components are dependent on temperature and time. Deformations in accordance with A and B are reversible, while those in accordance with C are irreversible. The lower viscosity of the liquid in the dampers at higher temperatures in the model provides a plausible explanation of the empirical dependence of viscoelasticity on temperature. Entropy-elastic and plastic components too then come to the fore. A considerable drop in the modulus of elasticity is the result.

The definition of the modulus of elasticity described so far fails to provide a complete description of the viscoelasticity of polymers. A more precise description splits viscoelastic behaviour into a viscous component and an elastic component in accordance with the *Burger* model and then describes it with appropriate individual moduli. At the complex number level the viscous components are identified by loss modulus E" which serves as imaginary part and the elastic components are identified by storage modulus E' acting as real part [3.4.35]. Addition then leads to the complex modulus of elasticity E*.

$$E^* = E' + i \cdot E''$$

The scalar value of the complex modulus can be calculated by obtaining the root of the sum of the squares of the individual modulus components. The quotient of the two moduli E"/E' or tan δ value also serves as the mechanical loss angle in assessments of a coating's mechanical behaviour. Owing to the dependence of E» on temperature the value of tan δ is also a function of temperature.

$$\tan\delta = \frac{E''}{E'}$$

For the engineer it is important to know that tan δ reaches a maximum in the region of the glass transition temperature. In this temperature range the coating thus offers the best compromise between energy dissipation and mechanical resistance – a fact which should be taken into account when impact-proof elastic coatings are being developed. Viscoelasticity is therefore always a complex combination of viscous and elastic behaviour and represents an intermediate stage to be quantified between two extremes of ideal elastic behaviour and ideal viscous behaviour.

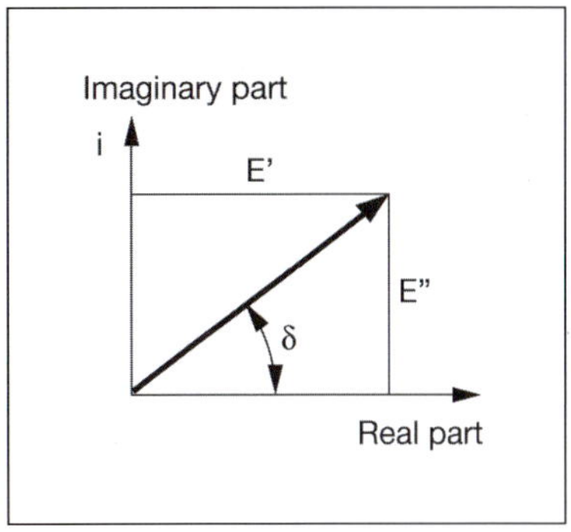

Figure 3.2.70: Complex modulus

Measurement Methods

Methods of measuring viscoelasticity can be divided into two different categories. The first seek to determine performance indicators by simulating the stresses that occur on coatings in practice and the second entail observing temporal phase shifts between a cyclic deformation force and deformation of the material. Owing to the time dependence of viscoelastic deformations it has proved expedient to split the individual test methods into procedures for determining **bending elasticity** and **impact elasticity**.

Testing of Bending Elasticity

All tests for characterising the bending behaviour of coatings are basically designed to reproduce practical stresses under defined conditions. In the **mandrel bending test** in accordance with ISO 1519 (cylindrical mandrel) sample strips coated at a defined velocity are bent around cylinders of different diameters. The diameter at which the coating cracks or flakes off after stress serves as an indicator of the coating's elasticity.

For many applications the test is conducted with a tapered (conical) mandrel as per ISO 6860. This has the advantage that a whole range of different bending radii can be covered in one test. In the conical mandrel bending tester the mandrel is a truncated cone approximately 20 cm long and decreasing in diameter from 38 mm to 3.2 mm.

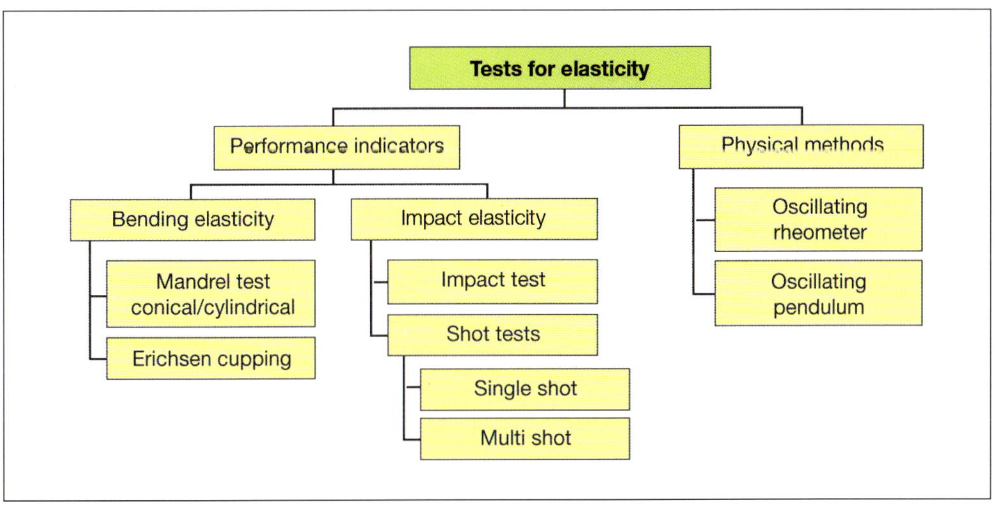

Figure 3.2.71: Overview of elasticity tests

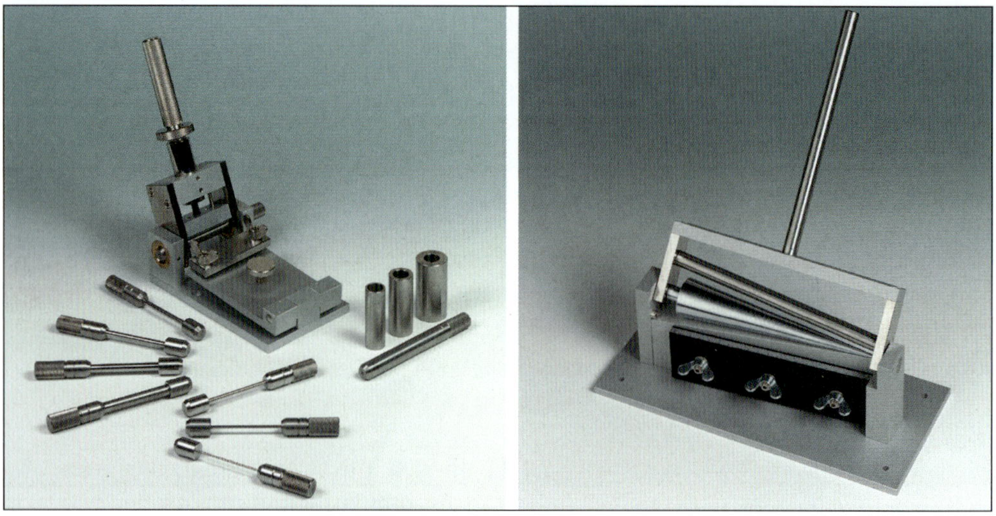

Figure 3.2.72: Cylindrical mandrel test as per ISO 1519 and the conical mandrel test as per ISO 6860
Source: Byk Gardner

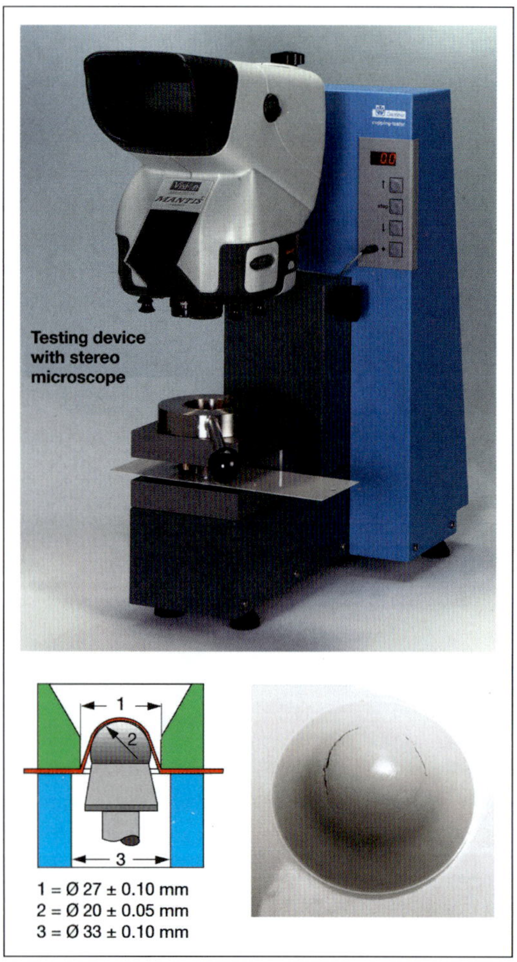

Figure 3.2.73: Erichsen cupping test instrument as per ISO 1520 *Source: Byk Gardner*

The test panel is covered with a sheet of paper, clamped firmly and bent through 180° at a constant rate in approximately 10 s. The paper is then removed and the point at which the cracking stops is marked. The diameter of the mandrel at which the coating cracks or detaches is stated. A different method of evaluation is to state the percentage elongation due to bending.

Apart from one-dimensional elongation to the point of cracking, two-dimensional elongations are also used to describe the elasticity of coatings. In the **Erichsen cupping test** a plunger is used to plunge a hardened, polished steel ball of 20 mm diameter into the back of the test panel at a uniform feed rate of about 0.2 mm/s until the first crack appears in the surface of the coating. The cupping value read off on the measuring device represents the distance travelled by the plunger. The cupping test places not only the film of paint under deformation stress, but also the entire test panel. For this reason, the substrate for the sample coating should only involve the same materials and pretreatment methods as are used in practice or have been standardised for development purposes.

All the tests presented so far merely provide information about the maximum extensibility of the polymeric film. Reproducible results can only be obtained if the film thickness of the substrate in particular, along with the film thickness of the coating and the test temperature, conform precisely to the values agreed in the specifications.

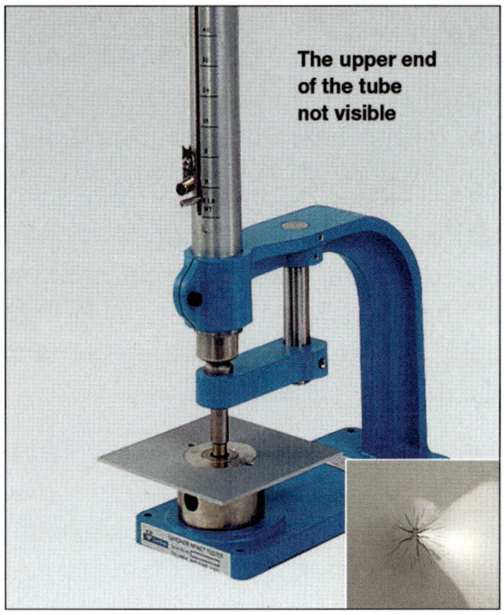

Figure 3.2.74: Single-impact test device according to ASTM B 2794 *Source: Byk Gardner*

Shock and Impact Elasticity

Dynamic deformation under impact stress is assessed with the ball impact tester as per ASTM D 2794. The tester consists of a base plate with a die and a solid stand with a holding arm for a drop pipe. On the drop pipe there is a height scale and a device for setting the drop weight. The falling body itself must be defined with regard to geometric design and mass.

The aim of the impact tester is to test the resistance of a film of paint to cracking or delamination from a metal substrate when a falling body strikes the coating. Depending on the objective, the body strikes the coating from the front side or from the reverse side. This causes compression or elongation of the coating film (impact or reverse impact). The test has to be repeated at increasing or decreasing drop heights until the first signs of cracking or delamination appear.

The special devices for testing the resistance of coatings to stone impact can be divided into **single-impact devices** and **multi-impact devices**. In the case of single-impact devices a defined impact is applied to the coated area. The test itself can be performed

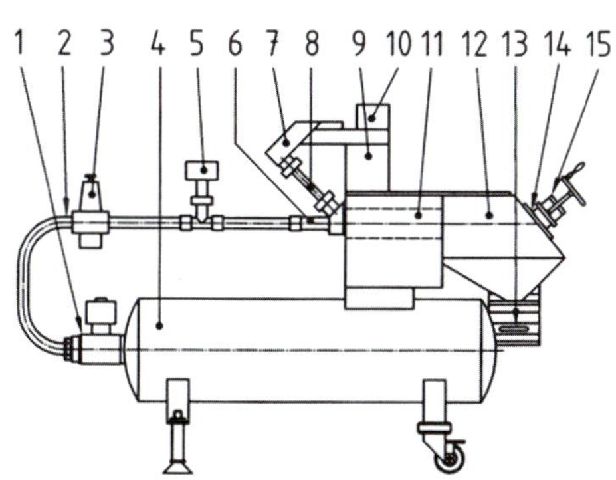

Figure 3.2.75: Schematic drawing of a multi-impact tester as per ISO 20567-1

in two ways. The first uses a hemispherical impact body with a diameter of 2 mm. The impact energy is generated by a rapid-start electric motor which projects the body onto the object. The impact energy is governed by the nominal rotational speed, the power and the mechanical time constant of the motor.

Single-impact tests conforming to ISO 20567-2 have established themselves especially in the automotive industry where they are also known as mono-impact tests. Here, a defined body is projected onto the test coating by a pneumatic device at a defined velocity and a defined angle of impact. The stone-chip resistance is assessed in both methods by determining the damaged area in square millimetres and interpreting the damage pattern.

The test for stone-chip resistance with multi-impact devices as per ISO 20567-1 seeks to simulate the damage done to the outer skin of vehicle bodies by a large number of small, often sharp-edged impact bodies which strike the surface simultaneously and in rapid succession. The standardised device consists of a steel pipe of standardised size, at one end of which the test panel is held in place by a holder at a defined projection angle. The feed chute for the projectiles is located at the other end. The projectiles, consisting of steel shot, steel balls or road chippings of various grain sizes, are projected onto the test panel by compressed air. If the effect of impact stress on the corrosion resistance of a coating is to be assessed, the test panel can be subjected to corrosion stress before and after blasting. The test conditions and total duration of corrosion stress vary with the test method.

Scientifical Methods of Measurement

The methods of measurement for describing the viscoelastic behaviour of coatings are essentially aimed at directly or indirectly recording the loss and storage moduli (see above) as an individual parameter of the viscoelastic material property. Dynamic mechanical measurements with an oscillating rheometer or a **torsion pendulum** make it possible to determine viscoelasticity from the behaviour of cyclically stressed paint films. Temporal phase shifts between a deforming force and deformation on one hand and damping of pendulum oscillations on the other are the measurable effects which can serve in the calculation of individual moduli as well as relaxation or retardation times.

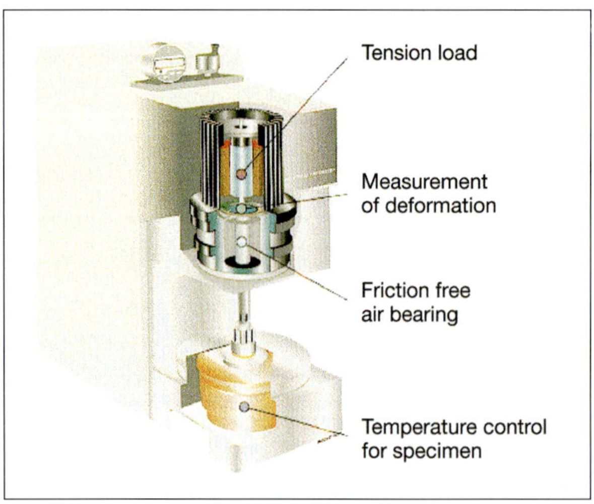

Figure 3.2.76: Principle behind an oscillating rheometer

Source: Rheometrics

Scientific physical methods are becoming more and more popular for characterising physical/technological properties although a correlation between the data found and performance is not always evident. The principle behind dynamic stressing with an oscillating rheometer is shown in Figure 3.2.76.

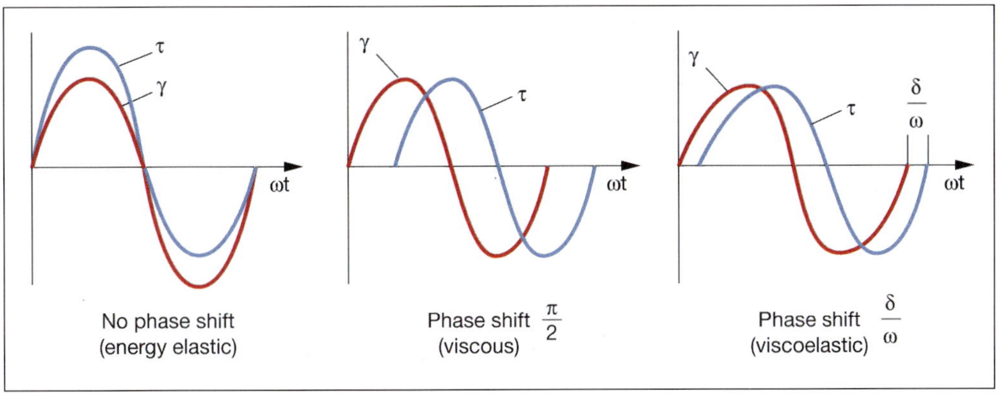

Figure 3.2.77: Stress-strain curves for energy-elastic, viscous and viscoelastic materials

An oscillating drive subjects a coating to cyclical shearing. The instrument is designed so that the shear angle γ follows a time-dependent sine-wave function. The oscillation which is transmitted to the bottom plate is delayed by a shift in phase angle, the extent of the delay varying with the elastic properties of the coating. If the shift angle $\delta = 0$, the body is an **ideal-elasticity** body, and if the shift is 90°, it is an **ideal-viscosity** body. For viscoelastic substances subjected to a specified deformation γ, the phase shift of the transferred shearing stress τ lies between $\delta = 0$ and $\delta = \pi/2$, with the exact Figure depending on the components of the two properties, i.e. the purely elastic and purely viscous properties are shifted by 90° relative to one another.

$$\gamma(t) = \gamma_0 \cdot \sin(\omega \cdot t)$$
$$\tau(t) = \tau_0 \cdot \sin(\omega \cdot t - \delta)$$

The smaller the phase shift δ, the more pronounced are the energy-elastic properties. The shift angle δ can be split up by mathematical treatment into a component which is in phase ($\delta = 0°$) and a component which is shifted by $\pi/2$ ($\delta = 90°$). Both parameters have to be converted into loss and storage moduli. The resulting quotient is the mechanical loss angle tan δ.

Another way of obtaining information about the elastic properties of coatings is to measure the damping of artificially created oscillations, without input of further energy.

Free films, frequently reinforced by backing films, are affixed to

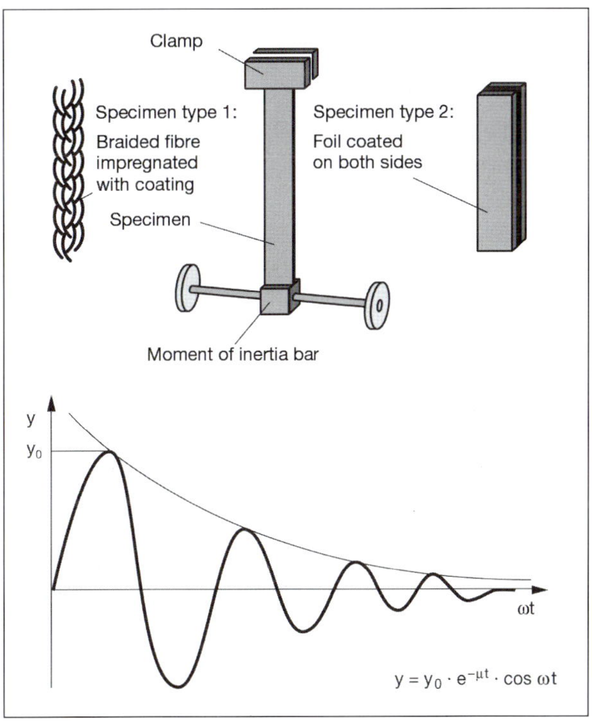

Figure 3.2.78: Schematic drawing of a torsional pendulum and a curve showing damping of the oscillations

one side and, after attachment of a flywheel, caused to oscillate by torsion. The oscillation time and damping can then be used to calculate the viscoelastic parameters. The pendulum oscillation y at maximum deflection y_0 can be described as follows.

$$y = y_0 \cdot e^{-\mu \cdot t} \cdot \cos(\omega \cdot t)$$

The smaller the damping factor μ, the less the degree of damping and the larger is the energy-elastic component of the test piece. Conversion into mechanical loss modulus tan δ is possible.

3.2.5.3 Hardness

To offer good performance, coatings must have not only high adhesion and elasticity but also a level of hardness that is fit for purpose. Hardness is a property which defies any attempts to define it with physical precision. It is variously described as a material property characterised by low extensibility, i.e. a high modulus of elasticity, high resistance and cohesion or resistance to abrasive effects. A definition originating from Swiss experts deems it to be a quasi-static resistance of a body to strictly local, inhomogeneous deformations subjected to the action of punctiform or linear force centres [3.4.36]. This definition indicates a relationship with elasticity and establishes a reciprocal correlation between it and the physical parameter of hardness. However, equating a high modulus of elasticity with high hardness only partially chimes with the generally held view of the term hardness.

Hardness scale	Reference material
1	Talc
2	Gypsum
3	Calcite
4	Fluorite
5	Apatite
6	Feldspar
7	Quartz
8	Topaz
9	Corundum
10	Diamond

Figure 3.2.79: Mohs' scale of hardness dating from 1822

The definition of hardness is extended by the **hardness scale** published in 1822 by *Mohs* who defined a total of 10 degrees of hardness by reference to hard minerals. On this

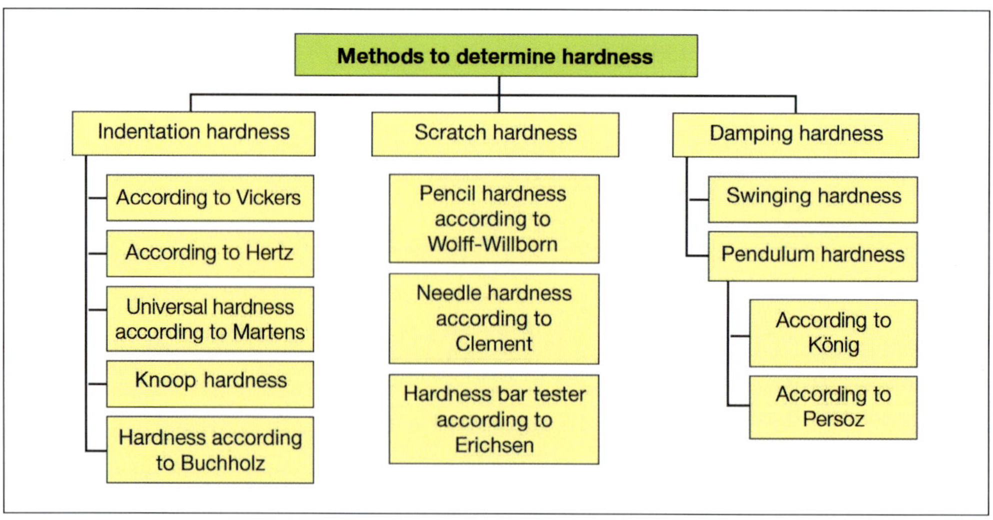

Figure 3.2.80: Methods for determining the hardness of coatings

scale, a hard material is one that will scratch a different one, without itself being scratched. Such a definition of hardness relates less to deformability and resistance to the action of force than to the internal resistance and hence to the cohesive forces of the material being assessed. Unfortunately, the *Mohs* hardness scale [3.4.37] does not contain any polymers as reference substances. It is therefore not very suitable for assessing organic coatings.

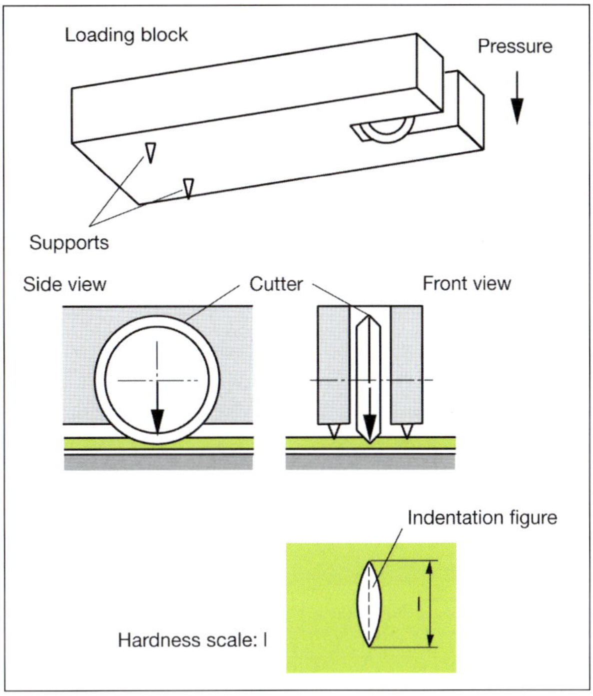

Figure 3.2.81: Indentation hardness as per Buchholz and ISO 2815

The definition in ISO 4618, in which hardness is described as the ability of a dry film to resist indentation of penetration by a solid object, is a broader one and more suitable for polymers, but also takes into account the response to pressure, friction or scratching. If it is further added that the result of stressing is assessed with regard not only to deformation but also to possible damage in the form of fracture or other irreversible changes, the outcome is a much closer idea of what is generally understood by the term hardness.

Misunderstandings can be avoided by not talking merely of absolute hardness but by also incorporating the principal specific characteristics into the definition. It should thus be clear at this stage that the term hardness involves various properties which are covered by various test methods and seldom fully correlate with each other.

Test Methods

Various instruments are available for determining the hardness of coatings by measuring the indentation resistance, the damping of oscillating free films or the resistance to scratching tools.

The most common method of determining **indentation hardness** is the *Buchholz* test described in ISO 2815. The indenter is a double truncated steel cone, which, after placement on the test coating, acts on it at right angles with a force of 5N.

After 30 seconds the cone is removed, the coating is allowed 35 seconds to regenerate and then the residual damage is measured. As set out in ISO 2815, the reported measured variable is the indentation length in millimetres.

The reciprocal value of the indentation length multiplied by 100, formerly known as Buchholz hardness, will be replaced by this in the future. The practicability of the test method is contingent on a sufficient film thickness (see below).

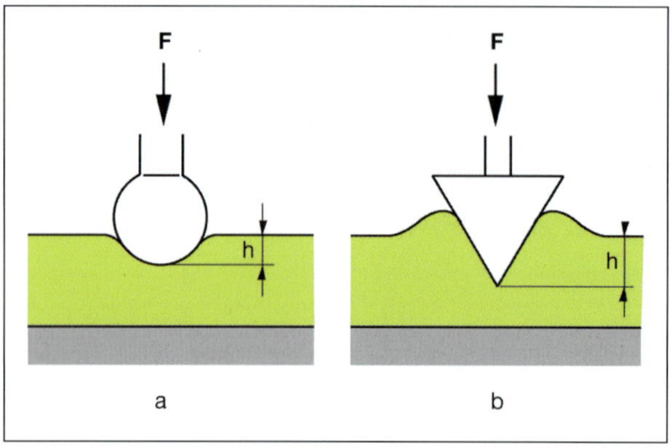

Figure 3.2.82: Test specimens for measuring (a) Hertz and (b) Vickers indentation hardness

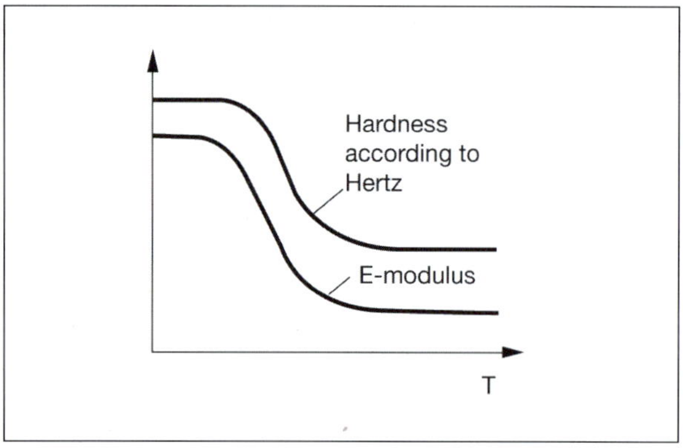

Figure 3.2.83: Correlation between Hertz indentation hardness as a function of temperature and the modulus of elasticity for coatings

Experience has shown that a short indentation length is associated with high mechanical strengths and hence good performance properties. The question as to which physical variables are measured in which bonding is difficult to answer and in terms of a correlation with performance properties it is ultimately of little consequence.

A rarely practised method of hardness testing is the *Knoop* hardness test. The apparatus consists of a device for applying a load of 250 mN to the coating via the tip of a conical diamond. The resulting deformation, reported in terms of indentation depth or length, serves as the measure of hardness.

Also rare in the paint and coatings industry is the penetration hardness determined under load with reference to defined physical parameters. The most commonly employed indenters here are the *Hertz* ball and the four-sided *Vickers* pyramid.

The ratio of the force F acting on the test body to the resulting penetration depth h is used to calculate hardness H_K and hardness H_v. The calculation utilises the radius R of the penetration ball in the case of *Hertz* hardness and the angle of inclination of the pyramid in the case of *Vickers* hardness.

$$H_K = \frac{3 \cdot F}{4 \cdot \sqrt{R} \cdot h^{\frac{3}{2}}} \qquad H_V = 3{,}8 \cdot 10^{-2} \cdot \frac{F}{h^2}$$

Both methods yield information about the hardness of coatings. However, the numerical values are not interchangeable. If there is no plasticity, deformation of the film is completely reversible according to the *Hertz* method. The hardness values largely correlate with the modulus of elasticity E. Figure 3.2.83 shows good agreement between the temperature dependency of the modulus of elasticity and *Hertz* hardness values.

The interaction of the *Vickers* pyramid with the polymeric film is different. The load area, which is initially very small, causes permanent damage owing to the sharp edges of the indenter. Consequently, there is no direct relationship between *Vickers* hardness and modulus of elasticity.

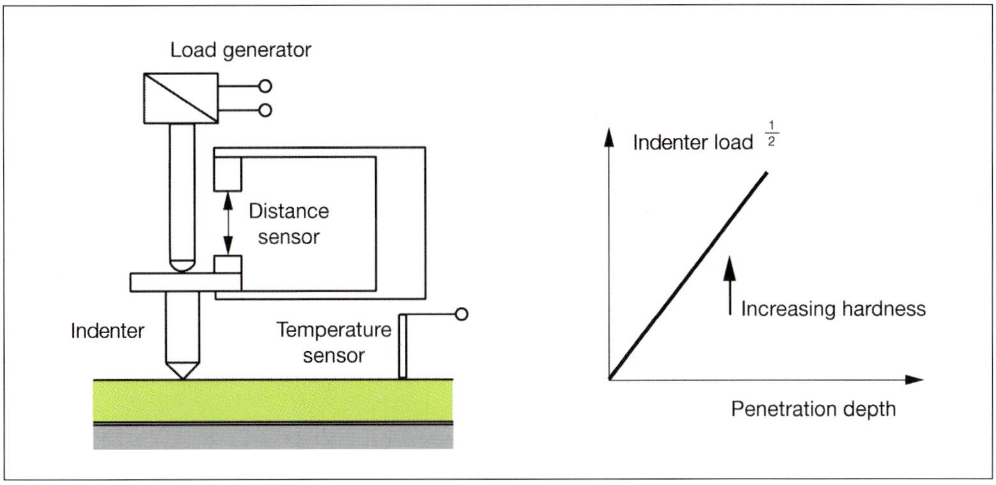

Figure 3.2.84: Measuring principle behind the determination of Martens universal hardness as per ISO 14577

Comprehensive studies incorporating all the important penetration hardness tests have repeatedly demonstrated that reproducible measurements are possible only if deformation of the film occurs only over a small fraction of the film thickness. *Van Laar* specifies that for reproducible results the penetration depth should not exceed 1/12th of the film thickness. At film thicknesses of 60 µm deformations of > 5 µm are thus inadmissible [3.4.38].

To satisfy these requirements two other test techniques have been developed. Either very light weights are used for the indenter or the deformations are determined under declining load, with extrapolation of the resultant hardness curve by regression to load 0. The latter method thus determines the gradient at the origin of the force-compression diagram.

The former measurement method determines ***Martens* universal hardness**. As described in ISO 14577-1 to 3 the test instrument measures the penetration depth of a pyramid-shaped diamond with a square base and a 136° angle of inclination. The test load is divided up into ranges from 0.4 to 25.6 mN, from 4 to 256 mN and from 16 to 1024 mN or according to additional agreement. The measurements are obtained by increasing a test force F quadratically at defined intervals so that the resulting penetration depth h can be recorded immediately. The universal hardness H_n is then calculated from

$$H_n = \frac{F}{26.43 \cdot h^2}$$

Apart from successive increases in force, other load processes can be set [3.4.39]. Thus test series performed under declining load can supply additional information about the time dependence of restoration. The relative depth resilience can be calculated from the penetration depth h_2 measured under the test load after a specified time and the residual penetration depth h_3 immediately after the load has been removed. Relative depth resilience t_r is then:

$$t_r = \frac{h_2 - h_3}{h_2} \cdot 100$$

The time dependence of penetration depth under constant test load yields supplementary data about creep characteristics, i.e. the irreversible deformation of coatings.

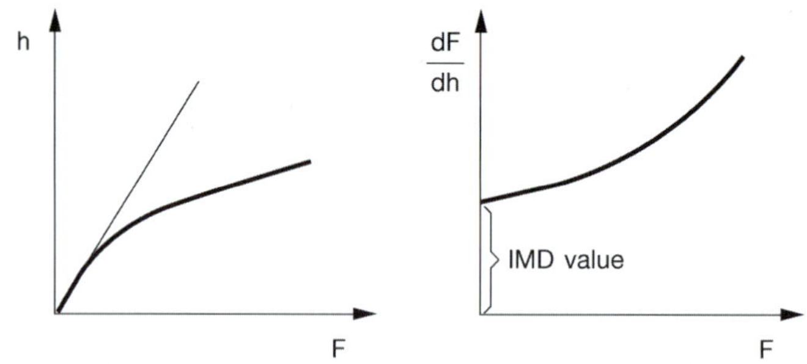

Figure 3.2.85: Diagrams for determining the infinitesimal modulus (IMD) according to Oesterle

One device suitable for determining *Martens* hardness is the "Fischerscope". This device makes it possible to not only record traditional penetration hardness but also viscoelasticity and plasticity.

Another method of eliminating errors due to substrate influences was developed by *Oesterle* [3.4.40]. The measuring principle behind the infinitesimal modulus determination (IMD) is similar to that of the *Hertz* and *Fischer* methods. Penetration depths h are measured as a function of increasing test force F. The increasing compression increasingly alters the elasticity of the thin test specimen. To eliminate the resulting errors, the slope in the force-compression diagram is determined as a function of the test force and extrapolated to the slope at the origin. The results yielded by the IMD method are independent of the shape of the test specimen. Owing to the extrapolation procedure and the resulting sensitivity very small differences in pigment distribution can be identified, as can fluctuations in the physical/technological properties at different levels of relative humidity in the environment.

This critical discussion of penetration hardness makes it clear that the *Buchholz, Vickers, Hertz, Oesterle* and *Martens* methods ultimately tend to be used for measuring viscoelasticity and plasticity, with it being just as impossible to make distinct attributions to the basic physical variables as it is to performance properties.

Test methods which take into account the influence of coatings on the damping of vibrations have little in common with penetration hardness tests. *König* **pendulum hardness** as described in ISO 1522 is a method frequently encountered in practice. The device itself consists of a pendulum, with two agate balls attached to the support.

The two 5 mm balls of the pendulum are placed on the coating and the pendulum is moved 6° away from the rest position. The pendulum, which weighs 200 g, is released and the oscillations are recorded by a counter. The damping is measured in terms of the number of oscillations or the time in seconds which elapses until the amplitude of the pendulum's swing has dropped from 6° to 3°. For use in accordance with the standard the device, an adjusting screw on the fitted standard plate is adjusted in such a way that for an oscillation period of 1.4 s a total of 179 oscillations have taken place by the time half the amplitude has been reached.

A slightly different technique for measuring damping hardness is that proposed by *Persoz*. The design and weight of the ***Persoz pendulum*** differ from those of the *König* pendulum and so it produces different measurements. For example, the 8 mm balls are much larger in diameter than the pendulum balls. The weight of the pendulum, 500 g, is also greater. The hardness is given by the number of swings made by the pendulum as the amplitude is damped from 12° to 4°.

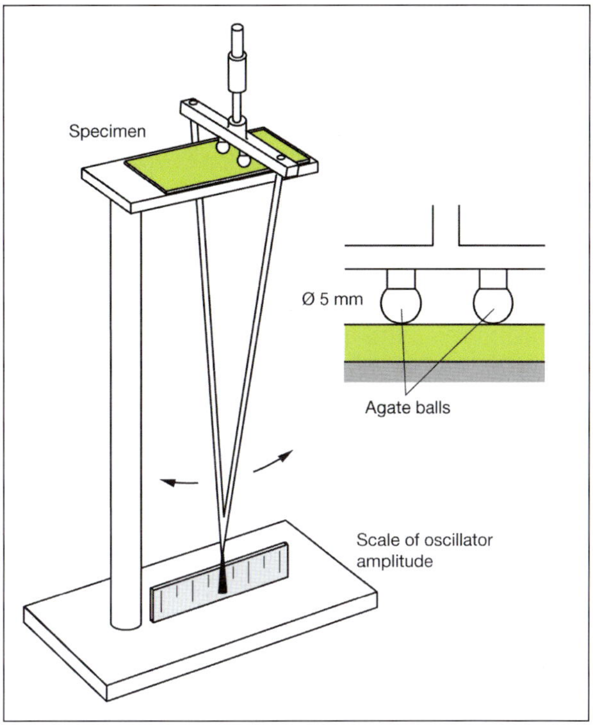

Figure 3.2.86: Principle and device for determining the König pendulum hardness as per ISO 1522

The damping effect stems from the periodic change in deformation induced by the balls. The interaction between the pendulum and the coating varies with the viscoelastic properties, especially the retardation behaviour, and the stored deformation energy is not fully recovered on account of the time delay in restoration.

The measured half-life $t_{1/2}$, oscillation time T_p, the length of the pendulum l and the radius of the balls r make it possible to calculate t_{ret}. To simplify comparative measurements of coating materials, $t_{1/2}$ is used.

$$t_{\frac{1}{2}} = \frac{T_P^2 \cdot l \cdot \ln 2}{\pi^2 \cdot r \cdot t_{ret}}$$

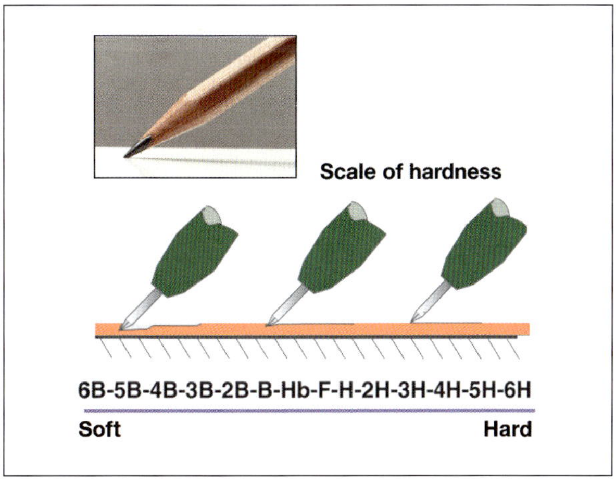

Figure 3.2.87: Device for determining the Wölff-Willborn pencil hardness *Source: Byk Gardner*

Damping hardness records the viscoelastic properties in a similar way to penetration hardness.

Measuring techniques that damage the coating with a scorer or suitably shaped test bodies come much closer to the definitions of hardness provided by ISO 4618 and *Mohs*. These methods of determining **scratch hardness** employ two principles:

- Either several test specimens of different hardness under the same weight or
- A single test specimen under different weights is moved horizontally over the coating.

The first is the *Wolff* and *Willborn* hardness test method. Also known as the **pencil hardness test** in ISO 15184, it utilises 17 different pencils ranging in hardness from 6B to 9H.

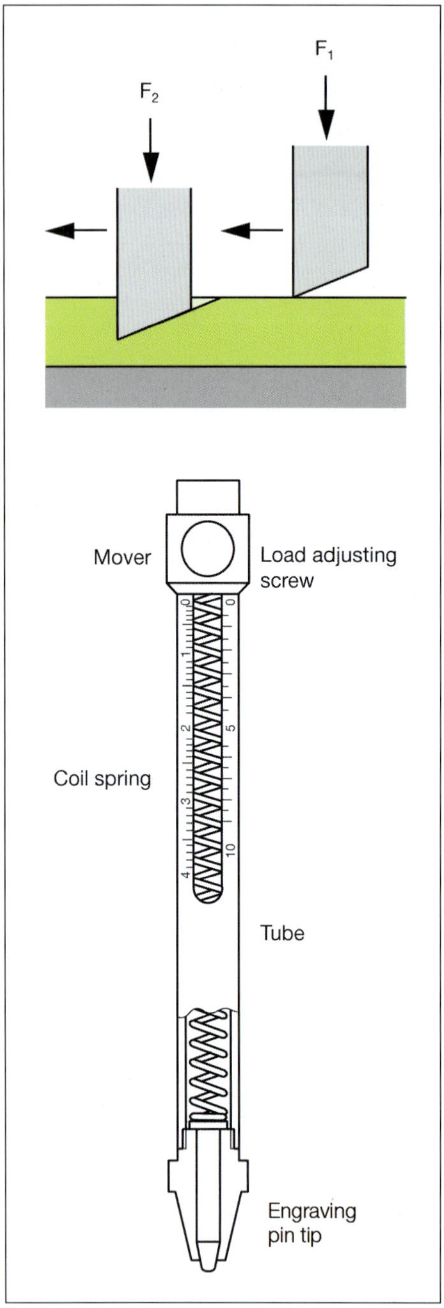

Figure 3.2.88: Principle behind scratch hardness determinations

The pencils, starting with 6B, are moved over the coating by an appropriate device under a constant weight of 300 g and at an angle of 30 to 40°. The hardness is given by that pencil which causes permanent damage to the coating.

To eliminate sources of error special attention has to be paid to pencil preparation and ensuring a dust free test coating.

The second way to obtain data on the hardness of coatings is to determine the resistance of a scorer pulled over the coating under different weights. The hardness is given by the applied force under which the scorer starts cutting into the coating. Both fracture mechanics and delamination of the coating by peeling play a physical role here. The amount of damage done by each influence is also governed by the velocity at which it is done.

The scoring tools used are chisel-shaped tungsten-carbide cutters (Clement method) or needles with spherical steel heads.

The measuring procedure has to be repeated under systematic load increases until signs of scratching appear in the coating. The test result for a given weight is rated on a scale of 1 to 10. Starting with 50 g the weights are increased to a maximum of 500 g.

Measurements with a spherical engraving tip are performed with the **Erichsen test bar**. The tester consists of a cylindrical tube into which a load-adjustable coil spring has been inserted. The adjustable spring tension acts on an engraving pin, the tip of which projects from the bottom of the tube. The test is performed by placing the tip of the hardness bar at right angles to the coating, applying sufficient force to ensure that it is pressed into the tube and then moving the bar over the coating for a distance of 5 to 10 mm at a constant velocity of approx. 10 mm/s. The hardness is given by that spring tension which is just sufficient to cause the engraving tip to leave a visible trace.

3.2.5.4 Abrasion Resistance and Mar Resistance

Abrasion resistance provides information about the ability of coatings to withstand without damage the influence of abrasive materials such as sand, dirt, chippings, and corundum, as well as scrubbing brushes. As with all physical/technological properties described so far, the background is extremely complex. Without a doubt the abrasion resistance is functionally linked both to the elastic properties and to the cohesive forces characterised by molecular entanglement and polar interaction. A key factor here is also the surface friction of the film, especially in the case of relatively weak abrasion stress. This is frequently known as **mar resistance** [3.4.41]. Owing to the complexity of this property set, test methods for quantifying abrasion resistance do not aim to determine defined physical indicators. Basically they are nothing more than a simulation of practical stresses.

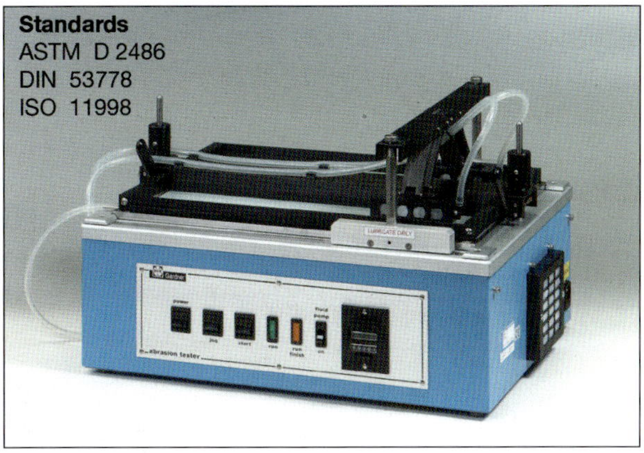

Figure 3.2.89: Device for determining scrubbing resistance as per ISO 11998 *Source: Byk Gardner*

High abrasion and mar resistance are particularly essential in the automotive, furniture and parquet flooring industries, many coil coating sectors and the paints industry. The last of these classifies its product range according to scrubbing and wash resistance. In emulsion paints these are usually determined in accordance with ISO 11998. The *Gardner* device employed here stresses the sample, which has been moistened with water and wetting agents, by moving a scrubbing brush back and forth over the surface under a defined force. Wash-resistant emulsions have to withstand 1,000 cycles with virtually no damage and wear-resistant ones, more than 5,000 cycles. For high scrubbing resistance in this product class the surface hardness of the coating is the dominant factor.

This contrasts with the other sectors mentioned above, where all potential influential forces play a role. Numerous test methods have therefore arisen.

The ***Taber abraser*** as per ISO 7784 (ASTM D 1044) is a long established method. The test consists in "sanding" a test panel with rotating frictional rollers under a defined force. The coating is subjected to frictional rollers of different roughness and hardness, with the choice depending on the hardness or abrasion resistance (see Figure 3.2.90). The abrasion resistance is given by the measured weight loss of the test panels as a consequence of abrasion over a specified test period. In other test methods different materials are drawn periodically across test panels under defined weights in simple apparatuses. The scratches created are usually measured as haze or reduced gloss. The type and shape of scratch depend heavily on the materials used [3.4.42].

The automotive industry is extremely interested in coatings that offer permanent scratch resistance. This interest is due on one hand to the huge popularity of dark colours for premium vehicles which make scratches much more easy to see and on the other to the sometimes wide differences in abrasion resistance of polymer coatings and body paints in the first few months following coating application. The minimum goal here is to preserve the abrasion resistance of the entire body until the vehicle reaches the dealer's showroom.

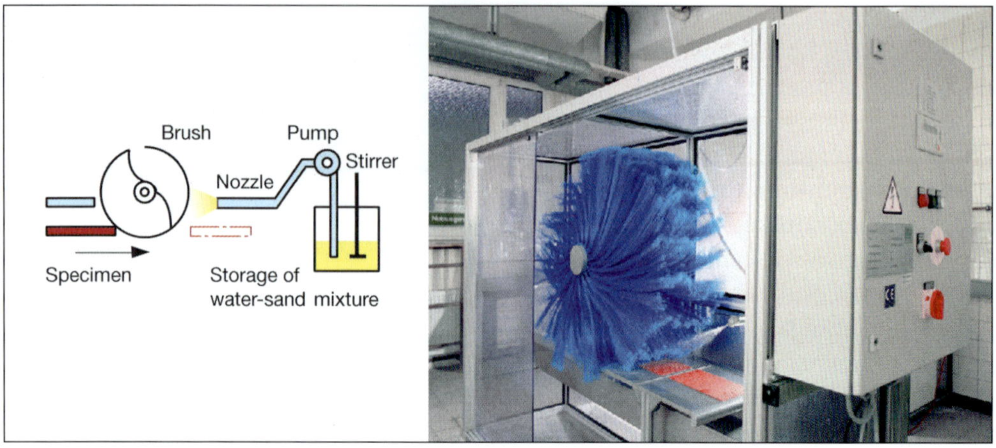

Figure 3.2.90: Principle and device for determining the mar resistance of automotive coatings as per ISO 20566

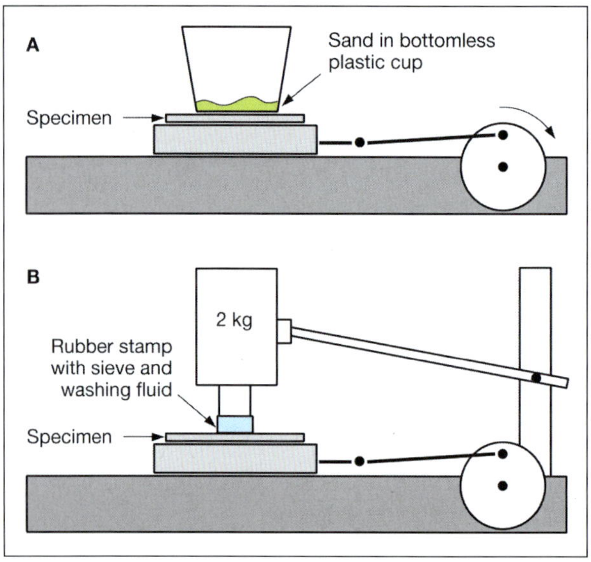

Figure 3.2.91: Simple devices for determining scratch resistance

In real life, carwashes are a major contributor to scratches. Reliable tests have been sought in this regard for years. However, it has still not proved possible to find the ideal test that shows an absolute correlation with real-life behaviour [3.4.43].

One standardised test method specified in ISO 20566 tries to reproduce the conditions obtaining in carwashes. Test panels are stressed with polyether or polypropylene brushes which rotate in opposite directions to the movement of the test bench as aqueous silica sand slurries (1.5 g/l) are sprayed on continuously. The tests are assessed by measuring gloss on the unstressed and stressed samples at right angles to the direction of abrasion. The reproducibility of the test conditions and that of the test device itself is still problematic, though.

A further method gaining acceptance for coatings is the Crockmeter which is used to measure colourfastness in the textiles industry [3.4.43]. This device moves different grinding papers back and forth across the coating surface under different pressures. The abrasion resistance is given by the loss of gloss.

Yet another method utilises the nanoscratcher. A single scratch is made by drawing a needle over the coating under increasing force. The method also relies on AFM (atomic force microscopy) for analysis [3.4.44].

Coatings whose surfaces offer permanent scratch resistance are extremely hard to achieve. The first step is to define the type of scratch, i.e. as merely deformation or as deformation and paint removal.

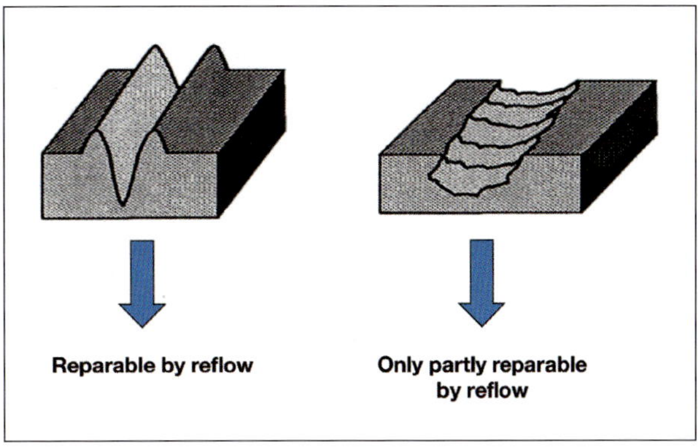

Figure 3.2.92: Characterisation of scratches: deformation and deformation with paint removal

The usual approach to producing high surface hardness, e.g. as in the case of glass or even diamond, cannot be adopted for an automotive OEM finish, for example, owing to the comprehensive property profile which it must possess. Ease of repair and elasticity would be ruled out. Very soft surfaces could also be deemed to possess scratch resistance where the deformed coating readily flows back together again to yield a smooth surface. This is an advantage of coatings for polymers which have been rendered "soft" so as not to overly impair the polymers' properties. The corollary here is the frequently observed phenomenon that car bumper coatings are very scratch resistant during the initial period of use because when the temperature rises slightly they allow the deformations and small scratches on the surface to flow back together. However, this property is lost over a period of 6 to 12 months as the film ages. Added to which, such coatings are not very resistant to chemicals.

All these different aspects have since found their way into recent improvements which avoid scratches in the first place by producing a hard, very thin layer of nanoparticles on the surface [3.4.45] or by generating a nanotextured film matrix with the aid of silanes which crosslink to form hard nanosegments within the film (see Chapter 2.1.5) [3.4.46]. Optimisation of the restoring forces, e.g. through the use of polyurethane structures, also helps to improve scratch resistance, at least where surface deformation is concerned [3.4.47].

3.2.5.5 Other Tests

Apart from the visual and traditional functional properties such as adhesion, elasticity and hardness, surface properties are increasingly being demanded which indicate not only visual attractiveness and physical functionality but also tactile effects.

This has led to the development of **soft touch coatings**. This property, also known as haptics, is intended to convey the look and feel of velvet and leather. Soft touch coatings are increasingly finding application in the automotive, phonographic and computer industries.

Haptics is therefore an area of tribology, the theory of friction. This addresses the contact characteristics of surfaces moving relative to one another, i.e. interactions of solid surfaces at the molecular level, influences exerted by the surface texture, and deformation behaviour. In the case of haptics one of the interacting surfaces is the human hand, where contact effects are transformed into a sensory impression by sensory cells.

Haptic properties are assessed more or less subjectively as a combination of measurements of the **coefficient of friction**, standardised in ISO 8295 for polymer films, of scratch resistance, of gloss and simply of the feeling obtained when touching a surface or running a fingernail over it. Plenty of development work still needs to be done in the future with regard to objective quality assessments of this complex of properties.

Another important characteristic of good long term resistance is the freedom of coatings from **internal stresses** and strains. Excessively rapid shrinkage during the film forming process can cause thermodynamically unstable molecular structures to be frozen inside. Subsequent external stresses or other mechanical or chemical influences can cause them to manifest in the form of cracks.

One simple method of visualising internal stresses and strains is to apply paint to thin substrates. If the coatings experience internal stresses and strains during film formation, the substrates yield and become deformed.

A more elegant but also more elaborate method of detecting stresses and strains in polymers employs stress optics techniques. Measurements of birefringence of polarised light provide important information for investigating stresses and strains in polymeric films.

3.2.6 Summary

Coatings for commodities have to meet many different requirements which can be grouped under the terms of function and colour. The first step in meeting all user criteria is to apply uniform, defect free layers in sufficient thickness. Film thickness measurements therefore form an integral part of coatings characterisation. Depending on the substrate and where necessitated by the coated item, sophisticated instruments can measure film thicknesses in fractions of seconds, without making contact with or causing damage to the surface.

The coating itself must meet not only visual requirements such as gloss, colour and effect but also physical and technological ones. These include good adhesion, elasticity and hardness. Any tests aimed at determining the mechanical properties must distinguish between short-term stress and continuous stress. The normal test methods for quantitatively describing the properties of coatings either simulate practical stresses and are therefore performance tests or they return defined physical indicators of limited reliability as regards applicability in practical situations.

Sophisticated metrology allows colour and optical effects to be determined objectively for accurate control over colour and gloss. For describing gloss, not only are optical methods used but also mechanical scanning processes for characterising surface texture.

3.3 Durability of Coatings

3.3.1 Fundamental Features of Ageing

The quality of coatings is determined primarily by film forming agents. But a crucial role is also played by pigments, extenders, additives, and the fine distribution produced by the dispersion process. The properties do not occur by chance though. They are always the outcome of a judicious combination of experience and scientific knowledge.

The aim of coating technology is not only to create quality but, equally, to also ensure that the degree of protection and the aesthetics of the coating remain intact for the agreed period of time. Ageing in polymeric materials has two different causes: internal causes, such as internal stresses, and external causes [3.4.48]. The latter are the most important and can vary enormously: sunlight containing aggressive ultraviolet radiation in conjunction with atmospheric oxygen, heat, cold, moisture in the form of water vapour, rain, dew, frost or snow, with and without salty impurities, all in virtually any combination, and in the form of cyclical or continuous stress. Industrial emissions of paint-damaging chemicals such as SO_2, volatile organic acids and alkalis in some regions are also worth listing in this context. Nor should natural active substances such as bird droppings, tree resins, bacteria and fungi be forgotten. Contamination of automotive finishes by road tar, spilled fuel and battery acid are also active substances that should be mentioned. All these factors combine with mechanical strains acting on the coated items such as vibration, and deformation due to tension, pressure, compression and shear, and so lead to a wide variety of stresses on the coating.

Atmospheric factors	Consequences
• Photochemical reactions • Hydrolytic degradation • Thermal decomposition • Volume change through swelling • Volume change due to temperature change	• Molecular degradation • Post-curing • Release of volatile cleavage products • Leaching of degradation products

Figure 3.3.1: Overview of atmospheric factors and their impact on coatings

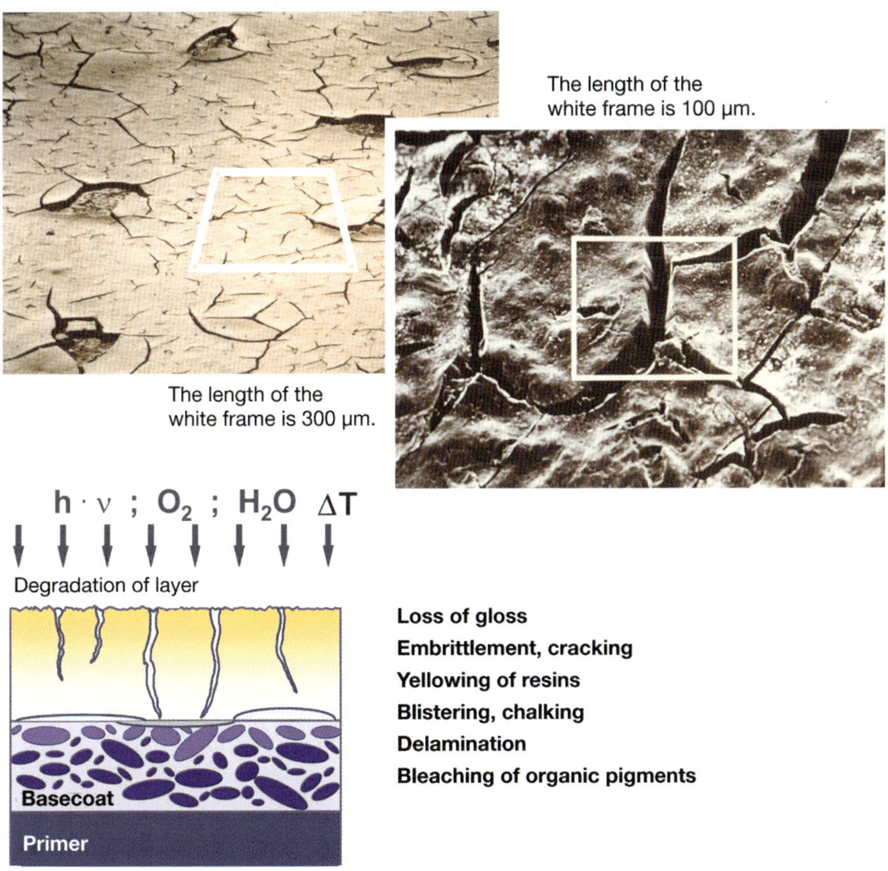

Figure 3.3.2: Photochemical and wet-chemical reactions on coatings and typical results

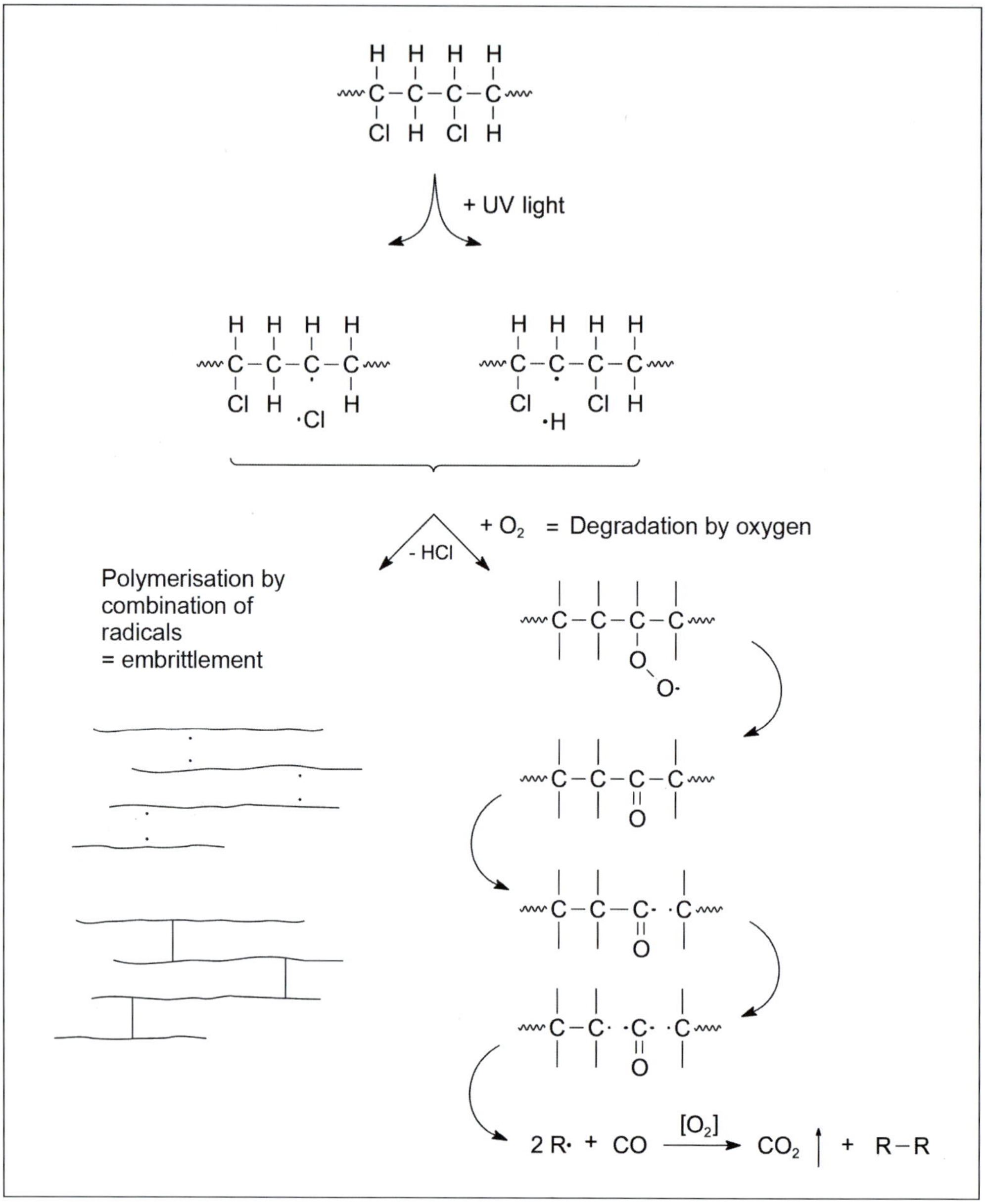

Figure 3.3.3: Photochemical degradation of PVC

The physical, chemical and biological factors, with the exception of the mechanical ones, can be grouped under the term **atmospheric factors**. DIN standards 50010-1 and -2 use the term climate to cover all the atmospheric conditions occurring at a location, including daily and seasonal changes.

The type of climate determines the extent of the changes induced in coatings. These are photochemical reactions initiated by ultraviolet light, hydrolytic degradation due to aqueous agents [3.4.49], thermal degradation by pyrolysis as well as swelling and shrinkage due to cyclical influences of moisture and changes in temperature. The rate at which performance deteriorates consequently depends on the intensity of the various

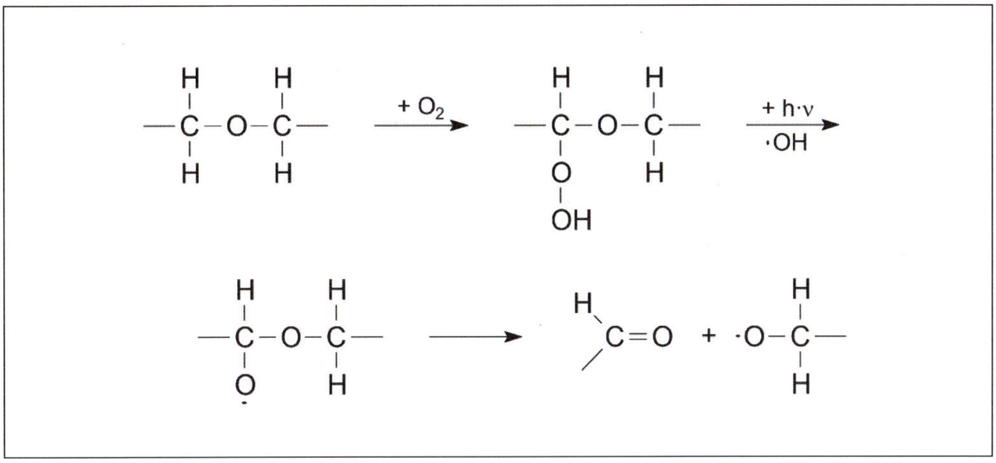

Figure 3.3.4: Photochemical degradation of polymers with ether groups in the presence of atmospheric oxygen

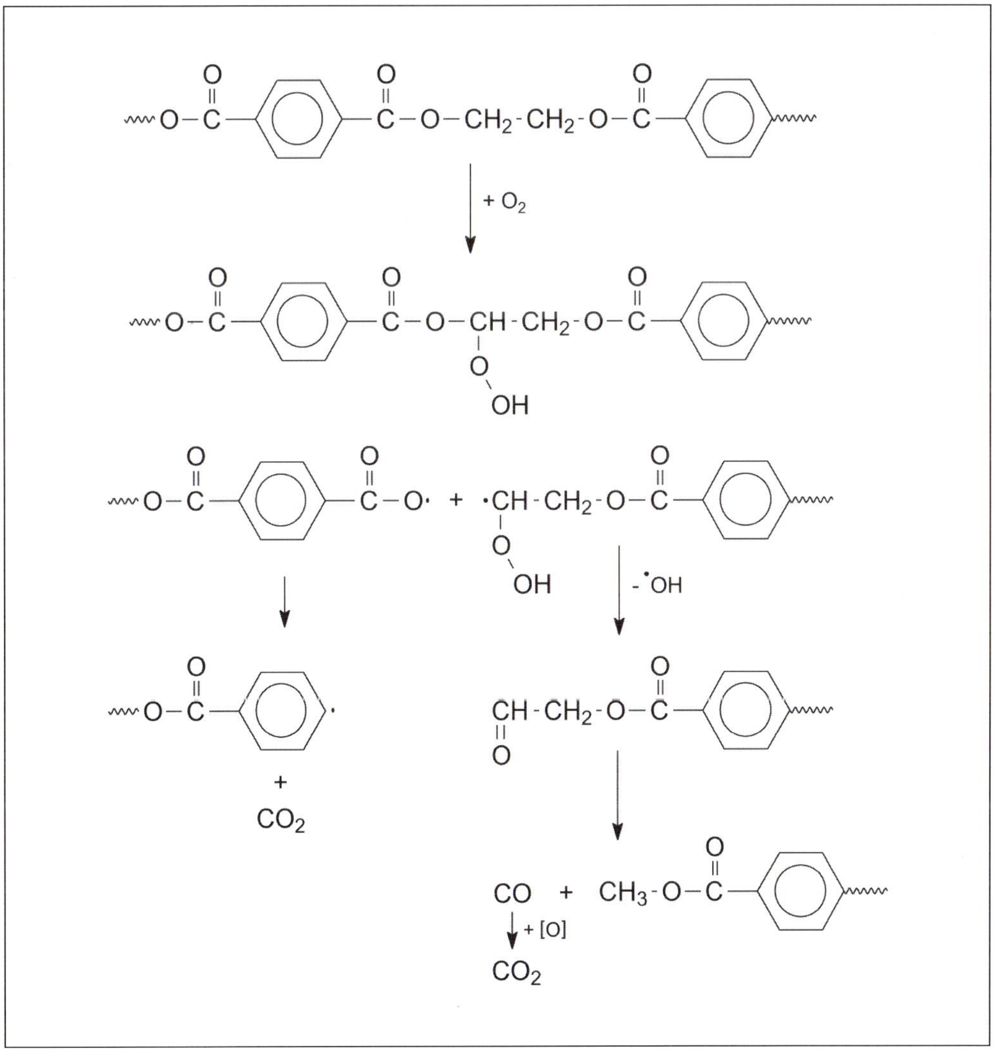

Figure 3.3.5: Degradation of polyesters due to the influence of light and atmospheric oxygen

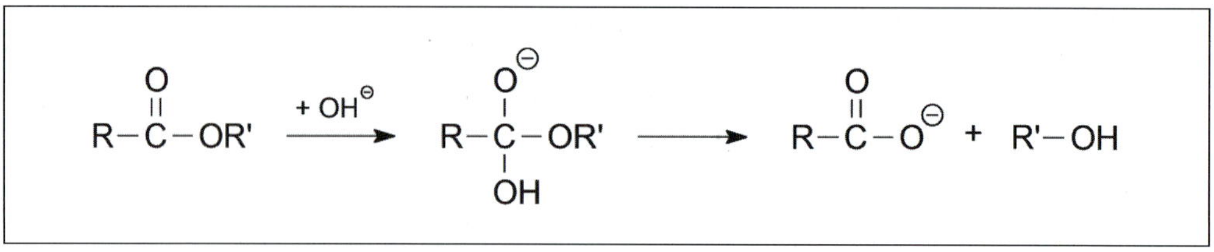

Figure 3.3.6: Wet-chemical degradation of polyesters, as exemplified by alkaline saponification

factors and the nature of the polymers (see Figures 3.3.3 to 3.3.12). Chain shortening induced by ultraviolet light on one hand and subsequent postcuring on the other lead to impaired elasticity and a decline in internal resistance. Where changes in temperature and humidity occur, increasing stresses and strains will cause physical destruction of the coating. This first becomes apparent as degradation with loss of gloss, accompanied by cracking, and, in extreme circumstances, flaking of the entire coating. That in turn completely destroys the protection afforded to the surface.

The chemical reactions responsible for the various **stages of ageing** through to complete degradation of the coating vary enormously. Degradation by UV light, hydrolysis and thermolysis, always in the presence of atmospheric oxygen, induce all kinds of changes in all constituents of the coating. If surface protection is to be retained the focus unquestionably needs to be placed on macromolecular film forming agents. The most important ageing reactions of paint-specific polymers, as illustrated by different examples, are therefore presented below.

Homolysis of chemical bonds by UV rays in the presence of atmospheric oxygen can shorten the polymer chains in PVC. On the other hand, new crosslinks occur as a result of the intermediate peroxide stage. Crosslinking and simultaneous chain reduction are hence the cause of embrittlement and weakening of internal strength.

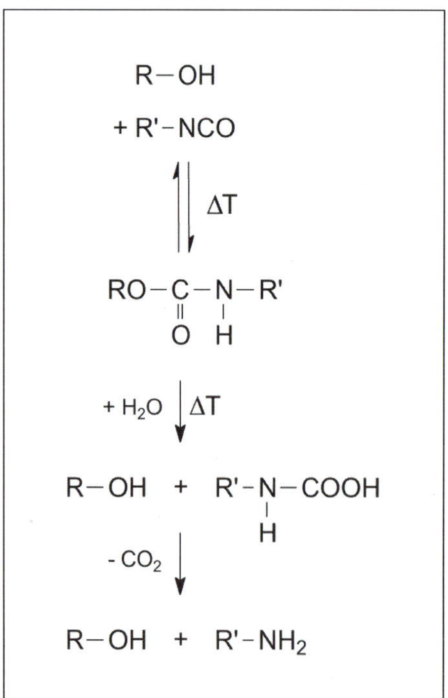

Figure 3.3.7: Thermal hydrolysis of polyurethanes

The ageing process commences with the formation of radicals which ultimately leads to the generation of hydrochloric acid and CO_2.

Degradation of ether groups in polyethers or epoxy resins takes place via the intermediate stage of peroxides whose molecular chains are then cleaved by ultraviolet light.

Under extreme exposure to ultraviolet light even saturated polyesters known to be stable in light undergo similar attack, albeit to a much lesser extent. Hydroperoxide formation, chain cleavage and in some cases oxidative degradation through to CO_2 can all be detected upon exposure to radiation of sufficient intensity and duration.

Apart from **radiation-induced degradation reactions** the atmospheric factors include **wet**

chemical degradation processes in the form of saponification reactions. Polyesters and alkyd resins are examples of film forming agents which, when subjected to prolonged intense interaction with acidic or alkaline electrolytes, can undergo chain degradation to an extent depending on the type and strength of the acid formed.

Thermal Influences on Polymers

Apart from degradation of the macromolecules by photochemical, oxidative or hydrolytic cleavage, it may also be necessary to take account of thermal influences, where pres-ent. These frequently lead to secondary reactions and therefore make ageing processes extremely complex. Consequently, raising the temperature affords a way of detecting not only the expected acceleration in esterification and oxidation reactions but also direct cleavage resulting from depolymerisation and random opening of the base chain.

One example of intelligible and in this case desirable thermal cleavage accompanied by secondary reactions and subsequent reprocessing is the **thermolysis of polyurethanes**.

Thermolysis of polycondensates is more complex. Normal film forming agents such as alkyd resins, amino resins, phenolic resins and polyamides yield a plethora of

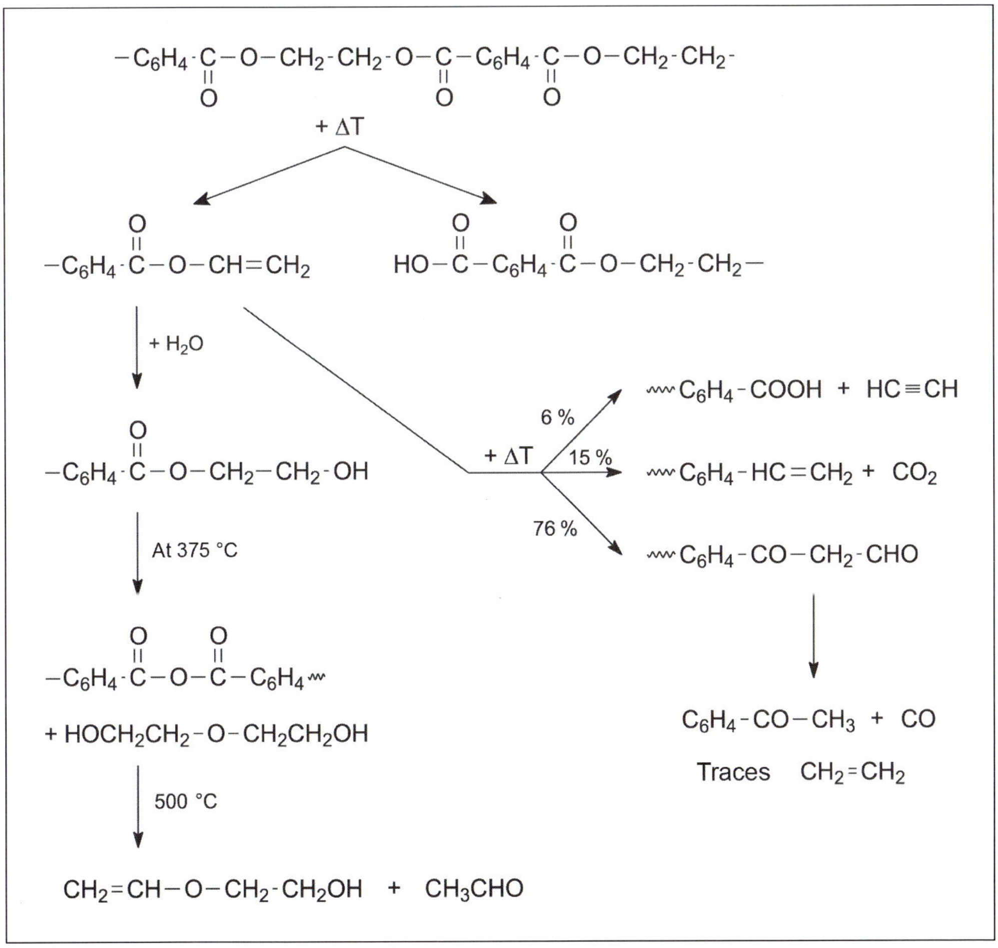

Figure 3.3.8: Thermolysis of polyester resins

Coating Film

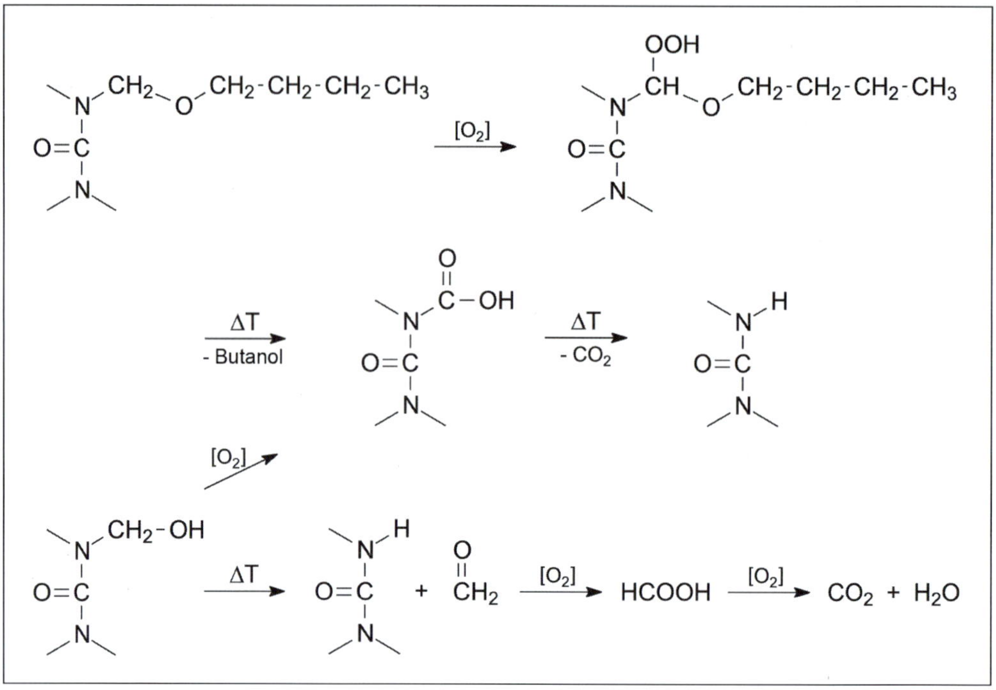

Figure 3.3.9: Thermal degradation of urea resins

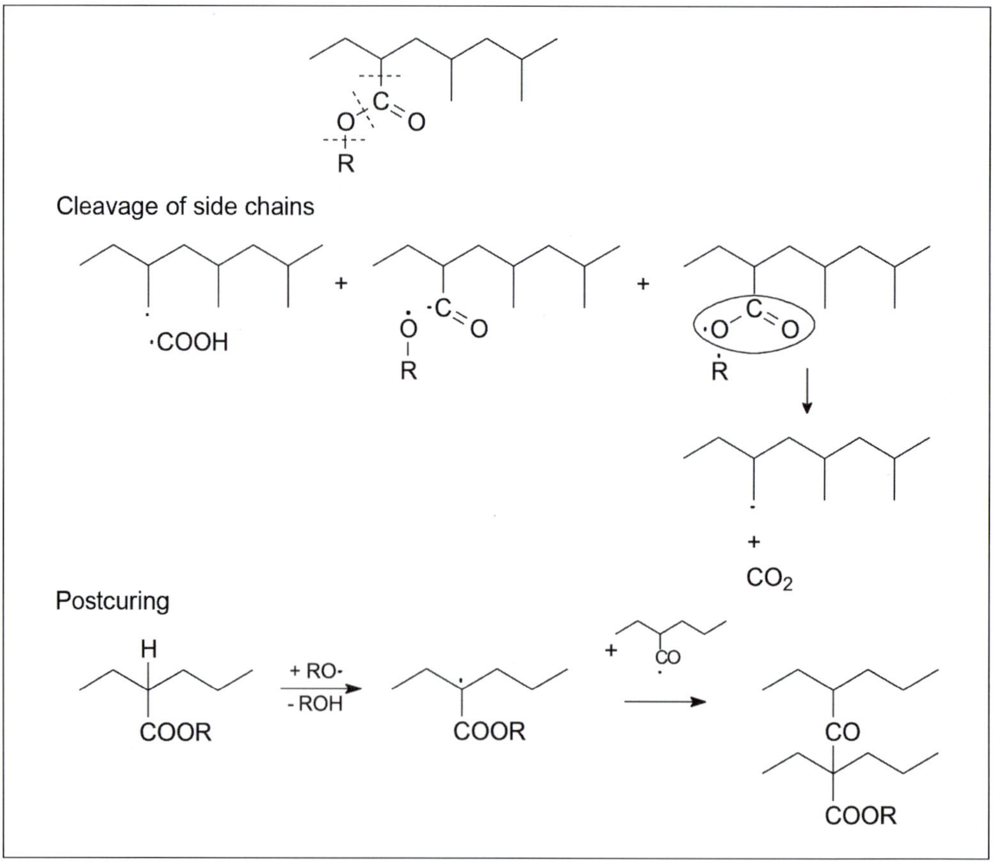

Figure 3.3.10: UV-induced post-curing of polyacrylates

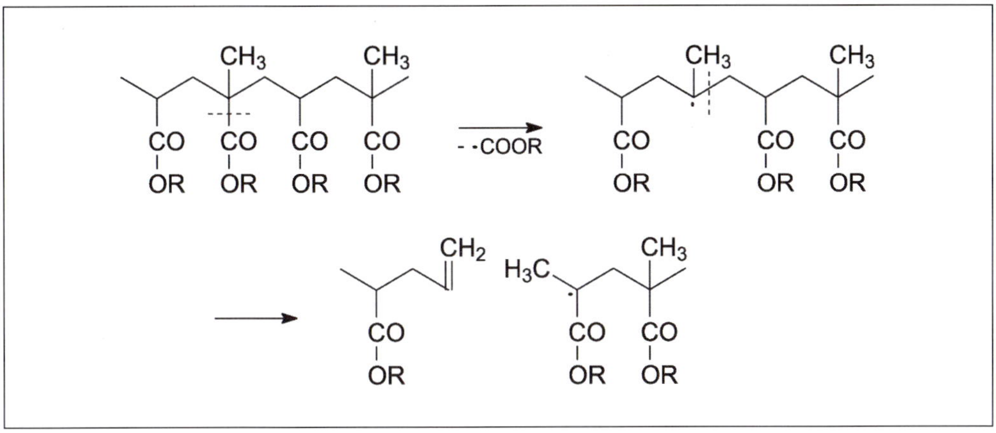

Figure 3.3.11: UV-induced cleavage of the basic chains of acrylic resins

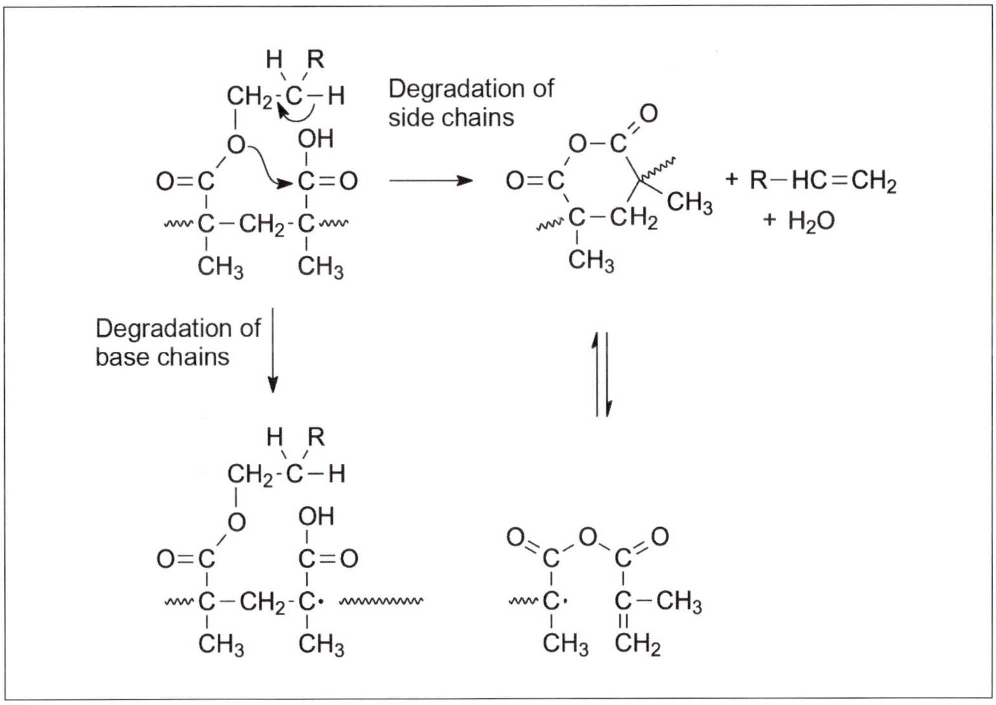

Figure 3.3.12: Thermal degradation of polymethyl methacrylates

secondary products. The most important degradation reactions will be illustrated by taking a linear saturated polyester (see Figure 3.3.8) and a urea resin (see Figure 3.3.9) as examples.

With polyester resins the chain undergoes degradation and double bonds occur which continue to react in a variety of ways.

With urea resins peroxide formation at the hydroxymethylene group gives rise to free carbamic acid derivatives which, after decarboxylation, are transformed into urea groups.

Thermal degradation of polymerisates is fundamentally simpler. Figures 3.3.10 to 3.3.12 show the most important ageing reactions of thermoplastic acrylic resins in the presence of ultraviolet light.

Side chains are cleaved homolytically by the ultraviolet light of the sun. The opening of C-C and C-O bonds gives rise to various radicals which become stabilised through hydrogen abstraction or combination with other radicals. The forming of new bonds involves crosslinking. Embrittlement of the coating is the result.

The ultraviolet rays are also capable of cleaving base chains. Acrylic resins containing methacrylic acid esters illustrate how radicals cleaved from the side chains cause a break in the base chain, forming double bonds and simultaneously transferring hydrogen. This also involves a reduction in molecular weight and hence in the internal strength of the coating.

Ageing Processes in Pigments, Fillers and Additives

The paint-damaging influences are not restricted to the degradation and embrittlement of polymeric film forming agents. The pigments, extenders and additives are also involved in the ageing process. Ultraviolet light or chemical influence can induce colour changes and even pigment-induced degradation of the polymer matrix. Comprehensive studies of multilayer coatings are aimed at allowing early evaluation of the durability of top coats [3.4.50].

New methods like atomic force microscopy (AFM) will help to foster a deeper understanding [3.4.51].

For multilayer paint systems with electrocoat paints acting as primer it is essential that the least-possible degradation by UV occur in order that delamination may be avoided [3.4.52].

A not unsubstantial ageing effect of coatings stems from the composition of the components themselves. Thus high boiling solvents or plasticisers that evaporate slowly cause a significant change, e.g. in the glass transition temperature and other mechanical properties such as elasticity and hardness of the films. Additionally, the crosslink density rises in the first weeks to months due to slow post-curing, even in the absence of weathering [3.4.53]. For this reason, tests are frequently conducted only after artificial ageing in the form of heat stress.

3.3.2 Ageing Tests

The ageing processes which have been described for coatings lead to two different types of damage. The first type comprises colour shifts, reduced gloss, chalking, i.e. **surface changes** which are detrimental to visual attractiveness. The second type comprises a diminution of **protection** as a result of blistering and cracking or complete delamination with subsequent damage to the substrate.

The first type of property change is normally covered in testing by determining the weather resistance and lightproofness as well as lightfastness whilst the second is dealt with by special tests for studying the protective function under the influence of damaging agents such as salts, alkalis, acids and natural substances such as tree resin and bird excrement. Many tests combine the two types of stress.

3.3.2.1 Methods for Testing Surface Durability

Weather Resistance

It has already been pointed out that weather is determined by sunlight, atmospheric oxygen, humidity, changes in moisture and temperature as well as all the natural and anthropogenic atmospheric features. The influences exerted on coatings by these factors vary with the geographic location. Temporary sunshine on the coast in conjunction with salt spray impacts a coating in a different manner to extreme, long-lasting exposure to light in the desert where substantial changes in temperature occur.

The conditions in an industrial atmosphere bear hardly any comparison with those in the mountains or in the tropics.

It is not possible to have one standard test capable of simulating how all the variables are affected by natural environmental factors. Either separate tests have to be performed in the different climatic zones or artificial climatic conditions must be created in which all the factors can be simulated. The former are known as natural weathering tests whilst the latter are called accelerated weathering tests [3.4.54]. Studies of the different tests fail to reveal a good correlation with the behaviour of the coatings in real life and corroborate the incomplete conditions obtaining for all test methods [3.4.55].

So if the intention is to predict the long term behaviour of coatings on the basis of testing, two requirements first have to be met:

1. Highly accurate knowledge of the conditions in the individual climatic zones must be available.
2. For an informative quality inspection the test must be measurable and proportionable in order that the test result may then be wholly reliably correlated with the change in quality under real life conditions.

Natural weathering tests nevertheless play an important role in testing the long term resistance of many coatings, especially as users of coatings frequently look for the results of such tests. It must be borne in mind that the weather is a factor which fluctuates considerably with time and place. Solar radiation, particularly, is sensitive to these parameters. For the paint and raw material manufacturer the intensity and distribution of the radiation are therefore important starting points for forecasting the ageing behaviour of coatings.

The sun with its surface temperature of 5900 K emits a continuous spectrum of electromagnetic waves from 200 nm in the short-wave ultraviolet range to several centimetres in the infrared range (A). On passing through the earth's atmosphere components in the ultraviolet and infrared ranges are selectively removed from the radiation by ozone and water vapour. The spectrum particularly loses radiation up to 300 nm in the short-wave range on account of

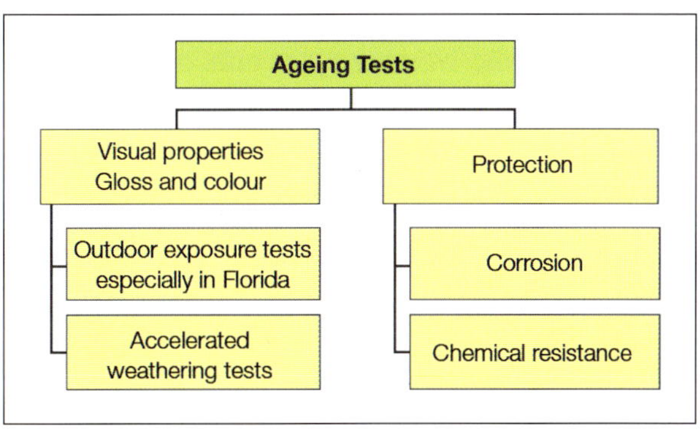

Figure 3.3.13: Overview of ageing tests

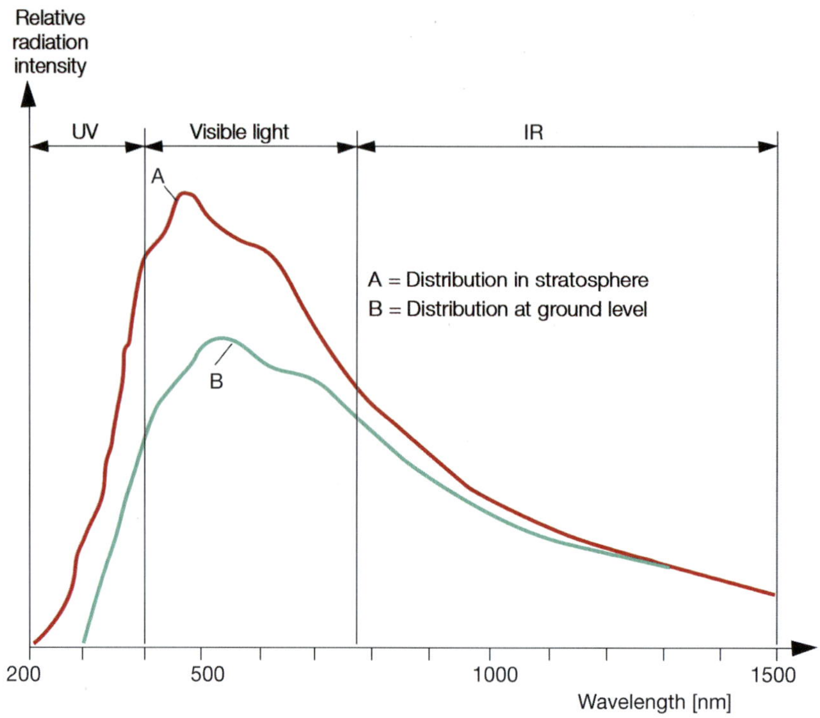

Figure 3.3.14: Change in the spectral components of sunlight as it passes through the atmosphere

the influence exerted by the ozone belt. In the longer-wave range selective weakening of radiation is caused by the excitation of molecular vibrations. Apart from water vapour, atmospheric oxygen and, to a lesser extent, carbon dioxide also play a role. At the same time declines in intensity can be seen over the entire spectrum due to scattering and absorption by atmospheric aerosols such as smoke and dust particles. This phenomenon, which is known as ***Rayleigh scattering*** and ***Mie scattering***, is ultimately the reason that the remaining radiation intensity, naturally varying with time and place, has on average lost approximately 50 % or 0.7 kW/m^2 of the energy originally emitted (B).

Given the temporal and spatial differences in radiation distribution it may be necessary in individual instances to conduct accurate measurements of the **radiation distribution** in order that the weathering results may be interpreted. Pyrheliometers are suitable devices for this. However, they only record the total radiation and are therefore unable to provide information on radiation distribution. The principle of measurement is the differential warming of white and black coated test panels under the influence of sunlight. Both panels are fitted with thermocouples which record any difference in temperature as a function of the intensity of the solar radiation.

The total intensity measured on the ground in the case of vertical solar radiation under a cloudless sky is, as already mentioned, 0.7 kW/m^2 or 0.07 W/cm^2. As a result, the energy for a solar minute is 4.2 Ws/cm^2 or 4.2 J/cm^2. This variable has become established as a customary unit of measure for the **radiation energy** per unit area (i.e.

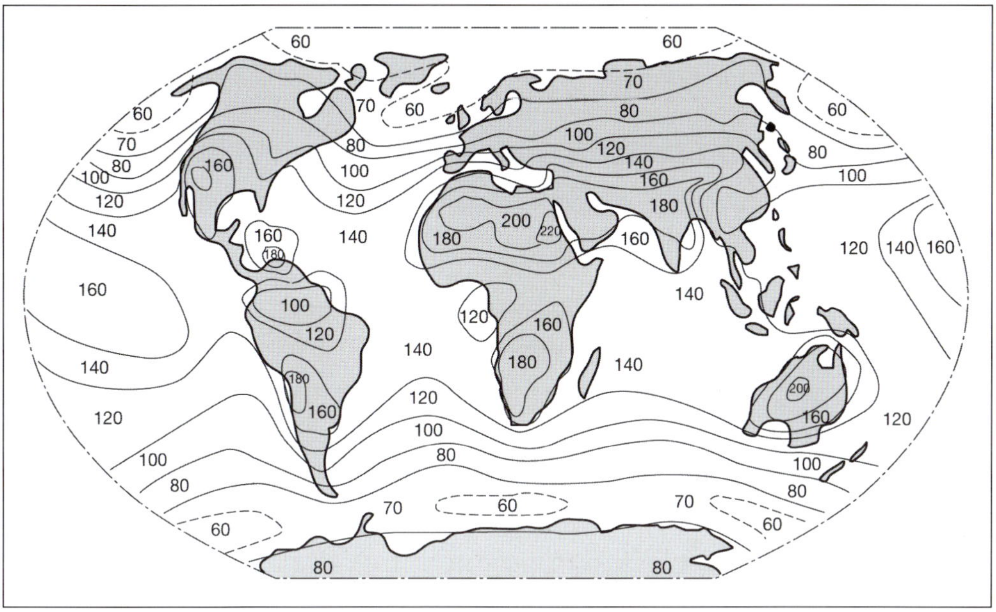

Figure 3.3.15: Map of the world showing lines of same light intensity in 1000s of Langley units

Langley). Figure 3.3.15 illustrates the annual global total radiation per unit area for different re-gions of the world, stated in units of Langley.

The highest values of over 200,000 Langley are recorded in desert regions only. Nevertheless, weathering tests are not conducted exclusively in such regions. The reason for this is that the ageing of coatings is not just of a photochemical nature. Degradation processes are also accelerated by moisture and heat. This explains why regions with a high level of solar radiation such as North Africa, Arizona or Australia with their lack of moisture do not rank among the most aggressive regions in the world for coatings. Regions with high humidity are far more damaging to coatings. The simultaneous action of radiation, moisture and temperature changes induces chemical and physical ageing processes which, to an extent depending on the chemical structure of the film forming agent and the pigmentation, cause the visual and physical/technological properties to change more or less rapidly. **Florida** and **Arizona** are therefore reliable regions for conducting natural weathering tests (see Figure 3.3.16), Florida especially is characterised by a higher level of moisture than Arizona and Germany. This is easily seen from the partial pressures of natural water vapour as illustrated in Figure 3.3.17.

Broad guidelines have been laid down in ISO 2810 concerning the performance of tests and evaluation of the results. According to these the term **long term outdoor exposure** includes all changes to the coatings which are caused by the interaction of all the physical and chemical factors contained in the weather. Testing for long term outdoor exposure by natural weathering therefore consists in stress with exposure to light, water and temperature changes as well as all types of industrial exhaust emissions and atmospheric features such as salt, soot, etc. The test covers the full range of changes in the coating.

The paint systems to be tested in **Florida** are laid out at the weathering stations either on rigid racks with a constant exposure angle or on swivelling panel supports. The exposure angle can thus be adjusted to match the position of the sun. On account of the

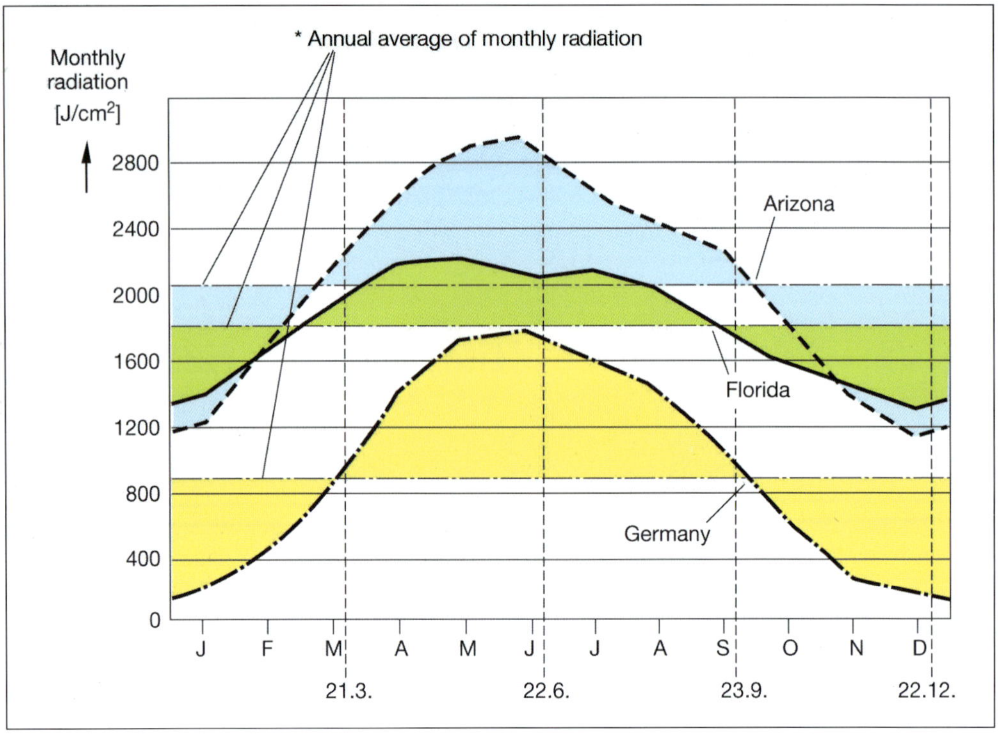

Figure 3.3.16: Monthly average radiation intensity of different regions in J/cm²

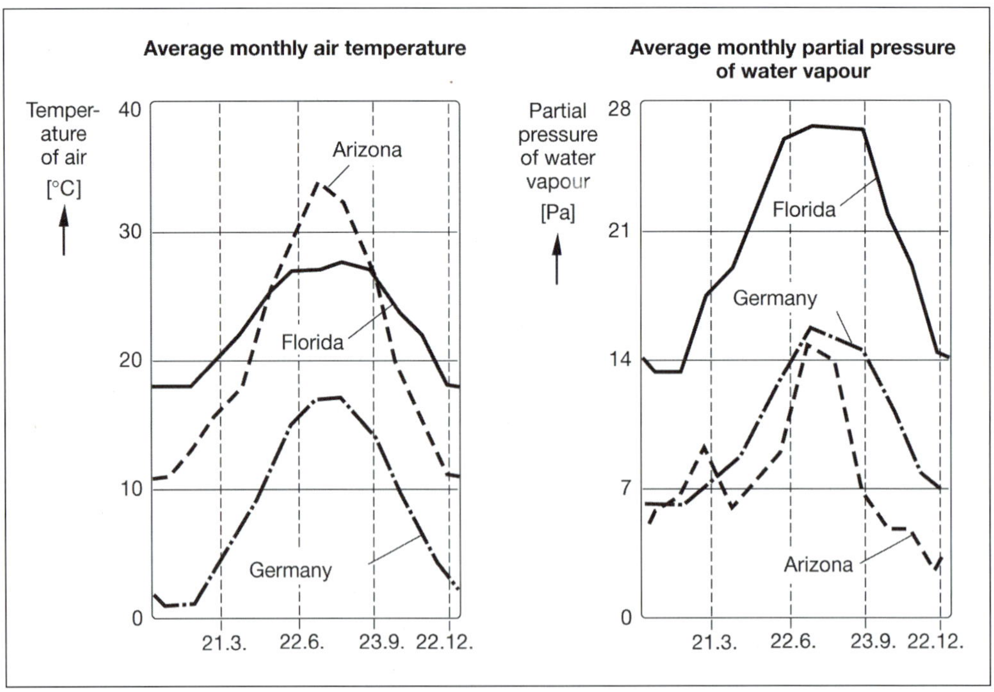

Figure 3.3.17: Local and temporal fluctuations of air temperature and moisture in Florida, Arizona and Germany

latitude of 25° N and hence the high position of the sun the test panels are chiefly exposed to the weather at an angle of 5°. However, the slight angle of such a rack prevents particles of dirt from being washed off by natural rainfall. The test panels therefore have to be cleaned at regular intervals. The 5° southern exposure arrangement is particularly specified by vehicle manufacturers.

Since solar radiation in Florida is more intense than in Germany, the humidity higher, the quantities of precipitation larger and the average temperature much higher, the ageing processes set in after a much shorter time and occur more rapidly than in the Central European climate. For some colours, peak temperatures can be as high as 60 °C. The test in Florida takes place in roughly half the time needed for the same degree of damage to occur in Germany.

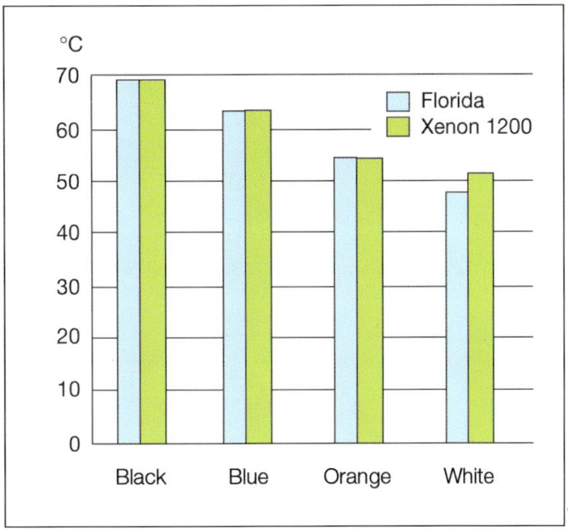

Figure 3.3.18: Peak temperatures of test panels in Florida as a function of colour at an exposure angle of 5° compared with the Xenon test

For more intensive stress the test panels are exposed to the weather in black boxes. These are made of sheet metal and have been painted black. The only opening in the top is covered with the test panels. Heated up by the solar radiation, the black boxes radiate additional heat onto the test panels and speed up ageing at a faster rate than would otherwise occur on exposed open racks.

Figure 3.3.19: Weathering stations in Florida with exposure angle of 5° south, and a heated black box

Source: Atlas (top), BASF Coatings (bottom)

The heated black boxes (HBB) are also fitted with a heater, a thermostat and a fan, so that a uniform temperature of 50 to 55 °C can be achieved in the walls of the box, even if solar radiation is obscured by clouds.

Weathering stations have also been set up in the desert climate of **Arizona**. Despite the lack of moisture, special devices called **EMMAs** (Equatorial Mount with Mirrors for Acceleration) have been developed which increase the already high radiation intensity again

Figure 3.3.20: EMMAQUA equipment for accelerated weather tests in Arizona *Source: Atlas*

by a factor of 8 through the use of aluminium mirrors. The racks track the sun with the aid of a motor and suitable sensors. Sample warming is limited by fan cooling. In view of the lack of moisture EMMA devices are only suitable for testing the lightfastness of pigments. Weak pigments manifest themselves after only a few months in the form of colour changes and changes in colour intensity.

Owing to their ability to generate high radiation loads, there have been plenty of attempts to convert these devices so that they can also be used for long term outdoor exposure tests. The resultant test devices known by the name of **EMMAQUA** are therefore equipped with spraying apparatus for wetting the test panels artificially. Since the test panels in these devices remain hot despite the cooling system, the water sprayed onto the panels evaporates very quickly. Consequently, the water does not exert the same influence necessary for film degradation as it does in Florida.

The **acid etch test** featuring a combination of weathering and acid resistance is an important test in North America. It is a natural weathering test conducted, e.g., on Blount Island near Jacksonville, North Florida. There, the deposition of industrial dust from the local concrete industry, brief rain-showers and heat can lead to punctiform marks and etching of the clearcoats [3.4.56]. The usual exposure duration is 14 weeks from June to September. An accelerated laboratory test has been developed [3.4.57]. The key factors are brief showers of acid rain, deposition of industrial dust and rapid drying phases.

Weathering tests are also performed **in the Central European region**. Since the sun is in a different position, the test panels are arranged at an angle of 45° facing south. There is only a marginal reduction in testing time relative to the service life of the coating.

Owing to the still relatively longer exposure time of several years, even in climatically extreme regions, attempts are being made to further shorten the testing time by conducting **accelerated weathering tests** with artificial light sources of high light intensity, large temperature fluctuations and shorter intervals between wetting and drying.

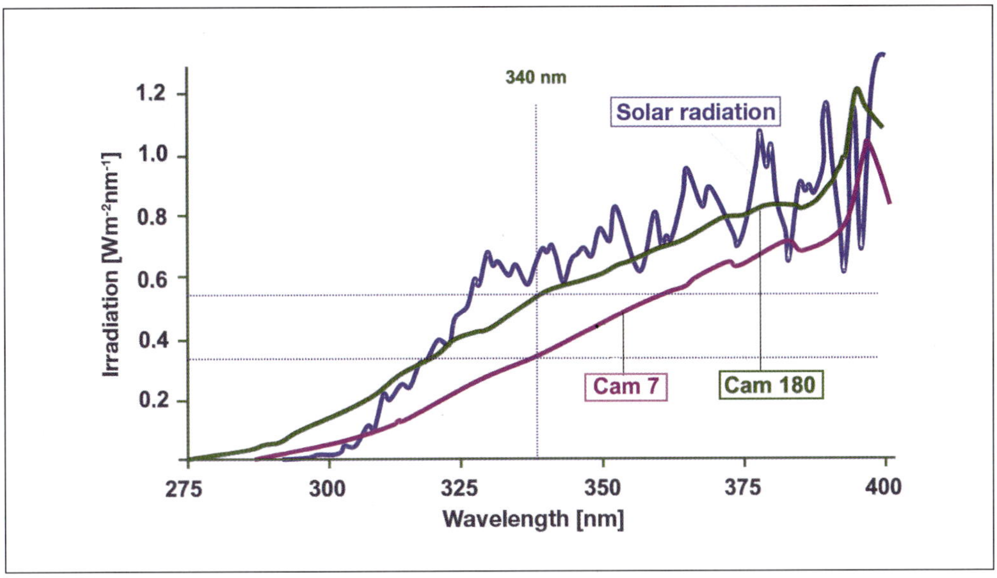

Figure 3.3.21: Spectral distribution of the emitters in weather-o-meter test equipment Cam 180-Test, Cam 7-Test

Source: Clariant

With the aid of suitable light sources such as carbon arc lamps and xenon lamps, the use of water sprays to mimic artificial rain and adjustment to a specified black panel temperature used for verification, it is possible to simulate natural conditions to a large extent. Black panel temperature is the temperature of a black reference panel which becomes established in the test device under the accelerated weathering conditions.

The **spectral distribution of the radiation sources** is crucial to the usability of the test results. For every radiation source there are sets of filters which, on one hand, reproduce the intensity distribution of sunlight and, on the other, essentially eliminate those short-wave ultraviolet components of light sources which are not present in the solar spectrum.

The best-known devices and ones which are used most frequently in the paint and coatings industry are the "Weather-o-meter" (WOM) manufactured by Atlas and the QUV weathering tester or Q-Sun Xenon test chamber made by Q-Lab. These devices essentially operate on the same principle and they satisfy the standards ISO 11341 and ISO 11507 for coatings. There is a lamp fitted to the centre of an enclosed test chamber. At a fixed distance from it is a sample drum rotating through 360° once a minute. Between the radiation source and the sample drum are one or more spraying units which spray the test samples with distilled water at certain intervals. Radiation sources in the Weather-o-meter devices can be carbon arc lamps, also referred to as sunshine lamps, or xenon high-pressure lamps. The carbon arc lamps are surrounded by a filter frame which serves as an air duct and as a support for eight filter glasses. Since during continuous operation the permeability of the filters to light declines quickly below the wavelength of 350 nm, the filter glasses are cleaned once a day and replaced every 2,000 operating hours. The xenon high-pressure lamps are generally surrounded by borosilicate (Pyrex) filters. The highly aggressive ultraviolet radiation below 290 nm is filtered out because it is not present in sunlight close to the earth. Through the use of different filters and combinations of filters it is thus possible to alter the spectral distribution of energy in the artificial radiation and thus adapt to the test conditions required.

Cam 180	Cam 7
• Quartz and borosilicate filters • 2 hours' illumination • 1 hour dark phase with watering	• Pyrex filter • 2 hours' permanent exposure, 18 minutes with watering
Recommended by VDA for	
• Testing crack formation	• Testing weathering resistance

Figure 3.3.22: Test cycles and test chamber of a Weather-o-meter
Source: BASF Coatings

The Xenon tester 1200 is fitted with 3 xenon high-pressure lamps. To ensure that the radiation energy applied to the test samples is as uniform as possible, each lamp is replaced in turn every 500 hours. In the first row of filters the xenon lamps are surrounded by a common quartz cylinder. That is followed by Pyrex filters, consisting of three 1/3 dishes. Dissipation of excess heat from the Weather-o-meter is achieved by air and water cooling. The test chamber is also fitted with a test drum, which can be run synchronously with samples on one side only or asynchronously with samples on both sides, whereby the sample supports are then rotated when they have passed through 180°.

Since the presence of water plays an important role in accelerated degradation, the test panels in all devices are sprayed with doubly-distilled water in certain cycles. Minute traces of silicates in the water would otherwise cover the test panels with a UV-absorbent protective layer over the course of time and thus impair the damaging effect of radiation in an uncontrolled manner.

The proven and therefore now customary test cycles of the "Weather-o-meter" are summarised in Figure 3.3.22.

Under various names the market offers further useful weathering devices which are moreover inexpensive due to their design. They were developed using new lamps and are based on the fact that it is ultimately only the short-wave ultraviolet component which is responsible for the light-induced ageing processes. Their use is preferred wherever intense solar radiation has to be simulated, i.e. North America and Asia.

Figure 3.3.23 shows a cross section through a weathering device. The fixed samples are exposed to radiation from light emitted by eight fluorescent tubes. The manufacturers offer lamps of different radiation distribution. The UV-A 340 lamp starts at 300 nm and is approximately equivalent to the short-wave component of sunlight whilst the UV 313 lamp at 280 nm emits a much more energy-rich radiation.

In contrast to the spraying units in the Atlas equipment the wetting necessary for speeding up ageing is achieved not by spraying but rather by evaporation and subsequent condensation on the sample panels, which are cooler than the sample chamber. This prevents UV-absorbent silicates from being deposited. The entire test cycle takes

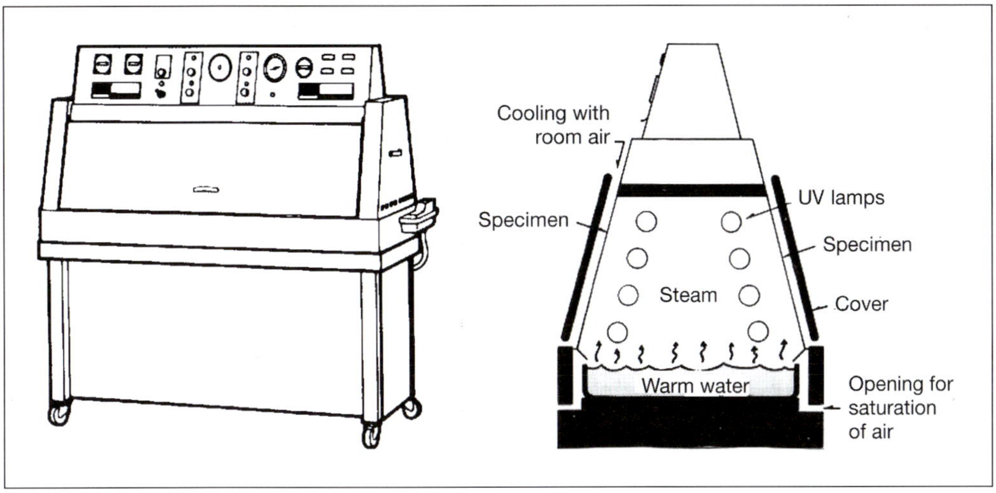

Figure 3.3.23: Overall view and schematic drawing of the test chamber in a weathering device *Source: Q-Panel*

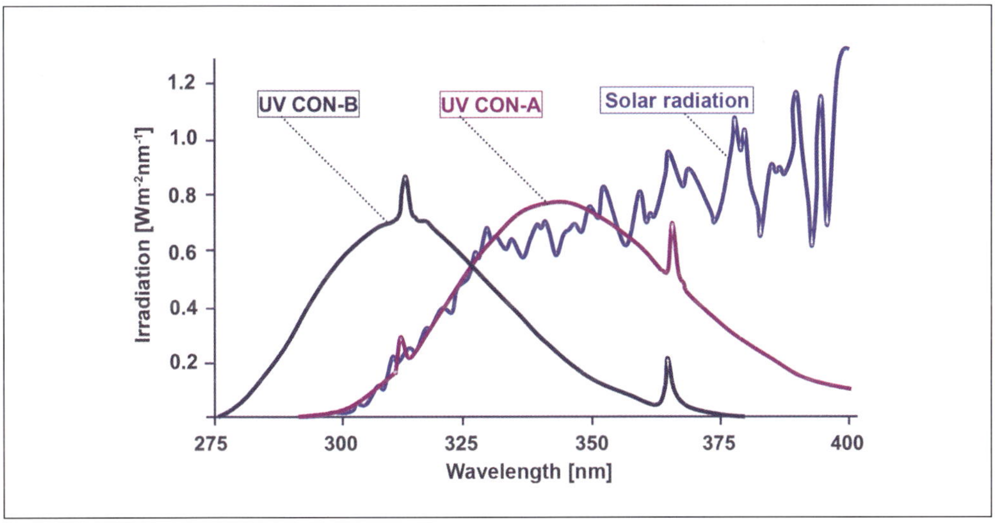

Figure 3.3.24: Spectral distribution of UV-A 340 and UV-B 313 compared with that of sunlight *Source: Clariant*

12 hours and includes eight hours of exposure to radiation at 70 °C, followed by a 4-hour condensation phase at 50 °C, also with exposure to radiation. The test result is normally assessed after a stressing time of 2,000 hours.

Assessment and Evaluation of Weathering Results

The experience gained with these weathering units indicates that the light resistance of solid-colour coatings does not correlate very well with the results of long term weathering in Florida, although the cracking tendency of 2-layer metallic coatings does correlate very well.

The assessment and evaluation of weathering results is contingent on an accurate test method for which all items have been agreed. The field of application of the coating materials determines how the test samples are treated for evaluation purposes. Coatings for facade elements are assessed visually without any further sample preparation.

Automotive coatings, by contrast, are washed in some cases or even polished and then the treated areas are assessed. Depending on the problem at hand the test panels are examined for **gloss retention, chalking** and **colour resistance**.

Other changes in the coatings such as **cracking** and **blistering, surface** and **edge corrosion, adhesive strength** and **flaking** are normally assessed visually as well.

As all the data are recorded in chronological sequence, it is also possible to make statements about the dynamics of the degradation reactions.

When accelerated weathering tests are being developed it must be borne in mind that although unrealistic conditions for the parameters can lead to shorter testing times the ability to make reliable predictions about behaviour in practice is considerably impaired [3.4.58]. Many studies are examining the testing cycle as a route to improving the correlation with natural weathering [3.4.59].

The experience gained from comparisons of Florida weathering and accelerated testing indicates an acceleration factor of 5 to 6. Consequently, the minimum testing time for accelerated weathering tests on automotive coatings is four to six months.

The test devices are also used for polymers, interior coatings, seals and other materials exposed to weathering.

3.3.2.2 Corrosion Protection Tests

Like weathering tests, functional tests, i.e. specific tests for certain levels of protection afforded by the coating, can take the form of natural weathering and accelerated testing. In this context, changes due to ageing have less to do with appearance than with functionality.

In effect this means considering how the coating's resistance responds to temperature and moisture, to changes in temperature and moisture, and to a wide variety of salts and chemicals in the form of gases and liquids. The assessment primarily focuses on damage in the form of blisters and delamination, as well as consequential damage resulting from damage to the coatings.

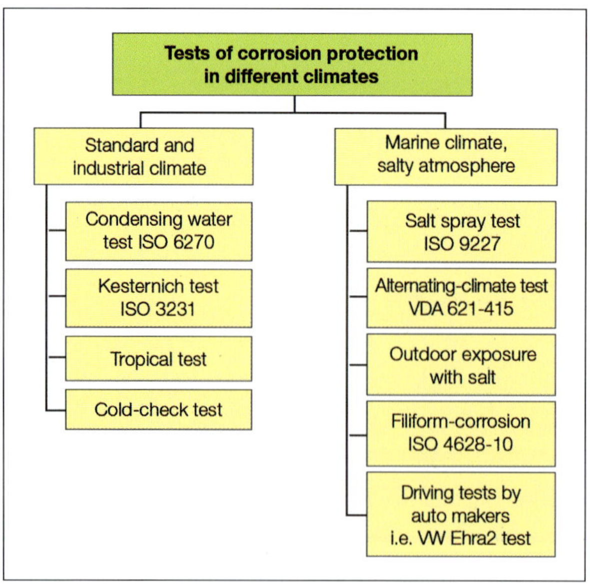

Figure 3.3.25 Overview of tests for assessing corrosion protection

From an economics point of view, one of the chief tasks of coatings on metal substrates is to protect against corrosion. Despite the enormous efforts that have been undertake in this area, experts estimate that 3 to 5 % of the gross national product of the industrialised nations is lost to corrosion every year. Protection against corrosion is therefore an important economic goal [3.4.60].

The term corrosion is used below in accordance with ISO 12944-1 and ISO 8044 to cover all the changes in the metal coated object caused by the influence of oxygen, water and catalytic electrolytes. From a chemical point of view, it involves the emission of electrons from the metal materials in the presence of water and oxygen. The outcome is the formation of an insoluble metal oxide layer (see Chapter 4.1).

The rust formed on iron is porous and so it is incapable of retarding the progressive destruction of the metal. Effective protection against corrosion is therefore necessary. This can be achieved with organic coatings, for example, frequently in conjunction with corrosion-inhibiting pigments.

Standard Reference Atmosphere and Industrial Atmosphere

The specific conditions obtaining in tropical regions, as well as in wet rooms such as indoor swimming baths, cold stores or kitchens, are simulated in so-called **condensing water alternating-climate tests** and **condensing water constant-climate tests** in which high humidity in conjunction with temperature and/or changes in temperature create either a consistently high or an alternating water vapour concentration. In all cases the samples in the test apparatuses are exposed to the test conditions specified in the standards.

ISO 6270 contains all the relevant information required for conducting the tests. The standard distinguishes between three variants of the test:

- condensing water, alternating humidity (and temperature)
- condensing water, alternating temperature
- condensing water, constant humidity

One variant which is now rarely specified is the alternating condensing-water/alternating humidity test in SO_2 atmosphere. In accordance with the agreement reached, either 0.2 litres or 2 litres of SO_2 gas are admitted to the 300 litre test chamber in order to simulate industrial atmosphere conditions. This test, which is standardized under ISO 3231 and goes by the name of the **Kesternich test** is unrealistic on account of the large spread of results and excessively high concentration of SO_2 compared with the stress encountered in practice and so is only used for very specific test objectives. The individual test conditions are all listed in Figure 3.3.26. The test is terminated when the testing time agreed with the customer has elapsed or when visible changes occur such as loss of gloss, blistering or corrosion.

Apart from the tests standardized in ISO standards there are numerous special-purpose tests which are agreed upon between customers and suppliers. The simulation of high but also substantially fluctuating temperatures in the tropics and the correspondingly low but also widely fluctuating temperatures in arctic regions should also be mentioned: the **tropics test** and the **cold check test**.

Marine Climate

The most widespread corrosion protection test is the **salt spray test** as per ISO 9227 and ASTM B117. It is designed to simulate the salty and corrosive atmosphere of coastal regions and the effects of road salt. At a test chamber temperature of 35 °C a 5 % sodium chloride solution is continuously sprayed with the aid of pre-warmed, humidified compressed air. Before exposure a scorer is used to damage the test panels parallel with their narrow edge and as far as the substrate. Completion of testing is followed by

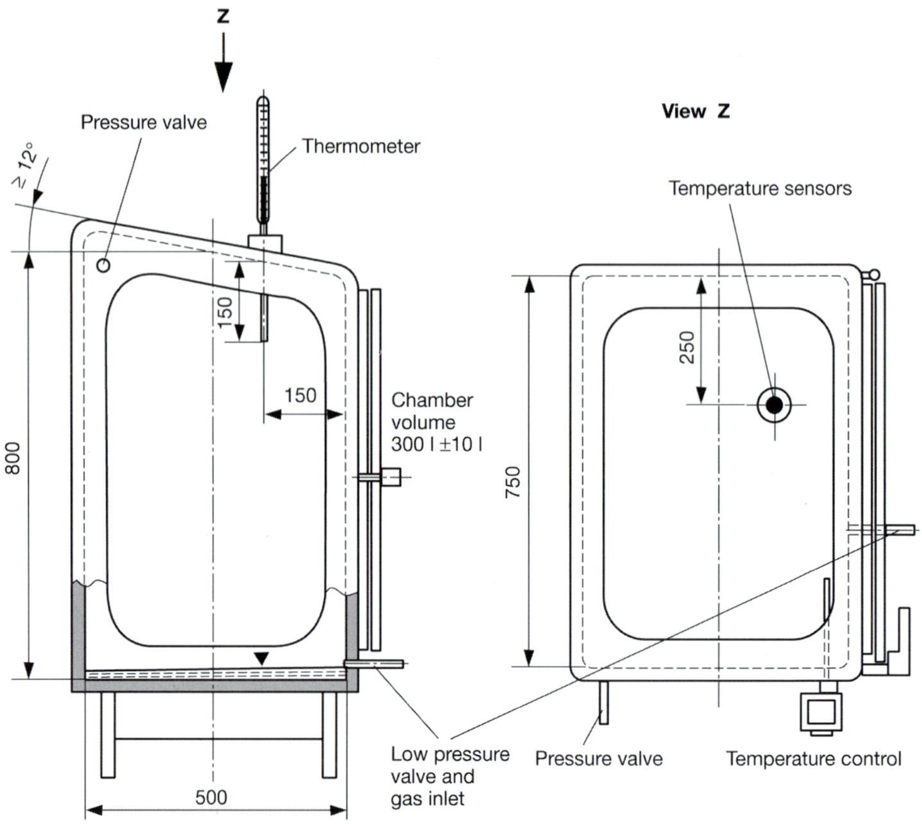

Type	German abbreviation	Duration of cycle hours	Duration of phases hours	Temperature in chamber [°C]	Relative humidity [%]
Condensing water, alternating humidity	SFW	24	8 Including preheating	40±3	100
			16 Including cooling (chamber open)	Room-temperature [2]	Under 75
	STW	24 [1]	8 Including preheating	40±3	100
			16 Including cooling	Room-temperature [2]	100
Condensing water, constant humidity	SK	–	From preheating until test finished	40±3	100

[1] Standard cycle duration; see also Chapter 2.2 [2] See DIN 50 014 for more on the term room temperature

Figure 3.3.26: Schematic drawing of a test chamber and conditions used in condensing water tests in accordance with ISO 6270

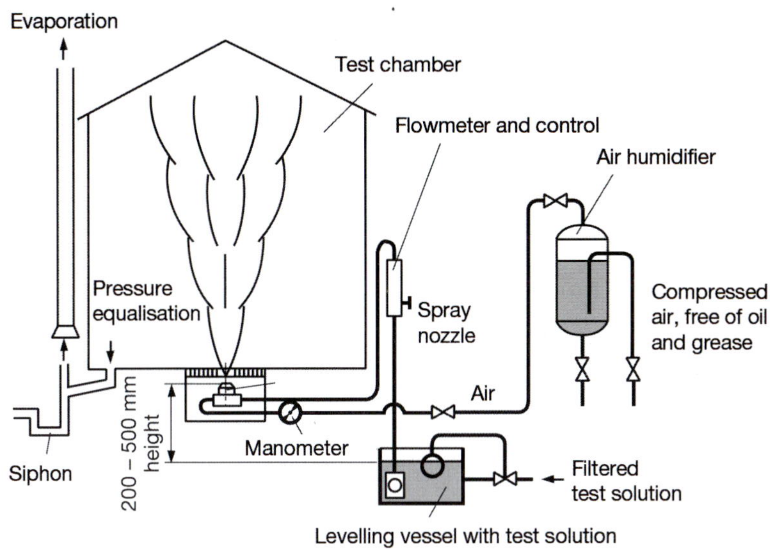

	Type control	Salt spray test ISO 9227
Spray solution	Sodium chloride in [g/l]	50±10
	Further additions	None
Collected solution	Sodium chloride content in g/l	50±10
	pH	6,5 to 7,2 at 25 °C ±2
	Temperature in chamber	35 °C ±2

Figure 3.3.27: Schematic drawing of test chamber and test conditions used in the salt spray test as per ISO 9227 and ASTM B 117

assessment of the width of creepback originating from the score as well as corrosion at the edges and on the surface.

The test is conducted in test chambers with capacities ranging from 300 litres to several cubic metres, in which it is possible to test entire vehicle bodies.

Preparation of the test panels and evaluation of the corrosion pattern are conducted in accordance with ISO 4628. Standard damage patterns to help make evaluation more objective are available (see Figure 3.3.28). The pictures shown in DIN standards serve as the basis for a scale of grades ranging from 0 to 5. The degree of damage can be established by comparing the sample with the standards. If blisters form, the number m and size g are evaluated or, if extensive corrosion occurs, the corroded area. Ri is the internationally agreed symbol for the degree of corrosion and in conjunction with a grade from 0 to 5 it indicates the degree of extensive corrosion. Digital image analysis in accordance with ISO 21227-3 is possible.

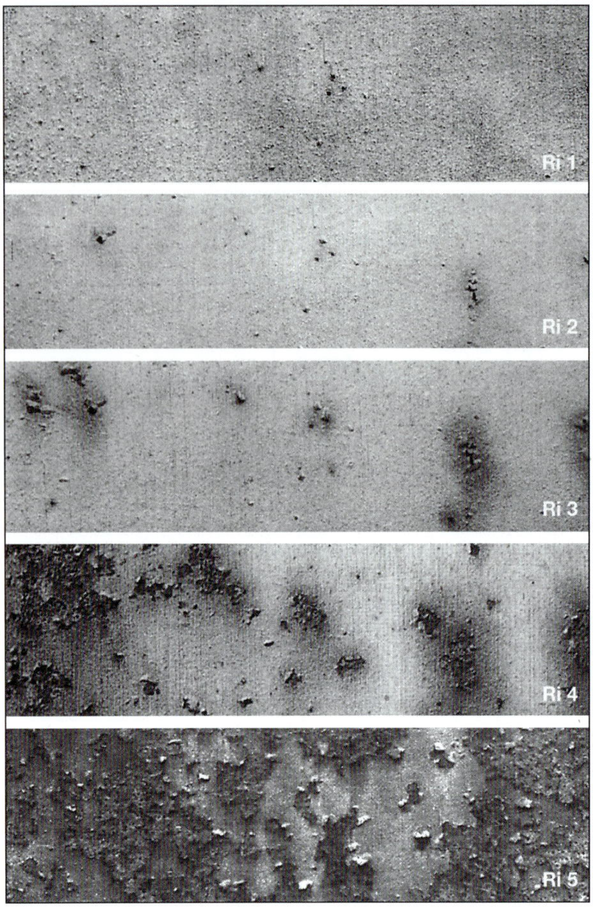

Figure 3.3.28: Standards for objective evaluation of surface corrosion; Ri = degree of rust

Despite its widespread use special care must be taken when the results of the salt spray test are applied to long term corrosion protection. In the literature more and more critical voices are being raised which liken the accuracy of the salt spray test to that of a weather prediction by an experienced shepherd. Even the information content of comparative tests on known corrosion protection systems is being questioned. Such opinions are confirmed when the corrosion pattern created by the salt spray test is analysed. Despite the loss of all protection, areas of extensive **delamination** often show no signs of damage due to corrosion as opposed to corrosion due to weathering. Especially on galvanised steel it is impossible to make any reliable statements based on test results about behaviour under practical conditions.

One example of a type of corrosion which cannot be simulated by the salt spray test is **filiform corrosion**, which is found not so much in the European climate but more frequently in tropical regions. Originating in mechanical damage to the coating this type of corrosion takes the form of thread-shaped creepback. As this form of corrosion cannot be simulated in a salt spray test, different tests have been developed specially for filiform corrosion, e.g. ISO 4623. The procedure consists in immersing the panels prepared for the salt spray test in 0.1 % sodium chloride solution for 30 to 60 s, drying them and then exposing them to 40 °C and 80 % relative humidity in a climatic test cabinet. Any channel-shaped corrosion creepback produced stems from the formation of so-called oxygen cells. On account of the higher O_2 concentration at the exit of the resulting channels cathodic and anodic areas are formed in conjunction with the creepback areas which have less O_2 (Figure 3.3.29). The aluminium employed in the aviation industry is subjected instead to DIN EN 3665, which uses a different type of scoring and evaluation from that of the method described.

In view of the dubious reliability of the salt spray test, efforts have been undertaken to develop better corrosion protection tests. The automotive industry, in particular, has a pressing need for absolutely reliable forecasts of long term corrosion protection. Various tests, usually combinations of test methods which were already known but were not very informative on their own, were therefore checked for correlation with practical conditions [3.4.61].

The **cyclical climate test** described in VDA 621-415 is hugely important because it is recognized by the automotive industry. This test is a combination of the salt spray test and the condensing water test. The corrosion media are a 5 % saline solution and deionised water used in alternation. A cycle lasts one week and consists of 24 hours of salt spray test, 96 hours of condensing water and humidity in alternating atmosphere and 48 hours' regeneration time at room temperature at a humidity of < 75 %. The total testing time is 10 or 12 weeks and is roughly equivalent to natural weathering involving artificial salt spray lasting 1 year (see below). As in most corrosion tests the coated test panels are damaged with a scorer down to the bare metal before being subjected to stress.

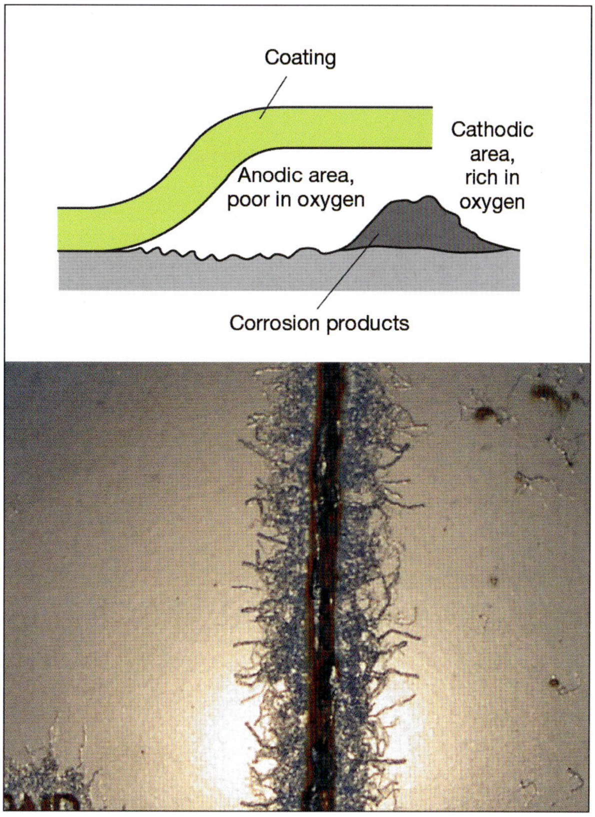

Figure 3.3.29: Underlying mechanism and example of filiform corrosion *Source: Streitberger*

Whilst with the pure salt spray test there is only corrosion creepback and hence a loss of adhesion, with the alternating atmospheric test corrosion products develop which are identical to the corrosion pattern in reality. If the corrosion resistance of the coating is inferior, perforation of the kind observed in practice occurs after a testing time of >12 cycles. The extent of it depends on the type and thickness of the metal substrate.

Natural weathering with salt, already mentioned, is also a method of accelerated testing for corrosion resistance. In this method the test panels, which are oriented southwards at angles of 5° and 45° to the horizontal, are usually positioned on sites belonging to the paint-manufacturing industry or user industries where they are weathered for a period of 6 months or more. Unlike the case for normal weathering, the test panels are sprayed with a 3 % saline solution once a week to accelerate the corrosion process. Also in this test the test bodies are scored or damaged by blasting steel chippings at them prior to exposure. This increases the rate of corrosion by a factor of 3 to 4 compared with normal weathering.

Despite a considerable improvement in correlation between the test result and corrosion pattern with the findings from practical stressing in the cyclical climate test, vehicle manufacturers have developed more test methods, some of which are elaborate. These involve exposing entire cars to fixed, detailed test cycles. As an example, consider the **VW Ehra2 test** (see Figure 3.3.30):

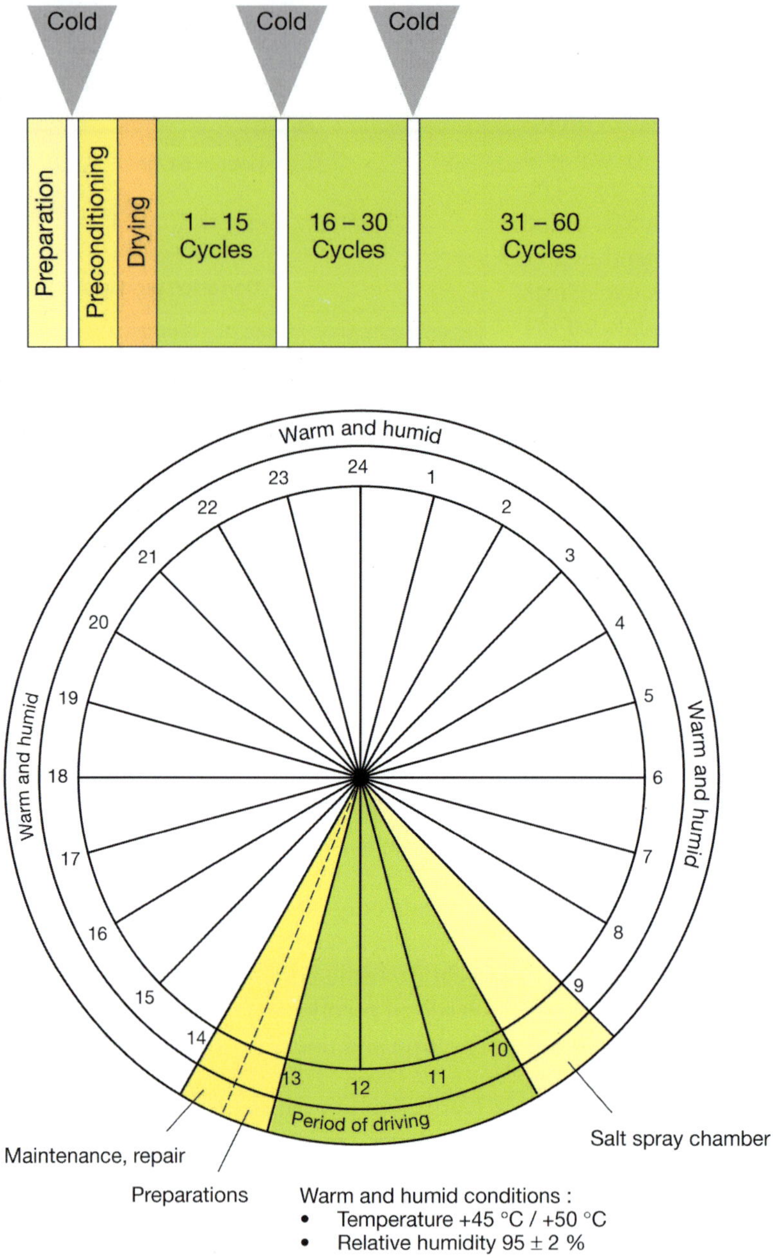

Figure 3.3.30: Overall procedure and individual cycles in the VW Ehra2 vehicle corrosion test

Sample cars selected from normal production are first cooled down to -35 °C in cold chambers and subjected to torsion by means of car jacks to simulate mechanical stress caused by driving. Any resulting cracks in the coating of the cavities would thus lead to weak points in the corrosion protection. For further preconditioning the test car is driven on road chippings and salty sludge road sections for 4 days. After cleaning and seven days of storage the corrosion test proper can begin. Over a total of 60 cycles with two interruptions for repeating the torsional stress in the cold chamber and subsequent salty sludge treatment the test cars are exposed to a humid heat chamber at 45 °C and 95 % relative humidity every 24 hours, initially for 19 hours. That is followed by a one-hour salt spray test, a three-hour driving stage, maintenance and possibly repair work, and finally repetition of this cycle another 59 times.

Figure 3.3.31: Weathering station in Münster, Germany

Source: BASF Coatings

These corrosion protection tests with their extremely corrosive environment are not equally useful for all substrates. For example, the salt spray test is unable to predict corrosion protection performance on galvanised steel [3.4.62].

Further, Indirect Test Methods

Further, indirect test methods are designed to provide information about corrosion and corrosion protection by analysing the electrochemical processes taking place. Unfortunately, the enormous reduction in testing time and the advantage of obtaining physical indicators are again partially offset by insufficient correlations with behaviour in practice. Out of the large number of tests described in the literature it is worth drawing attention to electrochemical impedance spectroscopy (EIS) [3.4.63]. Impedance, the resistance to alternating current presented by a stressed and nonstressed coating as a function of measuring frequency, supplies data which provides information about the level of protection against corrosion. However, there is no doubt that experience to date is not extensive to base a quality assurance system on it [3.4.60]. The measurement of **dielectric constants, corrosion potentials** and the plotting of **polarisation curves** are examples of corrosion protection tests whose application is chiefly limited to comparative testing for development purposes.

Very interesting in this respect is the use of a *Kelvin* probe for early indication of adhesion losses of organic layers on steel [3.4.64]

3.3.2.3 Chemical Resistance

Tests for chemical resistance are aimed at obtaining an overall performance profile, consisting of visual properties and protective functions. It has already been pointed out that ageing processes are accelerated by aggressive gases and liquids. **Sulphur dioxide** tops the list of agents which damage paint on exposure to outdoor weathering. It exerts a highly reducing effect and can therefore damage pigments and film forming agents in the same way. Other pollutants are found in industrial regions apart from SO_2. The type

Coating Film

of industry and the region determine the extent to which the **pH of rain** is influenced by acids, which can be substantial in some cases. Overall though, the problems caused by industrial emissions have been alleviated considerably as a result of numerous measures such as desulphurisation at power stations and catalytic converters on vehicles.

In addition to the gaseous and liquid **anthropogenic pollutants**, nature also makes its contribution by producing paint-damaging substances. **Bird droppings** and **tree resins** do not harm the entire surface of a coated object but they are able to cause considerable spot damage to the coating. Physical swelling, chemical reactions of an inorganic, organic and even enzymatic nature can take place simultaneously or successively and destroy the coating. By swelling is meant the absorption of liquids due to diffusion, which then leads to an increase in volume and softening of the surface. It also involves a reduction in **adhesion** and **resistance**. Such purely physical processes are reversible if the active substance is sufficiently volatile. High-boiling terpenes and natural resins, as occur in conifers, are withheld in the film of paint because of their inadequate volatility. In that event regeneration of the damaged area is no longer possible. One absolutely irreversible process is chemical attack by reactive natural substances. There are numerous natural resins and solvents which, on account of their reactive functional groups, not only penetrate the coating but also initiate degradation processes in polymers. Salts dissolved in the substances mentioned above are also damaging in that they penetrate the coating along with the water and remain in the film after the water has evaporated. It is rarely possible to divide the effect of paint-damaging atmospheric features into physical and chemical influences.

A whole host of **test substances** are deployed in the furniture, can coating and coil coating industries. Substances such as cola, coffee, tomato juice, etc. can attack and discolour unsuitable coatings extensively.

For practical tests it is necessary to select active substances which are appropriate to the particular application. These are delivered in droplets from a pipette and allowed to act on the coating for a specified time. To achieve maximum reliability in such a test the test panels are hardened in a gradient bar oven (see Chapter 3.1.5). The setting of a certain temperature gradient makes it possible to reproducibly create inadequate and excessive crosslinking on a test panel. In addition, the gradient bar permits tests to be conducted on identically stoved test panels under various temperature conditions. The use of temperatures above and below the glass transition temperatures of the coatings extends the breadth of information produced by the test results.

The actual test and the evaluation of the results should take place in accordance with ISO 2812-1 in conjunction with agreements on the choice of **test chemicals** between paint manufacturers and paint users. Important test solutions are fuels, lubricants, brake fluids, antifreeze, cold cleaners, preservatives and aqueous substances such as hydrochloric acid, sulphuric acid, sodium hydroxide solution, as well as oil soot from furnaces. Even water serves as a test substance. In certain industries the use of special test substances such as tar solutions, tar removers, insect juices, bird droppings, and tree resins is common. Household chemicals and foods are also used for testing the surfaces of furniture. Judicious combinations of the test substances and test conditions permit most of the damage encountered in practice such as discoloration, swelling and etching to be simulated. In the test for microorganisms the coating substances are exposed to **certain fungus or algae cultures** under various environmental and atmospheric conditions.

When it came to testing coatings **for oil soot resistance**, initially samples of soot were taken from furnaces to serve as test substances. The discovery that the damaging agents were mainly acidic in nature prompted the VDA to recommend a list of substances especial for the automotive industry. The test procedure consists in applying the test solution to the coating, which is then covered with a clock glass. The test piece is then exposed to stress for either 1 hour at 70 °C or 48 hours at 23 °C.

The standards specifically developed for the **furniture industry**, DIN EN 12720 and particularly DIN 68861-1, provide a classification of coatings according to stress categories 1 A to 1 F: 26 testing agents, including acetic acid, alcohol, white wine, red wine, cola, coffee, mustard and ballpoint pen paste, are specified as test substances.

3.3.3 Summary

Coating Film

Coatings, which are compositions of polymeric film forming agents, pigments and additives, can be exposed to considerable stresses during the service life of the substrate. The ultraviolet light of the sun, temperature fluctuations, water in all physical forms and chemicals of various composition are the most important factors affecting quality. The rate at which the resulting ageing reactions occur depends on the chemical structure of the individual constituents of the coating. Chain cleavage and subsequent formation of new crosslinking points impairs the cohesion and elasticity of the protective film. This may be accompanied by a direct attack on the substrate. In the case of metal materials, the loss of protection is manifested in the formation of corrosion products. Ageing reactions are not restricted to the polymeric film forming agents. With pigments they lead to more or less extensive colour shifts and loss of gloss and in many cases to unacceptable changes in visual properties.

For absolutely reliable forecasts of the long term behaviour of coatings, tests are conducted under stringent conditions. This is achieved by exposing sample panels in extreme climate regions or in weathering apparatus under defined aggressive atmospheric conditions. With a view to predicting the protective functions coatings are subjected to special tests involving common salt and other chemicals. To improve correlation with the results gained in practical use, elaborate systematic empirical tests are underway, particularly in the automotive industry.

3.4 Literature

References

3.4.1 S. E. Orchard, Appl. Sci. Res., Sect. A, 11, p 451 (1962)

3.4.2 A. Mercurio, S. Lewis, J. Paint Techn., 47, 607, p 37 (1975)

3.4.3 P. Grassmann, Wiss. Grundlagen der Anstrichstoffe u. Anstrichtechnik, Band 1,: Eigenverlag des Verb. der Schweizerischen Lack- und Farbenfabrikanten, 1960

3.4.4 K. H. Hornung, U. Biethan, F+L 76, p 461 (1970)

3.4.5 WO 1994009913 A I (1994) Ferro Corp.

3.4.6 R. Mattern, Dissertation Uni Paderborn, Aug. 1998

3.4.7 W. S. Overdiep, Progr. Org. Coat. 14 (2), p 159 (1986)

3.4.8 W. Schlesing, M. Buhk, M. Osterhold, Progr. Org. Coat. 49, p 197 (2004)

3.4.9 J.-C. Castaing, J.-N. Bousseau, P. Beurdeley, ECJ 11/2005, p 28

3.4.10 Anonymus, MO 66(6), p 40 (2012)

3.4.11 H. Petry, Technisches Messen 65 (11), p 396 (1998)

3.4.12 H. Schulz, DFO-Tagungsband 4100-030, p 3–9 (1996)

3.4.13 K. Lex, Welt d. Farben 03/2006, p 14
3.4.14 Champeney, Fourier Transformations and their Physical Applikations, Academic Press: London (1973)
3.4.15 G. Kigle-Böckler in Automotive Paints and Coatings, 2.ed. (Hrsg.Streitberger/Dössel), p 389, Wiley-VCH, Weinheim, 2008
3.4.16 T. Dauser, K. Lex, JOT 49 (3), p 24 (2009)
3.4.17 S. Georgen, Welt d. Farben, 09/2009, p 16
3.4.18 E. Lübbe, Messung der Farbsättigung und des Farbabstandes auf der Basis der Farbsättigung, Proceedings 78. Lacktagung, p 42, Schwäbisch-Gmünd, 2013
3.4.19 G. Kigle-Böckler in Automotive Paints and Coatings, 2.ed. (Hrsg.Streitberger/Dössel), p 402, Wiley-VCH, Weinheim, 2008
3.4.20 F. Schmollgruber, M. Adby, ECJ 05/2011, p 18
3.4.21 G. Pfaff, Welt d. Farben 09/2005, p 13; W. R. Cramer, F+L 118(1), p 34 (2012)
3.4.22 U. Schröder, Kunststoffe 98 (1), p 50 (2008)
3.4.23 W. D. Karl, Phänomen Farbe 22 (4), p 30 (2002); E. Bendin, Phänomen Farbe 22 (9), p 18 (2002)
3.4.24 W. Brockmann, Adhäsion 13, 44, 1969 a. 52, 1970
3.4.25 F. M. Fowkes, R. L. Patrick, Treatise on Adhesion a. Adhesives, Vol. 1, M. Dekker Inc., N.Y., 1967
3.4.26 W. A. Zisman, Advances in Chemistry, Series 43, American Chemical Society,: Washington 1964, p 1–51
3.4.27 R. J. Good, L. A. Girifalco, Treatise on Adhesion and Adhesives, Vol. 1, p 9–68, M. Dekker Verlag, N.Y., 1967
3.4.28 H. Potente, R. Krüger, Adhäsion (12), p 381 (1979)
3.4.29 D. K. Owens, R. C. Wendt, J. Appl. Poymer Sci. 13, p 1741 (1969)
3.4.30 U. Zorll, E. Klinke, F+L, 77 (2), p 125 (1971)
3.4.31 S. R. Plischki, Dissertation Universität Stuttgart 1997
3.4.32 U. Zorll, Goldschmidt informiert 1, Nr. 66, Lackadditive, p 54–67 (1989)
3.4.33 M. Kempe et al., F+L 102 (2), p 36 (1996)
3.4.34 Ferry, Viscoelastic Properties of Polymers, John Wiley & Sons, New York 1980
3.4.35 A. Zosel, Lack- und Polymerfilme, p 21–26, Vincentz Verlag, Hanover, 1996
3.4.36 D. Braun, Schweiz, Archiv für angew. Wiss. u. Techn. 14, p 106 (1958)
3.4.37 C. Gerthsen, Physik, p 86, Springer Verlag, Berlin, 1960
3.4.38 J. A. van Laar, II. Fatipec Kongress 1953, Congress paper, p 194 (1953)
3.4.39 H. H. Behnke, W. Weiler, Materialprüfung, 7–8, p 1030 (1988)
3.4.40 K. M. Oesterle, F+L 80 (1), p 9 (1974)
3.4.41 W. Shen, JCT Coat.Tech. 3 (3), p 54 (2006)
3.4.42 S. Rossi, F. Deflorian, A. Cambruzzi, L. Fontanari, L. Fredrizzi, P. L. Bonora, Pitt.e Vernici Eur. Coat. 04/2004, p 7 (2004)
3.4.43 C. M. Seubert, M. E. Nichols, JCT Res. 4(1), p 21 (2007)
3.4.44 U. Meier-Westhues, T. Klimmasch a. J. Tillack, ECJ 09/2002, p 28; L. Mi, H. Ling, W. Shen, R. Ryntz, B. Wichteran, A. Scholten, JCT Res. 3 (4), p 249 (2006)
3.4.45 EP 1204701, PPG (2005)
3.4.46 N. Laryea, S. Sepeur, C. Thurn, M. Mönkemeyer, F+L 116(7), p 23 (2010)
3.4.47 O. Fleck, S. Nowak, U. Meier-Westhues, ECJ 04/2009, p 106
3.4.48 V. Wachtendord, A. Geburtig, Nachr. Chemie 60, p 1018 (2012)
3.4.49 K. Henderson, R. Hunt, J. Boiseau, L. Pattison, JCT Coat.Tech. 5 (9), p 50 (2008)
3.4.50 J. L. Gerlock, A. V. Kucherov, M. E. Nichols, JCT 73 (918), p 45 (2001); D. R. Bauer, JCT 74 (924), p 33 (2002)
3.4.51 S. Biggs et al., Progr.Org.Coat. 42, p 49 (2001)
3.4.52 M. Entenmann, D. Koch, H. Greisinger, T. Schauer, ECJ 01/2013, p 32
3.4.53 W. Schlesing, M. Buhk, M. Osterhold, Progr.Org.Coat. 49, p 197 (2004)
3.4.54 A. Zielnik, PPCJ 202(4574), p 16 (2012)
3.4.55 M. Hoeflaak, B. de Ruiter, J. H. Maas, ECJ 03/2006, p 30
3.4.56 K. Henderson, R. Hunt, K. Spitler, J. Boisseau, JCT Coat.Tech. 2(18), p 38 (2005)
3.4.57 J. Boisseau, D. Campbel, J. Quill, W. Wurst, P. J. Brennan, PCI 21(6), p 33 (2005)
3.4.58 A. Riedl, Welt d. Farben 2/2006, p 16 (2006)
3.4.59 J. W. Chin, T. Nguyen, X. Gu, E. Byrd, J. Martin, JCT Coat. Tech. 3 (2), p 20 (2006); L. Podgorski, M. Arnold, G. Hora, Coat.World 8 (2), p 39 (2003)
3.4.60 P. A. Sorensen, S. Kiil, K. Dam-Johansen, C. E. Weinell, JCT Res. 6(2), p 135 (2009)
3.4.61 S. Pietsch, W.-D. Kaiser, F+L 108 (8), p 18 (2002)
3.4.62 J. Kittel et al., Progr. Org. Coat. 41, p 93 (2001)
3.4.63 T. Gichuhi, A. Balgeman, S. Prince, C. Wagner, S. O'Brian, A. Adams, JCT Coat.Tech. 8 (6), p 32 (2011)
3.4.64 M. Stratmann et al, Progr. Org. Coat 27, p 261 (1996)

Bibliography

H.-J. Adler, K. Potje-Kamloh (publisher), Quo Vadis – Coatings, XXVI Fatipec Congress, Dresden, Wiley-VCH Verlagsges., Weinheim, 2002
D. Brune, R. Hellberg, H. J. Whitlow, O. Hunderi (publisher), Surface Characterization, Wiley-VCH Verlagsges., Weinheim, 1997
H. G. Voelz, Industrielle Farbprüfung, Wiley-VCH, Weinheim, 2nd revised edition, 2001
D. G. Weldon, Failure Analysis of Paints and Coatings, Wiley-VCH Verlagsges., Weinheim, 2001
U. Schulz, Accelerated Testing, Vincentz, Hanover, 2009
W. Possart (publisher), Adhesion, Wiley-VCH, Weinheim, 2005
J. Sander (publisher), Anticorrosive Coatings, Vincentz, Hanover, 2010
Streitberger/Dössel (publisher), Automotive Paints and Coatings, Wiley-VCH, Weinheim, 2008
D. Satas, A. A. Tracton (publisher), Coatings Technology Handbook (2nd ed.), M. Dekker, New York, 2001
G. Kreysa, M. Schulz (publisher), Corrosion Handbook Vol 6 (2nd ed.), Wiley-VCH, Weinheim, 2004
P. Mischke, Film Formation, Vincentz, Hanover, 2009
H. G. Völz, Industrielle Farbprüfung (2nd ed.), Vieweg, Wiesbaden, 2003
R. Dietrich, Paint Analysis, Vincentz, Hanover, 2009
Kittel, Lehrbuch der Lacke und Beschichtungen (publisher. H.-J. Streitberger), Vol. 5 (2nd ed.), Hirzel, Stuttgart, 2004
H. Bubert, H. Jenett (publisher), Surface and Thin Film Analysis, Wiley, New York, 2002
Z. W. Wicks, F. N. Jones, S. P. Pappas, D. A. Wicks, Organic Coatings (3rd ed.), Wiley&Sons, Hoboken, 2007
J. Prieto, J. Kiene, Holzbeschichtung, Vincentz Network, Hanover, 2007
R. A. Wolf, Plastic Surface Modification, Hanser, Hanover, 2010
J. L. Keddie, A. Ronth, Fundamentals of Latex Film Formation (H. Pasch, ed.), Springer, Dordrecht, 2010
Brock/Groteklaes/Mischke, Lehrbuch der Lacktechnologie, 4th ed., Vincentz Network, Hanover, 2012
W. Kettler et al., Colour Technology of Coatings, Vincentz Network, Hanover, 2016
E. Lübbe, Farbempfindung, Farbbeschreibung und Farbmessung, Springer, Heidelberg – Berlin, 2012
Informationsserie des Fonds der chemischen Industrie Vol. 28 Lacke und Farben, Deutsches Lackinstitut, Frankfurt, 2003

4 Coating Technology

The term "coating technology" will be used to cover all the engineering processes which are necessary to transform the coating material described in the previous Chapter into a functional coating. The stages of pretreatment, application and drying or curing, along with their peripheral equipment in the form of air, energy and material supply systems, conveyors and systems for disposal or the recovery of overspray and gaseous emissions from the waste air together with paint-removal systems for grilles and conveyor chains represent the major system requirements.

The decision in favour of a particular coating technology and the desired quality of the coating will yield the optimum result only if the specific properties of the substrate are taken into account. The chemical nature and the morphology of the material surface are as important in this respect as the design, joining method and construction of the object to be protected. The object, its pretreatment and the actual painting process itself have to be optimally coordinated with each other. Coordination here does not mean the matching of a specific coating technology to an already specified object, but rather a concept which duly addresses the issues of pretreatment and construction such as sharp edges, panel overlaps and hard-to-reach cavities.

The objects to be coated pass through numerous stations on their way to acquiring a permanently decorative and protective skin. First, suitable pretreatment methods must be applied to the untreated to render them paintable. The associated process stages focus on physical and chemical cleaning methods, on smoothing processes and on the application of thin, inorganic conversion layers.

The pretreatment line is followed by the application, drying and curing of the coating material. The application lines consist of units which are charged with conditioned fresh air and material in order that the liquid or, in the case of powder coatings, solid coating material may be transformed there into a uniformly thin, mechanically strong film. Application can be effected by brush, roller, dip, flow, curtain or spray. The overspray and the gaseous solvents which occur or escape during spray coating must be extracted from the waste air.

The process-engineering operations in the oven concentrate on removing escaping solvents on one hand and initiating or accelerating chemical crosslinking reactions on the other.

All the individual processes mentioned, such as paint and air supply, application, drying, curing, as well as conveying and all the environment-related supplementary measures for treating exhaust gas, waste water and overspray, together with the necessary process control, can be broken down into individual elements that correspond to fundamental process-engineering operations. As a result, all the complex processes needed for a specific individual

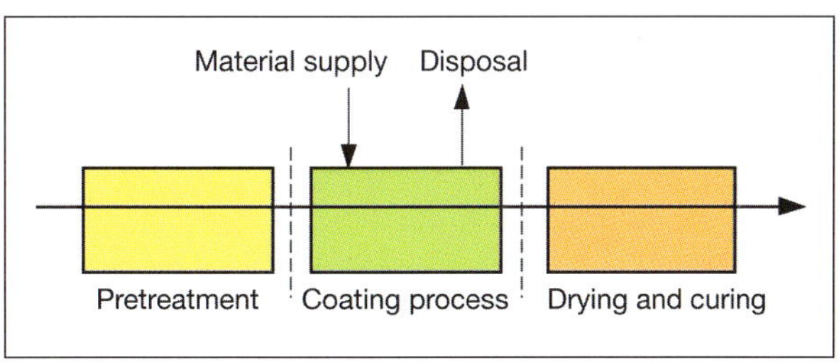

Figure 4.0.1: Process stages in a coating line

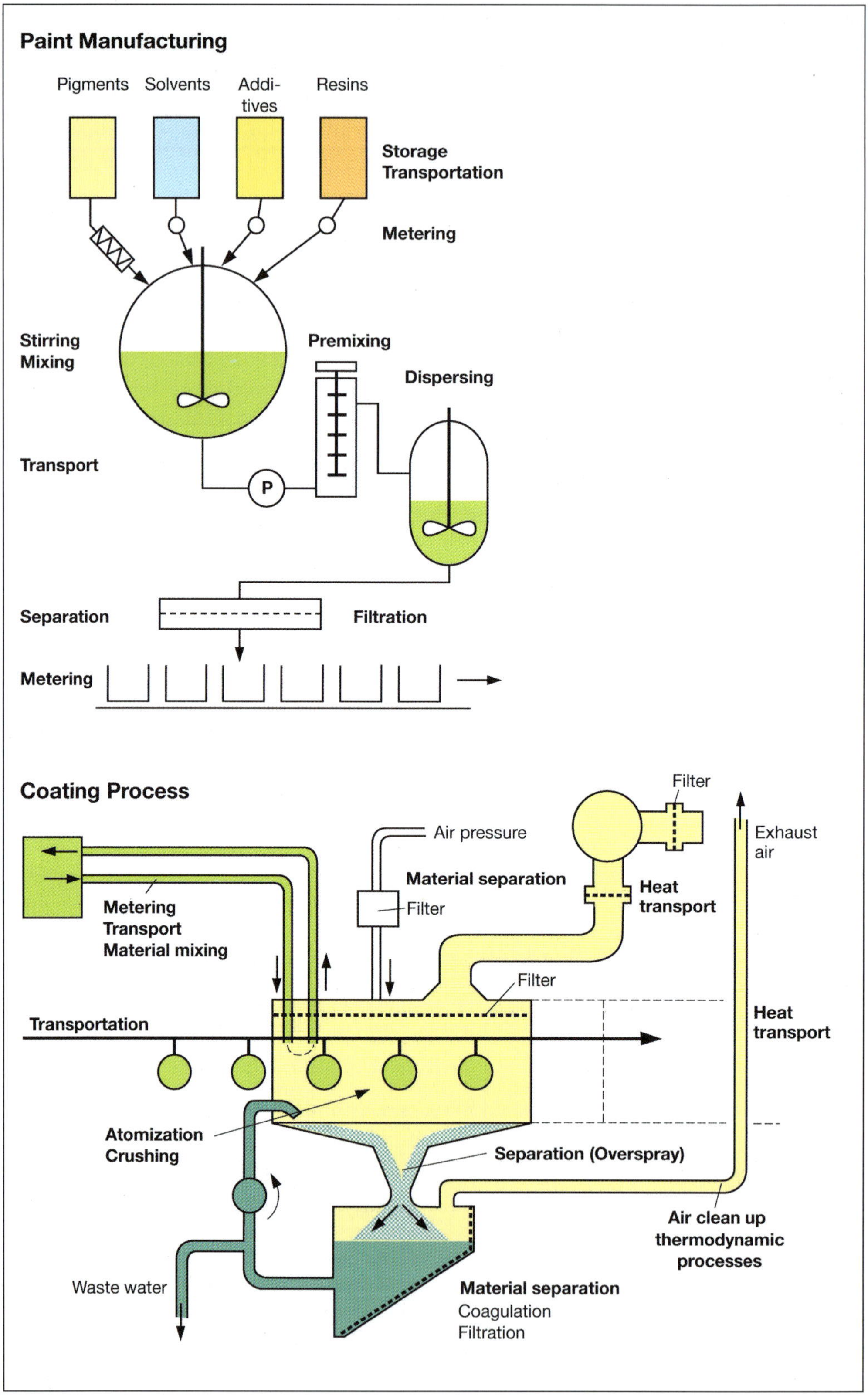

Figure 4.0.2: Basic operations in the paint manufacturing and spray coating processes

case can be assembled by simply joining a relatively small number of basic operations (individual modules) together.

The major basic processes of importance to coating technology are:

- Mass transfer
- Heat transfer
- Dispensing (metering)
- Material mixing
- Crushing
- Material separation
- Thermodynamic processes

The large number of different systems in existence is therefore the outcome of assembling the various modules in different ways. This also applies to both the paint manufacturing process and, for example, the spray coating process shown in Figure 4.0.2.

4.1 Influence of the Substrate

The coating material and the processing method play an important role in the desired surface finish. Paintability and design are still very closely interlinked and should never be discussed independently of each other.

The aim of the following section is to examine the influences exerted by the substrate to ensure that these are given due consideration when a painting process is being designed.

One main category of substrates is represented by natural materials, which may be subdivided into inorganic materials, such as concrete and mortar, and organic materials, such as wood, paper, cardboard, leather and rubber. Synthetic materials, the other main category, also have to be coated. Notable members here are metals, such as iron, alu-

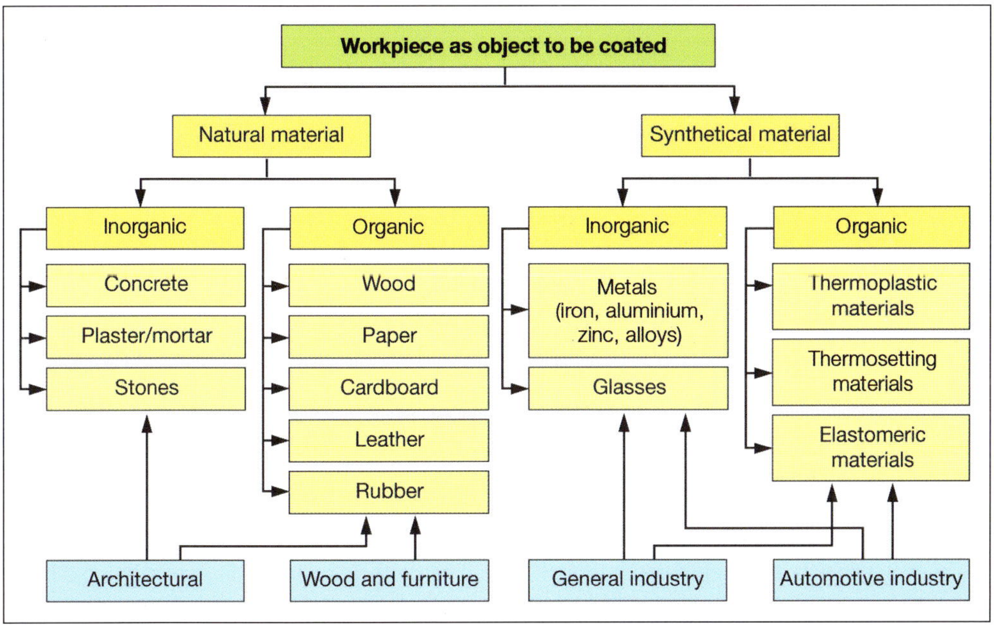

Figure 4.1.1: Overview of the most important substrates for coating processes and their applications

minium, zinc, magnesium and their alloys, as well as glass, ceramics and thermoplastic, thermosetting and elastomeric organic polymers. Metals are thermally stable yet susceptible to corrosion, whereas wood, by contrast, is unaffected by corrosion but can be damaged by micro-organisms.

The materials can be soft, as in the case of leather and rubber, but also extremely hard. Their surfaces can be rough or smooth, neutral or alkaline. The upshot of all this is that there are a great many various properties which need to be given due account when the material and process for surface treatment and coating are being selected.

4.1.1 Wood and Wood Materials

4.1.1.1 Wood as a Material

Wood, as the main constituent of the trunks, branches and roots of plants, is a complex organic material made of cells and therefore has a range of special features. On account of the way it has grown it exhibits **nonhomogeneous structures**. Wooden items sawn from trunks, such as planks or boards, are therefore nonhomogeneous as far as their expansion properties are concerned. Wood can absorb up to 30 % water, the exact amount depending on the relative humidity of the ambient air. And the type of wood determines by how much the volume increases: up to 10 % in some cases. Lightweight wood which has grown quickly absorbs greater volumes of water more quickly, whereas the hard, slow-growing types have a correspondingly slower water exchange process. Owing to the different alignment of the cells, **swelling** resulting from absorption is not the same in every direction. Wood swells and contracts 10 times more in the radial direction and 17 times more in the tangential direction than it does longitudinally. The swelling increases linearly by 0.2 % vol/vol for every percentage point increase in wood moisture content. At a relative humidity of approx. 50 %, which is the mean value for central European climatic conditions, the moisture content in wood reaches equilibrium at 8 to 10 %. Ideally, then, painting processes should take place under these conditions [4.4.1].

At the same time the **influence of water** brings about a significant change in density and hardness. Completely dry, compressed wood can have a density of up to 1.56 g/cm^3. The hardness can increase twelve-fold during compression.

Sawn timber also has a characteristic surface structure. It is composed of the annual rings or medullary rays, i.e. areas of different-sized cells and pores. The annual rings comprise two different elements, namely the lighter, outer, still living parts which are softer and known as the sap wood, and the darker heart wood or duramen, which forms during periods of slow growth. The latter merely provides mechanical support. It is protected against oxidative decomposition by tannic acid.

As a result of their cell structure, i.e. of the chemical secondary and tertiary structures, all types of wood are nonhomogeneous in terms of their properties and are therefore anisotropic materials.

Their chemical structure too is complex. Cellulose, which makes up 40 to 50 %, is the main constituent of wood (see Chapter 2.1.1). Although it swells in water in its 1,4-β-D-poly-glucosan form, it is completely insoluble and therefore constitutes the fundamental provider of mechanical strength and resistance to external influences. Hemicellulose

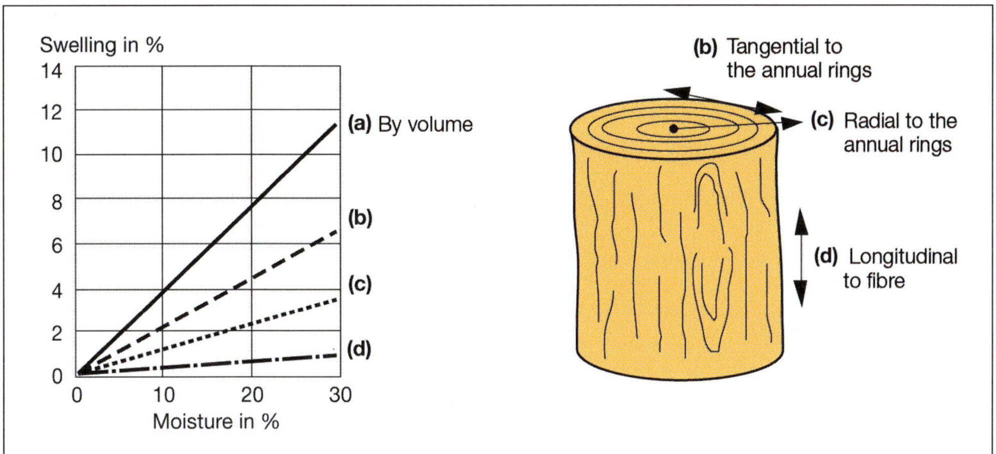

Figure 4.1.2: Swelling properties of wood, as illustrated by pine wood

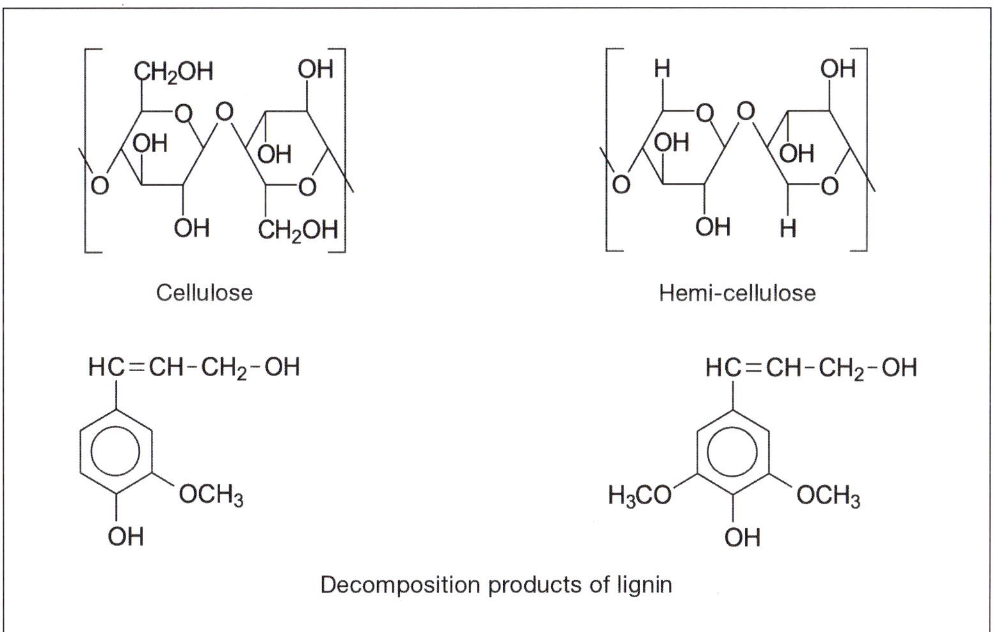

Figure 4.1.3: Chemical components and decomposition products of wood

deposits, which represent 15 to 20 %, are responsible for elasticisation (xylan), while lignin at 20 to 25 % operates as a supplementary stabiliser.

Whereas hemicelluloses such as cellulose are oligomeric sugar derivatives (main constituent: xylose), the lignin cells have a completely different chemical structure. Their decomposition products reveal that polyvalent phenol bodies, in some cases etherified with methanol, are present in a complex combination with allyl alcohol (see Figure 4.1.3).

In addition to the basic substances specified above, wood sometimes contains soluble substances in a very wide variety of types and quantities. Mono- and diterpenes, waxes and oils are present, along with wood-specific substances which can cause problems for surface treatment because they are soluble in paint solvents [4.4.2].

Paper and cardboard are also cellulose-containing materials and therefore wood derivatives. These are flexible sheet- or film-like materials for writing or printing on and for packaging. They mainly consist of plant fibres which are felted, pressed and bonded together. The fibrous materials are derived almost exclusively from wood. The purely mechanical processing of the wood yields wood pulp, a short-fibred mixture of cellulose and lignin. Combined mechanical and chemical pulping of wood, which largely breaks down the less resistant lignin without degrading the cellulose fibres, affords long-fibred chemical pulp consisting of up to 85 to 95 % cellulose.

Increasingly, low-quality **paper** is being made from recycled paper. Apart from the fibres, fillers such as kaolin, barium sulphate, calcium and magnesium carbonate, talcum and titanium dioxide are key basic materials used in paper. The fillers produce smoother surfaces and brighter colours. The addition of natural soaps (colophony) reduces the paper's ability to absorb writing and printing inks. Synthetic-resin glues and waxes serve to increase the wet strength, starch and starch derivatives the tensile and folding strength, and water glass the hardness.

4.1.1.2 Wood as a Workpiece

The simplest form of woodworking involves the production of solid sawn timber in the form of laths, planks and beams. **Solid timber** is mainly processed in the construction and window industries. In the furniture industry the use of solid timber requires highly developed craft skills and is one of the most interesting forms of joinery. Wall-mounted shelving, mouldings and furniture assembled from frames and panels are made with planks and boards. Up to 90 % of the timber used in the manufacture of chairs and tables is solid wood.

The natural working of solid timber as a result of shrinkage, swelling and the associated cracking and warping make special demands on coatings applied subsequently. Solid timber therefore requires very careful processing.

Owing to the changes and warping induced by anisotropic swelling and shrinking, numerous steps are taken to provide solid wood with better dimensional stability. The wood is rendered hydrophobic and thus largely impervious to the effects of water in the environment by means of mechanical compression on one hand and by chemical means on the other, including impregnation with polymerising monomers in pressure chambers and subsequent polymerisation.

The property profiles of veneers, which are only fractions of a millimetre thick, are similar to those of solid wood from a coating technology perspective. They are obtained from the solid wood by peeling (rotary-cut veneers) or cutting (sliced veneers) after the solid wood has been suitably softened and elasticised by the application of steam. As the veneers are so thin, they need to be stabilised with a suitable substrate. There is always the possibility that during the production process the substrate or the glue will affect the painting process via pores or flaws that extend as far as the substrate. Pretreatment steps are therefore often necessary (see below).

Rotary-cut veneers mainly serve as cross-band veneers. Sliced veneers are used to decorate surfaces. High-grade decorative woods such as walnut, mahogany, oak, elm, ash or rosewood are employed for this purpose. The substrate may be wood core plywood, plywood, chipboard and hardboard panels.

Battens which have been cut in different directions from solid wood and bonded together to form panels with covers of different thickness are termed plywood or wood core plywood. Wood core plywood panels are plywood panels which consist of a centre ply and at least one veneer ply on each side. The type of centre ply determines whether these are termed laminboard, blockboard or battenboard. Wood core plywood panels manufactured in this way are better able than planks or boards to withstand the effects of moisture. The danger of warping, distortion and cracking is minimised. The centre plies of wood core plywood panels are usually made of pine, fir and obeche. The cross-band and face veneers are bonded to the battens on both sides at right angles to the grain of the wood.

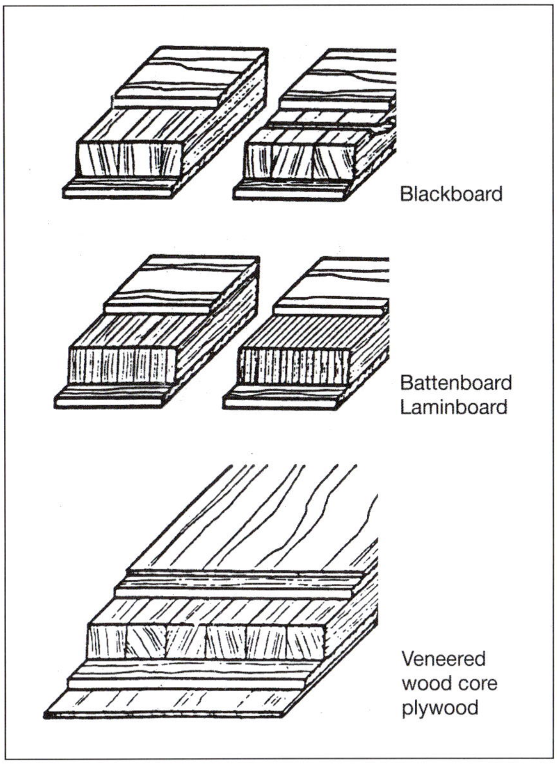

Figure 4.1.4: Various types of plywood as wooden workpieces

Another method of achieving wood materials with isotropic properties which is now in widespread use for furniture manufacturing consists in compressing comminuted wood particles. A wide variety of fibre-board and **chipboard** (also known as particleboard) products are now used. Fibred or shredded wood in an extremely wide variety of sizes and shapes is compressed with urea, melamine or phenol resins and thermally cured. These boards are variously classified as softboard, medium-density fibreboard (MDF) and hardboard to reflect the applied pressure of up to 2000 N/m^2. These in turn are used in furniture manufacturing with and without veneer to suit the particular requirements. These materials can be compressed in all three directions. Special tubular boards

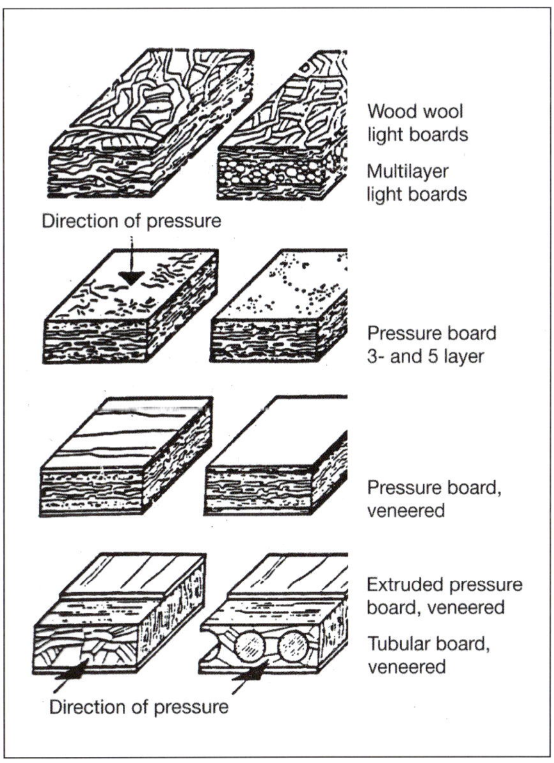

Figure 4.1.5: Types of fibreboard and chipboard

are also manufactured, not only for their high strength, but also because they are lightweight and relatively inexpensive.

4.1.1.3 Pretreatment of Wood

Before solid wood or wood materials can be subjected to surface treatment that will produce a surface layer, the workpiece must pass through a number of pretreatment stages. The most important stages before the actual painting process are wetting, cleaning, sanding, deresinification, removal of adhesive bleed-through and, if appropriate, bleaching and staining.

In craft processes the woodworking machines often leave behind visible pressure or drag marks which impair the attractiveness of the surfaces. These are eliminated by treating the parts in warm water and then drying them for 1 to 2 days at 18 to 20 °C.

Sanding machines have largely superseded surface cleaning in the grain direction which used to be done with short double planes or scrapers ahead of the **sanding process**. Sanding as a prerequisite for a perfect surface to which to apply paint must be tailored to each individual type of wood. In wet sanding, solvents, unlike water, do not change the volume of the wood, a fact which should be borne in mind when wood is sanded. Whereas water enlarges the cells through osmotic effects and raises the fibres of the cut surfaces, a similar effect does not occur when, for example, gasoline is used. It takes thorough sanding to satisfy the preconditions for uniform application of the coating materials.

Application

Corundum or silicon carbide particles of different levels of coarseness serve as **abrasives**. The various types of abrasive are attached to a backing fabric with synthetic resins, usually phenol resins, and nonstick coatings. Additional antistatic treatment prevents clogging with dust, and ensures that no problems occur when sanding belts are used.

The **sanding process** itself is electronically controlled by machine nowadays. Computers apply infinitely variable pressure control and so enable uniform sanding even at the edges of the workpieces. A combination of several sanding processes is the best preparation for the subsequent surface treatment. With wood, too, the time and effort spent on preparing the surface pay dividends in the quality of the paint finish [4.4.3].

Should it be necessary to remove any resins or glues from the surface to be coated, this can be done with solvents or soft soap in conjunction with ammonia. For light-coloured woods oxalic acid or a 30 % H_2O_2 solution is used. The wood should be scrubbed in the direction of the grain until all the glue residues have been removed.

The **bleaching** which occurs as a side effect of the treatment with H_2O_2 is carried out deliberately on certain wood types to achieve uniform lightening. Beech, oak, but also wood types which discolour in daylight such as maple, birch, and pear, are brightened using concentrated ammonia and subsequent treatment with a 30 % H_2O_2 solution. The oxygen-containing foam which forms on the surface is spread with a brush and then thoroughly rinsed off.

The **character of the wood** can be enhanced by exploiting the different properties of the sap wood and heart wood sections, i.e. by dyeing them differently. Dyes dissolved in water or solvents, smoke mordants, metal salt or double stains should be used. Metal complex dyes are dissolved in alcohols to act as spirit stains, soluble anionic azo dyes in water and conventional azo dyes in gasoline or waxes to act as wax or gloss stains. Metal complex dyes feature good fastness to light. Although the browning of heart wood which is achieved by smoking with ammonia in enclosed smoke chambers is smudge-proof and optically attractive, it is unfortunately not waterproof or light-resistant.

Metal stains are better because they are more durable. Wood containing tannic acid forms coloured complexes with various metal salts such as those of iron, cobalt, and copper. If there is no tannic acid in the wood to be treated, double staining can yield the same effect. This consists in soaking the wood in solutions of tannin, pyrogallol or hydroquinone followed by treatment with metal salts. The stains are applied with a brush or sponge and spread at right angles to the grain. Stains should be protected with a clearcoat to improve their resistance.

As the residue of a former living organism, wood is vulnerable to attack by **micro-organisms** which decompose it unless appropriate protective action is taken. In accordance with the degree and type of protection necessary, biocidal agents are dissolved to this end in solvents in the form of impregnations or in film forming primers or transparent paints. Fungicidal, bactericidal, insecticidal or algicidal active ingredients should be used as required (see Chapter 2.1.4).

In the battle against cellulose-degrading brown rot and lignin-degrading white rot, both of which are caused by fungi, it was standard practice in the past to use organo-metallic compounds or halogenated phenol bodies as effective general-purpose biocides. However, the potential danger to humans as a result of the sometimes high doses and the cumulative effect in the environment have led to the withdrawal of these substances and their replacement by modern wood preservatives which specifically target the micro-organisms' metabolisms (see Chapters 2.1.4 and 5).

4.1.1.4 Surface Finishing of Wood Panels

The desire to lend the appearance of a high-quality natural wood to chipboard used for furniture production and other purposes in general led to the development of grain printing and other surface finishing processes, even though face veneers were scarce and increasing in price. Grain printing enables an image very closely resembling natural wood to be printed in any colour onto chipboard, hardboard, paper and other materials. Before printing, the panels must be properly filled, ideally with the aid of filler machines which smooth against the grain. **Printing** itself is generally conducted in a combined rotogravure/offset process with printing rollers. The printing of pore formers with a paint-repelling printing ink creates open-pore effects (synchronised pores) when the panel is later coated.

The printing images are produced photochemically. This entails photographing a well prepared face veneer which is as flawless as possible. The image is adapted to the roller so that it is endless on the roller, generating a continuous pattern which is not interrupted at any point when printed.

Impregnated papers are also employed in the furniture and wood panel industry. The shortage and rising price of genuine veneers has contributed greatly to the greater popularity of such paper films. A clearcoat must be applied to the films after they have been pressed onto the chipboard and hardboard panels, as otherwise their surface protection is inadequate.

Moves to simplify the production sequence have led to the creation of **finish foils**. These are impregnated or printed paper films which are given a clearcoat by the film manufacturer. Once the foils have been pressed onto panels by the processor, no further finishing of the surface is necessary. Endless edges, which serve as roll stock in the furniture and wood panel industry for the continuous coating of edges, are also given this surface finish by the foil manufacturer (see Chapter 7.10).

The manufacture of such finish foils and endless edges is carried out in special impregnation-coating systems. The original paper films are mainly impregnated with urea and melamine resins, but also with acrylic and polyester resins and thereby so hardened that they do not delaminate or crack during later furniture production, i.e. during sawing, drilling and milling of the film-coated panel. The impregnated papers are dried in float driers. Further treatment of the dried paper films to produce the finish foil takes place either directly following the impregnation process or in a separate stage. Directly coupling the impregnation process to the application of the topcoat after an interim drying stage avoids the wasting of time and material which would otherwise happen if the material were to be rolled up and unrolled again.

The manufacturing process for the **production of endless edges** is analogous to that used for the production of finish foils. The papers used in edge production must be substantially heavier than those used for the panel faces in order that the chipboard edge, particularly the relatively uncompacted chipboard centre ply, may be satisfactorily bridged.

White or natural-colour base papers are chosen for decorating the edges in the desired edge colour. The manufacturer provides the papers with a smooth surface on one side, and thus little material and just one working stage are required when the topcoat is applied (see Chapter 7.10).

4.1.2 Metallic Materials

4.1.2.1 Properties of Metals

Despite the development of numerous attractive high-performance plastics, most consumer goods are still made of metallic materials. The most important materials are iron, aluminium, zinc, magnesium and their alloys.

These metals are chemical elements which, with the exception of rare meteorite finds, do not occur naturally in the pure form since they are thermodynamically unstable as metals. They therefore occur as oxides, hydroxides, carbonates and other stable compounds.

Metal	Melting point
Iron	1530 °C
Copper	1083 °C
Aluminium	658 °C
Magnesium	650 °C
Zinc	419 °C
Tin	232 °C

Figure 4.1.6: Melting points of commonly coated metals

Oxidation of metals often leads to oxide layers which may be either porous or continuous. Zinc and aluminium form stable, continuous oxide layers in certain pH ranges and are therefore passivated against further action of atmospheric oxygen, whereas iron is converted into porous layers. As the latter are permeable to oxygen and water they do not offer any protection to the metal.

Since the metals described are not present as chemical elements in nature, they must be manufactured artificially and later protected against decomposition by means of protective coatings. Their properties depend to a great degree on the composition of the metals. Their susceptibility to corrosion varies with their position in the electrochemical series and with their degree of purity. The melting points can be influenced by combining different metals in the form of alloys.

Of particular interest are the metals' **surface properties** which influence the painting process and thus the quality of the coating. The surfaces of the workpieces are impacted both chemically and mechanically by the manner in which they are produced, e.g. by rolling or casting. These impacts include mill scale and a degree of surface roughness which varies with the manufacturing process. Even traces of corrosion products can significantly impair the protective effect afforded by coatings. It is therefore absolutely essential that all corrosion products be completely removed prior to coating. This is the only way to guarantee effective anticorrosive protection. Surface treatment methods are greatly facilitated by the fact that metals are completely insoluble in all organic solvents and have high **melting temperatures**. Metallic materials do not usually impose any restrictions on the stoving temperature of coating materials. Exceptions here are certain alloys, e.g. soft solder, whose melting points of approx. 180 °C may well lie within the range of the stoving temperatures of industrial paints. Some aluminium alloys are also sensitive to such stoving temperatures.

Application

The use of coatings to provide **anticorrosive protection** as a means of preserving value simultaneously conserves resources, saves energy and so constitutes environmental protection. Every effort should therefore be made to avoid the destruction of material by corrosion. It is estimated that one-fifth of all manufactured steel is lost to corrosion. By corrosion is meant attack on the material as a result of a reaction with the environment and a subsequent change in the material properties. The term corrosion, which is derived from the Latin "corrodere" (to gnaw), refers to both the process and the resulting damage.

The causes of corrosion lie in the formation of electrochemical potentials caused by unequal charge distributions in the material. These can stem from inhomogeneities in the material itself (local cells) or from different surface states arising from the external action of oxygen. This gives rise to cathodic and anodic areas which, in the presence of water and oxygen, give rise to corrosion products when the iron releases electrons which are then absorbed by the oxygen. The magnitude of this potential difference between the anodic and cathodic areas can be quantified with the aid of the ***Nernst* equation**. This describes the correlation between the single potential φ of a metal electrode and the activity a of the cations in the solution. F represents *Faraday's* constant and φ_0 the reference potential.

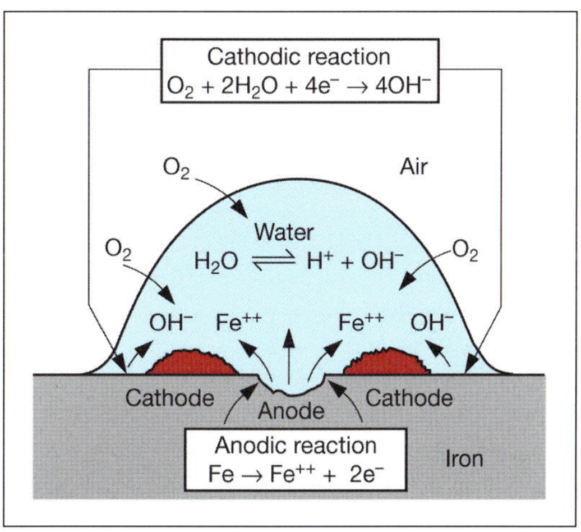

Figure 4.1.7: Corrosion of iron in an aeration cell

$$\varphi = \varphi_0 + \frac{RT}{nF} \cdot \ln a$$

The formation of corrosion products will now be illustrated in detail with the example of a water droplet. The potential here is determined by differences in the concentration of the oxygen dissolved in water.

Whereas the iron in the low-oxygen area goes into solution at the anode, OH ions are generated in the cathodic area from H_2O, oxygen and the electrons emitted by the iron. These initially form insoluble iron (II) hydroxide at the edge of the water droplet. The oxygen dissolved in the water then oxidises the iron (II) hydroxide to hydrated iron (III) oxide which dehydrates to become stable haematite.

$$\left.\begin{array}{l} 2\,Fe \xrightarrow[-4\,e^-]{} 2\,Fe^{2+} \\ 2\,H_2O + O_2 \xrightarrow{+4\,e^-} 4\,OH^- \end{array}\right\} 2\,Fe^{2+} + 4\,OH^- \longrightarrow 2\,Fe(OH)_2$$

After oxydation by oxygen

$$\longrightarrow Fe(OH)_3$$

$$2\,Fe(OH)_3 \xrightarrow[-2\,H_2O]{} 2\,FeOOH \xrightarrow[-H_2O]{} Fe_2O_3$$

(stable haematite)

Corrosion products are formed in this way over the entire surface of the metal. As they are formed, the atmospheric oxygen is increasingly prevented from passing through them, with the result that the rate of corrosion slows down. The critical relative humidity for triggering corrosion is around 60 %, meaning that corrosion products cannot form at a lower relative humidity and in the absence of liquid water. In dry areas, such as desert regions, corrosion problems are therefore virtually unknown. Factors such as airborne pollutants and the pH of the environment also play a decisive role. Airborne pollutants such as sulphur dioxide are major catalysts of the corrosion process. Salty sea air or acidic soot deposits can also promote the formation of galvanic cells and thus of corrosion. Oxidative decomposition occurs particularly quickly in the acid range, and slows down with increase in pH value. The corrosion of iron slows down more and more between pH 4 and pH 10.5 and practically comes to a standstill above pH 10.5. This fact is of consequence for the stability of steel in the alkaline environment of concrete. Where local cells are absent because the material is particularly pure, there is also little tendency to form corrosion products.

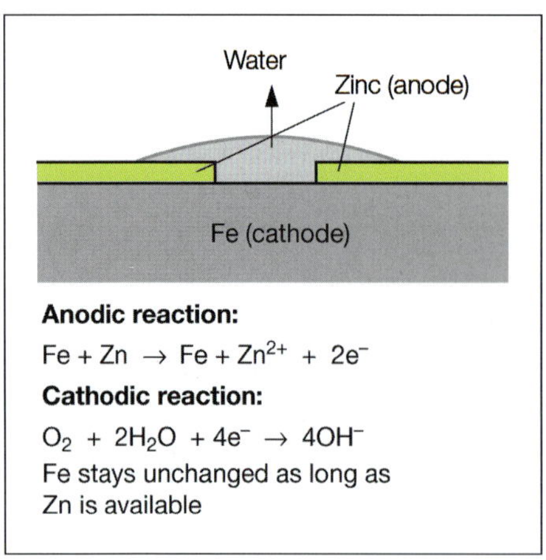

Figure 4.1.8: Cathodic protection of iron by zinc

The situation is different, however, for the amphoteric elements aluminium and zinc. Unlike iron, they are passivated in the mid-pH range due to the formation of a continuous oxide layer, but are susceptible to corrosion in the acidic and alkaline ranges because the protective layer dissolves.

They are increasingly being used as metallic materials because of their passivation in the neutral range: zinc primarily as a coating on steel, and aluminium in numerous alloys as a uniform material. Despite having a more basic character than iron, zinc has proved successful as a protective coating for iron. Even if the coatings suffer local mechanical damage, protection is maintained as a result of electrochemical processes. The protection provided by the more basic zinc consists in the transfer of electrons to iron to prevent the formation of iron ions during corrosive attack. In so acting as an anode the zinc provides cathodic protection for the steel. The damaged location fills up with zinc oxide, also known as white rust. The still intact zinc provides full surface protection for the steel against corrosion until it has been completely used up.

Purity	Amount dissolved in g/m²/day
99.999 %	6
99.990 %	112
99.970 %	6500
99.880 %	36000

Figure 4.1.9: Stability of aluminium to corrosive influences as a function of purity

The zinc coatings needed for protecting the steel can be applied in various ways. The important processes include **hot-dip galvanising**, in which the objects or coils are immersed in a zinc bath heated to 450 °C. The disadvantages of this process are the very thick coatings which it produces and the structures, known as flowers of zinc or zinc bloom, which occur during cooling. Variations in the configuration of the crystalline matrix create irregularities which are difficult to hide by coating.

As far as the attractiveness of the coating is concerned, hot-dip galvanising coupled with rolling (**Sendzimir process**), spray-galvanising or electrochemical galvanising are much more suitable. Spray-galvanising with flame cleaning consists in melting zinc and spraying it onto the surface as a liquid with the aid of compressed air. Galvanising by means of electrolytic deposition of zinc salts yields coatings which are even more uniform and at the same time extremely thin (2 to 10 µm).

Sherardising is an effective way to galvanise small parts. The objects are cleaned with acids and then heated for several hours at 400 °C in rotating drums that also contain zinc and sand.

Another method of providing cathodic protection for steel by zinc involves the application of zinc-rich paints. A high concentration of zinc powder is distributed in organic film forming agent solutions such that after application to the substrate and subsequent drying the individual zinc particles are in contact with each other. However, the protection they offer is not as good as that provided by continuous zinc layers.

Initially, simple zinc-rich paints could not be used to pre-coat sheets in the coil coating process, as they soiled the pressing tools while they were being stamped. The breakthrough came when the pre-coated sheets were treated with a clearcoat that had a low coefficient of friction (see Chapter 7.4).

Another metal frequently used in the production of goods is aluminium. It is extremely base. Despite this, aluminium is highly resistant to atmospheric oxygen because the oxide skin prevents the passage of oxygen. As mentioned above, however, the oxide layers are not stable in acidic and alkaline environments. The oxide layer also fails to protect against chloride ions, as it then dissolves to form chloro-complexes. The range of stable passivation depends heavily on the purity of the metal.

Of the various oxide layers, some of which can be quite thick, the skin which forms the oxygen barrier directly on the metal surface is extremely thin at 0.005 μm. The rest is porous, impairs electrical conductivity and the adhesion of coatings, and therefore must be removed. To promote the stability of aluminium, therefore, defined oxide layers are created by electrochemical oxidation and are known as anodised coatings.

Aluminium is highly valued as a material because of its low density of 2.70 g/cm^3 and its excellent physical properties, but is only rarely used as a pure metal.

It is alloyed with magnesium, silicon, copper and zinc to increase its dimensional stability and hardness. Aluminium alloys are classified as either wrought alloys or casting alloys in line with how they are manufactured and the form in which they are delivered. The materials for wrought alloys are processed by casting into ingots which are then drawn wrought by rolling and converted into coils, while the materials for casting alloys are shaped by pressure die casting, gravity die casting and sand casting.

In aluminium alloying, magnesium promotes strength and corrosion resistance, silicon lowers the coefficient of thermal expansion and the density, and other alloying metals such as copper and zinc increase its strength. Either aluminium/zinc/magnesium/copper or aluminium/silicon alloys are therefore used as befits the particular requirements. The compositions and designation of important alloys are specified in DIN EN 576 and DIN EN 573-1 and -4.

The first stage in the manufacture of aluminium coils and sheets is the casting of aluminium ingots. These are rolled to a certain interim thickness at 400 to 500 °C and then cold-rolled to the required final thickness. As happens with rolled steel, thermal removal of drawing greases can give rise to accumulations of crack products in the form of carbon which adversely effect the corrosion resistance at the interface.

In recent years, magnesium-containing materials have also come to prominence, particularly for automotive engineering applications. Like aluminium, magnesium is self-passivating, and is also unstable at high and low pH values and when exposed to chloride and similar ions. Although it has excellent deformation properties in the pure form, it is often unable to meet all the requirements as a material because of its inad-equate hardness. Like aluminium it has to be alloyed with other metals. By comparison with aluminium, whose density is only 1/3 that of steel, magnesium at 1/4 offers even greater potential weight reductions, provided the same strength can be achieved as with steel. At 5 kg/car body, the use of magnesium in carmaking is still modest. The corresponding figure for aluminium is already 100 kg. Greater usage of magnesium will also be contingent on its paintability.

The low-iron, low-copper and low-nickel or iron-free, copper-free and nickel-free alloys which have recently been developed feature improved corrosion resistance, and an increase in the proportion of magnesium used as a material in future can be expected.

As each of the metallic materials has specific properties, several different materials are often combined so that the broad property profile specified for consumer goods can be met. It should be borne in mind that, wherever a conductive connection is created between different metals, local cells form which significantly accelerate oxidative decomposition through galvanic corrosion. In metal-processing industries, such as in the production of cars, buses or commercial vehicles, galvanised components are often combined with aluminium and steel components. In such hybrid construction methods, lasting anticorrosive protection is only achieved if the design specifies that the different metals be separated from each other by nonconductors. Where contact is unavoidable, the baser anodic areas should be as small as possible by comparison with the cathodic areas since the corrosion current density varies with the relative sizes of the parts in contact. The practical implication, therefore, is that aluminium screws should never be used to attach sheet steel.

4.1.2.2 From the Material to the Workpiece

Not only the chemical features of the material, but also the surface structure of the workpiece affects the optical and functional properties of the coating which is later applied. The manufacturing process for the material and the specifications relating to deformability impart a more or less pronounced texture to surfaces. To this must be added changes in surface roughness due to upstream cleaning process of brushing, sanding or blasting with sand or steel shot. For example, steel castings and sheet steel which has been shot-blasted can have very rough surfaces with roughnesses of 15 to 30 µm (see Chapter 3.2.4).

Bodywork panels, on the other hand, have peak-to-valley heights R_a of only a few micrometres which vary with the degree of deformation that necessarily occurs in the stamping presses.

Since the optical attractiveness and the protective effect of the coating to be applied later depend on the way in which the surface structure, the paint system and the application process are matched, the mean peak-to-valley height R_a and the difference in height between the highest and lowest peaks R_t must be measured and factored into the substrate specification. As the roughness of the substrate increases, so the mean peak-to-valley height of the coating also increases. The influence of the substrate can still be seen in the topcoat even when several coats are applied.

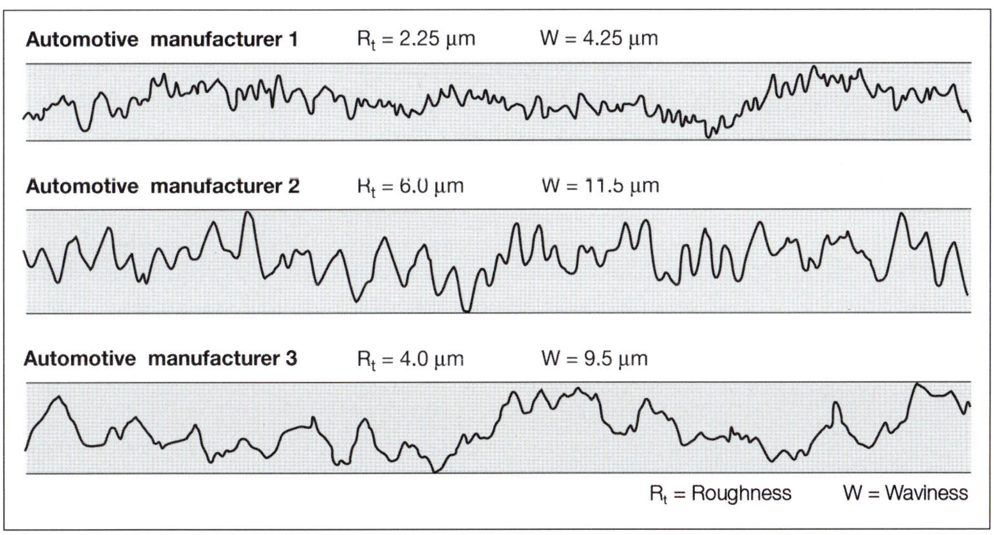

Figure 4.1.10: Roughness profiles of carbody steels

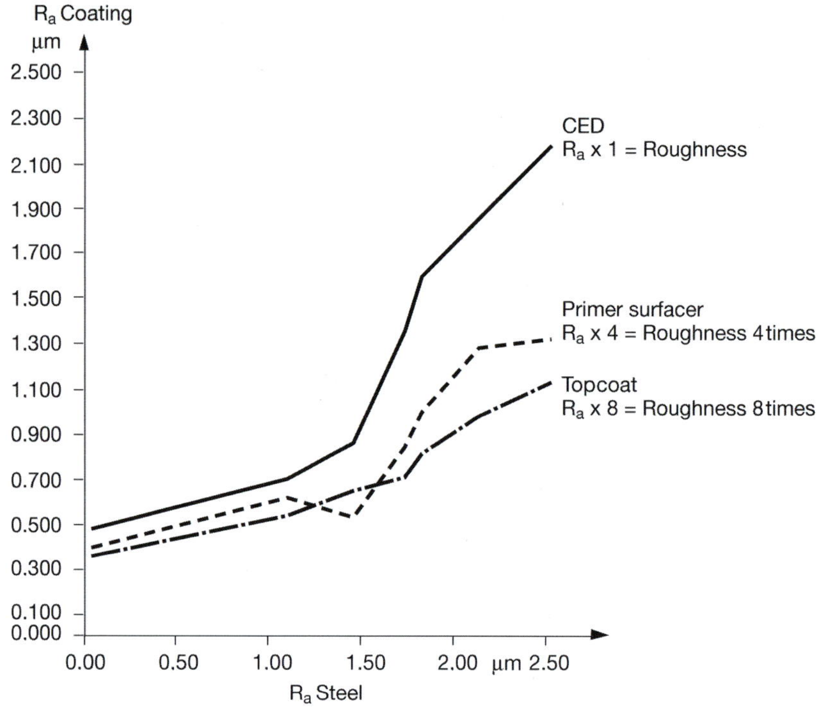

Figure 4.1.11: Correlation of R_a of substrate and a multilayer coating

4.1.2.3 Pretreatments

As oxide layers form fairly rapidly, bare sheets are coated with anticorrosive oils by way of temporary **anticorrosive protection** immediately after manufacture. The objects must therefore be cleaned and pretreated appropriately before the actual painting process. Impurities arising from the production process such as mill scale and oxide layers or anticorrosive oils from the manufacturing process for steel coils and drawing greases from the deforming process must be removed from the untreated objects. In many cases, mechanical cleaning by means of brushing, sanding or blasting is carried out. In any event, organic impurities must be removed from technical surfaces by solvents or surfactant-containing aqueous agents. As opposed to these physical methods it is also expedient to carry out chemical pretreatment in the form of pickling or, in the case of coatings subject to severe stresses, to carry out a film forming pretreatment before the actual painting process. Phosphating or chromating are processes which optimise the structure and conductivity of the surface and thus the anticorrosive protection in various ways. The greatest contribution in this connection stems from the improved adhesion of the organic anticorrosive coating to the substrate [4.4.4].

Physical Methods

A characteristic of mechanical cleaning processes is that in removing contaminants they create new surfaces with a modified roughness profile. **Brushing and sanding** remove corrosion products, contaminants or old coatings through the action of oscillating or rotating grinding heads. Inevitably some of the metallic substrate will also

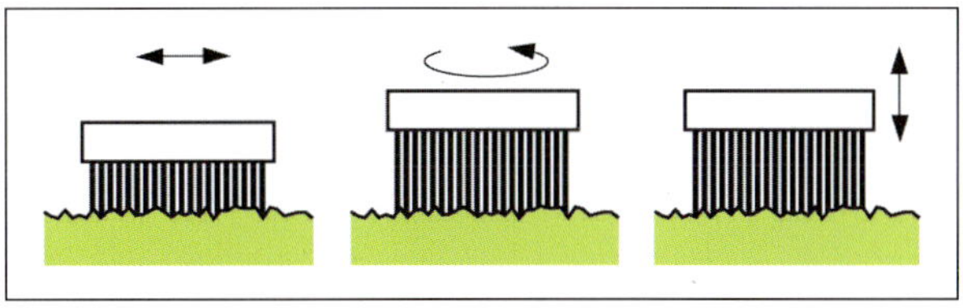

Figure 4.1.12: Brushes for mechanically cleaning metal surfaces

be removed. The changes in the roughness profile vary with the type of abrasive, the brush design and the amount of pressure exerted. Sanding tools which work only horizontally are not so suitable for pretreating heavily textured surfaces. Rotating brushes which oscillate vertically in addition to their horizontal movements ensure a much better cleaning outcome.

A number of electric and pneumatic sanding tools are also available for wet and dry mechanical rust removal and cleaning which also work with rotating or vibrating discs. The abrasives are synthetic mats which have been coated with silicon carbide or corundum.

Blasting is an effective way to clean metallic surfaces of all contaminants such as rust, mill scale, and old coatings. Essentially, this consists in projecting jagged quartz sand, corundum (Al_2O_3), glass beads or steel shot of different sizes from an air-powered pressure gun against the surface to be treated. The particles impact the surface with a high degree of kinetic energy, causing partial abrasion. This simultaneously increases the roughness which in turn increases the adhesion of the coating to be applied. If the roughness increases too much, however, peaks of the substrate can stick out from the coating, acting as weak points in the anticorrosive protection which it provides. Soft metals are therefore unsuitable for sand blasting.

On account of the fact that some of the metal substrate is abraded, the blasting medium becomes contaminated with metal particles. This means that, where different metallic materials are treated, traces of metals can be transferred from the blasting medium to the surface. The consequence is the formation of local cells and increasing susceptibility to corrosion. This is particularly the case when steel shot of a different chemical composition serves as the blasting medium. Such problems do not arise if the blasting medium and the substrate are chemically identical.

The **cryo-clean process** poses no problems whatsoever with regard to galvanic corrosion and nor does it cause metal abrasion. The blasting medium in this case is solid carbon dioxide at temperatures below -78 °C. In addition to the mechanical effect, the object is also cooled down. As a result, embrittlement and cracking occur in the layer to be removed, with partial delamination of the surface coating. Any CO_2 which penetrates the cracks and then evaporates blasts away the contaminants without attacking the metallic surface. Added to which, there is no waste blasting medium [4.4.5]. Bridges and steel structures are among the objects which lend themselves to this process. Substrates treated in this way make excellent surfaces for coating.

General industrial companies and the automotive industry also resort to **sand- and steel-shot blasting** if surfaces capable of being painted or phosphated have to be created on castings and other heavy-section steel components. However, the components must be thoroughly cleaned and degreased before blasting, which would otherwise press the contamination into the substrate.

Process	Application	Parameter	Process data: Removal of material after 40 s	Process data: Change in roughness R_t of steel – Before	Process data: Change in roughness R_t of steel – After
Wet blasting with Al_2O_3	Removal of rust from surfaces; preparation for coating with mild metal abrasion	Blasting agent: Al_2O_3 d = 90 – 150 µm/H_2O Pressure: 300 – 600 kPa	2 µm	3.0	5.5
Wet blasting with glass beads	Removal of thin layers; cleaning and polishing with mildest-possible metal abrasion	Blasting agent: glass d = 150 – 200 µm/H_2O Pressure: 300 – 600 kPa	2 µm	0.9	1.7
Dry blasting with Al_2O_3	Removal of corrosion products; roughening for better coating adhesion	Blasting agent: Al_2O_3 d = 50 – 150 µm Pressure: 50 – 600 kPa	≈ 50 µm	4.5	7.5
Dry blasting with steel shot	Removal of rust and old coatings with significant metal abrasion	Blasting agent: sharp-edged steel shot d = 600 – 1000 µm Pressure: 200 – 300 kPa	≈ 25 µm	3.5	8.5

Figure 4.1.13: Objectives and operating parameters of the various blasting processes

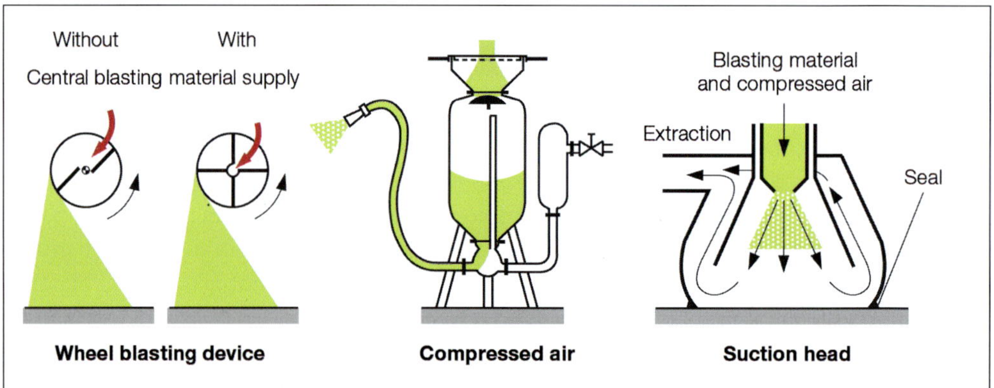

Figure 4.1.14: Schematic drawings of blasting equipment

Even better cleaning is achieved by **steam and wet blasting** with added alumina or silicon carbide particles, pumice powder or glass beads. These are admixed to the water in advance and blasted onto the workpiece with steam or high-pressure water. Mechanical blasting is so effective that this process is mainly reserved for removing firmly attached scale and firmly adhering old coatings.

The machines themselves operate on various particle-acceleration principles. Both shot-blasting wheels and compressed-air systems are used. Acceleration of the medium is effected by throw-shovels in shot-blasting wheels and by compressed air in compression systems. These systems pollute the environment with the blasting agent. However, if they are also fitted with an additional suction head to collect the blasting agent, this reduces the pollution while at the same time permitting reuse of the blasting medium. The suction head has to be positioned close to the surface of the workpiece and so suction-head systems are ineffective for intricately shaped parts.

There have been many attempts to provide immediate anticorrosion protection for bare metallic surfaces produced by blasting. The object can be sealed with an organic paint coat during the actual cleaning process either by means of paint-impregnated blasting media or by the simultaneous spray-coating of a primer in accordance with the ***Gritcote* method**.

Rust, all contaminants and even old coatings can be removed in a single step with the aid of **flame-cleaning blowers**. Bare metal surfaces can be produced with a hot reducing acetylene-oxygen flame and then coated with a primer while still warm and, in all probability, dry.

Physical cleaning with liquid agents serves to remove surface dirt and organic, artificially applied anticorrosive oils and drawing greases.

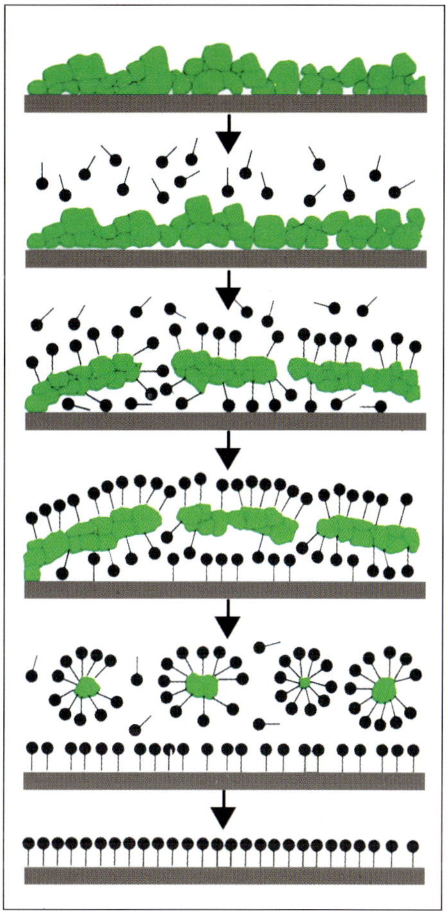

Figure 4.1.15: How surfactants in aqueous cleaning agents work

Modern solutions of aqueous cleaning agents contain surfactants, builders and complex agents. Apart from the classic anionic types based on carboxylates and esters of sulphuric acid the surfactants in use include sulphonates. Cationic and nonionic products are also growing in importance (see Chapter 2.1.4). Their task is to wet and emulsify the hydrophobic oils. Surfactants are either alkaline, acidic or neutral in accordance with their chemical structure. Alkaline cleaning agents have pH values between 9 and 12. They are particularly active and, used in conjunction with phosphates, initiate an additional pickling action when aluminium is being cleaned. Acidic cleaning agents contain acidic phosphates in addition to surfactants, with the result that a thin phosphate film is formed when steel is being pickled. The mildest cleaning agents are the neutral agents with pH values of 4 to 10. They are therefore equally suitable for the physical cleaning of steel, aluminium and zinc. All surfactants have a polar and thus hydrophilic moiety and a nonpolar, hydrophobic moiety. They connect the aqueous phase to the surface of the nonpolar greases and oils and promote emulsification upon suitable artificial agitation of the liquid. Surfactants and builders enclose the dirt and oil particles and stabilise them by forming forces of repulsion.

Builders are inorganic compounds which help surfactants to emulsify the greases to be removed. They are so strongly alkaline that they promote saponification of the greases and oils and in so doing enhance their emulsifiability. However, their use is limited to steel. They attack amphoteric aluminium and zinc too aggressively. Phosphates, borates or carbonates are instead more useful. In order that the effect of anionic surfactants may

Classification	pH	Ingredients	Application
Strongly alkaline	10.5 – 13	• Alkalis • Silicates • Phosphates • Complex formers • Surfactant	• Steel • Heavy contamination • High cleaning requirements
Weakly alkaline	8 – 10	• Phosphates • Borates • Carbonates • Activating agents	• Steel, zinc, aluminium and alloys • High cleaning requirements (spray application)
Neutral	7 – 9.5	• Surfactants • Corrosion inhibitors • Phosphates • Solubilisers	• Steel and zinc • Precleaning and anticorrosive protection
Weakly acidic	3.5 – 5.5	• Acidic alkali phosphates • Surfactants • Accelerators	• Cleaning and phosphating of steel and zinc • Cleaning of aluminium

Figure 4.1.16: Composition and applications of aqueous cleaning agents

not be impeded, it is essential that complexing agents be used to inactivate calcium and magnesium cations, which act as interferences.

The degree and rate of cleaning depend on proper selection of surfactants, complexing agents and builders, as well as on the temperature, duration and recirculation of the cleaning solution.

In view of the high salt concentration, surfaces treated with aqueous cleaning agents must be thoroughly rinsed with deionised water afterwards. Rinsing zones in spray processes offer advantages in respect of the emulsification rate. However, if cavities in intricately shaped components have to be protected, it is better to use the slower dip coating method for them. In smaller systems ultrasound has an accelerating effect.

Environmental and occupational health and safety legislation has substantially restricted the range of permitted solvents and thus the importance of solvent degreasing. The chlorinated hydrocarbons such as 1,1,1-trichloroethane, trichloroethylene and perchloroethylene, which were successfully used for de-greasing tasks in the past, may now be used either not at all or only in emissions-free, fully enclosed systems and subject to special conditions. Their use in coating processes is waning and they will therefore not be described in further detail.

Hydrocarbons, alcohols, ketones, esters and glycol ethers such as methoxypropanol (PM) and dipropylene glycol monomethylether (DPM) are still used because of their cost-effectiveness, though only under stringent occupational health and safety and environmental conditions. The solvent power of hydrocarbons is determined by their aromatic content. As a result of the restrictions surrounding the use of aromatic hydrocarbons, low-aromatic or aromatic-free hydrocarbons are mixed with esters or ethers to increase their solvent power (see Chapter 2.1.2).

Hydrocarbons with narrow boiling ranges prevent high-boiling residues from being left on the objects to be cleaned. Among the oxygen-containing cleaning agents, esters of lactic and succinic acids are of increasing interest in addition to the alkoxypropanols mentioned. Both classes of substances have relatively high flashpoints and good solvent power for oils and greases.

For environmental and economic reasons, the solvents are recirculated during cleaning. It is important to ensure that no dirt particles remain on the surfaces. Consequently, the solvents have to be periodically cleaned by distillation before they are reused. By contrast with the aqueous cleaning agents, organic solvents are generally used in dip or immersion systems. The individual sub-units must be enclosed in booths and equipped with air locks.

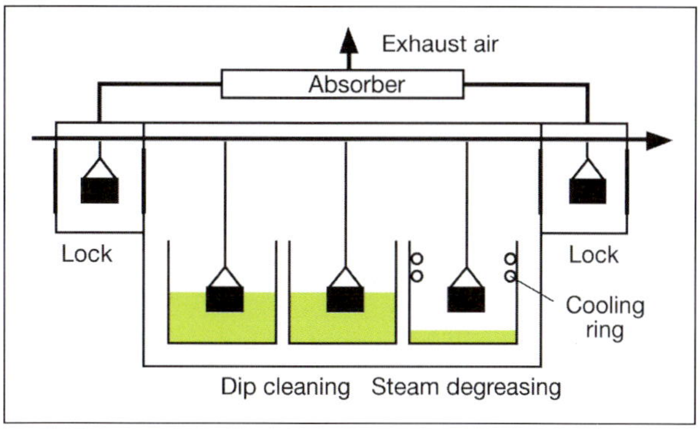

Figure 4.1.17: Schematic drawing of a process for combined dip and steam degreasing

Owing to the risk of an explosion spray processes may only be used with inert gas and are therefore only seldom encountered.

The cleaning process for bare metallic unfinished parts is made all the more difficult by the occurrence of microscopic cracks in the surface during deformation in the stamping presses, into which drawing greases can be pressed by the high stamping forces. To this must be added the fact that anticorrosive oils containing carbon double bonds resinify when stored for extended periods, which makes degreasing more difficult.

Application

Chemical Pretreatment

The chemical pretreatment methods can be divided into film forming and non-film types. The first group contains the already mentioned flame cleaning as a method of thermal reduction of oxides, and the pickling of metal surfaces with acids and bases. The second includes those processes which form inorganic conversion layers. By this is meant the various methods of phosphating, chromating and conversion based on titanium and zirconium compounds.

Pickling solutions are active substances dissolved in water which chemically attack the oxide and the material from the surface to remove or dissolve the oxide and contaminants. The type of material to be cleaned determines whether the chosen pickling solutions are acidic or alkaline. Proven acids for cleaning metals include sulphuric acid, hydrochloric acid and phosphoric acid. Caustic soda has proved effective for alkaline cleaning of amphoteric materials such as aluminium in pickling baths.

The objects are pickled by immersing them in the aqueous solutions of the acids or bases. With the exception of hydrochloric acid solutions, the baths are heated to 65 to 80 °C to accelerate the pickling process. The formation of hydrogen due to the pickling action and the associated embrittlement of the metal can be controlled by organic inhibitors such as quinoline, thioaldehyde, and pyridine.

The film forming **phosphating and chromating processes** are integrated after cleaning and before coating, especially where stringent requirements relating to anticorrosive protection have been specified. Such processes have been used very successfully for many decades. Even before 1930 phosphating was a familiar means of improving anticorrosive protection under the trade names Parker, Bonder, and Granodine processes. The chemical processes were very slow back then but have since been accelerated by means of catalysts and optimisation measures, and their quality has been greatly improved.

The prerequisite for a continuous inorganic conversion layer is thorough preliminary cleaning and the complete removal of all oxidic and organic residues.

The simplest method for forming a protective coating for steel consists in generating an iron phosphate coating. This involves giving the objects appropriate preliminary cleaning, followed by dip or spray coating with an acidic sodium phosphate solution. This pickling reaction generates hydrogen and acidic iron hydrogen phosphate, which is converted by atmospheric oxygen into insoluble, neutral iron phosphate.

$$Fe + 4\ NaH_2PO_4 \longrightarrow Fe(H_2PO_4)_2 + 2\ Na_2HPO_4 + H_2 \uparrow$$

$$2\ Fe(H_2PO_4)_2 + 2\ Na_2HPO_4 + {}^1/_2\ O_2 \longrightarrow 2\ FePO_4 \downarrow + 4\ NaH_2PO_4 + H_2O$$

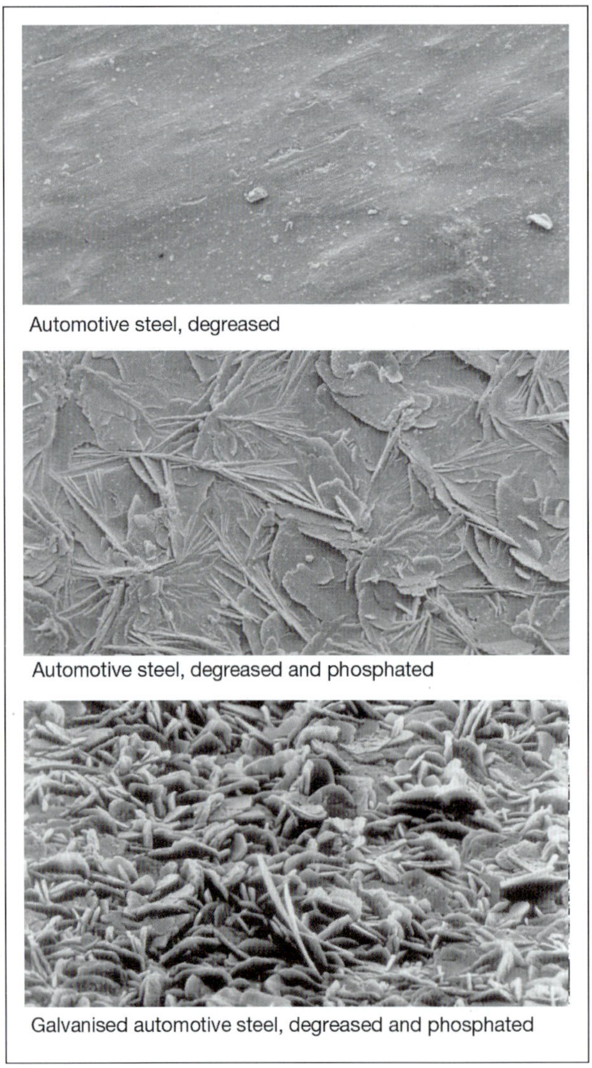

Figure 4.1.18: Electron micrographs of degreased, phosphated surfaces of carbody steels

The resulting coating is extremely thin (coating weight of 0.3 to 1.0 g/m^2 or less than 1 µm film thickness) which is beneficial when the product is subjected to tough loading.

Owing to the formation of iron phosphate, the term **iron phosphating** is used for this type of conversion coat. The term "alkali phosphating" is used as a synonym because of the acidic alkali phosphate acting as active substance. However, the term "non-film-forming phosphating" is misleading since measurable film thicknesses are verifiably produced. The explanation for this confusing term is that the layers are substantially thinner than in other phosphating processes and the fact that part of the film stems from the iron substrate.

A far greater level of anticorrosive protection is provided by **zinc phosphating** which takes longer to carry out and is more expensive due to higher material consumption. Zinc phosphating nowadays comes in numerous variants because the phosphating solutions need to be adapted to the various substrates, substrate mixtures and hybrid construction methods. After a degreasing step, zinc phosphating is performed either by spray coating or dip coating, the choice depending on the object and specifications. Combined spray and dip coating has proved successful in carmaking because of the improved anticorrosive protection provided for the cavities (see Chapter 7.1). The phosphating bath consists of an aqueous solution of phosphoric acid, acidic zinc phosphates and special additives. At equilibrium, the water-soluble hydrogen phosphates of the zinc and the insoluble tertiary zinc phosphate are present in a fixed ratio.

$$3\ Zn^{2+} + 6\ H_2PO_4^- \longrightarrow Zn_3(PO_4)_2\downarrow + 4\ H_3PO_4$$

Phosphoric acid is consumed during **phosphating** and so the equilibrium in the equation above is shifted to the right. More and more tertiary zinc phosphate is generated and is precipitated as a fine crystalline protective coating on the metal, which was etched to bare metal and roughened in the upstream acid action. Unfortunately, the hydrogen gas formed by the pickling action inhibits formation of the phosphate coating.

$$Fe + 2\ H^+ \longrightarrow Fe^{2+} + H_2\uparrow$$

To eliminate the hydrogen, sodium nitrate and nitrite, but also hydrogen peroxide or hydroxylamine, are added to the phosphating solution as accelerators. These oxidise the iron to trivalent iron ions while forming water at the same time.

The accelerators are also used in combination. As a result, dense fine-crystalline phosphate coatings with film weights of 1 to 3 g/m^2 are produced within 60 to 180 seconds. Without accelerating additives, the phosphating times would be considerably longer, i.e. 30 to 60 minutes at bath temperatures of 90 to 95 °C.

$$3\ Fe + NaClO_3 + 6\ H^+ \longrightarrow 3\ Fe^{2+} + NaCl + 3\ H_2O$$

$$6\ Fe^{2+} + NaClO_3 + 6\ H^+ \longrightarrow 6\ Fe^{3+} + NaCl + 3\ H_2O$$

The iron (III) ions contained in the phosphating solution combine with phosphate ions to form iron phosphate which is very difficult to dissolve. It is removed by sedimentation. This precipitation of the iron ions is necessary since large quantities of iron ions would impair the anticorrosive properties of the zinc phosphate coating by forming mixed crystals.

In addition to pure zinc phosphate coatings, phosphating treatments that employ several cations are also proving increasingly successful. Zinc-calcium and zinc-manganese phosphates may be cited as examples. The microcrystalline, firmly anchored mixed crystals offer excellent anticorrosive protection. Low-zinc phosphate treatments have also been specially developed for cathodic electrocoating. The baths have a high phosphoric acid content. This enables iron ions formed by the pickling reaction to be incorporated specifically into the phosphate coating.

$$Zn_2Fe(PO_4)_2 \cdot 4\ H_2O \quad \text{Phosphophyllite}$$

The resulting **phosphophyllite** exhibits good anticorrosive properties when combined with cathodically deposited primers. Furthermore, in the case of hybrid steel and zinc items, both materials are provided with an equal level of protection because of the more powerful pickling action of the low-zinc process. While the iron-containing phosphophyllite is formed on steel, tertiary zinc phosphate is deposited on the zinc as hopeite.

$Zn_3(PO_4)_2 \cdot 4\ H_2O$	Hopeite	
$Zn_2Mn(PO_4)_2 \cdot 4\ H_2O$	Manganese-containing hopeite	Tricationic phosphating
$Zn_2Ni(PO_4)_2 \cdot 4\ H_2O$	Nickel-containing hopeite	Tricationic phosphating
$Zn_2Ca(PO_4)_2 \cdot 4\ H_2O$	Scholzite	

The phosphophyllite-forming **low-zinc process** has been modified to improve the anticorrosive protection still further. Additives of manganese or nickel ions are incorporated into the phosphate coating, as in the zinc-calcium phosphating process.

The uniformity of the film forming process is promoted by pretreatment with titanium phosphate dispersions. The manganese ions support the pickling action and therefore contribute to the uniformity of the film forming process. Nickel is becoming less important because of physiological concerns.

Application

Apart from steel and zinc, aluminium is an important material. If aluminium is the only material in use, chromating is the usual pretreatment process (see below). If, however, a phosphate coat is to be applied to aluminium, the classic methods fail because of excessive pickling action and the formation of soluble aluminium ions. Interference by the soluble aluminium ions in the equilibrium reaction is eliminated by converting them into cryolite complexes with sodium fluoride. Silicates act as inhibitors to ensure a moderate pickling action. In this way it is possible to achieve a uniform conversion layer when steel, zinc and aluminium are used together [4.4.6].

The quality of the phosphating process also depends on the size and shape of the crystals which are firmly attached to the substrate. On account of their porous structure, coarse crystalline layers not only impair the anticorrosive protection, they also have a negative effect on the quality of the topcoat. Activators influence the fineness and thus the closed structure of the crystal matrix. The positive effect of manganese ions and titanium activators has already been mentioned above.

It is possible, with a view to improving the quality of phosphating coats, to use rinsing solutions to compact the coat and so to reinforce passivation. Chromic acid was used in the past but has been superseded by alternatives because of the hazard it poses to watercourses and because of the carcinogenicity of chromium (VI) ions demonstrated in laboratory animals. Zirconium fluoride and titanates have shown themselves to be particularly effective. Polyphosphates also seal the pores of the still free steel surface and therefore prevent the formation of microscopic local cells [4.4.7]. The rate of corrosion at damaged locations or pores is also a function of the solubility of those cations which have an inhibiting effect. OH-ion concentration decreases with increase in numbers of manganese ions, particularly in the alkaline surroundings of the areas acting as cathodes, and with it the progress of the corrosion.

The influence of the surface texture of the metallic substrate on the uniformity of the phosphate coat should also be noted. Roughness profiles of rolled sheets for automotive engineering are therefore selected with great care or are generated by post-treatment with laser beams that have defined abrading properties. This profile is significantly changed by the phosphate coat. An example is shown in Figure 4.1.19. Although the

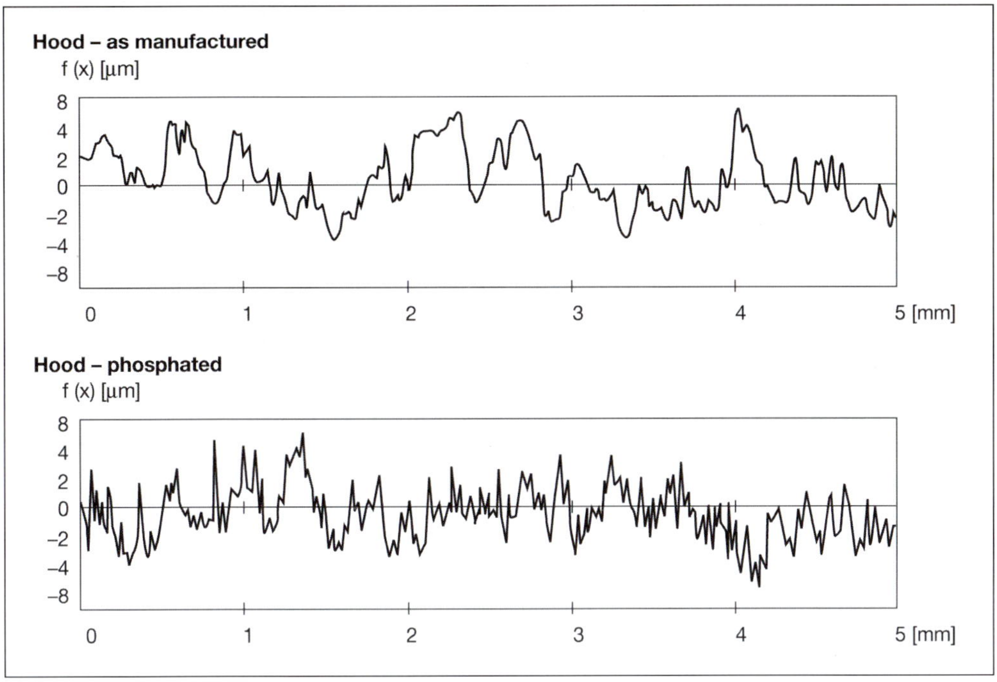

Figure 4.1.19: Changes induced in the surface roughness of steel surfaces by phosphating

Application

roughness profile does not increase in terms of peak-to-valley depth compared with the untreated sheet, it does increase in terms of the number of peaks. This is also accompanied by an improvement in wetting and, due to the increase in the surface area, mechanical fixation and thus adhesion of the primers.

The film thickness also has a decisive impact on the **quality** of the phosphate layer. Excessive film thickness impairs the elasticity and thus all the mechanical/technical properties. On the other hand, phosphate layers are electrical insulators and thus chemical inhibitors which slow down the action of corrosive agents on the metal.

One issue which is frequently discussed, but to which a generally applicable and binding answer is never given, concerns the purpose or effect of downstream drying of nonpassivated phosphate layers. The findings of a large number of trials are that the effects on the anticorrosive protection are system-dependent and a decision therefore has to be taken on individual merit as to whether drying is meaningful or not.

The development of **environmentally friendly pretreatment processes** has been influenced by the significant quantities of sludge produced and its handling, the large quantities of water involved, the energy needed for the conversion reaction at >40 °C, rising prices for zinc and phosphate, and the use of toxic chemicals such as nickel salts. A new generation of conversion coatings based on zirconium fluorides and adhesion promoters such as silanes is becoming increasingly important [4.4.8]. With the aid of the zirconium fluorides and adhesion promoters, a very thin, dense zirconium dioxide layer of about 20 to 50 nm is deposited on the metal surface by increasing the pH value. Attainable savings are up to 60 % in water consumption, 80 % in waste generation and up to 60 % in heating energy costs [4.4.9]. This pretreatment is often referred to as a **nanoceramic conversion layer**.

Pretreatment processes other than phosphating have also proved effective, especially for aluminium with its dense, firmly adhering oxide layer which has poor electrical conductivity, and occasionally also for zinc and magnesium alloys. The **chromating processes** mentioned earlier have the greatest practical significance. These are divided into green and yellow types in line with the composition of the baths. Both pretreatment methods form excellent substrates for the subsequent coating when subjected for a few minutes to either spray coating or dip coating.

The main constituents of **green-chromating baths** for the pretreatment of aluminium are chromic acid or chromates plus fluorides and phosphates. When zinc surfaces are being treated, chlorides and sulphates should also be used to supplement the main constituents.

$$2\,Al + 4\,H_3PO_4 + 2\,CrO_3 \longrightarrow 2\,AlPO_4 + \underbrace{2\,CrPO_4}_{\text{Green}} + 6\,H_2O$$

Aluminium and chromium phosphates form on the surface of the aluminium. As in phosphating, the fluoride serves as a complexing agent for the aluminium ions. The green colouration comes from the trivalent chromium in the chromium phosphate. Although hexavalent chromium is present in the bath, green chromating is free of hexavalent chromium. Solutions of chromic acid or chromates with fluorides are also used in **yellow chromating**. The first stage in forming a layer is a pickling reaction in which aluminium ions pass into solution. At the same time hexavalent chromate is reduced proportionately to chromium (III) cations by the hydrogen which is formed. As a result, cover coatings consisting essentially of oxides and hydrated oxides of chromium and aluminium are formed on the aluminium. In addition to chromium (III) compounds, varying quantities of the yellow chromate are also incorporated into this layer.

$$2\,Al + 3\,CrO_3 + 6\,H_2O \longrightarrow 2\,Al(OH)_3 + Cr(OH)_3 + \underbrace{Cr(OH)CrO_4 \cdot H_2O}_{\text{Yellow}}$$

The coating, which is initially still jelly-like in its softness, must be solidified by the action of heat. Since the heat resistance is limited to 150 °C, the chromate layers should only be heated to 70 °C, although 100 °C is acceptable in the event of downstream coating. This helps to avoid cracking or chipping. The specific weights of the chromate coats are range from 0.1 to 0.5 g/m^2.

The toxicity of hexavalent chromium compounds attempts has inspired much research into **chromium-free alternatives**. Ever more stringent environmental protection provisions are also adding to the demand for chromium-free pretreatment methods for aluminium. Chromium-free layers based on compounds that contain titanium, zinc or zirconium can be achieved by pickling in acidic phosphoric acid-containing baths with simultaneous complexation of the aluminium ions by sodium fluoride. After that the titanium and zirconium compounds are deposited in the form of phosphates or oxides. The coatings, with a film weight of only 0.01 to 0.02 $g/m^{2,}$ are extremely thin. Nonetheless, the protective effect in combination with a subsequent coating is equivalent to that of chromate coats.

An alternative to titanium and zirconium pretreatment is the tried-and-tested method of electrochemical oxidation. In this process, which is known as **anodisation**, the aluminium parts are connected up as the anode in dilute sulphuric acid and are coated with a thin, but pore-free barrier layer by application of a direct voltage of 20 volts.

Whether painted or unpainted, parts treated in this way offer a level of anticorrosive protection which depends on the type and quantity of the alloying components. Unfortunately, the process is more expensive than chromating because of the electrical energy required.

The pretreatment plants commonly used in industrial practice differ in the quality requirements imposed on the conversion layer. Regardless of whether they are operated as spraying, dipping or rotary systems, there are usually at least five individual stations for cleaning, phosphating, post-compaction, intermediate rinsing and post-rinsing (see Chapter 7.1). It is essential that the workpieces have been quantitatively freed from all bath components upon leaving the pretreatment plant. Even traces of electrolytes will form a weak point in the permanent anticorrosive protection due to the osmotic effects of any salts trapped inside the coating.

4.1.3 Plastics

4.1.3.1 Types of Plastics and their Properties

Plastics are artificially manufactured polymers. They can be obtained by synthesising low-molecular structural components or by modifying natural polymer materials. The processes employed in their manufacture are **polycondensation, polyaddition** and **polymerisation** (see Chapter 2.1.1) [4.4.10]. They are sub-divided into thermoplastics, thermosets and elastomers in accordance with the functionality of the basic structural elements or the form of the molecules. **Thermoplastics** consist of long, linear or branched molecular chains which occur as entangled individuals.

Thermoplastics have the specific property of liquefying when heated, owing to the increasing mobility of molecules, of solidifying when cooled down again, and of reversibly going into solution when interacting with certain solvents.

If, en route from monomers to solid plastic, the forming polymer molecules can be induced to crosslink, the materials can no longer be liquefied by heat action nor be dissolved by solvents. Such materials remain viscoelastic even under extreme conditions (see Chapter 3.2.5). They are, therefore, the materials of choice where resistance to solvents and heat are specified.

Elastomers, which are only slightly crosslinked and consist at the same time of movable basic chains, have a very specific property profile. They should always be considered potential solutions when extreme demands are imposed on extensibility or reversible deformability. Elastic sealing compounds for joining dimensionally non-stable objects, e.g. for window or concrete frames, or energy-absorbing front and tail components on cars are just two examples of the importance of highly elastic plastics.

Although the number of monomeric structural elements available is relatively low, the number of technically relevant plastics is extraordinarily high. The reasons for this are that not only do molecules come in so many shapes and sizes, but it is also possible to combine several types of monomers to make copolymers and to create physical blends of already finished polymers. Add to this the fact that copolymerisation can be manipulated

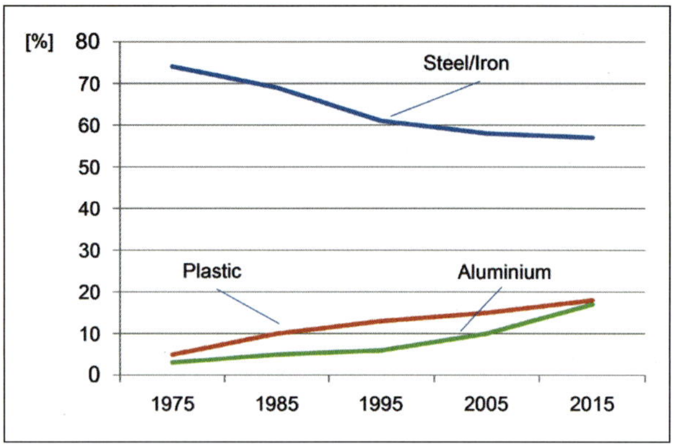

Figure 4.1.20: The growing use of plastic in passenger cars (in weight percent) *Source: authors' estimate*

such that the monomers can be arranged in random, alternating or block order within the copolymers and it is easy to see why the range of plastic types is so great.

The mechanical properties of plastics are often modified with fillers to influence their hardness or with fibres to enhance their cohesion. The embedding of reinforcing materials in the polymer matrix significantly improves the tensile and breaking strengths and the impact and notched impact strengths. Glass fibres are the most widely used reinforcing materials. Where a specification calls for high-strength, high-performance plastics with a low weight and low thermal expansion, carbon fibres also make suitable auxiliary materials.

Ongoing improvements in plastics are opening up more and more new application fields, both for everyday products and for industrial purposes. The major advantages of plastics over many other materials lie in their specific properties, such as low weight, deformability, corrosion resistance, toughness and elasticity on one hand, hardness and internal strength on the other, and high resistance to the action of solvents and chemicals.

In addition to their traditional applications in large areas of the packaging and construction industries and in electrical engineering, they are now also being used increasingly in automotive engineering, primarily because of their versatility in meeting specific requirements such as design options, weight reduction, anticorrosive properties, impact elasticity, and not least price.

It may sound paradoxical but plastics, made from crude oil, become cheaper than metallic materials when the price of energy rises. It is often forgotten in comparative studies that large amounts of energy are also needed to manufacture metallic materials from their oxides and to convert them into the workpiece by melting or by drawing and pressing processes. For example, 1 kg plastic is manufactured from 2.5 kg crude oil, after allowance for energy and material losses, whereas 4 kg crude oil equivalents are needed in the manufacture of steel, 12 kg for copper and as much as 15 kg for aluminium. These figures do not contain the energy for deformation processes required in the manufacture of the workpieces.

To this must also be added the simpler manufacturing process for intricately-shaped parts and components without lavish welding and joining methods. Solid or expanded-plastic components, which are lighter and therefore being increasingly used in aerospace and automotive engineering, can be manufactured quickly and economically by injection-moulding or extrusion methods. In similar fashion liquid feedstocks can be converted by chemical reactions into thermosetting plastics and closed- or open-cell foam plastics in every hardness grade and density variant. One long-established method is the **reaction injection moulding process** (RIM) for polyurethane plastics and unsaturated polyester resins. This involves injecting the reaction mixture into a

Plastic	Abbreviation (ISO 1043-1)	Application
Polyurethane	PUR • PUR rigid foam • PUR semi-rigid foam • PUR soft foam	 Window frames, furniture Spoilers, bumpers, steering wheels Spoilers, headrests, protective trim
Polycarbonate	PC • Pure • Modified	 Crash helmets, lamps, bottles, windows Front spoilers, bumpers, front ends
Polyamide	PA	Wheel covers, front spoilers, garden furniture, pipes
Polyethylene terephthalate	PET	Cooler grilles, front ends, petrol caps, bottles
Polypropylene-ethylene diene moulding	PP-EPDM	Bumpers, front and rear spoilers, wheel expansions
Acrylonitrile-butadiene styrene copolymers	ABS	Cooler grilles, rear spoilers, toys, computer housings
Polyphenylene oxide	PPO	Instrument panels, televisions, pipes, containers
Polystyrene	PS	Flower pots, televisions, radios, computer housings, toys, household goods
Polyvinyl chloride	PVC	Pipes, films, window frames, balcony balustrades, toys
Polymethyl methacrylate	PMMA	Taillights, windows, household goods, audio equipment
Glass-fibre reinforced	UP-GFK	Front spoilers, bumpers, sun roofs, boats
Unsaturated polyesters Epoxy resins	EP	Tennis racquets, gliders

Figure 4.1.21: Overview of commonly painted plastics

mould, where the materials react with each other to produce a solid or an expanded material. Where fibres are added for reinforcement purposes the process is termed **reinforced reaction injection moulding** (RRIM). Another method of producing thermoset mouldings is the compression moulding of semi-finished products, known as prepregs. These are fibre laminates impregnated with synthetic resin, generally unsaturated polyester resins (UP), which are only cured once they have been moulded into a workpiece in a die. This process is known as sheet moulding compound (SMC).

The selection of the plastic is therefore determined by the specific requirements it has to meet. The mechanical, chemical, thermal and electrical properties have an important role to play, as do factors such as transparency or pigmentability, resistance to longterm outdoor exposure, processability and the price.

Figure 4.1.21 shows an overview of the thermoplastic and thermoset products commonly used for general industrial purposes, together with their abbreviations in ISO 1043-1.

Because of their naturally smooth and relatively resistant surfaces plastics and plastic objects can certainly be used in unpainted form. Since most can be pigmented during their manufacture, the desire for colour can be satisfied without the need for a painting process which would involve both time and costs. Despite the inherent benefits of pigmented plastics, the process of painting plastics has nonetheless proved itself in many applications. The reasons for this may be of a cosmetic or functional nature. Improved resistance to light by means of UV-protective coatings, improved resistance to longterm outdoor exposure, waterproofing, solvent or chemical resistance through the use of saponifying, solid, highly crosslinked protective coats merit special mention. Effects on flame resistance, electrical conductivity, abrasion resistance and cleanability are further important arguments with regard to the functionality of the polymer workpiece.

From an economic perspective, **effect coatings** should particularly be borne in mind. A coating with a thin coloured polymer skin consisting of expensive effect materials and pigments is often more cost-effective than solid-body pigmentation of what are sometimes very large components, despite the additional surface treatment involved. Where hybrid construction methods are used, it is not possible to achieve an optically identical appearance for the pigmented plastics and the painted metal components. The different surface textures of metal and plastic surfaces require surface treatment with paint if a visual match is to be obtained.

Sometimes plastic elements such as instrument panel components and trim strips on motor vehicles are required to have a metallic appearance. This is achieved by **vapour-deposition of metals** such as aluminium, copper, silver or even gold. Since plastic surfaces are generally not smooth enough to permit the production of a perfect metal mirror surface they must be painted with a primer before metallisation. Furthermore, the vapour-deposited metal layer has to be coated to protect it against mechanical damage and corrosive influences.

If plastics are to be painted successfully, it is important to have precise information on the specific properties of the materials and workpieces. Thermal properties, solubilities, structure and polarity, as well as production-induced properties such as frozen-in strain, nonhomogeneous discharge of additives, release agents, sink marks, bond lines, blisters below the surface, visible or even projecting glass fibres, contaminants from the moulds or dies, electrostatic charges and particularly wetting and adhesion problems are the major causes of coating problems to mention in this context.

4.1.3.2 Pretreatments

The quality of coatings on plastics depends considerably on the correct choice of paint system and proper preparation of the surface. The **adhesion** which is necessary for lasting protection can generally only be achieved by applying a material-specific pretreatment. The pretreatment includes the removal of release agent residues and contaminants, the dissipation of electric charges and the elimination or levelling of production flaws.

When plastic parts are made by injection-moulding of thermoplastic materials, compression-moulding of thermosets or reaction injection moulding of polyurethanes, the surfaces come into contact with **release agents** which facilitate removal of the moulding. Such release agents generally consist of oils, wax-like substances or metal soaps. Silicone-containing release agents should not be used as they are difficult to remove and there is a high risk of contamination of the coating. Greasy soiling marks caused in transit, hand sweat and dust, which also impair adhesion and often the appearance of the paint coating as well, have just as adverse an effect as residues of release agent.

Wiping down with dry woollen and synthetic fibre cloths is not recommended because this generates a **static charge in the plastic** components. On account of plastics' low conductivity the electrical charges remain where they were generated and attract dust particles of the opposite charge to them. This effect can be prevented or mitigated by increasing the conductivity. Thus plastics can either be antistatically equipped during manufacture or subsequently treated with antistatic agents. However, these **antistatic agents** only unfold their full effectiveness above a certain relative humidity. Furthermore, they have a certain level of volatility, as a result of which their effect gradually diminishes. One method which has proved successful, therefore, is to blast the component with ionised air directly before it is coated. The alternating-current ioniser first neutralises the fixed electric charges and then blows the adhering dust from the surface by means of an adjustable air current.

Relatively simple processes such as **washing** with surfactant-containing aqueous domestic or industrial cleaning agents are often suitable means of thorough direct cleaning. The effect can be substantially enhanced by heating to approx. 50 °C and power washing. Any organic solvents used for cleaning purposes must be very carefully matched to the plastic to be coated in order that damage to the surface of the plastic may be avoided. Furthermore, solvents can diminish the internal strength because of their ability to penetrate the plastics.

Excessively rapid cooling of the plastic workpieces during manufacture leads to tensions or strain being frozen in, and these can cause cracks under the action of solvents. With many thermoplastics, therefore, particularly polystyrene, polycarbonate and polymethyl methacrylate, a few minutes of **tempering** should always be interposed before the painting process. This entails heating the materials to just below their softening point. The resulting increased mobility of the molecular segments cancels out the internal tensions, and the macromolecules assume their most thermodynamically favourable position.

For good **adhesion**, mechanical pretreatment operations to consider include roughening (satinising) and sanding. Regrettably blasting and sanding processes are difficult to automate and therefore costly. Care must be taken not to use excessively coarse abrasives or shot, or to sand too hard, particularly with glass-fibre-reinforced materials, as new problems will be created thereby because of surface faults. For these reasons mechanical processes are little used.

Where reliable adhesion between the cleaned plastic surface and the coating cannot be provided by special bonding agents or the use of solvents to effect slight dissolution or swelling of the primer, recourse is made to oxidising chemical processes. These include the action of fluorine or flame-treatment with gas burners adjusted to ensure oxidation. **Surface treatment with plasma** has proved particularly effective. Gas discharges are induced in vacuum chambers at pressures of approx. 1 mbar by applying a high direct voltage across two electrodes to ionise the gas molecules. The plasma is then introduced into the test chamber where it acts on the surface of the plastic parts and chemically changes them. Plasma consisting of oxygen or of oxygen/nitrogen mixtures is ideal. However, the effectiveness of the process must be weighed against the cost of the equipment.

One process operating under normal pressure uses high-frequency alternating voltages to generate free electrons which are also capable of chemically changing the surface of the

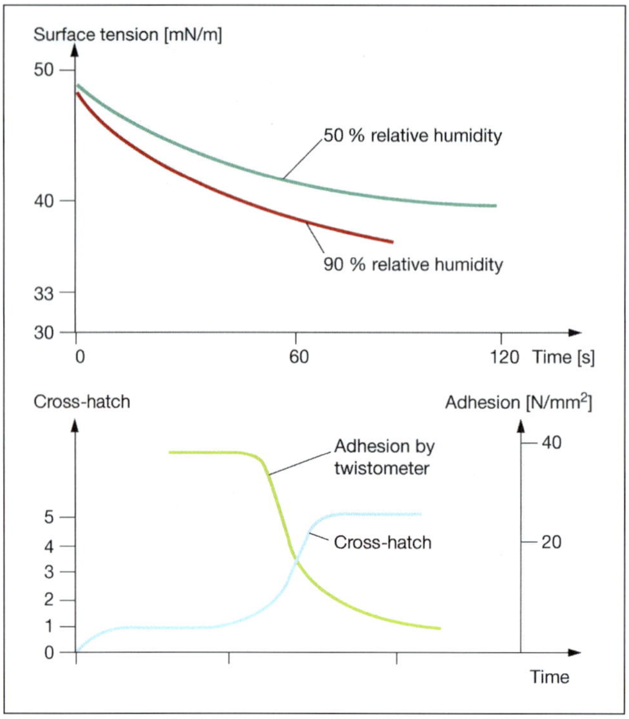

Figure 4.1.22: Change in surface tension and adhesion of corona-treated plastic surfaces over time

plastic. The process is based on a gas discharge induced by electrodes designed as contact elements under an applied alternating voltage of 10 to 20 kV at 20 to 40 kHz. The **corona discharges** generate atomic oxygen and extremely reactive atomic nitrogen. Both gases are adsorptively bound to the workpiece where they polarise the surface. The polarisation leads to an increase in the surface tension, and improves wettability and also intercoat adhesion. The volumes of ozone generated during operation of the systems (0.05 ppm up to limit values of 0.1 ppm) are within the permissible range for manual application. Infrared spectroscopy can identify generated peroxides and polar products, such as aldehydes, ketones, carboxylic acids and hydroxyl groups on the surface of the plastics. Compared with that of the untreated plastic, the surface tension can be temporarily raised by more than 10 mN/m.

This effect gradually diminishes again as a function of the relative humidity of the air. Some of the generated polar groups are evidently only physically bound. It is important, therefore, for the subsequent paint coating to be applied within one hour, if possible, wherever corona-discharge equipment is used in conjunction with a high relative humidity.

If the recommended times are observed, consistently good **intercoat adhesion** is achieved. However, the process only works with parts of simple geometric shape because of the very short range (only half a centimetre) of the activated gases.

Pretreatment also includes eradicating defects on the objects to be coated. Even under the most careful production conditions surface defects can occur in plastic components which seriously impair the outcome of painting operations. Eradication of flow lines, joint lines or differences in gloss levels requires high input in terms of polishing or buffing. A simple process is available specifically for the joint lines. Sprayable **joint line liquids** can be used to treat parts made of impact-resistant polystyrene (PS), styrene acrylonitrile (SAN), acrylonitrile-butadiene-styrene (ABS), and polycarbonate (PC). These joint line liquids are solvents which are matched to the plastic under treatment. They cause slight swelling and separation of the thin polymer skins without damaging the solid plastic. They not only cover the production faults, they also improve the general usability properties and the antistatic effect of the plastics.

Concealed **blisters** are a particular problem. The air inclusions in the liquid plastic of the reaction materials in the closed moulds or the gases which occur during processing enter the interface between the plastic component and the mould. A thin polymer film then often remains between the blister and the mould wall.

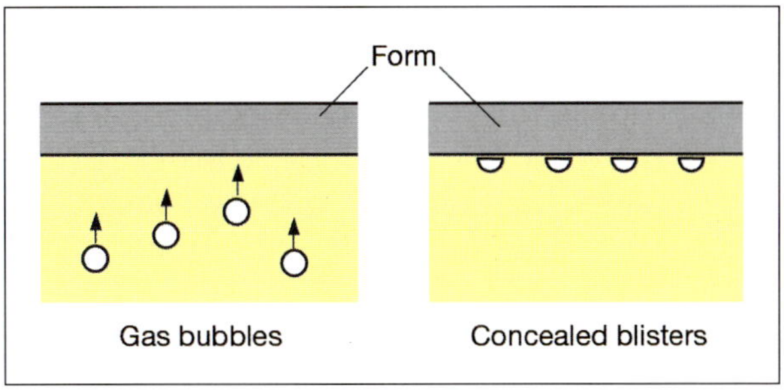

Figure 4.1.23: Concealed blisters before and after coating

After release from the mould these flaws are not visible to the eye initially. During downstream painting the covering skins are destroyed either by solvents or thermal expansion in the oven as a result of which the coating material can penetrate the blisters. This causes significant faults in the painted surface. In such cases it is sensible to puncture the blisters deliberately by thermal pretreatment and then to level the resulting indentations with filler.

The pretreatment methods described thus far do not raise the poor electrical conductivity of usual plastics sufficiently to permit the electrostatically assisted application of paints to be carried out. This would require the electrical surface resistance to be reduced from its original 10^{12} to 10^{14} Ωcm by several orders of magnitude. This can be done by adding conductive carbon blacks to the plastic material during compounding or by coating already manufactured components with electrically conductive paints.

One process which is specially designed for coating plastics involves the application of a primer or single-coat finish to parts while they are still in the mould. In this process, known as **in-mould coating** (IMC), the mould in which the plastic is pressed or cured in conventional methods is opened by a few millimetres to permit the primer material to be injected. When the mould is closed, this material is uniformly distributed over the surface of the plastic and then pressed and cured.

Because of the low resistance of thermoplastics to heat and solvents the standard paints for plastics are either water-borne, air-curing dispersions, oxidatively curing alkyd systems or force-dried 2-component polyurethane paints (see Chapter 7.3) [4.4.11].

4.1.4 Mineral Substrates

4.1.4.1 Concrete, Mortar, Plaster

Concrete is one of the most important construction materials. It is used in civil engineering projects, including the building of roads, bridges, industrial buildings and houses. According to ISO 22965-1 concrete is a synthetically manufactured stone material made of a mixture of cement, aggregates such as gravel and sand, other additives and water. It is sub-divided into reinforced concrete, prestressed concrete and simple concrete. Set concrete is classified on the basis of density as lightweight concrete with a maximum density of 2000 kg/m^3, normal concrete with a density in the range 2000 to 2800 kg/m^3 and heavy concrete with a density above 2800 kg/m^3.

Concrete is extremely resistant to ageing and normally requires no further protection for it to withstand weathering over the long term without damage. For that reason, it was not standard practice in the past to apply a surface protection to concrete. Coatings were applied solely for cosmetic purposes. By the time of the collapse of the prestressed concrete roof of the congress hall building in Berlin, however, it was abundantly clear that **carbonation** of the alkaline concrete can cause the pH to fall below 10, which means that the increasingly common **reinforcement steel** becomes susceptible to corrosion, even if it is fully enclosed by concrete. Since then, CO_2-barrier coatings have become the norm for functional reasons to protect reinforced concrete structures. It should be noted that concrete is a substrate whose absorbency varies as a function of its density and which requires pretreatment that depends on the admixtures and additives used in its manufacture.

The surface roughness of concrete varies with the raw material mixture, how it is processed, and the smoothness of the moulds or formwork. This confers a greater or lesser ability to absorb water via cavities. To this must be added the influences of the diverse range of additives and release agents. These include special surfactants to influence the rheological properties of the fresh concrete and hydrophobic compaction agents to increase its impermeability to water. Additives which retard or accelerate the setting process are also important formulation components in this context.

The requirements for specific concrete surface are specified in detail in ISO 22966 and DIN EN 206-1. It should be solid and free from loose layers and bloom. Sand-blasting of the surface may be advisable before an organic coating is applied. Any cracks present should be sealed with a resilient filler.

The release agents employed between the formwork and the concrete surface must also be taken into account in any later surface treatment or pretreatment. Mineral oils, water/oil or oil/water emulsions and waxes serve as release agents, and all residues of them must be removed before concrete can be painted. Otherwise either patches will form on the coating or the coating will peel off. Such removal is not easy, precisely because release agents are required to be water-insoluble and to remain effective over an extended period. Under no circumstances should organic solvents be used to remove release agents as these can penetrate the concrete.

A **concrete coating** must have a high diffusion resistance to CO_2, so that it can effectively prevent the carbonation mentioned above. Since concrete is always alkaline, longterm protection can only be provided by coatings which are resistant to saponification. Aqueous acrylic resin dispersions or styrene copolymers such as butadiene-styrene and acrylonitrile-butadiene-styrene, organically dissolved epoxy resins, and moisture-curing polyurethanes are established choices for this application. Chlorinated and cyclo rubber paints as modified natural materials also lend themselves readily to the protection of alkaline mineral substrates [4.4.12].

Mortar and **plaster** in all their numerous variants are similar to concrete in the conditions they present to coating materials. Plaster on external or internal walls or ceilings is a special mortar layer applied in a particular thickness. Whereas concrete is used solely as a base material, mortar provides the joints in masonry and plaster provides cosmetic cover on walls and ceilings. Unlike concrete, mortar and plaster contain various additives plus lime and, in special cases, even gypsum (calcium sulphate). Curing with water is accompanied by the slow conversion of the lime with CO_2 to calcium carbonate over the course of several weeks.

Given their alkalinity, the same rules apply to the surface treatment of plastered masonry surfaces as to that of concrete surfaces. Less preparatory cleaning work is needed because of the lack of contamination by release agents. It should be noted that H_2O is formed during curing of the lime component, as a result of which plastered surfaces release water for several months and therefore only coatings which are permeable to water vapour may be used.

4.1.4.2 Glass

Glass is an amorphous mixture of glass-forming substances such as SiO_2, B_2O_3 and P_2O_5, alkali metal salts to lower the melting point and stabilisers such as carbonates of calcium, barium, lead, and zinc. An inorganic transparent material with a softening point of approx. 1500 °C, glass is seldom surface-treated. If, however, an inscription is to be made or a particularly decorative appearance needs to be achieved, glass products are also provided with organic coatings.

Glass is a poor conductor of heat and electricity. As a result, internal tensions are quickly generated when it is heated. Glass is not absorbent and is extremely smooth. Coatings with a good finish quality are therefore easily achieved. Consequently, attention is mainly focused on good adhesion between the coating and the surface of the glass.

Glass for coating should be thoroughly prepared with aqueous or organic cleaning agents and, if necessary, also etched with aggressive acids, such as hydrofluoric acid. The paints are then selected to suit the requirements of the particular product.

4.1.5 Other Substrates

4.1.5.1 Rubber

The raw material for the production of rubber is obtained from hevea plants in the form of 1,4-cis-polyisoprene, or **caoutchouc**, in the form of a latex dispersion. The elastomeric rubber is produced by macroporous crosslinking with sulphur bridges.

The hydrophobic effect of aqueous dispersions of caoutchouc has been known for centuries. South American Indians impregnated their clothing with the extract from the hevea plant. Only later was the method for crosslinking crude rubber by means of vulcanisation discovered. The experience of *Goodyear* and *Parker* enabled car tyres to be made permanently resilient and wear-resistant. Rubber is a versatile material nowadays. Rubber soles, shoes, boots, balls, countless rugged parts in consumer goods and resilient sealers are based on this material.

The special properties of rubber impose particular demands on the coating materials used to paint it. In preparation for coating rubber articles, any impurities on the surface, particularly release agents from the manufacturing process, have to be removed. This is done by simple washing with solvents such as gasoline, alcohols or alkaline aqueous cleaning agents.

Consumer goods made of rubber have huge **extensibility** and the coatings used must be correspondingly elastic. Oil paints and alkyd resins which were used in the past because of their good adhesion and were often applied before vulcanisation to create a better bond, suffered from the disadvantage of embrittlement and therefore had only a limited lifetime.

More suitable coating materials are based on solutions and dispersions of elastic chloroprenes, chlorosulphonated polyethylene products, silicone resins and resilient chloroprenes. Polyurethanes have now become the most important coating materials for elastomers because of their resistance, low wear and longterm resilience. Their use is not limited to classic natural rubber but extends in the same way to numerous synthetic elastomers such as EPDM.

4.1.5.2 Leather

Leather is processed animal skin which has been bated and tanned to make it supple and stable. To this end the leather, which consists of proteins, is treated with tanning agents and inorganic complexing agents.

It is rare for a colour-imparting **surface treatment** not to be carried out. Measures to affect the appearance and the properties represent two different goals. By this is meant the colour, gloss and effect of the leather on one hand and protection against water, improved cleaning properties and constant elasticity on the other. Apart from impregnations to make the leather hydrophobic, primers are applied which have the effect of levelling pronounced grain in the leather. Finally, leather paints are applied for decorative purposes. These can be pigmented opaque materials or also transparent products. Candidate coating materials include conventional, solvent-borne, water-soluble and water-dilutable materials.

Solvent-borne **leather paints** contain cellulose nitrate, cellulose acetobutyrate and cellulose acetate, as well as polyurethanes and acrylic resin dispersions. To meet the demanding elasticity specifications, elastic polyurethanes and cellulose derivatives appropriately regulated with plasticisers should be used.

4.1.6 Workpiece Design and Coatability

In addition to the properties of the material, its specific surface structures and its pretreatment, the impact of the **design** and the **joining method** on paint workflows and on attainable coating quality should not be underestimated. Designers of consumer goods have to bear certain basic rules in mind if the desired quality expectations of the object to be painted are to be met in the long term. Specific design-related factors have to be addressed for the most commonly employed materials today, i.e. wood and composite materials, steel and other metals, and plastics.

In particular, measures have to be taken for finished components made of **wooden materials** to avoid stresses arising from longterm moisture action. Furthermore, conditions have to be created in the design which enable a coating to be uniformly applied.

The correlation between design and durability of the coating is especially evident with windows. The frame profile should be designed to ensure that water can drain off. All joints should be as flush as possible, using tenon-and-slot mortises or double tenons. All edges should be rounded to reduce the effect of dewetting at edges and thus to guarantee adequate film thickness in all areas of the coating. Effective seals must be provided between the casements and the frame. These seals must have sufficient longterm elasticity to cope with the permanent tensile, compressive and shear loads. This is the only way in which the necessary protection can be guaranteed at all connecting points between the window and the glass and masonry [4.4.13].

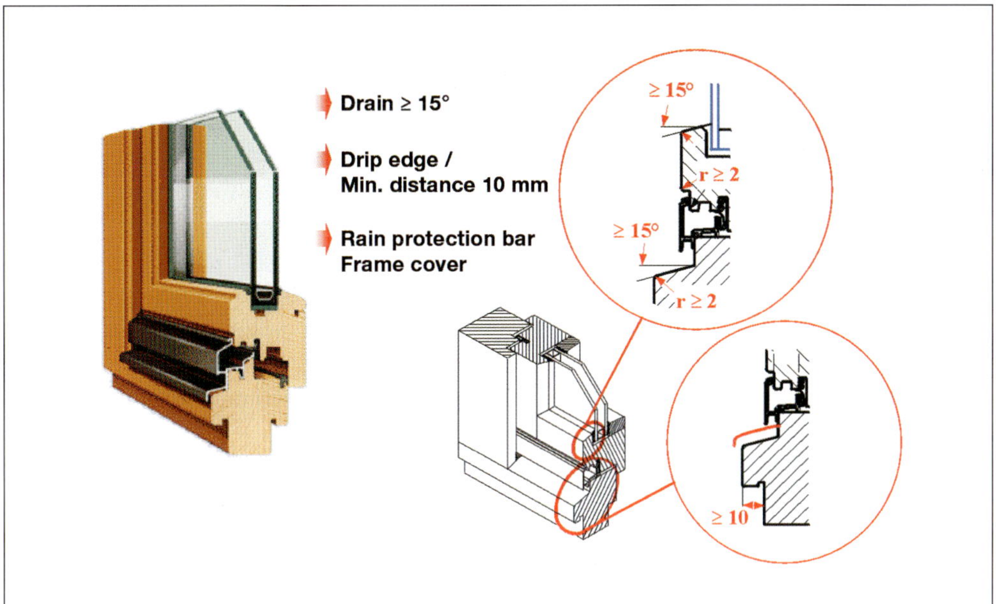

Figure 4.1.24: A window properly designed for coating

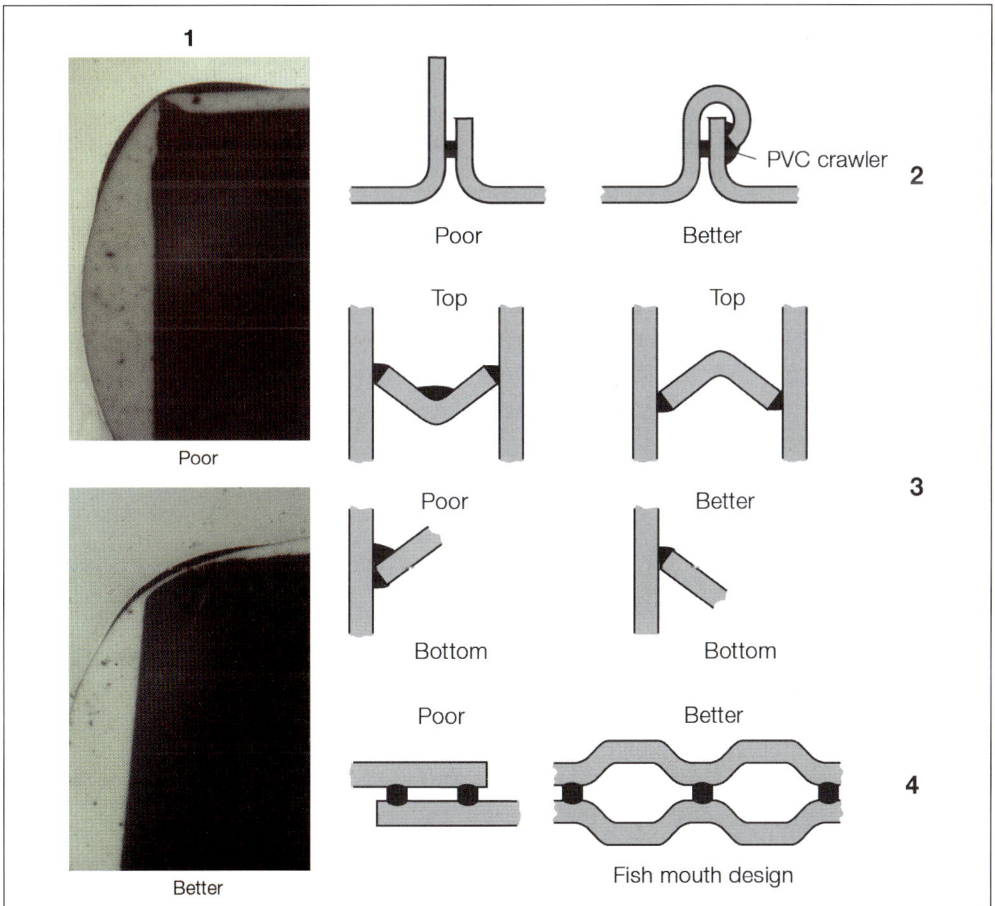

Figure 4.1.25: Construction elements properly designed for coating (1 – 4)

Application

Leaving aside the coating of planar endless coils in the coil coating process, metallic objects which have to be painted are often of extremely complex shape and include **welded and riveted joints** or **bonded connections** with cavities to stabilise the products. Beads and sharp cut edges make the geometry of the object to be coated even more complex. Cavities which are closed or are difficult to access must either be effectively and permanently sealed or made accessible to coating materials; then, however, good ventilation of the cavities must be provided, and with it the necessary conditions for longterm **anticorrosive protection**. Capillaries, narrow gaps and large overlaps must be avoided. Where dip coating is employed it is important to avoid having scoop-like parts in the object. Rounding of cut edges and dispensing with sharp-edged beads are further sensible requirements. Burr on edges causes the still-liquid material of fresh coatings to pull away from those edges. Sharp edges should therefore be rounded or protected by means of PVC strips after beading.

Places where water can accumulate must also be avoided. In addition, numerous **welding regulations** must be observed: sheet metal overlaps should only touch at the welds. There must be an adequate gap between all other areas so that they can be coated (fish mouth design).

Other potential weak points must be eradicated by the design. For example, the front of a car is particularly at risk from stone-chipping. Raked shapes mitigate this problem.

A passive measure which can be implemented is the correct choice of material. Corrosion-resistant metals, precoated sheets or plastic parts are therefore often integrated into the otherwise steel workpiece. The hybrid construction technique enables the individual parts of the object to be better adapted to practical requirements. Where different metals are used, the avoidance of local cells is of primary importance for good anticorrosion protection.

The fundamental rules are: avoid permanent damp spots by ensuring that water can drain properly, and do not create areas which are inaccessible to paint as a result of punched holes or similar measures. Furthermore, sharp edges should be eliminated by rounding, metal beads should be protected by plastic strips, suitable materials should be selected, and galvanic corrosion caused by contact between different metallic materials should be avoided.

4.1.7 Summary

For protection, the vast majority of all commodities and consumer goods need to be painted. Due account must be taken of the specific **material properties** of the objects to be painted. Wood, plastics, metals, mineral substrates and natural materials such as leather and rubber are different materials in many respects. The thermally stable metals possess only minor extensibility, but are susceptible to corrosion. Wood is a thermally unstable, absorbent material which degrades on longterm outdoor exposure and exhibits properties during the painting process which differ from those of plastics, which are sensitive to solvents and often thermally unstable.

Other factors which influence the selection of and the method for processing the coating material are the design and construction of the workpieces. Cavities in which moisture accumulates, edges from which the paint pulls away or overlaps of sheets which cannot be reached by the coating materials are weaknesses when it comes to the production of a lasting coating.

In all cases a perfect coating is only possible if the surfaces have been properly pretreated for the subsequent painting process. This involves removing all contaminants such as dirt, corrosion products, oils, and drawing greases from the surface. Metal objects to be painted often require supplementary chemical operations, such as film-forming phosphating or chromating, while plastics need oxidation reactions to increase the polarity and prepare the surface for good wetting and adhesion strength. Only when pretreatment which is appropriate to the process and has been adapted to the material or workpiece has been carried out have all the conditions been met for a lasting coating.

4.2 Paint Processing

A very wide range of processes is available to convert the paint into the end product, the coating. Successful implementation of the associated process stages is contingent, on one hand, on matching them to the nature, size, shape and number of objects to be painted per unit time and, on the other, on making due allowance for the properties of the paint to be used and the painting sequence for a multicoat paint system. All industrial painting processes share a division into two successive sub-stages.

The first process stage, namely application, consists in applying a generally organic coating material to the pretreated surfaces (see Figure 4.2.1). Methods here range from manual application, which requires skill and dexterity on the part of the user, to large-scale systems with high-output, process-controlled automatic systems and robots.

The second process stage consists in drying and curing the applied paints to convert them into the usable coating. During this, overlapping occurs between the physical process of solvent release and the chemical process of crosslinking (see Chapter 3.1). The energy required for this is transmitted by convective heat exchange, or by infrared, microwave, or UV or electron radiation.

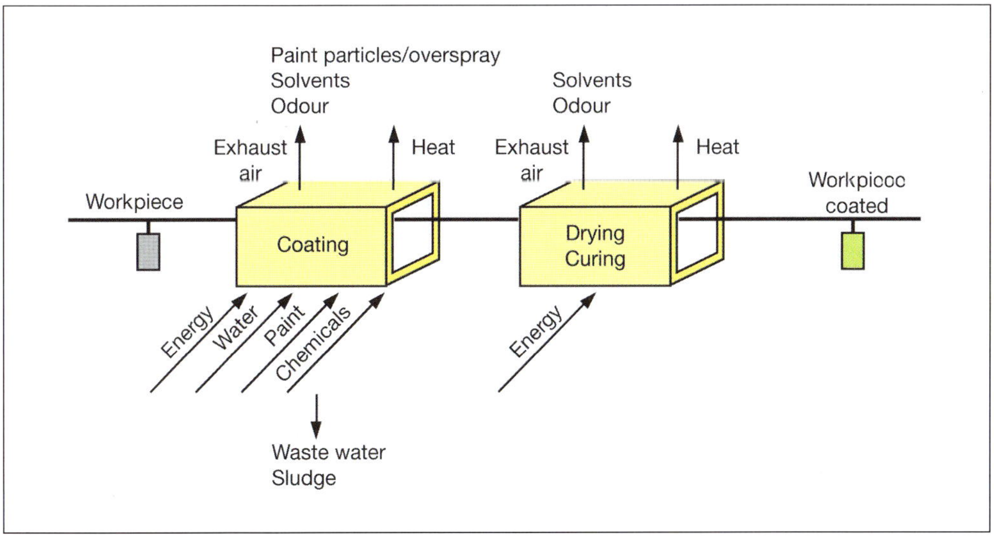

Figure 4.2.1: Application, drying and hardening of paints

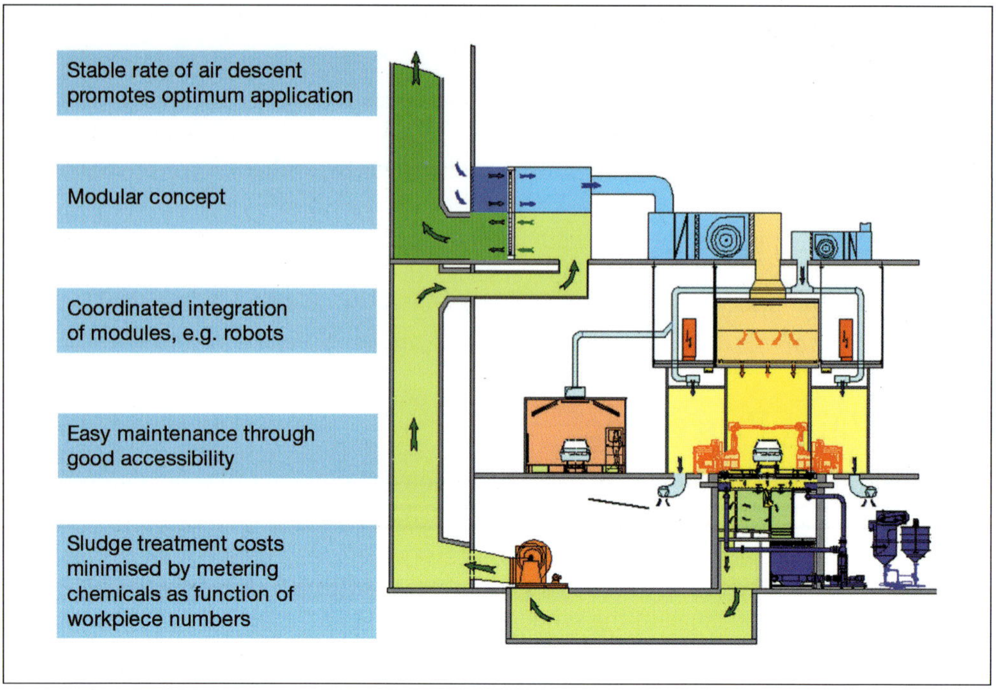

Figure 4.2.2: Cross-section of a paint line for spray application, with wet separation of overspray *Source: Dürr*

A key element of any industrial paint processing is monitoring and disposing of both the air in the booth and the overspray (see Chapter 4.2.1.5).

4.2.1 Processing of Wet Paints

Figure 4.2.3 shows a simple, intuitive classification system for painting processes.

In conventional dip coating processes or those assisted by electrochemical deposition the object is transported to the paint. Such processes include fluidised-bed coating for the processing of powder coatings.

In all other processing methods, the paint is taken to the object. Application is done directly by means of manual or mechanical roller, curtain coating or flow coating with only low material spill, or indirectly by means of atomisation and spray coating, albeit with considerably lower material yields in some cases. Despite the low-yield disadvantage the last of these processes predominates in industry. Tried-and-tested application methods include high-pressure pneumatic spraying, hydraulic airless spraying, and purely electrostatic and electrostatically assisted methods such as high-speed rotary atomisation.

Object to paint	Paint to object
Conventional dipping • Solvent-borne paints • Water-borne paints	**Direct processes** • Brushing • Rolling, scrolling • Flow-coating, pouring
Electrodeposition • Anodic • Cathodic	**Indirect processes** • High pressure atomisation • Airless atomisation • High-speed rotary bell • Electrostatic atomisation

Figure 4.2.3: Overview of important coating processes

4.2.1.1 Dip Coating (Object Taken to Paint)

Conventional Dip Coating

Dip coating is one of the simplest and oldest methods of painting industrial goods. It is easy to automate, has a much higher material yield than spray coating and has low capital costs. The process makes use of the adhesion properties of viscous paints on the objects to be painted. Its disadvantages are the poor processing window with regard to film thicknesses and the associated defects, such as runs and blisters, the cost and the need for special precautionary measures to avoid explosions or fires when solvent-borne paints are used. As a result of these dangers and corresponding incidents practically only water-borne dip coats are now employed in the few systems still operating.

Attempts to replace water by trichloroethylene failed to overcome all the difficulties associated with the process. Even though the tendency to run was reduced and greater uniformity of film thickness was achieved, there was the question of the fire risk, and thus it was the aforementioned occupational health and safety reasons which ensured that this very useful and sometimes successful process was relegated to secondary status again.

In dip coating the workpiece is immersed, in either a batch or a continuous process, in the relatively low-viscous paint, removed, drained, and conveyed to the oven. Draining, combined with evaporation of the solvents, has to last long enough to ensure that no surface defects in the form of blisters, pinholing or runs occur. Water-borne dip paints also require a sufficiently long blow-drying zone, which moreover is usually heated.

Despite all the optimisation measures implemented, it is not possible to achieve uniform film thicknesses on every part of the object. Furthermore, the paint still tends to run and pull away from edges, and these are further specific drawbacks of conventional dip coating with water-borne paints.

The disadvantages of conventional dip coating become particularly apparent with intricate objects containing cavities. During the evaporation phase and the subsequent heating in the oven a saturated vapour atmosphere can build up in those cavities through which air cannot pass. As a result, it is possible for the fresh paint film to be washed off again by condensation because of the slower heating of cooler parts of the object.

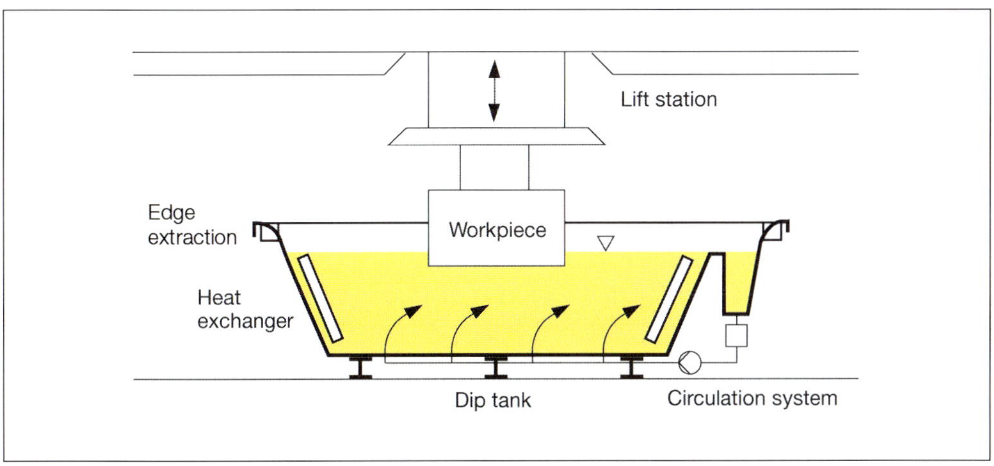

Figure 4.2.4: Schematic drawing of the conventional dip coating process

Application

The consequence of this is an incalculable weakening of the cavity protection which is impossible to monitor.

Electrocoating

More uniform coating of all parts of intricate objects without the risk of being washed out of cavities can be achieved on electrically conductive substrates by electrochemical deposition of ionic, aqueous secondary dispersions (see Chapter 2.1). If a direct voltage is connected across two electrodes, the charged paint micelles or dispersion particles in a diffusion-controlled interface move to the electrode of opposite charge, where they are deposited because of the change in pH value brought about by the electrolysis of the water. If the surface charge on the particles of the dispersion is positive, deposition takes place at the cathode, and if it is negative at the anode.

Although many studies and patents at the start of the 20th century were devoted to the subject of coating metals with resins, intensive development of commercial electrocoat paints really only got underway in the late 1950s when Ford started working on anodically deposited paints. The first plant for coating car bodies went on stream in USA in 1963, followed a short time later by one in Europe [4.4.14].

The development of cathodically deposited paints did not start until the mid-1960s [4.4.15]. These quickly proved to be high-quality primers and superseded anodic electrocoat paints in the mid-1970s, first in the USA and later in Europe and the rest of the world. Nowadays this system is exclusively used on automotive paint lines and is widely employed for other industrial applications.

Its advantages over its various predecessors such as spraying and dip coating are environmental compatibility by virtue of low emissions and, following the introduction of ultrafiltration, a high material efficiency > 99 % [4.4.16]. In addition, thanks to its good throwing power, it offers a winning combination of uniform coating of the interiors and exteriors of even intricate parts, and high process reliability.

The worldwide market for electrocoat paints in 2012 was estimated at 1400 kt. Some 60 kt were produced in Germany in 2012. They accounted for approx. 4 % of the total electrocoatings market.

Physico-Chemical Principles

Electrocoat paints are aqueous compositions of film forming agents, additives, pigments and extenders. The dispersion prepared from these typical paint ingredients is made to deposit an organic paint film on a conductive, generally metallic substrate by means of a direct current. Dispersions stabilised with a negative charge are deposited on the anode (anodic electrocoating or electrodeposition, AED). If they are stabilised with a positive charge, they are deposited at the cathode (cathodic electrocoating or electrodeposition, CED). To enable the paint to be deposited, the object to be painted is connected up as an electrode and is dipped along with the counter-electrode into the aqueous coating material. The paint is circulated to ensure a continuous supply of material for the interface, to dissipate the heat generated by the flow of current and ultimately also to prevent the generally pigmented dispersions from settling.

Figure 4.2.5 shows the chemical equations underlying dispersion in water and precipitation. Anodically deposited electrocoat paints require anionic carboxylic acid groups and

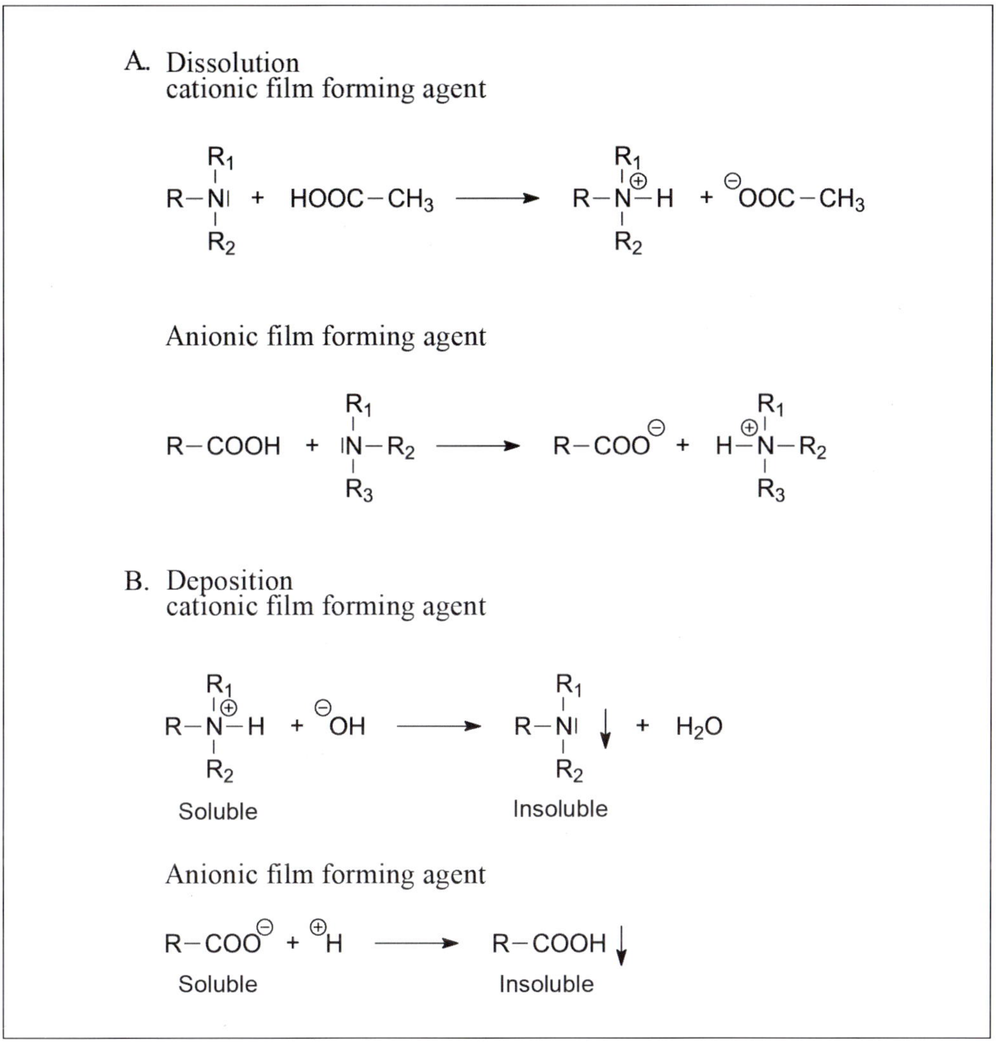

Figure 4.2.5: Dissolution and deposition reactions of electrodeposition resins

their reaction with amines or other bases, while their cathodically deposited counterparts need cationic groups such as amino groups and their reaction with acids. Film forming agents bearing different functional groups such as sulphonium or phosphonium are used only occasionally [4.4.17].

In accordance with the chemical reactions precipitation of the electrocoat paint takes place by means of deprotonation for cathodic deposition and protonation for anodic deposition. This occurs under the physical conditions of electrolysis of water in a diffusion-controlled interface on the substrate to be coated.

The presence of and necessity for a diffusion-controlled interface for successful deposition was demonstrated by Beck in experiments involving rotating disc electrodes [4.4.18]. For each paint system it is possible to identify a rotational velocity at which deposition of the electrocoat paint no longer takes place, irrespective of the voltage and current flow. To view it a different way, precipitation or deposition is the outcome of the forced migration of the paint dispersion particles into an area of greatly changed pH, as

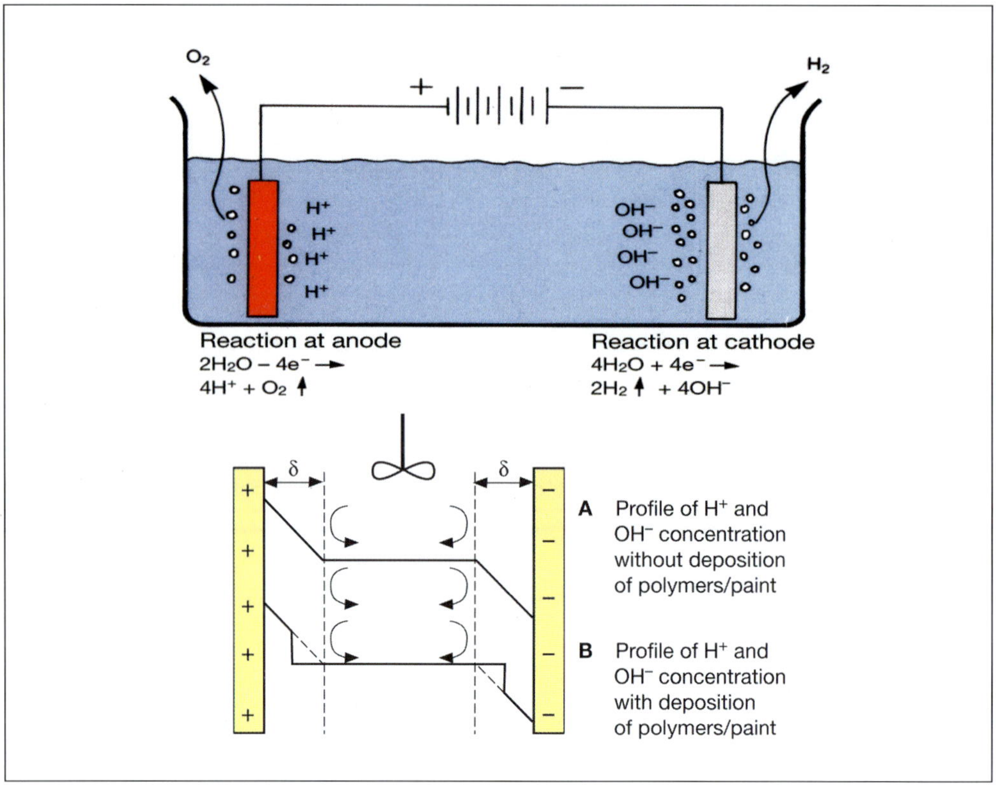

Figure 4.2.6: Electrode processes involved in electrocoating

shown in Figure 4.2.6 (A). In the case of coagulation the concentration curve of H^+ ions changes as shown in (B).

The pH at the electrodes can be estimated with the aid of simple assessments based on diffusion constants and the electrical current density in accordance with the *Sand* or *Cottrell* equations as at anodic deposition [4.4.19]:

$$C_{0_{H^+}} = \frac{j \cdot \delta}{2 \cdot D \cdot F}$$

j = Current density $(A \cdot cm^{-2})$
D = Diffusion constant of H^+-ions
F = Faraday constant
δ = Thickness of diffusion controlled layer

When appropriate data for the technical processes are substituted, the pH values at the electrode surface are found to be 2 for anodic and 12 for cathodic deposition.

As a result of the interplay between electrical attraction and retarding friction in the electric field of the diffusion-controlled interface, the disperse particles of the electrocoat paint, acting as charge carriers, migrate towards the substrate surface (electrophoresis) at a uniform rate. The film of water which adheres to the surface of the migrating particles and which contains some of the free-moving and retarding counter-ions partially shields the charges of the film forming agent particles. The still effective residual potential, known as the ζ potential, generates a migration rate u of 10^{-4} to 10^{-5} m/s. This is independent of the particle size and sufficient, given a correspondingly low viscosity, to generate film thicknesses of 15 to 20 μm in 1 minute.

$$u = \frac{\varepsilon \cdot \zeta \cdot E}{k \cdot \eta}$$

E = Dielectric field
η = Viscosity
ε = Dielectric constant
k = Constant
ζ = Effective residual potential

Owing to the insulating effect of the film being deposited, the applied deposition potential shifts to this film. Electric fields of approx. 10^5 V/cm form inside the deposited film because of the low film thickness [4.4.20].

Electro-Osmosis

The dispersion particles are neutralised before they even reach the electrodes and are thus precipitated. Accordingly, direct charge transfer does not take place at the electrodes. The originally mobile particles halt when they encounter the object to be painted. The electromotive force acting in the electric field causes the water to drift away, with the result that the transition from electrophoresis to electro-osmosis gives rise to dewatering and thus to a solid content in the film of more than 90 % directly on the substrate surface. As already mentioned, the film resistance increases as a result and the growth in film thickness slows.

The resultant film adheres firmly to the substrate, can no longer be washed off and has a characteristic morphology which is initially marked by blisters and pores. This morphology typifies electrocoat formulations and is heavily dependent on the viscosity properties [4.4.21].

Cathodically deposited films in particular exhibit these structures since twice as much gas is evolved at the cathode as at the anode.

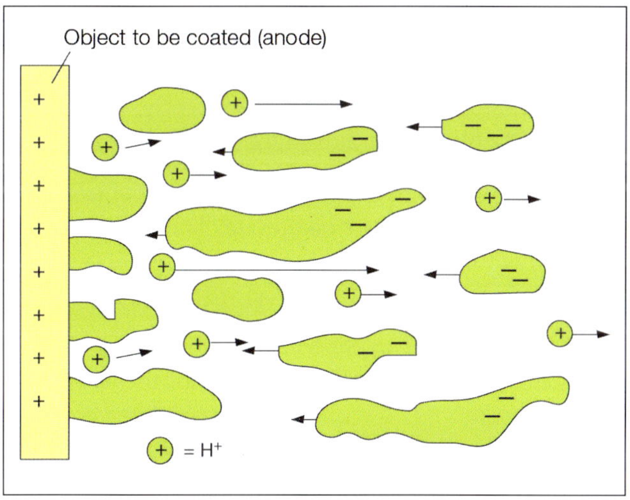

Figure 4.2.7: Dewatering of the electro-deposited coating by electro-osmosis

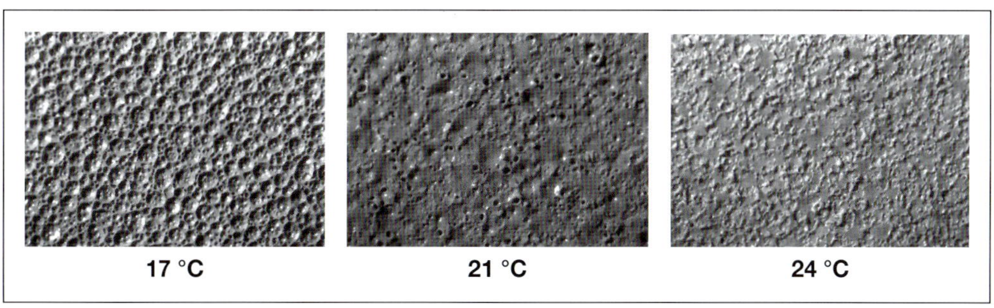

Figure 4.2.8: Morphology of cathodically deposited electrocoat films at different bath temperatures

Source: BASF Coatings

Secondary Reactions

A few secondary reactions take place at the electrodes during the electrochemical deposition of the polymer ions, and it is necessary to know what these are in order that practical problems with electrocoat tanks may be resolved. One electrode reaction worth noting is the dissolving of metals at the anode. Both the deposition of iron ions and the dissolving of amphoteric metals such as aluminium take place at the cathode, in accordance with the substrate material.

If metallic objects made of iron are being coated by the anodic process, it must therefore be expected that the object temporarily in the dipping bath will undergo some dissolution. Evidence of the incorporation of metal ions into paint films has been found. Conversely, in the case of cathodic deposition, the counter-electrode (anode) now permanently in the bath must be prevented from dissolving. However, it has also been found that measurable quantities of aluminium or zinc can pass into solution as anions in the strongly alkaline region of the cathode. Admittedly, the concentrations are significantly lower than those of the dissolved iron ions during anodic deposition [4.4.22].

Application

Anode: $Fe - 2e^- \longrightarrow Fe^{2+}$

Cathode: $Fe^{2+} + 2e^- \longrightarrow Fe\downarrow$

Mathematical Models of Deposition and Growth in Film Thickness

The dynamics of deposition are sufficiently well known as to permit the development of mathematical models for calculating the deposition properties of electrocoat paints, even on complex substrates [4.4.23, 4.4.24].

A certain time elapses until the OH^- concentration is reached at which the cathodic electrocoat paint starts to coagulate. The product of the current density j and the square root of this time t_i is called the specific paint constant k:

$$k = j \cdot \sqrt{t_i}$$

This leads to the following condition for the start of coating:

$$t_i = \frac{k^2}{j^2}$$

The ensuing phase of growth in film thickness ΔS_D can be calculated from the following equation:

$$\Delta S_D = \frac{1}{\ddot{A}} \cdot (j_t - j_d) \cdot \Delta t$$

ΔS_D = Film thickness (μm)
j_t = Current density at time t $(A \cdot cm^{-2})$
j_d = Lowest current density $(A \cdot cm^{-2})$
Ä = Electrochemical equivalence $(A \cdot s \cdot cm^{-3})$
t = Time (s)

If the current density j_t falls to values below j_d due to the increasing resistance of the depositing film, no further deposition takes place.

$$\frac{\Delta S_D}{\Delta t} = 0 \quad \text{at} \quad j_t < j_d$$

The current density j_t can be expressed by *Ohm's* law

$$j_t = (\kappa \cdot E)_t$$

where E is the electric field strength in V/cm and κ is the specific conductivity in S/cm. This yields a general equation for the growth in film thickness of

$$\int \frac{dS}{dt} = \frac{1}{\ddot{A}} \cdot ((\kappa \cdot E)_t - j_d)$$

In this equation the specific conductivity is composed of the conductivities of the bath and the film which is forming. Whereas the first is constant and easy to measure, the conductivity of the film is included as a function of the film thickness and the current density j_t. For calculations this is usually determined by measurements [4.4.24].

Another approach to calculating the growth in film thickness after *Machu* [4.4.25] is based on the simple assumption that the reduction in the free surface area dA is associated with the electric charge dQ. The quotient a of the surface area and electric charge is defined by *Machu* as the "electrophoretic equivalent."

$$a = -\frac{dA}{dQ} \left[\frac{cm^2}{A \cdot s}\right]$$

Since dQ = I dt and, according to *Ohm's* law U = I · R and R = ρ · l, it follows that:

$$\frac{dA}{dt} = -a \cdot \frac{U \cdot A}{\rho \cdot \ell}$$

$$\int_{A_0}^{A} \frac{dA}{A} = -\frac{a \cdot U}{\rho \cdot \ell} \int_0^t dt$$

ρ = Specific film resistance

ℓ = Distance between electrodes

Given the proportionality of the current I to the area A, the solution to this differential equation is:

$$I = I_0 \cdot e^{-\frac{a \cdot U}{\rho \cdot \ell} \cdot t}$$

The resultant current/time curve for the deposition of an electrocoat paint applies only to monomolecular films, but includes the major variables which determine film thickness, viz. film resistance, voltage and electric charge.

Physical Characteristics of Electrocoats

Standard deposition experiments in the laboratory are tracked via the current/time curve, which is then used to determine the integral characteristics. Schematically, all depositions more or less follow the dynamics shown in Figure 4.2.9.

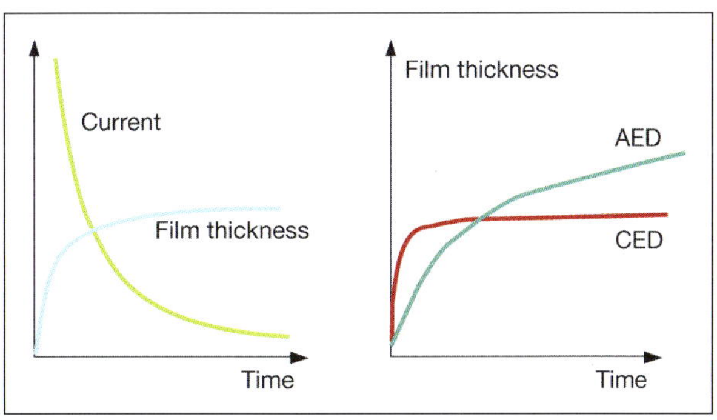

Figure 4.2.9: Schematic current/time and film-thickness/time curves of cathodically and anodically deposited electrocoats

The rise in the resistance associated with the build-up of the film and the reduction in the current flow mean that electrocoat paints of relatively uniform film thickness can even be achieved on intricately shaped objects. The ratio of the film thickness of the internal to the external areas of intricate bodies is termed the throwing power. Throwing power properties are studied in special test configurations [4.4.26] such as the BMW throwing power system to VDA 621-180.

By way of a guideline electrochemical characteristic the electrochemical equivalence Ä is determined in an electrical measuring system as the ratio of consumed electric charge to the mass of deposited electrocoat paint:

$$\text{Ä} = \frac{Q}{m} \left[\frac{A \cdot s}{g}\right]$$

The electrical energy W, e.g. for coating a car body, can be calculated with the aid of the electrochemical equivalence and the electrical conditions. Usual values are a voltage U of 350 V, a mass m of 2000 g for the electrocoat film and an electrochemical equivalence Ä of 40 C/g. Thus:

$$W = U \cdot I \cdot t \quad \text{if} \quad \text{Ä} = \frac{I \cdot t}{m} \quad \text{then} \quad W = U \cdot \text{Ä} \cdot m \quad [kWh]$$

Application

Substituting the given values yields a figure of 7.8 kWh. This fits in well with the actual figure of 5 to 8 kWh/body for most coating lines.

From the results of the laboratory-scale experiments described earlier the film resistance is determined at the end of the deposition process by substituting the measured bath resistance ρ_b into the corresponding equation based on *Ohm's* law as described:

$$I_E = \frac{U_E}{(\rho_b + \rho_s) \cdot S}$$

The specific resistance ρ_s of the coating S of the stoved film gives an indication of the potential throwing power of the electrocoat paints.

The most important physical parameters governing the deposition properties of electrocoat paints, apart from the applied voltage and the current density, are the bath temper-ature and the solid content. As far as the voltage and the current density are concerned, there are limit values at which deposition no longer occurs homogeneously over the entire surface area of the object; instead breakthroughs occur in the forming film as a result of the high temperature generated. These breakthroughs can adversely affect the state of the electrocoat paint in the form of both coagulation and destruction of the dispersion particles, with the result that deposition does not yield smooth and

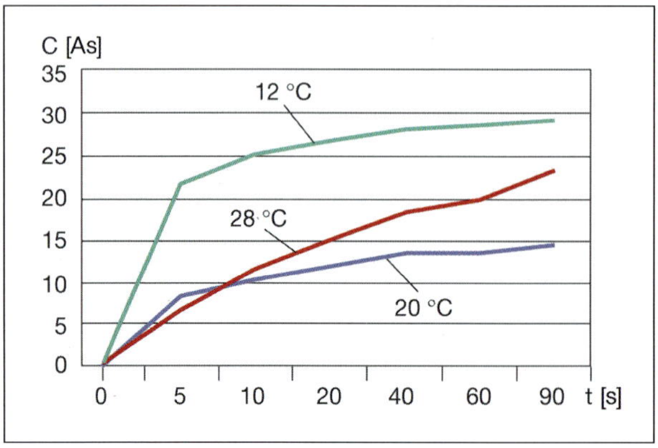

Figure 4.2.10: Coulomb/time curve of a cathodic electrodeposition coating at different bath temperatures

usable films. Experience shows that voltages determined in the laboratory should be 20 % lower in practice, as otherwise the defects just described may occur. The dependence of the coating process on the bath temperature is very high [4.4.21]. Figure 4.2.10 illustrates how the electric charge in coulombs varies with the coating time at different temperatures. The consumed charge C is proportional to the growth in film thickness. In this example the optimum bath temperature is around 20 °C: below that, the film thickness is too high; above it, the throwing power deteriorates. The specific paint formulation is responsible for this. Recent studies confirm that the glass transition temperature T_g of the relevant resin (see Chapter 3.2) plays an important role in the deposition mechanism of cathodically deposited electrocoat paints [4.4.27].

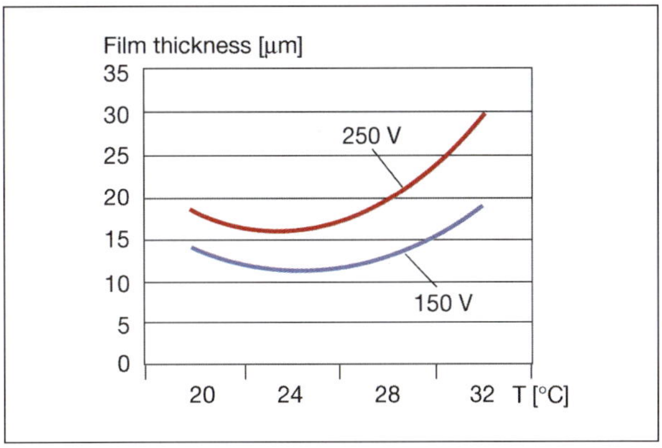

Figure 4.2.11: Change in film thickness as a function of solid content of a cathodic electrodeposition coating

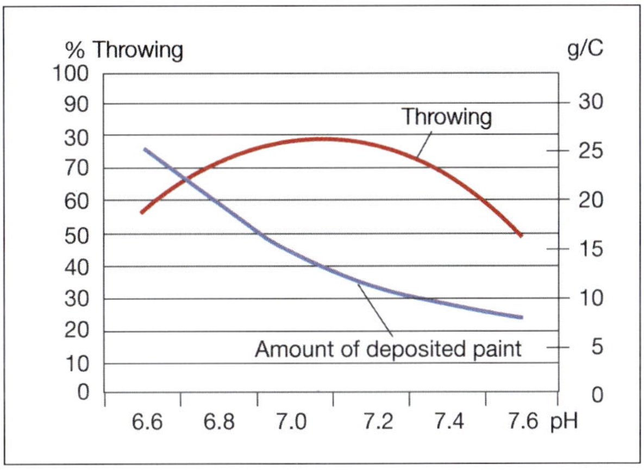

Figure 4.2.12: Change in deposited mass (g/C) and in throwing power versus degree of neutralisation (pH) of an anionic electrodeposition coating

Apart from the process-engineering variables, the composition of the electrocoat bath plays a key role in the deposition prop-erties. This concerns mainly the solid content, the neutralisation level of the dispersion, the specific bath conductivity and the solvent content.

Electrocoat paints with a low solvent content exhibit only a slight dependence of the growth in film thickness on the solid content. The throwing power is always improved by an increase in bath conductivity. Limitations on increasing the solid content result from the high sensitivity of the ultrafiltration performance of all commercial membranes to excessive solid contents in terms of both output and service life.

Raising only the neutralisation level of the electrocoat paints reduces the film thickness for the same application voltage [4.4.28]. The voltage increase which is therefore needed if the standard film thickness is to be achieved increases the throwing power. This is additionally boosted by the higher bath conductivity. Limiting factors here are increased evolution of gas and heat which can lead to electrical breakthroughs and a lower electrochemical yield, which is indicated by the increase in the deposition equivalent Ä in C/cm^3.

Parameter	CED	AED
Solid content (2 h/130 °C) in [%]	15 – 22	8 – 15
Pigment content (450 °C) in [%]	1 – 5	2 – 7
Solvent content (450 °C) in [%]	0.5 – 3.0	4 – 10
pH	5 – 7	7 – 9
Spec. conductivity (mS/cm)	1.5 – 2.5	2 – 3

Figure 4.2.13: Typical bath parameters for electrodeposition coatings (CED = cathodic; AED = anodic)

In the early days of anodic electrocoating technology only single-component paints were used, but today's electrocoating systems, mostly of the cathodic type, employ two components.

To maintain a uniform coating result, it is necessary to replenish the material consumed in coating the object with new material. There are two basic processes:

- Replenishment with a generally one-component, underneutralised electrocoat paint
- Replenishment with a generally two-component, neutralised electrocoat paint

In the first case the replenishment material is dispersed in the electrocoat paint by means of dispersers in a sufficiently large premixing vessel. The excess neutralising

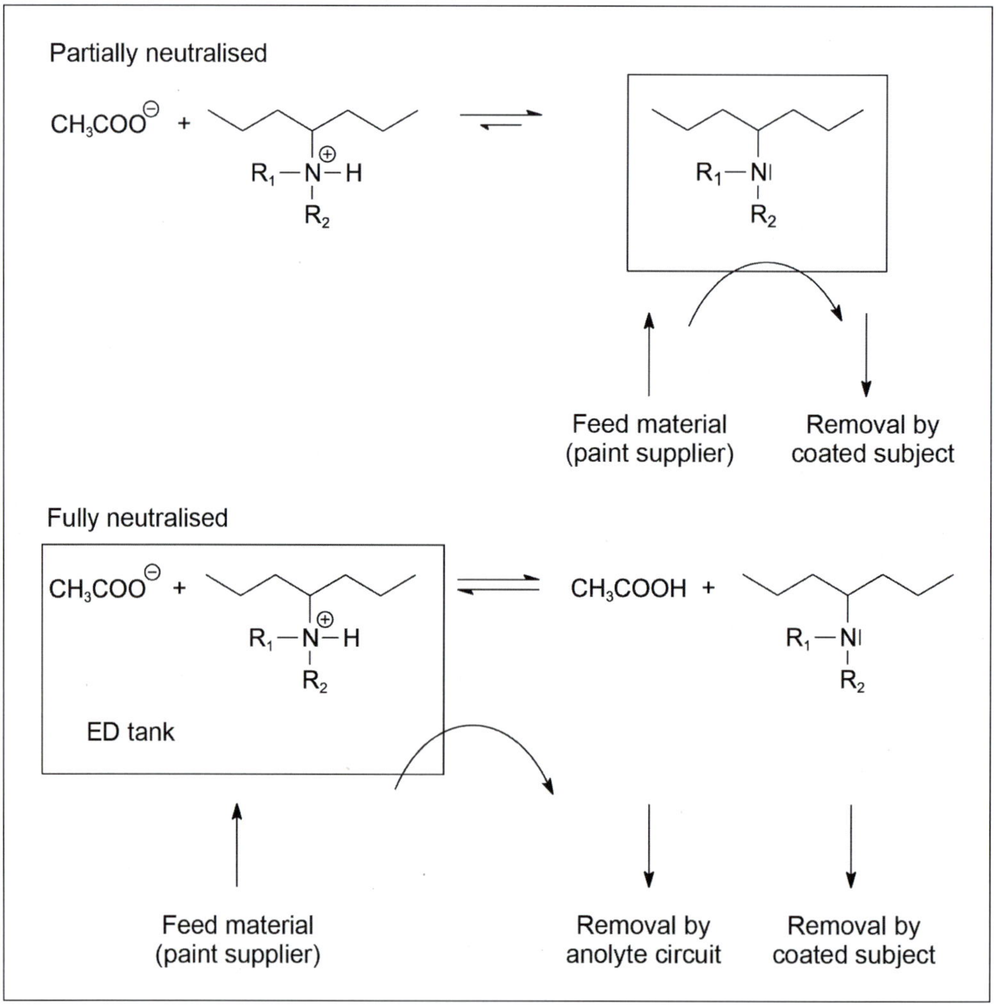

Figure 4.2.14: Acid-base equilibrium of the various replenishment methods

agent from the bath is generally sufficient to achieve this. The disadvantage of this process is that there is no simple way of monitoring the level of dispersion of the replenishment material. This replenishment method run smoothly if empirically determined process conditions are maintained.

The second replenishment method enables a considerably more stable and defined state to be achieved in the electrocoat tanks. This entails separate additions of a fully neutralised binder dispersion and a paste dispersion to the bath at an appropriate place, and does not require a dispersion step. A number of advantages result: only very minor fluctuations in the state of the bath since predispersed material is added, as well as automated, continuous dispensing as a function of electrical power consumption. Furthermore, a low-solvent replenishment material can be added which has only approx. 20 % of the solvents usually contained in 1-component replenishment materials.

However, the replenishment method with fully neutralised material requires a further circuit for extracting the excess neutralising agent which forms during deposition from the paint system. This circuit is known as the anolyte circuit in the case of cathodically deposited electrocoat paints and as the catholyte circuit for the anodic types. Furthermore, the replenishment method with fully neutralised material requires a large supply of material.

The volumes consumed determine the mode of transport and delivery: road tankers or, as is usual in the case of automotive paint lines, special 1-tonne steel vessels or 200-litre drums. Delivery by tanker requires the provision of correspondingly large storage tanks at the electrocoating lines.

Plant Engineering

Electrocoat paints are used in tanks ranging from 0.1 to 500 m^3 in volume. The largest tanks are found in the commercial vehicle industry, the smallest at job coaters or companies which have to coat small components and quantities using manual systems. The largest tanks are continuous systems, the most productive of which can handle 70 car bodies per hour (see Chapter 7.1).

Batch systems, on the other hand, handle about 15 units per hour, though are only approx. 30 % of the tank size.

Tank size plays an important role in the rate at which the bath material is replenished. A single theoretical replenishment is termed a turnover. For CED this should not be substantially longer than 12 months since the quality of the coating is no longer guaranteed for reasons of declining stability. This period is usually shorter for anodically deposited paints.

Figure 4.2.15: Examples of electrocoated industrial goods

Source: BASF Coatings

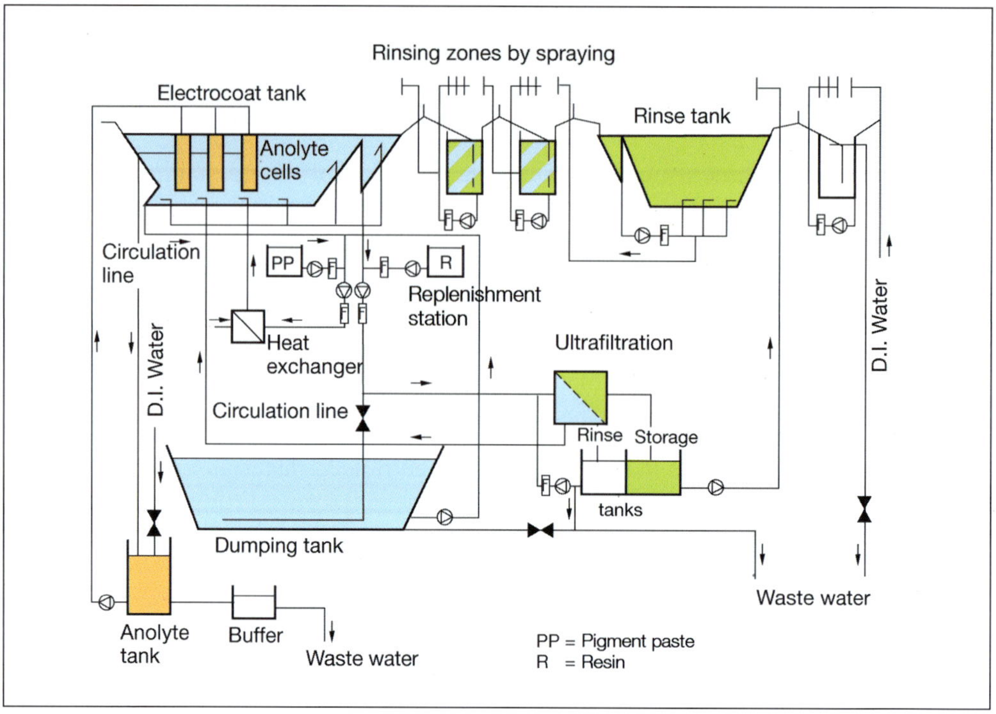

Figure 4.2.16: Schematic drawing of the cathodic electrodeposition process

Special steel or plastic materials are used for CED systems. Cold rolled steel can be used for the slightly alkaline anodic electrocoat paints.

Apart from the tank made of plastic-coated steel and the paint-supply vessel an electrocoating system for cathodic deposition requires the following plant elements:

- Power supply
- Heat exchanger
- Ultrafiltration system
- Filters
- Paint-dispensing (replenishment) system
- Anolyte circuit
- Rinsing zones

Deposition of electrocoat paints requires rectified **electric current**. The voltage is programmed to avoid excessive current densities which can cause coating defects and can even ruin the paint (see Chapter 6.2.2). The programme reduces current peaks at the start of the deposition process by setting the voltage to a lower level than in the later stages of the coating process. In batch systems this can easily be achieved by programming the voltage in one rectifier. Continuous systems, where there are several objects in electrocoat tanks at the same time, require at least a second rectifier. This enables a separate voltage to be variably programmed for the entry area. In less demanding processes it is possible to work with one rectifier and electrode-free entry zones because the greater bath resistance which can be achieved in this way can reduce the current peaks.

In the case of cathodic electrocoating the workpiece is connected up as the cathode and is at zero potential, as shown in example A of Figure 4.2.17, i.e. it is earthed. The anodes are connected up to the plus potential. The voltage programme with at least 2, generally 3, rectifiers is run via the anode potential. Example B shows a sectionalised rail where the voltage is generated via the cathode potential.

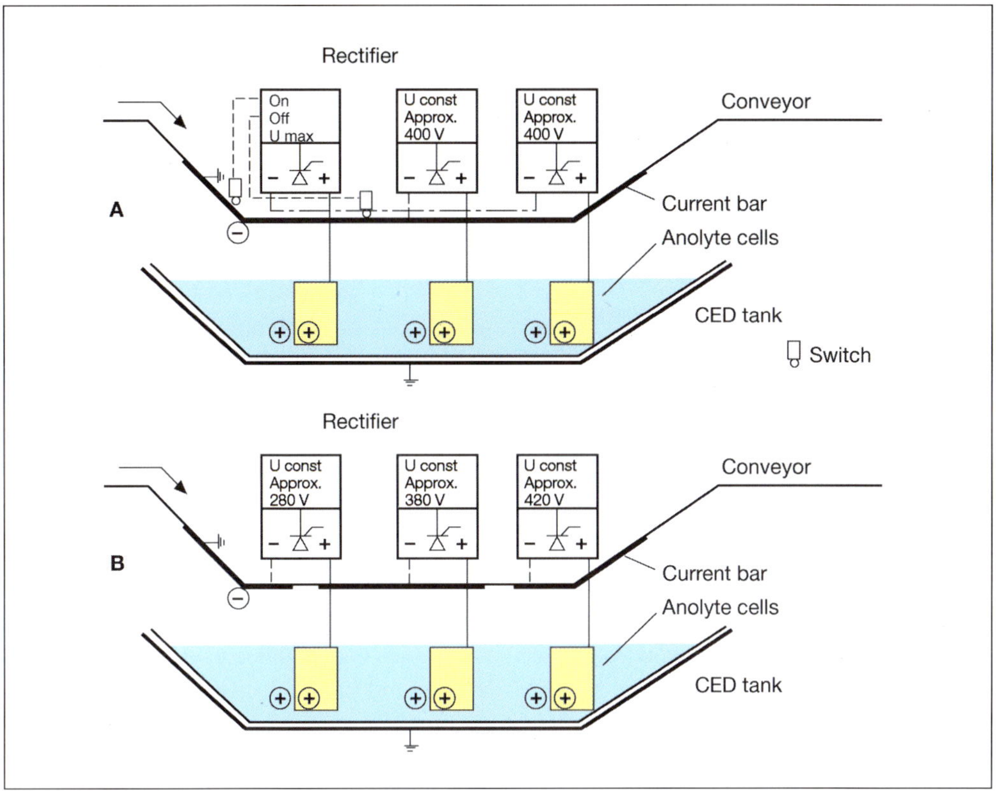

Figure 4.2.17: Principles behind the electric circuit for a continuous cathodic electrodeposition system

In industrial systems the direct current is usually supplied by thyristor rectifiers. For a good coating result, the residual ripple of the generated direct voltage should not exceed 5 %. The coating voltages are between 300 and 450 V for a coating time of between 2 and 4 minutes. The bath contents are at a temperature between 25 and 35 °C.

Several circuits with pumps are set up to operate the **heat exchanger**, the ultrafiltration system and the filters. These pumps maintain flow in the bath at the same time, thereby ensuring that sedimentation of the electrocoat paint is avoided. The jet nozzles are so arranged in the bath that no flow dead spots occur. The mean flow rate relative to the free cross-section of the bath should not be less than 0.3 m/s. This means that the tank volume is circulated 4 to 6 times per hour. Normally the direction of flow at the surface of the bath is in the direction in which the objects pass through the bath. The bath level may not be less than 20 cm above the objects in order that the necessary current flow may be guaranteed. The flow rate in pipes, as in the bath, should not be less than 0.3 m/s.

The pumps for conveying the paint should be fitted with double-acting mechanical seals which are feeded with the ultrafiltrate (see below) generated from the paint as the sealing liquid. Immersion pumps can also be used. Tungsten carbide/tungsten carbide is a suitable combination of materials for the seal surfaces. Filters should be installed upstream of the pumps in all paint circuits.

Part of effective filtration is careful selection of the extraction locations in the electrocoat tanks. It is definitely recommended that a partial flow be extracted from the contaminated entry area.

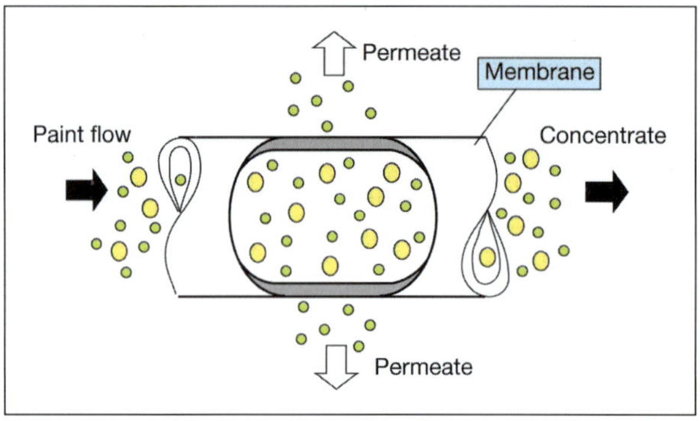

Figure 4.2.18: Schematic drawing of the ultrafiltration process

The bath temperature (see above) plays an important role in the application process. The bath quickly heats up because of the electrical energy input during the deposition process and the frictional heat of the circulating pumps and therefore has to be cooled down. Deviations of more than 1 °C from the standard temperature specified by the paint manufacturer have an impact on the film thickness and the surface quality of the stoved film. The tempering system must be designed to maintain the recommended bath temperature even under the most unfavourable conditions of a hot summer's day and also in winter.

The high material yields of the electrocoating process are achieved by **ultrafiltration**. This is a separation process by means of which low-molecular constituents are removed from a dispersion or colloidal solution. For electrocoat paints these constituents are, of course, water, organic solvents, electrolytes and low-molecular fractions of the binder. The process involves having the electrocoat paint pass a suitable membrane at high velocity. The concentration gradient between the inside and the outside and a pressure gradient of several thousand hPa are responsible for causing separation at the ultrafiltration module membrane. The separation threshold, expressed in terms of the molecular weights of dissolved constituents and the binder dispersion, varies from 0.0015 to 1.0 μm in accordance with the membrane system.

The flow of liquid which passes through the membrane is termed the permeate. The specific yield (or flux) is expressed in terms of the surface area and membrane and given in $l/(m^2\ h)$. The flux rate is a function of the membrane surface area, pore size, flow rate at the membrane surface, concentration gradient and pressure difference.

The two most important variables for a high yield of permeate are the membrane surface area and the flow rate of the electrocoat paint at the membrane surface. On the one hand, they maintain the concentration gradient at high throughputs and on the other hand, they retard the formation by paint deposits of a diffusion-inhibiting interface at the membrane surface. As the membrane slowly becomes clogged with dirt, unstable dispersion particles and pigments, the process becomes less regular, and a slow reduction in the flux rate is normal. The performance can be restored to almost its original level by cleaning with a suitable rinsing agent.

The ultrafiltrate generated is fed as rinsing agent on the counter-current principle into the last spray head or the rinsing tank. The discharging rinsing agent is generally returned to

Value	Dimension	CED	Ultrafiltrate
Appearance		Opaque	Clear
Solid content (2 h/180 °C)	%	20 – 22	0.3 – 0.6
Spec. conductivity (20 °C)	mS/cm	2.0	1.8

Figure 4.2.19: Physical data of the ultrafiltrate compared with the coating

the bath material via two to three further rinsing zones. The efficiency of the paint material depends, of course, on the amount of ultrafiltrate input into the rinsing zone. In rare cases, e.g. in the event of excessive contamination of the electrocoat paint by electrolytes, the ultrafiltrate is discarded and replaced by demineralised water.

Ultrafiltration systems were developed with the aim of obtaining the largest volume of ultrafiltrate per hour from the largest possible membrane surface area in the smallest possible space and with the smallest possible quantities of paint. The membrane modules play an important role in this. These can be classified as simple tubular modules, hollow-fibre modules, spiral modules and plate modules. Tubular modules have a diameter of several centimetres. The permeate is collected between the membrane and a plastic housing and is discharged to the outside. The advantages of such a module lie in high flux rates and simple mechanical cleaning. Hollow-fibre modules consist of a bundle of thin, tube-like membranes, numbering up to 90 units in a single housing. The internal diameter of the tubes is only 1.0 to 1.5 mm. Here, too, the resulting permeate is collected by a plastic sleeve and discharged to the outside. The advantages are a high membrane surface area per unit volume.

Spiral modules are designed such that two flat membranes are welded on three sides like an envelope. Owing to this design, they are also called winder modules. A porous medium between them absorbs the permeate. The end of this carrier material is fastened to a perforated manifold. The entire membrane/carrier system, along with a spacer, is wound up. The paint flows between the membranes which are separated by the spacers. The permeate passes through the membrane into the carrier material which drains tangentially into the perforated manifold. As with hollow-fibre modules, the advantages are a high membrane surface area per unit volume.

In the plate modules a stack of porous plates coated with a membrane are mounted in a frame. The paint flows through the interstices while the permeate passes into the porous plates and is discharged from there to the outside. The advantages of this module are low replacement costs for the membranes and low maintenance costs.

The paint to be filtered must not contain any contaminants whose diameter is the same or greater than the film thickness if defect-free electrocoat surfaces are to be achieved at a film thickness of 20 to 25 μm. As the paint ingredients are up to 5 μm in size, the pores in the **filter** bags must meet certain size specifications. These are usually in the order of 25 to 50 μm. The preferred filter material is polypropylene in a needle-felt finish. The filter bags are placed in special steel baskets, and these in turn are placed in special-steel filter vessels holding 2 to 8 such baskets. The electrocoat paint flows through the filter system from top to bottom. As a rule, all circuits undergo filtration. Correct operation of the filters can be verified by monitoring the pressure difference between the input and output pressures of the filter vessels. This difference should not exceed 1.5 bar. To ensure that the filter bags do not split, a pressure-relief system is needed for bypassing individual filter vessels in the event of a sudden rise in pressure. These bypasses must be rinsable, as the remaining electrocoat paint would sediment in the event of a fault. Under normal operating conditions a filter change should be carried out approximately every week as part of the general maintenance schedule.

Causes of contaminants in electrocoat paints can include:

- Entrainment of dirt from the unpainted body
- Dirt from the periphery of the CED system
- Abrasion and dirt from the conveyor system
- Dried-on paint from the hanger
- Coagulation and deposits of the electrocoat paint
- Electrical breakthroughs/high peak voltages
- Contaminated demineralised water
- Formation of coagulate due to shear forces, e.g. in defective pumps

Small quantities of specific surfactants make for an extremely unpleasant type of contaminant. To an extent depending on their hydrophobic nature (HLB value) and the surface tension, silicone oils and perfluorinated hydrocarbons at a concentration of ppb to ppm can cause cratering either in the electrocoat film itself or in the primer surface in a downstream coating stage. These contaminants are removed with the aid of special surfactant filters. Owing to the weak hydrophobic interaction of the filter medium with the contaminants, filtration under minimum pressure is the most effective removal method. Such filtration should be operated outside the normal circuit.

Application

An **anolyte circuit** is needed for cathodic electrocoat paints and replenishment with two-component neutralised material in order that the acid released during deposition may be removed from the paint bath. Otherwise the acid/base equilibrium would be lost. The resulting low pH leads to poor paint deposition, one of the reasons being the rise in the specific electrical conductivity.

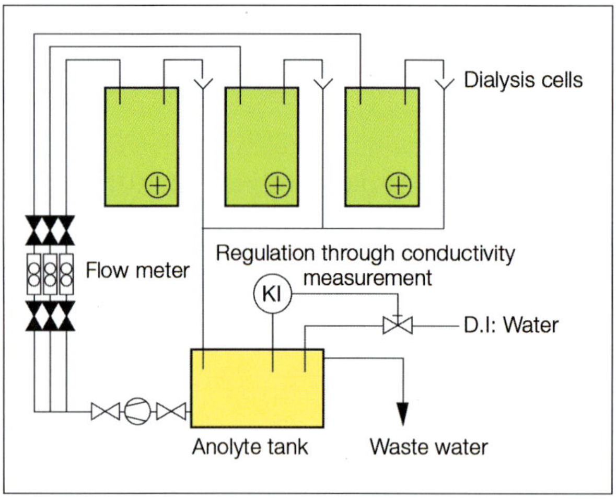

Figure 4.2.20: Schematic drawing of the anolyte circuit

Property	Dimension	Value
pH		2.2 – 2.8
Spec. conductivity (20 °C)	mS/cm	0.3 – 2.0
Solid content (2 h/130 °C)	%	< 0.2
Area ratio: Anode:cathode		1:4 – 6
Flow rate in anode cell	l/min	4 – 15

Figure 4.2.21: Process data for an anolyte circuit

There are two characteristic features of the anolyte circuit: first, the semi-permeable anion exchange membrane in front of the anode and, second, the water supply which is controlled via a conductivity measurement system.

The anion exchange membrane separates the box in which the anode is housed from the electrocoat tank. It ensures that only the anions necessary for ensuring electrical neutrality can pass through the membrane in the direction of the anode when current is flowing and the electrocoat paint is being deposited. If no voltage is connected, it acts as an electrolyte barrier.

The acid concentration is controlled by means of a conductivity measurement system and a control system which lowers the

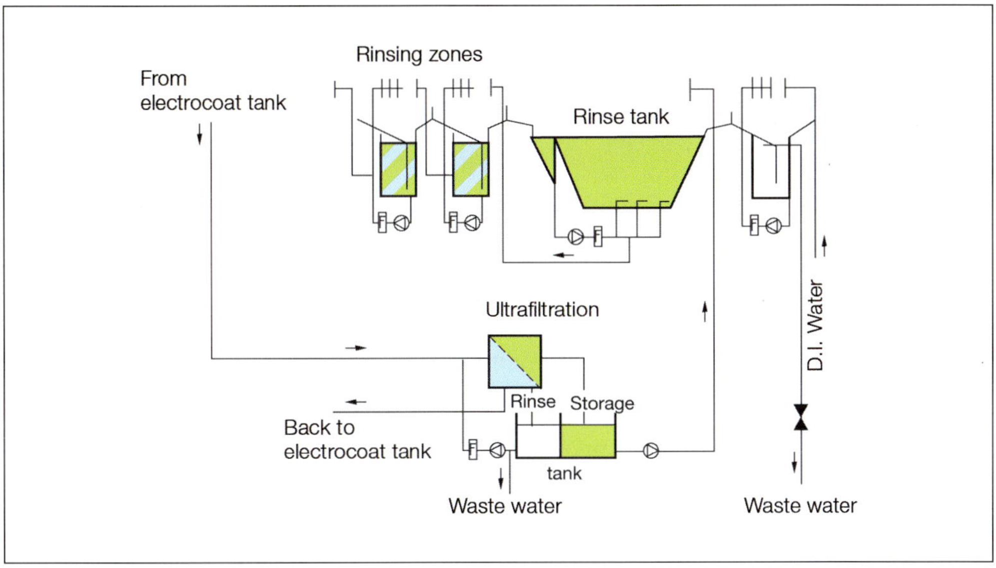

Figure 4.2.22: Schematic drawing of a rinsing zone with a rinse tank

concentration by feeding demineralised water to the anolyte tank. Concentrated acid is discharged via an overflow. In most cases this is acetic acid, and occasionally lactic and formic acid, which are readied for disposal by simple neutralisation.

The anolyte circuit is made of plastic, usually PVC, because the anolyte liquid has a low pH. The return lines of the anode cells should have a transparent section so that flow can be checked. The circulation pump must be made of acid-resistant special steel, and the pipework must be earthed.

A special type of stainless steel which is resistant to anodic dissolution in the acidic, slightly chloridic medium is required for the anode panels within the anode cell. Suitable types are special steels to DIN 1.4401 and DIN 1.4439. The anodes lose weight as a function of the electric charge passing through them. The usual service life in automotive systems varies with the position in the bath but is between two and four years.

Where paint enters the anolyte circuit due to leaks, material is deposited on the inside of the membrane and significantly impedes the current flow. This manifests itself in a reduction first in throwing power and then in the thickness of the deposited film. Production must then be halted immediately, the leak repaired and new membranes fitted, a fact which obviously causes serious production shortfalls. For this reason, turbidity sensors need to be installed in the control circuit for the sake of increasing operational reliability.

The electrocoat paint deposited on the object removed from the bath adheres firmly to the substrate, has a marked texture or is porous or both (see Figure 4.2.8). On its surface is a liquid film consisting of bath material and entrained gas bubbles. Unless this film is promptly removed, serious surface defects will occur after initial drying and cannot be eradicated during stoving. Furthermore, in the event of the complete failure of a rinsing zone, up to 60 % of the material would be lost (expressed in terms of the deposited electrocoat paint). The **rinsing zone** thus serves to optimise the surface quality and the material transfer efficiency and is charged with ultrafiltrate in accordance with the counter-current principle.

The rinsing zone generally has 5 or 6 individual units:

1. Rinsing of the surface shortly after removal from the bath
2. Zone I - Spraying
3. Zone II - Spraying
4. Zone III - Dipping
5. Spraying with pure ultrafiltrate

Optional: 6. Spraying with demineralised water

The last unit, spraying with demineralised water, which was still in use in the 1990s, has now been largely eliminated, as the ultrafiltrate quality has been improved so much through optimisation of the membranes and the electrocoat paint constituents that it very closely approximates the state of demineralised water. This permits costs to be cut.

The spraying process in unit 1 is important for ensuring a defect-free, smooth surface in the electrocoat paint. It must be carried out less than one minute after removal from the bath. This is done with spray nozzles which are installed above and to the sides of the object to be painted while it is still above the outlet area of the electrocoat bath.

Application

Then come three rinsing zones which have different paint concentrations to reflect the input of the electrocoat paint adhering to the object and are connected in a cascade towards the electrocoat bath.

The pure ultrafiltrate is fed into the last rinsing zone before the demineralised-water spray head. In some cases, part of the pure ultrafiltrate is used for the first rinsing which takes place while the object is still above the electrocoat bath.

The solid content in the rinsing zones varies with the quantity of ultrafiltrate supplied and the coating quantity, but is usually 1 to 2 % in rinsing zone I and 0.5 to 1.0 % in rinsing zone III. The rule of thumb for designing the ultrafiltration system is that there should be 0.8 to 1.5 l ultrafiltrate per m^2 surface to be coated. The contents of the individual rinsing stages are termed Recirculate I (Zone I) and Recirculate III (Zone III) in accordance with their position in the rinsing process.

The automotive industry now uses dip tanks almost only for rinsing zones II and III because of the rinsing process is more complete. If a spray head or tank with demineralised water is installed as the last unit, this material is not fed to the electrocoat tank via the recirculate stages. It must be disposed of. This is generally performed with the aid of precipitation agents in the systems themselves. Systems designed with such cascade rinsing units reach a material transfer efficiency of greater than 98 %.

Process Parameters

The process is monitored by periodically determining both analytical bath data and application figures. The cycle of different tests performed depends on the paint quality and the throughput rate (see Chapter 7.1).

The key analytical data include the solid content, the binder/pigment ratio or the pigment concentration as ash content, the pH, the specific conductivity and the solvent content of the bath material. These figures are primarily used for quality control of the replenishment material. The measurement methods are usually based on the pertinent ISO or ASTM standards (see Chapter 2.3).

Daily checks on the bath contents are limited to the solid content, pH value and specific bath conductivity. The physical deposition data determined under defined physical conditions confirm the constancy of the bath state on one hand and the deposition properties such as the throwing power on the other. These figures are more sensitive to fluctuations in the bath state than are the analytical data.

Overview of Binders and Paint Formulations

To make it possible to deposit film forming agents **anodically**, it is necessary as described above to incorporate negatively charged ionic groups into the matrix of the film forming agent. The carboxylic acid group is most frequently used for this. Conversion into the salt form with alkaline or amines (see Figure 4.2.14) can render the film forming agent dispersible in water. The desired properties of the coating such as anticorrosive protection, smooth surfaces and gloss are essentially determined by the chemical structure of the film forming agent and the crosslinkers. Polyacrylics and polyesters are the preferred film forming agents for top coats. Oils, alkyd resins, polybutadiene resins, phenol resins and epoxy resins serve as the base material for primers [4.4.29]. The first anodic electrocoat paints of significance were unsaturated oils which were made to react with maleic anhydride (see Chapter 2.1) [4.4.30]. The main raw material is linseed oil which is converted with 10 to 30 % maleic anhydride.

Application

The modified resins used in electrocoat paints have molecular weights of 1,000 to 10,000 g/mol. They are crosslinked at temperatures of 140 to 200 °C. Crosslinking is generally effected by esterification with phenol resins or resins rich in OH groups. Oxygen-induced polymerisation of the carbon double bonds for crosslinking the unsaturated film forming agents also plays a role. In the past this reaction was catalysed by the addition of lead compounds. To improve the anticorrosive protection, crosslinking is reinforced with phenol resin precondensates [4.4.31].

Oligobutadienes, also termed polymer oils, superseded the natural raw materials in the mid-1970s. They are easier to tailor optimally to the intended application by means of targeted synthesis processes, and they are resistant to saponification. The use of oligomers with cis double bonds is preferred because of their greater reactivity during maleinisation [4.4.32]. Apart from 1,4-butadiene, copolymers of various dienes such as isoprene, cyclopentadiene and 1,3-pentadiene, sometimes in combination with natural oils, are used to a lesser extent [4.4.33].

Unsaturated polyesters and alkyd resins can be made water-soluble by means of high acid values. They only find limited application nowadays. Saturated, acid polyesters have a certain importance for the steel furniture industry where they act as particularly smooth direct-gloss electrocoat paints. They are accordingly easy to manufacture and to crosslink with resins rich in OH groups.

Epoxy resins can serve as diols in the synthesis of polyesters or be converted directly as difunctional epoxides with polyfunctional carboxylic acids. The carboxylic acid function is introduced by making unsaturated fatty acids react with maleic anhydride [4.4.34]. Such resin-based electrocoat paints offer good anticorrosive protection. They have therefore been in use in the automotive industry for some time.

One simple means of synthesising anodic resins is to copolymerise acrylate, methacrylate or styrene with free acrylic or methacrylic acid. The latter must be present in a proportion of 5 to 20 % in the polymer. Crosslinking can be carried out via amino resins or phenol resin modifications if OH groups are present in the acrylic resin chain. Alternatively, self-crosslinking systems can be formulated by incorporating methylol groups into the acrylic chain [4.4.35]. Transesterification is another important crosslinking mechanism [4.4.36].

In general, acrylic resins are suitable for topcoats that offer acceptable fastness to light. Their anticorrosive protection is significantly inferior to that of electrocoat paints based on epoxy resins.

Anodic electrocoat paints based on polyamides, polyimides and polyester imides are also described in the literature, though they have never achieved practical importance [4.4.37].

Cathodic electrocoat paints are formulated on the basis of film forming agents which are modified by the incorporation of positively charged functional groups. The most important is the amino group which can be primary, secondary, tertiary and quaternary. Phosphonium, isothiouronium and sulphonium groups in resins have also been investigated. In principle they yield depositable coating materials. With the exception of the sulphonium group, however, they have not attained any practical prominence [4.4.38, 4.4.39].

Amine-modified epoxy resins, which deliver the best anticorrosive protection and excellent adhesion to metals, are important in practical applications. Over 90 % of them serve as primers in the automotive industry (see Chapter 7.1). The amino function can be achieved by converting free epoxy groups with amine or ammonium salts, starting from pure or polyester-modified epoxy resins [4.4.40]. Subsequent neutralisation of the amino groups with organic acids such as formic, acetic or lactic acids makes the resins dispersible in water. The commercially important electrocoat paints manufactured on this basis contain dispersed crosslinking agents derived from capped isocyanates [4.4.41]. For a while, an epoxy-modified phenol resin which was converted into water-dispersible, self-crosslinking resins in a *Mannich* reaction with secondary amine and formaldehyde enjoyed some importance. The *Mannich* group separates thermally from the amino group at approx. 190 °C and thus leads to crosslinking [4.4.42]. Such paints enable good anticorrosive protection on bare steel.

As with the anodic paints based on acrylic resin, cathodic acrylic resins can be manufactured with acrylate monomers containing amino groups, e.g. dimethylaminoethyl methacrylate, as polymerisation components. In addition, monomers containing hydroxyl groups or epoxy groups can be modified. Crosslinking, mostly to form direct-gloss, relatively weather-resistant coatings, is generally achieved with capped, multifunctional isocyanates [4.4.43]. Binders based on alkyd resins can either be functionalised to epoxy groups with the aid of peroxides or can be rendered amenable to cathodic deposition by making maleic anhydride react with diamines. These resins have not achieved industrial importance.

Variants on Electrocoating

In addition to straightforward electrocoating, painting processes have been available since the 1970s in which electrocoating is used to provide a complete paint system. A rinsed, as yet unstoved electrocoating can have a second electrocoat of the same or a different paint applied to it. The result is a total film thickness of up to 60 µm. This process is used in some industrial-scale applications.

Stoved CED films can also be coated a second time if, following suitable pigmentation after stoving, they possess the right film conductivity [4.4.44]. Compared with the first variant, this is a more reliable process since it is not necessary to pay attention to the compatibility of the two paints. The second electrocoat paint can be specifically formulated to the requirements of the particular user. This has been investigated in the automotive industry for the application of primer surfacers [4.4.45]. For process-engineering reasons and because of the high capital costs and a loss of leeway with colours, such coating processes have so far failed to gain acceptance.

Further examples of special electrocoating processes include continuous systems for the electrophoretic coating of endless coils or wires and flow-coating systems for steel pipes. In the former case the dip bath is provided with openings at the side through which the coils to be coated can be fed. The paint which escapes through the openings is collected in an overflow tray and then returned to the coating bath.

Electrophoretic flow-coating systems for coating metal pipes work in similar fashion. The paint flows out of the opened perforated base of a storage vessel which is connected up as an electrode and onto the rotating metal pipe which is connected up as the anode. After filtration, the excess material is also returned to the feed bath so that it can then be recycled into the coating process.

Electro powder coating (EPC) was developed as an electrocoating variant for the automotive sector with the specific aim of achieving a high film thickness that would render an additional primer surfacer superfluous [4.4.46]. The bath material consisted of suspensions of finely ground powder particles with particle sizes of < 5 µm in cathodic electrocoats.

Film formation occurs along the same principles as described for the electrocoat paints. The incorporation of paint-enclosed powder particles meant that films of thickness greater than 50 µm were achieved within short deposition times of 30 to 60 s. The extremely poor throwing power proved to be a disadvantage. In order that the anticorrosive protection necessary for intricate parts could be obtained nonetheless, it was essential to include a second coating tank so that the necessary cavity protection could be provided after interim drying at temperatures of approx. 110 °C. No aesthetic requirements were imposed on the material in the second bath as this was purely for cavity protection purposes.

EPC failed to deliver the required optical quality. In addition, with its need for an extra tank the EPC process was so costly in terms of plant and space requirements that before it even reached industrial-scale use it had to make way for the high film-build electrocoating process, which was developed at the same time. The high-film-build electrocoat can be deposited in one bath up to a film thickness of 35 µm and offers economic advantages because of the potential elimination of the filler line.

Another electrocoat variant which no longer exists was the reverse process. In this the outer skin of the object was primed with a conventional wet paint or powder coating. When this first film had been stoved onto the outer skin an electrocoat primer provided the necessary coating inside. The development of such coating materials was exclusively aimed at improving the anticorrosive protection. Although this process was used in the automotive industry for some years, it no longer has any practical importance.

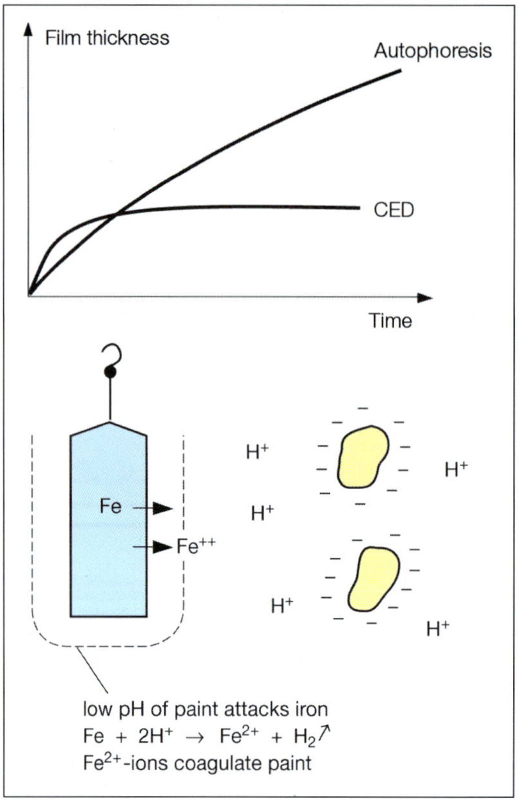

Figure 4.2.23: Paint deposition in the autophoresis process

Autophoresis

Autophoresis is a special variant of dip coating in which the object to be painted is completely immersed either in a batch or a continuous tank, as in straightforward dip-coating processes. The aggressiveness of the acidic and low-viscous autophoresis paint on steel cause the aqueous dispersion to coagulate on the steel surface [4.4.47]. The iron ions formed by the action on the steel surface precipitate the paint. The thickness of the deposited film is mainly determined by the object's dwell time in the bath and the bath temperature. The adhering film still contains over 50 % water and is therefore still very soft. The pre-drying stage which lasts 5 to 10 min at 100 °C solidifies the film sufficiently it to be crosslinked at 130 to 160 °C. The dissolution of the iron and its embedding in the film limits the use of the process considerably. On one hand, no light films can be produced and, on the other, the anticorrosive protection of such coatings is weaker than that of electrocoat paint films.

For a reliable coating to be obtain-ed appropriate process-engineering measures to produce a defect-free substrate surface need to be taken.

Until just a few years ago this process had failed to gain widespread acceptance and therefore remained a niche application. Recently, however, improved formulations have been introduced. The reasons for the renewed interest in this coating technology lie in the elimination of inorganic pretreatment (see Chapter 4.1) that is otherwise usual for dip coating and electrocoating, the opportunity to dispense with electrical energy for the deposition process, and complete coating of inner sections [4.4.48]. An example of applications for which this process is suitable is car seat frames which are generally painted with a black coating. For such an object, where practically none of the surface is visible in the finished seat, the requirements imposed on both anticorrosive protection and surface finish are low.

4.2.1.2 Brushing, Manual and Mechanical Roller, Flow and Curtain Coating Methods (Paint Applied Directly to the Object)

If the coating is brought to the object, two fundamentally different processing principles are commonly applied. The direct and technically straightforward processes include brushing, manual and mechanical roller, curtain, and flow coating. In all cases the coating material is transferred from the application device to the object to be painted without any detours and with the advantage of a high material efficiency.

In the other, indirect processes the material is first broken down into tiny particles and only then transported to the object to be painted where it is converted into a continuous film. This group of application employs different atomisation methods that lead to very different material yields. Whereas material yields of up to 98 % are achieved with purely electrostatic spray coating, pneumatic atomisation processes with application rates of sometimes less than 40 % are far from efficient.

Brushing, Manual Roller Coating

The oldest processes, which are primarily used for housepainting, are brush and manual roller application. The advantages of the brush and manual roller methods are universal applicability to intricate components, uniformly good wetting and distribution of the paint, and the excellent material yield of practically 100 %. The disadvantages are the low speed and the necessary skill and experience on the part of the painter in producing smooth and uniformly thick coatings. Brush coating is therefore not used as an industrial-scale application method.

Mechanical Roller, Flow and Curtain Coating

The manual roller method can, however, be scaled up. The application of paint, albeit limited to planar surfaces, is possible if the lambswool or foam rubber roller is replaced by rubber rolls. Continuous, large-scale application becomes feasible when the appropriate ancillary equipment and paint-supply systems are used. Automated roller-coating systems have been in operation in Germany since 1928 and are suitable, long-established paint-processing machines for coating planar workpieces such as wood panels, hardboard, plastic film, cardboard, paper, and metal coils.

A conveyor transports the workpieces underneath the fixed transfer roll which transfers the paint uniformly to the workpieces in an amount determined by the set nip. To this end, a rotating pick-up roll takes coating material from a storage tray. This polished roll has a hardened surface and transfers the paint to the transfer roll which is also rotating and which transfers the paint to the workpiece. **Three-roll systems** have a smoothing roll, called the doctor roll, located between the pick-up and the transfer rolls. The purpose of the doctor roll is to help to provide better conditions for obtaining a smooth film. Each roll is driven separately and can therefore be regulated independently of the speed of the other rolls.

The transfer roll, which consists of a metallic core with a rubber coating, can turn either in the same direction as the conveyor (forward roller-coating) or in the opposite direction (reverse roller-coating). As the forward process is limited in the amount which can be applied in one pass and with good levelling, thick coats are achieved by using rolls rotating counter to the direction in which the workpieces are moving. Such

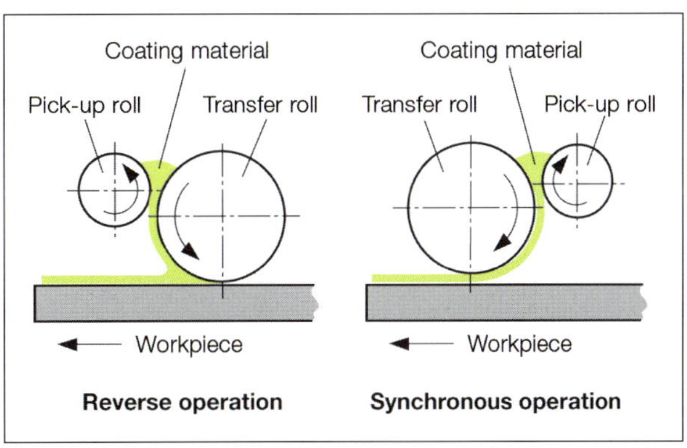

Figure 4.2.24: Schematic drawing of a two-roll applicator for the reverse and forward processes

Figure 4.2.25: Picture of a roll coater *Source: VAW*

reverse roller-coating allows a greater range of film thicknesses to be achieved with higher material viscosities and thus lower solvent emissions at the same time. With processing viscosities of up to 600 mPa s and 100 % material yield, roller-coating methods can produce texture-free films that have a uniform film thickness profile. Whereas the forward process is only capable of ap-plying adequately smooth coats 10 to 15 µm thick, films up to 100 µm thick can be applied by the reverse process. Highly thixotropic paints can also be processed because of the forced distribution of the paint by the rolls.

Direct transfer without the detour of interim enlargement of the paint surface area via at-omisation coupled with the short dwell time between application and the stoving oven offers the environmental advantage of low solvent emissions during processing. The solvent component of the paint is mainly given off in the oven where it can be eliminated or reclaimed without difficulty using already available technology, viz. incineration or adsorption methods.

Advanced systems running at line speeds of up to 200 m/min are not only environmentally sound, however. They also offer the most efficient processing methods from an economic perspective. A further technical advantage of the roller-coating process is that the planar objects can be coated on both sides when the rolls are configured appropriately.

This process imposes certain special demands on the coating materials. High stoving temperatures are required because of the high line speeds and the need to limit the length of the oven. Oil-free, heat-resistant, polyester-modified or silicone-resin-modified acrylic resins in conjunction with highly etherified HMMM resins have proved successful for this (see Chapter 2.1).

Electron-beam-curable and UV paints are an obvious choice for use with the high-speed roller-coating method because of their short curing times. These systems are used as liquid coatings and now also as powder by spraying and are cured in fractions of a second after they have been melted by IR radiation.

Particular demands are placed on the solvents of roller-coated paints. Apart from having a high evaporation number and a good solvent power for the film forming agents, they must not cause the rubberised transfer rolls to swell. Good results have been achieved with a combination of rolls sheathed in butyl rubber and solvents such as methyl glycol acetate, butyl glycol acetate and butyl diglycol acetate. Natural rubber jackets are more suitable for transfer rolls when aromatic-rich solvents are used.

A uniform coating result can only be guaranteed if the rheological properties of the paint have been adjusted and the contact pressure is appropriate for the line speed. The wet-film thickness must therefore be continuously measured and corrected, if necessary. In view of the high line speed (see Chapter 3.2.1), control actions must be implemented so that wastage may be limited.

The roller-coating units can be parts of continuous coating systems. In the **coil coating process** sheet metal coils are unrolled, degreased, pretreated, coated, cured and rewound continuously. To ensure a continuous process, coil reserves are stored in loop towers, and these reserves are then used whenever a new coil has to be welded to the end of the previous one (see Chapter 7.4). The coils coated in this way are then formed into a very wide range of consumer goods. Facade elements, garage doors, fridges, steel furniture, car body parts, cans and lighting fixtures are just some of the objects which are manufactured from these precoated coils. This process will become increasingly prominent in future because of the cost-effectiveness of the coating systems, the low emissions in the application area, the good material efficiency and the coating's excellent optical and mechanical/technical properties.

Another direct paint-processing method which is also suitable for planar parts is **curtain coating**. The paint is applied to the objects as they move by on conveyor belts, in similar fashion to the roller-coating method described above. By contrast with roll-er coating, the liquid coating material is allowed to fall onto the object. To ensure a uniform coat, the coating material is converted into a "curtain of paint" which flows permanently from a pouring head. The latter consists of a storage reservoir with a slot. The pouring head can be open or closed at the top. In the latter case different pressures can be set which provide an additional variable for the discharged quantity and thus for the film thickness.

As the object passes through the paint curtain, its surface is uniformly covered with paint. The film thickness is defined by the viscosity of the paint, the width of the slot in the pouring head and the speed of the object. If the curtain is wider than the object or if no objects are coated when the paint is flowing, the material is collect-ed and returned to the pouring head after filtration (see Figure 4.2.26). To avoid problems during the coating process the material is not fed directly to the pouring head. Air bubbles entrap-ped in the paint must be allowed to escape in a settling chamber. This is the only way to produce a continuous and uniformly discharging paint curtain. For better maintenance and trouble-free operation of curtain-coating systems, it must be possible to open the pouring head so that the discharge area can be cleaned properly.

The process permits a coating of 50 to 500 g/m^2 to be applied and therefore, unlike the roller-coating process which can only apply 10 to 100 g/m^2, is particularly suitable for high build coatings. Curtain coating machines have proved particularly useful in the woodworking and furniture sectors. They are used in these industries just as equally for high-build coatings with unsaturated polyesters or polymerisable polyacrylates as for thin-film, open-pore coatings with cellulose nitrate lacquers.

When the curtain-coating machines have only one pouring head the use of 2-component paints causes problems because of the limited pot life. In such cases multihead machines are used which apply several individual curtains of the individual components to the object. Additional mixing is effected by diffusion and flow in the wet film. The outcome is uniform curing.

Further variants are possible when unsaturated polyesters are used. The unsaturated polyester is dissolved in a nonreactive solvent and applied to the wood substrate

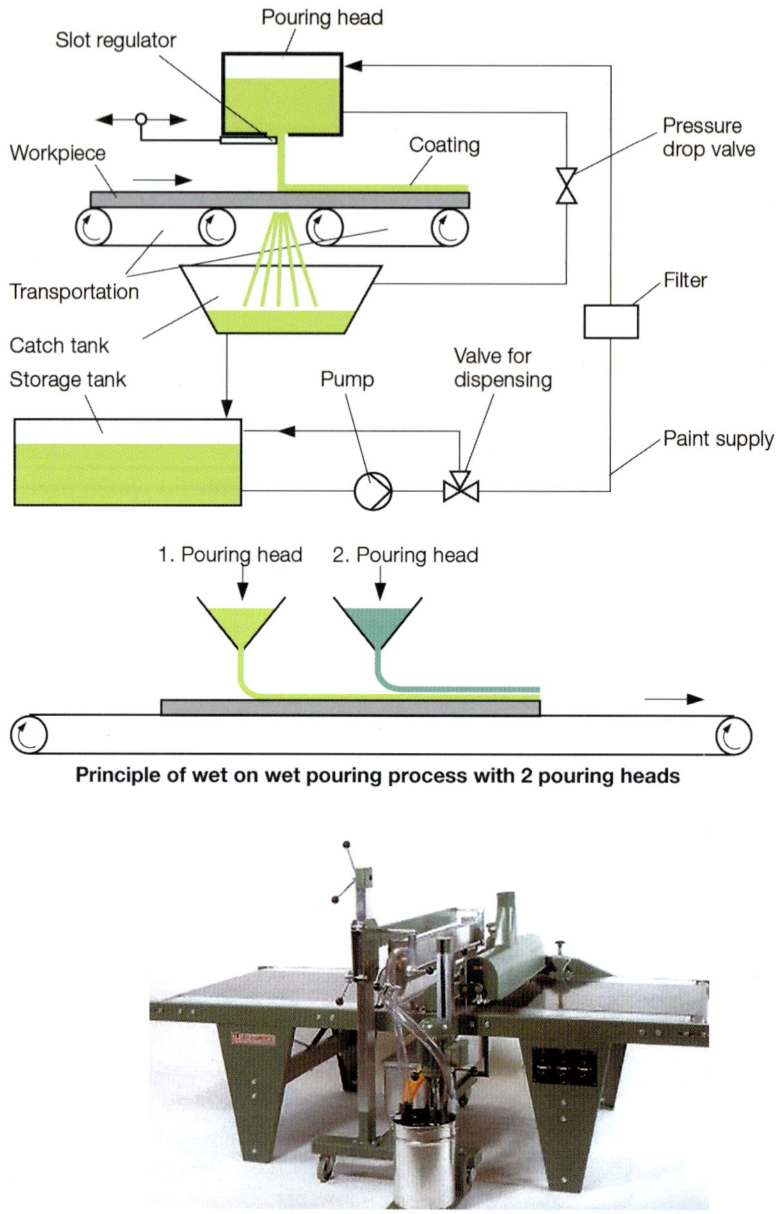

Figure 4.2.26: Schematic drawing and picture of a curtain-coating machine

together with peroxide. The substrate absorbs the solvent, with the result that the styrene-containing unsaturated polyester applied subsequently together with the primer accelerator is dissolved again such that chemical curing is initiated. In rare cases three-head curtain-coating machines are also used to ensure even better mixing of the components.

The curtain-coating process becomes particularly attractive when the sometimes very thick films are cured in seconds with UV or electron beams.

The roller/curtain-coating method is a technical hybrid between curtain coating and roller coating. Where low-solvent or solvent-free and highly viscous coating materials are processed, defined quantities of the film are transferred via a pick-up roll to a transfer roll. Here it is removed by an applicator (scraper or coating knife) and then poured onto the surface to be painted in a precisely metered quantity. The roller/curtain-coating method permits processing viscosities of up to 800 mPa s with shear of 100 s^{-1}.

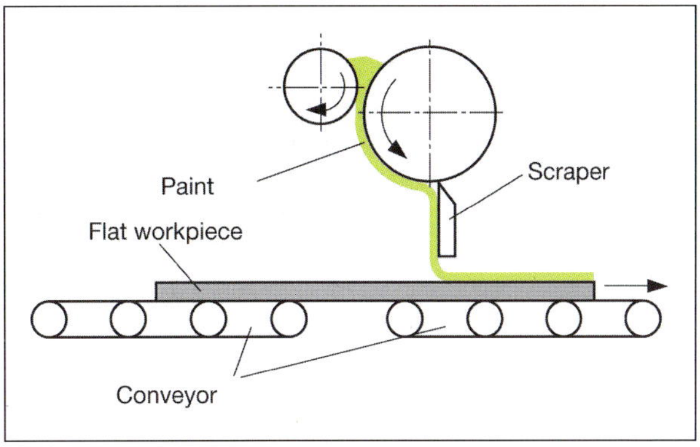

Figure 4.2.27: Schematic drawing of a roller/curtain coating system

Flow coating is similar to the dip-coating process described earlier. In this the paint is applied to the workpieces as they move past nozzles. The excess paint is collected, filtered and fed back to the nozzles.

Flow-coating processes have only limited use due to nonuniform film thickness and the formation of unavoidable runs. Even if the systems are optimised and the components are also rotated during the painting process, the surfaces are not attractive. Flow-coating has therefore only achieved a certain importance for coating engine blocks, central heating radiators, pipe racks or other bulky components for trucks or agricultural machinery.

As a relatively high amount of solvent evaporates during flow coating, the materials now used in open systems are mainly limited to water-borne paints.

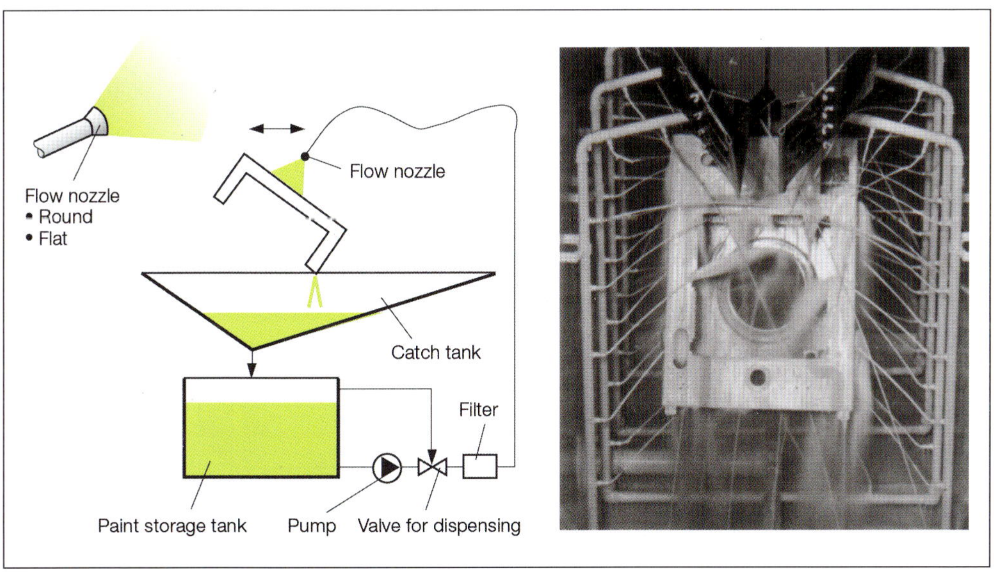

Figure 4.2.28: Schematic diagram and picture of a flood-coating machine

Application

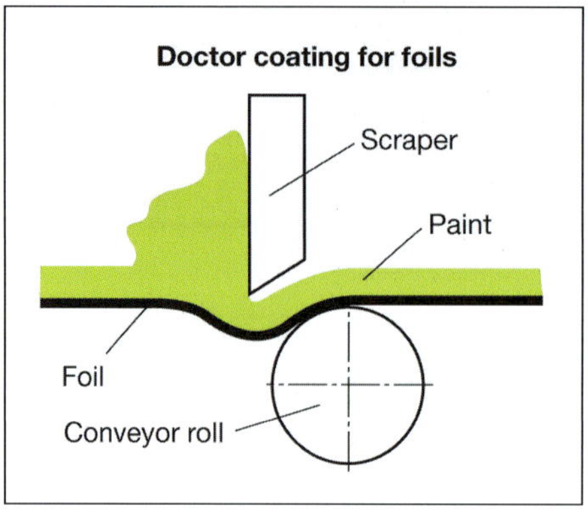

Figure 4.2.29: Schematic drawing of knife coating (doctor coating)

Other direct processes are tumbling, knife coating (doctor coating) and filling. The first of these is useful for coating small parts. The parts are shaken or rotated in a barrel or cage together with the paint and then, after the excess paint has drained off, the films are dried or cured by heat action.

In knife coating a coating knife (scraper) presses the highly viscous paste onto the paper or textile web as it runs through the machine, thereby guaranteeing uniform film thickness. Knife coating is an accepted industrial method for processing highly viscous paints in the textile and paper industries, but filling with non-free-flowing coating materials is limited to manual application.

4.2.1.3 Spray Process: Paint Applied Indirectly to the Object

Although indirect application processes are by far the most important for industrial paint processing (see Figure 2.3.61), they give the most cause for concern at the same time from an environmental perspective. Atomisation methods for spray coating have the disadvantage of sometimes extremely poor material yields combined with accumulation of paint waste, but they still dominate industrial-scale coating technology whenever the coating must meet demanding specifications on optical attractiveness. They en-able filigree, intricate parts to be given a uniform coating. If a wide range of colours is demanded and pronounced effects have to be achieved, there is practically no alternative to spray application.

The technical principle behind this process is the production of droplets which are transported to the object and then converted back into a continuous, smooth film. Modern

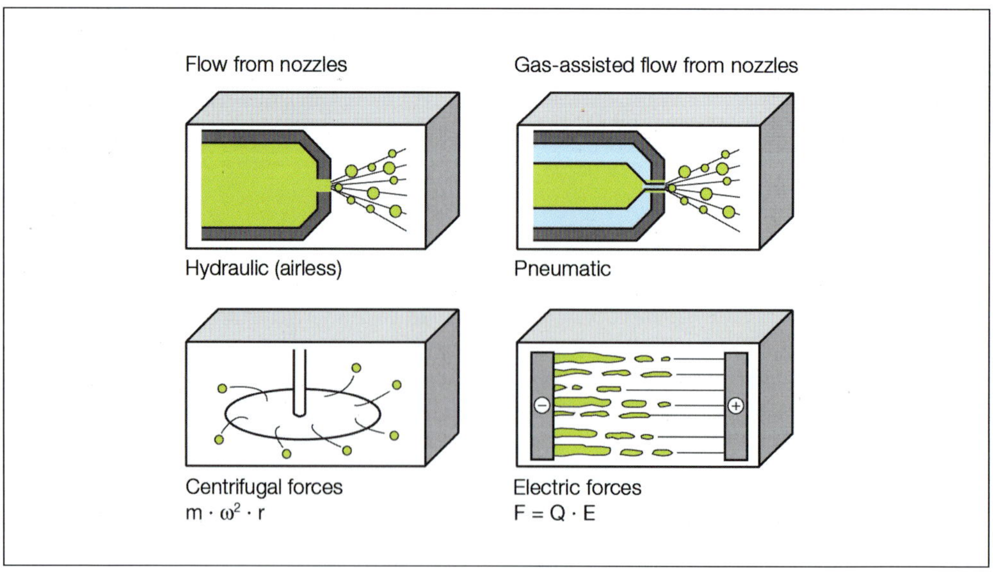

Figure 4.2.30: Methods of atomising

methods for producing the droplets make use of electrical forces, centrifugal forces, the energy of accelerated air or that of the highly accelerated paint itself to provide the necessary free surface energy. Transport to the object mostly is assisted by a flow of shaping air.

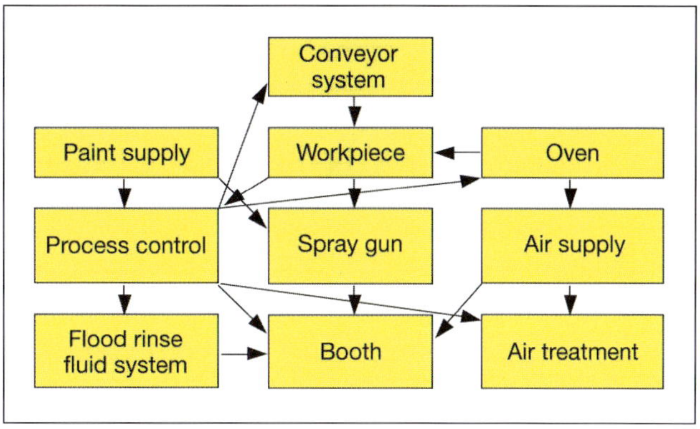

Figure 4.2.31: Elements of a spray coating facility

Spray processes require a booth in addition to the actual spray head itself. The spray booths must be provided with suitable inlet air, air-extraction, flushing and separation systems for dealing with overspray, and with paint-supply and conveying equipment (see Figure 4.2.2).

A specific droplet size spectrum is required in order that a particular coating result may be achieved. The droplet size spectrum is crucial for optically attractive films which are both smooth and thin and for the reproduction of colours, though certain constraints have to be observed. If the droplets become too small, the material spill increases as a result of increased overspray being the part of the spray cloud is not deposited on the object. In addition, more solvents evaporate because of the larger paint surface area. The consequence is high emissions and paint waste from spray booths, and poor levelling due to the viscosity. On the other hand, excessively large droplets yield an unacceptable surface property profile (see Chapter 3.1). In view of the contrary influences of the droplet size on the quality of the coating and environmental pollution the best compromise is usually average droplet sizes ranging from 30 µm to 35 µm.

Atomisation Through Nozzles

The free surface energy which has to be exerted to overcome the cohesive forces can be generated by flow in nozzles. A nozzle is a constriction at the end of a pipe for converting hydrostatic pressure into kinetic energy in fluids. Single- and multicomponent nozzles are usual for the industrial-scale application of this principle. The former are used in hydraulic atomisation. High material acceleration, resulting from high material pressure and narrow nozzles, causes the accelerated coating material exiting the nozzle to be broken down into tiny droplets by interaction with the ambient air. Multicomponent nozzles cause atomisation by the additional interaction of the exiting material with greatly accelerated air.

Droplet formation by liquids exiting nozzles is influenced by their velocity relative to the ambient air (see Figure 4.2.32). The conditions obtaining determine whether (1) dripping, (2) oscillation, (3) wavy sheet disintegration or (4) the desired spraying is achieved upon exiting. If a liquid is forced very slowly out of a nozzle, it initially forms a continuous filament. Eventually, having reached a certain minimum length, it undergoes necking until it reverts to the most thermodynamically stable state by forming individual droplets. Under these conditions, monodisperse distributions are obtained but they contain only a few large droplets.

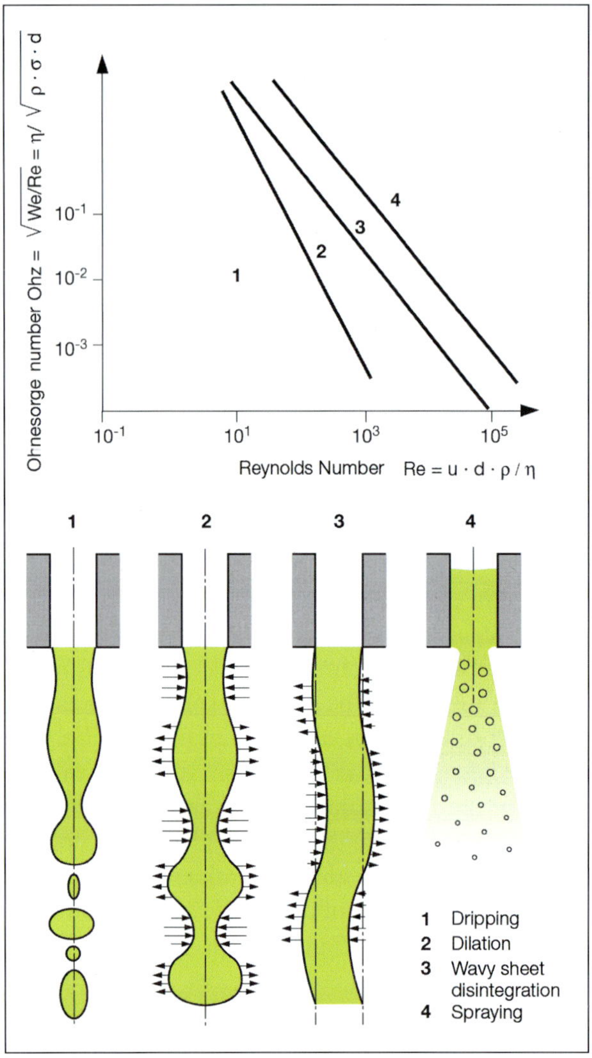

Figure 4.2.32: Mechanisms of droplet generation and working ranges during spraying

If the discharge rate is increased, longitudinal oscillations are induced in the filament as a result of friction with the resting ambient air, with the result that areas of dilation and constriction occur, creating the conditions for dripping. Again, the spectrum of droplet sizes is uniform. The droplet size is a function of nozzle width, surface tension and density.

A further rise in velocity leads to an increase in air friction and conversion of the liquid filament into transverse waves. This wavy sheet disintegration also produces large droplets with a narrow size spectrum.

At even greater pressure the exit velocities are so high that further, secondary atomisation is superimposed on the primary effects described as a result of collisions between the primary droplets and the now significantly slower-moving air. The dynamic pressure of the ambient air deforms the primary droplets, then causes necking and division. Further deformation leads to ever-smaller droplets until the deforming dynamic pressure is in equilibrium with the resisting cohesive pressure. This is the principle behind hydraulic atomisation in the form of airless atomisation.

The generation of tiny droplets which are of very different size can also be achieved by pneumatic atomisation involving a combination of greatly accelerated air and slow-moving liquid.

The operating ranges of the individual atomisation types can be shown graphically by combining the variables which characterise the flow state such as density ρ, flow rate u, viscosity η and surface tension σ to form dimensionless numbers which are then plotted mean-ingfully against each other. The numbers relevant to the present case are the *Reynolds* number Re, the *Weber* number We and the *Ohnesorge* number Ohz, which is the quotient of the square root of the *Weber* number and the *Reynolds* number.

$$Re = \frac{u \cdot d \cdot \rho}{\eta} \qquad We = \frac{u^2 \cdot d \cdot \rho}{\sigma} \qquad Ohz = \frac{\sqrt{We}}{Re}$$

If a liquid is present in the operating range of an atomiser for long enough, the droplet diameter d will, as described above, be reduced further and further until the cohesive pressure corresponds to the deforming dynamic pressure. Where there are no other variables influencing this size-reduction process, the droplet size can be calculated at the steady state. According to *Bernoulli* dynamic pressure p_{St} is given by:

$$p_{St} = \frac{\rho}{2} \cdot u^2$$

The cohesive pressure p_{Ko} in droplets with radius r is given by:

$$p_{Ko} = \frac{2 \cdot \sigma}{r}$$

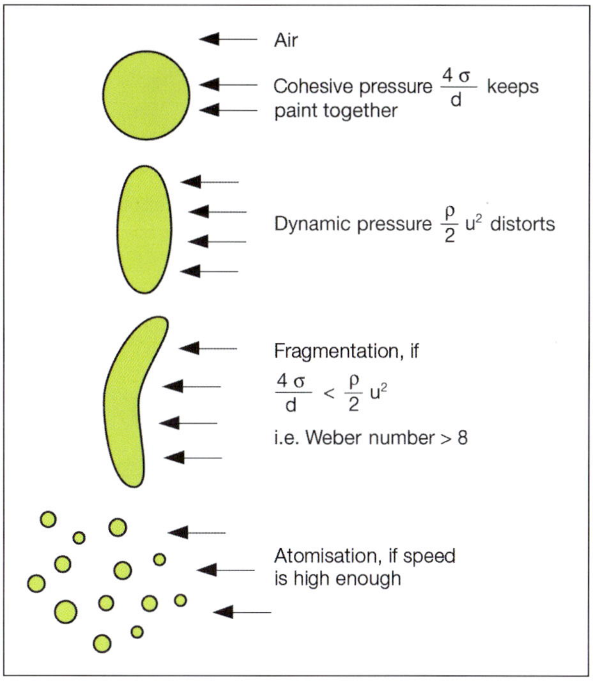

Figure 4.2.33: Mechanism behind droplet formation during atomisation

Size reduction therefore always takes place when

$$\frac{\rho}{2} \cdot u^2 > \frac{2 \cdot \sigma}{r} \quad \text{respective} \quad \frac{4 \cdot \sigma}{d}$$

or, solved differently,

$$\frac{\rho \cdot u^2 \cdot d}{\sigma} > 8 \quad \text{because of} \quad \frac{\rho \cdot u^2 \cdot d}{\sigma} = We \quad \text{follows}$$

atomisation if

$$We > 8$$

Consequently, the conditions for conversion into finer droplets are always right if the *Weber* number > 8. This shows an excellent fit for liquids such as water. For coating materials containing higher-molecular polymer solutions, however, certain amendments or corrections are needed.

Various methods are employed to bring about the high velocity differences between the substances to be sprayed and the air. As explained above, in the pneumatic process the supply of atomising air must attain a high velocity. The air meets the material either before exiting the atomisation device (internal mixing) or only after exiting the nozzle (external mixing), which is the usual method employed in coating technology. With external mixing, therefore, the paint droplets are generated not inside but outside the spray head.

Since external mixing is of far greater importance for pneumatic processing in coating technology, the following descriptions concentrate on this technology.

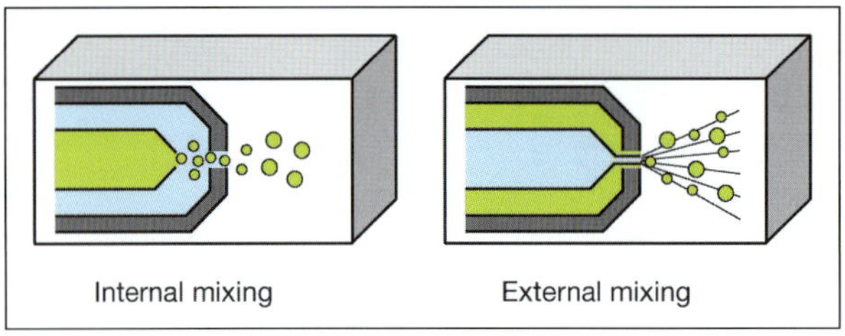

Figure 4.2.34: Pneumatic spraying with internal and external mixing

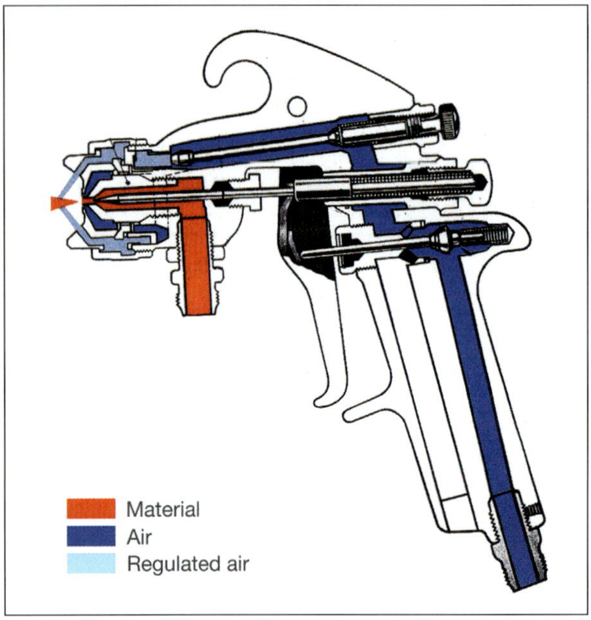

Figure 4.2.35: Cross-section of a spray gun

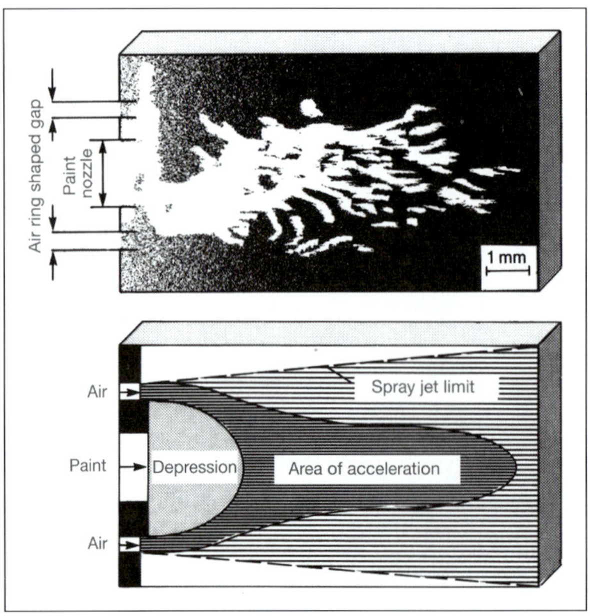

Figure 4.2.36: Picture and schematic drawing of the atomisation process

Pneumatic atomisation is subdivided on the basis of the effective air pressure into low-pressure (0.5 to 1 bar) and high-pressure atomisation (4 to 8 bar), with the latter the more important variant in industrial processing. As a result of the high air pressure the exiting air attains high velocities. Provided there is a sufficient quantity of air, up to 1 litre of paint per minute can be atomised, enough to meet the needs of high-performance industrial processes.

The spray heads or guns are based on the invention by the Dutch doctor *DeVilbiss* who registered a patent for a process for manufacturing aerosols for medical purposes in 1887. The principle behind such a gun, which has been constantly improved to the present day, is shown in the form of a cross-sectional drawing in Figure 4.2.35. The paint which is fed to the nozzle is initially prevented from leaving the gun by the nozzle pin. Once the trigger guard has been opened, only the atomising air which is under a pressure of 5 to 8 bar is discharged from an annular gap configured concentrically with the nozzle. As a result of the highly accelerated atomising air an area of underpressure is created at the nozzle aperture. After the nozzle has been opened, the coating material is entrained, deformed by the dynamic pressure acting in the acceleration zone of the spray air and ultimately converted into smaller droplets. The air can reach a velocity of up to 400 m/s.

The attainable discharge rates depend on the supply method for the paint. **Siphon feed spray guns** achieve only low material discharge rates. In more powerful spray guns, therefore, either gravity in the form of a gravity feed spray gun or additional material pressure ensures an effective increase in the material discharge rate. In the latter case the supply of material is no longer provided by storage reservoirs attached to the gun. It is handled instead by separate pressure vessels or circulation systems.

It should be noted that the resulting air velocity does not rise linearly with increase in air pressure. It reaches a plateau at air pressures of less than 8 bar, with the result that any further rise merely increases air consumption without significantly influencing the droplet size spectrum. Furthermore, increasing air volume leads to excessive energy consumption and a deterioration in material efficiency due to increased overspray. Wastage can be as much as 50 % of the material used. The causes are the large volumes of rapidly moving spray air which carry the small and less sluggish droplets past the object. In line with the shape of the object, the 1000 litres of accelerated air necessary for atomising 1000 g paint, along with the spray cloud, either is routed past the object to be painted or else rebounds from it. The rebound effect causes an undesirable distribution of the spray cloud in the vicinity of the workpiece. If the droplet size has to be reduced at high air pressures nonetheless, other measures must be taken, such as lowering the material output by reducing the material pressure or using smaller nozzles.

Figure 4.2.37: Different types of material feed for pneumatic spray guns *Source: ITW*

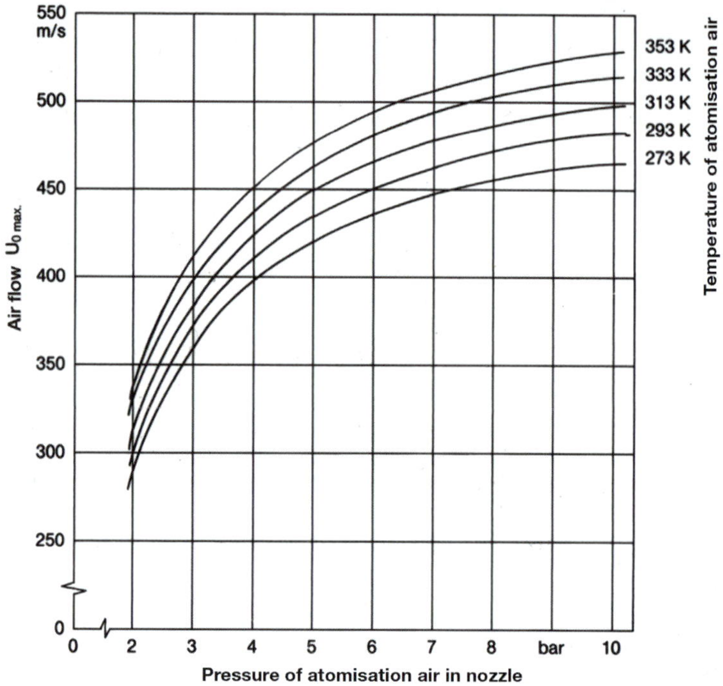

Figure 4.2.38: Velocity of atomisation air as a function of air pressure

There are limits on the rates of material discharge in pneumatic atomisation. At elevated rates the droplets increase in size. The reason for this is braking of the spray air upon interaction with the paint, and a resulting decrease in dynamic pressure. It is therefore clear that the formation of the droplet size spectra depends not only on the air velocity, but also on the air volume. The latter is determined by the surface area of the annular gap in addition to the pressure of the spraying air. The influence of the spraying air pressure or air volume on droplet size is indirectly proportional to the material discharge rate and thus to the material pressure at a constant specified nozzle surface area. Where a consistent spray pattern is required, changes in the material output will always have a corresponding effect on the volume and pressure of the spraying air. However, it only makes sense to make these changes if it is still possible to accelerate the air effectively (see Figure 4.2.38).

The volumes of atomisation air should be limited, if possible, for cost reasons. Approximately three-quarters of the total energy required for droplet formation is expended on accelerating the spraying air, while the rest, which is for providing friction and acceleration of the paint and the actual free surface energy, plays only a subordinate role [4.4.49].

There are other variables for influencing the droplet size spectrum on the paint property side, of which the viscosity and rheological properties of the coating material to be atomised are the critical ones. Assuming forced feed systems are not used for dispensing the material, the discharge rate is influenced in direct proportion by the material pressure and in indirect proportion by the viscosity in accordance with the *Hagen-Poiseuille* equation (see also Chapter 2.4). Since the viscosity of pseudoplastic coating materials falls not only because of the shear load but also, as with all liquid materials, by approx. 5 % per °C temperature increase, it is clear that fluctuations in ambient or material temperatures will lead to fluctuations in discharge rates. These in turn are reflected in different spray patterns and, associated with this, also in different film thickness, and will ultimately give rise to levelling defects in the form of, for example, orange peel or runs. Temperature constancy of material or forced delivery systems help fluctuations in quality to be avoided.

Independently of the indirect influence of viscosity on the discharge rate and thus on the droplet size spectrum, the rheological properties have a direct impact on the droplet size and especially the droplet size distribution. On account of the relatively short dwell time in the droplet-forming acceleration zone, the elastic components of the coating material play a role in addition to that of viscosity. Paint with a high elastic component requires a longer period in the shear zone to reach equilibrium. If it is also assumed that the dwell time needed in the shear zone for attainment of

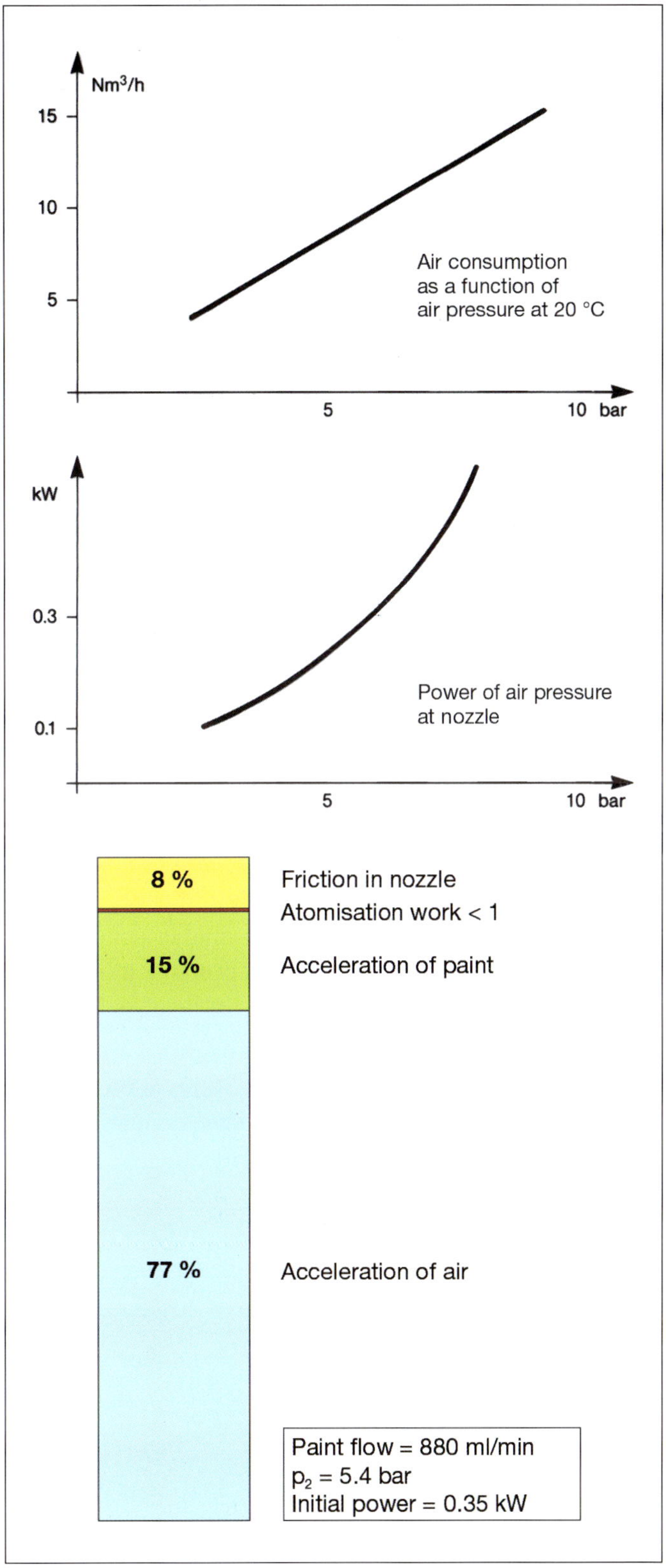

Figure 4.2.39: Air consumption, power consumption and energy used in pneumatic spray application

Application

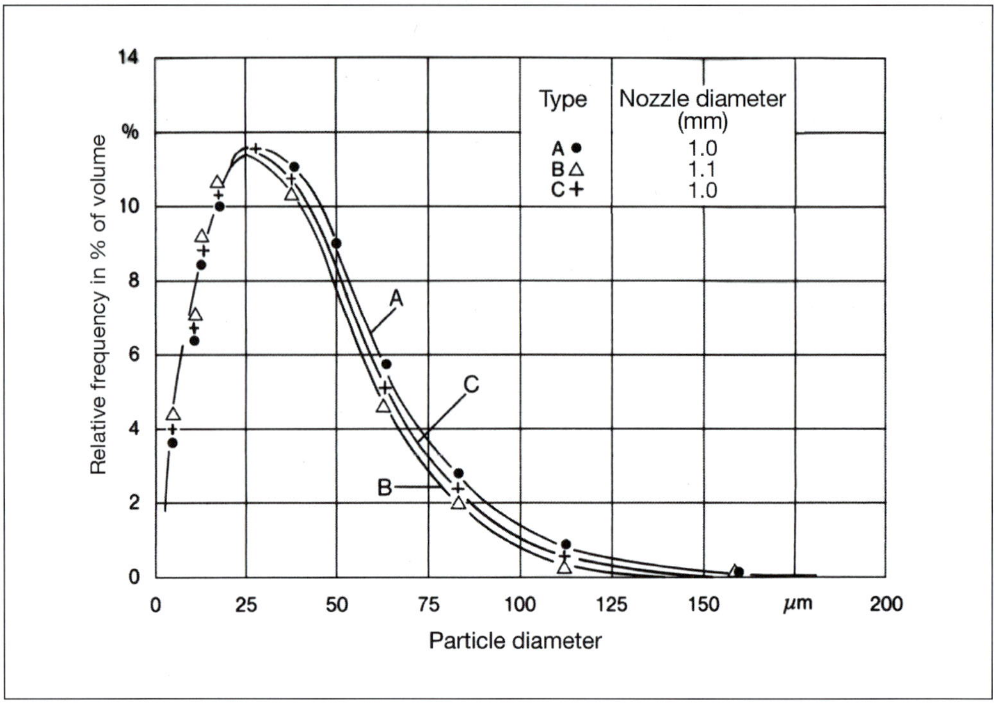

Figure 4.2.40: Typical spectrum of droplet sizes produced by high-pressure guns

equilibrium is not the same for all droplets and if, furthermore, account is taken of the fact that the shear zone is not homogeneous either, this provides an explanation of the broad spectrum of droplet sizes.

Further physical paint variables which influence the droplet size are surface tension and density. Low surface tensions lead to smaller cohesive pressure in the droplets and thus facilitate atomisation. High densities promote atomisation, since the dynamic pressure decreases more slowly due to the greater inertia.

There are numerous references in the literature to pneumatic atomisation and the calculation of droplet size [4.4.50–4.4.53]. Many publications have discussed ways to calculate the droplet size spectrum but *Walzel's* equation merits particular mention because it is simple and straightforward [4.4.54].

$$d_{32} = 51 \cdot Re^{-0.39} \cdot We^{-0.18} \cdot \left[\frac{\dot{m}_{Paint}}{\dot{m}_{Air}}\right]^{0.29} \cdot d_D$$

Here, d_{32} is the Sauter diameter or mean droplet size (see Chapter 2.3.3).

The *Reynolds* number Re and the *Weber* number We contain the relative velocities of air to paint droplets, viscosity, density and surface tension. These variables are supplemented by the ratio of the mass flow rates $\dot{m}_{paint}$ to $\dot{m}_{air}$ and by the diameter of the material nozzle d_D. This mathematical relation enables the effects of specific changes in the variables on the droplet size spectrum to be calculated with a good fit to measurements from practical applications.

Figure 4.2.41 shows the calculated effects of the material, equipment and process variables on the droplet size spectrum. The steeper the individual curves, the greater is the

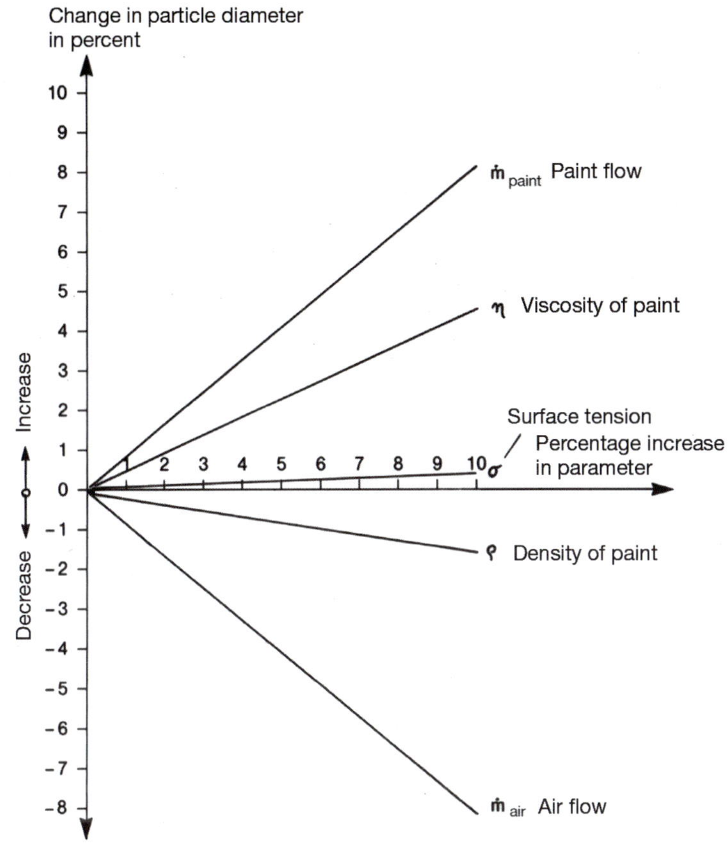

Figure 4.2.41: Sensitivity diagram of the variables that determine the spray pattern

influence on the spray pattern. A positive incline represents an increase in droplet size and a negative incline a decrease.

Consequently, the fineness of the atomisation deteriorates significantly with increase in paint throughput $\dot{m}_{paint}$ unless the air throughput $\dot{m}_{air}$ is increased correspondingly at the same time. Calculations also show a higher sensitivity to viscosity compared with surface tension and density, which also fits in with practical experience. Atomisation is therefore determined far less by the gun type than by the gun setting or by the ratio of paint throughput to air velocity. It should be noted in this regard that the same pressure gauge settings for air pressure under otherwise apparently identical conditions do not necessarily translate to the same droplet size. Different cross-sections and lengths of hoses give rise to different drops in pressure in the gun, despite the same pressure gauge setting. If reliable forecasts of the spray pattern are required, such factors have to be taken into account or the pressure gauges have to be positioned as close as possible to the spray gun.

Absolute calculations of droplet sizes to evaluate resulting levelling properties are made more difficult by the fact that the velocity of the individual droplets varies with their size. Collisions en route to the object cause droplet contamination which changes the primary droplet size spectrum. To this must be added the fact that the generated spectrum

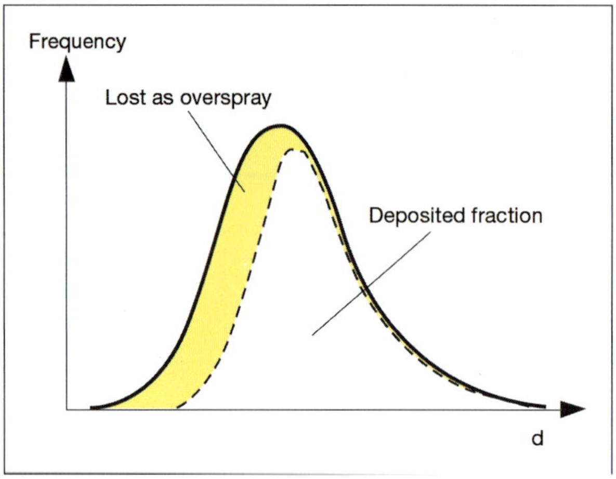

Figure 4.2.42: Difference between the generated and deposited droplet size spectrum in pneumatic atomisation

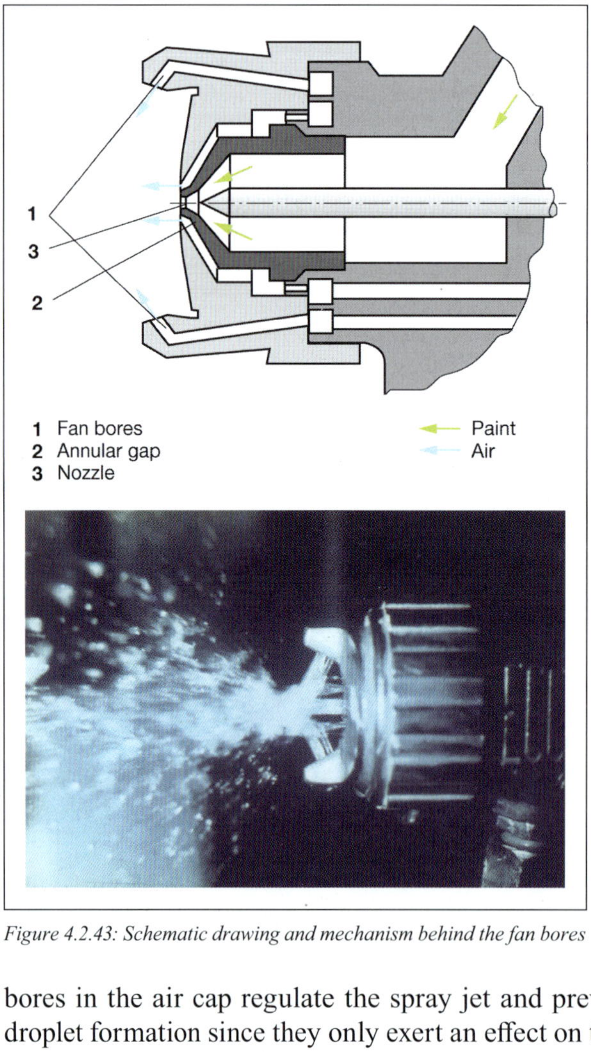

Figure 4.2.43: Schematic drawing and mechanism behind the fan bores

does not match the deposited droplets, as the finer particles are also less likely to be deposited on the object because of their greater mobility. They accumulate in the air carried past the object in the form of overspray and therefore change the deposited spectrum once more.

The scope of application for pneumatic high-pressure atomisation can be widened still further by attaching accessories to **high-performance atomisers** which permit the spray jet to be matched to the shape of the objects and for special effects to be set. It makes sense to adapt the spray pattern to the specific shape of the workpieces by expanding or constricting the spray cone to match the shape of the object. The normally round spray cone can be flattened to a greater or lesser extent by means of a separate supply of a partial flow of the atomising air, the fan air, to the fan bores to suit the particular task at hand.

Pneumatically operated spray guns are fitted with supplementary air outlet bores around the annular gap to ensure that the underpressure in the vicinity of the atomising nozzle does not cause any deposition of overspray in the atomising area of the spray gun. This prevents return flow of the overspray and thus soiling of the gun.

The cross-flows from the fan bores and the supplementary bores in the air cap regulate the spray jet and prevent deposits but do not influence droplet formation since they only exert an effect on the spray pattern after formation of the droplet size spectrum.

Application

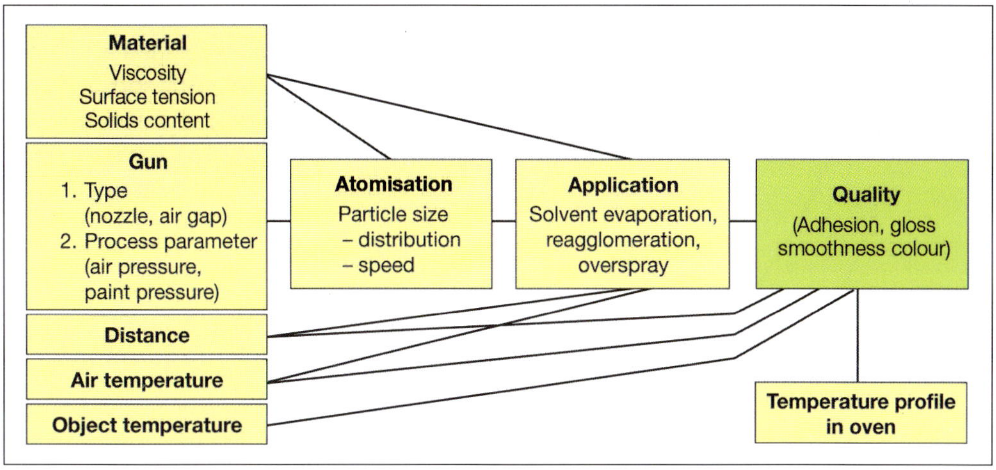

Figure 4.2.44: Factors determining the coating quality obtainable with high-pressure pneumatic atomisation

Pneumatic high-pressure atomisation has evolved into a complex process which is only manageable and capable of producing an optimum coating result if all the variables are known to those involved and are correspondingly addressed when the process is being implemented (see Figure 1.1.7). The viscosity, surface tension and density, as well as the solid content are relevant material variables; nozzle, annular gap, air cap and fan bores represent the equipment parameters; and air and material pressure are the adjustable process-engineering variables. The viscosity affects atomisation directly and also indirectly by influencing the discharge rate.

The air and object temperatures and especially the distance from the object are other significant variables affecting product quality in terms of gloss, colour, effect, levelling and, in certain cases, even adhesion. The average flight time of the spray cloud and thus the solvent content of the resulting film are determined by the distance from gun to object. A short distance leads to wet and therefore low-viscosity paint droplets. Long distances, on the other hand, allow the droplets to become too dry because they are in flight for too long. The outcomes of this are variations in levelling or running, as well as fluctuations in the formation of effects.

In practice, a distance of 25 to 30 cm from the object is generally the best compromise, as it takes account of the evaporation properties and the rising amount of overspray with increase in distance. In manual application, distances to the object are sometimes deliberately varied to compensate for different ambient temperatures and to influence effect formation.

Hot Spray Coating

It is possible to reduce emissions and overspray by heating the coating material during spray application. The lower viscosity resulting from the temperature increase opens up new means of influencing the processing method. Hot spraying affords a way of either raising the solid content or lowering the spraying air pressure. In the former case the higher viscosity resulting from the decision not to use solvents can be restored to its former level by means of heat action. The spray spray pattern obtained in this way is similar, despite the higher solid content. In the second case, a lower viscosity combined with a lower spraying air pressure can reduce the degree of overspray, without any

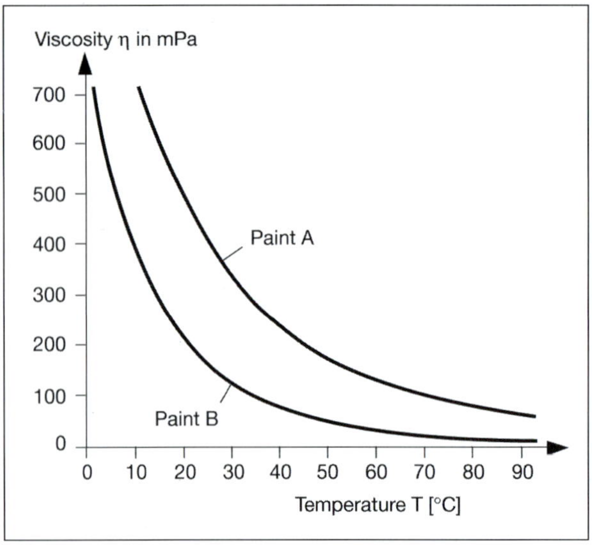

Figure 4.2.45: Spray viscosity as a function of paint temperature

Figure 4.2.46: Example of a high-volume, low-pressure gun

Source: ITW

deterioration in the spray pattern. Material consumption is reduced and so are emissions. Figure 4.2.45 illustrates the pronounced dependence of viscosity on temperature for the example of two paints. Experience shows that heating the material to 60 to 70 °C enables either the solid content to be increased by 4 to 6 % or the spraying air pressure to be lowered by 1 to 1.5 bar. The latter is equivalent to a 10 to 15 % reduction in overspray.

The material is heated indirectly by means of heat exchange with a heat transfer liquid. As the flash point of the solvents is generally exceeded here, the systems must be explosion-proof. The energy consumed in heating the paint is more than compensated for by savings in compressed air consumption.

Low-Pressure Process

The classic low-pressure process is only of importance for the manual sector. Only small quantities of material can be atomised in this process because the pressure and therefore the air volumes are low. Furthermore, it is not possible to achieve optimum droplet fineness, and so the coating result does not always meet very demand-ing surface quality requirements.

Interest in the low-pressure process initially was rekindled in the USA in response to more stringent environmental regulations. The statutory requirement to reduce wastage during spray coating to below 35 % and to reduce emissions spurred on further development of the low-pressure process to the extent that it now better meets the demands of industrial processing. This development has spawned a low-overspray variant of pneumatic atomisation, known as the **high-volume low-pressure (HVLP) process** [4.4.55]. It offers a compromise between an optically attractive coating and reduced overspray formation.

The atomisation principle corresponds to that of the pneumatic process, though it reduces the air velocity through a combination of lower air pressure, higher air volume and

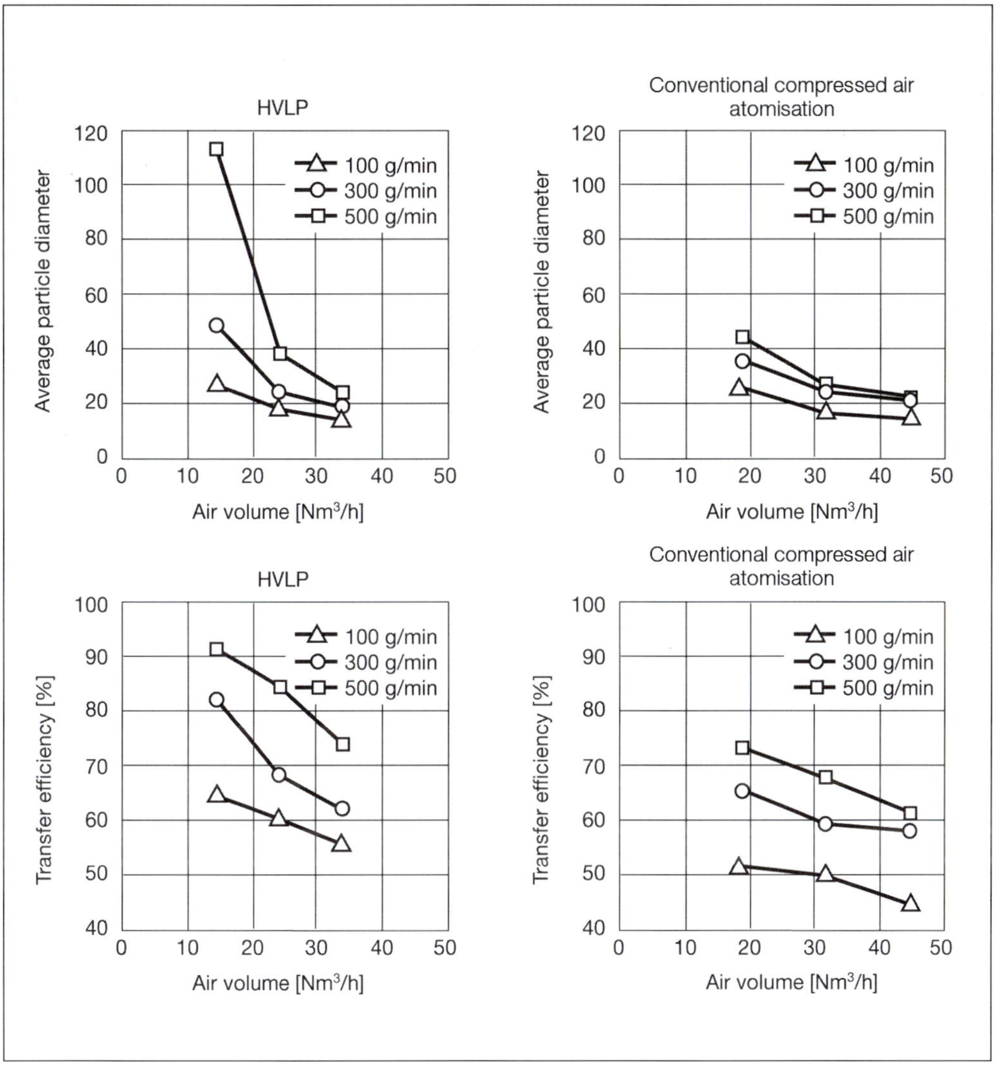

Figure 4.2.47: Droplet sizes and transfer efficiencies in the HVLP process compared with standard spray application, and their dependence on material flow and air flow (Nm^3 = standardised cubic metre)

a wider annular gap. This means that, despite the lower initial velocity of the spraying air because of the significantly greater air volume available, the air is also retarded less during atomisation. The positive side effect is a lower proportion of fine droplets in the droplet size spectrum.

The optimum spraying air pressure for HVLP guns is only 0.8 bar and less, although the gun is initially supplied with a pressure of 5 bar. The lower air pressure is achieved by expanding the air before it leaves the gun, and this greatly increases its volume. Thus, at a reduced discharge rate of 150 to 250 ml/min a droplet size spectrum similar to that of the conventional atomisation process is achieved along with a simultaneous reduction in overspray of 15 to 20 %. Figure 4.2.47 illustrates the sensitivity of droplet size in the low-pressure process to the volume of material to be atomised. It also shows the effect on transfer efficiency.

Hydraulic or Airless Atomisation

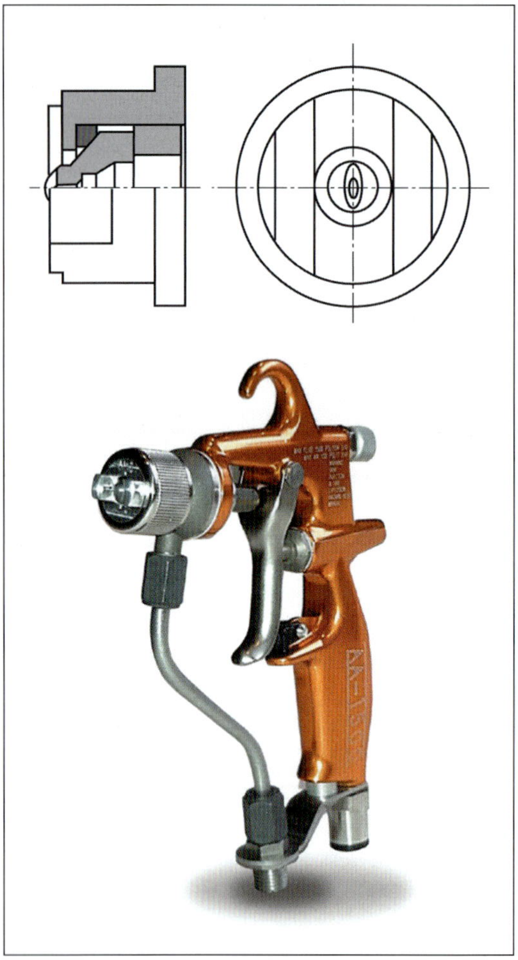

Figure 4.2.48: Schematic drawing of an airless nozzle and picture of an airless gun *Source: ITW*

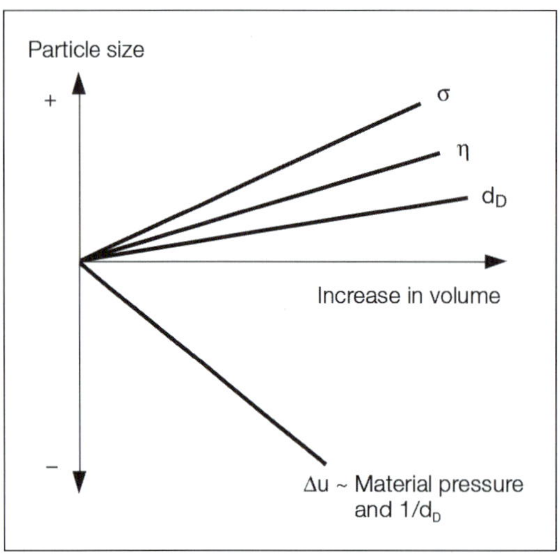

Figure 4.2.49: Sensitivity chart for airless application

The hydraulic atomisation process, also known as **airless** atomisation, brings about particle formation of the liquid coating material purely by high material pressure. The energy required for droplet formation is provided by the material pressure of 50 to 400 bar. Upon exiting the fine nozzles of diameter < 0.5 mm the paint is accelerated to more than 150 m/s and atomised by its spontaneous expansion and additional interaction with the stationary ambient air. The paint flow disintegrates immediately after exiting the nozzle because of the high degree of turbulence which is well above the critical *Reynolds'* number of 2000. The discharge aperture can be designed in the form of a slot nozzle to produce a shaped flat spray. The degree of turbulence can be increased even further by means of special atomising nozzles which form inlet and discharge eddies and thus produce even finer atomisation.

The variables affecting the spray pattern can be summarised according to *Walzel* as follows [4.4.56]:

$$d_{32} = We^{-0.5} \cdot \left[\frac{\rho_{paint}}{\rho_{air}}\right]^{-0.25} \cdot (1 + 3.3\ Ohz) \cdot d_D$$

A high particle velocity, concealed in the *Weber* number We, leads to a lower particle diameter. A higher viscosity, concealed in the *Weber* and *Ohnesorge* numbers, and a larger nozzle diameter d_D increase the droplet size. The discharge rate can be influenced by changing the material pressure or the nozzle diameter. It should be noted that the paint velocity u contained in the *Ohnesorge* and *Weber* numbers is defined by the material pressure and the nozzle width. Any change in d_D necessarily leads to a change in u. By contrast with the pneumatic process,

Application

the paint viscosity has only a minor influence on the droplet size spectrum, while the surface tension and density have a somewhat greater effect. The sensitivity chart in Figure 4.2.49 shows the characteristic dependencies in qualitative terms.

Hydraulic atomisers have no means for adjusting the paint quantities and are either open or closed. Any change in throughput would lead to changes in droplet formation.

The marked stress to which the nozzles are exposed has a detrimental effect on trouble-free longterm operation. Owing to the high velocity of the paint relative to the inside of the nozzle, material wear in the nozzle is high, especially when abrasive pigments are used. Even minor nozzle wear leads to increased film thicknesses and thus to runs and craters because the discharge rate increases to the power of four of the nozzle radius in accordance with the *Hagen-Poiseuille* equation. On the other hand, if the nozzles are made with hard metals there are also, of course, difficulties in producing fine bores. Satisfactory processing of hard nozzle materials only became possible with the advent of laser drilling.

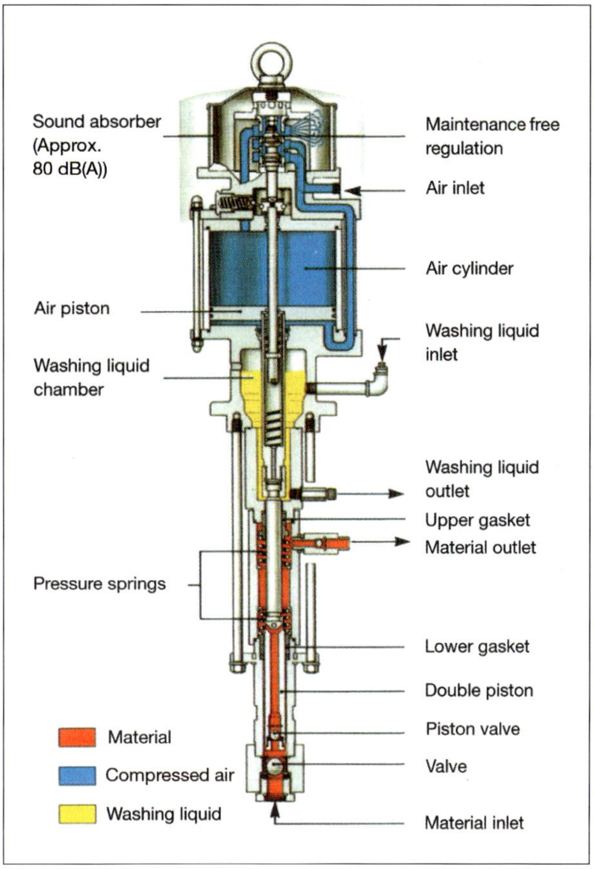

Figure 4.2.50: Schematic drawing of a pump for an airless atomiser
Source: ITW

Apart from a gun, the nozzle and the necessary safety features, a hydraulic atomisation system contains a means of generating the required paint pressure of up to 400 to 500 bar. The high material pressure is produced by pneumatically- or electrically-powered piston pumps capable of trouble-free local operation. Pneumatic double-piston pumps are special in that the high pressures are achieved merely by means of large differences in the cross-sections of the pistons. A surface area ratio of 1:10 to 1:20 enables the normal output pressure of 12 bar to be raised to 120 to 240 bar without difficulty. Approx. 200 litres of air expanded to normal pressure are required per litre of paint to be atomised.

The airless coating process permits the atomisation of relatively large volumes of paint of up to 2 l/min and lends itself to large surface areas. Airless atomisation has established itself in industrial coating applications, particularly for large objects such as ships, cranes, and industrial plant. It offers the advantage of a high material yield compared with the pneumatic process because it dispenses with the use of atomising air.

As spraying air is not used, the rebound effect familiar from pneumatic atomisation is absent. The consequence is lower overspray losses. Low overspray airless atomisation may therefore be used advantageously for paint spraying in enclosed spaces. Less solvent is evaporated in the spray booth because of the lower air volume, with the result that the bulk of the emissions are shifted to the drier or oven.

The 0.1 to 0.5 mm fine nozzles with their extremely small free nozzle surface area are unsuitable for effect paints as they become blocked by the large aluminium flakes (see Chapter 3.2.3). Numerous attempts have been made to extend the universality of airless guns by influencing the spray jet after the atomisation process. One solution has been to deform the spray cone by means of additional air flows, without detriment to the atomisation process. Such air-assisted hydraulic guns are operated at material pressures of less than 200 bar and air pressures of 1 to 4 bar. They are also suitable for more difficult and intricate components. Such processes are known by the name **Airmix**. They combine the benefits of pneumatic and hydraulic spraying.

Airless or Airmix systems are occasionally found to form relatively large quantities of overspray, which settles primarily on earthed parts in the vicinity. Tests have shown that the extensive friction experienced in the nozzles of airless guns by the poorly conducting paints and solvents causes substantial quantities of electrical charge to be transferred to the paint. Unearthed or poorly earthed parts then become charged relatively quickly during coating, as a result of which any subsequent incoming paint of the same charge polarity is repelled and deposited in the vicinity [4.4.57]. For hydraulic atomisation, therefore, it is recommended that the objects to be painted, together with conveyors, be earthed.

In closing, it should be noted that hydraulic paint atomisation can be optimised by processing heated material. This offers the same advantages as those described under pneumatic application.

Electrostatic Spray Coating

Electrical energy can also provide the energy required for spray coating. If charged liquids of charge Q are brought into an electric field of strength E, a force F_1 acts on them. The material is consequently attracted by the electrode of opposite charge.

$$F_1 = Q \cdot E$$

The electrical force is directly proportional to the charge on the liquid and the electric field strength. As frictional forces F_2 act in the opposite direction

$$F_2 = 6 \cdot \pi \cdot \eta \cdot r \cdot u$$

a steady state is soon established in accordance with Stokes' law. The electrically charged material then moves at a velocity of

$$u = \frac{Q \cdot E}{6 \cdot \pi \cdot \eta \cdot r}$$

along the field lines.

Such electrical phenomena can be exploited for coating purposes by designing equipment which can transfer charges to the liquids to be atomised and so create the conditions for formation and transport of droplets to the counter-electrode.

To this end a sufficiently strong electric field has to be generated by connecting a direct voltage between the atomiser and the object to be painted. The whole process of electrostatic paint application can therefore be broken down into generation of an electric field, charging of the paint, formation of paint droplets, transport to the object and electrical discharging of the film as it forms.

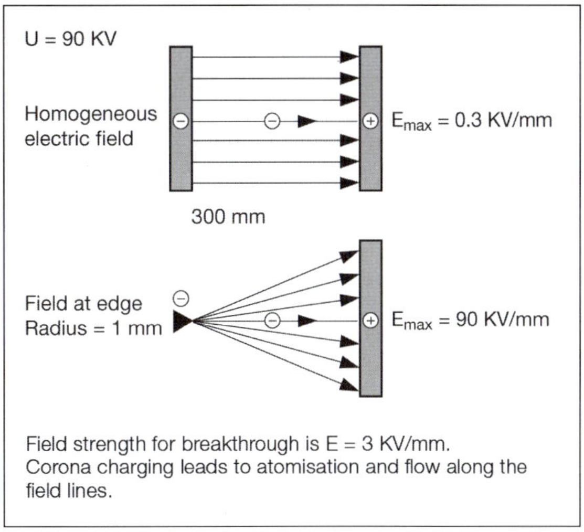

Figure 4.2.51: Homogeneous and non-homogeneous electric fields

An electric field strong enough to transfer charges is created by generating a negative potential in an electrode which is highly curved and which therefore has a greater field line density. The field strength E_S attainable in the range of such pointed electrodes can be approximated by $E_S = U/r$, where r describes the real rounding of such points. Calculation of the non-homogeneous field in the region of edge-shaped spray electrodes is somewhat more complicated. Here the field strength E(x) at location x obeys the following rule:

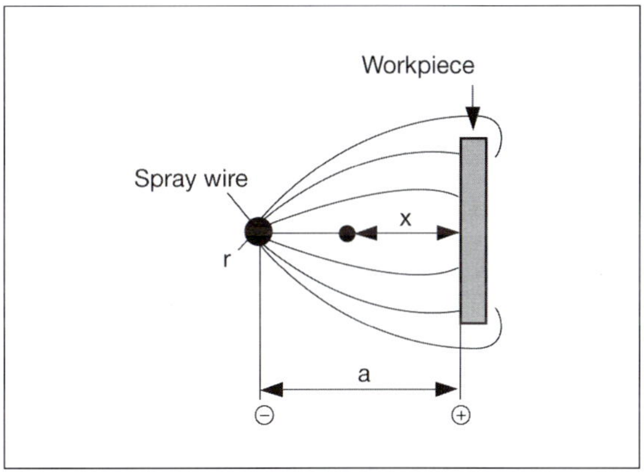

Figure 4.2.52: Calculation of the electric field strength in the spraying region

$$E(x) = \frac{2 \cdot a \cdot U}{(a^2 - x^2) \cdot \ln \frac{2 \cdot a}{r}}$$

here a is the distance between the two electrodes and r is the radius of the edge. Thus the maximum field strength E_{max} of the non-homogeneous field at the point x = a - r is:

$$E_{max} = \frac{U}{r \cdot \ln \frac{2 \cdot a}{r}}$$

With real spray edge radii of approx. 0.5 mm, the usual voltages of 90 kV generate field strengths capable of exceeding the breakdown field strength of approx. 3 kV/mm. This enables the electrons of the metal in the spray edge to leave the metal and to ionise the surrounding material.

Any paint in the region of the spray edge usually becomes negatively charged and is attracted by the object, which is connected up as the counter-electrode, and is thus pulled apart in the form of filaments. It is easy to estimate that the force F acting under the electrostatic atomisation conditions described is capable of performing the necessary work

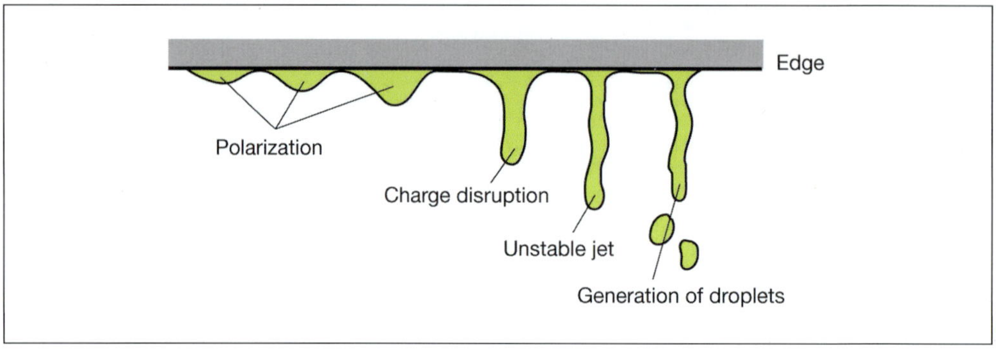

Figure 4.2.53: Generation of droplets by electric forces

to enlarge the surface area. It is influenced by the electric field strength and also by the dielectric constant ε_{paint} (see below).

$$F = E^2 \cdot \frac{\varepsilon_{Paint} - 1}{2 \cdot \varepsilon_{Paint}}$$

Application

Once the filaments reach a sufficient length (see Figures 4.2.32 and 4.2.53), necking occurs under the acting surface tension, and electrically charged droplets form. High-speed photography of charged liquids exiting nozzles show that the forming filaments do not take the shortest path to the counter-electrode as their surface charge is the same as that of the counter-electrode and thus repulsion occurs. Slowed down by the air resistance, wavy electrical counterforces which can be recognised by their meandering shape overlap en route to the counter-electrode. The filament then becomes excessively extended and thus thinner [4.4.58].

The droplet size in the case of electrostatic atomisation is estimated in the same way that the droplet size spectrum is calculated for pneumatic or hydraulic paint application. Particle formation is again influenced by two surface-related forces acting in opposite directions. In this case the repelling electrical forces act on the paint surface against the cohesive pressure p_{Ko} (see above).

$$p_{Ko} = \frac{2 \cdot \sigma}{r}$$

The electric counter-pressure p_{El} is given by

$$p_{El} = \frac{\pi}{2} \cdot \Psi^2 \cdot r$$

where Ψ is the charge per unit area.

At equilibrium both forces are equal, so at $2r = d_m$

$$\frac{2 \cdot \sigma}{r} = \frac{\pi}{2} \cdot \Psi^2 \cdot r \quad \text{or} \quad d_m = k \cdot \sqrt{\frac{\sigma}{\Psi^2}}$$

Since the charge per unit area Ψ is a function of the conductivity λ, the voltage U, the dielectric constant ε and the resulting mass flow rate $\dot{m}$, which is determined by the design, the charge per unit area which occurs during electrostatic spray coating can be calculated as follows:

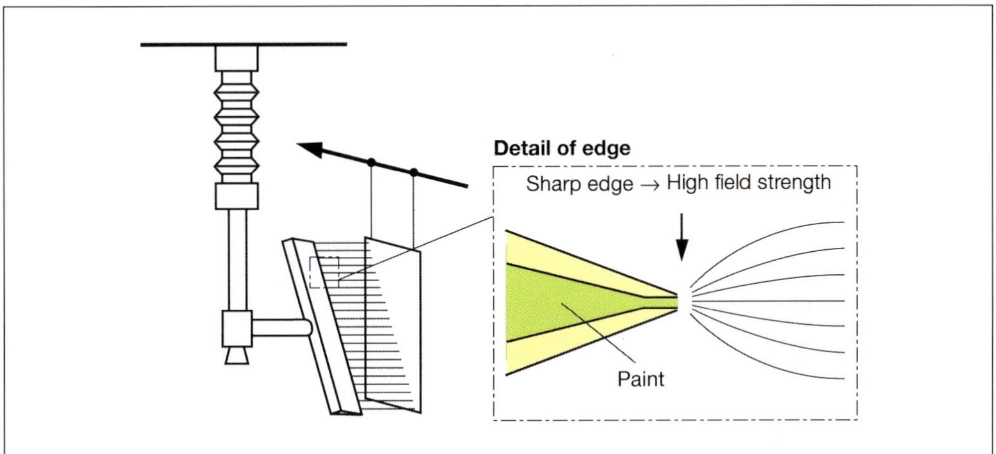

Figure 4.2.54: Schematic drawing of the spray gap for electrostatic paint application

$$\Psi = \frac{k_2 \cdot \lambda \cdot U}{\varepsilon \cdot \dot{m}}$$

The droplet diameter d_m is therefore given by:

$$d_m = \frac{k \cdot \sqrt{\sigma} \cdot \varepsilon \cdot \dot{m}}{\lambda \cdot U}$$

From this equation it is clear that in the case of purely electrostatic atomisation the droplet size is directly proportional to the material properties of surface tension and dielectric constant and indirectly proportional to the conductivity.

The technical equipment employed for electrostatic atomisation consists in its simplest form of a metal bar fitted with a sharp edge. The paint is supplied via an aperture in the area of the spray edge behind which the paint that is to be sprayed is passed when the equipment is in operation.

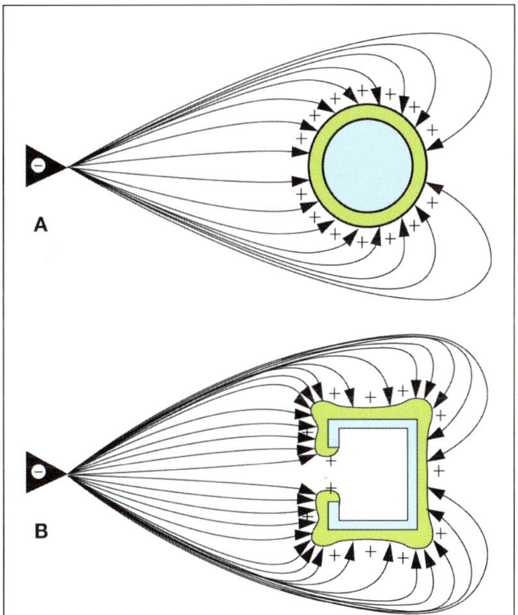

Figure 4.2.55: Film thickness distribution of electrostatically coated parts

The spray aperture (spray gap) shown in Figure 4.2.54 is only suitable for stationary systems because of its large mass which is under a high voltage of up to 160 kV [4.4.59].

When an earthed workpiece is brought into the vicinity of the aperture on a conveyor, an electric field is generated if a voltage is applied. The paint is then drawn out of the aperture, converted into droplets and transported along the field lines to the object to be painted. A particular feature of this process is that, despite a permanently connected voltage, atomisation only takes place if earthed objects, i.e. connected up as a

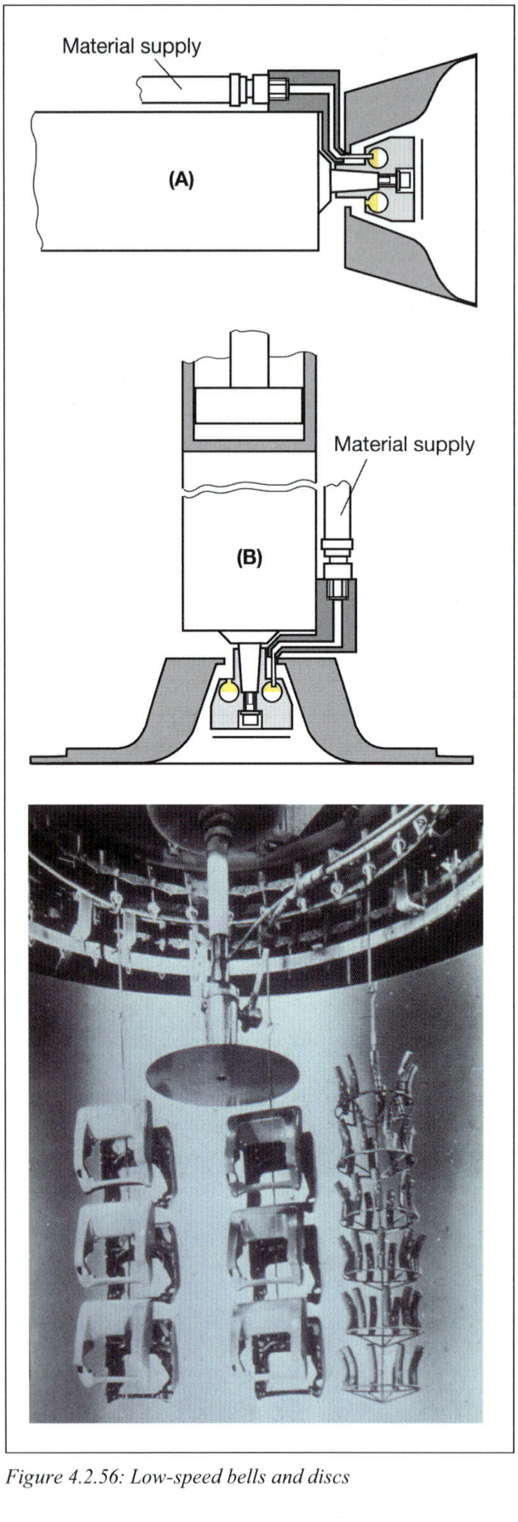

Figure 4.2.56: Low-speed bells and discs

counter-electrode, pass close enough to the spray head. Since the paint is transported only along the field lines to the object, the material yield is extremely high at approx. 98 %.

The painting width is essentially determined by the length of the spray aperture. Adjustable auxiliary electrodes are located at the ends of the aperture to optimise the spraying width. These enable the spray edge to be covered locally. The quantity of paint which can be atomised for a given spray aperture length varies with the set voltage. If this quantity of paint is insufficient to produce the desired film thickness, the aperture can be turned at a diagonal to shorten the coating width and thus increase the film thickness.

The thickness distribution of the resulting film is determined by the field density. If this is the same at all points of the object, extraordinarily uniform film thicknesses are achieved. Pipes and balls, i.e. objects without corners and edges, are optimally coated but satisfactory results are not obtained for angular objects or those containing cavities. Overlapping takes place on corners and edges, and can lead to faults in the form of runs during the subsequent film forming process. Cavities form *Faraday* cages, as a result of which no coating takes place at all because of the lack of field lines. It has already been shown that the droplet size is a function of the mass flow rate, the specific electrical properties of the paint and the applied voltage. A material already adjusted for optimum electrical conductivity will, given the dependence of the mass flow on the voltage, be influenced only by the voltage. Consequently, influences arising from fluctuations in materials under different ambient conditions of temperature or relative humidity are hard to compensate for. A further disadvantage is the only moderate output performance of approx. 1.5 g/min per 1 cm edge length.

To exploit the advantages of high paint yield at higher material output, equipment for electrostatic coating was developed which regulated the process via material dispensing. A more adaptable solution and one that simultaneously offers better performance is low-speed **spray bells**, which also atomise solely by electrostatic means. At a speed of < 3000 rpm droplet formation occurs purely in accordance with the laws of electrostatics described above and without mechanical assistance. The acting centrifugal forces merely ensure sufficient, velocity-dependent transport of the paint to the atomising edge.

The discharge rates can be adjusted within certain limits by the speed of rotation. Feed and discharge at the edge must remain in equilibrium. The disc speed imparts additional kinetic energy to the droplets, with the result that *Faraday* cages can be coated to an extent.

Like the spray aperture, the spray bell, which rotates at 900 to 2000 rpm under a voltage of 90 to 160 kV, has a design which lends itself only to automated processing systems. Atomisation of the paint occurs at the rotating, sharp bell edge which is under high voltage. The paint is fed to the centre of the bell on the inside, accelerated by the bell's rotation and transported to the edge of the bell by centrifugal forces. At the bell edge the paint forms a bead, from where it is then electrostatically entrained droplet by droplet. At approx. 3 g/min per cm edge length the maximum spray quantity is double that of the spray aperture.

Larger than the spray bell is the spray disc, which has a diameter ranging from 200 to 600 mm and is widely used for electrostatic coating of simple mass-produced goods. Atomisation takes place, as with the spray bell, at a rotating, sharp disc edge. Generally, the paint is applied to the underside of the disc, accelerated and transported to the disc edge via centrifugal forces. The remainder of the atomisation and paint droplet transport sequence is the same as that of the spray bell.

The advantages of this spray system come to bear when the workpieces travel around the horizontally configured disc in the form of an Ω (omega loop). As the disc sprays in one plane only, it must be moved vertically by a lifting cylinder to ensure sufficiently uniform coating (see Figure 4.2.56).

It has already been mentioned that electrostatic atomisation is a function of the conductivity of the coating material. The electrical conductivity is responsible for the rapid electric charging during droplet formation and for correspondingly rapid electrical discharging during film formation. Only if rapid discharging due to good earthing occurs it can be guaranteed that subsequent incoming spray mist will be deposited, and good levelling ensured. The latter is a fundamental problem in the coating of wood and plastics. If discharge occurs too slowly, irrespective of whether it the paint or the substrate has insufficient conductivity, levelling flaws are inevitable due to the presence of charges of the same polarity and thus repulsion.

The optimum conditions for electrostatic processing exist if the resistances of the coating material are between 10^6 and 10^{10} Ω cm or the conductivity ranges from 10^{-6} to 10^{-10} S/cm. Such values are easily achieved for solvent-borne paints, e.g. by means of blends of xylene ($R = 10^{10}$ Ω cm) and butanol ($R = 10^6$ Ω cm). If polar solvents cannot be used in the necessary quantity, the conductivity can be regulated additionally through the use of appropriate additives (see Chapter 2.1.4). Water-borne paints cannot be processed with this method because of their high conductivity.

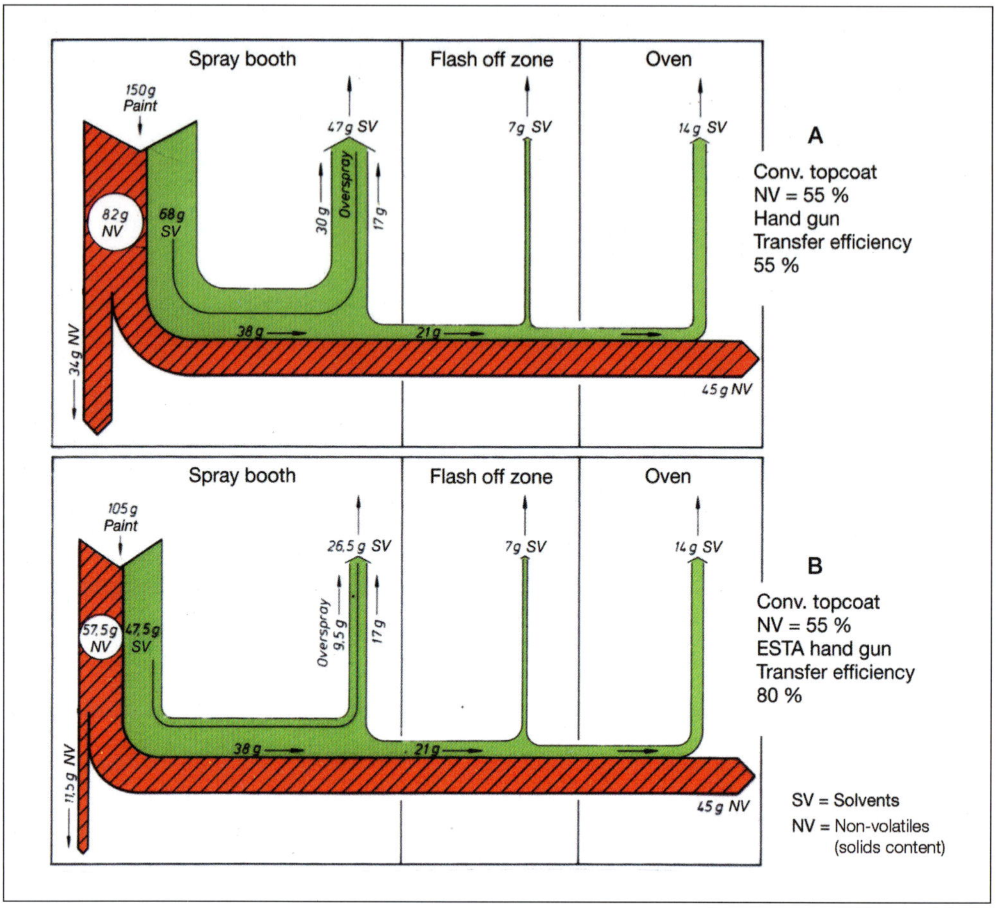

Figure 4.2.57: Yields of conventional and electrostatically assisted high pressure atomisation

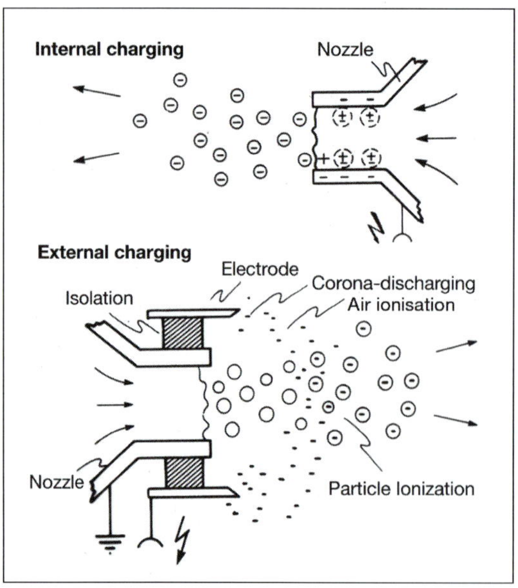

Figure 4.2.58: Schematic drawing of the two charging mechanisms for spray guns

Electrostatically Assisted Spray Processes

Pneumatic and Hydraulic Application

Given the good material yield in electrostatic application processes there have been plenty of attempts at combining pneumatic and hydraulic atomisation processes (see above) with electrostatic techniques. The achievable increase in the material yield has an impact on the total material flow in the spray booth, the evaporation zone and the oven. The following example provides detailed information on the advantages (see Figure 4.2.57). To achieve a coating of 45 g solid content pneumatic atomisation (A) requires

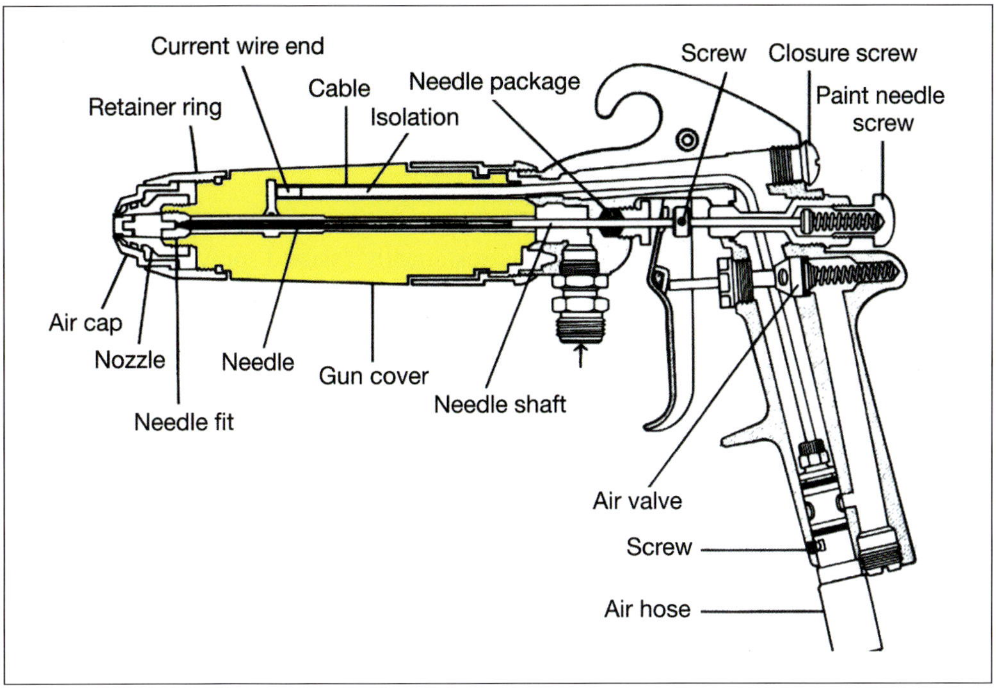

Figure 4.2.59: Cross-section of an electrostatically assisted high pressure gun

150 g paint of 55 % solid content, whereas the electrostatically assisted process (B) needs just 105 g paint.

Two different types of charging process have been shown to be effective, the choice between them depending on the paint. In accordance with the paint's conductivity, charging takes place either by means of **internal charging** before atomisation or by means of **external charging** after the paint has exited the spray head and droplet formation is complete. In the case of pneumatic application only the internal charging process is used in daily operations. In this the paint passes a high-voltage electrode in the electrically insulated gun attachment where it picks up the charges. Since the droplets which are later generated already carry a charge before they are formed, the instantaneous surface tension is reduced due to the repelling electrical forces. The outcome is better atomisation. This means that atomisation can be carried out at a higher spray viscosity or lower spraying-air pressure. The first case leads to even lower emissions due to the higher solid content, while the second leads to even less material lost to overspray. Under ideal conditions the material yield can be raised from 50 % for conventional processing to 80 %.

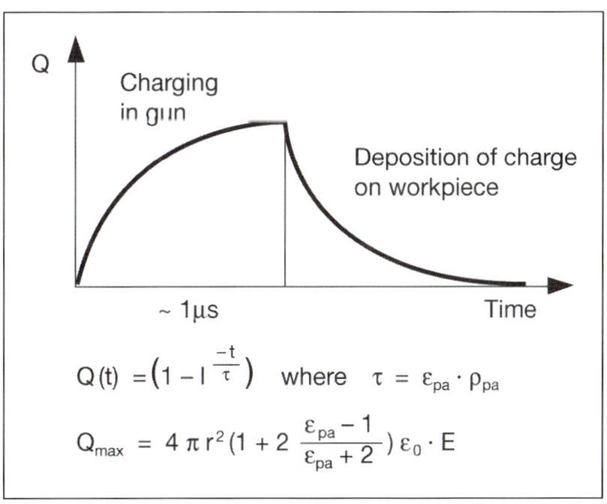

Figure 4.2.60: Charge uptake and maximum charging of coating materials as a function of processing time

However, internal charging is only possible if the specific resistances of the coating material lie within the already specified limits of 10^6 to 10^{10} Ω cm. Otherwise a considerable outlay would be required for safety reasons to insulate the paint-feed system. Furthermore, it should be noted that the charging rate depends directly on the dielectric constant ε_{pa} and the specific resistance δ_{pa} and therefore does not permit excessively high electrical resistances.

High-Speed Rotary Atomisation

The most advanced and most effective electrostatically assisted spray process is atomisation with **high-speed rotary bells**. The bells, which have diameters of 60 to 100 mm, rotate extremely fast at up to 80,000 rpm. Consequently, the atomisation mechanism is different from that of the low-speed bells and discs. Unlike exclusively electrostatic atomisation, high-speed rotary atomisation relies solely on centrifugal forces. The supplementary electrostatic charging of droplets merely serves to aim the spray jet towards the earthed object and thus to increase the material yield. This can reach extremely high levels of 85 to 90 %.

Unlike spray discs, high-speed rotary bells do not spray the object radially as it passes by. The spray jet is diverted through 90° by means of additional shaping air ("air shower") and thus directed onto the object to be painted which is under the bell. In this regard control over the shaping air is a key factor in the transfer efficiency of the bells [4.4.61].

High-speed rotary bells, with a material output of 15 to 20 g/min per cm edge length, are much more productive than purely electrostatic bells. In addition, the droplet size spectrum is narrower and easier to optimise to the particular requirement by means of machinery adjustments. The excellent visual quality of the coatings which can be achieved with this method has made high-speed rotary systems very popular.

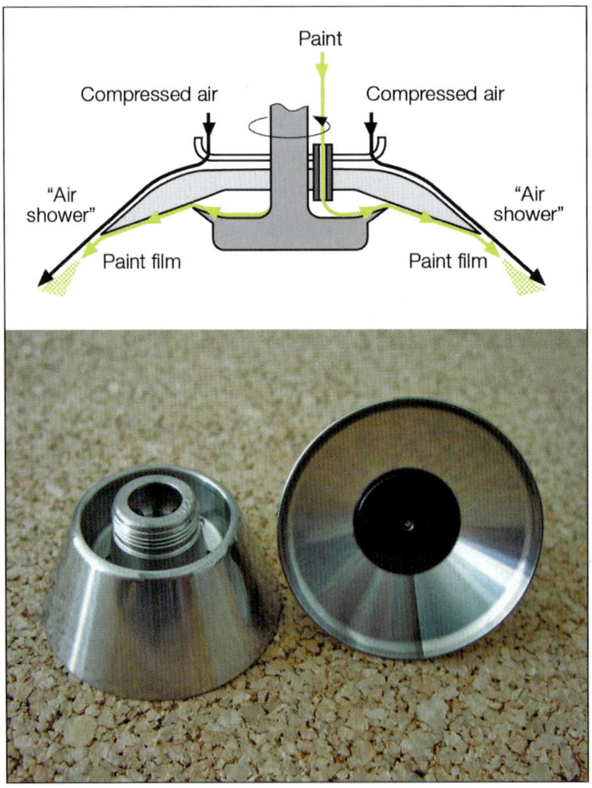

Figure 4.2.61: Schematic drawing and picture of a high-speed rotary bell *Source: BASF Coatings*

Having been fed to the inside of the bell, the coating material is transported by centrifugal forces to the bell rim where, in accordance with the operating conditions, it is atomised by various mechanisms. If the material feed is low at normal rotation speed, a bead forms at the bell rim from which individual filaments are drawn under the effect of the high centrifugal forces and then stabilise into droplets (see Figure 4.2.62, A). At a higher material feed the number of filaments increases. Once these reach a certain

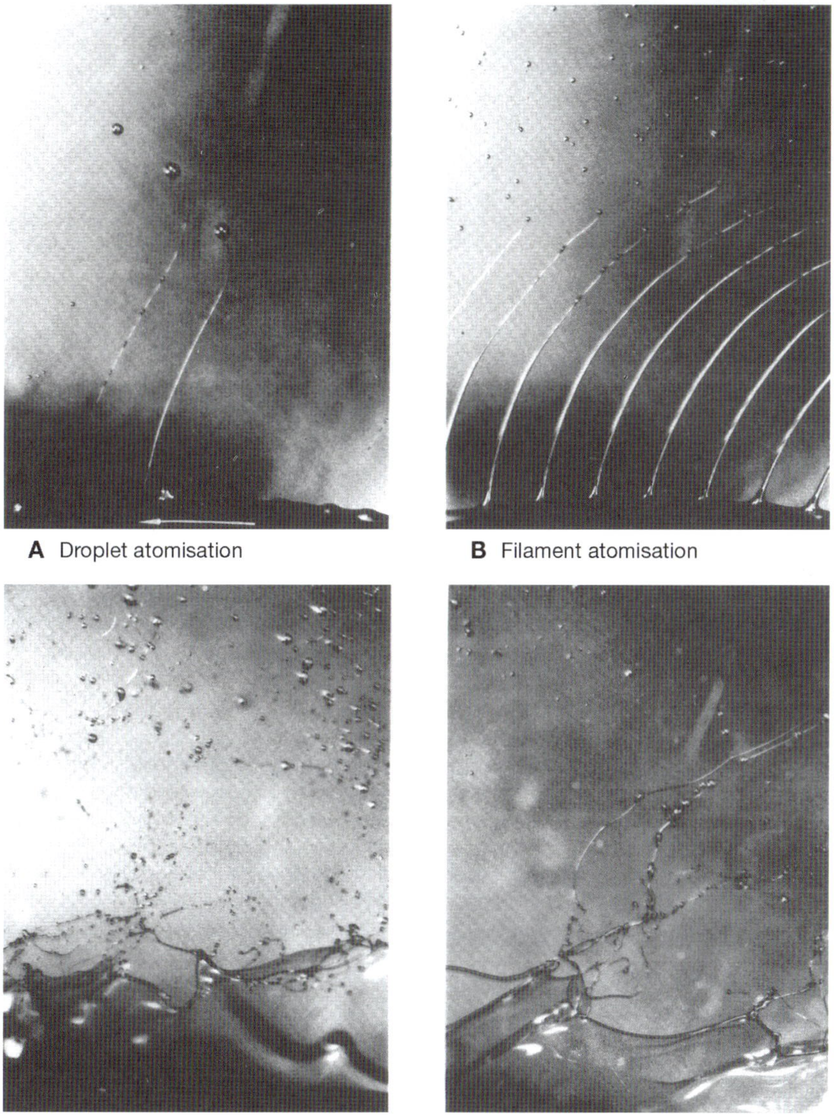

Figure 4.2.62: High-speed photographs of droplet formation at high-speed rotary bells

length, particle formation occurs due to the effect of the surface tension and the electrical forces, yielding a uniformly narrow droplet size spectrum of filament atomisation (B). This constitutes the best atomisation method for many applications. Further increase causes the supply of paint to become so great that the filaments are converted into lamellae. In aerodynamic lamellar atomisation (C_1) the paint bead is initially drawn out, without formation of filaments, to create a paint curtain which is made to oscillate by the ambient air. This process, which resembles a flag fluttering in the wind, also leads to droplet formation in the second stage. The droplet size spectrum generated by the lamellar atomisation is considerably less regular and is broader than that

Figure 4.2.63: Smooth and textured rims of bell cups
Source: BASF Coatings

created by filament atomisation. If the speed is increased, the lamellar motion changes to turbulence, with the result that regular disturbing fronts migrate from the edge to the end of the lamella where they have an impact on droplet formation (C_2). As a result of the high centrifugal forces an area of underpressure occurs in the central region of the bell (see Figure 4.2.61).

The bell rim is the most sensitive part of the bell design as regards fineness of atomisation. The bevel and texture (in the form of knurling) of the bell rim impact the droplet size spectrum and thus also the quality of the coating. In line with the aim of achieving the narrowest and finest possible droplet spectra with high rim coverage and thus high material output, the paint beads are converted into filaments by the grooves in the bell edges.

The relevant atomisation mechanism and thus also the droplet size spectrum depend on the material feed $\dot{V}$, rim coverage K_b, angular velocity ω, bell diameter D and coating material variables of surface tension σ, viscosity η and density ρ. The operating ranges just described can be represented graphically by calculating atomisation coefficients. The most favourable conditions can be described by plotting $We_2^{1/2} \cdot Kb^{5/6} \cdot Ohz_2^{10/36}$ against Ohz.

$$We_2^{\frac{1}{2}} \cdot Kb^{\frac{5}{6}} \cdot Ohz_2^{\frac{10}{36}} \quad = \text{Process number}$$

$$Ohz_2 = \frac{\eta}{\sqrt{\rho \cdot \sigma \cdot D}} \quad = \text{Paint number}$$

$$We_2 = \omega^2 \cdot D^3 \cdot \frac{\rho}{\sigma}$$

$$K_b = \dot{V}^2 \cdot \frac{\rho}{\sigma \cdot D^3}$$

The boundary lines in Figure 4.2.64 indicate the transition areas from one atomisation mechanism to another. The relevant working area is marked as a hatched surface.

The expected droplet size yielded by high-speed rotary atomisation can be somewhat forecast in much the same way as the droplet sizes for pneumatic and hydraulic atomisation. The results published in the literature do not present a coherent picture [4.4.62, 4.4.63]. The tests there were carried out with different liquids. Water, glycerine or paints have different atomisation properties and cannot therefore be described by a single mathematical equation either. A common feature of all the results, however, is the marked dependence of the droplet size spectrum on the surface tension, angular velocity and thus the rota-tional speed and size of the bell. The minor influence exerted by the viscosity over large operating ranges is also worth noting. This fact is confirmed in practice.

The heavy dependence of the atomisation quality or droplet size spectrum on the surface tension is striking, particularly for the processing of water-borne paints. In comparison

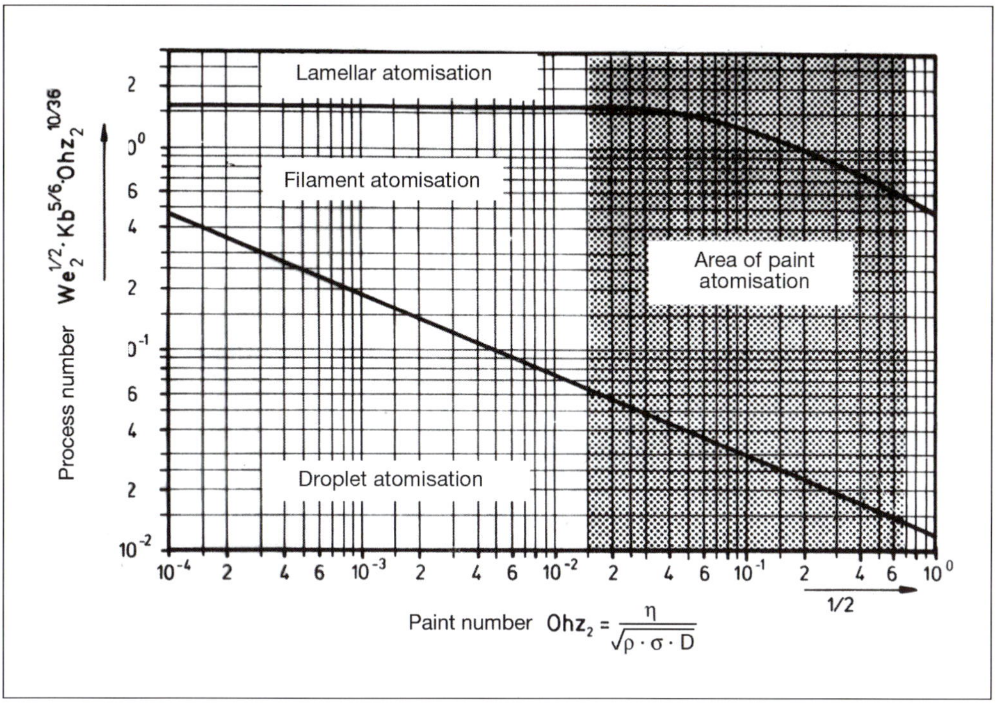

Figure 4.2.64: Working ranges for high-speed rotary atomisation

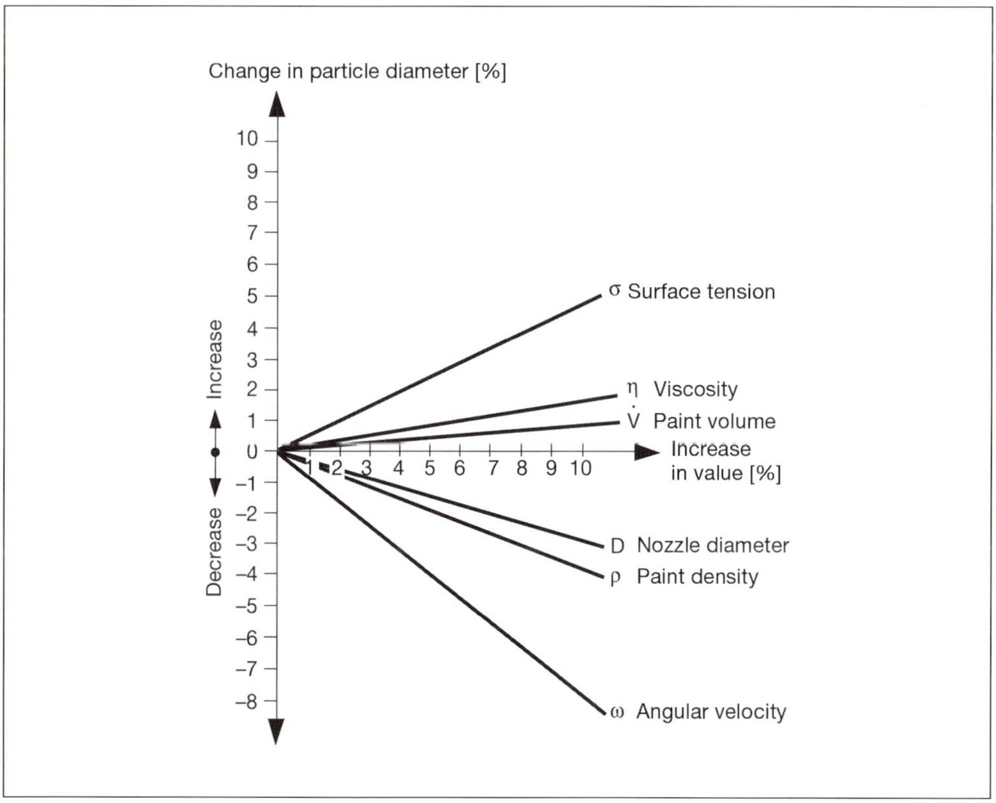

Figure 4.2.65: Sensitivity chart for high-speed rotary atomisation

Application

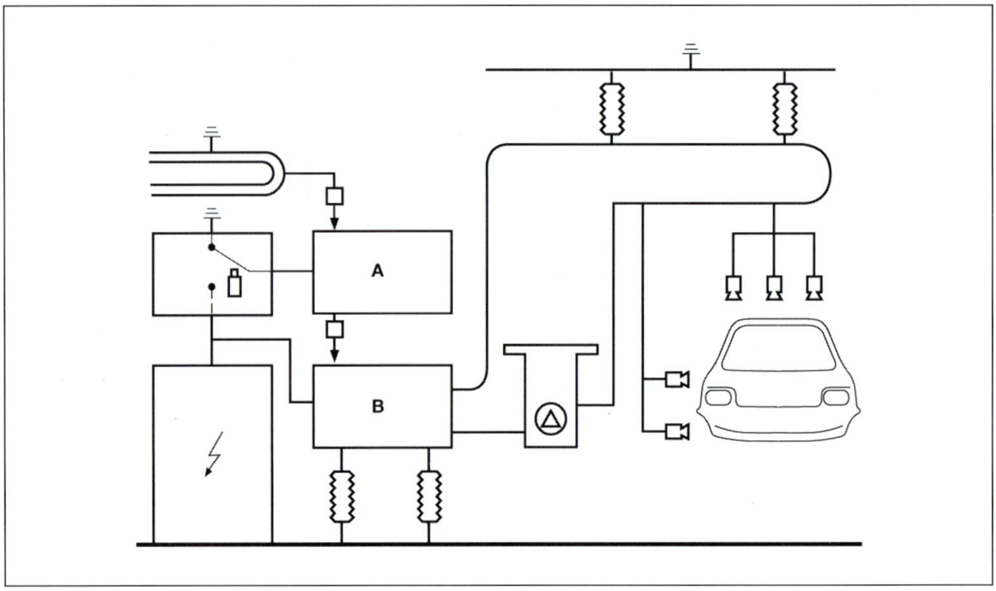

Figure 4.2.66: Processing of water-borne paints with internal charging by two separate feed tanks (A,B)

with solvent-borne paints, speeds up to 1000 rpm higher have to be set if the coarser droplet size spectrum of water-borne paints, caused by the high surface tension (see Chapter 5.6.2), is to achieve the same level of fineness as is attained with solvent-borne paints under the same conditions. The influence of paint throughput is also lower than in the case of pneumatic atomisation.

The feed and distribution of the material on the bell also require special attention. Deposition of overspray on the bell cup can be avoided by splitting the material flow. While the bulk of the material is discharged in the outer area of the cup, a smaller portion is fed to the cup through the centre bore. In this way overspray sucked back into the aforesaid area of underpressure at the bell is prevented from being deposited there and from later giving rise to quality defects in the form of pinholing on the object to be painted (see Chapter 6.2).

Rapid colour changes needed for many painting tasks can be achieved with only little material loss by circulating a rinsing agent. The rinsing process, including colour change, can be completed in just a few seconds.

As in the case of internal charging of spray guns described earlier, charging of the coating material in the high-speed rotary bell proceeds by subjecting the material feed to a voltage. This applies in particular to solvent-borne paints of sufficiently high electrical resistance (see above). For processing water-borne paints the entire system must be insulated. The paint-feed system (A) in particular must feature insulation of the working and top-up dispensing vessels (B) to prevent voltages or current flows in any part of the circulation system (see Figure 4.2.66). Colour changes in this system can only be achieved through corresponding investment.

The method of external charging mentioned earlier enjoys greater practical importance. In principle, it can be used with any type of paint. It leads to charging of the already formed droplets and is easy to implement in high-speed rotary coating processes. To bring about external charging, electrons have to be released which ionise the air molecules in the atomisation area in front of the material nozzle. The freshly created droplets adsorb part of the charge carriers as they fly through this ion cloud, becoming electrically charged particles in the process. A disadvantage is the high atmospheric ion concentration, which shields the object to be painted and also leads to a less uniform droplet charge compared with internal charging. Associated with this is the fact that material losses are up to 5 % greater than for internal charging.

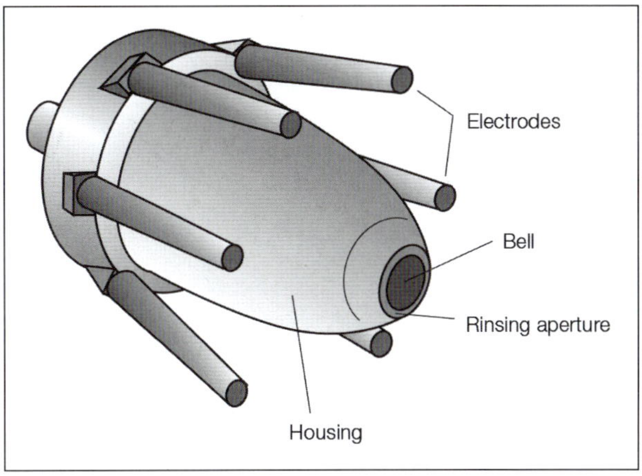

Figure 4.2.67: High-speed rotary atomiser for external charging

Optimised bell shapes were made possible by studies of the flow conditions in the spray cone with the aid of laser cutting techniques [4.4.64]. Some of the effort expended on developing high-speed rotary atomisation is being directed at effect coating. The droplet size spectra involved here are different from those of pneumatic coating. The latter leads to a measurable increase in the proportion of effect material in the deposited coating film. The shaping air of bell atomisation is less adept at capturing the large droplets of the spectrum than the small ones because of the acting centrifugal forces and their greater inertia, with the result that the profile of the deposited droplet size spectrum is shifted in the direction of the smaller and medium sized particles. This is the exact opposite of the situation for pneumatic application [4.4.65].

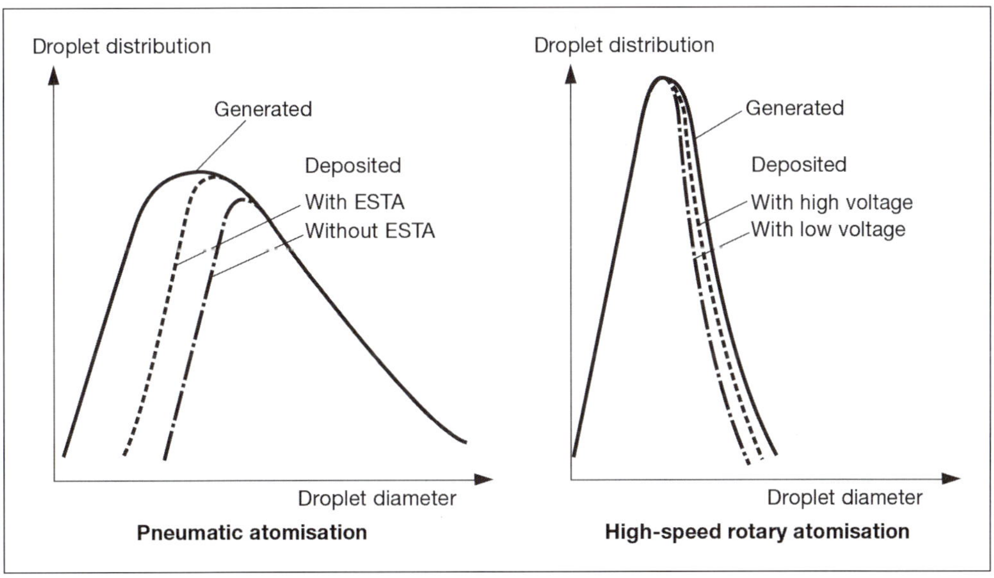

Figure 4.2.68: Differences in the generated and deposited droplet size distributions for pneumatic and high-speed rotary atomisation (ESTA = electrostatic application)

Application

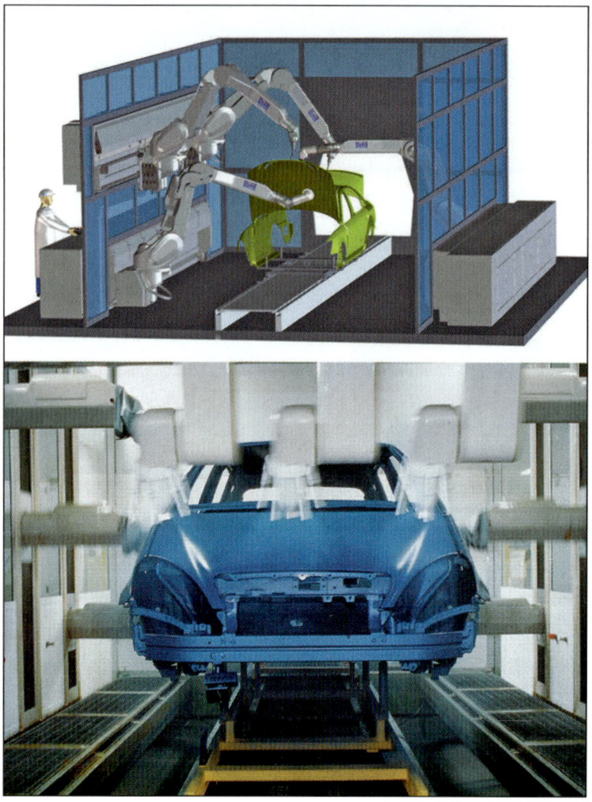

Figure 4.2.69: Schematic drawing and picture of a high-speed rotary bell application unit for coating cars

Sources: above: Dürr, below: BASF Coatings

On the other hand, high-speed rotary atomisation leads to separation and concentration of effect substances, particularly metal bronzes, in the droplets and thus to a nonhomogeneous distribution combined with poorer alignment in the deposited coating. This centrifugal separation is also induced on the bell cup and by rotation of the resulting droplets.

In order that all areas of intricate parts (such as car bodies) may be coated with a uniform layer while achieving good flow at the same time, the spray heads in the automated painting systems are configured so as to maintain consistent distances from the contours of the object to be painted. The advantages of electrostatic assistance in general are the good throwing power and edge expansion of the spray jet. This translates to gentle transition zones and a more uniform film thickness distribution on the back of the painted objects as well.

Thanks to ongoing refinements in the systems and atomiser bells, high-speed rotary atomisation is today the most advanced, cost-effective and, from an environmental perspective too, attractive atomising technology. The bells, which originally had diameters of 70 to 80 mm and speeds of 25,000 to 30,000 rpm, have become ever smaller and faster over the years. Bell cups with a diameter of 40 to 60 mm and speeds of up to 70,000 rpm are now standard. Such bells are fitted with shafts mounted on air-bearings for trouble-free operation. The smaller bells, which are at a shorter distance from the object, have the additional advantage of even better material yield. Owing to their lower moment of inertia they can follow complex contours and are therefore suitable for application by robots (see Chapter 4.2.1.6) [4.4.66].

Ultrasonic Atomisation

Any comprehensive picture of the basic principles of the atomisation of liquid coating materials must not neglect interesting atomisation techniques which employ ultrasound, but which are encountered only in other application areas. In capillary-wave atomisers thin liquid films are excited on ultrasonic oscillating plates to form waves. At a sufficiently high amplitude of the wave peaks, the tips start to neck and produce droplets. The size of the droplets is determined by the wavelength λ and thus by the frequency f of the exciting wave.

The unit itself consists of an upper stepped tube and a conical plate. This plate is excited to high-frequency oscillations by a piezo ceramic unit mounted on the underside, with the result that the liquid which is fed to the cone forms a thin film which is then converted into droplets.

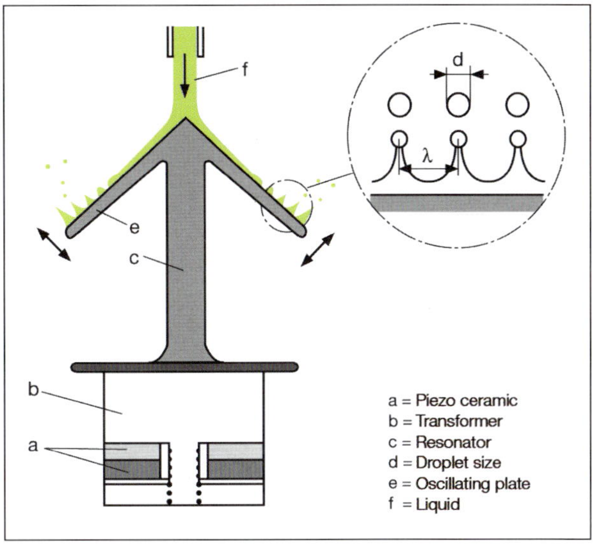

Figure 4.2.70: Ultrasonic capillary wave atomiser

The size of the droplets is a function of the frequency of the oscillation. The wavelength λ of the resulting wave formation by the film in turn depends on the surface tension σ and density ρ of the material to be atomised.

$$\lambda = 2 \cdot \left(\frac{\pi \cdot \sigma}{\rho \cdot f^2} \right)^{\frac{1}{3}}$$

As the particle diameter d is approx. 1/4 of the wavelength, the *Sauter* diameter d_{32} is given by:

$$d_{32} = 0.5 \cdot \left(\frac{\pi \cdot \sigma}{\rho \cdot f^2} \right)^{\frac{1}{3}}$$

Compared with standard atomisers capillary-wave atomisers have the disadvantage of poor deposition of the generated droplets, yet also offer a number of advantages. The most important of these are non-pressurised liquid feed, silent atomisation and energy savings.

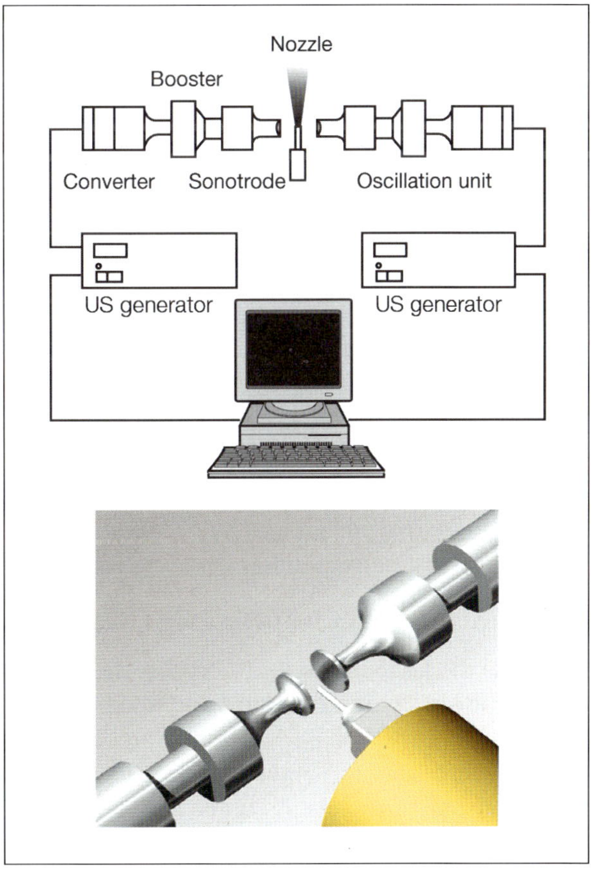

Figure 4.2.71: Scheme and model of an ultrasonic standing-wave atomiser

Another atomisation method, which is of interest for higher-viscosity fluids and thus also for melts, is ultrasonic standing-wave atomisation. This technique utilises a generator to produce the high-frequency energy for particle formation. A converter transforms the elec-

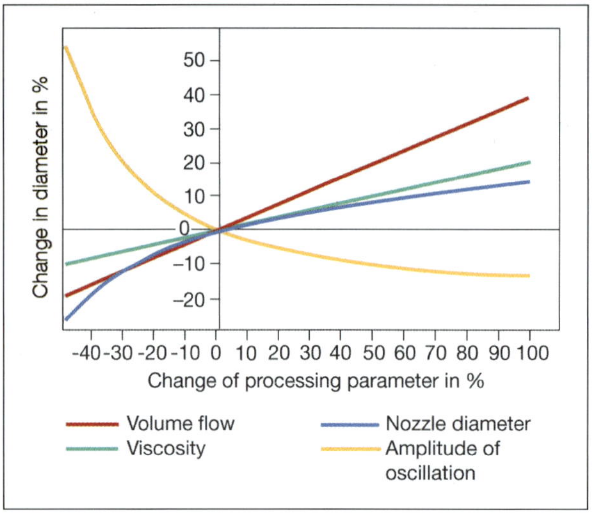

Figure 4.2.72: Sensitivity chart for ultrasonic standing-wave atomisation

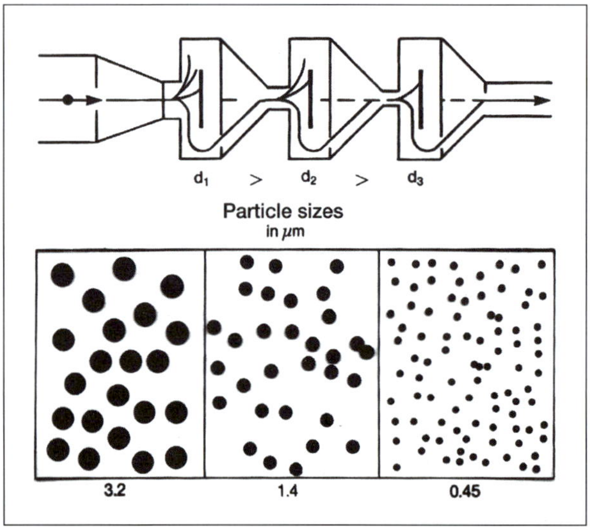

Figure 4.2.73: Principle behind spray droplet deposition in a cascade impactor

trical energy by means of piezo ceramics into mechanical oscillations of the same frequency.

Via a booster the oscillations reach the sonotrode which serves as the oscillating body. This emits the energy in the form of longitudinal ultrasonic waves of a correspondingly high intensity. If the sound waves hit a reflector or a second sonotrode, a standing wave with spatially fixed oscillation nodes is formed at certain distances. At the same time the oscillation nodes are pressure antinodes of the sound pressure of the standing wave.

The fluid to be atomised is introduced into the sound-pressure nodes through one or more feed tubes. The fineness of the generated droplets is determined to a significant extent by the mass flow rate and the positioning of the feed nozzle. On the material side, the surface tension and the viscosity are the most sensitive variables influencing droplet formation [4.4.67]. Figure 4.2.72 shows the percentage changes in the average particle diameters of an atomised polymer melt as a function of the different atomisation parameters.

Despite numerous advantages over the established atomisation techniques, ultrasonic standing-wave atomisation has yet to find application in coating technology.

Special Methods for Characterising Overspray

Owing to the large influence of droplet size on the visual and functional properties of coatings it is very important to be able to measure and describe drop-let sizes and distributions. Information on aerosols is important from an occupational health and safety perspective, too (see Chapter 5). The tests involved have been specially developed for measuring paint droplets. A simple method of estimating droplet sizes during spraying consists in passing plastic films through the spray jet and measuring the deformation of the captured droplets using image analysis. In another process the overspray is briefly directed to a nonmiscible liquid of the same density. The now nondeformed particles

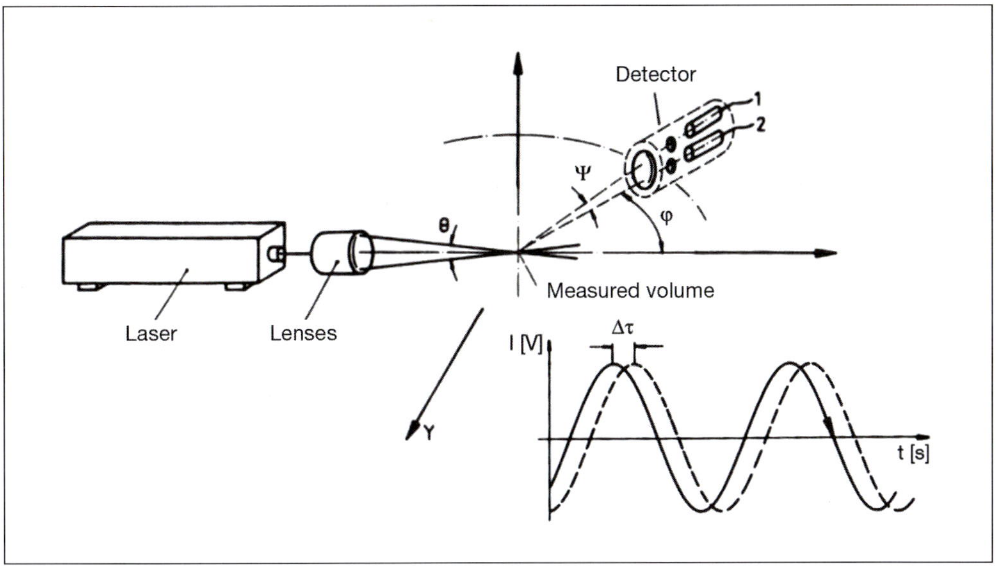

Figure 4.2.74: Phase Doppler anemometer

can be documented optically and again characterised with the aid of digital image analysis.

The cascade impactor, which is frequently used for health and safety tests, works on a completely different principle. This consists of a tube with several different constrictions and impact plates positioned after them. The aerosol droplets of variable size are introduced into the cascade impactor on an air current for the purpose of analysis. Since the air velocity increases with decrease in diameter of the constrictions, only the coarse droplets are deposited on the first plate, while the finer droplets are then deposited on the subsequent impact plates. Evaluation again proceeds by means of measurements or digital image analysis.

Recourse is also made to the methods of light scattering and light refraction for measuring the droplets in overspray (see Chapter 2.3.2). Phase *Doppler* anemometry (PDA) is of great importance in characterising a spray jet.

This measurement method is interesting because it provides information simultaneously on the average particle size, particle size distribution and particle velocity. To this end a measurement volume is generated at the intersection of two laser beams in the spray cone of the paint. Interference occurs in the overlap area of the two jets as spray droplets pass through. Passing paint droplets scatter the light as a function of the velocity and size of the particles [4.4.68]. Apart from the velocity and size of spray particles, local flow conditions between the spray head and the object to be painted are also of interest. To visualise the particle flows, the laser beams are expanded into planes by cylindrical lenses. The paths of the individual light points can be detected and calculated by recording them with a digital camera capable of processing electrical and optical signals (CCD camera).

4.2.1.4 Paint Supply Systems

Where manual painting of small areas or repair work only calls for a small quantity of paint, the material can be supplied from small containers via siphon-fed or gravity-fed devices integrated into the spray head (see above). Pressurised containers with a capacity of a few litres are frequently adequate if the quantities to be processed are not too large and the coating equipment is only operated temporarily. Material feed is effected with overpressure of 0.5 to 1.5 bar. However, if continuous coating is necessary, separate supply systems in the form of circulation systems and paint-supply systems have to be used for the material feed.

Circulation Lines

Circulation lines usually have to be installed where material throughputs exceed 200 kg per day. For greater safety and economical operation, the collection systems for spray heads are supplied with material by circulation systems. The characteristic feature of circulation systems is that the pipes coming out of the storage containers also return to them. To avoid sedimentation of pigments in circulation systems, it is absolutely essential to exceed flow rates of 0.2 to 0.3 m/s in the pipes. The circulation volume should be significantly greater than the drawn-off volume. This is the only way in which a relatively large number of spray booths can be supplied with sufficient material without a reduced discharge rate.

The coating materials, which are kept in a homogeneous state in the storage containers (A) with the aid of agitators (B), are brought to a material pressure of 10 to 20 bar by means of a paint circulation pump (C) and supplied to the individual paint draw-off points (N) through the long pipes (F) which can be up to several hundred metres in length. After passing through a return check-valve (G) the excess material, which is now depressurised again, returns to the storage containers (A) via valve (H). Such systems enable paint consumers to make centralised adjustments to the material to reflect the processing conditions. Several processing units in one or more booths can be supplied at the same time. This ensures the best conditions for obtaining identical colours on a workpiece, even if it has to be coated at several different units on the painting line.

Paint draw-off usually occurs in branch lines via material pressure regulators (I). These depressurise the material to a specified pressure of 1 to 2 bar and guarantee a uniform

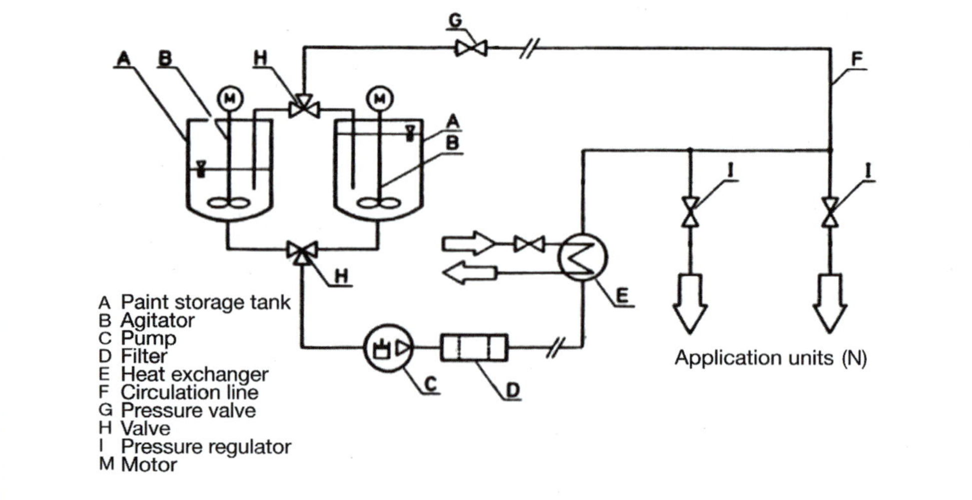

Figure 4.2.75: Schematic drawing and components of circulation systems for spray coating

discharge rate at a constant viscosity, even where there are several draw-off points in a circulation system upstream of the regulator. Sedimentation of pigments in the branch lines can be prevented, even when the gun is closed, by means of supplementary return lines from the material draw-off point.

Constant discharge rates and thus the same film thickness at constant material pressure are ensured by installing an additional thermostat (E in Figure 4.2.75) for the circulating material. In many cases it is better, and it is therefore becoming more and more common, to replace the material pressure regulators by metering pumps. Defects in the coating due to dirt are largely avoided by the installation of dirt filters (D in Figure 4.2.75).

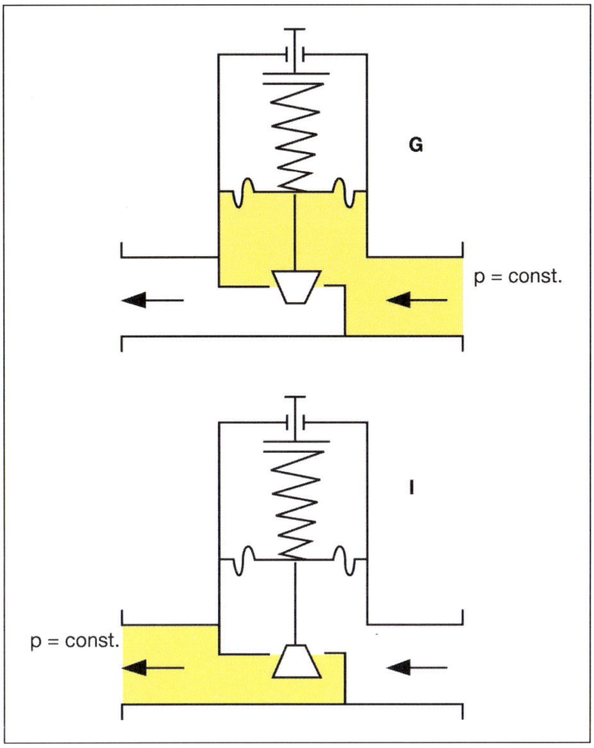

Figure 4.2.76: Schematic drawing of a pressure regulator (I) and return check-valve (G)

The pumps necessary for generating pressure and thus for conveying the coating materials should impart as little shear stress as possible to the material. Despite uniform pressure build-up and uniform material conveying, gear pumps are therefore not very suitable in view of the high shear stress imposed on the material. Secondary dispersing of the pigments in solid colours and creasing of the thin platelets of the effect substances in metallic paints can lead to shifts in colour and effect. Pneumatic piston pumps are therefore widely used for material transport in circulation systems. They switch off automatically on reaching the required pressure and switch back on automatically if the pressure falls below a specified set-point. Pulsation during conveying can be significantly reduced by means of double piston pumps and the additional installation of pressure-equalising containers.

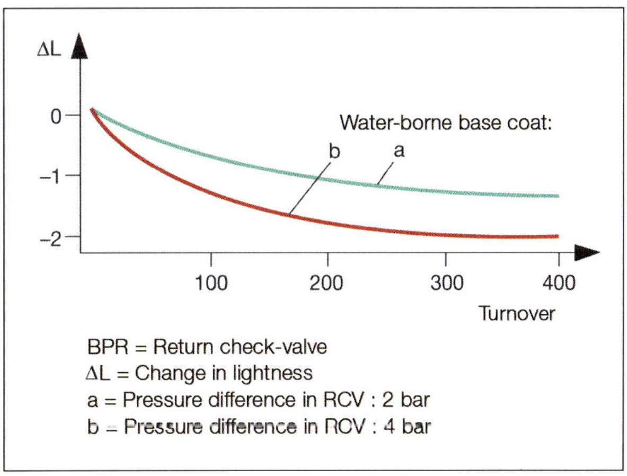

Figure 4.2.77: Decrease in lightness ΔL of metallic effect paints due to shearing in circulation lines

Figure 4.2.77 shows how the return check-valve can give rise to colour shift after an extended load period and low material draw-off.

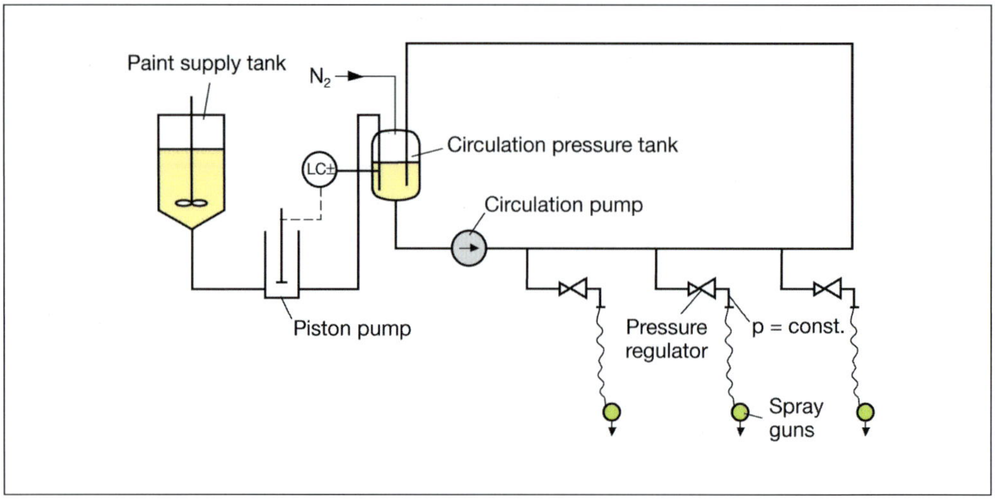

Figure 4.2.78: Circulation line without any shearing return check-valves

The shear load in circulation systems can be gradually reduced by ensuring that the coating material in the circuit is not depressurised when it returns. This can be achieved by installing circulation containers which are permanently pressurised. Since the return check-valve is not needed, shear forces are then transmitted in the pumps only. The paint supply container which is under normal pressure provides the requisite feed of material to the circulation container discontinuously in accordance with the fill level. Such systems have failed to gain acceptance because of the amount of equipment needed.

The constituent material of the circulation system too must suit the coating material circulating within it. The standard materials are iron for solvent-borne paints and stainless steel or plastic for water-borne paints. Polyamide should not be used if polyurethane paints or water-borne paints are being processed as it absorbs up to 5 % water from the air. This would induce moisture-curing of isocyanates while they are still in the circulation system.

Principles Behind Flow in Pipes

Accurate knowledge of flow properties in pipes is essential for checking volumetric flow rates of liquids and designing liquid-supply systems. The flow rate u in pipes is given by the ratio of the volumetric flow rate to the area of the cross-section, $\dot{V}/A$.

$$u = \frac{\dot{V}}{A} \quad \text{or} \quad \dot{V} = u \cdot A$$

The volumetric flow rate can be converted into a mass flow rate by replacing $\dot{V}$ with $\dot{m}/\rho$:

$$\text{because of} \quad \dot{V} = \frac{\dot{m}}{\rho} \quad \text{follows} \quad \dot{m} = \rho \cdot A \cdot u$$

For incompressible liquids, the following applies everywhere within the pipe, irrespective of diameter, when flow is stationary:

$$\dot{m}_1 = \dot{m}_2 = \dot{m}_i = k \quad \text{or} \quad \rho \cdot A_1 \cdot u_1 = \rho \cdot A_2 \cdot u_2 = \rho \cdot A_i \cdot u_i = k$$

$$A_i \cdot u_i = k$$

Incompressible liquids have different flow rates and thus different kinetic energies in pipes of different diameter. Disregarding gravity effects and internal friction initially, the following energy balance for the individual flow sections is obtained.

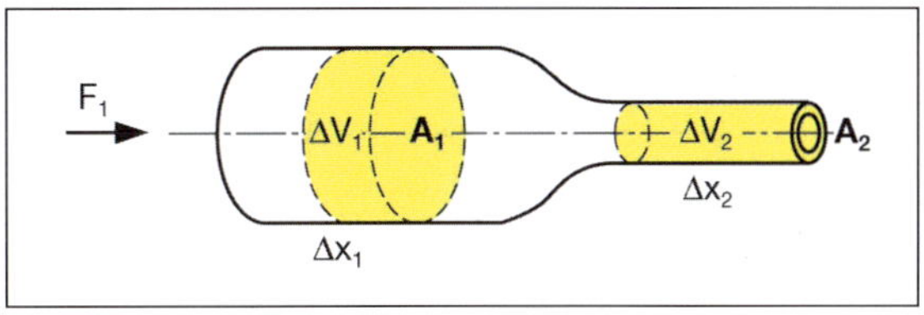

Figure 4.2.79: Flow in pipes of different diameter

By virtue of the incompressibility the transported volume is the same in all sections of the pipe, irrespective of the pressure:

$$\Delta V_1 = \Delta V_2 = K \quad \text{or} \quad A_1 \Delta X_1 = A_2 \Delta X_2$$

The volumetric work performed in section 1 $\Delta W_1 = p_1\ \Delta V_1$ as part of the total energy is not equal to the volumetric work ΔW_2 in section 2 because the flow rate is different.

$$\Delta W_1 = \Delta x_1 \cdot A_1 \cdot p_1 \quad \text{and} \quad \Delta W_2 = \Delta x_2 \cdot A_2 \cdot p_2$$

p_1 and p_2 represent the pressures acting in the sections.

Owing to the higher velocity u_2 in section 2 (right) the kinetic energy is also greater than in section 1. From the law of conservation of energy, the difference in kinetic energy ½ mu^2 must be equal to the difference between ΔW_1 and ΔW_2, i.e.

$$\Delta W_1 - \Delta W_2 = \frac{m}{2} \cdot u_2^2 - \frac{m}{2} \cdot u_1^2$$

$$p_1 \cdot A_1 \cdot \Delta x_1 - p_2 \cdot A_2 \cdot \Delta x_2 = \frac{\rho \cdot A_2 \cdot \Delta x_2}{2} \cdot u_2^2 - \frac{\rho \cdot A_1 \cdot \Delta x_1}{2} \cdot u_1^2$$

because of $A_1 \cdot \Delta x_1 = A_2 \cdot \Delta x_2$ follows

$$p_1 - p_2 = \frac{\rho}{2} \cdot u_2^2 - \frac{\rho}{2} \cdot u_1^2 \quad \text{or}$$

$$p_i + \frac{\rho}{2} \cdot u_i^2 = P$$

Here P is the total pressure which, under ideal liquid flow, i.e. excluding internal friction and gravity, is composed of the static pressure p and the dynamic pressure ½ ρu^2. Flow rates can be calculated from this relationship which was derived by *Bernoulli.*

Any consideration of flow properties in pipes must also take account of energy losses arising from the exchange of momentum in the form of internal friction. Since the liquids in the pipes adhere to the walls, i.e. show a flow rate of 0 m/s at the interface, different flow rates must exist as a function of the distance from the wall when liquids are conveyed. The individual zones cause mutual interference as a result of the exchange of momentum. According to *Newton* the following applies for parallel displacement of liquid layers, viz. laminar flow (see Chapter 2.3.2):

$$\tau = \eta \cdot D$$

If *Newton's* equation is applied to flow in pipes, the flow profile u(r), i.e. the flow rate distribution about the pipe radius R at distances r from the centre point, is given by:

$$u = \frac{\Delta p}{4 \cdot \eta \cdot \ell} \cdot (R^2 - r^2)$$

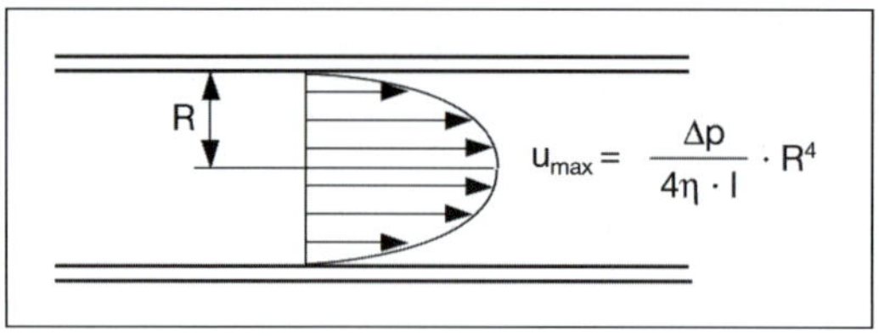

Figure 4.2.80: Velocity profile for laminar flow in pipes

The result is a parabolic function for the flow rate profile for laminar flow.

This states that all liquid molecules move parallel with each other in the direction of flow with a velocity vector. The *Hagen-Poiseuille* equation applies to the volumetric flow rate under laminar flow (see also Chapter 2.3.2).

$$\dot{V} = \frac{\Delta p \cdot \pi \cdot R^4}{8 \cdot \eta \cdot \ell}$$

To calculate the flow rate in pipes as a function of the pressure difference at a constant pipe diameter and a specified pipe length, it is recommended that the *Hagen-Poiseuille* equation be transformed by logarithmisation to simplify demarcations from other flow states and to simplify classification.

$$\ln \dot{V} = \ln \Delta p + \ln\left(\frac{\pi \cdot R^4}{8 \cdot \eta \cdot \ell}\right) \qquad \text{laminar flow}$$

A plot of ln $\dot{V}$ against ln Δp yields a linear relation of slope 1 and the y-intercept ln b, where b = ln (π R^4/8ηl).

The experiment shows that the *Hagen-Poiseuille* equation does not hold when $\dot{V}$ rises due to an increase in pressure after a transitional state which cannot be clearly described mathematically is exceeded. Although the discharge rate increases linearly again at higher pressures, it is considerably lower, with a slope of 1/2. Extrapolation of this line to the y-axis gives a new intercept, ln b'.

$$\ln \dot{V} = \frac{1}{2} \cdot \ln \Delta p + \ln \sqrt{\frac{\pi^2 \cdot R^5}{2 \cdot \eta \cdot \ell \cdot \rho}} \qquad \text{turbulent flow}$$

The reason for the significant change in flow behaviour is the transition from laminar to turbulent flow. The characteristic features of turbulent flow are that the parallel displacement of liquid films is overlapped by vortices and that no laminar flow elements remain if the flow rate is further increased. Since the average distance travelled by the moving molecules therefore increases, the discharge rate also rises more slowly. Turbulence can be recognised by the fact that the velocity profile is no longer parabolic. The velocity vectors are of almost the same magnitude across the entire pipe cross-section before falling steeply in the vicinity of the wall.

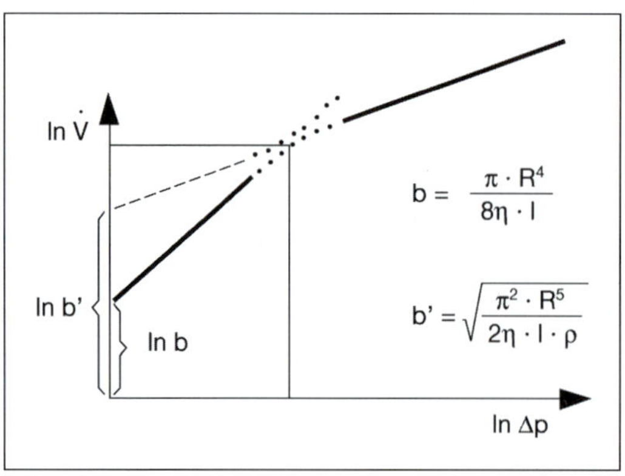

Figure 4.2.81: Flow rate as a function of pressure

The transition from laminar to turbulent flow is characterised by a critical flow rate u_{crit}. The transition point is also transferable to other technical

Application

equipment if all the variables associated with the flow state can be mathematically linked with each other in a dimensionless form to yield the *Reynolds* number. The transition from laminar to turbulent flow in pipes can thus be predicted by determining the critical *Reynolds* number. This is approx. 2000 for flowing water.

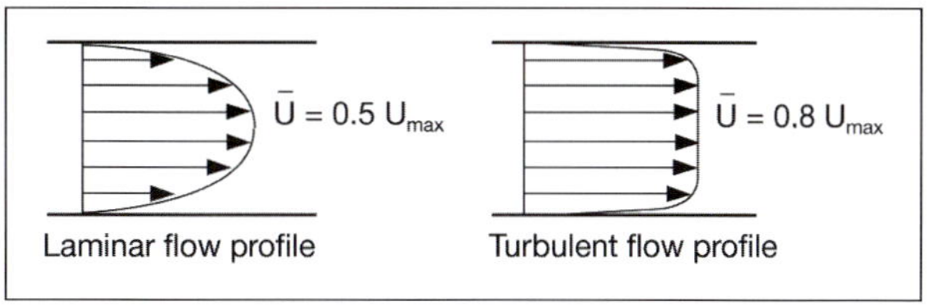

Figure 4.2.82: Profile of F turbulent flow in pipes

The calculation of the *Reynolds* number is of economic interest because the flow rate under laminar flow considerably exceeds that under turbulent flow, despite the same pressure difference. This is factored into the design of circulation systems: the usual flow rates of 0.2 to 0.3 m/s generate laminar flow in the pipes, without the sedimentation of paint constituents becoming an issue.

The energy losses incurred by moving liquids are a function of viscosity and flow rate, i.e. internal friction, and manifest themselves in the form of a pressure drop along the pipe. This pressure drop can be quantified by reformulating the *Hagen-Poiseuille* equation. If the *Hagen-Poiseuille* equation is solved for Δp:

$$\Delta p = \frac{8 \cdot \ell^2 \cdot \eta}{R^2 \cdot t}$$

and if the equation is expanded with u · ρ, it follows that:

$$\Delta p = \frac{4 \cdot 8 \cdot \ell^2 \cdot \eta \cdot u \cdot \rho}{D^2 \cdot t \cdot u \cdot \rho} \quad \text{or} \quad \Delta p = \left(\frac{64}{Re}\right) \cdot \frac{\rho}{2} \cdot u^2 \cdot \frac{\ell}{D}$$

This last equation applies to laminar flow only. To transform it into a universal form, i.e. one which is also applicable to turbulent flow, the term 64/Re must be replaced by a variable λ which is a function of Re. If this dependence is known, the pressure drop in pipes can be calculated for all flow conditions.

$$\Delta p = \lambda(Re) \cdot \frac{\rho}{2} \cdot u^2 \cdot \frac{\ell}{D}$$

So if log λ is plotted against log Re, different linear functions are obtained for laminar and turbulent flow whose y-intercepts are K and K', respectively.

$$\lambda = \frac{K}{Re} \qquad \lg \lambda = \lg K - \lg Re$$

If K or K' is known, it is possible to calculate λ, and thus the pressure drop or the flow rates in smooth pipes for all flow states, i.e. all *Reynolds* numbers.

Interestingly, the pressure drop is independent of the roughness of the pipe walls in the laminar flow range. The laminar interface "masks" the roughness. The turbulent range, on the other hand, is largely independent of Re, but dependent on the roughness of the pipe walls.

The artificial movement of liquids also leads to resistance when solid bodies are encountered because of the kinetic energy acting during flow. *Stokes'* law quantifies the force F_w

Application

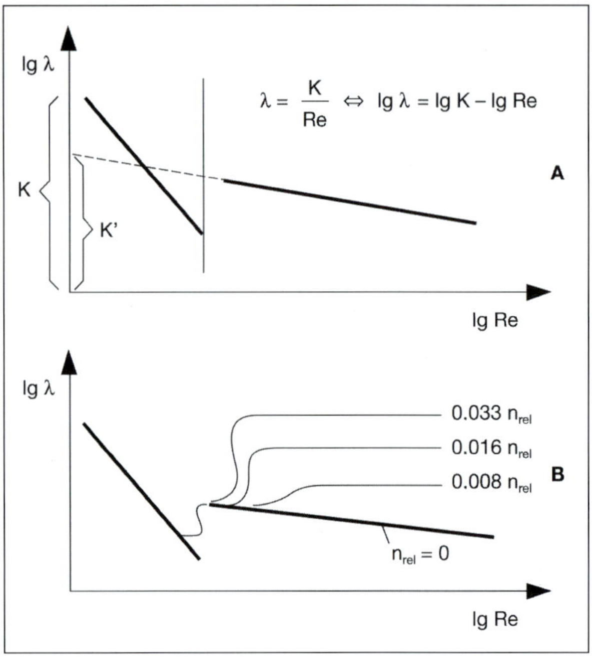

Figure 4.2.83: Dependence of λ(Re) for (A) smooth and (B) rough pipe walls

transmissible to a fixed sphere of radius r by a liquid at a velocity u in the laminar flow range.

$$F_W = 6 \cdot \pi \cdot \eta \cdot r \cdot u$$

In order for *Stokes'* law to be generalised for all flow states and also for nonspherical bodies, the dependence of the resistance on the flow state and the shape of the body against which flow occurs must also be included. To this end *Stokes'* law is expanded and transformed with the aid of the *Reynolds* number:

$$F_W = \frac{6 \cdot \pi \cdot \eta \cdot r \cdot u^2 \cdot \rho \cdot d}{u \cdot \rho \cdot d}$$

$$F_W = C_w(\text{Re}) \cdot \frac{\rho}{2} \cdot u^2 \cdot A$$

Applicability to all flow states and shapes of the body in question is achieved by replacing 24/Re by the C_w value which is also a function of the *Reynolds* number.

The movement of real liquids can only be guaranteed through the application of a permanent pressure difference because of energy consumption due to internal friction. This pressure difference is generated by means of pumps, which thereby create the right conditions for the continuous transport of liquids in pipes.

Pumps

Paint manufacturers and processors use numerous pumps which, on closer analysis, fall into two types of design. The essential difference between the types lies in the constancy of the output. A distinction is therefore drawn between positive-displacement pumps and other pumps. The first group includes those which move defined volumes, largely irrespective of the viscosity and the pressure difference. Important representatives in this category are piston pumps, gear pumps, fan pumps and roots pumps. Important types which do not feature positive displacement are centrifugal pumps, airlift pumps and jet pumps.

Despite the disadvantage of pulsating pressure build-up, piston pumps have proved extremely effective at conveying rheologically complex paints in pipes because the shear load is low. Together with gear pumps, which are also of the positive-displacement type, and centrifugal pumps, whose output is a function of pressure, these are the most important types for conveying paint material.

Piston pumps consist of a cylinder which is closed by valves and contains a movable piston. Through coordinated opening and closing of the cylinder chamber, the valves create the necessary conditions for transporting the liquid independently of the back-pressure. The principle therefore consists in autonomous intake and filling of the chambers, closing and subsequent emptying. Piston pumps are extremely efficient as the amount of slippage, i.e. the proportion of the volume not transported during a stroke, is of the order of 5 to 10 %.

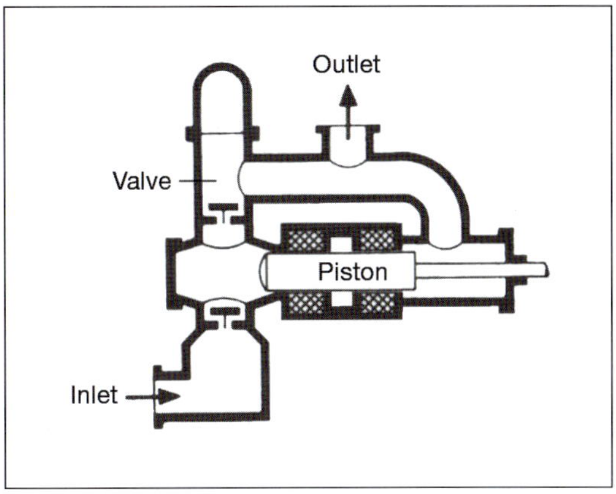

Figure 4.2.84: Cross-section of a piston pump

The pump's design enables the volume of liquid, which is determined by the piston stroke, the cross-section and the number of strokes, to be transported without difficulty in viscosity ranges from 10 mPa s to 50 Pa s and at pressures of up to 100 bar. Piston pumps play an extremely important role in moving coating materials in circulation systems, not least because of their smooth handling of pigmented coating materials. The disadvantages of valve wear arising from the transport of coating materials containing abrasive pigments and extenders are accepted in much the same way as the pulsating pressure build-up. To cancel out pressure peaks, therefore, pressure-equalising containers must be integrated into the system. The pressure pattern is smoothed by means of an air chamber separated by a rubber membrane which reduces the peaks and compensates for the periodic pressure drop in the pump by means of a phase shift.

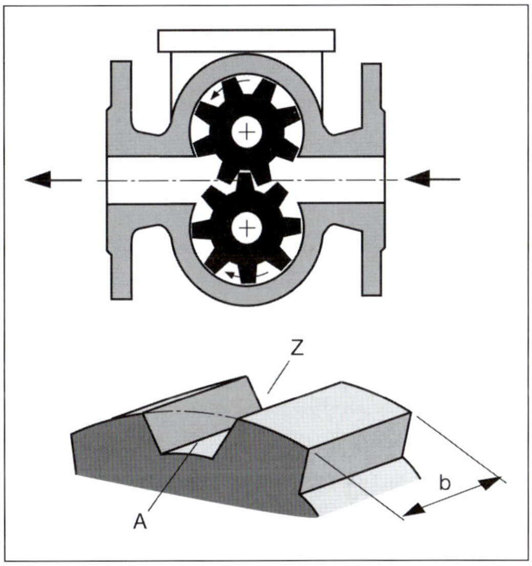

Figure 4.2.85: Cross-section of a gear pump

Gear pumps, which are also from the positive-displacement pump family, are rotary pumps and therefore work without valves. They convey liquids having viscosities of up to 10 Pa s in a nonpulsating manner at back-pressures of up to 20 bar.

Two volumes b · A are transported in the free tooth spaces z in the upper and lower parts of the pump as a function of the velocity n in the direction of flow. These volumes are greater than those in the central return-flow area. The return volume is substantially smaller than the transported $\dot{V}$ because of the meshing of the teeth in the central area. It varies with the pressure difference Δp and the viscosity η of the transported liquids due to incomplete sealing in the return-flow area. γ is a factor which has to be determined experimentally.

$$\dot{V} = 2 \cdot z \cdot A \cdot b \cdot n - \frac{\Delta p}{\eta} \cdot \gamma$$

Gear pumps are simple and reliable conveying devices, though they bring high shear forces to bear on the paint as it enters the pump. This causes excessive abrasion and

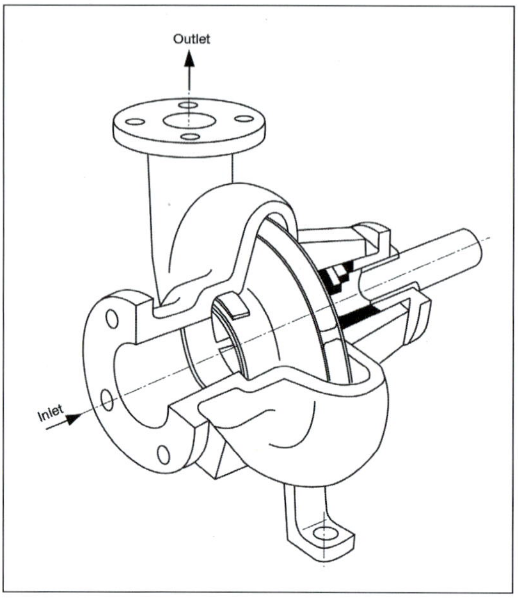

Figure 4.2.86: Cross-section of a centrifugal pump

creasing or breaking of the thin aluminium flakes in metal effect paints. Gear pumps are therefore not recommended for transporting every type of metallic paint in circulation systems. Like piston pumps, gear pumps, being positive-displacement pumps, should always have a bypass which serves as a safety feature and is opened by an adjustable minimum pressure in the event of pipe blockages.

Of the non-positive-displacement pumps only the **centrifugal pump** is worth mention-ing in the context of paint production and processing. It is a particularly simple and "gentle" conveying device, but is nowhere near as important in the paint and coatings industry as in other industrial sectors because of the frequently excessively high viscosities of the material to be conveyed. It is extremely difficult to transport media with viscosities in excess of 200 mPa · s.

The centrifugal pump works on the same principle as a propeller. Moving vanes draw the material in axially and discharge it radially. The material is therefore moved without pulsation, though not in defined volumes. This pump design has the further disadvantage of being unable to draw in liquids autonomously and being capable of overcoming only certain heights. On the other hand, the drawback of decrease in output with increase in back-pressure also means self-regulation or restriction of the pressure build-up in the event of closed or blocked pipes. The continuous circulation of the viscous liquid causes the temperature to rise, and so the pump will shut down automatically on reaching a setpoint temperature if it is fitted with a thermostatic switch.

Measuring, Dispensing and Homogenising

For many applications, such as the processing of two-component paints, material flows have to be dispensed, flow rates measured and corrected, and material homogenised. This is the only way to guarantee constant discharge rates or constant mixing proportions for the individual components and thus constant coat-ing quality. The numerous methods for determining flow rates are based on different measuring methods:

- Volumetric measurement methods
- Differential-pressure methods
- Pressure drop in pipes
- Variable-area methods
- Magnetic methods
- Incubation methods
- Thermal methods
- Ultrasound
- Mass flow measurement with gyroscopic methods

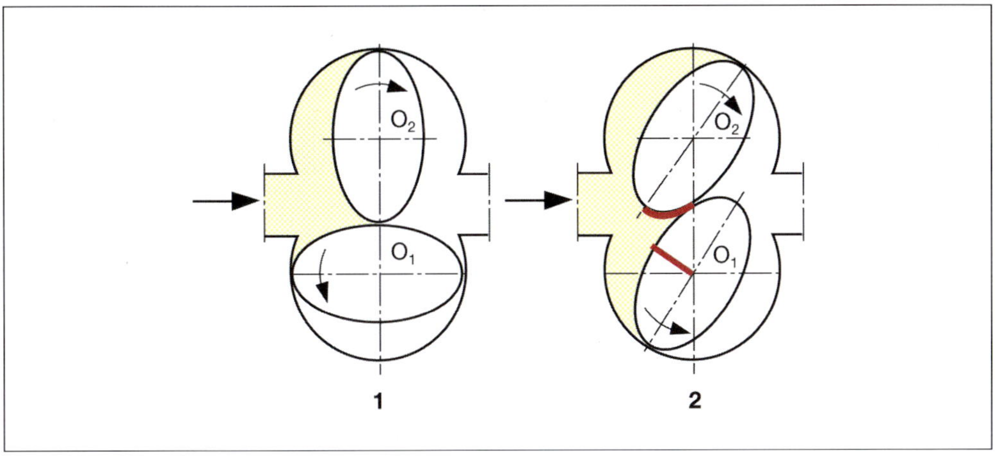

Figure 4.2.87: Principle behind the oval-wheel flowmeter

Whereas the volumetric methods give a direct indication of the material flow, all the other methods first yield physical variables which are then converted into volumetric or mass flow rates by mathematical linking with other variables.

The simplest methods of measuring material flows are **volumetric**. The equipment used here includes positive-displacement meters, in which category positive-displacement pumps also belong. These are the measuring instruments of choice if the pressure necessary to induce flow is generated by other units. The number of strokes executed by a piston pump acts as a measure of the flow if the volume displaced per stroke is known. Gear pumps can also serve as measuring instruments for the reproducible processing of industrial paints. They serve to monitor the discharge rate for 1-component paints and to maintain constant mixing ratios for 2-component systems. Oval-wheel flowmeters work on a similar principle. Gear pump meters and oval-wheel flowmeters have measurement errors of approx. 1 %.

The turbine meter employs a different volumetric measuring principle. This type of measurement basically involves repurposing a centrifugal pump to determine the flow rate. The velocity resulting from the flow of the liquid can be measured either mechanically or as a voltage induced by the movement of the turbine in an alternating magnetic field. Corresponding calibration curves are recorded to determine the throughputs as a function of the pressure difference and the viscosity of the liquid. This measuring principle therefore falls into the category of indirect processes.

Differential-pressure methods are based on pressure measurements and exploit the influence exerted by moving liquids on the distribution of the total pressure across the dynamic and static pressures. In accordance with the *Bernoulli* equation for ideal liquids the sum of the potential and kinetic energy is constant in the direction of flow, provided that internal friction is rendered negligible through close positioning of the measuring sensors. The different pipe cross-sections of this measuring system induce changes in the dynamic pressure which are measured and serve to calculate the volumetric flow rates (see above). A variant of this differential-pressure method involves measuring the pressure drop in pipes. For this, the pressure gauges are positioned at relatively large distances from each other over a constant pipe cross-section. If $\lambda(Re)$ and the density

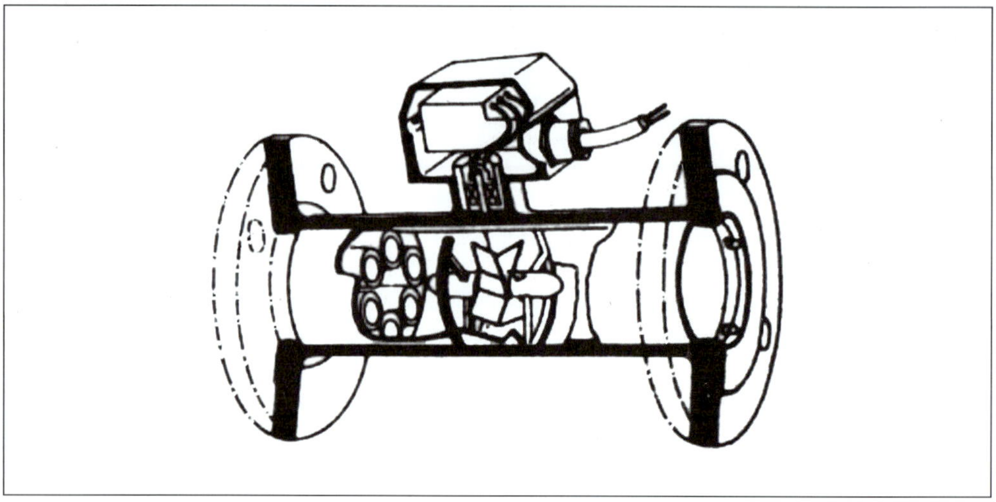

Figure 4.2.88: Turbine meter with induction measurement

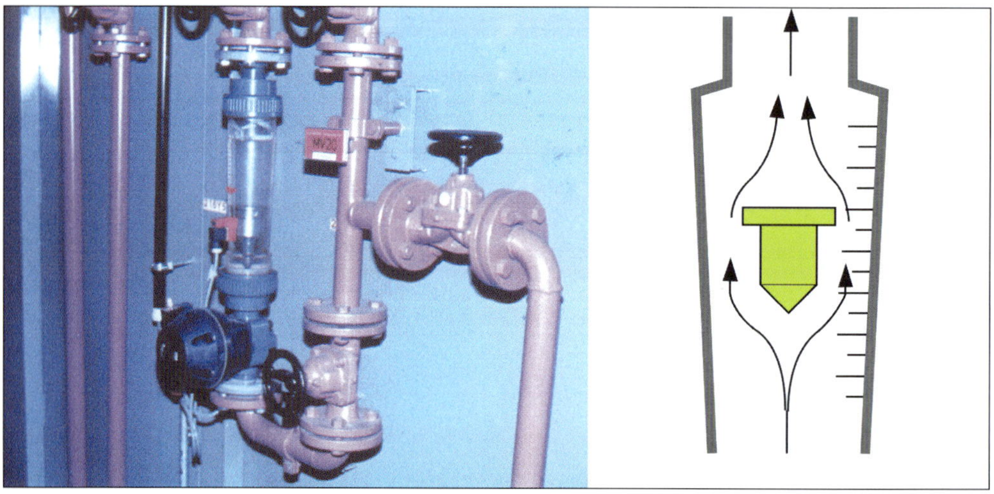

Figure 4.2.89: Photograph and schematic drawing of a volumetric flowmeter working on the float principle

of the liquid are known, the flow rate can be calculated by measuring the pressure drop across the measuring sensors and, if the cross-section is known, the volumetric flow rate (see above).

Orifice flowmeters determine the flow rate via the force transmitted to a float in a flowing liquid and are based on *Stokes'* resistance law (see above). The underlying principle utilises the counteraction of gravity and drag. The liquid to be measured flows from bottom to top in a vertical tube.

Installing a slightly conical float allows flows to be measured if different flow rates also give rise to different flow resistance C_w values. However, the measuring system must first be calibrated with the specific liquids to be measured.

In the **magnetic method** electrically conducting liquids are moved in a tube of diameter D under an applied magnetic field of magnetic flux B which in-duces a voltage proportional to the volumetric flow rate $\dot{V}$. This can be measured directly if insulated pipes are used.

$$\dot{V} = K \cdot \frac{D \cdot U}{B}$$

This is the preferred measurement method for water-borne paints. Calibration for the various paints is necessary.

Flow can be measured by the **seeding method** in which a heterogeneity is briefly induced in the flowing system at regular intervals at a specific location in the pipe. This can be done, for example, by means of a heat pulse or by seeding with a salt solution. The localised change in temperature or the increase in electrical conductivity can then be monitored at another location. The flow rates are calculated on the basis of the time delay between seeding and measurement of the seeding signal.

Seeding methods are very accurate, but can only be used for continuous flow. Interruptions will prevent the pulses from reaching the sensor and render calculation impossible. Flow rate measurements by the seeding method cannot be used to determine discharge rates for manual spray application because of the frequent interruptions.

The principle behind **hot-wire anemometry** is based on the velocity-dependent cooling of wires passing an electric current. The resistance of an electrically heated wire in a measuring cell is determined with the aid of a *Wheatstone* bridge. At a given voltage U the galvanometer of the calibrated bridge shows a deflection (A in Figure 4.2.92) proportional to the flow.

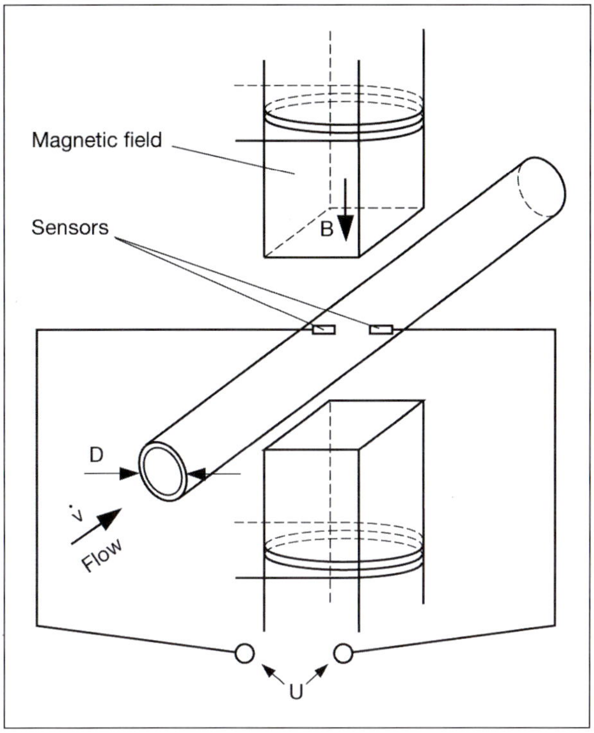

Figure 4.2.90: Principle behind volumetric flow by magnetic induction

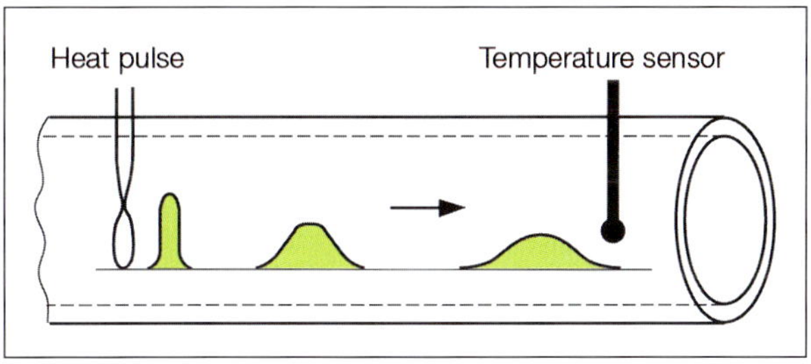

Figure 4.2.91: Principle behind volumetric flow measurement by seeding with heat pulses

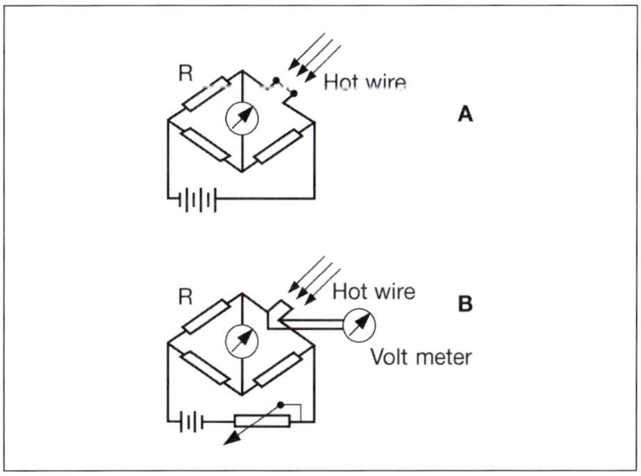

Figure 4.2.92: Hot-wire anemometer by evaluation method (A) and (B)

Application

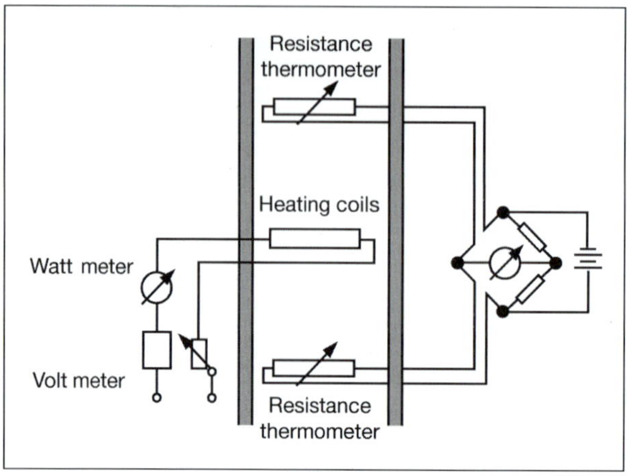

Figure 4.2.93: Volumetric flow measurement based on temperature rise following defined energy input

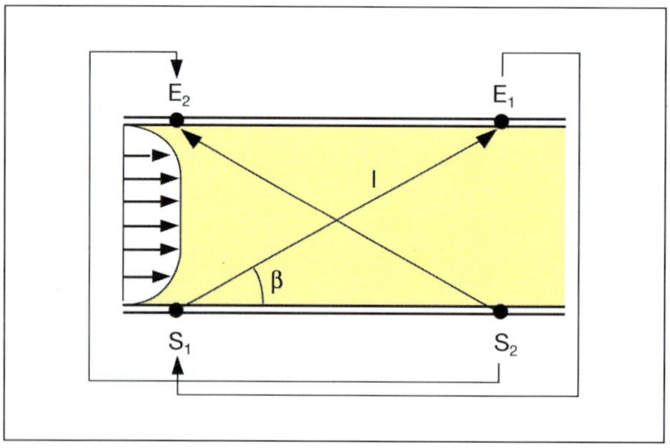

Figure 4.2.94: Ultrasonic measurement of volumetric flow

This stems from flow-dependent cooling of one leg of the circuit previously heated by the electric current and from the associated change in resistance.

As this measurement method is highly sensitive, it is only suitable for low flow velocities. If high flow rates have to be measured, calibration can be achieved by increasing the bridge voltage (B in Figure 4.2.92). The rise in the potential difference be-tween the input and output of the measurement section gives the flow velocity.

Sensors which detect **temperature increases** in the volumetric flow $\dot{V}$ after the input of a defined amount of electrical energy have also proved successful. By contrast with the pulse method, heat energy $\dot{q}$ is continu-ously supplied to the liquid via a hot wire. The temperature rise ΔT is then a function of both the material being heated (with parameters density ρ and heat capacity c) and the volumetric flow $\dot{V}$.

$$\dot{V} = \frac{\dot{q}}{\rho \cdot c \cdot \Delta T}$$

The principle behind **ultrasonic flow measurement** is the dependence of the velocity of sound on the velocity of the flowing liquid. The flow velocity determines how fast a sound emitted by a transmitter reaches a receiver. If the receiver and transmitter are positioned on opposite sides of the measuring circuit, the beam passes through the entire flow profile and determines the flow velocity as a mean value. Positioning the receiver and transmitter exactly perpendicular to the direction of flow does not provide any useful readings (see Figure 4.2.94). The accuracy of ultrasonic flow measurement is enhanced in the *sing-around* method. Measuring setup 1 is working in the direction of flow, while 2 is working against the direction of flow, with the transmitter and receiver positioned at a distance 1 from the tube. The transmitters deliver pulses only if the sound has not yet arrived at the receiver. When the sound reaches the receiver, the transmitter is switched off, before being switched back on again at the end of the sound information. Thus two different pulse sequences of durations t_1 and t_2 are received. The speed of sound c in system 1 is increased by the flow rate ü as a function of the incident angle β and is reduced in setup 2. The flow rate is calculated by mathematical treatment of the electronically measured time differences.

$$c + \bar{u} \cdot \cos\beta = \frac{\ell}{t_1} \quad \text{and} \quad c - \bar{u} \cdot \cos\beta = \frac{\ell}{t_2}$$

$$\bar{u} = \frac{1}{2 \cdot \cos\beta} \cdot \left(\frac{\ell}{t_2} - \frac{\ell}{t_1} \right)$$

The most accurate method for measuring mass flow rates utilises a measuring instrument working on the **gyroscopic principle**. It consists of a U-tube, through which the liquid to be measured flows after it has been made to oscillate electronically. A velocity-dependent *Coriolis* force, which twists the U-tube with a radius of curvature r out of its horizontal position, then acts on the mass of particles moving within the tube. The torsional angle α, which is measured at a specified oscillating frequency f, is proportional to the mass flow rate $\dot{m}$. This method has numerous advantages over all the others. Its accuracy is excellent, with an error rate of ± 0.3 %. The results are independent of the liquid's rheological properties and are unaffected by fluctuations in material pressure or different surface tension and density.

$$\dot{m} = k \cdot \frac{\alpha}{f \cdot r}$$

The gyroscopic method can also be used to measure pulsating flow and two-phase mixtures. The latter is important where dispersions are metered [4.4.69].

The reason for the large number of measurement methods presented is that no single method meets all requirements, e.g. for the transport of paints. This is because the accuracy of the measurement methods varies with the flow. For example, the differential-pressure method is only sufficiently accurate if large volumes are being transported, whereas the seeding method is only suitable for small volumet-ric flows (see Figure 4.2.96).

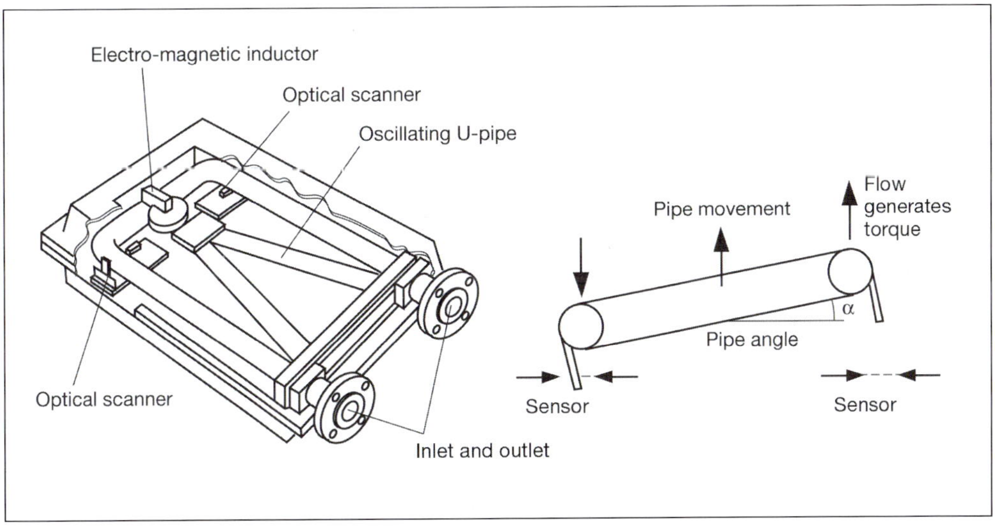

Figure 4.2.95: Principle behind mass flow measurement by the gyroscopic method

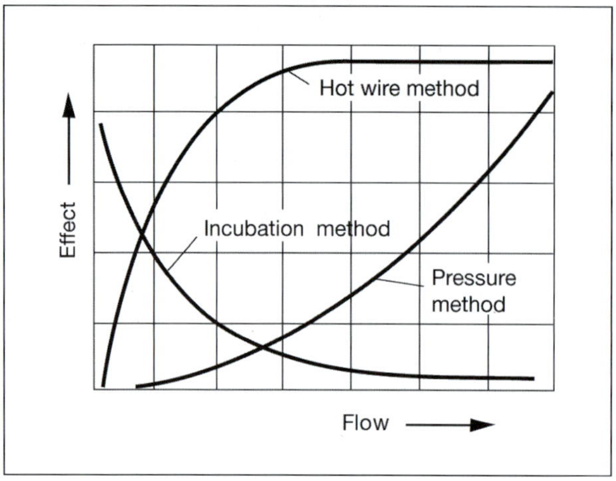

Figure 4.2.96: Sensitivity of different volumetric flow measurement methods

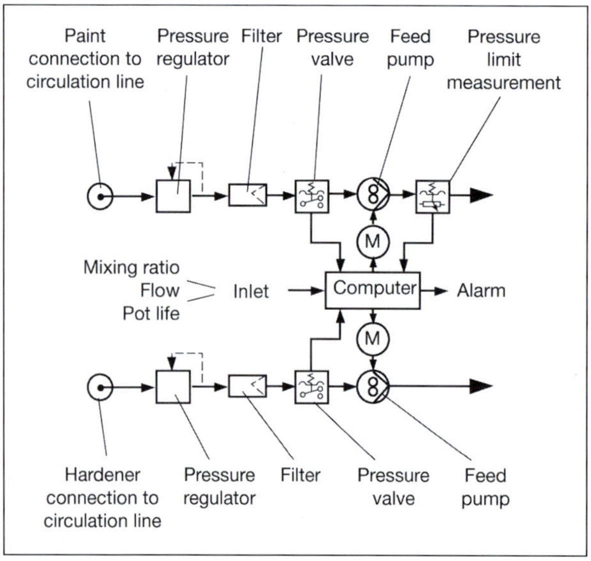

Figure 4.2.97: Electronic control of 2-component metering systems

Homogenisation in 2-Component Application Systems

In the industrial processing of 2-component paints, consisting of base and hardener with a pot life of just a few minutes, it is important, apart from precise metering of the volumetric flows, that the two components be homogenised thoroughly and rapidly for the sake of uniformly good coating. Small mixing volumes should make colour changes economically viable. In addition, a circulation system has to be installed for transporting the hardener and must be purged with dry nitrogen where isocyanate-containing materials are used.

The standard metering systems used in the past with their two mechanically linked piston pumps have largely been superseded by computer-controlled gear pumps. This has eliminated the problems of pulsating pressure build-up and bestowed greater flexibility over the setting of different mixing ratios.

Two process variants have proved successful with regard to electronically controlled metering systems. In the first the base quantity is determined by paint pressure regulators, and the volumetric flow rate is measured continuously with a flowmeter. This rate is transmitted to a computer which then actuates the stepper motor of a gearwheel meter for the hardener circuit. The second design dispenses with paint pressure regulators and meters the volumetric flows with computer-controlled gearwheel meters in order to achieve constant discharge rates, constant mixing ratios and greater flexibility over the setting of different mixing ratios. This method is the more important one.

Homogenisation of the volumetric flows is important for the quality of the coating produced by 2-component paints. It should be done as shortly before application as possible. Static mixers have proved successful for this and other reasons, with the *Kenics* mixer offering the best compromise of all requirements relating to paint processing.

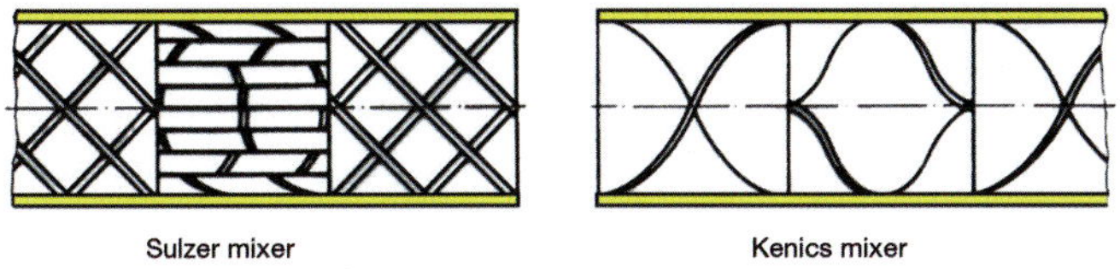

Figure 4.2.98: Schematic drawing and functional principle behind the Sulzer and Kenics mixers

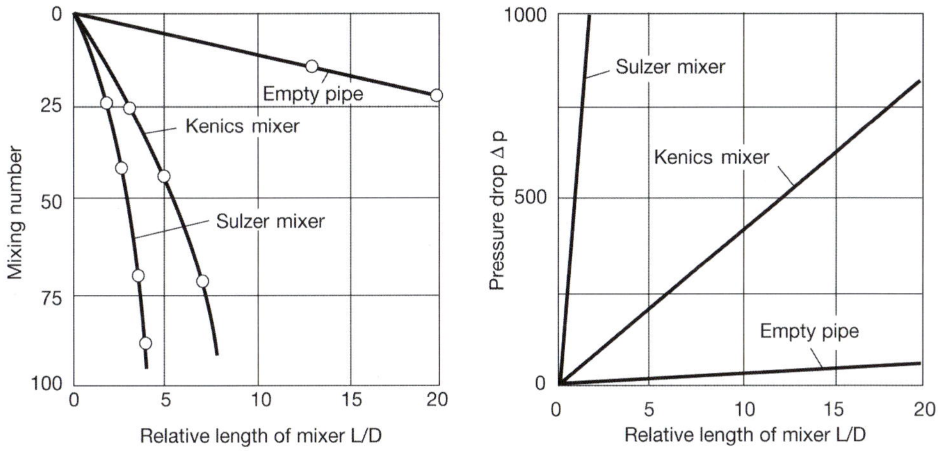

Figure 4.2.99: Diagrams of the mixing quality and pressure drop in Kenics and Sulzer mixers

Static mixers are devices in which two product flows are homogenised using the flow energy [4.4.70]. According to this definition any simple empty tube into which the two components are fed is, in principle, a static mixer. In such a tube, homogenisation is only achieved under turbulent flow if the length of the tube section is approx. 90 times its diameter. For that reason, the empty tube is fitted with baffles which considerably improve the ratio of length to cross-section. In all cases the improvements are associated with an increase in pressure drop. Of the many static mixers available, the *Kenics* mixer offers the best compromise between low pressure drop and high mixing quality.

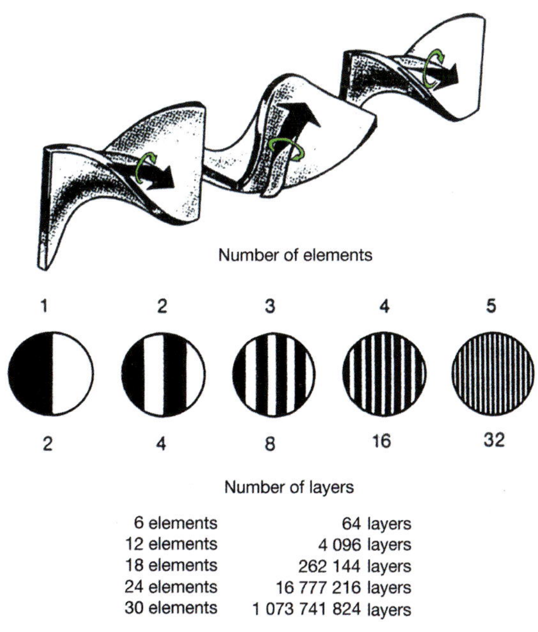

Figure 4.2.100: Design and mixing number of Kenics mixers

It consists of alternating left- and right-hand elements twisted at an angle of 180°. The individual elements are arranged at 90° behind each other in an empty tube. In this way laminar flow effects are used to split the flow into two halves.

The number of pitch increments as a function of the number of elements n is given by 2^n. For 24 elements, this is equivalent to no fewer than 16,777,216 sub-layers. For optimum mixing it is important during material selection that the viscosity differences in the flows for homogenisation be as small as possible. Static mixers, and *Kenics* mixers in particular, have become firmly established for the industrial processing of 2-component paints because of their good mixing quality, the low stresses imparted to the product, continuous operation, low space requirement and low capital costs. Their low space requirement even enables them to be integrated into spray heads.

Water-borne 2-component polyurethane paints, which have become more and more popular in recent years, require continuous extremely fine emulsification of the hardener components into the aqueous phase of the base dispersion. This imposes even greater demands on the quality of homogenisation. To this end the organic solution containing the isocyanate groups is injected through a nozzle into the aqueous polyol phase in droplets of 5 to 20 µm before it is broken down further into particle sizes of 0.5 µm in jet dispersers.

Application

4.2.1.5 Booth Conditioning and Overspray Elimination

A unique feature of indirect processing of paint by atomisation is the spray cloud. Associated with it are emissions of solvents to the environment and overspray. Action must therefore be taken in the interests of occupational health and safety and environmental protection to reclaim or dispose of the not inconsiderable quantities of solvents and overspray that do not strike the workpiece. The overspray has unwanted side effects on the process itself. Straying overspray can contaminate already painted workpieces if it deposits on fresh, still viscous coating. Film defects in the form of craters are often the consequence (see Chapter 6.2).

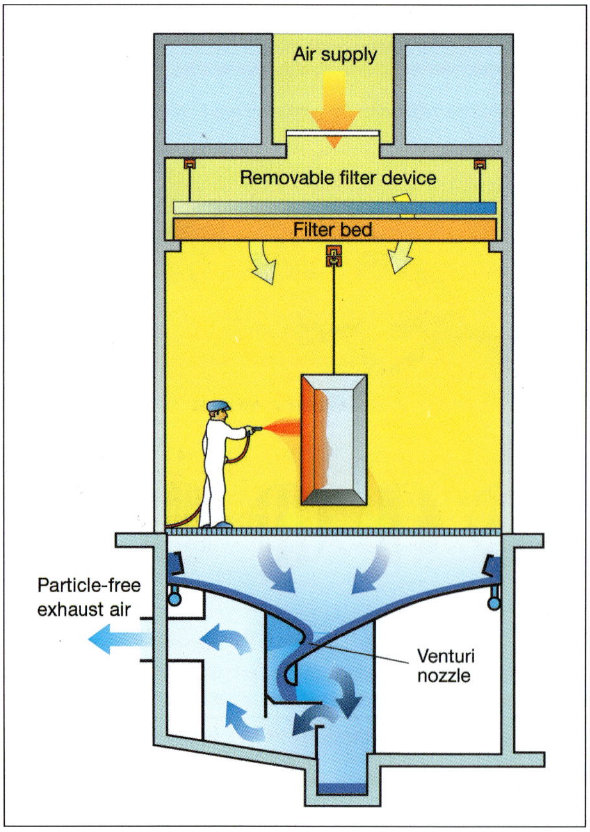

Figure 4.2.101: Schematic drawing of a spray booth *Source: Eisenmann*

It is therefore necessary to keep overspray and evaporating solvents under control. Bodyshops have small spray booths which accomplish this by extracting and filtering solvents and overspray by routing air through water curtains behind the spray booth.

In industry the spray equipment is enclosed in large booths. There the evaporated solvents and the overspray are removed by providing a supply of fresh, conditioned air which descends from top to bottom as laminarly as possible. If the air descends at a rate of 0.2 to 0.5 m/s, then 0.2 to 0.5 m^3/s or 720 to 1800 m^3/h of air is supplied for every m^2 of floor area. A floor area of 200 m^2 will therefore require up to 400,000 m^3 of purified, conditioned fresh air every hour. This is the way to avoid paint defects and comply with the regulatory MAC respective AGW values.

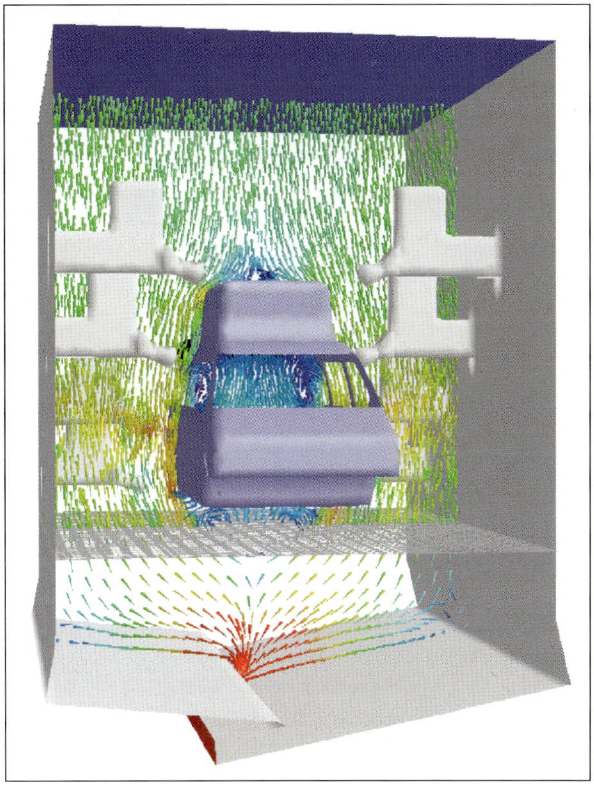

Figure 4.2.102: Air flow in a spray booth *Source: Dürr*

The supplied air must first be purified, tempered and, where water-borne paints are processed, also adjusted to the right relative humidity. This is achieved by means of fans, humidifiers, heat exchangers and upstream and downstream micro-filter units with pore sizes ranging from 10 to 0.1 µm.

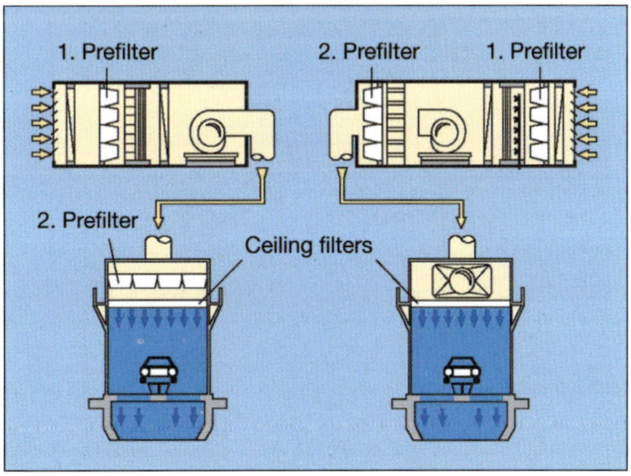

Figure 4.2.103: Conditioning of the air supply for a spray booth *Source: Dürr*

Although the amount of foreign matter in outside air is low at 0.1 to 0.3 mg/m^3, the number of actual particles is extremely large (1000 to 7000). Uncleaned fresh air supplied at a rate of 300,000 m^3 per hour would therefore introduce approx. 100 g dirt or more than 1 billion particles into the booth. Since dirt particles with sizes of more than 10 µm can give rise to paint defects on freshly coated workpieces, elimination of dirt is essential.

In view of the extremely small size of the particles, deep-bed filtration is the only method that can be used out of the usual process-engineering methods for separating particles. Wet-cleaning methods are unsuitable because they increase the air humidity excessively, and cyclones have inadequate separating action. Filter bags are used for microfiltration because they provide the largest possible filter surface area in a small

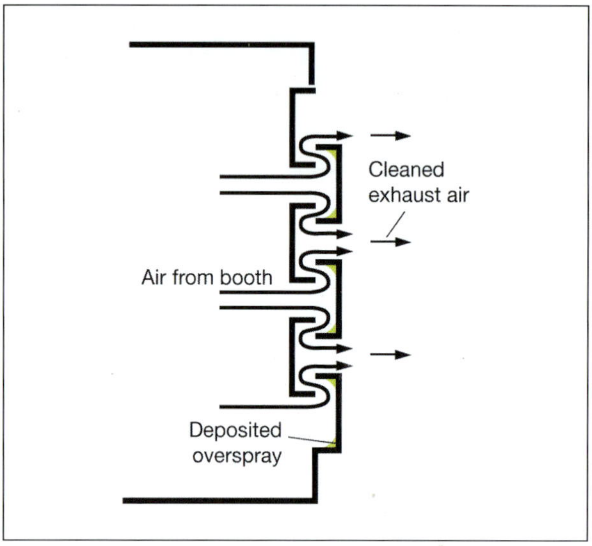

Figure 4.2.104: Overspray deposition on impact plates

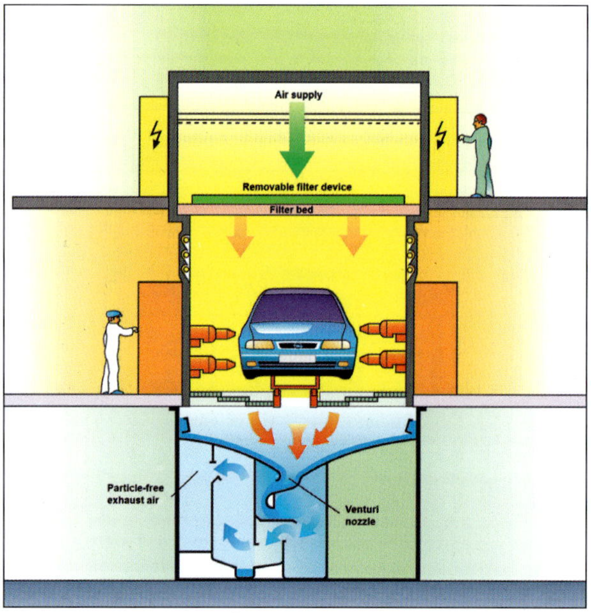

Figure 4.2.105: Wet deposition of overspray by the Venturi principle
Source: Eisenmann

space. The ceiling of the booth is fitted with large removable microfilter mats to distribute the air and ensure laminar flow.

The fresh air absorbs solvents which evaporate during the painting process and transports these and the overspray to the overspray traps. If only a little paint is being processed or if the processing method only generates a little overspray, as in the purely electrostatic process, filter mats can be employed over extended periods to separate the overspray.

Where large volumes of paint are processed and there is extensive overspray, impact plates can be fitted upstream of the dry filter mats to extend their service life.

When the air loaded with overspray is forced to change direction, inertia causes a large proportion of the droplets to be separated in advance. The overspray is then disposed of by removing the paint from the paint-covered impact plates and replacing the filter mats. The type of paint determines whether the paint-contaminated filter mats must be disposed of as domestic or special waste.

Different technical conditions must be provided for continuous coating systems. Assuming a material consumption of 7 kg paint per car and material spill of 20 % during daily production of 1000 cars, the daily overspray amounts to approx. 1.4 t. Dry filter systems have a short service life and so large-scale industrial coating systems are fitted exclusively with wet scrubbers. These are located below the booth. Their task is to ensure the most intensive interaction possible between the overspray droplets and the flood sheet fluid and thus to transfer the overspray to the aqueous phase. The overspray is then separated from the flood sheet fluid again and disposed of as paint sludge.

Rapid, complete absorption of overspray is effected by providing constrictions in the extraction devices to greatly accelerate the overspray before it is absorbed into the aqueous phase at a relatively high speed. Diffusers based on the *Venturi* principle work particularly well in this regard. After acceleration of the overspray-charged exhaust air, the widening of the aperture cross-section of the nozzles causes pronounced eddies. The resulting vacuum draws the flood sheet fluid into the air current, leading to the capture of the overspray.

The degree of separation achieved by such systems for average droplet sizes of 25 to 30 µm is extremely good at 99.8 to 99.9 %. As a rule, statutory regulations relating to aerosol residues in the spray booth exhaust air are met.

Surfactants and special coagulating agents are added to the flood sheet fluid to ensure better absorption of the overspray and, in particular, better separation from the fluid, which is recirculated.

Jet, cascade and vortex scrubbers are also encountered in practice. Jet scrubbing is the longest-established system. In this the scrubbing liquid is atomised via a series of nozzles in the exhaust air stream. Under optimum settings, good efficiencies of up to 99.5 % are achieved. Jet scrubbing is a relatively high-maintenance process due to soiling of the nozzles.

Cascade and vortex scrubbers are better in this regard. In cascade scrubbers the flood sheet fluid runs in a continuous film from an overflow container through a cascade of several semi-circular curved impact plates. These are configured such that the water entrained upwards with the air forms curtains of water as it flows back down and absorbs the overspray.

In vortex scrubbers the exhaust air is sucked through a narrow gap between the liquid and a limiter plate. The geometry of the gap and the rapidly moving exhaust air cause the flood sheet fluid to be atomised, in a manner similar to the scrubbing methods already described, before capturing the paint droplets.

The energy input required to separate the overspray is provided in jet scrubbers by the artificially moved flood sheet fluid and in all other systems by the kinetic energy of the exhaust air stream.

All overspray treatment processes include a water treatment stage which ensures that the paint sludge is separated and that the water treated in this way can be recycled to the flood sheet fluid again.

The systems described are unsuitable for handling only a small amount of overspray. Under such circumstances, electric filters are particularly attractive because of their superior efficiency. Their use is always recommended if complete removal of the overspray is required. This is the case, for example, where air is recirculated in spray booths (see Chapter 5).

As already described in the section on electrostatic atomisation of paints, such a high field strength E is generated at a filament-shaped cathode that the air becomes ionised and the overspray therefore becomes charged. The velocity u_{St}, at which the aerosols of diameter D migrate to the inner wall of the anode, down which water flows, is given by

$$u_{st} = \frac{Q \cdot E}{3 \cdot \pi \cdot \eta \cdot D}$$

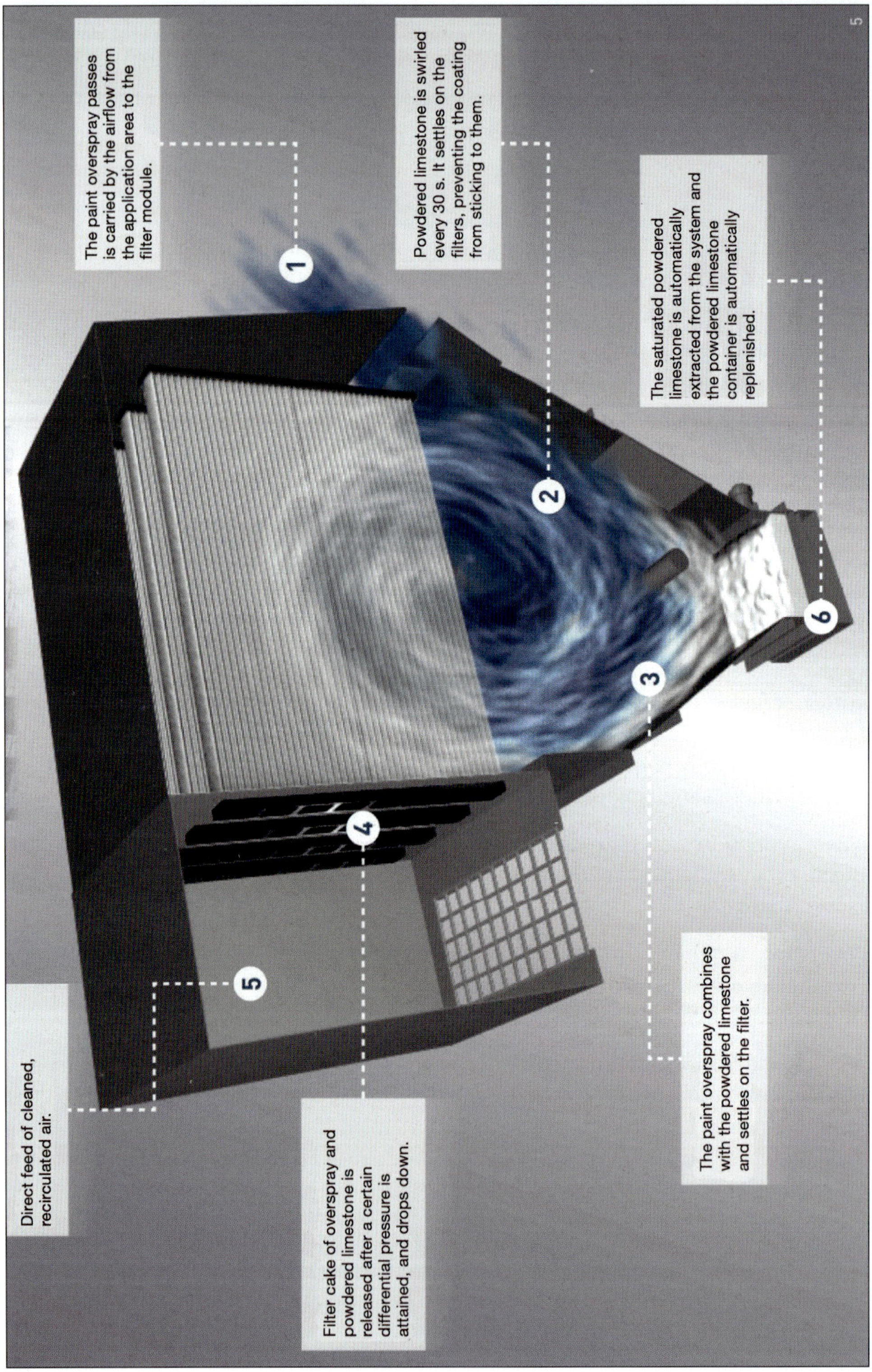

Figure 4.2.106: Dry deposition method employed in the automotive industry

Source: Dürr

Application

If account is also taken of the fact that the charged particles must cover the distance R, the air velocity may not exceed u_{air} in the electric filter.

$$u_{Air} = u_{st} \cdot \frac{h}{R}$$

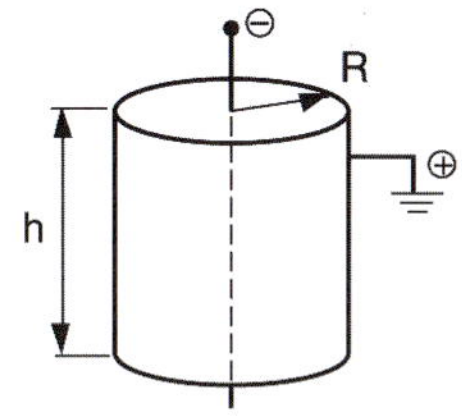

It therefore follows that

$$u_{Air} \leq \frac{h}{R} \cdot \frac{Q \cdot E}{3 \cdot \pi \cdot \eta \cdot D}$$

The charged overspray is separated at the pipes or planar walls, which are connected up as the anode and down which water trickles, and is then captured by the water.

With a view to reducing waste and increasing efficiency, modern methods of overspray separation have been developed and will be used more extensively in the future. One method adds powdered limestone to the air flow from a type of fluidised bed in the base of the booth to bind the paint particles in the booth air and then separate them collectively with filters. Consequently, there is no paint sludge that needs to be treated and disposed of with chemicals. The powdered limestone is sent for processing in the cement industry. In addition, the filtered booth air is recirculated, which slashes energy consumption substantially and the fresh air requirement to < 20 % [4.4.71] (see Figure 4.2.106).

Another process charges the paint particles electrostatically in an air stream and separates them on earthed separation plates. With the aid of a release agent on the surface of the plates, the separated material is bound, removed and collected in a tray. Some of the release agent is recovered and returned to the circuit (see Figures 4.2.106 and 4.2.107).

Disposal of the Paint Sludge and Booth Air

While filter mats charged with overspray must be landfilled or incinerated, the paint overspray has to be chemically and physically modified in order that it may be removed effectively from the circulating flood sheet fluid. Rendering the overspray droplets nonadhesive prevents pipes and pumps from becoming clogged and at the same time creates better conditions for separating the organic overspray from the aqueous phase. Wet separation is therefore accompanied by various methods of paint coagulation. Tried and tested processes can be divided into chemical and physical methods.

Figure 4.2.107: Electrostatic-assisted deposition method for overspray in the automotive industry *Source: Eisenmann*

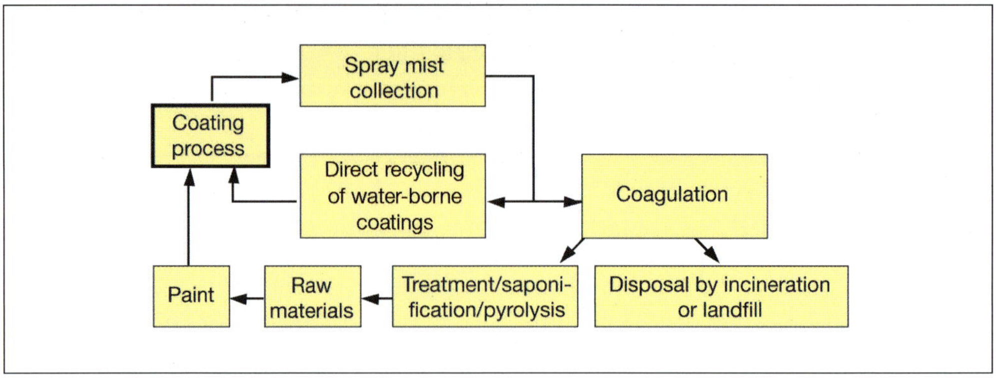

Figure 4.2.108: Traditional methods for wet treatment of overspray

Application

Saponification of the organic components of the paint sludge with caustic soda is no longer state of the art. A better method, because it is more universal and less aggressive, involves treatment with less basic metal hydroxides. In this relatively mild environment at pH 9 the paint droplets become destabilised and coated with colloidal aluminium or iron hydroxide. The destabilisation of the pigment dispersion leads to the formation of relatively large agglomerates, which is a desired side effect. The use of wax dispersions also renders the paint sludge nonadhesive by adsorption. One effective adsorption method exploits the physical interaction of alumina with the overspray droplets in the circulation water. Alumina initially acts as a substrate for the paint droplets and then, in the second step, extracts the organic solvents still remaining in them, leaving behind a nonadhesive coagulate. For the longest-possible service life, foam inhibitors should be added alongside the actual coagulation agents, and biocides should be incorporated to prevent bacterial attack.

In all cases steps must be taken to ensure that the coagulated paint sludge is scraped off or decanted after it has settled or after it has floated to the surface as a result of CO_2 formation, as happens in the case of polyurethane paints, so that it can be sent for further processing.

In some booths water-soluble paints are captured in the flood sheet fluid before being concentrated again by ultrafiltration or electrochemical coagulation and returned to the process (see Chapter 5). This is only useful and cost-effective wherever few or, preferably, no colour changes occur, e.g. when primer surfacer paints or primer material are processed [4.4.72].

For various reasons the booth air generated by treatment of the overspray is for the most part not recycled but instead undergoes waste-air treatment, usually involving adsorption on activated carbon, and is then discharged to the atmosphere (see Chapter 5). These reasons include, on one hand, compliance with the MAC values for solvent concentrations in booths to enable entry into the booths, if necessary, while painting is in progress, and, on the other, the cost of processing the humid exhaust air arising from the water-borne paints which are generally used in the automotive industry.

4.2.1.6 Automated Systems and Robots for Paint Processing

Industrial-scale application of uniform paint films by spray coating requires spray heads which, under constant conditions, pass by the workpiece, always at the same distance, and

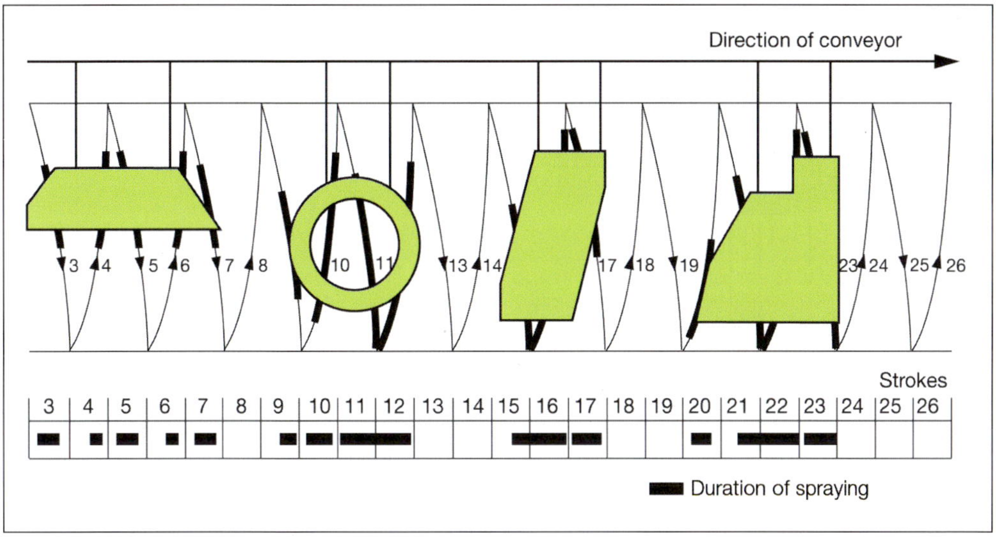

Figure 4.2.109: Example of programming paint output from reciprocators for different parts

in predetermined, coordinated trajectories. Traditional paint processing by pneumatic or hydraulic atomisation, mainly done manually in the past, is now being superseded increasingly by automated painting systems (reciprocators) and robots in industrial coating technology. Manual application is limited nowadays to areas which are hard to access and to occasional reworking during or upon completion of coating work.

As purely electrostatic spraying or high-speed rotary atomisation cannot be carried out manually, it is even more understandable that industrial paint processing is moving towards automated processing technology. Savings on staff, increases in material and energy yields and consistently good quality are just some of the conspicuous advantages.

The use of automated systems means that the experience of the painter can no longer be called on to compensate for fluctuations in material, process and ambient parameters, and so all the processes which influence application, equipment and material variables must be regularly, ideally continuously, measured and maintained at a constant level. Where such conditions are guaranteed, paint application can, in principle, be automated. The paint can then be applied by appropriate automated painting systems or robots, the choice depending on the complexity of the product to be painted. Such machines, as described below, can be fitted with almost all types of spray heads.

The simplest form of automation involves **reciprocators**. The workpiece is transported past the vertically moving spray head, with the result that the spray cone can coat the entire surface of the workpiece. It is necessary in this process to coordinate the forward speed and the lifting speed of the automated system's arm. If this is done, the outcome is a sawtooth spray pattern created by the regular lateral motion of the lifting arm in conjunction with the direction of movement of the workpiece.

The spray heads have to be switched on and off at specified, precisely determined intervals in accordance with the shape of the workpieces. In addition, the width of the spray jet must be matched to the feed rate to ensure uniform film thickness where the paint paths overlap (see Figure 4.2.109).

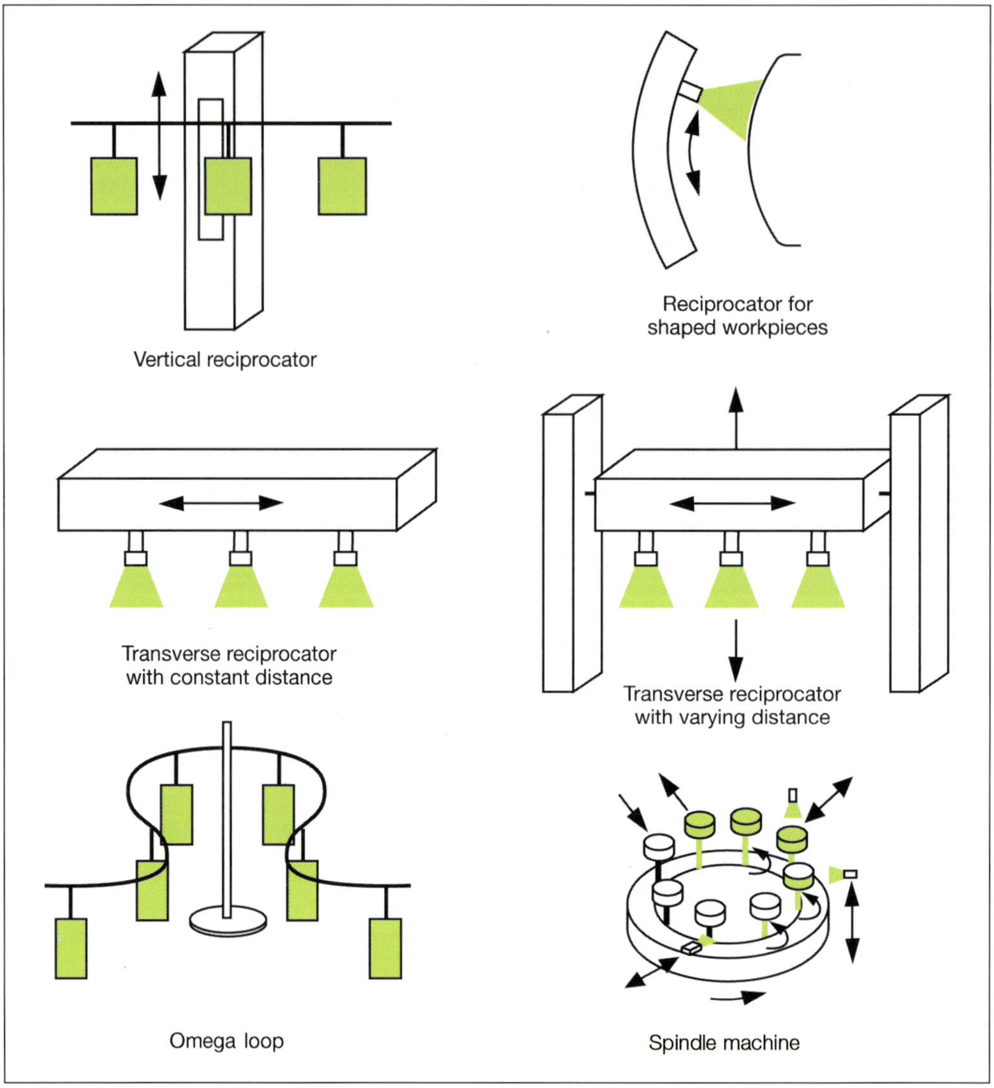

Figure 4.2.110: Reciprocator types and other machines for spray application

Not only the automated systems described, but also **transverse reciprocators** with an appropriate horizontal direction of movement are used for coating horizontal surfaces. For special applications transverse and rotary movements can be executed in addition to the horizontal movement. This makes it possible for the workpiece contour to be covered uniformly at a constant distance.

In the automotive industry, side and ceiling machines with several degrees of freedom are state of the art in high-speed rotary bell coating.

Manual application at the front and rear parts of car bodies can now be eliminated entirely due to the availability of rotatable, horizontal, automated systems. However, these machines cannot be used to coat concealed items such as lips or flanges, door entries and cavities.

The solution to this problem is offered by **painting robots** which can simulate the movements of the human arm. They consist of a manipulator, an electrical actuator unit and an electronic control unit. The manipulator emulates all the movements of a human sprayer by means of five or six continuously operating electric drives. The actuator unit contains all the drive systems along with sensors for monitoring movements. The electronic controller contains the robot's memory in the form of a computer program which enables a movement, once specified, to be repeated identically as often as required

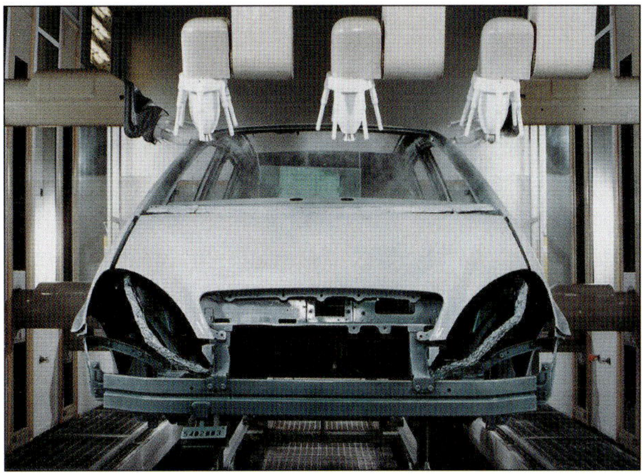

Figure 4.2.111: Side and roof reciprocators in automotive coating *Source: BASF Coatings*

This programming or "teach-in" can be carried out indirectly or directly. In the former case a separate manual device is moved so that the electronic data can be captured and then transferred to a robot movement program. In the direct teach-in method, the robot arm is moved by the programmer. The path movements are stored and then copied identically by the robot [4.4.73].

Figure 4.2.112: Painting robots at work *Source: Dürr*

Application

Additional process parameters such as the quantity of paint and the atomising air must be programmed separately. The most important thing is extremely precise timing of the desired air and paint quantities. Delays of the order of mere milliseconds can lead to paint defects or an inferior paint finish.

In the past the use of robots was problematic due to limitations on the masses that needed to be moved. Robots were unable to move heavy guns or high-inertia, high-speed rotary

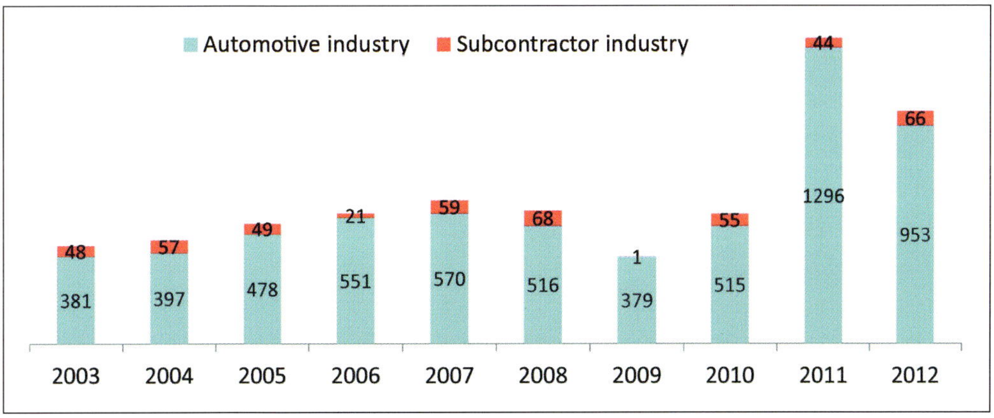

Figure 4.2.113: Sales of one manufacturer's robots for use in industrial coating technology

bells with sufficient accuracy and reproducibility. New generations of robots and smaller bells now make processing with high-speed rotary bells on robot arms easier [4.4.74].

Robots can deliver very high coating quality, provided they are properly programmed. However, this is contingent on constant material and ambient conditions. The advantages that can be achieved with robots are substantial. To constant film thicknesses, less paint spill and lower material consumption can be added the opportunity to replace the painter completely. The uptime of painting robots is now so high that the high capital costs are quickly offset by the advantages and savings. New painting lines for coating mass-produced goods now only feature robots.

4.2.1.7 Conveyors

Apart from application units, systems to eradicate overspray and systems to supply the coating materials, effective painting lines also require conveyors. Their job is to convey the workpieces to the individual surface treatment units cost-effectively and reliably. The flexibility of the painting line and the type of parts to be painted determine whether simple floor-based circular conveyors or overhead power-and-free conveyors are employed.

Application

Circular conveyors (see Figure 4.2.114) consist of a rail system, on which an endless chain fitted with hangers collects the workpieces, generally by means of a manual bracket, and then releases them again by the same method after they leave the final stage of the coating line. The chains in the rails run on idlers and are driven by infinitely adjustable motors. Conveying speeds vary from system to system but can reach up to 15 m/min.

Figure 4.2.114: Schematic drawing of a circular conveyor
Source: Eisenmann

The situation with **power-and-free conveyors** is different. These consist of two conveyor rails located either underneath or next to each other. The power rail (forming a circular conveyor) which supplements the carrier rail (free line) enables workpieces to be discharged from or added to the material flow by means of appropriate drivers without the need for stopping the conveyors. In this way buffer zones can be formed, and products returned for reworking can be diverted to another carrier rail. Furthermore, power-and-free conveyors make it possible to switch between continous and batch operation [4.4.75]. A further advantage is that loading and unloading of the conveyors can take place after discharge while stationary or on a side path. In addition, the individual conveying processes can be automated with programmable logic controllers.

From the painter's point of view, conveyors are a potential hazard if the workpieces are conveyed in suspended form due to possible contamination of the still fresh coating. Paint defects in the form of craters or impaired wetting can be caused by grease or oil particles falling from the lubrication of the conveyor rollers. Such defects can be avoided by recourse to crater-free lubricants or the installation of guards.

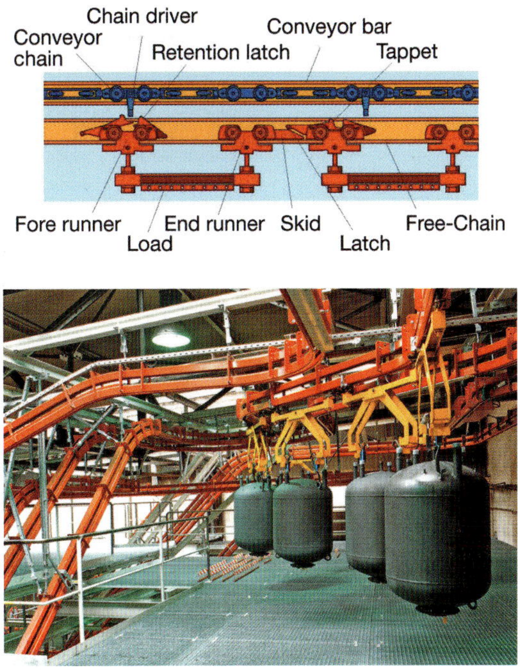

Figure 4.2.115: Power-and-free conveyor *Source, below: Eisenmann*

As mentioned above, aside from the usual overhead conveyors, floor conveyors can also be used in line with the specified task and the workpiece. These are either chains running in rail sections in the floor or induction loops which feed the generally large workpiece to be painted on modified carriages, termed skids, to the individual units. If only planar parts have to be conveyed, as in the woodworking and furniture industry, simple conveyors employing rubber, synthetic or metal belts are adequate.

Special conveyor systems have been developed for the pretreatment and electrocoating of car bodies and are increasingly being used (see Chapter 7.1).

4.2.1.8 Paint Removal

A specific problem with spray application is unwanted coating of conveyors, booth walls and the grilles in the booth floor. Since the conveyor plus workpiece always enter the downstream oven of the application system together, paint deposits arising from straying overspray are also stoved. This process repeats itself regularly because the conveyors are circulating continuously. For trouble-free operation, therefore, such paint has to be removed from all parts of equipment in the booths at regular intervals. Otherwise the paint film becomes so thick that it interferes with the operation of the conveyor system or pieces of film flake off because they have been embrittled by repeated stoving and so are a source of contamination.

Cold removal of paint with the aid of chlorinated hydrocarbon solvents, as frequently employed in the past, is seldom encountered nowadays. It is a different story for hot removal of paint with soda or caustic potash solution plus organic swelling agents. Given

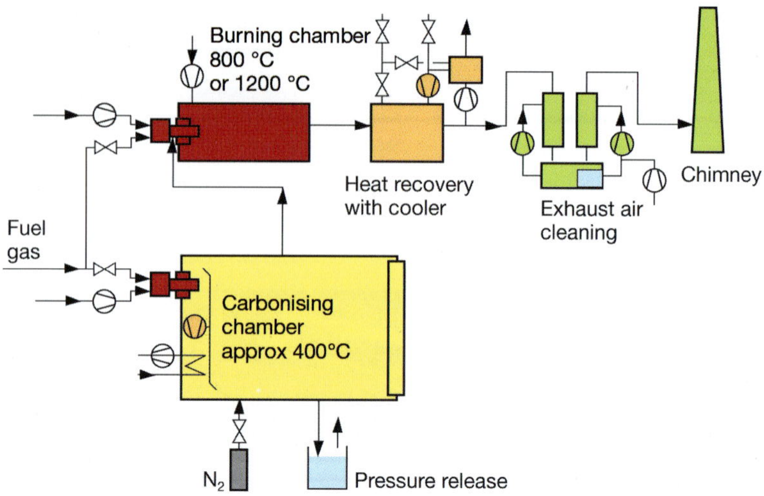

Figure 4.2.116: Schematic drawing of a paint removal process with carbonising chamber

the differences in chemical resistance of coatings, those containing amino groups in particular, such as cathodic electrocoats, are treated in acid baths, in due compliance with environmental legislation, while alkaline paint-removal agents are more suitable for stoving enamels and 2-component polyurethane systems.

Furthermore, high-pressure paint-removal and pyrolysis (thermal paint removal) processes are also employed. Removal of paint by high pressure requires specialist equipment which accelerates heated water at pressures of up to 1400 bar through fine nozzles to effect mechanical separation of the coating.

Treating the parts from which paint is to be removed with special primers in advance makes subsequent paint removal significantly easier. Sometimes such primers have different coefficients of thermal expansion which give rise to stress cracking once the coating cools down. Additional deployment of ultrasound usually makes the coating fall off automatically. The cryo-clean process described earlier (see Chapter 4.1.2) can also be used for final paint removal.

Other primers contain water-soluble film forming agents or carbonate-containing extenders. The former dissolve readily in aqueous paint-removal processes, while the latter foam when heated, thereby assisting with paint removal.

Thermal paint removal, which is the most common process in industry, consists in inducing pyrolysis at temperatures of 400 to 500 °C in chambers or fluidised-bed containers filled with aluminium oxide. The resulting pyrolysis products are burned in separate combustion chambers fitted with heat-recovery systems and fed to a downstream exhaust air cleaning system in compliance with environmental legislation. The recovered thermal energy is used for technical purposes. The advantages of thermal paint removal are that it can be used for almost all paint systems and that it is at the same time inexpensive and environment-friendly.

4.2.2 Processing of Powder Coatings

Unlike liquid coating materials, powder coatings are processed as solids before being melted in the stoving oven to form films. As there are no solvents involved, the application of powder coatings generates almost no emissions. Furthermore, any overspray consisting of powder particles can be recovered and recycled to the process. From an environmental perspective, therefore, powder coatings are the most attractive alternative to many solvent-borne wet paints (see Chapter 5.6.4).

There are several ways to transfer the powder coating particles to the workpiece. One is to heat the workpieces above the melting temperature of the powder coating and then dip them into a fluidised bed of the powder coating, in the manner of dip coating. In electrostatic powder coating an electric charge is imparted to the particles and causes them to fly along the field lines of a nonhomogeneous electric field to the earthed workpiece. Methods of processing the powder coating with the aid of a liquid medium have some practical experience. Aqueous dispersions (slurries) and electrocoat paints (EPC) are two examples of this (see Chapter 5.6).

Application

Despite offering major environmental advantages over other paint systems, powder coatings have failed to gain access to all areas of industrial coating. Excessively high stoving temperatures render powder coating technology unsuitable for thermally unstable substrates such as plastics and wood. However, the results of new research give hope for a breakthrough in these sectors too, and so an additional increase in the already high growth rates can be anticipated in future [4.4.76, 4.4.77].

4.2.2.1 Retrospective

Powder coating technology originated in the fluidised-bed method of applying thermoplastic powder coatings was invented in Germany in 1952 [4.4.78]. Fluidised-bed coating failed to become a major process because of the excessive film thickness and because a very competitive technology for electrostatic application of powder coatings was developed a short time afterwards in the USA. However, the launch of reactive powder coatings by Shell and further optimisation in application and fluidisation technology rekindled interest in Europe. As a result, industrial applications increased, slowly at first but then steadily. It was not until 1969 that companies in the Far East began to look into the question of expensive polyacryl-ics under licence from various American paint companies. When US companies turned to epoxy resins and polyesters in the early 1970s, powder coating technology was already well established in Italy and Germany. Acidic polyester, epoxy polyester and pure epoxy powder coatings were being processed successfully on an industrial scale. The technology was deemed so mature at that

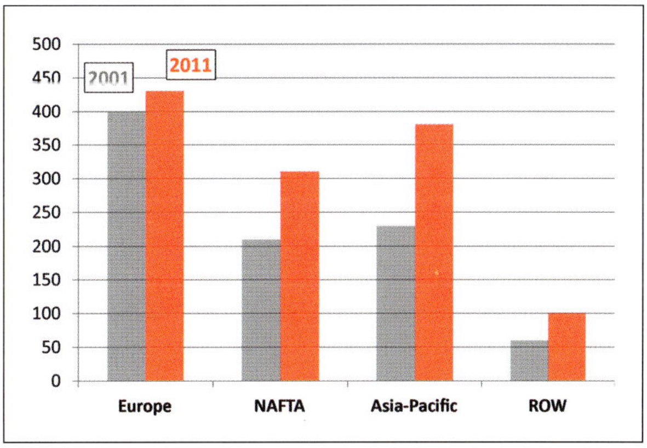

Figure 4.2.117: Expected regional growth in the powder coatings market

stage that new applications for it were being sought even in the automotive industry. It was thought that conventional powder coatings, slurries and EPC paints would, if used as primers, make the so-called reverse process attractive (see Chapter 4.2.1). These methods found application for a while but fell out of favour again after just a few years.

Very little growth occurred in the powder coatings market until 1980. Up to that time application equipment was too expensive and film thickness too high to be economically viable. Problems with colour changes and the high stoving temperatures significantly limited the range of colours, effects and substrates. However, as environmental awareness grew, the application of powder coatings grew more than that of conventional paints. Today powder coatings have a global market share of almost 10 %, expressed in terms of solid content.

New fields of use and new powder coatings were discovered by intensifying and coordinating research activities for powder coating technology in Germany and Europe as a whole as well as in the USA. The emergence of new powder coatings that could be cured with radiation meant that coating of heat-sensitive materials such as wood and timber-derived substances became both technically and financially feasible [4.4.77]. Better powder-metering technology, new knowledge concerning friction charging (tribo-charging) and better booth designs resulted from such insights. Powder coatings have therefore become more widely used. In the automotive industry they are already state of the art for primer surfacers and for limited time for clearcoats [4.4.79].

4.2.2.2 Plant and Equipment Details

Fluidisability is a fundamental prerequisite for handling powder coatings and is what makes it possible to meter powders with pneumatic pumps and to convey them in pipes. Fluidisation is also, of course, an important condition for technical application processes.

Fluidisation

Fluidisation is a process whereby the powder is converted from a static state to a dynamic fluid-like state by means of compressed air. The resulting powder/air mixture can be

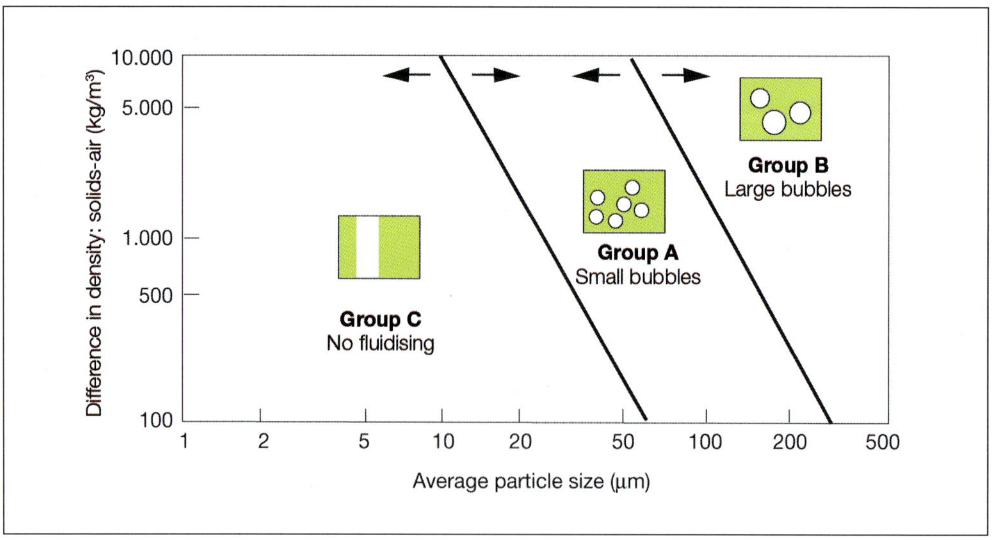

Figure 4.2.118: Product characteristics of bulk goods according to Geldart/Molerus

transported like a fluid. The powder is processed in hoppers which have floors which are permeable to air but not to powder. Certain special powder grades require horizontal vibration in the hoppers to ensure that fluidisation can occur smoothly.

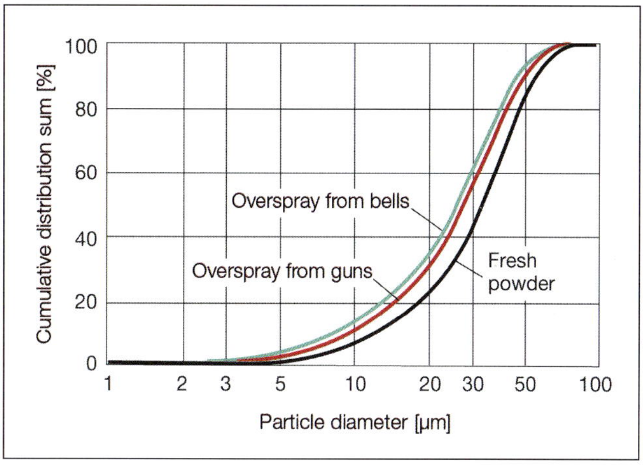

Figure 4.2.119: Typical particle size distribution of industrial powder coatings and related overspray

Powder coatings can be divided into three fluidisability categories on the basis of the essential physical properties of specific gravity and particle size. Group A is easily fluidisable, group C exhibits channelling and group B forms large bubbles when air is introduced.

Powders suitable for use in coating technology are mainly in the transition area between group A and C and usually have particle sizes of between 20 and 45 µm. The particle size distribution is also crucial for smooth handling and ensuring that the required range of film thickness can be achieved. The rule here is that the narrower the distribution, the better the handling. Fractions with particle sizes of less than 10 µm are a negative factor in many cases. Such fractions occur disproportionately in overspray, and these shift the material's fluidisability towards the unfavourable channelling range during recycling. Furthermore, they have a marked tendency to agglomerate.

Application

Powder Transport and Conveying

Powder coatings consist of solid particles and obey different laws from liquid paints in transit because of the need for fluidisation. Different pipes and pumps from normal are therefore used for transporting the powder to the spray heads and for returning and recycling the excess powder [4.4.80].

As the requirements relating to precision differ when powder coatings are transported, a distinction is drawn between precision conveyors for feeding the powder to the guns and bulk conveyors for transporting it between containers.

The entire powder-metering system for precision conveying consists of the storage container, a metering screw and the injectors. The powder injectors are located in front of the spray gun and convey the powder coating at a constant rate from the storage container to the gun where it is then mixed in a defined manner with further volumes of air.

Material is aspirated by means of a vacuum which is generated in a nozzle by accelerated conveying air. The aspirated powder coating is blown into the pipe leading to the gun in an amount depending on the air velocity. For uniform powder discharge, the conveying stream must not pulsate when the powder is being aspirated. The amount of air needed to ensure a uniform feed of the powder coating particles to the spray head varies with the quantity conveyed and so supplementary air may be needed. The length and

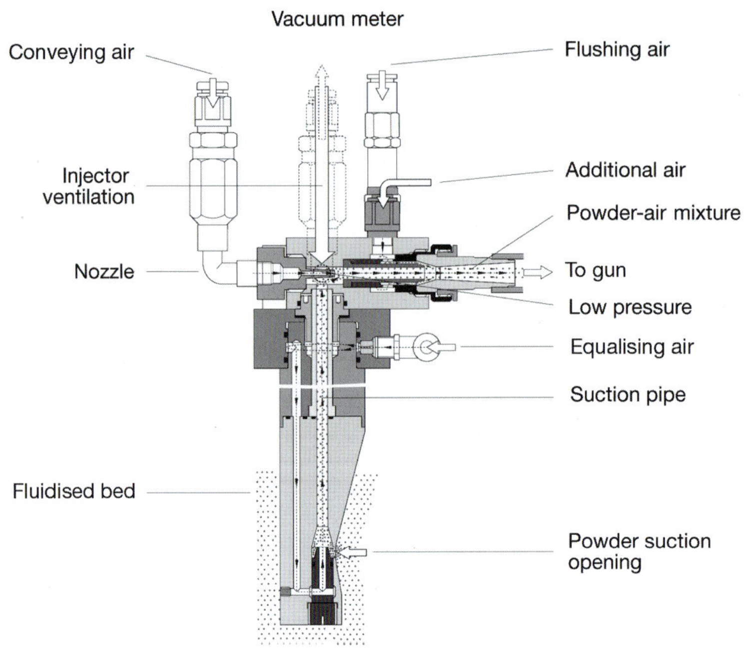

Figure 4.2.120: Schematic drawing of a powder injector

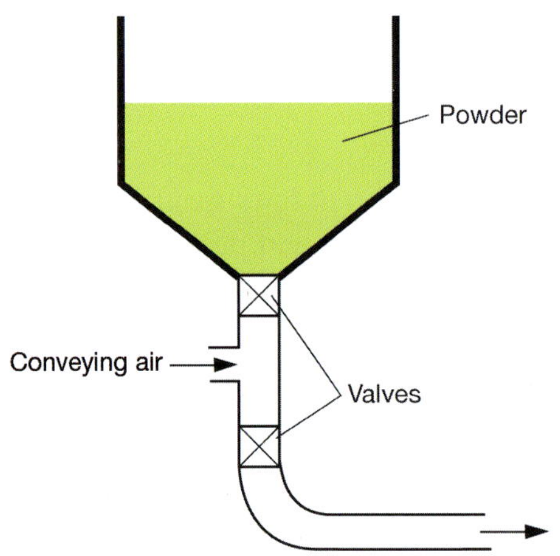

Figure 4.2.121: Schematic drawing of a pulsating powder pump

diameter of the conveyor hoses, their positioning and the discharge height also have a significant impact on powder transport.

Only 100 to 500 g/min has to be transported in precision conveying, but bulk conveyors must cope with up to 5 kg/min, so precision metering of the powder coating flow plays a subordinate role in bulk conveying. Economical operation and gentle handling of the powder are more important. Screw conveyors, oscillating conveyors, as well as high-performance injectors or powder pumps are employed.

In a plug conveyor the powder drops out of a storage container into a pipe after the valve has been opened. The valve closes, a second valve in the line opens and the material is then blown by a short blast of pressure lasting approx. 0.5 s. The valve is closed once more so that the process can start all over again. The plug method requires little air and so no additional filters or cyclones have to be installed at the end of the pipe.

When being conveyed through hose lines, the powder can become electrostatically charged as a result of friction. The main determining factors here are the constituent

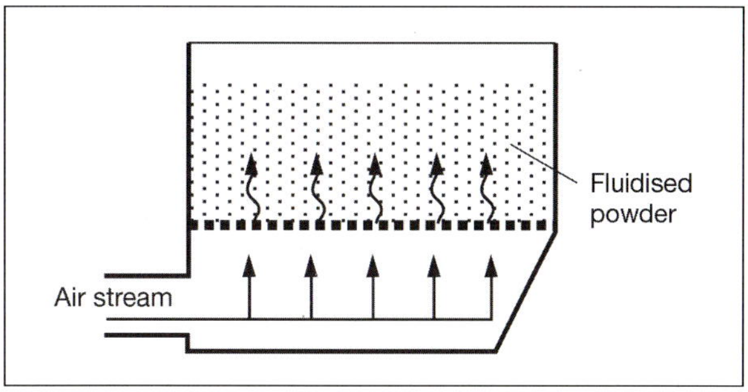

Figure 4.2.122: Principle behind fluidised-bed coating

materials of the hose lines and the hose cross-sections. Small cross-sections require high flow rates, at which powder particles tend to cake onto the line because of friction and impact. With large cross-sections the air flow is often insufficient to maintain the powder in a fluid state along the full length of long hoses. This leads to pulsating powder flows which naturally give rise to variations in film thickness.

Conveying technology based on digital dense phase conveying (DPC) significantly increases the precision of the powder quantities [4.4.81]. In this process a powder chamber is filled from a powder reservoir under vacuum and, after filling, emptied towards the spray gun with compressed air. The valves are controlled electronically (digitally). Applications are mainly found in the automotive and supplier industry [4.4.82].

Application of Powder Coatings

Despite the very high film thickness of the mainly thermoplastic coatings and despite the introduction of other processing methods for particular applications, **fluidised-bed coating** is firmly entrenched as a niche application. It is a dip-coating method in which workpieces that have been carefully pretreated and heated in advance are dipped in a bath of fluidised coating powder. Air flowing into the bath from below via a screen fluidises the powder. The fluidised powder comes into contact with the surface of the preheated parts where it melts if the workpiece temperature is high enough. The thickness of the resultant film depends on the dipping time; after being conveyed to the stoving oven, this film melts and flows due to the further increase in temperature and can also become crosslinked if the constituent powder is of a reactive type.

Electrostatic coating with powder coatings is achieved by conveying electrically charged powder particles in an electric field, followed by deposition on the workpiece by electric forces.

The equipment required for this type of powder coating includes storage containers, supply systems, equipment for charging the powder, spray systems and booths with units for extracting and recovering the overspray.

Two mechanisms exist for electrically charging the powder coatings. The first involves generating air ions, which subsequently attach themselves to the powder particles in an electric field by means of a corona discharge at pointed electrodes. The charged powder particles migrate along the field lines to the workpiece which has the opposite electrical charge.

Most guns work on the principle of external charging. The workpiece alone is the counter-electrode. The result is a high flow of electrons or ions to the workpiece. Since the powder particles only become charged when they leave the gun, there are only minor problems with soiling of the gun.

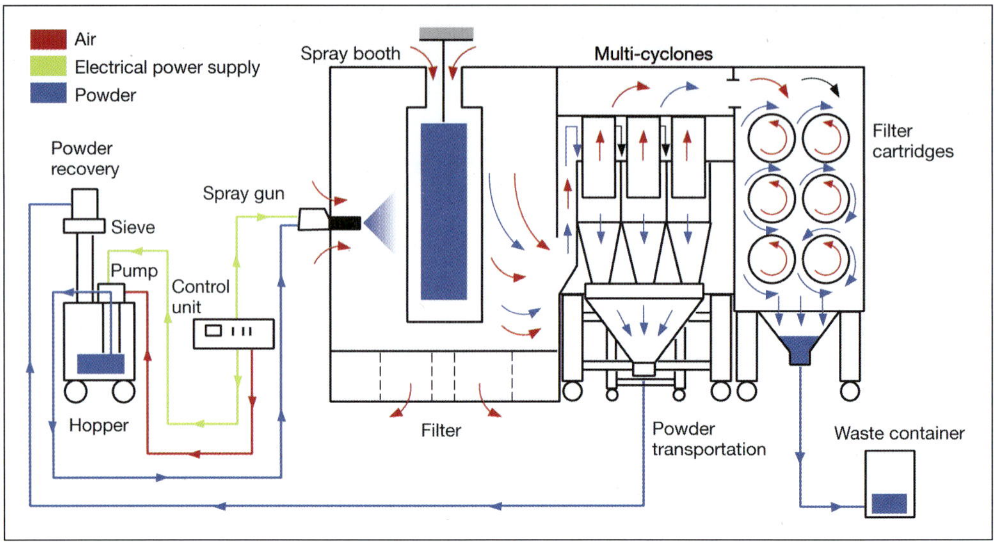

Figure 4.2.123: Schematic drawing of a powder coating unit featuring electrostatic charging

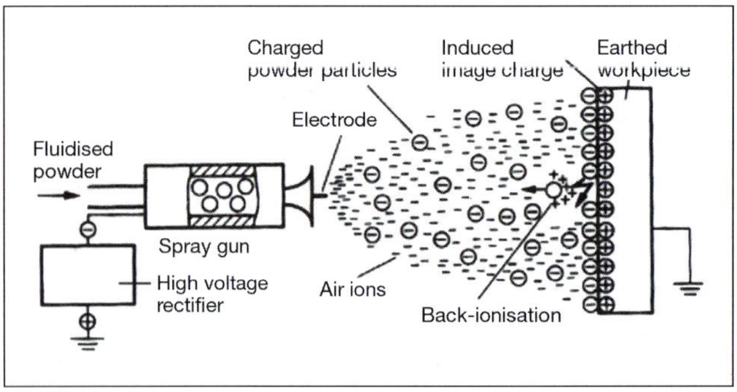

Figure 4.2.124: Schematic drawing of spray guns for external charging

Internal charging utilises a charging electrode and also an earthing point installed within the gun. Consequently, there are substantially fewer electrons available for the screening space charge. This leads to better cavity coating. However, there are problems with soiling of the earthed electrode inside the gun.

The size and shape of the powder coating particles are also crucial to the charging process. In theory spherical coating powder particles offer the best conditions for a uniform charge. The bizarrely and irregularly shaped particles produced by today's grinding technology therefore necessarily give rise to non-uniform charges on the individual particles and thus to different deposition properties. The charge in the case of corona charging is of the order of 2 to 12 μC/g.

The second way to electrically charge the powder particles utilises friction on suitable materials; no electric field is then required. In this **tribo-charging** by friction between the powder particles and a nonconductor of different electronegativity, electrons are exchanged between the two materials. The chemical nature of the paint resin and the plastic wall in contact with the particles determines whether they become positively or negatively charged. The functionality of this tribological method depends to a great extent on the nature of the film forming agent. The tendency to attract electrons increases with increase in electron work function (PMMA, PVC, PTFE). By contrast, if the ionisation energy is low (PA, PPO), the material emits electrons more easily. The most important powder formulations are given a positive charge in tribo-application devices. In tribo-electric spray heads, the powder particles flow through a spiral insulation duct. As a result,

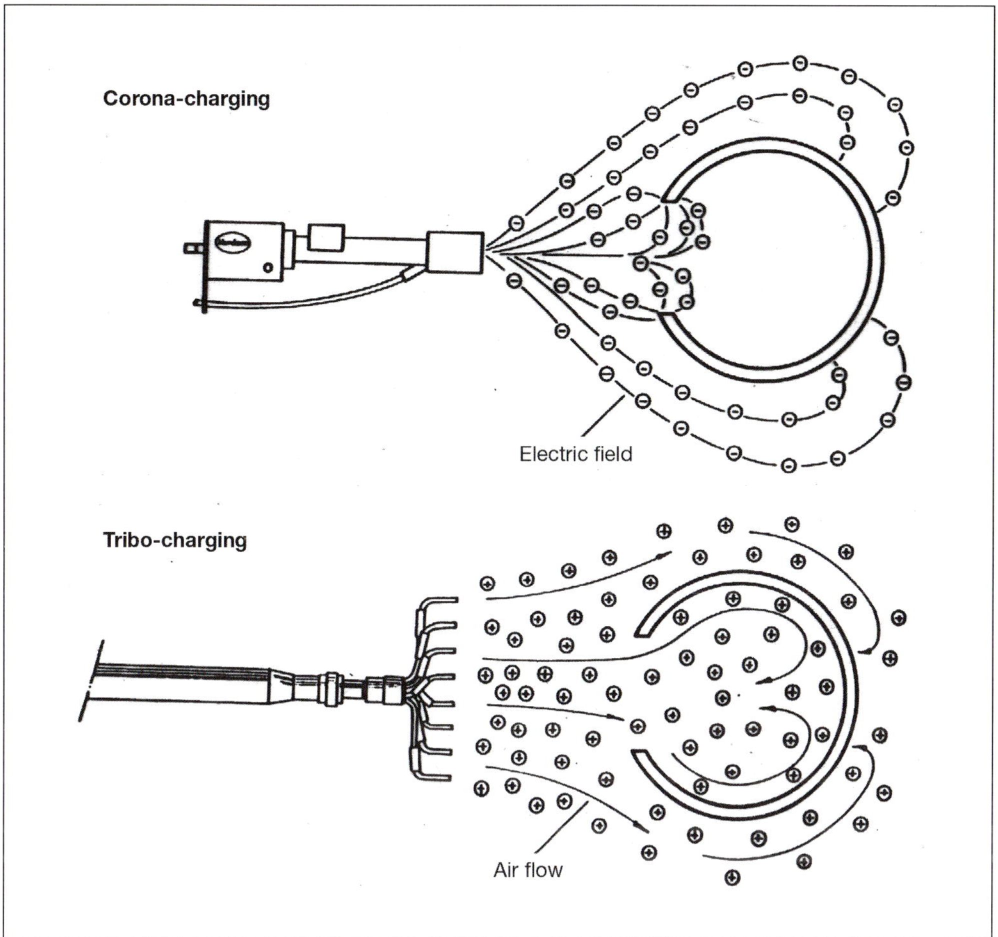

Figure 4.2.125: Deposition of the powder particles in corona and tribo-charging

the coating powder particles are pressed by centrifugal forces against the walls where they pick up their charge in direct contact with the duct wall. If the thus charged particles are then brought into the vicinity of an earthed workpiece by means of suitable spray heads, they form a discrete electric field with the workpiece. If a charged particle is already in a cavity, it will also form a field there. By contrast with external charging of powder coatings, particle deposition is therefore also possible in cavities.

Widespread application is hampered by the limited availability of tribo-capable film forming agents. However, the ease of tribo-charging can be improved by the use of additives such as aluminium oxide. With such formulations the dependence of the charging level on the relative humidity and the ambient temperature must be noted should conditions be extreme. Tribo-charging can produce charges of 2 to 10 μC/g on powder particles.

There are a number of reasons why tribo-application has only achieved about 15 % market penetration to date. One is the limited range of suitable powder resins plus the low performance resulting from the low material output. Another is that small particles are difficult to charge. Recyclability is therefore limited. Since the particles are larger than in the electrostatic method good levelling cannot be achieved in all applications.

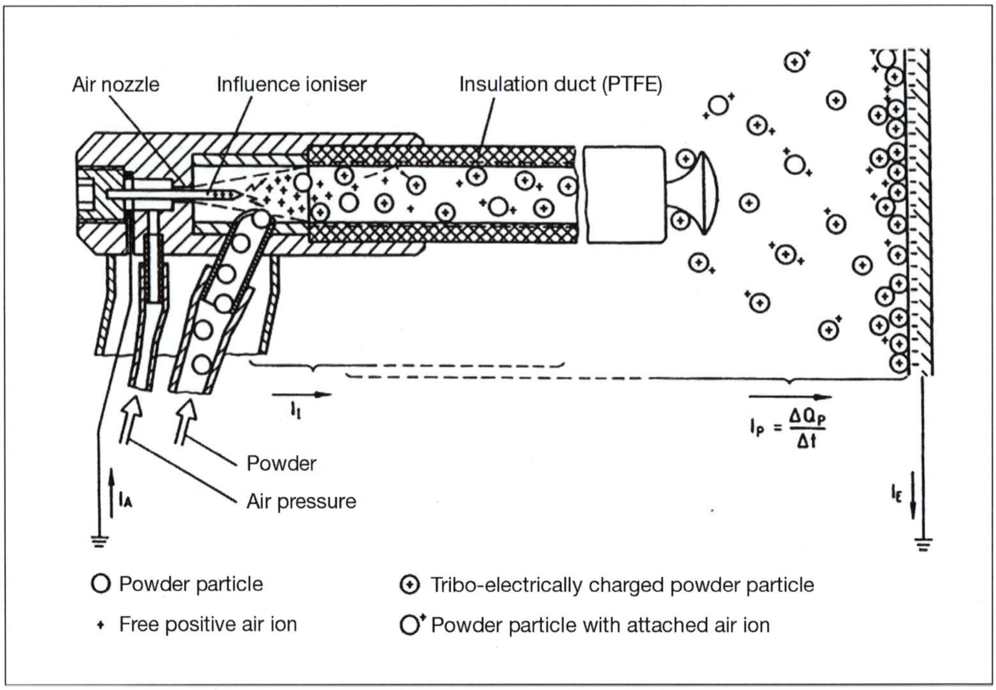

Figure 4.2.126: Schematic drawing of a gun for tribo-charging

Atomisation of the powder coating is carried out in the usual spray guns by fanning the air current out with the aid of mouth pieces. These either contain jet impact plate systems with free swirl flow or are fan jet or finger nozzles. The use of bells rotating at 2,000 to 10,000 rpm is relatively new. These create a broad spray cloud and are therefore ideal for large workpieces. The output rate is extremely high at up to 700 g/min.

Spinning discs in conjunction with an omega loop conveyor for the workpieces are a variant of the spray head design. The powder particles which are distributed horizontally to all sides form uniformly thick coats on the workpieces which are moved in a loop and also, if necessary, rotated about their own axis.

Unlike the case for guns, the atomisation parameters of spinning discs or rotary bells can be modified via the speed during operation.

While the charged particles in the electric field are **en route to the workpiece** electric forces, gravity and air resistance act on them. The electric forces must be greater than gravity so that the particles can largely follow the field lines to the workpiece. These conditions obtain for particle sizes of less than 100 µm. As a result of air friction the particles quickly attain a steady state. The migration rate is around 1 to 2 m/s.

Smaller particles deviate from their path sooner than larger ones owing to their lower overall charge. They tend to be more readily collected by the recovery systems and do not reach the workpiece. As a result of this enrichment with the fine components, the recovered powder has a different composition from the sprayed material (see Figure 4.2.119). To ensure a constant coating quality, therefore, the distribution profile of the particle sizes in the replenishment material is frequently modified.

Figure 4.2.127: Application example of powder coating atomisation by rotational bells *Source : Dürr*

A consequence of the high electrical resistance of the powder particles which reach the workpiece during **deposition** is that they only transmit a small portion of their charge to the conductive, earthed workpiece. The remainder of the charge remains on the particles until they enter the oven. After deposition of the particles on the workpiece, this charge induces opposite charges on the electrically conductive material surface. As a result, *Coulomb* forces in the form of image forces come into play and will even hold the powder on vertical surfaces.

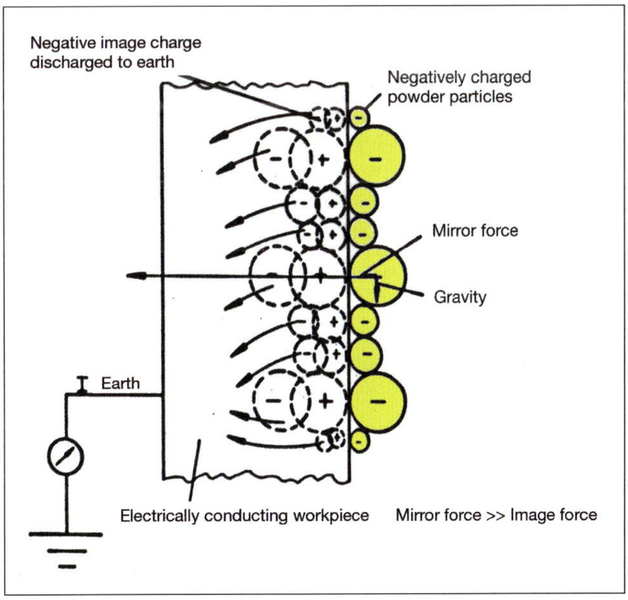

Figure 4.2.128: Powder adhesion to conductive vertical substrates

Where very thick powder coats, generally in excess of 150 µm, are applied by corona charging, effects occur which are attributable to the fact that a large excess of ionised air accumulates at the interface between the applied paint and the spray cloud and thus repels the powder particles of same electric charge. The quantity of air ions can be reduced by means of a supplementary electrode mounted on the atomiser. This colle-

Application process	Current
Counter corona with 60 kV	7 – 8 µA
Corona with 60 kV	29 µA
Counter corona with 90 kV	11 µA
Corona with 90 kV	40 µA

Figure 4.2.129: Effect of counter corona in the case of a polyester-epoxy hybrid powder coating

cts the free charge carriers and so causes low-ion spraying. The action of this component, termed a counter corona or super corona, can be verified via the current flowing from the substrate. It is smaller by a factor of 3 to 4. The powder charge itself is unaffected, as is evident from the same coating efficiencies of both processes. Surface finish is also considerably improved. Naturally, the re-ionisation process just described does not occur with tribo-application.

Recovery

Given that the initial coating efficiency of powder coatings at 30 to 50 % is low by comparison with liquid paints, reprocessing of the overspray, which therefore represents 50 to 70 % of the paint used, is necessary both economically and for the sake of the environment. There is no problem with this in terms of the properties of the powder material. Technically reliable recovery systems are available.

Application

The overspray is routed out of the booths by gravity and by downdraft air flow. Cyclones and plate, cartridge or bag filters are used for recovery. In **cyclones** the air flow charged with powder coating material is made to rotate tangentially. The rotation of the air can be enhanced by vanes. Centrifugal forces move the powder particles to the wall of the cyclone where they fall downwards under gravity and are then collected in a container.

The degree of powder separation depends on the particle size. Fine components in the exhaust air stream are not recovered in full by cyclones. Separation efficiency decreases with increase in cyclone diameter and so the air flow is divided into several individual flows in so-called multicyclones. Horizontally configured multicyclones have recently been incorporated directly into the booth wall.

Other methods of powder recovery are based on the use of plate filters. The plate material consists of sintered polyethylene with a porous polytetrafluoroethylene coating for easy separation of the powder from the filter surface. Filter processes are discontinuous. The filters have to be cleaned once a certain amount of material has accumulated. Under a high pressure drop separation levels of 99.99 % can be achieved even though throughput is high. The very long service life compared with other filter media must, however, be weighed up against a relatively high price. Cheaper cartridge or bag filters not only have shorter service lives, they are also much less effective, with efficiencies of only 99.0 to 99.9 %. Furthermore, they are more difficult to clean.

Trouble-free operation of the painting line is contingent on removal of agglomerates, foreign matter (dirt) and, if appropriate, the fine-grain fraction from the powder coating material recovered by cyclones or filters. If there is no foreign matter, simple vibrating screens are often sufficient, frequently in conjunction with cyclones. Turbo screens work on a different principle. The powder/air mixture is drawn through the screen by suction, and dirt and fibres are retained on the screen surface and have to be removed manually. Turbo screens, vibrating screens and ultrasonic screens are suitable for operations involving colour changes because they are easy to clean.

Figure 4.2.130: Multicyclone for powder recovery *Source: ITW*

Figure 4.2.131: Ultrasonic screen *Source: ITW*

Centrifugal and tumbling screens are suitable for continuous operation. In the former, the powder coating particles are routed to the cylindrical screen jacket by means of rotation. Dirt and screen oversize are continuously removed from the screen by a device and collected at the end. A tumbling screen operates on similar principles. Several screens of different mesh size move in such a way that the powder tumbles through them. The individual fractions gather at the edges of the individual screens in corresponding collecting channels.

Powder Booths

Trouble-free powder-coating processing in compliance with occupational health and safety legislation calls for feed, application and powder recovery units, as well as a booth operated with underpressure. The last of these primarily serves to prevent powder from escaping into the environment and enables controlled air feed for recovery. It also prevents the flammability limits for the powder/air mixture from being exceeded. When powder coatings are processed under high voltage supplementary conditions have to be met owing to the risk of explosions and fire (see Chapter 5.2.4).

Figure 4.2.132: Picture of a powder application unit for industrial coating *Source: BASF Coatings*

The booth walls are made of steel, stainless steel or plastic. Steel booths are inexpensive and easy to assemble and meet all safety regulations without difficulty. The disadvantages are susceptibility to soiling and the cost of cleaning after every colour change. Advantages in this respect provides smooth, joint-free plastic walls. On account of their poor electrical conductivity these repel the powder. The use of plastics as housing material makes compliance with the relevant safety regulations more difficult.

The same considerations apply to pipes as for the booth material. Plastic pipes must be earthed as well because of the risk of brush discharges.

For technical and safety reasons, manually operated equipment is only rarely found in industrial powder-coating booths. The guns are introduced into the booth through slits on powered arms and are operated from outside. The powered arms also ensure uniform distribution of the powder over the workpiece. At the same time the number of guns necessary for uniform film thickness is lower than in a static configuration. Vertical movement of the powered arms should not exceed 0.5 m/s in order that coating defects arising from turbulent airflow may be prevented.

Colour changes in powder-coating booths entails much more work than in wet-paint processing. It is not only the booths and pipes which have to be designed for easy cleaning; the powder-supply centre and the filters or cyclones for powder recovery must also be of a suitable design.

Many of the cleaning stages can only be carried out manually. The booth walls are air-blasted in two stages for periods of 5 to 6 minutes and then wiped down with sponges. The guns and hoses too must be cleaned and the immersed pipes and injectors purged. The control devices for the application equipment are usually located outside the booths, and so they do not have to be cleaned in the event of a colour change. The outcome is that colour changes can be performed in less than 20 minutes nowadays.

With appropriate investment in several interchangeable booths, colour changes can be carried out without any production downtime [4.4.85].

4.2.2.3 Powder Coating Materials

Powder coatings consist of thermoplastic or reactive polymers which are solid at room temperature. Thermoplastic powders nowadays are used to produce thick coatings by the fluidised-bed method, but only to a very small extent.

Most powders are reactive one-component systems which liquefy, flow and then cross-link as a result of treatment with heat. The most common film forming agents are epoxy resins, epoxy resin-modified polyesters, pure polyesters with hardeners containing epoxy groups or blocked isocyanates and acrylic resins with various crosslinking agents. Epoxy resins are used for the formulation of primer powders. Such powder coatings feature excellent anticorrosive protection and good mechanical/technical properties. They can be cured with dicyandiamide or polycarboxylic acids (see Chapter 2.1.1). The marked tendency of pure epoxy resin powders to undergo yellowing and their low resistance to longterm outdoor exposure are disadvantageous. The group of epoxy-polyester powders, often called hybrid powder coatings, is a mixture of epoxy resin and acidic polyester resin. The mixing ratio can range from 1:1 to 1:9 in favour of the polyester. Hybrid powder coatings feature excellent mechanical/technical properties. They have less tendency to undergo yellowing and chalking than pure epoxy resin powders. The hybrid powders are inadequate for demanding longterm outdoor exposure specifications. In such cases recourse should be made to polyester or polyurethane powders. Polyester powder coatings consist of acid polyesters together with a low-molecular hardener (see Chapter 2.1.1). Triglycidyl isocyanurate (TGIC) has been replaced by toxicologically safe alternatives because of suspected mutagenic effects. Glycidic esters of isophthalic acid or β-hydroxyalkylamides of adipic acid can serve as replacements for TGIC. If weather-resistant, mattable surfaces are required, polyurethane powders based on OH-functional polyesters and capped isocyanates are suitable systems. The hardeners, which are capped with ε-aminocaprolactam, release the capping agent in the stoving oven and contaminate the exhaust air from the oven, so the exhaust air has to be treated. Uretdiones do not have the disadvantage of forming cleavage products and are therefore gaining in prominence.

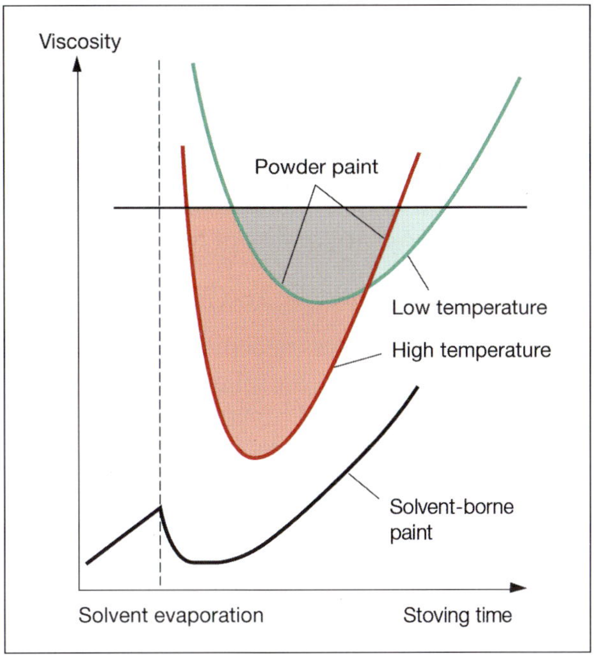

Figure 4.2.133: Schematic viscosity profile of powder coatings as a function of temperature rise in stoving ovens

Acrylic powders were unable to establish any major application fields in the past despite their excellent resistance to longterm outdoor exposure and their good levelling. It was only when clearcoats became established for sealing metallic basecoats in the automotive sector that acrylic powder coatings began to compete with the other powder types. The acrylic resins used in those formulations contain glycidyl methacrylate and react with carboxyl-functional crosslinkers.

Powder coatings have been processed in the automotive industry in the form of slurries which have the same chemical basis but are dispersed as microfine powders in water. Their curing temperature is rather low at 140 °C (see Chapter 5.6.3).

The rheological properties of powder coatings can be influenced to a limited extent only, due to the lack of solvents. The processing properties are therefore largely determined by the film forming agents. The molecular weight, molecular shape, polarity and reactivity determine the viscosity, and shape the entire levelling phase. Temperature control in the oven in conjunction with the viscosity profile of the powder coating are key to ensuring a smooth, sag-free coating. In the interests of good levelling the temperature-induced reduction in viscosity should reach its minimum as quickly as possible and should remain at this level for a sufficient time. The minimum viscosity level and the curve of the viscosity rise initiated by crosslinking are crucial for good levelling.

The practice of using the time which a powder melt spends below the target viscosity as a criterion for the levelling properties has proved useful. The area marked in Figure

4.2.133 is termed the fluidity integral and is a characteristic feature of the levelling of powder coatings (see Chapter 3.1.4).

Powder coatings trap large volumes of air during application. This air has to escape during melting and before crosslinking. This can be accomplished with the aid of benzoin and its derivatives. The flowability needed for trouble-free processing is maintained and adjusted with aluminium oxides or pyrogenic silica. Aluminium oxides also have an impact on the chargeability of the powder particles.

4.2.2.4 Powder-Specific Test Methods

Powder coatings possess specific properties which rule out many of the proven test methods for wet paints. As solid powders, they obey different laws of physics during processing (see above). Properties such as gel time, flowability, fluidisability, bulk density and bulk volume are significant physical variables for storage, transport and processing. Since these properties, just like the levelling of the already applied material, are also a function of the particle size or particle size distribution, these variables need to be factored into any description of the application properties of powder coatings (see Chapter 2.3).

Gel time and T_g

The specified conditions for storage and transport of powder coatings should be observed. Temperatures above 30 °C cause agglomeration of the powder particles and preliminary chemical reactions. Low temperatures prevent this and therefore ensure stable levelling properties over a period of 6 to 12 months, with the exact duration depending on the powder type.

The extent of preliminary crosslinking can be tested by reference to the gel time, the measurement method for which is described in ISO 8130-6. However, preliminary reactions which may have taken place during manufacture or storage can also be quantified by thermal analysis methods such as DTA or DSC. Shifts in the glass transition temperature T_g and a reduction in the enthalpy of reaction are regarded as indicators of improper treatment of the powder coatings during manufacture and storage.

Flowability

Just as in viscosity measurements of liquid paints, a certain quantity is allowed to pass a defined aperture in a fluidisation hopper. The flow-out time serves as an indicator of the powder's flowability. The volume required to fill the hopper is weighed in advance and is converted into the bulk density to act as an additional indicator (see Figure 4.2.134).

Fluidisability

Fluidisability is the most important quality-determining variable for evaluating processing properties, especially with regard to transport, dispensing and fluidised-bed coating. An amount of the powder coating commensurate with the load requirements is fluidised in a glass cylinder by an air current (determined by the air pressure) flowing through a porous base plate. The resulting depth of the fluidised powder, relative to the original depth h_0, is then a measure of its fluidisability. Since the results depend to a great degree on the equipment and conditions, precise and detailed agreements must be drawn up between customers and suppliers.

Good online characterisation of the fluidisation state for process control is possible by measuring pressure fluctuations in the fluidised bed. The occurrence of gas bubbles creates pressure fluctuations at a piezoelectric pressure transducer which are converted into a standard deviation which yields a characteristic value for the degree of fluidisation. Different curves are produced for the different levels of fluidisability as a function of the velocity of the air flow.

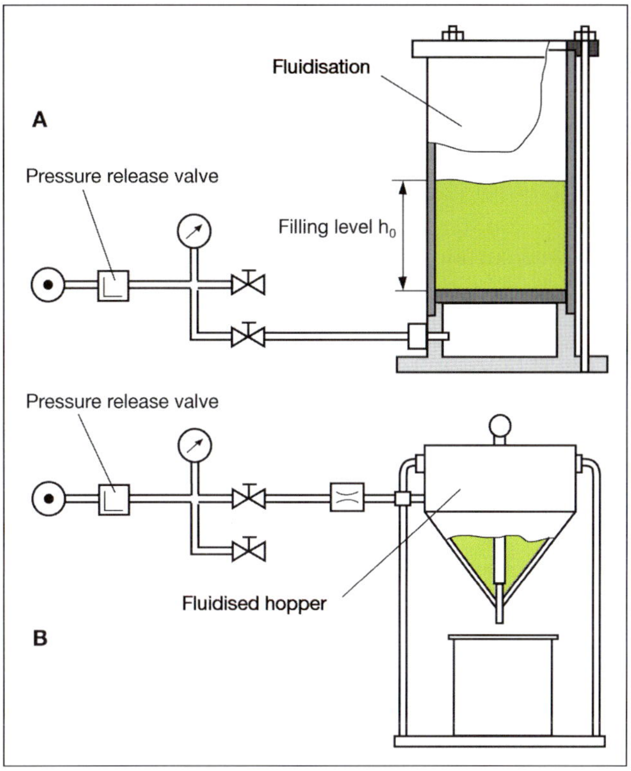

Figure 4.2.134: Methods for measuring the processing properties of powder coatings: (A) fluidisability and (B) flowability

Chargeability

Time-dependent pick-up and release of electrical charges are properties specific to electrostatic powder processing. Here it is necessary to determine chargeability under high voltage and tribo-chargeability. The electrical resistance of the powder material is related to its chargeability and so this easily measured variable serves as an indicator. A resistance-measuring cell, a high-resistance ohm meter and a plotter are required for the measurement. The resistance cell consists of an insulated inner cylinder with an electrode. The electrically insulated outer sleeve contains the second electrode. The powder is poured into the spaces between the electrodes, a potential difference of 1000 volts is applied and the resistance is read or recorded at various times. The final value after 5 minutes is taken as the measured reading.

Powder coatings with a specific resistance of more than 10^{14} Ω cm are regarded as highly suitable for electrostatic processing. At low resistances the coatings are thinner, and the influence of the relative humidity diminishes the stability of the image forces. Powder coatings with specific resistances of less than 10^{12} Ω cm are therefore unsuitable for electrostatic processing.

Tribo-chargeability is measured with apparatus similar to that employed for measuring flowability. The charge Q is determined as the product of the current I and time t. If current greater than 1.5 μA flows, the powder coating lends itself readily to tribo-charging. The transmissible charge per unit mass Q/m serves as a further characteristic. For good tribo-charging properties the charge transition per unit mass should always be in excess of 0.5 μC/g.

4.2.3 Summary

A number of methods have proved successful for processing coating materials. The size and nature of the workpieces, the coating speed and the desired quality of the coating determine whether dip, curtain, flow, roller or spray coating should be used.

Among dip-coating methods, electrocoating has gained acceptance worldwide for metallic workpieces because of its environmental and economic advantages. Whereas flow- and curtain-coating systems are encountered in only a few areas of coating technology, spray systems are ubiquitous. The advantages of this application method are its ability to coat intricate workpieces and the possibility of rapid colour changes. However, the advantages of the optically attractive coatings with various effects must be weighed against the disadvantage of high material spill. Spray systems employ pneumatic, hydraulic, electrical or rotation-induced centrifugal forces to atomise the paint. Spray systems for industrial-scale paint processing require a broad range of ancillary systems to supply material, energy and air and the appropriate equipment for disposing of the overspray and scrubbing the exhaust air.

Roller coating has a particular attraction because of the opportunity to precoat planar coiled sheet metal which is formed into consumer goods only at a later stage. The tendency to concentrate surface processes in just a few large high-speed coating systems that generate little environmental pollution is making coil coating increasingly important.

Besides the various application processes for liquid paints there are methods for applying coating materials in the form of solvent-free solid powders. The nonstick particles can be recovered without difficulty if they miss the workpiece and can be returned to the process. To date, powder coating technology has only proved successful for thermally stable painting substrates but from an environmental perspective it is the most attractive alternative to conventional technology involving solvent-borne liquid coating materials. New research into the rheological properties of powder-coating melts and new plant-engineering knowledge has made powder-coating technology an interesting proposition for thermally unstable materials such as wood and plastic. These new application areas have the potential to achieve above-average growth rates. If the film thickness, which is still too high for good levelling properties, can be reduced without any loss of aesthetic properties, the overall economic balance too will compare favourably with all other conventional coating technology alternatives.

4.3 Drying and Curing

The physical and chemical principles behind film formation were introduced and explained in Chapter 3.1. Film formation is a mass-transfer process in which solvents are emitted and, in the case of crosslinking systems, chemical reactions occur. Both processes are accelerated or initiated by heat input. Where liquid paints containing organic solvents or water are cured it is necessary to ensure that removal of the volatile components is as complete as possible and that the polyaddition or polycondensation reaction is accelerated by thermal initiation. The necessary heat transfer is achieved with heated air in the case of convection drying or by irradiation with electromagnetic waves in the IR range.

Heat transfer is not needed for paints which cure by means of polymerisation reactions. There, short-wave UV or electron-beam radiation is suitable for initiating chemical cross-linking. Were the paints formulated for radiation curing to be heated, the mostly volatile polymerisable reactants would be emitted and completely different coatings would result.

If coating is carried out in systems in which thermal energy is transferred by means of IR radiation, conduction or convection in combination with physical drying, such systems are ovens. If heat or short-wave radiation is not accompanied by the emission of solvents, as in the case of radiation curing, for example, and is the sole cause of film formation, these are curing systems.

The key variable for thermal curing is heat and mass transfer whereas for radiation curing it is the formation of initiator radicals which are produced by means of high-energy radiation and which excite the double bonds in the film forming agents or reactive diluents.

4.3.1 Film Formation by Heat Transfer

Heat transfer to organic substances such as paints causes the molecules of the paint components to absorb energy by means of oscillation, stretching or rotation. Given sufficient energy input, they can cross the boundary separating the solid, liquid and gaseous states. Furthermore, excitation of the molecules accelerates chemical reactions.

4.3.1.1 Theory

Since the aim is for mass and heat transfer to be economical and environmentally compatible, there are certain requirements governing the design and operation of ovens. Heat exchange must be particularly effective and accommodate the different specific heats and thermal conductivities of the individual materials and the design of the workpieces in such a way that the painted workpiece is heated uniformly and that uniform coatings are possible.

Mass and heat transfer can only be calculated if certain constraints are imposed. As heat and mass transfer in the ovens are dynamic processes, the adopted approach is either to study only short time periods in which there are no significant shifts in concentration and temperature or to create a calculable steady state by permanently loading a continuous-flow oven. This affords a way of determining overall balances via mass and heat transfer without the need to address the changing conditions in individual workpieces.

Mass transfer for solvents is calculated by determining the rate of emission $\dot{m}$ over short periods via the difference in concentration in the vapour immediately at the coating surface c_0 and the turbulent environment in the convective vapour c_∞. The mass flow $\dot{m}$ per unit area A is defined as:

$$\dot{m} = \frac{dm}{dt} \cdot \frac{1}{A}$$

This mass flow per unit area $\dot{m}$ is proportional to the difference $c_0 - c_\infty = \Delta c$. Here, c_0 is particularly high if high temperatures increase the kinetic energy of the solvent molecules to such an extent that a large number of molecules overcome the enthalpy of evaporation. The proportionality factor between $\dot{m}$ and Δc is defined as the mass-transfer coefficient β.

$$\dot{m} = \beta \cdot \Delta c$$

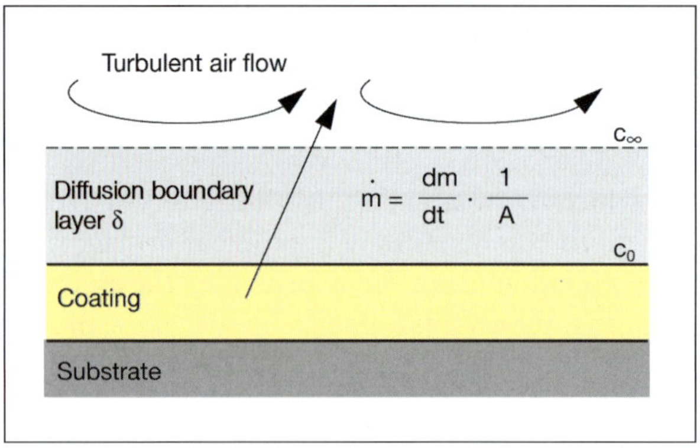

Figure 4.3.1: Exchange of material by diffusion

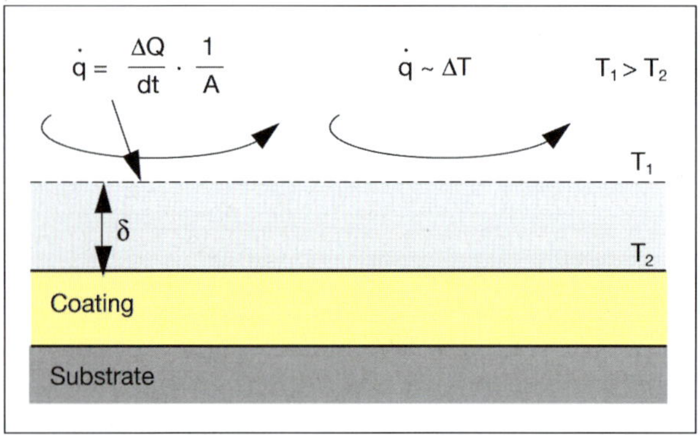

Figure 4.3.2: Mechanism behind heat transfer in convection ovens

Comparison with *Fick's* first law explains the physical importance of the mass-transfer coefficient. It is the quotient of the diffusion coefficients of the solvent D and the thickness of the laminar diffusion boundary layer δ. This variable, which determines the rate of mass transfer in convection drying, can be reduced to a thickness of just a few μm by artificially moving the air in the vapour space.

$$\dot{m} = \frac{D}{\delta} \cdot \Delta c$$

Large mass flows can therefore be set by means of large differences in concentration Δc and a thin laminar boundary layer. This is practically the case which obtains when high air velocities and high temperatures are present at the same time. The former lowers the thickness of the laminar boundary layer. The latter increases the vapour pressure and thus the concentration of the solvents at the interface. For the reasons cited, nozzles and air showers are installed in technical ovens with a view to meeting the target air velocities of up to 15 m/s.

Although the temperature and the velocity of the air determine the drying process, their settings are limited by potential film defects such as dirt transfer, matting and pinholing (see Chapter 6.2). Excessively rapid film hardening during drying and crosslinking makes it harder for solvents to move to the interface. Low film thicknesses in the coating are therefore conducive to ensuring complete mass transfer (see Chapter 3.1).

The rapid heating also initiates or accelerates the rate of chemical reactions. Film forming processes are therefore not only enhanced physically by the input of thermal energy. Efforts should also be made to achieve rapid **heat transfer** in the oven.

Heat transfer to paint films can be achieved in three different ways. The usual, indirect method consists in heating gases, generally air. The heated air transfers only thermal energy to the colder workpiece. The air cools down in the process and, to ensure ongoing heat exchange, has to be constantly reheated. Ovens which use artificially circulated air as the carrier gas are termed **convection ovens**. If a solid with a higher temperature is brought into direct contact with a colder solid such that heat can flow directly into the colder workpiece via conduction, this is called heat transfer by **conduction**. Furthermore, molecules in the coating material or in the substrate can be induced to oscillate by means of **infrared rays**, which are electromagnetic waves, and to transfer thermal energy without contact or interaction with any heated material.

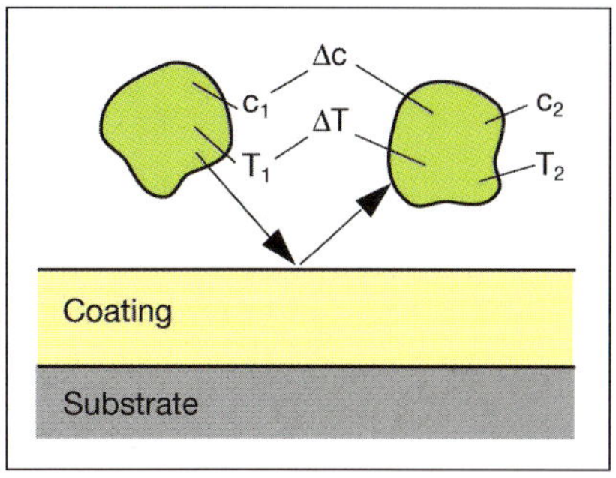

Figure 4.3.3: Mass and heat transport during convection drying

The consequence of turbulent air circulation in convection ovens is that the temperature gradient necessary for heat transfer is only found at the extremely thin laminar boundary layer between the workpiece and the surrounding gas space. Here, by contrast with the adjacent turbulent air flow, heat is transferred from molecule to molecule by molecular collision. The thickness of the laminar boundary layer therefore also determines the rate of energy flow. The aim for heat transfer, as for mass transfer, must be to keep this layer particularly thin. The heat flow per unit area over time $\dot{q} = dQ/dt/A$ is proportional to the temperature difference ΔT:

$$\dot{q} = \alpha \cdot \Delta T$$

where $\alpha = \lambda/\delta$. For its part the heat-transfer coefficient α contains the thermal conductivity λ of the air and again the film thickness δ which cannot be determined directly by experiment. As already mentioned, the thickness of this boundary layer can be influenced by the velocity of the carrier gas.

Application

When ovens are being designed, knowledge of the correlation between the mass-transfer coefficient β and the heat-transfer coefficient α is important. The following model can serve as a starting point for a mechanistic view of the transfers. When the turbulent air meets the coating numerous tiny vortices interact with the coating surface, transfer their heat energy and remove the solvents from the diffusion boundary layer.

The concentration of the solvents in the oven air corresponds to the solvent mass included in the volumetric flow rate:

$$\Delta c = \frac{\dot{m}}{\dot{V}} \quad \text{or} \quad \dot{m} = \dot{V} \cdot \Delta c$$

Since $\dot{m} = \beta\,\Delta c$ also, it follows in the present case that β is identical with $\dot{V}$. Furthermore, the heat transfer $\dot{q} = \alpha$. As ΔT can also be calculated from the volume of circulated air, it follows that:

$$\alpha \cdot \Delta T = \dot{V} \cdot \rho \cdot c_p \cdot \Delta T \quad \text{or} \quad \alpha = \dot{V} \cdot \rho \cdot c_p$$

$$\text{or} \quad \alpha = \beta \cdot \rho \cdot c_p, \text{ because} \quad \dot{V} = \beta$$

Mass and heat transfers are thus directly proportional, as expected.

The thermal energy transferred to the surface of the painted workpiece leads to surface heating and thus to a temperature gradient relative to those parts of the workpiece which are inaccessible to heat transfer. This triggers heat flow in the workpiece itself which is proportional to the temperature gradient ΔT within the workpiece.

$$\dot{q} = \frac{\lambda}{\Delta x} \cdot \Delta T$$

where λ is the thermal conductivity of the solid to be heated.

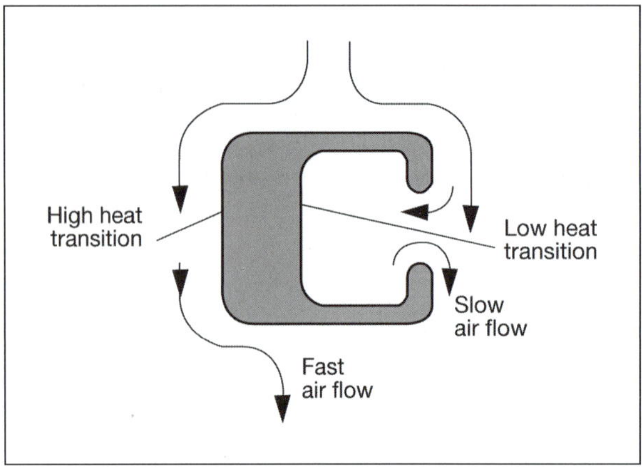

Figure 4.3.4: Causes of temperature differences in intricately shaped parts

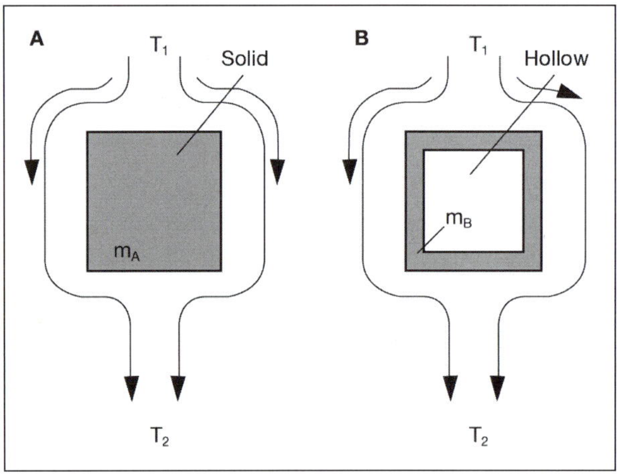

Figure 4.3.5: Heating of same-shaped parts A and B of different masses

If the thermal conductivity is high, as in all metals, the thermal energy flows relatively rapidly from the transfer point to the colder parts of the painted workpiece. The heat flow decreases with increase in distance Δx.

The situation is different for plastics. Their considerably inferior thermal conductivity leads to rapid heating of the outer areas. Continuation inwards is hindered by the low thermal conductivity. The risk of overstoving the outer part of the coating with simultaneous undercuring in the internal areas cannot be discounted.

Differences in the temperature profile of intricately shaped workpieces occur not only as a result of differences in heat transfer and thermal conductivities. Even in the event of rapid heat flow and uniform energy distribution, differences in workpiece temperatures are also bound to occur if there are differences in the masses that have to be heated up in individual areas of intricately shaped parts. Rocker panels of cars have a greater mass than the thinner panels used for the roof, for instance.

Since the quantity of heat transferred in the oven by the air

$$\dot{q} = \dot{V}_{air} \cdot c_p \cdot (T_1 - T_2)$$

is equal in both cases, the following applies to part A in Figure 4.3.5

$$\dot{q} = m_A \cdot c \cdot \Delta T_A$$

and to part B

$$\dot{q} = m_B \cdot c \cdot \Delta T_B$$

As masses m_A and m_B are different so the temperature difference ΔT_A is different from ΔT_B.

Irrespective of the heat-transfer method and despite great efforts, it has still not proved possible, given the sometimes intricate designs of the workpiece and the related differences

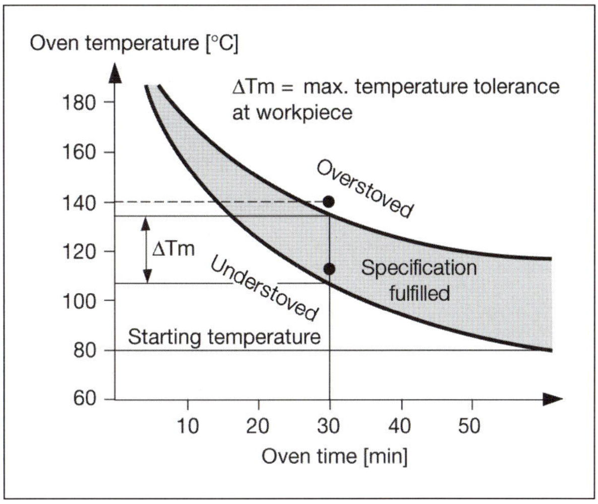

Figure 4.3.6: Typical curve of the processing window for stoving enamels in an oven

in energy exchange, to produce identical temperature profiles in all sections of the forming film during the dwell time in the oven. Nonetheless, no or merely negligible differences are permitted in the specified coating quality under those conditions. To put it another way: temporal and local fluctuations in process conditions may not exceed an agreed tolerance.

An important term in the context of drying is the process window for the coating materials. This indicates the processing tolerances within which the specifications can be met. A large process window for a coating means that constant quality can be achieved despite fluctuations in the oven conditions. The process window is the "specification fulfilled" band shown in the temperature/time chart for the oven.

Since understoving leads to poor hardness and scratch resistance (i.e. the degree of crosslinking is too low) and overstoving impairs elasticity (mesh density in the highly crosslinked resin molecules is excessively narrow), the limiting curves for the permissible drying conditions can be determined experimentally by specifying the minimum acceptable hardness and elasticity. The tolerance band between overstoved and understoved films narrows with decrease in drying time. Shortening the drying time by raising the temperature is then no longer acceptable as the spread of temperatures on the workpiece exceeds the permitted temperature range. The outcome is that the workpiece exhibits areas of both understoving and overstoving.

The temperature tolerances for the permitted range of drying conditions are determined not only by the paint systems, but also by the different colours. The pigment type and, particularly, the pigment quantity have an impact on hardness, adhesion and elasticity. Knowledge of the processing window enables predictions to be made about the reliability and economics of processing a new colour before the paint is even used (see Chapter 6.3).

Where it becomes necessary to change the stoving conditions, the time/temperature curve which is usually determined empirically for the purpose of optimum crosslinking can be recalculated for other stoving times or temperatures if the activation energy ΔE_A is known. If, for example, the optimum stoving temperature T_1 is known for a particular stoving time t_1 which is determined by the throughput time in the oven, it is possible to calculate the target stoving temperature T_x that will produce an identical degree of crosslinking or an identical coating quality if the stoving time is shortened to t_x:

$$\ln\left(\frac{t_x}{t_1}\right) = \frac{\Delta E_A}{R} \cdot \left(\frac{1}{T_x} - \frac{1}{T_1}\right)$$

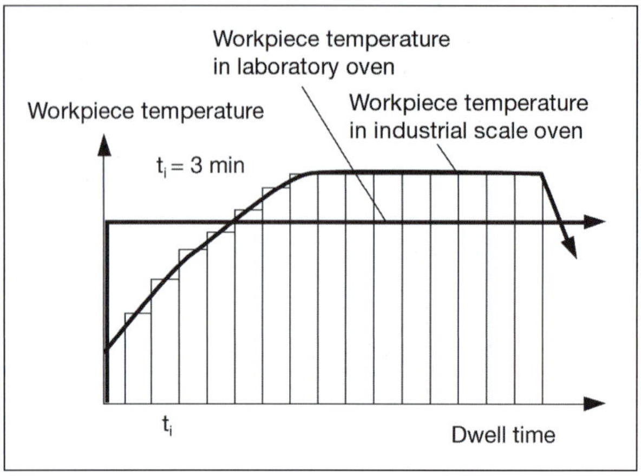

Figure 4.3.7: Principle behind conversion of the workpiece temperature in an industrial oven to the workpiece temperature in a laboratory oven

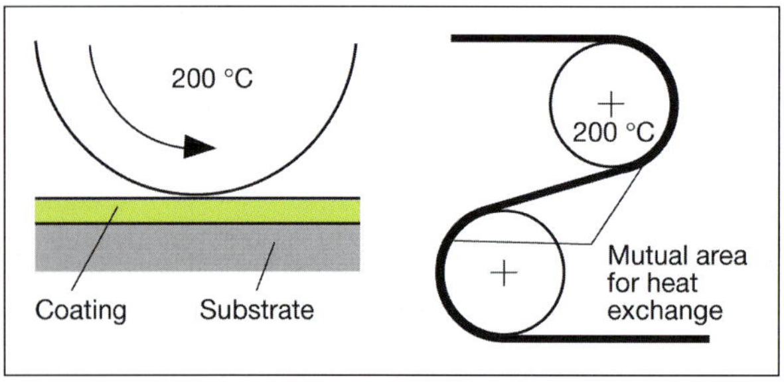

Figure 4.3.8: Heat transfer by conduction

Application

The equation above can also be used wherever curves which have been obtained for industrial ovens and which are sometimes complex have to be converted for use with laboratory ovens. This is done by dividing the oven curve into a number of segments of equal dwell time. Thus 10 segments of 3 minutes each are usual for total dwell times of 30 min.

The drying times in the laboratory oven corresponding to the separate three-minute time intervals t_0 are calculated from

$$\ln\left(\frac{t_i}{t_0}\right) = \frac{\Delta E_A}{R} \cdot \left(\frac{1}{T_i} - \frac{1}{T_0}\right)$$

where T_i is the average temperature within each individual segment of the industrial oven. The sum of all t_i then yields the total drying time in the laboratory oven for a specified constant stoving temperature T_0.

The workpiece temperature increases with length of dwell time in the oven as a function of the rates of heat transfer and of heat flow in the workpiece. The thus permanently decreasing ΔT between the circulating air and the workpiece causes retardation of the rate of heat transfer. In order for the required workpiece temperatures to be attained in an appropriate time nonetheless, the circulating air temperatures in the initial section of the oven should be higher than the desired maximum temperature of the workpiece, with the actual temperature difference depending on the paint system concerned. Such operating conditions are only feasible if no faults occur in a continuous system. Were the conveyor to come to a stop, overstoving would occur.

Another way of transferring thermal energy to the coating requires direct contact with heated solids, i.e. **conduction**. Heated rolls brought into contact with a coating will cause film solidification through energy transfer. Satisfactory heat transfer requires high temperature differences because the contact area between the rolls and a planar workpiece is small. In practice, therefore, contact drying has only proved successful with flexible painting substrates. Paper, cardboard and plastic film are suitable candidates for uniform conductive heat transfer.

Contact drying by conduction offers an additional advantage for the processing of waterborne coatings on wood. Wood fibres which have swollen as a result of osmotic processes tend to protrude after the coating process and would give rise to surface defects if they

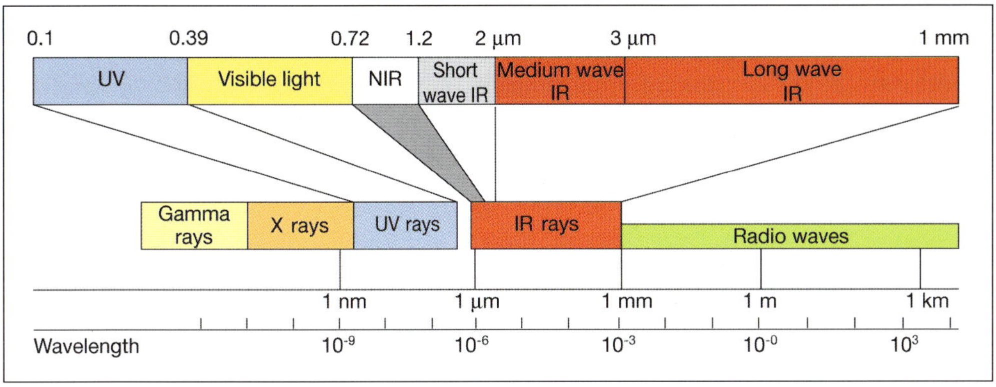

Figure 4.3.9: Wavelength spectrum of electromagnetic radiation

were not pressed back into the coating by the rolls. This eliminates the need for a sanding stage.

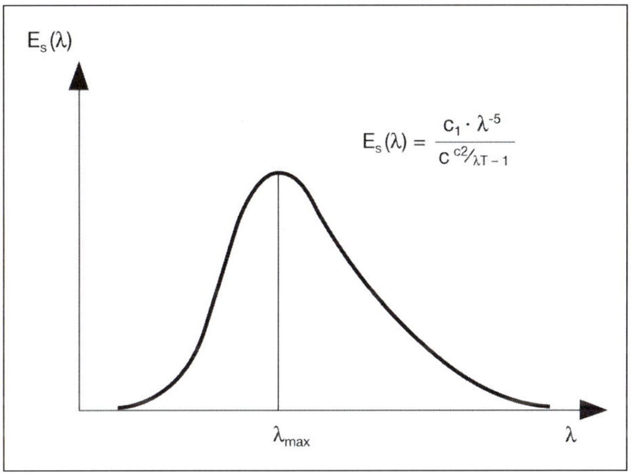

Figure 4.3.10: Energy distribution of an IR emitter as a function of wavelength in accordance with Planck's law

An alternative to the low efficient method of indirect heat transport via heat transfer from gaseous or solid materials is afforded by the direct method of electromagnetic waves in the form of **IR radiation** of suitable wavelength. If the wavelength of a natural emitter is such that the radiated energy quanta are capable of exciting molecular oscillations in a coating material, then heat energy can be transferred without the need for a carrier material. The optimum wavelength for heat transfer ranges from 0.8 to 6.0 µm, i.e. is in the longwave region beyond visible light.

The principle behind an IR emitter is that a material such as tungsten below is heated until it oscillates. This leads to periodic movements of electrical charges and the emission of electromagnetic waves from the material. The wavelength of the radiation decreases as the temperature rises. The wavelength range available for absorption by paints can therefore be regulated via the temperature of the emitter.

It should be noted that the molecular interaction of the charge movements prevents fixed-point radiators from emitting discrete wavelengths. A wavelength spectrum is produced whose radiation intensities vary with the wavelength. The quantitative dependencies are summarized in *Planck's* law of radiation [4.4.86].

$$E_s(\lambda) = \frac{c_1 \cdot \lambda^{-5}}{c^{\frac{c_2}{\lambda \cdot T - 1}}}$$

The *Stefan-Boltzmann* law states that the total power radiated per unit area may be found by integrating over the wavelength. *Stefan* and *Boltzmann* predicted and experimentally confirmed that the radiated power depends heavily on the temperature of the emitter. The radiated power E_s per unit area is proportional to the fourth power of the absolute temperature T. The constant of proportionality c_s for an ideal black body has been calculated as 5.67 10^{-8} $W/m^2/K^4$.

$$E_s(\text{black}) = c_s \cdot \left(\frac{T}{100}\right)^4$$

Real materials radiate energy at a fraction of black-body energy levels and are called grey bodies. The ratio of the constants of proportionality for a grey body to an ideal black body c_{grey}/c_s is a measure of ability to emit infrared radiation. This ratio is called the emissivity or ε value and lies between 0 and 1 for grey bodies.

The emitted energy is therefore determined by the ε value and the temperature.

$$E_s(\text{real}) = \varepsilon \cdot c_s \cdot \left(\frac{T}{100}\right)^4$$

Application

It has already been pointed out that raising the temperature of the emitter not only increases the energy output. At the same time the maximum on the radiation curve shifts to shorter wavelengths in accordance with *Wien's* displacement law.

$$\lambda_{max} \cdot T = 2897.8 \ \text{K} \cdot \mu\text{m}$$

Any quantitative description of energy transfer by infrared radiation must take into account the properties of the medium and the receiver in addition to the properties of the radiator of temperature T_1. It must be borne in mind that the receiver too emits radiation in accordance with its temperature T_2. The thermal energy $\dot{q}$ which can be effectively transmitted is therefore a function of the surface areas A of the radiator and receiver, the proportionality constants c_1 and c_2 and the difference between the outward and return flows of the radiation energy E_{eff}. If $\dot{q} = E_{eff} \cdot A$, it follows that

Material	ε
Ideal black body	1
Cu, polished	0.03
Al, polished	0.05
Cu, oxidised	0.6 – 0.9
Cast steel	0.65
Steel, oxidised	0.8
Clay brick	0.9
Porcelain	0.9
White coating	0.9
Water vapour	0.8
Flames	0.5
CO_2 gas	0.1

Figure 4.3.11a: ε values of different materials

Surface temperature		E_s	λ_{max}
[°C]	[K]	[kW/m²]	[nm]
200	473	2.84	6125
400	673	11.63	4305
600	873	32.90	3318
800	1073	75.10	2700
1000	1273	148.80	2276
1200	1473	266.80	1967
1500	1773	560.00	1634
2000	2273	1513.00	1275
2200	2475	2120.00	1171
2500	2773	3350.00	1045

Figure 4.3.11b: Radiant power E_s as a function of ideal black radiator temperature

$$E_{eff} = \frac{1}{\frac{1}{C_1} + \frac{1}{C_2} + \frac{1}{C_s}} \cdot \left[\left(\frac{T_1}{100} \right)^4 - \left(\frac{T_2}{100} \right)^4 \right]$$

For an understanding of radiation curing, it is necessary furthermore to note the interactions between the radiation and the receiver. A proportion E_R of the radiation E_{tot} striking the painted workpiece is reflected, while the rest penetrates the coating. As it passes through the coating, a further portion E_A is absorbed in the coating and converted into heat. The rest E_D is allowed to pass through to the substrate where it either is absorbed or leaves the painted workpiece unused.

$$E_{tot} = E_R + E_A + E_D$$

The magnitudes of the individual portions depend heavily on the wavelength of the radiation. The ratio of absorbed to unusable portions is additionally determined by the film thickness. If effective use is to be made of infrared energy for the drying and curing of coating materials, the specific absorption and reflection characteristics of paints must be known.

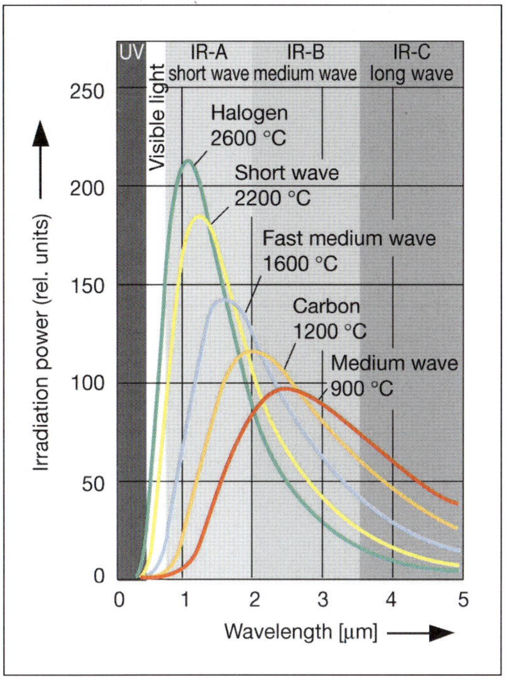

Figure 4.3.12: Effect of radiator temperature on radiant energy of IR emitters according to the Stefan-Boltzmann law

Absorption and reflection spectra in the wavelength range of interest for practical operations lie between 0.8 and 6 µm.

Radiators with a maximum wavelength of more than 6 µm emit far too little energy at a radiator temperature of approx. 200 °C for infrared radiation to be used meaningfully for rapid drying. It is only from wavelengths of 5.5 to 4 µm at about 400 °C that radiators with outputs of up to 20 kW/m^2 begin to become technically usable. These are termed dull radiators since they do not emit any visible radiation fractions below 700 °C. However, almost all coating materials absorb practically nothing in the range 4 to 5.5 µm. In the case of radiation-impermeable metallic substrates the energy fraction which is not used for conversion into heat for the coating is not lost. It is converted into heat in accordance with the properties of the substrate, i.e. it heats the substrate. This therefore offers the possibility of heating the coating from underneath, which often proves more advantageous than convection drying in which the heat is transferred from above. The risk of film defects in the form of pinholing is lowered as a consequence. For this reason, dull radiators have become an accepted means of drying thick paint films despite their low energy output.

If the radiator temperature is increased to 600 to 1000 °C, maximum wavelengths of 3.5 to 2.5 µm can be calculated in accordance with *Wien's* displacement law. The advantage of the substantially higher energy output is supplemented by marked substance-specific absorption of CH, OH or NH groups in the film forming agent. On account of their higher temperature, radiators in this medium-wave range also emit in the visible part

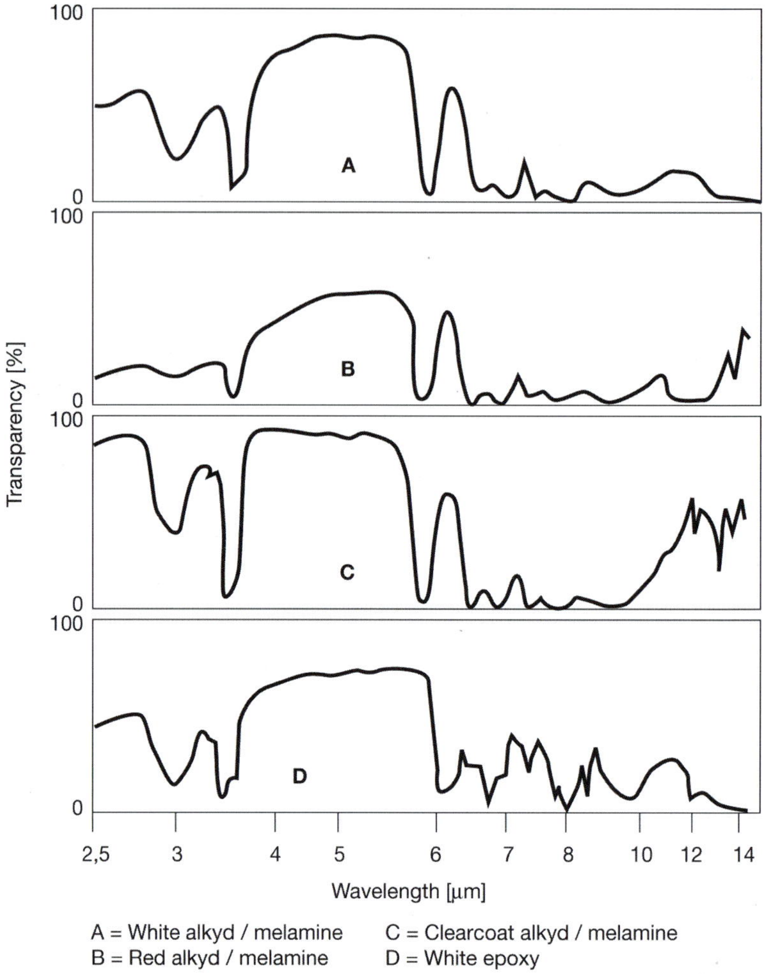

Figure 4.3.13: Absorption spectra of coating materials in the IR range of electromagnetic waves

of the spectrum. These are therefore known as bright emitters and can output radiation energy as high as 50 kW/m^2.

Radiators at 2000 to 2500 °C produce energy in the short-wave IR region with wavelengths of 1.6 to 1.2 µm. The substantially higher output of radiation energy of more than 100 kW/m^2 alters the absorption and reflection characteristics of the coating materials.

Type of radiation source	Surface temperature [°C]	Wavelength of maximum energy λ_{max} [nm]	Technically attainable output [kW/m^2]
Short wave	2000 to 2300	1300 to 1100	20 to 100
Medium wave	800 to 1000	2700 to 2300	8 to 50
Long wave	300 to 500	5000 to 3800	Up to 20

Figure 4.3.14: Classification of technical IR sources

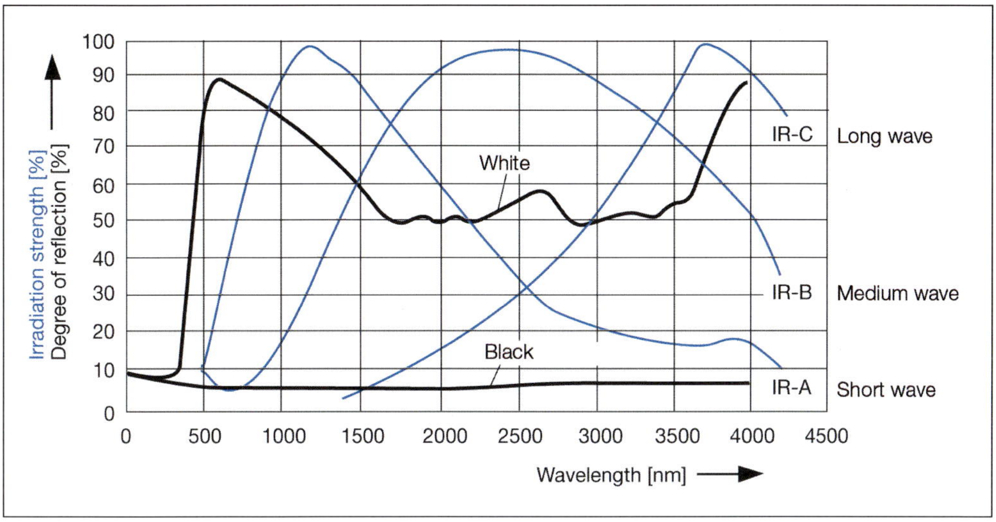

Figure 4.3.15: Degree of reflection of black and white coatings in the region of short-wave, medium-wave and long-wave IR radiation

The reflection characteristics of coatings in the short-wave range vary extensively with the type and quantity of colour pigments and effect substances. Whereas white paints pigmented with TiO_2 can have reflectance values in extreme cases of over 80 %, black paints pigmented with carbon black exhibit almost total absorption, irrespective of the wavelength.

In light pigments, short-wave radiators with a high energy output are incapable of effectively transforming the energy output into heat due to increasing reflection. It should be noted, furthermore, that absorption is colour-specific and thus heating of the coating is also colour-specific.

Metallic coatings heat up more slowly than solid colours because the aluminium flakes are strong reflectors. Given the same energy output, workpiece temperatures will differ from colour to colour and from one degree of lightness to another.

Consequently, for uniform crosslinking of coatings the energy intensities which have to be delivered vary with the colour. Contactless temperature measurement is needed for the energy-transfer process to ensure that the entire palette of colours and effects will possess the same service properties. This type of control technology has led to the development of methods which use even higher radiator temperatures to advance into even shorter-wavelength regions, despite the sensitive colour dependence and the greater risk of overcuring.

At 0.8 to 1.2 µm, radiation in the short-wave or near-infrared (NIR) region has the shortest wavelength and is thus particularly high in energy. The specific advantages of NIR stem from the fact that the electromagnetic waves can penetrate as far as the substrate. Radiation fractions reflected from there can even leave the film again and be reflected once more back to the coating by the radiator's reflectors. This multiple reflection, as it were, causes the energy input to be distributed homogeneously across the entire film. This uniform distribution and, ultimately, total absorption of the energy enables curing to take place in a matter of seconds [4.4.87]. NIR will even melt and harden powder coatings within just a few seconds. The short energy-transfer times also per-

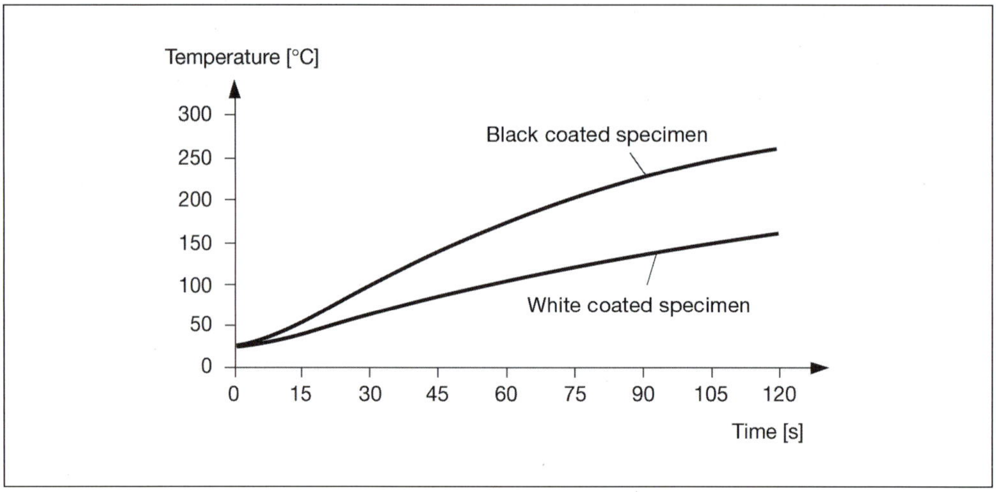

Figure 4.3.16: Heating curves of black and white coatings at the same radiation energy of 10 kW/m²

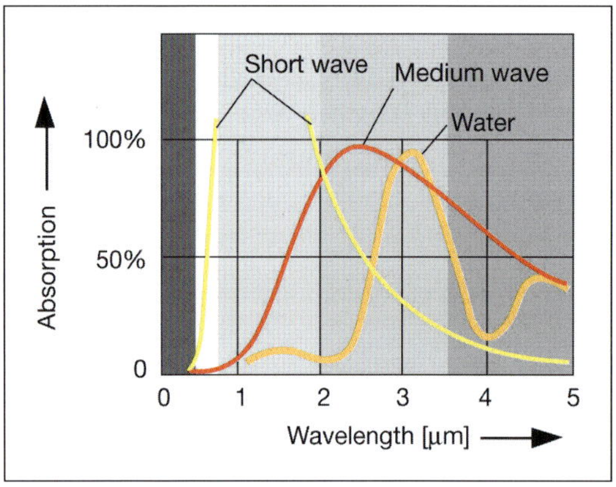

Figure 4.3.17: Absorption spectrum of water vapour

mit thermally unstable substrates such as wood and plastic to be coated with powder coatings and cured, provided the substrate has correspondingly poor thermal conductivity [4.4.88].

No description of the energy transfer from radiator to receiver would be complete without mentioning the medium. Absorption of the radiation by the ambient gas would cause the gas to heat up and thus impair the energy transfer. Carbon monoxide, carbon dioxide and water vapour are powerful absorbers of IR in the wavelength ranges of 2 to 5 µm and 12 to 16 µm. In view of the low concentration of CO_2 in the atmosphere, this disadvantage has little impact. Dry air has little tendency to absorb and is thus a good medium.

The processing or drying of water-borne paints is a different proposition. Medium-wave IR is extensively absorbed by water and would therefore be lost to the evaporating water if used for heating water-borne coatings. Only short-wave radiation and NIR penetrate the water vapour and thus heat the coating.

As already described, the energy output of radiators emitting wavelengths above 6 µm is too low and so they are not used. If longwave radiators had higher energy output, they would be attractive because absorption is very substantial and, at the same time, is not colour-specific or material-specific. This can be achieved with laser beams. The CO_2 laser has turned out to be usable for coating purposes. At 10.6 µm its emissions are in the extremely long-wave radiation region where high energy utilisation by all coating materials is assured. The high energy densities of IR lasers (IRASER) require that the radiation be scanned (deflected) with an oscillating mirror. Contactless measurement

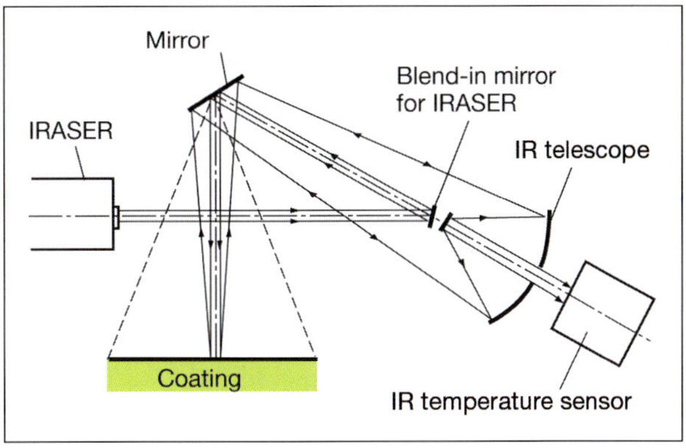

Figure 4.3.18: Radiation curing with laser beams

of the workpiece temperature is necessary for rapid control of the intensity of the laser beam (see Chapter 3.2.1).

The use of the even longer-wave **microwaves** to heat paint was successfully tested in the 1960s. Systems were installed specifically for the printing ink sector, first in the UK and Denmark, and later in Germany. The usual radiation frequency is 2.5 and 5.8 GHz. A functional correlation exists between the radiation to be absorbed to heat up the paint films and the dielectric constants or dielectric dissipation factor. The substrates too influence the outcome. Wooden substrates and aqueous dispersions have proved ideal for microwave drying [4.4.89, 4.4.90].

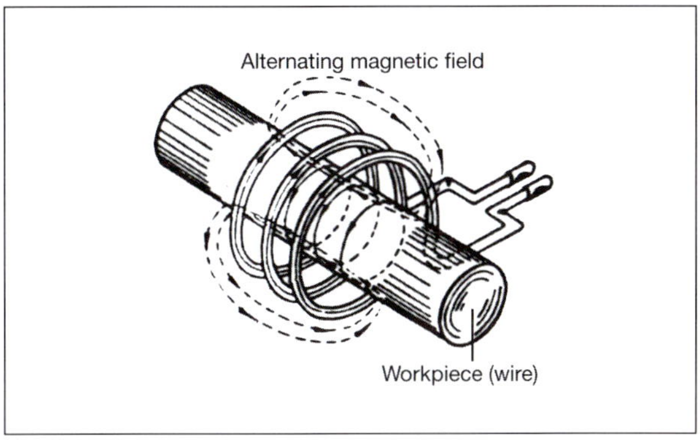

Figure 4.3.19: Induction heater

Other Methods of Heat Transfer

Other electrical phenomena can be used to generate heat for special applications. To paint wire, for example, an **electric current** passing through freshly painted wires develops sufficient heat to cure the paint. **Induction** of eddy currents in metallic materials can also be used to heat metals and consequently crosslink paints thermally.

If solid parts have to be coated and cured solely by thermal methods, eddy-currents induced by a primary voltage of sufficiently high frequency offer the opportunity to heat up only the surface. Curing of adhesives containing magnetic pigments such as γ-iron oxide is another method worth mentioning in this context. The thermal energy is generated by magnetic reversal in an alternating electromagnetic field.

4.3.1.2 Oven Design

Technical transfer of the thermal energy needed for accelerated film formation requires that the workpieces pass through a housing which is insulated against heat dissipation. They are conveyed in either batch or continuous mode in accordance with their shape, size and quantity. In the former case the painted workpiece is conveyed into the oven which is fitted with doors where, after the doors have been closed, it is heated over an appropriate period until it has reached the temperature required for crosslinking. In continuous ovens the tunnels, which are equipped with a conveyor, are open at their entrances and exits. This enables the painted workpieces to pass through the oven line continuously. For energy saving the ovens are mostly designed as so-called A-ovens to reduce the loss of hot air (see Figure 4.3.22).

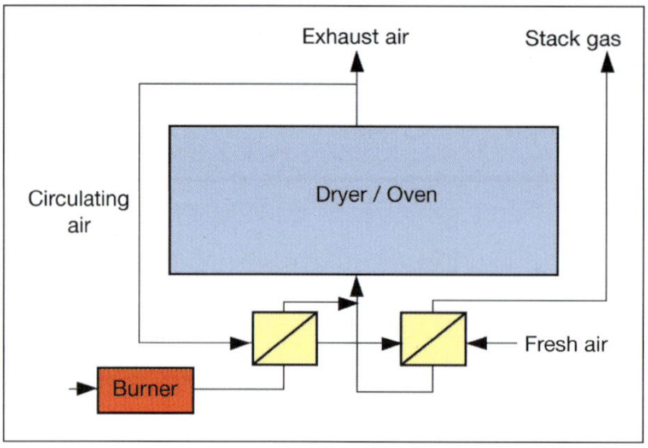

Figure 4.3.20: Components of an indirectly heated oven

Gas- or oil-fired boilers are used for heat generation and transfer in convection ovens. The thermal energy contained in the flue gas is transmitted to the oven air via heat exchangers.

If pure natural gas is used, it is also possible to dispense with the heat exchanger and to use the fuel gases direct. However, the advantage of the lower energy loss must be weighed against the risks of contamination of the still fresh coating surface by soot or SO_2. In rare cases the circulating air is also heated by means of superheated steam or electric heaters.

The air circulation needed for rapidly transferring the heat to the workpiece is provided by fans in conjunction with flow apertures, whereby air velocities of 2 to 5 m/s are attained. In individual cases the air velocity in jet ovens can be considerably higher if more rapid heat transfer is needed.

The distance which the heat carrier, air, has to cover in the oven varies with the design of the oven. Transverse or longitudinal ventilation can be applied to the chambers in the oven as befits the workpiece and paint system and thus the quantity of escaping solvents and cleavage products. Since the ambient air temperature falls as heat is transferred to the painted workpieces, in the event of uneven mass distribution, the circulating air should first flow over the solid parts which are more difficult to heat.

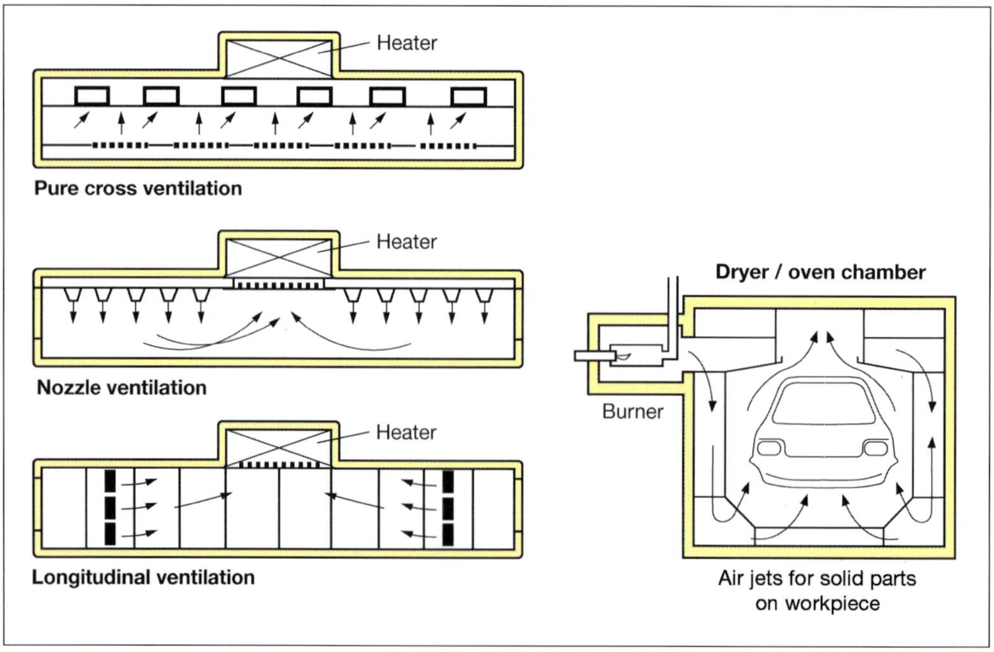

Figure 4.3.21: Air flows in ovens

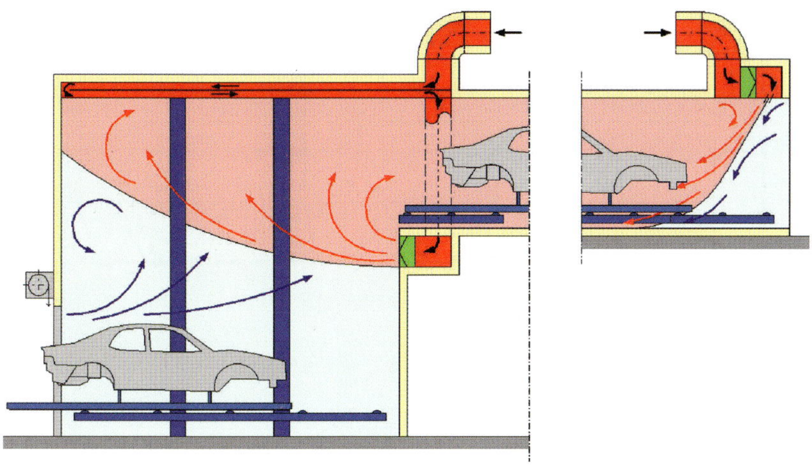

Figure 4.3.22: Schematic drawing of an A oven, with and without air locks *Source: Dürr*

For safety reasons, the oven air has to be diluted with fresh air in an amount proportional to the solvent content and the quantity and type of cleavage products. Operating in pure recirculation mode would continuously increase the concentration of flammable fractions and consequently possibly exceed the lower explosion limit. In addition, the risk of condensation would increase. Due to the explosion hazard, uniform statutory regulations relating to the circulating air concentration in ovens have been drawn up. In Germany this is in the form of orders BGV D25 and BGV D24 issued by the employers' liability insurance associations. BGV D25 specifies 0.8 % vol/vol as the upper threshold for flammable organic fractions in the oven air. The necessary volumes of fresh air can be calculated if the volatile organic components released during film formation are known.

The feed of fresh air leads of necessity to the same volume of solvent-charged, still hot exhaust air. This calls for secondary measures aimed at heat recovery and a reduction in the gaseous emissions from the exhaust air. Limit values for this are specified in Germany's Clean Air Guidelines and the VOC directive (see Chapter 5.2). To reduce the process energy, the entrances and exits of continuous ovens are lowered or air locks are formed by transverse flows.

There is no point in equipping an entire oven with a single heating and air circulation system as the workpiece is constantly undergoing changes in film states while it is passing through the oven; these changes concern the workpiece temperature, the degree of crosslinking and the emission of solvents and cleavage products. It is usual to divide the oven into 3 to 5 individual sections, each with a separate energy supply. Separate control of the circulating air/fresh air ratio must also be possible in the individual sections.

In the first zone, the heating zone, the workpiece temperatures are still low and the solvent content is correspondingly high. In the following holding zone, the coating undergoes major changes in terms of levelling and solidification as it dries and cures. The main portion of cleavage products leaves the film in this zone. Solvent emissions are completed in the subsequent holding zones. Hardly any cleavage products leave the

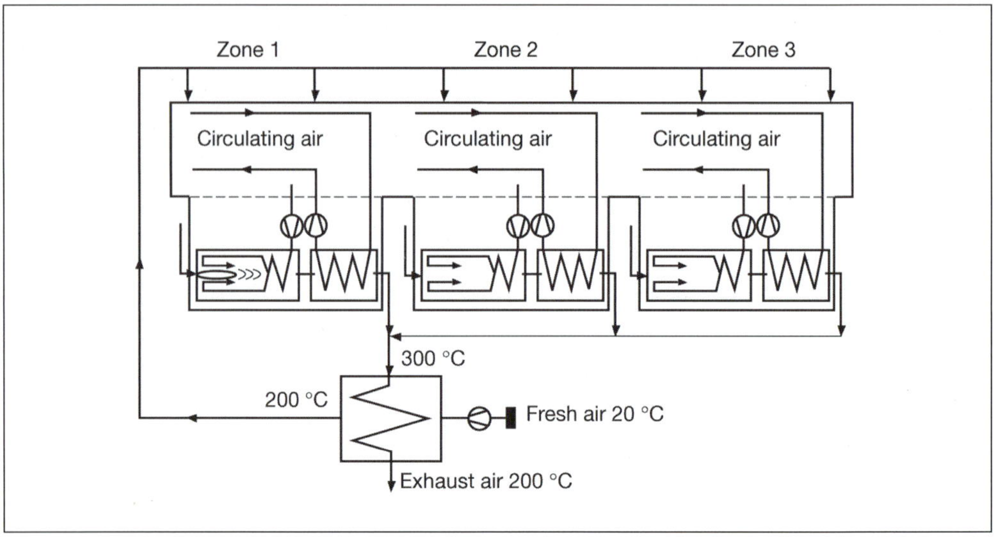

Figure 4.3.23: Schematic drawing of a multizone oven

Figure 4.3.24: IR ovens *Source: ITW*

coating film any more. Despite only further slight rises in the degree of crosslinking, the qualitative changes in the coating in these zones are still substantial. The film properties only reach their target values in the last section of the oven.

For economical oven operation the circulating air fractions should be kept as high as possible and the specified threshold concentrations complied with for safety reasons. This leads to the concentrations of organic emissions in the oven air being of the order of several g/m^3 and thus well above the limit values specified in the Clean Air Guidelines and VOC directive. The exhaust air from the oven must therefore be processed in secondary cleaning systems so that permissible limit values can be met. Various methods are available, of which thermal incineration and material recovery by adsorption on activated carbon are the most common (see Chapter 5.3). Since the thermal energy generated in the post-combustion process can also sensibly be used to heat the ovens, thermal incineration systems are often an integral part of drying systems.

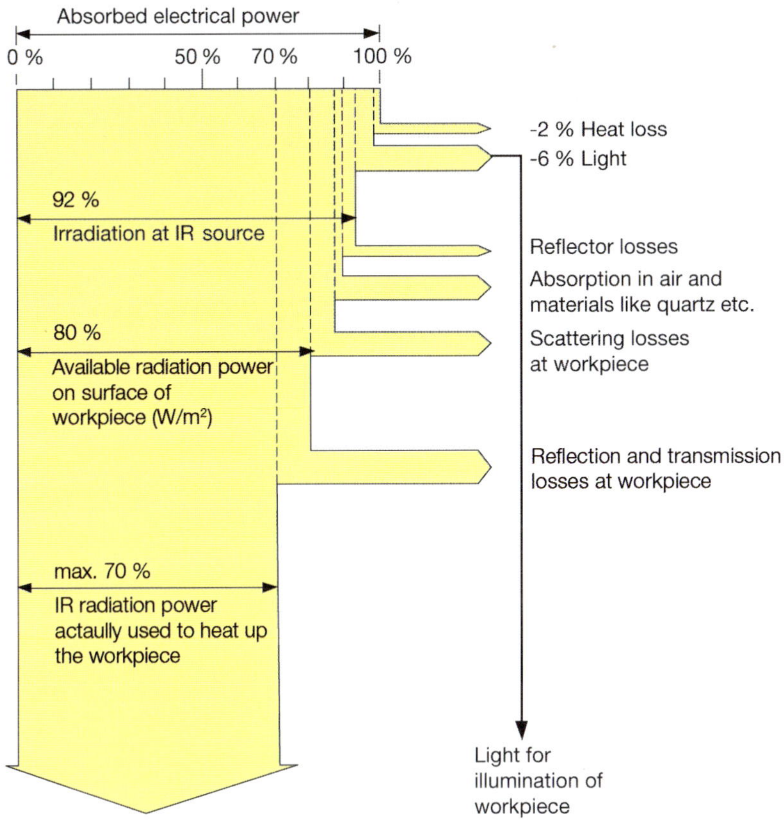

Figure 4.3.25: Yield of short-wave IR radiation

The heating of painted workpieces is frequently accelerated by installing infrared radiators in the first oven zone to supplement the heated circulating air. Dull or bright radiators are used in accordance with the particular task (see above). Dull radiators consist of metal or ceramic pockets through which the heated air is routed. The walls of the radiator pockets are heated to approx. 300 °C. Owing to the large masses of the radiator pockets these dull radiators have a very sluggish control response. They assist in the heating of solid parts.

Short-wave bright radiators consist of chromium-nickel or tungsten coils melted into quartz glass. If a sufficiently high electrical voltage is connected, they reach temperatures of up to 3000 K.

Reflectors should be employed to make effective use of the radiation energy emitted by the radiator. The IR rays are focused and thus intensified in different ways to match the design. Materials with small ε values (see above) should be used as the reflector material. Gold or aluminium reflectors with ε values of 0.02 to 0.04 have proved to be effective materials, the choice between them depending on the wavelength of the radiation. Despite the unusable visible fractions, reflector losses, absorption on the quartz sheath, as well as losses due to scattering, reflection and transmission, some 70 % of the installed power of short-wave IR radiators is still available for heating the workpiece.

Energy Balance of a Convection Oven

Energy-saving steps should not be determined simply on the basis of overall balances. It is more important to have a breakdown of what the energy is specifically used for and where it is lost in the process.

The total energy $\dot{Q}_{tot}$ input in the drier is calculated from the volume of circulating air $\dot{V}_{circ}$ and the difference in temperature between the entrance and the exit of the oven.

$$\dot{Q}_{tot} = \dot{V}_{circ} \cdot c_p \cdot \rho \cdot \Delta T$$

Unfortunately, this amount is not used solely to heat the coating. The sum of the individual energy fractions $\dot{Q}_{tot}$ to be provided per unit time corresponds to the following individual items:

$$\dot{Q}_{tot} = \dot{Q}_{pa} + \dot{Q}_s + \dot{Q}_{tr} + \dot{Q}_{surf} + \dot{Q}_o + \dot{Q}_{em}$$

$\dot{Q}$ is given in kJ/h. The following applies to the calculation of the exhaust air energy $\dot{Q}_{em}$:

$$\dot{Q}_{em} = \dot{V}_{em} \cdot c_p \cdot \rho \cdot (T_{oven} - T_{environment})$$

Application

The limit value for the concentration of the organic fraction of max. 0.8 % vol/vol as specified in BGV D25 serves as the basis for the calculation of $\dot{V}_{em}$:

$$\frac{V_{so}}{V_{em}} < 0.008$$

If the volume of solvents emitted in the oven is known, the fraction of exhaust air energy can be calculated. In practice the exhaust air volume calculated in this way is increased by a factor of 3 to 6, the precise figure depending on the paint system, to prevent the formation of condensation. The energy requirement $\dot{Q}_s$ to heat the workpieces (substrate) is calculated from

$$\dot{Q}_s = m_s \cdot c_s \cdot (T_s - T_{env})$$

The energy requirement $\dot{Q}_{tr}$ to heat the conveyor (transportation) is calculated from

$$\dot{Q}_{tr} = m_{tr} \cdot c_{tr} \cdot (T_{tr} - T_{env})$$

where in most cases $c_{tr} = c_s$ and $T_{tr} = T_s$. Energy losses due to the oven design must also be determined. The volumes of air escaping from the openings $\dot{V}_0$ can only be estimated.

$$\dot{V}_o = \frac{2}{3} \cdot A \cdot u \cdot z$$

A = Cross-section of oven entry

u = Stream of leaving oven air (for orientation: 0.2 m · s^{-1}

z = Opening factor of oven doors (z = 1, if always open^{-1})

In ovens with raised entrances and exits (A ovens; see Figure 4.3.22) air losses are negligible. Where the losses $\dot{V}$ due to the openings are known, the heat loss $\dot{Q}$ can be calculated as

$$\dot{Q}_o = \dot{V}_o \cdot \rho_{air} \cdot c_{pair} \cdot (T_{circ} - T_{env})$$

In addition, energy is lost due to heat exchange between the oven walls and the ambient air.

$$\dot{Q}_{surf} = \dot{V}_O \cdot K \cdot A \cdot (T_{circ} - T_{env})$$

K = Coefficient of heat transition

A = Outer surface of oven

The coefficient of heat transmission K is a function of, among other things, the oven insulation. A value of 0.5 kJ/m^{2}hK can be used in calculations.

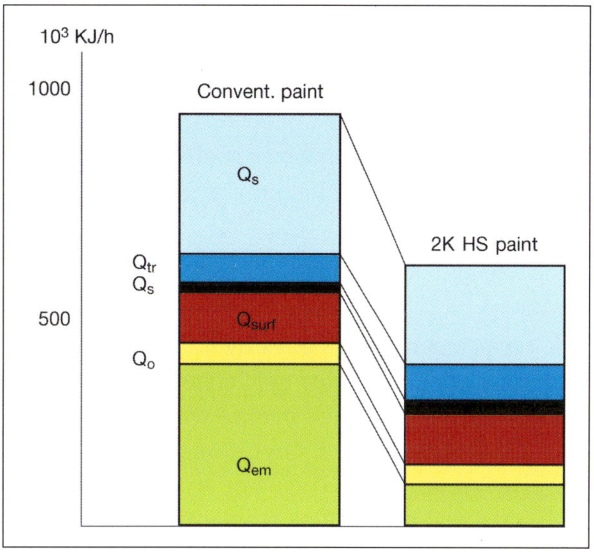

Figure 4.3.26: Energy consumption in the oven

The energy requirement $\dot{Q}_{pa}$ to heat the paint film and to evaporate the residual solvents is negligible in comparison with the other fractions. Although the heat of evaporation of water at approx. 2000 kJ/kg significantly exceeds that of organic solvents at 300 to 500 kJ/kg, as a proportion of the overall consumption it is marginal at only 1 to 2 %.

Experience shows that for the various items above the greatest energy inputs in the coating of industrial goods such as car bodies and refrigerators are due to exhaust air losses and heating of the workpieces.

An energy balance of this kind also yields information on the ratios of input to lost heat energy and indicates where potential savings may be made. Since the exhaust air volume is determined by the solvents and cleavage products escaping in the oven, energy savings can be achieved at the same stoving temperature by using high solids and poly-addition paints which do not form any cleavage products and so require less fresh air. It has been shown in practical operations that raising the solid content from 50 % to 70 % and replacing a solvent-borne stoving enamel by a 2-component polyurethane system enables the exhaust air energy losses to be reduced to approx. 50 % of the original value at the same stoving temperature of 130 °C.

4.3.2 Curing by UV and Electron-Beam Radiation

The possibility of initiating rapid, energy-saving chemical crosslinking reactions without heating the paint and substrate and with only low emissions was the motivation for applying photochemical reactions to the specific requirements of coating technology in the 1950s and 1960s.

It was known that the high-energy short-wave fractions of solar radiation can trigger chemical reactions by means of photochemical radical formation and without heat action. This type of reaction quickly found a technical application in polymer chemistry. Unsaturated polyesters which until that point could only be cured thermally or by the catalytic cleavage of peroxides were now curable in seconds by the action of radiation without any heat input or any pot-life restrictions. This “cold” UV curing was especially useful for planar and thermally unstable wooden materials in the furniture industry. Initially used only for the crosslinking of clearcoats and later also for thin non-hiding pigmented coatings, radiation curable paints subsequently conquered the printing inks sector and the printed-circuit board industry. Acrylic-acid-modified resins specially developed for curing have partially superseded the traditional unsaturated polyester resins (UP), despite being more expensive, because of their higher reactivity (see Chapter 2.1.1). For successful technical application, photoinitiators are absolutely essential for initiating polymerisation with UV-light. Most of them split homolytically into radicals under the influence of UV rays in the wavelength range 350 to 370 nm.

The replacement of UV rays by accelerated electrons (electron beams) opened up new routes for even faster curing of coating materials, again in the form of cold curing, without the inhibition which occurs with UV-absorbing paint components like pigments. Despite the qual-itative, environmental and energetic advantages, large-scale systems for electron-beam (EB) curing are relatively rare. High capital costs, the formation of X-rays and qualitative problems in the coating of metals have delayed the wider acceptance of EB curing to date. However, the few industrial-scale systems in operation are successfully coating wood in the form of doors, wood cement panels and parquet flooring, and curing printing inks.

The reasons for the sluggish success of the EB curing system lie in the refinements made to the technically simpler UV technology. New photoinitiators enabled the wavelength of the molecule-splitting radiation to be shifted to the short-wave visible range. It was then possible with the aid of modified radiators to match the maximum emissions to the absorption characteristics of the new photoinitiators, with the result that optically hiding paints pigmented with titanium dioxide could now be cured by means of electromagnetic radiation.

It should be noted in conclusion that cationic polymerisation can also be initiated by UV radiation, in addition to radical polymerisation. Strong acids released by UV rays initiate ring-opening polymerisation reactions of the kind familiar, for example, from the curing of epoxy resins with boron fluoride (see Chapter 2.1.1). Advantages of UV-initiated cationic polymerisation over radical polymerisation consist in complete curing upon merely brief exposure to UV rays and in the formation of OH groups, leading to better adhesion on metallic substrates.

Chemical Principle Behind UV Curing

The double bonds which are necessary for the radical growth reaction can be found in both the resins and the reactive solvents, known as reactive diluents. The role of the latter is to liquefy the paint and thus make it processable. By contrast with conventional solvents they do not leave the solidifying film, but become linked to the film forming agents through copolymerisation (see Chapter 2.1.4).

To initiate a polymerisation reaction, UV rays must act on the coating material to be cured. Although short-wave UV rays with a wavelength of 280 nm are theoretically capable of directly exciting the carbon double bond in the film forming agent chains, this method is unusual in technical applications because of the cost of designing the appropriate radiation sources. It is simpler to use long-wave and technically less complex mercury-vapour lamps. These can be employed in conjunction with additives capable

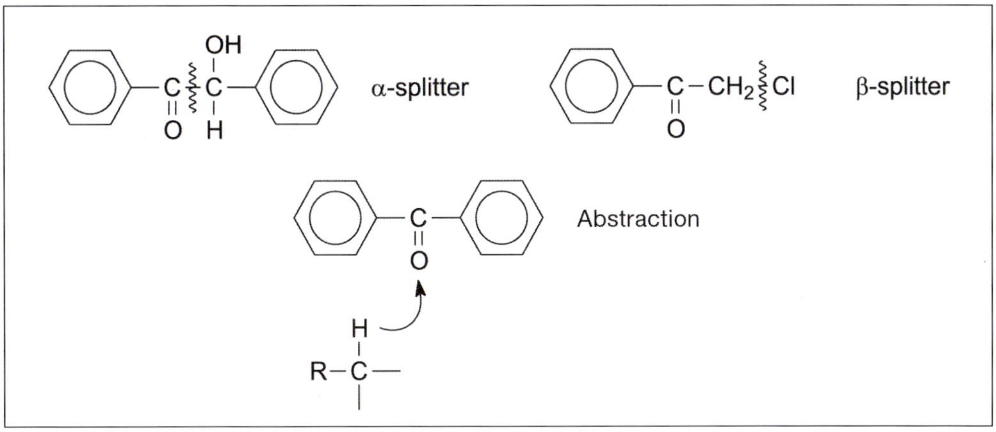

Figure 4.3.27: Classification of photoinitiators

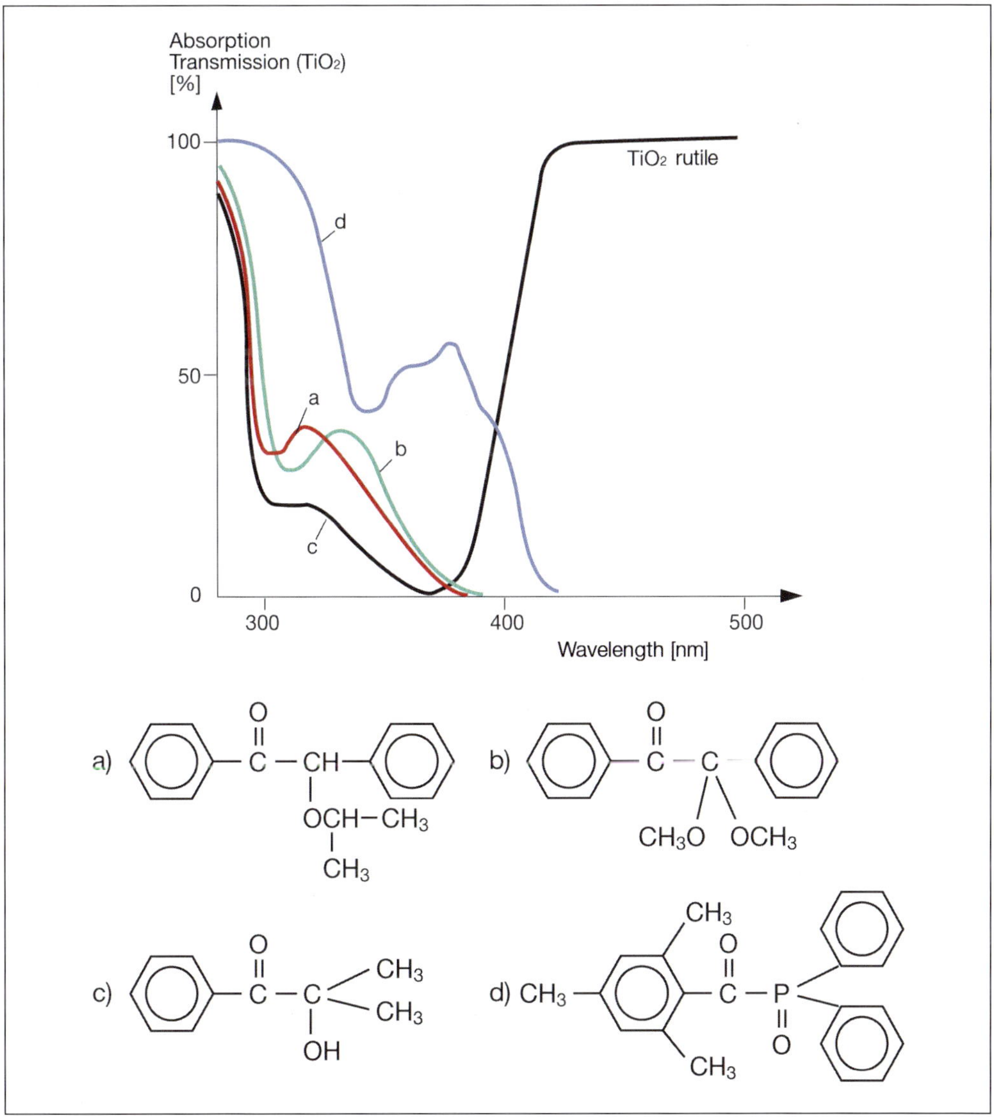

Figure 4.3.28: Absorption spectra of photoinitiators and transmission data for TiO_2

Application

of forming radicals at the relatively long-wave radiation fractions of 365 nm. Such photoinitiators or sensitisers absorb the UV rays in this wavelength range, cleaving carbon bonds to produce two radicals in each case (see Chapter 2.1.4).

Absorption of energy by the photoinitiators is impaired by the use of light-protection agents and in the presence of pigments unless their absorption characteristics have been matched accordingly [4.4.91]. It is therefore important that UV absorbers and pigments be used which do not absorb in the photoinitiator's absorption range. Zinc sulphide makes a suitable white pigment for radiation curable paints because of its favourable absorption characteristics. If it is desirable or necessary to use other pigments, for example titanium dioxide, there is no scope for the above-mentioned matching with traditional photoinitiators.

This problem was only resolved when new process materials were developed. The acylphosphine oxide group (see Chapter 2.1.4), when used in conjunction with UV rays in the wavelength range above 370 nm, is capable of initiating curing of coating materials pigmented with titanium dioxide which do not absorb in this range (see Figure 4.3.28).

The process of UV curing is accompanied by undesirable secondary reactions with the ambient air. The chain-terminating effect of the oxygen biradical during crosslinking leads to oxygen inhibition. This can be combated by chemical (see Chapter 2.1.1) or plant-engineering methods.

Equipment for UV Curing

The radiators used to produce a radical concentration sufficient for crosslinking need to be appropriately designed. Mercury-vapour lamps emit the necessary energy quanta for splitting the photoinitiators in the range 320 to 380 nm in several bands.

At 1 W/cm length of rod-shaped radiator, low-pressure mercury-vapour lamps have a low output in the UV range by comparison with high-pressure radiators which are considerably more powerful at up to 50 W/cm. The correspondingly slower curing induced by low-pressure radiators gives paraffin in the relevant coating materials sufficient time to float into the interface. The resultant diffusion barrier for oxygen prevents inhibition. The slow curing reduces the risk of "freezing" the internal strains. Furthermore, the low operating temperature reduces the loss of volatile reactive diluents. Low-pressure lamps are mostly used to effect preliminary gelling before the subsequent main crosslinking is carried out with high-pressure radiators.

The spectrum of mercury-vapour lamps particularly in the range above 400 nm can be expanded to include new emission bands by doping with gallium, indium or iron. Such lamps are particularly used for curing pigmented paints.

As with IR drying, recourse should be made to reflectors to make the maximum possible use of the radiation energy of the UV lamps for paint curing. The reflectance of aluminium reflectors is sufficiently high at over 90 %, yet they have failed to gain acceptance for high-pressure radiators. The reason for this is the high reflection of the IR fractions present in the mercury-vapour lamps' spectrum which heat the resulting coating excessively and thus lead to unwanted emissions of the reactive diluent. Reflectors made of highly heat-resistant specialty glass are more suitable because they are permeable to visible and infrared radiation but not to UV rays. As with IR radiators, mirror designs also exist for UV radiators which scatter or focus the light.

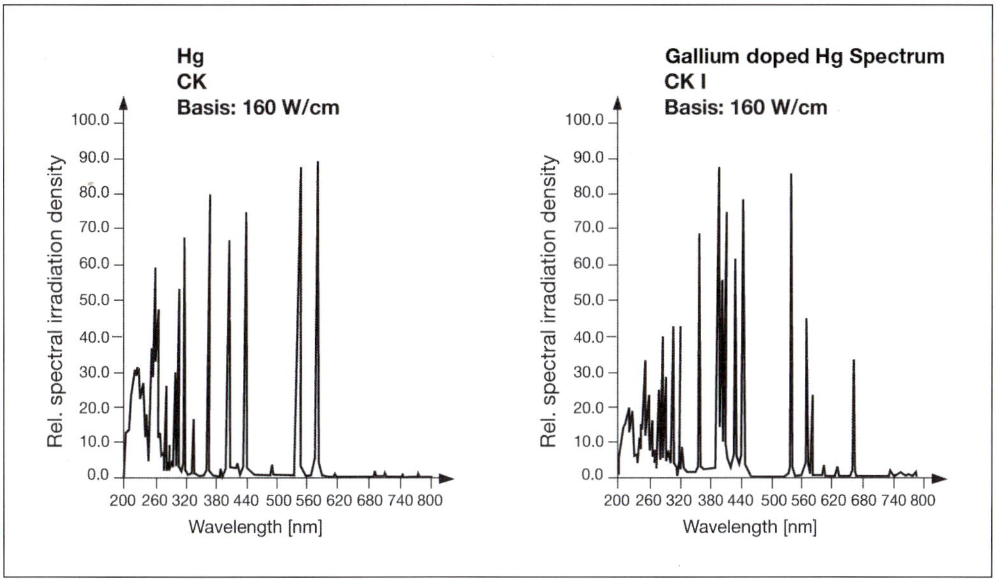

Figure 4.3.29: Emission spectra of standard and gallium-doped mercury-vapour lamps

To ensure a sufficiently high and at the same time constant energy input the radiators are regulated either in several increments or infinitely variably, and their energy output is monitored. If minimum values are not met, an acoustic or optical signal is triggered. Modern systems enable the UV levels to be measured continuously and, in the event of fluctuating conveying speeds, the setpoint values to be readjusted via closed-loop control circuits. The infrared rays emitted by the UV lamps require that the radiators be intensively cooled to permit constant operation over the long term.

The ozone generated by UV light makes it necessary to house UV radiators in ducts and to supply fresh air to ensure that the MAC value of 0.1 ppm is not exceeded.

Engineering measures for preventing oxygen-induced inhibition of crosslinking near the surface consist in purging the systems with nitrogen by way of inert gas. This stops the formation of ozone at the same time.

Since uniform crosslinking is only possible with a constant energy input, the use of UV-curing systems has so far been limited in practice to the coating of planar parts. The numerous advantages of rapid and intensive crosslinking and the associated high hardness and scratch-resistance of UV-cured coatings mean that this technology is also of interest for highly stressed surfaces such as those of cars. The three-dimensional form of the surfaces to be coated and cured require the permanently installed radiators in normal continuous systems to be replaced by movable radiator arms [4.4.92]. Program-controlled robot arms follow the contours of the surfaces at a constant distance and thus ensure consistent quality. The improved scratch-resistance and environmental benefits make the use of this technology in automotive coating probable [4.4.93].

Electron-Beam Curing

In the early 1970s attempts were made to progress from UV curing and explore other forms of radiation curing. It was known that accelerated electrons in the form of β-rays can trigger photochemical reactions. The effect is based on the fact that molecules are directly excited or ionised by the β-rays before decaying into radicals or radical ions. The use of such a technology renders photoinitiators superfluous. The accelerated electrons induce extremely rapid crosslinking of film forming agents or reactive diluents by polymerisation. The reactions are complete in fractions of a second, permit high coating speeds and offer the ideal conditions for economical coating. A further advantage of electron-beam curing is the ability of the accelerated electrons to penetrate deep into pigmented films and to ensure full curing of all pigmented coats.

As with UV coating, it is recommended with electron-beam curing, too, that the influence of atmospheric oxygen be counteracted by purging with inert gas.

All the film forming agents which have been described as suitable for UV curing lend themselves in the same way to electron-beam curing. The only difference from UV curing is the mechanism behind radical formation. Given sufficient kinetic energy, accelerated electrons can release bonding electrons from the film forming agents, giving rise to ionised molecules and radicals or radical ions.

$$AB \longrightarrow AB^{\oplus} + e^{\ominus}$$

$$AB^{\oplus} \longrightarrow A\cdot + B^{\oplus}$$

Application

If the energy of the exciting electrons is too low to release electrons, excitation of the resin molecules (AB*) may be possible. Radical formation can take place by means of subsequent homolytic decomposition.

$$AB \longrightarrow AB^{*} \longrightarrow A\cdot + B\cdot$$

Electrons with even less kinetic energy are captured and, as anions, are transformed into radicals and radical ions. Most interactions between the electrons and the film forming agents ultimately lead to the formation of radicals and thus to initiation of radical polymerisation.

The energy of the accelerated primary electrons is so high that they are capable of penetrating films several hundred micrometres thick. They generate slower secondary electrons which usher in rapid and complete curing in accordance with the mechanism described above [4.4.94].

Plant Engineering

The free electrons necessary for curing are released by electron accelerators, also known as electron guns. The devices, which are similar to a cathode-ray tube, contain a hot cathode, an anode with a punctiform opening, and a device for focusing the electrons which are released and accelerated by the applied high voltage.

The force F from the product of the elementary charge e and the electric field strength E which acts on the electrons causes an acceleration which imparts the kinetic energy $1/2m\ u^2$ to the electron. This corresponds to the work F · s and consequently:

$$\frac{m}{2} \cdot u^2 = F \cdot s = Q \cdot E \cdot s = Q \cdot U$$

$$\text{or} \quad u = \sqrt{\frac{2 \cdot Q \cdot U}{m}}$$

Q = Elementary charge
m = Mass of electron

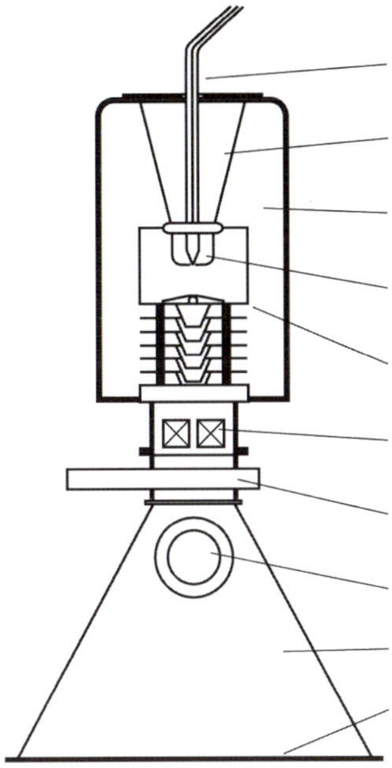

Figure 4.3.30: Schematic drawing of an electron accelerator

The electron of mass m and charge e, whose velocity u is a function solely of the applied high voltage U, passes through a 10 to 15 µm titanium window and thus reaches the atmosphere. The window, which absorbs 15 % of the energy and converts it to heat, has to be cooled to relieve the stress on it. The high intensity of the electron beam requires it to be permanently scanned in two dimensions.

The kinetic energy of the electrons produced by the electron gun is given in electron-volts (eV). 1 eV is the kinetic energy which an electron gains when it moves across a potential difference of 1 volt. 1 eV thus corresponds to $6.25 \cdot 10^{-18}$ Ws. In the past, it was standard practice to state the radiation dose as the energy in rad relative to the irradiated quantity. Since 1 rad at $6.25 \cdot 10^{13}$ eV/g is a very small unit, the radiation dose was converted to Megarad: 1 Megarad = $6.15 \cdot 10^{19}$ eV/g = 10 Ws/g = 10 Joule/g or, since recently, 10 kgy (kilogrey). With current paint systems, 50 J/g or 50 kgy are usual for ensuring adequate curing of the coating. Another important characteristic is the dose rate per unit time which is given in kgy/min.

Generally, 150 to 300 kV are applied to accelerate the electrons or, in rare cases, up to 500 kV. The highly accelerated electrons exiting the titanium window interact with the air and undergo substantial braking as a result. Electron beams therefore have a limited range. Furthermore, the electrons are scattered by interaction with the paint and consequently form new, secondary electrons. Only these are slow enough to be able to target molecules for ionisation and to form radicals. The braking of the primary electrons also gives rise to short-wave X-ray bremsstrahlung. Electron beam systems are therefore subject to radiological protection legislation in most countries and must be screened by a 12-mm-thick lead lining or 50-cm-thick concrete walls to prevent the emission of X-rays.

Amongst commercially available electron guns a distinction is drawn between devices with punctiform electron sources whose beam has to be deflected, i.e. scanned, to ensure its trouble-free use, linear accelerators, and guns with plate cathodes. Linear accelerators have a rod-shaped hot cathode which extends across the entire width of the coating to be cured. The linear accelerators in turn are divided into guns with and without grille scattering. In the former case the tungsten cathode is surrounded by a grille acting as a counter-electrode. As a result, an electron cloud forms. A second grille transforms the electrons into an electron cloud aimed at the workpiece so that it contains its final energy

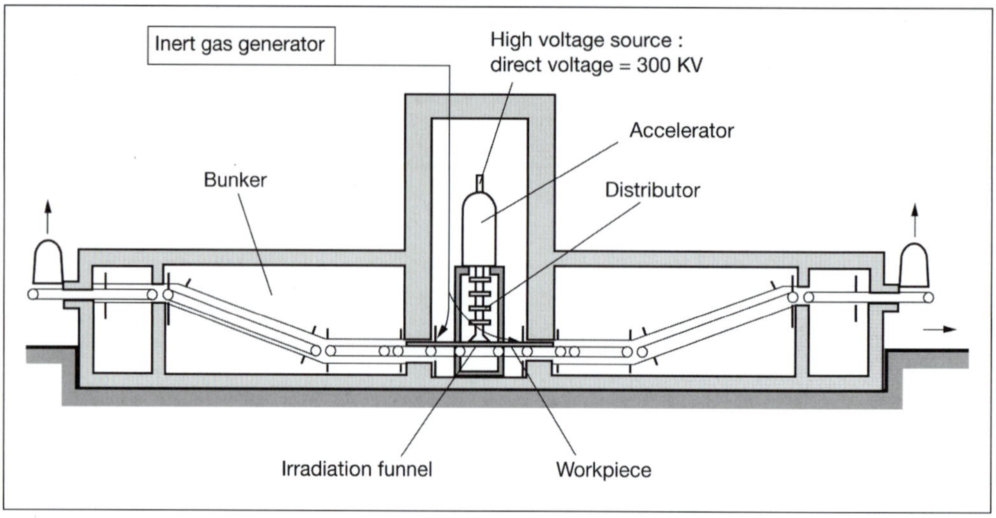

Figure 4.3.31: Schematic drawing of electron beam curing equipment

before it exits the titanium window. A curtain of accelerated electrons generated in this way is wide enough and therefore requires no additional scanning.

Electron guns without grille scattering produce direct acceleration of the electrons emitted by the tungsten wire.

Apart from linear accelerators, plate-type guns have proved successful. Several linear cathodes release electrons which are accelerated in two dimensions through a grille-type counter-electrode and towards the discharge window. Although the radiation intensity is not distributed as uniformly as that of linear accelerators, the cathode has a better service life.

Curing of coatings by means of electron beams is extremely attractive with regard to energy consumption. A simple energy calculation clearly illustrates the positive energy balance by comparison with convection drying. For a throughput time of 30 minutes and a stoving temperature of 150 °C the consumption of a convection oven with a temperature difference of 130 °C, a specific heat capacity of the paint of 2 J/g, and an energy ratio for the paint to be heated of 1 % of the oven's total energy consumption is: $130 \cdot 2 \cdot 100 = 26{,}000$ J/g (see Chapter 4.3.1).

Electron-beam curing, by contrast, assuming 100 % efficiency, is significantly cheaper at only 50 J/g coating material, i.e. its energy consumption is 500 times lower than that of convection drying. Even when the real efficiency is taken into account, the necessary energy requirement is approx. 10 % of that needed for drying by heat transfer.

4.3.3 Summary

Drying and curing systems are mainly used in industrial painting processes for film-formation purposes. They accelerate the physical drying process by emitting organic solvents or water and activate the crosslinking reactions via the transfer of thermal or radiation energy. Heat transfer takes place by convection with the aid of a carrier gas, by conduction or indirectly by IR and microwave radiation.

The "cold" curing method entailing the use of UV and electron beams is an interesting alternative to conventional methods from an environmental and economic perspective. The increased interest in UV-curing has been triggered by the new options provided by computer simulation and the use of robots which are making it possible to even coat and cure workpieces that have intricate three-dimensional shapes.

The geometry and materials of the substrates to be coated determine the choice of methods to be employed.

4.4 Literature

References

4.4.1 R. S. Williams, C. Jourdain, G. I. Daisey and R. W. Springate, JCT 72 (902), p. 35 (2000)
4.4.2 Wagenführ, Holzatlas, p. 53, Fachbuch Verlag, Leipzig, 1996
4.4.3 H. Pecina, Lack auf Holz, p. 153, Vincentz Verlag, Hanover, 1995
4.4.4 J. B. Bajat, J. P. Popic, V. B. Miskovic-Stankovic, Progr. Org. Coat. 69, p. 316 (2010)
4.4.5 R. Gassner, besser lackieren, 2 (4), p. 9 (1999)
4.4.6 W. A. Roland, P. Drouion, Metal Finishing 91 (12), p. 57 (1993)
4.4.7 J. Halbartschlager, I-Lack, 66, p. 672ff (1998)
4.4.8 T. Körner, JOT 48 (9), p. 52 (2008)
4.4.9 H. Gehmecker, M. Hueber, Proceedings Eurocar, Barcelona, 2005
4.4.10 M. Ottenberger, J. Feßmann, H. Sommer, JOT 41 (10), p. 68 (2001)
4.4.11 C. Kobusch et al., ECJ 12/2000, p. 46
4.4.12 F. M. Depke, Bitumen- und Teerlacke, p. 175, Colomb Verlag, Stuttgart, 1970
4.4.13 M. Bablik, Malerblatt 8, p. 28–31 and 9, p. 36–41 (1997)
4.4.14 G. L. Burnside, G. E. F. Brewer, J. Paint Techn. 38, p. 96 (1966)
4.4.15 DE-PS 1276260 (1965) BASF AG; US-PS 3962165 (1968) PPG Inc.
4.4.16 H.-J. Streitberger, Automotive Paints and Coatings (2nd ed.), p. 114, Wiley-VCH, Weinheim, 2008
4.4.17 US-PS 3955106 (1978) PPG Inc.; US-PS 4066592 (1978) PPG Inc.
4.4.18 F. Beck, F+L 72 (3), p. 218 (1966)
4.4.19 F. Beck, Progr. Org. Coat. 4, p. 1 (1976)
4.4.20 A. Doroszkowski, M. A. Toynton, Progr. Org. Coat. 17, p. 191 (1989)
4.4.21 H.-J. Streitberger et al., JOCCA 73, p. 454 (1990)
4.4.22 H. Spoor, U. Schenk, F+L 88, p. 94 (1982)
4.4.23 K. Rother, JOT 46 (3), p. 26 (2006); M. Schifko, MO 64 (9), p. 32 (2010)
4.4.24 D. W. Boyd, R. R. Zwack, Progr. Org. Coat. 27, p. 25 (1996)
4.4.25 U. Heimann, T. Dirking, H.-J. Streitberger, Numerische Berechnung des Umgriffverhaltens von Elektrotauchlacken, Proceedings Surtec, Berlin (1985)
4.4.26 A. Fieberg, E. Ehmann, F+L 107 (6), p. 66 (2001)
4.4.27 Y.-.I. Suzuki et al., Progr. Org. Coat. 42, p. 209 (2001)
4.4.28 I. Krylova, Progr. Org. Coat. 42, p. 119 (2001)
4.4.29 H. U. Schenk, H. Spoor, Progr. Org. Coat. 7, p. 1 (1979)
4.4.30 FR 1394007 (1962) PPG
4.4.31 DE-OS 1929523 (1968) Vianova
4.4.32 R. Kita, A. Kumi, J. Paint Techn. 48, p. 55 (1976)
4.4.33 DE-OS 2411419(1974) Hoechst AG; DE-OS 2411349 (1974) Hoechst AG
4.4.34 DE-OS 1645194 (1963) Hoechst AG
4.4.35 M. Marx, Makromol. Chemie 93, p. 126 (1966)
4.4.36 US-PS 584654 (1966) DeSoto
4.4.37 E. M. Boldebuch et al., Proceedings Fatipec, 1976
4.4.38 US PS 3955106 (1976) PPG Inc.
4.4.39 DE PS 150994 (1978) Bayer AG
4.4.40 US PS 3947339 (1975) PPG Inc.
4.4.41 US PS 3962165 (1968) PPG Inc.
4.4.42 E. Kempter, E. Gulbins, F. Caesar, Defazet 32, p. 412 (1978)
4.4.43 US PS 3883483 (1975) PPG Inc.
4.4.44 H. Guder, F. Beck, F+L 93, p. 539 (1986)
4.4.45 A. Schnell, Proceedings 7th DFO-Automobiltagung, Conference Proceedings p. 77, Straßburg, 2000

4.4.46 D. Saatweber, Defazet 33 (9), p. 296 (1976)
4.4.47 B. Pfeiffer. J. W. Schulze, J. Appl. Electrochem. 1991, p. 877; W. p. Hall, J. Coat. Techn. 52, p. 72 (1980)
4.4.48 T. Jones, Autodeposition of Organic Coatings, Metal Finishing: Organic Finishing Guidebook and Directory Issue 1994, p. 136, Elsevier Science, Inc., Hackensack, NJ, 1994
4.4.49 K. W. Thomer, Maßnahmen zur Verbesserung der pneumatischen Zerstäubung – Teilchengrößenbestimmung im Spritzstrahl, p. 111ff, Springer Verlag, Berlin, 1982
4.4.50 K. W. Thomer, Dissertation Uni Stuttgart 1982
4.4.51 A. Lefebvre, Atomisation and Sprays, Hemisphere Publishing, N.Y., 1989
4.4.52 K. Bauckhage, Chem.-Ing. Tech. 62, 888, p. 613–625 (1990)
4.4.53 T. Richter, S. Wilhelm, Maschinenmarkt 11, p. 26 (1991)
4.4.54 P. Walzel, Chem. Ing. Tech. 62 (12), p. 983 (1990)
4.4.55 Cudazzo and Strobeck, Automations- und Applikationstechnik, Hanover, DFO-Tagungsband, p. 33–44, DFO, Düsseldorf, 1992; W. Haferkorn, B. Keller, JOT 45(2), p. 40 (2005)
4.4.56 P. Walzel, Chem. Ing. Techn. 62 (12), p. 983 (1990)
4.4.57 H. Haagen, I-Lack 45 (3), p. 89 (1977)
4.4.58 W. Kleber, I-Lack 55 (3), p. 85 (1987), H. Schene I-Lack 34, p. 431 and p. 471 (1966)
4.4.59 R. A. Meisner, I-Lack 2, 55, p. 49ff (1987)
4.4.60 H. P. Fritsche, JOT 46 (7), p. 22 (2009)
4.4.61 Anonymous, MO 62 (11), p. 34 (2008)
4.4.62 M. Topp, Diplomarbeit FB Chemie, CTB, Universität Paderborn, p. 43 (2002)
4.4.63 P. Svejda, Dissertation Universität Stuttgart 1998
4.4.64 G. Brenn, J. Domnik, A. Lindenthal, M. Krüger, 8th International Coating Process Science and Technology Symposium, New Orleans, Proceedings p. 25–29 (1996)
4.4.65 R. Hruschka, MO 55 (11), p. 48 (2001)
4.4.66 M. Kohler, JOT 49 (9), p. 14 (2009)
4.4.67 J. Domnick, A. Scheibe, O. Ye, MO 55 (9), p. 41 (2001)
4.4.68 J. Vestweber, Dissertation Uni Paderborn 2002
4.4.69 J. Domnick, DFO Tagungsband, p. 157, Münster, 1991; K. Bauckhage, A. Schöne, Part. Charact. 2, p. 113 (1985)
4.4.70 J. Hoffmann (publisher), Handbuch der Messtechnik, p. 196, C. Hanser Verlag, Munich, 1999;W. Drahm and A. Rieder, Sensortechnik – Handbuch für Praxis und Wirtschaft (publisher H.-R. Tränkler and U. E. Obermeier, Chapter 13.13, p. 900, Springer Verlag, Heidelberg, 1998
4.4.71 P. Svejda, JOT 50 (9), p. 28 (2010)
4.4.72 U. Hoffmann, JOT 49 (3), p. 10 (2009)
4.4.73 R. Tacke, JOT 39 (5), p. 26 (1999)
4.4.74 Anonymous, JOT 41 (7), p. 12 (2001); P. Svejda, JOT 47 (3), p. 26 (2007)
4.4.75 Anonymous, MO 66 (9), p. 46 (2012)
4.4.76 Anonymous, JOT 46 (11), p. 32 (2006)
4.4.77 Anonymous, Coat. World 16 (5), p. 44 (2011)
4.4.78 R. A. Bayards, M. M. G. Antonisse, ECJ 05/2001, p. 16
4.4.79 G. Bell, PPCJ 200 (4545), p. 34 (2010)
4.4.80 E. Könneke, JOT 42 (3), p. 64 (2002)
4.4.81 J. Moser, F. F. Dietrich, JOT 41 (10), p. 56 (2001)
4.4.82 H.-G. Fritz, H. Voigtländer, MO 61 (5), p. 36 (2007)
4.4.83 R. Lüscher, JOT 43 (6), p. 32 (2003)
4.4.84 Peter G. de Lange, Powder Coatings and Technology (2nd ed.), p.341, Vincentz, Hanover (2004)
4.4.85 J. Zimmerhackel et al., JOT 50 (5), p. 28 (2010)
4.4.86 M. Planck, Ann. Phys. 4, p. 553 (1901)
4.4.87 M.-I. Bobb, JOT 45 (10), p. 80 (2005); J. Simon, JOT 42 (7), p. 30 (2002)
4.4.88 F. Zimmermann, MO 55 (1), p. 41 (2001)
4.4.89 Anonymous, JOT 49 (1), p. 26 (2009)
4.4.90 J. Schleusener, F+L 105 (7), p. 111 and 116 (1999)
4.4.91 M. Harsch, V. Bräutigam, P. Schüller, MO 56 (3), p. 42 (2002)
4.4.92 D. Skinner, PPCJ 192 (7), p. 14 (2002)
4.4.93 R. Schwalm, E. Beck, A. Pfau, ECJ 1–2/2003, p. 39
4.4.94 O. Röder, R. Bartel, G. Mattauch, Praxis-Forum 6, p. 103, Verlag Technik+Kommunikation, Berlin, 2000

General literature

H.-J. Adler, K. Potje-Kamloth, Quo Vadis – Coatings?, XXVI Fatipec Congress, Dresden, Wiley-VCH Verlagsges., Weinheim, 2002
Baehr, Stephan, Wärme- und Stoffübergang, Springer Verlag, Berlin, 1996
T. C. Dickenson (publisher), Valves, Piping and Pipeline Handbook, Elsevier Science, N.Y., 1999
J. Eisenlohr, J. Feßmann, Aufgaben und Verfahren der Oberflächenbehandlung, Expert Verlag, Renningen, 2000
Jahrbuch besser lackieren, Vincentz Verlag, Hanover, annually
R. Schwalm, UV Coatings, Elsevier, Oxford, 2007
P. Svejda, Prozesse und Applikationsverfahren (publisher U. Zorll, D. Ondratschek), Hanser, Munich, 2005
Emmanouil Spyrou, Powder Coatings – Chemistry and Technology (3rd ed.), Vincentz, Hanover, 2012
T. Brock, M. Groteklaes, P. Mischke, European Coatings Handbook (3rd ed.), Vincentz, Hanover, 2010
G. Wilke, J. Ortmeier, Coatings for Plastics, Vincentz, Hanover, 2012
Kittel: Lehrbuch der Lacke und Beschichtungen, Vol. 9: Verarbeitung von Lacken und Beschichtungsstoffen (2nd ed.), Hirzel, Stuttgart, 2004
Kittel: Lehrbuch der Lacke und Beschichtungen, Vol. 6: Anwendung von Lacken und sonstigen Beschichtungen (2nd ed.), Hirzel, Stuttgart, 2008
J. Prieto, J. Kiene, Holzbeschichtung, Vincentz, Hanover, 2004
B. Meuthen, A.-S. Jandel, Coil Coating (3rd ed.), Vieweg, Wiesbaden, 2013
Streitberger/Dössel (eds.), Automotive Paints and Coatings (2nd ed.), Wiley-VCH, Weinheim, 2008
P. Glöckner et al., Radiation Curing, Vincentz, Hanover, 2008
U. Schumacher, Th. Feist, D. Lehmann, Das Lernbuch der Lackiertechnik, 4th ed., Vincentz Network, Hanover, 2013

5 Safety, Environmental Protection and Health

Responsible care for the environment in the handling of raw materials and products, from manufacture and utilisation to recycling or disposal, is today a generally accepted principle. One of the manifestations of this is the fact that many software companies have now started to tackle the task of gathering and processing the abundance of data related to this problem (see Chapter 5.5).

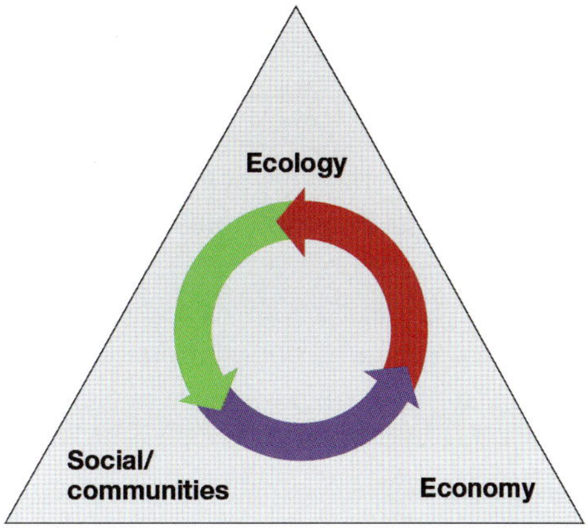

Figure 5.0.1: The three dimensions of sustainable development for the future
Sustainable, future-oriented development is a guiding principle according to which humanity must shape its future responsibly. The guiding principle has ecological, economic and societal dimensions.

For some time now, the development and manufacture of paints and coatings have been governed by different guidelines on "responsible care" and "sustainable development" issued first by the chemical industry associations and quickly followed by the paints and coatings industry around the world. The underlying principle, which was developed back in 1987 by the Brundtland Commission of the UN, is that the industry should only place as much strain on the environment as it can withstand through natural regeneration. In other words, we in the present undertake to leave future generations a world that is fit to live in.

In terms of the development of paints and coatings, both "responsible care" and "sustainable development" demand a comprehensive risk assessment from the points of view of the protection of health, environmental compatibility and other risks inherent to the handling of paints and coatings or the coated and cured object. This demand has increasingly repercussions on the composition of the products, the choice of application processes and the recycling of all substances involved after their use.

This voluntary undertaking by the paint and coatings industry is not sufficient for lawmakers in European nations and the USA, and so a full range of legislative provisions have been drawn up to regulate paints and coatings systems. In the last three decades, this has led to the creation of extensive legal requirements for painting technology, which are often characterised by the fact that they are constantly being updated to reflect technical developments.

5.1 Legislative Framework

The production and handling of paints and coatings has been in the sights of lawmakers all over the world for several decades. Outside Europe, for example in USA, these include the Clean Air Act (CAA), and the Toxic Substances Control Act (TSCA) which

Environment

regulates substances hazardous to health and the environment (HAPS or hazardous air-polluting substances), and the VOC regulations which have also been in force in China since 2010.

In Europe, Germany, together with the Scandinavian countries, stands out as a particularly regulation intensive country. In the past, this discrepancy has led to widely divergent situations inside the European countries in their handling of paints and coatings. Within the context of the Europeanisation and the harmonisation of national laws, the first task was to safeguard free trade in products, plants, appliances and waste materials. Consequently, the first areas of European regulation were the law on hazardous substances, the law on hazardous goods, the law on protection at machines and the law on waste management. Since the 1970s, environment-related laws have moved into the foreground, and they now play a major role, although they sometimes are highly controversial. These include the VOC Directive (see Chapter 2.3.2), the Biocidal Products Directive [5.8.1] and in particular the Chemicals Act (EC 1970/2006) on the registration, risk assessment and authorisation of chemicals during their life cycles (REACH), which has been in the implementation phase since 2007 [5.8.2] (see Figure 5.1.1). The associated registration periods for the products ends in mid-2018. It replaces the Europe-wide EINECS (European Inventory of Existing Chemicals) list and also constitutes comprehensive chemicals legislation for the European Community, bringing together the many laws existing in the various countries. A European Chemicals Agency (ECHA) has been set up to gather the extensive data on all products. Similar laws have now been introduced or are being implemented in the Asian region [5.8.3].

The basis for the production, storage, transport and application of coatings is therefore being increasingly influenced by Europe-wide guidelines. These represent the minimum requirements for laws, regulations and regulations to be introduced in the various countries.

The basic principles of the legislation described below are based on German law, which in many cases is analogous to that in other countries of the European Community. The main focus of the legislation and its implementing provisions for the paint industry can be divided into three areas of risk:

1. Handling of paints and coatings (→ Safety)
2. Environmental pollution caused by emissions to air, waste water and landfill (→ Environment)
3. Health risks (→ Health)

Date	Action
1.6.2007	REACH becomes law
1.6.2008	Start of the pre-registration period
1.12.2008	End of the pre-registration period
30.11.2010	Expiry of the registration period for a) Substances > 1000 t/yr b) Substances > 100 t/yr with environmentally hazardous properties c) Substances > 1 t/yr cmr-classified substances
31.5.2013	Expiry of registration period for substances >100 and < 1000 t/yr
31.5.2018	Expiry of registration period for substances >1 and <100 t/yr

Figure 5.1.1: REACH deadlines

This classification is based on the hazard characteristics of the coating materials (see Figure 5.1.2).

The associated laws, provisions and technical regulations often overlap in complex interrelationships, so that the handling of paints and coatings is covered by legal requirements on a number of fronts. Together with Europeanisation, which intervenes at different levels of German legislation, this complexity is one reason why considerable expertise is now necessary in the production and application of paints and coatings if all legal requirements are to be satisfied. Assistance is provided by the relevant industry associations, such as, in Germany, the Verband der deutschen Lackindustrie (VdL – Association of the German Paint and Coatings Industry), a part of the Verband der chemischen Industrie (VCI – Association of the German Chemicals Industry), with head office in Frankfurt. Additionally, many software companies offer update services on the legal requirements, often on a world-wide basis, together with the associated consequences for labelling, the compilation of material safety data sheets, storage provisions, transport provisions and other questions affecting the operations of the paints and coatings industry.

In Germany, the basis for the handling of coating raw materials is the "Chemikalienrecht" (Chemicals Act) along with the integrated REACH regulation. Important subsidiary laws include the "Gefahrstoffverordnung" (GefStV – Directive on Hazardous Substances), the "Bundesimmissionsschutzgesetz" (BimSchG – Federal Immission Control Act), the "Wasserhaushaltsgesetz" (WHG – Water Resources Act) and the "Kreislaufwirtschafts- und Abfallgesetz" (KrW-/AbfG – Waste Management Act), and the "Abwasserabgabengesetz" (Waste Water Levy Act). The last of these regulates the discharge of polluted waste water into sewage treatment plants or receiving waters. Of importance to coatings and coating materials in their capacity as preparations are the European CLP (Classification, Labelling and Packaging of Substances and Mixtures) Directive, the Federal Immission Control Act and the Preparations Directive based on EC Regulation 199/45/EC. Many other directives and administrative provisions are derived from these in Germany.

One important and long established instrument that provides the cornerstone for the safe handling of paints and coatings is the material safety data sheet (MSDS). The legal obligation to provide MSDS in the individual European countries was established by EU Directive 88/379/EEC, which was later superseded by 1999/45/EC (new directive on chemical preparations). Under this Directive, the manufacturer of paints and coatings is

Physical characteristic	Health characteristic	Environmental characteristic
Explosive Oxidising Highly flammable Readily flammable Flammable	Highly toxic Toxic Harmful to health Corrosive Irritating Carcinogenic (c) Mutagenic (m) Toxic to reproduction (r) Sensitising	Environmentally hazardous

Figure 5.1.2: Hazard characteristics

obliged to present the MSDS at the latest on first delivery of the product and must update the MSDS, with the obligation to provide proof thereof, in the event of any modification of the product. The format specified in EU Safety Directive 91/155/EEC has been further developed by the CEPE, the European umbrella organisation for national associations of the paints and coatings industry, in collaboration with the VdL, and has now been introduced in TRGS 220 ("Technische Regeln zur Gefahrstoffverordnung" – Technical Regulations Relating to the Directive on Hazardous Substances). This format covers all information concerning product composition, potential hazards of a technical nature and hazards to health, and it explains the necessary protective measures and the remedial action to be taken in the event of accidents and incorrect handling of paints and coatings. These aspects have been divided at European level into 16 headings, as shown in Figure 5.1.3. Findings from REACH notifications in the form of risk assessments and toxicological data are now being added to the sheets as dossiers.

Moreover, the identification of the products on packaging has been supplemented to enable rapid evaluation of the potential dangers, on the basis of visible and easily understood symbols. Substance-related data for the preparations of coating materials is often added electronically on the basis of standard wording in R (risk) phrases and S (safety) phrases. The R phrases describe the potential risks, whilst the S phrases contain safety instructions. These phrases are then included in the material safety data sheet and product labelling. On 1st October 2015 the phrases for mixtures, i.e. paints and coatings, were replaced by the H (Hazard) and P (Precautionary) statements of the GHS (Globally Harmonised System) system set out in Regulation EC 1272/2008, which is based on the proposal of the UN Commission of 2002. This is the case for chemicals since 2010. GHS is being introduced worldwide, especially in the European region with the exception of some Eastern European countries, in North and South America (Brazil and Argentina) and in South-East Asia with the important exception of India. Africa and the Middle East have not yet joined [5.8.3].

EU Safety Data Sheet guidelines	
1	Identification of the substance/preparation and of the company/undertaking
2	Composition/information on ingredients
3	Hazards identification
4	First-aid measures
5	Fire-fighting measures
6	Accidental release measures
7	Handling and storage
8	Exposure controls/personal protection
9	Physical and chemical properties
10	Stability and reactivity
11	Toxicological information
12	Ecological information
13	Disposal considerations
14	Transport information
15	Regulatory information
16	Other information

Figure 5.1.3: The 16 obligatory headings mandated by Directive 91/155/EEC

In Germany, assistance and state-of-the-art updates on safety questions are provided by, among others, the "Berufsgenossenschaften" (Occupation Cooperative Societies), which conduct safety inspections with the relevant companies in the event of accidents, inform the sectors concerned and issue corresponding workplace regulations.

The hazard symbols refer to the condition of the paint before and during application. Due to their polymeric character, applied and hardened coats are generally uncrit-

Figure 5.1.4: The most important hazard symbols according to the old Hazardous Substances Directive (above) and the GHS (Globally Harmonised System) (below)

ical in industrial use unless they contain chemicals classified as hazardous. However, their disposal must always be more closely evaluated and is increas-ingly becoming the focus of legislation in the context of life cycle analyses. A striking example is the end-of-life vehicle directive in Germany, based on the European Directive 2000/53/EC, which stipulated a recycling rate of > 95 % of the mass of all cars by 2015 in two stages.

Legislation in Europe is not only changing fast as a result of Europeanisation, but also, given the dynamisation clauses frequently built into the laws of individual countries, it is following the technical state of the art and the latest findings in the fields of safety, environmental protection and health.

From today's point of view, one problem affecting further legal developments is the compartmental approach, i.e. provisions are updated solely in respect of air, water or soil, so that there is practically no overall co-ordinated optimisation. Also, European environmental legislation does not necessarily mean harmonisation, as the member states can impose additional requirements when enacting the Directives.

In the chapters that follow, important features of the current legislative position are explained on the basis of the main focuses of the legal texts covering the specific fields of technical safety in the handling of paints and coatings (Chapter 5.2), environmental protection of air and water (Chapter 5.3) and the protection of human health (Chapter 5.4). The state of the art which is necessary for meeting the legal specifications in Germany concerning the manufacture, storage, transport and application of paints and coatings can only be attained through careful study of the laws in collaboration with the authorities and can only be outlined roughly in this chapter. It now requires considerable specialist personnel within companies.

5.2 Safety

Both the application and production lines for paints and coatings are at risk of explosion and fire due to the solvent content of solvent-borne and, in some cases, water-borne paints and coatings, and also due to potential particulate emissions from powder coatings. The

Directive on Industrial Safety and Health (BetriebSichV) together with the Directive on Hazardous Substances (GefStoffV) of 2010 and the Technical Rules for Hazardous Substances (TRGS) regulate the necessary information on the hazard potential and the provision of adequate work equipment for the safe handling of substances in systems requiring monitoring. These usually include manufacturing and industrial applications. The regulations on occupational safety technology are designed to avoid potential hazards and to eliminate known hazards. Comprehensive explosion protection concepts are offered, among others, by specialist safety companies [5.8.4].

One particular hazard in this respect is posed by electrostatic paint application (see Chapter 5.2.4).

Moreover, additional laws and provisions regulate the handling of paints and coatings as a function of the potential dangers of the substances within the process chain between production and application, i.e. during storage and transport. In some cases, these regulations also include emissions-related health protection, which will be discussed separately at a later stage in this Chapter (see Chapter 5.4).

For all stages in the production and application of coating materials, legislation requires a complete set of operating instructions, which must be made known to all employees concerned and must serve as the guiding basis for their working procedures.

The company's safety management is supported by the employers' liability insurance associations of the chemical industry (Occupation Cooperative Society) and the German statutory accident insurance association (DGUV) through brochures, such as "Work&Health".

In the annual statistics issued by the Occupation Cooperative Societies in Germany, the chemicals industry, which includes the paints and coatings industry, tops the list as the "safest" industry segment.

5.2.1 Manufacturing

The legal provisions and standards for protection against dangerous working substances during the planning and operation of a production site for liquid paints and coatings include the "Regeln der Berufsgenossenschaft" (BG-Regeln – Occupation Cooperative Society Regulations), formerly termed "Unfallverhütungsvorschriften" (UVV – Accident Prevention Regulations) and the "BG-Richtlinien" (BG Directives). These provisions are continually updated on the basis of analyses of actual accidents.

From the point of view of safety, one of the key legal texts governing the use of flammable paints and coatings containing solvents is the "Explosionsschutzrichtlinie" (EX-RL – Explosion Protection Directive), which provides comprehensive information for the evaluation of explosion risks as a function of the substances to be observed and their flash points (see Figure 2.1.156) and provisions for safety measures, including structural and technical precautions for the prevention of explosions or deflagrations. One essential section of this Directive is the description of precautions for preventing explosive air mixtures. This section classifies sites into different zones on the basis of risk. It also describes precautions and safety measures for repair work.

In general, the law stipulates that electrical appliances in explosion hazard rooms must conform to the Explosion Protection Directives. For example, switches designed to meet this provision prevent any sparks induced by the switching operation from entering into contact with explosive vapour/air mixtures.

The assessment of the required scope of further protective measures is based on the risk assessment of the relevant production areas, which are summarised in Figure 5.2.1.

For the application of flammable paints and coatings (see Chapters 5.2.2 and 5.2.4), separate rooms must be provided and must be isolated by fireproof means from other rooms and buildings. The same applies to the store, generally situated next to the application line, for supplying coating materials to the application installations. All these rooms must be marked with the notice: "Fire, open light and smoking prohibited". Floors and walls must be noncombustible and must be easy to clean. The doors have to open outwards, must be closed during normal operation and must shut automatically in the event of a fire. Rescue routes must be designed and marked to ensure that the paint room can be evacuated by the shortest possible route. A minimum number of ABC fire extinguishers is stipulated: 2 for a floor area up to 50 m^2 and a further 2 for every additional 200 m^2, each containing 12 kg of extinguishing powder. In addition, fire blankets and showers for extinguishing burning clothing must be installed in a visible position in the rooms.

The application installations and the workstations handling paints and coatings, e.g. measuring stations, transfer filling installations etc. must be fitted with air extraction systems that prevent the formation of explosive solvent/air mixtures. After the relevant equipment has been switched off, the extraction systems must remain in operation until the formation of explosive or ignitable air/solvent mixtures can be safely ruled out. In the event of failure of the extraction systems, the application equipment and its downstream plant components must switch off automatically.

All conductive equipment and building parts, e.g. walls, ceilings, floors, conveyers, dispersing machines and containers must be earthed to prevent the generation of a static charge that can be discharged by sparking, in particular during transfer, dispersion and mixing.

Zone	Risk assessment
Zone 0	Rooms/areas where an explosive atmosphere is present in permanence or for long periods.
Zone 1	Rooms/areas where it must be assumed that an explosive atmosphere can occur occasionally during normal operation.
Zone 2	Rooms/areas where it can be assumed that an explosive atmosphere does not occur during normal operation. If it does occur, it does so only for a short period.
Zone 20	Rooms/areas where an explosive atmosphere is present in permanence or for long periods or frequently in the form of a dust cloud in the air.
Zone 21	Rooms/areas where an explosive atmosphere can occur occasionally during normal operation in the form of a dust cloud in the air. Deposits or layers of combustible dust are usually present.
Zone 22	Rooms/areas where it can be assumed that an explosive atmosphere does not occur during normal operation in the form of a dust cloud in the air. If such an atmosphere does occur, it does so only for a short period or while accumulations or layers of combustible dust are present.

Figure 5 2.1: Zone-based risk assessmant of rooms or areas

Particular attention must be paid to the floors. They must be electrostatically conductive, with a maximum permissible leakage resistance of 1 MΩ. In this respect, it must be borne in mind that accumulations or deposits of paints and coatings residues can increase the earth leakage resistance. Consequently, the floor must be cleaned regularly.

5.2.2 Storage

In late 2002 the "Betriebs-Sicherheits-Verordnung"(BetrSichV – Directive on Industrial Safety and Health) superseded the "Verordnung über brennbare Flüssigkeiten" (VbF – Directive on Combustible Liquids) concerning storage sites for supplying the application equipment and every storage site used during the manufacture and application of paints and coatings. These provisions apply to products classified in specific hazard classes.

The technical and physical criteria for safe storage include the flash points and water solubility of the coating materials. The flash point is the lowest temperature at which a solvent or a preparation such as paint under specified conditions releases such large quantities of vapour into the air that a flammable mixture is generated. This characteristic value is measured in technically standardised apparatus (see Chapter 2.3.2). The substances are classified into different "hazard classes" according to the position of their flash point and water solubility (see Figure 5.2.2).

In accordance with the quantities stored and the location, storage sites are classified as storage not subject to compulsory display notice and permission and storage subject to compulsory display notice and permission. Here, limit values are laid down as a function of the hazard class, the type of container and the type and size of the rooms.

The storage of combustible liquids and paints is prohibited in passageways, thoroughfares, stairwells and generally accessible corridors, on the roofs of residential buildings, hospitals, office blocks and similar buildings and their penthouses, in guest rooms and canteens and also in workrooms. Moreover, no emptied containers may be stored in these places if they were previously used to store liquids from hazard classes F+ and F. This requirement must be strictly observed, because even small quantities of combustible liquids constitute an explosion hazard.

Class	Characteristics
F+	High combustible with flash points < 0 °C
F	Combustible with flash points > 0 °C < 21 °C
R 10	Combustible with flash points > 21 °C < 55 °C
No differentiation of water soluble or water mixible solvents	

Figure 5 2.2: Classification of combustible liquids according to the "Betriebssicherheits-Verordung" (Directive on Industrial Safety and Health)

This directive does not apply to workplaces where combustible liquids are stored for only short periods and in small quantities to ensure uninterrupted continuation of work in progress.

5.2.3 Transportation

At European level, the transportation of dangerous goods is governed by the European Agreement on the International Carriage of Dangerous Goods on Road, the ADR (Accord européen relatif au transport international des marchandise dangereures par route). The latest amended version of this agreement entered into force on 1.7.2001 and had to be adopted into the laws of the individual member states by the end of 2002.

The corresponding laws in Germany are the Gefahrgutverordnung Straße/Schiene (GGVSE – Directive on the Transportation of Dangerous Goods by Road/Rail) and Gefahrgutverordnung See (GGVSee – Directive on the Transportation of Dangerous Goods by Sea).

The principal elements of these laws are a classification of dangerous goods specifically for transport, the corresponding labelling and provisions specifying the capacity and size of containers and transportation system. The classification is harmonised under the CLP directive.

All flammable paints and coatings are governed by these regulations. The regulations also apply to water-borne paints due to the risk of groundwater contamination.

5.2.4 Application

The laws and provisions governing technical safety in the handling of paints and coatings in paintshops are described in Chapters 5.2.1, 5.2.2, and 5.2.3. A European standard has been devised for spray booths that addresses all the safety aspects [5.8.5].

A special feature is **electrostatic spray application**, as it involves the processing of solvent-borne or water-borne paints and powder coatings at high voltages of up to 100 kV. The associated moments of danger are, first, in the ignition of solvent-borne and water-borne paints having high solvent contents (< 55 % water) and of their solvent/air mixtures and, second, in the possibility of incorrect operation of the devices by the operators. The risk of ignition also exists with powder/air mixtures and must therefore also be taken into account during powder coating.

Protection of Persons

The safest and simplest protection against contact with voltage-conducting system components consists in enclosing these components in closed booths. The necessary access points in the form of doors or gates must be integrated into safety circuits, so that when they are opened the voltage is automatically switched off. Additionally, the system components must be earthed, and in general an alarm must be triggered before contact can occur.

The design of the electric circuit for electrostatically assisted hand-held devices must ensure that any contact with live parts has no effect on the relevant persons and prevent a short circuit if the gap between the gun and the workpiece is too small. Paints and coatings with flash points < 21 °C must not be applied by handheld devices, and voltage sources and their connections must not be installed in the booth.

Conductive, usually water-borne paints and coatings require insulation of the feed tanks, because there is a danger of short circuits across the metering lines. These metering lines must also be insulated.

Tribo-electric spray devices for powder coatings do not need high voltage electrodes. They generate the necessary voltage by friction of the powder at the spray units (see Chapter 4.2.2). The currents generated during application are in the µA range and are non-critical. However, extensive charging may occur if all live parts are not earthed. The same also applies to the operators who must therefore wear conductive footwear and operate on conductive floors.

Explosion and Fire Protection

There is a risk of explosion in electrostatic coating lines if the following combination of conditions occurs:

1. an explosive air mixture is present,
2. an ignition source with a higher energy content than the minimum ignition energy of the explosive air mixture is present and
3. a spark discharge is generated from the ignition source.

The aim of the safety standards and therefore of all coating plants and application personnel is to eliminate at least one of these conditions with as high a probability of success as possible. Here, the avoidance of explosive air mixtures takes top priority. This can be accomplished relatively simply with a suitable air intake and its monitoring in the paint booths. For example, when the air supply falls below a specified value, the high voltage can be automatically switched off.

For powder coatings, this precaution is also significant but it cannot be safely employed in practice due to the accumulation of powder particles and the swirling of these particles in the air line. Ignitable mixtures occur at levels greater than 20 g powder/m^3 air.

However, for both liquid and powder coatings, these measures are ineffective in the spray jet itself. For this reason, precautions must address the ignition sources.

First, all ignition sources other than the application devices themselves must be absolutely prohibited. Moreover, the spray guns, as a source of ignition, must have an energy-limiting system and a mechanism to prevent spark discharge. Proof of the technical safety of the various types of spray system is provided by tests conducted by State test bodies such as the Technische Überwachungsvereine (TÜV – Technical Inspectorates) or the Physikalisch-Technische-Bundesanstalt (PTB – Federal Physical and Technical Institute) in Braunschweig. Air/gas mixtures of known ignition energies are used for these tests. Guns for the tribo-electric charging of powder coatings are officially approved if their charge currents do not exceed a maximum of 15 μA.

Appropriate earthing must ensure a grounded state in less than 2 seconds after the high voltage is switched off, so that the system components can be touched without danger.

The following requirements apply to the earthing:

- Grip components of hand-held guns must have an electrically conductive and earthed contact surface of at least 20 cm^2.
- All conductive parts, such as booths, operating devices and other objects must be fully earthed.
- Operators must wear electrically conductive footwear.
- The leakage resistance between the operator's footwear, the floor and earth must not exceed 1 MΩ.
- Booths for applying liquid coatings must be made of electrically conductive material and must be earthed.

- The resistance at points of contact between each workpiece and the earthed pole of the high voltage source must not exceed 1 MΩ during the coating process.

This last requirement can only be met by careful and frequent cleaning of all suspended devices. Also, in this context, it is highly advisable to test that suspended devices entering the coating process are properly earthed.

5.3 Environmental Protection

5.3.1 Exhaust Air

The agreement reached by the international community on 12.11.2001 in Marrakesh under the auspices of the UN laid down concrete measures for the implementation of the Kyoto Protocol from the previous year. The aims of this climate protection agreement was to reduce the emission of greenhouse gases, including carbon dioxide, by an average of 5.2 % by 2012, as agreed in a subsequent undertaking [5.8.6]. The Federal Republic of Germany set itself the target of a 21 % reduction. Further-reaching targets are contained in the 5th and 6th European Union Environment Programme and the United Nations Environment Programme. These targets form part of a package that also includes improvement in the quality of life and a reduction in the ozone concentration at near-ground level. It is precisely here, in the lower atmosphere, that ozone is generated by hydrocarbon emissions, e.g. from cleaning systems, paint shops and automobile emissions in combination with oxygen and UV radiation.

Environment

The share of the paint and coatings industry is roughly a third of the "solvent use" segment in Figure 5.3.1, the most important sector in which is the cleaning industry. The German Federal Environment Agency (Umweltbundesamt) expected the target for 2012 to be just about reached [5.8.7]. However, it can be noticed that the measures in the field of solvent use, i.e. including paint production and processing, has not lead to a significant reduction in solvent emissions by 2010 and account for about 70 % of total emissions. The need for action to attain the targets set for Germany is much greater for the application of paints and coatings than for their production, which today takes place almost exclusively in closed systems. For this reason, the main focus of the descriptions that follow will be on paint and coating application.

Solvent emissions from the production and application of paints and coatings in industrial plants is regulated by the "31st Bundesimmissionsschutzverordnung" (BImSchV – Federal Directive on Immission Control), which is the adopting directive for incorporating the European VOC Directive of 1999 into German law [5.8.8].

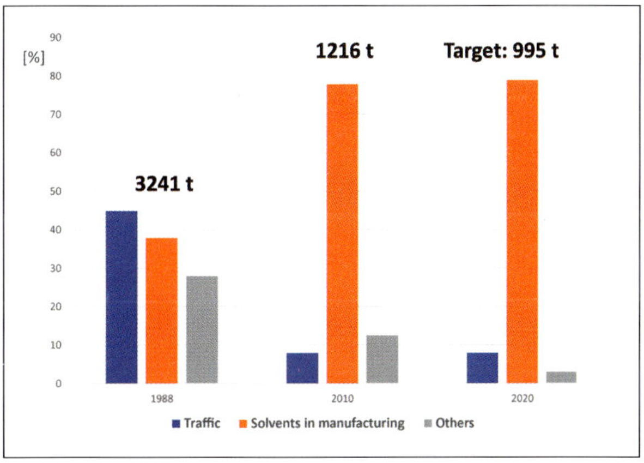

Figure 5.3.1: Contributions by organic substances to emissions in Germany *Source: VdL; Bundesumweltamt*

Paragraphs 1 and 22 of BImSchG lay down the **qualitative targets for industrial plants**:

§1: "The aim of this Act is to protect humans and also animals, plants and other things from harmful environmental impacts and also, in the case of industrial plants subject to permission, from dangers, significant disadvantages and significant pollution ... and to prevent the generation of harmful environmental impacts."

§22: "Industrial plants not subject to permission must be installed and operated in such manner that

1. harmful environmental impacts that are avoidable according to the state of the art are prevented,
2. environmental impacts that are unavoidable according to the state of the art are limited to a minimum and
3. the waste materials generated by the operation of the industrial plants can be removed in accordance with regulations."

Coating lines and paintshops are subject to permission if they use more than 25 kg of organic solvents per hour and/or if their annual consumption of organic solvents is greater than or equal to 15 t.

Emissions limits and in some cases the permissible limits vary from industry to industry, e.g. for automotive paintshops, package coating lines, coil coating lines, and other coating lines for mass goods. The 31st BImSchV permits solutions based on low emission paints and coatings in the form of reduction plans as well as solutions based on the aftertreatment of emissions. This regulation also regulates smaller paintshops that do not require permission.

The airborne pollution limits laid down for all industrial plants under the 31st BImSchV vary with the pollutant class of the polluting components.

The legislative provisions are based on the principle of dynamisation. For example, the limits should be considered simply as minimum requirements that must be attained within specified deadlines in the case of older plants. By contrast, new plants are assessed according to the current state of the art, which is considered to mean the development and testing of advanced processes as set out under environment legislation.

Emission limits as per BImSchV	
Ventilation systems with exhaust gas treatment	
Solvent consumption > 5 t/yr to 15 t/yr	100 mg/m^3
Solvent consumption> 15 t/yr	50 mg/m^3
with thermal afterburning	20 mg/m^3
Ventilation systems without exhaust air treatment and with diffuse emissions	
Solvent consumption > 5 t/yr to 15 t/yr	25 mg/m^3
Solvent consumption > 15 t/yr	20 mg/m^3
Automotive industry	
All installations*	35 g/m^2
**Valid from 2007 in mgC/m^2 coated aerea*	

Figure 5.3.2: General emission limits laid down in the 31st BImSchV for paintshops and paint manufacturing in mgC/m^3 air

The "TA Luft" technical regulation for the attainment of emission limits, dating from 1986, was reformulated for this purpose and the new version came into force on October 2002. However, "TA-Luft" now only regulates particulate or dust emissions from the production and application of paints and coatings. It comes into play for VOC emissions if the installation is not covered by the 31st BImSchV. This is the case, for example, with resin production in the coatings industry.

Product-Related Emissions Aspects

The progress made in reducing emissions through the development of new coatings and processes can be illustrated by the example of the automotive coating process.

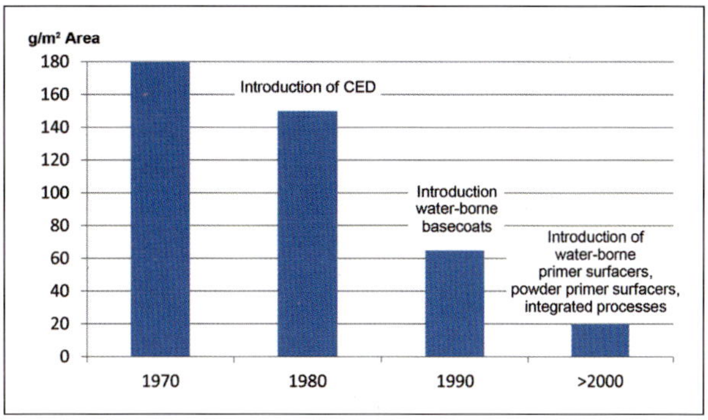

Figure 5.3.3: Reductions in emissions from automotive OEM coatings, expressed in terms of m^2 coated area

These low-solvent or solvent-free paints and coatings have also found applications throughout industry. The adoption of the European VOC Directive in the form of the 31st "Bundesimmissionsschutzverordnung" (BImSchV – Federal Directive on Immission Control) not only includes large industrial plants but also to a significant extent small-and-medium sized paintshops and other production sites that apply solvents. In most cases, the threshold solvent consumption triggering the obligation to obtain permission from the relevant authorities is reduced by a third to 5 t/yr. In the case of automotive paintshops the threshold is 15 t/yr for large-scale mass production, as mentioned above.

One important instrument of this directive, mainly for the small-and-medium-sized paintshops, is the Solvent Management Plan (SMP). This plan calculates a solvent balance between the purchase and consumption of solvents in all used products, with allowance for assumed diffuse solvent losses in the application department.

Irrespective of this management plan, obligations are laid down for the reduction and monitoring of emission sources. Here, paintshops are given the option of either adhering to the specified limits in mgC/m^3 (see Figure 5.3.2) or presenting a reduction plan based on the solvent content per paint-solid content. These reduction targets must be based on the current industrial standard and must be verifiably attained in accordance with target deadlines to be agreed within the specifications laid down by law. These specified values vary with the field of application but lie between 150 l and 320 l of solvent emission per 400 to 200 kg of paint solids consumed.

A key paints and coatings metric for drawing up an SMP is the VOC value (volatile organic compounds). This value applies to paints and coatings ready for application and specifies the content considered to be volatile and therefore a potential source of emissions. The VOC level is determined in accordance with ISO 11890 Part 1 (> 15 % solvent) or Part 2 (0.1 to 15 % solvent) and is calculated from the following equation (see Chapter 2.3.2):

$$\text{VOC (g/l)} = \frac{(\text{Mass of volatile contents [g]} - \text{Mass of water [g]})}{(\text{Volume of coating material [l]} - \text{Volume of water [l]})}$$

The distinction between volatile and nonvolatile (solid) for an organic compound is defined as follows:

Environment

> ***Definition:*** *An organic compound is volatile if at 293.15 K it has a vapour pressure of 0.01 KPa or more, or if it has equivalent volatility under the relevant operating conditions. The creosote content that exceeds this vapour pressure at 293.15 K is considered a volatile organic compound.*

The Industrial Emissions Directive (IED) 2010/75/EU applies to solvent-processing companies that deal with surface processes. These include companies that process more than 150 kg/h or 200 t/yr of solvents. Countries were given until 2013 to transpose the IED.

The Decopaint Directive (2004/42/EC) and its counterpart in Germany (ChemVOC-FarbV from 2004) have been adopted with a view to reducing exposure of tradesmen, including automotive refinishers and DIY enthusiasts, to VOC emissions. Based directly on the coating materials, it imposed maximum levels of volatile organic compounds on all paints and coatings used in the building sector and for automotive repairs in a first stage from 2007 on, with further reductions from 2010 on. For this purpose, the products used were divided into different categories and corresponding limit values specified. Figure 5.3.4 shows some selected values [5.8.8].

A further criterion for specifying emission limits is a toxicological assessment of the extent to which organic substances are carcinogenic, mutagenic or toxic for reproduction. These substances are known as "cmr substances" and have much lower emissions limits than other organic substances (see Figures 5.3.2 and 5.3.5).

In Europe, paint makers are allowed to emit 2 % of solvents consumed in production, while 1 % is allowed in Germany.

Product category	Paint type	VOC [g/l] from 1.1.2007 on	VOC [g/l] from 1.1.2010 on
Matt coating materials for interior walls and ceilings	W S	75 400	30 30
Glossy coating materials for interior walls and ceilings	W S	150 400	100 100
Coating materials for exterior walls made from mineral building materials	W S	75 450	40 430
Clearcoats and glazes for building structures and building elements	W S	150 500	130 400
One-component specialty coatings for buildings	W S	140 600	140 500
Two-component specialty coatings for buildings	W S	140 550	140 500
Primer surfacers for automotive refinishes		540	
Topcoats for automotive refinishes		420	

W = Water-thinnable paint
S = Solvent-borne paint

Figure 5.3.4: Limit values for selected paints as per the Decopaint directive

Types of Emission	Phrases	Area	Emissions limits	Specified in
Total dust		Application	20 mg org. C	TA-Luft
		Spray application	3 mg org. C	TA-Luft
		Paint production	10 mg org. C	TA-Luft
cmr substances*	Old: R 40, 45, 46, 49, 60, 61 New: H 340, 350, 350i, 360d, 360f	Application	50 mg org. C	BImSchV
		Spray application	1 mg org. C	BImSchV
		Manufacturing	20 mg org. C	BImSchV

** See Chapter 5.1 for more about R and H phrases*

Figure 5.3.5: Limit values for volatile cmr and other substances per standard m³ exhaust air from paintshops in Germany, as specified in the 31st BImSchV and TA Luft.

One characteristic that is not directly related to harmfulness is the odour of the emitted substances. The BImSchG also lays down that any odour nuisance to the residents near an industrial plant must be prevented. The appropriate technical measures for the plant are listed. The emission-limiting requirements can be quantified on the basis of olfactometrically measurable values in accordance DIN EN 13725. To this end, the odour thresholds of solvents and other substances are determined by trained test persons. Olfactometers are used to help gauge the odour detection threshold of odour-charged test air which is increasingly diluted with odourless air; the results then serve as a basis for calculating metrics. The volume of the sample which has been diluted until it is odourless (odour threshold) is expressed as a ratio of the volume of the original sample and stated in odour units (OU). As a result, it is possible to quantify the success of exhaust air and respiratory air purification measures by calculating odour reduction numbers (ORN). The odour reduction number is defined as: ORN = $(1 - OU_A/OU_B)$, where OU_B and OU_A are the odour units before and after air purification. Here, it should be noted that substances such as hydrogen sulphide can already be smelled before they reach a harmful concentration. On the other hand, there are also substances which are odourless or which can act on the organism in harmful concentrations before they have a perceptible smell. Within the framework of industrial plants subject to permission, the "Geruchsimmissionsrichtlinie" (Odour Emissions Directive) must be observed.

Process Engineering Possibilities for Emissions Reduction

Current legislation in Europe and other regions of the world such as North America permits process engineering solutions for reducing emissions to the prescribed limits. In that event, it is the pollution level in the exhaust air which must be monitored and measured on site.

In paintshops and coating lines, solvent emissions can occur at three points, namely paint supply/paint storage, paint application including transportation and curing/oven.

Air contamination levels vary considerably in accordance with the coating process, coating type and emissions point. However, it can be assumed that, where the volume of paint processed is much greater than the legal threshold of 5 t/yr of solvents, the exhaust air must be cleaned if the requirements are not met by the use of low-emissions coating materials. Various processes have proved successful for measuring high and low contaminant levels in the exhaust air.

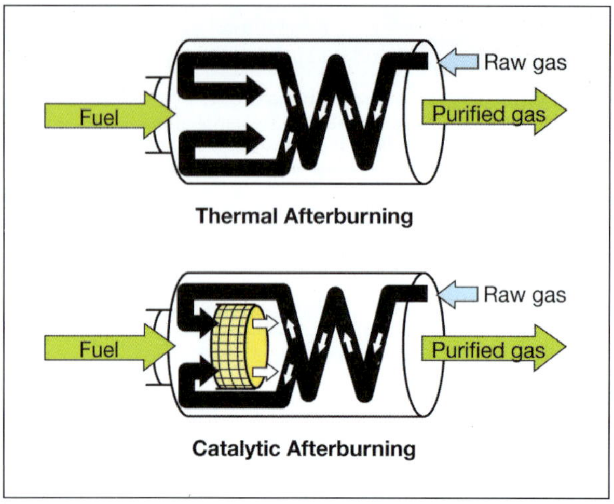

Figure 5.3.6: Schematic drawing of thermal afterburning equipment

Thermal afterburning is the method of choice for the relatively high contaminant levels in the exhaust air from the stoving oven compared to the spray booth, whilst the adsorption method has proved best for lower contaminant levels at the paint supply and application stages. Large users of coatings such as the automotive industry combine the exhaust air flows, thereby meeting the emissions limits by means of one central afterburning system for all emissions sources.

Thermal afterburning is based on the complete combustion of the organic substances to form water, carbon dioxide and nitrogen oxides. The last of these originate partly from the nitrogen content of the coatings, e.g. PUR coatings, and partly from the oxidation of atmospheric nitrogen at temperatures of 700 to 900 °C.

Environment

The temperatures necessary for complete thermal afterburning are attained by supplying additional fuels to the combustion chamber. In addition to the temperature, other factors crucial for high efficiency are the dwell time of the exhaust air and thorough mixing of the added fuel with the exhaust air.

The longer the dwell time in the combustion chamber, the better the cleaning effect of the system. A combustion temperature limit < 1000 °C is necessary, because formation of nitrogen oxides increases considerably above 1000 °C and can exceed the permissible limits. On the other hand, the highest-possible temperature is better for reducing the carbon monoxide content. Modern systems can be set to the optimum conditions for the coating materials concerned and so achieve high exhaust air cleaning efficiencies.

The thermal energy is mostly used for the production process, for preheating of the exhaust air flow or for other heat requirements. Such regenerative thermal oxidisers (RTO) for purifying exhaust air [5.8.10] operate discontinuously and, at a solvent concentration of about 1.2 g/m^3 and below in the exhaust air, autothermally and highly economically.

As already mentioned, all methods of thermally treating exhaust air experience problems due to the generation of secondary emissions. For example, if exhaust air with a low solvent content has to be aftercleaned to a level below 20 mg C/m^3, up to 2 to 3 tonnes of carbon monoxide, up to 2 to 3 tonnes of nitrogen oxides and up to 1700 tonnes of carbon dioxide can be emitted per tonne of emitted VOC, the precise figures depending on the untreated gas content. Since carbon monoxide and nitrogen oxides participate in the photochemical process that generates ozone in the lower atmosphere, this aftertreatment can be counter-productive.

Apart from thermal afterburning, recourse is also made in isolated cases to catalytic afterburning. Like thermal afterburning, the catalytic method oxidises the organic substances to water and carbon dioxide. In this method, the activation energy for oxidation is reduced to 350 to 500 °C by suitable catalysts. Despite the energy advantages over thermal afterburning, especially when integrated operation is impossible, this method has failed to find general favour because the efficient catalysts become contaminated by the substances contained in the coating lines and are thus rendered inactive [5.8.11].

Adsorption methods are more significant, especially when the exhaust air is contaminated by low levels of organic substances. In these processes, the organic substances in the vapour phase are deposited on a solid surface termed the "adsorbent" and are thereby removed from the air flow. The bond with the adsorbent is primarily physical in nature and can therefore be reversed under certain conditions. One adsorbent of very high surface activity is activated carbon, which has also been found effective in plants for the paint and coatings industry. For problem-free operation, the exhaust air must be free of particulate matter. Activated carbon was previously used in fixed bed systems that operated discontinuously with a through-flow of exhaust gas. Now, it is moved continuously over perforated filters in form of a fluidised bed, so that the adsorbent can remove the solvents evenly from the exhaust gas. With a specific surface area of more than 1000 m^2/g, the highly porous activated carbon is charged with solvents from the exhaust air and then channelled to a desorption system to be freed of solvent either by overheated steam at 150 °C or by low-oxygen air (< 10 %) at 600 °C The solvents are transformed into liquids by cooling to room temperature.

The disadvantages of this process are the problems inherent in treating solvents with hot steam. If water-miscible substances are contained in the solvent mixture, they will be present not in the organic but rather in the aqueous phase. The water must therefore be treated. Desorption with inert gas or low-oxygen air under conditions of thermal treatment at 600 °C induces chemical changes in the solvents, the extent of which varies with the type of solvents, and in the catalytically active surface of the activated carbon. Regular partial replacement of the adsorbent is therefore necessary. Nor should it be overlooked that the adsorption rate of the solvents depends on their volatility and polarity. Consequently, low-polarity solvents that have a low boiling point are more difficult to separate than the other components of the solvent mixture in the exhaust air. The regeneration process is very energy-intensive.

Emissions from Coating Films

For furniture or furniture surfaces and automotive interiors, it is evaporation of odour-intensive or even physiologically harmful components of the coating, wood substrate or plastics which needs to be addressed. For example, the "Gefahrstoffverordnung" (Hazardous Substances Directive) specifies limits on furniture emissions in closed rooms. For formaldehyde, the maximum permitted value is 0.1 ppm (ml/m^3). Any solvents remaining in the cured coating can also constitute a health hazard or add to the pollution level in closed rooms. If necessary, control tests must be conducted under defined conditions (ISO 16000-28) in the normal environment or in a special test chamber.

Emissions of all kinds from the many of plastic parts installed in automotive interiors are increasingly being targeted by consumers.

5.3.2 Wastewater

The "Wasserhaushaltsgesetz" (WHG – Water Resources Act), in conjunction with the corresponding Water Acts of the various German states, details the rules for the handling of groundwater and surface water. This Act applies to all industrial plants that either wish to or have to discharge contaminated process water into watercourses. The current version of the WHG has adopted the principle of prevention, i.e. the avoidance of wastewater, and the dynamisation clauses of the Federal Immission Control Act. The federal structure of the "Wasserrecht" (Law relating to Water) requires the adoption of these new provisions into the water laws of the various states in order to ensure their implementation throughout the Federal Republic.

Specifically, for their own needs, the states have passed directives on indirect discharging sources, laying down that substances hazardous to water must be discharged into main collectors. Direct discharge means that process wastewater is discharged directly into surface waters, e.g. streams or rivers.

Today, the general consensus of legislation is that the discharge of toxic metals and chlorohydrocarbons is prohibited. Approvals for these plants can be granted only if state-of-the-art wastewater treatment is provided.

The "Abwasserverordnung" (Wastewater Directive) describes the industry-specific requirements regarding the pollution limits of various substances for direct and indirect discharging sources. Further-reaching requirements, e.g. fees based on pollution levels (see below), may be imposed by local government statutes.

In order to ensure that the specific limits for contamination of wastewater are met, the lawmakers consequently demand measures to increase the efficiency of the process water within the processes. These measures include reductions in dredging losses from process baths, recycling of rinsing bath contents, minimisation of rinsing bath discharges by means of closed loops, longer operating times for process baths, such as floor washers in spray booths and other measures conforming to the state of the art.

In paint shops, the biggest sources of wastewater requiring treatment are the spray booth flood sheet fluid systems. Other important sources are wastewater from the electrocoating process in the form of diluted acetic or similar acids. However, these acids are nonhazardous, and are discharged simply on the basis of COD values and in line with the capacity of the wastewater treatment plants. Other forms of wastewater can be discharged from other units necessary for operation, e.g. cooling water. These discharges have nothing to do with the coating materials.

Substance	Appendix 40 of German Wastewater Directive	Requirements of communities
Aluminium	3 mg/l	
Iron	3 mg/l	
Lead	0.5 mg/l	1 – 2 mg/l
Phosphorus	2 mg/l	
Hydrocarbon	10 mg/l	10 – 20 mg/l
COD	300 mg/l	Approx. 600 mg/l

Figure 5.3.7: Threshold values for environmentally hazardous substances at OEMs, according to the German Wastewater Directive

The legal requirement for high-quality recycling of waste has transformed the treatment of wastewater from paint booths. For example, the water from the flood sheet fluid systems in spray booths for water-borne paints is sometimes not coagulated but instead is post-stabilised to allow an ultrafiltration to recover the coating components without any problems, i.e. with longer times for extended filtration cycles. The circulating volume should be kept as low as possible to ensure the efficiency of ultrafiltration. This drastically reduces the quantity of wastewater arising from the application of water-borne paints and coatings.

If recovery of solvent-borne paints and coatings or their components (see Chapter 5.3.3) is required, the coagulating agents must be adapted to this process. Aggressive coagulating agents should not be used. The coagulants must effect rapid phase separation and should form a coagulate with a low water content [5.8.12].

The generated wastewater is treated by established processing stages to ensure compliance with the limits on heavy metals and other substances. These stages are stripping, detoxification, flocculation, sedimentation and neutralisation. The chemicals required are specific to each coating [5.8.13].

In the automotive industry, the normal procedure is to channel all wastewater from electrocoating, overspray treatment and the cleaning operations into a single collector for appropriate treatment. Here, the chemical oxygen demand (COD) can be so high that the water/solvent mixture may have to be pretreated in the paintshop before the collecting stage. Established treatment processes are available for this purpose, the choice of which depends on the composition, quantity and location of the wastewater. These processes include biological decomposition similar to the process used in sewage plants, adsorption on activated carbon or adsorption on molecular filters, reverse osmosis and chemical oxidation [5.8.14].

The coating materials, like their raw materials, are classified into three water-hazard classes: slightly hazardous to water (WGK 1), hazardous to water (WGK 2) and very hazardous to water (WGK 3). The relevant class is specified in the material safety data sheet. Under the "Verwaltungsvorschrift Wassergefährdende Stoffe" (Administrative Provision on Water-Hazardous Substances), the classification criterion is the toxicity of the substances to aquatic organisms, starting with algae and rising to miniature life forms such as waterfleas and culminating with fish. The most critical value obtained from all the tests is adopted. However, ready biological decomposition can lead to a lower hazard classification, whereas poor decomposition capacity can lead to a higher one. These last substances, termed persistent organic pollutants (POPs) are currently the focus of considerable attention. The discharge quantity and the water-hazard class (WGK) are combined to generate risk levels (A to D) which impose different basic requirements on the relevant plants [5.8.15].

5.3.3 Recycling and Disposal

In Germany, this environmentally important topic is regulated by the "Kreislaufwirtschafts- und Abfallgesetz" (KrW-/AbfG – Waste Management Act). This act requires firstly the avoidance of waste, then its reutilisation by direct or indirect recycling and finally, if unavoidable, its conditioning for disposal.

For purely economic reasons, considerable importance is already attached to high utilisation or efficiency during both production and application of paints and coatings. As a result, today's established coating processes such as coil coating, electrocoating and

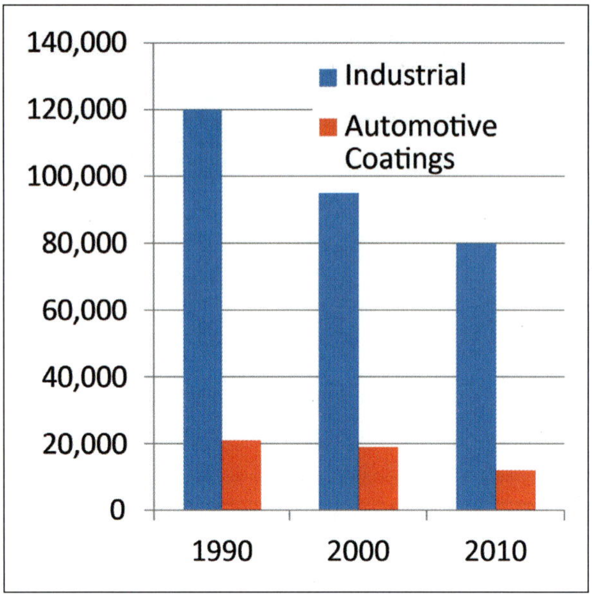

Figure 5.3.8: Amounts of coatings sludge in Germany
Source: authors' estimate

powder coating attain material efficiency levels of more than 98 %. Electrocoating and powder coating require special processing measures, such as ultrafiltration and powder recovery (see Chapters 4.2.1 and 4.2.2).

Waste cannot be totally prevented, especially where coloured coatings are produced by spray application. The overspray occurring during these processes cannot be recirculated to the coating process direct. First, frequent colour changes make recovery uneconomical, and, second, the substantially modified material properties of the dried coatings, frequently irreversible as in the case of 2-component systems, do not permit direct recovery.

The paint sludge levels generated by different treatment methods in the industrial segment in Germany amounted to approximately 80,000 tonnes in 2010. This figure represents 5 % of the overall quantity of 1.6 million tonnes of applied coatings. Related to industrially applied coating materials, the figure therefore represents more than 10 %.

In this case, despite the high proportion of spray coatings, the automotive industry is much more efficient (see Figure 5.3.8).

A great deal of effort is constantly devoted to the avoidance of waste in coating processes. Especially in the last few years, considerable progress has been made in the process engineering of spray booths [5.8.9] and especially in the improvement of application conditions and more fault-free coating processes [5.8.17].

Efforts to achieve the direct recycling of coating wastes in coating processes have led to some solutions and approaches [5.8.18], but have been redirected as described to the optimisation of processes with a view to increasing the coating efficiency and developing shorter, more efficient processes. The reasons are that the various projects, e.g. for the treatment of water-borne primer surfacers or the separation of coating wastes into solvent, pigment and film forming agent fractions cannot be transposed into everyday industrial operation. Generally, this has been due not to a lack of technical feasibility but rather to cost estimates and the legislative position that promotes avoidance over recycling. Recycling of the waste materials that are nevertheless generated takes place at a low level of material utilisation. For example, waste materials are channelled to steel furnaces as carbon donors, are used as fuel, are processed in the cement industry (see Chapters 7.1 and 4.2.1.5) or are converted to synthetic gas (see Figure 5.3.9) [5.8.19].

In the field of powder coating, the situation is much more favourable. Since the efficiency of first application is only about 50 to 60 % and the associated loss of coating solid is

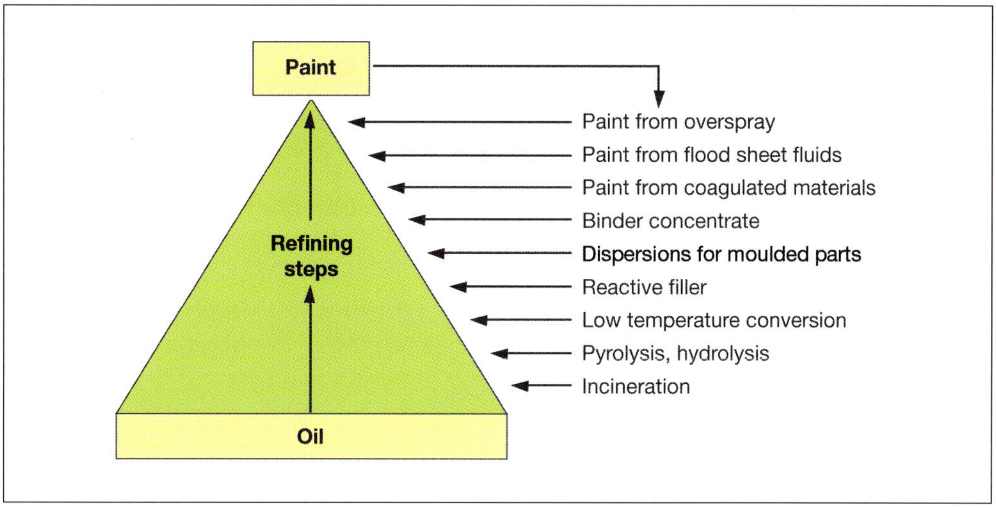

Figure 5.3.9: Scope for recycling organic paint waste along the raw materials chain

very high, and in addition powder as a solid is easy to handle, direct recycling of the overspray is now standard practice. Otherwise, powder coating would lose its competitiveness against liquid coatings (see Chapters 4.2.2 and 5.7).

Recycling is also an important theme in the packaging of paints and coatings. If there is a sufficiently good separation between the packaging and the filled product, i.e. the coating material, disposable packaging is channelled to the relevant material recycling system. In other words, cans are routed to steel extraction, whilst wood, cardboard and bags are generally incinerated. In the industrial field, it is standard practice to employ returnable containers that can be reused after cleaning.

The landfilling of surface treatment waste is prohibited for most substances, because they have a loss on ignition > 3 % or can release soluble substances into leachate.

5.4 Health

The development, manufacture and application of coating materials in Germany are governed by the “Chemikaliengesetz” (Chemicals Act) together with the “Gefahrstoffverordnung” (Hazardous Substances Directive), the “Technische Regeln für Gefahrstoffe” (TRGS – Technical Regulations relating to Hazardous Substances) and the “Unfallverhütungsvorschriften” (UVV – Accident Prevention Regulations). Most paints and coatings are deemed to be hazardous substances under the law. Moreover, the principles of “Responsible Care” adopted by the chemicals industry are also morally binding on the paint and coatings industry. This obligation is expressed in the “Leitlinien zur Sicherheit, Gesundheits- und Umweltschutz” (Guidelines on Safety and the Protection of Health and the Environment) issued by the Verband der deutschen Lackindustrie (VdL – Association of the German Paints and Coatings Industry).

The European Directives that were initially passed as framework laws are increasingly intervening directly in the laws of the individual member states. For example, one of the most recent directives of the “Gefahrstoffrecht” (Law on Hazardous Substances), contains references to the EU Directives, with the result that EU law will also be directly

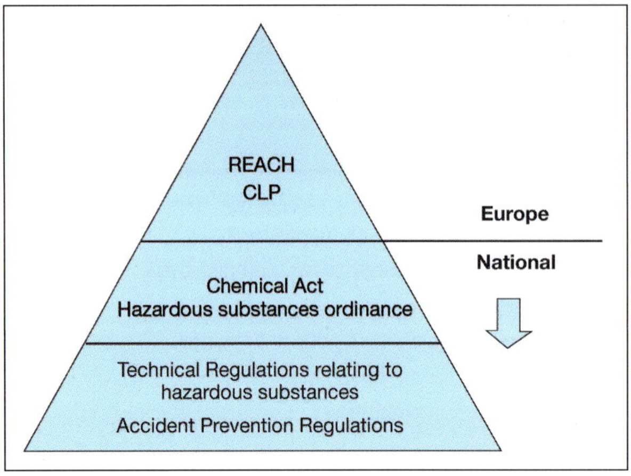

Figure 5.4.1: Structure of European Directives on hazardous substances

applicable in Germany after the normal periods allowed for the adoption of EU legislation into national law have elapsed.

The recently revised "Gefahrstoffverordnung" (Hazardous Substances Directive) has now also added the "Stoff-Richtlinie" (Materials Directive), the "Zubereitungsrichtlinie" (Preparations Directive) and the "Sicherheitsdatenblattrichtlinie" the Safety Data Sheet Directive as a compendium.

The most recent noteworthy amendment at European level, viz. Directive 1272/2008/EC (CLP Directive), has specified the classification of preparations as dangerous for the environment and the identification of sensitising substances. The German version on classification and labelling can be found in TRG 200.

Environment

Acute toxicity values (see Figure 2.1.156) have been determined for the various raw materials, such as the solvents of coating materials (see Chapter 2.1.2). These values, based on Directive OECD 401, are determined orally (LD or lethal dose), dermally (LD) and on inhalation (LC or lethal concentration) in accordance with the nature of the substance. Generally, the specified values are the doses found to be lethal to 50 % of the test animals, usually rats and mice, within 14 days. The substances are then classified by the Senate Committee into highly toxic, toxic and harmful to health.

The aims of legislation and the principles of "Responsible Care" are, first, to ensure safe handling and the prevention of health risks of all kinds and, second, to provide rapid and accurate emergency information for public authorities, doctors and of course customers. The basis for this is the material safety data sheet mentioned above (see Chapter 5.1). The MSDS describes the risks in standardised "R phrases" and the preventive measures in "S phrases" and also recommends the remedial action to be taken in the event of accidents involving the relevant product. These recommendations essentially apply to acute accidents and their risks (see Chapter 5.1).

During handling of paints and coatings in development, production and application, contamination of the skin and eyes and exposure and inhalation are particular potential dangers. Contamination can in most cases be avoided by personal protective equipment

	LD_{50}, oral	LD_{50}, dermal	LC_{50}, inhalation
Highly toxic	< 25 mg/kg	< 50 mg/kg	< 0.5 mg/l x 4 h
Toxic	25 – 200 mg/kg	50 – 400 mg/kg	0.5 – 2.0 mg/l x 4 h
Harmful to health	200 – 2000 mg/kg	400 – 2000 mg/kg	2.0 – 20 mg/l x 4 h

Figure 5.4.2: Classification of substances according to the Hazardous Substances Directive, in mg/kg body weight and mg/l concentration in air

such as overalls, safety shoes, protective gloves, protective goggles, breathing masks and if necessary full-body protective safety suits. This equipment must meet the requirements of the relevant industry associations.

Where products classified as hazardous to health are regularly handled, regular preventive medical check-ups by doctors qualified in industrial medicine are also specified.

On the basis of studies and risk assessments of the relevant substances, the exposure of employees is limited under the "Gefahrstoffverordnung" by the AGW levels (occupational exposure limit value in former times: MAC = maximum allowable concentration at workplace) to a level considered uncritical to health. The AGW level of a substance is its concentration in the air, in the form of gas, vapour or suspended particles, which according to present knowledge does not adversely affect health upon exposure of 8 hours per day and 40 working hours per week. This level must be regularly monitored by measurements at the workplace conforming to specified methods and it may be necessary to adopt technical or organisational measures to limit exposure on the basis of exposure scenarios under REACH. To complete the risk assessment, the characteristic values in the working environment, in the form of the described AGW and TRK values, are supplemented by limits on the quantity of substances or derived products in the human organism. On the basis of preventive inspections under industrial law, "BAT" values ("Biologische Arbeitsplatztoleranz" – biological tolerance in the workplace) for specific substances are laid down as threshold concentrations in the urine or blood. If these values are exceeded, additional safety measures must be taken.

The legislature, for its part, is pursuing a strategy of the avoidance of risky formulations. The aim of the relevant laws is to lower the risk to the health of the population from products or processes as far as possible.

The two most important laws that have a significant impact on the raw materials portfolio are "Regulation (EC) No 1907/2006 of the European Parliament and of the Council of 18th December 2006 on the Registration, Evaluation, Authorisation and Restriction of Chemicals" (REACH), which came into force in mid-2007, and Directive 98/8/EC of 16th February 1998 on the placing of biocidal products on the market, for which a legal basis was put in place in mid-June 2002 within the framework of the new version of the Chemicals Act in Germany.

REACH applies not only to chemicals and substances, but also to their mixtures such as coatings and their application scenarios, i.e. application and film formation. REACH involves costs, the impacts of which vary widely, especially for small quantities of products [5.8.20]. The registration phase for all substances produced and imported in Europe in a market volume of 1 t/a and above was completed in 2013 (see Figure 5.1.1). Manufacturers and importers now have to demonstrate, to an extent depending on volume classes, that their products can be handled safely throughout the entire process chain [5.8.21].

Due to the bioreactive nature of biocides, which serve as wood preservatives, disinfectants, preservatives, insecticides, rodenticides etc., the adoption of the Biocidal Products Directive (BPD) 98/8/EC after 2006 has placed products with clearly toxic effects for humans on a list of prohibited products. In addition, all products that were used before 2000 have been registered and, on the basis of appropriate documents, must be assessed to determine whether they can remain on the market [5.8.22]. This has already had a significant impact on the range of raw materials. A revision of the Directive came into force on 1st September 2013 in the guise of Regulation 2012/528/EU.

5.5 Ecobalances

Up to about 1960, the main focus for the development of paints and coatings was on achieving technical progress in the functionality of the coatings and coating processes (see Chapter 1.2). Later, considerations relating to environmental protection gained in importance. The pendulum temporarily swung disproportionately towards the ecological side. Legal interventions were responsible for this swing, with the core emphasis on emissions into the air and water. Since the early 1990s, there has been a coming together of the economic and ecological objectives for the benefit of technical and social progress as well as the environment. This is reflected in numerous publications which also highlight the penetration of the term "sustainability" within the sector during the development of new products by all kinds of manufacturers of coating materials [5.8.23, 5.8.24, 5.8.25].

Various evaluation models have been developed for this purpose, and are summed up under the terms sustainability, life cycle analysis (LCA), eco-efficiency and life cycle assessments [5.8.26]. Standardisation has also addressed this topic and created standards under ISO 14040 and 14044, which have now been accepted (see Chapter 6.5). The essential, new element for developers, machine makers and marketers is that records are kept and evaluations conducted of the entire product-manufacturing process, from raw material consumption to energy and environmental impact from production through to use and disposal. The same principle applies to life cycle engineering (LCE), in which the main focus is on manufacturing processes [5.8.27].

Environment

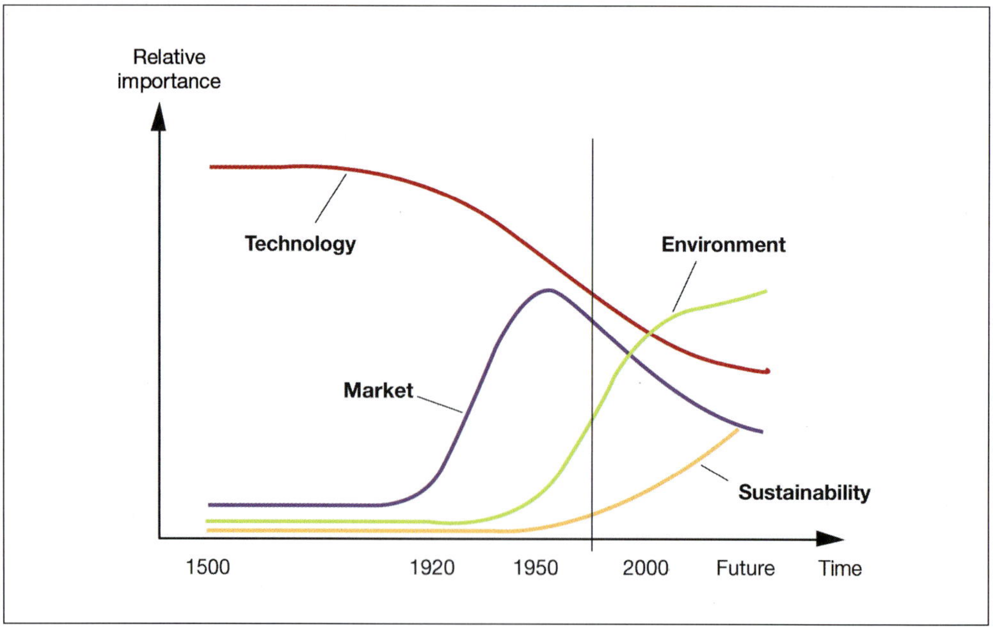

Figure 5.5.1: Development of the relative influence of technology, economy, environmental protection and sustainability on the coatings industry according to S. Bancken

The concept of "sustainability" extends the assessment of a commercial activity to include the aspect of social responsibility (see above). This is implemented, among other things, within the BASF Group's "SEE Balance", which is certified by various organisations and is based on the company's already established eco-efficiency analysis and includes the social responsibility and sustainability of the entire value-added chain under the heading "Social Responsibility" [5.8.28]. The criteria considered and evaluated from quality and quantitative aspects are product safety, health protection, cooperation between management and employees, e.g. through training and development programmes, as well as corporate social responsibility. For example, raw material suppliers are classified according to these criteria, with potential risks recorded and avoided. A "social fingerprint" allows a simple visual comparison of the investigated products and processes.

Legislative strategies at European level are giving considerable backing to this approach. For example, the directive passed by the EU in October 2000 for the disposal of end-of-life vehicles stipulated recycling rates of 85 % for the year 2006. This figure applied to all materials used in the car and not just steel as the most significant material by weight. In 2015, the recycling rate was raised to 95 %. In this context, there is frequent debate concerning the stage at which to recycle the materials (see Figure 5.3.9). For example, thermoplastic polymers can be returned to the same production stage albeit with some reduction in quality, whereas thermosetting polymers and their associated coatings must instead be transformed at the upstream stage at the level of the raw materials, e.g. as synthesis gas.

The coating process often represents the largest emissions source for environmental pollution in many manufacturing processes for products such as trains and aircraft. Coating also involves considerable energy and investment costs, which vary greatly in scale as a function of the type of coating and application process. Clearly only a comprehensive assessment of all these factors can permit a meaningful comparison of the production process in terms of resource conservation and environmental pollution (see Figure 5.5.2). These two aspects have emerged as the most important criteria for assessing suitably comparable processes.

Environmental pollution in turn is composed of the elements of emissions into the air, water and soil, toxicological hazards to humans and animals, and recyclability.

At present, an absolute assessment of all components of a "life cycle analysis" is still highly complex and costly, because the data structure is to a large extent only now in the process of being developed. Comparability is also limited by the fact that the system limits have not yet been standardised and are defined differently.

A simple but effective way out of this dilemma is provided for example by the BASF Group's eco-efficiency analysis [5.8.28]. The heart of this analysis is a product's practical utility to the customer. In general, customers can choose between several products or processes to satisfy their requirements. The eco-efficiency analysis compares the economic and ecological advantages and disadvantages of these different problem solutions. In other words, it is not products which are to one another as such, but instead their applications such as the coating of a square metre of the front of a piece of furniture. Problem solutions are adjudged to be eco-efficient if they provide distinct customer benefits at lower cost and with less environmental pollution than other solutions. Lower cost, in this case, is synonymous with greater conserving of resources. Here, the analysis is based on all stages of the value added cycle (see Figure 5.5.3). The costs and environmental

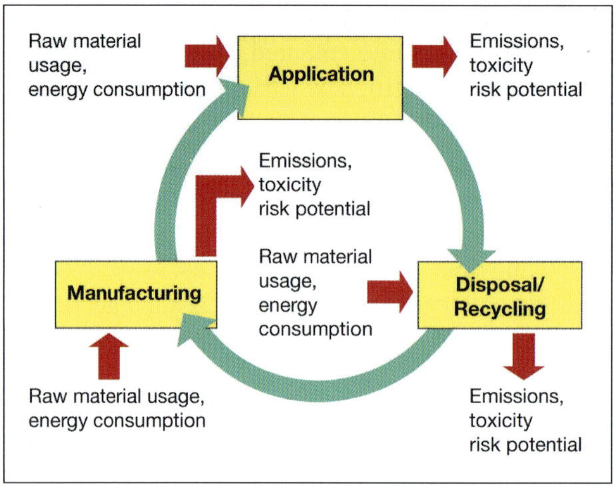

Figure 5.5.2: Value added cycle

pollution of all processing stages, i.e. production, application and disposal or recycling, are measured. The processing stages before or after production are also included in the calculation. The result of the analysis is summarised in an eco-efficiency portfolio.

Costs and environmental pollution form the two axes. The scales are arranged so that the best solution is positioned on the right or at the top. The relative positions of all the products tested are then added. Products or processes positioned in the right hand upper field are eco-efficient. Anticipated future product improvements can also be determined and entered in the portfolio for visualisation purposes.

The environmental impact important to this assessment is described on the basis of five categories, which together constitute the "ecological fingerprint" (1 = Good; 0 = Poor):

Environment

- Consumption of raw materials
- Use or consumption of land area
- Consumption of energy
- Emissions into the air, water and soil (waste)

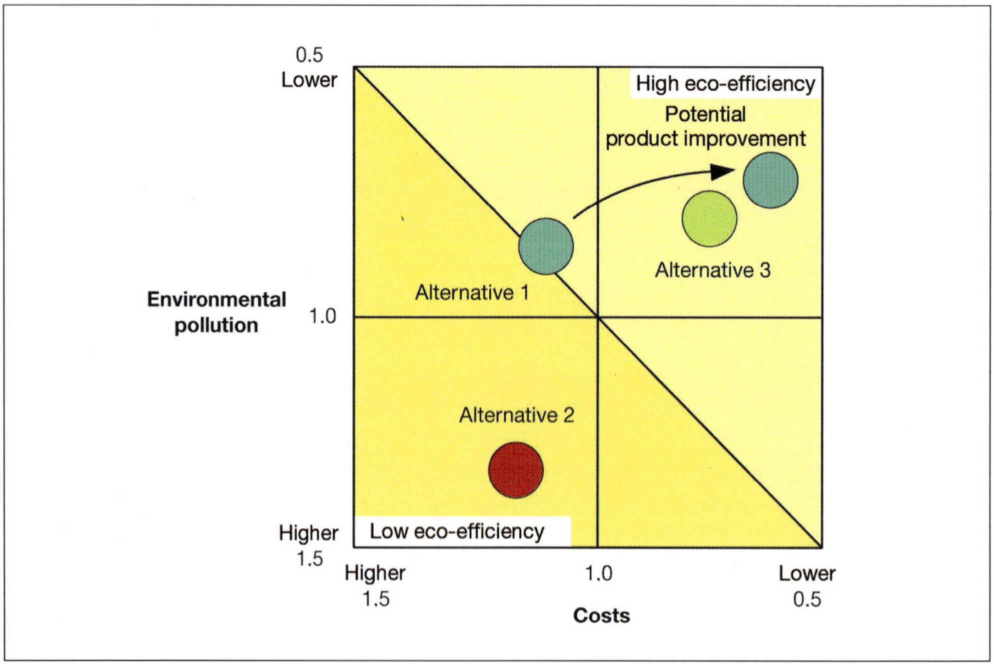

Figure 5.5.3: Eco-efficiency portfolio

- Toxicity potential of the used and released raw materials
- Potential misuse and risk

Each of these six categories conceals a mass of detailed individual criteria which are grouped together to determine the overall environmental impact of a product or process. The significance and scope of the properties at the various stages in the life cycle are taken into account by "relevance factors". These factors specify the extent to which individual criteria are represented in the overall environmental impact, for example how the potential for ozone destruction is weighted against the greenhouse potential. In the case of consumption of raw materials, the availability, natural occurrence and associated different sustainability ranges are taken into account. The weighting can also make allowance for the importance that society attaches to the different forms of environmental pollution.

The other axis of the portfolio is used to plot the economic data. Here, the material and energy flow of the compared processes and their resulting products are considered, including all relevant secondary flows. One important factor here is the definition of observation limits. Almost all clearly definable systems can be comparatively assessed by this method.

The costs and work involved in an analysis of this kind are considerable. However, experience shows that processes found to be eco-efficient by this method, together with their resulting products, have prevailed on the market.

For this reason, the results of the eco-efficiency analysis flow into the development of product range strategies. Together with additional information from company divisions such as marketing, research, development and application, strategic orientations can then be defined. In particular, the analysis, extended with the social component, helps to put the debates between industry and social forces about the need for and sustainability of products on a factual and sound basis.

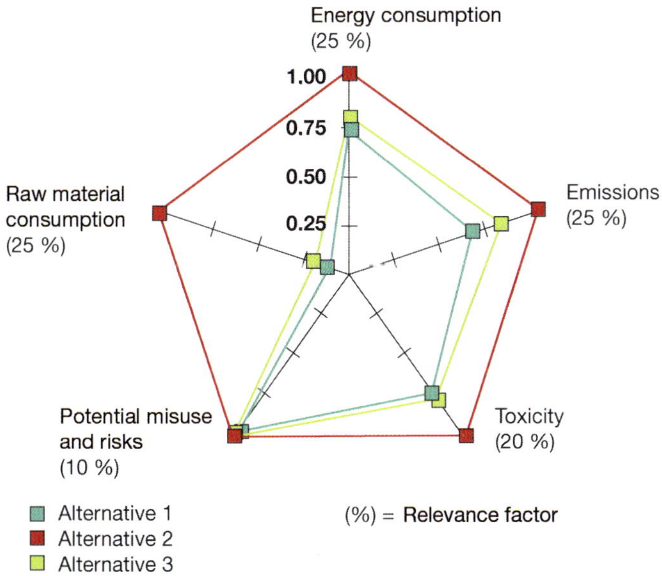

Figure 5.5.4: Ecological fingerprint

5.6 Environment-Compatible Paints and Coatings

The term "environmentally friendly", i.e. environment-compatible, has become established for certain types of paints and coatings in recent decades. In this context it should be noted that "environmentally friendly" can be defined in different ways and with different emphases. The most popular for paints and coatings places the primary emphasis on emissions into air, water and soil and the secondary emphasis on the toxicological component. The main spotlight is therefore on emissions, because the coating process generally requires a transfer agent, the solvent, which enables the paint or coating to be applied to the surface of the substrate. The reference point for evaluating environmental compatibility is conventional solvent-borne paint, which has a solid content of between 10 and 50 % and uses organic solvents as transfer agents.

From this point of view, the following paints and coatings are generally classified as "environmentally friendly" in popular usage:

1. Paints and coatings with a high solid content, termed "high solids"
2. Water-borne spray paints
3. Electrocoat paints
4. Powder coatings
5. Certain radiation-curable paints.

In recent years, the looming problems of crude oil availability and prices have led to a growing focus on coating systems that increasingly use biobased raw materials. In principle, these also belong to the group of environmentally friendly coating materials, but they are seldom referred to as "environmentally friendly paints".

A closer look at the low-solvent coating systems, which are commonly referred to as environmentally friendly, reveals that along the entire process chain from production, distribution to application there are further factors which testify to their environmental compatibility in the sense of resource conservation.

It is not possible to give relative rankings in product classes like water-borne coatings or powder, as environment-compatible coating systems cannot be universally used to meet all paint and coating requirements.

Indisputably, after powder coatings, electrocoatings and special radiation-curable coatings stand out as being particularly environment-compatible. The last of these have only limited application possibilities, because complex shaped parts cannot be uni-

Coatings	Solvent emissions	Transfer efficiency	Energy consumption	Manufacturing/transportation	Toxicology	Raw material consumption
High solids	+	o	+	+	–	o
Water-borne	++	+	–	–	+	o
Electrocoat	++	++	–	–	+	+
Powder	+++	++	+	+	++	–
Radiation-curable	++	o	+++	o	–	o

Figure 5.6.1: Qualitative evaluation of coatings as regards environmental compatibility (excl. bio-based) compared to conventional solvent-borne systems (o = comparable, + = better, - = worse)

formly cured by this technology. Moreover, not all formulations of such coating systems are solventfree. The low molecular, reactive compounds must also be subjected to careful toxicological assessment and selection.

A more precise comparative assessment of coating systems, taking into account the necessary processes, can be provided for specific applications by the eco-efficiency analysis described above (see Chapter 5.5.).

Conversions to environment-compatible coating systems almost invariably require plant modifications and extra investment in the coating lines. The most important aspects will be examined below.

5.6.1 Low-Emissions Coatings

5.6.1.1 High Solids

There is no unequivocal technical definition of high solid paints and views differ considerably between Europe and North America. In addition, it should be noted that the term "high solids" is relative, given that metallic base coats, for example, are already "high solid" coatings by virtue of their 30 % solid content.

In Europe, paints with a solid content > 50 % are generally classified as high solids, whereas in North America the requirement is > 55 %. One proposal for classification defines a coating with < 420 g/l VOC, > 60 % solid by weight or 80 % solid by volume as a high-solids coating [5.8.29].

Two formulation principles are used for these paints, which are mainly developed in North America:

1. Use film forming agents of low molar mass and high crosslink density
2. Choose solvents of high solution power

Upon delivery, high solid paints have the same level of viscosity as conventional solventborne paints. In other words, virtually no investment in the application line is necessary for a conversion to high solids. This is the decisive reason for the widespread use of this technology in North America.

The second principle mentioned above for the formulation of high solid paints has led to the increased use of ketones as solvents in North America. These solvents tend to be less popular in Europe where they are considered undesirable due to their relatively intense smell.

	Data/unit	Physically drying	Low solids	High solids
Molar mass	Mw	> 100,000	10,000	4,000
Functionality	Mol/g	0	300	600
As delivered	NV* [%]	25 %	55 %	75 %
	Viscosity [mPa s at 20 °C]	2,000	2,000	1,500

* *Solid content (2 h/130 °C)*

Figure 5.6.2: Typical technical data of resins for different coating systems

Environment

The formulation principles mentioned above offer very little scope, if any, for raising further the solid content. The only safe way to stay below the threshold of 3.0 lbs/gal (370 g/l) of solvents for liquid paints is to use water-borne or radiation curable coatings. Nevertheless, further development in North America has led to so-called "ultra-high solid" coatings, which when applied contain more than 80 % solid. In Europe, many paint manufacturers also offer systems containing < 250 g/l VOCs for various applications under this term.

The focus of further developments to enhance environmental compatibility is on the toxicology of the solvents and raw materials.

5.6.1.2 Water-Borne Paints

The term water-borne paints and coatings includes coating materials that are mainly formulated on the basis of dispersions. Here, a distinction is drawn between primary and secondary dispersions (see Chapter 2.1.1.2). The former are produced by means of emulsion polymerisation and are fine-particle dispersions of high molecular binders which can be air-dried, as in the case of decorative paints, or crosslinked as in the case of water-borne primer surfacers or water-borne basecoats in automotive finishing. Secondary dispersions consist of film forming agents of medium molecular weight of around 10,000 M_n. They are either self-crosslinking or need to be crosslinked with the aid of a reaction partner. They are manufactured in two production stages: resin synthesis in conventional solvents followed by dispersing in water. In some cases, the solvents are then removed by distillation at slightly elevated temperature and reduced pressures. The technical term for this process is "stripping" and it reduces the VOC levels considerably. In this way, almost all resin systems such as polyesters, epoxy resins [5.8.30], alkyd resins and most importantly acrylates [5.8.31] are accessible.

Stabilisation of both dispersion types may be effected by steric or electrostatic means, and the latter can be anionic or cationic (see Chapter 2.1.1.2). A typical example of the latter product class is the electrocoat, which was first introduced back in 1965 (see Chapter 4.2.2). Today's systems, mostly cathodic (see Chapter 7.1.2), are characterised by very low solvent contents, lead-free composition and almost entirely closed circuit processes that now generate virtually no wastewater or waste materials.

Only a few coating systems are based on film forming agents that are dissolved in water.

Type of dispersion	Particle size [nm]	Solid content [%]	VOC [g/l]	M_w
Secondary	100 – 400	20 – 40	50 – 250	< 10,000
Primary	1 – 200	25 – 45	0 – 100	> 100,000

Figure 5.6.3: Typical data for water-borne dispersions

	pH	NV (1)	% solid content/NV (2)	Pb	Biocides
AED 1970	8.5	10	55	Yes	Yes
CED 2002	6.0	18	5	No	No

(1) Bath
(2) As delivered

Figure 5.6.4: Typical bath parameters of today's (CED) and former (AED) electrodeposition coatings

Compared with organic solvents, water has a relatively low boiling point of 100 °C at 1013 hPa, but it has a very high evaporation energy of 2.26 J/g at the same temperature, a high evaporation value of 80 compared with diethyl ether, a high density of 1.00 g/ml at 4 °C, a high surface tension of 72.2 mN/m at 20 °C and a very low specific electrical resistance of 10 kΩcm. These properties have an influence on the formulation of paints for spray application and on the right choice of raw materials (see Chapter 2.1.1). A conversion from solvent-borne to water-borne coatings also incurs expenses and conversion costs due to the following physical and chemical properties of these coatings:

- Corrosiveness
- Evaporation characteristics of water
- Electrical resistance

The first aspect requires a change from plain steel to stainless steel for the walls, pumps and other devices that come into contact with the water-borne coating. If coating formulations are low in chlorine, one of the simpler stainless steel types such as 4.4301 is sufficient.

Specifically, for water-borne products, the quality of the coating materials used can be monitored via the pH or by determining the acid and base levels (see Chapter 2.3). The specific conductivity also serves as an indicator of constant quality.

In solvent-borne coatings, the important rheological behaviour during application and film formation is determined by the evaporation characteristics of the solvent. In this respect, these coating materials often exhibit Newtonian characteristics in their shear/viscosity function. This relationship must be different i.e. pseudoplastic or thixotropic for water-borne coatings, because water has a much lower evaporation rate (see Figure 5.6.5) [5.8.32]. Moreover, the viscosity curve as a function of the solid content features a "water mountain". This viscosity peak occurs at a solid content at which a water-in-oil dispersion

Environment

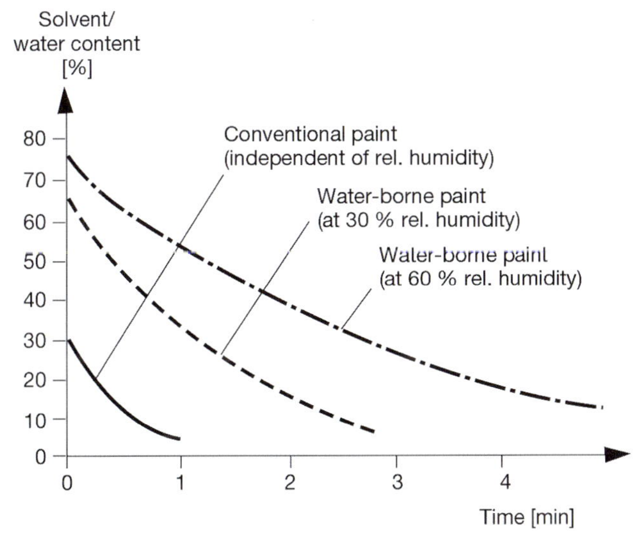

Figure 5.6.5: Typical evaporation behaviour of water at different relative humidity values and in comparison to conventional coatings

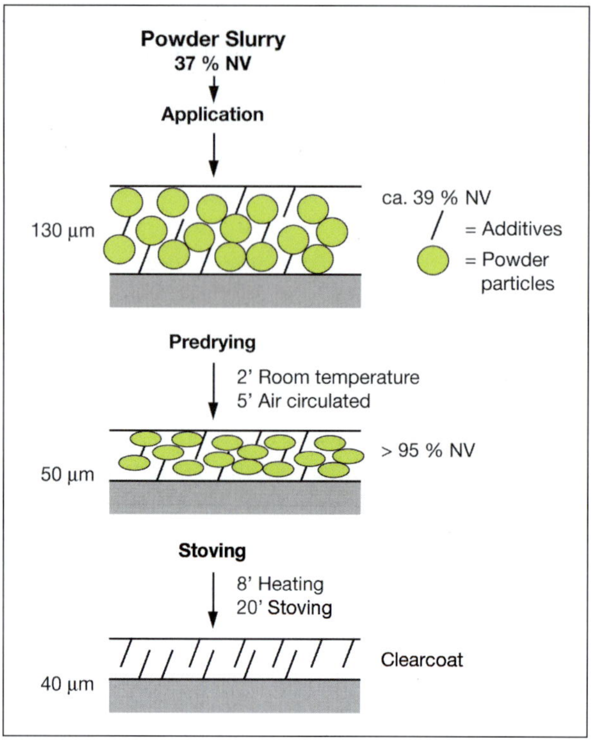

Figure 5.6.6: From slurry to clearcoat film

turns into an oil-in-water dispersion. As the addition of water does not generally effect any improvement in the application characteristics of technically demanding coatings, and the solid content plays a decisive role in the economics of the paint systems, water-borne paints are generally delivered ready for application. For this purpose, the formulation and composition are varied to imbue the coating materials with either pseudoplastic or thixotropic properties (see Chapter 2.3.3). Carefully adjusted rheology is key to storage properties and spray application. However, each type of application demands a specifically adjusted viscosity profile.

Consequently, quality control of the technical application properties with the aid of the viscosity profile needs to be specified more carefully and can generally no longer be based on measurement in the ISO flow-out cup (see Chapter 2.3).

Centrifugal pumps and similar pumps can only meter the material with sufficient precision if the viscosity obeys the principles of Newtonian fluids. Consequently, in many cases, volumetric or positive-displacement pumps are used for water-borne paints to meter constant quantities under non-Newtonian conditions. These pumps include gear pumps.

In the case of multilayer coating systems in industrial mass production, allowance must be made for the high evaporation energy and necessary evaporation time when flash-off zones between the individual application stages are being designed. These zones can often be shortened by recourse to IR radiation. The temperature profile in the stoving oven should be specifically designed for water-borne coating materials, because rapid heating can lead to pinholing in the film (see Chapter 6.2).

High electrical conductivity and frequent colour changes and associated product changes have a major impact on the supply of coating materials to electrostatic spraying apparatus. Insulation is also needed between the application device and the coating supply line, and the supply system must be properly earthed (see Chapter 4.2.1).

With organic solvent levels of < 5 % in water-borne paints, consideration needs to be given to the risk of bacterial growth (see Chapter 2.1.4).

One particular variant of water-borne spray paints is "powder slurry". This is a solvent-free dispersion of powder coatings in water; it combines the advantages of powder technology with those of a metering and application system for liquids. In addition, this technology enables ultrafine powder particles with a diameter of about 3 μm to be stably dispersed. This fineness is necessary for applications that require films just 35 to 45 μm thick and good levelling properties. This thickness is enough to produce the necessary surface and service properties and the required weathering resistance in automotive engineering. This coating system was used for clearcoats in automotive painting [5.8.34, 5.8.35].

5.6.1.3 Powder Coatings

Powder coating (see Chapters 2.2.4 and 4.2.2) is still a relatively new coating process. It was first used in the mid-1960s. In 2010, global demand for powder was 1.2 million tonnes.

The main focus of the application of powder coatings is in industrial painting and is attributed to the following segments worldwide:

Architecture:	20 %
Heating/radiators:	14 %
Metal furniture:	11 %
Car accessories:	12 %
White goods:	10 %
Machines/heavy vehicles:	13 %
Other:	20 %

Regional priorities are very different. For example, architecture is much bigger in Europe than in North America. In the saturated regions, annual growth in the powder market is around 1.5 %, but is much higher in RoW and Asia-Pacific.

In addition to the classic applications in the fields of domestic appliances and architecture, powder coatings are increasingly being used for add-ons and small parts in the automotive sector, such as wheels, axle and chassis components and also, over the last few years, automotive OEM finishes. Consequently, powder coatings are estimated to account for approximately 5 % of automotive primers and clearcoats today, predominantly in the USA.

Even though, at a rough estimate, it takes 1.3 times as much powder coating to coat the same area as a liquid coating (see Figure 5.7.1), the commercial use of powder coatings is constantly increasing. This fact can be explained by the following key characteristics of powder coatings:

1. No solvents are used
2. Legislation treats powder coatings essentially as non-hazardous material in respect of storage, transportation and waste
3. High material transfer efficiency > 98 % due to recycling
4. Economical application of high film build, where required
5. Competitive investment and operating costs
6. Low paint costs per NV

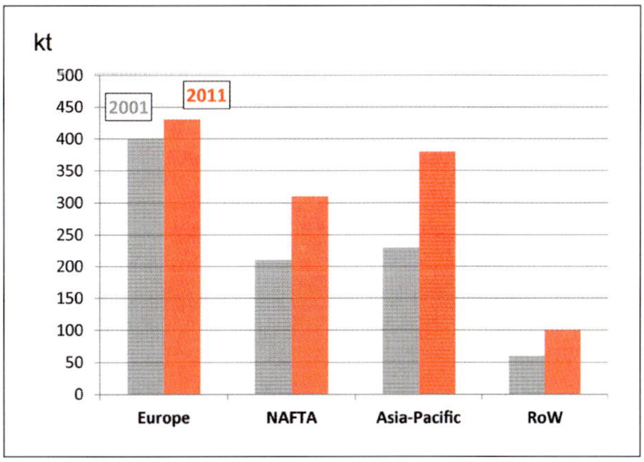

Figure 5.6.7: Growth in the global powder market

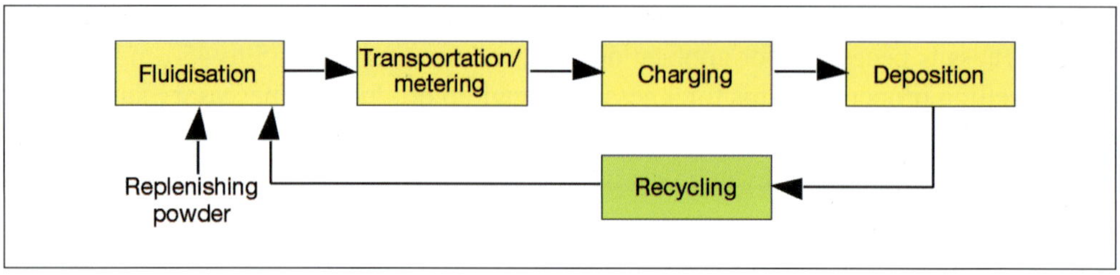

Figure 5.6.8: The process chain for powder application

When powder coatings are being handled, it is essential to bear in mind that the material properties are different from those of liquid coatings. As far as the technical aspects of powder coating systems are concerned, the application process can be broken down into a number of subprocesses that are defined by product and process parameters.

Key elements for mastering powder technology are the particle size and particle size distribution [5.8.36], the associated fluidisability of the powder, the metering system, the application equipment and material recovery (see Chapter 4.2.2) [5.8.37]. Meanwhile, its ecological appeal stems from the recovery and simple recycling of the overspray, combined with the absence of solvents.

The disadvantages are, first, the high film thickness of 70 to 100 µm required for good levelling, second, the still awkward and time consuming procedure for colour changes and, third, the high stoving temperatures (see Chapter 4.2.2). Developments aimed at radiation curable powder coatings (see Chapter 5.6.5) using UV and near-infrared (NIR) radiation suggest that its application is set to increase. The heat generated in the powder film by NIR radiation (see Chapter 4.3.1) effects melting and crosslinking within a few seconds [5.8.38]. Stoving temperatures of 140 °C are attainable [5.8.39].

Environment

An overall assessment based on a life cycle analysis or eco-efficiency analysis (see Chapter 5.5) shows that the clearcoat film thickness of > 60 µm necessary for attractive surfaces is unfavourable. The reason that film build is still high nowadays is the need for average particle sizes of 30 to 40 µm, as they are the only way to ensure safe handling. To reduce film build with a view to obtaining more attractive surfaces would require the use of finer powder particles, and that still raises problems of stabilisation and handling [5.8.40]. This problem has inspired numerous projects such as standing wave atomisation (see Chapter 4.2.2), powder production with supercritical CO_2 and powder slurry (see above) [5.8.41, 5.8.42].

5.6.1.4 Radiation Curable Paints and Coatings

Radiation curable paints and coatings (see Chapter 4.3.2), which have been in industrial use since the 1960s, gained their "environment-compatible" credentials mainly for two reasons (see Figure 5.6.1). First, they can be formulated for low emissions as 100 % non-volatiles (solid) products through the use of curable diluents or as aqueous products and, second, they conserve resources because coatings which are curable by UV radiation or electron beam require very little energy for film formation and crosslinking. Their range of uses was expanded when it proved possible to combine photoinitiators with the UV absorbers needed by clearcoats [5.8.43]. The application technology for UV coatings does not differ from that for conventional solvent-borne coatings. Also worth mentioning are electron-beam-curable coatings and UV-initiated cationic curing. Although these play a very small role in the overall radiation curing market, but the cationic-cure coatings do offer much better adhesion to metallic substrates.

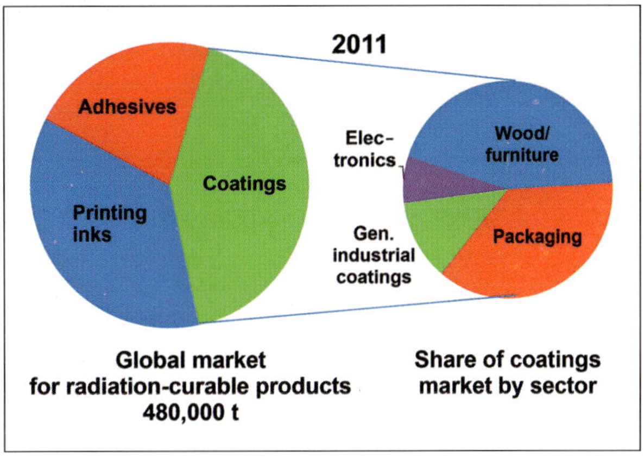

Figure 5.6.9: Global applications of radiation-curable coating materials and their share of the various coatings sectors

In 2001, radiation curable coatings had a global volume of around 350,000 tonnes. of which, 44 % was accounted for by coatings. The main application area is the wood and furniture industry [5.8.44] mainly with flooring coatings [5.6.45], followed by the packaging industry (see Figure 5.6.9). In industry, applications can be found in polycarbonate headlamp housings, automotive side and rear windows, as well as truck axles. Planters made of plastic are treated with UV-curable coatings to produce high-quality products [5.8.46]. Applications in the electronic field also include the coating and printing of CDs and DVDs, as well as coating of glassfibre. All product examples are characterised by simple geometries.

Electron beam curing (EBC) accounts for only a very small share of radiation-curable systems in the furniture industry due to the high safety requirements. An advantage of this technology is that the type and level of pigmentation exert a far lower influence on full-hardening.

Resins suitable for curing by UV radiation are unsaturated polyesters, still on a small scale, but these are losing share to acrylics and modified acrylics [5.8.47]. At 80 %, they now form the basis of radiation curable coatings. In particular, urethane acrylates are gaining in importance due to their balanced profile of film properties [5.8.48].

Pure or modified acrylic resins predominate among film forming agents, with a share of 75 %. Cationic curing is insignificant, with a share of 1 % by volume [5.8.49].

Radiation curable coatings consist of very low molecular film forming agents with a high proportion of polymerisable C-C double bonds, which are diluted in the liquid state with reactive monomers or solvents or, in the form of aqueous dispersions [5.8.50] are combined with photoinitiators and other components to produce a coating formulation for the intended application. The photoinitiators are activated by UV radiation and trigger a radical polymerisation process. This phenomenon occurs in fractions of a second under the usual energy input of approx. 1 J/cm^2 in the UVA radiation range. Pigmented systems of sufficient transparency require roughly double that amount of energy (see Chapter 4.3.2). Predrying to release the solvent or water at 70 to 90 °C is beneficial, the precise temperature depending on the formulation of the UV curable coating. The outcome is smooth, defect-free surfaces [5.8.51].

The most important classes of film forming agents used industrially are unsaturated polyesters and specially designed acrylic resins (see Chapter 2.1.1). The first generation of UV-curable wood coatings was limited to a coating line speed of 5 to 10 m/min

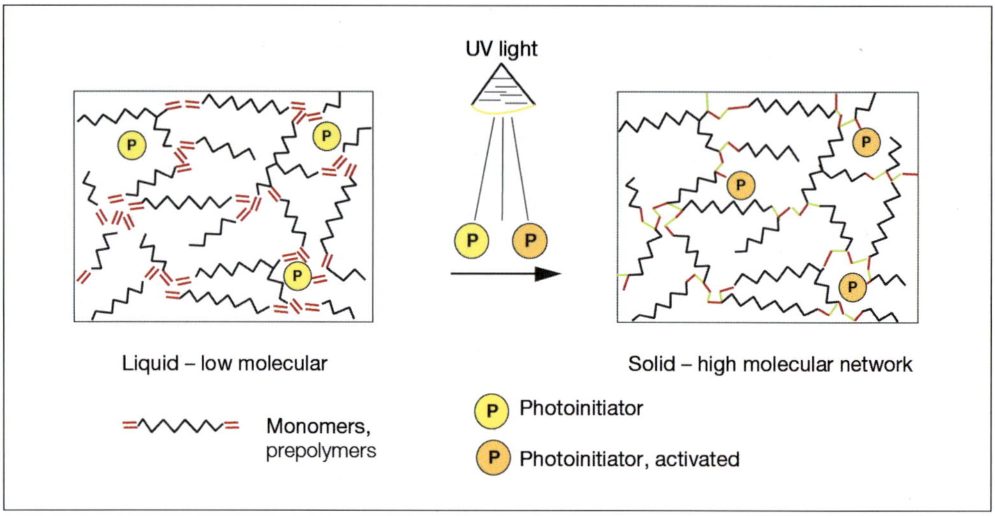

Figure 5.6.10: Principle behind UV curing

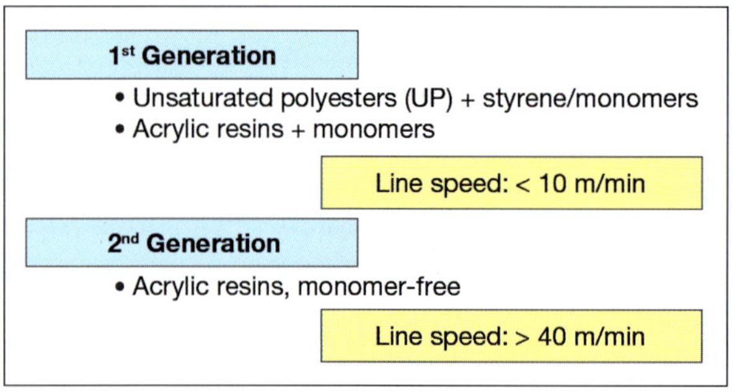

Figure 5.6.11: Generations of UV curable foil coatings for wood

due to the use of unsaturated polyester resins, and styrene as reactive diluent. Monomer-free acrylic resins offer greater reactivity and permit coating lines to operate up to 4 times faster. Even-higher production and coating speeds are attained in foil production, e.g. for the furniture industry in a roll-to-roll production process similar to that of coil coating. Full-hardening at a line speed of 140 m/min can only be guaranteed by UV-curable coatings [5.8.52].

In all cases, the coating lines are much shorter than in thermal curing. Energy consumption of UV curing is only about 10 % that of thermal curing, partly because the generation of UV radiation takes less energy and partly because the energy is only needed to crosslink the coating (see Chapter 4.3.3). The substrate does not heat up.

Although attractive in this respect, radiation curing plays only a secondary role in the overall paint and coatings market, with a share of less than 1 %. There are several reasons for this. First, the curing of intricately shaped parts is problematic and, second, not all pigmented coatings can be cured with UV radiation, as they do not have the necessary UV transparency. However, proposals are being made for a metallic paint finish [5.8.53]. Adhesion to metals is still not good enough, and so UV curable coatings do not attain the level of conventional primers. The coatings lack sufficient polar functional groups which are necessary for anchoring on the metal surface, i.e. essentially OH, NH, CO, CONH groups etc. are missing. However, due to their very high crosslinking density, especially as clearcoats, UV-curable coatings yield very hard coatings during film formation and have come into focus in response to increased demands on scratch resistance [5.8.54].

The coating manufacturers are currently developing so-called hybrid coatings, which are crosslinked by both radiation curing and thermal curing. These systems offer the advantages of high hardness and scratch resistance on the outside of the formed parts, with the inside receiving the requisite film properties from thermal curing, generally at 80 °C, via conventional crosslinking reactions. The high crosslink density of UV-cured hybrid systems is also utilised for coating plastic SMC parts, as a way of preventing recurrent outgassing from SMC products [5.8.55].

The disadvantages of the low penetration depth of UV light in pigmented coatings are overcome by curing pigmented coatings with electron beams (see Chapter 4.3.2).

Recently, UV powder technology has gained a certain position in the USA and Italy for coating MDF boards for the furniture industry The combination of powder coatings and UV curing is understandably very attractive due to its ecological and economic advantages. A major role here is also played by the shortness of the line and the throughput speed [5.8.56].

Developments in sensitisers and corresponding radical formers for NIR (near infrared) radiation have led to radical-curing products that penetrate to a much further depth than is the case for UV-curable coatings.

Further activities aimed at expanding the application potential of radiation-curable coatings are to be found in plant engineering [5.8.57]. The potential of UV curing by LED lamps is worth mentioning as they offer advantages in terms of environmental pollution and a more efficient wavelength spectrum compared to that of mercury lamps [5.8.58]. The combination of UV curing and coil coating is also an interesting development area [5.8.59].

5.6.1.5 Other Coating Systems

Coatings using supercritical CO_2 as solvent substitute have been on the market for a while [5.8.60]. CO_2 can be brought to the supercritical phase at temperatures above 31 °C and pressures above 7.8 MPa. It then has a specific gravity of 0.464 g/ml. As the physical parameters are close to those of conventional spray conditions in industrial applications, it has proved possible to develop coating systems that have a high viscosity and low solvent content and are brought to the necessary spray viscosity by means of a mixing system and supercritical CO_2. A 30 to 60 % reduction in solvent content can be attained by this method.

The cost of the equipment and excessively rapid drying of the applied film have for the most part produced unsatisfactory coating results, so that this technology is now only used in niche applications.

In principle, the coil coating process too can be classified as environment-compatible. The paint composition is of secondary importance in this connection. The application process has an efficiency of more than 95 % due to the high solid content of the coatings and the low material loss during the rolling process. The combination of the rolling process and the short distance between application and stoving oven means that, for a low investment, solvent emissions can be almost entirely shifted to the stoving oven, and thus can be almost entirely eliminated by means of thermal treatment (see Chapter 7.4).

5.6.2 Biobased Coating Systems

In recent years, the availability and price of crude oil have caused the focus to shift in favour of coating systems that are increasingly derived from biobased raw materials, i.e. from renewable sources. This is nothing new as far as coating materials are concerned because as described in Chapter 2.1.1.1, alkyd resins are formulated from natural fatty acids, colophony is still an important resin for printing inks, as well as cellulose and natural raw material based polyester building blocks used in various coatings. According to our own estimates, some 5 to 10 % of these raw materials are used in the global coatings market.

New approaches are going further and using fatty acids and oils as building blocks for polyols. Such polyols made from soy oil and fatty acid dimers are used for epoxy crosslinking [5.8.61], isocyanate crosslinking [5.8.62] and also for water-borne polyurethane dispersions [5.8.63]. In most cases, the corresponding "bio" polyol serves as a substitute for the previously used petroleum-based polyol. However, some raw materials for resin synthesis are obtained from biomass. For example, substitutes for pentaerythritol or dipentaerythritol (see Chapter 2.1) have been synthesised which are obtained from starch [5.8.64]. Itaconic acid obtained from sugar acts as a less toxic building block than methacrylate for UV-curable polyurethane acrylates and can be used as a polyurethane itaconate [5.8.65].

The technical objective in all cases is to retain the properties of the synthetic, quality-determining coating materials and paints produced from crude oil. Coating systems predominantly made with biobased raw materials currently fail to deliver the requisite quality profile, especially as regards the fast curing needed for industrial coating processes. For this reason, they have not yet found widespread use in industrial coating technology.

Additives too are being formulated with biobased raw materials.

Projects and studies are underway to expand the conventional sources of raw materials for fats and oils. Experiments are being conducted on algae, which are capable of yielding oils and fatty acids, as well as polysaccharides and basic substances for acrylic resins [5.8.66].

The entire value-added chain in the chemical industry, which has so far been based on crude oil, is also being examined for ease of conversion to renewable raw materials. Of huge importance in this regard are the availability and cultivation of the bio-raw materials, as it is still unclear how they can be cultivated in a sustainable and ecologically defensible manner [5.8.67]. Furthermore, the question as to whether land should be given over to food production or industrial is problematic, as is highlighted by the debate surrounding "food versus fuel". There is no doubt that the area under cultivation is increasing worldwide, with a clear focus currently on the USA, Brazil, Argentina and Canada. Soybeans and corn make up almost 80 % of this cultivated area. While there are still no signs of a clear strategy for a biobased raw materials base in the chemicals industry, proposals have been made [5.8.68]. The major chemical companies prefer cellulose to avoid the already mentioned discussion about „food versus fuel“ in view of the growing world population.

Biobased raw materials can contribute to sustainability if carefully selected, as has already been found in simple life cycle assessments [5.8.69]. Current evaluations are usually carried out via a carbon footprint analysis or eco-efficiency analysis (see Chapter 5.5) [5.8.28].

Figure 5.6.12: Manual foil coating of vehicles

5.6.3 Foil Coating

Foils have long played a certain role in the lettering of painted surfaces, especially on automobiles [5.8.70]. The aim here is to manually apply a coloured film, which has been either pigmented, coated or, as in the wood sector (see Chapter 7.10), created with printed finish foils, primarily to smooth or slightly curved surfaces. The foils provide temporary advertising or give the vehicle a new colour, but they must also be easy to remove again. To this end, solid-colour foils or white foils are printed with advertising messages or to other specifications.

The manual process is difficult to automate. For this reason, this type of foil coating is still a niche area and is barely used at all in industrial coating. Nevertheless, it is increasingly superseding conventional painted lettering and full-body repainting of vehicles [5.8.71]. Prime examples of foil wrapped vehicles are taxis and police cars, which are easier to sell in their original colour after the foil has been removed.

Since the mid-1990s, the potential of in-mould coatings has been explored extensively for use with plastic surfaces [5.8.72]. In addition to reel to reel film coating, which is very environmentally friendly, in-mould coating makes sense in the context of SMC (sheet moulding compound) plastics, the coating of which is not always free of defects due to gas emissions in cavities (see Chapter 7.3). For add-on parts, e.g. in vehicle production, acceptance of the technology is contingent on the attainable colour match with the original finish. UV-curable clearcoats have been mentioned in connection with proposed methods [5.8.73]. Up to now, technical implementation has been restricted to pilot plants and small runs.

5.7 Economics of Coating Processes

As described in the discussion of the eco-efficiency analysis (see Chapter 5.5), a company wishing to install a coating line first has to calculate the economics of the coating process before it can take the final decision on an investment. The economic aspects include investment costs, amortization, running costs for the coating line and the costs of the coating materials.

The running costs in turn consist of a multitude of cost blocks that are highly specific to the application process and the coating system used. These cost blocks include the disposal or utilisation of overspray, the cost of the air supply, the energy required for application and stoving or crosslinking, the cleaning of the suspended apparatus and of the coating line and all personnel costs for the paint line operators.

To make it easier to compare the coating cost factor, i.e. the cost of the coating materials, it has become common practice to employ the standard unit "costs/m^2 of coated surface area". This factor needs to be calculated carefully but it is not always easy to determine for every coating process.

In addition to the price per kg of delivered coating material, other factors enter into consideration, such as the content of nonvolatile components in the coating, the transfer efficiency of application, the film thickness and the specific gravity of the film.

This comparative calculation can be illustrated by the example of coating 100 m^2 of metal workpieces by three different methods – electrocoating, liquid spray coating and powder coating.

The equations used for calculating the economics (7) as shown in Figure 5.7.1 are as follows:

$$\frac{(1) \cdot (2)}{10} = (3)$$

$$\frac{(3) \cdot 100}{(4)} \cdot \frac{(5) \cdot 100}{(6)} = (7) = \text{Coating costs/100 m}^2$$

This simple approach aids decisions in favour of particular coating lines. However, it must always be remembered that (7) only compares the coating costs. It does not take

Type of coatings	Film thickness [µm]	Specific gravity of solids [g/cm^3]	Solids needed [kg]	Transfer efficiency	Price[1] [Euro/kg]	NV of replenished paint [%]	Costs for coating [Euro/100 m^2]
	(1)	(2)	(3)	(4)	(5)	(6)	(7)
Powder	80	1.60	12.8	98 %	3.00	100	39.2
Liquid	30	1.50	4.5	80 %	3.80	50	42.8
CED	22	1.45	3.2	95 %	4.00	42	32.1

[1] *about as delivered today*

Figure 5.7.1: Economics of three different coating systems in EUR/100 m^2

into account ecological aspects, capital costs or running costs (see Chapter 5.5). Moreover, clear agreements must be reached between supplier and customer on the determination of the solid content in order that material losses in the drying and curing process may be kept comparable between the application and the lab.

A widely used parameter for comparing coating costs is the theoretical **yield E** of a coating material in m^2/kg. The yield is defined as the quotient of the paintable surface area A and the necessary mass of coating material m_{Pa} required to coat this area:

$$E = \frac{A}{m_{Pa}} = \frac{NV}{\rho_T \cdot S_D \cdot 100}$$

m_{Pa} is calculated in kg from the film thickness S_D , the density of the dry film ρ_T (see Chapter 3.2.1) and the nonvolatile content of the coating material NV, using the following equation:

$$m_{Pa} = \frac{(\rho_T \cdot S_D \cdot A \cdot 100)}{NV}$$

Consequently, coatings are adjudged to have a particularly high yield when a high nonvolatile concentration (NV) is accompanied by a low dry film density and a low covering film thickness, i.e. a high hiding power (see Chapter 2.3.3). Here, it is assumed that the protective effect and function of the coating depend not on its mass but essentially on its volume or film thickness. This is not always the case.

Environment

The mathematical relationship for calculating the yield can also be used to estimate the solvent emission quantities L. Replacing m_{Pa} by $m_{So} + m_{dr}$, i.e. the solvent content and the proportion of dry film residue, and substituting this sum into the yield equation:

$$\frac{A}{(m_{So} + m_{dr})} = \frac{NV}{(\rho_T \cdot S_D \cdot 100)}$$

Replacing m_{dr} by $\rho_T \cdot S_D \cdot A$ and solving the equation for m_{So} (solvent emission L) yields:

$$L = \rho_T \cdot S_D \cdot A \cdot \left(\frac{100}{NV} - 1\right)$$

Therefore, if the film thickness to be applied S_D, the nonvolatile concentration of the coating material NV and the dry film density p_T are known, the solvent emissions can be estimated for coating a given area A. The result is only an estimate, in that this equation only holds to a degree for manual coating processes such as application by brush and roller, because it fails to take into account the transfer efficiency and application conditions in industrial processes and plants [5.8.74].

5.8 Literature

References

5.8.1 D. Leroy, ECJ 04/2006, p 16
5.8.2 M. Bross, JOT 47(/), p 8 (2007); A. Rommert, F+L 119 (9), p 50 (2013)
5.8.3 K.-F. Dössel, The World of Changing Regulations for Coating Materials and Paint Shops, Proceedings Surcar, Cannes, 2013; M. Cleuvers, Nachr. Chemie 62 (1), p 53 (2014)
5.8.4 M. Letzner, Verfahrenstechnik 05/2013, p 38
5.8.5 M. Riester, JOT 45 (4), p 8 (2005)
5.8.6 H. J. Drexler, J. Snell, ECJ 04/2002, p 24
5.8.7 H. Schrübbers, JOT 48 (2), p 30 (2008)
5.8.8 Dokumente zu Lacken und Farben 9 (pub.M. Bross), Deutsches Lackinstitut, Frankfurt
5.8.9 P. Svejda, JOT 50 (9), p 28 (2010)
5.8.10 S. Seibert, Welt d. Farben 9/2005, p 18
5.8.11 K. Wittstock, Dissertation, Universität Karlsruhe, 1995
5.8.12 D. Lippok, I-Lack 66 (11), p 624 (1998)
5.8.13 T. Schneider, JOT 38 (4), p 112 (1998)
5.8.14 D. Kargol, JOT 41 (1), p 38 (2001)
5.8.15 R. Möhlenbrock, JOT 41 (4), p 70 (2001)
5.8.16 P. Brodt, I-Lack 63 (11), p 388 (1995)
5.8.17 G.. Scheffels, JOT 50 (9), p 18 (2010)
5.8.18 W. Heine, JOT 41 (11), p 14 (2001); Anonymous, JOT 49 (3), p 10 (2009)
5.8.19 G. Leininger, MO 48, p 644 (1994)
5.8.20 A. Scott, Chem. Week 169 (11), p 17 (2007)
5.8.21 Anonymous, JOT 46 (10), p 96 (2006)
5.8.22 D. Leroy, ECJ 04/2006, p 16
5.8.23 C. Challener, JCT Coat. Tech 8 (10), p 32 (2011)
5.8.24 G. Kirchner, C. Kraft, ECJ 07–08/2011, p 22
5.8.25 D. Sykes, ECJ 02/2011, p 40
5.8.26 S. Papasavva et al., ECJ 01/2001, p 92; A. Koller, ECJ 04/2009, p 122
5.8.27 D. Piwowarczyk, MO 55 (2), p 26 (2001)
5.8.28 P. Saling, Int. J. Life Cycle Assessm. 10 (5), p 364 (2005)
5.8.29 M. Manea, High Solid Binders, p 118, Vincentz, Hanover, 2008
5.8.30 M. Tsang, M. Gelsberger, R. Grasböck, JCT Coat. Tech 7 (2), p 28 (2010)
5.8.31 R. Gertzmann, T. Irle, JOT 48(7), p38 (2008); M. Almato, M. Melchiors, E. Tejada, ECJ 07–08/2010, p 38; O. Groß, F+L 118 (2), p 32 (2012)
5.8.32 R. D. Hester, D. R. Squire, Jr., JCT 69 (864), p 109 (1997)
5.8.33 O. Sack, T. Mezger, Welt d. Farb. 3/2012, p 8
5.8.34 J. Woltering, Es muss nicht immer flüssig sein! – Vom Pulverlack bis zum Trockenmörtel, Vol. 2, Schriftenreihe VILF-Vorträge, November 1999
5.8.35 J. Woltering, JOT 38 (9), p 14 (1998)
5.8.36 D. Pugh, P. Kippax, S. Ward-Smith, PCI 23 (4), p 44 (2007)
5.8.37 R. Freeman, J. Cooke, PCI 23 (4), p 60 (2007)
5.8.38 F. Zimmermann, MO 55 (1), p 41 (2001)
5.8.39 Anonymous, JOT 48 (11), p 32 (2008)
5.8.40 R. Deane, F+L 107 (7), p 46 (2001)
5.8.41 E. Weidner et al., Chem. Eng.Techn. 24 (5), p 533 (2001)
5.8.42 J. Woltering, W. Kreis, H.-J.Streitberger, Powderclearcoat for Conventional Application Equipment, Proceedings XVII Surcar, 1997
5.8.43 A. Valet, C. Decker, Modern Paint & Coatings 91 (5), p 29 (2001)
5.8.44 K. Menzel, M. Biehler, M. Lokai, Pitt.e Vernici 2–3/2006, p26
5.8.45 J. A. Arcenaux, S. Smeets, J. C. Smith, G. Vanmeulder, JCT Coat. Tech. 10 (3), p 62 (2013)
5.8.46 Anonymous, MO 66 (10), p 16 (2012)
5.8.47 V. Shukla, M. Bajpai, D. K. Singh, M. Singh, R. Shukla, Pigm. & Res. Tech. 33 (5), p 272 (2004)
5.8.48 E. Spyrou, F+L 116 (11), p 20 (2010)
5.8.49 B. Balmer, ECJ 10/2001, p 16
5.8.50 J. Weikard, E. Lühmann, S. Sommer, ECJ 05/2008, p 36
5.8.51 R. Stephen, PPCJ 189 (11), p 10 (1999)
5.8.52 T. Guenther, J. Rad. Curing 17, p 4 (1990)
5.8.53 O. Kipfmüller, F+L 119 (2), p 22 (2013)

Environment

5.8.54 F. Köhn, MO 61 (3), p 12 (2007)
5.8.55 W. Kreis, MO 56 (3), p 38 (2002)
5.8.56 Anonymous, Coat. World 16 (5), p 44 (2011)
5.8.57 H. Friedrich, Welt d. Farben 5/2006, p 8
5.8.58 S. Jennings, PPCJ 202 (4571), p 20 (2012)
5.8.59 D. Skinner, ECJ 10/2009, p 18
5.8.60 J. N. Hay, K. Johns, JOCCA 83 (3), p 106 (2000)
5.8.61 E. Honcoop, W. McNamee, PCI 26 (11), p 24 (2010)
5.8.62 J. M. Shukla, M. R. Patel, K. H. Patel, N. K. Patel, ECJ 11/2006, p 38; G. Bourscheidt, T. Roloff, Welt d. Farben 4/2007, p 10
5.8.63 V. Mannari, S. Rengasamy, ECJ 03/2013, p 80
5.8.64 M. Tober, T. Biemans, J. Thiem, F+L 118 (9), p 16 (2012)
5.8.65 T. Biemann, K. Pohl, W. Wagner, S. Friebel, Neuartige zuckerbasierte UV-vernetzende Polyurethandispersionen und UV-vernetzende UV-Harze, Proceedings 78. Lacktagung, Schwäbisch Gmünd, 2013
5.8.66 J. van Haveren, R. Blaauw, D. van Es, J. Sanders, F+L 117 (7), p 18 (2011)
5.8.67 C. Janik, VAA Magazin 12/2012, p 6
5.8.68 B. Kamm, M. Gerhardt, S. Leiß, Nachr. Chemie 57 (12), p. 1182 (2009)
5.8.69 D. Kralisch, G. Kreisel, A. Dielmann, D. Preuße, W. Karl, F+L 110 (3), p 34 (2004); H. Kastien, W. Selter F+L 109 (1), p 75 (2003)
5.8.70 Anonymous, Lackiererblatt 12 (1), p 8 (2003)
5.8.71 D. Vaughan-Lee, PPCJ 194 (4478), p 24 (2004)
5.8.72 P. Bannwitz, H. Gleich, JOT 44 (9), p 40 (2004); G. Scheffels, JOT 47 (9), p 38 (2007)
5.8.73 D. Zulkarnian, U. Schlotterbeck, H. Gleich, Kunststoffe 95 (3), p 1213 (2005)
5.8.74 M. Scheithauer, I-Lack 66 (6), p 325 (1998)

General literature

K. Beritt, Umgang mit Gefahrstoffen, Ecomed Verlag, Landsberg, 4th edn 2002
E. Spyrou, Powder Coatings – Chemistry and Technology (3rd ed.), Vincentz, Hanover, 2012
R. Schwalm, UV-Coatings, Elsevier, Oxford, 2007
P. Glöckner, T. Jung, S. Struck, K. Studer, Radiation Curing, Vincentz, Hanover, 2008
G. Jonschker, Sol-Gel-Technology in Praxis, Vincentz, Hanover, 2014
B. Meuthen, A.-S. Jandel, Coil Coating (3rd ed.), Vieweg, Wiesbaden, 2013
T. Feist, D. Lehmann, Arbeits- und Aufgabenheft der Lackiertechnik, Vincentz, Hanover, 2010
Kittel – Lehrbuch der Lacke und Beschichtungen, Vol 3 (2nd edn.), Bindemittel für wasserverdünnbare Systeme (Hrsgb. K. D. Ledwoch), Hirzel, Stuttgart, 2003
Kittel – Lehrbuch der Lacke und Beschichtungen, Vol 8 (2nd edn.), Herstellung von Lacken und Beschichtungsstoffe, Arbeitssicherheit, Umweltschutz (pub. W. Ritter, W. Zöllner), Hirzel, Stuttgart, 2004
W. Jäger, Arbeitsschutzlexikon, Ecomed Verlag, Landsberg, 18th edn. 2001
M. Manea, High Solid Binders, Vincentz, Hanover, 2008
C. S. Chern, Principles and Application of Emulsion Polymerization, Wiley, New York, 2008
D. Urban, K. Takamura (pub.), Polymer Dispersions and Their Industrial Applications, Wiley-VCH Verlagsges., Weinheim, 2002
Die VOC-Richtlinie in Deutschland, Dokumente zu Lacken und Farben 8, Deutsches Lackinstitut, Frankfurt, 2002
Die Decopaint-Richtlinie in Deutschland, Dokumente zu Lacken und Farben 9, Deutsches Lackinstitut, Frankfurt, 2007
Ausgehverbot! Aktuelles zur Chemie der Farben und Lacke, HighChem hautnah, GDCh, Frankfurt, 2008
Handlungshilfe zur Minimierung der Lösemittelbelastung bei der Lackherstellung, VdL, Deutsches Lackinstitut, 2012

6 Principles of Quality Management

6.1 Evolution of Quality Concepts over Time

Industrial coatings are produced in large quantities and at high speeds, and generally to a high level of quality. As it only takes small quantities of paint to enhance the decorative and functional features of products, particular attention must be paid to ensuring a consistent level of quality that satisfies all the various requirements.

One significant aspect that distinguishes the paint and coatings industry from many other sectors is that it works with semi-finished products which are produced by the paints and coatings manufacturer with all due regard for quality but which are then processed and applied in a different location by the user (see Chapter 1). As a result, the conditions under which the manufacturer performs quality control measures are different from those of the customer. In addition, due to the highly technical nature of the often mass-processes, the raw materials used in paints and coatings are subject in some cases to variations, so that the manufacture of paints and coatings is not always a question of producing identical mixtures but of finding compositions that have identical application characteristics while maintaining coating properties which are identical with those of previous batches. Although measurements of key physical properties such as density, viscosity, surface tension and solid content are certainly important factors in testing and control, they are not always sufficient to guarantee quality. This fact is one of the reasons why the simulation of the practical stresses caused by bending, scribing, chipping or scratching and the visual assessment of test coatings are often adopted as additional criteria for evaluating quality (see Chapter 3.3).

In view of this situation, close contact and a sharing of ideas and experience between the supplier and the application site are clearly essential if the coating material is to be given the best foundations conducive to optimal quality before it takes shape in the finished coating.

Given that many paint shops procure their materials from different suppliers whilst insisting that the different products must be interchangeable in the application systems, it is also clear that not accurate agreements on the performance of tests and inspections will nor be sufficient. Consequently, standardisation bodies have been set up all around the world whose standards form the basis for agreements on the preparation of test samples and the performance and evaluation of tests (see Chapter 8).

In the first stage of this evolving process, the originally purely in-house "working standards" were supplemented – but unfortunately not fully replaced – by national, neutral test specifications. Then, progressive globalisation necessitated agreements that transcended national borders, with the result that nowadays test methods are well on the way to becoming harmonised at global level. ASTM, DIN, EN and ISO standards are increasingly becoming established and are gradually superseding working standards and national regulations. One example of efforts to harmonise test methods is the "VDA-Blätter 621", which are documents issued by the German Association of the Automotive Industry (VDA) that provide a snapshot of the test methods currently practised in the automotive industry [6.6.1].

Whereas standards simply contain methods for determining quality, a required minimum quality cannot be defined and met without agreement having been reached on the level of quality itself. Technical delivery conditions and/or specifications concerning the supply of paints and coatings therefore stipulate not only the test methods for quantifying quality but also the tolerances and the permissible measurement uncertainty (see Chapter 2.3.1).

Nevertheless, forecasts about the quality of the end product are beset with imponderables. One of the peculiarities of the industrial coatings sector is the fact that whilst considerable investment and work is needed on the part of the coating manufacturer to transform the coating materials into a coating that will suit the conditions of the user, after delivery to the customer the coating is not always applied under identical conditions.

This state of affairs has led paints and coatings manufacturers and their customers' application departments to engage more closely with each other and to view the creation of quality as a common goal. Quality concepts which have been developed over the last few years go far beyond mere product testing of the material and final coating. They include services such as safety, adherence to lead times, clear agreements on quality, and openness on the sharing of information. However, if quality is to be guaranteed, there is a need not only for reliable production systems, modern test equipment and controllable services but also for all company employees to adopt a corresponding mental attitude.

Quality management systems (QMS) must therefore instil an awareness of quality into the minds of all employees to pave the way for the introduction of defect prevention concepts so that faults are no longer only detected after production and corrected after the event. Total Quality Management (TQM) is one such concept [6.6.2]. To ensure that quality management systems or concepts are practicable and above all measurable in terms of their effectiveness, the principles and practical steps that lead towards TQM have been laid down in standards.

For example, the ISO 9000 ff. series of standards describe in detail the procedures for preventing defects (nonconforming products) and for ensuring reliable and punctual production and delivery. A quality standard specific to the automotive industry is ISO TSC 16949-2009.

The emphasis of testing and inspection is moving away from an exclusive focus on materials testing and towards process monitoring, with the simultaneous introduction of control loops. The target of zero defects in the production of coatings has still not been attained, however. Even in large paint shops such as those of the automotive industry, which pioneered quality and quality awareness, paint defects have still not been totally eliminated. However, thanks to intensive collaboration between the paints and coatings industry and its customers to control all process stages, including the quality strategy, clear progress has been made in both the quality level and the reproducibility of coatings.

The huge strides taken over the last few decades regarding the ability to analyse materials and coatings for defects has contributed significantly to this progress (see Chapter 2.3.2 and 3.2). The next section will therefore present a selection of the most common coating defects and how they may be identified.

6.2 Defects in Coating Processes and Applied Coatings

In the overall chain of production, sale and application of paints and coatings, a whole series of parameters, starting with the substrate or workpiece [6.6.3], moving on to the pretreatment process and extending to the paint properties and finally the application and curing conditions, have to be observed if the coating result is to satisfy the specifications. Nonconformances in one or more parameters can lead to a defective coating (see Figure 1.1.7).

These defects can include, e.g. a pronounced orange peel surface texture or runs and sags on vertical surfaces, inadequate stone-chip protection, a colour deviating from the approved sample, and localised or systematic surface defects of any kind. Expert application of carefully tested and approved coating materials to specified substrates almost invariably leads to defect-free coatings for the relevant purpose. On this basis, many nonconformances are remedied by careful analysis of the application parameters and inspection of the coating batches. Since the coating product and the application process are highly interdependent, it is necessary to decide whether optimisation of the application system or of the coating composition is the most economical, rapid and reliable starting point for solving a particular problem.

One example can be seen in the problem of runs. If there is enough scope within the key film properties to reduce the film thickness, optimising the application process is the simplest and most economical remedy. If not, more reliable sagging resistance must be provided by modifying the formulation of the paint.

For many years, the focus of optimising the reliability of coating processes has been on localised or systematic surface defects in industrial coatings. These defects seldom lead to any diminishment in terms of utility or "adequacy", but they considerably detract from the consumer's subjective appreciation of the product's value. Experience of attempts to classify the causes of these defects shows that the underlying causes are often found to be external disturbance factors in the substrates, their cleanness and pretreatment, as well as in the coating material and coating process [6.6.4].

Substrate and pretreatment play an important role in wetting and adhesion. Nonconformances and processes deviating from standard can lead to cosmetic defects in the form of wetting problems and even, in the case of inadequate adhesion, to serious deficiencies in anticorrosion protection. For all substances, it is essential to have complete mastery over the type of pretreatment and its practical implementation.

Modern environment-compatible paint and coating systems require greater care than their more robustly formulated solvent-borne counterparts, because being aqueous or powder products they are to a certain extent in a metastable state and are highly liable under certain conditions to suddenly revert to a thermodynamically more favourable state. This phenomenon can manifest itself as agglomeration, phase separation and the formation of micelles in individual coating components. The application of such nonhomogeneous coatings leads to a variety of surface defects, such as roughness and craters, and can also adversely affect the film properties.

In addition to these inherent sources of defect in the coating itself, a whole range of external factors also affect the outcome of the coating process. The most important problem in this respect will be familiar to most painters: it is dust, which is deposited, for example, during the slow drying of air-dried paint surfaces and may become trapped. Dust is often the main problem in industrial coating processes (see Chapter 7.1.6). However, not only dust but also auxiliary substances in the paint shops and the workpiece can give rise to visible defects in the paint systems. Particular mention should be made here to the widely encountered silicone-based oils and sealing compounds, and perfluoro compounds. These materials are characterised by very low surface tension values of < 20 mN/m. Due to their incompatibility, they cause surface defects in almost all coatings, especially water-borne and high solid types. In addition, the speed of the coating processes can create problems, e.g. due to insufficient evaporation of solvents, including water, in multiple layer paint systems. Correct calibration of spraying equipment is also necessary for achieving defect-free, smooth films (see Figure 1.1.7). The human eye is capable of discerning regular surface textures down to a depth of 0.5 µm.

Regardless of the cause, if the quality requirements imposed on these highly technical coating lines are very strict, any visible defects can incur enormous repair costs.

Where measurable defects occur in the coating process, rapid remedial action is needed. It is essential here to deploy the right analytical methods and to utilise the experience of application engineers to find the cause of the defect and remedy it by means of a targeted approach. In this context, four action parameters can be distinguished:

1. Careful analysis of the surface defect
2. Immediate recording of the defect statistics by time and location
3. Checking of all application parameters
4. Analysis of the quality data for the coating materials

To an extent depending on the severity of the incident, it may be expedient to adopt all four approaches simultaneously. The interplay of the human element, material properties and application parameters does not always lead to clear and rapid identification of the cause of a surface defect simply from its nature. What is needed is the ability of all involved to apply the methods employed by detectives, especially for multilayer paint systems, because localising the defect is a decisive step on the way to detecting the cause.

The chapters that follow describe the possibilities afforded by surface analysis and the form of the defects known to be caused by specific errors during application or the coating product itself.

6.2.1 Identifying Defects and their Causes

As already described, defects and failures in the coating of consumer goods can be divided into three classes, namely visible defects due to localised surface defects, visible defects due to large surface defects, and defects that impair the performance or functionality of the coating.

These classes of defect can occur both separately and together in the various coating processes.

Visible, localised surface defects are generally classified by their appearance. Thus people speak of runs, craters, solvent boils, blisters, pinholing, dirt and similar terms, which in many cases fail to provide a clear indication of the cause. For example, without closer inspection under a microscope, an apparent crater may in fact be a large solvent boil. Moreover, this terminology is not clearly defined, and that makes communication even between experts difficult. In contrast contamination of the coated surface by external dirt can generally be identified by analysing the defect under the microscope [6.6.5].

These kinds of surface defect can occur both sporadically and regularly in industrial coating processes. If they occur sporadically, the cost of repair is lower, but the time taken to locate the cause and remedy the problem is greater than if they happen regularly. Rapid action is imperative here, especially in the case of automotive coatings, can coating and coil coating.

The first step in remedying defects should always be the logging and visual classification of the surface defect at the production line, to serve as a reference base for cause analysis. This step includes a description of the defect, to the extent that it can be detected with the naked eye, localisation on the workpiece and a record of the chronological sequence. If the outcome of the visual assessment is unclear, a magnifying glass or portable microscope should be used on the production line to permit more extensive description of the defects and possible identification of the causes. If this analysis fails to yield a clear classification of the defect, the damaged area must be prepared for further investigation.

Here, in line with the substrate, three different processes may be used:

1. More detailed visual analysis of the imperfection, with the aid of a scalpel for preparation purposes and a stereomicroscope. The defect is cut into a wedge shape by creating two oblique scalpel cuts under the stereomicroscope, so that the defect is visible in the cross-section of the paint system. In multi-layer coatings, there is a high probability that this method will pinpoint the source of the defect in the affected layers. Once this technique has been mastered, this approach can be applied to all substrates.
2. Microtomy is a method of cutting thin sections off the coat in slices approx. 1 µm thick. These specimen slices can be examined under all types of microscope, in both reflected and transmitted light. In transmitted light, different contrast methods can be used to accentuate the effects. This method is reliable for plastic substrates.
3. In the paints and coatings industry, the most common method used is the cross-section preparation technique. Here, the defective point to be examined is cut out of the object and embedded in light curable or other curable plastics. This cast is then ground down to the damaged area, and a plan-view photograph of the defect point is taken under the microscope.

Defect analysis by one of these methods in combination with visual inspection usually enables engineers to classify the defect into dust contamination, craters, boils or pinholes, blisters and other defects. This classification gives pointers, but it is not sufficient to identify the cause or source of the defect.

Another degree of difficulty is added wherever multiple layer paint systems are examined, because successful fault remediation is contingent on knowing which layer the cause is located in.

Microscopy is usually highly adept at identifying defects caused by dirt or dust. Where visual, microscope-assisted analysis fails to localise and remedy the defect, recourse

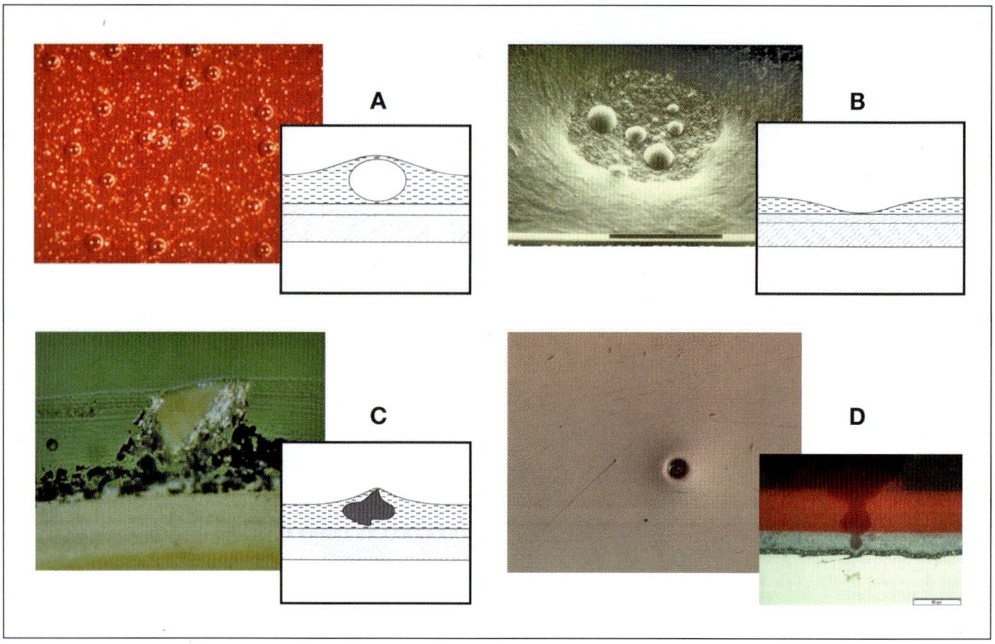

Figure 6.2.1: Typical defects in paint surfaces: (A) blisters, (B) craters, (C) dirt and (D) pinholing due to solvent boilers in a lower layer *Source: BASF Coatings*

is made to analytical methods, which provide the clearest possible clues to the type of substance at fault. One such method consists in the use of Fourier transform infrared spectroscopy (FTIR) to determine the structural features of the contaminating substance on micropellets or under an IR microscope if the concentration is high enough [6.6.6]. Energy-dispersive X-ray microanalysis (EDX) can also reveal structural elements in the defect area. The electron beam of a scanning electron microscope (SEM) stimulates emission of characteristic X-rays from the target sample in inverse proportion to the atomic number of the relevant element [6.6.7]. This method provides element-specific but not structure-specific information and its strength lies in the detection of inorganic components.

If coating defects are caused by boundary layer phenomena involving surface-active substances, their concentrations in the coating surface are usually so small that detection with IR spectroscopy or EDX is not feasible. Other, more sensitive processes such as electron spectroscopy for chemical analysis (ESCA), secondary ion mass spectroscopy (SIMS) and the laser microprobe mass analyser (LAMMA) are based on the method of bombarding the surface of the defect zone with a high energy beam or with mass particles in order to obtain spectroscopic or mass spectroscopic information from the reaction of the molecules. However, these processes are not sensitive enough to detect and measure impurities that are active at contaminant levels in the ppm and ppb range [6.6.6]. When it comes to the detection of surface contaminants, a low information depth or low degree of interaction with the bombarded surface equates to high detection sensitivity, as illustrated in Figure 6.2.2. Local resolution and the size of the bombarded area also play an important role.

The instruments developed by *Benninghofen* [6.6.9] on the basis of time-of-flight secondary ion mass spectroscopy (TOF-SIMS) offer the high sensitivity needed for the analy-

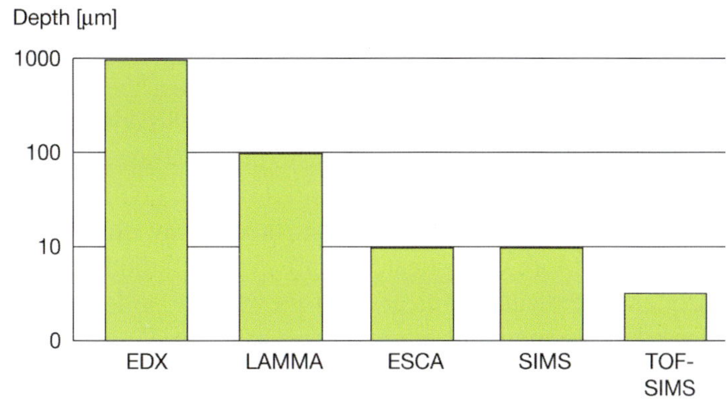

Figure 6.2.2: Depth of information for different methods of surface analysis

sis of chemical substances on the surface of coatings. These gently bombard the surface with argon or metal ions with a view to generating secondary ions from the coating. This method is well established today and has in many cases enabled engineers to identify the contaminating substances in defect zones.

Visible, wide area surface defects that can be rapidly and clearly analysed by the trained eye include orange peel texture, haze, mottling in metallic coatings, insufficient light-to-dark flop in metallic coatings and runs on vertical surfaces.

The procedure for locating the cause is exactly the same as for localised defects. In most cases, chemical analysis does not make sense on account of the macroscopic texture. The most frequently adopted approach for rectifying these defects is to modify the application method and/or the coating formulation.

The third class of defects, namely **adverse effect on film properties**, only occurs during the later life cycle of the coated product and is therefore not usually detectable during the coating process. Here, the most common defects are delamination from the substrate or between coating films, chalking or excessively rapid loss of gloss, blistering, discolouration and premature corrosion.

Defects in the performance of film properties must be avoided by deploying a quality management system (see Chapter 6.3).

As explained at the start of this chapter, although the phenomenological description of a defect does provide pointers to possible causes, it does not always yield unambiguous results. In every case of doubt, the substrate, pretreatment, coating process and coating material must be checked for possible deviations from the standard. The types of defect and causes described in the sections that follow are grouped according to probability of occurrence in these fields.

When it comes to repairing the surface defects, the previously described visual-inspection approach to classification will generally suffice to determine the necessary process, possibly in combination with light sanding (see Chapter 7.1.6).

6.2.2 The Most Frequent Causes of Surface Defects

Substrate and Pretreatment

Without exception, every substrate to be coated will require a specific form of pretreatment in order that a lasting coating may be obtained which conforms to specification (see Chapter 4.1). Pretreatment that is not specific to the substrate and not up to standard can lead to considerable deviations from the desired surface quality. The simplest form of pretreatment for the coating is to clean the surface of any dirt or other deposits or alterations of the surface. More sophisticated pretreatment methods consist in a cleaning stage with subsequent chemical modification of the surface, followed in turn by a further cleaning stage. For the most important classes of material, the search for potential causes of defects in the subsequent coating process will focus on certain specific areas.

Wood, being a natural raw material, is inhomogeneous and anisotropic. It therefore has irregular surface textures and pores which are due, for instance, to irregular growth. Moreover, to an extent depending on the type of wood and region it may be necessary to take various constituent substances such as resins, waxes and water into account. The most common pretreatment stages therefore include removing substances which are incompatible with the coatings, e.g. by means of soap solutions, ammonia and in some cases solvents, sanding in several stages with papers having grain sizes to > 400, and sealing and filling pores and cracks.

A multilayer coating system consisting of insulating primer, stain and several clearcoats helps to correct the substrate-induced defects of outgassing and moisture.

Medium density fibreboard (MDF), which is widely used in the furniture industry (see Chapter 4.1.1), is also coated in several layers comprising insulating primer, sometimes several applications of primer surfacer, basecoat and clearcoat. In this process, any bleeding of glue from the boards must be avoided. Where water-borne coatings are used, protruding wood fibres can be a problem. If necessary, to counter this problem, the wood surface is wetted with up to 50 g/m^2 water and then smoothed after drying [6.6.10].

Quality

Unlike wood, **metals** are isotropic materials. However, this is not wholly true of the surface, especially if sheet metal panels have passed through drawing and other forming processes. Also, spot welds in the production of formed parts can cause local problem points for pretreatment and coating. The auxiliary substances necessary for the manufacture of a metal part, e.g. anticorrosion greases for the storage and transport of coils or other sheetings, and the drawing greases used for the forming process are not always compatible with the coating materials. If they are not fully removed before the coating process, they have an adverse effect on the formation of the inorganic conversion layer. Unwanted surface effects can then occur in the electrocoatings in the form of wetting defects and bleeding (see below). On the other hand, if the anticorrosion oil is not evenly spread over the surface, flash rust or more severe rusting can occur, and although this surface rust can be removed by pretreatment such as sanding it will give rise to an uneven coating. Welding globules and dirt on and in the part are other sources of defects. Off-spec areas of roughness on the panel surface often lead to deviations in the surface quality of the entire paint system.

Plastic components do not have a homogeneous surface and, if produced by injection moulding of thermoplastics, often have flow lines. Mould release agents and exuding plasticisers and additives can cause differential wetting during the priming process, and so give rise to readily visible textures. The thermal characteristics and the internal stresses in the plastic must be taken into account when the coating materials are being formulated. During the application process, the low conductivity of plastics is responsible for high levels of dirt inclusion. Manual wiping and cleaning of the surfaces should therefore be avoided if possible. A source of defects in many expanded materials is the release of blowing agents during the coating process. This problem can occasionally be remedied by "tempering", i.e. briefly warming the plastic part until it is close to the deformation limit with a view to relieving internal stress.

Glassfibre-reinforced plastics such as sheet moulding compounds (SMCs), which are manufactured from unsaturated polyesters and monomeric components such as styrene and are then moulded into shape, constitute two major potential sources of defects for the coating. First, the orientation of the glass-fibre segments can cause them to protrude from the substrate, and second, during the coating process, cavities coated in a thin film can produce "voids" or, on drying, pores and holes in the coating.

Plastics of low surface tension must be carefully conditioned if they are to be homogeneously wetted by the coating material (see Chapter 4.1.3).

Defects Caused by the Process or Coating Product

Electrocoating can suffer from the same defects such as craters, pinholing and dust inclusion that occur in the other application methods and often with the same causes.

One significant source of defects specific to electrocoatings is the automotive body-in-white. While the causes of defects here vary, the most frequent are welding globules, metal particles and sweating of adhesives. The remedy is cleanness and optimisation of the cleaning processes in the body-in-white shop and directly before electrocoating.

Moreover, contamination of the tank via the overall production process must be avoided. Low-solvent, environment-compatible electrocoat paints can be contaminated by hydrophobic silicones in the ppm range or even by fluorinated hydrocarbons in the ppb range, to such an extent that craters can occur in the coating itself or even in the subsequent layer due to wetting defects. In addition, there are various highly specific defects that can have their causes in the process, the bath stage or the paint quality. The most important of these problems is the rupture voltage effect, which manifests itself as small spots or pinholes accompanied by centimetre-large defects in the form of quasi-fused, coagulated and localised areas of the coating. Such defects can already be detected in the wet film prior to the stoving oven. The causes can be excessively high voltage, bath temperature and bath conductivity during the process. Here, the appropriate remedial action is to lower the voltage or the bath temperature and to discard some of the ultrafiltrate and replace it with fully deionised water. As a rule of thumb, the voltage must be at least 10 % lower than the rupture voltage measurable in the lab. Another defect that can occur in continuous coating lines is the appearance of immersion marks that take the form of beads and pinholes on the immersed part. The causes here can be foam on the surface of the bath or excessive current density in the entry zone. Problems in the wetting of the substrate by the electrocoat paint can also lead to similar effects. During rinsing, water marks occur if rinsing in the exit zone is of poor quality and possibly occurs too late, or in general if insufficient

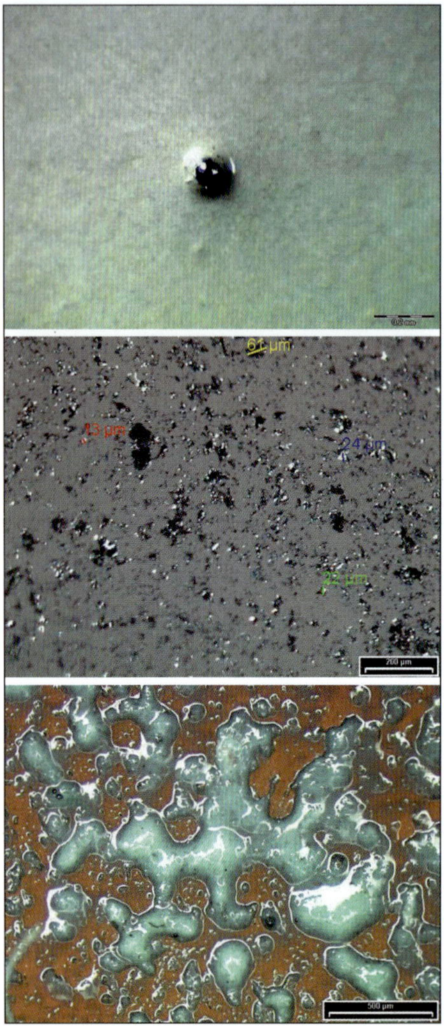

Figure 6.2.3: Welding globules, metal particles and body adhesives as contaminants in cathodic electrodeposition coatings (from top to bottom)
Source: BASF Coatings

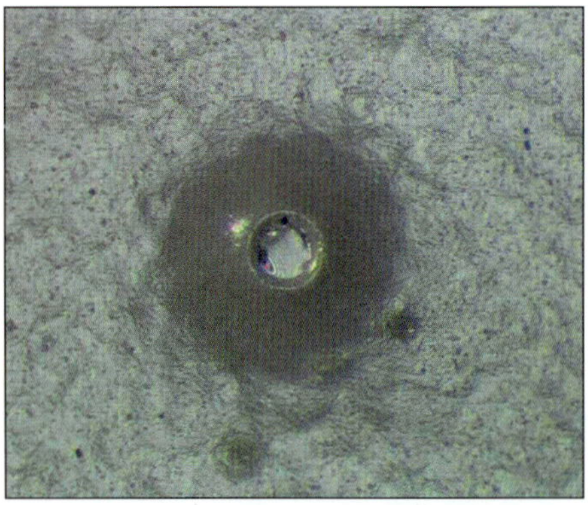

Figure 6.2.4: Pinholes in an electrodeposited coating that are caused by local electrical effects
Source: BASF Coatings

ultrafiltrate is generated for the entire rinsing zone. Pinholes in electrocoatings stem not only from the water or solvent content in the deposited film but also and more often from the bath composition, in particular too low a pigment-binder ratio or too low a bath temperature. Last but not least, large surface defects accompanied by the need for a significant change in application parameters just to maintain the same film thickness and throwing power are a sign of bacterial contamination. This problem can be bacteriologically analysed and combated with bactericides (see Chapter 2.1.4).

The specific defects associated with **spray paints** vary with the function of the paint. For example, primers must be totally free of surface defects if they are to perform their function of acting as a base for a topcoat which will possess high smoothness and gloss. The causes of defects here can be classified into several areas. For example, residues of dust from sanding operations can occur in cathodic electrocoatings, whilst residues of seam sealing compounds can also occur, especially if these are manually applied and if PVC serves as material. Spray mist incompatibilities at several stages in the application of the primer and finally dust inclusions are other notable sources of defects.

The spraying equipment can lead to defective coatings both in primer and topcoat application. These defects can be excessively thin films due to incorrect dispensing quantities, over-rough films due to excessive propellant air and "eggshells" due to deposits of coating material that accumulate on the spray heads and are then sporadically ejected onto the coated part.

The need for exact calibration of all parameters for a defect-free coating is therefore greatest in the case of topcoats, because they are responsible for the visible surface of the coated workpiece. Regardless of the type of topcoat, dust inclusions are the most frequent cause of defects (see below and Chapter 7.1).

In the case of basecoat and clearcoat systems, e.g. for metallic paints, specific defects can be an imperfect metallic effect, spray mist incompatibility and pinholing or solvent boils.

The metallic effect, especially for very light shades of colour, can be disturbed locally by black points. The defect is often caused by incorrect orientation of the aluminium flakes due to the inclusion of impurities in the basecoat or underlying layers.

Figure 6.2.5: Overlooked PVC residues from a PVC seam sealer on a car body surface

Source: BASF Coatings

The appearance of large zones of incorrectly oriented metallic paints is termed “mottling”. The immediate cause is differences in the evaporation and drying characteristics of the basecoat over significant areas. The ultimate cause must be sought both in the formulation of the relevant paint and in the application conditions. In addition to visual inspection, mottling can be detected by means of statistical measuring processes (see Chapter 6.3). Another large-area defect is known as the “salt and pepper” effect. This is caused by the formation of *Bénard’s* cells during the basecoat evaporation time (see Chapter 3.2.1).

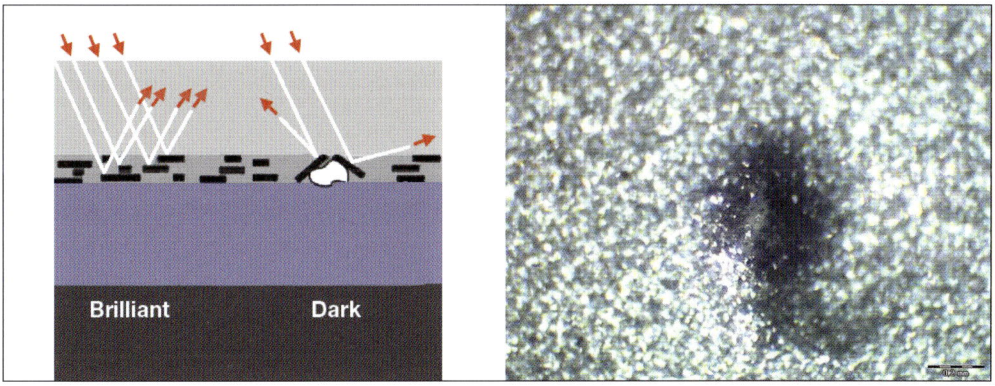

Figure 6.2.6: Schematic drawing and picture of a dark spot in a basecoat

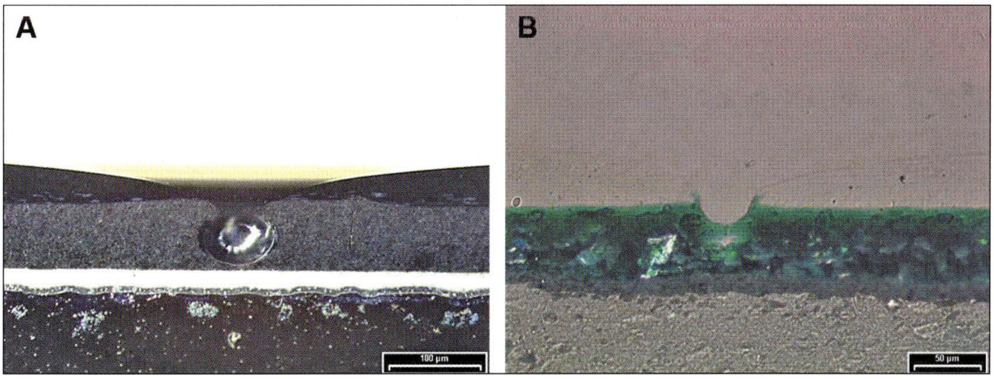

Figure 6.2.7: Pinholes caused by solvent boil in (A) the primer and (B) the basecoat

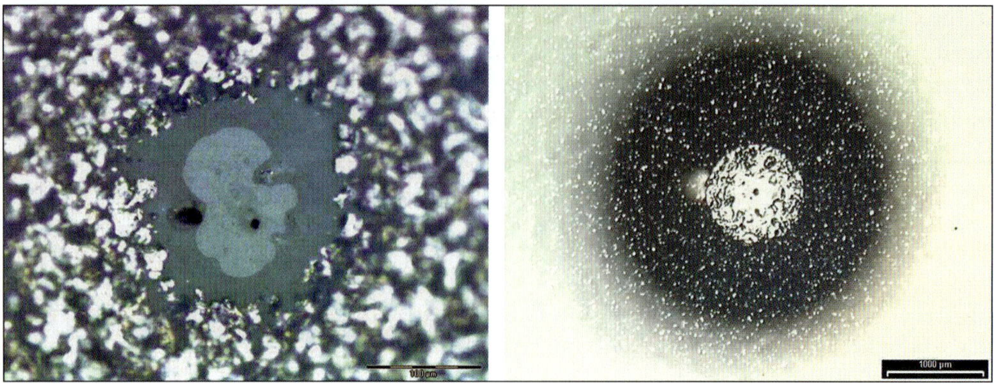

Figure 6.2.8: Craters in an automotive topcoat system, extending as far as the cathodic electrocoating and caused by surface-active contaminants *Source: BASF Coatings*

Spray mist incompatibility is evidenced by areas of very poor levelling. This phenomenon stems from the fact that previously coated areas, e.g. the door sills of a car body, are too dry upon reaching the next application stage, so that the fresh spray droplets are no longer accepted by this film and can fully deliquesce. Possible remedies are an optimisation of the spray zones and a "longer" setting by high boiling solvents of the coating material.

Pinholing, solvent boils and craters are defects that can be very small in diameter and so are difficult to distinguish from one another. The term "pinholing" is generally used when this defect is caused by blistering from underlying layers during wet-on-wet application of multiple layers.

The most frequent processing causes are firstly insufficient intermediate drying of the basecoat layer before application of the clearcoat and secondly air or solvent inclusions in the various layers.

Another major problem is inadequate matching of the formulations for the basecoat and the clearcoat. In this case, incipient dissolving of the basecoat by the solvents and other components of the clearcoat can create impressions of colour and effect that deviate from the specified standard.

All layers of a topcoat, including fillers, must have optimally calibrated wetting characteristics. Nevertheless, even very small-scale contamination by a surface-active substance of low surface tension can give rise to localised concentrations on a surface. In the next processing stage, these concentrations can lead to wetting defects which manifest themselves in the form of craters (see Chapter 2.1.4).

Surface defects in **powder coatings** are for the most part very specific due to the consistency of the coating material and to the application process. For example, the reliability of the corona spraying process is contingent on precision control over powder metering and powder recovery.

The most common problems with powder coatings and their causes include craters or blisters, excessively thin powder films and insufficient coverage of the substrate. The last of these can be caused by several reasons. First, the voltage at one of the electrodes of the spray unit may be too low. Here, the requisite action is to check the high-voltage source, the electrodes and the cables. In all cases, it is advisable to clean the electrodes of any

Figure 6.2.9: Eggshell – agglomerated powder coating particles *Source: BASF Coatings*

powder deposits. Second, the workpiece to be coated may be poorly earthed. Here, the earthing of the hangers and conveyer belt must be checked. The contacts must be free of powder deposits. If the powder flow rate is too high, sufficient electrostatic charging will not be possible. In that event, reducing the powder flow rate will lead to more efficient coating. If the humidity in the spray booth is too high, the charge on the powder particles will be broken down rapidly. The humidity should therefore be reduced to a lower setting. In the case of a very low-efficiency first application, the particle size of the recycled powder may shift towards the smaller range. This factor should be checked by measurement; if necessary, the ratio of new powder to recycled powder must be increased. Excessive air in the powder feed line can lead to less deposition of powder on the substrate. In this case, either the air flow must be decreased or the distance to the substrate increased.

Another very common defect is excessive orange peel. The back-spray effects of powder application very often lead to poor surface quality (see Chapter 4.2.2). The causes may be excessive voltage, an insufficient gap between the spray unit and the workpiece, poor earthing or films that are generally too thick. To remedy these defects, the voltage must be reduced, the gap increased and earthing must be checked and improved, or the powder flow rate must be reduced. In individual cases, the number of spray units can also be reduced.

Irregular and intermittent powder metering naturally also produces irregular surfaces. The cause here can be low pressure in the metering or conveying air, and so the air pressure and the airlines must be checked for kinks etc. Blockages in the application devices, lines or suction tubes may also be sources of defects. Regular cleaning is the recommended preventive measure.

Swirling of the powder in the metering container may be inadequate. In this case, the swirling air must be adjusted, and if necessary the tank must be cleaned and the powder in the tank changed.

Finally, powder agglomerates can spoil the surface with clearly visible defects. These "eggshells" are soft agglomerates, generally of fine powder particles, which can form either in the powder supply lines or in the application device. Their generation can be traced to thermal or electrostatic causes. The following factors influence the formation of agglomerates:

- A high proportion of regenerated powder in the mixing tank
- Lines not being cleaned frequently enough
- Excessive humidity in the compressed air.

As with all types of coating, localised surface defects have two main causes. Firstly, external contamination may occur during the application process and secondly the

Figure 6.2.10: Cross-section and plan view of dirt stemming from repeated coating of conveyor elements, such as hangers and spacers Source: BASF Coatings

powder coating itself may contain or generate interferencing substances. Contamination should be kept to a minimum by filtration during production and application. However, if a filter bag or strainer tears, the released dirt particles or powder agglomerates will produce defects in the deposited film (see above). Another place where powder agglomerates can form is in the lines and tanks. In this case, the dispensing system should regularly be blown clean as an important preventive measure. Excessive humidity in the compressed air can yield powder agglomerates, as can an excess of recycled powder due to its smaller particle size. Finally, it should be noted that powder coatings only provide the criteria for optimal quality if the workpieces have been properly pretreated. Whereas solvent-borne paints can easily cope with traces of grease or anticorrosion oils because the solvents dissolve grease, solvent-free powder coatings require absolutely clean, readily wetted substrates.

Craters and bubbles in powder coatings can also be caused by the circumstances and factors described above [6.6.11].

The high throughputs of coating materials in **coil coating** have specific potential for defects. For example, the stoving oven may become rapidly fouled by deposits if it is not cleaned periodically as it is designed to be. The outcome of poor maintenance is essentially surface defects in the form of dirt particles. Craters, pinholes, blisters, fish-eyes and boils are the most common visible signs of localised defects on the coil

Material type	Material	Number of free fibres/m²
Cotton	Cleaning rag	> 2,000,000
	T-shirt	1,000,000
	Denim jeans	800,000
	Floor mop	500,000
Synthetics	All-purpose cloth	43,000
	New overalls	25,000
	Cloth for cleaning PVC	25,000
	Synthetic cloth with hot-cut edges	0

Figure 6.2.11: Number of free fibres in various materials Source: BASF Coatings

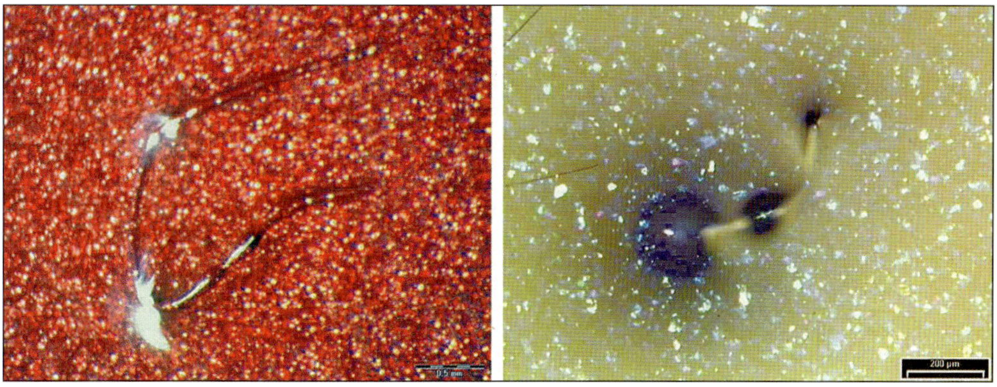

Figure 6.2.12: Examples of fibres in topcoats *Source: BASF Coatings*

coating sheet. This is particularly problematic for high-gloss coatings in the "white goods" sector where stringent demands are imposed on the surface finish. Inadequate degassing of the coatings, an overly steep oven curve and foam in the coatings can quickly lead to boils and blistering. Foam in a coating can specifically be reduced by a three-roller application process. If the transfer roller is worn, coating defects will take the form of score marks or streaks. In that event, the roller will have to be reground.

In general, the key factors governing surface finish are the coating line and the parameters of the delivered coating product. Remedial action at the end-user site is rarely possible.

Even though extensive automation and intensive work in the field of automotive coatings have by now largely brought contamination by dirt or dust under control, dirt is still the most important source of both defects and costs. The **dirt particles** that cause coating defects are not transported into the coating lines solely by the fresh air supplied from outside. Other dirt-bearing contaminants include particles borne into the booth by the operator, the compressed spraying air necessary to atomise the coating, the additional air required for metering in the case of powder coatings, the ambient air, and the workpiece [6.6.12]. Whilst dirt entrainment via the spraying air can be largely controlled through appropriate investment in air filters and oil separators, contamination by the workpiece and the handling system is still an unsolved problem in many cases. The outer skin of intricately shaped objects such as car bodies can be cleaned satisfactorily by cleaning systems fitted with emu feathers (see Chapter 7.1). However, this method fails to remove dirt particles in the cavities, of course. In the case of air-assisted paint atomising, the possibility cannot be ruled out that some of the dirt will be released inside cavities where it subsequently causes coating defects. Accordingly, a great deal of effort is being expended on eliminating these sources of defects through additional pre-cleaning or binding the cavity dirt entrained from the body-in-white shop.

In the case of manually operated application systems, the biggest problem is the contamination of coated surfaces with fibres introduced by personnel via

Size [mm]	Necessary repair work
< 0.1	None
0.1 – 0.3	Grinding and polishing
0.3 – 0.6	Spot repairs
> 0.6	Spot repairs or new coating

Figure 6.2.13: Sizes of dirt inclusions and repair work needed

their clothing or during cleaning operations. Here, the materials used for clothing and for cleaning the plant or workpiece can be ranked by the number of free fibres per m^2 of material. The surface defects generated by fibres are often visible and can be very easily identified under the microscope.

The clothing of coating line employees should be made only with fabrics having less than 100,000 free fibres per m^2, i.e. essentially only synthetic materials (see Figure 6.2.12).

The amount of work involved in repairing defects caused by particles, i.e. dirt or fibres, depends on the quality target defined for the coating, but can be roughly classified [6.6.11]. In all cases, repairs involve costs that can significantly contribute to the overall coating costs.

6.2.3 Summary

The causes of coating defects can vary extensively. The fact that they originate in the properties of the coating material, in the type of application, in the process and in human behaviour is all the more reason for supplier, plant manufacturer and customer to collaborate closely on preventing them. Today, analytical methods are available which can provide an explanation by chemical analysis for those defects which cannot be explained under the microscope.

6.3 Material Control

As indicated at the start of this chapter, quality assurance starts with monitoring the product so that it can be rejected if it does not conform to the tolerances agreed with the customer, and then seeking out the causes and taking remedial action where necessary. Initially, this was all about identifying quality characteristics that could then serve as criteria for assessing the ease of application, attractiveness and service life of the finished coating [6.6.14].

Subsequently, aspects of occupational health and safety and environmental compatibility were integrated into the monitoring system. The strategy therefore initially consisted in testing the properties during application and service while making allowance for the characteristics of the products during storage and transport.

There can be no question that the analytical determination of the composition of complex coatings is extremely useful, but it is equally indisputable that the analysis cannot serve as a criterion for exclusion should minor deviations arise. Determinations of solid content, density, viscosity and surface tension can without doubt act as important indicators that different batches are of identical quality, but they do not necessarily indicate quality failings when variations, within certain limits, arise. Therefore, the criteria that govern quality are physical and chemical characteristics which need to be defined numerically in the same manner as the application properties, which often have to be determined by simulating the application conditions and practical stresses.

For example, whilst modern oscillating rheometers can provide detailed information on molecular interactions, it is sometimes just enough to measure the flow-out time from an ISO cup to establish that two batches of the same product perform identically.

It is an accepted fact that it is still not possible in the majority of cases to wholly characterise the properties of coating materials and coatings by measuring their physical characteristics. Stresses that simulate the practical performance of a coating remain highly important test methods, and include cupping or bending on a conical or cylindrical mandrel, the cross-cut test and the *Buchholz* indentation test (see Chapter 3.2). Nevertheless, it must be noted that measurements of viscoelastic characteristics by means of an oscillating rheometer or thermomechanical analysis can yield data that are important in the assessment of quality [6.6.15]. Further measuring methods utilise the advantages of non-destructive and in some cases even contactless sensing of characteristic data as a means of determining the performance of coatings (see Chapter 3.3). Some pertinent examples are measurement of adhesion capability by ultrasonic reflection or thermal conduction, and measurements of film thickness by thermal waves or sound emission analysis. Changes in viscoelastic parameters, such as the modulus of elasticity, induced by means of artificial ageing make it possible to predict cracking tendencies long before the actual defect occurs (see Chapters 2.3 and 3.3).

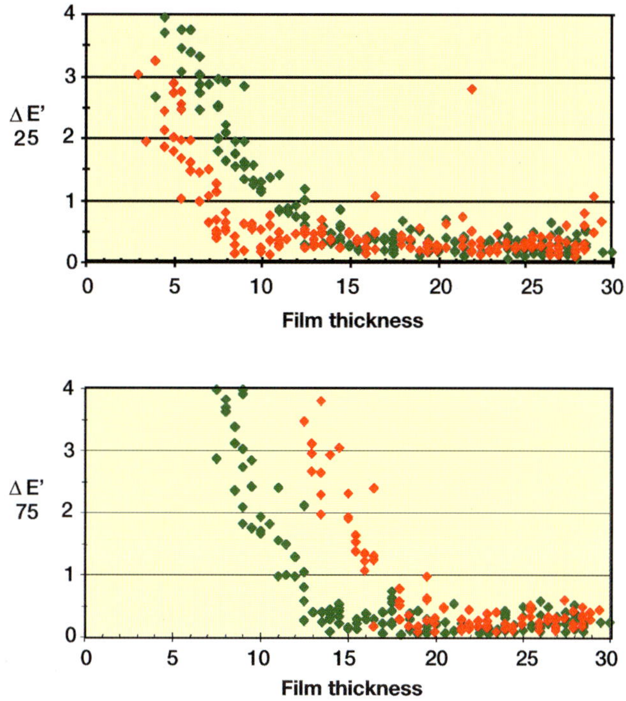

(A) Green: light colored primer surfacer; red: dark colored primer surfacer

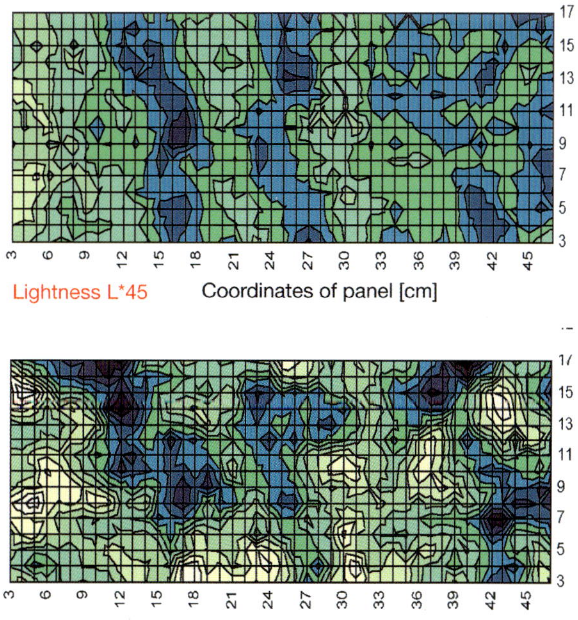

(B) Above: strong mottling; below: no mottling

Figure 6.3.1: Statistical analysis of coated test panels for determining (A) the hiding power using ΔE values and (B) mottling *Source: BASF Coatings*

All the tests mentioned so far relate to quality assessments of coating films under standardised conditions in a test lab or in the applications centre of a coatings manufacturer. Any test programme designed to determine the application properties of these coatings needs to be even larger and more varied. In view of the large number of different application systems used by paint shops and the undisputed fact that lab conditions are never identical with actual operating conditions, the coatings customer is often forced to retest new products or even individual batches under operating conditions, despite exhaustive prior testing by the coatings manufacturer.

Notable progress in the characterisation and optimisation of the application properties of coating materials has been achieved by means of statistical tests performed on metal panels coated with clearly defined film thickness profiles in a multilayer coating system consisting of basecoat and clearcoat [6.6.16]. The measurements are made at approximately 500 to 1000 points on each panel. The measured properties are film thickness, gloss, colour values and surface texture indicators (see Chapter 3.2). Statistical analysis yields quantitative descriptions of the application properties of the individual layers, e.g. in the case of the topcoat of a coating system, the hiding power, colour precision, gloss and surface requirements (see Figure 6.3.1). This method can be recommended for use in the development and the quality testing of the relevant coating materials.

The costs of the described conventional quality assurance procedures vary with customer requirements but can amount to 10,000 euros per batch.

6.4 Defect Prevention by Process Control and Control Loops

The coating result obtained in practice must be assessable and modifiable. For this purpose, additional plant-specific analyses of weak points and deficiencies must be conducted. Only in this way can the quality-sensitive points in the process be detected. These analyses include data acquisition, failure reports and of course the corresponding activities entailed in analysing and evaluating the mass of data, all aimed at the development of defect prevention strategies. Such strategies apply to the personnel performing the test as well as to the production equipment. In this context, qualified personnel [6.6.17] and constant process conditions are the most essential preconditions for ensuring constant quality.

Computers are essential for rapidly gathering data from the individual stations on the coating line and for evaluating the checklists. The most important application and coating line parameters for spray application are viscosity, flow rate and the spray air pressure, to name but three of the variables that affect quality.

There is more to modern quality assurance than the logging and describing of defects. Process optimisation across the entire supply and production process by means of control loops are the next step on the road to guaranteeing constant quality [6.6.18]. Spray coating is a longstanding practical example of how the monitoring and utilisation of process data facilitate the creation of process controls that eventually culminate in process automation. In this context **open-loop control** is a process by which one or more input values, such as viscosity, surface tension and temperature or the pressure of the spray air and material, influence the target variables of levelling, gloss and film thickness on the basis of known physical laws. Open-loop process control therefore has the function of adjusting processes and – if required – keeping them constant in accordance with a predefined program. In this case, the result is not monitored by the controller.

In contrast closed-loop control is a process by which the controlled variable is continuously monitored and compared to a setpoint value and then adjusted accordingly. The entire workflow therefore becomes a closed control loop, in which all disturbance variables are regularly detected and corrected [6.6.19]. Each type of open-loop and closed-loop control therefore begins with a metrological evaluation of the variables affecting the process. Due to the complexity of the interdependent relations (see Figure 1.1.7), the input variables are initially compared with the experimentally determined output data, so that binding calculation methods can then, where applicable, be found for the desired controller, e.g. with the aid of neural networks for the interconnections.

One example of successful process control with effective closed-loop control loops is the bath control employed for electrocoating baths. To ensure consistent coating quality, numerous bath parameters must be regularly measured and adjusted during operation of the electrocoating baths [6.6.20]. Electrocoating baths undergo continuous changes due to consumption of paint material as the workpieces are coated and also due to variations in ambient conditions (see Chapter 4.2.1). The consumption of material lowers the solid content of the bath, as a result of which the pigment-to-binder ratio may also change. Entrainment of pretreatment chemicals and addition of replenishment material can also modify the degree of neutralisation and therefore also the pH. Changes in the bath condition during replenishment must, if necessary, be offset by ultrafiltration and removal of liquid anolyte. At present this is mostly still a manual process, but there are some plants in which control parameters are captured by numerical control methods and used for process control. Since the solid content and pigment content cannot be measured directly, it is necessary to make recourse to other, indirect methods. Measurements of density and sound velocity have proved suitable in this regard. They can be detected precisely, rapidly and digitally and have a direct functional relationship with the desired variables of solid content and film forming agent concentration. Current consumption in electrocoatings is another basic variable that is used for, e.g., open-loop replenishment control (see Chapter 7.1).

It can be assumed that the experience gained with closed-loop process control for spray coating and the corresponding closed-loop electrocoating control will be utilised on a broader basis in future [6.6.21].

6.5 Quality Management

The traditional view of quality as the sum of the properties (quality criteria) of a product compared to the properties of other products for the same purpose has evolved towards a more comprehensive definition of quality as the regular fulfilment of the requirements and expectations of customers, with the aim of complete and sustained customer satisfaction. In this context, the product properties must also be extended to include environment-related parameters. The trend is now moving away from quality of product and towards quality of service.

Largely as a result of the globalisation of their activities, companies are facing more intense and geographically more widespread competition. Product quality as a criterion for assessing company expertise is now being supplemented by service, such as customer service, on-time-delivery and rapid response to problems at the customer's premises and to changes in customer requirements. Flexibility and an expanded vision of quality are the key foundations for meeting these criteria.

Through observation of the company's own products in use on the market and regular market analyses it is possible to anticipate customer requirements and to also beneficially steer these requirements towards a more ecologically sound or more economical alternative (see Chapter 5.5). These activities also form part of the foundations of a holistic view of quality and so serve as a basis for a quality management system. The aim of quality management is to recognise and properly control all the necessary quality-related measures over the product's life cycle. This process includes observations and findings made during product planning, production and utilisation on the market. The quality criteria in this case are not only the product properties but increasingly also service, adherence to lead times and competence in collaboration. They also have a major influence on the company's corporate image.

These criteria include measures to prevent defects. Experience has shown that the vast majority of quality defects can be traced back to the initial planning stage. These defects stem from latent weaknesses that only become detectable during production or utilisation on the market.

It should not be overlooked that defect prevention involves costs. The efficiency of defect prevention can only be calculated when the incurred defect costs can be set against the prevention costs.

As already mentioned, the modern interpretation of the term "quality" focuses not so much on the actual product as on satisfying the demands and expectations of the customer. ISO 9000 therefore defines quality as the totality of properties and characteristics of a product and service that are relevant to its suitability to satisfy predefined or assumed requirements. These characteristics include the functionality of the product, its durability and design as well as the services and ecological aspects relating to waste disposal and resource conservation. This concept also addresses the producer's claim to act in the interest of the customer by earning profits that will safeguard its future and enable it to engage in lasting cooperation.

The term quality therefore equally serves the interests of both manufacturers and customers. The definition of quality goes beyond product observation to the assessment of all the services provided by the company. Quality is therefore the regular fulfilment of the customers' requirements and expectations, to the customer's total and lasting satisfaction.

All departments in a company need to buy into this vision. Quality management is therefore a principle that suffuses the entire company and is ultimately the decision-making basis for all employees. In other words, quality management is a customer-oriented strategy of corporate management and must be deeply rooted in all units of the company.

This concept of quality-oriented companies is known as "Total Quality Management" (TQM) and is explained in detail in the aforementioned ISO 9000. In this standard, quality in the extended sense is the decisive variable to which all employees are committed, with the rider that a dynamic approach to quality implementation be adopted. The latter concept includes sustained efforts to improve the product, technical processes and services, always with the aim of providing even better fulfilment of the customer's expectations.

The instruments of quality management include quality control and quality assurance (QA) in conjunction with the holistic approach of quality management (QM) itself. Quality assurance is directed via the monitoring of processes and acceptance tests at determining the causes of deviations in the process with a view to eliminating them, if not quantitatively, then by means of control loops.

Numerous standards propose quality management systems and provide instructions for their implementation in the company. In addition, they define criteria for assessing the degree of implementation. Consequently, different quality management systems (QMS) from different companies can be compared and audited.

One of the most widely accepted quality management systems, covering all sectors of the economy, is described in the ISO 9000 ff. series of standards.

Whereas ISO 9000 is essentially an instruction manual for other standards, ISO 9001 deals with the installation of quality assurance systems and provides detailed system criteria that must be demonstrated. The standard also covers the company segments of development, design, production and customer service, concentrates on production and gives additional instructions on testing and inspection [6.6.22]. ISO 9001 also addresses in detail the tasks of management and personnel training. Other notable chapters include testing and inspection, statistical evaluation of test results, documentation, monitoring of test and inspection equipment, contract agreements and customer satisfaction. ISO 9004 provides information on the effectiveness and efficiency of the QM system during and after its introduction.

When introducing a quality assurance system, the company begins by implementing the basic principles of organisational structures and processes, with due allowance for operational interdependencies, and by defining responsibilities at the highest hierarchical level (creation of a quality assurance manual, see Figure 6.5.2). Only then are orders directed to the various divisions of the company with the aim of implementing the quality directives internally. Finally, at the third level, the tasks and activities in the work places are formulated and laid down as binding standard operating procedures. These procedures must cover workflows within departments and precise instructions on the handling of products, samples and the performance of tests and inspections.

The outcome is a self-contained quality assurance loop, which starts with market research, continues with product development and extends to the inspection of raw materials, the selection of production methods and final inspection. These provisions are supplemented by regulations governing sales and logistics activities, in combination with the methods of customer service and customer processing.

It should be noted that ISO 9000 was combined with the even older ISO 8402 to form the new ISO 9000:2000. The most recent version is DIN ISO 9000:2009. The former ISO 9001 to ISO 9003 have been combined into a uniform standard ISO 9001. ISO 9004 remains basically unchanged.

Over the last few years, the principles laid down in ISO 9000 ff. have been adopted and incorporated into the QA systems of numerous companies.

Includes	Distribution		Description
Companies Business units	**Internal:** Top management down to departmental managers **External:** In line with company policy	QM manual	Basic principles of organisational structures and processes, operational interdependencies, responsibilities, competencies: contains the company's organisational know-how. References to directives and standard operating procedures
Departments	**Internal only:** On a departmental basis	Directives	Detailed description of sub-areas of the quality system. Contains the company's organisational and technical know-how
Topics Individual activities	**External only:** In the work place	Standard operating procedures (SOPs)	Regulation of details. Comprehensive operating procedures, such as test specifications. Contains the company's technical know-how

Figure 6.5.1: Structure and description of quality assurance systems

Proof of conformity with ISO 9000 ff. must be provided in the form of certification issued by independent bodies. Acceptance of this new idea of quality assurance is gaining ground worldwide. In the year 2000 more than 32,000 companies were certified to ISO 9000 ff. in Europe, and worldwide the figure was more than 400,000.

Building on the ISO standards, the automotive industry in particular has produced its own specific regulations. For example, the major American automobile manufacturers have agreed on QS 9000 and now demand that their suppliers provide the relevant certifications [6.6.23]. In Europe, the French automobile manufacturers under the banner

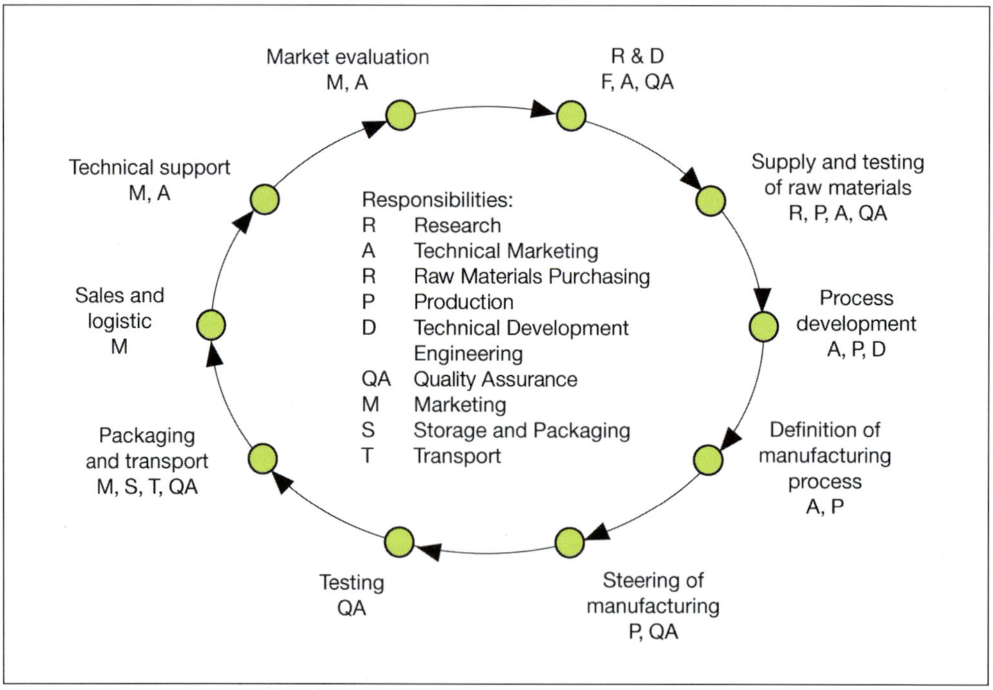

Figure 6.5.2: Example of a TQM-concept according to the Association of Chemical Industry (VCI)

ISO 9000:2005-12	ISO 9001:2008	ISO 9004:2009
Quality management systems: Principles and definitions	Quality management systems: Requirements for customer satisfaction	Quality management systems for organisation: Guidelines for quality improvement and their lasting success

Figure 6.5.3: ISO 9 000 series of standards

Standards	Description
ISO 14001	Environment management systems – specifications with guidance for use
ISO 14004	Environment management systems – general description of principles, systems and instruments for implementation
ISO 14010	Guidelines on eco audits – general principles
ISO 14011	Guidelines on eco audits – audit processes, audit of environment management system
ISO 14012	Guidelines on eco audits – criteria for qualification of auditors

Figure 6.5.4: Overview of eco standards

EAQF developed a set of regulations in the early 1990s. German manufacturers introduced a QM system slightly later under the auspices of the German Association of the Automotive Industry (VDA) dubbed VDA 6.1. In recent years, the globalisation of the automotive industry and the congruence that exists between the various regional QM systems, has increasingly led to the adoption of ISO TS 16949:2010 as a globally binding standard and as a replacement for the various regional standards. This standard concerns the entire supply chain, including the subcontractor industry and it has been accepted by the nine European and American automobile manufacturers which are members of the IATF (International Automotive Task Force). In contrast, Asian manufacturers still mostly have their own quality requirements. The aim of ISO TS 16949 is to improve supply chain processes in such a way that maximum customer satisfaction is achieved.

Quality

An important instrument for successful QM systems is the six sigma method introduced by Motorola in the 1980s and successfully used by General Electric. Its goal is to implement and monitor the process of continuous improvement throughout the company on the basis of sustainable data, standard deviations and quantitative targets for defined processes [6.6.24]. The balanced score card (BSC) has gained importance in the paint and coatings industry as well. The BSC takes into account financing and corporate goals as well as human resources, processes and customers, thus giving all instruments used, including the quality management system, the same strategic orientation. It is a concrete means of transforming the goals emerging from the company's vision and strategy into actual measures, an instrument suitable for communicating the strategy, and a tool for quantitatively tracking the measures. It entails determining key factors which are specific to the company and are known as key performance indicators (KPIs) [6.6.25].

The many quality and environmental requirements are met by making careful product development the focus of attention. The complexity of products, as a corollary of their numerous components, has given rise to the method of HTSC (high-throughput screening) [6.6.26]. This consists in using robots to automatically apply small samples of paint formulations on a micro-scale and to test for selected properties. Sample throughput and results are multiplied many times over compared to conventional laboratory

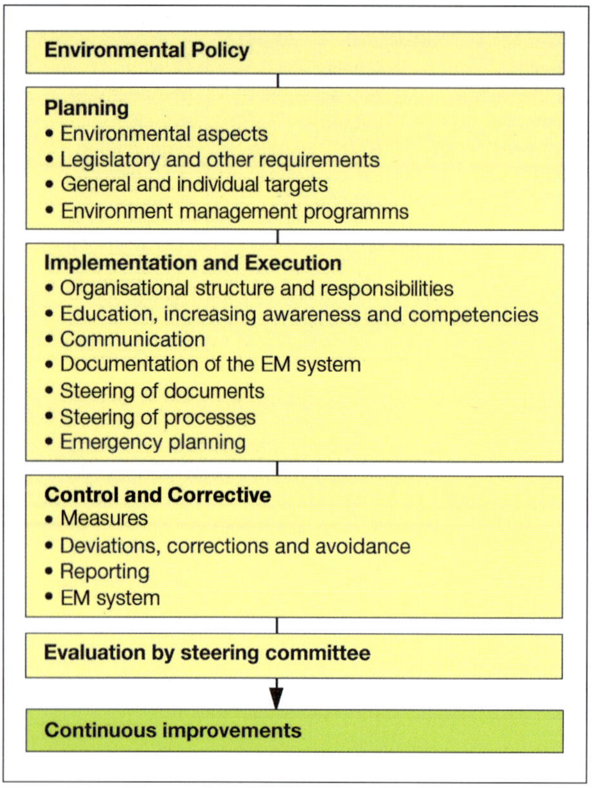

Figure 6.5.5: Basic elements of the environment management system as per ISO 14 001

methods. One factor underpinning the success of such systems is their flexibility with regard to application and measurements [6.6.27].

However, the company's goals are not just quality assurance. Pollution of the environment by industrial processes and the need for enhanced resource conservation are forcing industry to reduce emissions, waste and energy consumption.

Above and beyond the already existing laws or, more accurately, in order to implement the existing laws and to dynamise the implementation of ecological principles, environment management systems (EMS) that were initially developed in parallel with the QM systems are now being integrated into QMs. Whereas QM systems are now standard for modern companies, the implementation of EM systems is only in its infancy.

The first approaches can be traced back to the EU Eco Audit Regulation of 1993. The basic concepts have since been revised and were integrated in April 2001 into EU Regulation 761/2000. This regulation provides guidelines for the voluntary establishment of an environment management system with subsequent company audit. In the same way as quality management systems, the EU Eco Audit Regulation aims to achieve continuous, objectively assessed improvements in the company. A key characteristic is the duty to provide information to the public on the level of performance achieved and the active participation of all employees in the implementation of the set targets.

A similar goal, but standardised worldwide and therefore facilitating comparative assessments of companies on a global scale, is pursued by ISO 14000 ff., first published by ISO Technical Committee 207 in 1996. This norm comprises a total of 20 individual norms. The key norms here are ISO 14001:2004, ISO 14001:2009-11, ISO 14004, ISO 14010, ISO 14011 and ISO 14012. ISO 14004 describes the elements of the environment management system (EMS) in general without any requirement for an audit and can therefore be used by small and medium-sized enterprises (SMEs) in particular.

The basic philosophy behind these standards is to provide companies with assistance and incentives to achieve sustained improvements in environmental quality under their own responsibility. Unlike the EU Eco Audit Regulation, there is no obligation to provide information on individual activities to the public. Details on the establishment of an EMS are shown in Figure 6.5.5.

Whereas QM systems have been able to establish themselves worldwide, the introduction of EMS is only starting. By the end of 2004, some 4,300 companies had been certified under the EU Eco Audit Regulation, and about 90,000 companies worldwide under ISO 14001 ff.

Companies are therefore being confronted with the task of incorporating two different aims into their management. Embracing a quality mindset on one hand whilst endeavouring to act in an environmentally responsible manner is to pursue two goals at once, but the methods for introducing instruments and the principles required to attain them are similar. One thing is clear: the introduction of environmental activities is now a legal requirement. It is therefore a reasonable precaution, when the QM manual is being drawn up, to formulate environment-related concepts at the same time and to integrate them into the manual. Corporate strategies should harmonise the QM and EM systems with each other and also integrate workplace health and safety requirements into company guidelines.

QM and EM systems can therefore be introduced simultaneously or consecutively, in integrated or in separate form. The procedure which each company finds most expedient will have to be decided on the merits of the company's particular situation. In any event, the automotive industry today demands that its suppliers have an EM system certified to ISO 14001, and therefore many paint and coatings companies have already established the relevant systems by now.

6.6 Literature

References

6.6.1 H. Vesper, XXIV Fatipec, Interlaken, Congress paper, p. 163 (1998)
6.6.2 H. Rinne, H.-J. Mittag, Statistische Methoden der Qualitätssicherung, p. 14, Carl Hanser Verlag, Munich, 1995
6.6.3 M. Rochowitz, C. Ernst, JOT 46 (9), p. 60 (2009)
6.6.4 U. Vohrer, MO 61 (9), p. 40 (2007)
6.6.5 C. Schoff, JCT Coat. Tech. 3 (2), p. 36 (2006)
6.6.6 M. W. Urban, ECJ 7-8/2000, p. 58
6.6.7 L. Reimer, "Scanning Electron Microscopy", Springer Verlag, Berlin-Heidelberg-New York, 1985
6.6.8 R. Döring, U. Schernau, F+L 99, p. 321 (1993)
6.6.9 A. Benninghofen, J. Vac. Sci. Technol A3, p. 451 (1985)
6.6.10 U. Hoffmann, *besser lackieren!* Jahrbuch 2001, Vincentz Verlag, Hanover, p. 407ff
6.6.11 A. Grünberger, JOT 45 (10), p. 68 (2005)
6.6.12 J. Fichtner, JOT 42 (7), p. 38 (2002)
6.6.13 U. Hoffmann, *besser lackieren!*, Taschenbuch für Lackierbetriebe, p. 370, Vincentz Verlag, Hanover, 1993
6.6.14 U. Zorll, Prüftechnik bei Lackherstellung und Lackverarbeitung (Hrsg. O. Lückert), p. 3ff and p. 305ff, Vincentz Verlag, Hanover 1992
6.6.15 M. Osterhold, ECJ 04/2000, p. 18; B. Benzler, D. Neff, Kunststoffe 98 (3), p. 112 (2008)
6.6.16 C. Voye, F+L 106 (10), p. 34 (2002)
6.6.17 T. Feist, E.-H. Timmermann, *besser lackieren!*, Jahrbuch 2010, p. 382ff, Vincentz, Hanover
6.6.18 A. Riegel, K. Dekomien, ECJ 04/2013, p. 122
6.6.19 Anonymus, JOT 36 (9), p. 26 (1996)
6.6.20 W. M. Liu, I-Lack 66 (4), p. 222 (1998)
6.6.21 US PS 6073055 (2000) BASF Corp.
6.6.22 M. Hornung, C. Mai, *besser lackieren!*, Taschenbuch für Lackierbetriebe, p. 341ff, Vincentz Verlag, Hanover, 2002
6.6.23 R. Hoene, W. Kreis, F+L 104 (3), p. 108 (1998)

6.6.24 F.-D. Kopplitz, MO 59 (4), p. 38 (2005)
6.6.25 M. Schäfer, Nachr. Chem. 49 (5), p. 663 (2001)
6.6.26 S. Strazisar, M. Kendi, T. Fäcke, L. Hermans-Blackburn, X. S. Feng, JCT Res. 3 (4), p. 307 (2006); J. Anderson, R. Ortlz, JCT Coat. Tech. 10 (1), p. 30 (2013)
6.6.27 J. Ortmeier, S. Ansorge, H. Kost, F+L 113 (2), p. 35 (2007)

General literature

besser lackieren!, Taschenbuch für Lackierbetriebe, Vincentz Network, published annually
A. Rössler, Design of Experiments for Coatings, Vincentz Network, Hanover, 2014
E. Dietrich, A. Schulze, Statistische Verfahren zur Maschinen- und Prozessqualifikation (5th edn.), Hanser, Munich, 2005
B. Guttof, E. T. Cohen, Coatings and Drying Defects (2nd edn.), Wiley, 2006
D. G. Weldon (pub.), Failure Analysis of Paints and Coatings, Wiley, Chichester, 2009
V. Shah (pub.), Handbook of Plastics Testing and Failure Analysis, Wiley-VCH, Weinheim, 2002
R. Dietrich, Paint Analysis, Vincentz Network, Hanover, 2009
G. Meichsner, T. G. Mezger, J. Schröder, Lackeigenschaften messen und steuern, 2nd edn., Vincentz, Hanover, 2016
H. Bubert, H. Jenett (pub.), Surface and Thin Film Analysis, Wiley, New York, 2002
M. Finkbeiner, Umweltmanagement für kleine und mittlere Unternehmen – Die Normenreihe ISO 14000 und ihre Umsetzung, Beuth, Berlin (2012-03)

7 Coating Industries

The global paints and coatings market is not homogeneously structured by type of paint or coating technology. The regional associations define the segments in different ways. The market may be roughly divided up into decorative paints for buildings, automotive production paints and automotive refinish paints, general industrial coatings and printing inks (see Chapter 1.3). The largest segment is decorative paints for buildings, followed by general industrial coatings. Together with automotive OEM paints and automotive refinish paints this segment is also referred to as "industrial coatings". It is about 13.7 million tonnes in size and is technologically the most advanced. It includes industrial coating processes for mass-produced goods such as electrocoating, coil coating, spray coating with liquid or powder paints as well as radiation curing (see Chapters 4.2 and 4.3). The most important industries by sales, except decorative paints, in 2011 can be seen in Figure 7.0.1.

The manufacturers of commodities are increasingly resorting to different materials for the products they make. For example, in the automotive industry the production of vehicle bodies is always being subdivided into components which then can be manufactured from those materials which are best suited to meet the various requirements. A significant step forward in this regard was the manufacture of plastic bumpers to exploit the elastic properties in the event of collisions at low speeds. In addition, weight saving is a constant issue which has, in the meantime, led to the use of aluminium and magnesium and (in 2011) driven the proportion of weight accounted for by plastics up to 18 % (see Figure 4.1.20). Agricultural and construction plant contain more and more nonload-bearing parts (linings and covers) made of plastic. All materials have to look the same on the same product; this can be achieved with specially developed paints and coating processes.

The furniture industry is opting more and more for prefabricated and coated materials. Special mention should be made of the surface enhancement with foils, which are usually sealed with UV-curable clearcoats. Precoated parts are increasingly being used in shipbuilding, aircraft construction and heavy machinery construction [7.12.2]. New sectors, such as wind energy, are emerging and imposing their own requirements that the coating industry has to adjust to (see Chapter 7.11.8).

Coating Industry

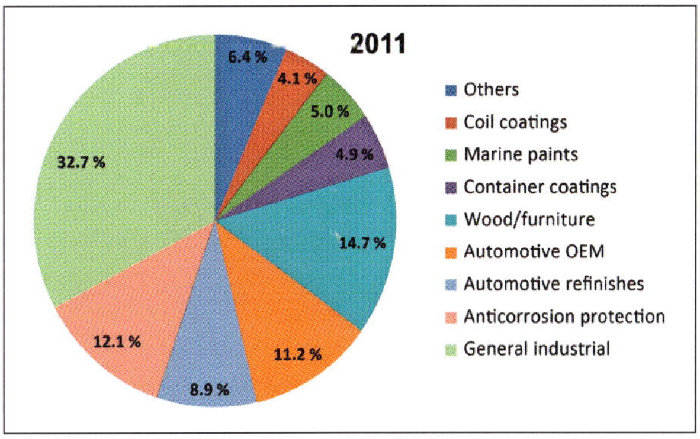

Figure 7.0.1: The most important branches for industrial coating on the global coatings market *Source: M. von Dungen, C. Maier [7.12.1]*

Technological trends in the various industries aside, there was a considerable shift in responsibilities for the workpiece during the 1990s. Owing to its high level of application know-how the coatings industry, supported by appropriate contracts with its customers on the basis of cost per m^2 or cost per coating unit, has assumed more and more responsibility for the results obtained in coating the workpiece. The pioneer in this area is the automotive industry, which restricts itself to planning, investing in and operating the coating lines. Technical implementation, quality monitoring and logistics of the lines are increasingly in the hands of the paint companies [7.12.3].

7.1 Automotive OEM Coating

In scarcely any other sector of surface coating have manual and lengthy coating processes evolved into such high-tech, high-speed industrial application processes as in the automotive industry. Whereas it used to take days or even weeks to coat vehicles, nowadays vehicle bodies need just 6 to 10 hours to pass through an entire coating line, complete with all the cleaning stages, pretreatment, electrocoating and the usual two to four stages needed for applying the topcoat.

This was also necessitated by the need to keep up with the continuous increase in the number of vehicles being manufactured. In 2012, annual global production of automobiles, including trucks, reached 9 million units, of which cars accounted for 72 million. In the last 10 years, there has been a significant shifting of production to Asia, where nearly half of all cars are now produced. German manufacturers achieved a market share of 19 % based on their global car production. Some 5.7 million units were produced in Germany. Roughly 180,000 tonnes of paint were consumed in their production. The global paint market for automotive coatings, including trucks, totals approx. 1 million tonnes.

The development and introduction of new technologies are initiated not only out of economic necessity. Driving forces behind the development of paints and processes for coating vehicles include reductions in environmental pollution as well as qualitative improvements, e.g. in protection against corrosion and in durability, not to mention appearance. Special requirements are imposed on resistance to chemical influences originating from aggressive airborne substances and salt as well as on physical properties such as adhesion and elasticity in conjunction with a high level of hardness.

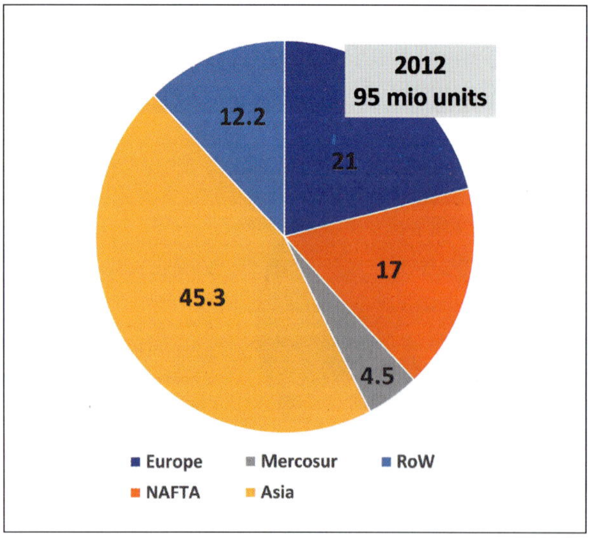

Figure 7.1.1: Global automobile production in 2012 *Source: OICA*

It is worth reflecting on the extreme stresses to which vehicles are exposed throughout their service life. High temperatures, e.g. up to 70 °C, for dark colours in Florida and other southern areas, as well as extremely low temperatures of down to -50 °C in the polar regions, constant temperature fluctuations of 10 to 20 °C a day, stone chipping on gravel and unfinished roads, high salt burdens in coastal regions and from road salt in winter months, high levels of ultraviolet radiation, the action of acidic and alkaline exhaust fumes as well as physical stress in car washes are all factors which constitute only some of the many different burdens and impacts which tend to damage coatings (see Chapter 3.3.2).

It is not hard to appreciate that such a catalogue of requirements cannot be met with a single coat of paint. For this reason, well-matched coatings are used which, when applied in several layers, perform various tasks from within the profile of requirements imposed on vehicles.

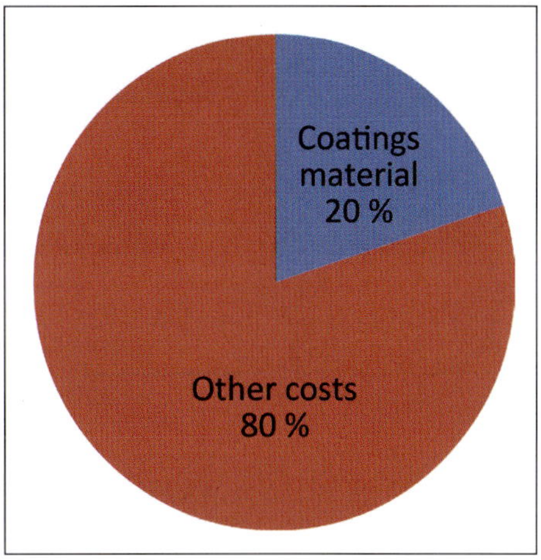

Figure 7.1.2: Cost of paint materials cost as a proportion of the total costs in automotive coating

Not only must a coating line meet the technical demands, it also has to increasingly contend with being the highest consumer of energy in the entire automotive production chain and with handling organisational aspects such as process sequences and the paint supplier's responsibility for the coating result [7.12.4]. This can be illustrated by a look at a cost analysis for the coating process. Of all the different types of cost such as capital investment, energy, staff, maintenance, and paint materials, the last of these only accounts for about 20 % of the total costs block on aver-age. When the cost of repairing defects or nonconforming surfaces is included in the overall analysis, crucial factors that contribute to success and help to ensure optimum coating results are on-time delivery of suitable quality paints to a well-designed coating line, and the application conditions. These considerations prompted the automotive industry, together with the paints and coatings industry, to develop and progressively implement exclusive supplier concepts and system supplier concepts [7.12.5].

The various substrates such as steel, galvanised steel and aluminium, as used nowadays in the manufacture of vehicle bodies, are first of all cleaned and then, in pretreatment, are provided with an inorganic conversion layer made of metal phosphates in a thickness

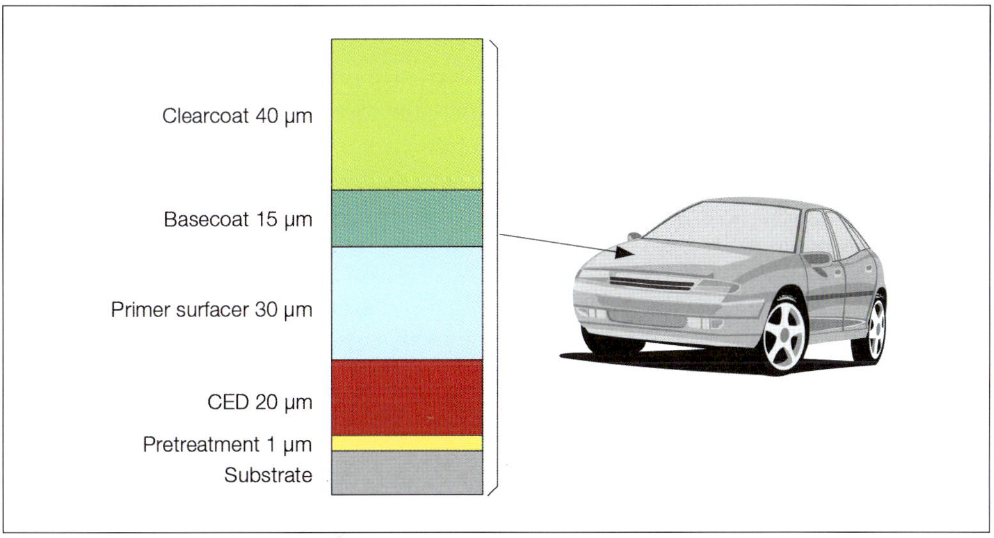

Figure 7.1.3: Schematic composition of an automotive coating

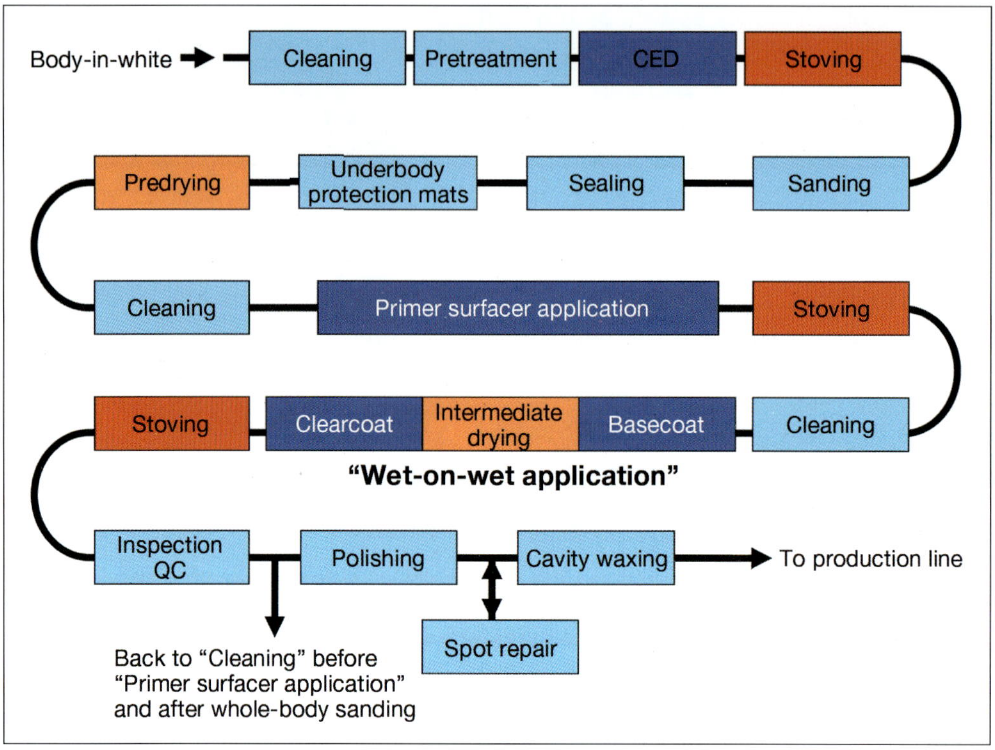

Figure 7.1.4: Typical process steps of a classic coating line in the automotive industry

of 1 to 2 µm or, increasingly, of zirconium dioxide (ZrO_2) in a thickness of just several nanometres. This step increases protection against corrosion and by enlarging the surface area it substantially improves the adhesion of the organic coating (see Chapter 4.1). Then comes the primer, usually applied by cathodic electrocoating and intended to afford passive protection against corrosion. The average film thickness is 20 to 22 µm on the outside and 5 to 18 µm on both the inside and in cavities. The primer surfacer that follows is designed to cover any uneveness as possible in the substrate and thus create such smooth surfaces that they can serve as a base for a bright topcoat without the need for sanding. At the same time the primer surfacer fulfils the function of protecting against stone chipping and has to provide anticorrosive properties if, although it is rare, the electrocoat primer has been sanded through to the metal. The coloured layer usually consists of a basecoat and clearcoat applied in two "wet on wet" steps. By this is meant that the two coatings are applied in rapid succession without stoving of the basecoat, and are then stoved and crosslinked together. This topcoat system must then provide the correct colour and a high appearance after interaction with the substrate and the other coating layers (see Chapter 3.2.4).

On average, from 15 to 20 kg paint is consumed in the electrocoating and topcoating of a standard vehicle with a total surface area of 70 m^2 and an external surface area of 22 m^2.

Since the end of the 1990s, systems for coating car bodies have been put into operation which initially deployed a "wet-on-wet-on-wet" process which eliminated stoving of the primer surfacer [7.12.6]. Today, the "integrated wet-on-wet" application of basecoat utilises a first coat comprising new types of basecoat which provide both the colouring and UV protection for the primer as well as stone chipping protection, while the second coat

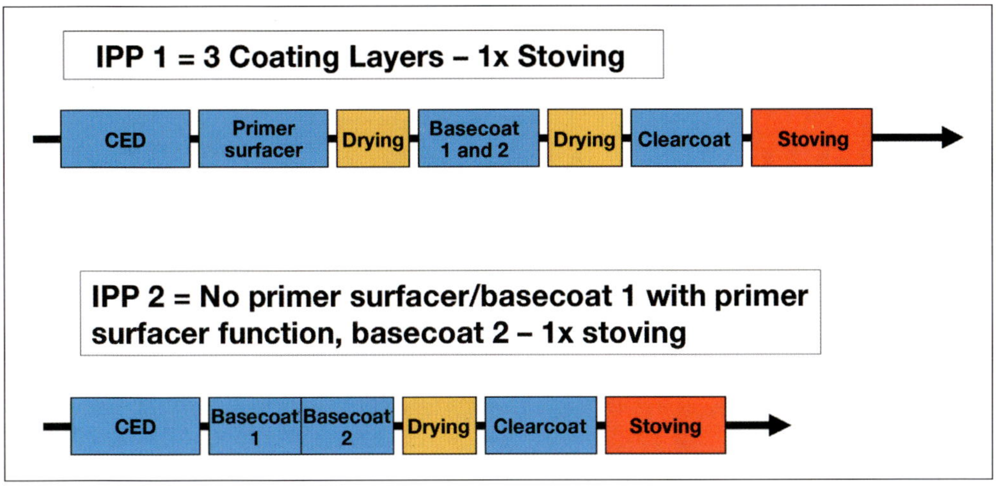

Figure 7.1.5: Process steps in the integrated paint process

provides the final colour followed by clear coat application [7.12.7]. This completely eliminates the need for primer surfacer application. Both processes are known as the integrated paint process (IPP). Occasionally, the term "IPP1" is used for the first process and "IPP2" for the second. Due to the significant reduction in energy and material costs of about 20 %, new coating lines are now being designed predominantly for the second process [7.12.8] mainly utilising water-borne paints, but also solvent-borne high solid coatings [7.12.9].

Application of cavity sealant, seam sealant and sound insulation is usually the responsibility of the paint shops. In the conventional process seam sealing occurs after electrocoating and exploits the curing of primer surfacers to achieve solidification whilst cavity sealant is applied after the completed coating process. These process steps are necessary to ensure excellent acoustic comfort and protection against corrosion but they are unpopular with the paint shops on account of soiling (see Chapter 6.2.7). In the newer IPP process, these steps are also carried out after electrocoating with a small downstream gelling oven and, in some cases, after topcoating.

The starting point for the quality of the coating in respect of all requirements imposed on anticorrosion protection and the attractiveness of the surface is the conceptual design of the vehicle body. Coatability depends just as much on the shape and design of a vehicle as on the chemical and physical properties of the materials to be coated. Anticorrosion protection and performance thus start on the design engineer's drawing board and with the selection of the materials to be used. It must be borne in mind that the surfaces of the various materials are not homogeneous substances. In addition, software programs are in use nowadays that can predict the paint results on car bodies without the need for trial coatings [7.12.10].

The steel panels from various pre-refinement stages used for the automotive industry, e.g. pure steel, electrolytically galvanised (EG), hot-dip galvanised (HDG), pre-phosphated, or organically precoated steel, as well as aluminium panels are subjected to various drawing processes and in some cases extreme deformation in the body-in-white shop, i.e. prior to the application of paint. It is therefore inevitable that there will be local variations in roughness. On the other hand, there is a directly proportional relationship between appearance, i.e. smoothness, build and gloss of the coating, and the roughness

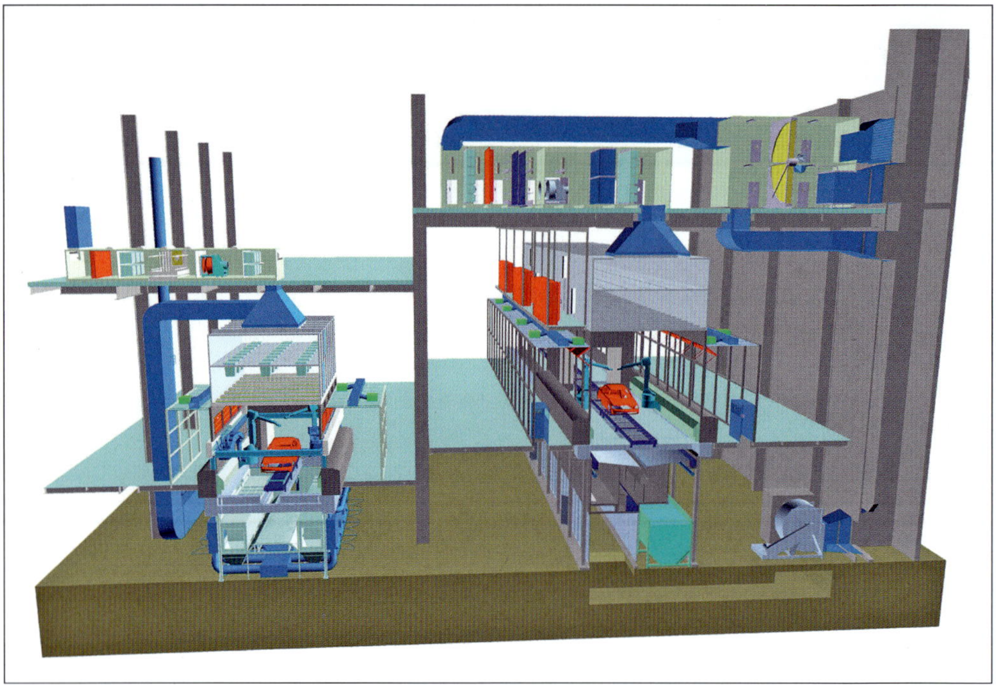

Figure 7.1.6: Comparison of plant concepts for recycling of overspray: left: dry; right: wet

of the metal panel (see Chapters 3.2.4 and 4.1.2). The substrate differences just mentioned can be compensated only to a certain extent by the paint materials and the coating process.

The length of a coating line, from transfer of the vehicle bodies from the body-in-white shop through to delivery to the final assembly stage, is usually between 2 and 3 km. It can be considerably longer if substantial effort is spent on making corrections to boost the "first run ok ratio" or if extreme line speeds need to be met.

"First run ok" means that a vehicle body passes through the paint shop without the need for any major corrective action and is accepted by final assembly as meeting the quality agreement. Given average speeds for all lines of between 3 and 6 m/min and cycle lengths of approx. 5 m per vehicle body, some 40 to 70 vehicle bodies are coated per hour. The dwell time in the paint shop is then 8 to 11 hours. However, American vehicle manufacturers often prefer faster lines operating at up to 12 m/min.

The environmental problems which have to be solved as coating lines operate more and more efficiently concern solvent emissions and waste, which consists of paint sludge and is due to overspray from application of primer surfacer and topcoat. In terms of solvent emissions, paints have undergone substantial improvements, particularly over the last 30 years (see Figure 5.3.3). This also applies to the quantities of waste generated in CED and overspray treatment [7.12.4].

Apart from the application units in all the stations, paint has to be supplied to a large number of delivery points via circulation systems. Here the paint is pumped out of a supply vessel of between 200 litres and 1,000 litres, circulated and thus fed to the spray guns or spray bells (see Chapter 4.2.1). The size of the supply vessels depends on the draw-off volume. For logistical reasons they are located together in one room, the "paint kitchen" or "paint mixing room", which has to meet regional storage and safety regulations (see Chapter 5.2.2). The tasks of the paint kitchen staff are not only to supply paints but also to perform quality controls, keep documentation and support the process of localising and avoiding defects.

To obtain a defect-free surface it is important to work under clean room conditions in the coating zones. Cleaning and corrective stages are also integrated into the workflow on the basis of experience. Preventive measures aimed at keeping vehicle bodies and ambient conditions clean in the paint shops are becoming increasingly prevalent [7.12.11].

The spraying units are chiefly electrostatic high-speed rotary atomisers which in some cases still use roof and side machines, the reciprocators, to apply the paint. According to estimates, robots now account for 90 % of spray-applied paint in Europe [7.12.12]. Occasionally, in special coating steps such as sill painting or coating of other areas that are difficult to access, recourse is made to the pneumatic spray atomiser on robots.

Ventilation of the spray booths and curing ovens incurs high energy costs. The use of robots has lowered the air requirement substantially. The overspray associated with spray application is diverted to the wash water underneath the systems by the air ducting in the spray booths. The wash water is reprocessed in appropriate cycles and freed from the paint materials which were coagulated by chemicals (see Chapter 4.2.1). New processes for separating dry overspray reduce the level of paint sludge, and substantially lower the energy consumption and the fresh air requirement of spray booths to < 20 % (see Chapter 4.2.1.5) [7.12.13].

Any necessary repairs of defective areas in the surface of the paint finish are identified in specially set-up inspection zones (see Figure 7.1.4) and are implemented by diverting the car body into a booth equipped with spot repair technology. Where surface defects are large or too numerous, the entire topcoat of the car body is usually repaired.

BASF's eco-efficiency analysis permits an ecological assessment of the entire automotive coating process to be carried out (see Chapter 5.5). This complex, but very systematic analysis of the processes, the application technology, the coating materials as well as the concepts for processing overspray, other waste and process chemicals reveals that the IPP processes, with their high solid or water-borne coatings, are the most environmentally friendly processes so far [7.12.14].

7.1.1 Pretreatment

In the body-in-white shop special attention is paid to the type of anticorrosion oils and drawing greases used to penetrate voids and grooves formed on the body panels by the drawing process. Their ease of removal in pretreatment baths coupled with a good level of compatibility with the paints substantially improve the efficiency of coating lines.

The pretreatment of vehicle bodies delivered from the body-in-white shop to the paint shop is organised by the various vehicle manufacturers in different ways. For some the cleaning process is the responsibility of the body-in-white shop and for others it is the full responsibility of the paint shops. In any event, pretreatment comprises the steps of cleaning, applying a conversion layer in the form of a phosphate coat 1 to 2 μm thick by means of a tri-cation phosphating process or, increasingly, a zirconium dioxide layer [7.12.15] a few nanometres thick, with subsequent rinsing and drying.

Cleaning and phosphating are mainly performed in dip baths suitably dimensioned to match the number of vehicle bodies/h. At rates of less than 15/h cycle tanks of ca. 60 m^3 are used, otherwise resort is made to continuous baths with a capacity of 150 to 450 m^3.

On continuous lines, bubbles of air in the interior and in cavities are avoided by routing the conveyor belt so as to execute tilting motions inside the baths. Use is increasingly being made of new concepts which rotate the vehicle bodies, such as the "Rodip", "VarioShuttle" and "E-Shuttle" methods, as they pass along the line (see Figure 7.1.8) [7.12.16]. These processes are also used in electrocoating.

Cleaning

In practice, a weakly alkaline degreasing solution of aqueous mixtures of salts, wetting agents and emulsifiers has become standard. The application temperatures are usually 40 °C, and up to 60 °C in individual cases (see Chapter 4.1.2). The process is usually a combination of spray and dip treatments followed by rinsing stages with deionised water to avoid salt residues. Dip treatment offers the best cleaning of drawing and anticorrosion oils as well as good cleaning of the internal areas but requires longer throughput times than spray cleaning.

The cleanliness of the entire vehicle body is key to ensuring that the coating lines are operating at high efficiency. By efficiency is meant the so-called "first run ok ratio" of a vehicle body passing through the entire line without the need for significant corrective action per number of total bodies.

Conversion Coatings

The most common pretreatment by conversion layers is tri-cation phosphating [7.12.17]. This type of pretreatment is an important step towards increasing the durability of a car body finish in terms of anticorrosion protection and adhesion of the subsequent electrocoating. The uniformity of the application of the conversion coat, which is 1 to 2 µm thick, is an important factor governing the quality of the coatings as a whole. It takes place in two stages, firstly activation of the surface and secondly build-up of the conversion coat in the form of phosphate crystals. Modern materials have the ability, albeit limited, to treat many different substrates such as steel [7.12.18], galvanised steel, aluminium and magnesium at the same time. In particular, the growing proportion of aluminium hang-on parts calls for greater attention. A share of 1.5 % was forecast for 2005 [7.12.19] and by some estimates this will grow to more than 2 % by 2013. Entire car bodies [7.12.20], hoods and boot

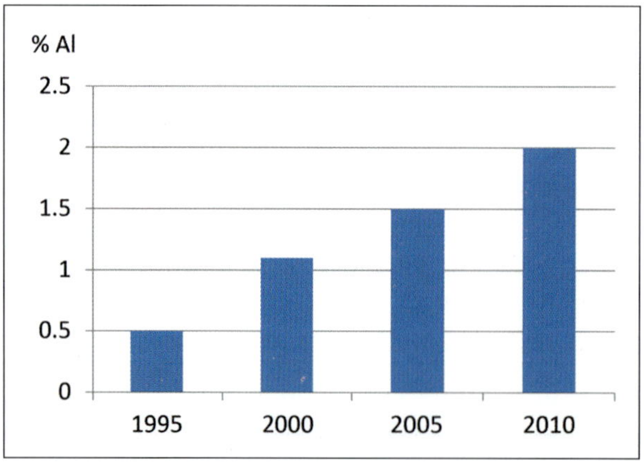

Figure 7.1.7: Growth in percentage share of aluminium in car bodies in western Europe

Coating Industry

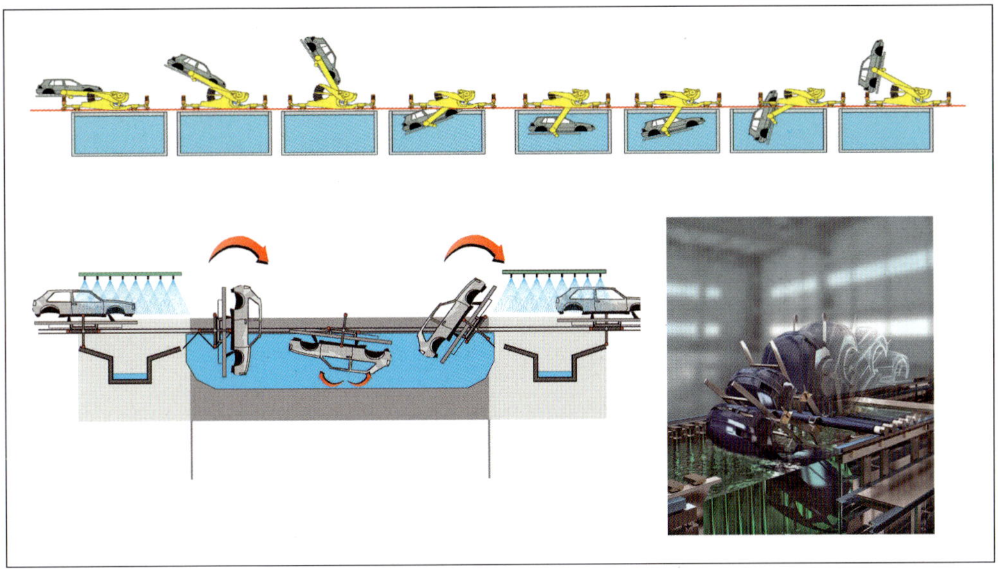

Figure 7.1.8: Rotary dip process for pretreatment and electrocoating of car bodies *Source: Eisenmann and Dürr*

lids account for most of it. Aluminium constituted 8 % of all the components in an automobile in 2010 [7.12.21].

On horizontal continuous lines, which are the most common type, the large amount of sludge generated during pretreatment of aluminium requires more stringent process control, more intensive rinsing at the bath exit and a specific adjustment of the pretreatment materials [7.12.22].

The state of the art includes zinc phosphate coatings containing manganese. These have superseded those containing nickel. Furthermore, the use of accelerators containing nitrates and nitrites to activate the surface is on the decline as they are being replaced by environmentally compatible active amino compounds as well as peroxides. In the case of aluminium substrates fluorides need to be used (see Chapter 4.1.2).

With regard to the following electrocoating process the rinsing zones downstream of the phosphating zones are sized so that as much of the phosphating solutions as possible is rinsed off. For one thing, that minimises contamination of the electrocoating bath by salts or other substances and, for another, flaws are avoided in the deposited electrocoating film which otherwise may be caused by salt residues on the surface of the workpiece. This is particularly necessary for surfaces where the appearance has to meet high standards.

In some cases a so-called passivation zone is located upstream of the rinsing zones. This improves the adhesion and anticorrosion protection properties of the electrocoat primer. The materials containing chromium that were used in the 1980s have nowadays been replaced by environmentally compatible, toxicologically safer chromium-free products such as titanates.

After the last rinsing cycle, which may be conducted with deionised water, it is advisable to install a hot blow zone with air temperatures of between 50 °C and 60 °C to completely dry the surface of the workpiece. Otherwise there is a risk of marks appearing during immersion into the electrocoating bath under tension, which would necessitate considerable sanding work on the coated workpiece.

Over the last few years, dip coating has become more common than spraying. The decisive factor here was the better cleaning and formation of conversion coats in cavities.

The "Vertak" process was used for some time because it took up less space. This entails treating the vehicle bodies by vertically dipping them into several cycle baths at each stage. "Rodip", "VarioShuttle" and "E-Shuttle" also occupy less space than continuous lines, rotating the vehicle body into the bath in order to finally pretreat it horizontally, albeit upside down. The advantage is that the visible surface to be coated is not contaminated by any sediment from the baths (see Figure 7.1.8 and Chapter 7.1.2).

As a result of the many treatment steps, not least on account of the high demands imposed on the cleanliness of the vehicle body when it is immersed in the downstream electrocoating bath, pretreatment lines are impressively long. Configurations vary, but can comprise tunnels up to 100 m long made of stainless steel sheet, through which the vehicle body has to pass. The effluent of a pretreatment line requires careful attention, although it can usually be disposed of in the normal sewage treatment plants.

The new pretreatment processes based on deposition of zirconium dioxide reduce outlay on pretreatment processes by up to 30 %. They are operated at room temperature, require fewer process steps and considerably reduce sludge formation (see Chapter 4.1.2) [7.12.23]. For car bodies made of aluminium and steel parts, a two-stage pretreatment of phosphating on steel and zirconium dioxide on aluminium is also carried out [7.12.21].

7.1.2 Electrocoating

Nowadays electrocoating is the most widespread process for evenly applying an anticorrosion primer to all metallic mass-produced products such as vehicles (see Chapter 4.2.1). Practically all vehicles throughout the world are primed by the electrocoating process.

Anodic electrocoating introduced at the end of the 1960s was a revolutionary process for intricate parts and offered major advantages in terms of enhanced utilisation of paint and

Parameter	Unit (conditions)	Anodic 1965	Anodic 1975	Cathodic ED 1980	Cathodic ED 2012
Solid content	(1 h/130 °C) %	8 – 15	12 – 18	15 – 20	15 – 20
Solvent content	%	10 – 15	5 – 10	3 – 5	< 1
Film thickness	µm	30 – 40	25 – 35	15 – 20	18 – 22
Lead/tin		Y/Y	Y/Y	Y/Y	N/N
pH		7 – 9	7 – 9	6 – 8	5 – 7
Bath temperature	°C	20 – 25	20 – 25	25 – 30	30 – 35
Deposition time	s	120 – 200	120 – 200	120 – 200	120 – 200
Voltage	V	150 – 250	150 – 250	250 – 350	350 – 500
Electr. equivalent	mg/C	60 – 80	50 – 70	30 – 50	30 – 40

Figure 7.1.9: Technical data of electrocoating generations (in the tank)

hence greater economy and environmental friendliness compared to the dip processes hitherto employed. The introduction of ultrafiltration boosted utilisation even further. The ultrafiltrate produced in ultrafiltration installations serves as a rinsing medium in a counter-current system for washing off electrocoat clinging to the surface and for returning it to the tank. In this way material efficiencies of > 98 % are achieved.

The chemical basis of the first anodic primers was maleinised linseed oil, which was later replaced by maleinised polybutadiene. These were self-crosslinking binders that were hardened with the aid of catalysts at approx. 160 °C. In the mid-1970s cathodic electrodeposition coatings replaced the anodic systems and have remained state of the art. The latest generations are characterised by very low solvent contents, the absence of lead and a high level of reliability.

The reasons for the rapid switch from anodic to cathodic electrocoats despite significant conversion were as follows:

- 3 to 4 times better protection against corrosion in conjunction with a suitable pretreatment on steel, prevention of iron dissolution and enhanced crosslinking through the use of new film forming agents
- Much-improved throwing power, i.e. better coating of cavities
- Greater process reliability due to replenishment with two components consisting of neutralised and pre-dispersed binder as well as a pigment concentrate and greater chemical stability of the components in water
- Smaller film thickness
- Greater alkaline resistance of the film at the points of contact in zinc/iron hybrid designs

The magnitude of these advantages varies from one industrial application to another. In all cases, however, process reliability is crucial because industrial mass production relies heavily on the homogeneity of the products employed. Since a coating line handles up to 1,500 vehicle bodies a day, each of which has a surface area of 60 to 90 m^2, the demands imposed on the formulation and production of the coating materials are high. This also applies to process control and plant engineering at the user. Both physical and analytical parameters serve as control criteria for maintaining the electrocoat tank at the same level of quality as regards the coating result. Nowadays there are control circuits which, for example, maintain the solid content in the bath via the power consumption. Another important contribution to process reliability has been made by the aforementioned post-addition of fully neutralised coating material and the splitting into two components. Post-addition of already dispersed material simplifies this process by reducing it to simple addition, without high risk of flocculation. However, the nondepos-ited acid builds up in the tank during the process and has to be withdrawn from the system via an anolyte circuit (see Chapter 4.2.1).

Property/test	Unit	Anodic 1975	Cathodic ED 2012
Film thickness	µm	30	20
Salt spray test ISO 9227 (480 h)	mm[1]	2	0
Cyclical climate test VDA 621 (10 cycles)	mm[1]	Destroyed	< 2

Figure 7.1.10: Comparison of anticorrosion protection of anodic versus cathodic electrodeposition coatings on zinc-phosphated steel panels as determined by rusting at the scribe[1]

Coating Industry

Figure 7.1.11: Dipping a car body into an electrocoating tank
Source: BASF Coatings

The improvement in anticorrosion protection afforded by cathodic electrocoats in comparison to their anodic counterparts can be verified by various tests (see Figure 7.1.10).

The hourly throughput determines, in a similar way to pretreatment, whether the tanks are operated in cyclic or continuous mode. The tank capacity of cyclic systems can be up to 100 m^3 compared with up to 450 m^3 for continuous systems.

The actual electrocoating process commences upon transfer from pretreatment [7.12.24]. The vehicle body can be wet or dry when it is dipped into the tank. Cyclic systems utilise one rectifier to set a variable voltage programme aimed at achiev-ing optimum coating results in terms of surface finish, film thickness and throwing power. In the case of continuous systems, the voltage programme is set with at least two rectifiers. The input rail is separated from the main rail for the central dipping area and the exit zone. This is necessary for reducing high current peaks which can cause surface defects through the development of high local heat and excessive gas (see Chapter 6.2.2). The new "Rodip", "VarioShuttle" and "E-Shuttle" systems can dip the car bodies faster and so they lower the risk of surface marks caused by excessively slow wetting of the surfaces by the electrocoating in the entrance area. They allow the car bodies to be coated upside down, i.e. reverse coated, so that sedimentation does not occur on the visible surfaces [7.12.25]. Another advantage is their smaller required space due to smaller tanks.

For consistent coating quality, the bath temperature must remain within ± 0.5 °C of the specified values. Heat exchangers must be designed in such a way that they can dissipate the heat generated by the flow of electric current. Since the largest currents occur in the dipping area, flow rates there should be high enough to dissipate the heat rapidly. For this reason and to prevent the pigmented electrocoats from settling, the bath contents must be circulated continuously. As a rough guide, the pumps should be of a size capable of circulating the bath about five times per hour (see Chapter 4.2.1).

The electrocoating process is so reliable that many automotive paint shops now have only one tank supplying a number of topcoat lines. This makes it possible to prime up to 1,500 units a day. Given this level of throughput, regular material replenishment is essential for ensuring a consistent level of quality. Automatic post-addition linked to power consumption is particularly expedient for this (see Chapter 4.2.1).

Upon exiting the tank, the film which has been compacted by electroosmotic effects during deposition and adheres firmly to the substrate still has paint material clinging to it that must be removed by rinsing. As this material can still be used, the vehicle body is rinsed by the counter-current method with the ultrafiltrate removed from the tank, and the rinsing water is returned to the coating bath (see Chapter 4.2.1). This takes place in at least two cascades. Modern electrocoats no longer require any further rinsing with wholly deionised water after they have been rinsed with pure ultrafiltrate.

The design of the rinsing zones must ensure that paint is prevented from drying on the coated workpiece. The vehicle body should reach the first rinsing process, which is usually conducted with a spray halo, in less than one minute after emerging from the tank. Once paint runs have dried on they can only be removed afterwards by sanding and they cause considerable disruption to the overall vehicle coating set-up. As in pretreatment, the main rinsing zones are configured as dip tanks. To the extent that the ultrafiltrate feed contains < 0.4 % solid, these baths will have solid contents of less than 1 % for the sake of efficient rinsing. Material efficiencies of over 99 % are obtained under such conditions.

Deposition of the electrocoat and the rinsing process are followed by a pass through the stoving oven. There the coating's residual water content of approx. 5 % is removed and good levelling along with crosslinking to yield the anticorrosion protection film are achieved. Stoving takes place at a workpiece temperature of between 160 °C and 190 °C for at least 15 minutes, with the actual duration depending on the type of electrocoat. The crosslinking process releases organic substances and this can lead to deposits in the oven. To comply with most legal regulations, these have to be sent for thermal afterburning. In view of this aspect an optimum supply and circulation of air must be set in the ovens, which are usually of type A design (see Chapter 4.3.1).

An electrocoating process can for the most part be operated in a closed loop without significant pollution of the environment with waste materials. The environmental pollution can have four sources of emissions:

1. Emission of volatile compounds in the tank and rinsing zone areas
2. Emission of organic compounds in the oven due to crosslinking reactions and elevated paint film temperature
3. Waste water in the form of ultrafiltrate or rinsing water
4. Anolyte circuit containing organic acids.

Coating Industry

The solvent content of the latest generation of cathodic bath materials is less than 0.5 %. Together with the usual air flow via the tank which is directed into the stoving ovens, this causes the concentration of organic compounds to be below the maximum allowable concentrations at the workplace. The minimally polluted air, together with the cleavage products from the crosslinking reaction in the stoving ovens, is then treated by thermal afterburning.

Any ultrafiltrate that occasionally has to be disposed of can be fed directly to the sewage treatment plants because it is wholly biodegradable under relatively short dwell times. The same applies to the substances in the anolyte circuit. It should be remembered that the pH there is about 4, whilst the ultrafiltrate is within the working range of the electrocoat, i.e. pH 5.8.

Before the primed vehicle body is transferred to the next station, the surface is inspected and repair steps may have to take place, usually in the form of sanding of defective surfaces. It is recommended that wet sanding be performed as that reduces the level of dirt and dust generated by such sanding operations. Apart from that, very careful cleaning prior to application of the primer surfacer is needed.

7.1.3 Seam Sealant and Underbody Protection

In most paint shops that do not employ IPP the electrocoated vehicle body is sealed with sealing materials and provided with underbody protection. In isolated cases soundproofing mats are also inserted at this stage of the coating process.

The electrocoat primer makes a suitable substrate for the plastisol materials (e.g. PVC or PUR). The sealing materials considerably improve the anticorrosion protection of cavities with poor accessibility for CED, whilst the underbody protection coating of approx. 1 mm film thickness protects against stone chipping. Underbody protection coating has now been widely superseded by plastic panels due to the substantial contamination of the coated surface and the decline in efficiency of the paint shop associated with application.

The seam sealant is applied at elevated temperature by robots using airless spray guns. Manual adjustments may still be necessary in isolated cases if the seams do not always have the same gap width and if the seam has to be coated where there are awkward connecting joints with several body parts.

The underbody protection coating too is applied by the airless method with the vehicle body jacked up. The spray jet offers greater precision and manual corrections are rarely necessary.

The pieces of soundproofing material are predominantly inserted by hand. They, along with the seam sealing and underbody protection materials, are thermally treated in the downstream primer surfacer oven. In the case of PVC, the emulsion particles will then dissolve in the plasticiser. The material cools to a gel, and forms a film which is highly resistant to physical stress caused by abrasion and stone chipping.

For some time, a number of automotive manufacturers also used to apply special stone chip protection coatings or plastisols in lower outer areas, e.g. door sills and front spoilers. Modern primer materials have now rendered the use of such materials superfluous.

7.1.4 Primer Surfacer

Unlike the electrocoating the primer surfacer is essentially spray-applied to the exterior, in the entrance area of vehicle bodies, and in the engine and luggage compartments. The corollary is that the vehicle body is thus given a second external coating, which not only provides extra anticorrosion protection and stone chipping protection and shields the electrocoating from possible attack by ultraviolet radiation but also serves as further preparation for a smooth surface and a bright top coating.

The multi-functionality of primer surfacers makes it difficult to formulate them. Nowadays the most popular film forming agents are polyurethane-modified polyesters in conjunction with epoxy resins. The corresponding coating formulations are rounded out with crosslinkers based on melamine or blocked isocyanates. In Europe they are predominantly water-borne coating materials generally characterised by a high level of pigmentation. Thus, the solid content is between 50 % and 65 % and the binder/pigment ratio is between 1 and 0.5. Some 10 % of all vehicle bodies throughout the world are coated with powder primer surfacers nowadays. With a layer thickness of 0.1 mm and a high material yield due to recycling on the line, they constitute an economical and ecological alternative to water-borne primer surfacers which are already fully functional in a thickness of about 25 µm. Crosslinking of most formulations is performed at a workpiece temperature > 150 °C for a period of 15 minutes, although around 130 °C is also employed.

The purpose of primer surfacers is to fill out and smooth any unevenness to level the substrate for the following layers of topcoat. Apart from good build, good sandability is an important quality criterion, even though the goal of modern coating lines is to make the sanding of primer surfacer superfluous. On the other hand, levelling of substrates has grown more important because increasingly disparate substrates such as steel, galvanised steel, aluminium and plastics are being used which have to produce an identical surface on the vehicle body despite the fact that some of them differ substantially in their surface structures.

Over the last few years, primer surfacers have received special attention in relation to the hiding power of the coloured topcoat. Consequently, the colour of the primer surfacer is key to minimising the thickness of the subsequent and more expensive basecoats for any given range of topcoat colours. Many vehicle companies have adopted the "colour keyed primer surfacer" concept. Here between 3 and 7 primer surfacer colours are supplied to the spray booths and matched to the colour of the basecoat to obtain maximum colour agreement and minimum basecoat film thickness. Some auto makers have elaborated on this to create a so-called "colour specific primer surfacer" concept in which each basecoat is based on a specific primer surfacer of largely identical colour.

Whereas the emphasis in these considerations is on the visual quality of the primer surfacer on one hand, this intermediate layer has to fulfil crucial tasks in the area of physical/technological properties such as stone chipping protection on the other. Wherever protection of the substrate against corrosion is missing, e.g. where the electrocoating has been sanded right through, the primer surfacer also has to fulfil anticorrosion protection functions.

In **preparation** for application of primer surfacers the vehicle body often undergoes a cleaning process. In most cases this is first performed manually and is often completed with an automatic process rather like a car wash station. Manual cleaning of the surface is done with special, lint-free wiping cloths that are usually damp. Any seam scalant resulting from improper application also has to be removed. The automatic cleaning systems have rotary brushes which are fitted with ostrich feathers to prevent any electrostatic charging and ensuing accumulation of dirt. All the measures serve to achieve the highest-possible "first run ok ratio".

On some coating lines the surface of the body is cleaned with air from air locks and precision air flow. The air flow is designed in such a way as to easily separate out the particles of dirt.

Liquid primer surfacer can be applied to the vehicle body via a spraying assembly and an appropriate number of supply lines, the circulation system pipes, and without any substantial loss of material. The circulation system is supplied from a central supply

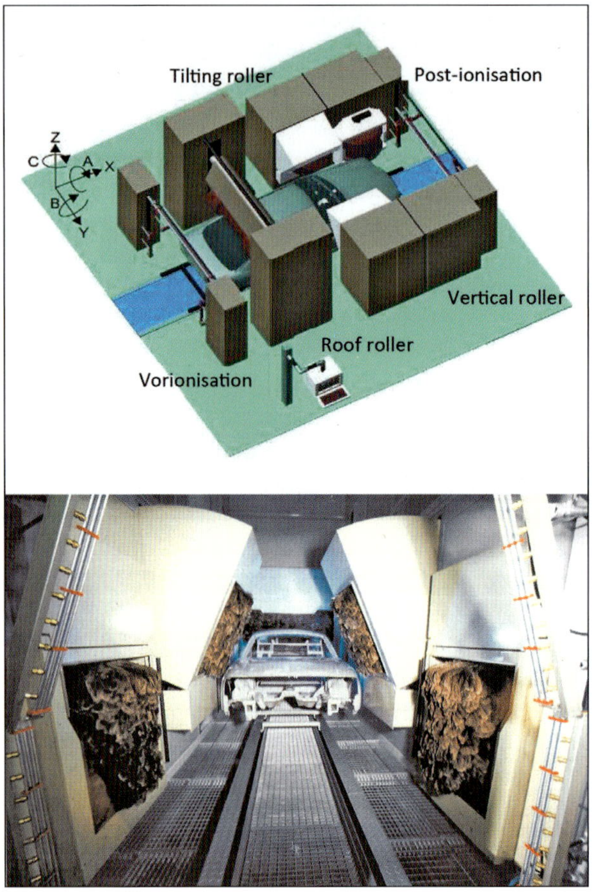

Figure 7.1.12: Schematic drawing and picture of an automated cleaning station for preparing defect-free coatings in automotive construction *Source: Dürr*

tank which usually has a refill tank. The coating material in the refill tank can be set to the required viscosity level. The circulation system pipes, which can be several hundred metres long, are fed from the supply tanks. Any unconsumed excess material flows back into the tank via a back-pressure regulating valve. At the outlets of the roughly 7 to 10 spraying units per coating line are pressure controllers which ensure consistent material pressure at the outlets for the sake of uniform film thickness. As water-borne coating materials are more rheologically demanding, positive-displacement systems have now largely superseded their pressure-controlled counterparts (see Chapter 4.2.1).

Whilst coatings used to be applied with compressed air, the state of the art is to use electrostatically assisted high-speed rotary bells. The requisite number of spray units largely depends on the line speed and the film thickness of the primer surfacers, usually water-borne, to be applied. To reach points with poor accessibility such as the door rocker panel in the interior, the first stage of the coating process usually features two robots with pneumatically atomising spray units. Each of these is followed by two side machines and two roof machines (reciprocators) for the overall coating. The spray jets of the latter can be programmed to follow the contours and ensure a uniform film thickness. When only one colour is used, internal charging is employed whilst, if more than one colour is employed, external charging is more favourable from the viewpoint of capital investment. The two methods have slightly different levels of transfer efficiency, with external charging being slightly inferior. Roof and side machines are increasingly being replaced by robots, which can be adjusted even more accurately. System availability in the event of robot failure is an important consideration in the design of such lines [7.12.26].

Cleaning of spray booths is a key factor for high "first run ok ratios". The frequency and the economics of overspray recovery from booth water [7.12.27] depend on the target transfer efficiency.

Environmentally-friendly, eco-efficient powder coating was first used in the early 1990s for primer surfacers, serving initially as stone chipping primer. The high film thickness for good stone chipping protection, absence of solvents, high level of transfer efficiency achieved by recycling the overspray directly in the application system, plus the lower investment and running costs were the main reasons for the spread of powder primer surfacers. Applications in the USA were additionally encouraged by legislation [7.12.28].

Figure 7.1.13: Application of powder primer surfacers in the automotive industry *Source: BASF Coatings*

An estimated 5 million vehicles were coated with powder primers in 2012. In the commercial vehicles sector some manufacturers have now switched to priming the all-galvanised cabs of commercial vehicles with powder only and eliminating the electrocoating.

Powder coating technology has been slow to penetrate the automotive industry up to now because of the need to change the metering equipment, the fact that powder coatings are handled differently from liquid coating materials in an industrial context, and problems with colour changes.

Application of powder primers requires that the car body be cleaned just as thoroughly as is the case for application of water-borne primer surfacers.

The low volumetric flow of powder and the high film thickness necessitate roughly twice as many spray units as are necessary for liquid paints. The highest delivery rates for applying the powder with the highest-possible transfer efficiency are provided by electrostatically assisted rotary atomisers. The initial transfer efficiency varies with the type of body but is at most 60 %. Consequently, the booth must be carefully designed for reusing the oversprayed powder. This calls for appropriate handling of the vehicle body, efficient air flow and a high degree of cleanliness. For reasons of quality, primer surfacers are frequently applied in high film thicknesses in the underbody area. That is where the recovered powder is mainly used. This can boost material transfer efficiencies to > 98 %. Powder layers 80 to 100 µm thick yield surfaces which are just as smooth as those obtained with liquid coatings, with the added advantage of exhibiting less "orange peel" on vertical surfaces.

In the case of water-borne or solvent-borne primer surfacers, after the body has left the application zone it passes through a flash-off zone before entering the stoving oven. The ovens are dimensioned to suit the type of coating applied, the workpiece and the line speed. The heating curve plays a major role in ensuring defect-free, smooth surfaces (see Chapter 4.3). After leaving the oven, passing through the sanding and correction zone and repeated cleaning of the surface, the vehicle body passes into the topcoat application zone.

The new basecoats employed in the IPP method are rendering water-borne and conventional primer surfacers superfluous (see Figure 7.1.5) [7.12.29].

7.1.5 Topcoat Application

High demands are made on the last coating layer because it determines the visual impression of the vehicle throughout its service life and is subjected to enormous stresses in all parts of the world during that period. By virtue of its interaction with all coating layers the topcoat helps to guarantee the level of physical properties afforded by the coating. Worthy of special mention in this regard are good adhesion between all coating layers and with the substrate, and a harmonised level of formability and elasticity in combination with a film surface of high hardness. Topcoats are furthermore expected to meet requirements for high gloss, good levelling, resistance to ultraviolet radiation in combination with weathering, and resistance to chemicals and organic solvents such as fuels (see Chapters 3.2 and 3.3). At the same time, it is the topcoat which provides the colour. Under the conditions described it must guarantee a high level of surface uniformity, lend itself to repair at the coating line and, during use of the vehicle and throughout its service life, consistency with regard to the visual and physical/technological properties.

Colour styling nowadays plays a much more important role than some 30 years ago. This is evident from the fact that there are now entire departments in automotive manufacturers and coating manufacturers dedicated to monitoring colour trends, developing new colours and special effects, and selecting them carefully for new models (see below) [7.12.30].

Figure 7.1.14: Example of a coating that supports the design
Source: BASF Coatings

The importance of colour styling is also supported technologically, on one hand by new application processes and on the other by new raw materials and pigments, which have created enormous scope for colour variations. In a reflection of the need to meet the described requirements more fully, coating materials and application processes have undergone substantial changes over the last 30 years. For instance, the topcoat nowadays is applied in two separate layers by the "wet on wet" process almost everywhere. Consequently, the two functions, namely colour and durability, have been separated and assigned to the first coat and the second coat (the clearcoat) respectively. The paint formulations for both coats have been optimised so much in respect of the described functions that durability levels, e.g. of gloss and colour, have been extended to 10 years and more. An important formulation step was not only the use of acrylic resins possessing greater UV resistance but also the use of UV absorbers and radical scavengers at the end of the 1970s as well as the development of pigments that have a higher level of lightfastness (see Chapter 2.1.3).

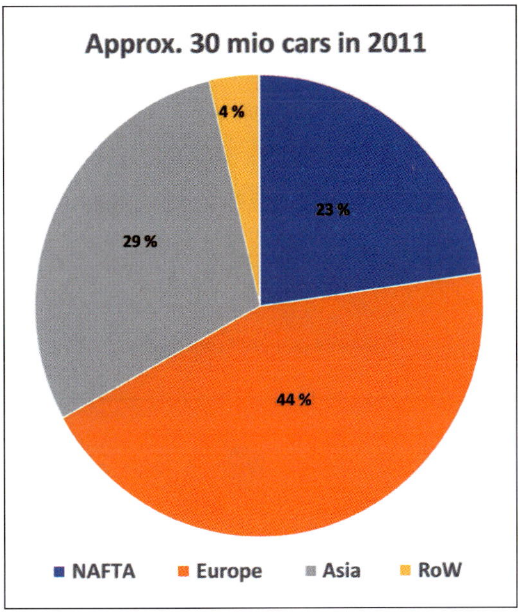

Figure 7.1.15: Regional distribution of automotive water-borne basecoat coating lines

The introduction of this two-layer topcoat also paved the way for much brighter metal effects. Thus, light-dark effects were obtained which varied with the angle of view and which were previously unknown from single-layer topcoats. These two-layer topcoats are of such good overall quality that solid colours, previously single-coat products, are now applied in two coats too. This has opened up scope for setting up standardised coating lines which can meet customers' wishes more easily, no matter which colour is currently popular.

Experts refer to the first layer as the basecoat (BC) and the next layer as the clearcoat (CC).

Basecoats

Solvent-borne basecoat is a low-solid paint which can contain 45 % solid for solid colours in North America and 35 % solid in other countries. The solid content in effect coatings is even lower. The associated high levels of solvent emissions have seen water-borne and high solid coating materials formulations become common worldwide. Compared with their solvent-borne counterparts, water-borne coatings only have a solvent content of 10 to 20 %. Since launch in 1986, they now account for about 40 % of passenger carsworldwide (see Figure 7.1.15). Basecoats are applied in a dry film thickness of 12 to 30 µm, with the precise thickness depending on the colour.

Before the vehicle body reaches the automatic coating machines any areas which are difficult to access are precoated by robots or by hand. These include the door sills and some zones in the engine compartment and the boot. In the case of robot application the vehicle body is usually stopped briefly in so-called "stop and go" mode and the hoods, trunks and doors are opened so that coating can take place automatically. This may have to be repeated for clearcoat application. There is no need for the "stop and go" mode in manual application.

Effect coatings are applied twice at line speeds of about 3 m/min and upwards because of their low solid content. Solid-colour coatings usually require only a single application with electrostatically assisted high-speed bells because they have a higher solid content.

		Solvent-Borne BC	Water-Borne BC
Metallic	NV	25 %	25 %
	Solvents/NV	3.0	0.5
Solid colour	NV	40 %	35 %
	Solvents/NV	1.5	0.3
Total [1)]	Solvents/NV	2.5	0.4

NV = nonvolatiles
[1)] Metallic : solid colour = 60 : 40

Figure 7.1.16: Reduction in solvent content due to water-borne basecoats (BC), expressed in terms of nonvolatiles (NV)

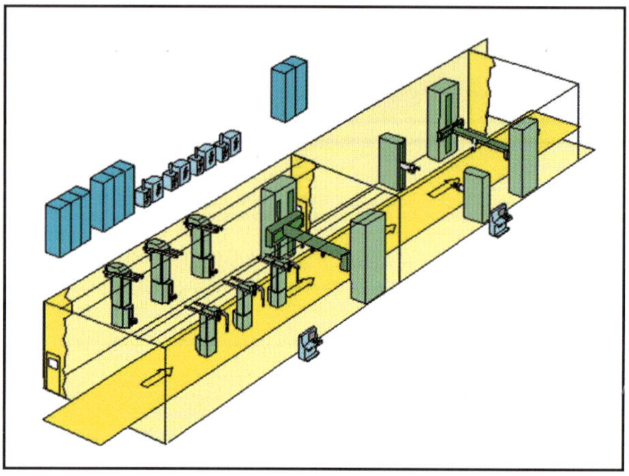

Figure 7.1.17: Typical arrangement of bells for roof and side machines
Source: Dürr

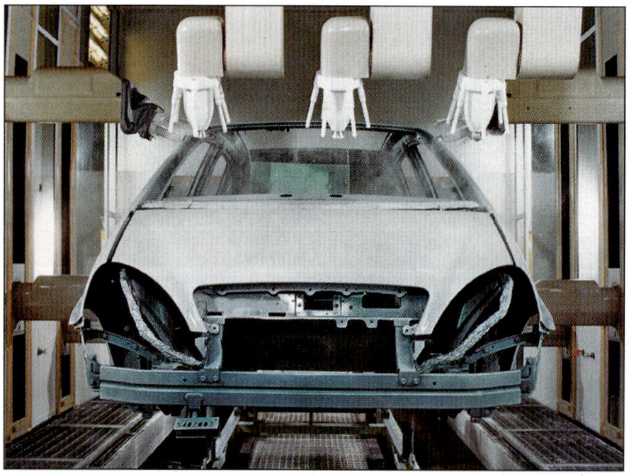

Figure 7.1.18: Application of automotive topcoats by high-speed rotary atomisers and external charging electrodes *Source: BASF Coatings*

Coating Industry

The first coat of effect coatings too is usually applied with electrostatically assisted high-speed bells. The second coat, comprising about 30 % of the requisite quantity of paint, is often still applied pneumatically so that the desired high orientation of the flake pigments may be obtained and for the sake of easy repair. To increase the transfer efficiency and lower the costs, high-speed bells, so-called micro-bells, are being used increasingly for the second layer of basecoat nowadays. The resultant coating effect is a very close match to that of pneumatic atomisation and facilitates manual repairs, mostly by pneumatic application.

The high conductivity of water-borne paints and the many colours involved in OEM coating, usually more than 12, means that electrostatic charging is of the external type. This largely eliminates the need for elaborate electrical isolation of the paint supply for the many colours. However, the development of the "cartridge" system is so far advanced that even direct charging systems technically configured to afford perfect potential isolation are used in production.

The bells are usually arranged so that the sides of the vehicle bodies are coated first and only then does the roof machine start up. Both types of machine are capable of following the contours of the vehicle bodies within certain limits. As the spray jet does not distribute the material evenly, the roof machine works at right angles to the movement of the vehicle body for greater coat uniformity (see Chapter 4.2.1).

A brief flash-off time of 1 to 2 minutes is provided between the two application stages of the effect coating. This prevents incipient dissolution of the first coat and thus loss of orientation of the effect particles. This is particularly important for ensuring reliable reproducibility of the flop effects and the associated perceived colour. The general rule is that after intermediate drying the basecoat layers should have a residual water content of less than 10 %. In the IPP process, the special coatings used for the first application take on the tasks that were previously provided by the primer surfacer, namely UV protection for the cathodic ED coating and stone chipping protection. The colour is identical with that of the second application (see Figure 7.1.5).

When such basecoats are being formulated, the composition of all the paint components should be optimised so as to ensure that the perceived colour is as uniform as possible for the given application conditions. The reason is that different shapes of vehicle body give rise to different distances and overlapping of the spray jets. Colours which are not very robust will elicit significant differences in colour perception and effect on the same vehicle body

Effect paints for automotive coating can be classified as metal effect paints with brightness flop ("flip-flop") [7.12.31], pearlescent effect paints with slightly changing colour flop, and colour flop paints with considerable colour change.

The perceived colours, which vary with the viewing angle, are the outcome of two physical principles at play. Metal effect paints reflect the incident light to the extent that the aluminium flakes are orientated and thus create a light-dark effect. Colour flop paints require flakes which are provided with two or more oxide coatings of different refractive index. In this case, it is the interference of certain wavelength components which gives rise to the angle-dependent colour. Pearlescent effect paints have a less pronounced effect which comes about in a similar way to that of natural mother of pearl (see Chapter 2.1.3). All the pigment categories can be combined to generate a whole host of new perceived colours (see Chapter 3.2.4).

Nowadays, the use of multi-angle optical instruments to perform the necessary quality control checks and to draw up the specification as precisely as possible is state of the art. Nevertheless, owing to the aforementioned problems in respect of the variations in shape of vehicle bodies and the vagaries of the application method, visual inspections are still performed at the vehicle manufacturer's (see Chapter 3.2.4).

Clearcoats

After application of the basecoats the vehicle body passes through a short intermediate drying zone termed "flash off". The conditions here are chosen such that downstream clearcoat application exerts little or no influence on the colouration or effects of basecoats. Usually the parameters are as follows: 50 to 70 °C workpiece temperature for 3 to 5 minutes. As is the case in most drying zones, infrared lamps are also used here to speed up heating and shorten the zones. Infrared lamps essentially heat up the coating, not the vehicle body, and so the cooling zones can also be smaller (see Chapter 4.3.1).

		North America	Europa
2-Component clearcoats	Solid content	65 %	55 %
	VOC	250 g/l	350 g/l
1-Component clearcoats	Solid content	60 %	50 %
	VOC	300 g/l	420 g/l
volatile organic compounds (VOCs)			

Figure 7.1.19: Solid content and VOC vales of common clearcoats

The clearcoats increasingly being used nowadays are two-component acrylic resin-borne bases which are cured with polyisocyanates (see Chapter 2.1.1). Mixing of the two components in the proportion of 3:1 to 5:1 is performed in the gun or bell just prior to atomisation. The clearcoat and dried basecoat are cured together at 130 °C to 150 °C to yield a durable film.

One-component clearcoats, which are used in addition to their two-component counterparts, are also acrylic resin-based and they contain melamine resins or blocked isocyanates by way of crosslinker. The stoving conditions are comparable to those for 2-component clearcoats.

The film thickness ranges from 40 to 50 µm. It holds the key to the brilliance, smoothness and appearance of the applied coating. In particular, it is important to observe the film thickness on vertical surfaces to avoid any sagging. This is essential for ensuring that the difference in orange peel between the horizontal surfaces (minor) and vertical surfaces (major) is not too striking. Thixotropy, or specific pseudoplastic behaviour, forms the basis for clearcoat formulations which can be applied reliably. This viscosity characteristic is achieved either with the aid of additives or special resin combinations (see Chapter 2.1.4).

The one-component and two-component types are dissolved in solvent and, if high solid systems, can have a solid content of up to 65 %, as is particularly the case in North America where this level is achieved by optimising the resin and improving the dissolving power of all the paint components (see Chapter 5.6.1). The North American coatings industry has been developing such high solid paints for approx. 25 years. By means of polymerisation technology it succeeded in lowering the molecular weight of the base resins and crosslinkers to such an extent that paint formulations with relatively high solids but identical viscosity can be achieved. The same applies to polyisocyanates and melamine crosslinkers.

The European coatings industry and paint-applying automotive industry opted for water-borne basecoats and so gained flexibility in terms of application reliability, performance profile and emissions from clearcoats. Water-borne, solvent-free products such as slurries and powders found application for a while [7.12.32, 7.12.33] but failed to gain acceptance due to problems with appearance and stone chipping resistance, with the result that only solvent-borne clearcoats are used around the world. Intensive work is being done on developing 2-component water-borne clearcoats [7.12.34].

Colour Trends and Colour Styling

The impression which a vehicle makes on a potential buyer depends very heavily on how it is colour-styled. Over the years, greater attention has been paid to regional aspects of automobile colours around the world. Substantial differences in the frequency of vehicle colours in North America, Asia and Europe (see Figures 2.1.159 and 7.1.20) are a reflection of different regional consumer tastes [7.12.35]. Automotive coatings manufacturers therefore maintain design departments tasked with harmonising the range of colours available for vehicles to prevailing colour trends. They do this by keeping a very close eye on trends in fashion, culture, art, furniture and social developments. Additionally, they need to make room for new colour effects aimed at creating greater market differentiation and attracting customers to those vehicles which are either bound for or already on the drawing board. At the same time, the design departments permanently liaise with the auto makers to agree on the optimum range of colours for new models or for special editions. Predictions of customers' taste in colour are important, but so too is the feasibility of such colours in terms of application reliability [7.12.3]. Over the last few years there has been a distinct trend towards achromatic colours. By this is meant black and grey shades as well as metallic ones with a slight tint of colour. This trend is most evident in Europe. These colours convey a neutral impression, yet emphasise the character of high-quality vehicles.

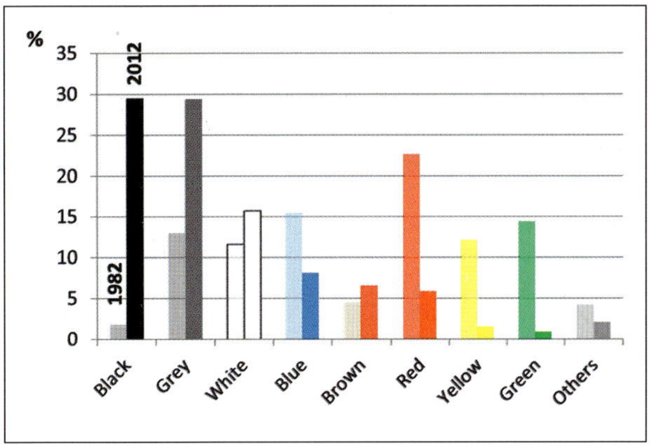

Figure 7.1.20: Popular trends in automotive colours
Source: Federal Motor Transport Authority, Germany

All in all, when customers purchase a new vehicle they have an ever widening selection of colours to choose from, provided they are prepared to pay more. Some manufacturers even supply so-called "tie colours", which means that the vehicles are coated and delivered in colours that match the customer's wishes, rather like choosing ties.

Matt finishes are currently one design element which has gained a certain degree of prominence. They are mostly achieved with foils and, more rarely, with matt clearcoats [7.12.36] in OEM coating. Effect pigments such as the "Xirallic" type, or so-called VMPs (vacuum metallised pigments) made of aluminium foils, which are being superseded by PVD-coated flakes [7.12.37], are opening up more and more new colour effect variants which designers are readily adopting and testing for their acceptance.

It takes an average of 1 to 2 years to launch new colours, due to the need to ensure safe handling and reliable quality in the paint shops and during repairs. Long-term tests such as exposure in Florida or accelerated tests in test chambers (see Chapter 3.3.2) are essential.

7.1.6 Repair in Automotive OEM Lines

Despite every precaution, sometimes even under clean room conditions, not all defects such as dirt inclusions, pinholes and other coating abnormalities can be completely avoided. Upon leaving the cooling zone of the clearcoat stoving oven the coated vehicle body is subjected to careful inspection. This is done at a dedicated work station under special illumination and by some 6 to 20 trained staff being able to detect surface defects quickly and reliably under those conditions (see Chapter 6.2).

Figure 7.1.21: Inspection line for coated car bodies

Source: BASF Coatings

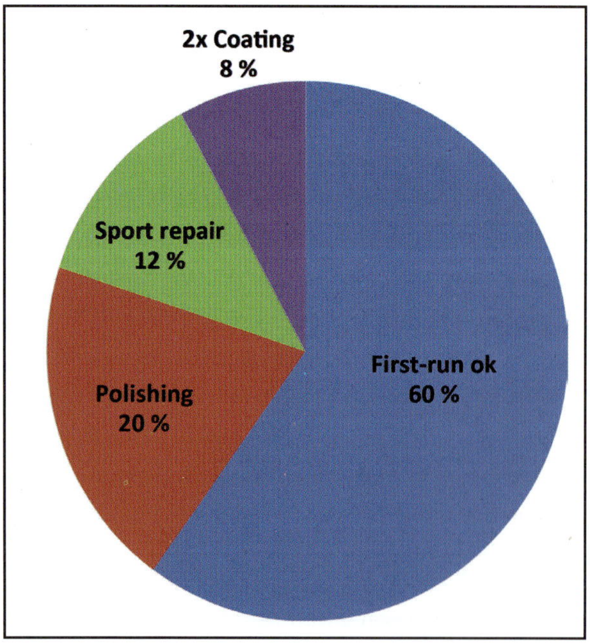

Figure 7.1.22: Common coating process data for troubleshooting in efficient automotive paint shops

Minor defects, especially in the clearcoat, are remedied by sanding and polishing, others call for either a spot repair or a whole body repair, whereby the vehicle body is prepared accordingly, usually with light sanding, and passes through the entire topcoat process a second time.

Spot repairs of rejected vehicle bodies are conducted manually in specially equipped booths. Special tools are used to prepare the small defects only by means of skilful light sanding, planing and cutting out and then refinishing with special spraying units chosen for the particular problem. The OEM paints can be used for this purpose. The necessary crosslinking temperature is created locally by infrared lamps at the point of repair. This manual activity calls for a high level of manual skill in handling special application units and on average it takes about one hour. The labour-intensive repairs of large patches that were common in the past have been eliminated as a result. When a whole body repair is performed the vehicle body passes through the normal coating process a second time after careful preparation of the defective areas.

In addition to the surface defects visible to customers, the inspection zone also captures data on the objective quality of the coating. Aside from measurements of film thickness, gloss, haze and surface smoothness, which are conducted offline with portable measuring instruments, gantries are now being deployed to carry out automated quality control. These perform remote measurements of defined quality characteristics online by video [7.12.38]. However, quality control has only limited scope to intervene and correct the process or the paint where nonconformities arise. The response time for intervention effectively depends on the stage at which intervention becomes necessary. Consequently, several hours can elapse, during which period a large number of vehicle bodies may escape the intervention.

Parameter	Unit	Value
Total film build	µm	90 – 130
Gloss (at 20°)	%	> 90
Haze		50 – 100
Wavescan		
Longwave		< 10 (H)/< 15 (V)
Shortwave		< 20 (H)/< 25 (V)
H = horizontal, V = vertical		

Figure 7.1.23: Common quality data for automotive OEM coatings

Optimisation of efficiency requires a holistic analysis of the coating line and paint quality, from conceptual design through to implementation. This parameter critically determines the capacity of the paint shop and the cost per vehicle body. It fluctuates considerably from line to line. The "first run ok ratio" is defined in a variety of ways. Coating lines designed to be a core process in which work is performed, for example, without any dirt-generating application of underbody protection, and carefully operated lines have "first run ok ratios" of about 92 % when first-time capability, polishing and spot repairs are viewed together.

Further coating work has to be performed in final assembly if carelessness or technical reasons give rise to defects in the coating of the already completed vehicle. These are then repaired with conventional automotive refinish paints and cured at temperatures of around 80 °C.

Some automotive manufacturers dip the finish-coated vehicle bodies half in baths of hot wax to flood and wet the cavities as a further step towards improving anticorrosion protection.

7.1.7 Outlook

The quality of automotive coating has been raised continuously over the last 30 years. The increasing use of galvanised and other types of precoated panels, enhancements in pretreatment, cathodic electrocoating, flooding with wax to protect the cavities, and design measures, has improved resistance to corrosion to such an extent that various automotive companies grant anticorrosion warranties of more than 12 years. This is no longer the preserve of expensive vehicles. The quality costs arising from customer complaints about the coating over the course of a vehicle's service life have been reduced to just a few euros per car.

Another remarkable advance in the durability of the automotive coating concerns the resistance to weathering that has resulted from the introduction of the basecoat and clearcoat systems. Nowadays, visible loss of gloss varies regionally but no longer occurs before the 8th to 10th year of a vehicle's life, depending on the region. This has been achieved, despite a decline in the consumption of paint following the introduction of more efficient application equipment and despite a 20 % reduction in film thickness.

The development potential of paint materials has still not been fully exploited. Ideas for lowering the interior temperature of primarily black car bodies in the sun through the use of IR-reflecting pigments are being studied. A desire to generate energy via the automotive coating is keeping research into photovoltaic coatings active. Today and in the future, however, efforts will concentrate on further optimising the entire coating process. A further focus of attention is on introducing toxicologically safe substances, now that substances such as cadmium, lead and chromium compounds, which are regarded as highly critical, and some glycol ethers have already been largely eliminated. For further energy savings, water-borne coatings, especially water-borne basecoats are required that will permit more cost-effective air-conditioning during application.

There is plenty of scope for optimisation within the various stages of the modern coating process. This is particularly the case where the body is not subjected to any further production steps after coating and the coating process has been designed as a core process.

For example, the primer surfacer can be replaced by a suitable electrocoating if the crosslinked electrodeposited films are sufficiently conductive. Likewise, an unstoved but dried electrocoat film can be recoated with an electrocoating or a water-borne spray paint and the two can then be stoved together. However, trials on such processes have so far failed to deliver compelling results. Eco-efficiency analyses are increasingly being used to evaluate and support the further development of coating processes from the aspect of sustainability (see Chapter 5.5) [7.12.14].

The appeal of radiation curable paints that stems from the speed at which they crosslink is supported by the high scratch resistance of the resultant clearcoats. The disadvantages of curing three-dimensional objects such as the vehicle body are being tackled with a great deal of creativity on the technical and material development sides [7.12.39, 7.12.40]. What all these concepts have in common is a desire for a more efficient coating process. Their success will be judged on the basis of the transfer efficiency of the application and the quality of the finish.

The rise in the number of mostly plastic hang-on parts for the vehicle's outer skin is leading to further considerations concerning colour management. Given that, as is normal nowadays, parts in the original vehicle colour are sourced from five or more suppliers and hence considerable quality assurance costs are incurred, ways of managing this complexity are being devised [7.12.41]. For instance, one proposal concerns a coating technology based on materials which cure at low temperatures of 80 °C and which make it possible for plastic hang-on parts on the vehicle bodies to be coated at the same time [7.12.42]. Another which has captured the imagination is to replace the coating with foils or coil coating. In that event, the use and application of conventional coating materials would no longer play a role in colour matching. Once the coating process has been eradicated, the auto makers' cost structure will shift much further into the variable costs.

The automotive industry has been monitoring developments in coil coating for many years now because this coating technology not only is the most efficient and fastest but it also obviates the need for a coating process during vehicle manufacture (see Chapter 7.4). The first organically precoated coils have now been used in the automotive industry. They eliminate the need for outlay on cavity sealing [7.12.43].

For decades, automotive manufacturers have been constantly researching the use of new, lighter materials for car body construction. Plastics reinforced with carbon fibres and glass fibres are coming increasingly to prominence because so much experience has been gained with handling them in aircraft construction. It has quickly been recognised that the current coatings have lower stoving temperatures (see above) and need to be optimised to provide the familiar, expected appearance. Lightweight magnesium, with a density of 1.74 kg/dm^3, is of great interest in this regard, but the concept of anticorrosion protection here needs to be clarified [7.12.44].

7.2 Automotive Refinishing

The automotive refinish coatings market has been waning for about 20 years in developed regions such as North America and Europe, but is growing significantly in other regions such as Asia. In 2010, the global market was about 700,000 tonnes. Thanks to the technology employed in automobiles, fewer accidents are occurring, especially those involving minor damage. Whereas some 20 years ago the customers of automotive refinish paints were small workshops with 1 to 5 employees, there is now a clear tendency towards large companies operating several larger paint shops. However, refinish paints as a class of product are growing slightly overall due to the fact that they are being used in many other sectors such as heavy vehicles, cranes, heavy machinery and trucks due to their good film properties at low baking temperatures.

The main difference between automotive refinishing and OEM coating at the vehicle manufacturing plant is that firstly coating is not conducted on a conveyor belt under industrial conditions and secondly it has to be performed on a finished vehicle with all its components. It consists in manual spray application by contractors under conditions of drying and curing which do not usually exceed 80 °C. Higher temperatures, even if they only occur locally, can cause plastic parts to deform or melt, tyres to deform or to irreparably damage a vehicle's electrics and electronic equipment. The auto makers work together with manufacturers of repair coatings to certify the repair systems for their own workshops or contract paint shops and to ensure that new colours are available for repairs.

From the technical requirements, it is easy to deduce that different paint materials are needed for the refinishing process than for OEM coating. Whilst physically drying or catalysed reactive paints are still in occasional use, two-component technology based on polyacrylates and on polyesters with isocyanate crosslinkers has become established worldwide due to the outstanding quality of the coating (see Chapter 2.1.1). Curing times for the 2-component systems are less than one hour under IR-assisted drying [7.12.45] and up to one day at room temperature. Water-borne products are gaining in importance. They crosslink according to the same principles as solvent-borne 2-component paints. UV-curable coatings are used for filler primers because they cure rapidly and can be sanded. Activities aimed at using them as clearcoats are ongoing, but have not yet become established. The "Decopaint" Directive in Europe (see Chapter 6) and, among other things, legal requirements in North America [7.12.46] have given a clear boost to the water-borne and high solid systems.

Refinishing is used for localised repair of defects in the coating but it can also be used to fully upgrade vehicles by providing more attractive coatings. The refinishing process

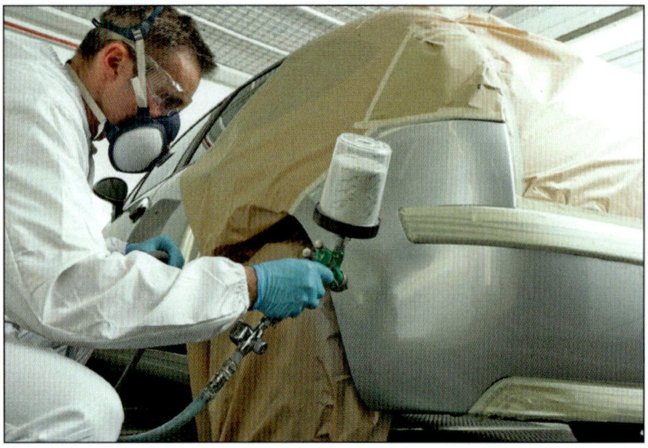

Figure 7.2.1: Automotive refinishing is craftsmanship
Source: BASF Coatings

depends on the type and amount of damage to the coating. Minor dents are smoothed with a coarse-grain body filler. A fine-grain body filler is usually then applied, followed by primer surfacer and topcoat. Once the filled areas have hardened they have to be carefully sanded.

Scratches and the like are repaired by sanding right down to the metal panel, if necessary, and applying first an anticorrosion or wash primer, and then a primer surfacer and topcoat. In many cases the topcoat only is lightly sanded and a topcoat composed usually of basecoat followed by clearcoat is applied to all of the affected area. The length of time needed for a refinish varies with the equipment on the coating lines and the extent of the repair, but is between one hour and one day.

The vehicles arriving into the workshop at an automotive refinisher are of all makes. They differ in age, structure of the OEM coating and the condition of the surface. It takes great manual skill and experience on the part of the refinisher to recognise the prior history and select the right process and work materials to effect the repair. For the customer, the degree to which the colour of the repair matches the rest of the vehicle is the most visible sign of quality and hence that is the most important goal to be achiev-ed. The increasing lifetime of OEM coatings is causing the number of colours to rise continuously. Nowadays it is estimated that there are more than 460,000 different colours adorning 1100 million vehicles worldwide. Rising vehicle exports are ensuring that all colours can be found everywhere. Such large numbers of colours spread across all regions can only be managed efficiently with the aid of sophisticated mixing systems and well-organised logistics.

Since refinishing is a highly skilled manual activity, coaters need to be provided with proper training. This is handled by the paint suppliers because, apart from the technical conditions, the paint quality exerts a major influence on the coating parameters and the quality of the work done by the coaters.

Specific factors influencing processing include the spray gun parameters, e.g. whether HVLP guns should be electrostatic or just pneumatic, the booth temperature and, in the case of water-borne products, the humidity.

Special manual skills, such as blending in, call for great dexterity in the use of the spray gun and determine the saleability of the refinish. By "blending in" is meant coating of a vehicle body part such as a wing, where the defective area is repaired in situ and the area surrounding the defect is coated with refinish and feathered out. This is highly recommended for mitigating colour discrepancies between the original paint and the refinish paint until a tolerable difference in colour has been achieved. This technique conceals any colour discrepancy at the margins of the repair.

One very important component for adjusting paints is the reducer or thinner. Usually there is a whole host of solvents available, which mainly influence the drying rate during and after application. The various products are dubbed "fast" if evaporation is rapid and "slow" if it is gradual. Portable IR lamps speed up curing and make for faster repairs.

Safe handling during storage and application of the materials is essential and users must be aware of potential hazards to people and the environment.

Figure 7.2.2: Examples of high-quality effect refinish coatings
Source: BASF Coatings

The quality properties of the various coatings and coating systems used in refinishing are defined by most vehicle manufacturers themselves and certified in the form of material releases. The most common materials nowadays, accounting for more than 75 % worldwide, are acrylic resins in combination with isocyanates in the form of 2-component systems. If the work is done expertly, the repairs will be a close match for the original coating in terms of film properties. In particular, the topcoats achieved the same level of weathering resistance as the OEM coating owing to the use of light stabilisers (see Chapter 2.1.4).

Determining the Repair Process

The average vehicle life has increased continuously to approx. 12 years. At the same time OEM coating processes have become more standardised worldwide. Nevertheless, in the case of vintage cars, for example, it may be necessary to characterise the original coating in order to define a suitable repair process. Figure 7.2.3 shows the most important types

	Original coating	Remarks/characteristics
A	**Solid colours and 1-coat metallics**	
	Stoving enamel (alkyd/melamine)	When cured: no problems
	Stoving enamel (acrylic/melamine)	When cured: no problems
	Thermoplastic acrylic	Sensitive to solvents: swelling
	Nitrocellulose coating	Sensitive to solvents: swelling
	PUR coating	When cured: no problems
	Oxidatively hardened alkyd paint	When cured: no problems
B	**2-Coat metallics**	
	Stoving enamel (acrylic/melamine)	When cured: no problems
	Thermoplastic acrylic	Sensitive to solvents: swelling

Figure 7.2.3: The most important types of OEM coatings prior to 1980 and their characteristics

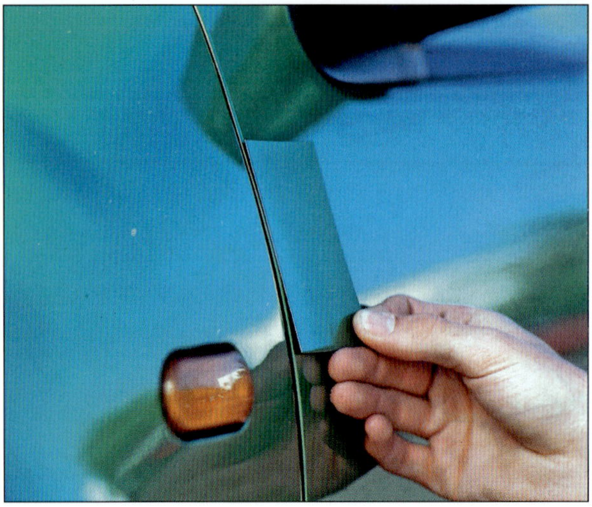

Figure 7.2.4: Colour matching during a refinish job
Source: BASF Coatings

of original coatings. It is essential to match the paint materials and sanding requirements to the vehicle type. This is a growing specialist market in the saturated regions of Europe and North America.

A simple process for identifying the type of paint used in the original coating exploits the tendency of the various paint films to dissolve and to swell. A white piece of cloth soaked in paint reducer or thinner, i.e. solvent, is rubbed over an inconspicuous part of the surface. Thermoplastic paints and nitrocellulose-based coatings will discolour the cloth or will swell visibly.

Colour Check

A colour check is another necessary preparatory step in vehicle refinishing. Every original coating changes colour as a function of type, age, exposure to stress, e.g. ultraviolet radiation, and the degree of care lavished on it by the owner (see Chapter 3.3.1). This change in colour can be largely eliminated by polishing. Nevertheless, the colour of the refinish paint may have to be adjusted. This is performed by a visual comparison of the original coating with a sample of refinish sprayed onto a separate piece of metal. The refinish is prepared according to the mixing formula for the vehicle's OEM colour code. If the comparison reveals the need for a colour adjustment, this is usually done via the components specified in the mixing formula. Colourimetric measurements to find the right colour formula are also in use.

Repair of effect coatings is more problematic because at every viewing angle the colour and intensity of the effect depend heavily on the application method and equipment parameters. The following merit particular mention:

1. Internal diameter of the nozzle and atomiser air pressure of the spray gun. Atomisation is finer if the internal diameter is small and the spray pressure is high; this makes the colour lighter.
2. Reduction and spray viscosity. A fast reducer can often make the silver effect brighter (i.e. the difference between light and dark is greater) than a slow reducer. The higher the viscosity, the coarser the particle distribution will be in the overspray and the greater will be the tendency to cause mottling (i.e. patches of discrepancies in light/dark properties).
3. Spraying. If, for example, fast reducers are used to spray dry, or if there is considerable overspray, the colour will appear lighter and more metallic. If, on the other hand, the paint is sprayed wet, the colour will be richer and darker. In between will be all the various shades of metallic effect and colour. Mottling is avoided by allowing time for evaporation.

4. Spray gun distance. If the distance is short the film applied will be wetter, i.e. darker and more chromatic, whereas if the distance is large the film will be drier and the colour will be lighter and more metallic.

Pretreatment

Once the preparations described have been completed, the repair process can commence. There are four different category of repair, each of which entails a different amount of work for achieving the right outcome or satisfying the customer's expectations:

1. New coating refinish. This is a repair to a coating on a new vehicle and so it must match the quality of the original coating in terms of colour, surface finish and adhesion to the substrate.
2. Market value coating. This is a refinish on a relatively old vehicle. The condition of the original coating determines whether pretreatment with body filler and sanding is required or not.
3. Saleable or used vehicle coating. Usually a whole-vehicle coating which makes an unattractive vehicle attractive again. This coating dispenses with elaborate sanding and body filling work. Cleaning and rust removal are followed by application of an adhesive primer and the topcoat in a wet on wet process, i.e. without any lengthy intermediate drying of the two coatings.
4. Blending in or spot repair (see Chapter 7.1). This is performed on minor damage that affects the topcoat only.

In preparation for the chosen type of refinish job, the surface areas not to be coated are masked and parts such as protective trims and spoilers are removed. Any necessary initial sanding work is also done. This is followed by giving the original coating a thorough wash. After drying, tar stains, residues of wax and silicone on the surface are removed with silicone remover. If these are not removed they will prevent wetting and adhesion of the refinish. In some instances, particularly in the case of a market value coating or a used vehicle coating, the original paint may have to be removed with paint stripper. The paint remover or stripper is usually applied with a brush and is allowed to soak in for a specified time. The original coating is scraped off with a scraping knife. This process is repeated a number of times until the original coating has been completely removed. The next step is thorough cleaning and careful removal of the stripper with mineral spirit or similar solvents.

Sanding

If the original coating has been cleaned but not stripped, the next stage consists in wet or dry sanding. Severely weathered or cracked coatings are sanded down until a perfect film has been reached, e.g. in the layer of primer or electrocoat. After sanding, the surfaces must be thoroughly cleaned of all dust. For wet sanding the surface is washed down with pure clean water afterwards and the edges and folds are rubbed completely dry. Sufficient time is left for the remaining moisture to evaporate. For dry sanding the surface is blown down with compressed air and then cleaned with a dust-binding cloth. Areas of rust are sanded down to the bare metal.

The sandpaper used in the stages described above and below should have the appropriate grit size. The rule is: the higher the number, the finer the sandpaper.

Application	Grit	Type of machine
Derusting, coarse sanding	P 16 to P 80	Angle grinder
Putty	P 40 to P 120	Vibrating grinder
	P 40 to P 80	Grinding stick
	P 80 to P 120	Eccentric grinder
Fine putty	P 24 to P 360	Manual sanding
	P 80 to P 120	Eccentric grinder
	P 80 to P 180	Vibrating grinder
Prime-filler	P 400 to P 800	Manual sanding
	P 120 to 180	Eccentric grinder
	P 120 to P 320	Vibrating grinder
Filler (primer surfacer)	P 400 to P 800	Manual sanding
	P 180 to P 220	Eccentric grinder
Topcoat	P 400 to P 800	Manual sanding
Original coating	P 180 to P 220	Eccentric grinder
2-Layer topcoats	P 600 to P 1200	Manual sanding

Figure 7.2.5: Grit sizes and machines for the various stages of a refinish job

The main types of abrasives are aluminium oxide and silicon carbide. Original coatings and hard substrates are chiefly treated with aluminium oxide paper whilst relatively soft substrates are sanded with silicon carbide paper.

Beads and edges should be sanded manually because with power tools it is too easy to sand all the way through to the substrate. Wet sanding with emery cloth is a fast way of working.

Applying Body Filler, i.e. Putties

Minor dents and irregularities are cleared of any rust and then corrected with body filler. The filler is allowed to harden and is then sanded to level off the defects. Application of a second layer of special fine body filler may be needed in order to achieve the good topcoat appearance of the original coating.

One-component body filler based on nitrocellulose/alkyd resin mixtures take about 2 hours to harden at room temperature whilst 2-component body fillers based on unsaturated polyester need only 15 minutes. The main ingredients in the hardener of a 2-component body filler are peroxides whose purpose is to accelerate radical curing of the unsaturated polyester.

Primers

In the case of high-quality coatings, once the surface of the body filler has been sanded and cleaned for the last time a primer is applied. This can be water-borne or solvent-borne and is usually a 2-component system. Drying takes about 2 hours at room temperature or 30 minutes at 60 °C. The recipes are usually formulated for smoothness, sandability and rate of drying.

Primers are also employed wherever sanding has gone through to the bare metal and no body filler is required. Such filler primers are usually 2-component systems based on epoxy resins with amine cross-linkers (see Chapter 2.1.1). They provide the best level of anticorrosion protection and good adhesion to metal substrates such as steel, galvanised steel and aluminium. If the main purpose of the primer is to provide stone chipping protection, 2-component polyacrylic or polyester-PUR systems are advantageous.

Figure 7.2.6: Samples of mixing colours for identifying a colour
Source: BASF Coatings

Topcoats

The topcoat system and the application procedure are dictated by the type of colour. Topcoats in solid colours make the least demands of paints and the coating process, metal effect paints make medium demands and pearlescent and colour flop paints make the highest demands (see above).

As in OEM coating, effect paints are applied in two coats wet on wet, i.e. a basecoat for colour and effect and a clearcoat.

The paint used for repairing is either the original refinish paint made on the basis of the mixing formula for the OEM colour or, usually in the case of relatively old coatings, paint which has been adjusted to the colour to be repaired by means of spray tests and appropriate corrections.

In most refinishing shops the original refinish paint is prepared with the aid of mixing paints mixed to recipes found by colour comparison or colourimetrc measurement. The mixing paints are located in mixing machines which stock 40 to 50 different colours. The paint is mixed either manually or even automatically using the components of the paint formula which has been predeveloped for it so as to provide a match to the vehicle's OEM colour code. These mixing formulas are worked out by coating manufacturers and managed electronically. They are an essential part of using the mixing machine.

As solvent-borne basecoats have high solvent contents, the purpose of which is to orient the aluminium particles to produce bright effects and which therefore pollute the environment, water-borne products are now in use around the world which have been adjusted pseudoplastically or thixotropically and which achieve the same effects, but with much lower solvent levels. These, in conjunction with spray guns operating on the HVLP (high volume, low pressure) principle (see Chapter 4.2.1.3), have brought about a decline in paint consumption because the paint transfer efficiency has been increased by 30 %.

Repairing Plastic Parts

Bumpers, spoilers, front ends, wings, boot lids and tailgates have been increasingly manufactured from plastics since the 1970s. On some models plastics in the outer skin now account for more than 30 % of the surface area. A whole host of different types of single-resin and blended polymers are used. They are tailored to the relevant requirements profile (see Chapters 4.1.3 and 7.3).

For the most part, the materials used are thermoplastic. Where large parts must have high rigidity and hardness, sheet moulding compounds (SMCs) are used. SMCs are thermosets based on unsaturated polyesters which are frequently reinforced with glass fibres. The term SMC is also the name of the process in which the plastic parts are manufactured by thermoforming and hardening reinforced prefabricated mats.

Of the various types of material, the most common are polypropylene (PP) thermoplastics. This is used for bumpers by almost all automotive manufacturers. In Europe it is pretreated for OEM coating by flaming the surface whereas in North America special adhesion promoters based on chlorinated polyolefins (CPOs) are employed. Such parts are refinished by possibly sanding and cleaning the PP surface and then applying a thin coat of a special adhesion primer. The parts are left for about 20 minutes at room temperature and subsequently treated as already described above for the standard refinish. The other thermoplastic and thermoset polymers can be easily coated in production after surface cleaning and possibly tempering at 60 °C, and then refinished accordingly. Thorough cleaning is essential for removing any surface-active release agents used in the manufacturing process for the plastic parts. Tempering is particularly necessary in the case of polyamide (PA), PMMA, PC and PUR polymers.

In certain cases, the topcoats may have to be provided with elasticising additives. Otherwise, the elasticity or notch impact resistance of the plastic may be affected by the coating. Generally speaking, it is important not to exceed a total film thickness of 100 µm in the refinish system.

7.3 Automotive Supply Industry

Over the last few years the automotive supply industry has gained substantially in importance. Global turnover came to 700 billion euros in 2012 as a result of the global strategy pursued by the automotive manufacturers to reduce the production depths of their products (cars) and shift the development of components and their assembly to the supply industry. In Germany, for example, production depths in passenger car construction were only around 30 % in 2000. In the USA some time ago, GM and Ford parted ways with their supplier subsidiaries Delphi and Visteon. At the same time the number of direct suppliers has fallen drastically, with the result that a pyramid of first, second and third tier suppliers has emerged in the automotive industry. The wave of consolidation which thus took place at the end of the 1990s has spawned major global suppliers with sales of up to 33 billion euros. The top 10 list is headed by two German companies: Continental and Bosch. Total sales of the German suppliers amounted to approx. 170 billion euros in 2012.

The diversity of products in this industrial segment is enormous. This inevitably means that practically all coating processes are used, including the full range of eco-efficient processes such as powder coating, electrocoating and coating processes based on radiation curable paints. This market segment is often where new application techniques and coating materials are used for the first time.

The industry can be divided up into the following segments:

- Exterior plastic hang-on parts
- Interior plastic hang-on parts
- Wheels
- Axles
- Engine blocks

Exterior Plastic Hang-on Parts

Among the most important exterior plastic hang-on parts on all car models nowadays are the bumpers, which are moulded from a very wide variety of thermoplastics and the bulk of which are coated. To these must be added tailgates, boot lids, sliding roofs, wings, trims and outside mirrors, which are already coated when delivered because they are fitted to the vehicle body later. Furthermore, these hang-on parts, such as the outside mirrors, are frequently assemblies and so cannot be coated at the car plant [7.12.47].

On some models more than 30 % of the area of the outer skin is now accounted for by plastic parts. Downward pressure to further reduce weight and costs has meant that the use of plastics continues to rise.

The polymers employed are primarily tailored to provide the necessary functionality on the vehicle body. Consequently, most bumpers are made of highly flexible, impact-resistant, low-cost polypropylene (PP) blends.

They are coated on specially designed coating lines. On many lines throughout Europe the procedure provides for a power wash with water, followed by flaming to improve adhesion, a water-borne prime coating, and application of water-borne basecoats and solvent-borne 2-component clearcoats. Between the primer and topcoat application there is a short intermediate drying process lasting 2 to 3 minutes. The crosslinking temperatures are 80 to 90 °C and they require dedicated paints. The trend towards eliminating the primer is on the increase because flaming removes the need for special-purpose adhesion primers. Moreover, there are substantial cost advantages to be gained from eliminating a coating stage. In North America and Asia flaming is often replaced by applying an adhesion primer (promoter), which is usually formulated with chlorinated polyolefins. The cross-

Figure 7.3.1: Coating of bumpers *Source: BASF Coatings*

Coating Industry

linking temperatures are then slightly higher at approx. 120 °C. Water-borne products are not yet widespread in these regions.

On many coating lines application is performed by robots and pneumatic spray guns, which achieve a maximum transfer efficiency of 40 %. For an electrostatically assisted coating, plastics have the crucial disadvantage of usually being nonconducting. This can cause electrostatic charges on the workpiece to repel the charged paint and impair the quality of the coating. This disadvantage can be eliminated by recourse to conductive primers. Proper earthing and the use of electrostatically assisted high-speed rotary atomisers then yield good results in the downstream application of basecoat and clearcoat [7.12.48]. Because they are highly conductive, water-borne paints are charged by means of external electrodes after spraying (see Chapter 4.2.1). Under these conditions the transfer efficiency for the coating materials is 60 % to 70 %.

The efficiency of the coating lines, as measured by "first run ok", is not as good as in automotive OEM coating because the low conductivity of the polymers harbours the risk of electrostatic charging, the consequence of which is that particles of dirt are attracted. Efficiency in the range 60 % to 70 % is a good level for a coating line handling plastic parts. The low material and production costs of bumpers mean that it is not necessary everywhere to rework all defective parts. Besides, it is actually more cost-effective to make new ones. What is more, the thermoplastic material can also be easily recycled [7.12.49].

On one hand, the success of the supplier hinges on having a colour specification that lays down the visual or colorimetric aspects. On the other, it is essential that bumper coatings can be corrected quickly because colour discrepancies can arise during original body coating or the coating may be outside the colour specification. Other concepts provide for parallel, coordinated paint development and production as well as intensive sharing of test results and empirical results [7.12.50].

Furthermore, the coating of polymers may exert only a minimum adverse effect on the mechanical properties of the substrate (e.g. impact resistance at -30 °C). For this reason, the paints used in automotive OEM coating do not always lend themselves to the coating of polymers. In particular, the primers and clearcoats for polymer coatings are formulated with highly flexible resins and isocyanate crosslinkers. Nevertheless, it has still not proved possible to ensure that the crosslinked coating will not adversely affect properties, especially the low-temperature impact resistance of the polymer.

Bumpers are made from all kinds of polymer blends to the auto makers' specifications. Usually they are simply primed at the manufacturer's and are then coated on the respective OEM coating line after electrocoating. The reason for this approach is that it greatly facilitates integration of the hang-on part into the vehicle body, which is assembled from the various materials. Bumpers not made of thermoplastic polyolefins (TPOs), e.g. of polyamide, merely require good surface cleaning, e.g. a "power wash" so that they can then be coated with a primer. The same applies to the outer skin of tailgates, which, on account of the rigidity requirements of such a hang-on part, are increasingly being made from SMCs, e.g. based on glass-fibre reinforced unsaturated polyesters (see Chapter 4.1.3). Special primers are needed for SMC parts that will prevent the effects of gas bubbles on the surface caused by outgassing or so-called voids. In-mould coating (IMC) in particular has proved successful here [7.12.51]. A paint is injected between the wall and the component in the mould and is then cured. This can be done with-out pressure beforehand and, after moulding, under pressure [7.12.52].

Interior Automotive Hang-on Parts

Figure 7.3.2: Various coated plastics parts inside a car
Source: Eisenhans,www.Fotolalia.com

Coating is often required when low-cost polymers are used to enhance the interior trim or the housing of the dashboard and many other parts of the vehicle's interior. The coating is designed not only to improve the surface visually but also to create a certain amount of grip. These coatings are termed "softfeel" paints.

Such modern coatings are mainly 2-component, polyurethane-based systems which are also available in water-borne form. The latter are starting to become more common. The usual film thickness of 20 to 40 µm also fulfils highly specific requirements, e.g. laser engravability [7.12.53], low surface reflection in the form of "velvet" coatings, high flexibility at low temperatures, e.g. for airbags, the triggering of which at low temperatures must not create any splinters. Avoidance of emissions from the plastic parts is also a requirement.

Lately, coloured coatings have been used for small vehicles as a way of following fashion trends. All in all, this market is growing very strongly because this coating has greatly enhanced the acceptance of plastic as a construction material and is helping to avoid its "cheap" image [7.12.54].

Application is performed on lines similar to those used for exterior coating. However, the polymers mainly employed in this segment, e.g. PA, PC, ABS, and polybutylene terephthalate (PBT) do not require pretreatment by flaming (see Chapter 4.1.3). Careful cleaning in the "power wash" process is usually sufficient.

Wheels

More than 500 million wheels are manufactured each year for the roughly 90 million vehicles that roll off the production lines annually around the world. Steel wheels still rank in first place. However, the proportion of aluminium wheels has been steadily increasing for years now, due to the weight savings and design options.

Wheels made of steel have become more and more standardised in design. They are adapted to the various types of vehicle with custom-coated plastic hub caps. The primary purpose of the coating has been to protect the metal against corrosion. Only a few steel wheels for cars are provided with topcoat, and then mostly in silver metallic. This does not apply to wheels for commercial vehicles, much higher percentages of which are given a topcoat.

Most steel wheels are cleaned, then pretreated, primed with electrocoat and possibly provided with a single layer of topcoat as described. As with vehicle coating, pretreatment

Figure 7.3.3: Steel wheels primed with a cathodic electrocoating *Source: BASF Coatings*

consists in zinc phosphating to confer maximum anticorrosion protection. The more environmentally friendly pretreatment based on zirconium dioxide deposition is also starting to be used here (see Chapter 7.1) [7.12.55]. State of the art in priming is cathodic electrodeposition coating. This is characterised by a high edge protection at a low layer thickness of 15 to 18 µm. Wheels in particular have a high edge-to-area ratio. Since the edge is frequently where most corrosion starts, it constitutes the most important determinant of the coating's quality. The importance of both anodic electrocoating with silver metallic effect, which was still standard only 20 years ago, and red cathodic electrocoating has declined.

The coating requirements are predominantly specified in anticorrosion tests such as the salt spray test and the VDA cyclical climate test (VDA 621-415), and adhesion and chemical resistance to brake fluids, fuel and engine oil (see Chapter 3.3.2).

For some time now, cost pressures in this business area have driven suppliers and customers to collaborate more closely in a spirit of partnership with a view to optimising the entire manufacturing process. As a result, in many cases nowadays supply contracts are negotiated in cost per square metre or per part. In some contracts further common savings potentials are agreed, which may stipulate reductions in process costs or paint costs.

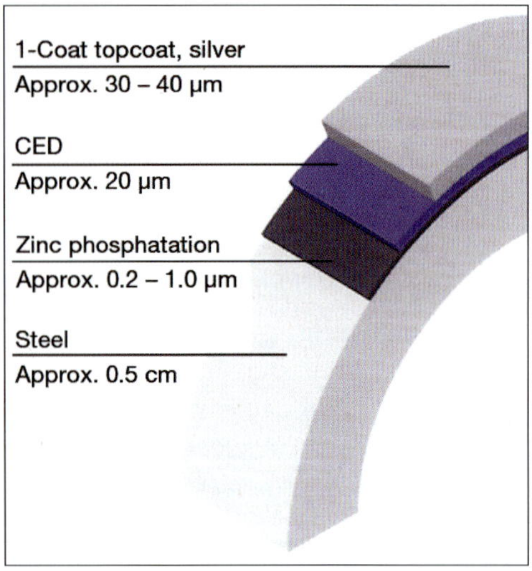

Figure 7.3.4: Multilayer structure of a steel-wheel coating

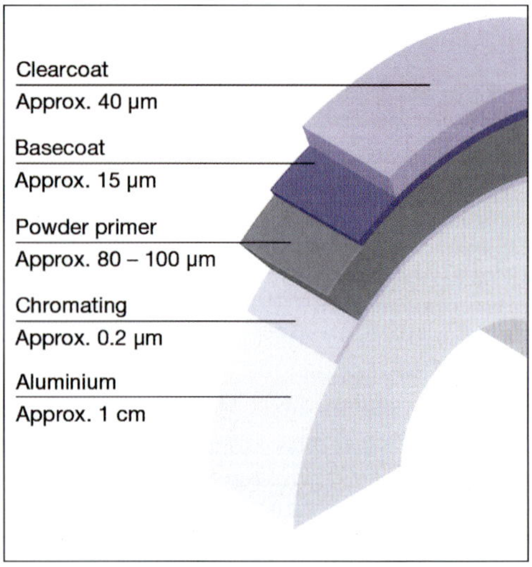

Figure 7.3.5: Multilayer structure of an aluminium-wheel coating

Coating Industry

Aluminium wheels have been accounting for an ever greater share of the total production volume of wheels for years. They are particularly important in the retail sector because of their attractive design and weight savings, as well as for OEM equipping of heavy passenger cars such as SUVs (sports utility vehicles) and luxury cars.

They are usually made by the die-casting method. All wheels are coated with a standard paint system, some are "turned off" for design reasons and then provided with clearcoat. By "turned off" is meant that the coating is milled off at certain points and the aluminium surface is sanded or treated specially.

Application of the standard paint system first requires the surface to be carefully degreased and chromated. The coating is an epoxy-based or epoxy-polyester-resin-based powder primer, which is crosslinked at temperatures of 160 to 190 °C. It is followed by application of topcoats comprising a basecoat and clearcoat. In Europe the two topcoats are solvent-borne systems, whilst in the USA the clearcoat is much more frequently a powder coating. The topcoats are crosslinked at 130 to 150 °C.

The specifications for the coating of aluminium wheels are based, amongst other things, on the CASS (Copper Accelerated Salt Spray) test in accordance with ISO 9227 or ASTM B368 as well as on tests for determining weathering resistance and chemical resistance, as are required in the case of steel wheel coatings.

Axles

Axles for cars and heavy trucks are being increasingly manufactured as components and delivered to the automotive industry as complete units for final vehicle assembly. Usually the parts required for this are primed by electrocoating prior to assembly. High demands in terms of anticorrosion protection means that pretreatments consist in zinc phosphating and almost exclusively cathodic electrocoating (see Chapter 4.2.1). Some automotive manufacturers specify high build electrocoats, which can be applied up to a film thickness of 35 µm.

For a time, the high amount of energy consumed in heating the heavy parts drove the development of electrocoats which crosslink at workpiece temperatures of less than 150 °C. However, the paints and coatings industry has so far failed to develop a product which offers the same level of anticorrosion protection at those temperatures as that conferred by the usual commercial products at 180 to 200 °C. In special cases the electrocoat primer is coated with powder, e.g. to create coloured coatings. There are also primers made of powder coatings which are applied after pretreatment.

Standard processes are used for these pretreatment and coating stages. The sizing of the coating line is essentially determined by the geometry of the parts and the throughput.

Engine Blocks

Engine blocks are cast from molten aluminium scrap and any necessary alloying constituents. After the mouldings have been cleaned of residues a matt black layer of powder paint is usually applied in a film thickness of 100 µm and cured at approx. 230 °C circulating air temperature. The powder paints are based on epoxy resins in order to ensure good anticorrosion protection. In addition, chemical resistance, good adhesion, high abrasion resistance and high heat resistance must also be achieved.

Mirrors

Mirrors are highly complex modules, the design and manufacture of which has now become the sole responsibility of the supply industry. Interior mirrors are not usually coated.

The exterior mirror, though, is made from thermoplastic of variable grades and mostly delivered in the vehicle's colour. A small percentage is supplied black and uncoated. The polymers vary from polyamides (mainly in North America) to ABS through to PC/ABS and PC/PBT blends in Europe.

After the housing has been cleaned, a primer is applied, followed by a basecoat and clearcoat wet on wet. The paints are predominantly solvent-borne systems. Only in the case of basecoats are water-borne products on the increase. The clearcoats are 2-component coating materials made of acrylic resins and isocyanate crosslinkers. The cross-linking temperatures for the primers and topcoats used in Europe are 80 to 90 °C whilst in North America they are 120 °C.

Application is pneumatic and is only electrostatically assisted in a few exceptional cases.

Headlight Systems

Most headlight systems were switched from steel and glass to plastics in the course of the 1990s. This step led to weight savings on one hand and offered new scope for design on the other. Headlight systems are entirely developed, assembled and delivered to the automotive industry as a finished component by the supply companies.

They have two main components: the reflector and the lens.

The reflector is increasingly being made of "blow moulding compounds" (BMCs) produced by the eponymous process [7.12.56] and is superseding the formed, die-cut metal panel that used to be cathodically electrocoated in advance of metallisation. Reflectors made of BMC do not require this step. Both types of reflector are spray-coated with pigmented, UV-curable primer in the next process step. UV-curable coatings have the advantage of not shrinking during curing and so they yield extremely smooth surfaces. This is essential for flawless formation of the reflector surface by downstream vapour-deposition of aluminium in autoclaves.

Figure 7.3.6: Typical headlight system made from plastic

Source: Streitberger

The lens is made of thermoplastic PMMA or PC. To enhance the weathering resistance, the lens is coated externally. There are currently two types of paint available for this, namely scratch-resistant clearcoats cured at 120 to 140 °C and UV-curable clearcoats. Both types are applied by spraying.

Coating Industry

Door Handles

Door handles are complete modules, the handle and handle recess of which are usually coated in the vehicle's colour for design reasons. Increasingly, the raw materials are trending away from cast aluminium and towards plastic.

The coating is applied on suitable lines, and is usually a basecoat-clearcoat system with brief intermediate flash-off. As is the case for the other coatings of plastic parts the use of water-borne paints is on the rise.

Other Parts

Other parts, most of which are coated, include trim strips, windscreen wipers, radiator grilles, heat exchangers, airbags and seat frames.

Electrocoating is the most important method for coating windscreen wipers, heat exchangers made of aluminium, and seat frames. Windscreen wipers are also given a topcoat to increase their weathering resistance. In the case of seat frames, recourse is occasionally made to autophoresis (see Chapter 4.2.1). Radiator grilles and airbags are made of plastic and are often spray-coated in the vehicle colour.

There are many job coaters active in this sector.

7.4 Coil Coating

The term coil coating refers to the continuous coating of metal coils, usually with liquid coating materials. Slightly more than 40 years ago this coating technology was implemented in a line 200 mm wide [7.12.57].

The importance of this efficient coating process has increased steadily ever since. Even from a modern-day perspective above-average growth can be expected over the next few decades. Coil coating usually enables the user to dispense entirely with coating lines in production and, through the reduction in fixed costs, to shift the cost structure from manufacture to the variable category.

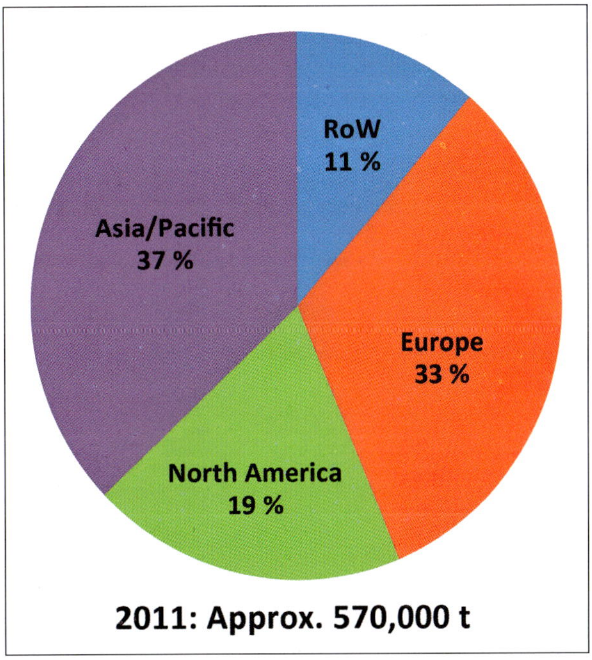

Figure 7.4.1: Regional breakdown of the worldwide coil coating market of approx. 570,000 tonnes in 2011 Source: M. von Dungen [7.12.58]

The most important substrates for coated coils are steel, galvanised steel and aluminium. At the moment aluminium coils are growing faster than steel in terms of surface area. In terms of the weight of the substrates, the ratio in Europe is about 1 to 6. A total of 17 million tonnes of steel (including galvanised steel) and 3 million tonnes

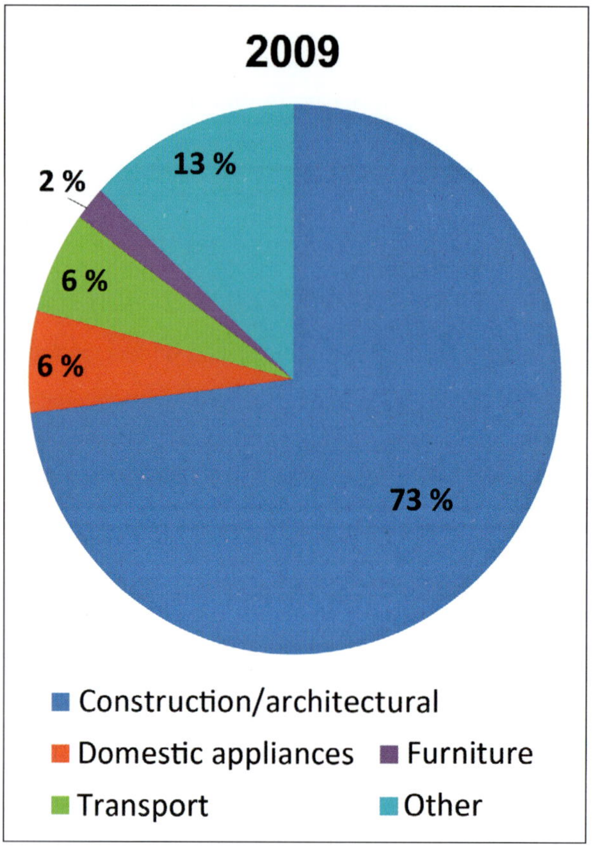

Figure 7.4.2: Breakdown of coil coating applications in industry in Europe in 2009 *Source: ECCA*

Figure 7.4.3: Storage of coated coils *Source: BASF Coatings*

Coating Industry

of aluminium are coil-coated worldwide. The corresponding Figure for Europe is roughly 5 million tonnes, equivalent to a coated area of 1.4 million m^2.

The numbers of coating systems and colours are rising substantially across all applications. Nearly all industrial sectors use coil coated panels nowadays. The main customers are the construction industry (internal and external cladding), the packaging (cans) industry, household appliances industry, the commercial vehicle sector including motor caravans and trailers, as well as industrial lighting systems and others. There are regional differences in the main application areas. For example, in Europe the construction sector accounted for nearly 73 % in the year 2009 and was thus above the world average of 54 %.

The roughly 60 coil coaters process metal coils of 0.2 to 2 mm thick. The materials consist of cold-rolled strips of soft or general structural steels, electrolytically galvanised steel sheet, hot-dip galvanised coil or aluminium and its wrought alloys The maximum coil widths are about 1,800 mm for steel coils and 2,000 mm for aluminium coils. Line speeds for steel coils can reach 200 m/min but average between 80 and 140 m/min. For aluminium coils, line speeds reach 150 m/min.

The lines consist of a coil feeder, the necessary coil storage unit (strip storage), a pretreatment zone, the first coating station

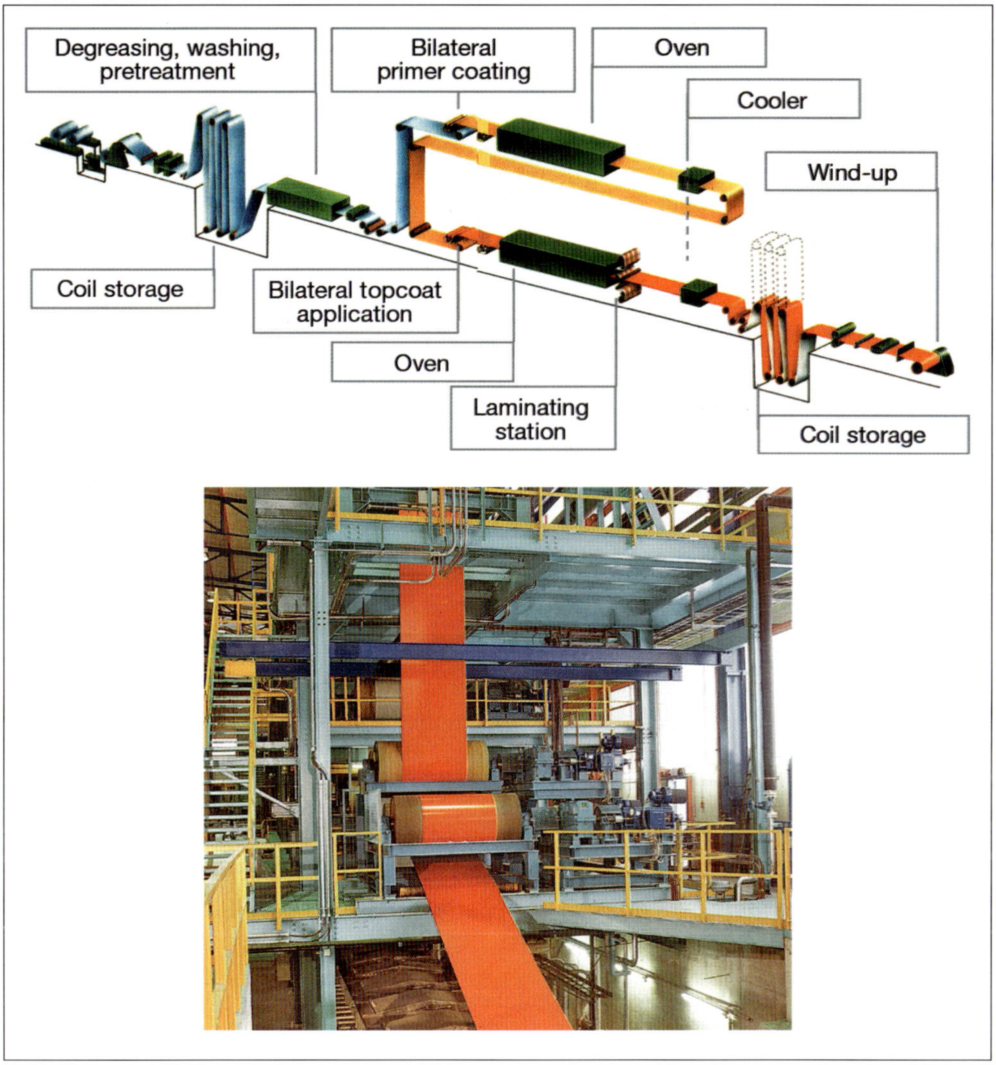

Figure 7.4.4: Process flow and picture of a coil coating line *Source: BASF Coatings*

Coating Industry

plus stoving oven and downstream cooling zone, a second coating station with oven, laminating station and cooling as well as a coil storage unit and coiler. Most lines are set up for two application steps, the first for applying a basecoat, and the second for applying a topcoat on the front and a protective coat on the rear, or a topcoat on both sides of the coil. This set-up requires two stoving ovens.

As with any industrial coating of metal substrates, high coating quality in coil coating is contingent on a perfect coil surface. This is achieved by chemical and physical cleaning steps. Nowadays such pretreatment is done by dip and spray methods, as well as by roller. Roller application, being a no-rinse process, is becoming more and more common on new lines. For many applications, chromium-based materials have been replaced by chromate-free formulations, both for pretreatment and for priming. Aluminium coils for the production of Venetian blinds can even be pretreated by the plasma process, albeit at low strip speeds of 40 m/min [7.12.59].

Coating Layer		Solid content [1]	Film thickness	Stoving conditions
Primer		55 %	5 – 8 µm	1 – 2 min/200 – 240 °C
Topcoat	polyester	45 – 55 %	20 – 25 µm	1 – 2 min/180 – 220 °C
	PVC plastisols	14 – 25 %	100 – 300 µm	1 – 2 min/180 –220 °C
Rear coating		50 – 60 %	8 – 10 µm	1 – 2 min/180 – 220 °C

[1] 30 min at 180 °C

Figure 7.4.5: Technical data of major coil coatings

Property	Unit	Total coating
Film thickness	µm	25 – 40
T-Bend		0
Hardness (Buchholz)		100
Gloss (20°)	%	70 – 90
Exposure in Florida (5°)	Years	< 3

Figure 7.4.6: Specific film properties of coil coatings

More than 98 % of primers and topcoats are applied by roller (rolling process). This offers a combination of high productivity and high transfer efficiency, with the result that the coil coating process is probably the most efficient coating method, expressed in terms of the amount of area coated per unit of time. In addition, the short time between application and crosslinking facilitates highly efficient reduction in solvent emissions in the form of thermal afterburning of the emission sources from the application zone and the oven. Merging of the two emissions sources yields such high concentrations that afterburning is self-sustaining in terms of energy.

The dry film thickness of the primer is 5 to 8 µm, that of the topcoat 15 to 25 µm and that of the rear coat 8 to 10 µm if crosslinkable coating materials are used. Where plastisols are employed the film thickness can be up to 300 µm.

The high coating speeds impose stringent demands on the rheological properties of the coating materials. Since the oven zone begins about 3 to 10 s after roller application, the applied coating has virtually no time to flash off. Therefore, for flawless surfaces, special knowledge about the composition of the solvents and additives is essential during formulation.

The curing or crosslinking of the organic coating materials takes place in continuous ovens divided into several zones. The ovens are of the induction type, most of which employ forced air circulation. Given the high line speed, they are extremely long so as to allow between one and two minutes of curing time at workpiece temperatures of up to 250 °C. Given the brevity of this time, it is vital that the heat be transferred as evenly as possible across the width of the coil.

Of the various test methods for coil coating, the cupping and T-bend tests are specifically designed to forecast whether the requirements will be met under practical conditions. The cupping test is a modification of the Erichsen cupping test and the T-bend test is a variant of the mandrel bending test. In accordance with the test method recommended by the European Coil Coatings Association (ECCA) under T 7 in 1996 the cracking and/or the pick-off of the organic coating of a coil is examined by bending the test panel parallel with the process direction. Various devices are allowed. Bending must always take place within 1 to 2 seconds. The reported value is the bending radius at which no cracks occur. T 0 therefore means that the smallest possible radius has been reached.

Owing to the versatility of coil coating in various industries almost all paint raw materials are used. For façade applications it is mainly plastisols made of PVC and polyvinylidene fluoride (PVDF) which are employed. Polyesters, acrylics, PUR- and silicone-modified polyacrylics and polyesters find application as topcoats. The primer is usually an epoxy-polyester mixture, chosen for its good adhesion and good anticorrosion protection. Carefully matched coating systems nowadays ensure coil formability, the limits of which are determined only by the steel itself. This means that no damage must be caused to the surface during die-cutting, forming, profiling, trimming or deep-drawing.

The last section of a coil coating line, the laminating station, can either involve embossing of the surface, application of laminates or application of protective plastic films. PVC coatings are embossed in the heated state by means of engraved rolls that create textured surfaces. Peel-off protective plastic films usually made of PE or PVC with a thickness of 70 to 150 μm protect the coated coil against soiling and damage to the surface during storage, transport and assembly. The adhesion of the plastic film to the coating must be adjusted such that on one hand further processing of the coated coils is not disrupted and on the other the plastic film can be easily peeled off the finished item.

Outlook

In Europe and Asia coil coating lines are chiefly located at the steel and aluminium manufacturers. This is a distinct forward-looking strategy in the value added chain of steel manufacturers. In the USA, on the other hand, the operators of coil coating lines are chiefly job coaters. It is understandable that especially in Europe steel makers have gone to enormous lengths over the years to promote the use of coil coated panels in manufacturing industry and particularly in the automotive industry. The associated concepts are gradually advancing with the introduction of "pre-sealed" coatings, which are weldable and for the moment only eliminate secondary anticorrosion protection measures such as cavity sealing and other sealing on vehicle bodies. The next, planned step is to entirely replace pretreatment and cathodic electrocoating in automotive paint shops. The requisite "pre-primed" materials will also be weldable. This implies that the task of protecting the cut edges resulting from coil processing will be achieved in a different manner. For the future, work is also being conducted on applying the coloured automotive topcoat to the coils and thus rendering the entire automotive coating line in its current form superfluous [7.12.60].

From the point of view of the current state of knowledge the idea that vehicle bodies can be manufactured by welding together parts stamped from such coils is illusory. Instead, a completely new vehicle production method consisting of clinching, riveting and adhesive bonding will have to be developed. This is currently the greatest challenge facing the successful introduction of coil coating to vehicle manufacture.

Coatings that create new surface effects will support the spread of coil coating into all kinds of industries. In particular, thick powder coatings offer scope for achieving such surfaces, but the high line speeds would have to be retained in that event. There are limits to the line speeds that conventional spraying units can work with. One variant of fluidised-bed coating, the "cloud chamber" process, is currently being investigated for its ability to handle very high line speeds [7.12.61]. Furthermore, a process involving electromagnetic brushes (EMBs), similar to those used in photocopying processes, is currently being developed [7.12.62]. It allows line speeds of up to 300 m/min for powder coating and is already finding application for small runs in the construction industry [7.12.63].

Metallic effects can further increase the share of coil coating the manufacture of household appliances [7.12.64].

Further growth for coil coating is expected in the market for house-roof structures. The Scandinavian and North American markets are serving as models for ecological building with coil coated panels in this area [7.12.65].

7.5 Commercial Vehicles

The term commercial vehicles covers a large number of industries extending from the manufacture of heavy trucks, buses, trailers, fork-lift trucks and agricultural machines to a whole host of one-off products such as special-purpose transporters and earthmoving vehicles.

The need to save weight so as to reduce fuel consumption has led to the widespread substitution of aluminium and plastics for traditional body materials such as steel and wood in superstructures [7.12.66].

From a coatings point of view, the best way to get an overview of this segment is to look at the production method and the size of the associated vehicle. This approach reveals two categories, namely mass production akin to OEM series production and manual production. Truck housings fall into the first category whilst bodies for silo or refuse vehicles, and huge earthmoving vehicles, belong in the second.

Despite the size of agricultural machines such as harvesters, OEM-like coating can be achieved by modifying the type of assembly operation, e.g. by coating the bulk of the components prior to assembly. An OEM coating set-up allows efficient coating processes such as electrocoating and robot coating to be deployed. If the parts are sorted accordingly high crosslinking temperatures of up to 180 °C can be used. By contrast, manual production processes and the size of the workpieces involved that need to be coated require coating systems which cure without the need for stoving f.e. refinish paints.

In many cases so few small parts have to be coated that this is done by job coaters, which use electrocoating lines for parts susceptible to corrosion, and flexible spraying lines, frequently powder lines, for hang-on parts that need to be coloured. Coil coating panels too are used in bus manufacture and in trailer production.

Coating Under Production Conditions

Where the quantity of products or components is high it is advisable to invest in a company paint shop. This would apply, for example, to production of trucks and buses.

The coating process starts with cleaning and pretreatment. For this purpose, the lines are configured to operate in a cycle. A continuous process would require a bath size that would require an excessively long dwell time for the coating materials. An alternative is to use the spray method. For the most part the pretreatment consists in environmentally-friendly zinc/manganese phosphating.

This is immediately followed by electrocoating with cathodic materials, which have become popular because of their excellent throwing power, the good anticorrosion protection which they afford even to the hybrid designs common in this industry, and their high level of stability. Stoving temperatures are 180 to 200 °C and the film thickness ranges from 20 to 25 μm.

In preparation for application of the usually water-borne primer surfacer, the body in white is sanded and body filler is applied to eliminate any defects. The primer surfacer is chiefly applied by stationary spraying robots or spraying machines. The film thickness ranges from 30 to 35 μm and curing takes place at 130 to 180 °C, as dictated by the specification.

The topcoat is generally a single-coat, alkyd-melamine-resin-based stoving enamel. It is applied in the same way as for primer surfacer application. Stoving is conducted for approx. 30 minutes at a workpiece temperature of 130 to 150 °C and the thickness ranges from 40 to 45 μm. Some companies have switched to high-quality 2-component or 3-component PUR topcoats because of their greater durability, [7.12.67]. These products may also be water-borne [7.12.68].

Powder coating is becoming more common where single-coat, coloured topcoats or primers need to be applied to galvanised steel [7.12.69]. The use of powder coatings obviates the need to apply primer because the high film thickness of the powder coating makes it possible to cover substrate defects much better than is the case for liquid topcoats [7.12.70].

Job Coaters

For very small runs it is unprofitable nowadays to have a paint shop which would be capable of coating all production parts. This is usually where job coaters step in.

Some manufacturers, however, employ a pretreatment and electrocoating process that is capable of priming all kinds of parts from the full range in an eco-efficient manner. The diversity of the materials is such that the chosen method of pretreatment needs to be able to provide them with a perfect conversion layer. In many cases this entails special degreasing and precleaning or stripping the substrate. Many of the hang-on parts are made of aluminium to save weight, whilst the bare skeleton and chassis are made of steel or galvanised steel. Wood was used for bodies up until some years ago and has all but disappeared now.

Aluminium needs to be pretreated with just as much care as steel. Although it protects itself against corrosion by quickly forming a passivation layer of oxide, this layer is not always conducive to good paint adhesion and this can be the cause of blistering and flaking. The whitish oxide layer, which is difficult to see, must be removed by physical or chemical pretreatment. After that the workpiece is sanded in order to increase the surface area. The sanding procedure and the sandpaper depend on the grade of aluminium. For example, hard grades of aluminium are used for road tanker bodies, medium ones for stake bodies and claddings, and soft to medium-hard ones for bodies. The fundamental rules are: for hard alloys use mechanically assisted sanding with a fine grit and for soft alloys use manual sanding with a coarse grit.

The application conditions of any downstream electrocoating must accommodate the various materials employed.

Figure 7.5.1: Example of manual application *Source: BASF Coatings*

Many manufacturers have manual coating stations in which primer surfacers and topcoats are applied either to the finished product or to selected hang-on parts. Spray application consists in pneumatic atomisation by gun, perhaps with electrostatic assistance [7.12.71].

All these skilled manual operations are increasingly being rendered superfluous by the use of precoated aluminium panels. However, even these panels are sanded and cleaned if they are going to be re-coated.

Galvanised panels too require very careful cleaning and pretreatment. True, the zinc layer is passivated by the formation of an oxide layer, as in the case of aluminium. However, as this readily undergoes chemical and physical attack, it does not offer long-lasting protection and must be removed before the coating process. Sanding must be avoided because the zinc layer is soft and is soon stripped off. This particularly applies to electrolytically galvanised panels, which has a very small zinc film thickness of just 8 µm.

Many of these vehicles have very large, bulky chassis. Sometimes they cannot be integrated into continuous processes. Coating therefore commences with manual spray cleaning. This is followed by rinsing and application of primer, also by spraying. The primers or filler primers used are 2-component epoxy-resin-based coating systems. They are cured at temperatures of 60 to 80 °C. After that a coloured topcoat can be applied. This is normally a 2-component polyacrylic coating with isocyanate crosslinker.

Apart from conventional gravity-fed and siphon-fed guns, other more powerful spraying techniques are used for spraying large vehicles (see Chapter 4.2.1).

Pressure vessels with capacities of up to 50 litres and paint and air hoses some 10 metres long are not uncommon. They have the advantage that the paint and spray pressures can be adjusted separately; this improves atomisation and ensures a better supply of paint. The siphon tube is immersed in the delivered vessel direct.

Airless spray guns assisted with shaping air are also used in this sector. These achieve fine atomisation and prevent the spray mist from being reflected from the coated workpiece. Overspray is reduced by approx. 50 %, although the spray pattern is not as good as with standard equipment. Suitability is restricted to large areas where the standard of visual attractiveness does not have to be particularly high.

Coatings for fire trucks and other rescue vehicles need a special paint system because the corresponding coating formulations for daylight fluorescent paints are not opaque. On top of the primer a white filler is applied, usually in the colour RAL 9010. Finally, the daylight fluorescent paint is applied in a film of at least 100 μm, allowed to dry, and covered with clearcoat. The colours appear slightly different when exposed to sunshine. However, this is tolerated on account of the high level of safety afforded.

Vehicle lettering nowadays is chiefly made by applying prefabricated foils. This method has almost entirely superseded the application of topcoats with the aid of templates (see Chapter 5.6.3).

The self-adhesive foils may only be applied to well cured, solvent-free topcoats. The degree of curing can be established by means of solvent resistance tests and scratch tests (see Chapters 3.1.5 and 3.2.5).

7.6 Mechanical Engineering

Mechanical engineering is a diversified industry involving so many different types of products that all kinds of coating materials are used. Textured and hammer finish effect paints round out the range available. The RAL colours which were customary in the past are nowadays often replaced by company-specific colours.

The materials for coating are chiefly metals such as steel, galvanised steel and cast iron. Light metals such as aluminium, along with plastics, usually in a pigmented form, are coming increasingly to the fore.

Pretreatment may terminate with solvent degreasing or flash-rust removal or may also include iron or zinc phosphating. Anodising may be sufficient in the case of aluminium. If necessary, a colour coating can be applied with the aid of 2-component PUR topcoats. With other metals pretreatment is followed by application of 2-component epoxy primers applied by spray or electrocoating. The topcoat colour nowadays is mostly produced by electrostatic spraying of powder coatings.

Some manufacturers divide coating quality into three classes on the basis of the expected level of demands and the functions of the parts, with the result that there are different grades of coating material within different product classes.

Parts too large for an oven often are coated with automotive refinish products.

Construction plant and similar equipment is subjected to such enormous stress that its coating is really only of temporarily decorative nature. For this reason, the workpieces undergo simple cleaning and are then given a single coat of powder paint based on polyester-epoxy resins (hybrid powders).

Machine tools and metalworking machines have to contend with so-called "wet runs" because they are operated with aggressive coolants, drilling oil emulsions as well as lubricating oils and hydraulic fluids. Only high-quality coatings based on 2-component PUR primers and topcoats are resistant to these materials. Hydraulic-driven machines also require a high-quality, highly crosslinked coating such as a 2-component PUR system. Although the coating softens when exposed to such oils, this process is reversible.

Coatings for machinery in the food industry must comply with regulations imposed by the authorities. The coating may not be attacked by greases, water, alcohol or fruit juices.

The manufacturers of printing presses also make special demands on the coating. For example, it must not be weakened by acids, alkalis, inks, printing ink, solvent or chrome cleaner.

7.7 White Goods

The market for all kinds of household appliances is still covered by the term "white goods". The term "brown goods", which was used for self-evident reasons to describe the segment of audio and TV industrial products, is no longer used nowadays. In the "white goods" segment, a distinction is made in Germany between large appliances such as refrigerators, freezers, dishwashers, washing machines, electric stoves, cooker hoods and cooking hobs on one hand. This market is worth 7.1 billion euros. On the other, the small appliances segment comprising vacuum cleaners, irons, kitchen devices and the like is worth 3.5 billion euros. For reasons of design and cost small devices tend to be coated less frequently. The market is shifting towards eastern Europe and Asia, both in terms of customers and production.

Although colours are playing a growing role in the design of household appliances [7.12.72], white still predominates because many household appliances are integrated into kitchen furniture and so are no longer directly visible.

The manufacturing process for refrigerators, freezers, washing machines, dryers, dishwashers, microwave ovens and cooker hoods has been undergoing upheaval for some years. For one thing, new production lines, especially for refrigerators, dryers and microwave ovens, have been designed for use with coil coated and precoated panels, with the result that there is no longer any need for a paint shop. For another, the share accounted for by liquid paints is on the decline and on traditional production lines they are being mostly replaced by powder coatings, for which the household appliances market, including radiators is the second-largest consumer segment at around 23 % worldwide (see Chapter 5.6.1.3).

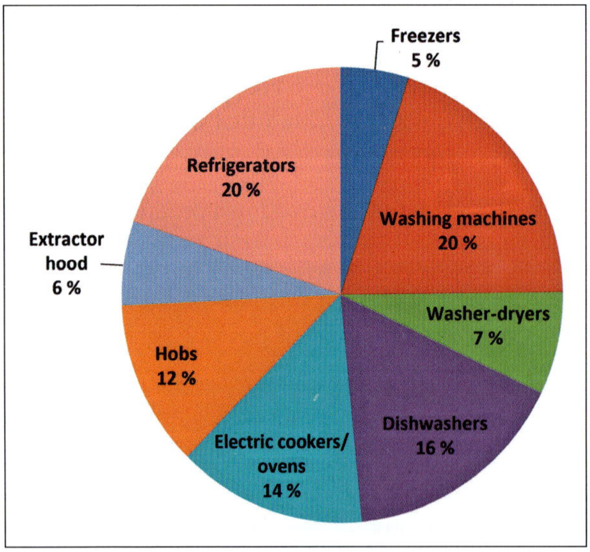

Figure 7.7.1: Market for household appliances in Germany in 2010
Source: ZVEI/GfK

The "wet" appliances segment, i.e. washing machines and dishwashers, employs conventional coating concept for good anti-corrosion protection on steel, namely inorganic pretreatment, followed by cathodic electrodeposition coating with subsequent powder coating (see Figure 7.7.2).

In addition to the housings and structural parts of household appliances, manufacturers are still making and coating various other components. These include evaporators, compressors, heat exchangers and motors. Such parts merely require protection against corrosion because they are not parts of the appliances which are visible to the user. In most cases anticorrosion protection is provided by an electrocoat primer or simple dipping in conjunction with iron or zinc phosphating.

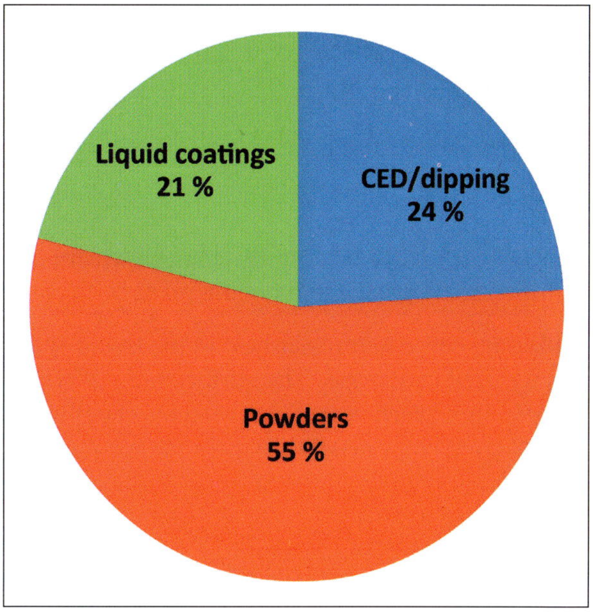

Figure 7.7.2: Breakdown of coating technologies in the European market for household appliances *Source: authors' estimate*

In terms of the coating technology the market for coating materials can therefore be divided into three product categories in addition to coil coating: electrocoats, liquid coatings and powder coatings. For the so-called "low-temperature" segments such as refrigerators and chest freezers a single-layer coating of the usually galvanised and pretreated parts is sufficient because there is less exposure to corrosion. This can be achieved with powder coatings or liquid coatings. The "wet" segment comprising dishwashers, washing machines and dryers is frequently given a coating consisting of a cathodic or anodic electrocoat and a topcoat of powder or liquid paint because of the greater exposure to corrosion. Apart from electrocoats there are also applications for water-borne dip coatings.

In the USA a substantial number of microwave ovens are provided with a single-layer electrocoat. The grades employed are based on acrylic resins and offer greater resistance to light than the traditional epoxy resin-based primers.

7.8 Building Supplies

In 2011, approx. 980,000 tonnes of coating materials were used in Germany to coat and renovate interior and exterior walls. Of this amount, 50 % was accounted for by emulsion paints, 25 % by silicate paints and 25 % by other coating materials. These activities are carried out by painters and consumers themselves (DIY).

A whole host of components are manufactured, coated and then supplied to building sites to be installed mainly in industrial buildings, as well as in the private house build-ing sector.

The most important of these included external and internal cladding, heating systems and radiators, industrial light fittings, doors, gates and frames, and aluminium windows.

These products are generally produced in series and coated by various processes, which are chosen firstly to match the geometry of the parts and secondly to meet the quality requirements imposed on the indoor and outdoor coatings. The focus here is mostly on colour diversity and weathering resistance.

Storefront elements are predominantly coil coated (see Chapter 7.4). So-called sheets or blanks are powder-coated to a lesser extent. This is usually done by a job coater.

The same applies to interior cladding such as ceilings, although the proportion of powder-coated sheets is much higher here.

The steel panels covering the burners and boilers of **heating systems** are coated. The panels are usually given an iron phosphate pretreatment, cleaned and then powder coated. Relatively small heating systems, which are usually fitted to the walls of kitchens or bathrooms, are increasingly being made with coil coated panels.

Radiators serve to transfer heat in the living and working areas of buildings. They are mainly used in Europe. The surface area of the steel used in the 60 million radiators manufactured each year is equivalent to roughly 10 % of the area of steel used in vehicle manufacture. This makes radiators the second largest steel surface area to be protected in any mass-product segment. Panel radiators fitted with convector plates are virtually the only type used nowadays and are mass-produced. Upon leaving the fully automated welding lines they are degreased, pretreated and then coated. The most commonly employed form of pretreatment is iron phosphating. Cathodic electrocoating, followed by powder coating, has become quite common, largely for quality reasons. In addition, there are coating lines featuring anodic electrocoating, particularly for aluminium radiators, and simple dip-coating lines for water-borne dip coatings (see Chapter 4.2.1). The continuity and reliability of the process is such that invoicing in this market segment is increasingly shifting to cost per square metre. Specially shaped designer radiators are primed with electrocoats and afterwards topcoated by spray coating in the colours requested by customers.

Protection against fire and burglary are the main reasons why **steel doors** are used especially in industrial construction. Fire doors and fire-retarding doors are made of sheet steel. The inside surface is coated with an anticorrosion protection primer, and a fire-retarding insulation panel is inserted and welded to another door panel. This is followed by spraying the exterior with air-curing anticorrosion primers. The workpieces must be stackable within a short time and they must be paintable with the usual coatings.

Other metal doors generally receive a silky-matt powder coating stoved at 150 to 170 °C. In preparation for this the doors are sprayed with a wet chemical pretreatment.

Door frames are mainly made of galvanised sheet steel, phosphated and then primed by electrocoating. Garage doors too are made of hot-dip galvanised steel and powder-primed. This priming has good adhesion to zinc, is resistant to weathering for several years and can be painted over with standard coatings. Coil coated panels are also used in this segment nowadays.

Like ceiling elements, industrial **extra-long light fittings** have a decorative role to play in addition to their real purpose. They are an integral part of modern ceiling cladding and are also predominantly made from coil coated panels. Apart from that, polyester powder coatings are used.

Outdoor light fittings must meet special requirements in terms of anticorrosion resistance. Suitable materials are aluminium alloys and galvanised steel. They are coated with powder paints.

Aluminium window frames are chiefly colour-treated by an inorganic process (anodisation), but are also coated to a certain extent. For this purpose, the frames are degreased, passivated or chromated and then given a single-layer colour coating of powder paint.

Specially designed components made of all kinds of materials are coated with powder paint at job coaters and are electrocoated beforehand if a high standard of anticorrosion protection is required.

Future requirements, which will especially affect exterior façades, are a further increase in weathering resistance in conjunction with dirt repellency, colour stability and anticorrosion protection. The air-purifying effect of buildings is also becoming increasingly popular for mega cities around the world (see Chapter 7.11.9).

7.9 Rail Vehicles

Rail vehicles include railway wagons and locomotives for trains, underground railways and trams. They are made from different materials. The cladding and structural parts of the vehicle body are mostly of aluminium to reduce weight. Aluminium has gained favour over steel, not least because it requires less pretreatment and rework. Furthermore, moulded parts are usually made with glass-fibre reinforced plastic while load-bearing components are made in steel.

The procedure for both original coating and whole-vehicle refinishing starts with descaling of the surface by sandblasting, usually by hand. Chemical pretreatment has not been performed for some time now, for reasons of environmental protection. The preparatory work involved in refinishing a locomotive, for example, entails manually sanding its entire surface. Further preparatory work includes applying body filler to level off relatively large cavities. The body filler compounds are based on unsaturated polyesters.

These types of rail vehicle are usually coated manually in suitably sized buildings using 2-component coatings which cure within a few hours at a temperature of 60 to 80 °C. Hang-on parts such as doors are coated separately.

In Germany water-borne 2-component primers have now become established whilst in other European countries recourse is still made to solvent-borne 2-component coating materials. These are applied within 1 to 2 hours and cured at 60 to 80 °C in appropriately sized chambers. The base film formers are combinations of epoxy resins and amine hardeners because of the good anticorrosion protection which they afford. The usual thickness of the primer is 50 µm for the outer skin and 300 µm for the floor area. Consequently, the paints employed need to vary in quality. Water-borne products are necessary on these coating lines because the size of the coating chamber makes emissions-reduction measures economically unviable.

The second stage of coating comprises application of a water-borne 2-component PUR primer surfacer onto the primer, especially in the case of trams and railways. It is applied in a thickness of 100 µm and then manually sanded down to 50 µm. This is necessary for meeting the high standards imposed on the appearance of municipal vehicles. This layer too is dried at temperatures of 60 to 80 °C and then the 2-component PUR topcoat is applied.

Figure 7.9.1: Coating of rail vehicles in huge, fully automated booths
Source: B&M

Application of primer and topcoat is performed with 2-component guns by the airless or airmix process (see Chapter 4.2.1), and in many cases is still done by coaters who are driven past the rail wagons or locomotives on mobile stages. Recently, robot-assisted coating lines have been put into operation [7.12.74]. Since a whole swathe of different colour shades are now being used, a short distance from the mixing block to the spray gun is essential for ensuring perfect colour change with as little rinsing agent as possible.

The entire coating system has to have a high level of resistance to acids and alkalis, particularly so that graffiti can be removed with suitably aggressive materials.

For this reason, the two-layer system comprising basecoat and clearcoat serves as topcoat, especially on railway wagons and underground railways.

In this connection, the anti-graffiti properties such as weak adhesion or poor wetting of the graffiti paints can be exploited through the use of specially formulated clearcoats. The standards imposed on appearance, weathering resistance and chemical resistance are similar to those required for automotive coating. It takes about a day to coat a locomotive or wagon by this process.

Locomotives and wagons frequently have service lives of more than 20 years. A new coating is applied at least once during their life cycle. The lines employed are similar to or identical with those of the original coating, the difference being that the process is preceded by a stripping stage. This is designed to completely remove all organic coatings.

7.10 Wood Coating

The term wood coating describes an extensive range of various coating measures and processes for the wood-processing industry. The processes can be divided into four segments:

- Impregnation of wooden parts for construction and outdoor facilities
- Direct coating of wooden products such as solid furniture, pianos, structural components, window frames
- Components made of wood, usually coated chipboards, e.g. for the furniture industry
- Floors

The global market for wood coatings in 2010 was approx. 2.030 million tonnes. In 2007, the roughly 600,000 employees in the entire wood and paper industry in Germany generated sales of 80 billion euros. Of these roughly 1/3 are carpenters and about

20 % are employed in the furniture industry. A wealth of products and processes are deployed in the protection of wood and for surface treatment. Spray application is still the most common. In terms of materials PUR coatings enjoy the lion's share, followed by nitrocellulose, and UV-curable, acid-curing and water-borne coatings in roughly equal proportions [7.12.75]. The breakdown of coating technology is quite different in North America [7.12.76]. The Decopaint Directive has been in force since 2007/2010 for coatings that are not processed in industrial plants (see Chapter 5). It has led to the greater adoption of water-borne and high solid coatings in the DIY and trade sectors.

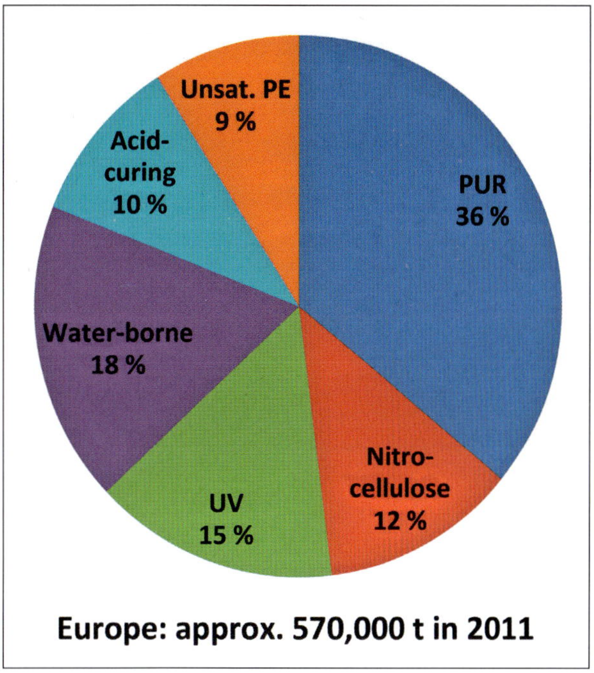

Figure 7.10.1: Breakdown of wood coatings by technology in Europe
Source: authors' estimate

Direct Processing of Wood

When it comes to the coating of wood the texture and properties of the various types play a major role. Wood is a naturally renewable material and its structure is not as standardised as that of metals or plastics for example. It is hard to surpass in type and diversity. Almost all types of deciduous and coniferous wood are used, such as pine, larch, fir, spruce, alder, birch, koto, gabun, cedar, anigre, cembra pine, wenge, oak, beech, teak, walnut, mahogany, macassar ebony, limba, yew, cherry, Indian rosewood, elm, makore, mansonia, ash, pear, abachi, afrormosia, zebrano and ramin.

The inhomogeneous texture of wood greatly affects its physical and technological properties. For example, "summer wood", which chiefly grows in late summer and autumn, has a much higher density than "spring wood", the main growth of which occurs in spring (e.g. pine trees). The reasons for this are the differences in xylem formation and the variation in cell wall thickness as a function of the growth period (see Chapter 4.1.1).

Building timber is pressure-treated with preservative at the manufacturer's, particularly for use outdoors. There are different classes of wood protection [7.12.77]. The European Biocidal Products Directive 2012/528/EU severely restricts the use of fungicides and algicides (see Chapter 5). Where they are to be coloured, pretreated parts can usually be easily overpainted without much effort. Untreated solid wood is used in the window industry and for high-quality furniture. Some 90 % of chairs and tables are made with solid wood.

The coating system for wooden surfaces is expected to cope with many different requirements that vary with the particular application. The coating's appearance now plays a major role, especially in kitchen furniture [7.12.78]. Generally speaking, it can only

Figure 7.10.2: Coated piano *Source: Steinway & Sons*

be achieved by a combination of different coats of paint. What all the coating systems have in common is that they must offer an attractive appearance, good adhesion to the substrate and good adhesion between the individual layers of the coating. For interior applications such as furniture and doors, chemical resistance to household substances such as coffee, red wine or detergents is also important. Exterior applications such as window coatings, on the other hand, call for a high level of weathering resistance, water resistance [7.12.79] and resistance to physical forces such as hail.

Coating frequently consists in pretreatment, staining, priming and applying the topcoat. Water-borne coatings are now available for all applications, and their market share is increasing at the expense of solvent-borne products. They are applied by spraying or flow coating.

Powder coating has become attractive for wood because of the powders' good filling power and the absence of solvent. Conductive primers are used to solve the problems of high electrical resistance associated with electrostatic application (see Chapter 4) [7.12.80].

The requirement for good adhesion to the wood is best fulfilled with a water-borne adhesion primer. The adhesion-enhancing effect is achieved by causing the wooden surface to swell and enabling it to be penetrated. The wood fibres stand up, are fixed during drying and are then sanded down. This not only guarantees optimal adhesion but yields very smooth surfaces.

High-quality finishes are achieved nowadays with 2-component PUR coatings which possess the optimum performance profile for weathering resistance, gloss and chemical resistance. Flat surfaces are increasingly being treated with UV-curable coatings.

Special effort is put into piano coating. Usually a three-coat system comprising UP filler, black PUR-based coat and UP clearcoat is applied by spraying. The individual coats, which can be up to 100 µm thick, are sanded carefully after each application and then polished. Musical instruments such as guitars are coated with water-soluble UV-curable coatings [7.12.81].

Manufacture of Components made of Wood Composites

Boards in the form of blockboard for high-quality furniture such as shop fittings and even pianos, and of chipboard and fibre board for all segments, are extremely important in the furniture industry (see Chapter 4.1.1).

Many of these boards are covered with foils which are made by printing wood patterns on impregnated paper and sealing with clearcoat. This is performed from roll to roll,

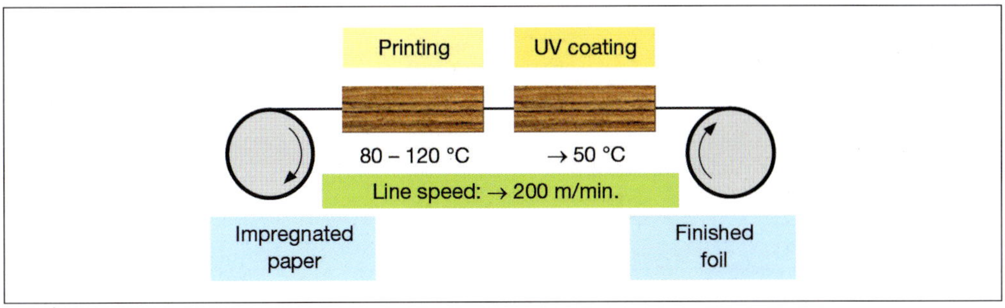

Figure 7.10.3: Schematic drawing of foil coating

i.e. by the coil process, at speeds of up to 200 m/s. The best clearcoats for this purpose are coatings which can be cured with ultraviolet radiation because they offer a good performance profile in terms of scratch resistance and chemical resistance at the high production speeds.

Solid or even flat wood veneer components for the furniture industry are also sealed with UV clearcoats. These may also be water-borne UV coatings wherever the particular application and performance profile allow.

Coating of wooden components for outdoor applications such as windows is gaining in importance because the use of more weatherproof tropical woods is declining. The service life of coated woods is about 5 years. To boost performance relative to aluminium and plastic windows UV-curable coatings are also recommended [7.12.82].

Other outdoor applications employ the traditional wood-coating systems, whereby the topcoat has a high level of weathering resistance.

Floors

Particularly high standards are imposed on the adhesion and abrasion resistance of coatings for industrially manufactured parquet and laminate. Parquet is coated in a multilayer structure, while MDF laminate is laminated with several layers of foils. The rapid production process and the geometrically straightforward parts soon led UV-curable coatings to become established in this product segment at a very early stage. They are applied by roller. Originally their formula was based on unsaturated polyesters (UP) with styrene as reactive diluent. In the meantime, they have been largely replaced in European clearcoats by unsaturated, urethane-modified acrylic systems [7.12.83]. Compared with paint formulations based on UP resins they offer the following advantages:

- Faster curing which allows higher line speeds
- Less inhibition on some wood substrates such as exotic woods
- No swelling of the rubber lining on the application rollers.

The coating of parquet can entail as many as three priming stages. These must be followed by sanding prior to application of clearcoats. The primers may still be based on unsaturated polyester resins, but there is a shift towards water-borne PUR or modified UV-curable coatings [7.12.84]. Abrasive strength is enhanced by adding approx. 10 % finely divided corundum to the clearcoat. The addition of silicone-modified silicate nanoparticles also enhances abrasive strength. The unsaturated acrylic resins are formulated as 100 % solid systems with reactive diluents such as hexanediol diacrylate

Figure 7.10.4: Coating of boards for the furniture industry
Source: BASF Coatings

(HDDA), tripropylene diacrylate (TPGDA) or dipropylene glycol diacrylate (DPGDA).

More attention needs to be paid to the smells of all indoor wood products owing to the substantial reduction in production and delivery times. Water-borne UV-curable coatings are advantageous in this regard [7.12.86] and dominate the market [7.12.87]. Solutions are also being developed for solvent-borne systems, e.g. with a view to lowering formaldehyde emissions [7.12.88]. Since 2011, parquet flooring in particular has been subject to the "EU Building Products Regulation (2011/305/EU)" which specifies the level of emissions during the first months of use on the basis of tests in accordance with EN 14342.

7.11 Other Fields of Application

The number of manufacturing industries is so large that no coating manufacturer can be equally successful in all of them. Many of the market segments call for substantial specialisation in coating and application technologies. To provide an overview of the most important segments, the following fields will be described briefly.

7.11.1 Protection of Structures

Coating Industry

Large steel structures have commensurately large surfaces which have to be protected against corrosion if their functionality is not to be impaired. They include road bridges, railway bridges, steel masts, wind turbines, all types of tank and tank facilities, cranes, exploration installations such as drilling rigs as well as buildings and production installations of all kinds.

For the most part, basic protection is provided by inorganic coatings such as galvanisation and zinc dust paints. Most structures are additionally coated in order to guarantee a service life of more than 30 years in many cases. Last but not least, coloured coatings are also applied for decoration. A recommendation on so-called "duplex systems" has been issued by the relevant associations in Germany [7.12.90]. Good protection of structures is also afforded by two-layer powder coating [7.12.91]. Legal requirements have also led to the development of special water-borne liquid coatings that offer high anticorrosion protection [7.12.92]. ISO 12944 describes the key factors for reliable coating of hot-dip galvanised substrates with organic paints that will yield a high, guaranteed level of durability.

Although a large proportion of structural components are delivered to the construction site already painted in a 2-component coating of zinc-rich primer and topcoat [7.12.93], nonetheless in many cases the topcoat has to be applied on site. The surfaces of the parts need careful treatment. Where the parts are new, the steel surface is often covered with firmly adhering scale. Efficient methods for removing it are sandblasting, flame cleaning, pickling and grinding, the choice depending on the size and accessibility of the object. The coatings consist of a thin zinc-rich primer approx. 100 µm thick, a 2-component intermediate layer also 100 µm thick and a coloured 2-component topcoat 50 – 70 µm thick [7.12.94].

Figure 7.11.1: Powder coating of pipeline *Source: BASF Coatings*

For commercial and ecological reasons these operations are frequently performed directly at the steel mill. Here the parts can be blasted fully automatically and the abrasive recycled. They are then treated with an antirust primer to prevent corrosion damage during transport to the construction site and in storage. The actual formulation varies with the intended subsequent use of the part. Thus, a high-build primer is used in connection with bolting while a weldable primer that offers only temporary protection against corrosion is employed in the case of welding. The parts should be further coated with primers and topcoats as soon as possible.

This also applies to the large structures of off-shore platforms which have to be dried as quickly as possible by special drying techniques, usually after degreasing and blasting with steam, to be treated with 2-component PUR coatings for an efficient coating sequence. Preconditioned air is often used for this [7.12.95].

One special application of structural protection is the coating of oil and gas pipelines. The exterior of pipelines laid in the ground or in the sea has to be prepared for a long service life. The bulk of them are coated with special powder paints to a film thickness of between 350 and 500 µm. This is conducted either on coating lines at the pipe manufacturer's or, in rarer cases, on site, as the standards imposed on the surface finish are understandably low. The pipe is heated to approx. 230 °C and the unusually thick coatings are produced by a sequence of several spray units. The residual heat is sufficient for curing all layers. In other cases, a powder coat 100 µm thick assumes the role of primer for a polyolefin coating. A service life of 50 years and more is achieved in operation by means of cathodic protection [7.12.96]. The epoxy powder coatings are specially optimised for resistance to alkalis.

7.11.2 Steel Furniture

Steel furniture chiefly comprises lockers, wardrobes, tool cabinets, office furniture and fittings, shelves for shop construction, vending machines, as well as furniture for doctors' surgeries and hospitals.

On account of the low exposure to corrosion indoors the materials employed are chiefly plain steel, frequently black plate and steel tubing. For greater protection against corrosion galvanised material or even aluminium is used.

As far as furniture for doctors' surgeries and hospitals is concerned, the requirements include resistance to disinfectant and hot steam sterilisation.

The type of pretreatment varies with the stresses to be expected. For example, lockers and cabinets are merely degreased whilst steel furniture for offices, doctors' surgeries and hospitals are also phosphated. The production programme determines whether pretreatment is done automatically in cycles or performed continuously.

Coating is almost always conducted with powder paints, which serve both as primer to protect against corrosion and as single-coat pigmented coatings to additionally provide colour. In some cases, such as garden furniture, the powder coating is applied by the fluidised bed method (see Chapter 4.2.2). Liquid coatings are used in isolated cases to produce high-quality, smooth and glossy surfaces.

In a few applications it can be expedient to additionally conduct an electrocoating process to meet the higher standard of anticorrosion protection required.

7.11.3 Aviation Industry

In this market segment there are three important areas:

- Commercial aircraft construction
- Military aircraft construction
- General aviation industry

The first segment is currently composed of Boeing and Airbus, the world's largest manufacturers of passengers and cargo aircraft. This segment has the largest potential of the global market for aircraft coatings, with a sales volume of approx. 200 million euros [7.12.97].

The technical requirements on coatings for military aircraft are regionally much more specific than those for commercial aircraft.

In all cases the usually two-layer coating must not only provide good anticorrosion protection on components made of aluminium, magnesium and titanium, their alloys and increasingly fibre composite materials, but also withstand extreme conditions and cope with the following stresses during the average 25 years` lifespan of commercial aircraft:

- Abrasiveness associated with high velocities
- Intense ultraviolet radiation at altitudes between 10 km and 14 km
- Extreme, frequent temperature fluctuations from 100 °C on the ground to -60 °C in flight
- Expansion of components at altitudes between 10 km and 14 km
- Resistance to
  - hydraulic fluid
  - de-icing agent
  - kerosene
- Easy paint removal using ordinary paint stripper

Figure 7.11.2: Aircraft exterior coatings involve laborious manual work *Source: Grandt; www.Fotolalia.com*

This must be guaranteed at minimum film thickness because weight is an extremely sensitive parameter in respect of utilisation and energy costs, particularly in aircraft construction.

The structural components for an aircraft are pretreated separately and in the case of aluminium they are anodised or chromated. In some cases, so called active primers, usually containing phosphoric acid, are applied in a film thickness of 6 to 8 μm after the cleaning stage. The anticorrosion protection primer is mostly based on an epoxy resin which cures with amines or isocyanates at room temperature. On account of the high safety standards imposed on component resistance, chromate-containing active pigments are still used today to prevent dangerous filiform corrosion. A 2-component PUR topcoat may also be applied to the part [7.12.98].

A chromate-free pretreatment and chromate-free primers are currently in an extensive test phase and are finding application in new aircraft models as well as for chrome-free electrocoatings [7.12.99].

The topcoat for the aircraft body is made with a 2-component PUR paint and possesses high gloss, colour fastness and weathering resistance. The two coats are applied in suitably large hangars by manual spraying. The often multi-coloured coatings are usually applied with the aid of templates. Curing then takes place at room temperature.

Aircraft require a complete overhaul, including a repaint of the outer skin, after approx. 5,000 flying hours on account of the high stresses which they experience. To this end they are manually stripped of their paint in the large hangars, cleaned, inspected for defects and fatigue, and finally recoated. As with the original coating this is usually done manually by conventional spraying.

The high emissions of solvents which are bound to occur where conventional exterior coatings are applied, coupled with legislative requirements, are leading to increased use of water-borne products. A further major challenge is presented by chromate-free formulations, which must not impair the important anticorrosion protection.

A large segment is accounted for not only by exterior coating, but also by interior coating. Although wall linings, window frames, doors and luggage shelves are partially made of pigmented plastics, these and the nonpigmented components can be provided with appropriate coatings to match the airlines' design specifications. The paints used for this purpose are characterised by good adhesion to plastics and by high flammability resistance to international FST (Fire, Smoke, Toxicity) standards. Again, 2-component PUR systems mainly are used for this segment. Water-borne systems are advancing as much in this area as in structural coating [7.12.100].

7.11.4 Electrically Insulating Coatings

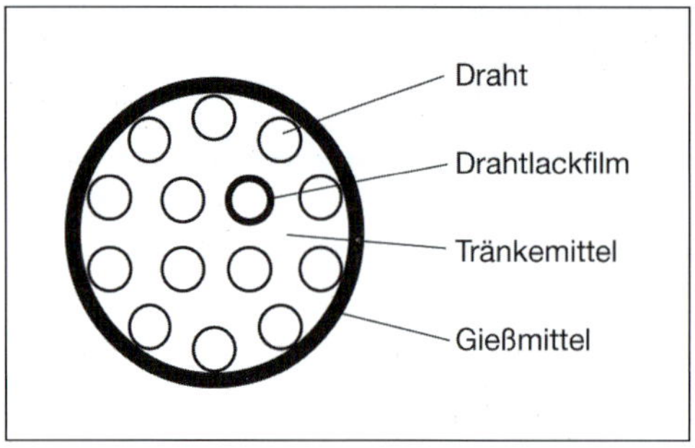

Figure 7.11.3: Various types of coatings and their functions, as illustrated by the example of electrical windings for electric motors

Electrically insulating coatings have the task of insulating electrically conducting substrates and so occupy a special position amongst the various coatings. Decoration plays a virtually negligible role, even though protection is increasingly desirable in the field of electronics parts and components.

For example, the items to be coated are either homogeneous materials as in the case of wire enamels, live parts of electric motors and transformers in the case of impregnating agents, or electronic parts and components in the case of casting materials.

With electric motors it is important to match all types of paint with one another to create the most effective electrically insulating system.

Electrical insulation is critical to ensuring that electric motors and transformers operate properly. Thermal resistance is also crucial to safe, continuous operation. For this reason, imide structures are contained in most of the coating systems described (see Chapter 2.1).

Wire Enamels

Wire enamels provide an insulating coat for wires used in electric motors, transformers and coils. These are not to be confused with the wire insulation for wires for electric cables.

The wire enamel is continuously dip-applied in several thin layers and stoved at temperatures of 400 to 500 °C. Solderable wire enamels permit direct tinning without the need for removing the coating. Other wire enamels feature high thermal resistance. The latter are formulated with polyesterimides.

Coating Industry

Impregnating Materials

The task of impregnating materials is to fill cavities in windings and strengthen them physically. They thus prevent penetration by harmful substances, e.g. water, improve heat dissipation and avoid the risk of wire breaks when being switched, as well as the occurrence of imbalance during rotation.

Impregnating materials are applied either by flow coating, dipping, vacuum impregnation or trickle impregnation with trickle coatings. The last of these processes prevents the housing from being coated by directing a thin jet of trickle coating at the end winding of the rotating workpiece. The coating then flows inside and is distributed inside the winding by centrifugal force.

The impregnating materials are generally stoved at 120 °C. The impregnating materials in question are based on unsaturated polyesters which are activated by peroxide hardeners. The solvents are styrene and similar unsaturated compounds. To a certain extent phenol-alkyd resin-based impregnating materials are still used. Modification with imide structures improves thermal resistance. However, long drying times are a drawback.

Casting Agents

The use of casting agents ranges from additional surface protection for electrically conducting winding parts to encapsulation of electrical components such as high-volt-age coils through to solid casting as an insulator and mechanical load-bearing structure.

The casting agents are based on polyester resins, polyurethane resins or epoxy resins to match the performance requirements. They are usually pigmented with extenders and activated by hardeners and so curing takes place rapidly at room temperature.

Application methods range from simple casting, as is usual with cable connection components, to the vacuum process, which optimally adjusts the electrical insulation by preventing cavities.

7.11.5 Communications Equipment

Apart from TV cabinets and computer housings, which are coated on conventional lines, the choice of which depends on the substrate, applications worth mentioning are the coating of printed circuit boards with photoresists, the sealing of CDs or DVDs and the coating of fibre-optic cables [7.12.101].

Cabinets for television sets, personal computers and peripheral equipment such as printers and scanners, as well as cameras and mobile phones are predominantly made of plastics. Many of these items are not coated, as they are manufactured with already pigmented plastic pellets. The reason for this is that they are not overly exposed to stress due to light, heat or moisture high. Cameras and mobile phones, on the other hand, are subjected to greater stress and ever more demanding design requirements and so they are often coated. The very wide range of polymer materials requires a specific pretreatment for each (see Chapter 4.1.1). They are spray coated on standard lines. The number of coats varies with the requirement for the surface finish of the item concerned. Two-layer topcoats are usual for effect coatings, e.g. on mobile phones.

The manufacture of printed circuit boards does not require durable coatings; instead temporary ones are cured selectively by ultraviolet radiation and are then washed, etched and metallised to generate appropriate structures on microscopic electrical conductors. These coatings can be both electrocoats and conventional solvent-borne coatings.

The high production speed of compact discs (CDs), digital versatile discs (DVDs) and Blu-Ray discs (BDs) based on PMMA or PC plastics [7.12.102] requires an equally fast sealing process for the aluminium film.

Figure 7.11.4: Examples of coated mobile phones

Source: Scanrail; www.Fotolalia.com

It is usually provided by UV-curable coatings which are applied by spin-coating owing to the circular shape of the information storage device. For this purpose, droplets of coating are applied in the centre whilst the disc rotates at an appropriate speed and the resulting film is cured by means of ultraviolet radiation.

To cover long transmission distances fibre-optic cables require an external coating with a different refractive index than that of the substrate. UV-curable clearcoats are used for this purpose also.

7.11.6 Road Marking paints

Road markings have to fulfil the following specific requirements:

- Daytime visibility with a durable, high degree of white
- Night-time visibility through the use of reflective elements
- Night-time visibility under wet conditions
- Good adhesion to road surfaces
- Slip resistance
- Permanent elasticity
- Abrasion resistance
- Rapid cure

These requirements stem from scientific studies into the safety aspects of road markings. In 1998 European Standard EN 1436 was introduced which has since led to a significant improvement in the quality of markings.

Easy application and very rapid solidification and curing are essential properties of the materials if the road marking procedure is to proceed smoothly without obstructing traffic excessively with road blocks.

Road marking materials can be divided up on the basis of requirements and composition into thin-layer and thick-layer types.

Thin-layer markings with a wet film thickness of 0.5 to 1.5 mm and applied as purely physically drying one-component coatings are predominantly solvent-borne systems because they offer a much shorter time between application and usability than water-borne types [7.12.103]. The resin base is usually a VC copolymer (see Chapter 2.1.1). Two-component systems cure very rapidly after the curing component is added directly before application. Despite the high cost of the raw material, isocyanate crosslinking has become established on account of its speedy curing. In North America, on the other hand, the federal authorities are demanding the application of water-borne marking materials to a new specification. Specially formulated acrylic dispersions which undergo a rapid reduction in pH after application are being proposed there [7.12.104].

Thick-layer markings are applied with hot-melt plastics, i.e. thermoplastics, and require heating to 180 °C for application. Once the markings have cooled down, vehicles can drive over them. The most important base resins in Europe are hydrocarbon resins and esters of colophony (see Chapter 2.1.1). Maleic-acid-modified natural resins account for two thirds of the total market in North America and are gradually gaining in Europe because of the higher pigmentability and wetting of glass beads [7.12.105].

Also used are so-called cold-setting plastics which are designed as solvent-free 2-component systems and also feature rapid curing. However, they are much more expensive than hot-melt plastics.

Construction sites are marked with adhesive foils which can be applied and removed afterwards by heating.

In Germany road markings are applied with materials which are approved in accordance with the regulations of the Federal Road Research Institute ("BASt"). The conditions and specifications are defined in the "TPM" or "Technical Provisions for the Testing of Marking Materials for Trunk Roads". These classify the materials by load capability and service life.

7.11.7 Shipbuilding

The coating of ships can be divided into in two segments:

- Commercial ships such as container vessels, tankers and cruisers
- Yachts and other leisure boats

The paints provided for this purpose are subdivided into OEM coatings (approx. 35 %) and repair and maintenance coatings (65 %).

Commercial ships are made of steel and need to be well protected by the coatings against corrosion because the constant effect of salt, particularly in marine transport, seriously increases steel corrosion. To assist with anticorrosion protection, particularly below the waterline, most ships are also cathodically protected by active anodes or stray current. The resulting hydroxyl anions create an alkaline environment, which requires the ship's coating to possess saponification resistance [7.12.106].

Furthermore, the paints and coatings must be resistant to algal growth and encrustation by other seawater organisms. Such growth not only contaminates the anticorrosive coating but also constitutes an economic problem. Clinging mussels roughen the otherwise smooth hulls, creating drag that reduces the speed of the vessels and increases diesel fuel consumption by several percent [7.12.107]. The use of resistant coatings extends the intervals between ship maintenance in dock by up to 30 %. Consequently, expressed in terms of the world fleet, some 1.5 billion US dollars could be saved every year by using such resistant coatings. Tin compounds contained in such formulations used to be state of the art when it came to the biocidal action of underwater coatings. However, such "antifouling" coatings release their

Figure 7.11.5: Ship coating *Source: Peter Bader; www.Fotolalia.com*

inorganic compounds fairly rapidly over the course of their lifetime and thus destroy the algae in the direct vicinity of the coating. The ban on organic tin compounds such as TBTO (tri-butyl tin oxide) and its now equally dubious replacements, namely copper oxides, in the wake of international legislation such as the EU's Biocidal Products Regulation (see Chapter 5) has spurred enormous development efforts aimed at discovering new substances and different solutions (see Chapter 2.1.4) [7.12.108]. Similar coatings are also employed in the growing market for drilling platforms and not inconsiderable quantities find application in yacht building.

The further requirements imposed on the coating of ships vary with the particular section of the ship to be treated. Upper deck coatings must withstand foot traffic and be slip-proof and hard-wearing. Ideal for this purpose are solvent-free 2-component liquid plastics. Deck superstructures such as command bridges, masts, container brackets, hatches, railings, etc. are given coatings of a quality which is equivalent to those used on the shell plating, often in different colours.

Coatings in the engine room must have a low level of flammability and conform to the specifications issued by the Seamen's Accident Prevention and Insurance Association. This also applies to the interior finish of passenger ships.

Special coatings based on epoxy resins and combinations of epoxy resins with hard coal tar are necessary for the oil bunkers in tankers. The same coatings, as well as 2-component epoxy primers, are suitable for tanks of exchanging oil and ballast water, and pure ballast water tanks. However, zinc silicate coatings are also used for transporting nonreactive hydrocarbons. In general, the coatings for tanks that transport chemicals are particularly critical. For example, they call for careful coordination of paint formulation, quality assurance test methods, manufacture, surface pretreatment and monitored application and inspection of the coating. The coatings are generally applied by the airless process. Poorly accessible surfaces are prepared manually and coated by brush.

For storage rooms 2-component systems based on epoxy resins and isocyanate cross-linkers have proved successful.

Today, large container vessels, for example, are usually constructed to a modular design. The anticorrosion protection may be applied during the production of these modules before final assembly [7.12.109].

The coating of yachts and other leisure boats is technically different because many boats from this sector have plastic hulls. However, wood is still encountered as a material, primarily in the luxury class. Coating is conducted with primers for the plastics and with coloured underwater paints that have an "antifouling" function.

7.11.8 Wind Energy

The search for new energy sources in conjunction with the abandoning of nuclear power and the combustion of fossil fuels has brought a surge in the global popularity of wind energy, as well as solar energy. This is especially true of Germany, which accounts for 28 % of the total wind energy harnessed today.

In 2008, some 1.5 % of the global energy requirement was already covered by wind power of 120 gigawatt, with the stated goal being to raise that percentage to 12 % by

Figure 7.11.6: Wind turbines and rotor blade painting *Source: B. Wylezich; www.Fotolia.com*

2020. If such ambitious forecasts are to be realised, future wind power plants will need substantial optimisation in terms of size and efficiency. At the same time, the turbines need to be relocated to windier offshore areas where the conditions for constant, stronger winds are more favourable.

With a tower height of up to 90 metres and rotor blade lengths of 20 metres, today's turbines are already capable of generating six megawatts per turbine. To really achieve the declared goal of a tenfold increase in the share of wind energy by 2020, turbines with up to 20 GW capacity and a rotor blade length of 90 metres are currently being developed [7.12.110].

The associated increase in size of the towers and rotors, coupled with an increase in rotary speed, will place ever-increasing demands on surface treatment. This also applies in the same way to towers made of steel or concrete and to the glass or carbon fibre reinforced plastic rotors. As a result of their rotation, the tips of the blades on current systems can reach speeds of up to 300 km/h. This generates extremely abrasive interaction and impact stress by atmospheric agents such as rain, hail, snow and dirt particles. The influence of UV rays from the sun and the salt burden in offshore plants also place high demands on the quality of the paint finish [7.12.111].

The rotor blades are manufactured in two halves (prepregs) by the vacuum infusion process. Epoxy resin impregnated glass or carbon fibre mats are inserted into two moulds which have been pretreated with release agents, the mould halves are sealed airtight and negative pressure is applied to release all the trapped air. The prepregs are hardened by heating to 70 °C and then the two separate rotor halves are bonded together to form the finished component.

It is vital that all release agents be removed before the specified coating is applied. Painting robots then apply a gelcoat in the first of a multi-stage coating process [7.12.112]. Wind turbines have an estimated economic lifetime of 20 years. The sensitivity of the underlying economics and thus the success of wind power, together with the extraordinary physical and weather-related loads to which the turbines are subjected, call for the use of high-quality materials which need to be continually enhanced to keep up with the demands of ever-larger wind turbines. In any case, special edge protection coatings [7.12.113] and repair systems for rotor blade coating are under development [7.12.114]. Water-borne fluoroacrylics are also being developed because of the good durability which they offer [7.12.115].

The gelcoat mentioned above is therefore designed for good adhesion to both the substrate and the subsequent intermediate layer. A special filler and additional edge protection are used to smooth uneven surfaces. The final topcoat comprises a high-grade 2-component PUR coating with a viscoelastic, scratch-resistant finish and special UV and anti-icing protection (see next chapter). All layers are applied in the airmix process by robots and then cured with infrared radiation. To avoid reflections and thus interference with air and road traffic, the topcoat is generally matt.

In view of the mechanical stress which is generated on the usually steel towers by the motion of the rotors in conjunction with the corrosive stress, mainly in the offshore area, extremely good anchoring in the ground is essential. The steel segments of the towers are subjected to wholesale, careful derusting, followed by galvanising and a powder or 2-component PUR coating to achieve 20 years' protection.

7.11.9 Smart Coatings

Coating materials that generate a novel function within the film have been around for some time now in the guise of smart coatings. This term is, of course, very broad. Some of the functions [7.12.116, 7.12.117, 7.12.118] in the corresponding films are:

- Special anticorrosion protection
- Self-healing films
- Colour matching to the environment
- Biocidal surfaces
- Catalytic surfaces for air regeneration
- Special optical effects and functional surfaces to lower resistance in aircraft building and shipbuilding
- Prevention of ice formation in wind turbines

A host of research activities in this field are aimed at generating new or improved functions and increasing added value through coatings [7.12.119]. Nanoproducts based on corresponding nanotechnology play an important role in this regard (see Chapter 3.1.5). The furthest advanced and in some cases most commercialised smart coatings include antibacterial or biocidal systems, catalytic systems in façade construction and systems for special anticorrosion protection.

The formulation principles underlying such smart coatings are straightforward. Only nanoparticles, e.g. with biocidal properties, are incorporated into the coating materials, where they either remain isotropically in the film matrix or migrate to the surface to generate the corresponding properties. These properties are usually characterised by ultrahydrophobic or ultrahydrophilic surfaces. They become particularly attractive if they are also switchable [7.12.120]. In the case of self-healing coatings, usually filled nanocapsules are added which, if they become damaged, physically compromised or if their pH changes, release polymerisable substances [7.12.121] or inhibitors [7.12.122] into the films.

Not necessarily linked to the use of nanoparticles is the production of coatings whose surface has a biocidal effect. Particularly in the finishing of façade coatings [7.12.123] and in the sanitary area, various additives are used to provide effective surface protection. In this connection, ionic, stable structures play an important role in the film [7.12.124]. However, the efficacy demonstrated in the various tests, e.g. ISO 846 or JIS Z 2801, is not always replicated in the real world [7.12.125].

A very interesting approach is that of surfaces which help to reduce nitrogen oxide pollution in metropolitan areas, for example, by acting as photocatalysts that cleave these airborne pollutants into nitrogen and oxygen [7.12.126].

7.11.10 Miscellaneous Coating Applications

A major segment for the application of coatings is the **packaging industry**. In this market segment a distinction is drawn between cardboard boxes and tubes, which are essentially served with printing inks, and consumer packaging. In the latter segment, beverage cans share the market with food cans. They are made of tin-plated steel (tin plate) or aluminium whilst the interior and exterior are coated. In addition, the exterior is printed with the necessary product information to comply with food legislation as well as for advertising purposes.

The per capita consumption of beverage cans varies substantially across the regions. For example, it is approx. 330 a year in the USA but in Germany only about 50 cans. The situation as regards food cans is similar. In North America almost only aluminium cans are used whilst in Europe tinplate still accounts for a considerable share. Beverage cans are made by deep-drawing the can body. It is then spray-coated inside and primed outside with rollers, usually in a white colour. The can body is then provided with a prefabricated lid in a combined forming and crimping process. This "2-piece" process has largely superseded the "3-piece" process, in which the bottom, body and lid are manufactured separately and then soldered together.

Beverage can coating especially is a high speed process which can handle up to 3,000 cans per minute. The interior lining applied by spray heads in an amount of 100 mg to 300 mg per can must be absolutely void-free. Either one or two coats are applied, the choice depending on the aggressiveness of the filling. The coatings used in North America are predominantly water-borne. By contrast, especially in Germany, solvent-borne coatings may be used if a thermal afterburner is installed. Curing takes only a few minutes at more than 180 °C.

Whilst beverage cans are made from a single sheet of metal, apart from the lid, a food can, especially where relatively large volumes are concerned, can only be made as a three-piece welded can. It is important to systematically protect the welded seams of the parts made of coated plate. This is predominantly done with the aid of powder coatings. In the past, the insides of the cans were mostly coated with a "gold" lacquer comprising an epoxy coating that turned yellow, but nowadays they are frequently provided with an opaque white coating for visual reasons.

Coatings for beverage cans and food cans require the approval of the food safety authorities. The American Food and Drug Association (FDA) is accepted throughout the world. The coatings are mainly based on epoxy resins.

Glass containers, e.g. for perfumes or similar products for personal care, are coated with powder or water-borne one-component PUR coatings [7.12.127].

Drum coating is another market segment in which some coating companies specialise. Epoxy/phenolic resins constitute the basis for these highly solvent-resistant coatings which are applied to the interiors in a thickness of 20 to 40 µm. Spray-applied water-borne coating materials are increasingly being used for this. The exterior coating is applied in one layer. Gas cylinders are protected on the outside with powder coatings [7.12.128].

Bicycles are mostly coated with powder paints. High-quality effect coatings are achieved by applying a powder primer followed by a topcoat comprising liquid basecoat/clearcoat layers. The same considerations apply to the coating of **motorcycles** as to automotive coating, with 2- to 3-coat systems being applied in line with the nature of the substrate. In the USA powder clearcoats are used on steel parts.

Spray can lacquers for DIY applications, which also serve as graffiti paints, have gained a certain significance. Based on mixed colour systems and similar to automotive refinishes, the formulations have to be compatible with the propane or butane propellant gas. Sprayability and hiding power are the essential quality criteria. The creation of special effects such as "granite" or hammer finish are important in this segment. Spray paints are increasingly being used in the automotive repair sector [7.12.129], with special spray heads to achieve a wider spray jet and to permit use of two-component coatings (see Chapter 7.2).

Various **sports articles** such as skis, snowboards, rollerblades and protective helmets are coated for design reasons. Articles of simple geometry are usually given a clearcoat based on UV-curable coatings which confer a high level of scratch resistance.

Lettering on vehicles, signs or other coated items has almost entirely switched to decal technology, i.e. the sticking on of prefabricated foils. Coating is no longer of any great significance (see Chapter 5.6.3).

Granite effects can be created on the surfaces through the single application of **multi-colour coatings**. Here, the paint system comprises deliberately separated micelles created from various inorganic components [7.12.130].

7.12 Literature

References

7.12.1 M. von Dungen, C. Maier, F+L 117 (11), p. 10 (2011)
7.12.2 Anonymous, JOT 45 (4), p. 46 (2005)
7.12.3 F. Bloser, Coat. World 10 (10), p. 52 (2005)
7.12.4 P. Svejda, Coat. World 16 (6), p. 34 (2011)
7.12.5 W. R. Cramer, Fahrzeug + Karosserie 58 (5), p. 34 (2005)
7.12.6 E. Wegner, Coat. World 9 (10), p. 44 (2004)
7.12.7 N. Löw, F. Hezel, F+L 118 (2), p. 15 (2012)
7.12.8 Anonymous, MO 66 (9), p. 26 (2012)
7.12.9 M. Golek, MO 63 (9), p. 28 (2009)
7.12.10 K. Rother, JOT 46 (3), p. 26 (2006); M. Schiffko, MO 64 (9), p. 22 (2010)
7.12.11 H.-F. Carstensen, MO 54 (4), p. 46 (2000)
7.12.12 G. Scheffels, JOT 50 (9), p. 18 (2010); Anonymous, JOT 49 (9), p. 14 (2009); P. Svejda, JOT 47 (3), p. 26 (2007)
7.12.13 P. Svejda, JOT 50 (9), p. 28 (2010); D. Wieland, JOT 49 (3), p. 14 (2009)
7.12.14 C. Hilger, H. Schumacher, Sustainability of Coatings and Coating Processes – Eco-Efficiency Analysis, Proceedings Surcar, Cannes, 2013
7.12.15 P. Kuhm, P. Droniou, M. Franks, ECJ 11/2007 "Automotive", p. 4; J. Rudolph, R. Moyle, E. Ehinger, M. Simpson, JOT 48 (4), p. 16 (2008); S. Winkels, JOT 49 (9), p. 70 (2009)
7.12.16 Anonymous, MO 63 (9), p. 11 (2009); P. Svejda, JOT 41 (9), p. 36 (2001)
7.12.17 H. Gehmecker, Automotive Paints and Coatings (publisher Streitberger/Dössel), p. 61, Wiley-VCH, Weinheim, 2007
7.12.18 Anonymous, Lackiererblatt 04/2006, p. 20
7.12.19 Anonymous, Automobil-Produktion, 06/2001, p. 126 and p. 136
7.12.20 Anonymous, Fahrzeug + Karosserie 62 (8), p. 72 (2009)
7.12.21 T. Weiler, JOT 50 (9), p. 32 (2010)
7.12.22 H. Gehmecker, JOT 41 (9), p. 70 (2001)
7.12.23 T. Körner, JOT 48 (9), p. 52 (2008)
7.12.24 K. Rapp, Proceed. 25. Tagung, Automobilkreis, Bad Nauheim, 2008
7.12.25 H.-J. Jüngling, JOT 44 (9), p. 18 (2004); Anonymous, JOT 46 (3), p. 22 (2006)
7.12.26 P. Svejda, JOT 49 (9), p. 16 (2009)
7.12.27 H. Krug, JOT 39 (4), p. 20 (1999)
7.12.28 H.-J. Streitberger, Weltweite Aktivitäten zur Anwendung von Pulverlacken in der Automobilindustrie, DFO-Tagung Pulver, Hamburg, 1995
7.12.29 K.-F. Dössel, MO 60 (6), p. 12 (2006)
7.12.30 Anonymous, Automobil Produktion 12/1999, p. 152; H. Beckersjürgen, JOT 41 (10), p. 88 (2001)
7.12.31 C. Vignolo, F+L, 119 (1), p. 34 (2013)
7.12.32 J. Klasing, Fahrzeug + Karosserie 53, p. 10 (2000)
7.12.33 T. Türk, JOT 41 (9), p. 16 (2001)
7.12.34 P. Barbeau, R. Klucker, J.-L. Loubet, S. Pavan, PCI 27 (2), p. 34 (2011)
7.12.35 Anonymous, Agenda Farbe 5 (5), p. 43 (2012)
7.12.36 WO 2012065828 (2012) BASF Coatings
7.12.37 F. J. Maile, G. Pfaff, P. Reynolds, Progr. Org. Coat. 54 (1), p. 150 (2005)
7.12.38 M. Große-Ophoff, JOT 41 (9), p. 82 (2001)
7.12.39 P. Burger, JOT 41 (9), p. 52 (2001)
7.12.40 R. Schwalm, F+L 106 (4), p. 58 (2000)
7.12.41 R. Edelberg, Phänomen Farbe 21 (10), p. 36 (2001)
7.12.42 D. Stranghoener, K.-F. Dössel, ECJ 06/2000, p. 39
7.12.43 G. Steinbeck, JOT 41 (9), p. 68 (2001)
7.12.44 C. K. Schoff, JCT Coat. Tech. 8 (4), p. 48 (2011)
7.12.45 M. Burkamp, JOT 38 (12), p. 26 (1998)
7.12.46 K. Pianoforte, Coat. World 17 (10), p. 34 (2012)
7.12.47 T. Grupp, U. Hoffmann, JOT 39 (12), p. 26 (1999)
7.12.48 M. Maß, MO 54, p. 42 (2000); T. Scholz, JOT 38 (7), p. 34 (1998)
7.12.49 H. Schiffer, W. Sinning, JOT 39 (7), p. 30 (1999)
7.12.50 F. Bloser, Coat. World, 10 (10), p. 52 (2005); U. Schröder, Kunststoffe 95 (1), p. 81 (2005)
7.12.51 Anonymous, MO 66 (11), p. 19 (2012)
7.12.52 G. Wilke, J. Ortmeier, Kunststoffbeschichtung, p. 92, Vincentz, Hanover, 2009

7.12.53 S. Jacob, C. Bernau, JOT 44 (3), p. 30 (2004)
7.12.54 Anonymous, MO 55 (3), p. 48 (2001)
7.12.55 S. Winkels, JOT 49 (9), p. 70 (2009)
7.12.56 R. Bluhm, A. Eippe, I. Fitta, Kunststoffe 98 (3), p. 61 (2008)
7.12.57 A. Jandel, MO 55 (2), p. 31 (2001)
7.12.58 M. von Dungen, ECJ 02/2012, p. 10
7.12.59 Anonymous, JOT 47 (7), p. 36 (2007)
7.12.60 M. Hickl, MO 60 (4), p. 44 (2006)
7.12.61 K. Mallet, ECJ 12/2001, p. 37; B. Meuthen, JOT 40 (2), p. 16 (2000)
7.12.62 P. H. G. Binda, PPCJ 192, p. 21 (2002)
7.12.63 Anonymous, MO 65 (12), p. 16 (2011)
7.12.64 Anonymous, MO 60 (9), p. 16 (2006)
7.12.65 Anonymous, JOT 47 (2), p. 32 (2007)
7.12.66 H.-P. Pfeiffer, H.-P. Hecker in Kittel – Lehrbuch der Lacke und Beschichtungen, vol. 6, 2nd ed.(publisher H.-J. Streitberger), Anwendung von Lacken und sonstigen Beschichtungen, p. 347, Hirzel, Stuttgart, 2008
7.12.67 H. Voigtländer, MO 60 (4), p. 22 (2006)
7.12.68 T. Barmbold, JOT 48 (3), p. 44 (2008)
7.12.69 V. Perk, JOT 47 (8), p. 28 (2007)
7.12.70 H.-G. Fritz, H. Voigtländer, MO 61 (5), p. 36 (2007)
7.12.71 H. Zander, MO 47 (12), p. 596 (1999)
7.12.72 Anonymous, MO 60 (9), p. 16 (2006)
7.12.73 W. Jäger, JOT 48 (10), p. 74 (2008)
7.12.74 S. Kramer, JOT 41 (7), p. 22 (2001)
7.12.75 J. Prieto, J. Kiene, Holzbeschichtung, p. 49, Vincentz, Hanover, 2007
7.12.76 K. Pianoforte, Coat. World, 16 (2), p. 20 (2011)
7.12.77 C. Körber, Welt d. Farben 09/2009, p. 10
7.12.78 A. Riegel, K. Dekomien, ECJ 04/2013, p. 122
7.12.79 N. Havaux, D. Vanaken, F. Simal, PCI 26 (10), p. 34 (2010)
7.12.80 C.-T. Lin, JCT Coat. Tech. 5 (3), p. 48 (2008)
7.12.81 Anonymous, MO 66 (7-8), p. 22 (2012)
7.12.82 S. Friebel, ECJ 1-2/2001, p. 68
7.12.83 M. Müller, F+L 107, p. 145 (2001)
7.12.84 J.-P Bleus, M. Decaux, G. Farnzolin, M. Philips, Pitt.e Vernici 09/2005, p. 5
7.12.85 K. Menzel, M. Biehler, M. Lokai, Pitt.e Vernici 2-3/2006, p. 26
7.12.86 J. Prieto, PPCJ 191 (7), p. 15 (2001)
7.12.87 J. Prieto, ECJ 04/2010, p. 20
7.12.88 J. C. Brogan, B. Equi, S. Quinn, ECJ 03/2013, p. 100
7.12.89 J. Rother, I-Lack 66 (4), p. 210 (1998)
7.12.90 R. Lüscher, JOT 42 (7), p. 32 (2002)
7.12.91 H. Kämpf, B. Drummer, J. Gehrke, MO 59 (11), p. 25 (2005)
7.12.92 O. Elizalde, S. Amthor, C. Moore, JCT Coat. Tech. 7 (9), p. 22 (2010)
7.12.93 H.-J. Bönstedt, JOT 45 (11), p. 28 (2005)
7.12.94 Q. Bell, Surf. Coat. Australia 30 (9), p. 14 (1993)
7.12.95 M. Virnau, MO 56 (3), p. 48 (2002)
7.12.96 Anonymous, MO 60 (10), p. 22 (2006)
7.12.97 M. Agosta, Coatings World 08/2002, p. 26
7.12.98 Anonymous, MO 60 (5), p. 14 (2006)
7.12.99 T. Wright, Coat. World 17 (8), p. 31 (2012)
7.12.100 S. Schröder, R. Stein (publisher H.-J. Streitberger) in Kittel – Lehrbuch der Lacke und Beschichtungen, vol. 6 (2nd ed.), Anwendung von Lacken und sonstigen Beschichtungen, p. 313, Hirzel, Stuttgart, 2008
7.12.101 G. Shen et al., JCT 71 (891), p. 53 (1999)
7.12.102 K. Roth, CHIUZ 41, p. 334 (2007)
7.12.103 B. Osterath, Nachr. Chemie 59, p. 1156 (2011)
7.12.104 K. B. Kosto, D. C. Schall, JCT Coat. Tech., 5 (7), p. 22 (2008)
7.12.105 S. Seeley, PPCJ 191 (11), p. 14 (2001)
7.12.106 I. Royston, JOCCA 82 (1), p. 186 (1999)
7.12.107 J. Brown, PPCJ 202 (4574), p. 16 (2012)
7.12.108 C. Challener, JCT Coat. Tech., 8 (4), p. 28 (2011)
7.12.109 Anonymous, JOT 45 (5), p. 46 (2005); Anonymous, JOT 46 (10), p. 108 (2006)
7.12.110 Anonymous, Agenda Farbe 5 (5), p. 6 (2012)
7.12.111 R. Stokes, PPCJ 202 (4570), p. 20 (2012)

7.12.112 Anonymous, JOT 05/2012, p. 36
7.12.113 Anonymous, MO 66 (5), p. 14 (2012)
7.12.114 Anonynmus, MO 67 (3), p. 17 (2013)
7.12.115 M. Durali, K. Wood, D. Battocchi, M. Hatzenbeller, ECJ 10/2011, p. 30
7.12.116 A. Stewart, PPCJ 202 (4571), p. 29 (2012)
7.12.117 S. Sell, N. Rehfeld, A. Kreider, A. Stake, V. Föste, V. Stenzel, MO 67 (3), p. 20 (2013)
7.12.118 S. Scharf, N. Rehfeld, V. Stenzel, F+L 118 (4), p. 15 (2012)
7.12.119 C. Challener, JCT Coat. Tech. 5 (8), p. 32 (2008)
7.12.120 P. Uhlmann, K. Oleschko, N. Laber, G. Hüttner, B. Lehmann, F+L 117 (9), p. 17 (2011)
7.12.121 S. Tillner, U. Mock, F+L 113 (10), p. 35 (2007)
7.12.122 S. J. Garcia; H. R. Fischer; P. A. White, J. Mardel, Y. Gonzalez-Garcia, J. M. C. Mol, A. E. Hughes, Progr. Org. Coatings, 70, p. 142 (2011)
7.12.123 S. Sauer, PPCJ 203 (4580), p. 37 (2013); E. Bagda, Welt d. Farben 03/2010, p. 11
7.12.124 Anonymous, JOT 04/2013, p. 50
7.12.125 T. Brenneer, M. Meder, F+L 116 (6), p. 25 (2010)
7.12.126 S. Fischer, MO 67 (4), p. 16 (2013)
7.12.127 R. Stewart, PPCJ 202 (4572), p. 20 (2012)
7.12.128 Anonymous, I-Lack 61 (9), p. 327 (1993)
7.12.129 Anonymous, Lackiererblatt 4/2006, p. 32
7.12.130 A. Lork, M. Möller, ECJ 05/2013, p. 24

General Literature

H.-J. Adler and K. Potje-Kamloth (publisher), Quo Vadis – Coatings?, XXVI Fatipec-Kongress, Dresden, Wiley-VCH Verlagsges., Weinheim, 2002
C. Hellio, D. Yebra (publisher), Advances in marine antifouling coatings and technologies, Woodhead Publ., Cambridge, 2009
Streitberger/Dössel (publisher), Automotive Paints and Coatings, Wiley-VCH, Weinheim, 2007
J. Sander (publisher), Korrosionsschutz durch Beschichtungen, Vincentz, Hanover, 2011
A.-S. Jandel, B. Meuthen, Coil Coating (3rd ed.), Springer-Vieweg, Wiesbaden, 2013
V. Stenzel, N. Rehfeld, Functional Coatings, Vincentz, Hanover, 2011
J. Prieto, J. Keine, Holzbeschichtung, Vincentz, Hanover, 2004
H.-J. Streitberger (publisher), Kittel – Lehrbuch der Lacke und Beschichtungen, vol.6 (2nd ed.), Anwendung von Lacken und sonstigen Beschichtungen, Hirzel, Stuttgart, 2008
E. Spyrou, Powder Coatings – Chemistry and Technology, (3rd ed.) Vincentz, Hanover, 2012
F. Bulian (publisher), Wood Coatings – Theory and Practice, Elsevier, Science and Technology, 2009
R. Schwalm, UV Coatings, Elsevier, Oxford, 2007
S. Sepeur et al., Nanotechnology, Vincentz, Hanover, 2008
P. Svejda, Prozesse und Applikationsverfahren, Hanser, Munich, 2007
M. Lake (publisher), Oberflächentechnik in der Kunststoffverarbeitung, Hanser, Munich, 2009
Kunststoffe im Automobilbau, VDI Verlag, Düsseldorf, 2011

8 Standards

8.1 General Information on Standardisation Work

The standardisation of products, processes and methods of measurement and testing promotes rationalisation and quality assurance in industry, technology, science and administrative circles. It serves to provide safety for people and property and to enhance quality in all spheres of life.

Standardisation is the systematic harmonisation of material and immaterial items, carried out jointly by interest groups, for the benefit of the general public. Internationally accepted industrial standards eliminate trade barriers and promote cross-border trade. They are therefore of major economic importance.

Industrial standards are designed to be introduced as accepted rules of engineering. They constitute a benchmark for quality and quality consistency. This benchmark is of importance within the framework of legal regulations. Standardisation also serves to regulate sensibly and provide information in the various technical fields.

The standardisation body in Germany is Deutsches Institut für Normung e.V. (DIN) in Berlin. It is an institution which is self-governed by all groups interested in standardisation. DIN is the round table at which manufacturers, trade, consumers, craftsmen, service companies, academic researchers, technical monitoring bodies, the government and anyone who has an interest in standardisation meet in order to determine the state of the art and place it on record in German standards. Its work is based on 10 basic policies:

1. Voluntary participation
2. Disclosure to the public
3. Participation by all interest groups
4. Unified approach and freedom from contradiction
5. Objectiveness
6. Consensus
7. Orientation towards the state of the art
8. Orientation towards economic circumstances
9. Orientation towards general good
10. Internationalism

Many German standards and draft standards are currently available for reading on the internet at www.din.de. All German, European and International Standards can be obtained from the publishers, Beuth-Verlag, which is associated with DIN. The Association of the German Paints and Coatings Industry (VdL) also publishes regular overviews concerning the current status of standards for members.

In the field of coating materials and coatings it is chiefly test methods and properties as well as technical terms which are in need of harmonisation and standardisation. This harmonisation is desirable because over the years a whole host of "company standards" have become established which have evolved from the mostly specific requirements of paint processors and the wealth of experience gained in dealing with these standards and test methods. The verification of standards and test methods, as required by quality management systems, has recently eliminated a large number of test methods because of their poor repeatability and measuring inaccuracies (see Chapter 6.5). Furthermore, standardisation regulates how paints are to be handled from a safety point of view. An essential trilingual guide to the terminology in German, French and English has since been published [8.4.1].

Standards dealing with coating materials and coatings are summarised in a DIN handbook entitled "Paints, Varnishes and Similar Coating Materials". This is a loose-leaf collection which is updated four times a year by the standardisation committee responsible, NAB (standardisation committee for coating materials and coatings). This committee celebrated its 60th anniversary in 2009.

Apart from this committee there are other standardisation committees which are of relevance to the paints and coatings industry. They include:

- Standardisation committee for principles of environmental protection (NAGUS)
- Standardisation committee for personal safety equipment (NPS)
- Standardisation committee for materials testing (NMP)
- Standardisation committee for pigments and extenders (NPF)
- Standardisation committee for quality management, statistics and certification principles (NQS).

A DIN standard is therefore a standard which has been drafted and issued by Deutsches Institut für Normung and which bears the DIN logo. This may take the form of a pre-standard which is still subject to reservations about application and which can be used on a trial basis for the time being. A supplement contains information about the standard but no additional standardised definitions. A draft standard is the provisionally completed result of standardisation work which is presented to the public for opinions.

Like industry, standardisation work is also subject to trends in globalisation. Consequently, efforts are being made to develop standards jointly for the paints and coatings industry in Germany, Europe and the rest of the world. DIN standards still exist, but now only country-specific standards are filed under this classification and these may not be accepted or have any significance at European or global level. European standards are filed under DIN EN if they have been introduced by Germany or other CEN (Comité Europeen de Normalisation or European Committee for Standardisation) member states and have passed a check for competitive advantages by that body.

CEN collaborates closely with ISO, the International Organisation for Standardisation which is based in Geneva. Accepted standards are given the designation of DIN EN ISO if they originate in Germany and have been passed in Europe.

Apart from that there is, relevant for the coatings industry, the ASTM, the American Society for Testing and Materials, in West Conshohocken, PA, which works on standards and standardisation processes for the North American market, more or less in parallel with ISO, CEN and DIN. Its renam-ing as ASTM International in 2001 demonstrates that this organisation is also giving due consideration to the globalisation of markets.

Ultimately paints and coatings companies with global operations currently have to negotiate their way about the respective and accepted standardisation systems in all markets because, for example, American manufacturers in Europe often specify ASTM standards.

Apart from the harmonisation of standardisation work on inspection and test methods, a further area of standardisation is that of quality management systems and environmental standards (see Chapter 6.3).

The management standards developed to date have found widespread acceptance and are extensively employed within the paints and coatings industry.

8.2 Selection of DIN Standards for the Paints and Coatings Industry and their Circle of Users

A selection of important standards for the coatings industry in Germany and, in some cases, in Europe as of 07/2013 is presented below in numerical order (List 8.2.1) and by keyword (List 8.2.2).

It should be noted that the standards are constantly being updated, on one hand with regard to the accuracy of the methods and on the other through extension or restriction of the application.

8.2.1 Standards by Numerical Order

8.2.1.1 Standards on Terminology

Standard	Brief description
DIN 55651	Solvents for paints and varnishes – Symbols
DIN 55655-1	Electrodeposition Paints – Terminology
DIN EN ISO 18451-1	Pigments, Dyestuff and Extenders – Terms and definitions
DIN 55945	Additional terms and definitions to DIN EN ISO 4618
DIN 55950	Symbols for the basis of the paint medium
DIN EN ISO 4618	Paints and varnishes – Terms and definitions
DIN EN ISO 8130-14	Coating powders – Terminology

8.2.1.2 Standards on Coating Materials

Standard	Brief description
DIN 53219	Volume of non-volatile matter
DIN 55655-2-9	Paints and varnishes – Electrodeposition paints
DIN 55659-1	pH Value – pH electrodes with glass membrane
DIN 55659-2	pH Value – pH electrodes with ISFET Technology
DIN 55685	Practical determination of non-volatile and volatile matter content during application
DIN EN 14210	Surface active agents – Determination of interfacial tension of solutions of surface active agents by the stirrup or ring method
ISO 3270	Temperatures and humidities for conditioning and testing
DIN EN ISO 1513	Preparation of test samples

Industrial Standards

Continuation

Standard	Brief description
DIN EN ISO 1516	Determination of flash/no flash – Closed cup equilibrium method
DIN EN ISO 1523	Determination of flash point – Closed cup equilibrium method
DIN EN ISO 1524	Fineness of grind
DIN EN ISO 2431	Flow time
DIN EN ISO 2811-1	Density – Pycnometer method
DIN EN ISO 2811-2	Density – Immersed body (plummet) method
DIN EN ISO 2811-3	Density – Oscillation method
DIN EN ISO 2811-4	Density – Pressure cup method
DIN EN ISO 2884-1	Determination of viscosity using rotary viscometers – Cone-and-plate viscometer operated at a high rate of shear
DIN EN ISO 2884-2	Determination of viscosity using rotary viscometers – Disc or ball viscometer operated at a single specified speed
DIN EN ISO 3219	Determination of viscosity using a rotational viscometer
DIN EN ISO 3233-1,-2,-3	Determination of the percentage volume of non-volatile matter – Method using a coated test panel to determine non-volatile matter and to determine dry film density by the Archimedes principle
DIN EN ISO 3679	Determination of flash point – Rapid equilibrium closed cup method
DIN EN ISO 9514	Pot life of multicomponent coating systems
DIN EN ISO 13736	Flash point – Abel closed-cup method
DIN EN ISO 14680-1	Pigment content – Centrifuge method
DIN EN ISO 14680-2	Pigment content – Ashing method
DIN EN ISO 14680-3	Pigment content – Filtration method
DIN EN ISO 15091	Electrical conductivity and resistance
DIN EN ISO 17895	VOC content in water-thinnable emulsion paints

8.2.1.3 Standards on Coatings

Standard	Brief description
DIN 50933	Measurement of coating thickness by differential measurement using a stylus instrument.
DIN 50986	Measurement of coating thickness; Wedge cut method for measuring the thickness of paints and related coatings
DIN 55660-1,-2,-3,-4	Wettability – Determination of the polar and dispersive fractions of the surface tension of liquids from an interfacial tension.
DIN 55660-5,-6,-7	Wettability – Determination of the polar and dispersive fractions of the surface tension of liquids from contact angles measurements on a solid with only a disperse contribution to its surface energy

Continuation

Standard	Brief description
DIN 55665	Natural weathering of coatings – Determination of corrosion protection performance
DIN 55670	Testing pores and cracks
DIN EN ISO 1514	Standard panels for testing
DIN EN ISO 1518-1	Scratch resistance – Constant-loading method
DIN EN ISO 1518-2	Scratch resistance – Continuously increasing load
DIN EN ISO 1519	Bend test (cylindrical mandrel)
DIN EN ISO 1520	Cupping test
DIN EN ISO 1522	Pendulum damping test
DIN EN ISO 2178 DIN 50981	Measurement of coating thickness – Magnetic method
DIN EN ISO 2360	Measurement of coating thickness – Amplitude-sensitive eddy current method
DIN EN ISO 2409	Cross-cut test
DIN EN ISO 2808	Determination of film thickness
DIN EN ISO 2810	Natural weathering
DIN EN ISO 2812-1	Resistance to liquids – Immersion in liquids other than water
DIN EN ISO 2812-2	Resistance to liquids – Water immersion method
DIN EN ISO 2812-3	Resistance to liquids – Method using an absorbent medium
DIN EN ISO 2812-4	Resistance to liquids – Spotting methods
DIN EN ISO 2812-5	Resistance to liquids – Temperature-gradient oven method
DIN EN ISO 2813	Gloss
DIN EN ISO 2815	Buchholz indentation test
DIN EN ISO 3231	Resistance to humid atmospheres containing sulphur dioxide
DIN EN ISO 3248	Determination of the effect of heat
DIN EN ISO 3497	Measurement of coating thickness – X-ray spectrometric methods
DIN EN ISO 3543	Measurement of thickness – Beta backscatter method
DIN EN ISO 3668	Visual comparison of the colour of paints
DIN EN ISO 4623-1	Filiform corrosion – Steel substrate
DIN EN ISO 4623-2	Filiform corrosion – Aluminium substrates
DIN EN ISO 4624	Pull-off test for adhesion
DIN EN ISO 4628-1	Evaluation of degradation of coatings – General introduction and designation system
DIN EN ISO 4628-2	Assessment of degree of blistering
DIN EN ISO 4628-3	Assessment of degree of rusting

Continuation

Standard	Brief description
DIN EN ISO 4628-4	Assessment of degree of cracking
DIN EN ISO 4628-5	Assessment of degree of flaking
DIN EN ISO 4628-6	Assessment of degree of chalking by tape method
DIN EN ISO 4628-7	Assessment of degree of chalking by velvet method
DIN EN ISO 4628-8	Assessment of degree of delamination and corrosion around a scribe
DIN EN ISO 4628-10	Assessment of degree of filiform corrosion
DIN EN ISO 6270-1	Resistance to humidity – Continuous condensation
DIN EN ISO 6270-2	Resistance to humidity – Procedure for exposing test specimens in condensation-water atmospheres
DIN EN ISO 6272-1	Impact resistance tests – Falling-weight test, large-area indenter
DIN EN ISO 6272-2	Impact resistance tests – Falling-weight test, small-area indenter
DIN EN ISO 6504-1	Hiding power – Kubelka-Munk method for white and light-coloured paints
DIN EN ISO 6504-3	Hiding power – Determination of contrast ratio
DIN EN ISO 6860	Bend test (conical mandrel)
DIN EN ISO 7784-1	Resistance to abrasion – Rotating abrasive-paper-covered wheel method
DIN EN ISO 7784-2	Resistance to abrasion – Rotating abrasive rubber wheel method
DIN EN ISO 7784-3	Resistance to abrasion – Reciprocating test panel method
DIN EN ISO 9117-1	Drying tests – Through-dry state and through-dry time
DIN EN ISO 9117-2	Pressure test for stackability
DIN EN ISO 9117-3	Surface-drying test using ballotini
DIN EN ISO 9117-4	Test using a mechanical recorder
DIN EN ISO 9117-5	Modified Bandow-Wolff test
DIN EN ISO 9117-6	Print-free test
DIN EN ISO 9227	Salt spray tests
DIN EN ISO 11890-1	Volatile organic compound (VOC) content – Difference method
DIN EN ISO 11890-2	Volatile organic compound (VOC) content – Gas-chromatographic method
DIN EN ISO 11997-1	Cyclic corrosion conditions – Wet (salt fog)/dry/humidity
DIN EN ISO 11997-2	Cyclic corrosion conditions – Wet (salt fog)/dry/humidity/UV light
DIN EN ISO 12137	Determination of mar resistance
DIN EN ISO 13076	Lighting and procedure for visual assessments of coatings
DIN EN ISO 13129	Electrochemical measurement of the protection provided to steel by paint coatings – Current interrupter (CI) technique, relaxation voltametry (RV) technique and DC transient (DCT) measurements

Continuation

Standard	Brief description
DIN EN ISO 13803	Determination of reflection haze on paint films at 20ø
DIN EN ISO 15110	Artificial weathering including acidic deposition
DIN EN ISO 15184	Determination of film hardness by pencil test
DIN EN ISO 15234	Formaldehyde emission
DIN EN ISO 15528	Sampling
DIN EN ISO 15710	Corrosion testing by alternate immersion in and removal from a buffered sodium chloride solution
DIN EN ISO 15711	Determination of resistance to cathodic disbonding of coatings exposed to sea water
DIN EN ISO 16474-1, -2, -3	Paints and Varnishes – Methods of exposure to laboratory light sources
DIN EN ISO 16773-1	Electrochemical impedance spectroscopy (EIS) on high-impedance coated specimens – Terms and definitions
DIN EN ISO 16773-2	EIS – Collection of data
DIN EN ISO 16773-3	EIS – Processing and analysis of data from dummy cells
DIN EN ISO 16773-4	EIS – Examples of spectra of polymer-coated specimens
DIN EN ISO 16862	Evaluation of sag resistance
DIN EN ISO 16925	Paints and Varnishes – Testing of resistance of coatings to pressure water-jetting
DIN EN ISO 17132	T-bend test
DIN EN ISO 17872	Scribe marks on metallic panels
DIN EN ISO 20566	Determination of the scratch resistance of a coating system using a laboratory-scale car-wash
DIN EN ISO 20567-1	Stone-chip resistance – Multi-impact testing
DIN EN ISO 20567-2	Single-impact test with a guided impact body
DIN EN ISO 20567-3	Single impact test with free flying impact body
DIN EN ISO 21227-1	Optical imaging – General guidance
DIN EN ISO 21227-2	Optical imaging – Evaluation procedure for multi-impact stone-chipping test
DIN EN ISO 21227-3	Optical imaging – Evaluation of delamination and corrosion around a scribe
DIN EN ISO 21227-4	Optical imaging – Evaluation of filiform corrosion
DIN EN ISO 28199-1	Fingerprint analysis – Relevant vocabulary and preparation of test panels
DIN EN ISO 28199-2	Fingerprint analysis – Colour stability, process hiding power, redissolving, overspray absorption, wetting, surface texture and mottling
DIN EN ISO 28199-3	Fingerprint analysis – Visual assessment of sagging, formation of bubbles, pinholing and hiding power

8.2.1.4 Standards on Aluminium

Standard	Brief description
DIN EN 12206-1	Coating of aluminium – Coatings prepared from coating powder

8.2.1.5 Standards on Coil Coatings

Standard	Brief description
DIN EN 1396	Coil coated aluminium sheet – Specifications
DIN EN 10169	Continuously coil coated steel flat products – Technical delivery conditions
DIN EN 13523-0	General introduction and list of test methods
DIN EN 13523-1	Film thickness
DIN EN 13523-2	Specular gloss
DIN EN 13523-3	Colour difference; instrumental comparison
DIN EN 13523-4	Pencil hardness
DIN EN 13523-5	Resistance to rapid deformation (impact test)
DIN EN 13523-6	Adhesion after indentation (cupping test)
DIN EN 13523-7	Resistance to cracking on bending (T-bend test)
DIN EN 13523-8	Resistance to salt spray (fog)
DIN EN 13523-9	Resistance to water immersion
DIN EN 13523-10	Resistance to fluorescent UV radiation and water condensation
DIN EN 13523-11	Resistance to solvents (rubbing test)
DIN EN 13523-12	Resistance to scratching
DIN EN 13523-13	Resistance to accelerated ageing by the use of heat
DIN EN 13523-14	Chalking (Helmen method)
DIN EN 13523-15	Metamerism
DIN EN 13523-16	Resistance to abrasion
DIN EN 13523-17	Adhesion of strippable films
DIN EN 13523-18	Resistance to staining
DIN EN 13523-19	Panel design and method of atmospheric exposure testing
DIN EN 13523-20	Foam adhesion
DIN EN 13523-21	Evaluation of outdoor exposed panels
DIN EN 13523-22	Colour difference – Visual comparison
DIN EN 13523-23	Colour stability in humid atmospheres containing sulphur dioxide
DIN EN 13523-24	Resistance to blocking and pressure marking
DIN EN 13523-25	Resistance to humidity
DIN EN 13523-26	Resistance to condensation of water

Continuation

Standard	Brief description
DIN EN 13523-27	Resistance to humid poultice (Cataplasm test)
DIN EN 13523-29	Resistance to environmental soiling (Dirt pick-up and striping)
DIN EN 16074	Determination of non-volatile-matter content and spreading rate of coil coating materials

8.2.1.6 Standards on Galvanised Steel

Standard	Brief description
DIN EN 13438	Powder organic coatings for galvanised or sherardised steel products for construction purposes
DIN EN 15773	Duplex systems – Specifications, recommendations and guidelines

8.2.1.7 Standards on Wood Outdoors

Standard	Brief description
DIN EN 927-1	Classification and selection
DIN EN 927-2	Performance specification
DIN EN 927-3	Natural weathering test
DIN EN 927-5	Assessment of the liquid water permeability
DIN EN 927-6	Exposure of wood coatings to artificial weathering using fluorescent UV lamps and water
DIN CEN/TS 16358 DIN SPEC 91029	Assessment of air inclusions/microfoam in coating films
DIN CEN/TS 16359 DIN SPEC 91030	Assessment of knot staining resistance of wood coatings
DIN CEN/TS 16360 DIN SPEC 91031	Assessment of film extensibility by indentation of a coating on a wooden substrate

8.2.1.8 Standards on Anticorrosion Protection of Steel Structures

Standard	Brief description
DIN 55634	Corrosion protection of supporting thin-walled building components made of steel
DIN EN ISO 12944-1	Corrosion protection of steel structures by protective paint systems – General introduction
DIN EN ISO 12944-2	Classification of environments
DIN EN ISO 12944-3	Design considerations
DIN EN ISO 12944-4	Types of surface and surface preparation
DIN EN ISO 12944-5	Protective paint systems
DIN EN ISO 12944-6	Laboratory performance test methods
DIN EN ISO 12944-7	Execution and supervision of paint work
DIN EN ISO 12944-8	Development of specifications for new work and maintenance

Continuation

Standard	Brief description
DIN EN ISO 16276-1	Assessment of, and acceptance criteria for, the adhesion/cohesion of a coating – Pull-off testing
DIN EN ISO 16276-2	Assessment of, and acceptance criteria for, the adhesion/cohesion of a coating – Cross-cut testing and X-cut testing
DIN EN ISO 29601	Assessment of porosity in a dry film

8.2.1.9 Standards on Nuclear Facilities

Standard	Brief description
DIN 53233	Abrasion behaviour of coatings
DIN 53781	Resistance of coatings to ionising radiation in nuclear plants
DIN 55991-1	Coatings for nuclear facilities – Requirements and test methods
DIN 55991-2	Coatings for nuclear facilities – Identity testing of coating materials
DIN 55991-3	Coatings in nuclear facilities – Coating work, planning and procedure

8.2.1.10 Standards on Aerospace

Standard	Brief description
DIN 29597	Application of liquid coating materials
DIN 65046-1 to -6	Methods of test for surface protective coatings; paints and varnishes
DIN EN 3840	Technical specification

8.2.1.11 Standards on Exterior Masonry and Concrete

Standard	Brief description
DIN EN 1062-1	Classification
DIN EN 1062-3	Liquid water permeability
DIN EN 1062-6	Carbon dioxide permeability
DIN EN 1062-7	Crack bridging properties
DIN EN 1062-11	Conditioning before testing
DIN EN 15457	Testing the efficacy of film preservatives in a coating against fungi
DIN EN 15458	Testing the efficacy of film preservatives in a coating against algae
DIN EN 16105	Release of substances in intermittent contact with water
DIN EN ISO 7783	Water-vapour transmission properties – Cup method

8.2.1.12 Standards on Shipbuilding

Standard	Brief description
DIN 80200	Symbols used for surface preparations, shop primers and priming coats
DIN EN ISO 10890	Modelling of biocide release rate from antifouling paints by mass-balance calculation
DIN EN ISO 15181-1	Determination of release rate of biocides from antifouling paints – General method for extraction of biocides
DIN EN ISO 15181-2	Determination of copper-ion concentration in the extract and calculation of the release rate
DIN EN ISO 15181-3	Calculation of the zinc ethylene-bis(dithiocarbamate) (Zineb) release rate by determination of the concentration of ethylenethiourea in the extract
DIN EN ISO 15181-4	Determination of pyridine-triphenylborane (PTPB) concentration in the extract and calcula-tion of the release rate
DIN EN ISO 15181-5	Calculation of the tolylfluanid and dichlofluanid release rate by determination of the concentration of dimethyltolylsulphamide (DMST) and dimethylphenylsulphamide (DMSA) in the extract
DIN EN ISO 15181-6	Determination of release rate of biocides from antifouling paints

8.2.1.13 Dispersion Coatings for Indoors

Standard	Brief description
DIN EN 13300	Classification
DIN EN ISO 11998	Wet-scrub resistance and cleanability of coatings

8.2.1.14 Synthetic Resin Plasters

Standard	Brief description
DIN 18558	Terminology, requirements, application
DIN EN 15824	Specifications for external renders and internal plasters based on organic binders

8.2.1.15 Powder Coatings

Standard	Brief description
DIN 55607	Dispersion of pigments in powder coatings and their colourimetric evaluation after application
DIN 55633	Corrosion protection of steel structures by powder coating systems
DIN EN 12206-1	Coating of aluminium – Coatings prepared from coating powder
DIN EN 13438	Powder organic coatings for galvanised or sherardised steel products for construction purposes
DIN EN 15773	Duplex systems – Specifications, recommendations and guidelines
DIN EN ISO 8130-1	Coating powders – Determination of particle size distribution by sieving

Continuation

Standard	Brief description
DIN EN ISO 8130-2	Determination of density by gas comparison pycnometer
DIN EN ISO 8130-3	Determination of density by liquid displacement pycnometer
DIN EN ISO 8130-4	Calculation of lower explosion limit
DIN EN ISO 8130-5	Determination of flow properties of a powder/air mixture
DIN EN ISO 8130-6	Determination of gel time of thermosetting coating powders at a given temperature
DIN EN ISO 8130-7	Determination of loss of mass on stoving
DIN EN ISO 8130-8	Assessment of the storage stability of thermosetting powders
DIN EN ISO 8130-9	Sampling
DIN EN ISO 8130-10	Determination of deposition efficiency
DIN EN ISO 8130-11	Inclined-plane flow test
DIN EN ISO 8130-12	Determination of compatibility
DIN EN ISO 8130-13	Particle size analysis by laser diffraction
DIN EN ISO 8130-14	Terminology

8.2.1.16 Road Marking Paints

Standard	Brief description
DIN EN 1436	Road marking performance for road users

8.2.1.17 Radiator Coatings

Standard	Brief description
DIN 55900-1	Primers and industrially applied priming coats
DIN 55900-2	Finishing paints and industrially applied finishing coats

8.2.1.18 Surface Preparation

Standard	Brief description
DIN EN ISO 8501-1 to -4	Visual assessment of surface cleanliness
DIN EN ISO 8502-2 to 6, 8, 9, 11, 12	Tests for the assessment of surface cleanliness
DIN EN ISO 8503-1 to -5	Surface roughness characteristics of blast-cleaned steel substrates
DIN EN ISO 8304-1 to -3	Surface preparation methods

8.2.1.19 External Thermal Insulation Composite Systems (ETICS)

Standard	Brief description
DIN 55699	Application of ETICS
DIN EN 13494	Determination of the tensile bond strength of the adhesive and of the base coat to the thermal insulation material
DIN EN 13495	Pull-off resistance of ETICS
DIN EN 13496	Mechanical properties of glass fibre meshes
DIN EN 13497	Resistance to impact of ETICS
DIN EN 13498	Resistance to penetration of ETICS
DIN EN 13499	ETICS based on expanded polystyrene
DIN EN 13500	ETICS based on mineral wool
DIN EN ISO 9229	Thermal insulation – Vocabulary

8.2.1.20 Colourimetry

Standard	Brief description
DIN 5033-1 to -9	Colourimetry
DIN 6164-1 und -2	DIN colour chart
DIN 6172	Metamerism index
DIN 6173-1 DIN 6173-2	Colour matching
DIN 6175-1	Tolerances for automotive paints – Uni paints
DIN 6175-2	Tolerances for automotive paints – Goniochromatic paints
DIN 6176	DIN99 formula
DIN EN ISO 11664-1	CIE standard Colourimetric observers
DIN EN ISO 11664-2	CIE standard illuminants
DIN EN ISO 11664-3	CIE tristimulus values
DIN EN ISO 11664-4	CIE 1976 L*a*b* – Colour space
DIN EN ISO 11664-5	CIE 1976 L*u*v* – Colour space
DIN EN ISO 11664-6	Colour difference formula

8.2.1.21 DIN Reports

Specification	Brief description
DIN SPEC 91063	Adhesion of coatings
DIN SPEC 91064	Hardness and wear resistance of coatings
DIN SPEC 91143-2	Modern rheological test methods – Thixotropy – Determination of the time-dependent structural change – Fundamentals and interlaboratory test

Continuation

Specification	Brief description
DIN-Reports 143	Modern rheological test methods – Determination of the yield point – Fundamentals and comparative testing methods
DIN-Reports 28	Corrosion protection of steel structures by protective paint systems – Testing of surfaces before application of coating materials for contain-ments that cannot be detected visually
DIN-Reports 49	Colour tolerances

8.2.2 Standards by Keyword

Keywords	Standard
Abrasion behaviour of decontaminable coatings – Method using a tube with falling abrasive	DIN 53233
Abrasion test – Reciprocating test panel method	DIN EN ISO 7784-3
Abrasion test – Rotating abrasive rubber wheel method	DIN EN ISO 7784-2
Abrasion test – Rotating abrasive-paper-covered wheel method	DIN EN ISO 7784-1
Abrasion: Resistance of coil coated metals to abrasion	DIN EN 13523-16
Adhesion of foam on coil coated metals	DIN EN 13523-20
Adhesion of strippable films – Coil coated metals	DIN EN 13523-17
Adhesion/cohesion (pull-off test) of a coating and acceptance criteria for steel structures	DIN EN ISO 16276-1
Aerospace: Processing of liquid coating materials – Aerospace	DIN 29597
Aerospace: Technical specifications for coating materials – Aerospace	DIN EN 3840
Antifouling: Copper-ion concentration in the extract of antifouling paints	DIN EN ISO 15181-2
Automotive: Tolerances for automotive paints – Goniochromatic paints	DIN 6175-2
Automotive: Tolerances for automotive paints – Uni paints	DIN 6175-1
Bandow-Wolff test – Modified	DIN EN ISO 9117-5
Bend test – Conical mandrel	DIN EN ISO 6860
Bend test – Cylindrical mandrel	DIN EN ISO 1519
Binder: Specifications for external renders and internal plasters based on organic binders	DIN EN 15824
Binders – Symbols	DIN 55950
Biocide: Determination of release rate of biocides from antifouling paints	DIN EN ISO 15181-6
Biocide: Modelling of biocide release rate from antifouling paints by mass-balance calculation	DIN EN ISO 10890
Bond: Tensile bond strength of the adhesive and of the base coat to the thermal insulation material	DIN EN 13494
Buchholz indentation test	DIN EN ISO 2815

Continuation

Keywords	Standard
Cathodic disbonding: Resistance to cathodic disbonding of coatings exposed to sea water	DIN EN ISO 15711
Chalking – Kempf method	DIN 53159
Chalking of coil coated panels – Helmen method	DIN EN 13523-14
Chalking: Degree of chalking – Tape – Assessment	DIN EN ISO 4628-6
Chalking: Degree of chalking – Velvet method – Assessment	DIN EN ISO 4628-7
Classification of environments in steel structures	DIN EN ISO 12944-2
Classification of exterior masonry and concrete	DIN EN 1062-1
Classification: Dispersion coatings for indoors – Classification	DIN EN 13300
Coating materials – Vocabulary	DIN EN ISO 4618
Coating materials – Vocabulary	DIN 55943
Coating materials – Vocabulary	DIN EN ISO 8130-14
Coating of aluminium – Coating powders	DIN EN 12206-1
Coating: Determination of coating thickness	DIN EN ISO 2808
Coatings for nuclear facilities – Coating work, planning and procedure	DIN 55991-3
Coatings for nuclear facilities – Identity testing of coating materials	DIN 55991-2
Coatings for nuclear facilities – Requirements and test methods	DIN 55991-1
Coatings materials – Classification	DIN 55944
Coats: Primers and industrially applied priming coats for radiators	DIN 55900-1
Coil coated aluminium sheet – Specifications	DIN EN 1396
Coil coated metals – General introduction and list of test methods	DIN EN 13523-0
Coil coated panels – Resistance to blocking and pressure marking	DIN EN 13523-24
Coil: Continuously coil coated steel flat products – Technical delivery conditions	DIN EN 10169
Coil: Evaluation of coil coated panels exposed outdoors	DIN EN 13523-21
Colour chart: DIN colour chart	DIN 6164-1 and -2
Colour comparison – Visual	DIN EN ISO 36681
Colour difference formula	DIN EN ISO 11664-6
Colour difference – Instrumental comparison of coil coated panels	DIN EN 13523-3
Colour difference – Visual comparison of coil coated panels	DIN EN 13523-22
Colour matching	DIN 6173-1 DIN 6173-2
Colour space – CIE 1976 L*a*b*	DIN EN ISO 11664-4
Colour space – CIE 1976 L*u*v*	DIN EN ISO 11664-5
Colourfastness of articles for common use – Test with artificial sweat	DIN 53160-2
Colourfastness of articles for common use – Test with saliva	DIN 53160-1

Continuation

Keywords	Standard
Colourimetric: Standard colourimetric observers – CIE	DIN EN ISO 11664-1
Colourimetry	DIN 5033-1 to -9
Colourimetry according DIN 99	DIN 6176
Combustibility: Sustained combustibility of liquids	DIN EN ISO 9038
Conditioning of samples before testing on exterior masonry and concrete	DIN EN 1062-11
content – Filtration method	DIN EN ISO 14680-3
Corrosion protection of steel structures by powder coating systems	DIN 55633
Corrosion protection of steel structures by protective paint systems – General introduction	DIN EN ISO 12944-1
Corrosion protection of supporting thin-walled building components made of steel	DIN 55634
Corrosion testing – Wet (salt fog)/dry/humidity	DIN EN ISO 11997-1
Corrosion testing by alternate immersion in and removal from a buffered sodium chloride solution	DIN EN ISO 15710
Corrosion testing conditions – Wet (salt fog)/dry/humidity/ UV light	DIN EN ISO 11997-2
Corrosion: Design considerations for corrosion protection of steel structures	DIN EN ISO 12944-3
Corrosion: Evaluation of corrosion protection of steel structures	DIN EN ISO 12944-6
Corrosion: Surfaces and surface preparation for corrosion protection of steel structures	DIN EN ISO 12944-4
Crack bridging properties of coating materials on exterior masonry and concrete	DIN EN 1062-7
Cracking: Degree of cracking – Assessment	DIN EN ISO 4628-4
Cross-cut test	DIN EN ISO 2409
Cross-cut: Adhesion/cohesion of a coating – Cross-cut testing and X-cut testing and acceptance criteria for steel structures	DIN EN ISO 16276-2
Degradation: Evaluation of degradation of coatings – General introduction and designation system	DIN EN ISO 4628-1
Degree of flaking – Assessment	DIN EN ISO 4628-5
Delamination and corrosion around a scribe – Assessment	DIN EN ISO 4628-8
Density – Immersed body method	DIN EN ISO 2811-2
Density – Oscillation method	DIN EN ISO 2811-3
Density – Pressure cup method	DIN EN ISO 2811-4
Density – Pycnometer method	DIN EN ISO 2811-1
Density of powder coatings using gas comparison pycnometer	DIN EN ISO 8130-2
Density of powder coatings using liquid displacement pycnometer	DIN EN ISO 8130-3
Deposition efficiency of coating powders	DIN EN ISO 8130-10

Continuation

Keywords	Standard
Dispersion of pigments in powder coatings and their colourimetric evaluation after application	DIN 55607
Drying recorder – Mechanical	DIN EN ISO 9117-4
Drying tests – Through-dry state and through-dry time	DIN EN ISO 9117-1
Duplex systems – Specifications, recommendations and guidelines	DIN EN 15773
EIS – Collection of data	DIN EN ISO 16773-2
EIS – Examples of spectra of polymer-coated specimens	DIN EN ISO 16773-4
EIS – Processing and analysis of data from dummy cells	DIN EN ISO 16773-3
EIS (electrochemical impedance spectroscopy) – Terms and definitions	DIN EN ISO 16773-1
Electrical conductivity and resistance	DIN EN ISO 15091
Electrochemical measurement: Protection provided to steel by paint coatings – Electrochemical measurement	DIN EN ISO 13129
Electrodeposition paints – Paints and varnishes	DIN 55655-2 to 9
Electrodeposition paints – Terminology	DIN 55655-1
Emissions of formaldehyde	DIN EN ISO 15234
ETICS based on expanded polystyrene	DIN EN 13499
ETICS based on mineral wool	DIN EN 13500
Explosion: Lower explosion limit – Coating powders	DIN EN ISO 8130-4
Exposure: Paints and varnishes – Methods of exposure to laboratory light sources	DIN EN ISO 16747-1, -2, -3
Filiform corrosion – Aluminium substrates	DIN EN ISO 4623-2
Filiform corrosion – Assessment	DIN EN ISO 4628-10
Filiform corrosion – Steel substrate	DIN EN ISO 4623-1
Film thickness – Overview of methods and execution	DIN EN ISO 2808
Fineness of grind	DIN EN ISO 1524
Fingerprint analysis – Colour stability, process hiding power, re-dissolving, overspray absorption, wetting, surface texture and mottling	DIN EN ISO 28199-2
Fingerprint analysis – Relevant vocabulary and preparation of test panels	DIN EN ISO 28199-1
Fingerprint analysis – Visual assessment of sagging, formation of bubbles, pinholing and hiding power	DIN EN ISO 28199-3
Flaking: Degree of flaking – Assessment	DIN EN ISO 4628-2
Flash point – Abel closed-cup method	DIN EN ISO 13736
Flash point determination – Closed cup equilibrium method	DIN EN ISO 1523
Flash point determination – Rapid equilibrium closed cup method	DIN EN ISO 3679
Flash point/no flash point determination – Closed cup equilibrium method	DIN EN ISO 1516
Flow properties of a powder/air mixture	DIN EN ISO 8130-5

Continuation

Keywords	Standard
Flow properties of coating powders – Incline plane	DIN EN ISO 8130-11
Gel time of powder coatings	DIN EN ISO 8130-6
Gloss of coil coating panels	DIN EN 13523-2
Gloss: Specular gloss	DIN EN ISO 2813
Hardness and wear resistance of coatings	DIN SPEC 91064
Haze	DIN EN ISO 13803
Hiding power – Determination of contrast ratio	DIN EN ISO 6504-3
Hiding power – Kubelka-Munk method for white and light-coloured paints	DIN EN ISO 6504-1
Hiding power – Tests of pigments, Determination of a hiding power	DIN 55601
Illuminants: Standard illuminants – CIE	DIN EN ISO 11664-2
Impact of heat	DIN EN ISO 3248
Impact resistance – Falling-weight test, large-area indenter	DIN EN ISO 6271-1
Impact resistance – Falling-weight test, small-area indenter	DIN EN ISO 6271-2
Impact resistance of coil coated panels	DIN EN 13523-5
Impact resistance of ETICS	DIN EN 13497
Impact: Single-impact test with a guided impact body	DIN EN ISO 20567-2
Impact: Single-impact test with free flying impact body	DIN EN ISO 20567-3
Impact: Stone-chip resistance – Multi-impact testing	DIN EN ISO 20567-1
Indentation test	DIN EN ISO 1520
Indentation test on coil coated panels	DIN EN 13523-6
Insulation: Thermal insulation – Vocabulary	DIN EN ISO 9229
Lighting: Visual assessments of coatings – Lighting and procedure	DIN EN ISO 13076
Mar resistance: Determination of mar resistance	DIN EN ISO 12137
Mechanical properties of glass fibre meshes for ETICS	DIN EN 13496
Metamerism index	DIN 6172
Metamerism on coil coated panels	DIN EN 13523-15
Non-volatile and volatile matter content during application	DIN 55685
Non-volatile: Percentage volume of non-volatile matter – Method using a coated test panel to determine non-volatile matter and to determine dry film density by the Archimedes principle	DIN EN ISO 3233-1
Non-volatile: Volume of non-volatile matter	DIN 53219; DIN EN ISO 23811
Non-volatile-matter content and spreading rate of coil coating materials	DIN EN 16074
Optical imaging – Assessment of filiform corrosion	DIN EN ISO 21227-4

Industrial Standards

Continuation

Keywords	Standard
Optical imaging – Evaluation of delamination and corrosion around a scribe	DIN EN ISO 21227-3
Optical imaging – Evaluation of multi-impact stone-chipping	DIN EN ISO 21227-2
Optical imaging – General guidance	DIN EN ISO 21227-1
Paint: Execution and supervision of paint work in steel structures	DIN EN ISO 12944-7
Particle size analysis of coating powders by laser diffraction	DIN EN ISO 8130-13
Particle size distribution of coating powders – Determination by sieving	DIN EN ISO 8130-1
Pencil hardness of coil coated panels	DIN EN 13523-4
Pencil test: Hardness of coatings – Pencil test	DIN EN ISO 15184
Pendulum damping test	DIN EN ISO 1522
Penetration resistance of ETICS	DIN EN 13498
Permeability: Carbon dioxide permeability of coating materials on exterior masonry	DIN EN 1062-6
pH Value – pH electrodes with ISFET Technology	DIN 55659-2
pH Value – pH-electrodes with glass membrane	DIN 55659-1
Pigment content – Ashing method	DIN EN ISO 14680-2
Pigment content – Centrifuge method	DIN EN ISO 14680-1
Porosity – Assessment	DIN 55670
Porosity assessment – Corrosion protection by protective paint systems	DIN EN ISO 29601
Pot life of multicomponent coating systems	DIN EN ISO 9514
Powder coatings – Loss of mass on stoving	DIN EN ISO 8130-7
Powder coatings: Determination of compatibility of powder coatings	DIN EN ISO 8130-12
Powder organic coatings for galvanised or sherardised steel products for construction purposes	DIN EN 13438
Powders: Coating powders – Terminology	DIN EN ISO 8130-14
Preservative: Efficacy of film preservatives in a coating on exterior masonry and concrete against fungi	DIN EN 15457
Preservative: Efficacy of film preservatives in a coating on exterior masonry and concrete against fungi	DIN EN 15458
Print-free test	DIN EN ISO 9117-6
Processing of ETICS	DIN 55699
Protective paint systems in steel construction	DIN EN ISO 12944-5
Pull-off resistance of ETICS	DIN EN 13495
Pull-off test for adhesion	DIN EN ISO 4624
Radiators: Finishing paints and industrially applied finishing coats for radiators	DIN 55900-2
Recoatability	DIN 53221

Industrial Standards

Continuation

Keywords	Standard
Release of substances from coatings on mineral substrates	DIN EN 16105
Release rate of biocides from antifouling paints – General method for extraction of biocides	DIN EN ISO 15181-1
Release rate of pyridine-triphenylborane (PTPB) in the extract of antifouling coatings	DIN EN ISO 15181-4
Release rate of tolylfluanid and dichlofluanid in the extract of antifouling coatings by determining the concentration of dimethyltolylsulphamide (DMST) and dimethylphenylsulphamide (DMSA)	DIN EN ISO 15181-5
Release rate of zinc ethylene-bis(dithiocarbamate) (zineb) by determination of the concentration of ethylenethiourea in the extract of antifouling paints	DIN EN ISO 15181-3
Resin: Synthetic resin plasters – Terminology, requirements, application	DIN 18558
Resistance of coatings to ionising radiation in nuclear plants	DIN 53781
Resistance of coil coated metals in humid atmospheres containing sulphur dioxide	DIN EN 13523-23
Resistance of coil coated metals to accelerated ageing by the use of heat	DIN EN 13523-13
Resistance of coil coated metals to fluorescent UV radiation and water condensation	DIN EN 13523-10
Resistance of coil coated metals to moisture	DIN EN 13523-25
Resistance of coil coated metals to soiling	DIN EN 13523-29
Resistance of coil coated metals to solvents (rubbing test)	DIN EN 13523-11
Resistance of coil coated metals to spotting	DIN EN 13523-18
Resistance of coil coated metals to water condensation	DIN EN 13523-26
Resistance of coil coated metals to water immersion	DIN EN 13523-9
Resistance of wood coatings to knot staining	DIN CEN/TS 16359 DIN SPEC 91030
Resistance: Paints and varnishes – Testing of resistance of coatings to pressure water-jetting	DIN EN ISO 16925
Resistance to coil coated metals to humid poultice (Cataplasm test)	DIN EN 13523-27
Resistance to humid atmospheres containing sulphur dioxide	DIN EN ISO 3231
Resistance to humidity – Continuous condensation	DIN EN ISO 6270-1
Resistance to humidity – Exposure in condensation-water atmospheres	DIN EN ISO 6270-2
Resistance to liquids – Immersion in liquids other than water	DIN EN ISO 2812-1
Resistance to liquids – Method using an absorbent medium	DIN EN ISO 2812-3
Resistance to liquids – Spotting methods	DIN EN ISO 2812-4
Resistance to liquids – Temperature-gradient oven	DIN EN ISO 2812-5
Resistance to liquids – Water immersion method	DIN EN ISO 2812-2
Road marking performance for road users	DIN EN 1436

Continuation

Keywords	Standard
Rusting: Degree of rusting – Assessment	DIN EN ISO 4628-3
Sagging: Resistance to sagging	DIN EN ISO 16862
Salt spray testing – Coil coated panels	DIN EN 13523-8
Salt spray tests	DIN EN ISO 9227
Sampling	DIN EN ISO 15528
Sampling of coating powders	DIN EN ISO 8130-9
Scratch resistance – Constant-loading method	DIN EN ISO 1518-1
Scratch resistance – Continuously increasing load	DIN EN ISO 1518-2
Scratch resistance of a coating system using a laboratory-scale car-wash	DIN EN ISO 20566
Scratching: Resistance to scratching (coil coating)	DIN EN 13523-12
Scribe marks on metallic panels	DIN EN ISO 17872
Shipbuilding: Steel Components for Shipbuilding – Symbols used for surface preparations	DIN 80200
Solvents – Symbols	DIN 55651
Stackability – Pressure test	DIN EN ISO 9117-2
Standard panels for testing	DIN EN ISO 1514
Steel structures – Development of specifications for new work and maintenance	DIN EN ISO 12944-8
Storage stability of coating powders	DIN EN ISO 8130-8
Surface cleanliness – Tests for assessment of	DIN EN ISO 8502-2 to -6, -8, -9, -11, -12
Surface cleanliness – Visual assessment	DIN EN ISO 8501-1 to -4
Surface drying with ballotini	DIN EN ISO 9117-3
Surface preparation methods	DIN EN ISO 8304-1 to -3
Surface roughness characteristics – Blast-cleaned steel substrates	DIN EN ISO 8503-1 to -5
Surface: Test methods for surface protective coatings – Aerospace	DIN 65046-1 to -6
T-bend test	DIN EN ISO 17132
T-Bend test on coil coated panels	DIN EN 13523-7
Tension: Interfacial tension by the stirrup or ring method	DIN EN 14210
Test samples: Preparation of test samples	DIN EN ISO 1513
Thickness: Coating thickness of coil coated panels	DIN EN 13523-1
Thickness: Measurement of coating thickness – Amplitude-sensitive eddy current method	DIN EN ISO 2360
Thickness: Measurement of coating thickness – Wedge cut method	DIN 50986
Thickness: Measurement of coating thickness by differential measurement using a stylus instrument	DIN 50933

Continuation

Keywords	Standard
Thickness: Measurement of thickness – Beta backscatter method	DIN EN ISO 3543
Thickness: Measurement of thickness – X-ray fluorescence	DIN EN ISO 3497
Thickness: Measurement of coating thickness – Magnetic method	DIN EN ISO 2178 DIN 50981
Tristimulus values – CIE	DIN EN ISO 11664-3
Viscosity determination – Cone-and-plate viscometer operated at a high rate of shear	DIN EN ISO 2884-1
Viscosity determination – Disc or ball viscometer operated at a single specified speed	DIN EN ISO 2884-2
Viscosity determination – Flow out time	DIN EN ISO 2431
Viscosity determination – Rotational viscometer	DIN EN ISO 3219
VOC content – Difference method	DIN EN ISO 11890-1
VOC content – Gas-chromatographic method	DIN EN ISO 11890-2
VOC content in water-thinnable emulsion paints	DIN EN ISO 17895
Water permeability: Liquid water permeability of coatings for exterior masonry and concrete	DIN EN 1062-3
Water permeability: Liquid water permeability of wood coatings	DIN EN 927-5
Water-vapour transmission properties of coatings for exterior masonry and concrete	DIN EN ISO 7783
Weathering: Artificial weathering including acidic deposition	DIN EN ISO 15110
Weathering: Artificial weathering of wood coatings	DIN EN 927-6
Weathering: Natural weathering of coatings – Corrosion protection performance	DIN 55665
Weathering: Natural weathering of coil coated panels	DIN EN 13523-19
Weathering: Natural weathering of wood coatings	DIN EN 927-3
Weathering: Natural weathering test	DIN EN ISO 2810
Wet-scrub resistance and cleanability of emulsion coatings for indoors	DIN EN ISO 11998
Wettability – Determination of the free surface energy of solid surfaces by measuring the contact angle	DIN 55660-2
Wettability – Determination of the polar and dispersive fractions of the surface tension of liquids from contact angles measurements on a solid with only a disperse contribution to its surface energy	DIN 55660-5
Wettability – Determination of the surface tension of liquids using the pendant drop method	DIN 55660-3
Wettability – Terminology and general principles	DIN 55660-1
Wood coatings – Classification and selection	DIN EN 927-1
Wood coatings – Performance specification	DIN EN 927-2

Continuation

Keywords	Standard
Wood: Air inclusions/microfoam in wood coatings	DIN CEN/TS 16358 DIN SPEC 91029
Wood: Extensibility of wood coatings	DIN CEN/TS 16360 DIN SPEC 91031

8.3 Literature

References

8.3.1 Beschichtungsstoffe (publisher DIN), Vincentz Verlag, Hanover, and Beuth Verlag, Berlin (2001)

Industrial Standards

9 Appendix

9.1 List of Physical Constants

9.1.1 Latin Symbols

A	Area, Hamaker Constant
a	Electrophoretic Equivalent
a*	Distance on Red-Green Axis from Zero in CIE-Lab System
Ä	Electrodeposition Equivalent
B	Magnetic Flux
b	Barometric Reading, Interference Factor in Sedimentation Equation
b*	Distance on Blue-Yellow Axis from Zero in CIE-Lab System
C	Relative Centrifugal Force, Capacity of a Condenser, Mass Concentration, Specific Heat Capacity, Velocity of Sound
C*	Chroma (Colourfulness) in CIE-Lab System
C_w (Re)	Friction Coefficient in the Flow Resistance Law
C_p	Specific Heat of Gases at Constant Pressure
C_v	Specific Heat of Gases at Constant Volume
C_s	Constant of Proportionality of an Ideal Black Source
C_{gr}	Constant of Proportionality of a Real (Grey) Source
D	Diameter, Cumulative Sum, Diffusion Coefficient, Directional Force, Dispersity, Velocity Gradient (Shear Rate)
D_v	Hiding Power, Expressed in Terms of Volume
D_m	Hiding Power, Expressed in Terms of Mass
d	Diameter
d_{32}	Sauter Diameter
d'	Average Particle Size According to RRSB Method
E	Modulus of Elasticity, Electric Field Strength, Penetration Coefficient, Initial Weight
E_s	Power Radiated, Expressed in Terms of Area
E'	Storage Modulus
E''	Loss Modulus
E_a	Activation Energy
ΔE^*_{ab}	Colour Distance in CIE-Lab System
e	Elementary Charge
e_a	Sum of Acid Equivalents in the Patton Equation
F	Force, Faraday Constant

Appendix

F_{rel} Relative Tinting Strength

F_w Mechanical Resistance Force

f Frequency, Form Factor

G Free Enthalpy, Shear Modulus

ΔG Change in Free Enthalpy

g Acceleration due to Gravity

H Enthalpy, Adhesive Power

H_K Hertz Hardness

H_V Vickers Hardness

Hn Martens Universal Hardness

h Height

h* Hue in CIE-L*C*h* System

I Ionic Strength, Intensity, Electrical Current Density

j Electrical Current Density

K Absorption Coefficient

k Boltzmann Constant, General Constant

L Power, Length

L* Lightness in CIE-Lab System

l Length

M Torque

M_n Number Average Molecular Weight

M_W Mass Average Molecular Weight

m Mass

m_0 Number of Moles in Patton Equation

m Volumetric Flow Rate

N Power Dissipation

N/V Energy Dissipation Density

Ne(Re) Newtonian Performance Index

n Refractive Index, Number of Moles, Rotational Speed, Number, Degree of Uniformity in the RRSB Method

O Surface Area

O_m Surface Area, Expressed in Terms of Mass

O_v Surface Area, Expressed in Terms of Volume

Ohz Ohnesorge Number

p Pressure, Conversion Rate (Here: Critical Conversion Rate According to Patton)

Q Electrical Charge, Thermal Energy

$\dot{Q}$ Heat Flow

$\dot{q}$ Heat Flow, Expressed in Terms of Area

R	Radius, General Gas Constant, Degree of Reflection, Electrical Resistance
R_s	Gloss Peak
Re	Reynolds Number
Re_M	Reynolds Number for Agitator Processes
r	Radius
R_a	Roughness
R_t	Peak-to-Valley Heights
S	Entropy, Scatter Coefficient, Film Thickness, Spreading Coefficient
S_D	Film Thickness
s	Standard Deviation of a Finite Number of Readings
T	Temperature, Oscillation Time
T_g	Glass Temperature (Glass Transition Temperature)
T_c	Ceiling Temperature
ΔT	Temperature Difference
t	Time
$t_{1/2}$	Half-Life
t_i	Induction Period in the Electrodeposition Process
t_{ret}	Retardation Time
t_{rel}	Relaxation Time
$t_{S,N}$	Student's t-Factor
t_{Fg}	Flash Point, Measured
t_{Fn}	Flash Point, Standardised
U	Electrical Voltage, Nonuniformity
u	Velocity
u_A	Peripheral Speed
V	Volume
V_T	Potential Surface Energy
V_A	Potential for Attraction
V_R	Potential for Repulsion
V_{Pi}	Pigment Volume
V_{Bi}	Binder Volume
$\dot{V}$	Volumetric Flow
W	Work, Probability
W_p	Adhesive Force in the Peel Test
W_k	Adhesive Force in the Wedge Feed Method (Hesiometer)
We	Weber Number
X,Y,Z	Standard Trichromatic Values
x,y,z	Standard Trichromatic Value Components
$\overline{x},\overline{y},\overline{z}$	Standard Spectral Values
$\overline{x}$	Arithmetic Mean
z	Valency

9.1.2 Greek Symbols

α	Contact Angle, Heat Transfer Coefficient, Polarisability, Correction Factor for Calculating Hiding Power
β	Mass Transfer Coefficient
γ	Shear Angle, Volumetric Dilation Coefficient
δ	Thickness of the Diffusion Boundary Layer, Loss Angle, Solubility Parameter
Δ	Difference
ε	Dielectric Constant, Relative Elongation, Relative Power Radiated
ζ	Zeta-Potential (Electrokinetic Potential)
η	Dynamic Viscosity
κ	Specific Electrical Conductivity
λ	Wavelength, Distance between Texture Peaks on Freshly Deposited Coating Films
$\lambda(Re)$	Specific Proportionality Factor for Pressure Drop in Pipes
μ	Dipole Moment, Poisson's Number, Damping Coefficient, Average Number in an Infinite Number of Readings (True Value)
ν	Kinematic Viscosity, Frequency, Specific Conductivity
ξ	Space Requirement of N_2 Molecule during Adsorption
ρ	Density, Specific Electrical Resistance
ρ_T	Dry Film Density
σ	Surface Tension
Σ	Sum
τ	Shear Stress
φ	Electrochemical Potential, Fluidity
φ_0	Electrochemical Reference Potential
Φ	Good Factor, Magnetic Flux
χ	Exponent in the Hückel Function
ψ	Charge Per Unit Area
ω	Angular Velocity, Monomolecular Surface Coverage in Gas Adsorption
Ω	Electrical Resistance

9.2 Internet Addresses of Interest to the Coatings World

9.2.1 Organisations/Associations/Authorities

American Society for Testing and Materials, USA	www.astm.org
Association for UV-/EB-Curing Technology, NL	www.radtech-europe.com
Brazilian Coatings Manufacturers Association (ABRAFATI), Sao Paulo, BR	www.abrafati.br
British Coatings Federation (BCF), Leatherhead, UK	www.coatings.org.uk
Bundesanstalt für Arbeitsschutz und Arbeitsmedizin, Dortmund, D	www.baua.de
Bundesanstalt für Materialprüfung, Berlin, D	www.bam.de
Danish Coatings and Adhesives Association – DFL Copenhagen, DK	www.fdlf.dk
Deutsche Farbwissenschaftliche Gesellschaft, Berlin, D	www.dfwg.de
Deutsche Forschungsgesellschaft für Oberflächenbehandlung e.V., Neuss, D	www.dfo-online.de
Deutsches Institut für Normung e.V., Berlin, D	www.din.de
Deutsches Lackinstitut (VdL), Frankfurt am Main, D	www.lacke-und-farben.de
ECCA – Sekretariat, Brussels, B	www.eccacoil.com
Europäische Gesellschaft für Lackiertechnik eV, Neuss, D	www.egl-online.de
European Commission for Standardization, Brussels, B	www.cen.eu
European Council of Paint, Printing Ink and Artist's Colours Industry (CEPE), Brussels, B	www.cepe.org
European Coil Coating Association, Salzgitter, D	www.ecca.de
Fachverband der chemischen Industrie Österreichs, Vienna, A	www.wk.or.at/fcio/
Federation of Societies for Coatings Technology, Whitpain, PA, USA	www.coatingstech.org
GDCh Fachgruppe Lackchemie, Frankfurt, D	www.gdch.de/index.php?id=110
Gesamtverband der Aluminiumindustrie, Düsseldorf, D	www.aluinfo.de
Industrieverband für Lacke und Beschichtungen (IVP), Brussels, B	www.fedichem.be
Informationsportal „REACH", Dessau-Roßlau, D	www.reach-info.de
Italian Association for Adhesives and sealants, paints and varnishes, printing and silk-screen inks (AVISA), Milan, I	www.federchimica.it
Japan Paint Manufacturers Association (JPMA), Tokyo, JP	www.toryo.or.jp
Canadian Association for Paints and Coatings, St. Laurent, CDN	www.cdnpaint.org
National Paint and Coating Association (NPCA), Washington, USA	www.paint.org

Swedish Association of Paint Manufacturers (SVEFF), Stockholm, SE	www.sveff.se
Stahlinformationszentrum, Düsseldorf, D	www.stahl-info.de
United States Environmental Protection Agency (EPA)	www.epa.gov
Verband der deutschen Lackindustrie, Frankfurt am Main, D	www.lackverband.de
Organisation Internationale des Onstructeurs d'Automobiles	www.oica.net
Federation of the Paints, Coatings and Adhesives Industry (FIPEC), Paris, France	www.fipec.org
Verband der Ingenieure des Lack- und Farbenfaches e.V., Düsseldorf, D	www.vilf.de
Association of Paint Manufacturers of the Czech Republic, Prague, CZ	www.avnh.cz
Verband der Mineralfarbenindustrie (VdMi), Frankfurt, D	www.vdmi.de
Spanish Association of Coatings and Printing Inks Manufacturers (ASEFAPI), Madrid, E	www.asefapi.es
Verband Schweizerischer Lack- und Farbenfabrikanten (VSLF), Zürich, CH	www.vslf.ch
Association of Paint and Printing Ink Manufacturers (VVVF), Leiden, NL	www.vvvf.nl
World Wind Energy Association (WWEA), Bonn, D	www.wwindea.org

9.2.2 Publications/Information

Automobil Produktion, mi-Verlag, Augsburg, D	www.mi-verlag.de
BNP Media, (Paint & Coatings Industry), Troy, MI, USA	www.pcimag.com
Chemical Week Associates, New York, USA	www.modernpaintandcoatings.com
Coatings World, USA	www.coatingsworld.com
Ecomed Verlag, Heidelberg, D	www.ecomed.de
Elsevier Science Verlag, (Progress in Organic Coatings) New York, USA	www.elsevier.com
Expert Verlag, Renningen, D	www.expertverlag.de
Fachinformationszentrum Technik, Frankfurt, D	www.fiz-technik.de
Carl Hanser Verlag, Munich, D	www.hanser.de
S. Hirzel Verlag, Stuttgart, D	www.Hirzel.de
Informationen zu Lackprodukten/Lieferanten	www.chemcompass.de
Information Platform of JPCL and PCE, Paintsquare Inc, USA	www.protectivecoatings.com
Information platform for industrial corrosion protection, USA	www.corrosion.org
Institute of Corrosion, Northampton, UK	www.icorr.org
Jedermann-Verlag, Heidelberg, D	www.jedermann.de
Journal für Oberflächentechnik (Vieweg + Teubner Springer), Wiesbaden, D	www.jot-oberflaeche.de

Journal of Coatings Technology, FSCT, Whitpain, PA, USA	www.coatingstech.org
Metalloberfläche (Organ der EGL), Neuss, D	www.egl-online.de
Products Finishing, Cincinnati, OH, USA	www.pfonline.com
Polymers Paint Colour Journal (PPCJ), Redhill, Surrey UK	www.polymerspaintcolourjournal.com/
Rodman Publishing, Ramsey, NJ, USA	www.paintandcoatings.com
Springer Vieweg, Heidelberg, D	www.jot-oberflaeche.de www.vieweg.de
Vincentz Network, (European Coatings Journal, European Coatings Library, FARBE UND LACK, FARBE UND LACK Bibliothek, besser lackieren) Hanover, D	www.european-coatings.com www.farbeundlack.de www.besserlackieren.de
WEKA Media, Kissing, D	www.weka.de
Welt der Farben (publisher: E. Berger), Cologne, D	www.welt-der-farben.de
Wiley-VCH Verlagsgesellschaft, Weinheim, D	www.wiley-vch.de

9.2.3 Institutes/Universities

Coatings Research Institute, Limelette, B	www.cori-coatings.be
European Centre for Coatings and Surface Technology, Enschede, NL	www.eccs-online.org
Fachhochschule Aalen, Aalen, D	www.fh-aalen.de
Fachhochschule für Technik, Esslingen, D	www.fht-esslingen.de
Hochschule Niederrhein, Krefeld, D	www.hs-niederrhein.de
Forschungsgesellschaft für Pigmente und Lacke, Stuttgart, D	www.fpl-ev.de
Fraunhofer Institut für Produktionstechnik und Automatisierung, Stuttgart, D	www.ipa.de
Fraunhofer Institute for Coating and Surface Technology, Braunschweig, D	www.ist.fhg.de
Fraunhofer Institute for Manufacturing Engineering and Applied Materials Research (IFAM), Bremen, Dresden, D	www.ifam.fraunhofer.de
Institute of Materials Science, New Paltz, NY, USA	www.ims-np.org
Institut für Fenstertechnik, Rosenheim, D	www.ift-rosenheim.de
Institut für Holztechnologie, Dresden, D	www.ihd-dresden.de
Institut für Korrosionsschutz, Dresden, D	www.iks-dresden.de
Institut für Kunststoffprüfung und Kunststoffkunde, Stuttgart, D	www.ikp.uni-stuttgart.de
Institut für Lacke und Farben, Magdeburg, D	www.lackinstitut-magdeburg.de
Institut für Polymerforschung, Dresden, D	www.ipfdd.de
Netherlands Organisation for Applied Scientific Research - TNO, Eindhoven, NL	www.tno.nl
Paint Research Association, Hampton, Middlesex, UK	www.pra-world.com
Paint and Coatings Resource Center (EPA), USA	www.paintcenter.org

Technische Akademie Wuppertal, Wuppertal, D	www.taw.de
Universität Bremen, FB4, Produktionstechnik, Bremen, D	www.uni-bremen.de
Universität Paderborn, Paderborn, D	www.uni-paderborn.de

9.2.4 Miscellaneous

Automotive Circle International	www.automobillackierung.de
Technik+Kommunikation Verlags GmbH, Berlin, D	www.karosserierohbau.de
www.body-in-white.de	
www.tuk-verlag.de	
Euroforum Conferences	www.euroforum.com
European Coatings Conference, Vincentz Network, Hanover, D	www.european-coatings.com /Events/(category)/1
GDCh Congresses	www.gdch.de/index.php?id=51
Farbmetrik Gall	www.farbmetrik-gall.de
Internationaler Rohbauexpertenkreis, Stuttgart, D	www.rohbau-expertenkreis.de
Portal for Surface Topics etc., USA	www.finishing.com
Technische Akademie Wuppertal, Weiterbildungszentrum, Wuppertal, D	www.taw.de

9.3 Acronyms/Abbreviations

AAS	Atomic Absorption Spectroscopy (analytical technique)
ABS	Acrylonitrile-butadiene-styrene copolymer (thermoplastic)
ADR	Accord Européen relatif au transport international des marchandise dangereuse par route
AED	Anodic Electrodeposition
AFM	Atomic Force Microscopy (Analytical Technique)
AGW	Maximum Workplace Concentration (Germany)
AGS	Committee for Hazardous Substances (Germany)
AQMB	Autospect. Quality Measurement System (Measurement Method)
ASTM	American Society for Testing and Materials, West Conshohocken, PA, USA
AV	Acid Value (Analysis)
BAT	Biological Workplace Tolerance (Germany)
BASt	Federal Highway Research Institute (German Authority)
BC	Basecoat (Pigmented Layer Underneath the Clearcoat)
GDP	Gross Domestic Product
BC	Base Coat
Blends	Mixture (e.g. of Different Polymer Types)
BMBF	Federal Ministry of Education and Research (Germany)

BMC	Blow Moulding Compound (Production of Plastic Parts by Blowing Process)
BONS	β-Oxy-naphthoic Acid
BPD	Biocidal Products Directive (EU)
BOD	Biological Oxygen Demand (Waste Water)
BSC	Balanced Score Card (Company Management)
CASS	Copper Accelerated Salt Spray (Corrosion Test)
CC	Clear Coat
CCD	Charge-Coupled-Device (Integrated Circuit for Processing Electrical and Optical Signals and Video Cameras)
CD	Compact Discs
CED	Cathodic Electrodeposition
CEN	Comité Européen De Normalisation, Brussels
CEPE	Conseil Européen De L' Industrie Des Peintures, Des Encres, D' Imprimerie Et Des Couleurs D' Art
CHIUZ	Chemie in unserer Zeit (Trade Journal, Wiley-VCH)
CIE	Comité Internationale D' Eclairage (International Lighting Commission)
CLP	Classification, Labelling and Packaging (EU Regulation)
CMC	Critical Micelle Concentration
CMI	Chloromethyl Isothiazolinone (Biocide)
CMR	Cancerogenic, Mutagenic, Reproduction-toxic
CNSL	Cashew Nut Shell Liquid (Natural Raw Material)
COD	Chemical Oxygen Demand
Coil	Wound Metal Strip of Different Widths (Steel or Aluminium)
COD	Chemical Oxygen Demand (Waste Water)
CPO	Chlorinated Polyolefins
CPU	Cost Per Unit
CPVC	Critical Pigment Volume Concentration
CVD	Chemical Vapour Deposition
DABCO	Diazabicyclooctane (Catalyst)
DBTL	Dibutyl Tin Laurate (Catalyst)
DDF	Digital Dense Phase Conveying (Powder)
DETA	Diethylene Triamine
DFG	Deutsche Forschungsgesellschaft (Semi-State Institution)
DFO	Deutsche Forschungsgesellschaft für Oberflächenbehandlung e.V., Düsseldorf
DGUV	Deutsche Gesetzliche Unfallversicherung E. V., Berlin (Safety)
DH	Dispersion Hardness (Test Variable)
DIN	Deutsches Institut für Normung e.V., Berlin

DIY	Do-It-Yourself
DLVO	Derjaguin, Landau, Verwey, Overbeek (Theory on the Stabilisation of Aqueous Dispersion)
DOA	Dioctyl adipate (Plasticiser)
DOI	Distinctness of Image
DOP	Dioctyl Phthalate (Plasticiser)
DPA	Diphenolic Acid
DPGDA	Dipropylene Glycol Diacrylate (Reactive Diluent)
DPM	Dipropylene Glycol Monomethyl Ether (Solvent)
DPP	Diketopyrrolopyrrole (Pigment)
DTA	Differential Thermal Analysis (Analytical Technique)
DSC	Differential Scanning Calorimetry (Analytical Technique)
DVD	Digital Versatile Disc (Communications Technology)
EAQF	Evaluation Aptitude Qualité Fournisseur (QM System)
EB	Electron Beam
ECCA	European Coil Coating Association
ECCS	European Centre for Coatings and Surface Technology (Institute)
ECHA	European Chemicals Agency (Headquarters: Helsinki, FIN)
ECJ	European Coatings Journal
EDA	Ethylene Diamine
EDP	Electrodeposition Coatings (Paints)
Effect Coatings	Coatings that Produce Different Brightness or Colour Impressions of the Applied Film at Different Viewing Angles
EG	Electrogalvanised (Electrolytically Deposited Zn Layer on Steel)
EIS	Electrochemical Impedance Spectroscopy (Testing)
EMB	Electromagnetic Brushing Process (Powder Application)
EMMA	Equatorial Mount with Mirrors for Acceleration (Testing)
EM	Environmental Management
EMS	Environmantel Managment System
EN	European Norm
EN	Evaporation Number (Analytical Technique)
EP	Epoxy (Resin)
EPA	Environmental Protection Agency (USA)
ESCA	Electron Spectroscopy for Chemical Analysis
ESH	Electron Beam Curing
ESTA	Electrostatic Application
EU	European Union
ETICS	External Thermal Insulation Composite Systems (Architecture)
EX-RL	Explosion Protection Directive (Germany)

FDA	Federal Drug Administration (USA)
NV	Solids (Non-Volatile) Content in the Coating Material
F+L	FARBE UND LACK (German Trade Journal)
Flash Off	Intermediate Drying of a Coating at Room Temperature or Slightly Elevated Temperature Prior to the Next Coating.
Flip-Flop	Different Brightness at Different Viewing Angles of the Painted Surface
FROK	First Run Ok = Measure of the Efficiency of Paint Shops
FST	Fire, Smoke, Toxicity
GC	Gas Chromatography (Analytical Technique)
GDCh	Gesellschaft Deutscher Chemiker (Association of German Chemists)
GDP	Gross Domestic Product (Economic Data)
GefStoffV	Hazardous Substances Ordinance (Germany)
GGVSEB	Dangerous Goods Ordinance Road/Rail/Waterways (Germany)
GGVSee	Dangerous Goods Ordinance Sea (Germany)
GHS	Globally Harmonised System (Hazardous Substances Labelling)
GPC	Gel Permeation Chromatography (Analytical Technique)
HALS	Hindered Amine Light Stabiliser
HDDA	Hexanediol Diacrylate (Reactive Diluent)
HDE	High Pressure Paint Stripping
HDG	Hot Dip Galvanised (Deposition of Zinc Layer by Dipping)
HDI	Hexamethylene Diisocyanate (Crosslinker)
HLB	Hydrophilic-Lipophilic Balance
HMDI	Hydrogenated Diphenylmethane Diisocyanate (Crosslinker)
HMMM	Hexamethoxymethyl Melamine (Triazine-Based Crosslinker)
HPLC	High Performance Liquid Chromatography (Analytical Technique)
HS	High Solids (Rich in Solids)
HVLP	High Volume Low Pressure (Spray Gun)
HTSC	High Throughput Screening (Method)
IATF	International Automotive Task Force (Project Group)
IC	Intumescent Coatings (Fire Safety Coating)
IED	Industrial Emissions Directive (Legislation)
ILB	Industrie-Lackier-Betrieb (Paint Journal)
IMC	In-mould Coating (Coating Process)
IR	Infrared
IPDI	Isophorone Diisocyanate (Crosslinker)
IPP	Integrated Paint Process (Coating Process Used in the Automotive Industry)
IRASER	Infrared Laser

ISO	International Organisation for Standardisation Based in the USA
JCT	Journal of Coatings Technology (Trade Journal)
JPCL	Journal of Protective Coatings & Linings (Trade Journal)
1K-	One-Component
2K-	Two-Component
SME	Small to Medium-Sized Enterprises (Classification)
KNV	Catalytic Afterburning
KOH	Potassium Hydroxide
KrW-/AbfG	Closed Substance Cycle and Waste Management Act (Germany)
KPGV	Critical Pigment Weight Concentration
KPI	Key Performance Indicator
LAMMA	Laser Microprobe Mass Analyser (Analytical Instrument)
LC	Liquid Crystal
	Lethal Concentration
LCA	Life Cycle Analysis
LCD	Liquid Crystal Display
LCE	Life Cycle Engineering
LD	Lethal Dose (Toxicology)
LEPC	Low Emissions Paint Consortium (USA)
MA	Maleic Anhydride (Intermediate)
MAC	Maximum Permissible Workplace Concentration
MDF	Medium Density Fibreboard (Medium Density Chipboard)
MDI	Diphenylmethane Diisocyanate (Crosslinker)
MEQ	Milliequivalent (Analysis)
Metallic Paints	Paints that Produce a Difference in Brightness at Different Viewing Angles of the Surface in the Applied Film
MF	Melamine Formaldehyde Resin (Film Former)
MFT	Minimum Film Forming Temperature
MIG	Multi-Stage Impulse Counter Agitator (Manufacturing)
MO	Metalloberfläche (Trade Journal)
MSDS	Material Safety Data Sheet (Safety Data Sheet)
NAB	Standards Committee For Coatings And Coatings (DIN)
NAD	Non Aqueous Dispersions
NAGUS	Standards Committee on the Fundamentals of Environmental Protection (DIN)
NAFTA	North American Free Trade Agreement (Canada, USA, Mexico)
NV	Non-Volatiles (DIN Term For Solids)
NIR	Near Infrared (Wavelength Range)
NMP	N-Methylpyrrolidone

	Standards Committee For Materials Testing (DIN)
NPF	Standards Committee For Pigments And Fillers (DIN)
NPG	Neopentyl Glycol (Intermediate For Resin Production)
NPS	Standards Committee on Personal Protective Equipment (DIN)
NQS	Standards Committee on Quality Management, Statistics and Certification Issues (DIN)
OEM	Original Equipment Manufacturer
OHV	OH Value (Analysis)
OICA	Organisation Internationale des Constructeurs d'Automobile
Orange Peel	Orange-Skin-Like Appearance of Surfaces, Visible Waviness in mm and cm Range
Organosol	See: Plastisol
Overspray	Common Term for Paint Particles that Do Not Hit the Object During Spraying
PA	Phthalic Anhydride (Intermediate for Synthetic Resins)
PA	Polyamide (Thermoplastic Polymer)
PBT	Polybutylene Terephthalate (Thermoplastic Polymer)
	Persistent, Bioaccumulative and Toxic
PC	Polycarbonate (thermoplastic polymer)
PCE	Powder Coating Europe
PCI	Paint & Coatings Industry (Trade Journal, USA)
PDA	Phase Doppler Anemometry (Analytical Technique)
PE	Polyester (Resin, Polymer)
PF	Phenol Formaldehyde Resin (Film Former)
pH	Negative Logarithm to Base 10 of The Hydrogen Ion Concentration
PIG	Paint Inspection Gauge (Measurement Technology)
PM	Methoxypropanol (Solvent)
PMMA	Polymethyl Methacrylate (Thermoplastic Polymer)
Plastisol	Dispersion of PVC Polymers In Plasticisers (Possibly with Hydrocarbons as Thinner = Organosols)
POP	Persistant Organic Pollutant
Potlife	Duration of the Workability of 2-Component Coatings after Mixing of Both Components
PP	Polypropylene (Thermoplastic Polymer)
ppb	Parts Per Billion (1 Part Per 1 Billion Parts)
PPCJ	Polymers, Paint & Colour Journal (Trade Journal, UK)
ppm	Parts Per Million (1 Part Per 1 Million Parts)
PPO	Polypropylene Oxide (Polymer)
PRA	Paint Research Organisation (UK)
PS	Polystyrene (Thermoplastic)

PTB	Physikalisch-technische Bundesanstalt, Braunschweig (German Authority)
PTFE	Polytetetrafluoroethylene (Raw Material)
PUD	Polyurethane Dispersions
PUR	Polyurethane (Film Former)
PVC	Pigment Volume Concentration (PVC)
PVC	Polyvinyl Chloride (Thermoplastic Polymer)
PVDF	Polyvinylidene Fluoride (Thermoplastic Polymer)
QA	Quality Assurance
QELS	Quasi-Elastic Light Scattering (Analytical Technique)
QMS	Quality Management System
RAL	German Colour Standards
RBN	Rub-Out Number (Measurement Technology)
REACH	Registration, Evaluation and Authorisation of Chemicals (EU Directive)
R&D	Research and Development
RIM	Reaction Injection Moulding (Process for Manufacturing Plastic Parts)
RRIM	Reaction Injection Moulding (Process for Manufacturing Plastic Parts)
RRSB Distribution	Graphic Representation of Particle Size
RTO	Regenerative Thermal Oxidiser
SAN	Styrene Acrylonitrile (Polymer)
SAM	Self-Assembling Molecules (Application)
SDS	Safety Data Sheet
SEA	South East Asia (Excluding Japan)
SEM	Scanning Electron Microscopy (Analytical Technique)
SIMS	Secondary Ion Mass Spectroscopy (Analytical Technique)
SMC	Sheet Moulding Compound (Process for Manufacturing Plastic Parts)
SMP	Solvent Management Plan (EU Regulation)
SN	Saponification Number (Analysis)
Solid Colours	Shades Without Effects
SOP	Standard Operating Procedure
Spot Repair	Small, Local Manual Repair of Paint Damage
TACT	Trisalkoxycarbonyl Aminotriazine (Crosslinker)
TA Luft	German Directive on Emissions to Air
TBTO	Tributyl Tin Oxide (Catalyst)
TDI	Toluene Diisocyanate (Crosslinker)
TE	Thermal Paint Stripping

TETA	Triethylene Tetramine (Crosslinker)
TG	Thermogravimetry (Analytical Technique)
TGIC	Triglycidyl Isocyanurate (Crosslinker)
TMP	Trimethylolpropane (Solvent)
TNO	Netherlandstoegepast Natuurwetenschappelijk Onderzoek (Dutch Institute)
TNV	Thermal Afterburning
TPGDA	Tripropylene Glycol Diacrylate (Film Former)
TPM	Technical Regulations for the Testing of Marking Materials (Germany)
TPO	2,4,6-Trimethylbenzoylphosphine Oxide
	Thermoplastic Polyolefins
TPM	Technical Specifications for Testing of Marking Materials
TRA	Thermally Regenerative Exhaust Air Purification
TRbF	Technical Rules for Flammable Liquids (Germany)
TRGS	Technical Rules for the Hazardous Substances Regulation (Germany)
TRK	Technical Guideline Concentration (Germany)
TMA	Trimellitic Anhydride (Film Former)
	Thermosetting Acrylics
TSCA	Toxic Substances Control Act (USA)
TÜV	Technical Inspectorate
UBS	Underbody Coating Protection
UF	Ultrafiltrate/Ultrafiltration
	Urea-Formaldehyde Resin (Film Former)
UP	Unsaturated Polyester
USCAR	United States Council For Automotive Research
UV	Ultraviolet
UVA	UV Radiation in the Wavelength Range from 360 to 400 Nm
UVV	Accident Prevention Regulations (Germany)
VbF	Ordinance on Flammable Liquids (Old German Law)
VC	Vinyl Chloride (Intermediate)
VCI	Association of the German Chemical Industry, Frankfurt
VDA	Association of the German Automotive Industry, Frankfurt
VdL	Association of the German Paints and Coatings Industry, Frankfurt
Vertak	Vertical Dipping Process for Car Bodies
VMP	Vacuum Metallised Pigments (Effect Pigments)
VOC	Volatile Organic Compounds
Wet on wet	Application of Two Coats of Paint by Spraying without Crosslinking; only Drying of the First Coat

WGK	Water Hazard Class (Germany)
WHG	Water Resources Act (Germany)
WOM	Weatherometer
XDI	Xylylene Diisocyanate (Crosslinker)
ZWM	Centrifugal Fluidised Bed Mill (Manufacturing)

Authors

Professor Dr Artur Goldschmidt, born in 1938, studied chemistry at the University of Münster, Germany and began working for BASF Coatings in 1969, where he was responsible for developing synthetic resins. In 1975, he was appointed head of the department of Automotive Coatings Development and Application Technology. He took up a professorship at the University of Paderborn, Germany in 1979, where until 2002 he was in charge of department of Chemistry and Technology of Coating Materials within the Faculty of Chemistry and Chemical Engineering. During his time at the University of Paderborn he also held the chair of the Deutsche Forschungsgesellschaft für Oberflächenbehandlung (German Research Society for Surface Treatment) between 1991 and 1997.

Dr Hans-Joachim Streitberger was born in 1945 and studied chemistry at the University of Münster, Germany before commencing his professional career at BASF Coatings in 1976. After holding various managerial positions in electrocoatings development and automotive coatings R&D in the USA and Europe, he was given responsibility for a business unit in the field of general industry. In 2003 he became an independent consultant to the coatings industry. During his career, he held positions on the supervisory board of BASF Coatings, the Verband der Lackindustrie (Association of the German Paints and Coatings Industry) and the Gesellschaft Deutscher Chemiker e.V. (German Chemical Society).

Index

Symbols

1,2,4-triazole 94
1,2-propanediol 68
1,3-propanediol 68
1,4-butanediol 68
1-component high solids coating 65
1-component paint 70
1-component polyurethane paints 94, 95
1-nitroso-2-naphthol 154
2,4,6-trimethylbenzoylphosphine oxide (TPO) 183
2-component high solids paints 337
2-component metering equipment 555
2-component paints 69, 101
2-methyl-2-aminopropanol 181
3,5-dimethyl-1,2-pyrazole 95
31. BImSchG 19
α-ethylhexanoic acid 57
α-halogen ketones 70
ß-hydroxy alkylamide 65
ß-oxy-naphthoic acid (BONS) 153
β-rays 608
ε-aminocaprolactam 94
ω-aminocaprolactam 72
ω-aminoundecanoic acid 72

A

Abbé refractometer 304
abbreviations 790
Abel-Pensky 265
abietic acid 40
abrasiveness 314
abrasives for wood 446
ABS 114, 721, 724
absorption, selective 363
accelerated sedimentation 250
accelerated weathering tests 422
aceteoacetic ester groups 62
acetoacetanilide 151
acetoacetic ester 94
acetone process 95
acetylacetone 62, 94
achromatic axis 369
achromatic point 367
acid anhydrides 106
acid etch test 422
acid polyesters 111
acid value 268
acronyms 790
acrylamide 123
acrylated alkyd resin 60
acrylic acid 65
acrylic acid-modified carbonic esters 125
acrylic acid-modified epoxy resins 124
acrylic polyols 98
acrylic powder coatings 581
acrylic resin 27, 34, 51, 67, 71, 77, 84, 99, 100, 118, 119, 120, 122, 123, 126, 201–203, 275, 335, 380, 416, 472, 474, 498, 502, 580, 581, 649, 650, 703, 706, 713, 724, 729, 735, 741
acrylic resin, dispersions 474
acrylic resin, fluorinated 124
acrylic resin, isocyanate-modified 126
acrylic resin, modified 649
acrylic resin, radiation-curable 123
acrylic resin, self-curing 122
acrylic system 741
acrylonitrile-butadiene-styrene copolymer 114
activated carbon 631
activation energy 85, 266, 335
additive 29
additive colour mixing 365
additives 27, 167, 171, 177, 178, 180–182, 184
adhesion 99, 101, 377, 381, 383, 386, 661
adhesion promoters 718
adhesion value 382
adhesive failure 383
adhesive forces 380
adhesive power 376
adhesive strength 376, 382
adipic acid 56, 64, 68
adipica acid 170
adsorbent 631
adsorption method 630, 631
aerosils 194
afterburning, catalytic 630
ageing 421
ageing process 412, 421, 424
agglomerate 143, 172, 176, 192, 209ff, 234, 236, 239ff, 247ff, 293, 311, 315, 322, 359, 562, 571
agglomerate separation 210, 214
agglomeration 661
aggregates 143, 208
agitation speed 233
agitator disc 236
agitator mills 213, 226, 239, 245
agitator mills, mixing time of 231
agitator shaft 231
agitator tools 242
agitators 228
agricultural machines 730
AGW levels 637
AGW values 267
air bubbles 177

air velocities 586
airbags 725
aircraft paints 93
airless application, sensitivity diagram for 520
airless atomisation 321, 508, 520–522
airless gun, stress on the nozzles of 521
airless guns, universal use of 522
airless process 738
airlift pumps 546
airmix 522
airmix process 738, 752
alcohols 137
aldehyde starch 46
aldimines 93, 107
algicides 189
aliphatic hydrocarbons 136
aliphatic polyesters 168
alkyd resin 45, 53, 60, 82, 84, 92, 116, 174, 183, 497
alkyd resin, acrylated 60
alkyd resin, styrene-modified 60
alkyd resin, thixotropic 60
alkyd resin, water-emulsifiable 62
alkyd resin, water-soluble 62
alkyd resins, metal-supported 62
alkylbenzenes 136
alkylphenol 80
allophanate 95, 98
aluminium 451, 725
aluminium flakes 669
aluminium materials, dimensional stability 452
aluminium oxide 167
aluminium silicate 147, 150
aluminium wheels 721
aluminium windows 735
aluminium, alloying of 452
aluminium, stability of 452
amber 22, 33, 40
amine crosslinkers 717
amine hardener 106
amino resins 78
aminoanthraquinone 156
aminosilanes
ammonium sulphate, ethoxylated 181
amphoteric surfactants 173
analysis of binders 278
analysis of defects 663
anchor mixer 230
Andrade equation 284
angle-dependent chroma 373
aniline 86
aniline processes 147
animal glue 49
anionic surfactants 173
annular gap 510
anodic electrocoat 39, 108, 114, 123, 480, 488, 490, 497, 498, 694, 722, 735, 736
anodisation 464
anodized coatings 452
anolyte circuit 494
anthanthrone 156
anthraquinone pigments 155, 156
anticorrosion greases 666
anticorrosion oil 454, 666, 672
anticorrosion protection 16, 33, 115, 580, 717
anti-fogging coatings 196
anti-fogging properties 194
antifouling 196
antifouling coatings 190, 749
anti-graffiti properties 738
anti-reflex coating 196
antirust primer 743
A oven 597, 599, 602
apparent or bulk density 144
application properties 676
application properties of coating materials 676
application system 200, 554, 567, 646, 659, 661, 673, 676, 701
application, ease of 674
application, hydraulic 528
application, pneumatic 528
aqueous dispersion 116, 119
architecture 647
areometer 301, 302
arithmetic mean 261
aromatic hydrocarbons 136
artificial copal 80
assessment, toxicological 628
associations 787
associative thickeners 182
atmospheric factors 410
atmospheric influences 16
atmospheric oxygen, inhibition by 71
atomisation 279, 292, 321, 328, 478, 501, 506, 514, 515, 529, 706, 714, 732
atomisation air, volumes of 512
atomisation, airless 321, 508, 520
atomisation by external charging 529
atomisation by internal charging 529
atomisation, mechanism 532
atomisation, electrostatic 523–525, 527
atomisation methods 321, 506
atomisation through nozzles 507
atomisation types 508
atomisation types, characteristic variables 508
atomisation, high-pressure 510, 517
atomisation, hydraulic 507, 508, 521, 522, 563
atomisation, low-pressure 510
atomisation, pneumatic 508, 510, 512, 514, 517, 528, 534, 563, 704, 732
atomisation, ultrasonic 536
atomising 506
atomising air 510

Index

attritor 240
authorities 787
automotive coating 24, 663, 673
automotive coating, durability of 709
automotive industry 634, 659
automotive interiors 631
automotive OEM coating, efficiency of 709
automotive OEM finishes 647
automotive refinish paints 93, 99
automotive refinishing paints 119, 711
automotive refinishing sector 111
automotive repair filler 68
automotive sector 65
automotive supply industry 718
automotive supply industry, global sales of 718
autophoresis 500
autophoresis paint 500
Autospect Quality Measurement System (AQMS) 363
auxochrome groups 150
availability 203
aviation industry 744
avoidance of waste 634
axles 723
azeotrope 133
azo groups 150
azo pigments 151
azopyrrolones 75

B

back-spray effects 671
bacterial contamination 668
bactericides 189
Baekeland 78
bag filters 253, 254, 257
Bakelite 78
balanced score card 681
balata 42
ball impact tester 395
ball mills 213, 214, 239, 240
ballast water tanks 750
balsam 40
barium sulphate 167
baryte 167
basecoat 203, 360, 581, 660, 669, 670, 676, 688, 699, 701, 703ff, 717, 720, 723ff, 738, 754
basecoat, incipient dissolving of 670
base-paints 92
basket mills 245
BAT values 637
batch production mode 205
bath material, rate of replenishment 489
bath temperature 492
bathochromic shift 150
Bayer, O. 87
beads 667
beater 230
bell diameter 532
bell rim 530, 532
Bénard cells 179, 186, 336, 669
bending elasticity 389, 391, 393
Bennet 144
bentonites 181, 182, 194
benzil ketals 70, 183
benzines 136
benzoguanamine 81, 85
benzoic acid 57
benzoin derivatives 70, 183
benzophenone 71, 183
benzotriazoles 188
benzyl cellulose 47
Bernoulli 286, 509
Bernoulli equation 549
BET method 144, 312
beverage cans 753
Beyer, Adolf von 78
bicycles 754
binders 27
Bingham liquid 283
bio raw materials 652
biobased coating systems 652
biobased raw materials 652
Biocidal Products Directive 190, 739
biocides 189, 190
Biocides Product Directive 637
biological decomposition 633
biological threshold limit values 264
bird droppings 434
bismuth molybdate 149
bismuth vanadate 149
bisphenol A 102
bitumen 114, 115
black body 592
black boxes 421
black panel temperature 423
black pigments 146
black spinel 148
blade agitator 229, 230
blanc fixe 167
blasting, cleaning technology 455
blending, with spray gun 712
blistering 416, 426, 427, 665, 670
blisters 131, 429, 471, 663, 672
blockboards 740
blocked isocyanate 66, 67, 95, 108, 123, 50, 698, 706
blocking 107
blocking agent 94, 95
blow molding compound 724
blown oils 38
BMW throwing power 486
BOD5 264, 267
body colour 364

body filler 737
body of standards 659
boils 663, 672
bond strength 376, 384
bonding agents 469
BONS pigments 153
boot lids 719
booth air, treatment of exhaust air 562
booth conditioning 556
borates 186
boundary layer phenomena 664
bridging flocculation 221, 257
bright emitters 594
bright radiators 601
brominated hydrocarbons 186
brown goods 734
brown iron oxide 147
brown rot 447
Brunauer 144, 312
brushability 319, 320
brushes, electromagnetic 730
brushing, method of paint application 501
bubble pressure method 299
bubble viscometer 291
Büche 231
Buchholz 399
Buchholz hardness 399
Buchholz indentation test 675
building supplies 735
buildings 742
bulk volume 225
Burger 392
Burger model 392
butanediol 56
butynediol derivatives 174

C

cadmium sulphide 141
cake filtration 252
calcite powders 166
calcium carbonate 166
calcium magnesium carbonate 166
calcium sulphate 167
calculating the yield 655
calculation of colour difference 370
calculation of emission quantities 655
can coating 663
can coating industry 434
cane sugar 45
caoutchouc 42
capacitative methods 349
capillary method 296
capillary viscometers 285
capillary wave atomisers 536, 537
capital costs 655
capped isocyanate hardeners 93
carbamate-modified 120
carbamic acid 89
carbamic acid esters 100
carbodiimides 111
carbohydrates 43
carbon blacks 146
carbonates 166
carbonation 472
carbonic esters 113
carbonyl pigments 154
cardanol 80
Carothers 50, 72
cartridge filters 254
cascade and vortex scrubbers 559
cascade method 246
cascade rinsing 496
casein 49
Casson liquid 283
casting materials 747
castor oil 35, 36, 59
catalysts 182
catalytic afterburning 630
cathode-ray tube 608
cathodic electrocoat 108, 110, 461, 480, 484, 490, 494, 498, 499, 568, 668, 688, 695, 696, 709, 722, 723, 729, 736
cathodic electrodeposition coating 722
cationic electrocoat paints 95
cationic surfactants 173
cavity protection 499
CDs 747
CED films, stoved 499
ceiling temperature 73, 390
cellulose 33, 45, 46
cellulose acetobutyrate 43, 48
cellulose ester 48
cellulose ethers 179
cellulose nitrate 33, 43, 47, 81, 116
cellulose paints 22
cellulose propionate 48
centrifugal and tumbling screens 579
centrifugal fluidized bed mill 244
centrifugal pump 546, 548
centrifuges 255
certification 680
chain degradation 415
chain mobility 168
chain reaction 50
chalk 166
chalk flint 166
chalking 416, 426, 665
chamber filter presses 253
chargeability 583
check-ups, medical 637
chemical resistance 94, 124, 127

China clay 165
chloride process 145
chlorinated hydrocarbons 137
chlorinated polyolefins 718
chlorinated rubber 43
chlorohydrocarbons 632
chloromethyl isothiazolinone (CMI) 189
chroma 138, 364, 369, 370
chroma, angle-dependent 373
chromating 163
chromating processes 464
chromium green 149
chromium oxide 141
chromium oxide green 148
chromium titanium yellow 148
chromium yellow 141
chromophore structures 150
CIE system 365, 367
CIELAB 368
circular conveyor 566
circulating air/fresh air ratio, control of 599
circulation lines 540
circulation systems 540
circulation systems, flow profile in 545
circulation water, bacterial attack 562
clarification 252, 253
cleaning of exhaust air 629
cleaning processes, mechanical 454
clearcoat 27–30, 95, 120, 138, 163, 187, 188, 192, 203, 206, 252–255, 257, 271, 422, 447–450, 570, 581, 604, 628, 646–650, 653, 666, 669, 670, 685, 688, 703, 705ff, 723, 725, 733, 738,740, 741, 748, 754
clearcoats, crosslinking of 604
cleavage products 65, 85, 95, 336, 337
Clement method 404
climate change test 431
closed agitator mills 241
closed-loop control 677
cloud chamber process 729
CLP directive 617, 623, 636
cmr substances 264, 628
coagulating agents 559, 562, 633
coarse powders 165
coat of paint 202
coating additives 171
coating, biobased 652
coating cost factor 654
coating line, process stages in 439
coating, market 17, 18, 125, 569, 570, 650, 652, 685, 711
coating material 16, 27, 665
coating materials, characterisation of 258
coating materials, supply of 646
coating process 665, 719
coating process, environmentally friendly 19, 23, 661
coating processes, defects in 661
coating quality 200
coating result 507
coating systems, large-scale industrial 558
coating technology 439
coating technology, basic processes of 441
coating technology, external mixing 509
coating technology, pneumatic 509
coating with high-speed rotary bells 564
coating, continuous 540
coating, defective 661
coating, droplet size spectrum of 535
coating, industrial 557
coating, mean peak-to-valley height 453
coating, throwing power of 536
coatings, defects in applied 661
coatings, water-borne 645
cobalt blue 148
cobalt green 148
cobalt salts 70
cobalt soaps 184
cochineal 141
coconut oil 59
COD 264, 267
coefficient of friction 407
coefficient of variation 261
coffee 740
cohesion 398
cohesion energy 129
cohesive energy density 129
cohesive failure 383
cohesive forces 294
cohesive pressure 297, 332
cohesive strength 166, 375, 385
coil coating 64, 67, 76, 94, 584, 626, 633, 650, 651, 663, 672, 685, 710, 725, 727, 728, 729, 730, 735, 736
coil coating industry 434
coil coating lines 726
coil coating process 503, 651, 672
coil coating sector 65
coil coating, commercial vehicle sector 726
coil coating, domestic appliances industry 726
coil coating, internal and external cladding 726
coil coating, line speed 726
coil coating, packaging industry 726
coil coating, test methods 728
coiling 127
cold check test 427
cold removal of paint 567
collaboration 678
collagen 49
collodion cotton 47
colloidal solutions 128
colloids 191
colophony 33, 40
colorant 138, 141
colorants, physical principles of 142
colour 142, 144, 365

colour adjustment of refinish paint 714
colour changes 580, 648
colour check 714
colour classification systems 374
colour concentrates 205
colour difference 367, 370
colour impression 365
colour index 144
colour match 653, 712
colour measurement 373
colour sample catalogues 374
colour specification 720
colour styling 702, 706
coloured pigments 147
coloured pigments, organic 150
commercial vehicles 730
communications equipment 747
company standards 759
compartment mills 244
completing mixers 228
components made of wood composites 740
compressors 735
computers, housings for 747
concrete 99
concrete coating 472
concrete, surface treatment 472
condensed water alternating climate tests 427
condensed water constant climate test 427
conduction 586, 590
conductive coating 16
conductive primers 720
conductivity, electrical 181
cone-and-plate rotary viscometers 290
cone-and-plate viscometers 288, 291
cones 364
consumer packaging 753
consumer's risk 262
contact drying 590
container vessels 749
contaminating substance 665
contamination 636, 664, 667, 671
contamination, bacterial 668
continuous heat resistance 77
continuous heat treatment 107
continuous processes 503
control loops 660, 676
convection drying 584
convection oven 586, 598
convection oven, energy balance of 602
conversion coatings 463, 654, 692
conversion layer, nanoceramic 463
conveyors, for painting line 566
conveyors, power-and-free 566
cooker hoods 734
copal 33, 40
copper oxides 750
copper phthalocyanines 154
corona charging 578
corona discharge 573
corona treatment 197
corroded area 429
corrosion 426, 427, 430, 665
corrosion control 202
corrosion of aluminium 450
corrosion of iron 450
corrosion of zinc 450
corrosion protection test 427, 722
corrosion, rate of 450
corrosiveness 645
cosolvents 65, 339
cost effectiveness 203
costs/m^2 654, 722
cottonseed oil 36
Couette viscometer 288
Coulomb forces 208, 572
Coulter counter 144, 309
counter corona 578
CPVC 225
crack resistance 120, 188
cracking 416, 426
cranes 742
crater 127, 131, 176, 179, 180, 254, 259, 292, 293, 335, 494, 521, 556, 567, 661ff, 667, 670, 672
crazing 203
creepback 429, 431
creeping flow 285
cresols 138
crimson 141
critical micelle concentration (CMC) 173
critical PVC (CPVC) 165
critical surface tension 379
crockmeter 406
cross hatch method 382, 383
cross-cut test 675
crosslink density 85
crosslinking 37, 330, 335, 412
crosslinking agents 78
crosslinking, degree of 71, 343
crosslinking, oxidative 34, 35
crosslinking, oxidative secondary 114
crosslinking reaction 111, 336
crosslinking resin 77
cross-section preparation technique 663
cross-sections 347, 352
cruisers 749
cryo-clean process 455, 568
cumene hydroperoxide 182
cumyl hydroperoxide 70
curing by electron beams 603
curing by UV radiation 603
curing of liquid paints 584
curing time 70

curing with laser beams 597
curing, oxidative 37, 183
current density, excessive 667
curtain coating systems 503
curves of changes in tinting strength 317
customer satisfaction 677
CVD process 164
cycle method 246
cycloaliphatic hydrocarbons 136
cyclones 578
cyclorubber 42
cylinder viscometer 288, 290

D

dammar 33, 40
damping 399
damping hardness 403
daylight fluorescent paints 733
deaerators 176
Debye 217
decahydronaphthalene (Dekalin) 136
decanter 255
decomposition, homolytic 608
Decopaint Directive 628, 711, 739
decorative paints 644
decorative surfaces 444
deep bed filter 257
deep bed filtration 252, 557
defect prevention 676, 678
defect prevention strategies 676
defective points 663
defects 653, 661, 663, 667, 671
defects, analysis of 663
defects, causes of 665
defects, cosmetic 661
defects, identifying 662
defects, through clothing 674
defects, types of 665
defects, visible 662
definition, tasks and economic importance 15
defoaming agent 175
deformation, types of 387
degradation process 419
degradation reactions 415
degree of corrosion 429
delamination 665
dendrimers 31
dendritic polyesters 126
dense phase conveying (DPC), digital 573
density 144, 301, 302, 304, 327, 353, 659, 674
density of solids 353
depletion flocculation 222
depolymerisation 413
deposition potential 483
depth resilience, relative 401
Derjaguin 215
design 474
detergents 740
deviations, random 260
deviations, systematic 260
dewetting from edges 178
dextrin 46
diastereoisomers 44
di-azo pigments 151
dibenzoyl peroxide 70, 182
dicyandiamide 106
dielectric constant 524
Diesbach 149
diethanolamine 181
diethyl malonate 95
diethylene glycol 68, 138
differential scanning calorimetry (DSC) 276
differential thermal analysis (DTA) 276
differential-pressure methods 549
diffusion barrier for oxygen 606
digital versatile discs 747
diglycidyl ethers 109
dihydroxycarboxylic acids 95
diisocyanate 60, 61, 90–92, 95, 104, 278
diketene 62
diketopyrolopyrazoles (DPP) 161
dilatancy 282, 288
dilatant flow 236
dimerised fatty acids 72
dimethyl-1,3-propanediol (neopentyl glycol) 56
dimethylamine 181
dimethylpyrazole 94
DIN colour charts 375
dioxazine pigments 159
dip coat, water-borne 479
dip coating, conventional 479
dip coating, ease of 319
dip coating, optimisation of 479
dipentaerythritol 56
diphenolic acid (DPA) 80
diphenylmethane diisocyanate (MDI) 91
dipole forces 294
dipping, ease of 320
direct discharge 632
direct dispersion 226
dirt 662, 663, 666, 674
dirt entrainment 673
dirt particles 673
dirt repellent 195
dirt, trapped 662
disc centrifuge 255, 308
discharging, electrical 527
discolouration 665
dishwashers 734
dispensing 548
disperse component 379

dispersers 234
dispersing 144, 243
dispersing of pigments 234
dispersing properties of pigments 206
dispersion 98, 113, 114, 127, 128, 200, 225, 236, 282, 331–333, 644, 649
dispersion coatings 113
dispersion forces 294, 377
dispersion hardness 317
dispersion process 209, 212, 215
dispersion rate 212, 244
dispersion time 223
dispersion, aqueous 116, 119
dispersion, theory of 206
dispersions, secondary 644
dispersions, solvent-free 646
dispersity 275
disposal of 558
dissolvers 213, 235, 245
distribution curve 261
diterpene 40
division of the market 18
DLVO theory 216, 217
doctor blade 238
doctors' surgeries, furniture for 743
dodecyl alcohol 180
DOI measurements 360
dolomite 166
domestic appliances 647
door handles 725
doors, gates and frames 735, 736
double screw kneaders 247
double trough kneaders 247
DPC, dense phase conveying 573
drawing greases 666
drilling rigs 742
drop method 296
droplet formation; in airless atomisation 520
droplet size 328, 509, 525
droplet size distribution 328, 511–514, 524, 530, 531
droplet size, influence of 538
drum coating 754
dry film density 304, 353, 655
dry film thicknesses 345, 728
dry filter mats 558
dry filter systems 558
dry sanding 715
dry sieve sets 307
dryers 734
drying 651
drying of paint, contamination during 598
drying process 586, 589
drying, physical 30, 329
drying, temperature tolerances for 589
dual-cure coatings 125
dull radiators 593, 601
duplex systems 742
dust 662
dust contamination 663, 667, 668
dust inclusions 668
duty to provide information 682
dynamic mechanical analysis 276, 340, 396
dynamic viscosity 281

E

earth pigments 147
earthing, poor 671
easy-to-clean 195
EBC process 651
eco audit 682
eco-efficiency analysis 639, 641, 643, 652
eco-efficiency audit 24
eco-efficiency portfolio 640
eco-efficiency 638
ecological fingerprint 640
eddy-current heating 597
eddy-current method 349
edge protection 722
edge protection coatings, for rotor blades 752
effect coatings 184, 468, 717
effect coatings, repair of 714
effect pigments 161, 373
effect, bactericidal 198
efficiency 203, 720
eggshells 668, 671
elastic coatings 473
elastic modulus 675
elasticity 388, 412, 718
elastomers, property 465
electric filters 559
electrical conductivity 181, 314, 321, 527, 646
electrical double layers 215
electrical insulation coatings 95, 746
electrical insulation properties 107
electro powder coating (EPC) 499
electrocoat paints, market for 480
electrochemical impedance spectroscopy 433
electrocoat 644, 735
electrocoat bath, composition of 487
electrocoat bath, electrode reaction in 484
electrocoat bath, leaks in 495
electrocoat bath, replenishment method for 489
electrocoat paint, bath temperature of 487
electrocoat paint, breakthroughs in 486
electrocoat paint, coagulation of 484
electrocoat paint, current/time curve for 485
electrocoat paint, deposition of 481
electrocoat paint, potential throwing power of 486
electrocoat paint, sedimentation of 491
electrocoat paint, surface defects in 495
electrocoat paints 110, 480, 497, 642

electrocoat paints, contaminants in 493
electrocoat paints, physico-chemical principles of 480
electrocoat primers 108
electrocoat tanks, extraction, 491
electrocoat, morphology of 483
electrocoat, reverse variant 499
electrocoat, surfaces 493
electrocoating 480, 633, 667, 677, 719
electrocoating baths 677
electrocoating process 632
electrocoating, anodic 114, 123, 480
electrocoating, application voltage 667
electrocoating, cathodic 461, 480, 490
electrocoating, electrode processes in 482
electrocoating, immersion marks 667
electrocoating, market 480
electrocoating, variants of 498
electrocoating, water marks in 667
electrodeposition, principles behind 491
electromagnetic brushes 730
electron cloud 609
electron guns 609
electron guns, energy consumption of 610
electron microscopy 311
electron spectroscopy 664
electron spectroscopy for chemical analysis 664
electron-beam 2019,
electron-beam curing 71, 607, 648, 649
electron-beam of the scanning electron microscope 664
electron-beam-curable coatings 183
electrons, accelerated 610
electro-osmosis 483
electrophoresis 482
electrostatic stabilisation 221
elementary charge 608
embrittlement 412
emission of solvent 337
emissions 22, 63, 105, 268, 638, 642, 655
emissions reduction 629
emitter power, dependence 592
EMMA test 422
EMMAQUA 422
Emmet 144, 312
emulsion 63, 95, 109, 113, 117, 127, 172, 192, 405, 472, 698, 733
emulsion paints 116, 128, 148, 165, 166, 235, 405, 735
emulsion polymerisation 116, 117, 118, 644
emulsion polymers 34, 51, 98, 113, 118
enantiomer 44
end groups, reactive 32
endomethylene tetrahydrophthalic acid 55, 72
energy consumption 19
energy dissipation 234, 239
energy elasticity 390
energy losses 602
energy-dispersive X-ray microanalysis 664
engine blocks 723
entropic stabilisation 219, 220
environment management systems (EMS) 682
environmental pollution 639, 640
environmental protection 638
environmentally compatible coating technologies 23, 99
EPC 499, 569, 570
epichlorohydrin 102, 104
epoxidised oil 170
epoxy acrylates 124
epoxy alkyd resin esters 108
epoxy-containing dispersions 109
epoxy group 101
epoxy novolaks 80
epoxy paints 85, 109
epoxy resin emulsions 109
epoxy resin esters 107
epoxy resin powders 580
epoxy resins 87, 100, 101, 107, 108, 202, 497, 580, 717, 737
epoxy resins, acrylic acid-modified 124
epoxy resins, amine-modified 498
epoxy resins, silicone-reinforced 77
epoxy-polyester 729
Epprecht twistometer 383
equipment, personal protective 636
Erichsen test bar 404
error, relative 261
errors, random 261
esterification 40, 47
esterimides 113
esters 137
etching, of clear coats 422
ethanediol (ethylene glycol) 56, 68, 138
etherification level 84
ethoxylated ammonium sulphate 181
ethyl acetoacetate 95
ethyl cellulose 47
EU Building Products Directive 742
EU Eco Audit Regulation 682
European Directives 635
Evans 306
evaporation 128, 631, 735
evaporation behaviour 333
evaporation number 132, 304
evaporation rate 333
excessive crosslinking 335
exchange of materials 333, 334
exhaust air energy 602
exhaust air purification, regenerative thermal oxidisers for 630
exhaust air treatment 630
exhaust air, contaminant levels in 630
exhaust air, overspray-charged 559
exploration installations 742
explosion limits 131, 264, 265, 266

explosive mixtures 264
exposure 636
extenders 27, 29, 138, 139, 165
exterior plastic hang-on parts 719
external cladding 735
external charging 529
extruder types 248

F

façade elements 736
failure reports 676
falling-ball viscometer 291
fan bores, mechanism behind 516
fan pumps 546
Fast-Fourier transformation 362
fatty acid dimers 652
fatty acid process 57
fatty acids, dimerised 72
fatty acids, saturated 35
fatty oils 34, 39
ferrocene 183
fibre composite materials 744
fibreglass 167
fibre-optic cables 747
fibres 673, 674
Fick's first law of diffusion 333
fields of application 742
filiform corrosion 430
fillers 113, 363
fillers, lamellar 167
filling 506, 716
film build 85
film defects 179, 586
film formation 27ff, 49, 55, 59, 60, 63, 95, 96, 98, 107, 109, 114, 117, 127, 128, 167, 172, 183, 192–195, 201, 215, 282, 319, 327ff, 378, 382, 408, 499, 527, 584, 585, 597, 599, 637, 645, 648, 650
film formation by heat transfer 585
film forming agent 27, 28, 30, 31, 34, 575
film forming process 127, 331–334, 586
film of water, adhering 482
film shrinkage 327, 337
film thickness 328, 345, 347, 708
film thickness distribution 526, 536
film thickness gauges 348
film thickness measuring instrument 352
film thickness, growth in 484
film thickness, increase in 526
films, self-healing 752
filter aid 252, 253
filter bags 254, 493
filter media 251
filtration 250, 251
filtration units 250
final inspections 263
fine dispersion 242
fine powders 165
finish foil coating 653
finish foils, for wood 448
first run ok 709
Fischer, Karl 277
Fischerscope 402
fish mouth design 476
fish oil 36
fish-eyes 672
flame cleaning 743
flame ionisation detector (FIDs) 268
flame retardant coating 16
flame treatment 197
flame-cleaning blowers 457
flameproofing agents 186
flaming 718, 719, 721
flammability 264
flash point 131, 134, 264
flash rust 666
flash-off zones 646
flavanthrone 156
flip-flop 161
floating effect 186, 336
flocculation 209, 222, 258, 318, 319, 336
flood sheet fluid systems 632
flood-coating systems, electrophoretic 499
floors 741
Florida weathering 426
flow 231
flow coating 740
flow coating, method of paint application 505
flow cups 288
flow curve 225, 283, 288
flow graph (τ/D) 281
flow in pipes 286
flow lines 666
flow measurement by seeding 551
flow point determination 225
flow profile 286, 543
flow properties in pipes 543
flow properties of powder coatings 340
flow rates 543
flow resistance law 231
flow states 546
flow, laminar 235
flowability 582
fluidisation 570, 582, 648
fluidisation, degree of 583
fluidised bed method 569, 744
fluidised-bed coating 573
fluidised-bed process 115
fluidity 281
fluorinated hydrocarbons 667
fluoroacrylates 752
fluorosurfactants 178

flushing 239
flux rate 492
foam 174
foam persistance 296
foil coatings 650, 653, 741
foils 710, 733, 740
food cans 753
foreign matter 257
form factors 309, 313
formaldehyde 78, 81, 85, 86, 189, 631
formaldehyde cleavage 85
formulation 199, 200, 205, 225, 226
formulations, general rules for 200
Fourier transform infrared spectroscopy (FTIR) 664
fractional analysis 304
freezers 734
Fresnel's equation 355
friction rolls 237
fungicides 189
furniture 631
furniture industry 434, 666, 739

G

galactose 45
galvanic cells, formation of 450
galvanised steel 692, 699, 717, 725
galvanising 451
gap flow 285
garage doors 736
Gardner 405
gas chromatography 274
gas cylinders 754
gas pipelines 743
gas-phase hydrolysis 193
Gaussian distribution curve 261
GC/MS 274
gear pumps 253–546, 547, 646
gel particles 179, 180
gel permeation chromatography 275
gel time 70, 582
gelatin 49
Gibbs-Helmholtz equation 215, 220
Gibbs-Marangoni effect 175
glass containers, coating of 754
glass transition temperature 119, 169, 202, 276, 390, 582
glass-fibre segments 667
global energy requirement 750
globalisation 24, 677
gloss 131, 184, 356, 359, 676, 708
gloss indicatrix 358
gloss peak 358, 363
gloss retention 188, 426
gloss, loss of 203
glucose 45
glue, bleeding of 666
glycerin triacrylate, propoxylated 125
glycerine 56, 64
glycerobenzyl resins 80
glycidyl acrylate 120
glycidyl ethers 102
glycidyl methacrylate 120
glycol ethers 137, 138
glycols 137
glycoluril 81, 85
glyptal 49, 51, 63
goniophotometer 358, 359, 363
Good's interaction parameter 379
Gouy 216
gradient bar oven 340, 434
graffiti 738
grain fineness 315
graphite 167
gravity sedimentation 251, 254, 308
green chromating 464
green earth 150
greenhouse potential 641
grille scattering 609
grinding 743
grinding balls 239
grinding media 240, 241, 242
grinding trough shape 243, 244
grindometer 315
grip 721
Gritcote method 457
groundnut oil 36
growth reaction, radical 604
gutta-percha 42

H

H_{12}MDI 91
Hagenbach correction 286
Hagen-Poiseuille equation 286, 512, 544,
HALS 188
hammer finishes 77
haptics 407
hard 195
hardener 92
hardening, chemical 30
hardness 399
hardness bar 404
harmonisation of test methods 659
harvesters 730
hazard symbols 266, 618, 619
hazardous materials 267
Hazardous Substances Ordinance 635, 636
hazardous to water 633
haze 665, 708
headlight systems 724
heat exchange, effective 585
heat exchangers 725, 735

heat flow in the workpiece 590
heat generation during flow 285
heat recovery 599
heat resistance 73, 76, 95, 104
heat transfer 586, 587
heat transfer by conduction 590
heat transfer coefficient 587
heat transfer, direct 586
heat transfer, indirect 586
heat transfer, methods of 597
heat transmission, coefficient of 603
heated black boxes (HBB) 421
heating systems 735, 736
height profile 360
Heller 221
Helmholtz 216
Hemicellulose 442
Hertz hardness 400
hexahydrophthalic acid 55
hexamethoxymethyl melamine (HMMM) 83, 100
hexamethylene diisocyanate (HDI) 91
hiding power 138, 142, 144, 146, 164, 197, 200, 207, 209, 317, 655
high build electrocoats 723
high film-build electrocoating process 499
high performance liquid chromatography 274
high solids 65, 199, 212, 253, 337, 642
high solids coating, classification of 643
high solid paints 19, 27, 28, 83, 128, 137, 337, 338, 603, 642–644, 651, 662, 689, 691, 706, 711, 739
high throughput screening 681
high volume low pressure (HVLP) process 518
high-boiling solvents 179
high-energy UV or electron radiation for curing 65
high-performance atomisers 516
high-pressure paint removal 568
high-rotation coating, droplet sizes in 532
high-rotation coating, influence of viscosity on 532
high-rotation coating, quality of the coating 532
high-speed rotary atomisation 530, 536
high-speed rotary atomisers 720
high-speed rotary bell, atomisation 532
high-speed rotary bell, atomisation mechanism of 532
high-speed rotary bells 530
high-speed rotary coating, material losses in 535
high-speed rotary coating, paint-feed system for 534
high-speed rotary systems, advantages of 530
hindered amine light stabilisers (HALS) 188
HLB value 174
HMMM resins 96
homogenisation 229, 237
homogenisation, 2-component application systems 554
homogenising 229, 548
Hooke's law 388
hopeite 462
hoppers 571
Höppler viscometer 291
hornblende 150, 166
hospital furniture 743
hot removal of paint 567
hot spray coating 517
hot-dip galvanising 451
hot-wire anemometry 551
household substances 740
HPLC 274
Hückel 217
hue 369, 370
humidity-hardening polyurethane coatings 93, 99
HVLP guns 712
HVLP process, atomisation principle behind 518
hybrid coatings 651
hybrid polymers 194
hybrid powder coatings 580
hybrid powders 110
hydantoin 104
hydrated aluminium oxides 186
hydrated chromium sesquioxide 148
hydraulic atomisation 520
hydraulic guns, air-assisted 522
hydrocarbons 458
hydrocarbons, aliphatic 136
hydrocarbons, aromatic 136
hydrocarbons, brominated 186
hydrocarbons, cycloaliphatic 136
hydrocarbons, fluorinated 667
hydrocyclones 255
hydrogen transfer 70
hydroperoxide 69, 182
hydrophilic-lipophilic balance (HLB) 174, 175, 494
hydrophobic 117, 172, 184, 198, 315, 444, 457, 472, 473, 494, 667
hydrophobing agents, nanoscale 195
hydroquinone 69
hydrosol 194
hydroxy-benzophenone 188
hydroxycyclohexyl phenyl ketone 183
hydroxyethyl cellulose 47
hypsochromic shift 150

I

ideal liquid flow 543
ideal-elastic bodies 388
identical quality 674
IG gauge 347
ignition point 131
ignition temperature 264, 266
ilmenite 145
imines 93
immersed balls 302
immersed bodies 301
immersion marks 667

immersion mills 245
impact elasticity 391, 395
impact load 239
impact resilience 389
impact resistance 720
impact tester 395
impedance spectroscopy 433
impregnating materials 746
impurities 664
in household appliances 730
incipient dissolving of the basecoat 670
indentation hardness 399
indigo 141
indirect discharge 632
induction 597
induction forces 377
industrial light fittings 735
inert gas for drying 607
infinitesimal modulus determination (IMD) 402
informative task 15
infrared rays 586
infrared spectroscopy 271
inhibition 606
inhibition by atmospheric oxygen 71
inhibitor 69, 70, 177
initiators 70, 182
injection moulding 666
in-mould coating 471, 653, 720
insufficient intermediate drying 670
integrated paint process (IPP) 270, 689, 691, 698, 701, 705
integrated "wet-on-wet-on-wet" 688
interaction parameter 379
intercoat adhesion 470
interface 300
interference pigments 163, 197
interior automotive hang-on parts 721
intermediate drying 670
intermediate drying, insufficient 670
internal cladding 735
internal charging 529
internal plasticisation 168
internal stresses 408
internal stresses in plastic 667
internal tension 391
intrinsic viscosity 177, 179
iodine value 36
ionic polymerisation 183
ionic surfactants 173
ioniser 469
IR drying, effective use of 593
IR drying, short-wave range 594
IR emitter, energy distribution of 591
IR emitters, yield from 601
IR lasers 596
IR oven 600
IR radiation 201, 272, 502, 585, 591, 595, 646
IR radiation, short wave 595, 601
IR radiation, medium wave 595, 596
IR radiation, long wave 595
IR radiators, temperature of 593
iron complex 154
iron oxide 147
iron oxide, black 146
iron oxide, lamellar 147
iron oxides 141
iron phosphating 460, 736
ISO 14000 ff. 682
ISO 9000 679
ISO 9000 ff. 679
ISO 9000 series 660
ISO TSC 16949-2009 660
isocyanate hardeners, capped 93
isocyanate crosslinkers 98, 711, 720, 724, 732, 750
isocyanate hardener 88, 89, 93, 95, 118, 120, 123
isocyanate resins 65, 78
isocyanate-free 99
isocyanate 32, 50, 65, 66, 77, 87, 89ff, 124, 126, 134, 182, 269, 272, 314, 336, 498, 542, 554, 556, 652, 713, 745
isocyanurates 89
isoelectric point 218
isoindolinones 155
isononanoic acid 57
isophorone diisocyanate 104
isophorone diisocyanate (IPDI) 91
isophthalic acid 54, 64
isophthalic acid phenyl ester 74
isoprene 42
isopropylbenzene (cumene) 137

J

jet dispersion 98, 556
jet pumps 546
jet scrubbing 559
job coater 730, 736
joining technique 474

K

kaolin 165
Kenics mixer 554
kermes 141
Kesternich test 427
ketimine 93, 107
ketones 137
key analytical data 496
key performance indicators 681
kieselgur 166
kinematic viscosity 281
kneader 238, 239, 246
knife coating, method of paint application 506

Knoop hardness 400
König 402
KrW-/AbfG – Waste Management Act 633
Kubelka 316
Kuhn 50

L

L* C* h* chromatic system 370
Lambert's law 310
lamellar atomisation 531
lamellar extenders 167
lamellar iron oxide 147
lamellar silicates 181
laminar flow 235, 543
laminar flow gradients 234
laminate 741
Landau 215
landfilling of waste 635
laser doppler anemometry 219
laser emitters, for drying 596
laser microprobe mass analyser method (LAMMA) 664
latex milk 33
latex paints 114
layer silicates 194
layer thickness measurements with ultrasound 350
layer thickness measurements with X-ray fluorescence 350
layer thickness measurements with β backscatter 350
LC 134
LC pigments 164
LD 134
LD_{50} 266
lead chromate pigments 148
lead times, adherence to 678
leafing 163
lecithin 34, 173
LED lamps 651
lens 724
letting down 223, 226
level of material utilisation 634
levelling 131, 178, 292, 328
levelling defects 513
levelling fault 179, 527
levelling properties 515
Leverkus 150
Lewis acid 104, 107
life cycle analysis (LCA) 638
life cycle engineering 638
light scattering 138, 142, 276, 309
light scattering pigment 207
light/dark effects 356
lightfastness 416
lightproofness 416
lightspar 167
light-to-dark flop 665
line speeds 502
linseed oil 22, 36
liquids, pseudoplastic 281
Lithol Red 151
lithopone 146
local cells 455, 476
local defects in the surface 662
lockers 743
locomotives 737
long term outdoor exposure 99, 419
long term weathering 425
loss angle 392
loss of gloss 427, 665
loss on ignition 314
lotus effect 196
lower explosion limit 266

M

MAC levels 134
MAC values 264
Mackor 219
macro-foaming 175
magnesium 711
magnesium, paintability of 452
maleic anhydride 39, 55, 67
maleic resins 40
maleinised 108, 110, 113, 278
maleinised linseed oil 695
maleinised polybutadiene 695
malonic acid esters 111
malonic ester 94
mandrel bending test 393
manganese soaps 184
Mannich reaction 498
mannose 45
manual painting 539
manual roller, method of paint application 501
manufacturer's risk 262
manufacturing instructions 226
mar resistance 405
Marcusson 265
market value coating 715
marking materials, water-borne 748
Martens universal hardness 401
mass flow measurement, based on gyroscopic principle 553
mass flow per unit area 585
mass spectroscopy 273
mass transfers 587
mass-transfer coefficient 587
material consumption 518
material control 674
material discharge rate 511
material efficiency levels 634
material feed 540
material flow 205

Index

material output 512
material safety data sheet 636
material to be dispersed 228, 247
material utilisation 634
material yield 479, 492, 501, 528
materials, combining of 228
matrix segments
matt black powder paint 723
mauveine 142
maximum admissible concentrations in the workplace (MAC) 134, 264, 267
MDF board 651, 666
MDF fibreboard 445
mean error 261
mean particle diameter 306
mean roughness value 360
measurement accuracy 260
measurement methods, volumetric 549
measurement of colours 365
measurement sedimentation 308
measurement, statistical 676
measuring the magnetic attractive force 348
measuring uncertainty 660
mechanical engineering 733
mechanical properties 99, 116, 119, 165, 276, 375, 408, 416, 466, 720
medium powders 165
Megele 306
melamine resins 78, 81, 83, 85, 117, 167, 186
melt process 95
melting temperatures, metals 449
membrane osmometry 276
MEQ value 270
MEQB value 270
MEQS 270
mercury vapour lamps 71, 606
metal complex pigments 154
metal particles 667
metal pigments 161
metallic coatings 159, 356, 665
metallic effect 360, 669
metallic finish 163
metallic paints 669
metallic soaps 38, 182
metal-supported alkyd resins 62
metamerism 371
methacrylic esters 119
methylcarbamate 99
methyl cellulose 47
methyl formamide 138
methyl isothiazolinone 189
Meyer, C. H. 78
mica 167
micaceous iron ore 148, 167
micelle concentration 182
micelles 117
micelles, formation of 661
Michael addition 111
microfilter mats 557, 558
microfoam 175
microorganism 434, 447
microscope 663, 674
microstructures 338
microtomy 663
microwave drying 597
microwave ovens 734
Mie scattering 418
mill base 225, 240, 245
mill base composition 222
mill base formulation 223, 225
mineral blue 149
mineral pigments 147
mineral spirits 136
minimum film forming temperature (MFT) 116, 119, 333
mirror forces 377
mirrors 724
miscellaneous coating applications 753
mixed production 228
mixed stabilisation 221
mixed-phase pigments 148
mixers, static 555
mixing coefficient 233
mixing formulations 226, 771
mixing machines 717
mixing paint principle 226
mixing paints 226
mixing quality 229, 555
mixing time 229, 232
mixing time of agitator mills 231
mixing, continuous 556
mixtures, explosive 264
modified acrylic resins 649
modified natural material 33
modified natural substance 28
modular design, container vessels 750
modulus of elasticity 388, 389, 390
Mohs 398
molecular enlargement 329
molecular weight 31
molecular weight distribution 31, 59, 275
molybdate orange 149
molybdate red 149
monoazo pigments 151
monobutyl phosphate 181
monomer 50
monosaccharide 43, 45
montmorillonites (bentonites) 179
mortar 472
motor cycles 754
motors 735
mottling 665, 669, 714
mould release agents 666

Index

Müller-Rochow synthesis 76
multicutter instrument 383
multi-impact devices 395
multistage impulse agitator 230
multizone oven 600
Munk 316
Munsell catalogue 374

N

Nanocapsules, filled 752
nanoceramic 463
nanoparticles 30, 191ff, 407, 741, 752
nanoparticles by hydrolysis 192
nanoparticles, silicon-containing 197
nanoproduct 191, 197, 198, 752
nanoscale 192
nanoscale dimensions 192
naphthol AS 151
natural material 33, 34
natural oils 34
natural resin 21, 40, 49, 51, 59, 279, 434, 748
natural rubber 42
natural substance, modified 28
natural weathering 417
natural weathering correlation 426
natural weathering with salt 431
near infrared radiation 648
NCO number 268
neopentyl glycol 64, 68
Nernst 449
N-ethylpyrrolidone 95, 138
neutral cleaning agents 457
neutralising agents 181
new coating refinish 715
Newton 543
Newtonian twin plate model 280
nickel titanium yellow 148
NIR radiation 648
nitrating acid 47
nitrogen oxides, formation of 630
N-methylolchloracetamide 189
N-methylpyrrolidone 95, 138
nonionic polyurethane dispersions 98
nonionic surfactants 174
non-leafing types 163
nonsolvent 131
nonuniformity 275
nonvolatile matter (NV) 279
nonylphenol-ethoxylate 109
no-rinse process 727
Nouy 295
novolak 78
nozzle wear 321
nuclear magnetic resonance 272
nuclear magnetic resonance spectroscopy 273
number average molecular weight 275

O

object to be painted 200
object to be painted, earthing of 522
occupational exposure limit 91
occupational health, safety and environmental protection regulations 203
odour 629
odour reduction number 629
odour threshold 134, 629
Oesterle 402
office furniture 743
offshore plants 751
OH value 269
Ohnesorge number 508
oil absorption value 144, 211
oil length 54, 59
oil number 315
oil paints 22
oil part 54
oil pipelines 743
oil, epoxidised 170
oils, blown 38
oils, fatty 34, 39
oils, natural 34
oiticica oil 36
olfactometry 134, 629
oligomeric polyurethanes 179
on-time delivery 678
open-loop control 676
o-phthalic acid 54
optical brighteners 186
optimum mill base composition 224
orange peel 513, 661, 671
orange peel texture 665
Orchard 328
Ordinance on hazardous substances 266
organic coloured pigments 150
organosol 113, 194
orientation forces 377
orifice flowmeters 550
original coatings 714
orthosilicates 193
oscillating rheometer 396
oscillation 399
oscillator 301, 304
Ostwald 283
outdoor light fittings 736
outside mirrors 719
oven air, threshold for 599
oven, air circulation in 598
oven, air velocities in 598
oven, design of 587, 597

Index

ovens, air movements in 598
ovens, economical operation of 600
ovens, operating conditions of 590
ovens, secondary cleaning system for 600
Overbeek 215
overcuring 595
overspray 5, 127, 179, 180, 439, 478, 507, 511, 516ff, 522, 529, 534, 538, 556, 558, 559, 561, 562, 633–635, 690, 691, 700, 701, 714, 732
overspray absorption 127, 558
overspray formation, reduced 518
overspray losses 522
overspray treatment 559, 562
overspray, characterisation of 538
overspray, elimination of 556
overstoving 588
Owens 380
oxalanilides 188
oxazolidine ring 93
oxazolidines 93
oxazolines 111
oxidative curing 37, 183
oximes 94, 177
oxirane group 101
oxygen cells 430
oxygen inhibition 55, 126, 606
ozone formation during drying 607

P

PA 721
paddle type kneaders 247
paint 20
paint and coatings market, global 17
paint atomisation 530
paint atomisation, electrostatic 523
paint defects 567
paint defects, avoiding of 557
paint factory, material flow in 204
paint finish 565
paint, high solids 337
paint, important solvents for 134
paint manufacture, separation processes in 250
paint manufacturer 20
paint material, variables 514
paint particles, separation of 561
paint pattern, variables affecting 517
paint, physically drying 330
paint plant, processing sequences 227
paint processing, automated 563
paint processing, elimination of dirt in 557
paint processing, indirect 556
paint processing, industrial 506
paint production 226
paint removal 567
paint removal with carbonising chamber 568
paint removal, high pressure 568
paint removal, thermal 568
paint resins 78
paint sludge 634
paint sludge, disposal of 558
paint structure 27
paint system, function 202
paint systems, multiple layer 663
paint throughput 515
paint waste 22
paint, water-borne 19, 27, 60, 64, 83, 107, 128, 144, 167, 171, 178, 181, 189, 237, 315, 339, 478, 479, 505, 532, 534, 542, 557, 562, 596, 619, 623, 632, 644, 646, 689, 720, 725
paint yield 527
paint, economic processing of 589
paint, environment-compatible 642
paint, thermal curing 119
paint-damaging substances 434
painting lines for mass-produced goods 566
painting process 558
painting process, industrial 477
painting robots 564
painting width 526
paint-removal process 568
paints 27
paint-supply systems 540
Palm kernel oil 36
palmitic acid 35
panel surface, roughness on 666
panels, coil coated 736
panels, precoated 711
parquet 741
particle counter 309
particle size 117, 142ff, 150, 165, 177, 191ff, 207, 208, 223, 225, 234, 248ff, 305ff, 332, 336, 339, 482, 499, 539, 556, 571, 575, 578, 582, 644, 648, 671, 672, 769
particle size distribution 144, 212, 250, 305, 322, 539, 571
particle size distribution, measurement of 539
particle velocity, measurement of 539
passivation 451, 462, 693, 731
pass method 245
paste 205
paste mixers 228
pastes, production of 226
PC 718, 721, 724
PC/ABS blends 724
PC/PBT blends 724
peak-to-valley heights 453
pearlescent pigments 163
peel test 385
peel-off method 384
pencil hardness 403
pending drop method 297
pendulum hardness 402
penetration coefficient 176

penetration hardness 400
Penniman process 147
Pensky-Martens 265
pentaerythritol 56, 64, 125
pentaerythritol, substitutes for 652
perfluoro compounds 662
performance index 232
perinone pigments 157
Perkin 142
peroxide 69, 183
persistant organic pollutants 633
personal protective equipment 636
Persoz 402
Persoz pendulum 402
perylene pigments 157
Peters chip test 386
petroleum benzines 136
pH of a pigment suspension 314
pH regulators 181
pH value 645, 677
phase Doppler anemometry 539
phase separation 661
phase shift angle 397
phenolic resins 49, 78, 79, 81, 107, 116, 497
phenols 94
phenoxy resins 103
phenylmethylpyrazolone 151
phosgene 89
phospating 461
phosphating and chromating processes 459
phosphating, quality of 463
phosphites 186
phosphophyllite 461
phosphoric acid 170
phosphoric esters 170
photoinitiators 69, 70, 124, 125, 183, 184, 604–608, 648, 649
photothermometry 350, 351
photovoltaic coatings 710
phthalate plasticisers 170
phthalic acid 170
phthalic esters 170
physical characteristics 675
physical constants 783
physical drying 30, 329, 330
pickling 743
pickling solutions 459
pigment 29
pigment absorption capability 211, 224
pigment agglomerates 234, 240
pigment dispersion 217, 218
pigment dispersion, stabilisation 215
pigment pastes 205, 223, 228
pigment shock 222, 226
pigment size description 305
pigment stabilisation 177
pigment surface area per unit volume 312
pigment surface area, specific 312
pigment volume concentration (PVC) 165, 208, 211
pigment wetting 212, 238
pigment-binder ratio 668
pigments 27, 29, 138, 140, 142, 144, 198, 200, 208, 314
pigments important to coating technology 145
pin mill 249
pinholes 180, 662, 663, 667, 668, 670, 672
pipes 542
piston pumps 253, 546
Planck's law of radiation 591
planetary kneaders 247
plasma pretreatment 468
plasma process 727
plaster 472
plastic components 666
plastic internal stresses in 667
plastic parts, repairing of 718
plastic surfaces, metallising of 468
plastic, conductivity of 469
plastic, glassfibre-reinforced 667
plasticisation 116, 169
plasticisation, internal 168
plasticiser 29, 116, 167, 168, 200, 331, 667
plasticity 282
plastics 465
plastics, adhesion of 469
plastics, conductivity of 471
plastics, modification of 466
plastics, reinforced with carbon fibres 711
plastics, tempering of 469
plastisols 113, 115, 333, 728, 729
plate filter 253
plate-and-frame 253
plate-type guns 610
plough mixers 248
plug conveyor 572
PMMA 718, 724
Poisson's ratio 388
polar component 379
polishing 709
polyacrylate 111, 114, 118, 119, 179, 711
polyacrylate, OH-functional 120
polyacrylate, thermoplastic 118
polyacrylic acids 182
polyaddition 50, 87, 335
polyaddition resins 87, 100
polyalkyl ether polyols 109
polyamides 72, 718, 724
polyamines 106
polyaminoamides 60, 72, 106
polybenzimidazole 72, 74
polybenzimidazopyrrolone 72
polybutadiene resin 114, 497
polybutadiene oil 114

polybutylene terephthalate (PBT) 721
polycondensation 50, 335, 336
polycondensation reaction 65
polycondensation resins 51
polycondensation curing 337
polyester acrylates 124
polyester polyols 98
polyester resins, unsaturated 67
polyesters 580, 711, 729
polyesters, aliphatic 168
polyesters, dendritic 126
polyesters, saturated 63
polyester, unsaturated 29, 55, 56, 67ff, 123, 125, 157, 177, 182, 183, 336, 466, 467, 497, 503, 504, 603, 604, 649, 650, 667, 716, 718, 720, 737, 741, 746
polyethers 95, 98, 101, 107, 109, 178, 182, 221, 406, 412
polyethylene glycol 107, 221
polyethylene wax dispersions 114
polyimidazopyrrolones 75
polyimides 72, 73, 75
polyisocyanate 65, 67, 120, 706
polyisoprene 42, 114
polymeric solutions 330
polymerisation 335
polymerisation reaction, start of 604
polymerisation, ionic 183
polymerisation, radical 608
polyolefin waxes 181
polyolefins 114
polyolefins, chlorinated 718
polypropylene blends 719
polysaccharid 43, 45
polysiloxane 76, 176, 178
polystyrol 114
polyurethane coatings 89, 92, 93,99, 336
polyurethane dispersions 34, 95ff, 98, 652
polyurethane paints 98, 99
polyurethane paints, dispersion 95
polyurethane paints, humidity-hardening 99
polyurethane paints, one-component 95
polyurethane powder coatings 581
polyurethanes 87, 202
polyurethanes, oligomeric 179
polyvinyl acetal 116
polyvinyl acetate 115
polyvinyl alcohol 116
polyvinyl butyral 116
polyvinyl chloride 115
polyvinyl compounds 114
polyvinyl esters 114
polyvinyl ethers 117
polyvinyl isobutyl ether 117
polyvinyl propionate 116
polyvinyl pyrrolidones 179
polyvinylacetate dispersions 113
polyvinylbutyral 81
polyvinylidene fluoride (PVDF) 729
poor earthing 671
pores 666, 667
portable 713
positive-displacement pumps 546, 646
post-chlorinating 114
postcuring 412
pot life 69, 107, 603
pouring heads 503
powder adhesion 577
powder agglomerates 671, 672
powder booths 579
powder coating 64, 99, 110, 113, 249, 634, 642, 647, 670, 672, 719, 723, 734, 736
powder coating particles, charging of 574
powder coating, atomisation of 576
powder coating, colour changes in booths 580
powder coating, degree of separation 578
powder coating, dip-coating method 573
powder coating, efficiency of 578
powder coating, electrostatic 569, 573, 574
powder coating, electrostatic unit for 574
powder coating, fluidisability of 648
powder coating, material recovery 648
powder coating, of cavities 575
powder coating, recovery of 578
powder coating, rotary bells for 576
powder coating, spray heads for 576
powder coatings, application 573, 647
powder coatings, constituent materials of 580
powder coatings, film forming agents for 580
powder coatings, fluidisability of 571
powder coatings, hybrid 580
powder coatings, market 569, 647
powder coatings, metering system for 571
powder coatings, one-component systems for 580
powder coatings, particle size distribution of 571
powder coatings, precision conveying of 571
powder coatings, processing of 569
powder coatings, processing properties of 581
powder coatings, reactive 569
powder coatings, risk of explosions and fire with 579
powder coatings, test methods for 582
powder coatings, transport and conveying of 571
powder coatings, tribocharging 574
powder dispensing 671
powder particles 647, 671
powder particles, agglomeration of 582
powder primer 699, 701, 723, 754
powder recovery 578, 580
powder slurry 646, 648
powder-coating technology, origins of 569
power consumption 231
power wash process 719
power-and-free conveyors 566
precoated panels 711

precondensation 77
preparation 663
prepolymer process 96
prepolymers 95
preprimed materials 729
presealed 729
preservation during storage 189
preservatives 189
press-stud theory 377
pressure-equalising containers, 547
pretreatment 661, 665, 666
pretreatment methods, chemical 459
pretreatment plants 465
pretreatment, chromate-free 745
pretreatment, environmentally friendly 463, 722
pretreatment, of plastic surfaces 468
preventive medical check-ups 637
primary dispersions 644
primary particles 143, 208, 209, 215, 234
primary wetting 220
primer surfacer 27, 81, 95, 98, 103, 108, 165, 166, 202, 499, 562, 570, 628, 634, 644, 666, 688, 689, 690, 698ff, 705, 710, 712, 716, 731, 732, 737
primer surfacer, colour specific 699
primer surfacer, water-borne 98, 634, 644, 699, 701, 731
primer, chromate-free 745
primer, for clearcoats 650
primer, hybrid coating 650
primers 110, 202, 338, 363, 668, 716
primers for UV curable coatings 650
primers, application of 728
primers, conductive 720
printed circuit boards, manufacture of 747
process control 676, 677
process monitoring 496
process optimisation 661
process water 632
processibility 319
processing performance 259
processing tolerance 589
product life cycle 678
product life cycle audit 24
product planning 678
production installations 742
production line 23
production of pastes 226
production resources 200
products, radiation-curable 649
profile depths 361
profilometer 338, 347
propanediol (propylene glycol) 56
propeller agitators 229
properties of alkyd resins 59
properties, application 676
propoxylated glycerin triacrylate 125
propylene carbonate 138
propylene glycol 138
protection against light 187
protection of structures 742
protective action 344
protective equipment 637
protective function 16
protein 49
Prussian blue 149
pseudoplastic materials 287
pseudoplastic liquids 281
pseudoplasticity 60, 282, 284, 288
pseudowetting 212
p-tert.-butyl benzoic acid 57
p-toluenesulphonic acid 182
PUDs 95
Pugh 221
pull-force method 295
pull-off method 382
pulverised powders 165
pump, airless atomiser for 521
pumps 546
pumps for airless atomisation 521
PUR coatings 718, 740
PUR primer surfacer 737
PUR topcoat 737
purple 141
purple straight 367
PUR-polyacrylics 729
putty 716
PVC 114, 729
PVC plastisols 115
pycnometer 301, 352
pyranthrone 156
pyrazoloquinazolone pigments 161
pyrex filters 424
pyrheliometers 418
pyrogenic silicas 179, 186
pyrolysis 568
pyromellitic acid 64, 74

Q

QM system 681
QS 9000 680
quality 15, 21, 24, 261, 659, 677, 678
quality agreements 660
quality assessments 676
quality assurance (QA) 21, 259, 260, 674, 676, 679
quality awareness 660
quality characteristics 674
quality concept 660
quality criteria 678
quality directives 679
quality level 659
quality management 659, 677, 678
quality management systems 660, 678, 679

quality tests 228
quality, constant 589
quality, definition of 678
quality, determination of 660
quality, term 678
quartz powder 166
quasi-elastic light scattering (QELS) 310
quinacridones 156
quinone 69
quinophthalones 161

R

R phrases 636
radiant power, total 591
radiation curable coatings 99, 125, 719
radiation curing, market 649
radiation dose 609
radiation intensities 591
radiation sources 423
radiation-curable acrylic resins 123
radiation-curable paints 642, 648, 710
radiator grilles 725
radiators 735, 736
radical formation, mechanism of 608
radical polymerisation 649
radical scavengers 187
radicals 71
rail vehicles 737
railway bridges 742
railway wagons 737
RAL charts (RAL 841 GL) 375
RAL colours 733
RAL register 8040 HR 374
Ramsay, W. 295
random deviations 260
random errors 261
raw materials 27
raw materials, biobased 642
raw materials, incoming goods inspections 263
raw materials, shortage of 28
Rayleigh scattering 418
REACH 20, 616, 617, 618, 637
reaction injection molding 466
reactive 96
reactive diluent 105, 124, 203, 741
reactive end groups 32
reactive polyurethane dispersions 96
reactive thinners 29
reagglomeration 209, 221, 318
rear coat 728
reciprocators for paint application 563
recovery 578, 633, 634, 648
recycling 634, 635
recycling of overspray 322, 690
recycling of waste 632
red iron oxide 147
red lead 146
red wine 740
reduced gloss 416
refinish paints, new applications for 711
reflection curve 364, 372
reflectometer 356, 357
reflector 724
refractive index 304
refrigerators 734
refuse vehicles 730
re-ionisation 578
reinforced reaction injection moulding 467
relative depth resilience 401
relative error 261
relative tinting strength 316
relaxation time 391
release agents 468
release agents, silicone-containing 468
Remmler 144, 305
repair finish, duration of 712
repair systems, certification 711
repair work 674
repulsion barrier 219
repulsive barrier 218
residence times 244
residual sum R 305
resins 27
resins by polymerisation 113
resins for anodic electrocoating 108
resins market, global 33
resistance to water 81
resites 79
resols 79
resource conservation 639
responsible care 24, 615, 635, 636
retardation 391
retardation time 391
retrospective 21
reverse osmosis 633
reverse process 570
reverse roll coating, method of paint application 502
Reynolds number 231, 285, 508, 545
Reynolds number, critical 545
rheogram 288
rheological behaviour 327
rheometers 293
rheopexy 283
ricinoleic acid 72
rinsing zone 458, 490, 493, 495, 496, 668, 693, 697
risk assessment 621, 637
risk classifications 264
road bridges 742
road marking paints 748, 749
robot 562, 564, 565, 611, 681, 691, 700, 730
rodip 692, 694, 696

rods 364
roller coating, method of paint application 501
roller-coated paints, properties of 502
rolling ball method 292
roots pumps 546
rosin 144, 305
rotary atomisation, high speed 478, 530, 535, 536, 563
rotary dip 693
rotary bells 530, 531, 534, 564, 566, 576
rotary systems 465
rotary viscometers 288
rotor blade lengths 751
rotor-stator agitators (inline dissolvers) 237
rotor-stator unit 214
roughness on panel surface 666
roughness profile 361, 463
RRSB 307
RRSB curve 305
rubber 42, 99
rub-out test 144, 223, 319
rules of engineering 759
running 178
runs 178, 513, 661, 662, 665
rupture voltage effect 667
rust removal, mechanical 455
rusting 666

S

S phrases 636
safety indicators 264
sag balance 341
sagging 292
sagging limit 341
sagging resistance 661
sagging tendency 341
saleable vehicle coating 715
salt and pepper effect 669
salt spray test 430, 433
salt spray test to ISO 9227 427
sand and steel-shot blasting 455
sandblasting 737, 743
sanding 446, 715
sanding dust 668
sanding dust, residues of 668
sandpaper 715
saponification value 268
sardine oil 36
Sato 220
saturated fatty acids 35
saturated polyesters 63, 64, 67, 83, 84
scanning electron microscope (SEM) 664
scattering coefficient 317
scattering power 142, 207, 316
Schlack 72
score marks 673
scratch and erasure methods 385
scratch hardness 403
scratch resistance 124, 179, 195, 407, 650, 710
screw type kneaders 247
scrubbing resistance 195
sealing compounds 662
seam sealing compounds 668
Searle viscometer 288
seat frames 725
sebacic acid 64, 68, 170
secondary dispersions 644
secondary emissions 630
secondary ion mass spectroscopy (SIMS) 664
sedimentation 144, 177, 179, 209, 215, 223, 250, 251, 254, 285, 307, 308, 314, 318, 321, 461, 491, 540, 541, 545, 633, 696
sedimentation inhibitors 177
sedimentation profiles 318
sedimentation rate 308
sedimentation, accelerated 250
sedimentation, tendency towards 318
SEE balance 639
selective absorption 363
self-cleaning method 196
self-condensation 79
self-curing acrylic resins 122
semi-finished products 659
sensitisers 70, 183
sensitising substances 636
separation by sieves 250
separation of 561
separation performance 251
separation processes 257
separator 255
sequential test strategy 263
services 678
settling, tendency towards 144
sewage plants 633
shaping air, control over 530
sharpness of reproduction 356
shear angle 388
shear energy 234
shear forces 213, 234
shear gradient 280, 284
shear load 238
shear modulus 388
shear stress 280, 281
sheet metal overlaps 476
sheet moulding compound 653, 667
shelf life 320
shellac 40, 78
shelves for shop construction 743
Shields, I. 295
shipbuilding 749
shop fittings 740
short-wave radiators 595

shot peening test 386
shot-blasting wheels 456
shrinkage 336
shuttle process 246
siccatives 38, 183
sieve bags 251
sieve centrifuge 255
sieve device 251
sieving 250
sieving processes 307
sieving units 250
significance threshold 262
silica, colloidal 193
silica, pyrogenic 179, 192
silicate 741
silicate paints 193
silicating 163
silicon carbide 166
silicone acrylates 77, 124
silicone alkyds 77
silicone oil 76, 77, 168, 171, 180, 494, 682
silicone pastes 77
silicone polyesters 77
silicone polyethers 128, 178
silicone resin 34, 62, 67, 75, 77, 196, 231, 468, 474, 502, 667, 715
silicone-modified acrylate 124, 502, 729
silicone-modified polyacrylics 729
silicone-reinforced epoxy resins 77
silicones 662, 667
silo vehicles 730
single-impact devices 395
single-impact tests 396
siphon feed spray guns 511
siphon-fed guns 732
six sigma 681
sizing process 250
sliding roofs 719
slurries 581
small angle scatter 358
smart coatings 752
SMC plastic parts 651
SMCs 720
Snell's law 305
social responsibility 639
soft resin 81
softening point 40
soft-touch coatings 721
solar radiation 417
sol-gel conversion 283
sol-gel products 197
sol-gel technology 192ff
solid bowl centrifuge 255
solid bowl centrifuge 255
solid wood, dimensional stability 444
solids content 279, 330, 659, 674
solubility parameters 130
solubility, theory of 129
solutions, colloidal 128
solvation 129, 331
solvent 19, 27, 29, 127, 131, 132
solvent boils 662, 670
solvent degreasing 458
solvent emissions 31, 330
solvent evaporation 505
solvent naphtha 136
solvent shock 222
solvent-free dispersion 646
solvents, chemical and physiological properties 134
solvents, physical properties of 131
sorbitol 45
sound emission analysis 387
soy oil 36
soya oil fatty acid 59
sparkle 374
special waste 558
specific conductivity 645
specific surface area 312, 313
specifications 259, 660
spectrophotometer 372
specular gloss 358
Sperling 144, 305
spin-coating process 748
spinning drop method 298
sports articles 754
spot repairs 709
spot welds 666
spots 667
spray application 506
spray application, unwanted coating 567
spray bells 527
spray booth exhaust air 559
spray booth, airflow in 557
spray can lacquers 754
spray can paints, for automotive repairs 754
spray cloud 511
spray coating, effect of temperature increase on 517
spray coating, electrostatic 522
spray distance from object 517
spray droplet deposition 538
spray droplets 670
spray gun, atomising area of 516
spray guns 510
spray mist incompatibility 668, 670
spray pattern 516, 517
spray processes, electrostatically assisted 528
sprayability 319, 321
spray-galvanising 451
spraying air pressure 517, 519
spread of the distribution 275
spreading coefficient 176
stabilisation 143, 209, 217, 218

stabilisation of nanoparticles 192
stabilisation, electrostatic 221
stabilisation, entropic 219
stand oils 38
standard chromaticity diagram 367
standard deviation 261, 262
standard light B 366
standard light D 65 366
standard light TL 84 366
standard observer 365, 366
standard operating procedures 679
standard spectral values 366
standard trichromatic value 367
standard trichromatic value components 366
standard trichromatic values 366
standardisation committee 760
standardisation work 759
standardlight A 366
standards 260, 660, 759
standing wave atomisation 648
starch 33, 45
starting radical 69
starting radicals 51
static mixers 554
statistical analysis 675, 676
statistical measurements 676
Staudinger 50
steam and wet blasting processes 456
stearic acid 35
steel doors 736
steel furniture 743
steel, galvanised 433, 687, 692, 699, 717, 725, 726, 731, 733, 736
steel masts 742
steel wheels 721
Stefan-Boltzmann law 591, 593
step reaction 50
stereomicroscope 663
steric stabilisation 219
Stern 216
Stokes' law 522, 545
stone chip protection 661
stone chipping, protection 202, 688, 689, 699, 701, 705, 717
stoving temperature 590
stoving times 589
streaks 673
stress-strain diagram 388, 389
stripping 644
structures, chromophore 150
styrene 67, 69
styrene-modified alkyd resin 60
substance, surface active 664
substances, sensitising 636
substances, water-hazardous 633
substrate, influence of 441
substrates 661, 665, 666
subtractive colour mixing 365
sucrose 45
sugar 44
sulphate process 145
sulphochlorination 114
sulphonamides 85
sulphonium salts (AB) 183
sunflower 36
sunflower oil 36
supercritical carbon dioxide 250
supercritical CO_2 651
supercritical phase 651
surface active substances 664
surface analysis 662, 665
surface area, specific 313
surface defect 175, 470, 479, 495, 590, 661ff, 691, 696, 707, 708
surface defects, causes of 666
surface defects, wide area 665
surface disturbances 357
surface filtration 252
surface finish 441
surface finishing processes 447
surface protection 17
surface quality 518
surface reflection 355
surface roughness 453
surface smoothness 708
surface tension 77, 127, 172, 175, 178, 179, 180, 210, 211, 212, 293, 295, 297, 299, 327, 328, 336, 378–380, 659, 662, 670, 674, 676
surface texture 347, 359, 361
surface treatment 473
surface treatment with plasma 469
surface, local defects in 662
surfaces, biocidal 752
surfaces, for air regeneration 752
surfaces, for lowering resistance 752
surfaces, soiled 195
surfactant-like 196
surfactants 172, 173, 301, 457
surfactants, amphoteric 173
surfactants, anionic 173
surfactants, cationic 173
surfactants, ionic 173
surfactants, nonionic 174
sustainability 638, 641, 652
sustainable development 24, 615
sweating of adhesives 667
swelling 434
swelling tendency 714
symbols 783
synergists 71, 126, 183
synthesis gas 639
synthetic monocarboxylic acids 56

synthetic resin putties 113
synthetic resins 48
systematic deviations 260

T

Taber-Abraser 405
TACT 100, 120
tailgates 719
talcum 166
tall oil 36
tall oil fatty acid 59
tamped density 313, 314
tamped volume 313
tank facilities 742
tankers 749
tanks 742
tar 115
T-bend test 728
t-butyl perbenzoate 70
technical terms and conditions 259
Teller 144, 312
temperature constancy, during paint application 513
temperature profile 588
temperature/time chart 589
tempering 667
tendency to dissolve 714
tendency to run 38
terephthalic acid 55, 64
terms and conditions, technical 259
terpene 40
terpene hydrocarbons 136
tertiarybutyl perbenzoate 182
test chemicals 434
test methods 659, 660
test methods, harmonisation of 659
test specification 226, 659
test strategy, sequential 263
testing and inspection 660
tetraalkylammonium sulphate 181
tetraaminobiphenyl 74
tetrachlorothioindigo 155
tetraethyl orthosilicate 193
tetrahydronaphthalene (Tetralin) 136
tetrahydrophthalic acid 55, 72
tetramethylolacetylene diurea (tetramethylol glycoluril) 189
TGIC 104
thermal afterburning 630
thermal analysis 276
thermal conductivity 588
thermal treatment of exhaust air 630
thermal wave technique 352
thermally curing alkyd resins 57
thermogravimetric analysis (TGA) 276
thermoplastic polyacrylates 118
thermoplastic polymer 580, 639
thermoplastics 465
thermoplastics, property of 465
thermosetting paints 119
thermosetting polymer 639
thickeners 181
thin-section microtomy 663
thioindigo 155
thixotropic agents 72
thixotropic alkyd resins 60
thixotroping 65
thixotroping agents 181
thixotropy 60, 177, 181, 282, 283
thorough moistening 210, 236
three roll mills 213, 214, 237
three-head curtain-coating machines 504
three-roll systems 501
throughput rate 496
throwing power 497, 499, 668
timber, solid 444
time/temperature curve 589
time-of-flight secondary ion mass spectroscopy (TOF-SIMS) 664
tin compounds, ban on organic 750
tin plate 753
tinting strength 144, 316, 336
tinting strength, curves of changes in 317
titanium dioxide 145, 148, 207
titanium dioxide, micronized 192, 197
tolerance threshold 263
toluene 136
toluene-p-sulphonamide 85
toluylene diisocyanate (TDI) 91
tool cabinets 743
topcoat 27, 95, 98, 102, 110, 115, 120, 123, 146, 147, 163, 165, 202, 239, 448, 453, 462, 498, 628, 668, 670, 673, 676, 686, 688, 690, 696, 699, 702ff, 712ff, 752, 754
topcoat, of rotor blades 752
torque 289
torsion pendulum 396
total quality management 660, 678
total radiation 418
toxicity 104
toxicity values 636
toxicology 19, 644
traditional urethanes 95
trams 737
transesterification 57
transesterification process 58
transfer efficiency 647, 654
transverse reciprocators 564
tree resins 434
triazine rings 86
triazines 188
tribo-application 575, 578
tribo-application devices 574

tribo-chargeability 583
tribo-charging 570, 574
triethylene glycol 68
triglycidyl isocyanurate 104
triglycidyl isocyanurate (TGIC) 65
trim strips 725
trimellitic acid 63, 64, 104
trimellitic anhydride 55
trimethylol propane 64
trimethylolethane 56
trimethylolpropane (TMP) 56, 91, 125
trims 719
triphenylmethane pigments 154
tris-(2-hydroxyethyl)iso-cyanatotriacrylate 125
trisalkoxycarbonylaminotriazine (see TACT)
TRK values 267, 637
tropics test 427
trough kneaders 247
tumble mixers 248
tumbling, method of paint application 506
tung oil 36
turbidity factor 310
turbidity sensors 495
turbo screens for powder coating recovery 578
turbulent flow 544
turpentine oil 22, 136
turpentine substitute 136
TV cabinets 747
twin screw extruders 248
twin-shaft dissolver 236
twistometer 383
two-component paints, processing of 548
two-layer topcoats 203
types of deformation 388

U

ultracentrifuge 276
ultrafiltrate 492
ultrafiltration 480, 487, 490ff, 496, 562, 633, 634, 677, 695
ultrafiltration membrane 492
ultrafiltration process 492
ultrafiltration systems 491, 493
ultra-high solids 644
ultramarine 141, 149
ultrasonic atomisation 250, 536
ultrasonic flow measurement 552
ultrasonic pulse echo method 387
ultrasonic standing-wave atomisation 537
ultrasound 350
underbody protection 115
undercuring 588
underground railways 737
Uniform Color Scale System of the Optical Society 375
unsaturated polyester resins 67, 69, 70, 183, 649, 741
upper explosion limit 265, 266
urea 83, 167, 186
urea resins 78, 81, 85, 116
uretdiones 95
urethane acrylates 124, 649
urethane alkyd 60
urethane oil 60
urethane resins 82
used vehicle coating 715
UV absorbers 187, 188, 197, 203, 278, 606, 648
UV coatings 648
UV curable coatings 650, 724
UV curing 650, 651
UV curing systems, 3D workpieces 607
UV curing, chemical principle behind 604
UV curing, cold 603
UV curing, equipment for 606
UV drying, continuous measurement of 606
UV powder technology 651
UV protection 194
UV radiation 201, 648
UV spectroscopy 270
UV spectrum 271
UV-curable clearcoats 741
UV-curable coatings 740, 741
UV-curable coatings for filler primers 711
UV-curable wood coatings 649

V

value added cycle 639
van der Waals forces 294, 377
van Laar 401
vanadium pentoxide 196
vapour deposition 468
vapour pressure osmometry 276
VarioShuttle 695, 696
VDA 6.1 681
vehicle, axles of 647
vehicle, chassis parts for 647
velocity gradient 281, 288
velvet coatings 721
vending machines 743
ventilation, longitudinal 598
ventilation, transverse 598
Venturi principle 559
versatic acid 57
Verwey 215
vibrating sieves 257
vintage cars 713
vinyl ketones 110
viscoelasticity 280, 389, 392
viscometers 291
viscosimetric methods 276
viscosity 211, 212, 213, 279, 280, 281, 283, 288, 290, 327, 328, 659, 674, 676

viscosity of a dissolved water-borne paint 339
viscosity profile 646
viscosity, dynamic 281
viscosity, kinematic 281
visible defects 662
visible spectrum 364
VOC 616, 625, 627, 628, 630
VOC directive 599, 600, 616, 625, 627
VOC values 268
void ratio 354
voids 667
volatile organic compounds (VOCs) 267
volume shrinkage 332
volumetric flow measurement methods, comparison of 554
volumetric flow rate 542
volumetric pumps 646
von Etvös, R. 295
VW Ehra2 test 431

W

wardrobes 743
wash primer 81, 117
washing machines 734
waste 634
waste air flows 268
waste materials 644
waste water 267, 632, 633, 644
waste, landfilling of 635
waste, recycling of 632
Wastewater Ordinance 632
water absorption value 144
water mountain 645
water repellent state 181
water, evaporation characteristics of 645
water, physical data of 644
water-borne basecoats 163, 644, 706, 710, 719
water-borne coatings 645
water-borne paints 64, 339, 646
water-borne polyurethane 34, 95, 96, 98, 652
water-borne primer surfacers 98, 634, 644, 699, 701, 731
water-borne spray paints 642
water-borne-2-component system 98
watercourses 632
water-emulsifiable alkyd resin 62
water-hazardous substances 633
water-soluble alkyd resin 62
water-thinnable epoxy resin 108
wavelength spectrum 591
wavescan 361, 362, 709
waviness 360
waviness filters 362
waviness spectra 362
wax dispersions 186
waxes 71, 666
wear resistance 99
weather resistance 416, 417
weathering stations 419
weathering tests 422
Weather-O-meter (WOM) 423, 424
Weber 207
Weber number 508
wedge applicators 341
wedge feed method (hesiometer) 384
wedge wire sieves 241
weight average molecular weight Mw 275
welding globules 666, 667
wet on wet 717, 724
wet sanding 715
wetting 178, 210, 212, 661, 667
wetting agents 178
wetting defects 666, 670
wetting point 212
wetting volume 211
wheatstone bridge 551
wheel gauge 346
wheels 721
WHG – Water Resources Act 632
white goods 673, 734
white lead (basic lead carbonate) 146
white pigments 145
white rot 447
wide angle scatter 358
wide area surface defects 665
Wien's displacement law 592
Wilhelmy 295
Willborn 403
wind energy 750
wind turbines 742
windscreen wipers 725
wings 719, 720
wire enamels 746
Wolff 292, 403
wood 666
wood and furniture sector 68
wood coating 738
wood coating system 739
wood fibres, projecting 666
wood industry 738
wood preservatives 447
wood, as material 442
wood, as workpiece 444
wood, bleaching 446
wood, expansion properties 442
wood, impregnation 447
wood, pretreatment of 446
wood, protective action 447
wood, surface finishing 447
wood, swelling of 442
wood, wet sanding of 446

wooden materials, design 474
workpiece temperature 588, 597

X

Xenon test chamber 423
Xenon tester 424
Xirallic 374
X-ray microanalysis (EDX) 664
X-rays 350
xylene 136
xylylene diisocyanate (XDI) 91

Y

yachts 749
yellow chromating 464
yellow iron oxide 147
yellow iron oxide pigments 147
yield 301, 318, 655
yield E 655
yield point 282, 283
yield point curve 224
yield point method 225
yield value 224
Young's equation 210

Z

Zeidler 292
Zein 49
zero defects 660
zeta potential 219, 221
zinc aluminium phosphates 146
zinc bloom 451
zinc calcium phosphating 462
zinc dust 200
zinc dust primers 147
zinc oxide 141
zinc phosphate 109, 146
zinc phosphate coatings 461
zinc phosphating 460
zinc soaps 184, 186
zinc sulphide 146
zinc white (zinc oxide) 146
zinc-rich paints 451, 742
zinc-rich primer 742
zirconium dioxide 463, 688, 692, 694, 722
zirconium fluorides 463
zirconium soaps 184
Zisman 379
ZnO, micronized 197